Mammals of North America - Volume 1

Sergio Ticul Álvarez-Castañeda

Mammals of North America - Volume 1

Systematics and Taxonomy

Springer

Sergio Ticul Álvarez-Castañeda
Centro de Investigaciones Biológicas del Noroeste
La Paz, Baja California Sur, México

ISBN 978-3-031-41663-7 ISBN 978-3-031-41661-3 (eBook)
https://doi.org/10.1007/978-3-031-41661-3

This Springer imprint is published by the registered company Springer Nature Switzerland AG
The registered company address is: Gewerbestrasse 11, 6330 Cham, Switzerland

Preface

The keys are presented in alphabetical rather than evolutionary order. A summary of the orders, families, subfamilies, genera, subgenera, and species information is included. This organization was selected to facilitate species identification, aiming to support non-specialists in the group. Presenting them in phylogenetic order implies a general knowledge of the group of mammals. The keys are arranged hierarchically, first into orders, then by family within each order and by subfamilies within each family, as needed. The arrangement of genera and species is based on the characteristics that support the most precise and didactic separation.

The book considers the species distributed in México, the United States (mainland), and Canada as of April 2021 and includes specific keys for each order. In total, 12 orders, 49 families, 48 subfamilies, 223 genera, and 782 species are presented. Preparing this version involved a compilation work and countless previous versions spanning more than 38 years, which have been reviewed and tested under different conditions.

How to Use the Keys It is convenient to have a measurement method suitable for the species to be identified. A vernier or caliper and a magnifying glass of 10× minimum are recommended to observe various characteristics of the skull of small mammals that are described; in the laboratory, a stereo microscope is recommended.

The keys have been written using a simple language deemed accessible for non-specialists. However, the use of technical words could not be avoided in some cases, and support material is then provided as illustrations based on photographs from specimens of the species of interest. The intention is to use morphological traits of organisms. However, sometimes these are difficult to appreciate by a non-specialist or in morphologically cryptic species whose differentiation is based on genetic analyses; in these cases, the geographic distribution was selected, as it is easier to determine and distinguish.

Each key contains external and cranial characteristics. For most species, only one of them can be used. However, users are encouraged to review both types of characteristics together to confirm the identification of species; therefore, it is convenient to combine the two options as many times as possible. The elaboration of keys by combining external and skull characters was complex since, on occasions, they are not fully compatible to build a logical identification key. Both types of characters are equally important in the keys; it was impossible to have both options in some cases, so the one considered most appropriate was used for differentiating between similar species.

When only the external or cranial characters are mentioned, it is due to the lack of diagnostic information for the other set of characteristics. The identification should proceed using both options found in the dichotomous key and analyzing which of the two yields the closest fit in the description and distribution. These issues will arise mainly in some subfamilies of bats, rodents, and shrews, where the differences between species are very subtle in many cases.

The identification guides are composed of dichotomous keys, with two options for identification, in which one of them is the best suits of the characteristics of the specimen to be identified. After one of the keys, the known distribution of the species is given, based on its general patterns. Subsequently, the valid scientific name of the species is given as of December 2022. For each species, a paragraph provides the description of the species, with its diagnostic characteristics highlighted in bold. The external characteristics are described in the first place, including the coloration pattern and morphological characters.

The coloration was described based on literature descriptions and a review of voucher specimens, highlighting the characteristics that allow for distinguishing the species. The coloration can vary greatly between regions, subspecies, and even seasons of the year, so the most common one is described. Secondly, the cranial and two length measurements define the specimen's size, one that can be recorded in the field when the specimen is alive, and the cranial one during laboratory work. In both cases, it should be considered that these measurements were compiled from intervals reported in the literature and others recorded directly from voucher specimens, which include the representativeness of the species. The aim has always been to consider the variation of specimens or populations. However, for several species, data are available from a single population only, so specimens from different populations may be slightly larger or smaller in size than the interval presented; the coloration may also show slight variations. In addition, as living specimens are somehow subject to selection processes, there may be some lying outside the range that do not fully match the characteristics described. Concerning dimensions, two measurements were used at the same time, expressed in millimeters in all cases, which may vary depending on their frequency of use in each group. The first is a somatic metric: total length or forearm length in the case of bats; the second is cranial: skull length or condylobasal length. The same set of measurements is always maintained between species that are comparable. A few measurements are used because the aim is to use them as a reference for size; there are several species that are morphologically similar, but with significant differences in meristic aspects. It is important to consider that the intervals given for the measurements are relative since there may be variations depending on the geographic region and subspecies.

Literature Used The literature used was varied and addressed aspects of taxonomy, nomenclature, and distribution, which are the axes of this work. In the keys of species, the section of the description compiles information from different sources cited in the file of each species. All the literature information describes the physical characteristics of the species and includes a comparison with other species sharing a potentially sympatric distribution. In many cases, aspects of the descriptions of genera, subgenera, and species are strongly influenced by the syntheses written in the early twentieth century, when the major taxonomic reviews of the species were carried out. In virtually all cases, a physical examination of specimens was conducted to verify the characteristics mentioned in the literature; additional information was recorded and mentioned since information on the description of several species is scarce and limited to the characteristics that allow distinguishing between species of the same genus that are found in the same or a close range.

Translation The work is presented in English, and the proper names of the locations are mentioned in the original language. In many cases, the difference in language is restricted to using a stress mark; for example, México (with a stress mark as it is the name of the country in Spanish) and New Mexico (no stress mark as it is the name of a state in the USA, in English). As a rule, the names of states are not translated as some lack a translation, such as Nuevo León. The distribution of subspecies is not translated either since subspecies are essentially named after the states where they are found.

Data Used for Maps The data used to elaborate the maps were gathered from the Vertnet database of different museums. (I accept that data from misidentifications may be included in the maps) For each species, the unique localities of specimens were obtained and sorted according to 25-year intervals. The localities were reviewed to confirm that they were within the theoretical distribution area of the species; localities outside the theoretical distribution range of the species were reviewed or sought to confirm that they were included in a publication that would allow for validating their identification. If this validation was not possible following either of these two approaches, these outlier localities were eliminated from the maps. This is significant for species from recent periods, especially those that have undergone taxonomic changes or considered cryptic species in some cases. An example is the division of *Peromyscus maniculatus* into different species, so the localities considered were adjusted to the theoretical distribution areas proposed by the authors for each of the new species. The distribution range may change over time and as detailed studies of these groups are conducted. For some species, some literature records have not been included in the databases of the museums consulted. In the map, these cases are marked with a box instead of a point, meaning that this location comes from the literature reviewed rather than the museum database.

In general, for old records, i.e., prior to 1980, the georeferences from the geographic location are calculated, so they are approximate. Thus, there may be minor variations in their actual location, considering the continental scale used for elaborating the maps. With the information gathered on the localities, ecology, types of associated vegetation, and physiography, a theoretical distribution area was drawn. In some cases, the range shown is far from being an accurate reflection of the current range, especially for large mammals. The new localities are shown in a darker color and the old ones in a lighter one, so an approximate variation of the distribution over time can be visualized, showing whether it has shrunk or expanded. Occasionally, tropical species—particularly those from southern México—have very few locations compared to those living in colder environments. In addition, these species are associated with tropical forests undergoing deforestation processes. Thus, a large part of their distribution could be considered as theoretical.

The delimitation of the geographic range of subspecies is an approximation because I did not have the opportunity of review all the material. The boundaries of the distribution between subspecies were based on previous publications, including publications of some species or genera, mammal species, and the publication of Hall (1981). For some species on which I have experience, the ecology or physiography across the distribution area can modify the distribution boundaries. The distribution of subspecies is an approximation and should be taken as such because only a detailed examination of specimens can define the assigned subspecies.

Photos All the photos were taken and edited by the author, some were used and modificated from "Key for the identification of the Mexican Mammals" (Álvarez-Castañeda et al. 2017a). Some of the photos edited to further emphasize the characteristic mentioned in the key. The photos were taken from museum specimens of different institutions which are mentioned in the acknowledgments.

Protection At present, it is hugely important to know the protection and risk categories of the species according to the laws and regulations in the country where you work; this is why it was decided to include this information. When carrying out fieldwork and *in situ* identification, it is essential to determine the protection and risk category of the species before proceeding further, because doing so may have legal implications for the collector.

The information obtained on the species listed as endangered is limited due to the scarce studies conducted on many of these species. The causes affecting a given species may include direct uses, habitat modification, or the introduction of exotic species, among others. This is why the early identification of the species allows us to obtain valuable information for their understanding and the implementation of conservation measures.

When working with species included in protection lists, it is highly recommended to collect the greatest amount of data possible, such as reproductive status, somatic measurements, sex, activity, behavior, ecological and niche aspects, associations with other animal and plant species, surrounding vegetation, degree of habitat alteration, and others. Taken together, this information can provide key points for understanding the species and subsequently support conservation actions and policies. Working with mammal species should always be supported by specific permits/licenses from the respective authority, particularly for species listed under a protection and risk category. On the other hand, the information generated can be used by the personnel in charge of wildlife protection for decision-making about any relevant actions for immediate implementation.

La Paz, Baja California Sur, México

Sergio Ticul Álvarez-Castañeda

Acknowledgments

A work of this type and scope needed a great work team who contributed throughout the various process stages. In most cases, the persons who supported me did so voluntarily and as a practice to gain knowledge about particular mammal taxa. The group of collaborators includes researchers from different institutions who provided advisory or reviewed specific parts of the text. Staff associated with my work group and students of different academic levels actively participated in the review of the keys and tested them with actual collection specimens. I received feedback from all of them, on parts that needed clarification or that should be illustrated for a better understanding of the work.

The idea to carry out this project was first proposed by Ticul Álvarez some 40 years ago as a "tropicalization" of the North American keys, aiming to improve the analysis and better differentiate the species in the same genus or similar genera with sympatric or close distribution. In this large-scale project, we first worked on the keys of Mexican bats, *Claves para los murciélagos de México* (Álvarez et al. 1994). Subsequently, given the significant changes in taxonomy and nomenclature, we decided to address all the mammals in México with the Keys for identifying Mexican Mammals (Álvarez-Castañeda et al. 2017). When this work was completed, I realized that one of the key reference books for North America—Mammals of North America (Hall 1981)—was already obsolete for many species. Thus, a more updated version was necessary to include new species and the recent taxonomic and distribution changes, but preserving the essence of being useful for the differentiation of species, especially in a world in which fieldwork and taxonomists have been reduced. That is why I continued with the initiative that emerged at the dawn of my scientific career in Mammalogy.

The second cornerstone in the development of this work was James Patton. I had the opportunity to learn from his extensive knowledge when I did a sabbatical stay at the Museum of Vertebrate Zoology at the University of California, Berkeley. He was then the driving force in one way or another taking on the task of preparing a synthetic work on the mammals of North America. I must confess that I was completely unfamiliar with several of the groups and species of the subcontinent at this point since my field of experience was centered in México. However, this challenge served as a personal learning project to broaden my knowledge of the group in at least this region of the world. The MVZ collection allowed me to start learning and understanding the different groups of mammals, in addition to other collections visited. Having access to Jim's huge knowledge was also pivotal in this respect. I am deeply grateful for his help in the revision of the language in some parts of the draft.

My very special thanks to Leticia Cab Sulub, who supported me with the database of localities for the elaboration of maps of the species distribution. This activity was a very arduous task because, besides the projection of localities and distribution areas, several versions had to be made for different scales, drawing types, and projections, in addition to working on the optimal design of the possible distribution areas. For most species, several versions of maps were drafted until I was convinced that the final map was the best theoretical distribution based on the sources of information consulted.

Alina Gabriela Monroy Gamboa and Patricia Cortés Calva provided continued and valuable support in the repeated reading of the different draft versions, always making highly pertinent observations that helped to clarify the writing and spotting errors in different parts. I am hugely grateful for the time spent by them in the detailed reviews. I should also highlight their great personal support. Thanks also to Consuelo Lorenzo, who reviewed the final version of the whole manuscript, contributing a great number of excellent recommendations and highlighting multiple areas that needed to be improved. To Carmen Izmene Gutierrez Rojas for its revision in detail of many specific task in the different drafts.

Noé González was key in many aspects of this process. Together, we discussed several approaches for the keys, the characteristics of the different taxa, and the reviews of several groups, particularly Chiroptera and Cricetidae. In this second work, he decided not to participate actively; however, I frequently consulted with him to resolve the keys, address specific characteristics, or discuss systematic changes, and occasionally talked about recently published literature.

Maria Elena Sánchez Salazar supported me in different aspects of the writing and editing of the manuscript, so it can be as clear as possible. I deeply thanks for all the support you have provided throughout my professional development, even when when I was a student of bachelor. Her recommendations improved the overall editorial quality of the text and have made the writing easier to read and understand.

This work involved the review of several scientific collections, which was possible thanks to the support of the personnel in charge of them and, in some cases, to the donation of material to the Collection of the Centro de Investigaciones Biológicas del Noroeste (CIB), which was used to observe details of the species and for testing the keys. The collections reviewed and the persons I wish to thank are the following. Museum of Vertebrate Zoology, University of California, Berkeley: James L. Patton, Michael Nachman, Cristopher Conroy, and Eileen Lacey. Smithsonian Institute: Alfred Gardner, Neal Woodman, Michael Carleton, Jesús Maldonado, Darrin Lunde, and Robert Fisher. University of New Mexico: Joseph Cook, Terry Yates, and Jonathan Dunnum. Brigham Young University: Duke Rogers. Field Museum: Bruce Patterson and Adam Ferguson. University of Kansas: Robert Timm and Maria Eifler. Museum of Texas Tech University; Robert Baker, Robert Bradley, and Heath Garner. American Museum of Natural History: Robert Voss. San Diego Natural History Museum: Philip Unitt and Scott Tremor. Angelo State University: Robert Dowler. Humboldt State University: Thorvald "Thor" Holmes. Escuela Nacional de Ciencias Biológicas del Instituto Politécnico Nacional: Ticul Álvarez, Aurelio Ocaña, and Juan Carlos López Vidal. Instituto de Biología del Universidad Nacional Autónoma de México: Oscar Sánchez, Fernando Cervantes, Julieta Vargas, and Yolanda Hortelano. El Colegio de la Frontera Sur, San Cristóbal de las Casas: Consuelo Lorenzo and Jorge Bolaños. Centro Interdisciplinario de Investigación para el Desarrollo Integral Regional, unidad Oaxaca del Instituto Politécnico Nacional: Miguel Briones and Natalia Martín Regalado. Instituto Nacional de Antropología e Historia: Joaquín Arroyo and Aurelio Ocaña. Universidad Autónoma Metropolitana, unidad Iztapalapa: José Ramírez Pulido and Noé González. Thanks also to the personnel in charge of the collection of Centro de Investigaciones Biológicas del Noroeste: Patricia Cortés Calva, Mayra de la Paz Cuevas, and Carmen Izmene Gutiérrez Rojas.

Parts, sections, or taxonomic groups in the manuscript were reviewed by different specialists in the groups that gave me their highly qualified opinion, which was extremely useful and was taken into account. In this respect, I am very grateful to James L. Patton, Robert Bradley, Noé González, Patricia Cortés, Lázaro Guevara, Consuelo Lorenzo, Gabriela Monroy, Sonia Gallina, Francisco X. González, and Elizabeth Arellano. Thanks also to Connor Burgin, Robert Timm, Ronald Pine, and Luis Ruedas for their multiple comments and points of view on the list of species included in this manuscript. The keys were also tested and evaluated by my students to identify the critical points that should be reviewed and addressed: Alina Gabriela Monroy Gamboa, Evelyn Patricia Ríos Mendoza, Ana Lilia Trujano Álvarez, Eduardo Aguilera Miller, Cintya A. Segura Trujillo, Leticia Cab Sulub, Horacio Cabrera Santiago, Issac Camargo Pérez, Gabriela Suárez Gracida, Luis Ernesto Pérez Montes, Lilia Isabel López Pérez, Erika Patricia González Quintero, Francisco Javier Navarro Frías, Hortensia Santillán Ortiz, Nansy Sánchez, Jorge Villalpando, Jacqueline Tun Balam, and Arturo Hernández Gutiérrez. Special thanks to Lia C. Méndez Rodríguez and Lia Montserrat Álvarez Méndez for the reviews to the different versions and their huge unconditional support at all times.

Taxonomic Criteria This work is a compilation of the existing information as of the date of publication. The nomenclature is based on the available literature. In cases when two different points of view are expressed, one is adopted but the other one is also mentioned; in these cases, the rationale for selecting a given author is explained. The nomenclatural review above the family level was not addressed in this work. The following two current classifications are taken for reference.

Contents

Taxonomic List and Nomenclature Used

Class Mammalia

Subclass Theria

Infraclass Marsupialia

Order Didelphimorphia

Family Didelphidae

Subfamily Caluromyinae

Caluromys (Mallodelphys) derbianus
Derby's Woolly Opossum, tlacuache lanudo

Subfamily Didelphinae

Tribe Didelphini

Chironectes minimus
Water opossum, tlacuache de agua
Didelphis marsupialis
Common opossum, tlacuache tropical
Didelphis virginiana
Virginia opossum, tlacuache
Philander vossi
Gray four-eyed opossums, cuatro ojos gris

Tribe Marmosini

Marmosa (Exulomarmosa) mexicana
Mexican mouse opossum, ratón tlacuache mexicano
Tlacuatzin balsasensis
Balsas mouse opossum, ratón tlacuache del Balsas
Tlacuatzin canescens
Gray mouse opossum, ratón tlacuache gris
Tlacuatzin gaumeri
Yucatán mouse opossum, ratón tlacuache de Yucatán
Tlacuatzin insularis
Tres Marías mouse opossum, ratón tlacuache de Tres Marías

S. T. Álvarez-Castañeda, *Mammals of North America - Volume 1*, https://doi.org/10.1007/978-3-031-41661-3_1

Tlacuatzin sinaloe
Northern mouse opossum, ratón tlacuache del norte

Tribe Metachirini

Metachirus myosuros
Brown four-eyed opossums, cuatro ojos pardo

Infraclass Eutheria

Order Artiodactyla

Suborder Rumiantia

Infraorder Pecora

Family Antilocapridae

Antilocapra americana
Pronghorn, berrendo

Family Bovidae

Subfamily Bovinae

Tribe Bovini

Bos bison
Bison, bisonte americano

Subfamily Antilopinae

Tribe Caprini

*Capra hircus**
Dometic goat, cabra
Oreamnos americanus
Mountain goat, cabra de montaña
Ovibos moschatus
Muskox, buey almizclero
Ovis canadensis
Bighorn sheep, borrego cimarrón
Ovis dalli
Dall's sheep, borrego cimarrón blanco

Family Cervidae

Subfamily Capreolinae

Tribe Alceini

Alces alces
Moose, alce

Tribe Rangiferini

Mazama pandora
Yucatán brown brocket deer, temazate café
*Mazama temama***
Central American red brocket deer, temazate rojo
Odocoileus hemionus
Mule deer, venado bura
Odocoileus virginianus
White-tailed deer, venado cola blanca
Rangifer tarandus
Reindeer, reno

Subfamily Cervinae

Tribe Cervini

*Axis axis**
Chital, ciervo moteado
Cervus canadiensis
Elk, ciervo rojo
*Dama dama**
Fallow deer, gamo común

Suborder Suina

Family Suidae

Subfamily Suinae

Tribe Suini

*Sus scrofa**
Wild boad, cerdo salvale

Family Tayassuidae

Subfamily Tayassuinae

Dicotyles tajacu
collared peccary, jabalí de collar
Tayassu pecari
White-lipped peccary, jabalí de labios blancos

Order Carnivora

Suborder Caniformia

Family Canidae

Alopex lagopus
Artic fox, zorra del ártico
Canis familiaris
Dog, perro

Canis latrans
Coyote, coyote
Canis lupus
Gray wolf, lobo gris
Urocyon cinereoargenteus
Gray fox, zorra gris
Urocyon littoralis
Island grey fox, zorra gris de isla Catalina
Vulpes macrotis
Kit fox, zorro del desierto norteño
Vulpes velox
Swift fox, zorro del desierto
Vulpes vulpes
Red fox, zorro rojo

Family Mephitidae

Subfamily Mephitinae

Conepatus leuconotus
Eastern hog-nosed skunk, zorrillo lomo blanca
Conepatus semistriatus
Striped hog-nosed skunk, zorrillo bilistado
Mephitis macroura
Hooded skunk, zorrillo rayado sureño
Mephitis mephitis
Striped skunk, zorrillo rayado norteño
Spilogale angustifrons
Southern spotted skunk, zorrillo manchado del sur
Spilogale gracilis
Western spotted skunk, zorrillo manchado oeste
Spilogale interrupta
Plains spotted skunk, zorrillo manchado de las planicies
Spilogale leucoparia
Desert spotted skunk, zorrillo manchado del desierto
Spilogale putorius
Eastern spotted skunk, zorrillo manchado este
Spilogale pygmaea
Pygmy spotted skunk, zorrillo pigmeo
Spilogale yucatanensis
Yucatán spotted skunk, zorrillo manchado de la península de Yucatán

Family Mustelidae

Subfamily Guloninae

Eira barbara
Tayra, cabeza de viejo
Gulo gulo
Wolverine, glotón
Martes (Martes) americana
American marten, marta americana
Martes (Martes) caulina
Pacific marten, marta del Pacífico
Pekania pennanti
Fisher, marta pescadora

Subfamily Ictonychinae

Galictis vittata
Greater grison, grisón

Subfamily Lutrinae

Enhydra lutris
Sea otter, nutria marina
Lontra canadensis
River otter, nutria de rio
Lontra longicaudis
Southern otter, nutria de río tropical

Subfamily Mustelinae

Mustela (Gale) nivalis
Least weasel, comadreja pequeña
Mustela (Mustela) erminea
Ermine, armiño
Mustela (Mustela) haidarum
Haida ermine, armiño del archipiélago de Haida Gwaii
Mustela (Putorius) nigripes
Black-footed ferret, hurón de patas negras
Mustela (Mustela) richardsonii
American ermine, armiño Americano
Neogale frenata
Long-tailed weasel, comadreja
Neogale vison
American mink, mink

Subfamily Taxidiinae

Taxidea taxus
American badger, tejón

Family Odobenidae

Odobenus rosmarus
Walrus, morsa

Family Otariidae

Arctocephalus townsendi
Guadalupe fur seal, lobo-fino de Guadalupe
Callorhinus ursinus
Northern fur seal, lobo marino del ártico

Subfamily Otariidae

Eumetopias jubatus
Steller sea lion, lobo marino de Steller
Zalophus californianus
California sea lion, lobo marino de California

Family Phocidae

Cystophora cristata
Hooded seal, foca capuchina
Erignathus barbatus
Bearded seal, foca barbuda
Halichoerus grypus
Gray seal, foca gris del Atlántico
Histriophoca fasciata
Ribbon seal, foca listada
Mirounga angustirostris
Northern elephant seal, elefante-marino del norte
Neomonachus tropicalis †
Caribbean monk seal, foca monje del Caribe
Pagophilus groenlandicus
Harp seal, foca de arpa
Phoca largha
Spotted seal, foca moteada
Phoca vitulina
Harbor seal, foca común
Pusa hispida
Ringed seal, foca anillada

Family Procyonidae

Bassariscus astutus
Ringtail cat, cacomixtle
Bassariscus sumichrasti
Cacomistle, cacomixtle tropical
Nasua narica
White-nosed coati, coatí
Potos flavus
Kinkajou, martucha
Procyon lotor
Raccoon, mapache
Procyon pygmaeus
Cozumel raccoon, mapache de Cozumel

Family Ursidae

Subfamily Ursinae

Ursus americanus
American black bear, oso negro
Ursus arctos
Grizzly bear, oso pardo
Ursus maritimus
Polar bear, oso polar

Suborder Feliformia

Family Felidae

Subfamily Felinae

Herpailurus yagouaroundi
Jaguarundi, Jaguarundi
Leopardus pardalis
Ocelot, ocelote
Leopardus wiedii
Margay, tigrillo
Lynx canadensis
Canada lynx, lince canadiense
Lynx rufus
Bobcat, gato montes
Puma concolor
Cougar, puma

Subfamily Pantherinae

Panthera onca
Jaguar, Jaguar

Order Chiroptera

Family Emballonuridae

Balantiopteryx io
Thomas's sac-winged bat, murciélago pardo de sacos en las alas
Balantiopteryx plicata
Gray sac-winged bat, murciélago gris de sacos en las alas
Centronycteris centralis
Thomas's shaggy bat, murciélago lanudo
Diclidurus (Diclidurus) albus
Northern ghost bat, murciélago blanco
Peropteryx (Peropteryx) kappleri
Greater dog-like bat, murciélago perro mayor
Peropteryx (Peropteryx) macrotis
Lesser dog-like bat, murciélago perro menor
Rhynchonycteris naso
Proboscis bat, murciélago narigón
Saccopteryx bilineata
Greater sac-winged bat, murciélago rayado mayor
Saccopteryx leptura
Lesser sac-winged bat, murciélago rayado menor

Family Molossidae

Subfamily Molossinae

Cynomops mexicanus
Mexican dog-faced bat, murciélago cara de perro mexicano
Eumops auripendulus
Black bonneted bat, murciélago negro de gorra

Eumops ferox
Central American bonneted bat, murciélago de gorra centroamericano
Eumops floridanus
Florida bonneted bat, murciélago de gorra de Florida
Eumops hansae
Sanborn's bonneted bat, murciélago de gorra pigmeo
Eumops nanus
Eastern lesser bonneted bat, murciélago de gorra enano
Eumops perotis
Greater bonneted bat, murciélago de gorra mayor del oeste
Eumops underwoodi
Underwood's bonneted bat, murciélago de gorra del Pacífico
Molossus alvarezi
Alvarez's mastiff bat, murciélago mastín de Yucatán
Molossus aztecus
Azteca mastiff bat, murciélago mastín azteca
Molossus coibensis
Coiban mastiff bat, murciélago mastín de Coiban
Molossus molossus
Pallas' mastiff bat, murciélago mastín peludo
Molossus nigricans
Black mastiff bat, murciélago mastín negro
Molossus sinaloae
Sinaloan mastiff bat, murciélago mastín de Sinaloa
Nyctinomops aurispinosus
Peale's free-tailed bat, murciélago de cola libre
Nyctinomops femorosaccus
Pocketed free-tailed bat, murciélago de cola libre de bolsa
Nyctinomops laticaudatus
Broad-eared free-tailed bat, murciélago de cola libre de orejas grandes
Nyctinomops macrotis
Big free-tailed bat, murciélago de cola libre mayor
Promops centralis
Big crested mastiff bat, gran murciélago mastín crestado
Tadarida brasiliensis
Brazilean free-tailed bat, murciélago de cola libre mexicano

Family Mormoopidae

Mormoops megalophylla
Ghost-faced bat, murciélago con cara arrugada
Pteronotus (Chilonycteris) personatus
Wagner's mustached bat, murciélago bigotudo menor
Pteronotus (Phyllodia) mesoamericanus
Common mustached bat, murciélago bigotudo mayor
Pteronotus (Phyllodia) mexicanus
Mexican mustached bat, murciélago bigotudo mexicano
Pteronotus (Pteronotus) davyi
Davy's naked-backed bat, murciélago de espalda desnuda menor
Pteronotus (Pteronotus) gymnonotus
Big naked-backed bat, murciélago de espalda desnuda mayor

Family Natalidae

Natalus lanatus
Haired funnel-eared bat, murciélago orejas de embudo peludo
Natalus mexicanus
Mexican funnel-eared bat, murciélago orejas de embudo mexicano

Family Noctilionidae

Noctilio (Dirias) albiventris
Lesser bulldog bat, murciélago pescador menor
Noctilio (Noctilio) leporinus
Great bulldog bat, murciélago pescador mayor

Family Phyllostomidae

Subfamily Carolliinae

Carollia perspicillata
Seba's short-tailed bat, murciélago colicorta
Carollia sowelli
Sowell's short-tailed bat, murciélago de colicorta pardo
Carollia subrufa
Gray short-tailed bat, murciélago colicorta gris

Subfamily Desmodontinae

Desmodus rotundus
Common vampire bat, vampiro común
Diaemus youngii
White-winged vampire bat, vampiro de alas blancas
Diphylla ecaudata
Hairy-legged vampire bat, vampiro de patas peludas

Subfamily Glossophaginae

Tribe Choeronycterini

Anoura geoffroyi
Geoffroy's tailless bat, murciélago sin cola gris
Choeroniscus godmani
Godman's long-tongued bat, murciélago de lengua larga
Choeronycteris mexicana
Mexican long-tongued bat, murciélago trompudo mexicano
Hylonycteris underwoodi
Underwood's long-tongued bat, murciélago de lengüilargo moreno
Lichonycteris obscura
Dark long-tongued bat, murciélago lengüilargo oscuro
Musonycteris harrisoni
Banana bat, murciélago hocicudo del Balsas

Tribe Glossophagini

Glossophaga commissarisi
Commissaris' long-tongued bats, murciélago lengüilargo pequeño
Glossophaga leachii
Gray's long-tongued bats, murciélago lengüilargo

Glossophaga morenoi
Western long-tongued bats, murciélago lengüilargo del oeste
Glossophaga mutica
Tres Marías Island long-tongued bat, murciélago lengüilargo de Islas Tres Marías
Leptonycteris nivalis
Mexican long-nosed bat, murciélago magueyero mayor
Leptonycteris yerbabuenae
Lesser long-nosed bat, murciélago magueyero menor

Subfamily Glyphonycterinae

Glyphonycteris sylvestris
Tri-colored big-eared bat, murciélago tricolor
Trinycteris nicefori
Niceforo's big-eared bat, murciélago orejón raro

Subfamily Lonchorhininae

Lonchorhina aurita
Common sword-nosed bat, murciélago nariz de espada

Subfamily Macrotinae

Macrotus californicus
California big-eared bats, murciélago orejón norteño
Macrotus waterhousii
Waterhous's big-eared bats, murciélago orejón sureño

Subfamily Micronycterinae

Lampronycteris brachyotis
Yellow-throated big-eared bat, murciélago orejón de garganta amarilla
Micronycteris (Micronycteris) microtis
Common little big-eared bat, murciélago orejón pequeño
Micronycteris (Schizonycteris) schmidtorum
Schmidts's little big-eared bat, murciélago orejón centroamericano

Subfamily Phyllostominae

Tribe Phyllostomini

Gardnerycteris keenani
Striped hairy-nosed bat, murciélago dorado nariz peluda
Lophostoma brasiliense
Pygmy round-eared bat, murciélago orejas redondas brasileño
Lophostoma evote
Davis's round-eared bat, murciélago orejas redondas mesoamericano
Macrophyllum macrophyllum
Long-legged bat, murciélago de patas largas
Phylloderma stenops
Pale-faced bat, murciélago cara pálida
Phyllostomus discolor
Pale spear-nosed bat, murciélago pálido de nariz de lanza
Tonatia bakeri
Stripe-headed round-eared bat, murciélago de orejas redondas
Trachops cirrhosus
Fringe-lipped bat, murciélago labio verrugoso

Tribe Vampyrini

Chrotopterus auritus
Woolly false vampyre bat, falso vampiro lanudo
Mimon cozumelae
Cozumelan golden bat, murciélago dorado de Cozumel
Vampyrum spectrum
Spectral bat, murciélago espectral

Subfamily Stenodermatinae

Tribe Stenodermatini

Artibeus hirsutus
Hairy fruit-eating bat, murciélago frutero de patas peludas
Artibeus jamaicensis
Jamaican fruit-eating bat, murciélago frutero común
Artibeus lituratus
Great fruit-eating bat, murciélago frutero gigante
Centurio senex
Wrinkle-faced bat, murciélago cara de viejo
Chiroderma salvini
Salvin's big-eyed bat, murciélago de grandes ojos pardo
Chiroderma scopaeum
Dwarf big-eyed bat, murciélago de grandes ojos enano
Chiroderma villosum
Hairy big-eyed bat, murciélago de grandes ojos güero
Dermanura azteca
Aztec fruit-eating bat, murciélago frutero menor de patas peludas
Dermanura phaeotis
Pygmy fruit-eating bat, murciélago frutero pigmeo
Dermanura tolteca
Toltec fruit-eating bat, murciélago frutero menor de patas desnudas
Dermanura watsoni
Thomas's fruit-eating bat, murciélago frutero menor centroamericano
Enchisthenes hartii
Velvety fruit-eating bat, murciélago frutero negro
Phyllops falcatus
Cuban fig-eating bat, murciélago higuero de Cuba
Platyrrhinus helleri
Heller's broad-nosed bat, murciélago listado nariz ancha
Uroderma convexum
Northern tent-making bat, murciélago acampador del norte
Uroderma davisi
Soconusco tent-making bat, murciélago acampador del Soconusco
Uroderma magnirostrum
Brown tent-making bat, murciélago acampador pálido
Vampyressa thyone
Northern little yellow-eared bat, murciélago cara pintada menor
Vampyrodes major
Great stripe-faced bat, murciélago cara rayada mayor

Tribe Sturnirini

Sturnira (Sturnira) hondurensis
Highland yellow-shouldered bat, murciélago de charreteras mayor
Sturnira (Sturnira) parvidens
Little yellow-shouldered bat, murciélago de charreteras menor

Family Thyropteridae

Thyroptera tricolor
Disk-winged bats, murciélagos con discos en las alas

Family Vespertilionidae

Subfamily Myotinae

Myotis albescens
Silver-tipped myotis, murciélago ratón plateado
Myotis auriculus
Southwestern myotis, murciélago ratón suroccidental
Myotis austroriparius
Southeastern myotis, murciélago ratón del sureste
Myotis californicus
California Myotis, murciélago ratón de California
Myotis ciliolabrum
Western small-footed myotis, murciélago ratón occidental
Myotis elegans
Elegant myotis, murciélago ratón elegante
Myotis evotis
Long-eared myotis, murciélago ratón de oeste
Myotis findleyi
Findley's myotis, murciélago ratón de las Islas Marías
Myotis fortidens
Cinnamon myotis, murciélago ratón canela
Myotis grisescens
Gray myotis, murciélago ratón gris
Myotis keenii
Keen's myotis, murciélago ratón de la Columbia Británica
Myotis leibii
Eastern small-footed myotis, murciélago ratón oriental
Myotis lucifugus
Little brown bat, pequeño murciélago ratón norteamericano
Myotis nigricans
Black myotis, murciélago ratón negro
Myotis occultus
Arizona myotis, murciélago ratón de Arizona
Myotis pilosatibialis
Haired legs myotis, murciélago ratón patas peludas
Myotis planiceps
Flat-headed myotis, murciélago ratón de cabeza plana
Myotis septentrionalis
Northern myotis, murciélago ratón grande del norte
Myotis sodalis
Indiana myotis, murciélago ratón de Indiana
Myotis thysanodes
Fringed myotis, murciélago ratón peludo

Myotis velifer
Cave myotis, murciélago ratón de las cuevas
Myotis vivesi
Fish-eating myotis, murciélago ratón pescador
Myotis volans
Long-legged myotis, murciélago ratón patas largas
Myotis yumanensis
Yuma myotis, murciélago ratón pequeño

Subfamily Vespertilioninae

Tribe Antrozoini

Antrozous pallidus
Pallid bat, murciélago pálido
Bauerus dubiaquercus
Van Gelder's bat, murciélago pálido tropical
Rhogeessa (Baeodon) alleni
Yellow bat, murciélago amarillo grande
Rhogeessa (Baeodon) gracilis
Slender yellow bat, murciélago amarillo delgado
Rhogeessa (Rhogeessa) aeneus
Yucatán yellow bat, murciélago amarillo de Yucatán
Rhogeessa (Rhogeessa) bickhami
Brickhami's yellow bat, murciélago amarillo de la costa de Chiapas
Rhogeessa (Rhogeessa) genowaysi
Genoways's yellow bat, murciélago amarillo de Chiapas
Rhogeessa (Rhogeessa) mira
Least yellow bat, murciélago amarillo pigmeo
Rhogeessa (Rhogeessa) parvula
Little yellow bat, murciélago amarillo pequeño
Rhogeessa (Rhogeessa) tumida
Black-winged little yellow bat, murciélago amarillo de alas negras

Tribe Eptesicini

Eptesicus (Eptesicus) brasiliensis
Brazilian brown bat, murciélago pardo de Brasil
Eptesicus (Eptesicus) furinalis
Argentine brown bat, murciélago pardo tropical
*Eptesicus (Eptesicus) fuscus***
Big brown bat, murciélago pardo grande
Lasionycteris noctivagans
Silver-haired bat, murciélago plateado

Tribe Lasiurini

Lasiurus borealis
Eastern red bat, murciélago rojo del este
Lasiurus cinereus
Hoary bat, murciélago canoso
Lasiurus ega
Southern yellow bat, murciélago amarillo del sur
Lasiurus frantzii
Western red bat, murciélago rojo del oeste

Lasiurus intermedius
Northern yellow bat, murciélago amarillo del norte
Lasiurus seminolus
Seminole bat, murciélago canoso de Florida
Lasiurus xanthinus
Western yellow bat, murciélago amarillo del oeste

Tribe Nycticeiini

Nycticeius humeralis
Evening bat, murciélago del anochecer

Tribe Perimyotini

Parastrellus hesperus
American pipistrelle, murciélago pequeño del oeste
Perimyotis subflavus
Eastern pipistrelle, murciélago pequeño del este

Tribe Plecotini

*Corynorhinus mexicanus***
Mexican big-eared bat, murciélago orejudo mexicano
Corynorhinus rafinesquii
Rafinesque's big-eared bat, murciélago orejudo norteamericano
Corynorhinus townsendii
Townsend's big-eared bat, murciélago orejudo pálido
Euderma maculatum
Spotted bat, murciélago orejudo manchado
Idionycteris phyllotis
Allen's big-eared bat, murciélago orejudo negro

Order Cingulata

Family Dasypodidae

Subfamily Dasypodinae

Dasypus (Dasypus) novemcinctus
Nine-banded armadillo, armadillo de nueve bandas

Family Chlamyphoridae

Subfamily Tolypeutinae

Cabassous centralis
Northern naked-tailed armadillo, armadillo de cola desnuda

Order Eulipotyphla

Family Soricidae

Subfamily Soricinae

Tribe Blarinini

Blarina brevicauda
Northern short-tailed shrew, musaraña de cola corta del norte
Blarina carolinensis
Southern short-tailed shrew, musaraña de cola corta del sur
Blarina hylophaga
Elliot's short-tailed shrew, musaraña de cola corta de las planicies
Blarina shermani
Sherman's short-tailed shrew, musaraña de cola corta de Florida
Cryptotis alticola
Merriam's small-eared shrew, musaraña tropical de montaña
Cryptotis berlandieri
Northeastern México least shrew, musaraña tropical del noreste de México
Cryptotis goldmani
Goldman's small-eared shrew, musaraña tropical de la Sierra Madre del Sur
*Cryptotis goodwini***
Goodwin's small-eared shrew, musaraña tropical de Chiapas
Cryptotis griseoventris
Guatemalan broad-clawed shrew, musaraña tropical de Guatemala
Cryptotis lacandonensis
Lacandona small-eared shrew, musaraña tropical de la Lacandona
Cryptotis magnus
Big small-eared shrew, musaraña tropical gigante
Cryptotis mayensis
Yucatán small-eared shrew, musaraña tropical Maya
Cryptotis merriami
Merriam's small-eared shrew, musaraña tropical de tierras altas
Cryptotis mexicanus
Mexican small-eared shrew, musaraña tropical mexicana
Cryptotis nelsoni
Nelson's small-eared shrew, musaraña tropical de Catemaco
Cryptotis obscurus
Grizzled Mexican small-eared shrew, musaraña tropical de Hidalgo
Cryptotis parvus
North American least shrew, musaraña tropical de Norteamérica
Cryptotis peregrinus
Oaxacan small-eared shrew, musaraña tropical de Oaxaca
Cryptotis phillipsii
Phillips' small-eared shrew, musaraña tropical la sierra de Oaxaca
Cryptotis pueblensis
Southern México least shrew, musaraña tropical del sur de México
Cryptotis soricinus
Valle de México least shrew, musaraña tropical del Valle de México
Cryptotis tropicalis
Tropical small-eared shrew, musaraña de tropical

Tribe Notiosoricini

Megasorex gigas
Mexican shrew, musaraña gigante
Notiosorex cockrumi
Cockrum's gray shrew, musaraña del desierto de Arizona
Notiosorex crawfordi
Crawford's gray shrew, musaraña del desierto
Notiosorex evotis
Large-eared gray shrew, musaraña del desierto del Pacífico
Notiosorex tataticuli
Ticul's gray shrew, musaraña de las Californias
Notiosorex villai
Villa's gray shrew, musaraña de Tamaulipas

Tribe Soricini

Sorex (Otiosorex) albibarbis
White-bearded water shrew, musaraña de agua de barba blanca
Sorex (Otiosorex) bendirii
Marsh shrew, musaraña de pantano
Sorex (Otiosorex) cinereus
Masked shrew, musaraña canosa
Sorex (Otiosorex) dispar
Long-tailed shrew, musaraña de cola larga
Sorex (Otiosorex) emarginatus
Zacatecas shrew, musaraña de Zacatecas
Sorex (Otiosorex) eximius
Western pygmy shrew, musaraña pigmea del oeste
Sorex (Otiosorex) fumeus
Smokey shrew, musaraña de las Smoky Mountains
Sorex (Otiosorex) haydeni
Prairie shrew, musaraña de las praderas
Sorex (Otiosorex) hoyi
American pygmy shrew, musaraña pigmea
Sorex (Otiosorex) ixtlanensis
Ixtlán shrew, musaraña de Ixtlán
Sorex (Otiosorex) jacksoni
Saint Lawrence Island shrew, musaraña de la isla Saint Lawrence
Sorex (Otiosorex) longirostris
Southeastern shrew, musaraña del suroeste
Sorex (Otiosorex) lyelli
Mount Lyell shrew, musaraña del monte Lyell
Sorex (Otiosorex) milleri
Carmen mountain shrew, musaraña de Sierra del Carmen
Sorex (Otiosorex) monticolus
Montane shrew, musaraña oscura
Sorex (Otiosorex) nanus
Dwarf shrew, musaraña enana
Sorex (Otiosorex) navigator
Western water shrew, musaraña de agua occidental
Sorex (Otiosorex) obscurus
Darker shrew, musaraña oscura
Sorex (Otiosorex) oreopolus
Mexican long-tailed shrew, musaraña mexicana

Sorex (Otiosorex) orizabae
Orizaba long-tailed shrew, musaraña de Orizaba
Sorex (Otiosorex) ornatus
Ornate shrew, musaraña ornamentada
Sorex (Otiosorex) pacificus
Pacific shrew, musaraña del Pacífico
Sorex (Otiosorex) palustris
Water shrew, musaraña de agua Americana
Sorex (Otiosorex) preblei
Preble's shrew, musaraña del noroeste
Sorex (Otiosorex) pribilofensis
Pribilof Island shrew, musaraña de la isla Pribilof
Sorex (Otiosorex) rohweri
Olympic shrew, musaraña de la península Olimpia
Sorex (Otiosorex) sonomae
Fog shrew, musaraña de la niebla
Sorex (Otiosorex) tenellus
Inyo shrew, musaraña de las montañas Inyo
Sorex (Otiosorex) ugyunak
Barren ground shrew, musaraña del Ártico
Sorex (Otiosorex) vagrans
Vagrant shrew, musaraña vagabunda
Sorex (Sorex) arcticus
Arctic shrew, musaraña del Ártico
Sorex (Sorex) maritimensis
Maritime shrew, musaraña de Nueva Escocia
Sorex (Sorex) minutissimus
Lesser shrew, Musaraña menor
Sorex (Sorex) tundrensis
Tundra shrew, musaraña de la tundra
Sorex (incertae sedis) arizonae
Arizona shrew, musaraña de Arizona
Sorex (incertae sedis) altoensis
Mexican highlands shrew, musaraña de las tierras altas de México
Sorex (incertae sedis) chiapensis
Chiapas shrew, musaraña de Chiapas
Sorex (incertae sedis) cristobalensis
Cristobal shrew, musaraña de Cristóbal
Sorex (incertae sedis) macrodon
Large-toothed shrew, musaraña de dientes grande
Sorex (incertae sedis) mediopua
Jalisco shrew, musaraña de Jalisco
Sorex (incertae sedis) merriami
Merriam's shrew, musaraña del Great Basin
Sorex (incertae sedis) mutabilis
Mutable shrew, musaraña de Oaxaca
Sorex (incertae sedis) salvini
Salvini's shrew, musaraña de la Sierra Madre del Sur
Sorex (incertae sedis) saussurei
Saussure's shrew, musaraña común de montaña
Sorex (incertae sedis) sclateri
Sclater's shrew, musaraña selvática
Sorex (incertae sedis) stizodon
San Cristobal shrew, musaraña de San Cristóbal de las Casas

Sorex (incertae sedis) trowbridgii
Trowbridge's shrew, musaraña de la costa oeste
Sorex (incertae sedis) ventralis
Chestnut-bellied shrew, musaraña de vientre castaño
Sorex (incertae sedis) veraecrucis
Veracruz shrew, musaraña de Veracruz

Family Talpidae

Subfamily Scalopinae

Tribe Condylurini

Condylura cristata
Star-nosed mole, topo nariz de estrella

Tribe Scalopini

Parascalops breweri
Hairy-tailed mole, topo de cola peluda
Scalopus aquaticus
Eastern mole, topo de la costa este
Scapanus anthonyi
Anthony's mexican mole, topo de San Pedro Mártir
Scapanus latimanus
Broad-footed mole, topo de patas anchas
Scapanus occultus
West coast mole, topo de la costa oeste
Scapanus orarius
Pacific mole, topo de la costa
Scapanus townsendii
Snow mole, topo de la nieve

Subfamily Talpinae

Tribe Neurotrichini

Neurotrichus gibbsii
Shrew mole, topo musaraña

Order Lagomorpha

Family Leporidae

*Lepus (Eulago) europaeus**
European hare, liebre europea
Lepus (Lepus) arcticus
Arctic hare, liebre del ártico
Lepus (Lepus) othus
Alaskan hare, liebre de Alaska
Lepus (Macrotolagus) alleni
Antelope jackrabbit, liebre antílope
Lepus (Proeulagus) altamirae
Tamaulipas jackrabbit, liebre de Tamaulipas

Lepus (Poecitolagus) americanus
Snowshoe hare, liebre de las nieves
Lepus (Proeulagus) californicus
Black-tailed jackrabbit, liebre de cola negra
Lepus (Proeulagus) callotis
White-sided jackrabbit, liebre de costados blancos
Lepus (Proeulagus) flavigularis
Tehuantepec jackrabbit, liebre de Tehuantepec
Lepus (Proeulagus) townsendii
White-tailed jackrabbit, liebre de cola blanca
Romerolagus diazi
Teporingo, zacatuche
Sylvilagus aquaticus
Swamp rabbit, conejo de los pantanos
Sylvilagus audubonii
Desert cottontail, conejo del desierto
Sylvilagus bachmani
Brush rabbit, conejo de los arbustos
Sylvilagus cognatus
Manzano mountain cottontail, conejo de las montañas Manzano
Sylvilagus cunicularius
Mexican cottontail, conejo mexicano
*Sylvilagus floridanus***
Eastern cottontail, conejo del este
Sylvilagus gabbi
Tropical cottontail, conejo tropical
Sylvilagus graysoni
Tres Marias islands cottontail, conejo de las Islas Marías
Sylvilagus idahoensis
Pygmy rabbit, conejo pigmeo
Sylvilagus insonus
Omilteme cottontail, conejo de Omilteme
Sylvilagus nuttallii
Mountain cottontail, conejo de las montañas
Sylvilagus obscurus
Appalachian cottontail, conejo de las montañas Apalaches
Sylvilagus palustris
Marsh rabbit, conejo de los humedales
Sylvilagus robustus
Robust cottontail, conejo del suroeste
Sylvilagus transitionalis
New England cottontail, conejo de Nueva Inglaterra

Family Ochotonidae

Ochotona (Pika) collaris
Collared pika, pica de collar
Ochotona (Pika) princeps
American pika, pica americana

Order Perissodactyla

Suborder Ceratomorpha

Family Tapiridae

Tapirus (Tapirella) bairdii
Baird's tapir, tapir Americano

Suborder Hippomorpha

Family Equidae

Tribe Equini

*Equus ferus**
Horse, caballo
*Equus africanus**
Donky, burro

Order Pilosa

Family Cyclopedidae

Cyclopes dorsalis
Silky anteater, osito hormiguero dorado

Family Myrmecophagidae

Tamandua mexicana
Northern tamandua, oso hormiguero

Order Primates

Suborder Haplorhini

Family Atelidae

Subfamily Alouattinae

Alouatta palliata
Brown howler, mono aullador café
Alouatta pigra
Black howler, mono aullador negro

Subfamily Atelinae

Ateles geoffroyi
Spider monkey, mono araña

Family Cercopithecidae

Tribe Papionini

*Macaca mulatta**
Rhesus monkey, mono resus

Order Rodentia

Suborder Eusciurida

Family Aplodontiidae

Aplodontia rufa
Sewellel, castor de montaña

Family Sciuridae

Subfamily Sciurinae

Tribe Pteromyini

Glaucomys oregonensis
Humboldt's flying squirrel, Ardilla voladora de Oregon
Glaucomys sabrinus
Northern flying squirrel; ardilla voladora del norte
Glaucomys volans
Southern flying squirrel, ardilla voladora del sur

Tribe Sciurini

Sciurus (Echinosciurus) aureogaster
Red-bellied squirrel, ardilla de vientre rojo
Sciurus (Echinosciurus) colliaei
Collie's squirrel, ardilla de la Sierra Madre Occidental
Sciurus (Echinosciurus) deppei
Deppe's squirrel, ardilla tropical
Sciurus (Echinosciurus) variegatoides
Variegated squirrel, ardilla centroamericana
Sciurus (Echinosciurus) yucatanensis
Yucatán squirrel, ardilla de Yucatán
Sciurus (Hesperosciurus) aberti
Abert's squirrel, ardilla orejas peludas
Sciurus (Hesperosciurus) griseus
Western gray squirrel, ardilla gris del oeste
Sciurus (Neosciurus) carolinensis
Eastern gray squirrel, ardilla gris del este
Sciurus (Parasciurus) alleni
Allen's squirrel, ardilla de la Sierra Madre Oriental
Sciurus (Parasciurus) arizonensis
Arizona gray squirrel, ardilla gris de Arizona
Sciurus (Parasciurus) nayaritensis
Mexican fox squirrel, ardilla de Nayarit
Sciurus (Parasciurus) niger
Eastern fox squirrel, ardilla zorro del este

Sciurus (Parasciurus) oculatus
Peters's squirrel, ardilla mexicana
Tamiasciurus douglasii
Douglas' squirrel, ardilla de la costa oeste
Tamiasciurus fremonti
Fremont's squirrel, ardilla roja de Nuevo Mexico
Tamiasciurus hudsonicus
Red squirrel, ardilla roja

Subfamily Xerinae

Tribe Marmotini

Ammospermophilus harrisii
Harris's antelope squirrel, ardilla listada sonorense
Ammospermophilus interpres
Texas antelope squirrel, ardilla listada texana
Ammospermophilus leucurus
White-tailed antelope squirrel, juancito de cola blanca
Ammospermophilus nelsoni
Nelson's antelope squirrel, ardilla listada del Valle de San Joaquín
Callospermophilus lateralis
Golden-mantled ground squirrel, ardilla de manto dorada
Callospermophilus madrensis
Sierra Madre ground squirrel, ardilla de manto de la Sierra Madre
Callospermophilus saturatus
Cascade golden-mantled ground squirrel, ardilla de manto de las montañas Cascadas
Cynomys (Cynomys) ludovicianus
Black-tailed prairie dog, perrito de las praderas colinegra
Cynomys (Cynomys) mexicanus
Mexican prairie dog, perrito de las praderas mexicano
Cynomys (Leucocrossuromys) gunnisoni
Gunnison's prairie dog, Perrito de las praderas de la planicie de Colorado
Cynomys (Leucocrossuromys) leucurus
White-tailed prairie dog, perrito de las praderas coliblanca
Cynomys (Leucocrossuromys) parvidens
Utah prairie dog, perrito de las praderas de Utah
Ictidomys mexicanus
Mexican ground ssquirrel, ardilla manchada mexicana
Ictidomys parvidens
Rio Grande ground squirrel, ardilla manchada texana
Ictidomys tridecemlineatus
Thirteen-lined ground squirrel, ardilla manchada de trece líneas
Marmota (Marmota) broweri
Alaska marmot, marmota de Alaska
Marmota (Marmota) monax
Woodchuck, marmota americana
Marmota (Petromarmota) caligata
Hoary marmot, marmota canosa
Marmota (Petromarmota) flaviventris
Yellow-bellied marmot, marmota de vientre amarillo
Marmota (Petromarmota) olympus
Olympic marmot, marmota de la península de Olimpia
Marmota (Petromarmota) vancouverensis
Vancouver island marmot, marmota de isla Vancouver

Notocitellus adocetus
Tropical ground squirrel, cuinique o juancito del Balsas
Notocitellus annulatus
Ring-tailed ground squirrel, ardilla de tierra cola anillada
Otospermophilus beecheyi
California rock squirrel, ardillón de California
Otospermophilus douglasii
Douglas's rock squirrel, ardillón del norte de California
Otospermophilus variegatus
Rock squirrel, ardillón común
Poliocitellus franklinii
Franklin's ground squirrel, ardilla de tierra gris
Urocitellus armatus
Uinta ground squirrel, ardilla de tierra de las montañas Uinta
Urocitellus beldingi
Belding's ground squirrel, ardilla de tierra
Urocitellus brunneus
Idaho ground squirrel, ardilla de tierra de Idaho
Urocitellus canus
Merriam's ground squirrel, ardilla de tierra canosa
Urocitellus columbianus
Columbian ground squirrel, ardilla de tierra de la Columbia Británica
Urocitellus elegans
Wyoming ground squirrel, ardilla de tierra de Wyoming
Urocitellus mollis
Piute ground squirrel, ardilla de tierra del pueblo Piute
Urocitellus parryii
Arctic ground squirrel, ardilla de tierra del ártico
Urocitellus richardsonii
Richardson's ground squirrel, ardilla de tierra
Urocitellus townsendii
Townsend's ground squirrel, ardilla de tierra del Río Columbia
Urocitellus washingtoni
Washington ground squirrel, ardilla de tierra de Washington
Xerospermophilus mohavensis
Mohave ground squirrel, ardilla de tierra de desierto del Mohave
Xerospermophilus perotensis
Perote ground squirrel, ardilla de tierra de Perote
Xerospermophilus spilosoma
Spotted ground squirrel, ardilla de tierra de desierto manchada
Xerospermophilus tereticaudus
Round-tailed ground squirrel, ardilla de tierra de desierto de cola redonda

Tribe Tamiini

Neotamias alpinus
Alpine chipmunk, chichimoco alpino
Neotamias amoenus
Yellow-pine chipmunk, chichimoco del pino ponderosa
Neotamias bulleri
Buller's chipmunk, chichimoco de Zacatecas
Neotamias canipes
Gray-footed chipmunk, chichimoco de patas grises
Neotamias cinereicollis
Gray-collared chipmunk, chichimoco de collar gris

Neotamias cratericus
Craters of the moon chipmunk, chichimoco de los cráteres de la luna
Neotamias dorsalis
Cliff chipmunk, chichimoco de los acantilados
Neotamias durangae
Durango chipmunk, chichimoco de Durango
Neotamias grisescens
Least gray chipmunk, chichimoco pigmeo gris
Neotamias merriami
Merriam's chipmunk, chichimoco del sur de California
Neotamias minimus
Least chipmunk, chichimoco pigmeo
Neotamias obscurus
California chipmunk, chichimoco de California
Neotamias ochrogenys
Yellow-cheeked chipmunk, chichimoco de cachetes amarillos
Neotamias palmeri
Palmer's chipmunk, chichimoco de Spring Mountains
Neotamias panamintinus
Panamint chipmunk, chichimoco del Valle Panamint
Neotamias quadrimaculatus
Long-eared chipmunk, chichimoco de orejas grandes
Neotamias quadrivittatus
Colorado chipmunk, chichimoco de Colorado
Neotamias ruficaudus
Red-tailed chipmunk, chichimoco de cola roja
Neotamias rufus
Hopi chipmunk, chichimoco de los Hopi
Neotamias senex
Allen's chipmunk, chichimoco de la Sierra Nevada
Neotamias siskiyou
Siskiyou chipmunk, chichimoco de las montañas Siskiyou
Neotamias solivagus
Howell's chipmunks, chichimoco de Coahuila
Neotamias sonomae
Sonoma chipmunk, chichimoco de Sonoma
Neotamias speciosus
Lodgepole chipmunk, chichimoco del pino
Neotamias townsendii
Townsend's chipmunk, chichimoco del noroeste
Neotamias umbrinus
Uinta chipmunk, chichimoco común
Tamias striatus
Eastern chipmunk, chichimoco del este

Suborder Ctenohystrica

Infraorder Hystricognathi

Family Agoutidae

Cuniculus paca
Lowland paca, tepezcuintle

Family Dasyproctidae

Dasyprocta mexicana
Mexican agouti, guaqueque negro
Dasyprocta punctata
Central American agouti, guaqueque

Family Echimyidae

Tribe Myocastorini

*Myocastor coypus**
Nutria, coipú

Family Erethizontidae

Subfamily Erethizontinae

Coendou (Coendou) mexicanus
Mexican hairy dwarf porcupine, puercoespín tropical
Erethizon dorsatum
North American porcupine, puercoespín norteamericano

Suborder Supramyomorpha

Family Castoridae

Castor canadensis
American beaver, castor

Family Cricetidae

Subfamily Arvicolinae

Tribe Arvicolini

Alexandromys oeconomus
Tundra vole, ratón colicorto nórdico
Lemmiscus curtatus
Sagebrush vole, chincolo de las artemisas
Microtus (Mynomes) canicaudus
Gray-tailed vole, ratón colicorto de cola gris
Microtus (Mynomes) drummondii
Drummond meadow vole, ratón colicorto del oeste
Microtus (Mynomes) dukecampbelli
Florida meadow vole, ratón colicorto de Florida
Microtus (Mynomes) montanus
Montane vole, ratón colicorto de las montañas
Microtus (Mynomes) ochrogaster
Prairie vole, ratón colicorto de las praderas
Microtus (Mynomes) oregoni
Creeping vole, ratón colicorto de Oregon
Microtus (Mynomes) pennsylvanicus
Meadow vole, ratón colicorto de los pastizales

Microtus (Mynomes) townsendii
Townsend's vole, ratón colicorto de la costa noroeste
Microtus (Pitymys) guatemalensis
Guatemalan vole, ratón colicorto de Guatemala
Microtus (Pitymys) oaxacensis
Tarabundí vole, ratón colicorto de Oaxaca
Microtus (Pitymys) pinetorum
Woodland vole, ratón colicorto de los bosques
Microtus (Pitymys) quasiater
Jalapan pine vole, ratón colicorto de Jalapa
Microtus (incertae sedis) abbreviatus
Insular vole, ratón colicorto de la Isla San Mateo
Microtus (incertae sedis) californicus
California vole, ratón colicorto de California
Microtus (incertae sedis) chrotorrhinus
Rock vole, ratón colicorto de las rocas
Microtus (incertae sedis) longicaudus
Long-tailed vole, ratón colicorto de cola larga
Microtus (incertae sedis) mexicanus
Mexican vole, ratón colicorto mexicano
Microtus (incertae sedis) mogollonensis
Mogollon vole, ratón colicorto de Mogollon
Microtus (incertae sedis) richardsoni
North American water vole, ratón colicorto de agua
Microtus (incertae sedis) umbrosus
Zempoaltepec vole, ratón colicorto del cerro Zempoaltepec
Microtus (incertae sedis) xanthognathus
Taiga vole, ratón colicorto de la taiga

Tribe Clethrionomyini

Clethrionomys californicus
Western red-backed vole, chincolo de lomo rojo del oeste
Clethrionomys gapperi
Southern red-backed vole, chincolo de lomo rojo del sur
Clethrionomys rutilus
Northern red-backed vole, chincolo de lomo rojo del norte

Tribe Dicrostonychini

Dicrostonyx (Dicrostonyx) groenlandicus
Northern collared lemming, leming de tundra común
Dicrostonyx (Dicrostonyx) hudsonius
Ungava collared lemming, leming de tundra de Ungava
Dicrostonyx (Dicrostonyx) nelsoni
Nelson's collared lemming, leming de tundra de Alaska
Dicrostonyx (Dicrostonyx) nunatakensis
Ogilvie Mountains collared lemming, leming de tundra del Yukon
Dicrostonyx (Dicrostonyx) richardsoni
Richardson's collared lemming, leming de tundra de la Bahía Hudson
Dicrostonyx (Dicrostonyx) unalascensis
Umnak island collared lemming, leming de tundra de la isla Umnak

Tribe Lemmini

Lemmus nigripes
Western brown lemming, leming pardo del oeste
Lemmus trimucronatus
brown lemming, leming pardo
Synaptomys (Mictomys) borealis
Northern bog lemming, leming de pantano del norte
Synaptomys (Synaptomys) cooperi
Southern bog lemming, leming de pantano del sur

Tribe Ondatrini

Neofiber alleni
Round-tailed muskrat, rata almizclera de cola redonda
Ondatra zibethicus
Muskrat, rata almizclera

Tribe Phenacomyini

Arborimus albipes
White-footed vole, chincolo arborícola de patas blancas
Arborimus longicaudus
Red tree vole, chincolo arborícola de las sequoyas
Arborimus pomo
Sonoma tree vole, chincolo arborícola de Sonoma
Phenacomys intermedius
Western heather vole, campañol del oeste
Phenacomys ungava
Eastern heather vole, campañol del este

Subfamily Neotominae

Tribe Baiomyini

Baiomys brunneus
Southern pygmy mouse, ratón pigmeo tropical
Baiomys musculus
Jalisco pygmy mouse, ratón pigmeo de Jalisco
Baiomys taylori
Northern pygmy mouse, ratón pigmeo norteño
Scotinomys teguina
Short-tailed singing mouse, ratón pigmeo negro

Tribe Neotomini

Habromys chinanteco
Chinanteco crested-tailed mouse, ratón arbóreo de Oaxaca
Habromys delicatulus
Delicate crested-tailed mouse, ratón arbóreo delicado
Habromys ixtlani
Ixtlán crested-tailed mouse, ratón arbóreo de Ixtlán
Habromys lepturus
Zempoaltépec crested-tailed mouse, ratón arbóreo de Zempoaltépec
Habromys lophurus
Crested-tailed mouse, ratón arbóreo de cola crestada
Habromys schmidlyi
Schmidly crested-tailed mouse, ratón arbóreo de Guerrero

Habromys simulatus
Jico crested-tailed mouse, ratón arbóreo de Jico
Hodomys alleni
Allen's woodrat, rata de campo del occidente
Megadontomys cryophilus
Oaxaca giant deermouse, ratón gigante de Oaxaca
Megadontomys nelsoni
Nelson's giant deermouse, ratón gigante de Veracruz
Megadontomys thomasi
Thomas's giant deermouse, ratón gigante de Guerrero
Nelsonia goldmani
Nelson and Goldman woodrat, rata enana de Michoacán
Nelsonia neotomodon
Diminutive woodrat, rata enana del oeste
Neotoma (Neotoma) albigula
White-throated woodrat, rata de campo de garganta blanca
Neotoma (Neotoma) angustapalata
Tamaulipan woodrat, rata de campo de Tamaulipas
Neotoma (Neotoma) bryanti
Bryant's woodrat, rata de campo de Baja California
Neotoma (Tenoma) cinerea
Bushy-Tailed woodrat, rata de campo de cola peluda
Neotoma (Neotoma) devia
Arizona woodrat, rata de campo de Arizona
Neotoma (Neotoma) ferruginea
Guatemala woodrat, rata de campo de Guatemala
Neotoma (Neotoma) floridana
Eastern woodrat, rata de campo del este
Neotoma (Neotoma) fuscipes
Dusky-footed woodrat, rata de campo de patas negras
Neotoma (Neotoma) goldmani
Goldman´s woodrat, rata de campo del altiplano
Neotoma (Neotoma) insularis
Ángel de la Guarda Island woodrat, rata de campo de Isla Ángel de la Guarda
Neotoma (Neotoma) lepida
Desert woodrat, rata de campo del desierto
Neotoma (Neotoma) leucodon
White-toothed woodrat, rata de campo de dientes blancos
Neotoma (Neotoma) macrotis
Big-eared woodrat, rata de campo de orejas grandes
Neotoma (Neotoma) magister
Appalachian woodrat, rata de campo de los Appalachian
Neotoma (Neotoma) melanura
Dark woodrat, rata de campo negruzca
Neotoma (Neotoma) mexicana
Mexican woodrat, rata magueyera mexicana
Neotoma (Neotoma) micropus
Southern Plain woodrat, rata de campo de las planicies del sur
Neotoma (Neotoma) nelsoni
Nelson's woodrat, rata de campo de Perote
Neotoma (Neotoma) palatina
Bolaños woodrat, rata de campo de Jalisco
Neotoma (Teanopus) phenax
Sonoran woodrat, rata de campo de Sonora
Neotoma (Neotoma) picta
Tehuantepec woodrat, rata de campo de Tehuantepec

Neotoma (Neotoma) stephensi
Stephen's woodrat, rata de campo de Arizona
Neotomodon alstoni
Mexican Volcano mouse, Ratón de los Volcanes
Onychomys arenicola
Chihuahua grasshopper mouse, ratón chapulinero de Chihuahua
Onychomys leucogaster
Northern grasshopper mouse, ratón chapulinero del norte
Onychomys torridus
Southern grasshopper mouse, ratón chapulinero del sur
Osgoodomys banderanus
Michoacán deermouse, ratón de Michoacán
Peromyscus (Haplomylomys) avius
Cerralvo island deermouse, ratón de isla Cerralvo
Peromyscus (Haplomylomys) californicus
California deermouse, ratón de California
Peromyscus (Haplomylomys) caniceps
Montserrat island deermouse, ratón de Isla Montserrat
Peromyscus (Haplomylomys) dickeyi
Tortuga deermouse, ratón de la Isla Tortuga
Peromyscus (Haplomylomys) eremicus
Cactus deermouse, ratón de los cactus
Peromyscus (Haplomylomys) eva
Baja California deermouse, ratón de Baja California
Peromyscus (Haplomylomys) fraterculus
Northern Baja deermouse, ratón de las Californias
Peromyscus (Haplomylomys) guardia
Ángel de la Guarda island deermouse, ratón de la isla Ángel de la Guarda
Peromyscus (Haplomylomys) hooperi
Hooper's deermouse, ratón de Coahuila
Peromyscus (Haplomylomys) merriami
Mesquite deermouse, ratón de los mezquites
Peromyscus (Haplomylomys) pembertoni †
San Pedro Nolasco deermouse, ratón de Isla San Pedro Nolasco
Peromyscus (Peromyscus) amplus
Highlands rock deermouse, ratón de las rocas de las tierras altas
Peromyscus (Peromyscus) angelensis
Angel deermouse, ratón de los ángeles
Peromyscus (Peromyscus) attwateri
Texas deermouse, ratón de Texas
Peromyscus (Peromyscus) aztecus
Aztec deermouse, ratón azteca
Peromyscus (Peromyscus) beatae
Orizaba deermouse, ratón de Orizaba
Peromyscus (Peromyscus) boylii
Brush deermouse, ratón de los matorrales
Peromyscus (Peromyscus) bullatus
Perote deermouse, ratón de Perote
Peromyscus (Peromyscus) carletoni
Carleton's deermouse, ratón de Nayarit
Peromyscus (Peromyscus) carolpattonae
Carol Patton's deermouse, ratón de El Triunfo
Peromyscus (Peromyscus) chrysopus
Coastal tawny deermouse, ratón leonado de la costa

Peromyscus (Peromyscus) collinus
Tamaulipas white-ankled deermouse, ratón de ancas blancas de Tamaulipas
Peromyscus (Peromyscus) cordillerae
Monte Cacaguatique white-footed Mouse, ratón del Monte Cacaguatique
Peromyscus (Peromyscus) crinitus
Canyon deermouse, ratón de las cañadas
Peromyscus (Peromyscus) difficilis
Southern rock deermouse, ratón de las rocas del sur
Peromyscus (Peromyscus) ensinki
Zinapécuaro deermouse, ratón de Zinapécuaro
Peromyscus (Peromyscus) felipensis
Oaxaca rock deermouse, ratón de las rocas de Oaxaca
Peromyscus (Peromyscus) furvus
Blackish deermouse, ratón negruzco
Peromyscus (Peromyscus) gambelii
Pacific common deermouse, ratón común del Pacífico
Peromyscus (Peromyscus) gossypinus
Cotton mouse, ratón del algodón
Peromyscus (Peromyscus) gratus
Tlalpan deermouse, ratón piñonero mexicano
Peromyscus (Peromyscus) greenbaumi
Coalcoman deermouse, ratón de Coalcomán
Peromyscus (Peromyscus) gymnotis
Naked-eared deermouse, ratón de orejas desnudas
Peromyscus (Peromyscus) hylocetes
Transvolcanic deermouse, ratón del Eje Volcánico
Peromyscus (Peromyscus) keeni
Northwestern deermouse, ratón del noroeste
Peromyscus (Peromyscus) kilpatricki
Kilpatrick's deermouse, ratón de Zitácuaro
Peromyscus (Peromyscus) labecula
Mexican common deermouse, ratón común de México
Peromyscus (Peromyscus) laceianus
Northern white-ankled deermouse, ratón de ancas blancas norteño
Peromyscus (Peromyscus) latirostris
Northern blackish deermouse, ratón negruzco del norte
Peromyscus (Peromyscus) leucopus
White-footed deermouse, ratón de patas blancas
Peromyscus (Peromyscus) leucurus
Tehuantepec deermouse, ratón de Tehuantepec
Peromyscus (Peromyscus) levipes
Nimble-footed deermouse, ratón de patas ágiles
Peromyscus (Peromyscus) madrensis
Tres Marías island deermouse, ratón de Isla Tres Marías
Peromyscus (Peromyscus) maniculatus
Northeastern common deermouse, ratón común del noreste
Peromyscus (Peromyscus) megalops
Brown deermouse, ratón de rostro ancho
Peromyscus (Peromyscus) mekisturus
Puebla deermouse, ratón de Puebla
Peromyscus (Peromyscus) melanocarpus
Zempoaltepec deermouse, ratón de Zempoaltepec
Peromyscus (Peromyscus) melanophrys
Black-eyed deermouse, ratón de máscara

Peromyscus (Peromyscus) melanotis
Black-eared deermouse, ratón de orejas negras
Peromyscus (Peromyscus) melanurus
Black-tailed deermouse, ratón de cola negra
Peromyscus (Peromyscus) mexicanus
Mexican deermouse, ratón mexicano
Peromyscus (Peromyscus) micropus
Small-footed deermouse, ratón de patas pequeñas
Peromyscus (Peromyscus) nasutus
Northern rock deermouse, ratón de las rocas del norte
Peromyscus (Peromyscus) ochraventer
Brown belly deermouse, ratón de vientre pardo
Peromyscus (Peromyscus) pectoralis
White-ankled deermouse, ratón de ancas blancas
Peromyscus (Peromyscus) perfulvus
Tawny deermouse, ratón leonado
Peromyscus (Peromyscus) polionotus
Beach mouse, ratón del sureste
Peromyscus (Peromyscus) polius
Chihuahuan mouse, ratón de Chihuahua
Peromyscus (Peromyscus) purepechus
Purepecha deermouse, ratón purepecha
Peromyscus (Peromyscus) sagax
Michoacán deermouse, ratón de Sahuayo
Peromyscus (Peromyscus) schmidlyi
Schmidly's deermouse, ratón de la Sierra Madre Occidental
Peromyscus (Peromyscus) sejugis
Santa Cruz Island deermouse, ratón de Isla Santa Cruz
Peromyscus (Peromyscus) simulus
Nayarit deermouse, ratón de Nayarit
Peromyscus (Peromyscus) slevini
Slevin's deermouse, ratón de la Isla Catalina
Peromyscus (Peromyscus) sonoriensis
North American common deermouse, ratón común de Norteamérica
Peromyscus (Peromyscus) spicilegus
Gleaning deermouse, ratón de las espigas
Peromyscus (Peromyscus) stephani
San Esteban island deermouse, ratón de Isla San Esteban
Peromyscus (Peromyscus) totontepecus
Totontepec deermouse, ratón de Totontepec
Peromyscus (Peromyscus) truei
Pinyon deermouse, ratón piñonero del norte
Peromyscus (Peromyscus) winkelmanni
Winkelmann's deermouse, ratón de Guerrero
Peromyscus (Peromyscus) yucatanicus
Yucatán deermouse, ratón de Yucatán
Peromyscus (Peromyscus) zamorae
Zamora deermouse, ratón de Zamora
Peromyscus (Peromyscus) zarhynchus
Chiapan deermouse, ratón de Chiapas
Podomys floridanus
Florida deermouse, ratón de Florida
Reithrodontomys (Aprodon) albilabris
Oaxaca small-toothed harvest mouse, ratón de dientes pequeños de Oaxaca
Reithrodontomys (Aprodon) bakeri
Baker's harvest mouse, ratón de las cosechas de Guerrero

Reithrodontomys (Aprodon) chrysopsis
Volcano harvest mouse, ratón de las cosechas de los volcanes
Reithrodontomys (Aprodon) gracilis
Slender harvest mouse, ratón de las cosechas esbelto
Reithrodontomys (Aprodon) mexicanus
Mexican harvest mouse, ratón de las cosechas mexicano
Reithrodontomys (Aprodon) microdon
Small-toothed harvest mouse, ratón de las cosechas de dientes pequeños
Reithrodontomys (Aprodon) spectabilis
Cozumel harvest mouse, ratón de Isla Cozumel
Reithrodontomys (Aprodon) tenuirostris
Narrow-nosed harvest mouse, ratón de las cosechas de Guatemala
Reithrodontomys (Aprodon) wagneri
Michoacán small-toothed harvest mouse, ratón de dientes pequeños de Michoacán
Reithrodontomys (Reithrodontomys) burti
Sonoran harvest mouse, ratón de las cosechas sonorense
Reithrodontomys (Reithrodontomys) fulvescens
Fulvous harvest mouse, ratón de las cosechas leonado
Reithrodontomys (Reithrodontomys) hirsutus
Hairy harvest mouse, ratón de las cosechas peludo
Reithrodontomys (Reithrodontomys) humulis
Eastern harvest mouse, ratón de las cosechas del este
Reithrodontomys (Reithrodontomys) megalotis
Western harvest mouse, ratón de las cosechas común
Reithrodontomys (Reithrodontomys) montanus
Plains harvest mouse, ratón de las cosechas de las planicies
Reithrodontomys (Reithrodontomys) raviventris
Salt-Marsh harvest mouse, ratón de las cosechas de la Bahía de San Francisco
Reithrodontomys (Reithrodontomys) sumichrasti
Sumichrast's harvest mouse, ratón de las cosechas de montaña
Reithrodontomys (Reithrodontomys) zacatecae
Zacatecas harvest mouse, ratón de las cosechas de Zacatecas
Xenomys nelsoni
Magdalena woodrat, rata de Magdalena

Tribe Ochrotomyini

Ochrotomys nuttalli
Golden mouse, ratón dorado

Subfamily Sigmodontinae

Tribe Ichthyomyini

Rheomys mexicanus
Mexican water mouse, ratita nadadora mexicana
Rheomys thomasi
Thomas's water mouse, ratita nadadora de Chiapas

Tribe Oryzomyini

Handleyomys alfaroi
Alfaro's rice rat, ratón de agua tropical
Handleyomys chapmani
Chapman's rice rat, ratón de agua de la Sierra Madre
Handleyomys guerrerensis
Guerrero rice rat, ratón de agua de Guerrero

Handleyomys melanotis
Black-eared rice rat, ratón de agua de orejas negras
Handleyomys rhabdops
Highland rice rat, ratón de agua de las tierras altas
Handleyomys rostratus
Long-nosed rice rat, ratón de agua de nariz larga
Handleyomys saturatior
Cloud forest rice rat, ratón de agua del bosque de niebla
Oligoryzomys fulvescens
Flavous pigmy rice rat, ratón de agua pigmeo
Oryzomys albiventer
White-bellied rice rat, rata arrocera de vientre blanco
Oryzomys couesi
Coues's rice rat, rata arrocera del Golfo
Oryzomys fulgens
Thomas's rice rat, rata arrocera del Valle de México
Oryzomys mexicanus
Mexican rice rat, rata arrocera mexicana
Oryzomys nelsoni
Nelson's rice rat, rata arrocera de las islas Tres Marías
Oryzomys palustris
Marsh rice rat, rata arrocera de los pantanos
Oryzomys peninsulae †
Lower California rice rat, rata arrocera de Baja California
Oryzomys texensis
Texas rice rat, rata arrocera de Texas

Tribe Sigmodontini

Sigmodon (Sigmodon) alleni
Allen's cotton rat, rata cañera del occidente
Sigmodon (Sigmodon) arizonae
Arizona cotton rat, rata cañera de Arizona
Sigmodon (Sigmodon) fulviventer
Tawny-bellied cotton rat, rata cañera de vientre amarillo
Sigmodon (Sigmodon) hirsutus
Southern cotton rat, rata cañera del sur
Sigmodon (Sigmodon) hispidus
Hispid cotton rat, rata cañera espinosa
Sigmodon (Sigmodon) leucotis
White-eared cotton rat, rata cañera orejas blancas
Sigmodon (Sigmodon) mascotensis
Jalisco cotton, rat, rata cañera de Jalisco
Sigmodon (Sigmodon) ochrognathus
Yellow-nosed cotton rat, rata cañera rostro amarillo
Sigmodon (Sigmodon) toltecus
Toltec cotton rat, rata cañera del Golfo

Subfamily Tylomyinae

Nyctomys sumichrasti
Sumichrast's vesper rat, rata del amanecer
Otonyctomys hatti
Yucatán vesper rat, rata orejuda trepadora de Yucatán
Ototylomys chiapensis
La Pera big-eared climbing rat, rata orejuda trepadora de Chiapas

Ototylomys phyllotis
Big-eared climbing rat, rata orejuda trepadora de común
Tylomys bullaris
Chiapan climbing rat, rata de árbol de Chiapas
Tylomys nudicaudus
Peters's climbing rat, rata de árbol común
Tylomys tumbalensis
Tumbalá climbing rat, rata de árbol de Tumbalá

Family Muridae

*Mus musculus**
House mouse, ratón de casa
*Rattus norvegicus**
Brown rat, rata café
*Rattus rattus**
House rat, rata gris

Family Geomyidae

Subfamily Geomyinae

Tribe Geomyni

Cratogeomys castanops
Yellow-faced pocket gopher, tuza de cara amarilla
Cratogeomys fulvescens
Oriental Basin pocket gopher, tuza de la Cuenca de Oriental
Cratogeomys fumosus
Smoky pocket gopher, tuza del centro de México
Cratogeomys goldmani
Goldman's pocket gopher, tuza del altiplano
Cratogeomys merriami
Merriam's pocket gopher, tuza del Valle de México
Cratogeomys perotensis
Perote pocket gopher, tuza de Perote
Cratogeomys planiceps
Toluca Volcano pocket gopher, tuza del Nevado de Toluca
Geomys arenarius
Desert pocket gopher, tuza del desierto
Geomys attwateri
Attwater's pocket gopher, tuza de Corpus Christi
Geomys breviceps
Baird's pocket gopher, tuza texana del este
Geomys bursarius
Plains pocket gopher, tuza de las planicies
Geomys jugossicularis
Hall's pocket gopher, tuza de las planicies centrales
Geomys knoxjonesi
Jones's pocket gopher, tuza texana del oeste
Geomys lutescens
Sandy Hill pocket gopher, tuza arenera de Sand Hills
Geomys personatus
Texas pocket gopher, tuza texana del sur

Geomys pinetis
Southeastern pocket gopher, tuza del sureste
Geomys streckeri
Strecker's pocket gopher, tuza texana del Carrizo
Geomys texensis
Llano pocket gopher, tuza del centro de Texas
Geomys tropicalis
Tropical pocket gopher, tuza tropical
Heterogeomys hispidus
Hispid pocket gopher, tuza gigante trópical
Heterogeomys lanius
Big pocket gopher, tuza gigante Veracruzana
Orthogeomys grandis
Giant pocket gopher, tuza gigante del pacífico
Pappogeomys bulleri
Buller's pocket gopher, tuza de Jalisco
Zygogeomys trichopus
Michoacán pocket gopher, tuza de Michoacán

Tribe Thomomyini

Thomomys (Megascapheus) atrovarius
Sinaloa pocket gopher, tuza de Sinaloa
Thomomys (Megascapheus) bottae
Botta's pocket gopher, tuza de abazones del norte
Thomomys (Megascapheus) bulbivorus
Camas pocket gopher, tuza del Valle de Camas
Thomomys (Megascapheus) fulvus
Fulvus pocket gopher, tuza del suroeste
Thomomys (Megascapheus) laticeps
Northern California pocket gopher, tuza del norte de California
Thomomys (Megascapheus) nayarensis
Nayar pocket gopher, tuza del Nayar
Thomomys (Megascapheus) nigricans
California pocket gopher, tuza de Baja California
Thomomys (Megascapheus) sheldoni
Sheldoni pocket gopher, tuza de la Sierra Madre
Thomomys (Megascapheus) townsendii
Townsend's pocket gopher, tuza de las montañas del oeste
Thomomys (Megascapheus) umbrinus
Southern pocket gopher, tuza mexicana
Thomomys (Thomomys) clusius
Wyoming pocket gopher, tuza de Wyoming
Thomomys (Thomomys) idahoensis
Idaho pocket gopher, tuza de Idaho
Thomomys (Thomomys) mazama
Western pocket gopher, tuza del oeste
Thomomys (Thomomys) monticola
Mountain pocket gopher, tuza de las montañas
Thomomys (Thomomys) talpoides
Northern pocket gopher, tuza del norte

Family Heteromyidae

Subfamily Dipodomyinae

Dipodomys agilis
Agile kangaroo rat, rata canguro del sur de California
Dipodomys californicus
California kangaroo rat, rata canguro de California
Dipodomys compactus
Gulf Coast kangaroo rat, rata canguro del Golfo
Dipodomys deserti
Desert kangaroo rats, rata canguro del desierto
Dipodomys elator
Texas kangaroo rat, rata canguro de Texas
Dipodomys gravipes
San Quintín kangaroo rat, rata canguro de San Quintín
Dipodomys heermanni
Morro Bay kangaroo rat, rata canguro de Morro Bay
Dipodomys ingens
Giant kangaroo rat, rata canguro gigante
Dipodomys merriami
Merriam's kangaroo rat, rata canguro común de cuatro dedos
Dipodomys microps
Chisel-toothed kangaroo rat, rata canguro con incisivos de cincel
Dipodomys nelsoni
Nelson's kangaroo rat, rata canguro de cola blanca
Dipodomys nitratoides
San Joaquín Valley kangaroo rat, rata canguro del Valle de San Joaquín
Dipodomys ordii
Ord's kangaroo rat, rata canguro común de 5 dedos
Dipodomys ornatus
Plateau kangaroo rat, rata canguro del Altiplano
Dipodomys panamintinus
Panamint kangaroo rat, rata canguro del valle Panamint
Dipodomys phillipsii
Phillips's kangaroo rat, rata canguro del Valle de México
Dipodomys simulans
Dulzura kangaroo rat, rata canguro de Baja California
Dipodomys spectabilis
Banner-tailed kangaroo rat, rata canguro de cola peluda
Dipodomys stephensi
Stephens' kangaroo rat, rata canguro de San Jacinto Valley
Dipodomys venustus
Narrow-faced kangaroo rat, rata canguro de rostro angosto
Microdipodops megacephalus
Dark kangaroo mouse, ratón canguro oscuro
Microdipodops pallidus
Pale kangaroo mouse, ratón canguro pálido

Subfamily Heteromyinae

Heteromys bulleri
Manantlán spiny pocket mouse, ratón espinoso de Manantlán
Heteromys desmarestianus
Desmarest's spiny pocket mouse, ratón espinoso común
Heteromys gaumeri
Gaumer's spiny pocket mouse, ratón espinoso de Yucatán

Heteromys goldmani
Goldman's spiny pocket mouse, ratón espinoso
Heteromys irroratus
Mexican spiny mouse, ratón espinoso mexicano
Heteromys nelsoni
Nelson's spiny pocket mouse, ratón espinoso de Chiapas
Heteromys pictus
Western spiny pocket mouse, ratón espinoso pintado
Heteromys salvini
Salvin's spiny pocket mouse, ratón espinoso de la costa
Heteromys spectabilis
Jaliscan spiny pocket mouse, ratón spinoso de Jalisco

Subfamily Perognathinae

Chaetodipus ammophilus
Dalquet's pocket mouse, ratón de bolsas del Cabo
Chaetodipus arenarius
Sand pocket mouse, ratón de bolsas de las arenas
Chaetodipus artus
Narrow-skulled pocket mouse, ratón de bolsas de cara delgada
Chaetodipus baileyi
Bailey's pocket mouse, ratón de bolsas gigante
Chaetodipus californicus
California pocket mouse, ratón de bolsas de California
Chaetodipus collis
Highland Coarse-haired pocket mouse, ratón de bolsas de las tierras altas
Chaetodipus durangae
Durango Coarse-haired pocket mouse, ratón de bolsas de Durango
Chaetodipus eremicus
Chihuahuan pocket mouse, ratón de bolsas de Chihuahua
Chaetodipus fallax
San Diego pocket mouse, ratón de bolsas de San Diego
Chaetodipus formosus
Long-tailed pocket mouse, ratón de bolsas de cola larga
Chaetodipus goldmani
Goldman's pocket mouse, ratón de bolsas Sonorense
Chaetodipus hispidus
Hispid pocket mouse, ratón de bolsas de los pastizales
Chaetodipus intermedius
Rock pocket mouse, ratón de bolsas de las rocas
Chaetodipus lineatus
Lined pocket mouse, ratón de bolsas Potosino
Chaetodipus nelsoni
Nelson's pocket mouse, ratón de bolsas del Altiplano
Chaetodipus penicillatus
Desert pocket mouse, ratón de bolsas del desierto
Chaetodipus pernix
Sinaloan pocket mouse, ratón de bolsas de Sinaloa
Chaetodipus rudinoris
Baja California pocket mouse, ratón de bolsas de Baja California
Chaetodipus siccus
Cerralvo Island pocket mouse, ratón de bolsas de Isla Cerralvo
Chaetodipus spinatus
Spiny pocket mouse, ratón de bolsas espinoso

Perognathus alticola
White-eared pocket mouse, ratón de abazones de orejas blancas
Perognathus amplus
Arizona pocket mouse, ratón de abazones de Arizona
Perognathus fasciatus
Olive-backed pocket mouse, ratón de abazones de coloración verdosa
Perognathus flavescens
Plains pocket mouse, ratón de abazones de las planicies
Perognathus flavus
Silky pocket mouse, ratón de abazones de pelo sedoso
Perognathus inornatus
San Joaquín pocket mouse, ratón de abazones de San Joaquín
Perognathus longimembris
Little pocket mouse, ratón de abazones pequeño
Perognathus merriami
Merriam's pocket mouse, ratón de abazones del este
Perognathus mollipilosus
Great Basin pocket mouse, ratón de abazones del Great Basin
Perognathus parvus
Columbia Plateau pocket mouse, ratón de abazones de Columbia

Family Zapodidae

Napaeozapus insignis
Woodland jumping mouse, ratón saltarín del bosque
Zapus hudsonius
Meadow jumping mouse, ratón saltarín de la pradera
Zapus princeps
Western jumping mouse, ratón saltarín del oeste
Zapus trinotatus
Pacific jumping mouse, ratón saltarín del Pacífico

Order Sirenia

Family Trichechidae

Trichechus manatus
West indian manatee, manatí

Family Dugongidae

Hydrodamalis gigas †
Steller's sea lion, vaca marina

*Introduced species. ** Species with recent taxonomic changes, see addendum

Class Mammalia and Order Didelphimorphia

Class Mammalia Linnaeus, 1758

The main characteristics used to define the members of the class Mammalia are: having mammary glands (mammae) that allow the female to nourish newborn young with milk; hair present, although confined to early development, even absent in the adult stage, as in some marine mammals; mandibular ramus of the lower jaw made up of a single bone (dentary); lower jaw articulating directly with the skull, without intervention of the quadrate bone; two exoccipital condyles that articulate with the atlas; a diaphragm dividing the pulmonary area; red blood corpuscles not nucleated; "warm blood"; complete double circulation; heart four-chambered; presence of amnion and allantois.

Subclass Theria Parker and Haswell, 1897

Young developing to a point of being recognizable as mammals in the reproductive tract of the female instead of developing in an egg outside the female; platelike teeth and triconodont teeth in general (talon and talonid absent or poorly developed), respectively.

1. Tail very long, generally bare; thumb opposed to the other toes; teeth 50 in total; upper incisors five and lower and upper molars four .... infraclass Metatheria; order Didelphimorphia (p. 40)

1a. Tail of variable length; thumb either opposed to the other toes or not; teeth less than 50 to none; upper incisors and lower and upper molars never more than three .... infraclass Eutheria (p. 57)

Infraclass Metatheria Huxley, 1880

Placenta absent or incomplete; epipubic bones present or rudiments in both sexes; skull with a small braincase and a large preorbital region; nasals expanded posteriorly; jugal bone contributing to the glenoid fossa; tympanic bone small, annular or tubular, rarely fused with other skull bones; auditory bullae, if present, formed entirely or principally from the alisphenoid bone; young born in a notably undeveloped condition and develop while firmly attached to mammae in an abdominal pouch; brain lacking corpus callosum.

S. T. Álvarez-Castañeda, *Mammals of North America - Volume 1*, https://doi.org/10.1007/978-3-031-41661-3_2

Order Didelphimorphia Gill, 1872

The Didelphimorphia includes opossum in North America, characterized by having five toes on each foot; tail more or less prehensile, in many species with less than the proximal half of the tail furred, the remainder naked; hallux of hindfoot clawless, opposable to the other digits; a great number of species with a pouch, others with two folds of skin; five upper subequal and four lower incisors; snout long and pointed; preorbital region long and pointed and braincase small; zygomatic arches large; hard palate with fenestrae near the molars; upper molars tritubercular; lower molars tritubercular with well-developed trigonids; epipubic bones present. Distribution is restricted to America. The only family is Didelphidae.

Family Didelphidae Gray, 1821

The family Didelphidae includes opossums that are characterized by a **marsupium opening anteriorly, if present; tail long, usually mostly naked, and prehensile; hallux well developed, opposable and without claw;** five digits in each limb, all sub-equal in length; **sagittal crest often well developed; lower incisors equal and not procumbent**; skull relatively long and slender; paroccipital process small; canines well developed; molar tubercular; auditory bullae relatively not inflated. Ranges exclusively in America with two subfamilies (Didelphinae and Caluromyinae).

1. Hair abundant and silky; proximal half of tail with hair; ears paler; a black line from nose to forehead; upper toothrow arched outward (Fig. 1; Veracruz and southern Yucatán Peninsula, México south through Honduras).. subfamily Caluromyinae; subgenus *Mallodelphys*; *Caluromys derbianus* (p. 30)

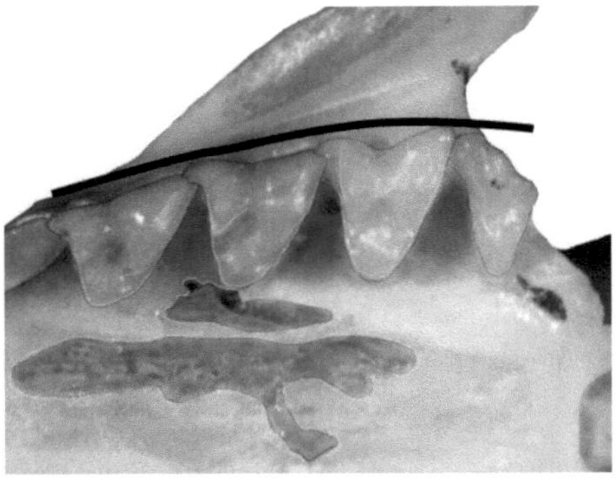

Fig. 1 The upper toothrow arched outward as in *Caluromys derbianus*

1a. Hair bristle in general; tail without hair over all; ears dark; no black line from nose to forehead; upper toothrow almost straight (Fig. 2) .... subfamily Didelphinae (p. 32)

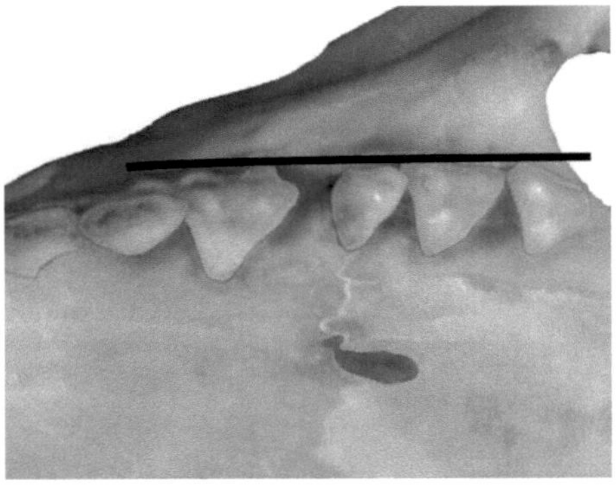

Fig. 2 The upper toothrow almost straight

Subfamily Caluromyinae Kirsch, 1977

Medium size; **tail long with the proximal half haired**; hallux without a claw, well developed, and opposable; five digits in each limb, all sub-equal in length; sagittal crest often well developed; lower incisors equal; skull relatively long and slender; paroccipital process small; canines well developed; **upper toothrow curved and converging anteriorly;** molar tubercular.

Genus ***Caluromys*** Allen, 1900

Medium sized; dorsal coloration reddish brown to pale gray; shoulders, forearms, and hindlegs in a different shade but paler; fur of underparts yellowish to orange, including a dark stripe on the face; pelage long, fine, and wooly; hair extending along almost all of, or more than, the basal third of the tail; tail longer than the head-and-body length; temporal ridges infrequently uniting to form a sagittal crest; postorbital processes well developed; **maxillary toothrows curved and converging anteriorly; third upper molars not larger than the first;** posterior palate lacking fenestrae.

Subgenus ***Mallodelphys*** Thomas, 1920

Tail haired 50% above and up to 20% below; ventral tail bases naked and covered with hard raised tubercles; pouch with lateral skin folds joined posteriorly and opening anteriorly.

Caluromys derbianus (Waterhouse, 1841)

Derby's woolly opossum, tlacuache lanudo

1. *C. d. aztecus* (Thomas, 1913). Puebla, Veracruz, and Oaxaca.
2. *C. d. fervidus* (Thomas, 1913). Campeche, Quintana Roo, Chiapas, and southeastern Tabasco south through Honduras.

Conservation: *Caluromys derbianus* is considered as Least Concern by the IUCN (Solari and Lew 2015).

Characteristics: Woolly opossum is medium sized; total length 600.0–700.0 mm, condylobasal length 60.2–62.5 mm. Dorsal pelage pale gray with **three spots that can even be in orange shades on shoulders, back, and hips, with a paler diamond-shaped spot on the back**; fur of underparts whitish cream; long and pinkish ears; a middle line dark brown on the rostrum; **tail very long, upperpart with hair of the same color as back, tip bare and white**; upperpart of forelimbs white; **upper toothrow arched outward**; last upper molars not bigger than first molars; supraorbital processes well developed; first molariforms very reduced, almost of the size of incisors; **more than two fenestra on the palatal**.

Comments: *Caluromys derbianus* is considered in the subfamily Caluromyinae, previously within the Didelphinae (Hall 1981). It ranges from southern Veracruz, Tabasco, northern Chiapas, southern Campeche, and southeastern Quintana Roo south through Honduras (Map 1). *C. derbianus* is sympatric, or nearly so, with *Chironectes minimus, Didelphis marsupialis, D. virginiana, Philander vossi, Marmosa mexicana, Tlacuatzin canescens*, and *T. gaumeri.* The genus *Caluromys* differs from other genera by having long, dense, and woolly hair with a dark, median, facial stripe often present, extending from the crown to the fleshy portion of the nose; dorsal pelage pale gray with three spots that can even be in orange shades on the shoulders and back. Hip with a paler diamond-shaped spot on the back; tail very long, about 58–67% of the total length and prehensile, upperpart with hair of the same color as the back, tip bare and white; well-developed postorbital processes; temporal ridges occasionally united to form a sagittal crest, possibly as a function of age; long slender canines and third upper molars not larger than the first molars; first upper premolars very small, upper toothrow arched outward and with more than two fenestrae on the palatal.

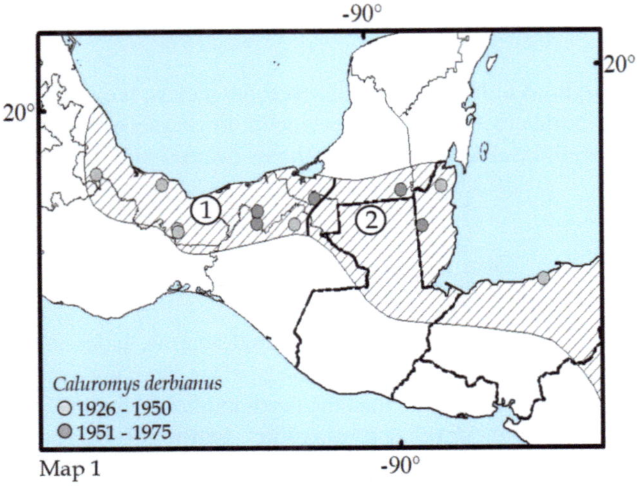

Map 1 Distribution map of *Caluromys derbianus*

Additional Literature: Goodwin (1942b, 1946), Hall and Kelson (1959), Hall and Dalquest (1963), Bucher and Hoffmann (1980), Gardner (2005a), Cáceres and Carmignotto (2006).

Subfamily Didelphinae Gray, 1821

Small to medium size; marsupium opening anteriorly, if present; tail long, naked, and prehensile; hallux without claw, well developed, and opposable; five digits in each limb, all sub-equal in length; sagittal crest often well developed; lower incisors equal size; skull relatively long and slender; paroccipital process small; canines well developed; **upper toothrow almost straight;** molar tubercular; auditory bullae relatively not inflated. The keys were elaborated based on the review of specimens and the following sources: Hall (1981), Jones and Manning (1992), and Álvarez-Castañeda et al. (2015, 2017a).

1. Eyes surrender by black spot, which contrast with head color; females without pouches; total length less than 400.0 mm; skull length less than 50.0 mm .... tribe Marmosini 2

1a. With or without dark spots around the eyes but not contrasting with head color, many times some pale supraocular spots; females with pouches; total length > 400.0 mm; skull length > 50.0 mm 3

2. Dorsal pelage grayish or slightly orange pelage, rarely mixed with cinnamon color; small pit between palatine and molar cleft (Fig. 3) .... *Tlacuatzin* (p. 51)

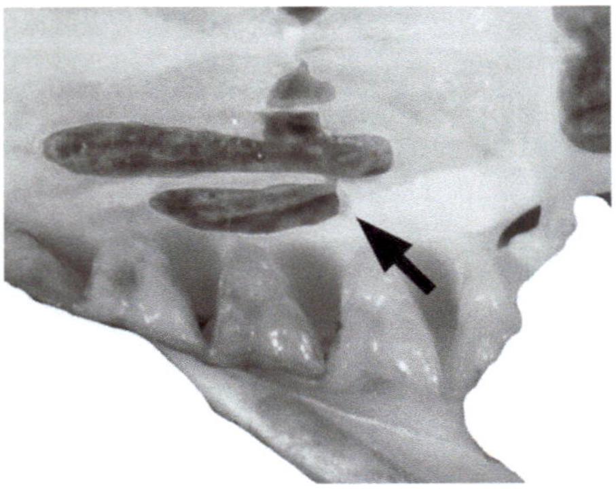

Fig. 3 Palatal with a small pit between palatine and molar cleft

2a. Dorsal pelage brown or cinnamon; no pit between palatine and molar cleft (Fig. 4; from Tamaulipas and Oaxaca, México south through Nicaragua)..*Marmosa mexicana* (p. 50)

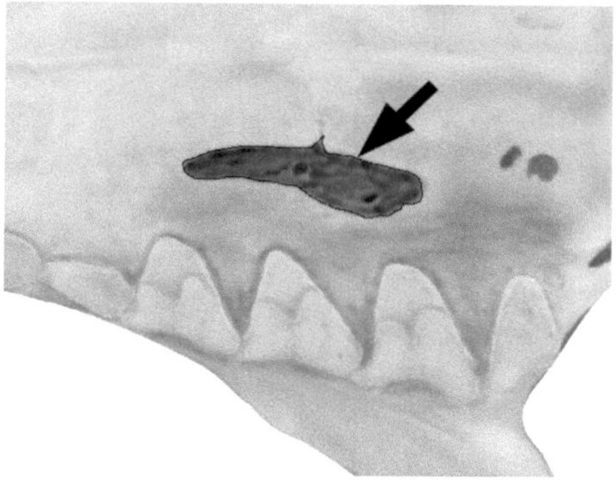

Fig. 4 Palatal without pit between palatine and molar cleft as in *Marmosa mexicana*

3. Dorsal coloration brown; tail without hair at base, all bicolor brown dorsally and white ventrally (from Chiapas, México south through South America) ....tribe Metachirini; *Metachirus myosuros* (p. 55)

3a. Dorsal coloration variable; tail with hair at base, not bicolor with dark dorsally and paler ventrally................................ .... Tribe Didelphini 4

4. Hindfoot toes webbed; dorsal pelage gray with four broad black bands across the back; zygomatic breadth >60% of basal length; single lacrimal foramen usually presents on each side (from eastern Oaxaca and southern Tabasco, México, south through Honduras and El Salvador) .... *Chironectes minimus* (p. 45)

4a. Hindfoot toes not webbed; dorsal pelage never with well-defined dorsal spots; zygomatic breadth less than 60% of basal length; two lacrimal foramina usually present on each side 5

5. Two types of long pelage; no spots above the eyes; preorbital process present (Fig. 5); skull length > 80.0 mm; several fenestrae on the palate not well defined (Fig. 6) .... *Didelphis* (p. 46)

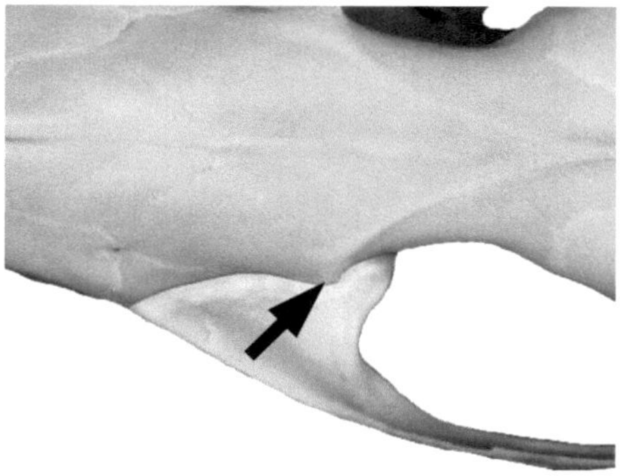

Fig. 5 The preorbital process present

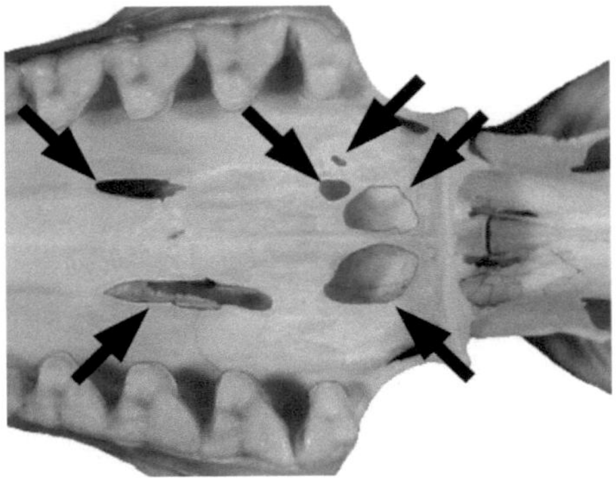

Fig. 6 With several fenestrae on the palate not well defined

5a. Pelage short; two white spots above the eyes; preorbital process absent (Fig. 7); skull length less than 80.0 mm; only four well-defined fenestrae on the posterior palate (Fig. 8; from Tamaulipas and eastern Oaxaca, México, south through Central America) .... *Philander vossi* (p. 48)

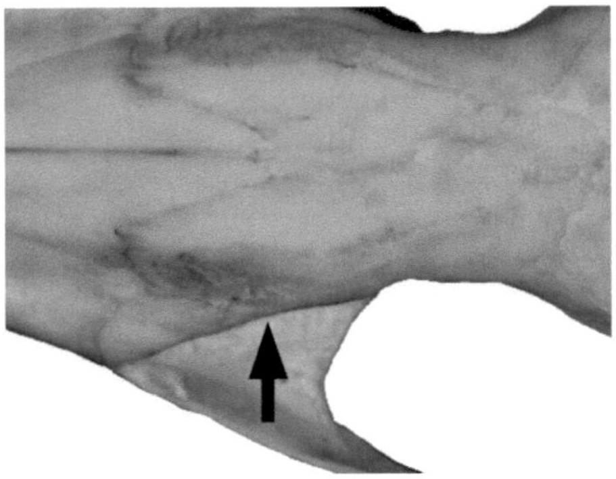

Fig. 7 The preorbital process absent as in *Philander vossi*

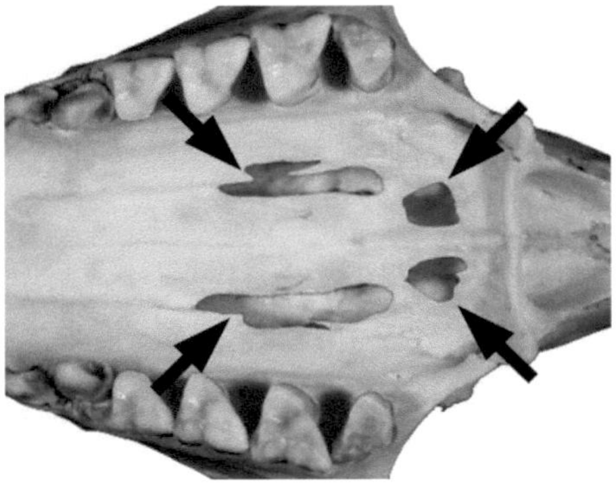

Fig. 8 With only four well-defined fenestrae on the posterior palate as in *Philander vossi*

Tribe Didelphini

Genus *Chironectes* Illiger, 1811

Medium sized; **color pattern marbled black and gray, rounded black areas confluent along the midline of the back; pelage relatively short, fine, and dense; a grayish-white band passing below the eyes from one ear to the other; underparts, chin, and chest pure white; rhinarium with a short backward extension on top of the muzzle**; ears moderately large, naked, and rounded, their anterior basal projections rudimentary; metatragus small; **toes webbed; pisiform area of the fore sixth digit**; hindfoot sole granular and without plantar tubercles; tail longer than the head-and-body length, round, and furred only at the base; marsupium present; **braincase notably broad; only two fenestrae in the bony palate**; sagittal and lambdoidal crests present in adults.

Chironectes minimus (Zimmermann, 1780)

Water opossum, tlacuache de agua

C. m. argyrodytes Dickey, 1928. Only subspecies in North America, recorded from Oaxaca and Chiapas.

Conservation: *Chironectes minimus* is listed as Least Concern by the IUCN (Pérez-Hernandez et al. 2016a).

Characteristics: Water opossum is small sized; total length 645.0–745.0 mm, skull length 68.2–81.0 mm. **Dorsal pelage pale gray with four broad black bands across the back ending in a semicircle,** coloration from brown chocolate to blackish brown; fur of underparts white; ears small, bare, and blackish; tail wide, with hair at the base, naked distally and white at the tip; **hindfoot toes webbed**; last upper molars almost of the same size as the first molar; supraorbital processes developed; first molariforms not reduced; two palatal fenestrae; braincase wide.

Comments: *Chironectes minimus* ranges from southern Oaxaca, southern Veracruz, and Tabasco south through Central America, including the lowlands of Chiapas (Map 2). This species is sympatric, or nearly so, with *Didelphis marsupialis, D. virginiana, Marmosa mexicana, Metachirus myosuros, Philander vossi,* and *Tlacuatzin canescens. C. minimus* differs from other genera by having dorsal pelage pale gray with four broad black bands across the back ending in a semicircle and hindfoot toes webbed; hair relatively short, dense, and with a few longer hairs intermixed with it; rhinarium with a short backward extension on top of the muzzle; ears moderately large, naked, and rounded; forefeet not webbed; fingers long, naked and tactile, and claws greatly reduced, but hindfeet webbed to the tips of toes, terminal toe-pads alone extending beyond webbing.

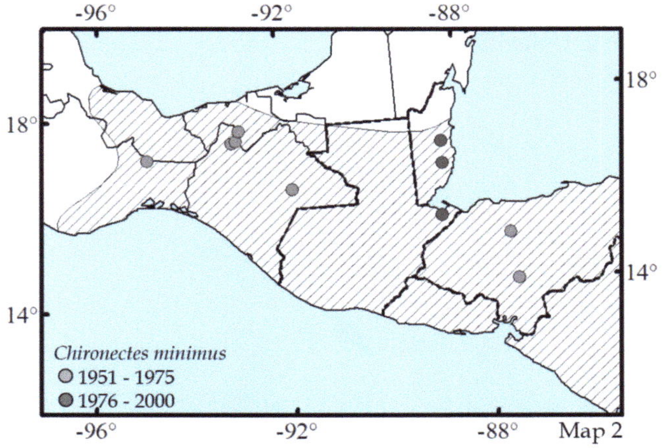

Map 2 Distribution map of *Chironectes minimus*

Additional Literature: Thomas (1888), Cabrera (1919), Goodwin (1942b), Oliver (1976), Larry (1978), Marshall (1978), Hall (1981).

Genus *Didelphis* Linnaeus, 1758

Dorsal pelage grayish, blackish, and occasionally whitish, coloration largely depending on length of hair coverage; **two types of long pelage**; fur of underparts yellow, cream, or orange; without spots on the head; ears bare and blackish; tail length varying in different species; **last upper molars almost of the same size as the first molars**; supraorbital processes developed; first molariforms not reduced; greater than two palatal fenestrae. The keys were elaborated based on the review of specimens and the following sources: Allen (1901), Gardner (1973), Hall (1981), Jones and Manning (1992), and Álvarez-Castañeda et al. (2017a).

1. Cheek pelage generally yellowish; posterior lacrimal extension pointed, sometimes square; dorsal-posterior palatine extension (forms part of the internal eye-socket wall) thin (Fig. 9; from Tamaulipas and Oaxaca, México, south through Nicaragua)..*Didelphis marsupialis* (p. 47)

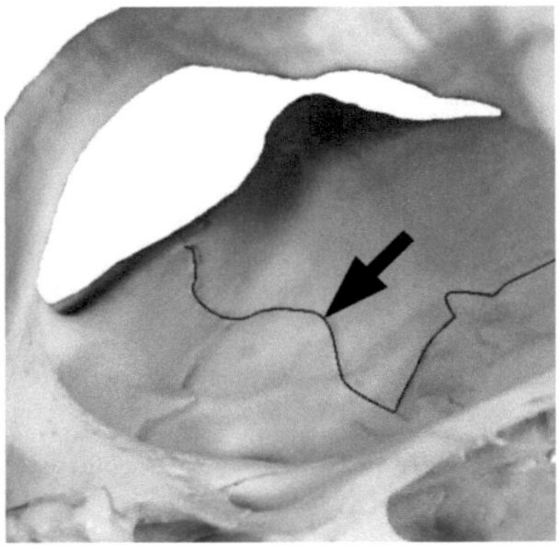

Fig. 9 The dorsal-posterior palatine extension is as thin as in *Didelphis marsupialis*

1a. Cheek pelage generally white; posterior lacrimal extension with a diamond-shaped border; dorsal-posterior palatine border (forming part of the internal eye-socket wall) wide (Fig. 10; Ontario and Québec, Canada, west through Nebraska south through Arizona, United States, and Sonora, México, to Costa Rica. An isolated population from British Columbia, Canada, south through California, United States) ....*Didelphis virginiana* (p. 47)

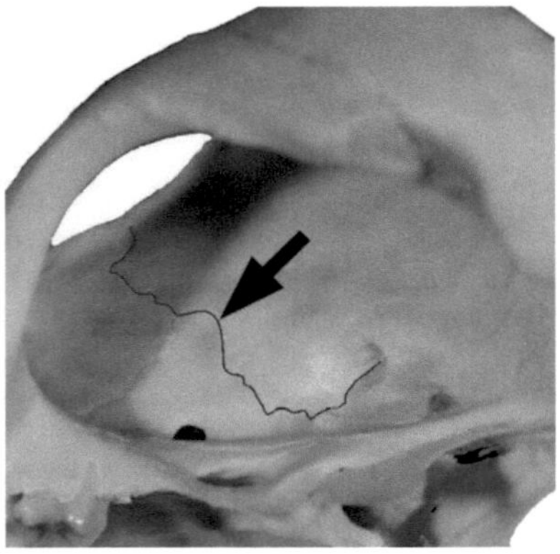

Fig. 10 The dorsal-posterior palatine border is as wide as in *Didelphis virginiana*

Didelphis marsupialis Linnaeus, 1758

Common opossum, tlacuache tropical

D. m. tabascensis Allen, 1901. Only subspecies in North America.

Conservation: *Didelphis marsupialis* is listed as Least Concern by the IUCN (Astúa de Moraes et al. 2016).

Characteristics: Common opossum is a medium-sized mammal; **total length 725.0–895.0 mm**, condylobasal length 85.5–123.0 mm in males and 82.5–114.4 mm in females. Dorsal pelage grayish, blackish, and occasionally whitish; fur of underparts yellow, cream, or orange; **cheeks yellow or yellowish cream**; all whiskers black; tail larger than the head-and-body length; **posterior lacrimal extension pointed, sometimes square; dorsal-posterior palatine extension thin**.

Comments: *Didelphis virginiana* was recognized as a distinct species (Gardner 1973), previously considered a subspecies of *D. marsupialis* (Hershkovitz 1951; Hall and Kelson 1959). It ranges from Tamaulipas and southern Oaxaca south through Nicaragua, including the Yucatán Peninsula (Map 3), and can be found in sympatry, or nearly so, with *Caluromys derbianus, Chironectes minimus, D. virginiana, Philander vossi, Marmosa mexicana, Metachirus myosuros, Tlacuatzin canescens,* and *T. gaumeri. D. marsupialis* differs from other genera by its larger size, greater than 725.0 mm, and skull length greater than 80.0 mm; two types of long pelage, no defined spots above the eyes, and several fenestrae on the palate not well defined. From *D. virginiana*, by its generally smaller size, different parts of the pelage yellow or yellowish cream, mainly on the cheeks; posterior lacrimal bone extension pointed, sometimes square, and dorsal-posterior palatine extension thin.

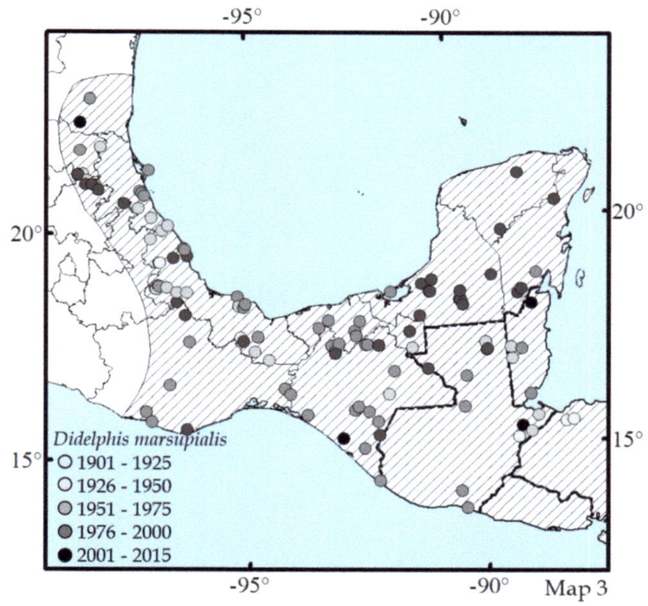

Map 3 Distribution map of *Didelphis marsupialis*

Additional Literature: Gardner (2005a), Cerqueira (1985).

Didelphis virginiana Kerr, 1792

Virginia opossum, tlacuache

1. *D. v. californica* Bennett, 1833. From Sonora and southern Texas south through Costa Rica.
2. *D. v. pigra* Bangs, 1898. From the southeastern coast of Texas to Georgia, including the Florida Peninsula.
3. *D. v. virginiana* Kerr, 1792. Western and eastern United States and southern Canada.
4. *D. v. yucatanensis* Allen, 1901. Yucatán Peninsula.

Conservation: *Didelphis virginiana* is listed as Least Concern by the IUCN (Pérez-Hernández et al. 2016b).

Characteristics: Virginia opossum is medium sized; **total length 900.0–970.0 mm**, females condylobasal length 77.1–114.5 mm; males 84.1–138.0 mm. Dorsal pelage grayish or whitish but rarely blackish; fur of underparts white, yellow, or cream; ears occasionally white at tip; **cheeks white**; whiskers black and white; tail equal to or shorter than the head-and-body length; **posterior lacrimal extension with diamond-shaped border; dorsal-posterior palatine border wide**.

Comments: *Didelphis virginiana* was recognized as a distinct species (Gardner 1973), previously regarded as a subspecies of *D. marsupialis* (Hershkovitz 1951; Hall and Kelson 1959). Its ranges from Ontario and New England in the east, including western Wisconsin, Nebraska, and from Arizona and Sonora, south through Costa Rica. An isolated population is found from British Columbia to California, Arizona, and Colorado. Not found in the Baja California Peninsula, highlands, and Sonora Desert (Map 4), and can be found in sympatry, or nearly so, with *Caluromys derbianus, Chironectes minimus, D. marsupialis, Philander vossi, Marmosa mexicana, Metachirus myosuros, Tlacuatzin canescens, T. gaumeri, T. balsasensis,* and *T. sinaloae. D. virginiana* differs from other genera by having larger size, greater than 900.0 mm and skull length greater than 83.0 mm, with two types of long pelage, no defined spots above the eyes, and several fenestrae on the palate not well defined. From *D. marsupialis* its generally larger size; different parts of the pelage whitish mainly on the cheeks; posterior lacrimal bone in a diamond-shaped border and dorsal-posterior palatine extension wide.

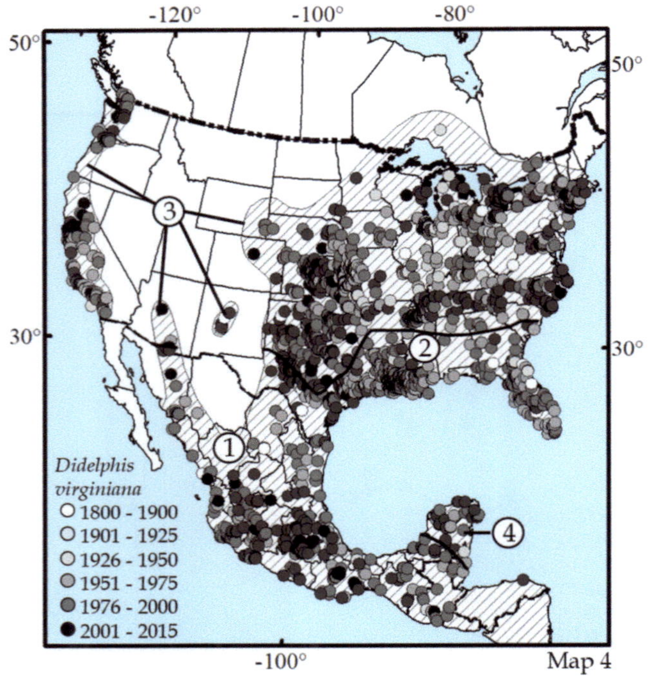

Map 4 Distribution map of *Didelphis virginiana*

Additional Literature: Gardner (1982, 2005a), McManus (1974).

Genus *Philander* Brisson, 1762

Medium size; dorsal coloration pale gray to black; face dark contrasts with white supraorbital spots above each eye; underparts, chin, and cheeks from yellowish-white to orange; chest pale gray; ears large and naked, rims black, and pale spot present on forehead at medial base; median black line runs along top of head; tail with hair at base and naked distally; marsupium present; four fenestrae in bony palate; sagittal and occipital crests well developed; auditory bullae small.

Philander vossi Gardner and Ramírez-Pulido, 2020

Gray four-eyed opossums, cuatro ojos gris

Monotypic.

Conservation: *Philander vossi* (as *Philander opossum*) is listed as Least Concern by the IUCN (de la Sancha et al. 2016)

Characteristics: Gray Four-eyed opossum is small sized; total length 500.0–610.0 mm, skull length 62.7–73.5 mm. Dorsal pelage dark grayish brown or blackish gray interspersed with white; **blackish head with two white spots over the eyes**; underparts and upperpart of the foot cream or yellowish; ears bare and blackish with a cream spot at the base; tail hairy at the base, posterior part black and distal third in contrasting white; last upper molars almost of the same size as the first molar; supraorbital processes developed; **no postorbital processes**; first molariforms not reduced; **four fenestrae on the palate**.

Comments: *Philander opossum* was previously considered *Metachirops opossum* (Pine 1973) and *P. pallidus* was elevated to full species, which was the subspecies with ranges in México (Voss et al. 2018). Gardner and Ramírez-Pulido (2020) replaced *P. pallidus* with the *nomen novum P. vossi* following the Article 60.3 of the ICZN 1999. *P. vossi* ranges from Tamaulipas and Oaxaca south through all Central America, including the Yucatán Peninsula (Map 5) and can be found in sympatry, or nearly so, with *Caluromys derbianus, Chironectes minimus, Didelphis marsupialis, D. virginiana, Marmosa mexicana, Metachirus myosuros, Tlacuatzin canescens,* and *T. gaumeri. P. vossi* differs from other genera by having the head blackish with two white spots over the eyes and four fenestrae on the palate. From *M. myosuros,* by having a gray dorsal coloration; tail with hair at the base, with the first half gray and the second half white.

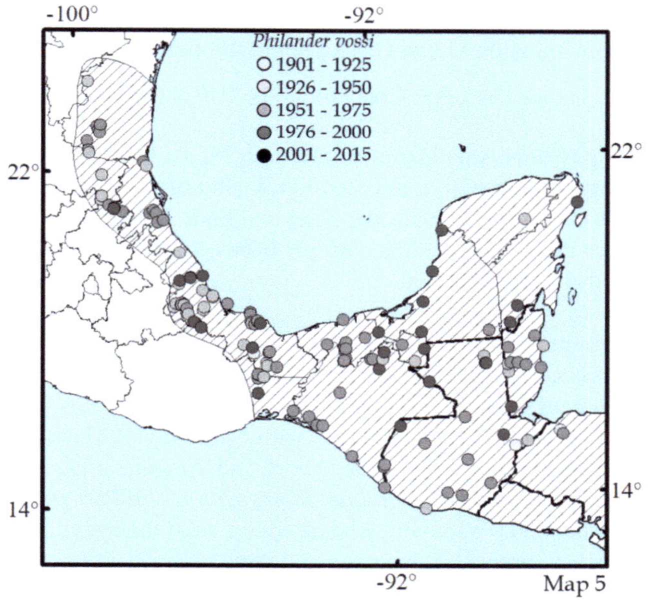

Map 5 Distribution map of *Philander vossi*

Additional Literature: Gardner and Patton (1972), Pine (1973), Hershkovitz (1981, 1997), Gardner (1993), Patton and Da Silva (1997), Castro-Arellano et al. (2000).

Tribe Marmosini

Genus *Marmosa* Gray, 1821

Small sized; **dorsal pelage pale brown or orange brown with a marked black ring around the eyes; tail longer than the head-and-body length, underside of tip modified for grasping; six plantar tubercles; marsupium absent; temporal ridges not united into a sagittal crest;** postorbital processes present or absent according to the species; toothrows convergent anteriorly; **third upper molars usually larger than the first molar; no pit between the palatine and the molar cleft.** Only one subgenus in North America: *Exulomarmosa* (Voss et al. 2014).

Subgenus *Exulomarmosa* Voss et al., 2014

Gular gland consistently present and well developed in adult males; fore claws small; medial and lateral carpal tubercles present; tail with rhomboidal or oblong scales below, usually in both spiral and annular series; tail prehensile with ventral surface not densely fringed with long hairs; postorbital processes absent; palatine fenestrae absent in North American species.

Marmosa mexicana Merriam, 1897

Mexican mouse opossum, ratón tlacuache mexicano

M. m. mayensis Osgood, 1913. Yucatán Peninsula.
M. m. mexicana Merriam, 1897. From Tamaulipas and Oaxaca south through Nicaragua.

Conservation: *Marmosa mexicana* is listed as Least Concern by the IUCN (Martin 2016).

Characteristics: Mexican mouse opossum is small sized; total length 282.0–386.0 mm, condylobasal length 32.0–40.6 mm. **Dorsal pelage pale brown or orange brown with a marked black ring around the eyes**; fur of underparts cream to yellow; ears bare and pale brown; tail brown longer than the head-and-body length; back might be slightly darker than the ventral side; last upper molars larger than the first molar; **no pit between the palatine and the molar cleft;** supraorbital processes developed and first molariforms not reduced.

Comments: *Marmosa mayensis* was previously considered a full species (Ramírez-Pulido et al. 2014) following a genetic analysis (Voss et al. 2014) where it was not considered a different species from *M. mexicana* (Gutierrez et al. 2010; Voss et al. 2014). *M. mexicana* ranges from Tamaulipas and southern Oaxaca south through Central America, including the Yucatán Peninsula (Map 6), and can be found in sympatry, or nearly so, with *Caluromys derbianus, Chironectes minimus, Didelphis marsupialis, D. virginiana, Metachirus myosuros, Philander vossi,* and *Tlacuatzin canescens.* The genus *Marmosa* differs from *C. derbianus, C. minimus, D. marsupialis, D. virginiana, M. myosuros,* and *P. vossi* by its smaller size, total length less than 400.0 mm and skull length less than 50.0 mm, with a black spot around the eyes contrasting with the head color. From *T. canescens,* by having the dorsal pelage brown or cinnamon and without pits between the palatine and the molar cleft.

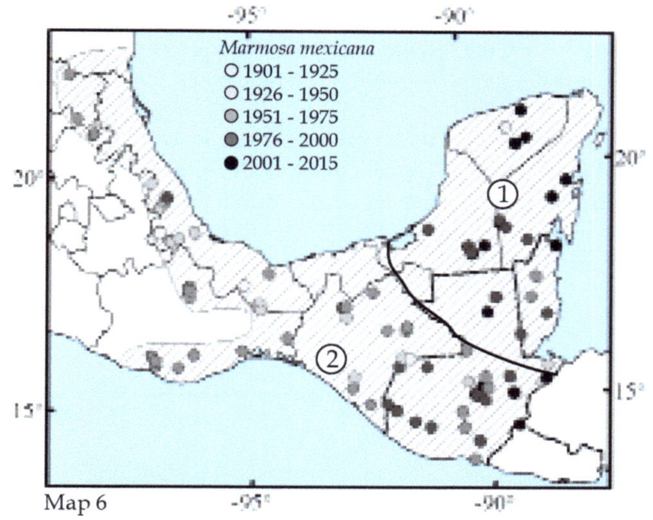

Map 6 Distribution map of *Marmosa mexicana*

Additional Literature: Osgood (1913), Cabrera (1919), Tate (1933), Jones and Álvarez (1964), Alonso-Mejía and Medellín (1992), Voss ad Jansa (2003), Rossi et al. (2010).

Genus ***Tlacuatzin*** Voss and Jansa, 2003

Dorsal **pelage pale gray** with a marked black ring around the eyes; fur of underparts cream-white; ears bare and pale brown; tail approximately of the same length as body, with back slightly darker and occasionally white at the tip; last upper molars larger than the first molar; supraorbital processes slightly developed; first molariforms not reduced; more than two fenestrae in the palate; **small pit between the palatine and the molar cleft**. The keys were elaborated based on the review of specimens and the following sources: Hall (1981), Álvarez-Castañeda et al. (2015, 2017), and Arcangeli et al. (2018).

1. Dorsal coloration with cinnamon shades; restricted to the Tres Marías Archipelago............ *Tlacuatzin insularis* (p. 54)

1a. Dorsal coloration without cinnamon shades; distributed in the mainland 2

2. Dorsal coloration in grayish shades; tail monocolor (from Jalisco south through western Chiapas, México).................. .... *Tlacuatzin canescens* (p. 52)

2a. Dorsal coloration in brownish shades; tail bicolored 3

3. Tail with a coloration gradient, from darker at the base to pale at the tip, no notorious differences between dorsal and ventral parts (from southern Sonora south through Jalisco, México) ....*Tlacuatzin sinaloae* (p. 54)

3a. Tail without a longitudinal coloration gradient. Dorsal coloration darker than fur of underparts 4

4. Dorsal coloration pale brown; larger rostrum and skull, condylobasal length > 28.5 mm, length from the first to the third upper molar equal to or >5.4 mm (from inland of Jalisco east through Oaxaca, México) ....*Tlacuatzin balsasensis* (p. 51)

4a. Dorsal coloration dark brown; skull and rostrum shorter, condylobasal length less than 28.5 mm; length from the first to the third upper molar less than 5.4 mm (Yucatán Peninsula, México) ....*Tlacuatzin gaumeri* (p. 53)

Tlacuatzin balsasensis Arcangeli et al., 2018

Balsas mouse opossum, ratón tlacuache del Balsas

Monotypic.

Conservation: *Tlacuatzin balsasensis* (as *Tlacuatzin canescens*) is listed as Least Concern by the IUCN (Martin 2017). However, with the taxonomical change, its conservation status needs to be re-evaluated.

Characteristics: Balsas mouse opossum is small sized; total length 205.0–302.0 mm, condylobasal length 28.6–37.4 mm. **Dorsal pelage brownish slightly gray** with a marked black ring around the eyes; fur of underparts **yellow buff**; feet yellowish; tail slightly bicolored, brownish dorsally and pale ventrally; palatine and nasals long and broad.

Comments: The genus *Tlacuatzin* was considered as *Marmosa* (Voss and Jansa 2003). *T. balsasensis* was previously as a part of *T. canescens* but was elevated to full species (Arcangeli et al. 2018) and ranges within the Balsas Basin, inland lowlands of Jalisco, Michoacán, Guerrero, Morelos, Puebla, and northwestern Oaxaca (Map 7). *T. balsasensis* cannot be found in sympatric, or nearly so, with other species of the genus *Tlacuatzin* but can be found with *Didelphis virginiana* from which differs by having total length less than 400.0 mm; skull length less than 50.0 mm, and with black spots around the eyes, contrasting with head coloration.

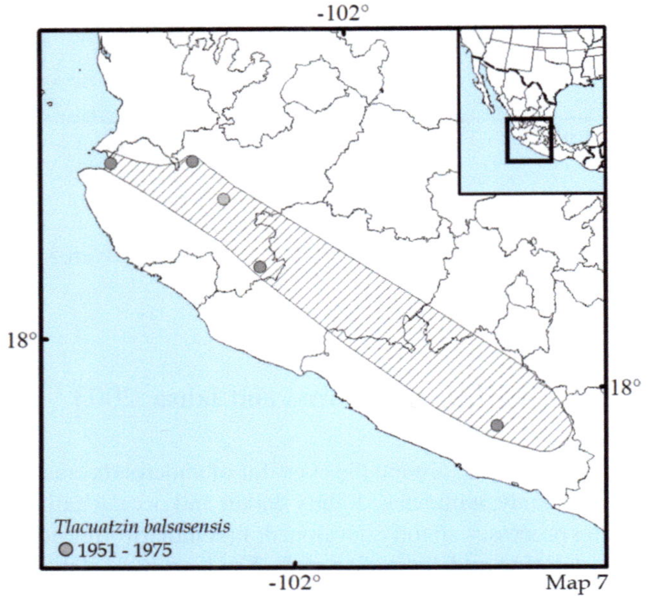

Map 7 Distribution map of *Tlacuatzin balsasensis*

Tlacuatzin canescens (Allen, 1893)

Gray mouse opossum, ratón tlacuache gris

Monotypic.

Conservation: *Tlacuatzin canescens* is listed as Least Concern by the IUCN (Martin 2017). However, with the taxonomical change the conservation status needs to be re-evaluated.

Characteristics: Gray mouse opossum is small sized; total length 172.0–281.0 mm, condylobasal length 27.0–35.8 mm. **Dorsal pelage pale gray** with a marked black ring around the eyes; fur of underparts **white**; feet grayish; tail pale brown; skull and palatine short and broad.

Comments: The genus *Tlacuatzin* was considered under *Marmosa* (Voss and Jansa 2003). The species of *T. balsasensis, T. insularis,* and *T. sinaloae* were previously subspecies of *T. canescens* (Arcangeli et al. 2018). *T. gaumeri* was previously considered a junior synonym of *T. c. canescens. T. canescens* ranges from Jalisco by the coast south through western Chiapas (Map 8), and is sympatric, or nearly so, with *Didelphis marsupialis, D. virginiana, Metachirus myosuros, Philander vossi,* and *Marmosa mexicana,* and it cannot be found in sympatric, or nearly so, with other species of the genus *Tlacuatzin. T. canescens* differs from other genera, by having total length less than 400.0 mm and with black spots around the eyes, contrasting with head coloration. From *M. mexicana,* by having dorsal pelage pale gray and a series of small pits between the palatine and molar cleft.

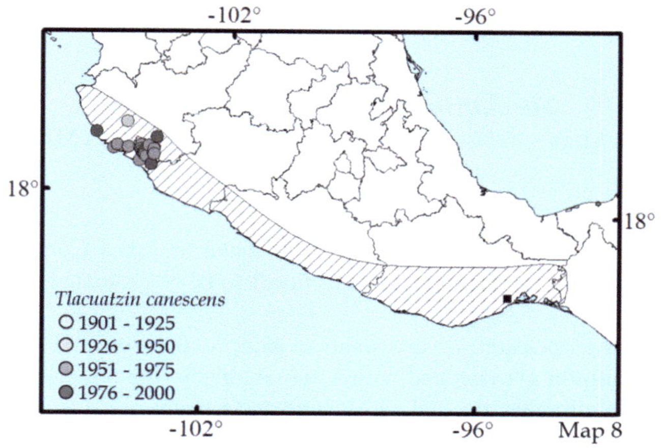

Map 8 Distribution map of *Tlacuatzin canescens*

Additional Literature: Tate (1933), Goodwin (1969), Reid (1997), Zarza et al. (2003).

Tlacuatzin gaumeri (Osgood, 1913)

Yucatán mouse opossum, ratón tlacuache de Yucatán

Monotypic.

Conservation: *Tlacuatzin gaumeri* (as *Tlacuatzin canescens*) is listed as Least Concern by the IUCN (Martin 2017). However, with the taxonomical change the conservation status needs to be re-evaluated.

Characteristics: Yucatán mouse opossum is small sized; total length 201.0–230.0 mm, condylobasal length 28.1–28.4 mm. **Dorsal pelage dark brownish gray** with a marked black ring around the eyes; **fur of underparts pale cream**; feet creamy white; tail brownish dorsally and pale ventrally; skull and nasals small and narrow.

Comments: The genus *Tlacuatzin* was considered under *Marmosa* (Voss and Jansa 2003). *Tlacuatzin gaumeri* was elevated to full species (Arcangeli et al. 2018), previously considered part of *T. canescens*. *T. gaumeri* ranges only in Campeche, Quintana Roo, and Yucatán (Map 9). It is sympatric, or nearly so, with *Didelphis marsupialis, D. virginiana, Philander vossi,* and *Marmosa mexicana*, but is mostly not sympatric with other species of the genus *Tlacuatzin*. *T. gaumeri* differs from other genera by having total length less than 400.0 mm and with black spots around the eyes, contrasting with the head coloration. From *M. mexicana,* by having the dorsal pelage pale gray and a series of small pits between the palatine and the molar cleft.

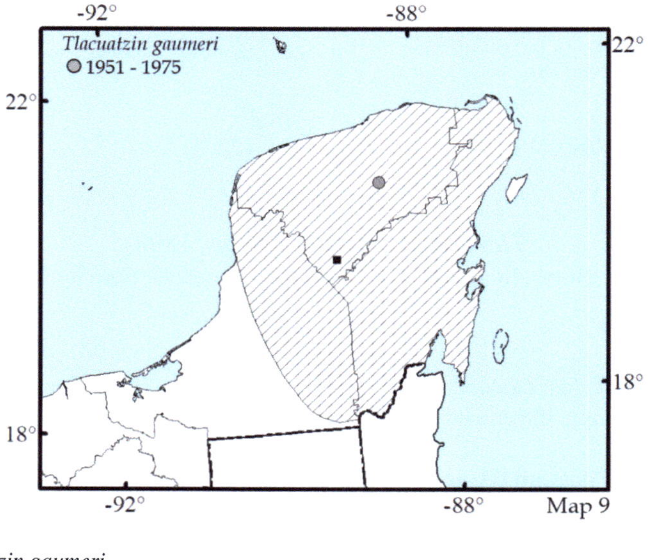

Map 9 Distribution map of *Tlacuatzin gaumeri*

Additional Literature: Osgood (1913), Tate (1933), Reid (1997), Zarza et al. (2003).

Tlacuatzin insularis (Merriam, 1898)

Tres Marías mouse opossum, ratón tlacuache de Tres Marías

Monotypic.

Conservation: *Tlacuatzin insularis* (as *Tlacuatzin canescens*) is listed as Least Concern by the IUCN (Martin 2017). However, with the taxonomical change, the conservation status needs to be re-evaluated.

Characteristics: Tres Marías mouse opossum is very small in size; total length 210.0–283.0 mm, condylobasal length 28.1–31.6 mm. **Dorsal pelage cinnamon grayish** and with a marked black ring around the eyes; fur of underparts **cream-buff**; feet buffy-brown; tail brown; skull, palatine, and nasals long and narrow.

Comments: The genus *Tlacuatzin* was considered under *Marmosa* (Voss and Jansa 2003). *Tlacuatzin insularis* was elevated to full species (Arcangeli et al. 2018), previously considered a subspecies of *T. canescens*. *T. insularis* is restricted to the María Madre, María Cleofas, and San Juanito Islands, Nayarit (Map 10), with no other Didelphimorphia found in sympatry.

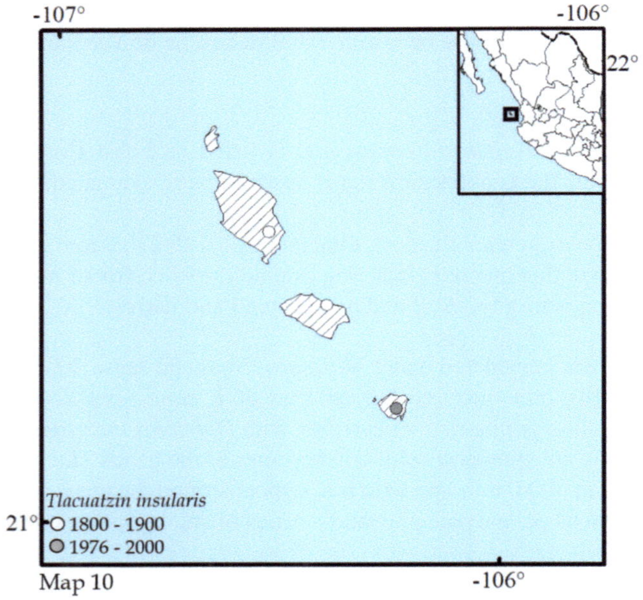

Map 10 Distribution map of *Tlacuatzin insularis*

Additional Literature: Nelson (1899), Wilson (1991), Zarza et al. (2003).

Tlacuatzin sinaloae (Allen, 1898)

Northern mouse opossum, ratón tlacuache norteño

Monotypic.

Conservation: *Tlacuatzin sinaloae* (as *Tlacuatzin canescens*) is listed as Least Concern by the IUCN (Martin 2017). However, with the taxonomical change, the conservation status needs to be re-evaluated.

Characteristics: Northern mouse opossum is small sized; total length 195.0–290.0 mm, condylobasal length 26.1–34.5 mm. **Dorsal pelage brownish gray** with a marked black ring around the eyes; fur of underparts **yellow**; feet creamy white; tail brownish at the base and pale at the tip; skull and palatine short and broad.

Comments: The genus *Tlacuatzin* was considered under *Marmosa* (Voss and Jansa 2003). *Tlacuatzin sinaloae* was elevated to full species (Arcangeli et al. 2018), previously regarded as a subspecies of *T. canescens*. Its ranges from southern Sonora, Sinaloa, Nayarit, and Jalisco (Map 11), and is mostly not sympatric with other species of the genus *Tlacuatzin,* only with *Didelphis virginiana,* from which it differs by having total length less than 400.0 mm and with black spots around the eyes, contrasting with the head coloration.

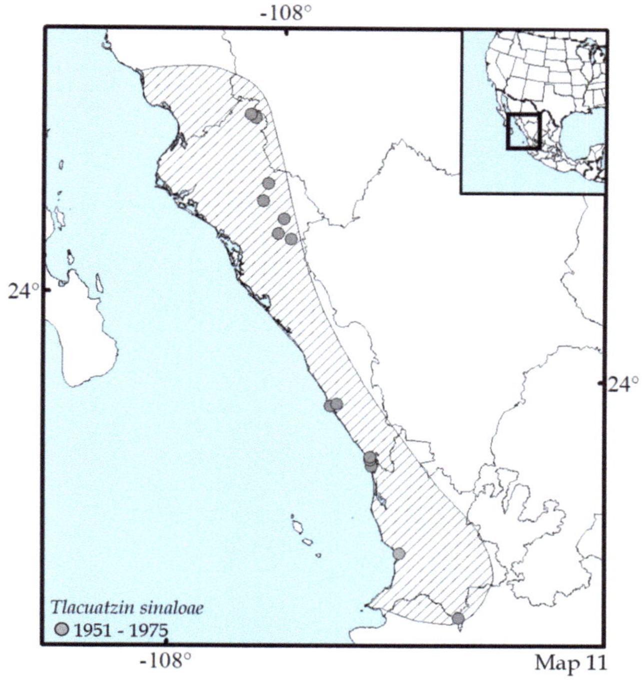

Map 11 Distribution map of *Tlacuatzin sinaloae*

Additional Literature: Tate (1933), Armstrong and Jones (1971a), Zarza et al. (2003).

Tribe Metachirini

Genus ***Metachirus*** Burmeister, 1854

Medium sized; **pelage short, dense and silky;** dorsally reddish to yellowish-brown and occasionally with thin pale streaks along the dorsum; often darkening as it approaches their rump and paler on their sides; **whitish spots above each eye; pale rostrum and cheeks**; ears brown; fur of underparts yellowish or buffy brown; tail sparsely haired and entirely bicolored, brown dorsally and with ventrally, white at the tip.

Metachirus myosuros (Temminck, 1824)

Brown four-eyed opossums, cuatro ojos pardo

Monotypic.

Conservation: *Metachirus myosuros* (as *M. nudicaudatus*) is listed as Least Concern by the IUCN (Brito et al. 2021).

Characteristics: Brown four-eyed opossum is small sized; total length 550.0–660.0 mm, condylobasal length 57.6–63.9 mm. Dorsal pelage brown and tan on the sides and below the ears; ears brown; **rostrum dark brown, with a dark stripe from the forehead to the nape; four spots cream-colored above the eyes and cheeks**; fur of underparts pale yellow; **tail sparsely haired, entirely bicolored, brown dorsally and white ventrally.**

Comments: *Metachirus myosuros* was previously assigned to *M. nudicaudatus* (Geoffroy, 1803), nowadays is considered a different species (Voss et al. 2019). *M. myosuros* has an isolated record from southern Chiapas in México and a continuous distribution from Nicaragua south through South America (Map 12). It can be found in sympatry, or nearly so, with *Caluromys derbianus, Chironectes minimus, Didelphis marsupialis, Philander vossi, Marmosa mexicana,* and *Tlacuatzin canescens.* It differs from *Caluromys derbianus, Chironectes minimus, Didelphis marsupialis* by its smaller size; skull length less than 80.0 mm; with two defined paler spots above the eyes; without a preorbital process, and the posterior palate with only four well-defined fenestrae. From *Marmosa* and *Tlacuatzin,* by having two defined paler spots above the eyes; without dark spots around the eyes contrasting with the head color; total length greater than 400.0 mm and skull length greater than 50.0 mm. From *P. vossi,* by having the dorsal coloration brown; tail without hair at the base and bicolored throughout, brown dorsally and white ventrally.

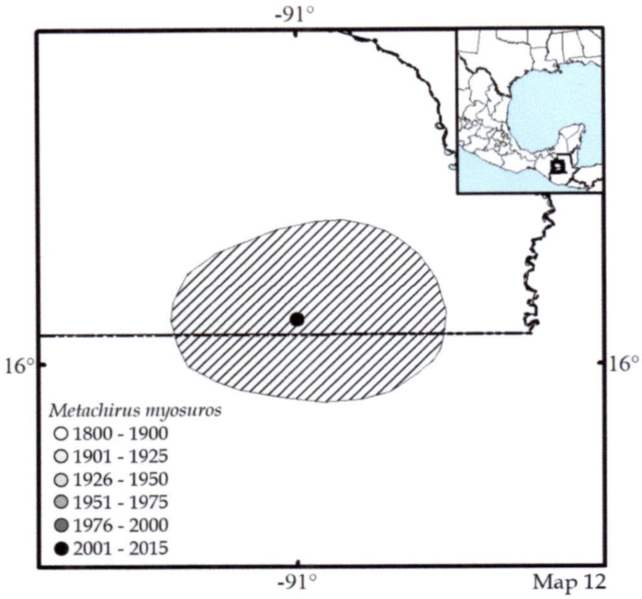

Map 12 Distribution map of *Metachirus myosuros*

Additional Literature: Medellín et al. (1992), Costa and Patton (2006).

Order Artiodactyla

Infraclass Eutheria Gill, 1872

Placentalia differs from other mammals in that the fetus is carried in the uterus of the mother until a relatively late development stage; there are no epipubic bones; malleolus is at the bottom of the fibula; and rearmost bones of the foot fit into a socket formed by the ends of the tibia and fibula, forming a complete mortise and tenon upper ankle joint (marine mammals not included in the book). The keys were elaborated based on the review of specimens and the following sources: Lawlor (1979), Hall (1981), Jones and Manning (1992), and Álvarez-Castañeda et al. (2015, 2017a).

1. Body fusiform, optimum for aquatic life; limbs paddle shaped; with dorsal fin; tail fin with two lobes; nasal conducts open in the dorsal part of the skull .... order Artiodactyla (p. 48), suborder Cetancodonta, not included in the book

1a. Body of variable shape, if fusiform and optimum for aquatic life, without a dorsal fin; if there is a fin in the tail, it has only one lobe; limbs with appendages or paddle shaped; nasal conducts open horizontally in the anterior part of the rostrum 2

2. Less than three digits and never covered by a hoof; body covered by a shell; no teeth, if present, all of similar shape and with a diastema between them 3

2a. Digits never with a sheath; if less than three, digits covered by a hoof; teeth at least of two different shapes 4

3. Body covered by hair; two or three digits; strong claws; no teeth or zygomatic arches .... order Pilosa (p. 563)

3a. Body covered by a shell; more than three digits; teeth and zygomatic arches present .... order Cingulata (p. 393)

4. Head less than 150.0 mm and without canine teeth; one lower incisor on each side, bevel shaped 5

4a. Head greater than 150.0 mm; if smaller, with canine teeth; more than one lower incisor on each side, not bevel shaped or without lower incisors 6

5. Tail small, tassel shaped; ears long and narrow; two upper incisors on each side, second incisor smaller and located on the back of the first incisor; skull not completely ossified, with fenestrae .... order Lagomorpha (p. 515)

5a. Tail large or short, never tassel shaped; ears equally long and wide; one upper incisor on each side; well-ossified skull, without fenestrae .... order Rodentia (Volumen 2)

6. Small specimens, less than 100 g in weight; skull length less than 50.0 mm; if larger, molars with the occlusal surface clearly W-shaped 7

6a. Large specimens, greater than 300 g in weight; skull length greater than 50.0 mm; if smaller, molar occlusal surface not W-shaped 8

S. T. Álvarez-Castañeda, *Mammals of North America - Volume 1*, https://doi.org/10.1007/978-3-031-41661-3_3

7. Forelimbs wing shaped; canines perfectly distinct from incisors and premolars; zygomatic bones present; if absent, teeth with a diastema....order Chiroptera (p. 199)

7a. Forelimbs not wing shaped; canines, although present, not differentiated from incisors and premolars; zygomatic bones present or absent .... order Eulipotyphla (p. 397)

8. Limbs mesaxonic, covered by hooves or paddle shaped; nasals very reduced, so that nostrils seem very big; molars with transverse lophos....9

8a. Limbs paraxonic; nasals long; teeth of different shapes, without transverse lophos....10

9. One to three digits in each limb, covered by hooves; eyes small relative to head size; three upper and lower incisors; canines present in old specimens....order Perissodactyla (p. 559)

9a. Limbs paddle shaped; eyes large relative to head size; incisors, when present, one upper and one lower; no canines.... order Sirenia (p. 573)

10. Limbs long and thin, same as with tail; temporal and orbital pits completely separated by a bone wall, in such way that eye sockets are directed frontward; foramen magnum situated in the ventro-posterior side of the skull........order Primates (p. 567)

10a. Limbs variable in length; temporal and orbital pits continuous, at most with a postero-external bar limiting them but with no bone wall separating them ....order Carnivora (p. 87).

Order Artiodactyla Owen, 1848

Artiodactyla includes antelopes, deer, musk deer, peccaries, pronghorns, and sheep. This order has no morphological characteristic that could be used as a diagnostic. Four suborders are recognized: Cetancodonta (Árnason et al. 2000; Spaulding et al. 2009; hippopotamus, whales, and dolphins), which are stout; naked-skinned; aquatic or semiaquatic. The other three suborders have hooves and horns or antlers in common; **frontal limbs have two or four digits and hooves**; when they have four digits, two are small and not touching the ground; the other two main digits almost equal in size but not symmetrical in shape (symmetrical axis between the third and fourth digits); femur without the third trochanter; **no alisphenoid canal**. These are Ruminantiamorpha (pronghorns, antelopes, sheep, deer, and musk deer), Suina (pigs and peccaries), and Tylopoda (camels). Distribution is nearly cosmopolitan. One suborder, Ruminantia, and four families occur in North America: Antilocapridae, Bovidae, Cervidae, and Tayassuidae. Artiodactyla was previously considered Cetoartiodactyla (Montgelard et al. 1997); however, a recent revision of the group considered cetaceans as belonging to Artiodactyla, so the name Artiodactyla needs to be used (Prothero et al. 2022).

1. Body extended; limbs proportionally long; hair short; upper incisors absent; upper canines absent; if present, not triangular in transverse section; molariforms selenodont....suborder Rumiantia (p. 59)

1a. Body compact and cylindrical; limbs proportionally short; hair long and bristle-like; upper incisors present; canines well developed, triangular in transverse section (Fig. 1); molariforms bunodont .... suborder Suina (p. 83)

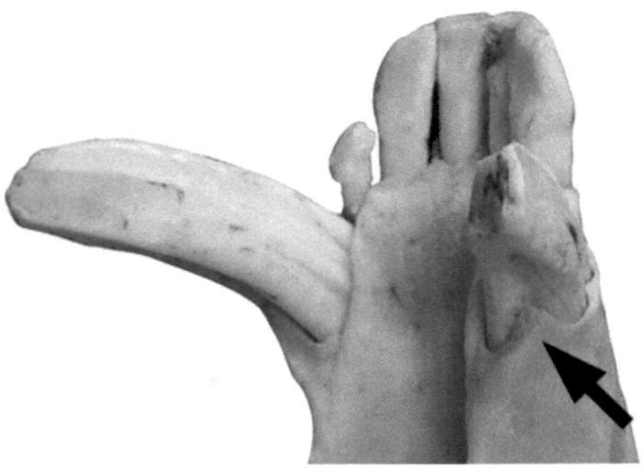

Fig. 1 The canines are well developed, triangular in transverse section

Suborder Rumiantia Scopoli, 1777

Metatarsal bones of the main pair of toes fused into a single cannon bone (same as in metacarpal bones); **metacarpals and metatarsals of lateral toes incomplete or absent; navicular and cuboid of the tarsus united; stomach with four chambers;** placenta cotyledonary; **horns or antlers usually present, at least in males; upper incisors absent; lower canines incisiform; molariforms selenodont.**

Infraorder Pecora Flower, 1883

Even-toed hoofed, ruminant digestion; a great number of species with cranial appendages projecting from their frontal bones. The keys were elaborated based on the review of specimens and the following sources: Lawlor (1979), Hall (1981), Jones and Manning (1992), and Álvarez-Castañeda et al. (2017a).

1\. Horns simple in both sexes; nasals articulating with lacrimals (Fig. 2) .... family Bovidae (p. 62)

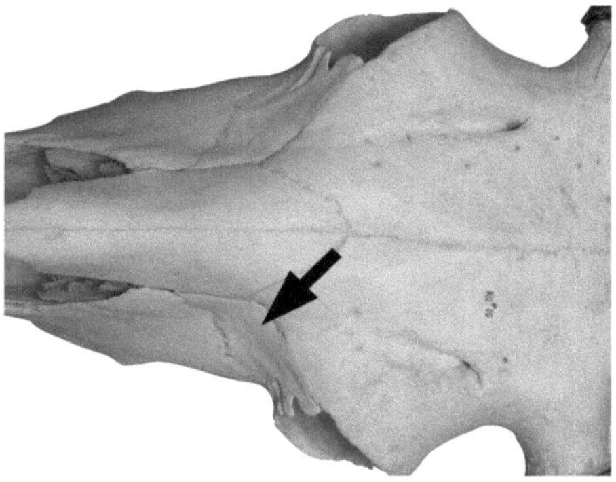

Fig. 2 The nasals bones articulating with lacrimals

1a. Antlers present, horn shaped or branched, less developed or absent in females; if horns are present, these are covered by a keratin sheath that sheds annually; nasals not articulating with lacrimals (Fig. 3)..2

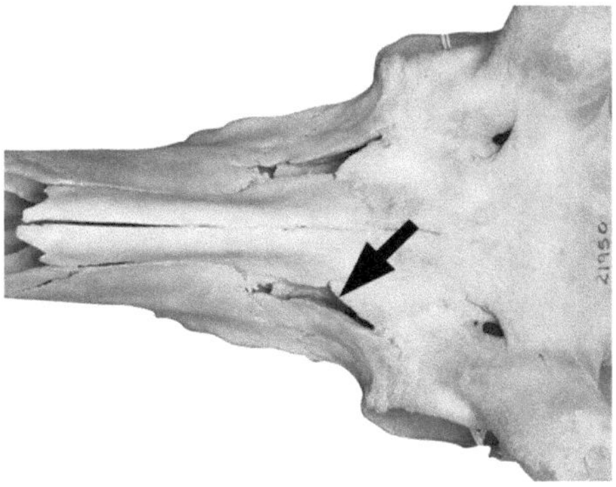

Fig. 3 The nasals bones not articulating with lacrimals

2\. Horns covered with a keratin sheath that sheds annually, present in males and females, growing anterior–posterior, anterior–posterior diameter larger than the transverse diameter, above the frontal bone and projecting perpendicularly; molariforms extended and hypsodont (Fig. 4) .... family Antilocapridae (p. 60)

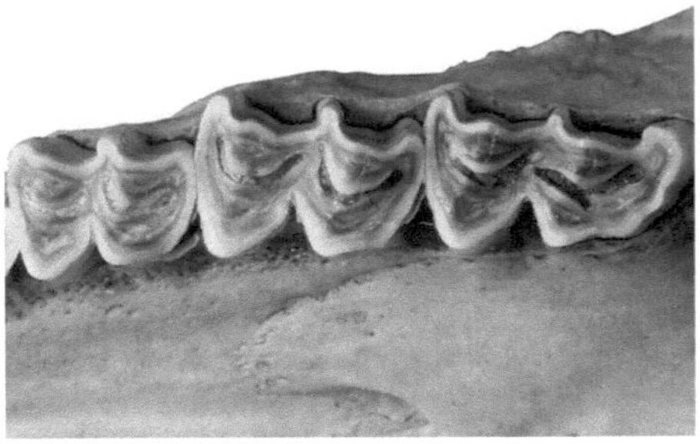

Fig. 4 The molariforms are extended and hypsodont as in *Antilocapra americana*

2a. Antlers instead of horns, usually absent in females, growing behind the orbital pits, with the main beam projecting toward the front and sides; anterior–posterior diameter almost equal to transverse one; square and brachydont molariforms (Fig. 5) .... family Cervidae (p. 58)

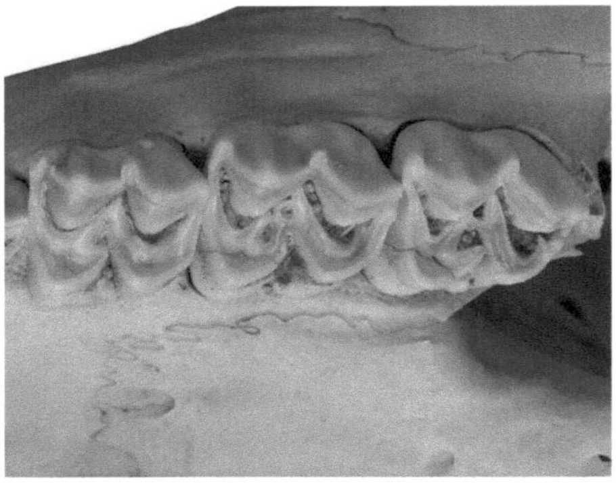

Fig. 5 The molariforms are square and brachydont

Family Antilocapridae Gray, 1866

This family includes only one species, the pronghorn or American antelope (*Antilocapra americana*), characterized by having horns instead of antlers; the **bone marrow is covered by a keratin sheath, forked in mature individuals; the sheath sheds annually** (a characteristic unique to the family); **both sexes have horns, but those of females are small and hood shaped;** tail short; limbs especially long and thin; **two lacrimal foramina inside the orbits; without upper canines; molar surface smooth in texture**. They are distributed exclusively in North America and belong to the order Artiodactyla, infraorder Pecora.

Genus ***Antilocapra*** (Ord, 1818)

Pronghorn is small sized; weight 47–70 kg; body colors contrasting from white, rusty brown to tan, black, and dark brown markings in the head and throat; eyelashes and mucous membranes of the nose and mouth coal black; brownish-black patches starting just below the ears in males and extending downward from 75 to 100 mm; mane hairs russet, with varying amounts of black hairs at the tip; **two rump patches of white erectile hairs, approximately 75 mm long, visible from great distances; males and females with horns compressed laterally, dark colored in males, whitish at the tip and anterior prongs, tips variable but normally inclined inward, backward, or both; horn sheaths of adult males from 33 to 50 cm in length and shed annually**; horn sheaths present or absent in females; head and eyes large; limbs and feet relatively long, slender, lacking lateral digits; **lacrimal duct opening into two orifices**.

Antilocapra americana (Ord, 1815)
Pronghorn, berrendo

1. *A. a. americana* (Ord, 1815). From California east through Manitoba to Oklahoma, and from Idaho to Manitoba south through Arizona to Oklahoma.
2. *A. a. mexicana* Merriam, 1901. From central Arizona, southwestern New Mexico, and Texas south through Sonora to San Luis Potosí.
3. *A. a. oregona* Bailey, 1932. Oregon and Washington.
4. *A. a. peninsularis* Nelson, 1912. Baja California Peninsula.
5. *A. a. sonoriensis* Goldman, 1945. Sonora and southern Arizona.

Conservation: *Antilocapra americana* is listed as Endangered in the Norma Oficial Mexicana (DOF 2019) and as Least Concern by the IUCN (IUCN SSC Antelope Specialist Group 2016).

Comments: *Antilocapra americana* historically ranges from Alberta and Manitoba south through Sonora, Chihuahua, Coahuila, Zacatecas, and San Luis Potosí, including the Baja California Peninsula, and from California and Washington west through the Dakotas and Texas. Now, it is patchily distributed (Map 1) and can be found in sympatry, or nearly so, with *Cervus canadensis, Odocoileus hemionus,* and *Odocoileus virginianus. A. americana* can be distinguished from the species of other genera by having small horns in both sexes, the bone marrow covered by a keratin sheath, forked in mature individuals, and with the sheath shedding annually.

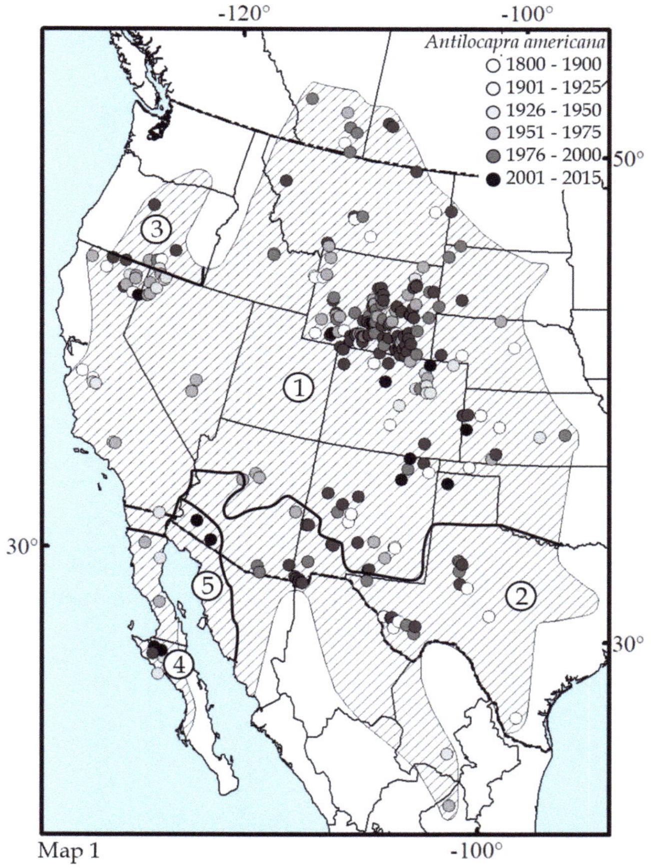

Map 1 Distribution map of *Antilocapra americana*

Additional Literature: Merriam (1901c), Nelson (1912), Goldman (1945), Hall (1946), Miller and Kellogg (1955), O'Gara (1969, 1978, 1990), Grubb (1993, 2005), and Hassanin and Douzery (2003).

Family Bovidae Gray, 1821

The family Bovidae includes antelopes, sheep, and goats characterized by small to large sizes (250.0–2,000.0 mm in height); **horns always present in males, variable in females, not branched and with a hardened keratinous sheath; sheath and bone marrow not deciduous**; tail large or short, not ending in a lock of hair; neck moderately long; **usually one lacrimal foramen and inside the orbit**; upper canines absent; molar surface smooth. Original distribution in North America, Eurasia, and Africa. Two subfamilies range in North America: Bovinae and Antilopinae.

Subfamily Bovinae Gray, 1821

The generic characteristics of the species include cloven hooves and usually at least one of the sexes of species having **true horns in lateral direction and upward**. The keys were elaborated based on the review of specimens and the following sources: Lawlor (1979) and Hall (1981).

1. Horns smooth, conical; accessory columns on the inner side of molars not reduced (Fig. 6); interorbital and mastoid width similar (from isolated areas along the central plains, Canada, and Alaska, south through Chihuahua and Coahuila, México) .... *Bos bison* (p. 63)

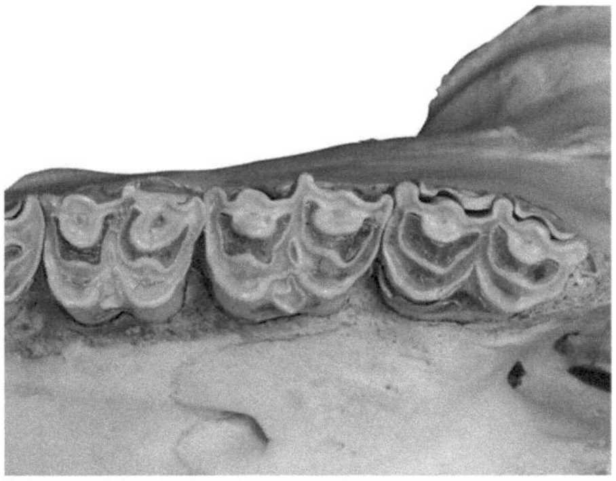

Fig. 6 The accessory columns on the inner side of molars are not reduced as in *Bos bison*

1a. Horns rugose, flat at the base, accessory columns on the inner side of molars reduced (Fig. 7); interorbital width greater than mastoid (only in Alaska and the Canadian tundra region and northernmost Minnesota, United States) .... *Ovibos moschatus* (p. 64)

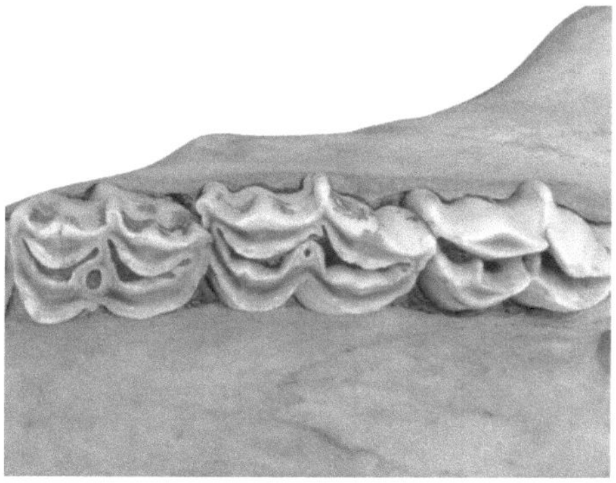

Fig. 7 The accessory columns on the inner side of molars are reduced as in *Ovibos moschatus*

Genus Tribe Bovini *Bos* Linnaeus, 1758

Bison has brown pelage, long over the forehead, neck, hump, and front quarters and short over the rear and tail; head large and carried low on the short and thick neck; horns black curved upward and inward and tapering to circular in both sexes; ears well buried in the long pelage of the head; tail relatively short and tufted at the tip; hooves black with a nearly circular print; frontal bone quadrate, flat to convex between the horns; parietal plane forming an obtuse angle with the occipital plane; occipitals broad; nasals broad, short, and triangular; premaxilla triangular.

Bos bison Linnaeus, 1758

American Bison, bisonte americano

1. *B. b. athabascae* (Rhoads, 1898). From Alaska and the Yukon south through central Alberta, British Columbia, and Saskatchewan.
2. *B. b. bison* Linnaeus, 1758. From central Alberta and Saskatchewan south through Chihuahua and Coahuila.

Conservation: *Bos bison* (as *Bison bison*) is listed as Endangered in the Norma Oficial Mexicana (DOF 2019). *B. b. athabascae* is listed as Threatened by the United States Endangered Species Act (FWS 2022); *B. b. bison,* as Threatened by the Canadian Endangered Species Protection Act; and *B. bison*, as Near Threatened (C2a (i): an estimated continuing decline in the percentage of mature individuals) by the IUCN (Aune et al. 2017).

Characteristics: Bison is very large in size, total length 3,042.0–3,803.0 mm in males and 2,132.0–2,896.0 mm in females. Dorsal pelage dark brown, head darker, long over the forehead, neck, hump, and front quarters, short over the rear and tail; with a notorious beard; hump very high; **horns small, conical, directed upwards; bone projections in frontal bones parallel to the frontal axes, with the horn cornea inserted and not covering all the parietal bones**.

Comments: Bisons are considered within the genus *Bos* (Groves 1981; Hartl et al. 1988), previously in *Bison*. Krumbiegel (1980) recognized *B. b. montanae* (northern plains bison) as a valid subspecies. *Bos bison* was found historically from Alaska and the Yukon south through Chihuahua and Coahuila and east through Ohio to northern Florida. Today, it can be found only in isolated populations in northern México and eastern United States and Canada (Map 2), in sympatry, or nearly so, with *Alce alces*, *Cervus canadensis, Odocoileus hemionus, Odocoileus virginianus,* and *Rangifer tarandus*. *B. bison* can be distinguished from all the species of other genera by having smooth conical horns in a lateral upward direction; a shoulder hump is formed by the elongation of the neural processes of thoracic vertebrae and shoulder musculature; with pelage differences between front and rear, hind quarters paler with pelage short over the rear and tail. Front quarters with pelage brown and long over the forehead, neck, and hump. Frontal bones quadrate, flat to convex between the horn cores; nasals broad, short, and triangular.

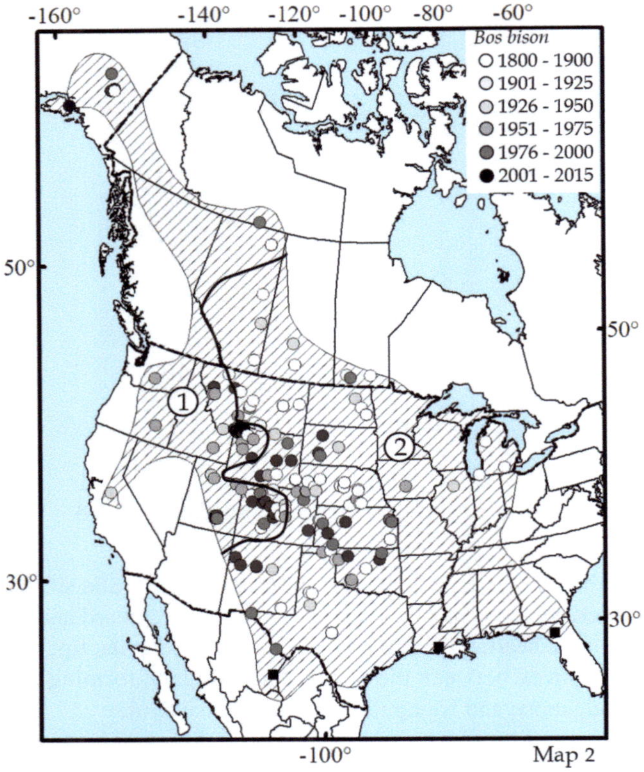

Map 2 Distribution map of *Bos bison*

Additional Literature: Skinner and Kaisen (1947), Reed (1955), Rostland (1960), Roe (1970), Christman (1971), McDonald (1981), Meagher (1986), Grubb (1995, 2005), Groves and Grubb (2011), and Wang et al. (2018).

Genus *Ovibos* Blainville, 1816

Muskox has long, **dark brown guard hairs extended almost to the ground and covering an extremely fine, cashmere-like underhair; limbs white to yellowish, contrasting with the dark coloration; horns deflected downward and slightly forward, then turning upward again**; in adult males, the horns come together in a massive boss; **preorbital gland well developed in both sexes** and with a lenticular structure lying in a shallow groove on the lacrimal bone; tail vestigial. **Skull massive with protruding orbits and upper braincase covered by horn bases.**

Ovibos moschatus (Zimmermann, 1780)

Muskox, buey almizclero

Monotypic.

Conservation: *Ovibos moschatus* is listed as Least Concern by the IUCN (Gunn and Forchhammer 2022).

Characteristics: Muskox is very large in size; total length 2,000.0–2,500.0 mm, skull length 422.0–516.0 mm in males and 385.0–430.0 mm in females. Dorsal pelage dark brown to dark gray, with long hair in all the body; upper part of the back higher than upper parts of the hips; **horns large, rugose, flattened at the base, and in** pale **color that contrasts with the body**; muzzle whitish; **horns flattened at the base, covering the parietal bones**; mastoid width similar in length to the width across the molars and less than the interorbital width.

Comments: Three subspecies have been considered for *Ovibos moschatus*: *O. m. moschatus, O. m. niphoecus,* and *O. m. wardi* (Allen 1913). However, in the latest review of the species, it was listed as monotypic (Tener 1965). *O. moschatus* ranges in the northernmost areas of the continent, from Alaska to northern Manitoba and the tundra islands of Canada and Minnesota in the United States (Map 3). It can be found in sympatry, or nearly so, with *Rangifer tarandus* and can be distinguished from it by having horns in a lateral upward direction, rugose and flat at the base, and accessory columns on the inner side of molars reduced.

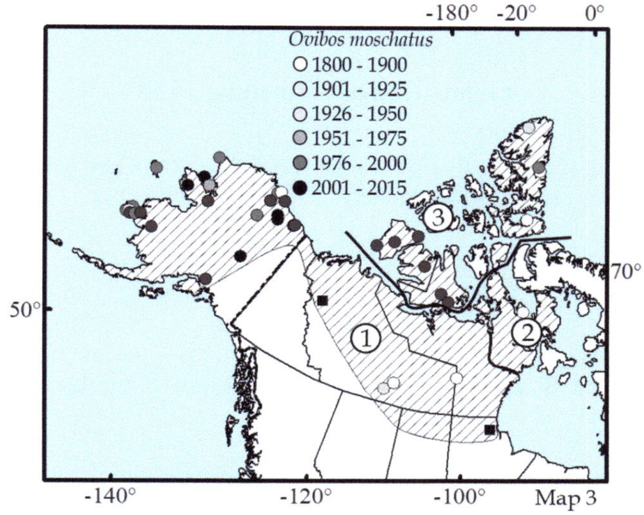

Map 3 Distribution map of *Ovibos moschatus*

Additional Literature: Zimmermann (1780), Blainville (1816), Miller et al. (1977), Lent (1988), Grubb (1995, 2005), and Groves and Grubb (2011).

Subfamily Caprinae Gray, 1821

This subfamily is characterized by dark brown, short, backward-curving, stiletto-like horns in both sexes; sexual dimorphism because males display the horn core and horn sheath; males of equal size as, or slightly larger than, females; scent glands varying in location and use. The keys were elaborated based on the review of specimens and the following sources: Lawlor (1979), Hall (1981), and Jones and Manning (1992).

1. Horn tips turning to the central line of the skull; lacrimal pits absent (from southeastern Alaska, Yukon, south through Washington, Idaho, and Montana. Populations have been introduced in other Midwest states)........................ .... *Oreamnos americanus* (p. 66)

1a. Horn tips turning outside of the central line of the skull; lacrimal pits present (Fig. 8)..2

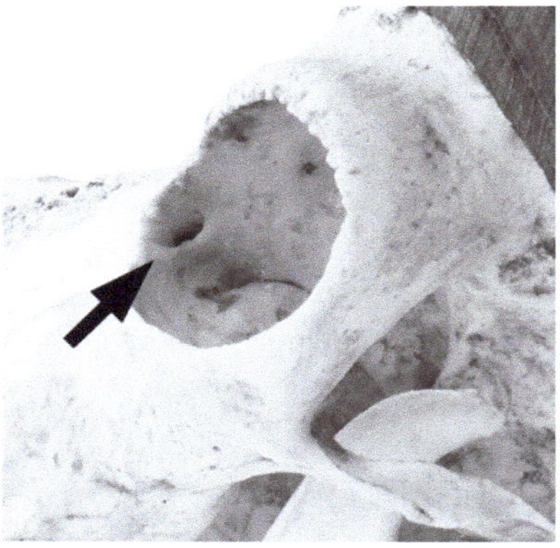

Fig. 8 Lacrimal pits present

2. Horns greater than 150.0 mm in circumference at the base; no scent glands close to the feet, in the groin, and anterior to the eyes; neither beard nor hairless calluses on the foreleg knees .... *Ovis* (p. 67)

2a. Horns less than 150.0 mm in circumference at the base; with scent glands close to the feet, in the groin, and anterior to the eyes; distinctive beard and hairless calluses on the foreleg knees .... *Capra hircus* (introduced species p. 66)

Tribe Caprini

Genus *Capra* Linnaeus, 1758

The genus is characterized by having scent glands close to the feet, in the groin, and anterior to the eyes, a distinctive beard in many species, and hairless calluses on the foreleg knees.

Capra hircus Linnaeus, 1758
Domestic goat, cabra

Conservation: *Capra hircus* is listed as an invasive species (Global Invasive Species Database 2022a) and not applicable by the IUCN (Giannatos et al. 2007).

Characteristics: The goat is a domesticated species derived from the wild goat (*Capra aegagrus*). It can be found as a feral species in many areas. The most important characteristics for identification are the presence of the beard, its small size, and hairless calluses on the foreleg knees, in addition to the scent glands.

Comments: *Capra hircus* can be found as a feral species in different parts of North America but is more common in the islands of northwestern México and isolated areas in Altiplano Mexicano.

Additional Literature: Gentry et al. (2004) and Kazanskaia et al. (2007).

Genus *Oreamnos* Rafinesque, 1817

The mountain goat has **a pelage consisting of white wool and guard hairs**, often with some dark brown hairs on the back and rump or defined by a dark brown line. In winter, it often appears yellowish; **beard pointed, approximately 100.0–135.0 mm in length; horns black, present in males and females, in posterior direction, and extremely developed in males (203.0–254.0 mm), circumference at the base less than 150.0 mm, and turning to the central line of the skull at the tip**; feet large, with oval hooves and prominent "dew claws." Skull light and fragile; lacrimal pits absent.

Oreamnos americanus (Blainville, 1816)
Mountain goat, cabra de montaña

1. *O. a. americanus* (Blainville, 1816). Restricted to the mountain areas of western British Columbia, Washington, and Oregon.
2. *O. a. columbiae* Hollister, 1912. Restricted to the mountain areas of northern-central British Columbia, Yukon, and Mackenzie.
3. *O. a. kennedyi* (Elliot, 1900). Restricted to the mountain areas of southeastern Alaska.
4. *O. a. missoulae* Allen, 1904. Restricted to the mountain areas of southeastern British Columbia, eastern Washington and Oregon, northern Idaho, and western Montana.

Conservation: *Oreamnos americanus* is listed as Least Concern by the IUCN (Festa-Bianchet 2022).

Characteristics: Mountain goats are large sized; total length 1,500–1,780 mm in males and 1,200.0–1,430.0 mm in females; basilar length is 270.0–293.0 mm in males and 230.0–255.0 mm in females. **The entire individuals are white; horns curved in backward direction, 150.0–280.0 mm in length and less than 150.0 mm in circumference;** lacrimal pits absent.

Comments: *Oreamnos americanus* ranges from the Rocky Mountains and other mountainous regions of the Western Cordillera of North America, southeastern Alaska, southern Yukon, and Alberta, south through Washington, Idaho, and Montana. Different populations have been introduced in other Midwest states (Map 4) and can be found in sympatry, or nearly so, with any other species of the suborder Rumiantia; however, it can be confused with young or females of *Ovis canadensis* and can be distinguished from other species by having black-colored horns. *O. americanus* differs from other genera by having its horns in a posterior direction, lock shaped, and extremely developed in males, less than 150.0 mm in circumference at the base and turning to the central line of the skull at the tip; lacrimal pits are absent.

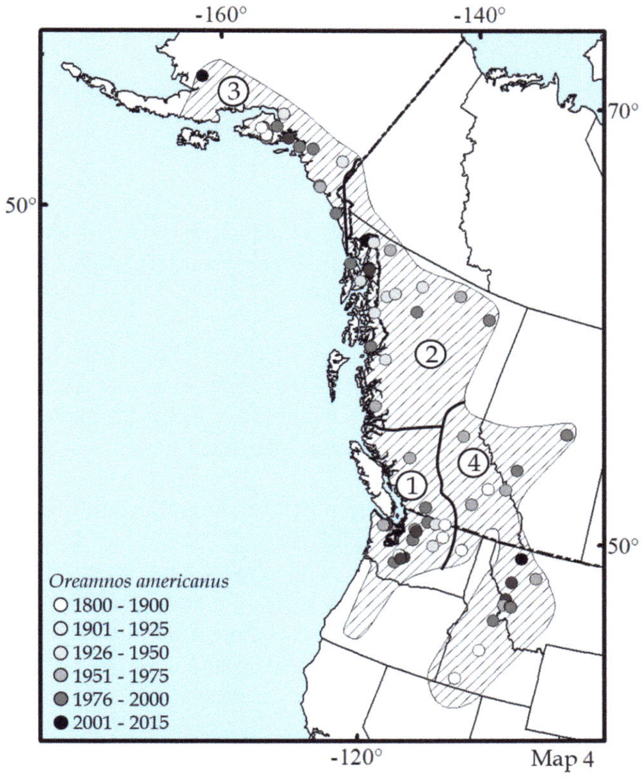

Map 4 Distribution map of *Oreamnos americanus*

Additional Literature: Allen (1904a), Seton (1929), Brandborg (1955), Hall and Kelson (1959), Dolan (1963), Cowan and McCrory (1970), Rideout and Hoffmann (1975), Grubb (1995, 2005), and Groves and Grubb (2011).

Genus *Ovis* Linnaeus, 1758

Bighorn sheep is medium sized, with **horns curved in a closed screw shape, more developed in males, barely reaching one-eighth of a turn in females; frontal bones with posterior projections are curved toward the back, where the corneal part of the horn is inserted**. The keys were elaborated based on the review of specimens and the following sources: Lawlor (1979), Hall (1981), and Jones and Manning (1992).

1. Upper parts brown; adult males with horns with more than three-fourths of a circle; palatal width more than 35% of its length; nasal length greater than 105.0 mm in males and 85.0 mm in females (from British Columbia and western Alberta south through the Baja California Peninsula, Sonora, Chihuahua, and Coahuila, and east through South Dakota)........ *Ovis canadensis* (p. 68)

1a. Upper parts brownish black or white; adult males with horns with less than three-fourths of a circle. Palatal narrowing less than 35% of its length; nasal length less than 105.0 mm in males and 85.0 mm in females (mountain ranges in Alaska, Yukon, Mackenzie, and British Columbia)........ *Ovis dalli* (p. 69).

Ovis canadensis Shaw, 1804

Bighorn Sheep, borrego cimarrón

1. *O. c. auduboni* Merriam, 1901. Western Dakotas and Nebraska.
2. *O. c. californiana* Douglas, 1829. From central British Columbia south through central California.
3. *O. c. canadensis* Shaw, 1804. From eastern British Columbia and western Alberta south through Nevada, Utah, and northern New Mexico.
4. *O. c. cremnobates* Elliot, 1904. Southern California and Baja California.
5. *O. c. mexicana* Merriam, 1901. Arizona, New Mexico, Texas, Sonora, Chihuahua, and Coahuila.
6. *O. c. nelsoni* Merriam, 1897. Southern California and western Nevada.
7. *O. c. weemsi* Goldman, 1937. Baja California Sur.

Conservation: *Ovis canadensis* is listed in México as Special Protection (DOF 2019). *O. c. nelsoni* and *O. c. sierrae* are listed as Endangered by the United States Endangered Species Act (FWS 2022), and *O. canadensis* as Least Concern by the IUCN (Festa-Bianchet 2020a).

Characteristics: Bighorn sheep is medium sized; total length 1166.0–1887.0 mm in females and 1326–1953.0 mm in males; basilar length 228.0–252.0 mm in females and 258.0–287.0 mm in males. **Dorsal pelage brown, with the snout whitish and the head of the same color as the body**; fur of underparts white; back part of the hips white; **nasal bones large**, greater than 105.0 mm in males and 85.0 mm in females; **horns large and heavy**.

Comments: *Ovis canadensis* inhabits mountain ranges from British Columbia and western Alberta south through the Baja California Peninsula, Sonora, Chihuahua, and Coahuila, including all the western United States, including New Mexico, Colorado, Nebraska, and South Dakota (Map 5). It can be found in sympatry, or nearly so, with *O. canadensis*, and can be distinguished from the species of other genera by having the horns in a posterior direction and lock shaped and extremely developed in males, greater than 150.0 mm in circumference at the base and turning outward from the central line of the skull at the tip; lacrimal pits are present. From *O. dalli,* the dorsal pelage is brown with the snout whitish and the head of the same color as the body, and the palatal width is more than 35% of its length.

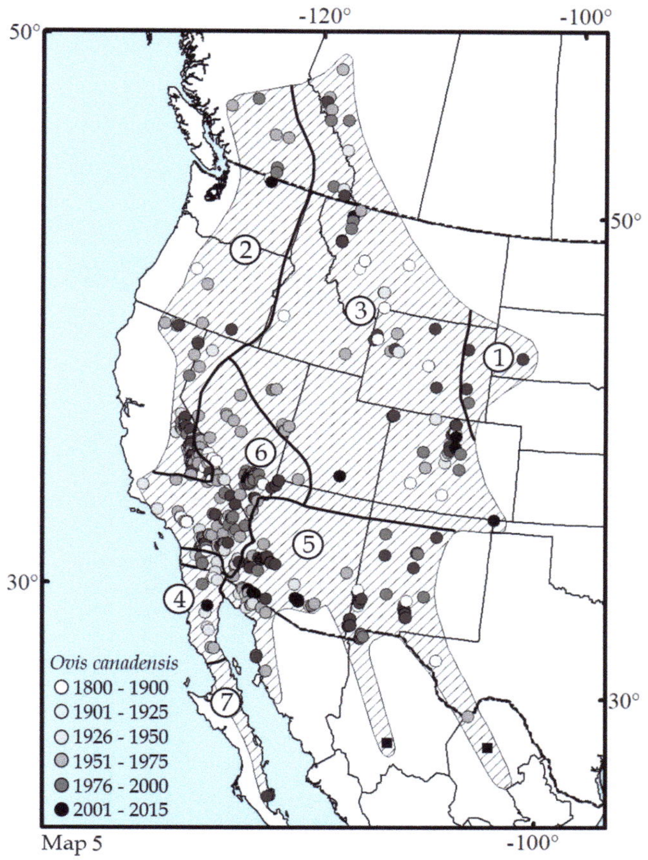

Map 5 Distribution map of *Ovis canadensis*

Additional Literature: Cowan (1940), Buechner (1960), Clark (1964), Baker and Bradley (1965), Bradley and Baker (1967), Geist (1971), Nadler et al. (1973), Trefethen (1975), Hansen (1980), Monson and Sumner (1980), Hall (1981), Shackleton (1985), and Groves and Grubb (2011).

Ovis dalli Nelson, 1884

Dall's sheep, borrego cimarrón blanco

1. *O. d. dalli* Nelson, 1884. Alaska Northern and southwestern Yukon, northwestern British Columbia, and western Makenzie.
2. *O. d. stonei* Allen, 1897. Central Yukon and northern British Columbia.

Conservation: *Ovis dalli* is listed as Least Concern by the IUCN (Festa-Bianchet 2020b).

Characteristics: Dall's sheep is medium sized; total length 1,300–1,800.0 mm in males and 1,320.0–1,620.0 mm in females; basilar length 235.0–274.0 mm in males and 215.0–238.0 mm in females. **Dorsal pelage white or brownish black; if brownish black, snout white and face white to dirty white,** paler **than the body color**; fur of underparts white; back part of the hips white; nasal bones small, less than 105.0 mm in males and 85.0 mm in females; horns relatively **small and** pale.

Comments: *Ovis dalli kenaiensis* is considered a junior synonym of *O. d. dalli* (Worley et al. 2004). It ranges in highlands of Alaska, Yukon, Mackenzie, and British Columbia (Map 6). *O. dalli* is almost never found in sympatry with any other species of the suborder Rumiantia, and can be distinguished from the species of other genera by having the horns in a posterior direction, lock-shaped and extremely developed in males, greater than 150.0 mm in circumference at the base and turning outward of the central line of the skull at the tip; lacrimal pits present. From *O. canadensis,* the dorsal pelage is white; if brown, snout is not whitish and head is darker than the body; palatal width is less than 35% of its length.

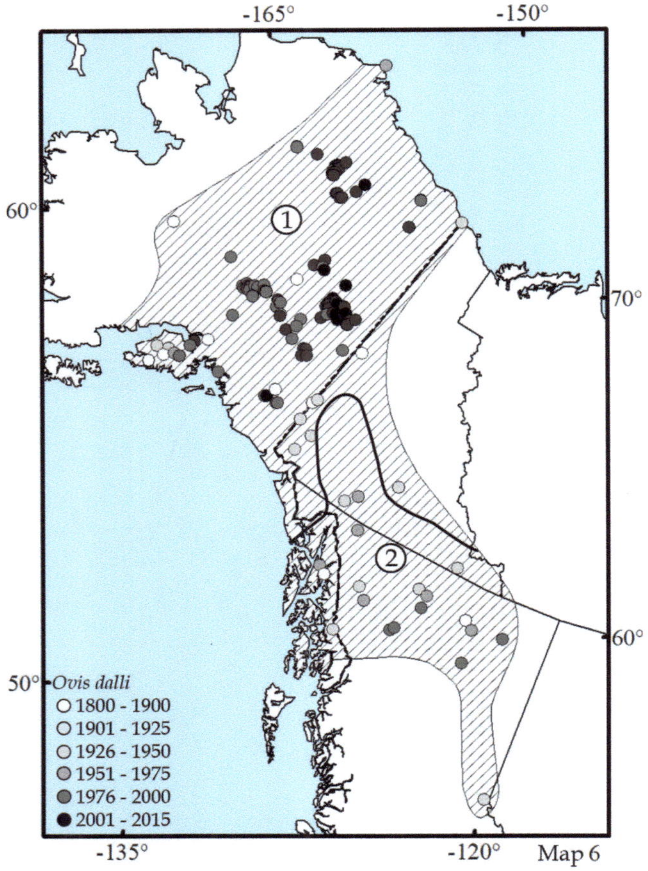

Map 6 Distribution map of *Ovis dalli*

Additional Literature: Cowan (1940), Clark (1964), Geist (1971), Nadler et al. (1973), Trefethen (1975), Hall (1981), Shackleton (1985), Bowyer and Leslie (1992), and Groves and Grubb (2011).

Family Cervidae Goldfuss, 1820

The family Cervidae includes deer, moose, and reindeer characterized by **lateral toes generally present, with constituent bones fully developed; lateral metacarpals lack either upper (telemetacarpalian) or lower (plesiometacarpalian) ends**; limbs powerful; eyes generally small and ears large; **facial glands present** (except in *Capreolus*); tarsal, metatarsal, and interdigital glands occasionally present; **sexual dimorphism**; ruminant species; placenta with few cotyledons; frontal appendages (when present) generally branched, bony, solid, deciduous antler tissue, supported on permanent skin-covered pedicels; **antlers shed and regrown each year**; upper canines long in antlerless species but absent or poorly developed in North America cervids, all of which have antlers in males at least; lower canines incisiform; molariforms selenodont, generally brachydont. The keys were elaborated based on the review of specimens and the following sources: Lawlor (1979), Hall (1981), and Jones and Manning (1992).

1. Antlers can be palmate; if not, a single main branch in the posterior or upper direction (Fig. 9); weight less than 150.0 kg; lateral metacarpals proximal (plesiometacarpal); postglenoid foramen entirely within the squamosal bone.... .... subfamily Capreolinae (p. 71)

1a. Antlers generally not palmate; if palmate, dorsal coloration completely covered with white spots; antlers with one main branch in the posterior or upper direction and one or two branches pointing forward (Fig. 9); weight greater than 150.0 kg; lateral metacarpals keep the distal ends (telemetacarpal); postglenoid foramen between the squamosal and petrous bone .... subfamily Cervinae (p. 81)

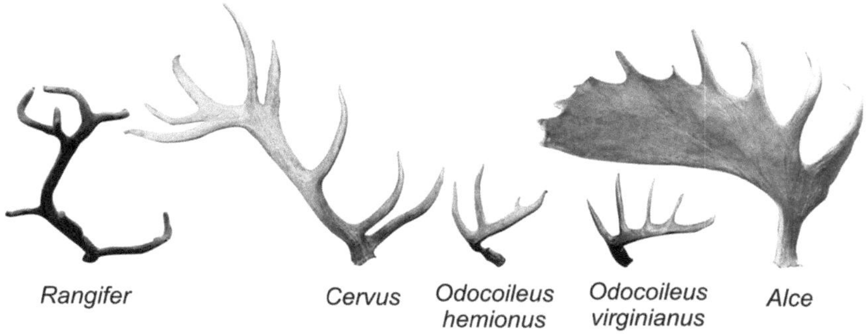

Fig. 9 Different antlers of Cervidae

Subfamily Capreolinae Brookes, 1828

This subfamily includes deer, caribou, and moose, ranging from small to large in size; **antlers instead of horns, mainly in males (except for the genus *Rangifer*),** generally with complex structures and a skin sheath named "velvet" when growing; **antlers generally shed each year**; stomach with four chambers, **lacrimal depression present; two lacrimal foramina present at the front edge of the orbit; lateral metacarpals keep the proximal (plesiometacarpal); postglenoid foramen entirely within the squamosal bone; all species present in North America without canines** (*Cervus* can have a small one). North American species are included in two tribes: Alceini (*Alces*) and Rangiferini (*Mazama, Odocoileus*, and *Rangifer*). The keys were elaborated based on the review of specimens and the following sources: Lawlor (1979), Hall (1981), and Jones and Manning (1992).

1. Antlers strongly palmate; upper part of the back notoriously over upper parts of the hips; skull greater than 500.0 mm (British Columbia to Nova Scotia and all states bordering Canada, in addition to Colorado, Utah, Wyoming, and Massachusetts) .... tribe Alceini *Alces alce* (p. 72)

1a. Antlers slightly palmate or not palmate; upper part of the back slightly or not over the hips at the same height, skull less than 500.0 mm .... tribe Rangiferini 2

2. Antlers somehow palmate (Fig. 9); hooves wider than the limbs or upper part of the back notoriously over upper part of the hips; nasal turbinate bones with increased surface area within the nostrils (British Columbia to Nova Scotia and all states of the United States bordering Canada) .... *Rangifer tarandus* (p. 80)

2a. Antlers not palmate, circular, or elliptical in cross-section (Fig. 9); hooves of similar width to limbs and upper part of the back at a similar level to upper part of the hips; nasal turbinate bones without increased surface area within the nostrils (Fig. 10) 3

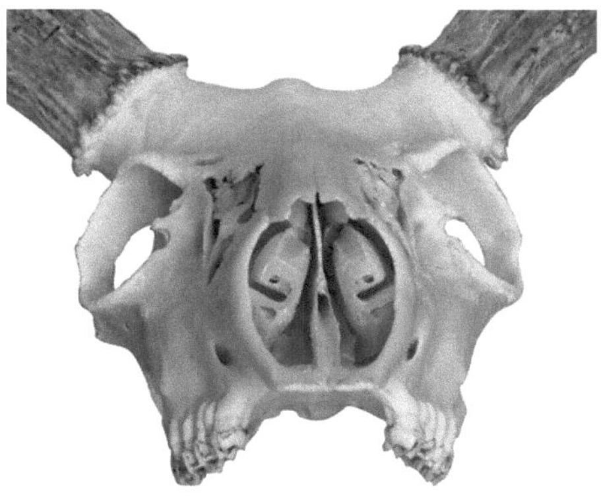

Fig. 10 The nasal turbinate bones without increased surface area within the nostrils

3. Fore shorter than hind limbs; antlers generally less than half of head length and unbranched, although in some cases a small fork is found close to the distal end (Fig. 11); auditory bullae lightly inflated; condylobasal length less than 230.0 mm .... *Mazama* (p. 73)

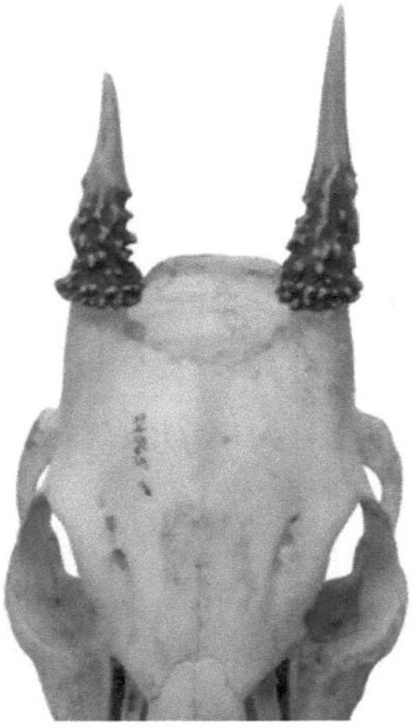

Fig. 11 The insertion of both antlers as parallel as in *Mazama temama*

3a. Limbs of equal size; antlers greater than half-length of the head, branched (Fig. 9); auditory bullae moderately inflated; condylobasal length greater than 230.0 mm .... *Odocoileus* (p. 76)

Tribe Alceini

Genus *Alces* Gray, 1821

Moose is the largest of cervids; the pelage consists of gray wool interspersed with pale brown to black guard hairs; guard hairs have a scale pattern distinguishable from that of other cervids; fur of underparts is generally pale-shaded; **with a "bell" (a pendulous dewlap of skin and hair at the neck) and a large, flexible, overhanging upper lip**; with lacrimal and small tarsal glands. **Skull greatly lengthened in the premaxillary region; sizable nasal aperture; distance from the front of nasals to the front of premaxilla about equal to that from the back of nasals to the back of occiput**; vomer lower posteriorly, not dividing the aperture of the posterior nares; lacrimal vacuity widely open; maxillary canines usually absent in both sexes and lower incisiforms; molars and premolars rather broad and low crowned; **antlers present only in males, large, with palmate blades and projecting tines.** The populations in America are considered at the subspecies level.

Alces alces (Linnaeus, 1758)
Moose, alce

1. *A. a. americana* (Clinton, 1822). From Michigan east through Newfoundland.
2. *A. a. andersoni* Peterson, 1950. From Yukon and British Columbia east through Minnesota and Ontario.
3. *A. a. gigas* Miller, 1899. Alaska and Yukon.
4. *A. a. shirasi* Nelson, 1914. From southern Alberta and Washington south through Idaho and Colorado.

Conservation: *Alces alces* is listed as Least Concern by the IUCN (Hundertmark 2016).

Characteristics: Moose is very large in size; total length 2,400.0–3,900.0 mm, skull length 538.0–686.0 mm. Dorsal pelage with a prevailing dark brown coloration, tail of the same color, short; massive body; short neck; **large head with an elongated snout; long dewlap from the neck known as "bell"**; limbs long and grayish; **shoulders higher than the rump;** hooves narrow; **long** limbs; nasal turbinate bones with increased surface area within the nostrils; **nasal region shortened with a sizable nasal aperture, distance from the front of nasals equal to the distance from the back of nasals to the back of the occipital**.

Comments: *Alces americanus* was considered a different species from *Alces alces* by its karyotype and morphology (Geist 1998; Boyeskorov 1999). Analysis of mtDNA revealed that the differences are not at the species level, and it is therefore considered the same species (Geist 1998; Hundertmark et al. 2002, 2003; Udina et al. 2002; Hundertmark and Bowyer 2004). There is still some debate about the taxonomy of the genus *Alces* at the specific level, so taxonomic changes are likely to occur in the future. *A. alces* ranges across Canada, Alaska, and northern North Dakota to New Brunswick and Nova Scotia, including from eastern Washington to western Montana south through Idaho and Colorado (Map 7). It can be found in sympatry, or nearly so, with *Cervus canadensis, Odocoileus hemionus, O. virginianus,* and *Rangifer tarandus.* It can be distinguished from these species by having the antlers strongly palmate; the upper part of the back is notoriously over the upper parts of the hips; hooves are wider than the limbs; the skull length is greater than 500.0 mm.

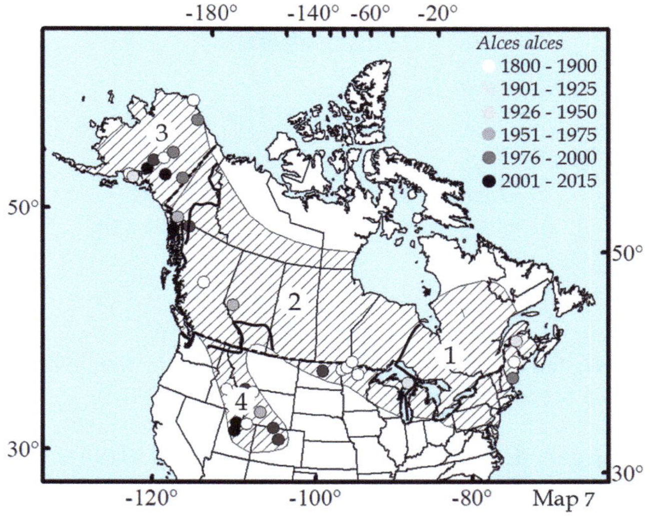

Map 7 Distribution map of *Alces alces*

Additional Literature: Peterson (1955), Hall (1981), Franzmann (1978, 1981), Franzmann et al. (1978), Groves and Grubb (2011), Randi et al. (1998), Pitra et al. (2004), Hassanin et al. (2012), and Heckeberg (2020).

Tribe Rangiferini

Genus *Mazama* Rafinesque, 1817

Temazates are small sized, throat pale but not white; tail relatively short, white ventrally; upperparts and limbs of the same color as the back; face elongated; ears in proportion to head size; facial glands small and exposed; tail short; toe-gland pits deep and triangular; skull small and back arched; **antlers conical and less than one-half of the head length, with only one beam heading back of the body**, occasionally with one branch; upper border of the orbits flat, nearly straight; auditory bullae slightly inflated; molars sometimes with supplementary columns; canines often present. The keys were elaborated based on the review of specimens and the following sources: Hall (1981), and Weber and Medellín (2010).

1. Pelage reddish; insertion of both antlers parallel (Fig. 11); internal distance between the antlers at the base equal to or less than occipital breadth (from Tamaulipas and Oaxaca south through Central America).. ....*Mazama temama* (p. 75)

1a. Pelage grayish-brown; divergent antler insertion (Fig. 12), internal distance between antlers at the base larger than occipital breadth (Yucatán Peninsula)..*Mazama pandora* (p. 74).

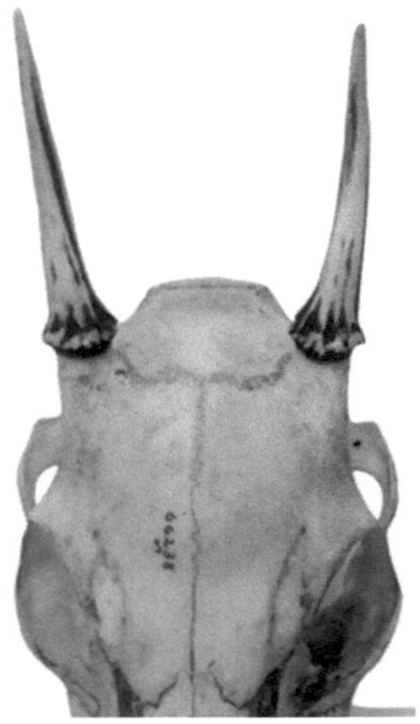

Fig. 12 The insertion of both antlers as divergent as in *Mazama pandora*

Mazama pandora Merriam, 1901

Yucatán Brown Brocket Deer, temazate café

Monotypic.

Conservation: *Mazama pandora* is listed as Vulnerable (A2c: population reduction in the last 10 years 30% or less and may not be reversible by decline in area of occupancy) by the IUCN (Weber et al. 2016).

Characteristics: Yucatán brown brocket deer is small sized; total length 1,750.0–1,850.0 mm, condylobasal length from 161.3 to 177.0 mm. **Dorsal pelage grayish brown including the whole snout;** fur of underparts brownish to whitish; **a tuft of stiff hair growing on the forehead; antlers with one or two points, may be curved inward; divergent antler insertion**.

Comments: The genus *Odocoileus* is used instead of *Mazama* (Gutiérrez et al. 2017). However, the change needs to be accepted by the community. On the other hand, given the commercial use of the species, I maintain the name *Mazama*. *Mazama pandora* was elevated to full species (Medellín et al. 1998), previously regarded as a subspecies of *M. americana pandora* (Genoways and Jones 1975) or as a subspecies of *M. gouazoubira* (Czernay 1987; Bisbal 1991). *M. pandora* ranges only in the Yucatán Peninsula, including Campeche, Yucatán, Quintana Roo, northern Guatemala, and northern Belize (Map 8). It can be found in sympatry, or nearly so, with *M. temama* only in the wettest southern areas of the Peninsula, and with *Odocoileus virginianus* in the entire Peninsula. *M. pandora* can be distinguished from *Odocoileus* by its smaller size; antlers are unbranched, a small fork is occasionally found in the distal end and less than the head length, and condylobasal length is less than 230.0 mm. From *M. temama*, by its grayish-brown coloration, antlers are more separated at the base than at the tip.

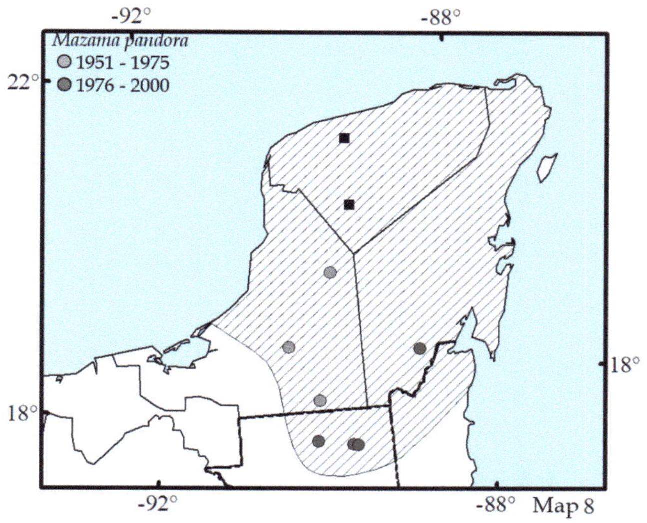

Map 8 Distribution map of *Mazama pandora*

Additional Literature: Merriam (1901d), Allen (1915b), Tate (1939), Goldman and Moore (1945), Hall (1981), Grubb (1995, 2005), Geist (1998), Pitra et al. (2004), Grubb (2005), Duarte et al. (2008), González et al. (2009), Abril et al. (2010), Weber and Medellín (2010), Groves and Grubb (2011), and Escobedo-Morales et al. (2016).

Mazama temama (Kerr, 1792)

Central American Red Brocket Deer, temazate rojo

M. t. temama (Kerr, 1792). It is the only subspecies in North America (*see addenda).

Conservation: *Mazama temama* is listed as Data Deficient by the IUCN (Bello et al. 2016).

Characteristics: Central American red brocket deer is small sized; total length 1,700.0–1,800.0 mm, condylobasal length 162.0–171.0 mm. **Dorsum reddish including all the snout, head, and neck; fur of underparts white; forelegs shorter than hind legs; antlers with one point; parallel insertion of antlers**.

Comments: *Mazama temama* is considered a different species from *M. americana* (Geist 1998). It ranges from Tamaulipas and Oaxaca south through Belize, including Campeche and Quintana Roo, but not the northern area of the Yucatán Peninsula (Map 9). It can be found in sympatry, or nearly so, with *M. pandora* only in the wettest areas of the southern part of the Peninsula and with *Odocoileus virginianus* across its range. *M. temama* can be distinguished from *Odocoileus* by its smaller size; antlers are unbranched; a small fork is occasionally found in the distal end and less than the head length, and condylobasal length is less than 230.0 mm. From *M. pandora,* it has reddish coloration and antler tips are at a similar distance than at the base. See adeendum.

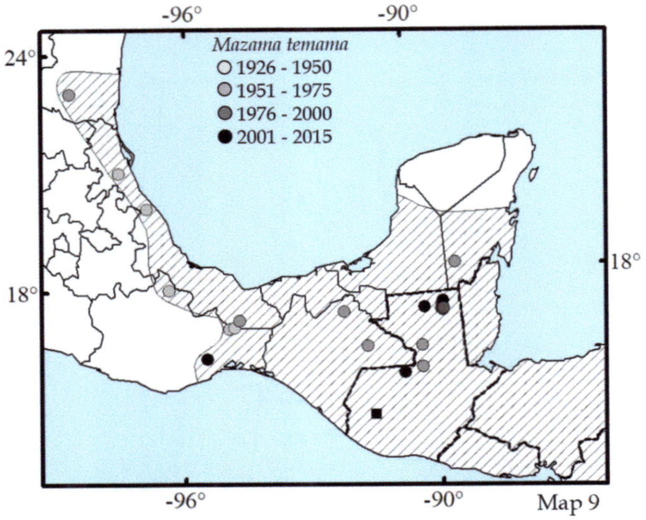

Map 9 Distribution map of *Mazama temama*

Additional Literature: Allen (1915b), Tate (1939), Goldman and Moore (1945), Genoways and Jones (1975), Hall (1981), Czernay (1987), Grubb (1995, 2005), Medellín et al. (1998), Pitra et al. (2004), Grubb (2005), Duarte et al. (2008), González et al. (2009), Abril et al. (2010), Bello-Gutiérrez et al. (2010), Groves and Grubb (2011), Gutiérrez et al. (2017), Heckeberg (2020), and Pino et al. (2022).

Genus *Odocoileus* Rafinesque, 1832

Dorsal pelage in different brown to grayish shades; **snout with black tip and upperpart pale; throat pale;** face long and narrow; rhinarium well developed; ears variable in size; tail long to medium; interdigital glands present on forefeet and hindfeet; **antlers conical and greater than one-half of the head length, branched in males, with more than one beam**; limbs of the same size; auditory bullae moderately inflated; vomer high, dividing the nostrils into two chambers posteriorly; lacrimal pits and preorbital glands small; lacrimal fossa shallow; upper canines usually absent. The keys were elaborated based on the review of specimens and the following sources: Kellogg (1956), Cowan (1956), Hall (1981), Jones and Manning (1992) and Álvarez-Castañeda et al. (2017a).

1. Tail dorsal hair brown with white edges; antlers with a main beam where points or tines grow (Fig. 9); ears approximately 50% of the head length; lacrimal pit shallow (from southern Canada south through South America, except for the Baja California Peninsula and southwestern United States) .... *Odocoileus virginianus* (p. 78)

1a. Tail dorsal hair black; antlers split progressively into two beams of equal diameter (Fig. 9); ears approximately 75% of the head length; lacrimal pit deep (from Alaska and Western Canada south through the Rocky Mountains and from the Western Plains states south through the Baja California Peninsula, Cedros, San José, Tiburon Islands to northwestern México) .... *Odocoileus hemionus* (p. 76).

Odocoileus hemionus (Rafinesque, 1817)

Mule deer, venado bura

1. *O. h. californicus* (Caton, 1876). Central California.
2. *O. h. cerrosensis* Merriam, 1898. Restricted to Cedros Island, Baja California.
3. *O. h. columbianus* (Richardson, 1829). Coastal areas from southern British Columbia to northern California.
4. *O. h. eremicus* (Mearns, 1897). Southeastern California, southern Arizona, and northwestern Sonora.
5. *O. h. fuliginatus* Cowan, 1933. Southern California and central northern Baja California.
6. *O. h. hemionus* (Rafinesque, 1817). From Yukon to Nevada east through Keewanti to Oklahoma, including Minnesota and Iowa.

7. *O. h. inyoensis* Cowan, 1933. Restricted to central-western California.
8. *O. h. peninsulae* (Lydekker, 1898). Restricted to the central Baja California Peninsula south through Baja California Sur.
9. *O. h. sheldoni* Goldman, 1939. Restricted to Tiburón Island, Sonora.
10. *O. h. sitkensis* Merriam, 1898. Coastal areas of British Columbia.

Conservation: *Odocoileus hemionus cerrosensis* is listed as Endangered; *O. h. sheldoni* as Threatened in the Norma Oficial Mexicana (DOF 2019); and *O. hemionus* as Least Concern by the IUCN (Sanchez-Rojas and Gallina-Tessaro 2016).

Characteristics: Mule deer is medium sized; total length 1,370.0 to 1,800.0 mm, condylobasal length 236.0 to 252.0 mm. Dorsal pelage brown; **tail short and narrow; fur of underparts white and black at the tip** and dorsal parts and limbs white; **ears large, around two-thirds of the face length; antlers without a main beam from where the beams grow**; antlers bifurcated with more than two points in adults; **lacrimal pit deep**.

Comments: Hall (1981) uses the genus *Dama* in place of *Odocoileus*, but this was not accepted following the International Commission on Zoological Nomenclature (opinion 581; Jones et al. 1982). The subspecies of *Odocoileus hemionus* can be partitioned into two divisions, Black-tailed deer (*O. h. columbianus* and *O. h. sitkensis*) and Mule deer (*O. h. californicus, O. h. cerrosensis, O. h. eremicus, O. h. fuliginatus, O. h. hemionus, O. h. inyoensis, O. h. peninsulae,* and *O. h. sheldoni*; Heffelfinger 2000; Latch et al. 2009). *O. h. crooki* was not previously considered a valid subspecies because it is a hybrid between *O. virginianus* and *O. hemionus*. *O. hemionus* ranges throughout western North America from Alaska and Western Canada through the Rocky Mountains and Western Plains States of the United States to the Dakotas and Texas south through to the Baja California Peninsula and Cedro, San José, and Tiburón Islands in northwestern México (Map 10). *O. hemionus* is sympatric, or nearly so, with *Alces alces, Cervus canadensis, O. virginianus,* and *Rangifer tarandus*. It can be distinguished from these species because its antlers split progressively in two beams of the same diameter, without forward beams, and ears are large in size, around two-thirds of the face length.

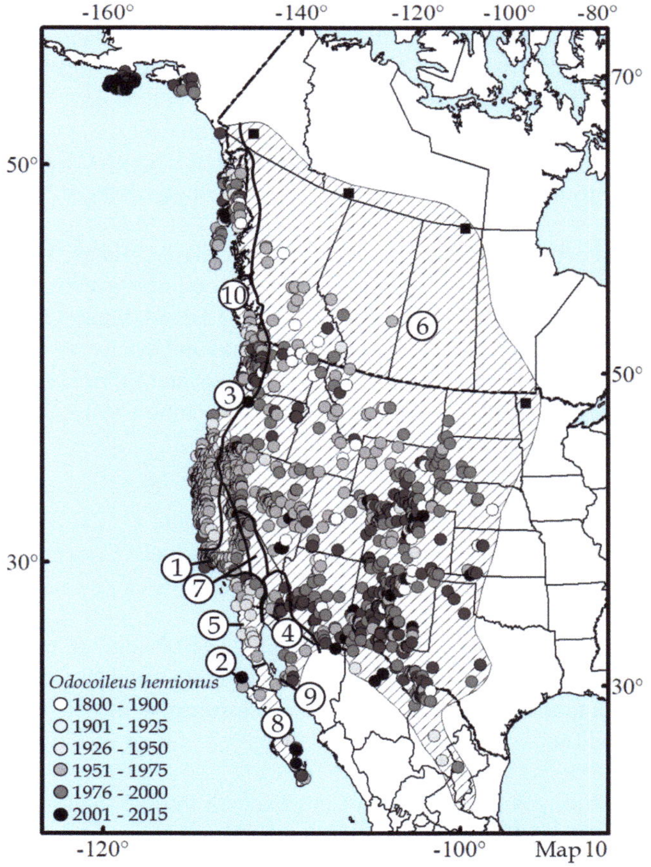

Map 10 Distribution map of *Odocoileus hemionus*

Additional Literature: Cowan (1936, 1956), Kellogg (1956), Geist (1981, 1998), Wallmo (1981a), Anderson and Wallmo (1984), Heffelfinger (2000), Pitra et al. (2004), González et al. 2009, Groves and Grubb (2011), Cortés-Calva et al. (2013), and Gutiérrez et al. (2017).

Odocoileus virginianus (Zimmermann, 1780)

White-Tailed Deer, venado cola blanca

1. *O. v. acapulcensis (*Caton, 1877). Coastal areas of Michoacán, Guerrero, and Oaxaca.
2. *O. v. borealis* Miller, 1900. From Ontario, central Minnesota Michigan, and Illinois eastward and from Indiana, Ohio, and Pennsylvania north through Labrador Peninsula and New Brunswick.
3. *O. v. carminis* Goldman and Kellogg, 1940. Coahuila and western Chihuahua.
4. *O. v. clavium* Barbour and Allen, 1922. Restricted to Key West, Florida.
5. *O. v. couesi* (Coues and Yarrow, 1875). From Arizona and New Mexico south through Sinaloa and northern Jalisco.
6. *O. v. dacotensis* Goldman and Kellogg, 1940. From Mackenzie River south through Minnesota, South Dakota, northern Colorado.
7. *O. v. hiltonensis* Goldman and Kellogg, 1940. Restricted to Hilton Head Island, South Carolina.
8. *O. v. leucurus* (Douglas, 1829). Coasts of Washington and Oregon.
9. *O. v. macrourus* (Rafinesque, 1817). From southern South Dakota and Minnesota south through Louisiana, including Nebraska.
10. *O. v. mcilhennyi* Miller, 1928. Coasts of Texas and Louisiana.
11. *O. v. mexicanus* (Gmelin, 1788). From Guanajuato and San Luis Potosí south through Guerrero, Puebla, and Oaxaca.
12. *O. v. miquihuanensis* Goldman and Kellogg, 1940. From southern Coahuila, Nuevo León, and Tamaulipas south through Zacatecas and San Luis Potosí.
13. *O. v. nelsoni* Merriam, 1898. Central Chiapas.
14. *O. v. nigribarbis* Goldman and Kellogg, 1940. Restricted to Blackbeard Island, Georgia.
15. *O. v. oaxacensis* Goldman and Kellogg, 1940. Restricted to central Oaxaca.
16. *O. v. ochrourus* Bailey, 1932. From British Columbia south through Oregon, Idaho, and Wyoming.
17. *O. v. osceola* (Bangs, 1896). Coastal Mississippi, Alabama, and northwestern Florida.
18. *O. v. seminolus* Goldman and Kellogg, 1940. Restricted to the Florida Peninsula.
19. *O. v. sinaloae* Allen, 1903. Sinaloa, Nayarit, Jalisco, and central Michoacán.
20. *O. v. taurinsulae* Goldman and Kellogg, 1940. Restricted to Bull Island, South Carolina.
21. *O. v. texanus* (Mearns, 1898). From Colorado and Kansas south through northern Sonora, Coahuila, Nuevo León, and Tamaulipas.
22. *O. v. thomasi* Merriam, 1898. The Tehuantepec isthmus, coasts of Chiapas, Tabasco, and southwestern Campeche.
23. *O. v. toltecus* (Saussure, 1860). Western Puebla, southern Veracruz, and northwestern Oaxaca.
24. *O. v. venatorius* Goldman and Kellogg, 1940. Restricted to Hunting Island, South Carolina.
25. *O. v. veraecrucis* Goldman and Kellogg, 1940. Southern Tamaulipas and Veracruz.
26. *O. v. virginianus* (Zimmermann, 1780). From Kentucky, West Virginia, and Virginia south through Mississippi, Alabama, and Georgia, and from Kentucky and Mississippi east through the Atlantic coast.
27. *O. v. yucatanensis* (Hays, 1872). Restricted to the Yucatán Peninsula.

Conservation: *Odocoileus virginianus leucurus* is listed as Threatened and *O. v. clavium* as Endangered by the United States Endangered Species Act (FWS 2022), and *O. virginianus* is listed as Least Concern by the IUCN (Gallina and López-Arévalo 2016).

Characteristics: White-tailed deer is medium sized; total length 1340.0–2062.0 mm, condylobasal length 170.0 to 246.5 mm. Dorsal pelage brown; **long tail white** ventrally and dorsally same as back; rump same as back; **ears in proportion to the head size; antlers with a main beam from where the beams grow, antler bifurcations**; antlers with more than two points in adults; lacrimal pit shallow.

Comments: *Odocoileus virginianus* ranges from southern Canada south through South America and in most of the United States and México. It is absent in western Colorado, Utah, northeastern New Mexico, northern Arizona, and the Baja California Peninsula (Map 11). It can be found in sympatry, or nearly so, with *Alces alces, Cervus canadensis, Mazama pandora, M. temama, O. hemionus,* and *Rangifer tarandus. O. virginianus* can be distinguished from the species of other genera by its long tail, fur of underparts white, and antlers with a main beam from where the beams grow, without forward beams.

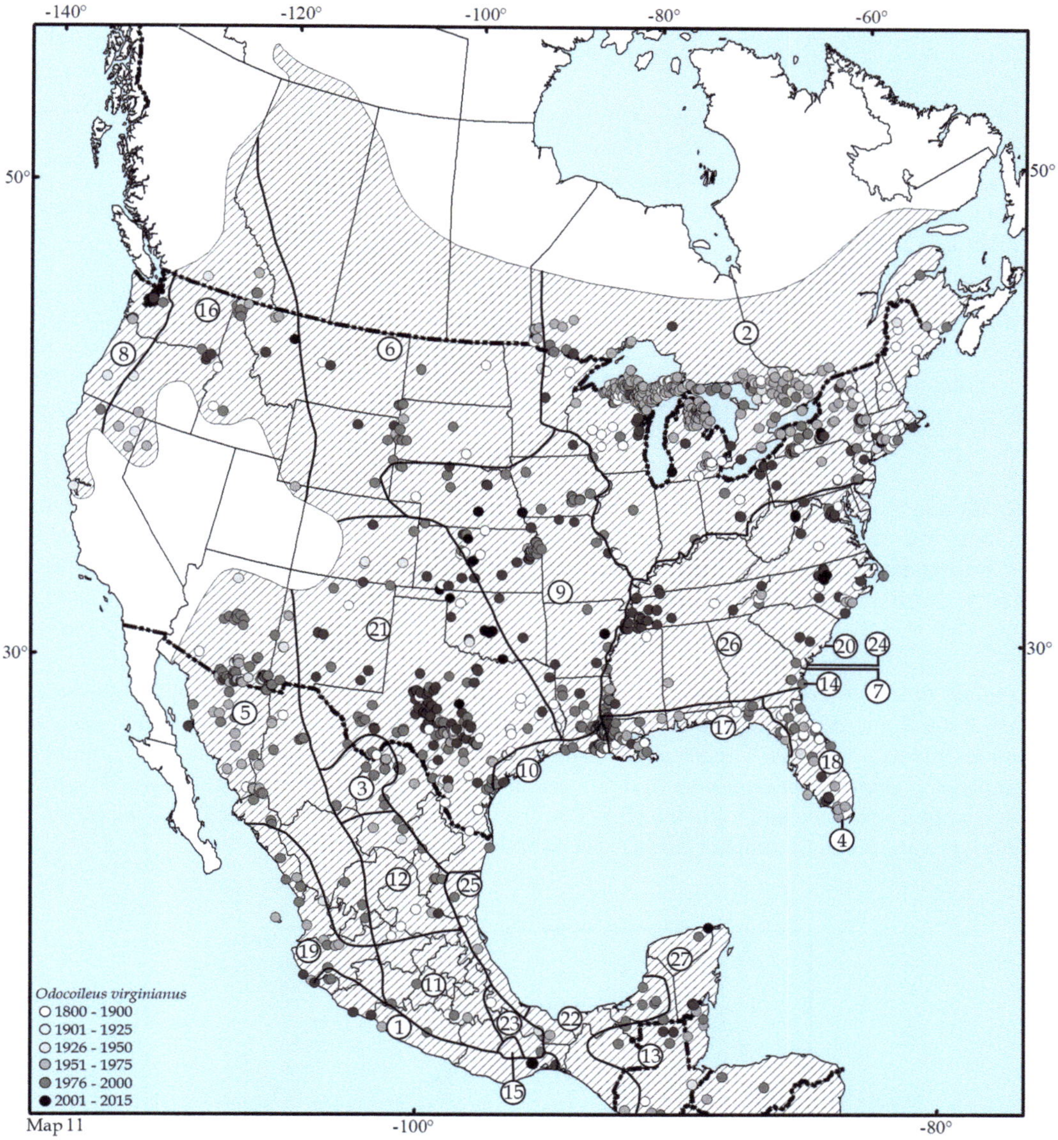

Map 11 Distribution map of *Odocoileus virginianus*

Additional Literature: Cowan (1936, 1956), Kellogg (1956), Taylor (1956), Teer et al. (1965), Halls (1978, 1984), Hall (1981), Baker (1984a), McCabe and McCabe (1984), Geist (1998), Smith (1991), Pitra et al. (2004), Groves and Grubb (2011), Hassanin et al. (2012), and Heckeberg (2020).

Genus *Rangifer* Smith, 1827

Caribou are of moderate size; pelage from pale gray or pale brown to dark brown or blackish; **mane on neck,** paler **grayish or creamy and with long hair in the ventral area; underpart usually darker than upperparts**; ears and tail short; muzzle wide and hairy; **lateral hoofs functional; main hoofs broad, deeply cleft, almost circular when held together**; **tarsal glands present and metatarsal glands absent; antlers large, variable, semipalmate distally, usually present in both sexes; lacrimal vacuity large**; vomer high posteriorly, completely dividing the posterior nares; orbital cavities not extended backward under the horn base pedicles; **lower incisors relatively small; upper canine usually present in both sexes**; posterior lobe of last molars small.

Rangifer tarandus (Linnaeus, 1758)

Caribou, Reindeer, reno, caribú

1. *R. t. caribou* (Gmelin, 1788). From southeastern Alaska, Yukon, and Mackenzie east through Newfoundland and New Brunswick, including the northern Idaho, North Dakota, Minnesota, Wisconsin, and Michigan.
2. *R. t. dawsoni* Thompson-Seton, 1900. From Queen Charlotte Island, British Columbia.
3. *R. t. groenlandicus* (Linnaeus, 1767). From Alaska eastward and north through Yukon, Makenzie, Keewatin, and Franklin.
4. *R. t. pearyi* Allen, 1902. Canadian Arctic Islands.

Conservation: *Rangifer tarandus dawsoni* from Queen Charlotte Island, British Columbia, is listed as extinct; *R. t. groenlandicus* as Special Concern; and *R. t. pearyi* as Threatened by the Canadian Endangered Species Protection Act. *R. t. caribou* is listed as Endangered by the United States Endangered Species Act, and *R. tarandus* is listed as Vulnerable (A2a; population reduction in the last 10 years 30% or less and may not be reversible by direct observation) by the IUCN (Gunn 2016).

Characteristics: Reindeer is medium sized; total length 1,700.0–2,100.0 mm, skull length 295.0–435.0 mm. Dorsal pelage varies among regions, sex, and season, from very dark brown in summer to nearly white in winter; white belly, neck, and above the hooves; **hooves large, wide, and concave; antlers candelabra-like in both sexes; neck and limbs long;** ears and tail short; males with a well developed neck mane; lacrimal vacuities large; vomer high posteriorly, completely dividing the posterior nares.

Comments: *Rangifer tarandus granti* is considered a junior synonym of *R. t. caribou*. *Rangifer tarandus* can be found historically mainly across Canada, Alaska, Washington, Idaho, Montana, Minnesota, Wisconsin, Michigan, Ohio, and Maine. Today, it is mainly restricted to northern Canada and Alaska (Map 12) and can be found in sympatry, or nearly so, with *Alces alces, Cervus canadensis, Odocoileus hemionus,* and *O. virginianus*. *R. tarandus* can be distinguished from species of other genera by its antlers slightly palmate and with one branch to the front; there is mane on neck; hooves are of similar width to the limbs; lateral hooves are functional; upper canines are usually present in both sexes.

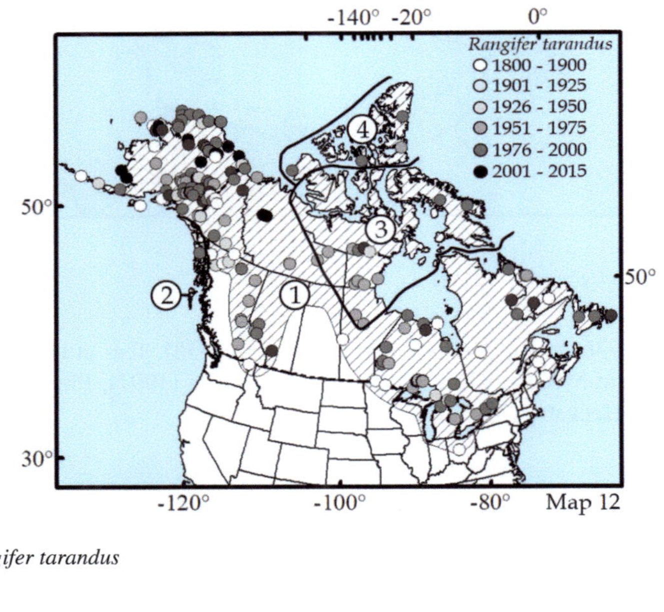

Map 12 Distribution map of *Rangifer tarandus*

Additional Literature: Banfield (1962a), Hall (1981), Thomas and Everson (1982), Røed et al. (1991), Cronin (1992), Geist (1998), Randi et al. (1998), Pitra et al. (2004), Grubb (2005), Gilbert et al. (2006), Groves and Grubb (2011), Hassanin et al. (2012), and Heckeberg (2020).

Subfamily Cervinae Godfuss, 1820

This subfamily includes Elk or Wapiti from America and introduced species; **antlers instead of horns, only found in males,** with complex structures and with a skin sheath named "velvet" when growing; antlers shed each year; stomach with four chambers; **lacrimal depression present; two lacrimal foramina present at the front edge of the orbit; lateral metacarpals keep the distal ends (telemetacarpal). Postglenoid foramen between squamosal and petrous bones; all species present in North America without canines**. Species present in North America are included in the tribe Cervini (*Axis, Cervus,* and *Dama*).

1. Antlers palmate; anal shield with a white background and framed by two black stripes that give it the shape of a broken heart, as it is crossed by the black tail (invasive species, British Columbia, California, Georgia, Kentucky, and Tennessee)........ *Dama dama* (introduced species p. 83)

1a. Antlers never palmate in any form; anal shield white but not framed by two black stripes........ *2*

2. Dorsal pelage from golden to rufous, completely covered in white spots; belly, rump, throat, insides of limbs, ears, and tail white throughout; tail long and marked with a dark stripe that stretches along its length; antlers with three prongs (invasive species, California, Florida, and Texas)........ *Axis axis* (introduced species p. 81)

2a. Dorsal pelage with white spots only in young individuals; chin, throat, and belly paler or grayish; tail medium to short; antlers subcylindrical or somewhat flattened (Texas, Colorado, Nebraska, Dakotas, Montana, and eastern........ *Cervus canadensis* (p. 81)

Tribe Cervini ***Axis axis*** (Erxleben, 1777)

Chital, ciervo moteado

Conservation: *Axis axis* was previously considered an invasive species (Global Invasive Database 2022b) and Least Concern by the IUCN (Masseti and Mertzanidou 2008).

Characteristics: Chital is medium sized; total length 1,500.0–1,800.0 mm. **Dorsal pelage from golden to rufous, completely covered in white spots;** underparts, rump, throat, insides of limbs, ears, and tail white throughout; **antlers with three prongs;** tail long and marked with a dark stripe that stretches along its length.

Comments: *Axis axis* is an invasive species in California, Florida, and Texas. *A. axis* is sympatric, or nearly so, with *Odocoileus hemionus* and *O. virginianus. A. axis* can be distinguished from *Odocoileus* by having the **dorsal pelage from golden to rufous, completely covered in white spots; antlers with three prongs**.

Additional Literature: Hanák and Mazák (1991), Nowak (1999), Grubb (2005), Raman (2013), and Duckworth et al. (2015).

Genus ***Cervus*** Linnaeus, 1758

Wapiti has a coat uniformly colored or spotted, young generally spotted; **chin, throat, and belly paler or grayish**; muzzle bare; tail medium to short; **main hooves long, pointed, hind pair united throughout most of the basal length by a deep interungual membrane; antlers subcylindrical or somewhat flattened**. Skull long and narrow; frontal-parietal region rounded; prelacrimal vacuity fairly shallow; **upper molars moderately short-crowned and wide; three crowns of lower incisors varying in size** and no upper incisors.

Cervus canadensis Erxleben, 1777

Elk, ciervo rojo

Monotypic.

Conservation: *Cervus canadensis* is considered Least Concern by the IUCN (Brook et al. 2018).

Characteristics: Elk is large sized; total length 130.0–160.0 mm**. Head and neck pelage dark brown**, dorsum paler; rump creamy white; **rump with a patch that widens laterally at the top, the spot may vary geographically**; fur of underparts darker than in the dorsum**. Antlers conical and pale or darker, with one main beam regularly with six smaller beams** of approximately two times the head length in males; females without antlers; limbs of equal size and darker than the body; auditory bullae moderately inflated.

Comments: Two species have been considered in the genus *Cervus*: *Cervus canadensis* and *C. elaphus* (Wilson and Reeder 2005; Zhang and Zhang 2012; Frey and Riede 2013). Lorenzini and Garofalo (2015) proposed using *C. elaphus* for North America. However, *C. canadensis* is still in use (Heckeberg et al. 2016; Meiri et al. 2018; Heckeberg 2020). The many differences between the species include the form of the patch of the upper rump and the color and form of the antler. In *C. canadensis*, the rump patch flares out laterally on the upper rump, then contrasts confined just to the inner margins of the upper thighs, and antlers are pale-hued with the distal part flattened. In *C. canadensis*, the patch does not flare on the upper rump and is more nearly parallel-sided to about halfway down the back of the thighs, and the antlers are darker hued with the distal part not flattened. The subspecies *C. e. manitobensis, C. e. merriami, C. e. nannodes, C. e. nelsoni,* and *C. e. roosevelti* are considered junior synonyms of *C. e. canadensis* (Lorenzini and Garofalo 2015). *C. canadensis* ranges from British Columbia and Alberta, south through Oregon, California, New Mexico, and Texas, and east through Québec, from New York and Pennsylvania to Georgia and across the southern United States; except for Alabama, Georgia, and Mississippi, all its range is found in isolated populations (Map 13). It can be found in sympatry, or nearly so, with *Alces alces, Odocoileus hemionus, O. virginianus,* and *Rangifer tarandus* and can be distinguished from these species by its antlers with one main branch in the posterior or upper direction; one or two branches forward; back of the nasal cavity not divided by the vomer.

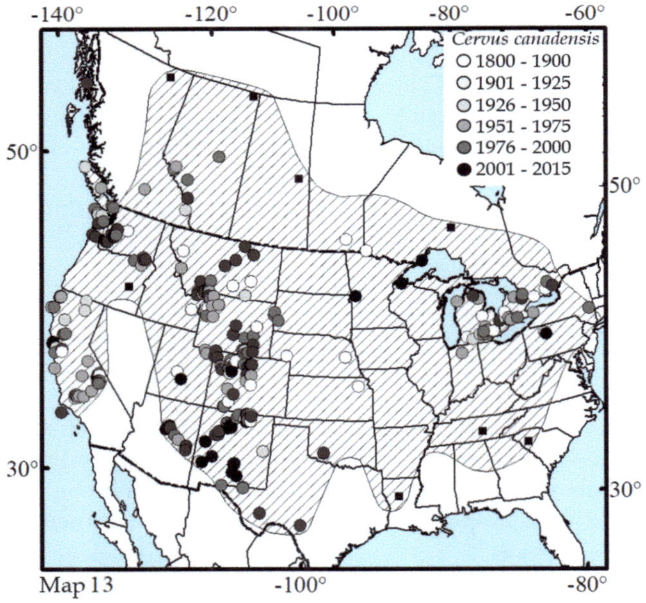

Map 13 Distribution map of *Cervus canadensis*

Additional Literature: Gustavsson and Sundt (1968), Geist (1998), Pitra et al. (2004), Groves (2005, 2006), Groves and Grubb (2011), Marcot (2007), Hassanin et al. (2012), Heckeberg et al. (2016), Kuwayama and Ozawa (2000), Polziehn and Strobeck (2002), and Heckeberg (2020).

Dama dama (Linnaeus, 1758)

Fallow Deer, gamo común

Conservation: *Dama dama is* considered an invasive species (Global Invasive Database 2022c) and Least Concern by the IUCN (Masseti and Mertzanidou 2008).

Characteristics: Fallow deer is medium sized; total length 1,180.0–1,550.0 mm. **Antlers palmate; dorsal and lateral coloration reddish-brown, hair dotted with white spots in spring and summer** (occasionally with a dark band on the back); **anal shield with a white background and framed by two black stripes that give it the shape of a broken heart, as it is crossed by the black tail.**

Comments: *Dama dama* is an invasive species in British Columbia, one barrier island in Georgia, western Kentucky, Tennessee, and different parts of California and can be found in sympatry, or nearly so, with *Alces alces, Cervus canadensis, Odocoileus hemionus,* and *O. virginianus. D. dama* can be distinguished from other genera by having antlers palmate; dorsal and lateral coloration is reddish-brown, hair dotted with white spots in spring and summer (occasionally with a dark band on the back); anal shield with a white background and framed by two black stripes that give it the shape of a broken heart, as it is crossed by the black tail.

Additional Literature: Blanco (1998), Apollonio (1999), Nowak (1999), and Grubb (2005).

Suborder Suina Gray, 1868

The species of the suborder are medium sized; eye small in relation to the head; snout ending in a cartilaginous disc; no horns or antlers; stomach complex, not ruminating, with three chambers; hindfoot with three or four digits; cannon bone lacking throughout the feet. Skull: upper canines directed downward or upward; two or three upper incisors; molars bunodont; molars with square quadricuspidate crowns.

1. Hindfoot with four digits; paroccipital process very large; upper canines directed outward and upward......................... .... Family Suidae; subfamily Suinae; tribe Suini; *Sus scrofa* (introduced species, p. 83)

1a. Hindfoot with three digits; paroccipital process small; upper canines directed downward... .... Family Tayassuidae; subfamily Tayassuidae (p. 84)

Family Suidae Gray, 1821

The species of the family are medium sized; eyes small in relation to the head; body sparsely haired or with coarse bristles; **snout elongated, mobile, flattened at the end;** stomach complex, not ruminating, with three chambers; **hindfoot with four digits; metacarpal and metatarsal not fused**; **upper canines directed outward and upward; molars forming a continuous series, decreasing gradually, with the largest in the middle of the molar series.**

Subfamily Suinae Gray, 1821

This subfamily includes all the living species within the family Suidae.

Tribe Suini

Sus scrofa Linnaeus, 1758
Wild Boar, cerdo salvaje

Conservation: *Sus scrofa* is listed as an invasive species (Global Invasive Database 2022d) and Least Concern by the IUCN (Keuling and Leus 2019).

Characteristics: Wild boar is medium sized; total length 1,100.0–2,000.0 mm, condylobasal length 221.0–270.0 mm. Body bulky, massively built, with limbs short and relatively thin; trunk short and robust, with hindquarters relatively underdeveloped; region behind the shoulder blades rising into a hump, neck short and thick to the point of being nearly immobile; head very large, taking up to one-third of the entire body length; **eyes small and deep-set; ears long and broad; canine teeth well developed, protruding from the mouth of adult males**.

Comments: *Sus scrofa* has been introduced in more than 61 states of Canada, México, and the United States. This species can be distinguished from all the native species, *Dicotyles* and *Tayassu*, by having the upper canines directed outward and upward and hindfoot with four digits.

Additional Literature: Lawlor (1979), Wood and Barrett (1979), Mayer and Brisbin (2009), and Sweeney et al. (2003).

Family Tayassuidae Palmer, 1897

The species in this family are medium sized; eye small relative to the head; snout ending in a cartilaginous disc; stomach complex, not with three chambers; hindfoot tridactyl. Skull: upper canines directed downward, not outward or upward, and short and straight; third upper incisors absent; the first upper and lower premolars and molars forming a continuous series, increasing gradually in size from first to last; fourth upper and lower premolars as complex as molars; molars with square crowns.

Subfamily Tayassuinae Palmer, 1897

This subfamily includes peccary, white-lipped peccary, wild hogs, and wild pigs or American wild boar. They are medium sized (height 75.0–105.0 cm); body covered with bristle hairs; three to four digits, only two functional in locomotion; snout extended, moveable, flattened anteriorly; two metatarsals and metacarpals fused together; upper and lower canines similar in size and with sharp cutting edges. Distributed from northern México to Argentina. The keys were elaborated based on the review of specimens and the following sources: Woodburne (1968), Hall (1981), and Álvarez-Castañeda et al. (2015).

1. Skin with or without a white or yellowish collar; upper lips not white. Maxilla not expanded laterally to the level of the molars; with a crest in the palate at the level of the canine and first premolars; skull length less than 260.0 mm (Texas and Arizona south through Central America) .... *Tayassu tajacu* (p. 85)

1a. Skin without a white or yellowish collar; upper lips white. Maxilla expanded laterally to the height of the molars; without a crest in the palate; skull length greater than 260.0 mm (from southern Veracruz and the southern region of the Yucatán Peninsula south through South America) .... *Tayassu pecari* (p. 86).

Genus *Tayassu* Fischer [von Waldheim], 1814

White-lipped jabalies or peccaries have the dorsum speckled from dark brown to black; **fur of underparts white in the inguinal and pelvic regions and on the throat and distal rostrum;** body robust and limbs proportionally short; head triangular; a single medial dewclaw on the posterior side of the hindfoot; **canines triangular**, with their length equal to or less than the distance of the diastema; sagittal crest absent in the posterior part of the skull or barely developed; **narial notch long, acuminate posteriorly; rostrum slightly concave, dorsum of rostrum flat; infraorbital foramen located more posteriorly under the zygomatic shelf, infraorbital opening narrow or slit-like**.

Tayassu tajacu (Linnaeus, 1758)

Collared peccary, jabalí de collar

1. *T. t. angulatus* (Cope, 1889). From Texas south through San Luis Potosí and Tamaulipas.
2. *T. t. crassus* Merriam, 1901. From southern Tamaulipas and central San Luis Potosí south to southern Veracruz.
3. *T. t. humeralis* Merriam, 1901. From Colima to Oaxaca.
4. *T. t. nanus* Merriam, 1901. Restricted to Cozumel Island.
5. *T. t. nelsoni* (Goldman, 1926). South and Central Chiapas and Quintana Roo south through Central America.
6. *T. t. sonoriensis* (Mearns, 1897). From Arizona south to Jalisco and Colima.
7. *T. t. yucatanensis* Merriam, 1901. Tabasco, Campeche, Yucatán, northern Chiapas, and Belize.

Conservation: *Tayassu tajacu* (as *Pecari tajacu*) is listed as Least Concern by the IUCN (Gongora et al. 2011).

Characteristics: Collared peccary is medium sized; total length 900.0–1,390.0 mm, condylobasal length 221.0–270.0 mm. **It has a distinguishable** paler **collar from the shoulders to the chest; hairs with a wider and darker band;** commonly showing a dorsal stripe with hairs longer and darker than in the lateral area.

Comments: The species of *Tayassu* were previously considered within the genus *Pecari* (Grubb 1993, 2005). *T. tajacu* was divided into two different species: *Dycotiles angulatus* and *D. crassus* following Woodburne (1968), Husson (1978), and Groves and Grubb (2011). However, the change is not accepted until more in-depth molecular and morphological studies are used to delimit species within the clade. I still use *T. tajacu,* which ranges in lowlands from Texas and Arizona south through Central America (Map 14), and can be found in sympatry, or nearly so, with *Tayassu pecari. T. tajacu* can be distinguished from *T. pecari* by having a white hair collar but no white spots in the inferior jaw, cheeks, and throat.

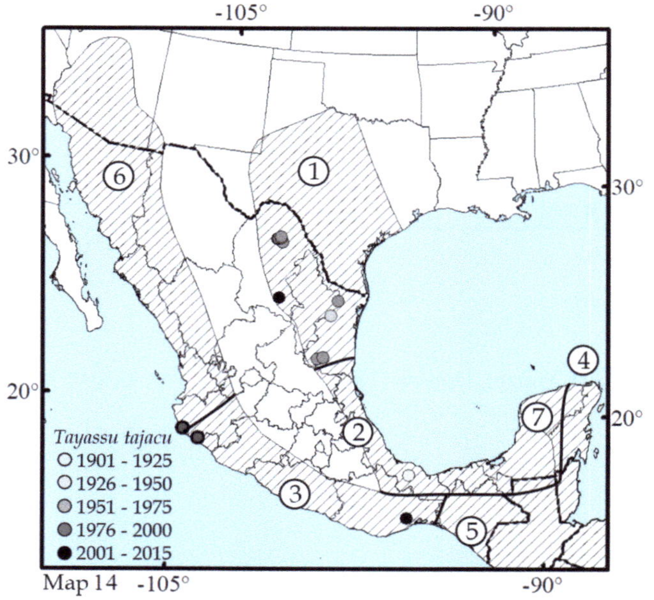

Map 14 Distribution map of *Tayassu tajacu*

Additional Literature: Miller (1914), Mayer and Wetzel (1987), and Wright (1989).

Tayassu pecari (Link, 1795)

White-lipped Peccary, jabalí de labios blancos

T. p. ringens Merriam, 1901. Only subspecies in North America.

Conservation: *Tayassu pecari* is listed as Endangered in the Norma Oficial Mexicana (DOF 2019) and as Vulnerable (A2bcde + 3bcde: population reduction in the last 10 years 30% or less and may not be reversible data obtained by the index of abundance, decline in the area of occupancy, actual levels of exploitation, and effects of introduced taxa + population reduction projected to be met in the future with bases on the index of abundance in the area of occupancy, actual levels of exploitation, and effects of introduced taxa) by the IUCN (Keuroghlian et al. 2013).

Characteristics: White-lipped peccary is medium sized; skull length 255.0–295.0 mm. Fur of dorsum and underparts is brown and speckled in dark grayish brown, varying broadly in between geographic areas, **with white spots in the inferior jaw, cheeks, and throat but without a white hair collar**; commonly with a dorsal stripe of longer and darker hairs than in the lateral area; robust body and limbs proportionally short; head triangular; canines triangular, length more than twice the diastema; sagittal crest very well developed in the posterior part of the skull.

Comments: The genus *Tayassu* is used according to Grubb (1993, 2005). *Dicotyles* was previously considered a junior synonym of *Tayassu* (Hershkovitz 1963); however, Woodburne (1968), Husson (1978), and Groves and Grubb (2011) accepted it as a full genus. *T. pecari* ranges from southern Veracruz and Oaxaca south through South America, including the southern Yucatán Peninsula, except for the coasts of Chiapas (Map 15). It can be found in sympatry, or nearly so, with *D. crassus* but not with *D. angulatus*. *T. pecari* can be distinguished from both species of *Dicotyles* by having white spots in the inferior jaw, cheeks, and throat, but without a white hair collar.

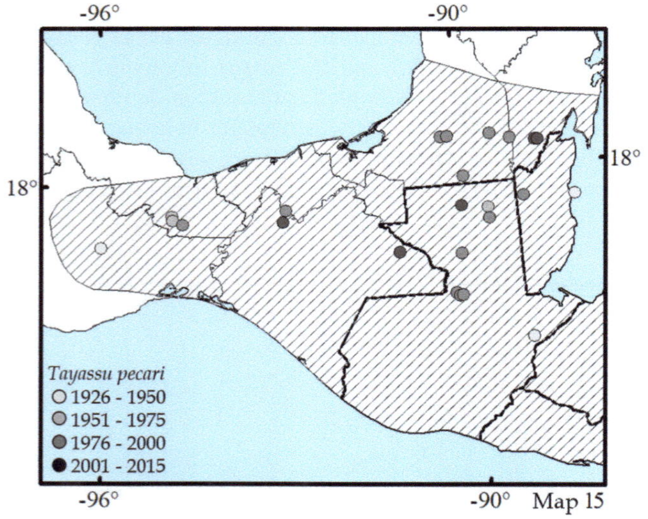

Map 15 Distribution map of *Tayassu pecari*

Additional Literature: Miller (1914), Wetzel, Donkin (1985), Mayer and Wetzel (1987), Wright (1989), and Grubb (2005).

Order Carnivora

Order Carnivora Bowdich, 1824

The Carnivora includes dogs, bears, cats, sea lions, seals, and others, characterized by **canines big, sharp, curved, semi or transversely circular, and carnassial**. Its distribution is nearly cosmopolitan. Two suborders, Caniform and Feliform, and eight families occur in North America: Felidae (Feliformia), and Canidae, Mephitidae, Mustelidae, Otariidae, Phocidae, Procyonidae, and Ursidae (Caniformia). The keys were elaborated based on the review of specimens and the following sources: Lawlor (1979), Hall (1981), and Jones and Manning (1992).

1. Forelimbs paddle-shaped or claws not flattened laterally; if sharp, not retractile or with fingernails; more than three lower molariforms; upper carnassial with a well-developed heel or squared-shaped (Fig. 1).. suborder Caniformia (p. 88)

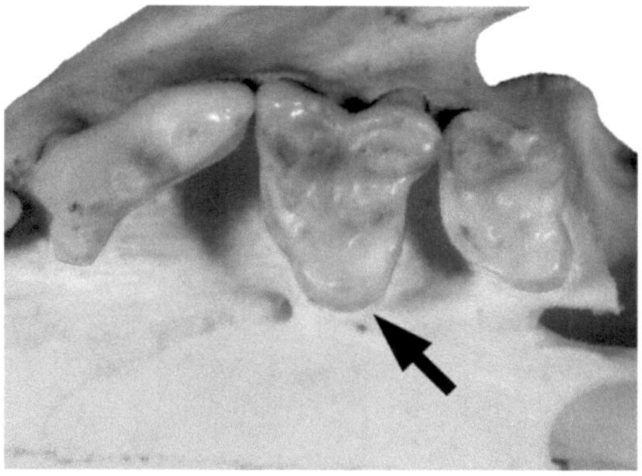

Fig. 1 The upper carnassial with a well-developed heel or squared shaped

2. Forelimbs not paddle-shaped and claws flattened laterally, sharp, and retractile; three lower molariforms; upper carnassial razor blade-shaped and without a heel (Fig. 2) .... suborder Feliformia (p. 185)

S. T. Álvarez-Castañeda, *Mammals of North America - Volume 1*, https://doi.org/10.1007/978-3-031-41661-3_4

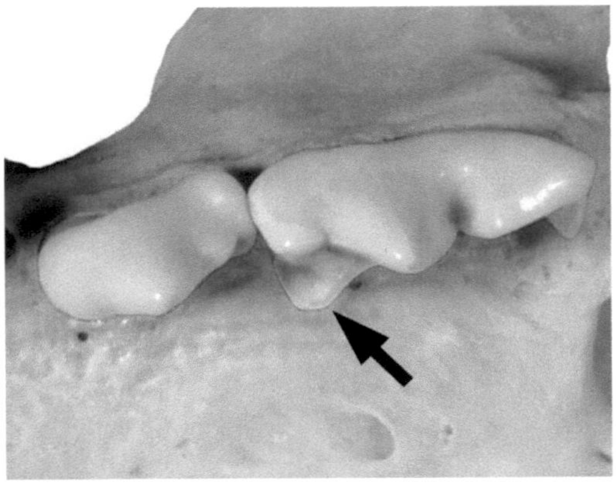

Fig. 2 The upper carnassial razor blade shaped and without a heel

Suborder Caniformia Kretzoi, 1943

The species within this suborder are generally plantigrade (except for the family Canidae); semiretractile claws (except for fisher, marten, sea otter [forefeet only], and ringtail); **bulbourethral glands and vesicula seminalis always absent**; baculum usually long; **rostrum long;** jaws long with more teeth than other suborders; **carnassial teeth less specialized;** tend to be omnivorous and opportunistic; **auditory bullae single-chambered or partially divided, composed of a single bone**. The keys were elaborated based on the review of specimens and the following sources: Lawlor (1979), Hall (1981), Jones and Manning (1992), and Álvarez-Castañeda et al. (2017a).

1. Forelimbs paddle-shaped; no tail; two lower incisors; all postcanine teeth of the same shape, generally conical........... 2

1a. Forelimbs not paddle shaped, with well-defined digits; with tail; three lower incisors; postcanine teeth differently shaped among themselves........... 4

2. Canines well developed in adults of both sexes, exposed even with the mouth closed; upper canines larger than the maxillar toothrow (northern Alaska, United States, and northern Quebec and Hudson Bay, Canada) .... family Odobenidae; *Odobenus rosmarus* (p. 133)

2a. Canines developed but not exposed with the mouth closed; upper canines shorter than the maxillary toothrow........... 3

3. Ears small; hind flippers directed frontward and functional for walking; total number of teeth 24 or less; with an alisphenoid channel (Fig. 3); teeth usually unicuspid; postorbital process well developed........... family Otariidae (p. 134)

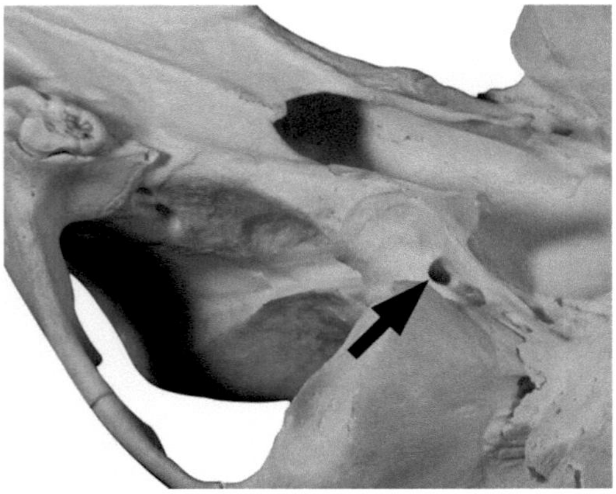

Fig. 3 The alisphenoid channel present

3a. Without ears; hind flippers directed backwards and not functional for walking; total number of teeth 26 or more; no alisphenoid channel (Fig. 4); teeth not unicuspid; postorbital process absent............................ family Phocidae (p. 141)

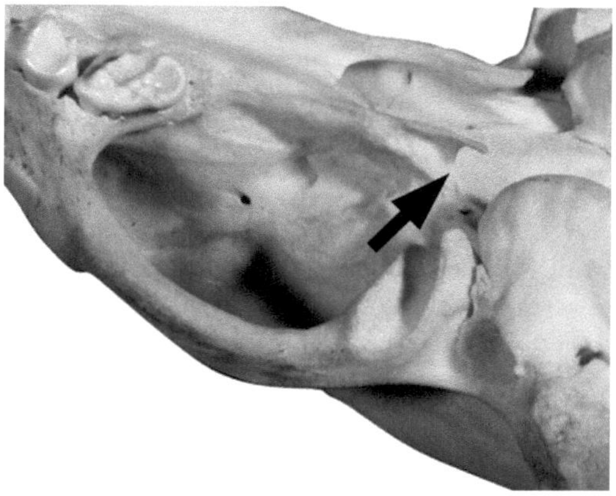

Fig. 4 The alisphenoid channel absent

4. Coloration homogeneous or grizzled but never with contrasting stripes, spots, or annulated color pattern, no hind digits webbed; tail not prehensile; posterior opening of the alisphenoid channel present toward the exterior (Fig. 3); total molariforms 26..........5

4a. Coloration with stripes, spots, or annulated color pattern or hind digits webbed or tail prehensile; posterior opening of the alisphenoid channel absent toward the exterior (Fig. 4); total molariforms 24 or less..........6

5. Medium sized; limbs slim, digits with claws not well developed; auditory bullae septum longitudinal; fourth upper molariforms sharp (in carnassial shape; Fig. 1); condylobasal length less than 280.0 mm............family Canidae (p. 78)

5a. Large sized; limbs robust with well-developed claws; auditory bullae septum not longitudinal; fourth upper molariforms flat with a small crest (not carnassial shaped; Fig. 5); condylobasal length greater than 280.0 mm.......... family Ursidae (p. 164)

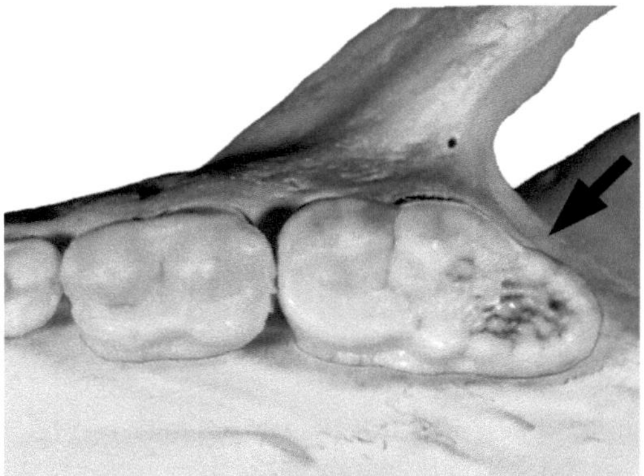

Fig. 5 The fourth upper molariforms flat or with a small crest

6. Tail with grizzly rings, dorsal stripes, and whitish (barely noticeable in some species), if rings are not present, body color old gold, tail prehensile; upper molars two or three; upper carnassial grinder............family Procyonidae (p. 155)

6a. Tail without grizzly rings in different colors, but may have two or more colors in longitudinal lines; one upper molar; cutting upper carnassial..........7

7. Specimens of different colors, never black with longitudinal white dorsal stripes; body extended; no odoriferous glands mixed with urine as a defense means .... family Mustelidae (p. 106)

7a. Specimens black, usually with white longitudinal dorsal stripes; body tends to be compact; odoriferous glands mixed with urine as a defense means .... family Mephitidae (p. 92)

Family Canidae Fisher, 1817

The family Canidae includes wolves, dogs, foxes, coyotes, and jackals characterized by well-developed claws, relatively straight and nonretractile; digitigrade; **tail long and bushy;** skull **elongated, rostrum relatively long and narrow**; **last upper molars relatively large and transversely elongated**; mastoid process smaller than paroccipital process; carnassial and canines well developed. Cosmopolitan distribution except for Madagascar, New Zealand, and the Philippine Islands. The keys were elaborated based on the review of specimens and the following sources: Lawlor (1979), Hall (1981), Jones and Manning (1992), and Álvarez-Castañeda et al. (2017a).

1. Color speckled in gray shades (combination of pale gray and black hairs) on the back and rostrum; rest of the body reddish cinnamon; temporal crest lyre-shaped (Fig. 6); frontal depression at the level of the postorbital processes; ventral border of jaw rising to the level of the angular process (forming a step) .... *Urocyon* (p. 86)

Fig. 6 The temporal crest lyre shaped

1a. Combination of different colors but different to those described in clause 1 above; temporal crest neither well-defined nor lyre-shaped; no frontal depression at the level of the postorbital process; ventral border of snout continuous throughout its length (not forming a step) 2

2a. Total length greater than 535.0 mm; hind limb length greater than 150.0 mm; condylobasal length greater than 125.0 mm; postorbital process slightly developed; postorbital breadth equal to or greater than the interorbital breadth .... *Canis* (p. 80)

2. Total length less than 535.0 mm; hind limb length less than 150.0 mm; condylobasal length less than 125.0 mm; postorbital process well developed; postorbital breadth notably less than interorbital breadth 3

3. Ears short and rounded; rostrum wide breadth at the level of the second upper premolars greater than 18% of the condylobasal length; first upper incisors with anteroposterior diameter greater than the distance between the upper incisors and canine alveoli (tundra regions of Canada and Alaska, United States) .... *Alopex lagopus* (p. 80)

3a. Ears large and pointed; rostrum narrow, breadth at the level of the second upper premolars less than 18% of the condylobasal length; first upper incisors with the anteroposterior diameter less than the distance between the upper incisors and canine alveoli .... *Vulpes* (p. 88)

Genus *Alopex* Kaup, 1829

The Arctic fox, the only species in the genus, has short and rounded ears; thickly haired feet and, usually, seasonal dichromatic, being brownish-gray in summer and white or "blue" in winter. *Alopex* and *Vulpes* were considered subgenera of *Canis* (Van Gelder 1978), but this was not accepted. *Fennecus* is a congeneric taxon of *Alopex* and, at the same time, a subgenus of *Vulpes* (McKenna and Bell 1997).

Alopex lagopus (Linnaeus, 1758)

Arctic Fox, zorra del ártico

1. *A. l. groenlandicus* (Bechstein, 1799). Restricted to Ellesmere Islands, Canada.
2. *A. l. hallensis* (Merriam, 1900). Restricted to San Matthew Island, Alaska.
3. *A. l. lagopus* (Linnaeus, 1758). From Alaska east through the border between Ontario and Québec.
4. *A. l. pribilofensis* (Merriam, 1902). Restricted to Saint Paul Island, Alaska.
5. *A. l. ungava* (Merriam, 1902). From border between Ontario and Québec east through Newfoundland.

Conservation: *Alopex lagopus* (as *Vulpes lagopus*) is listed as Least Concern by the IUCN (Angerbjörn and Tannerfeldt 2014).

Characteristics: Arctic fox is small sized; total length 350.0–550.0 mm, condylobasal length 109.3 a 132.5 mm. The coloration varies geographically and seasonally; however, **the dorsal coloration always homogeneous; in coastal areas, color ranging from bluish gray in winter to dark bluish gray to chocolate brown in summer;** inland specimens white in winter and chocolate brown in summer; ears rounded and tail of the same color as the body; **rostrum width at the second upper premolar greater than 18% of the condylobasal length; first upper incisor with anteroposterior diameter greater than the distance between the upper incisor alveoli and the canine**.

Comments: Similar genetic distances among *Vulpes macrotis, V. velox,* and *Alopex lagopus* are used to recognize the three at species level (Geffen et al. 1992; Mercure et al. 1993; Wozencraft 2005). I still recognized *Alopex* as a valid genus. *A. lagopus* has a restricted range to the tundra region northern Canada and Alaska (Map 1), and can be distinguished from other species by its absence of a gray speckled coloration at back and rostrum; temporal crest is not developed in lyre-shaped; total length less than 535.0 mm, and postorbital process well developed. From *Vulpes*, by having dorsal coloration white, bluish gray or chocolate brown; very short rostrum and limbs.

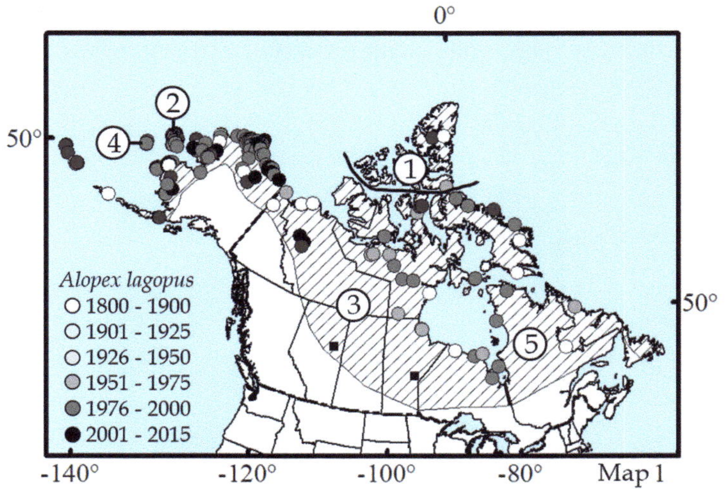

Map 1 Distribution map of *Alopex lagopus*

Additional Literature: Miller (1912b), Bobrinskii (1944, 1965), Prestrud (1991), Bininda-Emonds et al. (1999), Audet et al. (2002), Perini et al. (2010).

Genus *Canis* Linnaeus, 1758

Dorsal pelage speckled in yellowish-brown shades to dark grizzly brown, white, or black; fur of underparts creamy yellow or paler than in the dorsum; tail bushy; **limbs long; sagittal crest restricted to the central part of the braincase**; jaws without a "step" in the proximal part to the angular process. *Alopex* and *Vulpes* were considered subgenera of *Canis* (Van Gelder 1978), but this was not accepted. *Canis rufus* (red wolf) and *C. lycaon* (eastern wolf) have been considered hybrids between *C. latrans* and *C. lupus* based on the analyses of the whole genome, which makes them more similar to the coyote than to the wolf (vonHoldt et al. 2016). *Canis familiaris* had been considered by many authors as a subspecies of *Canis lupus*. I follow the Mammals Database of the American Society of Mammalogy. The keys were elaborated based on the review of specimens and the following sources: Jackson (1951), Lawlor (1979), Hall (1981), Jones and Manning (1992), and Álvarez-Castañeda et al. (2017a).

1a. Total length greater than 900.0 mm; ear length less than 45% of hind limb length; nose pad greater than 30.0 mm; anterior-posterior canine length greater than 11.0 mm; skull length greater than 210.0 mm; postorbital constrictions relatively narrow (Fig. 7; from Canada south through northern México and east through the Mississippi river, theoretical) .... *Canis lupus* (p. 95)

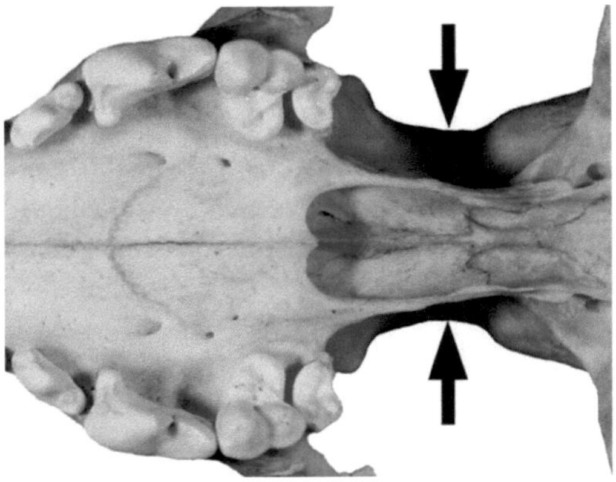

Fig. 7 The postorbital constrictions are relatively as narrow as in *Canis lupus*

1. Total length less than 900.0 mm; ear length greater than 60% of hind limb length; nose pad less than 30.0 mm; anterior-posterior canine length less than 11.0 mm; skull length less than 210.0 mm; postorbital constriction relatively wide (Fig. 8)........ 2

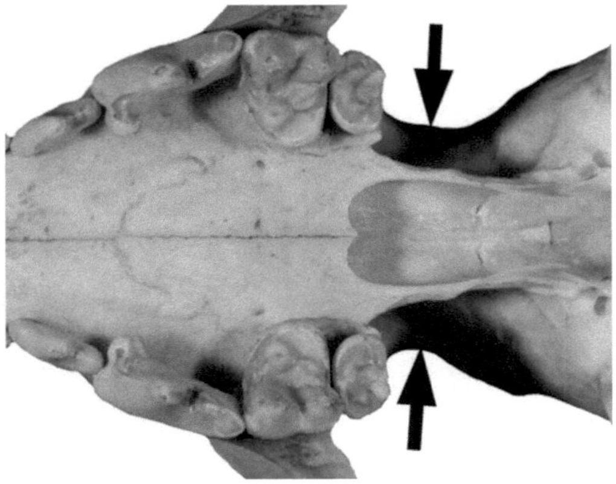

Fig. 8 The postorbital constriction relatively wide

2. Elbows lower than the sternum in lateral view; tail generally held down; track oval and claws in inward direction; palatal broad between the inner margins of the upper first premolar alveolus less than 33% of alveolus distance of upper tooth-row (throughout North America, except for northern and northwestern Canada) ....*Canis latrans* (p. 93)

2a. Dog larger or smaller than coyote, but both can be similar in size. Elbows higher than the sternum in lateral view; tail generally held up; track rounded and claws in outward direction; palatal broad between the inner margins of the upper first premolar alveolus greater than 33% of the alveolus distance of upper tooth-row..*Canis familiaris* (introduced species, p. 93)

Canis familiaris Linnaeus, 1758

Dog, perro

Conservation: *Canis familiaris* is not listed in any conservation list but is considered as invasive by the IUCN (as *Canis lupus*; Global Invasive Species Database 2022e).

Characteristics: Species historically subjected to a domestication process, which has caused a great variation in its morphology, so it is not possible to define a general morphology that can be used to define the species.

Comments: The dog was domesticated from *Canis familiaris* and has been recorded as wildlife only from different insolated areas in North America. *C. familiaris* differs from *C. latrans,* by having the rostrum shorter and wider; length between the inner margins of the upper first premolar alveoli divided by the length from the anterior margin of the first premolar alveolus to the posterior margin of the last molar alveolus greater than 33%. From *C. lupus*, by its small size in general.

Additional Literature: Howard (1949), Jackson (1951), Gentry et al. (2004).

Canis latrans Say, 1823

Coyote, coyote

1. *C. l. cagottis* (Hamilton-Smith, 1839). From southern Tamaulipas and central San Luis Potosí south through the Tehuantepec isthmus.
2. *C. l. clepticus* Elliot, 1903. From southern California to central Baja California.
3. *C. l. floridanus* Miller, 1912. From central Alabama, Tennessee, and Kentucky east through the Atlantic coast, including the whole Florida Peninsula.
4. *C. l. frustror* Woodhouse, 1851. From Arkansas and Kansas south through Texas and Louisiana.
5. *C. l. goldmani* Merriam, 1904. From the Tehuantepec isthmus south through Chiapas, including the Yucatán Peninsula.
6. *C. l. gregoryi* Goldman, 1937. From eastern Oklahoma and Texas east through central Alabama, Tennessee, and Kentucky
7. *C. l. impavidus* Allen, 1903. From southern Sinaloa and Sonora south through southern Zacatecas and Jalisco.
8. *C. l. incolatus* Hall, 1934. From Alaska, Yukon, and western Northwest Territories south through central British Columbia and Alberta.
9. *C. l. jamesi* Townsend, 1912. Restricted to Tiburón Island, Sonora.
10. *C. l. latrans* Say, 1823. From central Alberta and Saskatchewan south through northern Texas, Oklahoma, and central Kansas.
11. *C. l. lestes* Merriam, 1897. From central British Columbia and Alberta south through southern California to Colorado, and northern New Mexico.
12. *C. l. mearnsi* Merriam, 1897. From southern California to Colorado and western New Mexico, and from eastern Baja California south through northern San Luis Potosí and western Nuevo León.
13. *C. l. microdon* Merriam, 1897. Tamaulipas and Nuevo León.
14. *C. l. ochropus* Eschscholtz, 1829. Coastal California.
15. *C. l. peninsulae* Merriam, 1897. From central Baja California south through the Baja California Peninsula.
16. *C. l. texensis* Bailey, 1905. From northeastern New Mexico and northwestern Texas south through Nuevo León.
17. *C. l. thamnos* Jackson, 1949. From eastern Saskatchewan south to northwestern Kansas and from northern Arkansas east through New Brunswick.
18. *C. l. umpquensis* Jackson, 1949. Coastal Washington and Oregon.
19. *C. l. vigilis* Merriam, 1897. From central Jalisco along the coast south through the Tehuantepec Isthmus.

Conservation: *Canis latrans* is listed as Least Concern by the IUCN (Kays 2018).

Characteristics: Coyote is medium sized; **total length 1,000.0–1,350.0 mm, condylobasal length 160.2–203.5 mm**. Dorsal pelage speckled in yellowish brown to dark grizzly brown shades; fur of underparts creamy yellow; small body size; ear length greater than 60% of hind limb length; **canine anteroposterior diameter less than 11.0 mm**; length 215.0 mm in males and less than 205.0 in females.

Comments: *Canis rufus* has been considered a hybrid between *C. latrans* and *C. lupus* and is more similar to the coyote than to the wolf (vonHoldt et al. 2016); for that reason, the subspecies *C. rufus floridanus* and *C. r. gregoryi* are considered subspecies of *C. latrans*. *C. latrans* ranges from Alaska and central-northern Canada south through Central America. It is considered that the coyote expanded its ranges as consequence of the extirpation of wolf populations and modification of the landscape (Map 2). It can be found in sympatry, or nearly so, with *C. lupus, Urocyon cinereoargenteus, Vulpes macrotis, V. velox,* and *V. vulpes*. *C. latrans* can be distinguished from the species of *Vulpes* by its larger size in total, hind limb and condylobasal lengths greater than 535.0, 150.0, and 125.0 mm, respectively. From *C. lupus*, by its smaller size with a total length less than 900.0 mm; ears large, greater than 60% of hind limb length; nose pad less than 30.0 mm; skull length less than 210.0 mm, and canine anterior posterior diameter less than 11.0 mm.

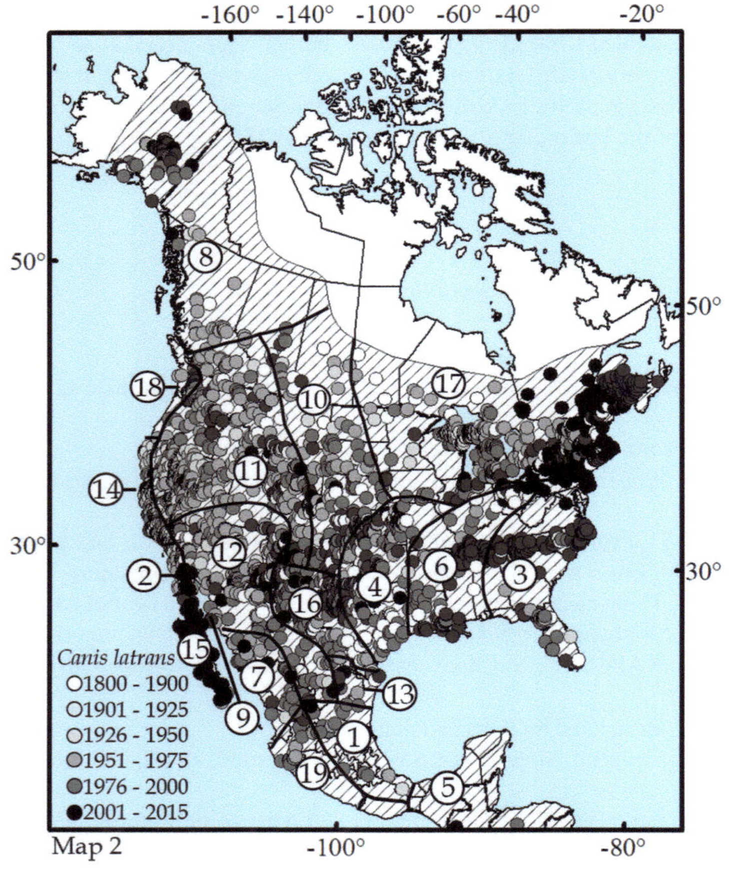

Map 2 Distribution map of *Canis latrans*

Additional Literature: Young and Jackson (1951), Jackson (1961), Lawrence and Bossert (1967), Mech (1974), Gier (1975), Riley and McBride (1975), Bekoff (1977), McKenna and Bell (1997), Hidalgo-Mihart et al. (2004), Wozencraft (2005).

Canis lupus Linnaeus, 1758

Gray Wolf, lobo gris

1. *C. l. alces* Goldman, 1941. Restricted to the Kenai Peninsula, Alaska.
2. *C. l. arctos* Pocock, 1935. Restricted to Banks, Ellesmere, and Sverdrup Islands.
3. *C. l. baileyi* Nelson and Goldman, 1929. From southern Arizona, New Mexico, and Texas south through Oaxaca.
4. *C. l. beothucus* Allen and Barbour, 1937. Restricted to Prince Edward Islands, Newfoundland.
5. *C. l. bernardi* Anderson, 1943. Restricted to Banks Island, Franklin.
6. *C. l. columbianus* Goldman, 1941. Southern Yukon and British Columbia.
7. *C. l. crassodon* Hall, 1932. Restricted to Vancouver Island.
8. *C. l. floridanus* Miller, 1912. From Ohio south through western Florida and east through the Atlantic coast, including the Florida Peninsula.
9. *C. l. fuscus* Richardson, 1839. Coastal of British Columbia south through northeastern California.
10. *C. l. gregoryi* Goldman, 1937. Eastern Texas and Oklahoma.
11. *C. l. griseoalbus* Baird, 1823. Southern Mackenzie, central Saskatchewan, and Manitoba.
12. *C. l. hudsonicus* Goldman, 1941. From western Mackenzie east through Keewatin, northern Manitoba, and western Ontario.
13. *C. l. irremotus* Goldman, 1937. From central Alberta south through southeastern Oregon to Wyoming.
14. *C. l. labradorius* Goldman, 1937. Central and northern Labrador Peninsula.
15. *C. l. ligoni* Goldman, 1937. Southeastern Alaska and north islands of British Columbia.
16. *C. l. lycaon* Schreber, 1775. From Manitoba to Illinois east through the Atlantic coast.
17. *C. l. mackenzii* Anderson, 1943. Northern Mackenzie.
18. *C. l. manningi* Anderson, 1943. Baffin Island, Franklin.
19. *C. l. mogollonensis* Goldman, 1937. Arizona and central-eastern New Mexico.
20. *C. l. monstrabilis* Goldman, 1937. Southwestern New Mexico, Texas, Coahuila, and Nuevo León.
21. *C. l. nubilus* Say, 1823. From central Saskatchewan and Manitoba south through northern Texas, Oklahoma, and Arkansas.
22. *C. l. occidentalis* Richardson, 1829. Central Makenzie, eastern Yukon, eastern British Columbia, and central-northern Saskatchewan.
23. *C. l. pambasileus* Elliot, 1905. Western Yukon and Alaska.
24. *C. l. rufus* Audubon and Bachman, 1851. From Arkansas and Illinois south through eastern Texas, Louisiana, and Mississippi.
25. *C. l. tundrarum* Miller, 1912a, b. Northern Alaska and northwestern Yukon.
26. *C. l. youngi* Goldman, 1937. California, Nevada, Utah, Colorado, and southern Wyoming.

Conservation: *Canis lupus baileyi* is listed as Endangered by the Norma Oficial Mexicana (DOF 2019). *C. l. baileyi* is listed as Endangered in the United States Endangered Species Act (FWS 2022), along with other subspecies in different regions of the country. *C. l. lycaon* is listed as Threatened by the Canadian Endangered Species Protection Act; *C. lupus*, as Least Concern by the IUCN (Boitani et al. 2018); *C. rufus*, as Critically Endangered (C2a (i, ii); D: small population size projected continuing decline with less than 50 mature individuals in each subpopulation) by the IUCN; (Phillips 2018). *C. lycaon* is not listed. In México, it was considered extinct in the wild, but now, with a reintroduction program in Chihuahua, it is considered as Endangered.

Characteristics: Gray wolf is medium to large sized; total length 1270.0–1640.0 mm in males and 1370.0–1520.0 mm in females; average skull length 215.0 mm in males and 210.0 mm in female. Dorsal pelage from white to black, including mottled and diverse brown and gray shades; underparts variable in relation to the dorsum; ear length smaller than 60% of the hind limb length; **canine anteroposterior diameter greater than 11.0 mm**.

Comments: Historically, *Canis lupus* has had a wide distribution range from 15 degrees north latitude south through northern México; it was present in much of the United States. Today, their distribution is restricted to the wilderness in remote areas of Canada and the United States (Map 3). *C. lupus* is sympatric, or nearly so, with *C. latrans, Urocyon cinereoargenteus, Vulpes macrotis, V. velox,* and *V. vulpes. Canis lupus* can be distinguished from the species of *Vulpes* by its large size, with total, hind limb, and condylobasal lengths greater than 535.0, 150.0, and 125.0 mm, respectively. From *C. latrans*, by its larger size, with total length greater than 900.0 mm; ears large but less than 60% of hind limb length; nose pad greater than 30.0 mm; skull length greater than 210.0 mm, and canine anterior-posterior diameter greater than 11.0 mm.

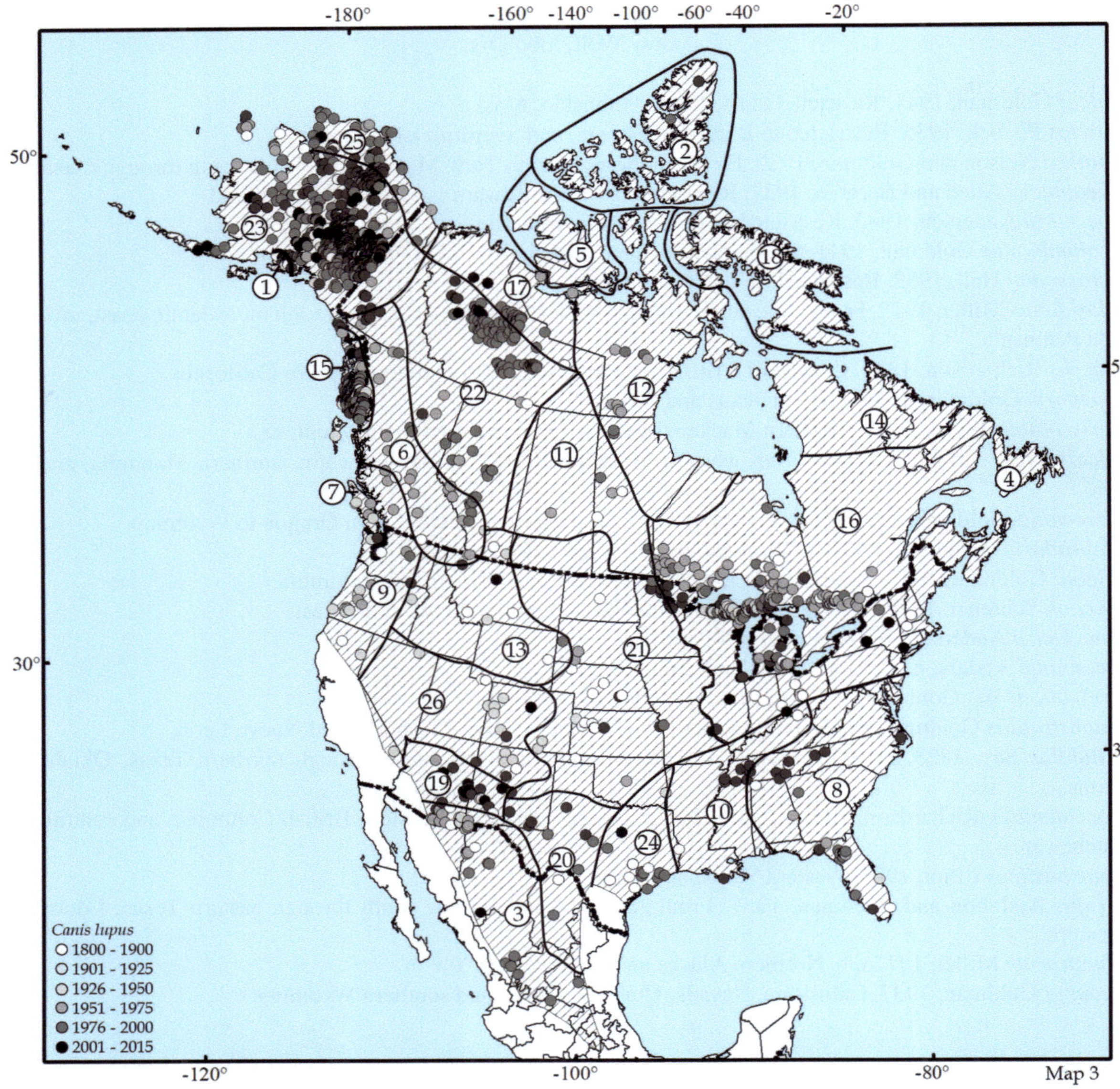

Map 3 Distribution map of *Canis lupus*

Additional Literature: Goldman (1937, 1944), Young and Goldman (1944), Young and Jackson (1951), Lawrence and Bossert (1967), Paradiso and Nowak (1972), Mech (1974), Wozencraft (2005).

Genus *Urocyon* Baird, 1857

Dorsal pelage speckled in gray shades; neck with conspicuous cinnamon-rusty color; chest and throat white; limbs, flanks, and ears reddish ochre and contrasting with dorsal; face with dark and white marks; **ears shorter than 50% of head length**; moderately short limbs; tail long, hairy, and blackish above; **sagittal crest is lyre-shaped**; section inside parietal bones same breath as length; **jaw with "step" in proximal part to angular process**. *Urocyon* was previously considered a subgenus of *Vulpes* (Clutton-Brock et al. 1976) and of *Canis* (Van Gelder 1978). The keys were elaborated based on the review of specimens and the following sources: Hall (1981).

1. Total length greater than 800.0 mm; condylobasal length greater than 106.0 mm (southern Canada, mainly throughout the United States and México, except for northwestern United States)........................*Urocyon cinereoargenteus* (p. 97)

1a. Total length less than 800.0 mm; condylobasal length less than 106.0 mm (Channel Islands, California, United States) ....*Urocyon littoralis* (p. 98)

Urocyon cinereoargenteus (Schreber, 1775)

Gray Fox, zorra gris

1. *U. c. borealis* Merriam, 1903. From New York and Massachusetts northwest through Maine.
2. *U. c. californicus* Mearns, 1897. Central southern California and northern Baja California.
3. *U. c. cinereoargenteus* (Schreber, 1775). From Illinois to Mississippi east through the Atlantic coast.
4. *U. c. floridanus* Rhoads, 1895. From western Texas to southeastern South Carolina, including the Florida Peninsula.
5. *U. c. fraterculus* Elliot, 1896. Yucatán Peninsula.
6. *U. c. guatemalae* Miller, 1899. Chiapas and southeastern Tabasco south through Central America.
7. *U. c. madrensis* Burt and Hooper, 1941. Southwestern Sonora, western Chihuahua, and Durango.
8. *U. c. nigrirostris* (Lichtenstein, 1850). From Sinaloa, Zacatecas, and San Luis Potosí south through Puebla and western Oaxaca.
9. *U. c. ocythous* Bangs, 1899. From southern Manitoba and southwestern Ontario south through Oklahoma and Arkansas.
10. *U. c. orinomus* Goldman, 1938. From Southern San Luis Potosí and central Veracruz south through western Chiapas.
11. *U. c. peninsularis* Huey, 1928. Baja California Peninsula.
12. *U. c. scottii* Mearns, 1891. From central Nevada to northern Texas south through central Sonora to central Veracruz.
13. *U. c. townsendi* Merriam, 1899. Oregon and central-northern California.

Conservation: *Urocyon cinereoargenteus* is listed as Threatened by the Canadian Endangered Species Protection Act and as Least Concern by the IUCN (Roemer et al. 2016).

Characteristics: Gray fox is small sized; **total length 800.0–1,125.0 mm, condylobasal length 110.0–130.0 mm**. Color and shape as described for the genus, similar to *U. littoralis* but larger.

Comments: *Urocyon cinereoargenteus* ranges from southern-central and eastern Canada and northeastern United States south through South America (Map 4). It can be found in sympatry, or nearly so, with *Canis latrans, C. lupus, Vulpes macrotis, V. velox, V. vulpes,* and *C. familiaris. U. cinereoargenteus* can be distinguished from these species by its speckled pelage in gray with shade coloration on the back and rostrum; other parts of body with reddish cinnamon hair; temporal crest in a very distinctive lyre shape. From *U. littoralis,* by its greater size.

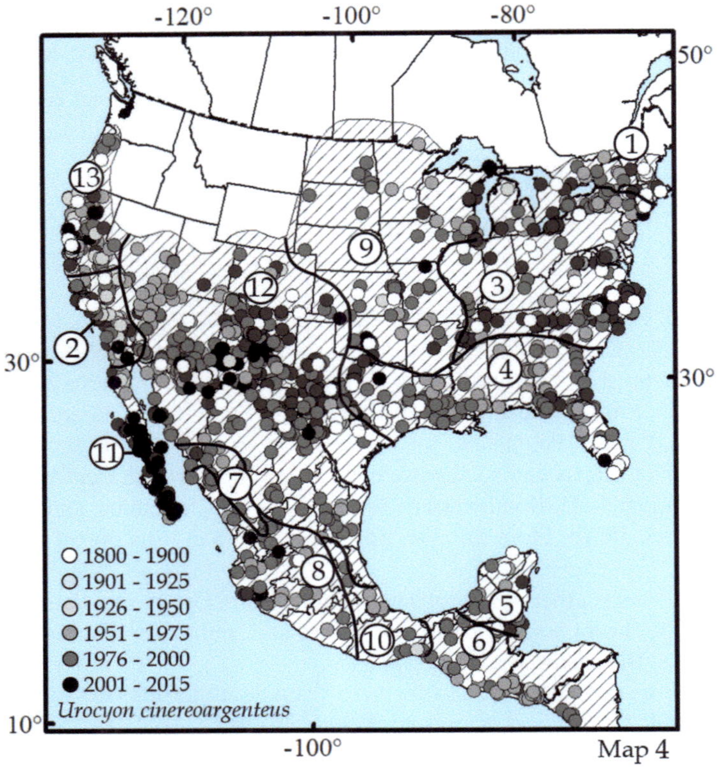

Map 4 Distribution map of *Urocyon cinereoargenteus*

Additional Literature: Grinnell et al. (1937a, b), Hildebrand (1954), Sullivan (1956), Jackson (1961), Wozencraft (2005).

Urocyon littoralis (Baird, 1858)

Island Gray Fox, zorra gris de isla Catalina

1. *U. l. catalinae* Merriam, 1903. Restricted to Santa Catarina Island, California.
2. *U. l. clementae* Merriam, 1903. Restricted to San Clemente Island, California.
3. *U. l. dickeyi* Grinnell and Linsdale, 1930. Restricted to San Nicolas Island, California.
4. *U. l. littoralis* (Baird, 1858). Restricted to San Miguel Island, California.
5. *U. l. santacruzae* Merriam, 1903. Restricted to Santa Cruz Island, California.
6. *U. l. santarosae* Grinnell and Linsdale, 1930. Restricted to Santa Rosa Island, California.

Conservation: *Urocyon littoralis catalinae* is listed as Endangered by the United States Endangered Species Act (FWS 2022) and *U. littoralis* as Near Threatened by the IUCN (Coonan et al. 2013).

Characteristics: Island gray fox is the smallest member of the genus; **total length 620.0–800.0 mm**, condylobasal length **92.0–105.0 mm**. Dorsal pelage and shape as described to the genus, similar to *U. cinereoargenteus* but smaller.

Comments: *Urocyon littoralis* range is restricted to the Channel Islands in California (Map 5). It is the only canid in the islands and can be differentiated from *U. cinereoargenteus* by its smaller size.

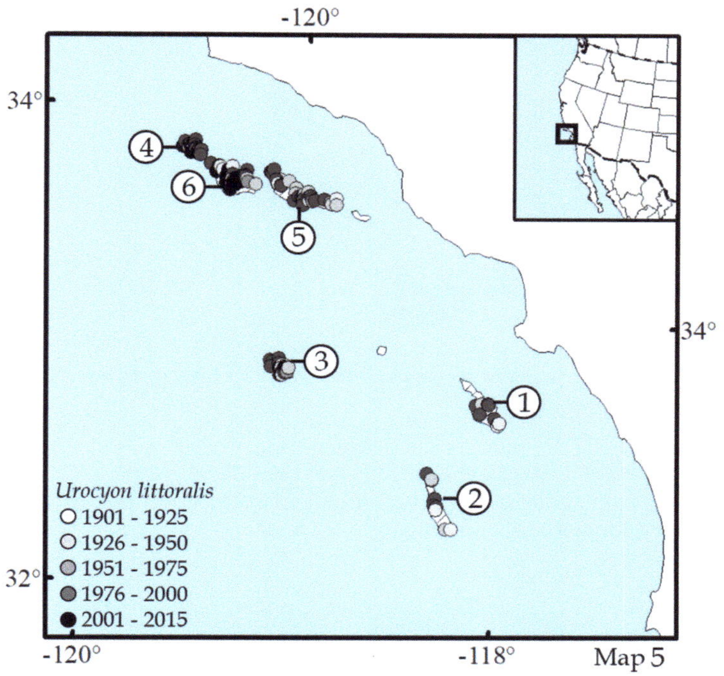

Map 5 Distribution map of *Urocyon littoralis*

Additional Literature: Grinnell et al. (1937a, b), Hildebrand (1952, 1954), Collins (1982, 1993), Gilbert et al. (1990), Wayne et al. (1991a, b), Moore and Collins (1995), Wozencraft (2005).

Genus *Vulpes* Frisch, 1775

Dorsal pelage with a homogeneous dorsal coloration in brown shades; limbs can be proportionally short in some species; **black triangular markings between the eyes and the nose;** tail often in different color at the tip; **skull small, sagittal crest not lyre-shaped**; jaw without a "step" in the proximal part to the angular process. *Alopex* and *Vulpes* were considered subgenera of *Canis* (Van Gelder 1978), but this was not accepted. The keys were elaborated based on the review of specimens and the following sources: Hall (1946, 1981).

1. Tail white at the tip; feet black; skull slender, mastoid width less than 34% of the total length (throughout Canada, eastern and central-western United States, except for desert areas) .... *Vulpes vulpes* (p. 103)

1a. Tail black at the tip; feet not black; skull not slender, mastoid width greater than 34% of the total length 2

2a. Tail greater than 69% of the total length and black at the tip; ears small, less than 80.0 mm; foramen magnum level at least 3.0 mm lower than the palatal; anterocondyloid foramen and lacerum posterior opening less than two times longer than wide (Fig. 9; desert areas of southwestern United States, northwestern and northern-central México).................... .... *Vulpes macrotis* (p. 100)

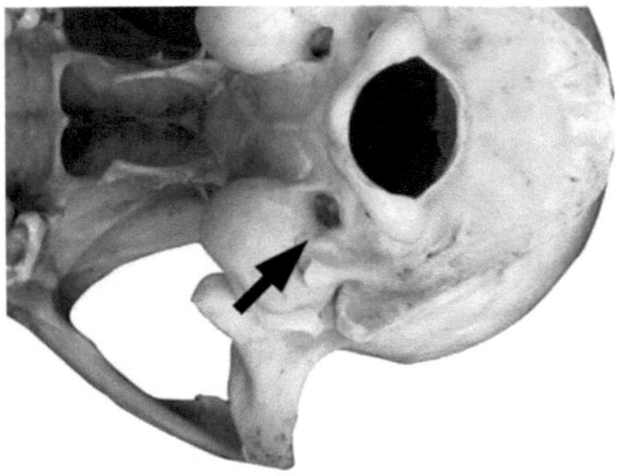

Fig. 9 The anterocondyloid foramen and lacerum posterior opening less than two times longer than wide like in *Vulpes macrotis*

2. Tail shorter, less than 69% of the total length and less black at the tip; ears large, greater than 80.0 mm; foramen magnum at the same level than the palatal; anterocondyloid foramen and lacerum posterior opening two times longer than wide (Fig. 10; Great Plains in United States and Canada).. *Vulpes velox* (p. 102)

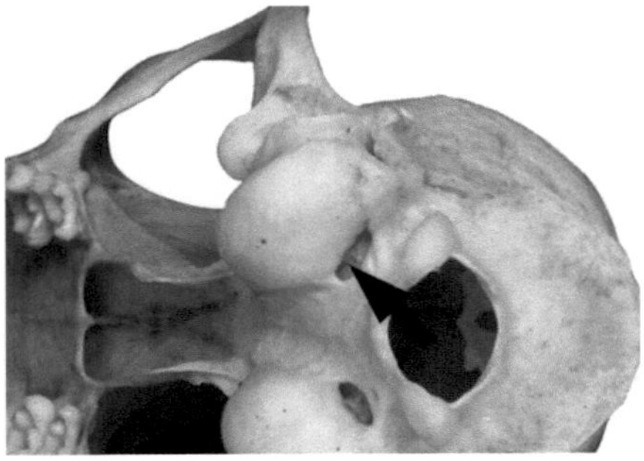

Fig. 10 The anterocondyloid foramen and lacerum posterior opening two times longer than wide like in *Vulpes velox*

Vulpes macrotis Merriam, 1888

Kit Fox, zorro del desierto norteño

1. *V. m. arsipus* Elliot, 1904. Southeastern California, southern Nevada, Arizona, and northern Sonora.
2. *V. m. devia* Nelson and Goldman, 1909. From central Baja California south through the Baja California Peninsula.
3. *V. m. macrotis* Merriam, 1888. Restricted to the San Jacinto Plains in Riverside, California.
4. *V. m. mutica* Merriam, 1902. Central Valley, California.
5. *V. m. neomexicanus* Merriam, 1902. From southwestern Colorado south through northern Chihuahua and southwestern Texas.
6. *V. m. nevadensis* Goldman, 1931. From southern Oregon and southwestern Idaho south through Nevada, including western Utah.
7. *V. m. tenuirostris* Nelson and Goldman, 1931. Restricted to the Valle de la Trinidad, Baja California.
8. *V. m. zinseri* Benson, 1938. From southern Chihuahua and Coahuila south through San Luis Potosí.

Conservation: *Vulpes macrotis mutica* is listed as Endangered by the United States Endangered Species Act (FWS 2022); *Vulpes macrotis*, as Threatened by the Norma Oficial Mexicana (DOF 2019) and as Least Concern by the IUCN (Cypher and List 2014).

Characteristics: Kit fox is small sized; total length 730.0–840.0 mm, condylobasal length 107.0–116.0 mm. Similar to *Vulpes velox* but with longer tail and ears, **without distinguished blackish nose patches; ears short, length less than 80.0 mm;** skull similar to *V. velox,* **but foramen magnum at a level at least 3.0 mm lower than the palatal; opening of the anterior condyloid and posterior lacerum foramens less than two times longer than wide**.

Comments: *Vulpes macrotis* and *V. velox* were considered conspecific (Blair et al 1968; Bueler 1973; Dragoo et al. 1990). Similar genetic distances among *Vulpes macrotis, V. velox,* and *Alopex lagopus* are used to recognize the three at the species level (Mercure et al. 1993, Wozencraft 2005). Subspecific designations for Kit foxes are not fully resolved; on the one hand, eight are recognized (McGrew 1979); on the other, none (Dragoo et al. 1990). *V. macrotis* ranges in the desert areas of southwestern United States, northwestern and northern-central México, including the Chihuahua, Sonora, and Baja Californian deserts, Great Basin, deserts of California, and the San Joaquin Valley (Map 6). The other two species of *Vulpes* have an allopatric distribution, mainly *V. velox. V. macrotis* can be found in sympatry, or nearly so, with *Canis latrans* and *Urocyon cinereoargenteus,* and can be distinguished from them by its absence of speckles in gray coloration on the back and rostrum; temporal crest not lyre shaped; total length less than 535.0 mm, and postorbital process well-developed. From *Alopex lagopus* and *V. vulpes*, by having dorsal coloration dark buffy gray with brownish sides; fur of underparts paler, including limbs and tail, and tail blackish at the tip, contrasting with the reddish coloration of *V. vulpes,* and the white, bluish gray, or chocolate brown coloration of *A. lagopus,* and a very short rostrum. From *V. macrotis,* by its tail greater than 69% of the total length and ears less than 80.0 mm.

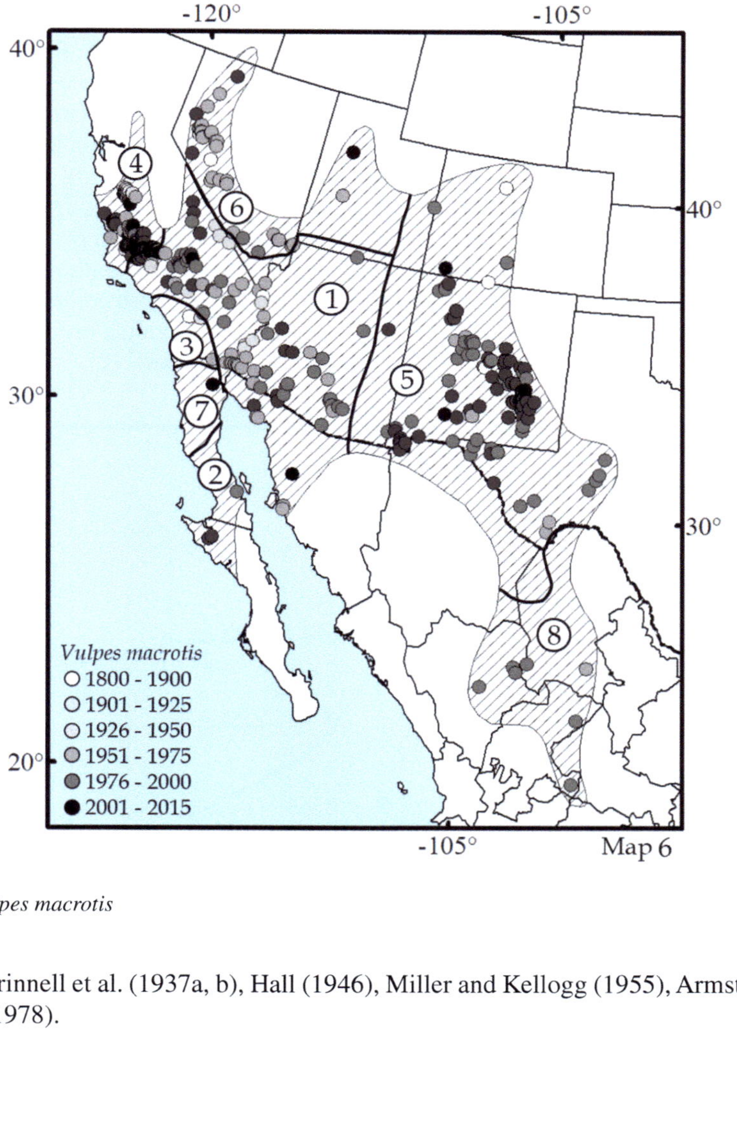

Map 6 Distribution map of *Vulpes macrotis*

Additional Literature: Grinnell et al. (1937a, b), Hall (1946), Miller and Kellogg (1955), Armstrong (1972), Thornton and Creel (1975), Van Gelder (1978).

Vulpes velox (Say, 1823)
Swift Fox, zorro del desierto

1. *V. v. hebes* Merriam, 1902. From Alberta southeastern to Minnesota and Iowa.
2. *V. v. velox* (Say, 1823). From Montana, South Dakota, and Southwestern Iowa south through western New Mexico and Northern Texas.

Conservation: *Vulpes velox* is listed as Endangered by the Canadian Endangered Species Protection Act and as Least Concern by the IUCN (Moehrenschlager and Sovada 2016).

Characteristics: Swift fox is small sized; total length 735.0–880.0 mm, condylobasal length 108.0–115.0 mm. Dorsal color dark buffy gray with brownish sides; fur of underparts paler including the limbs and tail; tail blackish at the tip, throat, chest, and belly buffy to white in winter and reddish gray in summer; ears triangular and of the same color than the back; **with blackish nose patches; ears long, greater than 80.0 mm;** sagittal crest curved, not well developed; **foramen magnum at the same level than the palatal; opening of the anterior condyloid and posterior lacerum of the foramen two times longer than wide**.

Comments: *Vulpes macrotis* and *V. velox* were considered conspecific (Blair et al 1968; Bueler 1973; Dragoo et al. 1990). The similar genetic distances among *Vulpes macrotis, V. velox,* and *Alopex lagopus* are used to recognize the three at the species level (Mercure et al. 1993, Wozencraft 2005). *Vulpes velox* is restricted to the central plains of the United States, Texas, New Mexico, Oklahoma, Colorado, Kansas, Nebraska, Wyoming, the Dakotas, Montana; in Canada, in Saskatchewan and Alberta (Map 7). The other two species of *Vulpes* are allopatric; mainly *V. macrotis*. *V. velox* can be found in sympatry, or nearly so, with *Canis latrans* and *Urocyon cinereoargenteus,* and can be distinguished from these two species by the rostrum speckled in gray; temporal crest lyre-shaped; total length less than 535.0 mm and postorbital process well developed. From *Alopex lagopus* and *V. vulpes*, by having the dorsal coloration dark buffy gray with brownish sides, fur of underparts paler, including the limbs and tail; tail blackish at the tip, contrasting with the reddish coloration of *V. vulpes,* and the white, bluish gray, or chocolate brown coloration of *A. lagopus.* From *A. lagopus*, by having a very short rostrum, and from *V. macrotis,* by its tail less than 69% of the total length and its ears greater than 80.0 mm.

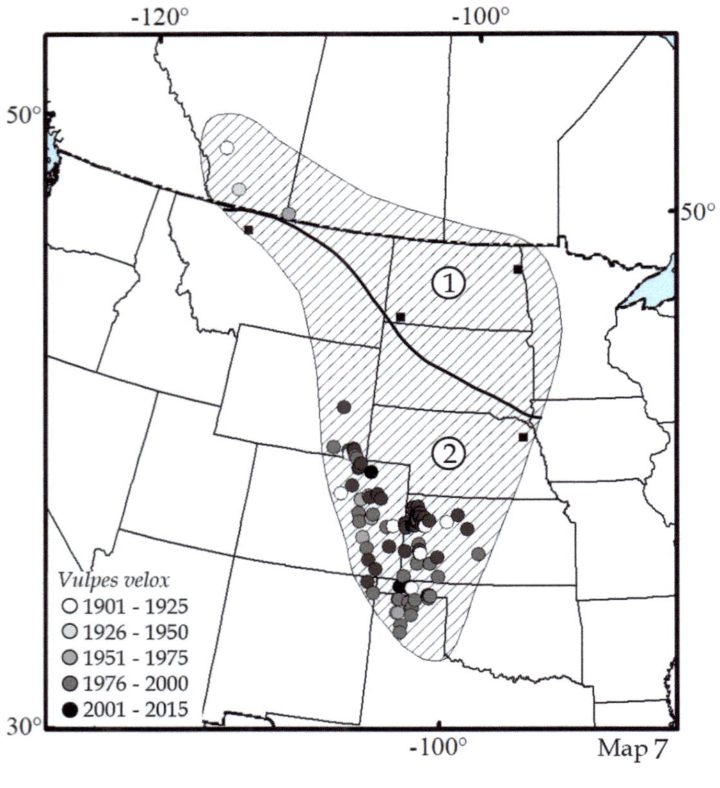

Map 7 Distribution map of *Vulpes velox*

Additional Literature: Thornton et al. (1971), Armstrong (1972), Thornton and Creel (1975), Van Gelder (1978), Egoscue (1979), McGrew (1979).

Vulpes vulpes (Linnaeus, 1758)
Red Fox, zorro rojo

1. *V. v. abietorum* Merriam, 1900. From southern Yukon, Mackenzie, and Nunavut to British Columbia, Alberta, and northwestern Saskatchewan.
2. *V. v. alascensis* Merriam, 1900. Alaska, Yukon, Mackenzie, and Nunavut.
3. *V. v. cascadensis* Merriam, 1900. From southern British Columbia to Oregon.
4. *V. v. fulvus* (Desmarest, 1820). From Colorado to Texas east through the Atlantic coast and north through Illinois, southern Ontario to Southwestern Maine.
5. *V. v. harrimani* Merriam, 1900. Restricted to Kodiak Island, Alaska.
6. *V. v. kenaiensis* Merriam, 1900. Restricted to Kenai Peninsula, Alaska.
7. *V. v. macroura* Baird, 1852. From southern British Columbia and western Alberta south through northern Arizona and New Mexico.
8. *V. v. necator* Merriam, 1900. California and Nevada.
9. *V. v. regalis* Merriam, 1900. From Nunavut and western Ontario south through Nebraska and Iowa.
10. *V. v. rubricosa* Bangs, 1898. From eastern Ontario and Baffin Island east through Newfoundland and Nova Scotia.

Conservation: *Vulpes vulpes* is listed as Least Concern by the IUCN (Hoffmann and Sillero-Zubiri 2021).

Characteristics: Red fox is medium sized; total length 900.0–1,050.0 mm, condylobasal length 139.0–150.0 mm. **Coloration homogeneous, red in most animals; fur of the throat, chest, and underparts white; tail unicolored, bushy, and white at the tip**; ears triangular with black in the dorsal part of the tip; **feet black; sagittal crest triangular**.

Comments: *Vulpes vulpes* from North America can be considered a different species (*V. fulva*; Statham et al. 2014). *V. vulpes* ranges include the whole Canada and central and eastern areas of the United States, except for desert areas and the western mountain range. Four isolated populations can be found in California (Map 8). *Vulpes vulpes* can be found in sympatry, or nearly so, with *Canis latrans, C. lupus,* and *Urocyon cinereoargenteus,* and can be distinguished from those species by its rostrum speckled in gray, and temporal crest lyre-shaped; total length less than 535.0 mm and postorbital process well developed. From the other *Vulpes*, by its homogeneous red coloration, including the throat and chest; belly white; tail unicolored, bushy, white at the tip; feet black; sagittal crest triangular.

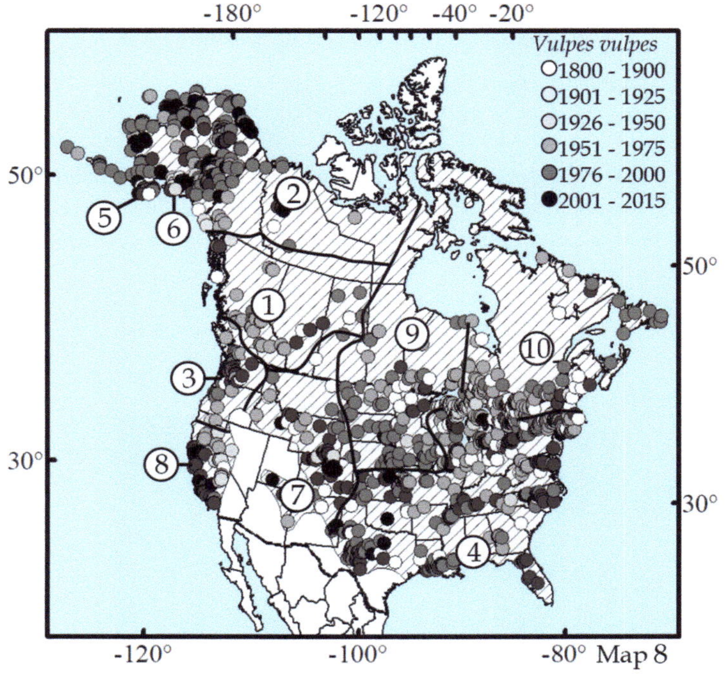

Map 8 Distribution map of *Vulpes vulpes*

Additional Literature: Jackson (1961), Ellerman and Morrison-Scott (1966), Hall (1981), Banfield (1987), Voigt (1987), Nowak (1991), Larivière and Pasitschniak-Arts (1996), Wozencraft (2005).

Family Mephitidae Bonaparte, 1845

The family Mephitidae includes skunks or stink badgers characterized by a small-to-medium size; **tail long, bushy, with black hairs; always in white-black coloration**; limb posture plantigrade; **glands producing strong and noxious odors; post-mandibular process frequently prominent and curved around the mandibular cavity**; **upper molars relatively large**; **skull blocky; rostrum short;** mastoid process absent or smaller than the paroccipital; carnassial usually well developed; no alisphenoid channel. Mephitidae differs from Mustelidae (Dragoo and Honeycutt 1997; Flynn et al. 2000; Wozencraft 2005). No subfamilies are recognized in this family.

1. Generally black, with one or two white stripes along the back; snout with no white spots; two upper premolars.......... *Conepatus* (p. 105)

1a. Color varied, never white along the back; snout with white spots; three upper premolars 2

2. Small sized, less than 500.0 mm; dorsal hair small, less than 20.0 mm; black pelage with many longitudinal and transverse white stripes or spots, giving it the appearance of a harlequin; palatine ending at the level or behind the last molar posterior border (Fig. 11); skull not arched in side view, with the highest part in the braincase .... *Spilogale* (p. 110)

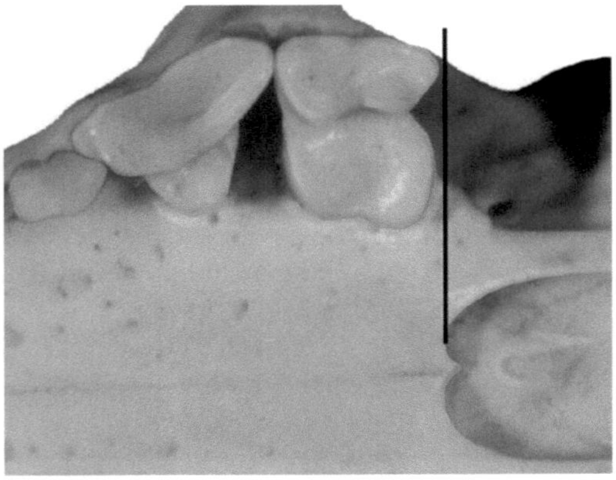

Fig. 11 The palatine ending at the level or behind the last molar posterior border

2a. Medium sized, greater than 500.0 mm; dorsal hair greater than 20.0 mm; black pelage, as much as two lateral dorsal whitish lines or spots; palatine ending in front of the last molar posterior border (Fig. 12); skull arched in side view, with the highest part in the interorbital area .... *Mephitis* (p. 107)

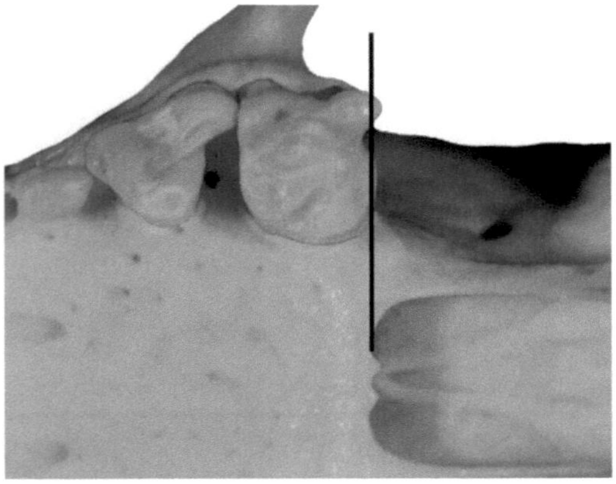

Fig. 12 The palatine ending in front of the last molar posterior border

Genus *Conepatus* Gray, 1837

Dorsal pelage black with one or two white stripes on back (no grizzle with black) running from forehead to back or hips with no white stripe on central rostrum; no transverse white lines to body; fur of underparts black; **short and white tail;** extended and snout bare pink; nose pad (ca. 20 mm width) width than on *Mephitis*; ears small and rounded; highest part of the skull at height of occipital area; three molariforms; last molariforms proportionally larger than next to last. The keys were elaborated based on the review of specimens and the following sources: Hall (1981), Dragoo et al. (2003), and Álvarez-Castañeda et al. (2017a).

1. Tail length less than one-third of the total length; back with two white stripes separated by a black stripe much narrower than any of the white ones; double or single anterior orbital foramen (Fig. 13; from Veracruz and the Yucatán Peninsula, México, south through South America) .... *Conepatus semistriatus* (p. 107)

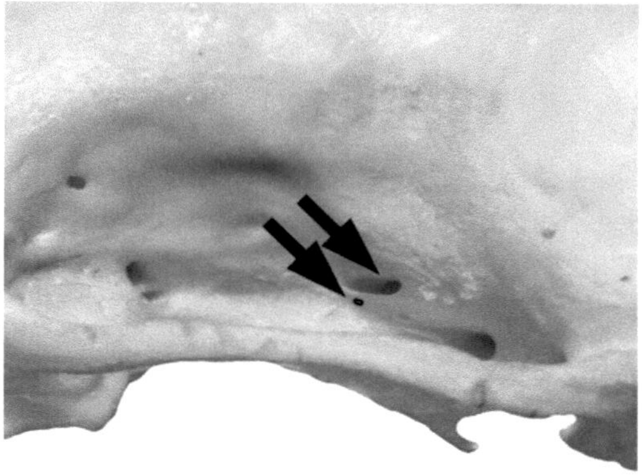

Fig. 13 Skull with one or two anterior orbital foramen present in *Conepatus semistriatus*

1a. Tail length greater than one-third of the total length; back with one white wide stripe; single anterior orbital foramen (Fig. 14; from Arizona, Colorado, and Texas, United States, south through Central America, except for the Yucatán Peninsula)... *Conepatus leuconotus* (p. 106)

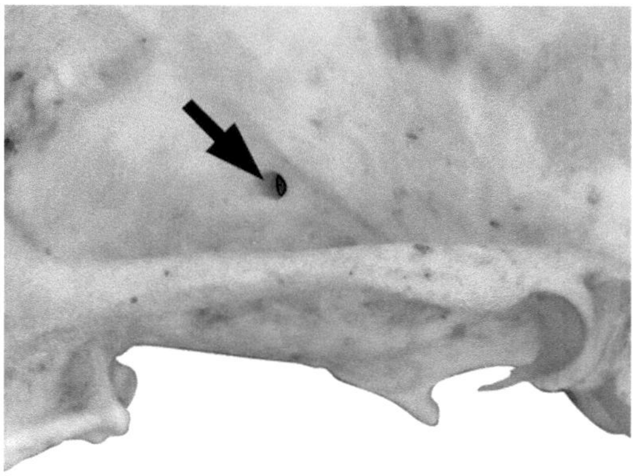

Fig. 14 Skull with only one anterior orbital foramen present in *Conepatus leuconotus*

Conepatus leuconotus (Lichtenstein, 1832)

Eastern Hog-Nosed Skunk, zorrillo de lomo blanco

1. *C. l. figginsi* Miller, 1925. Northwestern New Mexico and southeastern Colorado.
2. *C. l. filipensis* Merriam, 1902. Restricted to the highlands of central-northern Oaxaca.
3. *C. l. fremonti* Miller, 1933. Southeastern Colorado.
4. *C. l. leuconotus* (Lichtenstein, 1832). Veracruz.
5. *C. l. mearnsi* Merriam, 1902. From New Mexico and Texas south through Jalisco and San Luis Potosí.
6. *C. l. mesoleucus* (Lichtenstein, 1832). From Jalisco and San Luis Potosí south through Chiapas, including coastal Oaxaca.
7. *C. l. nelsoni* Goldman, 1922. Coastal areas of Colima, Michoacán, and Guerrero.
8. *C. l. sonoriensis* Merriam, 1902. Coastal areas from Sonora to Jalisco.
9. *C. l. telmalestes* Bailey, 1905. Restricted to eastern Texas.
10. *C. l. texensis* Merriam, 1902. Southwestern Texas, Tamaulipas, eastern Nuevo León, and northern Veracruz.
11. *C. l. venaticus* Goldman, 1922. Arizona, western New Mexico, eastern Sonora, and western Chihuahua.

Conservation: *Conepatus leuconotus* is listed as Least Concern by the IUCN (Helgen 2016).

Characteristics: Eastern hog-nosed skunk is the largest member of the family; total length 595.0–659.0 mm; condylobasal length 68.2–76.4 mm. Dorsal pelage black with **a medium white stripe; underparts at the tail base black or white**; **nose pad highly developed;** with a single anterior orbital foramen.

Comments: *Conepatus mesoleucus* is a junior synonym of *C. leuconotus* based on morphological and molecular data (Dragoo et al. 2003), previously considered a different species (Jones et al. 1992). It ranges from Texas, Arizona, New Mexico, and Colorado south through Central America, except for the Baja California Peninsula, western Sonora, Tabasco, Campeche, Yucatán, Quintana Roo, and Chiapas lowlands. The species is absent from hot deserts and tropical moist evergreen forests (Map 9) and can be found in sympatry, or nearly so, with *C. semistriatus, Mephitis macroura, M. mephitis,* and different species of *Spilogale.* It can be distinguished from the species of *Mephitis* and *Spilogale* by having a generally black coloration with one or two wide white stripes along the back and no white spots or stripes in the snout; presence of only two upper premolars. From *C. semistriatus*, by its larger tail, greater than one-third of the total length with only one white wide stripe; with a single anterior orbital foramen.

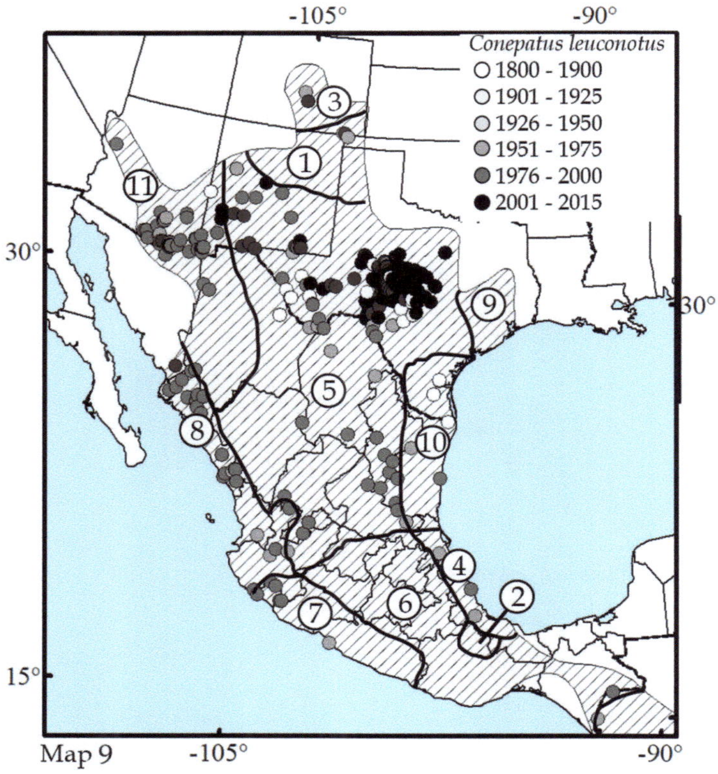

Map 9 Distribution map of *Conepatus leuconotus*

Additional Literature: Jones et al. (1992), Dragoo and Honeycutt (1997), Dragoo et al. (2003), Wozencraft (2005), Dragoo (2009), Dragoo and Sheffield (2009).

Conepatus semistriatus (Boddaert, 1785)
Striped Hog-Nosed Skunk, zorrillo bilistado

1. *C. s. conepalt* (Gmelin, 1788). From western Yucatán Peninsula west through central Veracruz.
2. *C. s. yucatanicus* Goldman, 1943. From central Yucatán Peninsula east and south through Central America.

Conservation: *Conepatus semistriatus conepalt* is listed as Subject to Special Protection in the Norma Oficial Mexicana (DOF 2019) and *C. semistriatus* as Least Concern by the IUCN (Cuarón et al. 2016f).

Characteristics: Striped hog-nosed skunk is medium sized; total length 533.0–688.0 mm; condylobasal length 78.0 mm. **Dorsal pelage black with two white stripes; fur of underparts black, hair short, and grizzle with white at the tail base;** with a double or single anterior orbital foramen.

Comments: *Conepatus semistriatus* ranges in southern México, from central Veracruz, including the whole Yucatán Peninsula south through South Central (Map 10), and can be found in sympatry, or nearly so, with *C. leuconotus,* from which it can be distinguished by its smaller tail, less than one-third of the total length; two white wide stripes separated by a black one, much narrower; with a single or double anterior orbital foramen. From species of *Mephitis* and *Spilogale,* by having a generally black coloration with one or two wide white stripes along the back and no white spots or stripes in the snout; only two upper premolars.

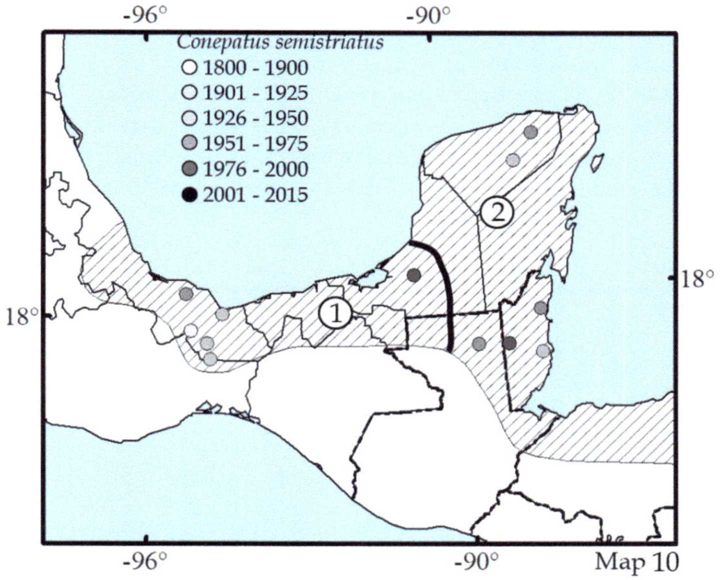

Map 10 Distribution map of *Conepatus semistriatus*

Additional Literature: Dragoo and Honeycutt (1997), Nowak (2005), Wozencraft (2005).

Genus *Mephitis* Geoffroy Saint-Hilaire and Cuvier, 1795

Dorsal and ventral coloration black with one or more white dorsal stripes running from the forehead to the hips, in some parts grizzled with black; with a white line on the middle of the rostrum; no transverse white lines in the body; fur of underparts black; tail long, with white hair grizzly with black; snout black; highest part of the skull at the level of the interorbital region; four molariforms; last molariforms proportionally equal to the next to last; palatine ending in front of the last molar posterior border. The keys were elaborated based on the review of specimens and the following sources: Howell (1901), Hall (1981), Dragoo (2009), and Álvarez-Castañeda et al. (2017a).

1. White stripes with no mixture of black hair; auditory bullae small, less than 12% of the condylobasal length (from southern Canada south through northern México, except for the United States southwestern deserts) .... *Mephitis mephitis* (p. 109)

1a. White stripes mixed with black hair; auditory bullae large, greater than 12% of the condylobasal length (Arizona, New Mexico, and Texas, United States, south through Central America, except for the Yucatán Peninsula) .... *Mephitis macroura* (p. 108)

Mephitis macroura Lichtenstein, 1832

Hooded Skunk, zorrillo rayado sureño

1. *M. m. eximius* Hall and Dalquest, 1950. Restricted to central Veracruz.
2. *M. m. macroura* Lichtenstein, 1832. From Nayarit, central Durango, northern Zacatecas, and Tamaulipas south through Chiapas.
3. *M. m. milleri* Mearns, 1897. From Arizona, New Mexico, and Texas south through southern Sinaloa, central Durango, and northern Zacatecas.

Conservation: *Mephitis macroura* is listed as Least Concern by the IUCN (Cuarón et al. 2016e).

Characteristics: Hooded skunk is medium sized; total length 598.0–678.0 mm, condylobasal length 59.0–63.0 mm. **Dorsal pelage black with white stripes in different shapes and forms; generally with a white stripe in the middle of the rostrum; fur of underparts black; auditory bullae large, greater than 12% of the condylobasal length**.

Comments: *Mephitis macroura* ranges from Arizona, New Mexico, and Texas south through Central America, except for the Yucatán Peninsula (Map 11), and can be found in sympatry, or nearly so, with *Conepatus semistriatus, C. leuconotus, M. mephitis,* and different species of *Spilogale. Mephitis macroura* can be distinguished from the species of *Conepatus* and *Spilogale* by having a generally black coloration, with one or two white stripes grizzled with black, not solid white; snout with a white spot; only three upper premolars, and skull length greater than 61.0 mm. From *M. mephitis*, by its dorsal stripes grizzled with black; auditory bullae small, less than 12% of the condylobasal length.

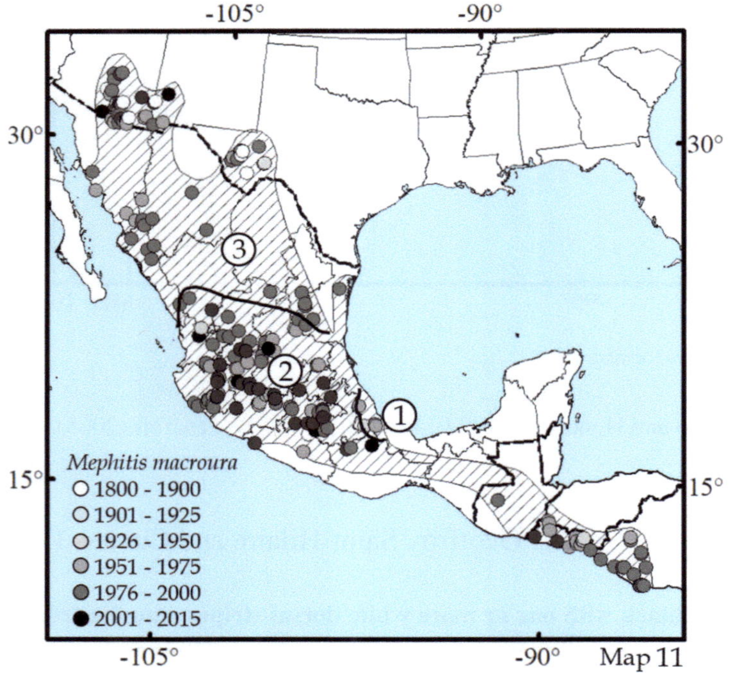

Map 11 Distribution map of *Mephitis macroura*

Additional Literature: Dragoo and Honeycutt (1997), Hwang and Larivière (2001), Wozencraft (2005), Dragoo (2009).

Mephitis mephitis (Schreber, 1776)
Striped Skunk, zorrillo rayado norteño

1. *M. m. avia* Bangs, 1898. Eastern Kansas, northern-central Missouri, and Illinois.
2. *M. m. elongata* Bangs, 1895. From Virginia south through Florida and west through the coastal plains to Louisiana.
3. *M. m. estor* Merriam, 1890a, b, c. From Utah and western Colorado south through northern Durango and western Chihuahua.
4. *M. m. holzneri* Mearns, 1897. From central and southern California north through Baja California.
5. *M. m. hudsonica* Richardson, 1829. From southern Mackenzie south through Colorado, and east through Wisconsin, including British Columbia.
6. *M. m. major* (Howell, 1901). Southeastern Washington, Oregon, western Idaho, Nevada, and western Utah.
7. *M. m. mephitis* (Schreber, 1776). From northwestern Manitoba east through Québec.
8. *M. m. mesomelas* Lichtenstein, 1832. From southern Kansas and southwestern Arkansas south through central Texas and Louisiana.
9. *M. m. nigra* (Peale and Palisot de Beauvois, 1796). From New Brunswick west through Michigan, Indiana, Missouri, Arkansas, and Mississippi south through southern Alabama and central Georgia.
10. *M. m. notata* (Howell, 1901). Washington and northern Oregon.
11. *M. m. occidentalis* Baird, 1858. From coastal Oregon to central California.
12. *M. m. spissigrada* Bangs, 1898. Coastal Washington.
13. *M. m. varians* Gray, 1837. From central Colorado and Kansas south through Chihuahua to Tamaulipas.

Conservation: *Mephitis mephitis* is listed as Least Concern by the IUCN (Helgen and Reid 2016b).

Characteristics: Striped skunk is medium sized; total length 575.0–800.0 mm, condylobasal length 62.0–74.0 mm. **Dorsal pelage black with two white dorsal stripes running from the forehead to the hips; auditory bullae small, less than 12% of the condylobasal length**.

Comments: *Mephitis mephitis* ranges throughout southern Canada, from British Columbia to Nova Scotia, mainly all United States south through northern México to Sonora, northern Durango, Coahuila, Nuevo León, and Tamaulipas (Map 12), and can be found in sympatry, or nearly so, with *Conepatus leuconotus, M. macroura,* and different species of *Spilogale,* from which it can be distinguished by its generally black coloration with one or two wide white stripes grizzled with black; if solid white, it does not cover all the back; snout with white spots; only three upper premolars, and skull length greater than 61.0 mm. From *M. macroura,* by its dorsal stripe less grizzled with black or solid white; however, if solid white, it does not cover all the back; auditory bullae small, greater than 12% of the condylobasal length.

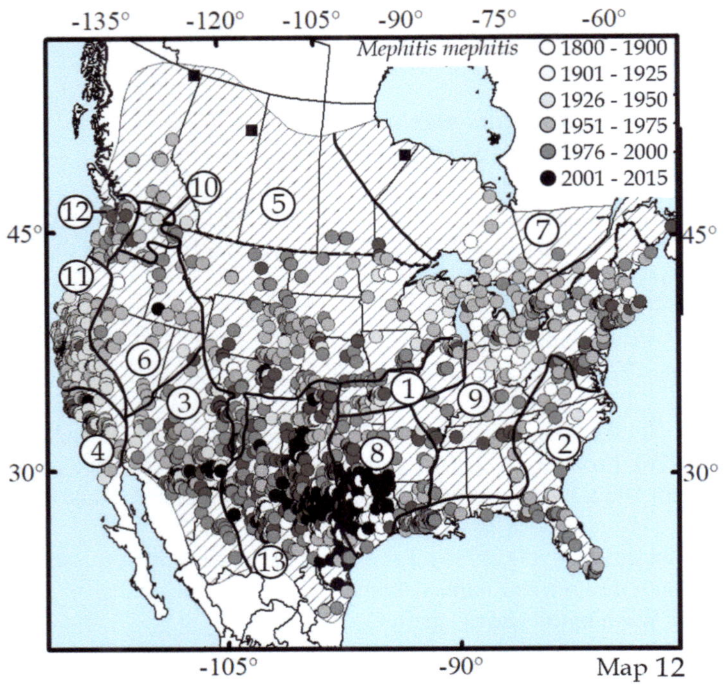

Map 12 Distribution map of *Mephitis mephitis*

Additional Literature: Dragoo and Honeycutt (1997), Dragoo (2009).

Genus *Spilogale* Gray, 1865

Dorsal and ventral coloration black with more than four white stripes on the back continuing toward the sides in transverse shape; general coloration and tail with white and black hairs, with the proportion of white-black color varying among the species; sagittal crest absent to well developed, and four molariforms. The keys were elaborated based on the review of specimens and the following sources Van Gelder (1959b), Hall (1981), Dragoo (2009), Álvarez-Castañeda et al. (2017a), McDonough et al. (2022).

1. Hind and forelimbs white; dorsal stripes black, continuous down to the hips; total length less than 290.0 mm; tail length less than 90.0 mm for both sexes; basal length less than 38.0 mm (coastal plains of Sinaloa south through Oaxaca, México) .... *Spilogale pygmaea* (p. 117)

1a. Hind and forelimbs black; dorsal stripes white not continuous to hips; total length greater than 290.0 mm; tail length always greater than 90.0 mm in both sexes; basal length greater than 38.0 mm 2

2. White dorsal stripes narrow, body more black than white; total length greater than 450.0 mm in males and greater than 400.0 mm in females; condylobasal length greater than 54.0 mm in males and greater than 50.0 mm in females; skull relatively broad and flat 3

2a. White dorsal stripes wider, body white-to-black ratio equal to or white dominant; total length less than 450.0 mm in males and less than 430.0 mm in females; condylobasal length less than 60.0 mm in males and less than 54.0 mm in females; skull relatively narrow and high 4

3. Tail prominently white at the tip; dorsal stripes wide and not interrupted; patch in front of the ears large, continuous with an outer dorsal stripe; mastoid capsules only slightly inflated, without well-developed processes; rostrum wide (eastern Mississippi River, Appalachian Mountains south through Louisiana to Florida, United States) .... *Spilogale putorius* (p. 116)

3a. Tail prominently black at the tip; dorsal stripes of variable width but on average narrower and frequently interrupted; patch in front of the ears markedly reduced, often absent, and usually not continuous with an outer dorsal stripe; mastoid capsules only slightly inflated, with well-developed processes; rostrum narrow (from western Mississippi River, Ontario, Canada and North Dakota, United States, south through Veracruz, México) .... *Spilogale interrupta* (p. 113)

4. In general, a greater proportion of black; total length greater than 360.0 mm in males and greater than 330.0 mm in females; skull relatively narrow and high 5

4a. In general, a greater proportion of white; total length less than 400.0 mm in males and less than 300.0 mm in females; skull relatively broad and flat 6

5. Mastoids not greatly inflated (From British Columbia, Canada, south through the Baja California Peninsula, México and east through Wyoming and Colorado, and eastern Idaho, including the Channel Islands, United States) .... *Spilogale gracilis* (p. 112)

5a. Mastoids greatly inflated (Chihuahuan and Sonoran Deserts south through the Altiplano Mexicano, including the Madrean Sky Islands and Texas) .... *Spilogale leucoparia* (p. 115)

6. Total length greater than 330.0 mm in males and greater than 300.0 in females; tail with two small spots at the base, not connected (from Jalisco, Durango, and Zacatecas south through Central America) .... *Spilogale angustifrons* (p. 111)

6a. Total length less than 330.0 mm in males and less than 300.0 in females; tail two small spots connected (horseshoe) at the base (Yucatán Peninsula, México) .... *Spilogale yucatanensis* (p. 118)

Spilogale angustifrons Howell, 1902

Southern Spotted Skunk, zorrillo manchado del sur

1. *S. a. angustifrons* Howell, 1902. From southern Durango, Zacatecas, and Guanajuato, south through Jalisco, Michoacán, and Estado de México.
2. *S. a. elata* Howell, 1906. Central Chiapas south through Central America.
3. *S. a. tropicalis* Howell, 1902. From Puebla and Morelos south through the coast of Chiapas and El Salvador.

Conservation: *Spilogale angustifrons* is listed as Least Concern by the IUCN (Helgen et al. 2016a).

Characteristics: Southern spotted skunk is small sized; total length; 325.0–350.0 mm in males and 310.0–340.0 mm in females; condylobasal length 50.0–52.2 mm in males and 45.0–47.0 mm in females. Dorsum **black with one white central stripe on the back, one lateral stripe running from the eye level to the middle of the back,** and some stripes running from the middle of the back to the limbs; **in general, a large proportion of white appearing white with black stripes**; tail black and white; sagittal crest well developed.

Comments: *Spilogale angustifrons* is considered a different species from *S. putorius* (Owen et al. 1996; Wozencraft 2005; Dragoo 2009) and *S. yucatanensis* from *Spilogale angutifrons* (McDonough et al. 2022). It ranges from Durango, Zacatecas, Jalisco, Aguascalientes, and San Luis Potosí south through Central America (Map 13), and can be found in sympatry, or nearly so, with *M. macroura. S. angustifrons* can be distinguished from the species of *Conepatus* and *Mephitis* by having the dorsal coloration black with many longitudinal and transverse white stripes or spots giving it the appearance of a harlequin; smaller size, less than 500.0 mm. From *S. gracilis* and S. *putorius,* by having a greater proportion of white pelage, appearing to be white with black stripes. From *S. pygmaea,* by its larger size, with total, tail, and basal lengths greater than 290.0, 90.0, and 38.0 mm, respectively; the two central lines on the back continue toward the hips and then to the forelimbs.

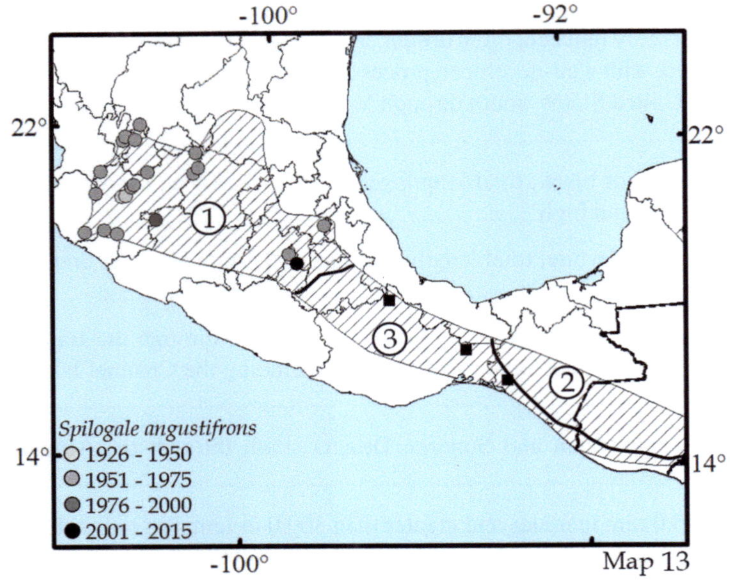

Map 13 Distribution map of *Spilogale angustifrons*

Additional Literature: Howell (1906), Van Gelder (1959b), Mead (1968), Dragoo et al. (1993), Dragoo and Honeycutt (1997).

Spilogale gracilis Merriam, 1890

Western Spotted Skunk, zorrillo manchado oeste

1. *S. g. amphiala* Dickey, 1929. Restricted to the Channel Islands, California.
2. *S. g. gracilis* Merriam, 1890. From central Washington to central Wyoming south through western California, Arizona, and New Mexico.
3. *S. g. latifrons* Merriam, 1890. Western British Columbia, Washington, and Oregon.
4. *S. g. lucasana* Merriam, 1890. Southern Baja California Peninsula.
5. *S. g. martirensis* Elliot, 1903. Baja California Peninsula to central Baja California Sur.
6. *S. g. phenax* Merriam, 1890. Coastal California.

Conservation: *Spilogale gracilis* is listed as Least Concern by the IUCN (Cuarón et al. 2016g).

Characteristics: Western spotted skunk is small sized; total length 360.0–457.0 mm in males and 330.0–431.0 mm in females; condylobasal length 49.9–59.7 mm in males and 47.2–53.8 mm in females. Dorsal pelage black with **very contrasting white stripes where black dominates,** with one central stripe on the back, one lateral stripe running from the eye level to the middle of the back, and some stripes running from the middle of the back to the limbs; **in general, a greater proportion of black, appearing as black with white stripes**; tail black and white; sagittal crest well developed.

Comments: *Spilogale gracilis* is considered a different species from *S. putorius* (Jones et al. 1992; Baker et al. 2003a; Wozencraft 2005). It ranges in western North America from British Columbia south through northern México, including the Channel Islands of California. The eastern records are from Wyoming, Colorado, and Idaho (Map 14). It can be found in sympatry, or nearly so, with *Conepatus leuconotus, Mephitis macroura,* and *M. mephitis,* from which it can be distinguished by its dorsal coloration black with many longitudinal and transverse white stripes or spots, giving it the appearance of a harlequin, and its smaller size, less than 500.0 mm. From *S. angustifrons,* by having a greater proportion of black pelage appearing black with wide white stripes, and from *S. putorius,* by having the white dorsal stripes narrow; total length less than 450.0 mm in males and less than 430.0 mm in females.

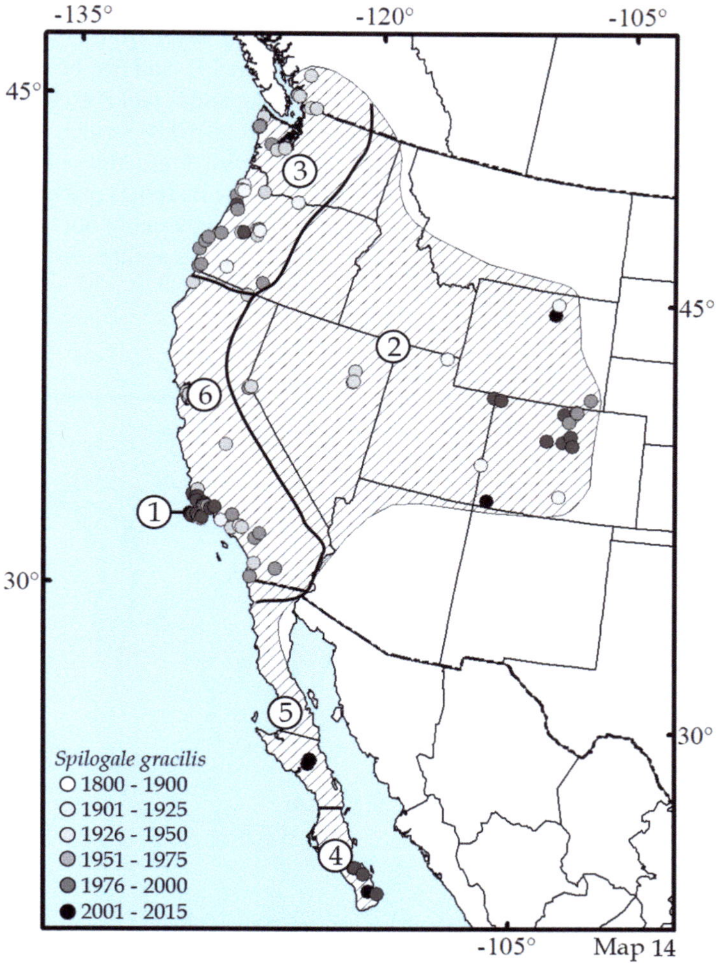

Map 14 Distribution map of *Spilogale gracilis*

Additional Literature: Howell (1906), Van Gelder (1959b), Mead (1968), Dragoo et al. (1993), Owen et al. (1996), Dragoo and Honeycutt (1997), Verts et al. (2001), Dragoo (2009), McDonough et al. (2022).

Spilogale interrupta (Rafinesque, 1820)

Plains Spotted Skunk, zorrillo manchado de las planicies

Monotypic.

Conservation: *Spilogale interrupta* (as *Spilogale putorius*) is listed as Vulnerable (A2abc + 3bc + 4abc: population reduction in the last 10 years 30% or less and may not be reversible by direct observation, index of abundance, and decline in area of occupancy + population reduction projected to be met in the future using index of abundance in area of occupancy + population reduction may not be reversible using direct observation, index of abundance, and decline in area of occupancy) by the IUCN (Gompper and Jachowski 2016) México.

Characteristics: Plain spotted skunk is small sized; total length 423.0–585.0 mm in males and 410.0–532.0 mm in females; condylobasal length 49.2–61.2 mm in males and 48.6–58.8 mm in females. Dorsal pelage black with one lateral stripe running from the eye level to the middle of the back and, in some individuals, from the middle of the back to the limbs; **dorsal stripes variable in width but on average narrower than in *S. putorius,* frequently interrupted (in about one-third of the specimens examined) and sometimes reduced to a few widely separated spots; a patch in front of the ears markedly reduced, often absent, usually not continuous with the outer dorsal stripe;** four or six white continuous dorsal lines from the back to the head; forehead with a white spot; **tail long, usually black throughout, occasionally with a very slender white hair at the tip, entirely surrounded by black hairs;** sagittal crest well developed; skull and teeth heavy.

Comments: *Spilogale interrupta* is considered a different species from *S. putorius* (McDonough et al. 2022). It is mostly restricted to the Great Plains of the United States, western of Mississippi River from Ontario Canada, North Dakota and Nebraska south through Texas, Louisiana, and Tamaulipas, México (Map 15), and can be found in sympatry, or nearly so, with *Conepatus leuconotus, Mephitis mephitis, S. angustifrons, S. gracilis,* and *S. putorius*. It can be distinguished from those species by its black dorsal coloration with many longitudinal and transverse white stripes or spots giving it the appearance of a harlequin and its smaller size, less than 500.0 mm. From *S. angustifrons, S. gracilis,* and *S. putorius,* by having a greater proportion of black pelage, appearing black with narrow white stripes; white markings markedly reduced; dorsal stripes variable in width; the patch in front of the ears markedly reduced, often absent, usually not continuous with the outer dorsal stripe; tail usually black throughout. From *S. putorius,* by having the mastoid capsules only slightly inflated but with well-developed processes; rostrum narrower.

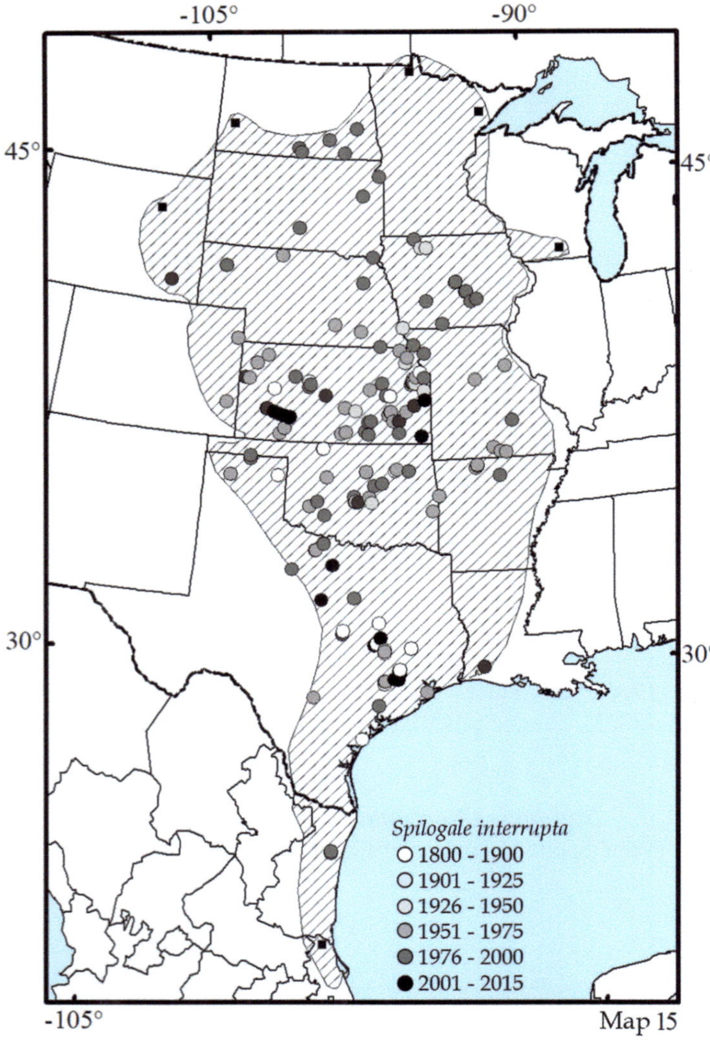

Map 15 Distribution map of *Spilogale interrupta*

Additional Literature: Howell (1906), Van Gelder (1959b), Mead (1968), Woodman and Ferguson (2021), McDonough et al. (2022).

Spilogale leucoparia Merriam, 1890
Desert Spotted Skunk, zorrillo manchado del desierto

Monotypic.

Conservation: *Spilogale leucoparia* (as *Spilogale gracilis*) is listed as Least Concern by the IUCN (Cuarón et al. 2016b).

Characteristics: Desert spotted skunk is small sized; total length 340.0–450.0 mm in males and 330.0–394.0 mm in females; condylobasal length 47.5–60.1 mm in males and 45.2–56.6 mm in females. Dorsal pelage white with **highly contrasting black stripes where white dominates,** one central stripe on the back, one lateral stripe running from the eye level to the middle of the back, and some stripes running from the middle of the back to the limbs; **stripes very broad, especially the lateral stripes and the outer pair of dorsal stripes; a frontal spot and a patch in front of the ears large; limbs not white; tail white in the distal one-fourth dorsally and the distal two-thirds ventrally;** sagittal crest well developed; **braincase broad and flat; mastoid capsules greatly inflated; auditory bullae large.**

Comments: *Spilogale leucoparia* is considered a different species from *S. gracilis* (McDonough et al. 2022). It ranges from Arizona, New Mexico, and Texas south through Durango, Zacatecas, and San Luis Potosí, México (Map 16), and can be found in sympatry, or nearly so, with *Conepatus leuconotus, Mephitis macroura, M. mephitis, S. angustifrons, S. gracilis,* and S. *putorius.* It can be distinguished from *C. leuconotus, M. macroura,* and *M. mephitis* by its black dorsal coloration with multiple longitudinal and transverse white stripes or spots giving it the appearance of a harlequin, and its smaller size, less than 500.0 mm. From *S. angustifrons*, by its greater proportion of black pelage, appearing black with wide white stripes. From *S. gracilis* and S. *putorius,* by having the white dorsal stripes narrow; total length less than 450.0 mm in males and less than 430.0 mm in females.

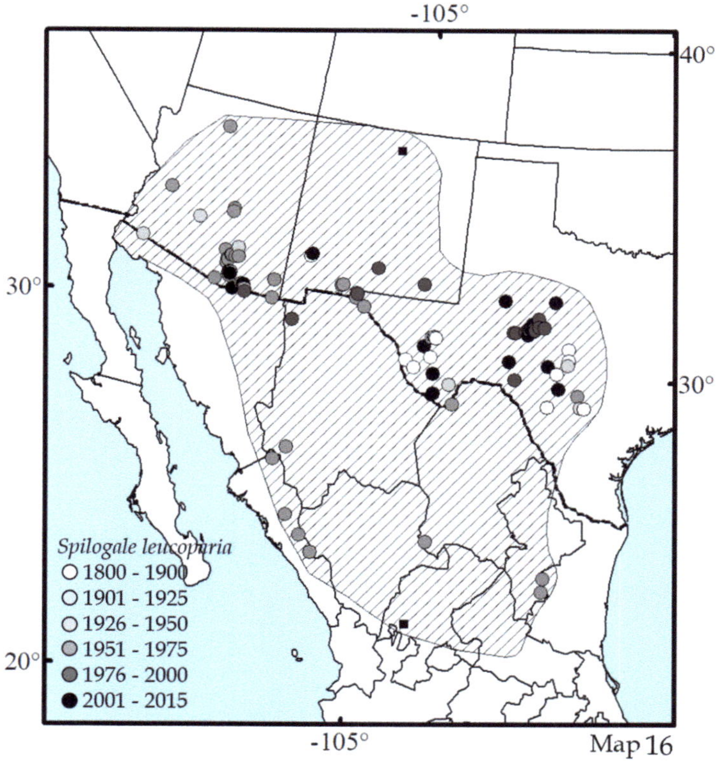

Map 16 Distribution map of *Spilogale leucoparia*

Additional Literature: Howell (1906), Van Gelder (b), Mead (1968).

Spilogale putorius (Linnaeus, 1758)

Eastern Spotted Skunk, zorrillo manchado este

1. *S. p. ambarvalis* Bangs, 1898. Florida Peninsula.
2. *S. p. putorius* (Linnaeus, 1758). From Illinois to Pennsylvania south through Gulf coast.

Conservation: *Spilogale putorius* is listed as Vulnerable (A2abc + 3bc + 4abc: population reduction in the last 10 years 30% or less and may not be reversible by direct observation, index of abundance, and decline in area of occupancy + population reduction projected to be met in the future using index of abundance in area of occupancy + population reduction may not be reversible using direct observation, index of abundance, and decline in area of occupancy) by the IUCN (Gompper and Jachowski 2016) and is not listed in the United States or México.

Characteristics: Eastern spotted skunk is small sized; total length 403.0–470.0 m in females, 453.0–610.0 mm in males; condylobasal length 51.2–56.4 mm in females; 53.6–61.9 mm in males. Dorsal pelage black with **one lateral stripe running from the eye level to the middle of the back** and some stripes running from the middle of the back to the limbs; **four or six white continuous dorsal lines from the back to the head; forehead with a white spot**; **tail black and white, long**; sagittal crest well developed; **skull and teeth heavy.**

Comments: *Spilogale putorius* is considered a different species from *S. gracilis* (Mead 1968; Dragoo et al. 1993). It ranges from eastern North America, eastern Mississippi River, Appalachian Mountains, and Illinois to Pennsylvania south through Louisiana to Florida (Map 17). It can be found in sympatry, or nearly so, with *Conepatus leuconotus, Mephitis mephitis, S. angustifrons, S. gracilis,* and *S. interrupta,* from which it can be distinguished by its black dorsal coloration with many longitudinal and transverse white stripes or spots, giving it the appearance of a harlequin, and its smaller size, less than 500.0 mm. From *S. angustifrons, S. gracilis,* and *S. interrupta,* by having a greater proportion of black pelage, appearing black with narrow white stripes; white markings markedly reduced; dorsal stripes variable in width; the patch in front of the ears markedly reduced, often absent, and usually not continuous with the outer dorsal stripe; tail usually black throughout. From *S. interrupta,* by its mastoid capsules only slightly inflated but with well-developed processes, and rostrum narrower.

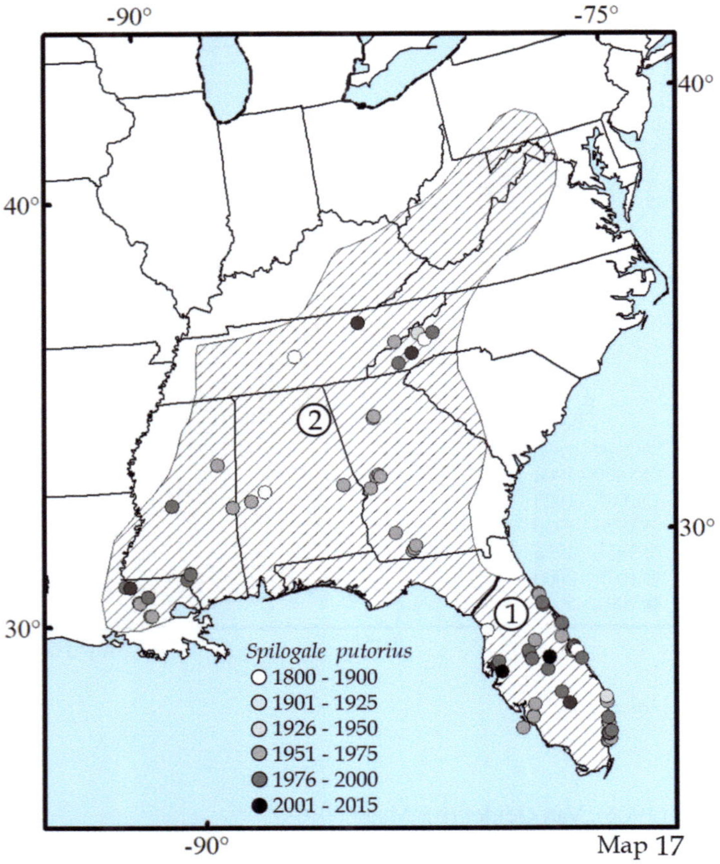

Map 17 Distribution map of *Spilogale putorius*

Additional Literature: Howell (1906), Van Gelder (b), Jones et al. (1992), Owen et al. (1996), Dragoo and Honeycutt (1997), Wozencraft (2005), Dragoo (2009), McDonough et al. (2022).

Spilogale pygmaea Thomas, 1898

Pygmy Spotted Skunk, zorrillo pigmeo

1. *S. p. australis* Hall, 1938. From the coast of Michoacán south through Oaxaca.
2. *S. p. intermedia* López-Forment and Urbano, 1979. Coast of Jalisco and Colima.
3. *S. p. pygmaea* Thomas, 1898. Coast of Sinaloa and Nayarit.

Conservation: *Spilogale pygmaea* is listed as Endangered in the Norma Oficial Mexicana (DOF 2019) and as Vulnerable (A2ce: population reduction in the last 10 years 30% or less and may not be reversible data obtained by decline in area of occupancy and effects of introduced taxa) by the IUCN (Helgen et al. 2016b).

Characteristics: Pygmy spotted skunk is the smallest member of the family; total length 248.0–271.0 mm in males and 235.0–265.0 mm in females; condylobasal length 41.3–43.2 mm in males and 36.0–37.0 mm in females. Dorsal pelage black with **two central lines on the back running toward the hips and then to the forelimbs,** one lateral stripe running from the eye level to the middle of the back, both joining on the forehead as a rostral line; **dorsal pelage of hind and forelimbs white; dorsal stripes continuous down to the hips;** tail white with black hair; sagittal crest barely developed or absent.

Comments: *Spilogale pygmaea* was previously considered a junior synonym of *S. putorius* (Ewer 1973). *S. pygmaea* is endemic to the tropical Pacific Coast of México, from Sinaloa south through Oaxaca; it is most common at lower altitudes (Map 18). It can be found in sympatry, or nearly so, with *Conepatus semistriatus* and *Mephitis macroura. S. pygmaea* can be distinguished from the species of *Conepatus* and *Mephitis* by its black dorsal coloration with multiple longitudinal and transverse white stripes or spots giving it the appearance of a harlequin and its smaller size, less than 500.0 mm. From *S. angustifrons, S. gracilis,* and *S. putorius* by its smaller size, with total, tail, and basal lengths less than 290.0, 90.0, and 38.0 mm, respectively.

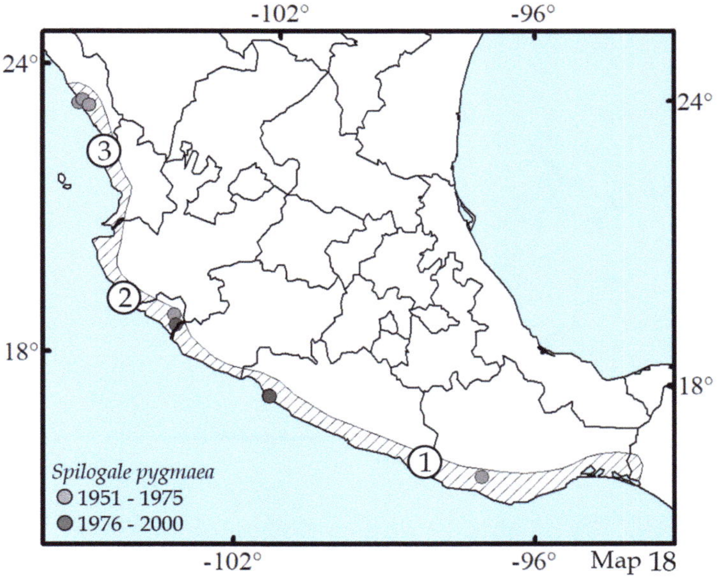

Map 18 Distribution map of *Spilogale pygmaea*

Additional Literature: Howell (1906), Van Gelder (b), Mead (1968), Medellin et al. (1989, 1998), Dragoo et al. (1993), Owen et al. (1996), Dragoo and Honeycutt (1997), Dragoo (2009), McDonough et al. (2022).

Spilogale yucatanensis Burt, 1938

Yucatán Spotted Skunk, zorrillo manchado de la península de Yucatán

Monotypic.

Conservation: *Spilogale yucatanensis* (as *Spilogale angustifrons*) is listed as Least Concern by the IUCN (Helgen et al. 2016b).

Characteristics: Yucatán spotted skunk is small sized; total length 260.0–280.0 mm in females, 290.0–320.0 mm in males; condylobasal length 40.0–40.5 mm in females, 25.0–30.0 mm in males. Dorsal pelage black with **central stripes on back no continuing toward the hips; no stripe to forelimbs,** one lateral stripe running from eye level to middle back not linking on the forehead spot; **no white dorsal pelage of hind and forelimbs;** tail with large part white at the tip; sagittal crest little developed or absent.

Comments: *Spilogale yucatanensis* is considered a different species from *S. angustifrons* (McDonough et al. 2022). *S. yucatanensis* is endemic to the Yucatán Peninsula including Belize and northeastern Guatemala (Map 19) and can be found in sympatry, or nearly so, with *Conepatus semistriatus,* from which it can be distinguished by its black dorsal coloration with many longitudinal and transverse white stripes or spots, giving it the appearance of a harlequin, and its smaller size, less than 500.0 mm.

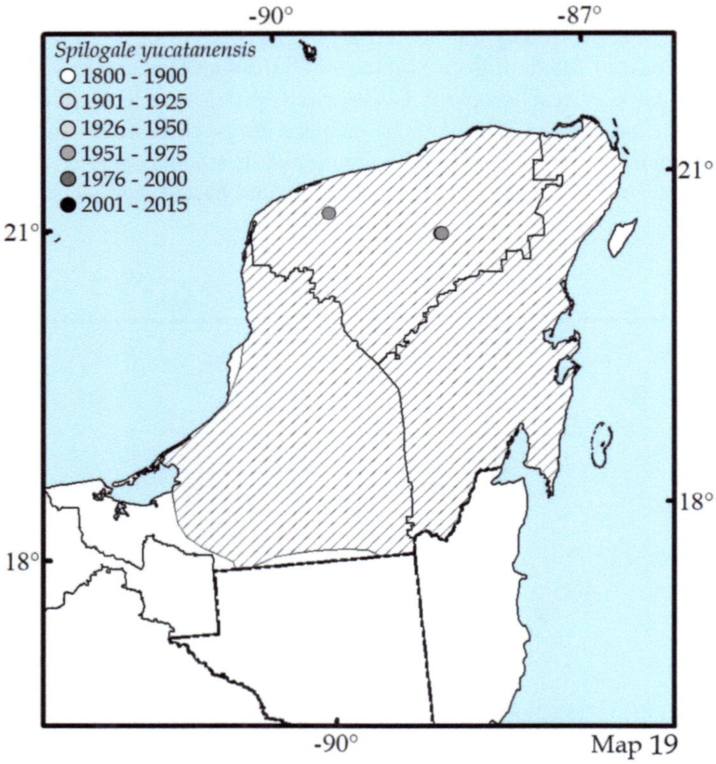

Map 19 Distribution map of *Spilogale yucatanensis*

Additional Literature: Van Gelder (1959a, b), Mead (1968).

Family Mustelidae Fischer, 1817

The family Mustelidae includes martens, otters, and weasels characterized by being small to medium sized; tail usually long but not bushy; limb posture plantigrade (badgers) to digitigrades (weasels); interdigital membranes in otters; glands producing odors; **post-mandibular process frequently prominent and curved around the mandibular fossa, often locking the lower jaw in place**; **upper molars relatively large, usually dumbbell-shaped or squarish**; **skull blocky, rostrum short, often flattened**; mastoid process absent or smaller than the paroccipital process; carnassial usually well developed; no alisphenoid canal. Cosmopolitan distribution, except for Oceania and Madagascar. It belongs to the superfamily Musteloidea. Two subfamilies range in North America: Lutrinae and Mustelinae. The keys were elaborated based on the review of specimens and the following sources: Lawlor (1979), Hall (1981), and Álvarez-Castañeda et al. (2015, 2017a).

1. Body very long with very short limbs; basal length less than 70.0 mm .... subfamily Mustelinae (p. 133)

1a. Body and limbs in the normal proportion; basal length greater than 70.0 mm 2

2. Tail base as wide as the hip area; hind digits webbed; last upper molariforms equal to or larger than the first molar ....subfamily Lutrinae (p. 129)

2a. Tail base slimmer than the hip area; hind digits not webbed; last upper molariforms smaller than the first molar (Fig. 15)..3

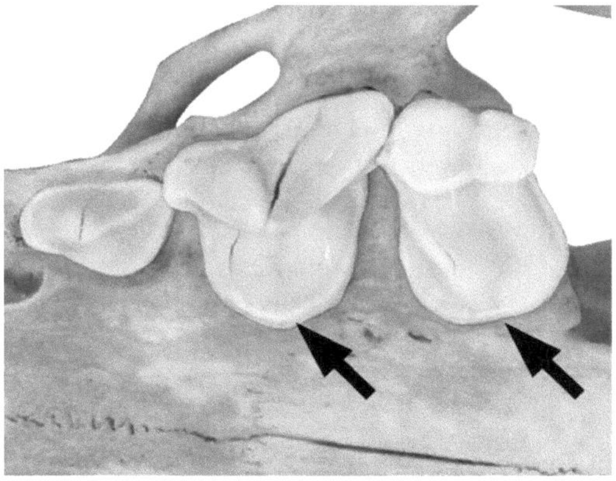

Fig. 15 Last upper molariforms smaller than the first molar

3. A pale-colored diagonal stripe on the head and shoulders; teeth large and strong in proportion to the skull; upper carnassial occupying nearly one-half of the inner surface of the tooth; first upper molars about twice wider than long (San Luis Potosí, Veracruz, and Chiapas including Yucatán Peninsula south through South America)........................... ....subfamily Ictonychinae; *Galictis vittata* (p. 128)

3. Without a pale-colored diagonal stripe on the head and shoulders; teeth of size proportional to the skull; upper carnassial occupying less than one-half of the tooth surface; first upper molars longer than wide 4

4. Brown pelage with a white line from the nose to at least the middle of the body; body appearing flat mainly because of very long lateral hair; tail short, less than 20% of the total length; skull triangular in dorsal view (Fig. 16); mastoid breadth greater than 65.0 mm (from central Canada south through Oaxaca, México, except for the eastern Mississippi River, United States) .... subfamily Taxidiinae; *Taxidea taxus* (p. 145)

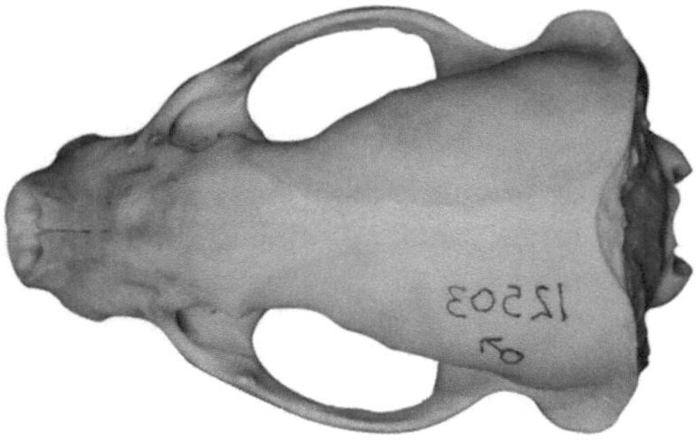

Fig. 16 Skull shape triangular in dorsal view

4a. Black or brown pelage but never with the white stripe described previously; body cylndrical; tail very long, greater than 20% of the total length; skull not triangular in dorsal view; mastoid breadth less than 65.0 mm ....subfamily Guloninae (p. 120)

Subfamily Guloninae Gray, 1825

This subfamily includes the tayra, wolverine, marten, and fisher, characterized by having a cylindrical body; tail very long, greater than 20% of the total length and bushy. The keys were elaborated based on the review of specimens and the following sources: Lawlor (1979), Hall (1981), Lariviere (1999a), and Álvarez-Castañeda et al. (2017a).

1. Body stocky; coloration generally dark brown with a blond band on the forehead and the flanks; four upper and lower premolars; skull length greater than 138.0 mm (throughout Alaska and Canada and the highlands of Washington, Oregon, California, Colorado, and Utah, and northeast of the United States).................................... *Gulo gulo* (p. 122)

1a. Body not stocky; body generally in different colors and without a blond band on the forehead and the flanks; less than four upper and lower premolars; skull length less than 138.0 mm 2

2. Dorsal and tail coloration black or grizzly; head coloration different from the back; anterior part of the orbital fossa at the level of the second upper premolars (Tamaulipas and Sinaloa, México south through South America)...................... ....*Eira barbara* (p. 120)

2a. Dorsal and tail coloration brown; head coloration not markedly different from the back; anterior part of the orbital fossa at the level of the third upper premolars 3

3. Tail less than 290.0 mm; ears pointed; head paler than the body; length of fourth upper premolars greater than 9.5 mm; length of first lower molars less than 11.0 mm ....*Martes* (p. 123)

3a. Tail greater than 290.0 mm; ears rounded; head and body of the same color; length of fourth upper premolars less than 9.5 mm; length of first lower molars greater than 11.0 mm (from Canada, except for the tundra region, south through northeastern United States, and Montana, Idaho, Wyoming, Utah, Oregon, and California).. .... *Pekania pennanti* (p. 126)

Genus *Eira* Smith, 1842

Body pelage black or dark brown with the head and neck gray or brown; fur of underparts yellow or a white spot on the chest; rhinarium grooved on the upper and anterior surfaces; body long and slender; tail long; neck long; head broad; limbs short; plantar pads separated from carpal and metacarpal pads by a deep and wide groove; pads confluent throughout their width; only a slight division between the two moieties of carpal pads, and metatarsal pad with no division; upper carnassial anterior in position, occupying less than one-half of the inner surface of the tooth; upper molars wider than long; talonid semibasined in the first lower molars.

Eira barbara (Linnaeus, 1758)

Tayra, cabeza de viejo

E. b. senex (Thomas, 1900). Only subspecies with range in North America.

Conservation: *Eira barbara* is listed as Endangered in the Norma Oficial Mexicana (DOF 2019) and as Least Concern by the IUCN (Cuarón et al. 2016d).

Characteristics: Tayra is medium sized; total length 920.0–1165.0 mm, condylobasal length 98.0–120.5 mm. **Dorsal pelage usually dark, head and neck in yellowish-golden shades contrasting with the body;** limbs long and thin, about one-fourth to one-fifth of the head-and-body length; long, tail dark brown or blackish, hairy; palate projecting behind the line of the last molars greater than the breadth of the last molars; last molars much wider than long; supraocular processes very sharp; **first lower molars (carnassial) not cutting**; skull breadth at the auditory bullae height not notably wider than the rest of the braincase.

Comments: *Eira barbara* ranges from Tamaulipas and Sinaloa south through South America (Map 20). It can be found in sympatry, or nearly so, with *Galictis vittata, Lontra longicaudis,* and *Neogale frenata*, from which it can be distinguished by its larger size, more than 1000.0 mm; dorsal pelage usually dark, head and neck in yellowish-golden shades contrasting with the body, and lower carnassial not cutting. From *Galictis*, by having the rhinarium grooved on both upper and anterior surfaces and nostrils larger and closer together. From *L. longicaudis*, by its narrow tail at the base; hind and fore digits not webbed, and last molariforms smaller than the first molariforms.

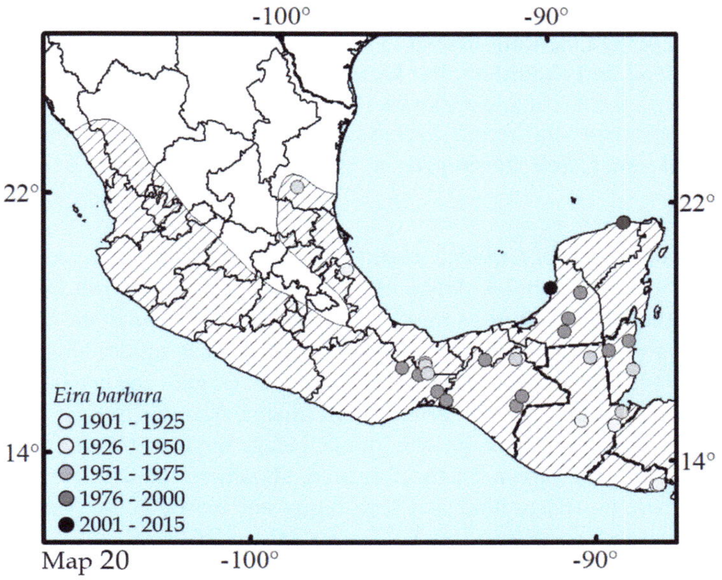

Map 20 Distribution map of *Eira barbara*

Additional Literature: Pocock (1921), Hall and Dalquest (1963), Hall (1981), Eisenberg (1989), Mares et al. (1989), Emmons and Freer (1990), Presley (2000).

Genus *Gulo* Pallas, 1780

Wolverines have a **general color blackish brown with a broad pale brown stripe beginning on each side of the neck and extending back alongside of the body to join its opposite stripe over and across the tail base**; **crown and cheeks paler and grayer**; fur of underparts whitish on the middle of the abdomen, throat and chest more or less spotted with yellowish white; **sagittal crest very well developed; zygomatic arches extraordinarily ponderous**; palate behind the upper molars; auditory bullae moderately inflated and separate from the paroccipital processes; mandibular and dental structures extremely strong; **lower first molar with a talonid longer than the trigonid, metaconid absent, talonid barely semitrenchant and with a single, median longitudinal ridge**.

Gulo gulo (Linnaeus, 1758)

Wolverine, glotón

1. *G. g. katschemakensis* Matschie, 1918. Restricted to the Kenai Peninsula, Alaska.
2. *G. g. luscus* (Linnaeus, 1758). Throughout Alaska, Canada, and a large part of northern United States.
3. *G. g. luteus* Elliot, 1904. From Washington south through California.
4. *G. g. vancouverensis* Goldman, 1935. Vancouver Island, British Columbia.

Conservation: The eastern population of *Gulo gulo* is listed as Endangered by the Canadian Endangered Species Protection Act and as Least Concern by the IUCN (Abramov 2016).

Characteristics: Wolverine is the largest mustelid in North America; total length 940.0–1080.0 mm; condylobasal length 138.0–160.2 mm. Dorsal pelage medium brown to black, except for the nape; two pale bands creamy, pale yellow, or pale brown on the forehead and shoulders, flanks, and above the tail at the base; fur of underparts similar to the nape in color, a pale whitish spot on the throat and chest; ears round and furred; head large; neck short and stout; limbs relatively short; feet large and with powerful semi-retractile claws; tail short, about one-fifth of the head-and-body length; skull very robust and broad; **sagittal crest well developed; zygomatic arches broad; dentals strong; four upper and lower molars**.

Comments: The North American species of *Gulo* have been considered a different species *G. luscus* (Hall 1981), but this was not accepted (Jones et al. 1992; Pasitschniak-Arts and Larivière 1995; Wozencraft 2005). *G. gulo* has a circumpolar distribution; in America, the southeastern limit of its range comprises all the States in the United States-Canada border and the southwestern through the highlands of Wyoming, Idaho, Montana, Washington, Oregon, California, and an isolated population in central-eastern Nevada and central-western Utah (Map 21). *G. gulo* and can be found in sympatry, or nearly so, with *Lontra canadensis, Martes americana, Mustela erminea, M. nivalis, Neogale frenata, N. vison,* and *Pekania pennanti;* it can be distinguished from these species by its large size; dorsal pelage medium brown to black except for the nape, with two bands creamy, paler yellow, or pale brown on the forehead, shoulders, flanks, and dorsally at the tail base. From *L. canadensis,* by having a narrow tail base, hind and fore digits not webbed; last molariforms smaller than the first molarifoms.

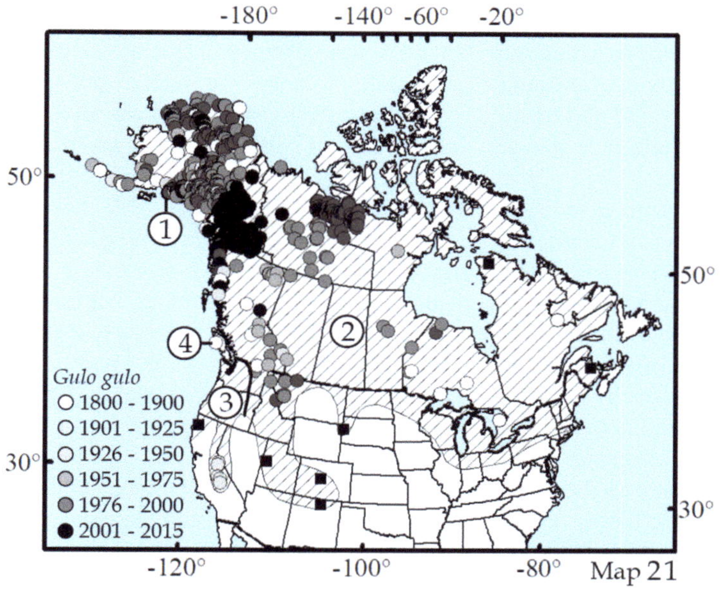

Map 21 Distribution map of *Gulo gulo*

Additional Literature: Jackson (1961), Wilson (1982), Nowak and Paradiso (1983), Banfield (1987).

Genus *Martes* Pinel, 1792

Dorsal and underparts pelage varying according to the species and season; **limbs short;** tail medium to long, bushy. Both species can be distinguished in skull shape and color patterns. The keys were elaborated based on the review of specimens and the following sources: Hall (1981), Powell (1981), and Clark et al. (1987).

1. Irregular markings on the throat and under surface generally whitish or yellowish; distance between the outer alveoli of the upper canines much less than the auditory bullae greatest length; rostrum laterally narrow and longer; mastoid breadth smaller (throughout Canada, except for the tundra region, south through northwestern United States, Washington, Idaho, Utah, and northwestern United States) ....*Martes americana* (p. 111)

1a. Irregular markings on the throat and under surface generally orange-red; distance between the outer alveoli of the upper canines equal to the auditory bullae greatest length; rostrum laterally dilated and shorter; mastoid breadth greater (from British Columbia, Canada, south through California and east through Colorado and New Mexico, United States)......... .... *Martes curina* (p. 112)

Martes americana (Turton, 1806)

American Marten, marta americana

1. *M. a. abieticola* (Preble, 1902). Saskatchewan and northwestern Manitoba.
2. *M. a. abietinoides* Gray, 1865. Central and southeastern British Columbia.
3. *M. a. actuosa* (Osgood, 1900). Alaska, Yukon, northern British Columbia, Alberta, and Mackenzie.
4. *M. a. americana* (Turton, 1806). From eastern Manitoba east through southeastern Québec and New Brunswick.
5. *M. a. atrata* (Bangs, 1897). Restricted to Prince Edward Island, Newfoundland.
6. *M. a. brumalis* (Bangs, 1898). Northern Québec and Newfoundland.
7. *M. a. kenaiensis* (Elliot, 1903). Southern coast of Alaska.

Conservation: *Martes americana atrata* is listed as Endangered Species in the Act of Canada and *M. americana* is listed as Least Concern by the IUCN (Helgen and Reid 2016a).

Characteristics: American marten is medium sized; total length 560.0–680.0 mm in males and 500.0–600.0 mm in females; condylobasal length 73.0–88.0 mm in males and 70.5–80.5 mm in females. Dorsal pelage yellow buff to tawny brown to almost black; head paler than the body; fur of underparts similar to dorsal fur, with **a spot vivid straw whitish or yellowish in the throat and chest;** limbs and tail darker; head triangular and nose sharp; ears rounded; limbs short and tail bushy; skull relatively long and narrow with slender zygomatic arches; auditory bullae elongated and relatively inflated.

Comments: *Martes caurina* was recognized as a distinct species (Carr and Hick 1997; McGowan et al. 1999; Dawson et al. 2017, Colella et al. 2018), previously regarded as a subspecies of *M. americana* (Wright 1953; Anderson 1970; Clark et al. 1987; Stone and Cook 2002). *M. americana* ranges in the higher-latitude taiga of northern North America, from southwestern Canada north through Alaska and west through the eastern coast, including Washington, Montana, Newfoundland, Michigan, Minnesota, Illinois, Ohio, and Maine (Map 22). It can be found in sympatry, or nearly so, with *Gulo gulo, Lontra canadensis, Mustela erminea, M. nivalis, Neogale frenata, N. vison,* and *Pekania pennanti,* and can be distinguished from all other species of *Mustela* and *N. vison* by its larger size, greater than 500.0 mm in total length. From *L. canadensis,* by having a narrow tail base; hind and fore digits not webbed; last molariforms smaller than the first molariforms. From *G. gulo,* by not having a stocky body and without a blond band on the forehead and flanks. From *P. pennanti,* by having a spot vivid straw whitish or yellowish on the throat and chest and head of paler color than the body. From *M. caurina,* by having a rostrum narrow and longer; auditory bullae longer and more inflated; palate shelf behind the plane at the last molar; first upper premolars larger and less crowded; last upper molars shorter and with a narrow saddle; distance between the upper canine outer alveoli much less than the greatest auditory bullae length; transverse diameter of the last upper molars less than the upper sectorial length; first lower molar length greater than the antero-posterior diameter at the last upper molars, exceeding the greatest breadth of the upper sectorial.

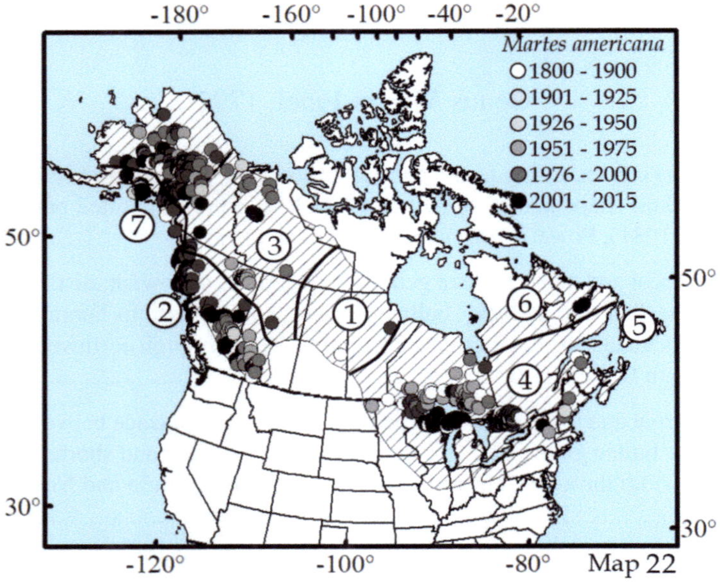

Map 22 Distribution map of *Martes americana*

Additional Literature: Merriam (1890c), Youngman (1982), Graham and Graham (1994), Colella et al. (2019).

Martes caurina (Merriam, 1890)

Pacific Marten, marta del Pacífico

1. *M. c. caurina* (Merriam, 1890). Central and southwestern British Columbia.
2. *M. c. humboldtensis* Grinnell and Dixon, 1926. Restricted to northwestern California.
3. *M. c. nesophila* (Osgood, 1901). Restricted to island of Graham, Queen Charlotte, Kuia, and Sewell Inlet.
4. *M. c. origenes* (Roads, 1902). Utah, Colorado, southern Wyoming, and northern New Mexico.
5. *M. c. sierrae* Grinnell and Storer, 1916. Restricted to California.
6. *M. c. vancouverensis* Grinnell and Dixon, 1926. Restricted to Vancouver Island, British Columbia.
7. *M. c. vulpina* (Rafinesque, 1819). Idaho, Montana, eastern Washington, Oregon, and western Wyoming.

Conservation: *Martes caurina sierrae* is considered a vulnerable species by NatureServe. *Martes caurina* (as *Martes americana*) is listed as Least Concern by the IUCN (Helgen and Reid 2016a).

Characteristics: Pacific marten is medium sized; total length 560.0–680.0 mm in males, 500.0–600.0 mm in females; condylobasal length 73.0–88.0 mm in males, 70.0–80.0 mm in females. Dorsal pelage from yellow buff to tawny brown to almost black; head paler than the body; fur of underparts similar to dorsal fur with **a spot vivid straw orange-red in the throat and chest;** limbs and tail darker; head triangular and nose sharp; ears rounded; limbs short and tail bushy; skull relatively long and narrow with slender zygomatic arches; auditory bullae elongated and relatively inflated.

Comments: *Martes caurina* was recognized as a distinct species (Carr and Hick 1997; McGowan et al. 1999; Dawson et al. 2017, Colella et al. 2018), previously regarded as a subspecies of *M. americana* (Wright 1953; Anderson 1970; Clark et al. 1987; Stone and Cook 2002). *M. caurina* ranges in temperate old-growth coastal forests, forests of the southern Rockies, the intermountain west, and southwestern sky islands from the Southern islands of Alaska (Admiralty, Kuiu, Graham, and Queen Charlotte), southwestern Canada (British Columbia and Alberta), and central-western United States (Washington, Montana, Idaho, Wyoming, Oregon, California, Utah, Colorado, and New Mexico (Map 23). It can be found in sympatry, or nearly so, with *Gulo gulo, Lontra canadensis, Mustela erminea, M. nivalis, Neogale frenata, N. vison,* and *Pekania pennanti. M. caurina* can be distinguished from all other species of *Mustela* and *N. vison* by having a larger size, greater than 500.0 mm in total length. From *L. canadensis,* by its narrow tail base, hind and fore digits not webbed; last molariforms smaller than the first molariforms. From *G. gulo,* by not having a stocky body and without a blond band on the forehead and flanks. From *P. pennanti,* by having a spot vivid straw orange-red on the throat and chest and head paler than the body. From *M. americana,* by having the rostrum broader and shorter; auditory bullae shorter and less inflated; palate shelf less produced behind the plane of the last molar; first upper premolars smaller and more crowded; last upper molars not only larger but with a much broader saddle; distance between upper canine outer alveoli equal to or greatest than auditory bullae length; transverse diameter of the last upper molars greater than upper sectorial length; first lower molar length less than last upper molar antero-posterior diameter and equal to the upper sectorial greatest breadth.

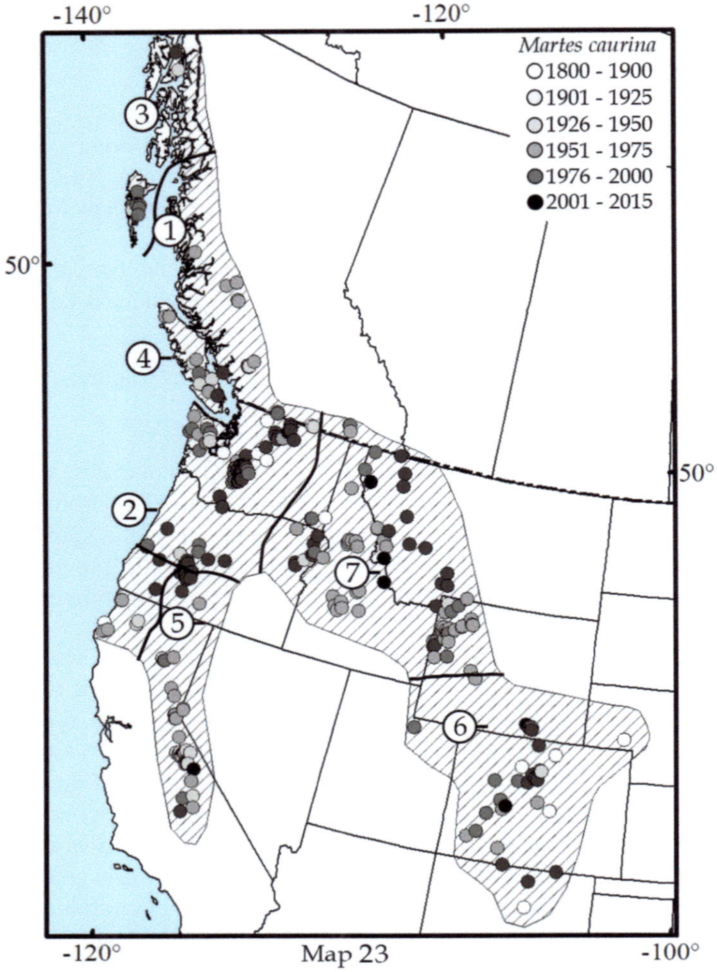

Map 23 Distribution map of *Martes caurina*

Additional Literature: Merriam (1890c), Youngman (1982), Graham and Graham (1994), Colella et al. (2019).

Pekania Gray, 1865

Dorsal and underpart pelage dark, varying according to the species, sex, and season; tail, rump, and legs black; face, neck, and shoulders hoary gold or silver; belly uniformly brown with white or cream patches; claws retractable; external median rootlet on the upper carnassial exposed.

Pekania pennanti (Erxleben, 1777)

Fisher, marta pescadora

1. *M. p. columbiana* Goldman, 1935. From southeastern Yukon and central western British Columbia east through central Manitoba, southern to Idaho and Utah.
2. *M. p. pacifica* (Rhoads, 1898). From the coast of British Columbia south through Central California.
3. *M. p. pennanti* (Erxleben, 1777). From Alberta and Montana east through the Atlantic coast in Québec to New York and inland to North Carolina.

Conservation: *Pekania pennanti* (as *Martes pennanti*) is listed as Least Concern by the IUCN (Helgen and Reid 2018).

Characteristics: Fisher is medium sized; total length 900.0–1,200.0 mm in males and 750.0–950.0 mm in females, condylobasal length 88.0–112.0 mm. **Dorsal pelage dark grizzle with cream;** face, neck, and shoulders honey gold to silver; pelage with three color bands; fur of underparts uniformly brown with white cream patches of different sizes on the chest, axilla, and genital area; limbs, rump, and tail black; ears round; **tail long, bushy**; skull relatively long and narrow with slender zygomatic arches; auditory bullae elongated and relatively inflated.

Comments: Koepfli et al. (2008) proposed using *Pekania* instead of *Martes*. *Pekania pennanti* has a discontinued distribution in four main ranges. The first includes central Canada, from the Pacific coast to the Atlantic coast, except for the Great Plains, and from the central part of the Appalachian Mountains north through Minnesota, Wisconsin, and Michigan. The second one comprises Idaho, Montana, Wyoming, and Utah. The third is in Washington, Oregon, and California. The fourth and last is in Sierra Nevada, California (Map 24). *P. pennanti* is sympatric, or nearly so, with *Gulo gulo, Lontra canadensis, Martes americana, Mustela erminea, M. nivalis, Neogale frenata,* and *N. vison.* It can be distinguished from all other species of *Mustela* and *N. vison* by its larger size, greater than 500.0 mm in total length. From *L. canadensis,* by having a tail narrow at the base; hind and fore digits not webbed; last molariforms smaller than the first molariforms. From *G. gulo,* by its body not stocky and without a blond band on the forehead and flanks. From *M. americana*, by having the fur of the throat, chest, and underparts brown with white creamy patches; head and body of the same color.

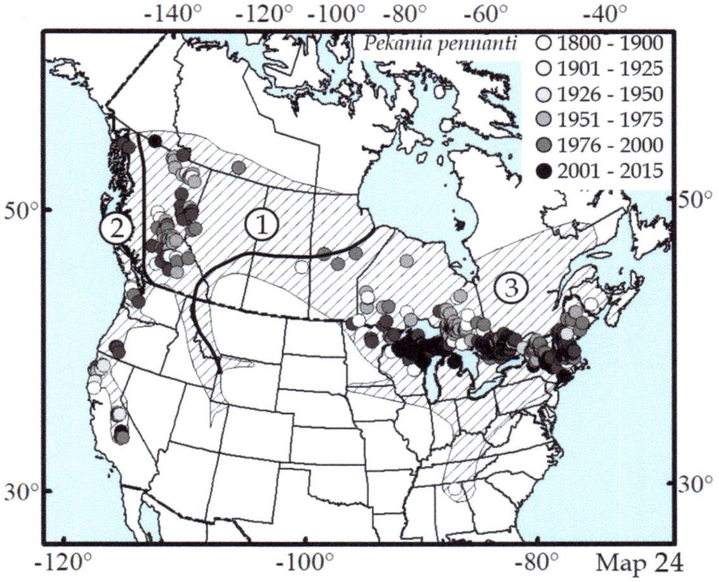

Map 24 Distribution map of *Pekania pennanti*

Additional Literature: Powell (1981, 1993), Strickland et al. (1982), Stone and Cook (2002), Drew et al. (2003).

Subfamily Ictonychinae Pocock, 1922

This subfamily includes the Greater Grison and is characterized by a pale-colored diagonal stripe on the head and shoulders, and on the back in some species; teeth large and strong; upper carnassial occupying nearly half of the inner surface of the tooth; first upper molars about twice wider than long.

Genus *Galictis* Bell, 1826

Greater Grison has the dorsal coloration black or brown, grizzled with white or yellow-tipped hairs; fur of underparts black or brown, with few white or yellow-tipped hairs; **pale-colored, diagonal stripe on the head and shoulders**; tail short; body slightly flattened dorsoventrally; limbs short; pinna of the ears well developed with a small bursa in front of the posterior margin; plantar pads separated from the carpal and metacarpal pads by a wide and deep groove; pads converging over their entire width; only a slight division between the two carpal pads and with no division on the metatarsal pad; skull strong, massive, with a dorsal profile relatively flat; **braincase broad, rounded, and V-shaped in the posterior area**; rostrum short; postorbital process short and pointed; zygomatic arches strongly developed and auditory bullae flattened; **teeth large and strong; upper carnassial occupying nearly one-half of the inner surface of the tooth; first upper molars about twice wider than long, with a lobe-shaped cingulum.**

Galictis vittata (Schreber, 1776)

Greater Grison, grisón

G. v. canaster Nelson, 1901. Only subspecies with range in North America.

Conservation: *Galictis vittata* is listed as Threatened in the Norma Oficial Mexicana (DOF 2019) and as Least Concern by the IUCN (Cuarón et al. 2016b).

Characteristics: Greater grison is medium sized; total length 600.0–760.0 mm, condylobasal length 80.3–97.9 mm. **Dorsal pelage generally blackish dark with a white stripe from the forehead to the shoulders, passing under the ears and above the eyes;** this line frequently divides the head into two shades, darker dorsally and white ventrally; fur of limbs and underparts black; limbs short, less than one-sixth of the head-and-body length; tail short; palate projecting behind the last molar line wider than large; supraocular processes slightly sharp; **first lower molars (carnassial) cutting**; skull breadth at the auditory bullae height notoriously wider than the rest of the braincase.

Comments: *Galictis vittata* ranges at lower elevations from San Luis Potosí, Veracruz, and Chiapas south through South America including Yucatán Peninsula (Map 25). It can be found in sympatry, or nearly so, with *Eira barbara, Lontra longicaudis,* and *Neogale frenata* and can be distinguished from *E. barbara* and *N. frenata* by having the dorsal pelage generally blackish dark with a white stripe from the forehead to the shoulders passing under the ears and above the eyes; this line frequently divides the head into two shades, darker below and white above. In addition, from *Eira*, by having the rhinarium grooved only at the lower half of the anterior surface and nostrils smaller and separated. From *L. longicaudis,* by having a narrow tail at the base; hind and fore digits not webbed; last molariforms smaller than the first molariforms.

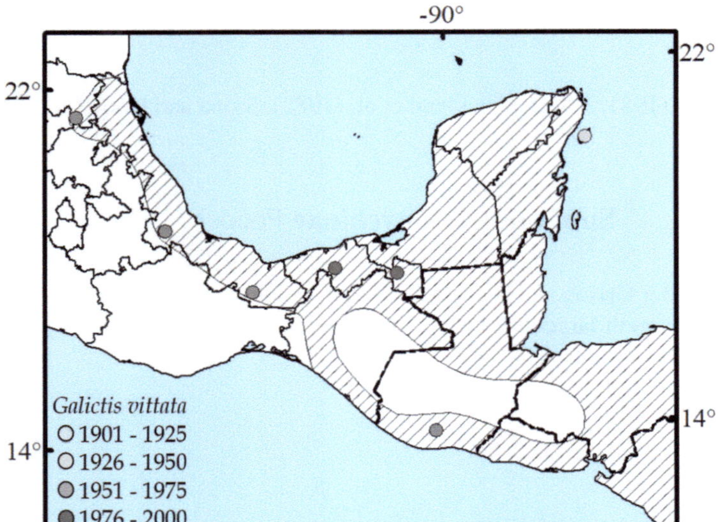

Map 25 Distribution map of *Galictis vittata*

Additional Literature: Pocock (1921), Dalquest and Roberts (1951), Goodwin (1969), Eisenberg (1989), Arita et al. (1990), Eisenberg and Redford (1999), Yensen and Tarifa (2003), Nowak (2005).

Subfamily Lutrinae Bonaparte, 1838

This subfamily includes otters and is characterized by long and slim bodies; limbs relatively short; webbed feet used to swim; ability of holding its breath underwater; underfur very soft and insulated, protected by an outer layer of long guard hairs; hair traps air, keeping them dry, warm, and somewhat buoyant underwater. The keys were elaborated based on the review of specimens and the following sources: Lawlor (1979), Hall (1981), Lariviere (1999b), and Álvarez-Castañeda et al. (2017a).

1. Head in pale golden shades contrasting with the body; two lower incisors on each side; large lacerated foramen close to 8.0 mm (Fig. 17; from Alaska, United States, south through Baja California, México) .... *Enhydra lutris* (p. 129)

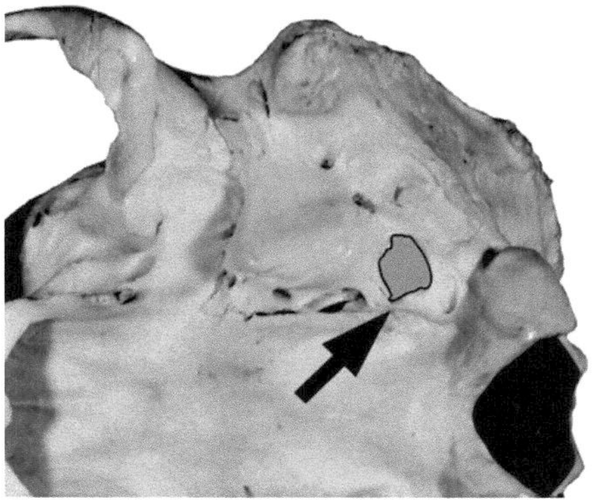

Fig. 17 Lacerated foramen large close to 8.0 mm

1a. Head and body of similar color; three lower incisors on each side; small lacerated foramen, generally inconspicuous .... *Lontra* (p. 131)

Genus *Enhydra* Fleming, 1822

Sea otters have a coloration varying from black to almost red, usually uniformly dark brown; fur of underparts grizzled; throat and chest paler than upperparts and head grizzled; head large and blunt; neck short and thick; ears short and pointed; limbs and tail short; **tail less than one-third of body length, somewhat flattened dorsoventrally**; **hindfeet webbed, flattened into broad flippers, furred dorsally and ventrally, fifth digit longest, claws retractile**; forefeet small, palms naked; skull high, inflated, **flattened dorsally, with the facial region broad and flat**; orbits relatively large; postorbital processes small; auditory bullae small; palate extending posteriorly to the first upper molar plane only a short distance; **two pairs of lower incisors**; upper premolars and first upper molars wider than long; **molars broad and flat with no cutting surfaces**.

Enhydra lutris (Linnaeus, 1758)
Sea Otter, nutria marina

E. l. lutris (Linnaeus, 1758). From the coast of Washington north through British Columbia and Alaska.
E. l. nereis (Merriam, 1904). From the coast of Washington south through the Baja California Peninsula.

Conservation: *Enhydra lutris kenyoni* and *E. l. nereis* are listed as Threatened by the United States Endangered Species Act (FWS 2022). *E. lutris nereis* is listed as Endangered in the Norma Oficial Mexicana (DOF 2019), as Threatened and as Special Concern in the Canadian Endangered Species Protection Act, and as Endangered (A2abe: population reduction in the last 10 years 50% or less and may not be reversible data obtained by direct observation, index of abundance, and effects of introduced taxa) by the IUCN (Doroff and Burdin 2015). The species was hunted since the 1700s getting to near extinction. Today isolated population can be found in Alaska, Canada, central California, and México.

Characteristics: Sea otter is medium sized; total length on average 1260.0–1450.0 mm in males; condylobasal length 115.0–136.0 mm in females; 125.0–144.0 mm in males and 1070.0–1400.0 mm in females. **Dorsal pelage generally dark brown, head pelage silvery-golden contrasting with the body; tail flattened dorsoventrally**; **hind digits webbed and three times larger than foredigits**; four large molariforms, the last molariforms almost as wide as the space between the two last molariforms; mastoid breadth almost equal to the cranial breadth at the auditory bullae level; **two lower incisors on each side; lacerated foramen diameter close to 8.0 mm**.

Comments: Historically, *Enhydra lutris* occurred across the North Pacific from the Aleutian Islands south through the Baja California Peninsula, México (Map 26, Kenyon 1969). *E. lutris* is not found in sympatric distribution, or almost never, with any other species of the family Mustelidae and other carnivores with four limbs. *E. lutris* can be distinguished from *Lontra canadensis* and *L. longicaudis* by being associated with seawater; head in pale golden shades, generally contrasting with the rest of the body; two lower incisors on each side and a larger lacerated foramen.

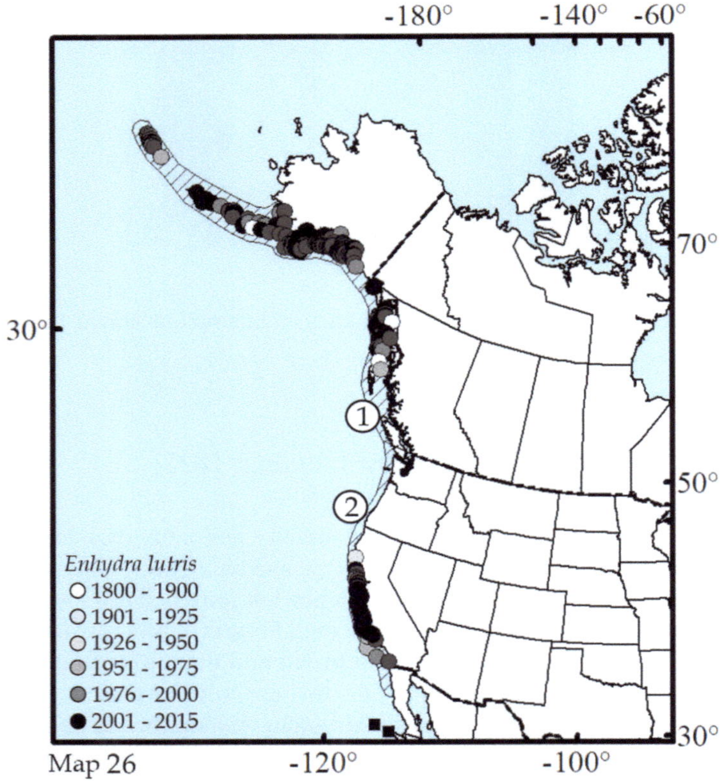

Map 26 Distribution map of *Enhydra lutris*

Additional Literature: Blair et al. (1957), van Zyll de Jong (1972), Roest (1973), Davis and Lidicker (1975), Estes (1980), Ralls et al. (1983), Wilson et al. (1991), Cronin et al. (1996), Burn and Doroff (2005).

Genus *Lontra* Gray, 1843

Dorsal pelage brown, not contrasting with the head color; hair short and very dense; tail flattened dorsoventrally; limbs short and stout; **hind and fore digits webbed and similar in size**; lateral rhinarium; ears short and small; five small molariforms; last molar width, almost twice the breadth between the two last molars; mastoid breadth much less than the skull breadth at the auditory bullae region; **three lower incisors on each side; lanceolate foramen small and generally inconspicuous**. The assignation of the genus *Lontra* has varied between authors; *Lontra* is considered a junior synonym of *Lutra* (McKenna and Bell 1997) and the species of the New World are within *Lontra,* different from those of the Old World, *Lutra* (Van Zyll de Joung 1972, 1987; Wozencraft 2005)

1. Foot sole naked; rhinarum furred laterally; skull convex dorsally (Fig. 18); southern of 28° N latitude (from warm-water rivers in México south through South America) .... *Lontra longicaudis* (p. 132)

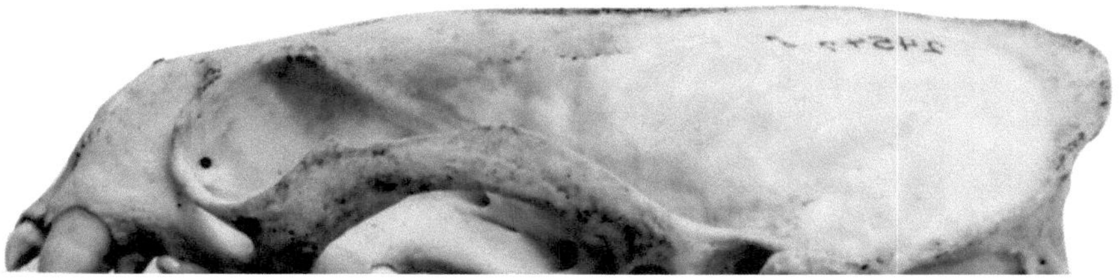

Fig. 18 Skull convex dorsally like in *Lontra longicaudis*

1a. Foot sole with a hair tuft under the toes; rhinarum naked; skull flat dorsally (Fig. 19); northern of 28° N latitude (rivers in the United States and Canada) .... *Lontra canadensis* (p. 131)

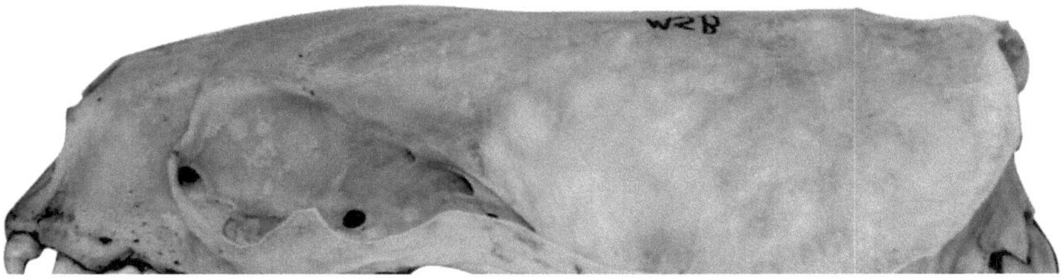

Fig. 19 Skull flat dorsally like in *Lontra canadensis*

Lontra canadensis (Schreber, 1776)

River Otter, nutria de río

1. *L. c. canadensis* (Schreber, 1776). From Ontario, northwestern Minnesota, Wisconsin, Michigan, and New York northwest through the Labrador Peninsula.
2. *L. c. kodiacensis* (Goldman, 1935). Restricted to Kodiak Island, Alaska.
3. *L. c. lataxina* (Cuvier, 1823). From New York to Nebraska south through the Gulf of México coast and from Nebraska to Texas east through the Atlantic Ocean.
4. *L. c. mira* (Goldman, 1935). Vancouver Island and coast of British Columbia.
5. *L. c. pacifica* (Rhoads, 1898). From Alaska south through California and Colorado, including eastern Wisconsin and western Ontario.
6. *L. c. periclyzomae* (Elliot, 1905). Restricted to Queen Charlotte Island, Alaska.
7. *L. c. sonora* (Rhoads, 1898). Colorado, Nevada, New Mexico, and eastern California.

Conservation: *Lontra canadensis* is listed as Least Concern by the IUCN (Serfass 2021).

Characteristics: River otter is medium sized; total length 900.0–1150.0 mm; basal length 106.0–115.0 mm. Dorsal pelage lustrous, pale gray to black; chin and upper lip grayer than the rest of the body; old individuals may become white-tipped; **foot sole with a hair tuft in the digits; skull flat dorsally**.

Comments: *Lontra canadensis* ranges throughout the United States and Canada (Map 27). Today, it is mainly absent in the Great Plains and the southwestern states. This species is not found in sympatry, or almost never, with any other species of the family Lutrinae. However, it can be distinguished from all other genera of the family Mustelidae by having the tail very wide at the base, as wide as the hips region; hind and fore digits webbed and similar in size; last molariforms equal to or larger than the first to the last molariforms. From *L. longicaudis,* by having the rhinarium naked and the foot sole with a hair tuft in the digits, and skull flat dorsally. From *Enhydra lutris,* by being associated with inland waters; head color generally not contrasting with the rest of the body; three lower incisors on each side and the lacerated foramen small.

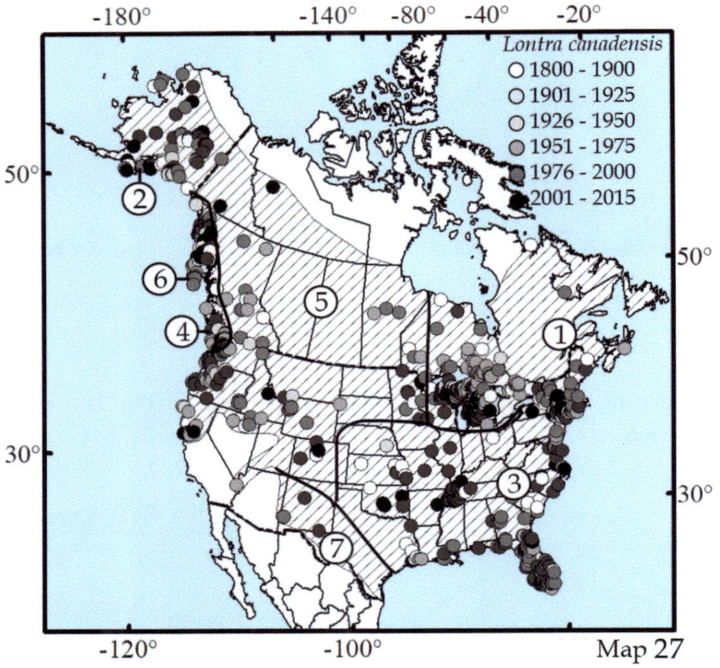

Map 27 Distribution map of *Lontra canadensis*

Additional Literature: Harris (1968), Kellnhauser (1983), Koepfi and Wayne (1998), Larivière and Walton (1998), Bininda-Emonds et al. (1999), Wozencraft (1993).

Lontra longicaudis (Olfers, 1818)

Southern otter, nutria de río tropical

L. l. annectens (Major, 1897). Only subspecies with range in North America.

Conservation: *Lontra longicaudis annectens* is listed as Threatened in the Norma Oficial Mexicana (DOF 2019) and as Near Threatened by the IUCN (Rheingantz et al. 2021).

Characteristics: Southern otter is medium sized; total length 890.0–1200.0 mm; basal length 92.6–101.0 mm in females. Dorsal pelage lustrous, grayish brown; hair of underparts and throat paler; chin and upper lip silver white to yellow; **foot sole without locks of hair in the digits; skull convex dorsally**.

Comments: *Lontra annectens* is a junior synonym of *L. longicaudis* (Van Zyll de Joung 1972; Wozencraft 2005); it was previously regarded as a different species. *L. longicaudis* ranges in the rivers from northwestern México south through South America (Map 28). It is not found in sympatry, or almost never, with any other species of the family Lutrinae; however, it can be found with species of the other genera of the family Mustelidae from which it can be distinguished by its tail very wide at the base, as wide as the hip region; hind and fore digits webbed and similar in size; last molariforms equal to or larger than the first to the last molariforms. From *L. canadensis,* by having the rhinarium furred laterally and the foot sole without a hair tuft in the digits, and skull convex dorsally. From *Enhydra lutris*, by being associated with inland waters; head color generally not contrasting with the rest of the body; three lower incisors on each side and the lacerated foramen small.

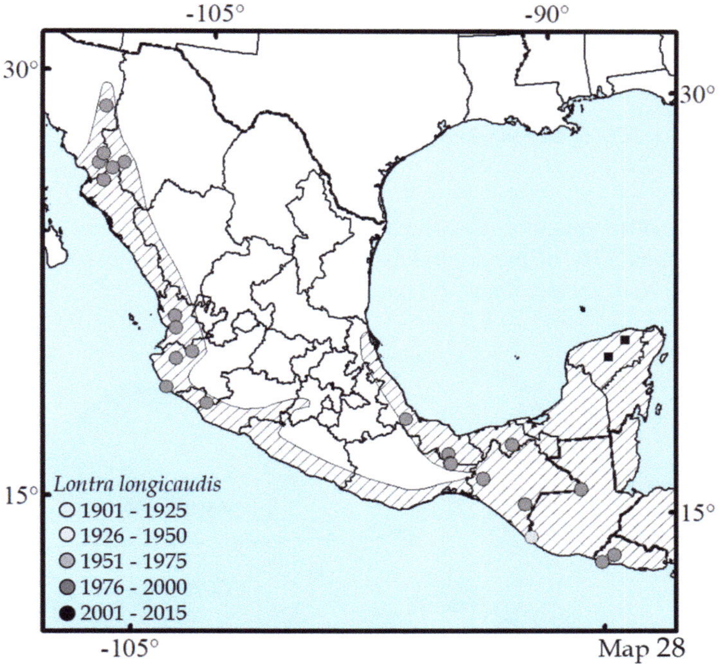

Map 28 Distribution map of *Lontra longicaudis*

Additional Literature: Harris (1968), Kellnhauser (1983), Koepfi and Wayne (1998), Bininda-Emonds et al. (1999), Larivière (1999b).

Subfamily Mustelinae Fisher, 1817

The ferret and weasels are among the smallest carnivores in North America, with the body long and slender; dorsum and underparts with color variations among species and seasons; limbs short, less than one-sixth of the head-and-body length; strong sexual dimorphism in size; inner moiety of the first upper molars larger than in the outer molars; fourth upper premolars simple; first lower molars sharp; auditory bullae greatly inflated. The keys were elaborated based on the review of specimens and the following sources: Hall (1981), Lariviere (1999a), and Álvarez-Castañeda et al. (2017a).

1. Total length greater than 300.0 mm; fur of the back and underparts brown chocolate; first lower molars with a suggestion of metaconid and trenchant; maxillary toothrow greater than 33% of the skull length............*Neogale* (part; p. 141)

1a. Total length less than 300.0 mm; fur of the back and of different colors but not brown chocolate; first lower molars lacking metaconid; maxillary toothrow less than 33% of the skull length..2

2. Feet, face, and body of the same color, with a yellow-brown mask; postglenoid length less than 47% of the condylobasal length (Fig. 20); upper tooth row less than 20.0 mm in males and less than 17.8 mm in females .... *Neogale* (part; p. 141)

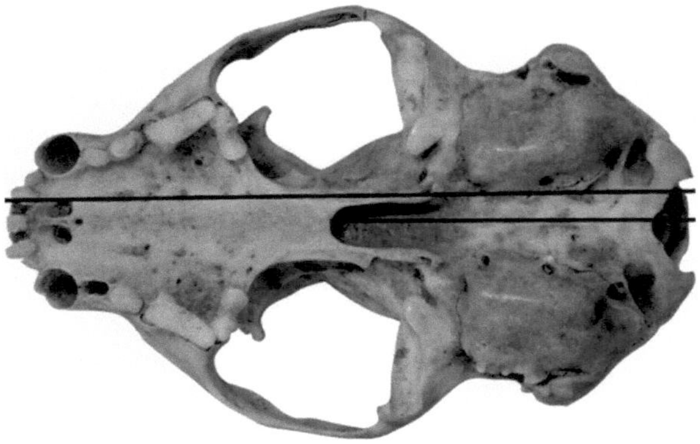

Fig. 20 The postglenoid length less than 47% of the condylobasal length

2a. Feet and face of the same or different color from the body coloration, with a yellow-brown or black mask; postglenoid length equal to or greater than 47% of the condylobasal length (Fig. 21); upper tooth row equal to or greater than 20.0 mm in males and equal to or greater than 17.8 mm in females .... *Mustela* (p. 134)

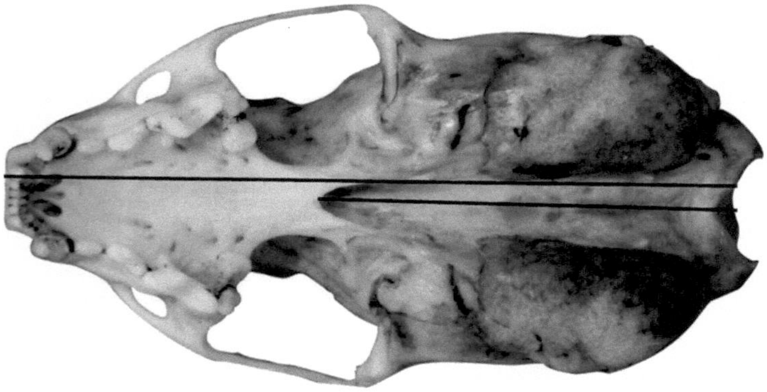

Fig. 21 The postglenoid length is equal to or greater than 47% of the condylobasal length

Genus *Mustela* Linnaeus, 1758

Dorsal and underparts fur with color variations among species and seasons; limbs short, less than one-sixth of the head-and-body length; strong sexual dimorphism in size; inner moiety of the first upper molars larger than in the outer molars; fourth upper premolars simple; first lower molar with a sharp talanoid; auditory bullae greatly inflated; posterior margin of the palatal notably behind the last upper molar. *Mustela* has three subgenera in North America: *Gale* (*M. nivalis*), *Mustela* (*M. erminea, M. haidarum, and M. richardsonii*), and *Putorius* (*M. nigripes*). The keys were elaborated based on the review of specimens and the following sources: Lawlor (1979), Hall (1951a, 1981), and Lariviere (1999a).

1. Feet and mask black with a sand-colored body; upper toothrow greater than 20.0 mm in males and greater than 17.8 mm in females (only in captivity; it has been reintroduced in the United States Great Plains and in Chihuahua, México) .... subgenus *Putorius*; *Mustela nigripes* (p. 140)

1a. Feet, face, and body of the same color; if a mask is present, yellow-brown, body color brown or white in winter; upper toothrow less than 20.0 mm in males and less than 17.8 mm in females 2

2. Tail distal third not black; baculum with a hook on the distal shaft end; skull with postorbital constriction short (throughout Canada and Alaska, except for tundra areas and northeastern Unites States and Appalachian Mountains)........................... .... subgenus *Gale*; *Mustela nivalis* (p. 135)

2a. Tail distal third black; baculum simple without the typical hook at the distal end; skull with the postorbital area short, rather abruptly constricted .... subgenus *Mustela* 3

3. Distributed only in the Haida Gwaii archipelago in the Graham and Moresby, Prince of Wales, and Suemez Islands, Canada..*Mustela haidarum* (p. 137)

3a. Distributed in other parts of Canada and the United States 4

4. Distributed only in Alaska and northern Yukon to Kewanee and the Kodiak and Admiralty Islands, Canada ....*Mustela erminea* (p. 136)

4a. Distributed in other parts of Canada and central and northern United States ....*Mustela richardsonii* (p. 138)

Subgenus ***Gale*** Wagner, 1841

Dorsum brown; fur of the throat and underparts cream to white; demarcation line between these colors strongly marked; **tail with distal third not black**; color in winter paler, only *M. nivalis* becomes completely white; plantar surfaces furred; baculum with a hook on the shaft distal end; size small; muzzle short; braincase long and narrow; skull without noticeable projecting mastoid processes; **postorbital constriction short**; auditory bullae large and inflated, bean-shaped, with inner margins nearly parallel; anterior margins with prominent inner corners; **foramen lacerum medium displaced to the anterior part of the auditory bullae;** dentition weaker than in the subgenus *Mustela*; second upper premolars with two roots.

Mustela nivalis Linnaeus, 1766
Least Weasel, comadreja pequeña

1. *M. n. allegheniensis* (Rhoads, 1901). From Wisconsin and Illinois east through Maine and south through Tennessee by the Appalachian Mountains.
2. *M. n. campestris* Jackson, 1913. South Dakota, Nebraska, Kansas, Iowa, and northern Arkansas.
3. *M. n. eskimo* (Stone, 1900). Alaska, Yukon, and northwestern Mackenzie.
4. *M. n. rixosa* (Bangs, 1896). From southeastern Yukon and southeastern Makenzie east through the Labrador Peninsula and south through British Columbia to Minnesota.

Conservation: *Mustela nivalis* is listed as Least Concern by the IUCN (McDonald et al. 2019).

Characteristics: Least weasel is small sized; total length 230.0–250.0 mm in males, 200.0–225.0 mm in females; condylobasal length 42.0–49.0 mm in males, 45.0–50.0 mm in females. Upper parts brown; fur of underparts whitish; **tail not black at the tip;** skull similar to *M. erminea*.

Comments: *Mustela nivalis* is probably a complex of many species (Reig 1997), in addition, it has been assigned to the subgenus *Mustela* (Youngman 1982) and *Gale* (Abramov 1999). It ranges throughout Canada, except for the tundra areas; in the United States, from the northern part of the Great Plains, Great Lakes, and Appalachian Mountains south through Wyoming, Oklahoma, Illinois, Indiana, and Tennessee (Map 29). It can be found in sympatry, or nearly so, with *Gulo gulo, Lontra canadensis, Martes americana, Mustela erminea, Neogale frenata, N. vison, Pekania pennanti,* and *Taxidea taxus. M. nivalis* can be distinguished from these species by its smaller size, total length less than 300.0 mm. From *N. frenata,* by having fur of underparts white and lacking a face mask. From *M. erminea* and *N. frenata,* by having a shorter tail, less than 30% of the head-and-body length, and tip with only one dorsal coloration during summer (brown coloration) and winter (white coloration)

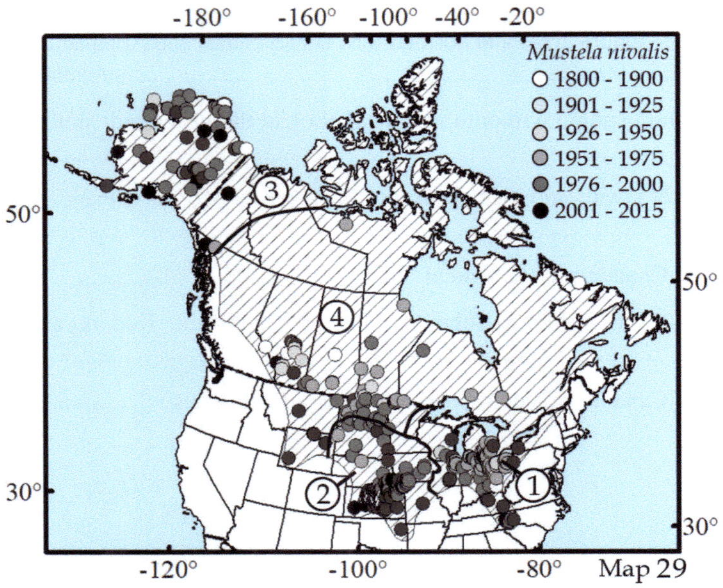

Map 29 Distribution map of *Mustela nivalis*

Additional Literature: Hall (1951a), King (1989), van Zyll de Jong (1993), Reig (1997), Nowak (2005), Wozencraft (2005).

Subgenus *Mustela* Linnaeus, 1758

Dorsum brownish; fur of the neck and underparts cream to white; demarcation line between these colors strongly marked; tail with distal third black; typically completely white during winter, except for the tail black at the tip; plantar surfaces furred; baculum is the simplest structure in the genus without the **typical hook at distal end;** size small; muzzle short, convex above; braincase long and narrow; mastoids in full-age sometimes prominent; **skull with a short, rather abruptly constricted postorbital area**; infraorbital foramen large; auditory bullae large and inflated; their inner margins closely arranged and nearly parallel; anterior margins without prominent inner corners; second upper premolars with two roots.

Mustela erminea Linnaeus, 1758
Ermine, armiño

1. *M. e. arctica* (Merriam, 1896). Alaska and northern Yukon to Kewanee, including central and western Franklin.
2. *M. e. kadiacensis* (Merriam, 1896). Restricted to Kodiak Island, Alaska.
3. *M. e. salva* Hall, 1944. Restricted to Admiralty Island, Alaska.

Conservation: *Mustela erminea* is listed as Least Concern by the IUCN (Reid et al. 2016a).

Characteristics: Ermine is small sized; total length of females 210.0–270.0 mm; males 290.0–340.0 mm; condylobasal length in females 30.4–35.6 mm; males 34.4–41.5 mm. Dorsal pelage brown; fur of underparts yellowish from the chin to the inguinal region; tail black at the tip; braincase long, mainly twice the rostrum length.

Comments: *Mustela haidarum* and *M. richardsonii* were recognized as distinct species (Colella et al. 2021), previously regarded as subspecies of *M. erminea* (Hall 1944, 1945). *M. erminea* ranges in Alaska and northern Yukon to Kewanee and the Kodiak and Admiralty Islands (Map 30) and can be found in sympatry, or nearly so, with *Gulo gulo, Lontra canadensis, Martes americana, Mustela nivalis,* and *Neogale vison. M. erminea* can be distinguished from these species by its smaller size, length less than 300.0 mm. From *M. nivalis,* by its larger tail, more than 30% of the head-and-body length, and black at the tip in summer (brown coloration) and winter (white coloration). From *M. haidarum* and *M. richardsonii,* only using molecular data or based on its distribution.

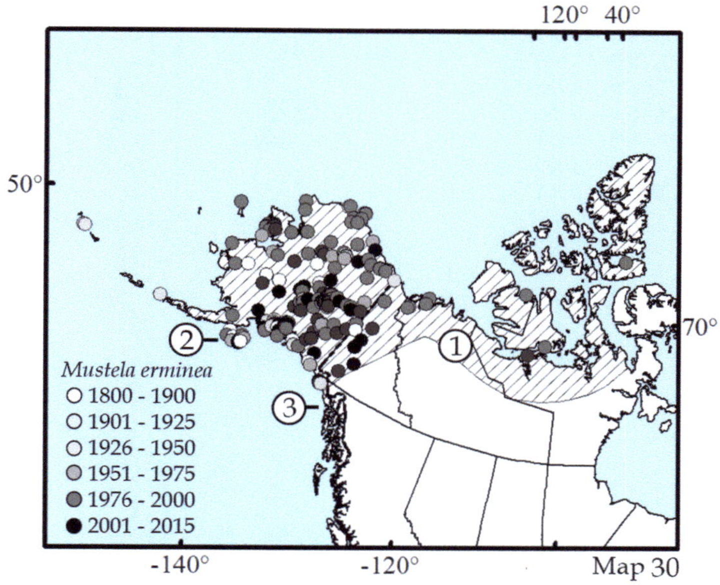

Map 30 Distribution map of *Mustela erminea*

Additional Literature: Merriam (1896), Hall (1951a), Ellerman and Morrison-Scott (1951), Youngman (1982), King (1983a), King (1989), Eger (1990), Sheffield and Thomas (1997), Abramov (1999), Nowak (2005), Wozencraft (2005), Colella et al. (2021).

Mustela haidarum (Preble, 1898)

Haida ermine, armiño del archipiélago de Haida Gwaii

1. *M. e. celenda* Hall, 1944. Restricted to Prince of Wales Island, British Columbia.
2. *M. e. haidarum* (Preble, 1898). Restricted to the Graham and Queen Charlotte Islands, British Columbia.
3. *M. e. seclusa* Hall, 1944. Restricted to Suemez Island, Alaska.

Conservation: *Mustela haidarum* is listed as Threatened by the Canadian Endangered Species Protection Act and (as *Mustela erminea*) as Least Concern by the IUCN (Reid et al. 2016a).

Characteristics: Haida ermine is small sized; total length 245.0–270.0 mm in males and 270.0–290.0 mm in females; condylobasal length 35.0–38.5 mm in males and 34.0–36.7 mm in females. Dorsal pelage chocolate brown; fur of underparts whitish from the chin to the inguinal region; tail black at the tip, completely white in winter; braincase long, mainly twice the rostrum length.

Comments: *Mustela haidarum* was recognized as a distinct species (Colella et al. 2021), previously regarded as a subspecies of *M. erminea* (Hall 1944). *M. haidarum* ranges in islands off the coast of British Columbia and southeast Alaska as Haida Gwaii archipelago in the Graham and Moresby Islands, Prince of Wales Island, and Suemez Island (Map 31), and can be found in sympatry, or nearly so, with *Gulo gulo, Lontra canadensis, Martes americana, Mustela nivalis,* and *Neogale vison. M. haidarum* can be distinguished from these species by its smaller size, total length less than 300.0 mm. From *Mustela nivalis,* by having larger tail, more than 30% of the head-and-body length, and black at the tip in summer (brown coloration) and winter (white coloration). From *M. erminea* and *M. richardsonii,* only using molecular data or by distribution.

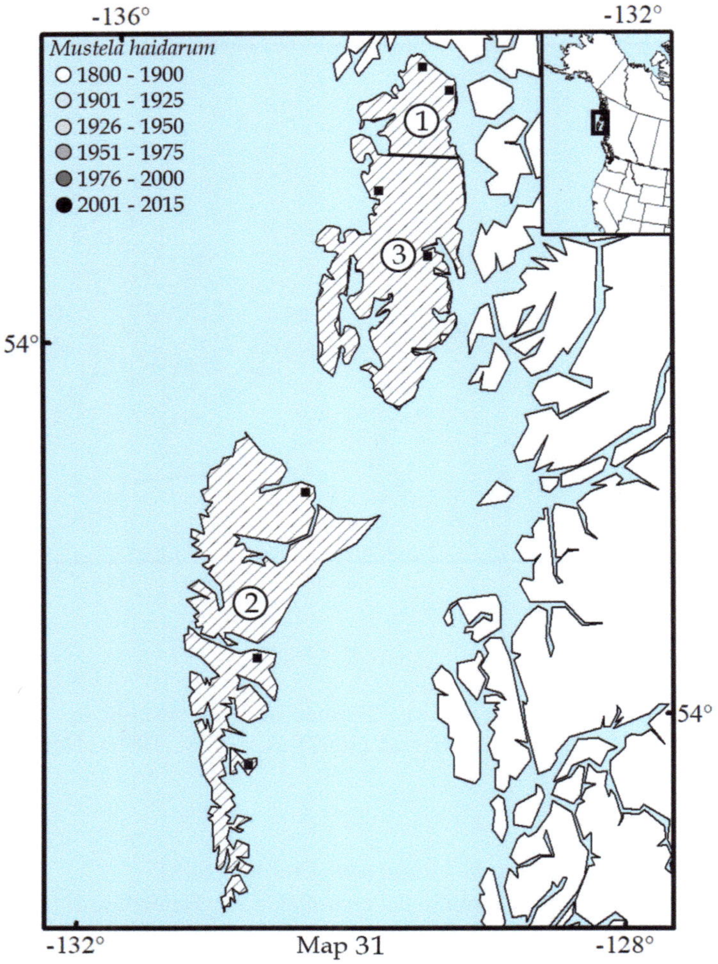

Map 31 Distribution map of *Mustela haidarum*

Additional Literature: Merriam (1896), Hall (1951a), Ellerman and Morrison-Scott (1951), Youngman (1982), King (1983a), King (1989), Eger (1990), Sheffield and Thomas (1997), Abramov (1999), Nowak (2005), Wozencraft (2005), Colella et al. (2021).

Mustela richardsonii Bonaparte, 1838

American Ermine, armiño americano

1. *M. e. alascensis* (Merriam, 1896). Central coast of British Columbia.
2. *M. e. anguinae* Hall, 1932. Restricted to Vancouver Island, British Columbia.
3. *M. e. bangsi* Hall, 1945. Southern Manitoba and Ontario, the Dakotas, Minnesota, Wisconsin, Iowa, and Michigan.
4. *M. e. cicognanii* Bonaparte, 1838. From southern Ontario and southwestern Québec south through Pennsylvania and New Jersey and north through Maine.
5. *M. e. fallenda* Hall, 1945. Central western British Columbia.
6. *M. e. gulosa* Hall, 1945. Central Washington.
7. *M. e. initis* Hall, 1944. Restricted to Baranof Island, Alaska.
8. *M. e. invicta* Hall, 1945. Southeastern British Columbia, southern Alberta, northern Washington, Idaho, and western Montana.
9. *M. e. muricus* (Bangs, 1899). From western Oregon and northwestern California east to western South Dakota to northern New Mexico, including southern Montana.

10. *M. e. olympica* Hall, 1945. Olympic Peninsula, Washington.
11. *M. e. richardsonii* Bonaparte, 1838. From British Columbia east through the Labrador Peninsula and from southern Yukon to central Keewatin south through central British Columbia to southwestern Québec.
12. *M. e. semplei* Sutton and Hamilton, 1932. Western Keewatin and western Franklin.
13. *M. e. streatori* (Merriam, 1896). Coast of Oregon and northwestern California.

Conservation: *Mustela richardsonii* (as *Mustela erminea*) is listed as Least Concern by the IUCN (Reid et al. 2016a).

Characteristics: Ermine is small sized; total length of females 202.0–250.0 mm; males 288.0–315.0 mm; condylobasal length in females 31.0–35.8 mm; males 38.4–45.7 mm. Dorsal pelage brown; fur of underparts whitish, from the chin to the inguinal region; tail black at the tip; braincase long, mainly twice the rostrum.

Comments: *Mustela richardsonii* was recognized as a distinct species (Colella et al. 2021), previously regarded as a subspecies of *M. erminea* (Hall 1945). *M. richardsonii* ranges in central and eastern Canada and in United States at northern of the Pacific Coast, Rocky Mountains, Great Plains, Great Lakes, and Appalachian Mountains, through California to New Mexico, Kansas, and east through Iowa to New Jersey (Map 32), and can be found in sympatry, or nearly so, with *Gulo gulo, Lontra canadensis, Martes americana, M. nivalis, Neogale frenata, N. vison,* and *Pekania pennanti. M. richardsonii* can be distinguished from these species by its smaller size, total length less than 300.0 mm. From *N. frenata,* by having the fur of underparts white and no face mask. From *M. nivalis,* by having the tail with the tip dark brown in summer and white in winter; tail larger, more than 30% of the head-and-body length. From *M. erminea* and *M. haidarum,* only using molecular data or based on its distribution.

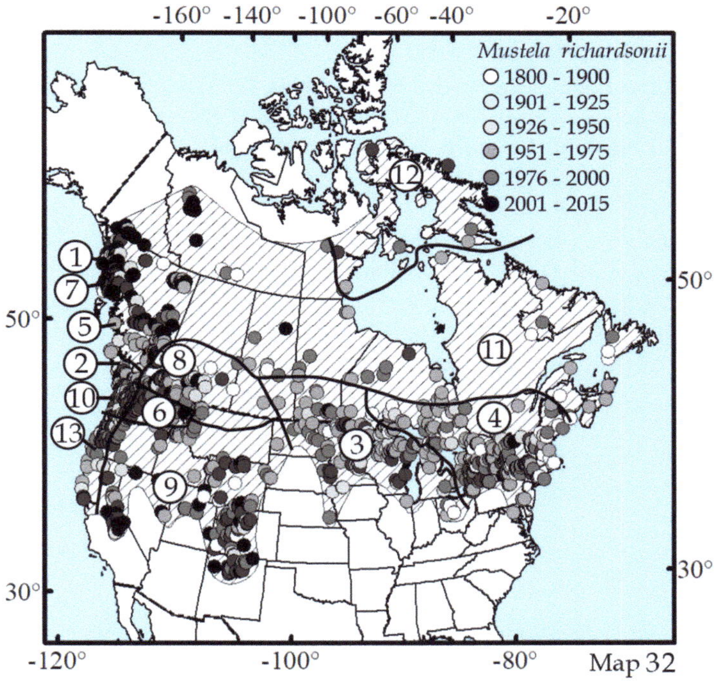

Map 32 Distribution map of *Mustela richardsonii*

Additional Literature: Merriam (1896), Hall (1951a), Ellerman and Morrison-Scott (1951), Youngman (1982), King (1983a, 1989), Eger (1990), Sheffield and Thomas (1997), Abramov (1999), Nowak (2005), Wozencraft (2005), Colella et al. (2021).

Subgenus ***Putorius*** Cuvier, 1817

Polecats differ from all the other species of *Mustela* in the coloration: fur of the feet, tail, and underparts (partly) blackish, the rest of the body with long black hairs producing a clouded effect over the whitish or yellowish underfur; brown "mask" on the muzzle; plantar surfaces furred; baculum with a hook on the shaft distal end; **size large**; **muzzle relatively long**; **braincase shorter than the muzzle**; skull robust and strongly built, rather flat, relatively short and broad; zygomatic arches widely arranged; **postorbital area prolonged and constriction strongly marked**; cranial crests well developed; **auditory bullae relatively very small, strongly flattened and distinctly triangular in outline,** inner margins not parallel and diverging posteriorly; anterior margins with prominent inner corners; **foramen lacerum medium distinctly displaced posterid; second upper premolars with one or two roots**.

Mustela nigripes (Audubon and Bachman, 1851)
Black-Footed Ferret, hurón de patas negras

Monotypic.

Conservation: *Mustela nigripes* is listed as Threatened by the United States Endangered Species Act (FWS 2022); listed as Extirpated by the Canadian Endangered Species Protection Act and as Endangered (C1 + 2a (i); D: small population size continuing decline in 20% per 5 years + estimated projected continuing decline in less than 250 of mature individuals in each subpopulation + with less than 250 mature individuals) by the IUCN (Belant et al. 2015). *M. nigripes* is associated with its prey, the prairie-dog *Cynomys*, so that when control programs extirpated prairie-dogs from most of its ranges, *M. nigripes* became endangered.

Characteristics: Black-footed ferret is small sized; total length 330.0–420.0 mm in males and 280.0–350.0 mm in females; condylobasal length 39.0–54.0 mm in males and 35.0–47.0 mm in females. **Dorsal and ventral pelage of sandy color; feet, limbs, and face mask black**; tail black at the tip; postorbital process well developed; mainly no diastema between the molariform teeth.

Comments: Historically, *Mustela nigripes* can be found throughout the Great Plains, from southern Canada to northern México (Map 33). Today, there are only a few isolated populations within the historical ranges, which can be found in sympatry, or nearly so, with *Lontra canadensis, Martes americana, Mustela erminea, Neogale frenata, N. vison, Pekania pennanti,* and *Taxidea taxus. M. nigripes* can be distinguished from these species by its smaller size, total length less than 300.0 mm. From *Mustela erminea* and *N. frenata,* by having the feet and mask black with a sandy-colored body.

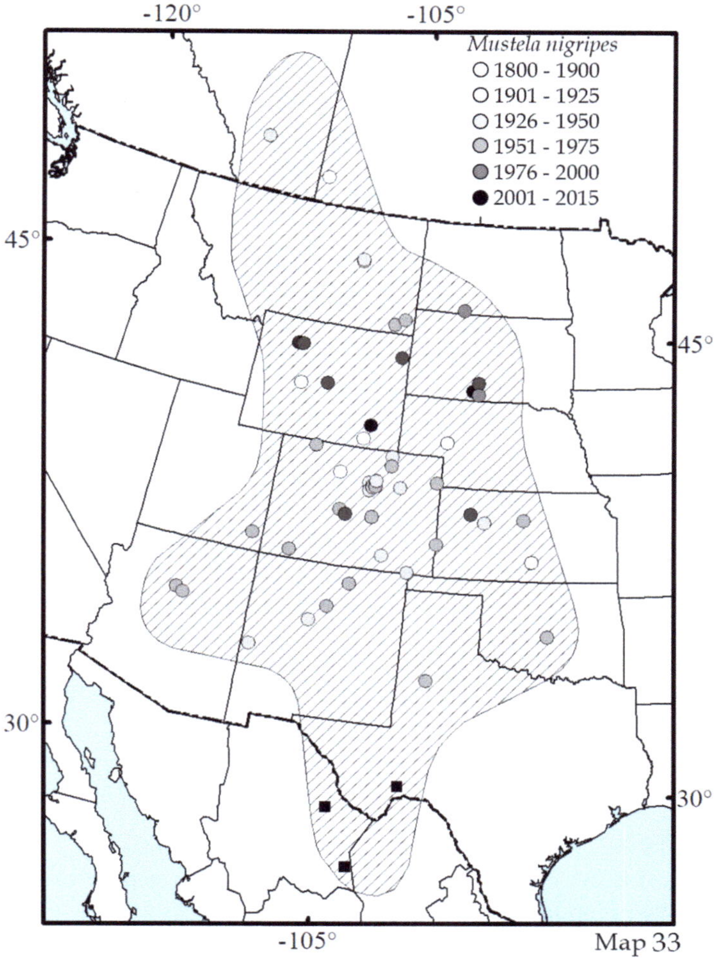

Map 33 Distribution map of *Mustela nigripes*

Additional Literature: Hall (1951a), Anderson (1977), Hillman and Clark (1980), Youngman (1982), King (1989), Abramov (1999), Nowak (2005), Wozencraft (2005).

Genus *Neogale* Gray, 1865

The genus *Neogale* was described as a subgenus of *Mustela* (Gray, 1865) with very ambiguous characteristics, such as face with a pale spot in front of the ears; back uniform; tail black at the tip. None of these characteristics are diagnostic for the genus, but a phylogenetic analysis shows that *Neogale* is independent from *Mustela* and needs to be considered a different genus (Harding and Smith 2009; Patterson et al. 2021). *Neogale vison* was considered within the genus *Neovison* (Harding and Smith 2009) and afterward within *Mustela* (Abramov 2000a, b; Kurose et al. 2008), but it was subsequently considered *Neogale* (Abramov 2000a, b; Harding and Smith 2009; Patterson et al. 2021). The keys were elaborated based on the review of specimens and the following sources: Hall (1951a, 1981).

1. Upper toothrow length less than 20.0 mm in males and less than 17.8 mm in females (from southern Canada south through South America, except for the southwestern desert, United States, and the Baja California Peninsula, México)............ *Neogale frenata* (p. 142)

1a. Upper toothrow length greater than 20.0 mm in males and greater than 17.8 mm in females (from Alaska and Canada, except for the northernmost part south through the United States, except for the desert)............ *Neogale vison* (p. 143)

Neogale frenata Lichtenstein, 1831

Long-Tailed Weasel, comadreja

1. *N. f. alleni* (Merriam, 1896). Western Wyoming and eastern South Dakota.
2. *N. f. altifrontalis* (Hall, 1936). Coast of Washington and Oregon.
3. *N. f. arizonensis* (Mearns, 1891). Central Arizona and central western New Mexico.
4. *N. f. arthuri* (Hall, 1927). Southeastern Texas, Louisiana, and southwestern Mississippi.
5. *N. f. effera* (Hall, 1936). Northern-central Oregon.
6. *N. f. frenata* (Lichtenstein, 1831). From southern Texas south through Tamaulipas and the highlands to Ciudad de México.
7. *N. f. goldmani* (Merriam, 1896). Coastal and Central Chiapas.
8. *N. f. inyoensis* (Hall, 1936). Restricted to Inyo County, California.
9. *N. f. latirostra* (Hall, 1936). Southwestern California and northwestern Baja California.
10. *N. f. leucoparia* (Merriam, 1896). From Nayarit to Oaxaca.
11. *N. f. longicauda* (Bonaparte, 1838). From central Alberta and southwestern Manitoba south through western Colorado and eastern Kansas.
12. *N. f. macrophonius* (Elliot, 1905). Western Oaxaca.
13. *N. f. munda* (Bangs, 1899). Central-western California, northern San Francisco Bay.
14. *N. f. neomexicana* (Barber and Cockerell, 1898). From southeastern Colorado and southwestern Kansas south through Nayarit, Jalisco, and western Zacatecas, including southeastern Arizona.
15. *N. f. nevadensis* (Hall, 1936). From Southern British Columbia and central-eastern Washington south through Nevada, Utah, and northern New Mexico.
16. *N. f. nigriauris* (Hall, 1936). Central western California, southern San Francisco Bay.
17. *N. f. noveboracensis* (Emmons, 1840). From Wisconsin to Tennessee east through the Atlantic coast, except for Maine to the north.
18. *N. f. occisor* (Bangs, 1899). Maine and western New Brunswick.
19. *N. f. olivacea* (Howell, 1913). From Mississippi east through North Carolina and south through northern Florida and the coast of Alabama.
20. *N. f. oregonensis* (Merriam, 1896). Coastal Oregon and northwestern California.
21. *N. f. oribasus* (Bangs, 1899). British Columbia, central western Alberta, western and southern Montana.
22. *N. f. peninsulae* (Rhoads, 1894). Florida Peninsula.
23. *N. f. perda* (Merriam, 1902). From Tabasco southern Veracruz to northern Chiapas, including the Yucatán Peninsula.
24. *N. f. perotae* (Hall, 1936). From Hidalgo, Tlaxcala, and Morelos south through northern Oaxaca.
25. *N. f. primulina* (Jackson, 1913). Central Iowa, western Nebraska, Kansas, and Oklahoma, northern Mississippi, and Arkansas and Missouri.
26. *N. f. pulchra* (Hall, 1936). Southern-central California.
27. *N. f. saturata* (Merriam, 1896). Northern-central California.
28. *N. f. spadix* (Bangs, 1896). Minnesota, western Dakotas, northeastern Nebraska, and central-northern Iowa.
29. *N. f. texensis* (Hall, 1936). Central Texas.
30. *N. f. tropicalis* (Merriam, 1896). Veracruz and eastern San Luis Potosí.
31. *N. f. washingtoni* (Merriam, 1896). Southern-central Washington.
32. *N. f. xanthogenys* (Gray, 1843). Central Valley of California.

Conservation: *Neogale frenata* (as *Mustela frenata*) is listed as Least Concern by the IUCN (Helgen and Reid 2016c).

Characteristics: Long-tailed weasel is small sized; total length 490.0–550.0 mm in males, 280.0–350.0 mm in females; condylobasal length 39.0–54.0 mm in males, 35.0–47.0 mm in females. **Dorsal pelage on the back cinnamon brown**; fur of underparts with ochre shades; **populations of México and Florida with a white spot on the forehead,** tail black at the tip; skull similar to *M. erminea*.

Comments: *Neogale frenata* has the broadest distribution of any species of the family Mustelidae in America, from southern Canada south through South America, except for the southwestern desert, United States, and the Baja California Peninsula, México (Map 34). It can be found in sympatry, or nearly so, with *Eira barbara, Galictis vittata, Gulo gulo, Lontra canadensis, L. longicaudis, Martes americana, Mustela erminea, M. nigripes, M. nivalis, Neogale vison, Pekania pennanti,* and *Taxidea taxus. N. frenata* can be distinguished from these species by its smaller size, total length less than 300.0 mm. From *Mustela erminea* and *M. nigripes,* by having the largest tail in the genus *Mustela*, its total length more than 44% of the head-and-body length, and a black face mask in most part of its ranges. From *M. nivalis*, by its brown or white coloration, never sandy in color, with feet and limbs of the same color.

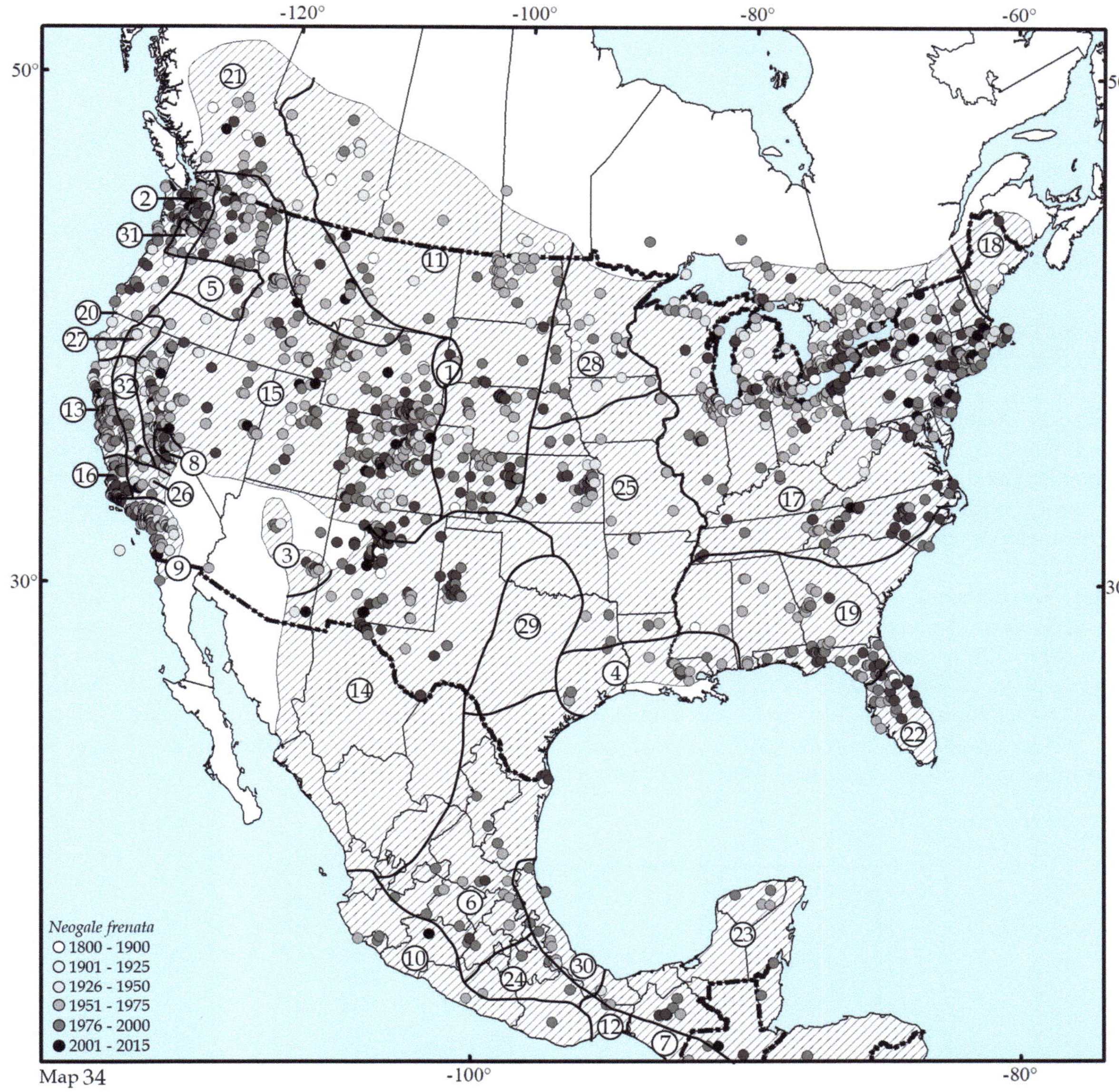

Map 34 Distribution map of *Neogale frenata*

Additional Literature: Hall (1951a), Youngman (1982), Eisenberg (1989), King (1989), Emmons and Feer (1990), Sheffield and Thomas (1997), Abramov (1999), Long (1999b), Nowak (2005).

Neogale vison (Schreber, 1777)

American Mink, mink

1. *N. v. aestuarina* (Grinnell, 1916). California.
2. *N. v. aniakensis* (Burns, 1964). Central western Alaska.
3. *N. v. energumenos* (Bangs, 1896). From Yukon south through Oregon and Arizona and New Mexico, including Alberta.
4. *N. v. evagor* (Hall, 1932). Restricted to Vancouver Island, British Columbia.
5. *N. v. evergladensis* (Hamilton, 1948). Southwestern Florida.
6. *N. v. ingens* (Osgood, 1900). Alaska and northern Yukon and Mackenzie.
7. *N. v. lacustris* (Preble, 1902). From Mackenzie to Central Ontario south through northern Montana to Wisconsin.
8. *N. v. letifera* (Hollister, 1913). From northern Montana to Wisconsin south through Oklahoma to Illinois.

9. *N. v. lowii* (Anderson, 1945). Central-northern Québec and Newfoundland.
10. *N. v. lutensis* (Bangs, 1898). Florida Peninsula and coast of Georgia and South Carolina.
11. *N. v. macrodon* (Prentiss, 1903). Restricted to the Gulf of Maine, New England area.
12. *N. v. melampeplus* (Elliot, 1903). Southern Alaska.
13. *N. v. mink* (Peal and Palisot de Beauvois, 1796). From Michigan and western New York throughout the southeastern United States, except for Louisiana, Arkansas, Florida, and the Appalachian Mountains.
14. *N. v. nesolestes* (Heller, 1909). Islands of British Columbia.
15. *N. v. vison* (Schreber, 1777). Southern Ontario, Québec, northwestern United States, and the Appalachian Mountains.
16. *N. v. vulgivaga* (Bangs, 1895). Louisiana and Arkansas.

Conservation: *Neogale vison macrodon* is listed as Extinct, *N. vison* (as *Neovison vison)* is listed as Least Concern by the IUCN (Reid et al. 2016d).

Characteristics: American mink is medium sized; total length 504.0–1805.0 mm in males, 488.0–810.0 mm in females; condylobasal length 65.3–76.3 mm in males, 58.5–70.0 mm in females. **Dorsal pelage uniformly dark brown, with tail close to black at the tip; chin white**, occasionally a mark in the throat, chest, or belly; ears short, scarcely projected above the hair; toes webbed at the base; mink farming can produce different pelage colors; skull dorsally flat with a short rostrum; lambdoid crest well developed; auditory bullae moderately inflated; palatal extending above the molars.

Comments: *Neogale macrodon* was previously a full species (Wozencraft 2005; Mead et al. 2000), but was proposed as a subspecies of *N. vison* (Graham 2001). *N. vison* ranges throughout Canada and the United States, except for desert and semi-desert areas (Map 35). It can be found in sympatry, or nearly so, with *Gulo gulo, Lontra canadensis, Martes americana, Mustela erminea, M. nivalis, Neogale frenata, Pekania pennanti,* and *Taxidea taxus. N. vison* can be distinguished from all other species by its dorsal pelage uniformly dark brown with the tail close to black at the tip and the chin white. From *G. gulo, L. canadensis, Martes americana, P. pennanti,* and *T. taxus,* by its smaller size, total length less than 900.0 mm. From *N. frenata* and *M. nivalis* by its larger size, 300.0 mm.

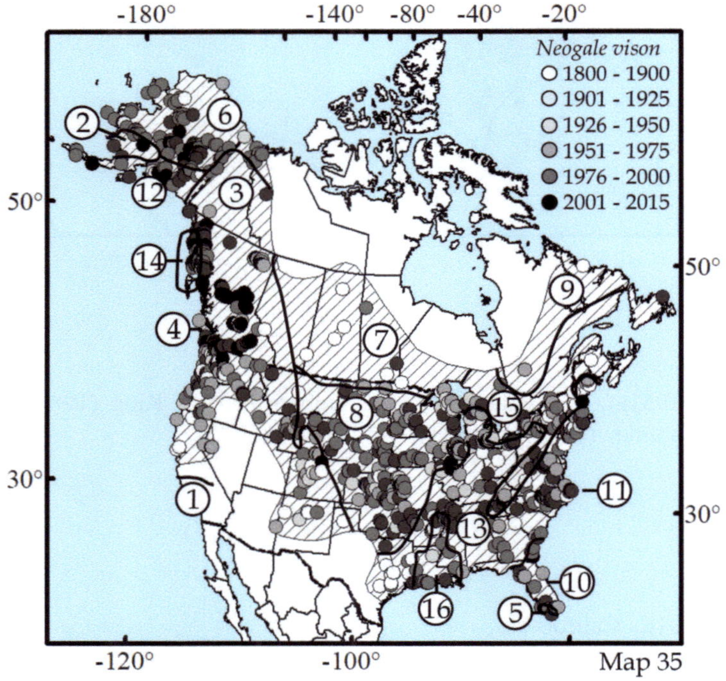

Map 35 Distribution map of *Neogale vison*

Additional Literature: Brinck et al. (1983), Taranin et al. (1991), Larivière (1999a), Kurose et al. (2008), Nowak (2005), Patterson et al. (2021).

Subfamily Taxidiinae Pocock, 1920

Genus ***Taxidea*** Waterhouse 1839

American badgers have a **pelage shaggy, grayish, or brownish, with a white medial stripe on the head, extending to the back; fur of underparts whitish or creamy; "badges" (spots) black on the face surrounded by white pelage**; feet black or dark brown, subdigitigrade, with long curved foreclaws and shovel-like hind claws; toes of forefeet partially webbed; limbs short; body depressed; **skull large and rugose; occiput depressed, facial angle steep**; palate extending posteriorly beyond the plane of the upper molars; auditory bullae highly inflated, not in contact with the paroccipital processes; **first upper molars triangular, often with cusps arranged in rows that are transverse to the long axis of the skull; carnassial premolars also subtriangular with longest side posteromedia**; trigonid of first lower molars longer than taloned; metaconid, hypoconid, and entoconid large.

Taxidea taxus (Schreber, 1777)
American Badger, tejón

1. *T. t. berlandieri* Baird, 1858. From central California to Oklahoma south through Puebla, including the Baja California Peninsula.
2. *T. t. jacksoni* Schantz, 1946. From eastern Manitoba and Wisconsin east through Ohio and southwestern Ontario.
3. *T. t. jeffersonii* (Harlan, 1825). From British Columbia to central Colorado west through California, including all the coast and northwestern Baja California.
4. *T. t. taxus* (Schreber, 1777). From Yukon to Manitoba south through Colorado to western Tennessee.

A Conservation: *Taxidea taxus jacksoni* and *T. t. jeffersonii* are listed as Endangered in the Canadian Endangered Species Protection Act, *T. taxus* as Threatened in the Norma Oficial Mexicana (DOF 2019), and as Least Concern by the IUCN (Helgen and Reid 2016d).

Characteristics: American badger is medium sized; total length 629.0–870.0 mm in males and 521.0–790.0 mm in females; average condylobasal 10.9–11.7 mm. **Dorsal pelage gray to brown, with a marked white line running from the nose tip to the back**; two white longitudinal spots at the eyes level; robust body tending to appear wider than tall; **not tall and slim, with well-developed short limbs less than one-sixth of the head-and-body length; tail short;** palate projected behind the last molar line greater than the width of the last molars, with almost the same wide as length; sharp supraocular processes; skull wide at the auditory bullae level, notoriously wider than the rest of the braincase.

Comments: *Taxidea taxus* ranges from southern Canada in British Columbia, Alberta, Saskatchewan, Manitoba, and southern Ontario south through the Baja California Peninsula and the Eje Neovolcánico in México, including most of the northern, western, and central United States (Map 36). It can be found in sympatry, or nearly so, with *Lontra canadensis, Martes americana, Mustela erminea, M. nigripes, M. nivalis,* and *Neogale frenata.* It can be distinguished from all other species by its dorsal pelage from gray to brown but with a marked white line running from the nose tip to the back; body slim with well-developed short limbs less than one-sixth of the head-and-body length and tail short.

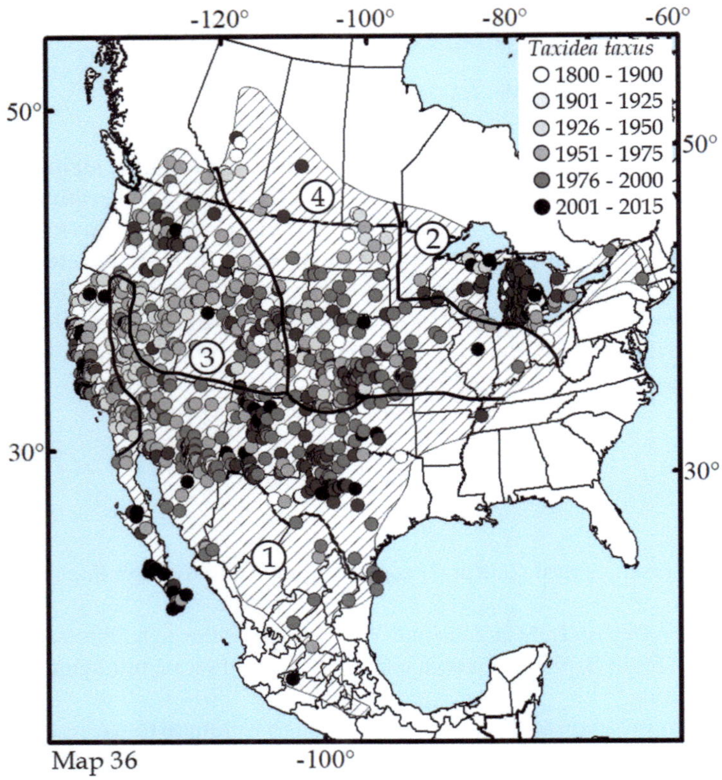

Map 36 Distribution map of *Taxidea taxus*

Additional Literature: Pocock (1920), Long and Long (1965), Long (1965, 1972, 1973, 1991b), Kyle et al. (2004), Nowak (2005), Wozencraft (2005).

Family Odobenidae Allen, 1880

The species in this family have the **body thick and heavy; head relatively short; muzzle blunt, broad, with coarse bristles; pinnae absent; anterior portion of the skull swollen; canines forming long tusks;** postorbital processes lacking; surface of scapula long, narrow; mandibular symphysis fused in adults; pars petrosa, dorsal to meatus acusticus internus.

Genus ***Odobenus*** Brisson, 1762

Body pelage yellowish-brown, becoming reddish-brown or chestnut on the belly and at the base of the appendages; **hair short, becoming sparse with age; skin thick and wrinkled; body thick and heavy**; head relatively short; muzzle blunt, broad, with coarse bristles; pinnae absent; fore flippers nearly as wide as long; **hind flippers approximately triangular, first and fifth toes much longer than the others; anterior portion of the skull swollen; upper canines large, elongate; ever-growing tusks in both sexes;** the other teeth simple, peg-like; postorbital processes lacking; crista tympani lacking overhang of free frontal edge; surface of scapula long and narrow; mandibular symphysis fused in adults.

Odobenus rosmarus (Linnaeus, 1758)

Walrus, morsa

1. *O. r. divergens* (Illiger, 1815). From Bering Sea and Chukchi Seas west through northern Russia.
2. *O. r. rosmarus* (Linnaeus, 1758). From eastern Canadian Arctic east through Greenland.

Conservation: *Odobenus rosmarus* is listed as Vulnerable (A3c: population reduction in the last 10 years 30% or less projected to be met in future data obtained by decline in area of occupancy) by the IUCN (Lowry 2016b). The United States Endangered Species Act (FWS 2022) and the Marine Mammal Protection Act only permit to hunt this species by Alaskan natives for subsistence or production of handicrafts. The northwestern Atlantic population is listed as Extirpated in the Canadian Endangered Species Protection Act.

Characteristics: Walrus is large sized; total length 2,500.0–3,500.0 mm in males and 2,300.0–3,100.0 mm in females; condylobasal length 335.0–387.0 mm in males and 322.0–373.0 mm in females. **General coloration pinkish, from whitish to brownish; pinna absent; snout broad and with many whiskers; skull semi-rectangular with the rostrum swollen in its anterior portion; canines larger than the maxillary toothrow in both sexes and forming long tusks**.

Comments: *Odobenus rosmarus* was divided into three subspecies (Rice 1998); the current treatment only accepts two (Fay 1982; Lindqvist et al. 2009). *O. rosmarus* ranges in North Pacific, North Atlantic, and Arctic oceans with a discontinuous circumpolar distribution. The Atlantic population ranges from the eastern Canadian Arctic west through Greenland, including southern of the Gulf of Saint Lawrence, but it was extirpated. The Pacific population ranges from the Bering Sea west and north through the Beaufort Sea (Map 37). *O. rosmarus* can be found in sympatry, or nearly so, with *Callorhinus ursinus, Erignathus barbatus, Histriophoca fasciata, Pagophilus groenlandicus, Phoca largha, P. vitulina,* and *Pusa hispida*. It can be distinguished from all other genera by its larger size; skin thick and wrinkled, mainly without hair; upper canines large and elongate, as tusks in both sexes.

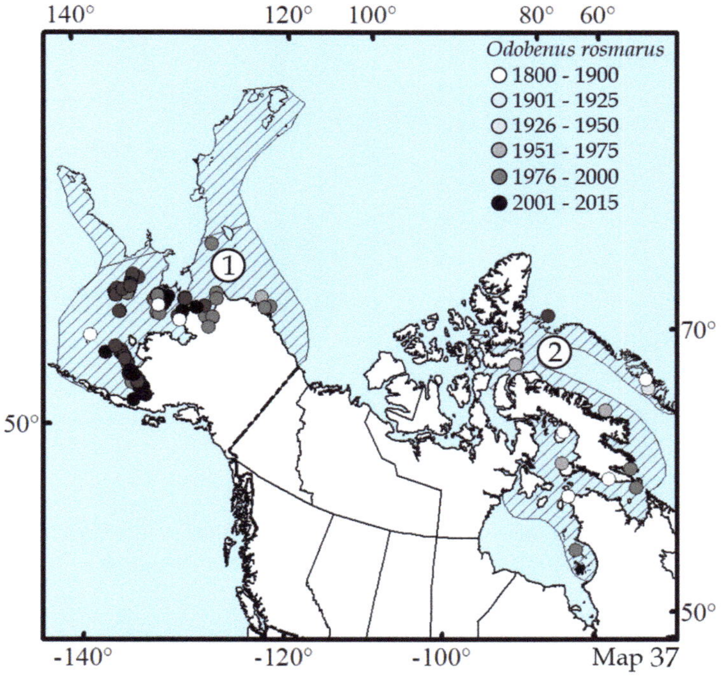

Map 37 Distribution map of *Odobenus rosmarus*

Additional Literature: Allen (1930), Scheffer (1958), Manville and Favour (1960), Harrington (1966), Burns (1981a), Fay (1981, 1985), King (1983a, b), McLeod et al. (2014), Stewart et al. (2014).

Family Otariidae Gray, 1825

The family Otariidae includes sea lions and fur seals, characterized by **ears visible; back limbs directed forward making them functional in terrestrial displacement; body coloration uniform, never spotted or banded; nails prominent on three middle digits of the hind flippers; testes scrotal; supraorbital process well developed; mastoid process separated from the auditory bullae; medial upper incisors with transverse grooves; cheek-teeth simple, usually unicuspid.** Cosmopolitan distribution in all temperate and cold coastlines. The species ranging in North America belong to two subfamilies: Arctocephalinae and Otariinae. Distinctions between subfamilies are not useful in the identification key, so they are not used.

1. Coloration pale brown; large animals, total length greater than 2,700.0 mm in males and greater than 1,900.0 mm in females; diastema between the last upper premolars and the molars equal to or greater than 5% of the skull length (from Alaska south through northern Baja California, México)... *Eumetopias jubatus* (p. 152)

1a. Different colorations; medium-sized animals, total length less than 2,700.0 mm in males and less than 1,900.0 mm in females; diastema between the last upper premolars and the molars less than 5% of the skull length............................2

2. Rostrum short, with the nose very sharply pointed; nasal length similar to its anterior width; rostral length less than 25% of the skull length (from Alaska south through California, United States) .... *Callorhinus ursinus* (p. 150)

2a. Rostrum size proportional to body size; nasal length greater than its anterior width; rostral length greater than 25% of the skull length..3

3. Total length less than 2,000.0 mm in males and 1,500.0 mm in females; palatine narrow, with parallel sides between the first and third upper premolars (Fig. 22); auditory bullae smooth and rounded (from Farallon Islands, United States, south through the Baja California Peninsula, México)..*Arctocephalus townsendi* (p. 149)

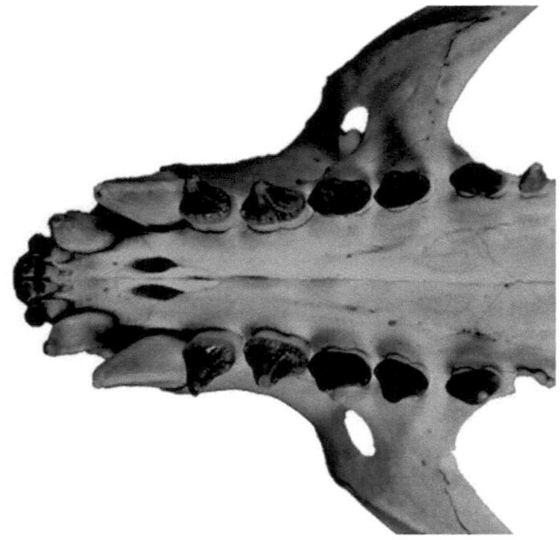

Fig. 22 The palatine narrow, with parallel sides between the first and third upper premolars like in *Arctocephalus*

3a. Total length greater than 2,000.0 mm in males and 1,500.0 mm in females; palatine width with sides slanting posteriorly between the first and third upper premolars (Fig. 23); auditory bullae irregular (from Alaska, United States, south through Central America) ....*Zalophus californianus* (p. 153)

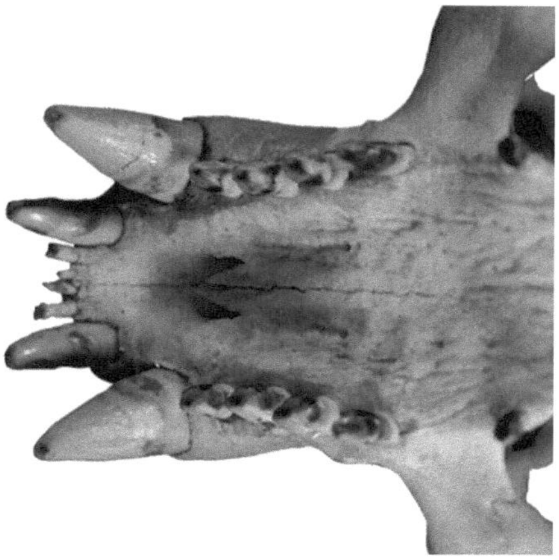

Fig. 23 The palatine width with sides slanting posteriorly between the first and third upper premolars like in *Zalophus californianus*

Subfamily Arctocephalinae Scheffer and Rice, 1963

The species within this family are named fur seals, having the following features in common: hair generally short, with denser underfur; farther and longer-lasting foraging trips; smaller and more abundant prey items; greater sexual dimorphism.

Genus ***Arctocephalus*** Geoffroy St.-Hilaire and Cuvier, 1826

Body pelage blackish-gray dorsally with the head and neck yellowish gray; fur of underparts **brownish-black; lips, chin, and proximal parts of the limbs rusty brown;** skull narrow; rostrum narrow and long, length averaging ~30% of the skull length; nasal bones long and slender; prolongations of palatine bones not thickened; zygomatic root of the maxillae broad; zygomatic arches narrow; postorbital processes triangular; teeth back of the canines lacking accessory cusps and widely spaced.

Arctocephalus townsendi Merriam, 1897
Guadalupe Fur Seal, lobo fino de Guadalupe

A. t. townsendi Merriam, 1897. From the Farallon Islands, California, south through the Guadalupe Islands, Baja California, including San Benito and Cedros Islands.

Conservation: *Arctocephalus townsendi* is listed as Least Concern by the IUCN (Aurioles-Gamboa 2015), as Threatened by the United States Endangered Species Act (FWS 2022), as Protected by the United States Marine Mammal Protection Act, and as Endangered by the Norma Oficial Mexicana (DOF 2019).

Characteristics: Guadalupe Fur seal is large sized; total length 1,900.0–2,400.0 mm in males and 1,300.0–1,500.0 mm in females; condylobasal length 235.0–255.0 mm in males and 195.0–217.0 mm in females. Dorsal pelage blackish gray with yellowish gray shades on the head and neck; fur of underparts blackish brown; males much larger than females; **first toe larger than the others in the fore flippers and all of similar size in the hind flippers**; ears small sized; hind flippers directed toward the front, functional for terrestrial displacement; **auditory bullae rounded**; **interorbital constriction well-defined, hourglass-shaped**; **cheek-teeth more closed in the middle part**; **sagittal crest low, highest point at braincase at the middle part**; postorbital process pointed; diastema between the last upper premolars and the molars less than 5% of the skull length.

Comments: *Arctocephalus* and *Zalophus* have been considered cogeneric (Van Gelder 1977). *Arctocephalus townsendi* was previously considered a subspecies of *A. philippii* (Scheffer 1958; Brunner 2004). They are currently treated as different species (Repenning et al. 1971; Wynen et al. 2001; Higdon et al. 2007). *A. townsendi* range is centered on the Guadalupe Island and can be found from the Gulf of California and the Pacific coast of the Baja California Peninsula north through the Farallon Islands, California (Map 38). It can be found in sympatry, or nearly so, with *Callorhinus ursinus* and *Mirounga angustirostris. A. townsendi* can be distinguished from *C. ursinus* by having a more elongated muzzle; lower forehead, and slope of anterodorsal aspect of the skull, facial angle always exceeding 125°; hair of fore flippers extends past the wrist. From *M. angustirostris,* lacking a lengthened snout in males; males and females less than 3000.0 mm in total length; skull with a thin rostrum and triangular.

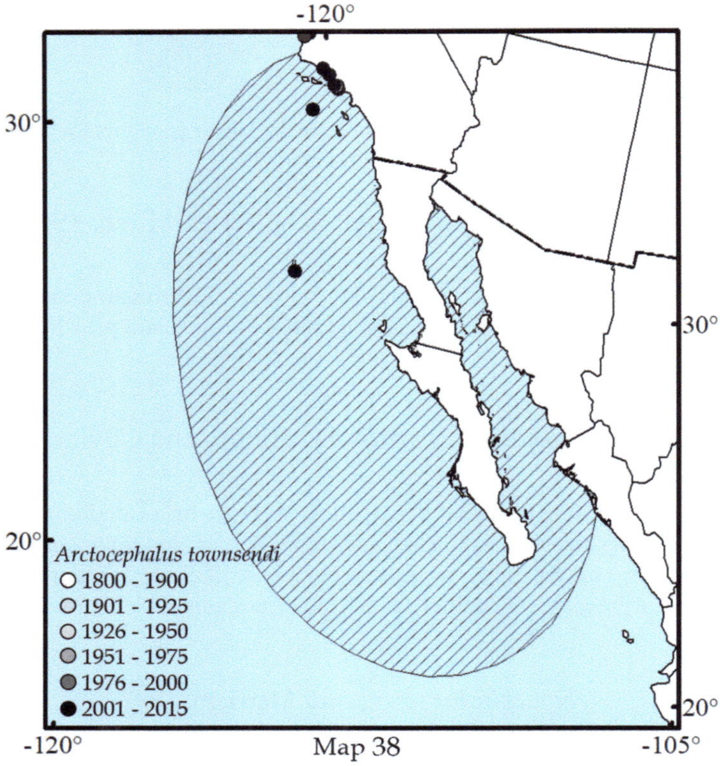

Map 38 Distribution map of *Arctocephalus townsendi*

Additional Literature: King (1954), Rice et al. (1965), Peterson et al. (1968), Hall (1981), Gallo-Reynoso (1994), Gallo-Reynoso and Figueroa-Carranza (1996), Belcher and Lee (2002), Aurioles-Gamboa et al. (2010).

Genus *Callorhinus* Gray, 1859

Body pelage with different patterns between sexes; in males, back from blackish to dark brownish, shoulders and neck gray; underparts and limbs reddish-brown. Females gray to grayish brown and underparts reddish-brown or gray; **muzzle small and forehead high, with a facial angle always less than 125°; hair of fore flippers ends abruptly at the wrist; nasals short and broad; rostrum short, broad, convex, only slightly depressed; postorbital processes triangular;** palate short, narrow, and deeply notched on the posterior margin.

Callorhinus ursinus (Linnaeus, 1758)

Northern Fur Seal, oso marino ártico

Monotypic.

Conservation: *Callorhinus ursinus* is listed as Vulnerable (A2b: population reduction in the last 10 years 30% or less and may not be reversible by index of abundance) by the IUCN (Gelatt et al. 2015) and in the United States Marine Mammal Protection Act.

Characteristics: Northern Fur seal is large sized; total length 1,900.0–2,100.0 mm in males and 1,200.0–1,500.0 mm in females; condylobasal length 220.0–275.0 mm in males and 165.0–200.0 mm in females. **Dorsal pelage dark brown to black; males with long coarse hairs in the neck, shoulders, and chest; rostrum short with the nose very sharply pointed; with a straight line at the fore flipper base**; **toes of fore and hind flipper of similar size**; ears small sized; hind flippers directed frontward and functional in terrestrial displacement; large body size; auditory bullae not rounded; **interorbital constriction well defined**; cheek-teeth parallel; sagittal crest absent or very small; postorbital process pointed; diastema between the last upper premolars and molars less than 5% of the skull length.

Comments: No subspecies are recognized (Taylor et al. 1955; Rice 1998). *Callorhinus ursinus* ranges in the North Pacific, including the coastal and pelagic waters, from northern Baja California north through the Bering Sea (Map 39). It can be found in sympatry, or nearly so, with *Arctocephalus townsendi, Erignathus barbatus, Eumetopias jubatus, Histriophoca fasciata, Mirounga angustirostris, Odobenus rosmarus, Phoca largha, P. vitulina, Pusa hispida,* and *Zalophus californianus. C. ursinus* can be distinguished from *A. townsendi* by having the muzzle less elongated; forehead higher; slope of the anterodorsal aspect of the skull; facial angle always less than 125°; hair of fore flippers ends abruptly at the wrist. From *E. barbatus,* by having small pinnae; hind flippers directed frontward and functional for walking; muzzle not broad, with vibrissae dense and straight. From *E. jubatus* and *Z. californianus,* by having the rostrum very short with the nose very sharply pointed; in addition, from *E. jubatus*, by its smaller size. From *M. angustirostris,* by not having a lengthened snout in males; smaller size, less than 3000.0 mm in both sexes; skull triangular and with a narrow rostrum. From *O. rosmarus,* by its smaller size with hair covering all the body, not naked; upper canines not large and elongated as tusks in both sexes. From *P. largha, P. vitulina,* and *P. hispida,* by its larger size, greater than 1700.0 mm; presence of small pinnae; hind flippers directed frontward and functional for walking.

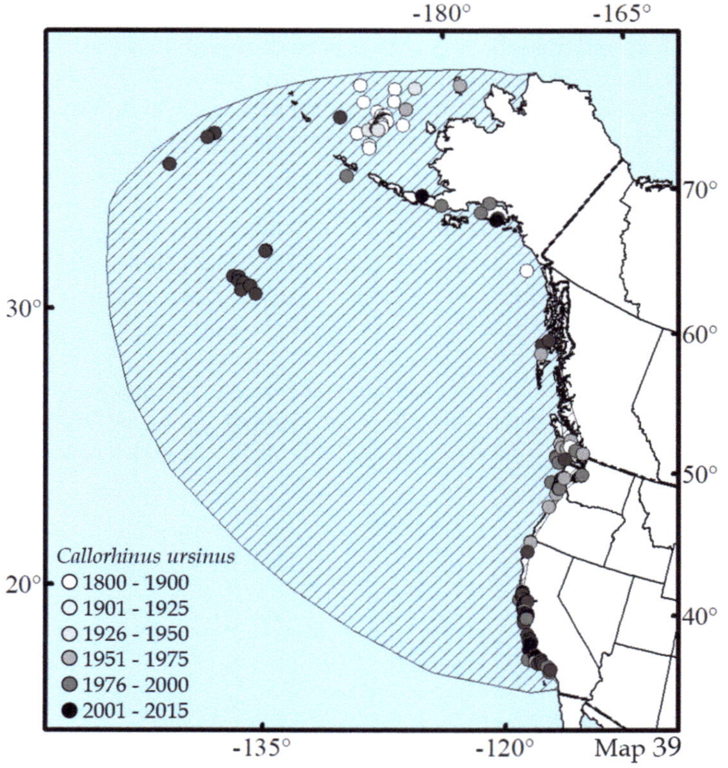

Map 39 Distribution map of *Callorhinus ursinus*

Additional Literature: Peterson et al. (1968), Repenning et al. (1971), Loughlin et al. (1994), Dickerson et al. (2010), Zeppelin and Orr (2010).

Subfamily Otariinae Gray, 1825

The species within this family are named sea lions and have the following features in common: hair generally long and with less dense underfur; short foraging trips; large prey items; less sexual dimorphism.

Genus ***Eumetopias*** Gill, 1866

Body pelage buff to reddish brown, paler immediately after molting; chest and abdomen slightly darker; naked parts of the skin black; **pelage harsh; neck with a mane in males; skull large; palate long, contracted and truncate posteriorly; postorbital processes quadrate and placed anteriorly**.

Eumetopias jubatus (Schreber, 1776)
Steller's Sea Lion, león marino de Steller

1. *E. j. jubatus* (Schreber, 1776). From Alaska west through southern Japan and Korea.
2. *E. j. monteriensis* (Gray, 1859). From Alaska east through Baja California.

Conservation: *Eumetopias jubatus* is listed as Near Threatened by the IUCN (Gelatt and Sweeney 2016), as Threatened in the United States Marine Mammal Protection Act, and as Special Concern in the Canadian Species Protection Act.

Characteristics: Steller sea lion is the largest seal of the family; total length 2,700.0–3,200.0 mm in males and 1,900.0–2,900.0 mm in females; condylobasal length 375.0–400.0 mm in males and 294.0–298.0 mm in females. Dorsal pelage brown to pale brown; hairs of the same color in underparts and the body; snout straight; males with long coarse hairs in the neck, shoulders, and chest; **fore and hind flippers with first toe larger than the others**; ears small sized; hind flippers directed frontward and functional in terrestrial displacement; large body size; auditory bullae not rounded; interorbital constriction marked but not hourglass-shaped; cheek-teeth parallel; **sagittal crest high, highest point at the parietal level; postorbital process quadrate; a large diastema between the last upper premolars and molars, greater than 5% of the skull length.**

Comments: *Eumetopias jubatus* ranges in the North Pacific Ocean, from northern Baja California north through the Bering Sea (Map 40). It can be found in sympatry, or nearly so, with *Callorhinus ursinus, Zalophus californianus, Erignathus barbatus, Histriophoca fasciata, Mirounga angustirostris, Phoca largha, P. vitulina,* and *Pusa hispida. E. jubatus* can be distinguished from *C. ursinus, H. fasciata, P. largha, P. vitulina, P. hispida,* and *Z. californianus* by its larger size, and 2700.0 mm in males and greater than 1900.0 mm in females. From *Z. califorianus,* by its paler coloration and larger size, weighing three times as much as adults, with females of *Eumetopias* as large as adult males of *Zalophus*; bearlike with a short straight nose; postcanine teeth not evenly spaced, with a diastema. From *M. angustirostris,* by lacking a lengthened snout in males; smaller size, less than 3000.0 mm in both sexes and skull triangular and with a narrow rostrum.

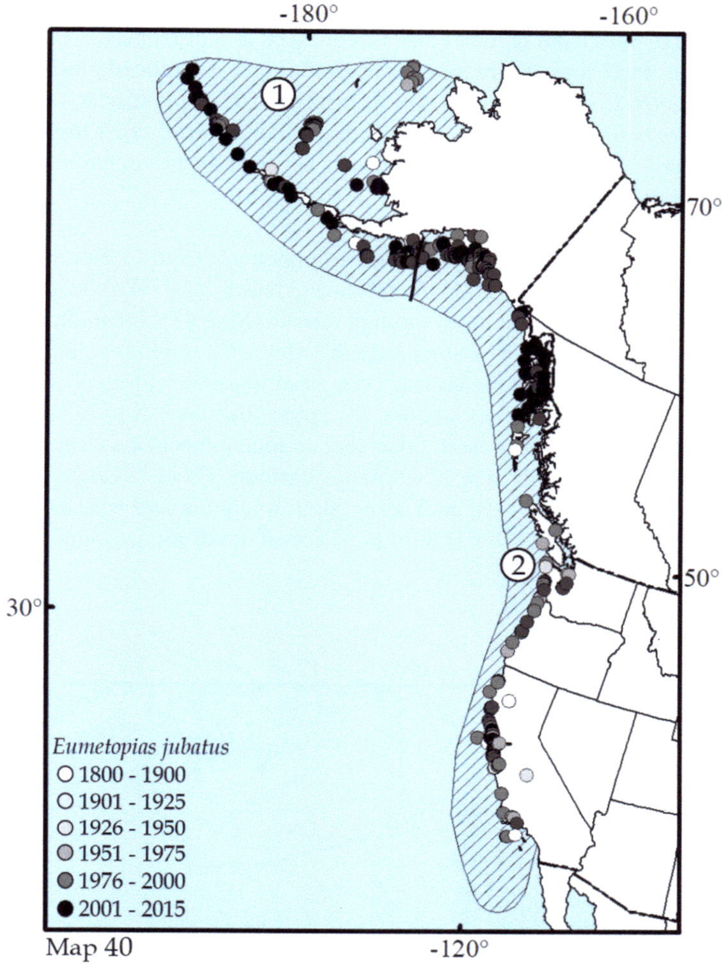

Map 40 Distribution map of *Eumetopias jubatus*

Additional Literature: King (1954), Itoo et al. (1977), Schusterman (1981), Loughlin (1997, 2009), Rice (1998), Loughlin et al. (2009), Phillips et al. (2009), Gallo-Reynoso et al. (2020).

Genus ***Zalophus*** Gill, 1866

Body pelage blackish to pale buff to deep sepia; **muzzle broad and heavy; males with a conspicuous crest on top of the head; males with a very high sagittal crest, rising steeply from the frontals in the supraorbital region, greatest height above the postorbital processes**; palate relatively short; anterior nares narrow; auditory bullae irregular in shape; zygomatic roots of the maxillae narrow.

Zalophus californianus (Lesson, 1828)

California Sea Lion, lobo marino de California

Monotypic.

Conservation: *Zalophus californianus* is listed as Least Concern by the IUCN (Aurioles-Gamboa and Hernández-Camacho 2015); listed as subject to Special Protection in the Norma Oficial Mexicana (DOF 2019) and by the United States Endangered Species Act (FWS 2022), and Marine Mammal Protection Act.

Characteristics: California sea lion is large sized; total length 2,000.0–2,500.0 mm in males and 1,600.0–1,800.0 mm in females; condylobasal length 250.0–330.0 mm in males and 195.0–250.0 mm in females. Dorsal pelage brown to dark brown; males bigger than females; dog-like face; **toes of fore flippers of similar size, first toe of hind flippers larger than the others**; ears small sized; hind flippers directed frontward and functional for terrestrial displacement; auditory bullae not rounded; interorbital constriction marked but not hourglass shaped; cheek-teeth parallel; **sagittal crest high, with the highest point at the postorbital process**; postorbital process pointed; with the diastema between the last upper premolars and molars less than 5% of the skull length.

Comments: *Zalophus* and *Arctocephalus* have been considered cogeneric (Van Gelder 1977). Nowadays three different species are within *Zalophus* (Rice 1998; Brunner 2004; Sakahira and Niimi 2007; Wolf et al. 2007). *Z. californianus* ranges in the North Pacific Ocean from Central America north through Alaska (Map 41). It can be found in sympatry, or nearly so, with *Callorhinus ursinus, Eumetopias jubatus, Mirounga angustirostris, Phoca vitulina,* and *Pusa hispida. Z. californianus* can be distinguished from *C. ursinus* by having the rostrum size in proportion with the body; without a straight line at the fore flipper base; toes of similar size in fore flippers and first toe larger than the others in hind flippers. From *E. jubatus,* by having a darker coloration; smaller size, weighing three times less as adults; therefore, females of *Eumetopias* are as large as adult males of *Zalophus*; postcanine teeth evenly spaced without a diastema. From *M. angustirostris,* by lacking a lengthened snout in males; smaller size, less than 3000.0 mm in both sexes; skull triangular and with a narrow rostrum. From *P. vitulina* and *P. hispida,* by its larger size, greater than 1700.0 mm; presence of small pinnae; hind flippers directed frontward and functional for walking.

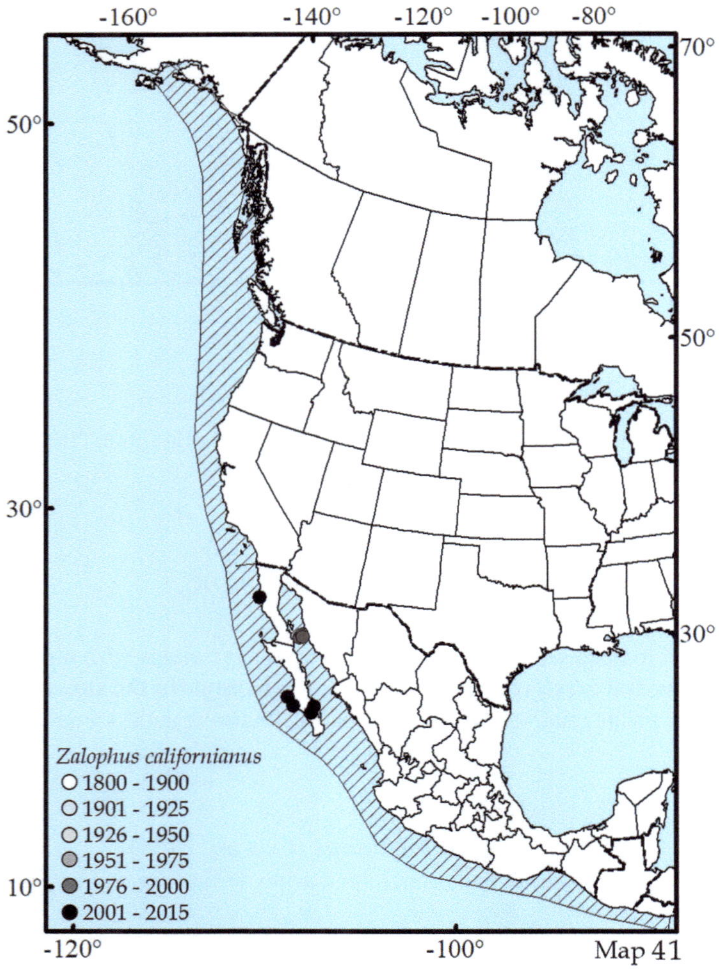

Map 41 Distribution map of *Zalophus californianus*

Additional Literature: Van Gelder (1960), Cowan and Guiguet (1965).

Family Phocidae Gray, 1821

The family Phocidae includes seals characterized by **non-visible ears; back limbs directed backward and not functional for terrestrial displacement; body coloration usually spotted or banded; claws present on the hind flippers; testes abdominal; supraorbital process rudimentary or absent; mastoid process bound to the auditory bullae;** medial upper incisors without transverse grooves; **cheek-teeth usually multicuspid**. Distributed throughout the coasts of temperate and cold seas. No subfamilies are recognized in this family.

1. Males with a proboscis; one lower incisor on each side; premaxilla not in contact with the nasal pits (Fig. 24)............2

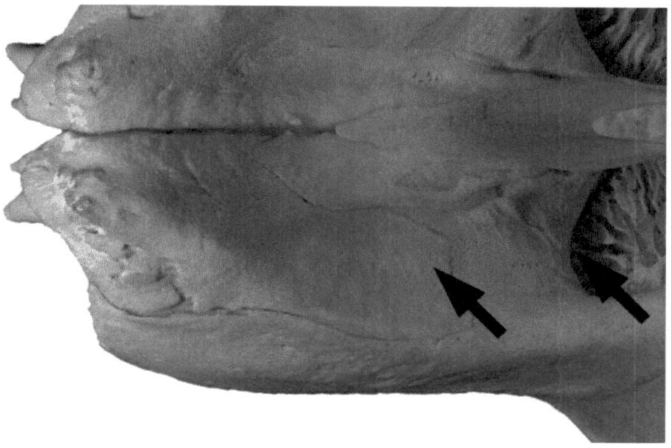

Fig. 24 The premaxilla bone is not in contact with the nasal pits

1a. Males without a proboscis; two lower incisors on each side; premaxilla in contact with the nasal pits (Pacific Ocean coast)............3

2. Males with an elongated inflatable tubular proboscis; nails small and rudimentary; auditory bullae concave on the anterior border; inner and outer incisors of similar size (from Alaska, United States, south through Bahia Magdalena, Baja California, México)............ *Mirounga angustirostris* (p. 164)

2a. Males with a skin pouch on the nose; nails well developed; auditory bullae straight on the anterior border; inner incisors less than one-half of the size of the outer incisors (from northern Newfoundland, Canada, south through the Gulf of Saint Lawrence, United States)............ *Cystophora cristata* (p. 158)

3. Rostrum straight from the frontal region to the nostrils; dorsal coloration gray, with irregular white spots; maxillary bone triangular on side view (Fig. 25); all molariforms unicuspid (from Newfoundland, Canada, south through New Jersey, United States)............*Halichoerus grypus* (p. 161)

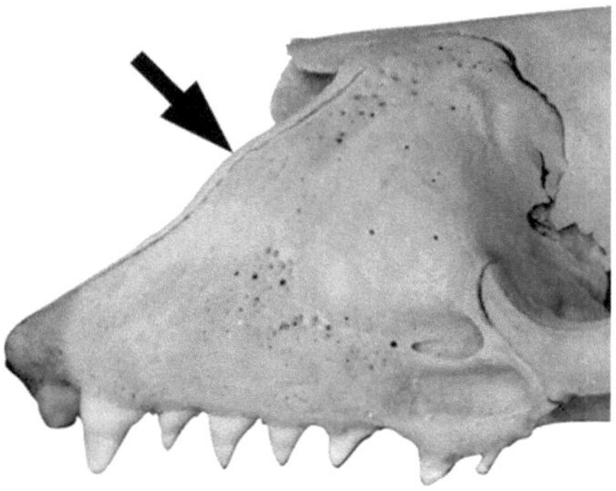

Fig. 25 The maxillary bone triangular on lateral view in *Halichoerus grypus*

3a. Rostrum not straight, with a clear curvature between the frontal region and the snout; dorsal coloration not gray, without irregular white spots: maxillary bone not triangular on side view (Fig. 26); not all molariforms unicuspid 4

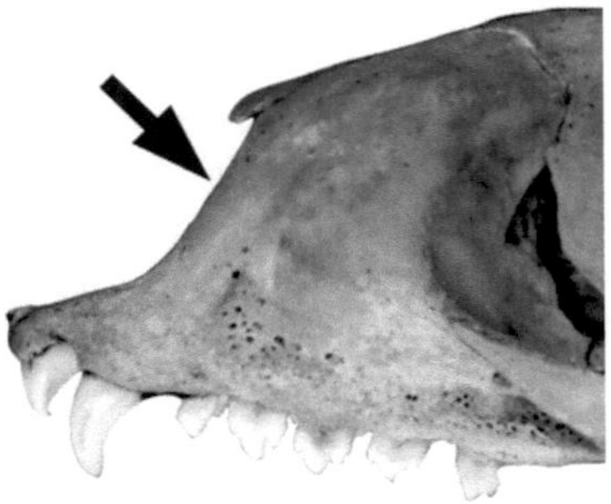

Fig. 26 Maxillary bone not triangular on lateral view

4. Dorsal coloration with pale rings on a black background; almost no neck and face short; upper angles of zygomatic with a plate-shaped expansion (Fig. 27; Arctic Ocean, from western Alaska to northwestern Newfoundland) .... *Pusa hispida* (p. 170)

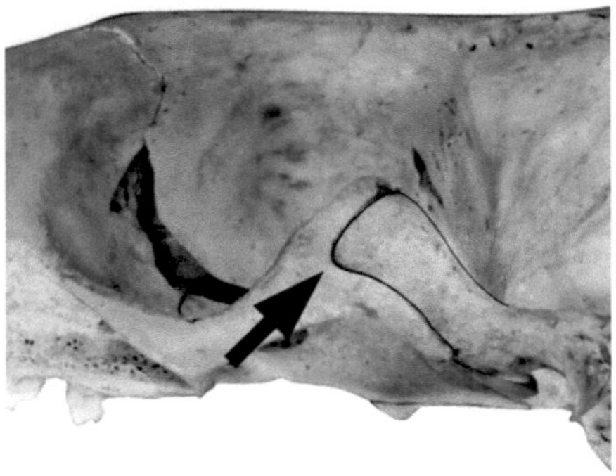

Fig. 27 The upper angles of zygomatic with a plate-shaped expansion

4a. Dorsal coloration with different patterns but no pale rings on a black background; neck distinguished and face not short; no upper angles of the zygomatic with a plate-shaped expansion (Fig. 28) 5

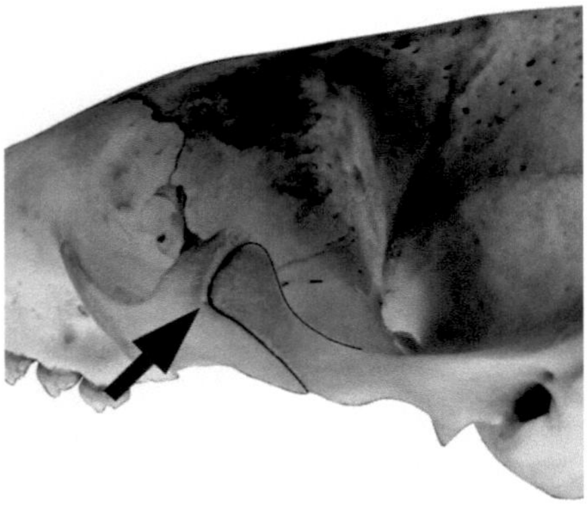

Fig. 28 No upper angles of the zygomatic with a plate-shaped expansion

5. Third digit of fore flippers larger that the first and second digits; jugal bond short, width more than one-half of length (from Alaska, United States, east through Newfoundland, Canada) .... *Erignathus barbatus* (p. 159)

5a. First and second digits of the fore fin larger; jugal bond short, width less than one-half of length (Fig. 29).................. 6

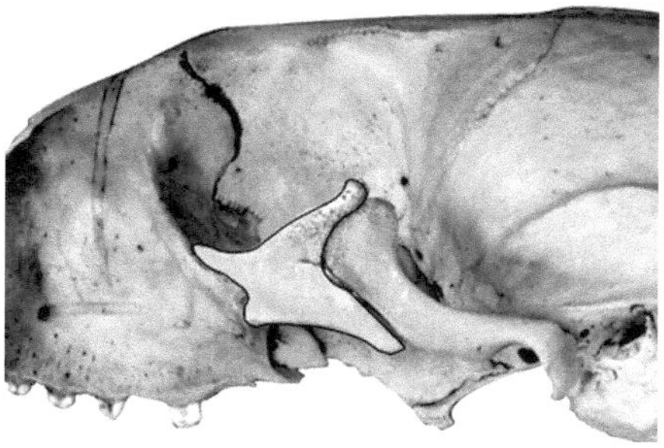

Fig. 29 The jugal bond is short, width less than one-half of length

6. Head coloration contrasting with the neck; posterior margin of palate broadly U shaped; nasal septum reaching or nearly reaching the posterior margin of the palate.. 7

6a. Head and neck coloration similar; posterior margin of the palate broadly V shaped; nasal septum not reaching the posterior margin of the palate (Fig. 30).. 8

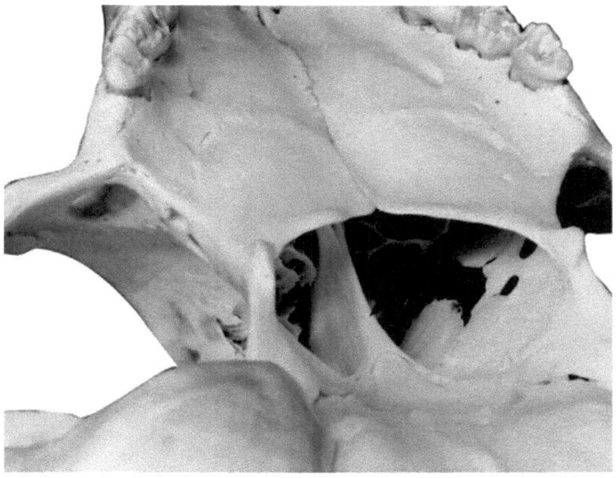

Fig. 30 The nasal septum not reaching the posterior margin of the palate

7. Brown with whitish ribbons around the head and flippers: cheek-teeth usually lacking an accessory cusp (Alaska, United States).. *Histriophoca fasciata* (p. 162)

7a. White with a black lyre-shaped marking on the back, face black; cheek-teeth with an accessory cusp (from Hudson Bay, Canada, south to the Gulf of Maine, United States) .... *Pagophilus groenlandicus* (p. 166)

8. Total length greater than 1800.0 mm; fingernails of flippers barely developed; two upper incisors (from Florida, United States, south through Quintana Roo, México, considered extinct)............................... *Neomonachus tropicalis* (p. 165)

8a. Total length less than 1900.0 mm; fingernails well developed in flippers; three upper incisors .... *Phoca* (p. 167)

Genus *Cystophora* Nilsson, 1820

Body pelage grayish to blackish dorsally and flanks paler spotted with white; **males with a bladder-like inflatable pouch of the skin on the nose;** hind flippers slightly emarginate on the distal borders; **toes of hind flippers subequal to or with the first and fifth toes much longer than the others and all digits with large claws;** palatines slightly emarginate on the posterior border; **auditory bullae square in the front; basisphenoid and basioccipital bones broad, flat, and not concave**; mandible wit upper notch deeply U shaped; **postcanine teeth single-rooted, except for the fifth upper** and occasionally the fourth upper postcanine teeth, which are double-rooted.

Cystophora cristata (Erxleben, 1777)

Hooded Seal, foca capuchina

Monotypic.

Conservation: *Cystophora cristata* is listed as Vulnerable (A3c: population reduction in the last 10 years 30% or less projected to be met in future by decline in area of occupancy) by the IUCN (Kovacs 2016b). Hunting was banned in Canada in 1972 and quotas were set on the rest of the Canadian harvest in 1974. *C. cristata* is listed among the top three Arctic marine mammal species in terms of sensitivity to climate change (Laidre et al. 2008).

Characteristics: Hooded seal is medium sized; total length 2,300.0–2,800.0 mm in males and 2,000.0–2,300.0 mm in females; condylobasal length 265.0–275.0 mm in males and 210.0–235.0 mm in females. **Dorsal pelage silver-gray with black splotches; fur of underparts paler; males with a hood on the top of the head that can be inflated, and a pouch of skin on the nose like a red balloon that can be inflated too;** no ears; hind flippers with well-developed fingernails; posterior palatine foramina posterior to the maxillopalatine suture, near the palatine center; posterior palate margin very broadly U shaped; cheek-teeth smaller, **with no accessory cusps, all of similar size and without a diastema**; space between auditory bullae narrower than the breadth; two upper incisors on each side, with the inner less than one-half of the size of the outer incisors.

Comments: No subspecies are recognized (Wozencraft 2005). *Cystophora cristata* ranges in the North Atlantic Ocean and adjoining waters of the Arctic Ocean, from the Gulf of Saint Lawrence to northern Newfoundland associated with pack ice (Map 42). It can be found in sympatry, or nearly so, with *Erignathus barbatus, Halichoerus grypus, Odobenus rosmarus, Pagophilus groenlandicus, Phoca vitulina,* and *Pusa hispida*; it can be distinguished from all other species by males having a hood on the top of the head that can be inflated and a pouch of skin on the nose like a red balloon that can be inflated too.

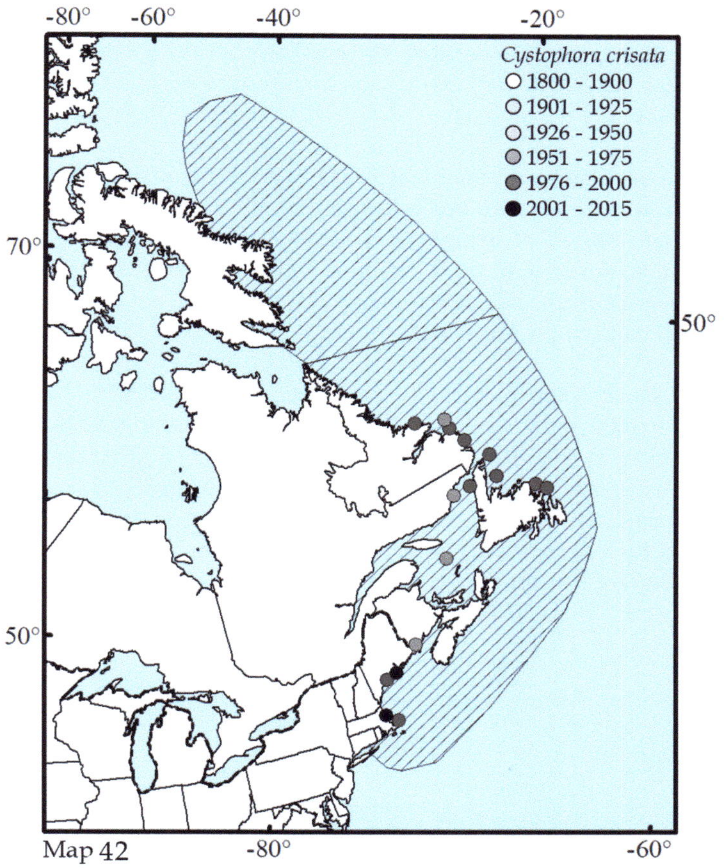

Map 42 Distribution map of *Cystophora cristata*

Additional Literature: King (1966), Wiig and Lie (1984), Kovacs and Lavigne (1986), Sundt et al. (1994), Coltman et al. (2007), Andersen et al. (2009), Folkow et al. (2010).

Genus *Erignathus* Gill, 1866

Body pelage grayish to yellowish, darker grayish on the back; **muzzle broad with vibrissae dense and straight; forehead high and convex; third digit of the fore flipper longest**; four nipples; **palatal area broad, oval shaped**; posterior border of the palate emarginate, not triangular; postorbital processes small; jugal short, jugal depth not less than one-half of its length; **narial septum incomplete**; posterior margin of the mandibular ramus forming a straight line between the angular and articular processes.

Erignathus barbatus (Erxleben, 1777)

Bearded Seal, foca barbuda

1. *E. b. barbatus* (Erxleben, 1777). North Atlantic Ocean and adjoining waters of the Arctic Ocean, from the central Canadian Arctic east through the Laptev Sea, Hudson Bay, east through Greenland and south through Newfoundland.
2. *E. b. nauticus* (Pallas, 1811). North Pacific Ocean and adjoining waters of the Arctic Ocean, from the central Canadian Arctic west through the Laptev Sea, all coasts of Banks Island and Yukon south through Alaska.

Conservation: *Erignathus barbatus* is listed as Threatened by the United States Endangered Species Act (FWS 2022); the Marine Mammal Protection Act only permits it to be hunted by Alaskan natives for subsistence or production of handicrafts. It is listed as Least Concern by the IUCN (Kovacs 2016a) subsistence hunting is accepted in Canada, comanaged by the Department of Fisheries and Oceans.

Characteristics: Bearded seal is medium sized; total length 2,000.0–2,600.0 mm, condylobasal length 208.0–257.0 mm. **Dorsal pelage gray or brownish, sometimes with red coloration on the face and flippers; fur of underparts of similar color; muzzle broad with vibrissae dense and straight;** fore flippers square; males without a proboscis; no ears; hind flippers with well-developed fingernails; posterior palatine foramina situated posterior to the maxillopalatine suture; posterior palate margin broadly U shaped; cheek-teeth smaller, with no accessory cusps, and of similar size; space between auditory bullae much greater than its breadth; three upper incisors.

Comments: Two subspecies within *Erignathus barbatus* have been recognized with a patchy circumpolar distribution, from the central Canadian Arctic to the Laptev Sea; *E. b. barbatus* ranges in the Atlantic Ocean and *E. b. nauticus* ranges in the Pacific Ocean (Rice 1998; Davis et al. 2008). However, some authors do not find a clear boundary between these subspecies, so these are not completely accepted (Burns 1981b; Risch et al. 2007; Map 43). *E barbatus* can be found in sympatry, or nearly so, with *Callorhinus ursinus, Eumetopias jubatus, Cystophora cristata, Histriophoca fasciata, Odobenus rosmarus, Phoca largha, P. vitulina,* and *Pusa hispida* and can be distinguished from all by having the muzzle broad with vibrissae dense and straight, and the third digit of fore flippers longest.

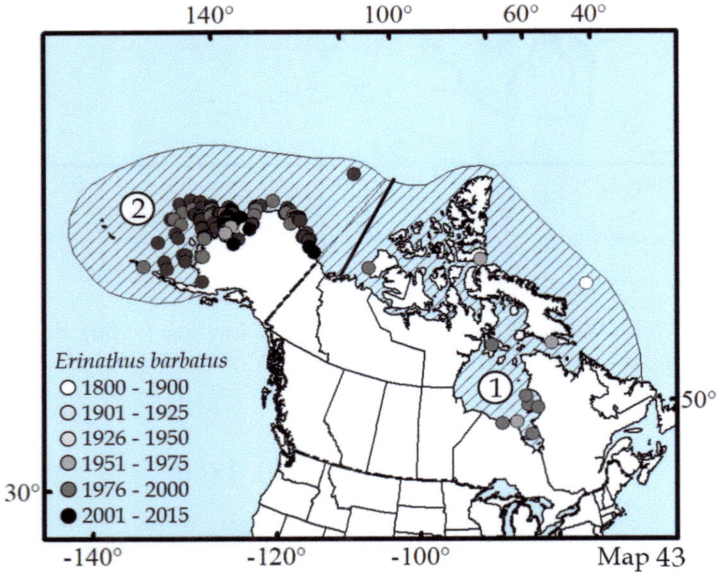

Map 43 Distribution map of *Erignathus barbatus*

Additional Literature: Gosselin and Boily (1994), Kovacs (2009).

Genus *Halichoerus* Nilsson, 1820

Body pelage silverish to dusky-gray with black spots on back and flanks; snout long, with distance between nose tip to eyes nearly twice than between eye and auditory meatus; muzzle broad; vibrissae crenulated; **skull arched dorsally from nasals to occipital;** upper postcanine teeth single-rooted, except for the last two and last lower tooth.

Halichoerus grypus (Fabricius, 1791)

Altantic Gray Seal, foca gris del Atlántico

H. g. grypus (Fabricius, 1791). North Atlantic Ocean and adjoining waters of the Arctic Ocean, from the Gulf of Saint Lawrence and Terranova Peninsula south through Massachusetts.

Conservation: *Halichoerus grypus* is listed as Least Concern by the IUCN (Bowen 2016).

Characteristics: Gray seal is medium sized; total length 2,000.0–2,700.0 mm in males and 1,600.0–2,200.0 mm in females; condylobasal length 285.0 a 325.0 mm in males and 295.0 mm in females. **Dorsal pelage black to gray-green, with irregular white spots, similar to W-shape;** fur of underparts paler; **rostrum straight from the frontal region to the nostrils, mainly with no angle**; males without a proboscis; no ears; hind flippers with well-developed fingernails; at least four foramina situated posterior to the maxillopalatine suture; **posterior palate margin broadly U shaped**; cheek-teeth smaller, with no accessory cusps, **second upper cheek-tooth largest;** space between auditory bullae similar to its breadth; three upper incisors.

Comments: Molecular data point to considering *Pusa* and *Halichoerus* as junior synonyms of *Phoca* (Fulton and Strobeck 2010; Nyakatura and Bininda-Emonds 2012). *Halichoerus grypus* has a confusing taxonomy; some authors have proposed the existence of three populations with no taxonomic level (Bonner 1981; Boskovic et al. 1996) or three subspecies: *H. g. grypus* in the western North Atlantic, *H. g. atlantica* in the eastern North Atlantic, and *H. g. baltica* in the Baltic Sea (Chapskii 1975), or only two subspecies: *H. g. grypus* in the western and eastern North Atlantic and *H. g. macrorynchus* in the Baltic Sea (Rice 1998). The Society for Marine Mammalogy recognizes only two subspecies. *H. grypus* ranges in cold temperature to subarctic North Atlantic waters over the continental shelf (Map 44). It can be found in sympatry, or nearly so, with *Cystophora cristata, Pagophilus groenlandicus,* and *Phoca vitulina*. *H. grypus* can be distinguished from all other species by its rostrum straight from the frontal region to the nostrils, mainly with no angle, with black spots on the back and flanks, and its general coloration in gray shades.

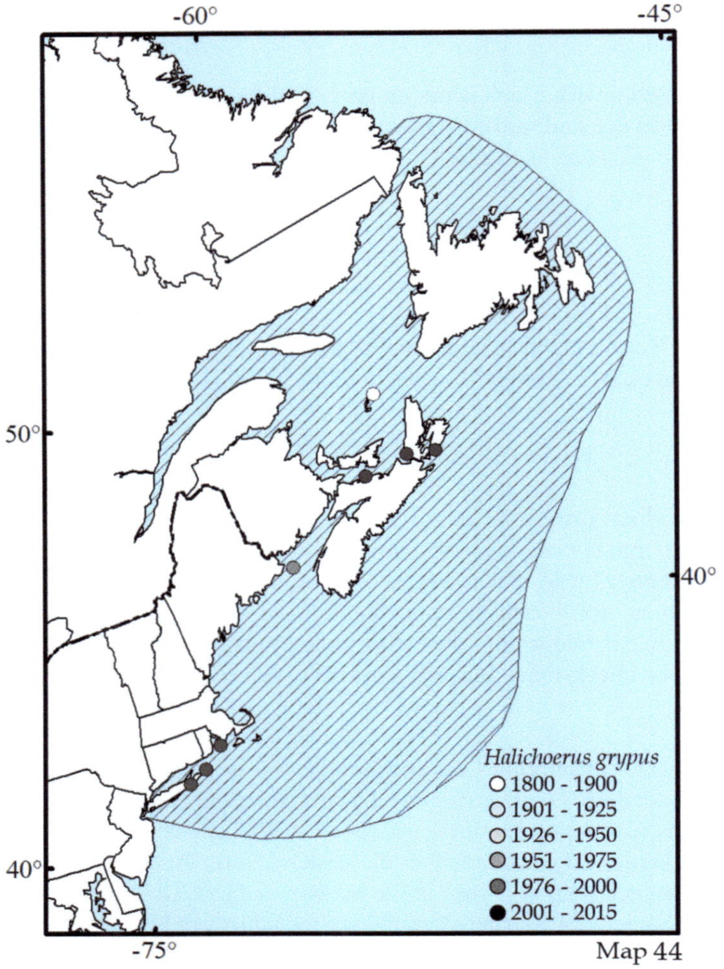

Map 44 Distribution map of *Halichoerus grypus*

Additional Literature: Rice (1998), Hall (2002).

Genus *Histriophoca* Gill, 1873

Body pelage color pattern differing between sexes, males dark brown with yellowish bands encircling the neck, rump and forelimbs; females grayish-yellow or grayish-brown with a faint indication of yellowish bands, the one encircling the rump most distinct; nasals narrow, tapering gradually to the posterior end; posterior margin of palate broadly U shaped; **auditory bullae large**; upper postcanine teeth small, usually lacking accessory cusps; posterior palatine foramina situated on, or posterior to, the maxillopalatine suture; mandible slender; mandibular symphysis long; masseteric fossa broad and deep; coronoid process large. *Histriophoca* has been considered a different genus (Scheffer 1958; Rice 1998).

Histriophoca fasciata (Zimmermann, 1783)
Ribbon Seal, foca listada

Monotypic.

Conservation: *Histriophoca fasciata* is listed as Least Concern by the IUCN (Lowry 2016a).

Characteristics: Ribbon seal is small sized; total length 1,500.0–1,600.0 mm; condylobasal length 195.0–223.0 mm. **Dorsal pelage brown with whitish ribbons around the head and flippers; fur of underparts with a similar pattern;** males without a proboscis; no ears; hind flippers with well-developed fingernails; posterior palatine foramina situated on or posterior to maxillopalatine suture; posterior palate margin broadly U shaped; **cheek-teeth large, with no accessory cusps**; third upper cheek-tooth largest; space between auditory bullae much greater than its breadth; three upper incisors.

Comments: *Histriophoca fasciata* ranges in the northern North Pacific Ocean and adjoining seas, but most commonly in the Bering and Okhotsk seas (Map 45). It can be found in sympatry, or nearly so, with *Callorhinus ursinus, Erignathus barbatus, Eumetopias jubatus, Mirounga angustirostris, Odobenus rosmarus, Phoca largha, P. vitulina,* and *Pusa hispida* and can be distinguished from these by its specific color pattern, with males dark brown with yellowish bands encircling the neck, rump, and forelimbs; females grayish-yellow or grayish-brown with faint indication of yellowish bands, the one encircling the rump most distinct.

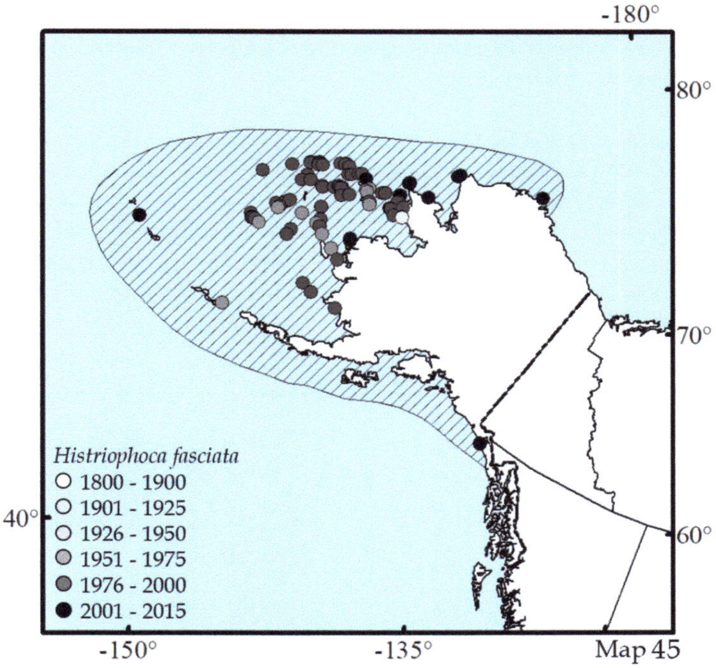

Map 45 Distribution map of *Histriophoca fasciata*

Additional Literature: Roest (1964), Burns and Fay (1972), Burns (1981b), Moore and Barrowclough (1984), Árnason et al. (1995), Higdon et al. (2007).

Genus *Mirounga* Gray, 1827

Body pelage dull yellowish-brown mixed with gray and darker on the back; fur of underparts yellowish; immature individuals grayish; **males with a proboscis-like inflatable snout; hind flippers bilobed; claws, if present, rudimentary**; palatine bones deeply emarginate on the posterior border; **auditory bullae concave anteriorly**; interbullar space narrow and deeply hollowed; superior notch of the mandible not deeply emarginate.

Mirounga angustirostris (Gill, 1866)

Northern Elephant Seal, elefante-marino del norte

Monotypic.

Conservation: *Mirounga angustirostris* is fully protected by the United State Marine Mammal Protection Act; it is listed as Threatened in the Norma Oficial Mexicana (DOF 2019) and as Low Concern by the IUCN (Hückstädt 2015).

Characteristics: Northern elephant seal is large sized; total length 3,600.0–4,200.0 mm in males and 2,200.0–3,000.0 mm in females; condylobasal length 480.0–510.0 mm in males and 310.0–345.0 mm in females. Dorsal pelage yellowish brown grizzly with gray, darker on the back; fur of underparts yellowish; males much bigger than females and **with a big proboscis; no ears; fingernails barely developed; nasal bones of small size**; posterior palatine foramen situated on or posterior to the maxillopalatine suture; posterior palate margin broadly U shaped; cheek-teeth small, with no accessory cusps, all of similar size; space between auditory bullae similar to its breadth; **two upper incisors on each side, inner incisors of similar size as outer incisors**.

Comments: *Mirounga angustirostris* ranges in the Pacific Ocean, from Baja California Sur north through the Aleutian Islands, Alaska, and can be found in pelagic areas as in the coast (Map 46). It can be found in sympatry, or nearly so, with *Arctocephalus townsendi, Callorhinus ursinus, Halichoerus grypus, Phoca largha, P. vitulina, Pusa hispida,* and *Zalophus californianus* and can be distinguished from these species by males having a lengthened snout; males and females greater than 3000.0 mm; skull with a wide square rostrum. From *A. townsendi* and *H. grypus,* by the overlapping size between the females of *Mirounga* with the males of the other two species and by not having ears; hind limbs directed backwards and not functional for walking.

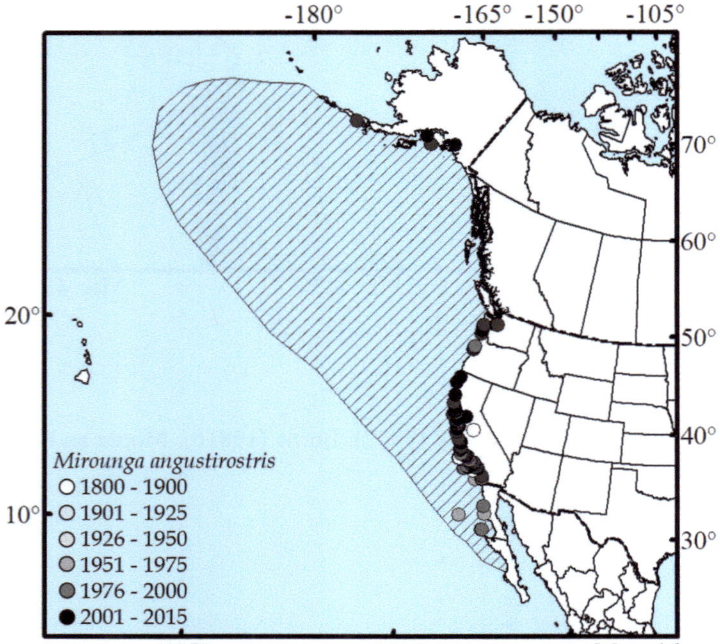

Map 46 Distribution map of *Mirounga angustirostris*

Additional Literature: Rice et al. (1965), King (1966), Briggs and Morejohn (1976), Ling and Bryden (1992), Stewart and Huber (1993), Le Boeuf et al. (2000), Robinson et al. (2012).

Genus *Neomonachus* Slater et al., 2014

West Indian Seal have the dorsal body pelage brownish-gray, without spots; fur of underparts white or yellowish-gray; **muzzle broad and slightly grooved between the nostrils; vibrissae short and not dense; nails well developed on the fore flippers and rudimentary on the hind flippers**; digits of hind flippers shorter proximally; **rostrum elongate and depressed**; palate notched posteriorly; canines large and conical; teeth posterior to canines conical, hollow, **with a strongly lobed cingulum on the proximal side and oblique relative to the median palatine line; all double-rooted, except for the two first.** *Neomonachus tropicalis* was considered in the genus *Monachus* (Rice 1998), but molecular and morphological analyses show differences from the Mediterranean monk seal (*Monachus monachus*); for this reason, *Neomonachus* is a full genus (Barnes et al. 1985; Scheel et al. 2014).

Neomonachus tropicalis (Gray, 1850)†
Caribbean Monk Seal, foca monje del Caribe

Monotypic.

Conservation: *Neomonachus tropicalis* (as *Monachus tropicalis*) is listed as Extinct in the Norma Oficial Mexicana (DOF 2019) and by the IUCN; has not been spotted since 1952 (Adam 2004).

Characteristics: Caribbean monk seal is small sized; total length 1,200.00–1,400.00 mm; condylobasal length 270.0–290.0 mm. Dorsal pelage grayish brown without spots; fur of underparts yellowish white; no strong difference in size between sexes; males without a proboscis; no ears; hind flippers with **fingernails barely developed;** posterior palatine foramina situated on or posterior to the maxillopalatine suture; posterior palate margin broadly V shaped; larger cheek-teeth, with accessory cusps, except for the tooth near the canines; space between auditory bullae much greater than its breadth; two upper incisors.

Comments: *Neomonachus tropicalis* ranges in the Caribbean Sea and Gulf of México south trough Honduras (Map 47).

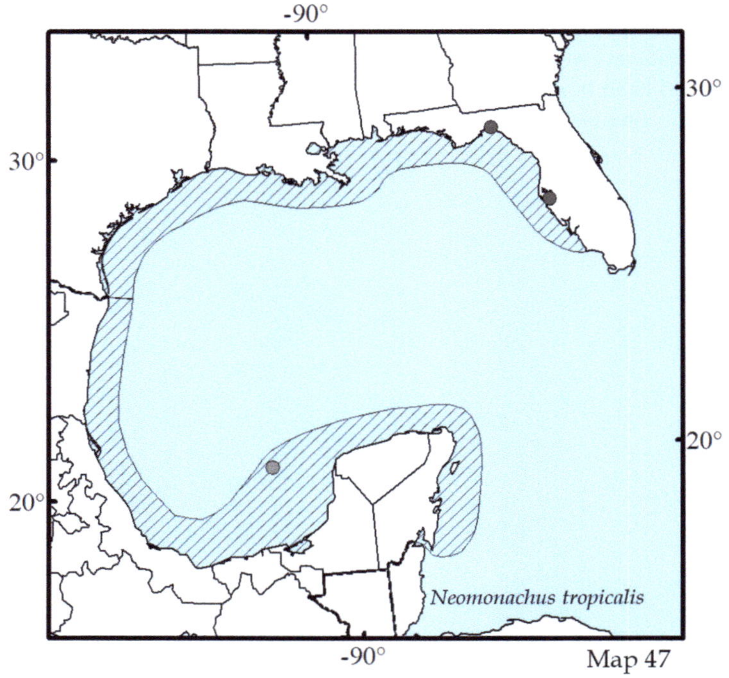

Map 47 Distribution map of *Neomonachus tropicalis*

Additional Literature: Allen (1942), Moore (1953), King (1956), Kenyon (1977), Le Boeuf et al. (1986), Allen (2004).

Genus *Pagophilus* Gray, 1844

Harp seal has a color pattern differing between sexes, **males with the back whitish to yellowish-white, face and head blackish, and a black elliptical band on the flanks, from the shoulders almost to the tail, lateral bands meeting broadly at the shoulders and barely meeting near the tail; hind limbs generally irregularly spotted with black. Females of similar color patterns as males but black markings absent, or, if present, irregular and indistinct**; nasals long, narrow, and tapering gradually to the posterior end; posterior margin of palate broadly U shaped; **teeth posterior to canines small, with accessory cusps, and the third upper and lower teeth largest;** posterior palatine foramina on, or posterior to, the maxillopalatine suture.

Pagophilus groenlandicus (Erxleben, 1777)

Harp Seal, foca de arpa

P. g. groenlandicus (Erxleben, 1777). North Atlantic Ocean and adjoining waters of the Arctic Ocean, from the Labrador Peninsula northeast through Greenland.

Conservation: *Pagophilus groenlandicus* is listed as Least Concern by the IUCN (Kovacs 2015).

Characteristics: Harp seal is small sized; total length 1,700.00–1,900.00 mm; condylobasal length 205.0–234.4 mm. **Dorsal pelage white with a black lyre-shaped mark on the back; face black; fur of underparts white;** males without a proboscis; no ears; hind flippers with well-developed fingernails; posterior palatine foramina situated on or posterior to the maxillopalatine suture; posterior palate margin broadly U shaped; cheek-teeth smaller, with accessory cusps; third upper cheek-tooth largest; space between auditory bullae much greater than its breadth; three upper incisors.

Comments: Harp seal has been included in the genera *Pagophoca, Pagophilus*, and *Phoca*. At species level between *P. groenlandicus* and *P. groenlandica*. The current name recognized by the Society of Marine Mammalogy (Committee on Taxonomy 2014) is *Pagophilus groenlandicus* (Berta and Churchill 2011). Two subspecies are recognized, *P. g. groenlandicus* and *P. g. oceanicus* (Rice 1998; Heptner et al. 1996), but three distinct populations are recognized (Lavigne 2002) and probably at subspecies level (Meisfjord and Sundt 1996; Perry et al. 2000). *P. groenlandicus* ranges in North Atlantic and Arctic Ocean, from Hudson Bay to the Gulf of Maine (Map 48). It can be found in sympatry, or nearly so, with *Cystophora cristata, Halichoerus grypus, Odobenus rosmarus, Phoca vitulina*, and *Pusa hispida. P. groenlandicus* can be distinguished from these species by its face and head blackish and a black elliptical band on the flanks, from the shoulders almost to the tail, the bands from each flank meeting broadly at the shoulder, and barely meeting near the tail.

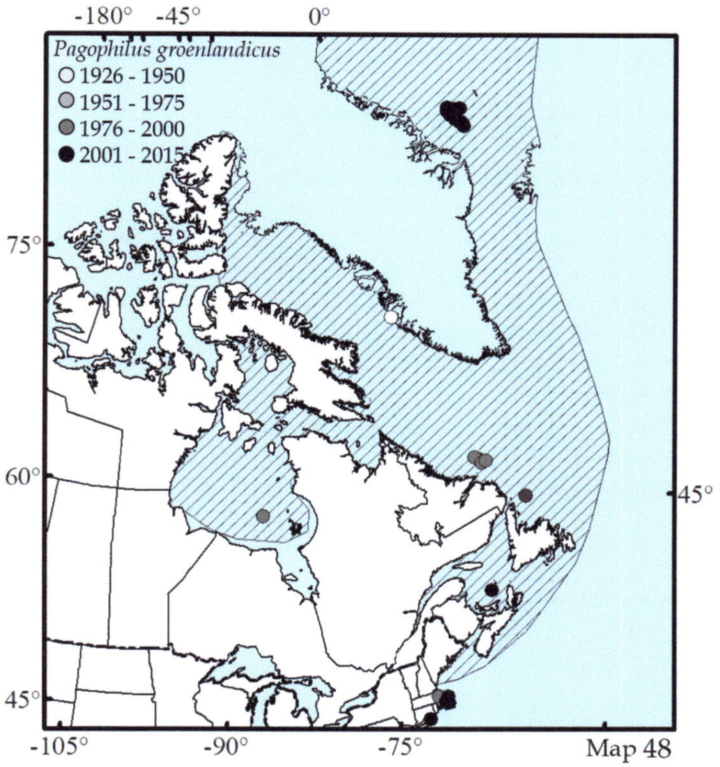

Map 48 Distribution map of *Pagophilus groenlandicus*

Additional Literature: Goodwin (1954), Lacoste and Stenson (2000), Harris et al. (2002), Committee on Taxonomy (2014).

Genus *Phoca* Linnaeus, 1758

Fur of underparts paler; no strong contrast in size between sexes; males without a proboscis; no ears; hind flippers directed backwards, not functional for terrestrial displacement; **well-developed fingernails;** posterior palatine foramina situated anterior to the maxillopalatine suture; posterior margin of the palate broadly V shaped; cheek-teeth large, with no accessory cusps; **third upper cheek-tooth largest; three upper incisors;** space between auditory bullae much greater than its breadth. The two species are hard to differentiate.

1. Grayish with dark splotches; on average, skull length greater than 205.0 mm in males and greater than 190.0 in females (western and northern Alaska, United States).. *Phoca largha* (p. 167).

1a. Grayish with irregular pale rings; on average, skull length less than 210.0 mm in males and less than 195.0 mm in females (from southern Alaska, United States, south through the Baja California Peninsula, México, and from Hudson Bay, Canada, south through Georgia, United States)...*Phoca vitulina* (p. 168)

Phoca largha Pallas, 1811
Spotted Seal, foca moteada

Monotypic.

Conservation: *Phoca largha* is listed as Least Concern by the IUCN (Boveng 2016) and as protected by the United States Marine Mammal Protection Act.

Characteristics: Spotted seal is small sized; total length 1,400.0–1,700.0 mm; condylobasal length 205.6–245.0 mm in males and 190.0–210.0 mm in females. Hard to differentiate from *P. vitulina.*

Comments: *Phoca largha* was recognized as a distinct species (Shaughessy and Fay 1977; King 1983a, b; Rice 1998), previously regarded as a subspecies of *P. vitulina* (Scheffer 1958). Genetic analyses show the possible population structure (Li et al. 2010). *P. largha* ranges in the North Pacific and Arctic Oceans, from the delta of the Mackenzie River to Bering Sea (Map 49). It can be found in sympatry, or nearly so, with *Callorhinus ursinus, Erignathus barbatus, Eumetopias jubatus, Erignathus barbatus, Histriophoca fasciata, Mirounga angustirostris, Odobenus rosmarus, Phoca vitulina,* and *Pusa hispida. P. largha* can be distinguished from *C. ursinus, E. barbatus, E. jubatus, M. angustirostris,* and *O. rosmarus* by its smaller size, less than 1,700.00 mm. From *C. ursinus* and *E. jubatus*, by the absence of small ears; hind flippers directing backward and not functional for walking. From *O. rosmarus,* by its smaller size; hair covering all the body, not naked; upper canines not large and elongate as tusks in both sexes. From *M. angustirostris, by* not having a lengthened snout in males; skull triangular and with a narrow rostrum. From *H. fasciata,* by not having the head contrasting with the neck; pattern of a paler wide band along and around the body. From *P. vitulina,* it is not easy to differentiate, the two characteristics that can be used are the restricted range of *P. largha* in the Arctic Ocean and Bering Sea, and its grayish coloration with dark splotches, while in *P. vitulina* the color pattern is highly variable. From *P. hispida*, by having a well defined neck and a snout proportional to the head size.

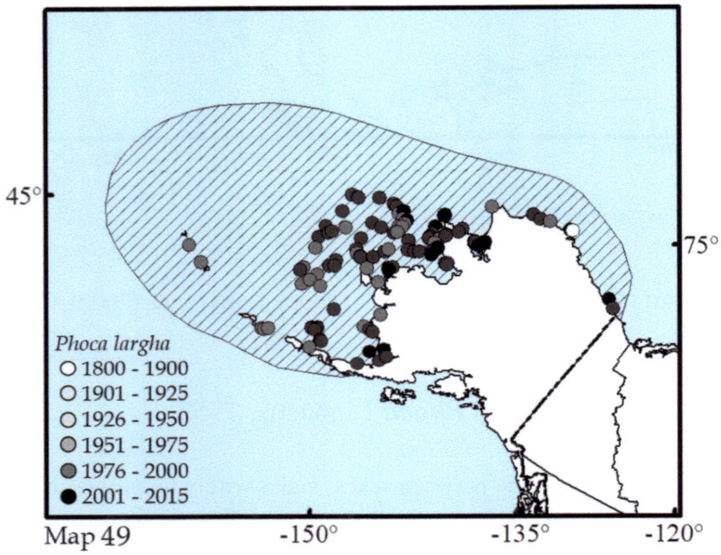

Map 49 Distribution map of *Phoca largha*

Additional Literature: Quakenbush (1988), O'Corry-Crowe and Bonin (2009).

Phoca vitulina Linnaeus, 1758

Harbor Seal, foca común

1. *P. v. concolor* De Kay, 1842. Atlantic coast, from Ellesmere Island, Nunavut south through Virginia.
2. *P. v. mellonae* Doutt, 1942. Atlantic coast, Restricted to the Ungava Peninsula, Québec.
3. *P. v. richardii* (Gray, 1864). Arctic and Pacific coasts, from the Aleutian Islands south through northern Baja California.

Conservation: *Phoca vitulina richardii* is listed as subject to Special Protection in the Norma Oficial Mexicana (DOF 2019), as Protected by the United States Marine Mammal Protection Act, and as Least Concern by the IUCN (Lowry 2016c).

Characteristics: Harbor seal is small sized; total length 1,400.0–1,900.0 mm; condylobasal length 192.6–210.0 mm in males and 180.0–195.0 mm in females. Hard to differentiate from *P. largha.*

Comments: *Phoca largha* was previously considered a subspecies of *P. vitulina* (Scheffer 1958). Of the three commonly recognized subspecies of *Phoca vitulina* for North America (Rice 1998), *P. v. concolor* was previously considered a junior synonym of *P. v. mellonae* (Berta and Churchill 2011). *P. vitulina* ranges in the Pacific and Atlantic oceans, from the Baja California Peninsula north through Alaska, including the Aleutian Islands, and from Georgia to the Melvile Peninsula and Hudson Bay (Map 50). *P. vitulina* can be found in sympatry, or nearly so, with *Callorhinus ursinus, Cystophora cristata, Erignathus barbatus, Eumetopias jubatus, Halichoerus grypus, Histriophoca fasciata, Mirounga angustirostris, Odobenus rosmarus, Pagophilus groenlandicus, Phoca largha, Pusa hispida,* and *Zalophus californianus. P. vitulina* can be distinguished from *C. ursinus, C. cristata, E. barbatus, E. jubatus, H. grypus, M. angustirostris, O. rosmarus,* and *Z. californianus* by its smaller size, less than 1900.0 mm. From *C. cristata*, by its males not having a hood on the head top that can be inflated, and a pouch of nose skin like a red balloon that can be inflated too. From *C. ursinus* and *E. jubatus*, by the absence of ears; hind flippers directed frontward and functional for walking. From *O. rosmarus,* by its smaller size with hair covering all the body, not naked, and upper canines not large and elongated as tusks in both sexes. From *M. angustirostris*, by its males without a lengthened snout and the skull triangular with a narrow rostrum. From *H. grypus*, by having a clear snout, with a distinctive angle between the frontal region and the nostrils. From *H. fasciata*, by not having the head color contrasting with the neck and a pattern of a paler wide band along and around the body. From *P. groenlandicus*, by not having a black elliptical band on the flanks from the shoulders almost to the tail, with the bands from each side meeting broadly at the shoulders and barely meeting near the tail. From *P. largha*, it is not easy to differentiate; the two characteristics that can be used are the restricted range to the Arctic Ocean and Bering Sea, and the grayish coloration with dark splotches. From *P. hispida*, by having a well-defined neck and the snout proportional to the head size.

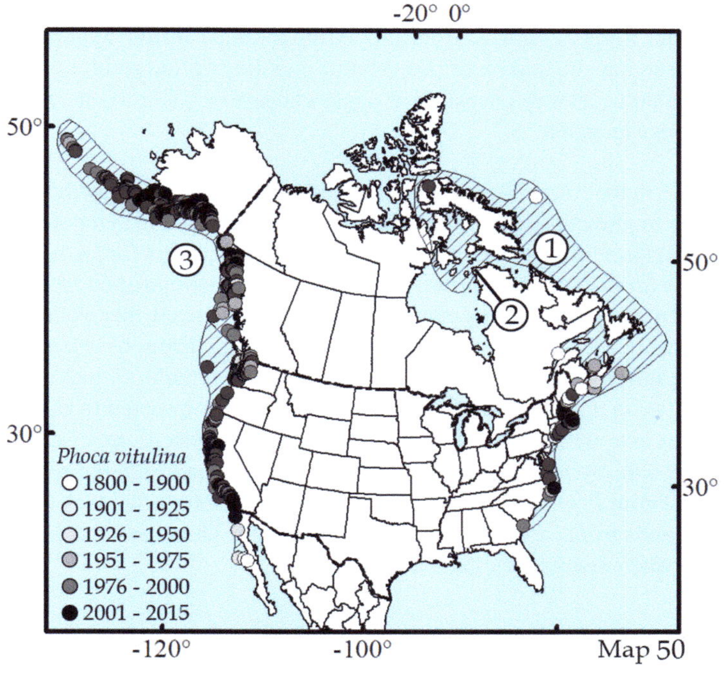

Map 50 Distribution map of *Phoca vitulina*

Additional Literature: Bartholomew and Boolootian (1960), Kenyon (1960), Caldwell (1961, 1969), Harper (1961), Edwards (1963).

Genus *Pusa* Scopoli, 1771

Ringed seal varies widely in color from white to silvery-white to yellowish; fur of underparts whitish to yellowish; **with a pattern of spots, streaked or marbled with black dorsally, sometimes having white spots with dark centers**; often patches of tawny or russet hairs around the lips, maxillae, and below the tail and hind limbs; **posterior margin of the palate V shaped; nasal septum not reaching the posterior margin of the palate; teeth posterior to the canines small, third upper and lower teeth largest**; posterior palatine foramina situated on, or posterior to, the maxillopalatine suture. *P. hispida* has been considered within the genera *Pusa* and *Phoca* (Rice 1998). Molecular data point to considering *Pusa* and *Halichoerus* as junior synonyms of *Phoca* (Fulton and Strobeck 2010).

Pusa hispida (Schreber, 1775)

Ringed Seal, foca anillada

P. h. hispida (Schreber, 1775). All the Artic Ocean of North America.

Conservation: *Pusa hispida* is listed as protected by the United States Marine Mammal Protection Act and can be harvested by Alaskan native hunters for subsistence purposes and is listed as Least Concern by the IUCN (Lowry 2016d).

Characteristics: Ringed seal is small sized; total length 1,500.0–1,600.0 mm; condylobasal length 150.1–175.3 mm. **Dorsal pelage with pale rings on a black background; fur of underparts paler; almost no neck and face short;** males without a proboscis; no ears; hind flippers with well-developed fingernails; **upper zygomatic angles with plate-shaped expansion;** posterior palatine foramina situated on or posterior to maxillopalatine suture; posterior palate margin broadly V shaped; cheek-teeth smaller with no accessory cusps; third upper cheek-tooth largest; space between auditory bullae much greater than its breadth; three upper incisors.

Comments: *Pusa hispida* ranges in the Arctic Ocean with a circumpolar distribution, with the southern distribution in western Alaska and northwestern Newfoundland (Rice 1998; Map 51). It can be found in sympatry, or nearly so, with *Callorhinus ursinus, Cystophora cristata, Erignathus barbatus, Eumetopias jubatus, Histriophoca fasciata, Odobenus rosmarus, Pagophilus groenlandicus, Phoca largha,* and *P. vitulina. P. hispida* can be distinguished from all these species by not having a well-defined neck, seen as absent, and snout very small proportional to the head size. From *C. ursinus, C. cristata, E. barbatus, E. jubatus,* and *O. rosmarus,* by its smaller size, less than 1700.0 mm. From *C. cristata*, by the absence in males of a hood on the head top that can be inflated, and absence of a pouch of skin on the nose like a red balloon that can be inflated too. From *C. ursinus* and *E. jubatus,* in the absence of small ears; hind flippers directed frontward and functional for walking. From *O. rosmarus,* by its smaller size with hair covering all the body, not naked; upper canines not large and elongate as tusks in both sexes. From *H. fasciata,* by not having the head color contrasting with the neck and a pattern of a paler wide band along and around the body. From *P. groenlandicus*, by not having the color of a black elliptical band on the flanks from the shoulders almost to the tail, bands from each side meeting broadly at the shoulders, and barely meeting near the tail. From *P. largha* and *P. vitulina,* see the main distinctive characteristics.

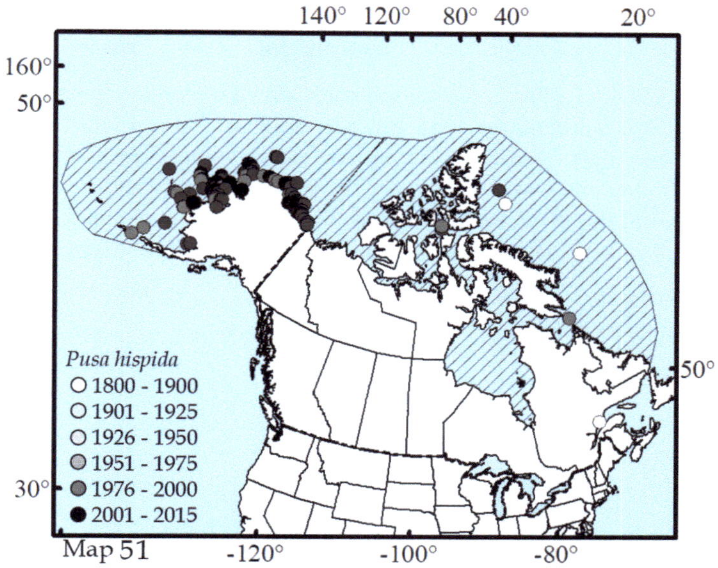

Map 51 Distribution map of *Pusa hispida*

Additional Literature: Nyakatura and Bininda-Emonds (2012).

Family Procyonidae Gray, 1825

The family Procyonidae includes raccoons, cacomixtles, and coatis, among others, characterized by their medium size (600.0–1350.0 mm); **tail long, generally ringed with alternating black and pale colored stripes**; limb posture plantigrade or semiplantigrade; claws prominent and nonretractile, some species semiretractile; **mastoid process equal to or larger than the paroccipital process; carnassial poorly developed** (except for *Bassariscus*); **molars broad; upper molars relatively large and rounded**, generally with hypocones. Distributed only in America. No subfamilies are recognized in this family. The keys were elaborated based on the review of specimens and the following sources: Hollister (1921); Lawlor (1979), Hall (1981), Jones and Manning (1992), and Álvarez-Castañeda et al. (2017a).

1. Rostrum less than 45.0 mm; molars mainly flat; canines with a groove 2

1a. Rostrum greater than 45.0 mm; molars with a well defined crest; canines without a groove 3

2. Tail old-gold colored, never marked with rings, prehensile; breadth of first upper molars less than 135% of its length (from Tamaulipas and Michoacán south through South America) .... *Potos flavus* (p. 175)

2a. Tail very clearly marked with alternating black and white rings; never prehensile; breadth of first upper molars greater than 135% of its length .... *Bassariscus* (p. 172)

3. Tail very long greater than 50% of the head-and-body length; rhinarium extended; rostrum extended and laterally compressed; rostrum breadth at the canine level less than the interorbital breadth; diastema between the canines and the first upper premolars (from Arizona and New Mexico south through South America, except for Altiplano Mexicano and Baja California Peninsula, México) .... *Nasua narica* (p. 176)

3a. Tail short, approximately 40% of the head-and-body length; rhinarium not extended; rostrum not laterally compressed; rostrum breadth at the canine level greater than the interorbital breadth; no diastema between the canines and the first upper premolars .... *Procyon* (p. 178)

Genus *Bassariscus* Coues, 1887

Dorsal pelage grayish to grayish brown, darker in the central back; fur of underparts white to pale yellow; body extended; rostrum black with marked white spots around the eyes; **tail very long and clearly marked with alternating black and white rings, never prehensile;** three upper and lower premolars; arched molariform line; if the following condition is not met, first molar almost as wide as long and diastema between the premolars and rostrum breadth at the premolars level less than canine breadth; **first and second upper incisors bicuspid**. The keys were elaborated based on the review of specimens and the following sources: Hall (1981), Poglayen-Neuwall and Toweill (1988), and Álvarez-Castañeda et al. (2017a).

1. Claws not retractile; ears sharp; upper carnassial triangular (Fig. 31); first and second upper incisors bicuspid (from Guerrero and Veracruz south through Central America)..*Bassariscus sumichrasti* (p. 174)

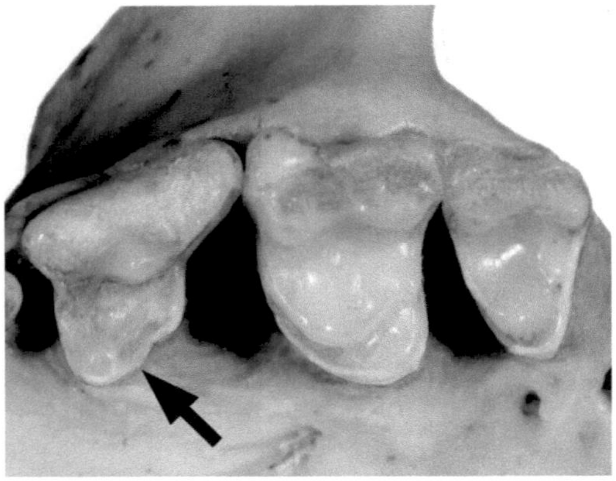

Fig. 31 Upper carnassial triangular like in *Bassariscus sumichrasti*

1a. Claws semi-retractile; ears rounded; upper carnassial irregular (Fig. 32); first and second upper incisors unicuspid (from Oregon east through Missouri and Louisiana, United States, south through to Oaxaca, México)...*Bassariscus astutus* (p. 173)

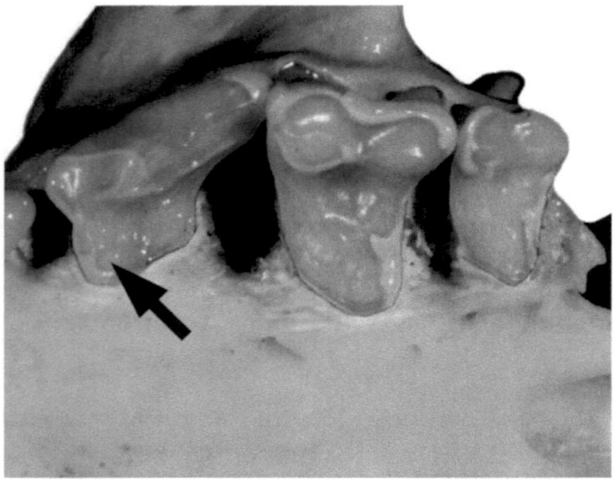

Fig. 32 Upper carnassial irregular like in *Bassariscus astutus*

Bassariscus astutus (Lichtenstein, 1830)

Ringtail Cat, cacomixtle

1. *B. a. arizonensis* Goldman, 1932. Southern Wyoming, eastern Utah, western Colorado, New Mexico, and Arizona.
2. *B. a. astutus* (Lichtenstein, 1830). From Tamaulipas and Zacatecas south through northern Oaxaca.
3. *B. a. bolei* Goldman, 1945. From Estado de México and Guerrero east through Oaxaca.
4. *B. a. consitus* Nelson and Goldman, 1932. From Sinaloa south through Estado de México and Guerrero.
5. *B. a. flavus* Rhoads, 1894. From Colorado, Kansas, Arkansas, and Louisiana south through Sonora to Tamaulipas.
6. *B. a. insulicola* Nelson and Goldman, 1909. Restricted to San José Island, Baja California Sur.
7. *B. a. macdougalli* Goodwin, 1956. Restricted to the central-eastern Oaxaca.
8. *B. a. nevadensis* Miller, 1913. Nevada, eastern California, and central-western Utah.
9. *B. a. octavus* Hall, 1926. Restricted to southern California.
10. *B. a. palmarius* Nelson and Goldman, 1909. Baja California Peninsula.
11. *B. a. raptor* (Baird, 1859). Southern Oregon and central-southern California.
12. *B. a. saxicola* Merriam, 1897. Restricted to Espíritu Santo Island, Baja California Sur.
13. *B. a. willetti* Stager, 1950. Restricted to southeastern California.
14. *B. a. yumanensis* Huey, 1937. Restricted to western Arizona.

Conservation: *Bassariscus astutus insulicola* and *B. a. saxicola* are listed as Threatened in the Norma Oficial Mexicana (DOF 2019) and *B. astutus* as Least Concern by the IUCN (Reid et al. 2016e).

Characteristics: Ringtail cat is small sized; total length 616.0–811.0 mm; condylobasal length 81.0–85.2 mm in males and 76.4–78.1 mm in females. **Dorsal pelage grayish;** fur of underparts cream to white; **claws semi-retractile**; **ears rounded; first and second upper incisors unicuspid; upper carnassial irregular**.

Comments: *Bassariscus astutus* is widely distributed from southern Oregon and California throughout the south-western states to Missouri and Louisiana and south through Oaxaca and central-eastern Veracruz, including the Tiburón, Espíritu Santo, and San José Islands in the Gulf of California (Map 52). It can be found in sympatry, or nearly so, with *Procyon lotor,* from which it can be distinguished by its dorsal pelage grayish to grayish brown; tail very long, similar to the body length and clearly marked with alternating black and white rings; rhinarium not extended or mobile, and from *B. sumichrasti,* by having the dorsal pelage grayish, semi-retractile claws, and ears rounded.

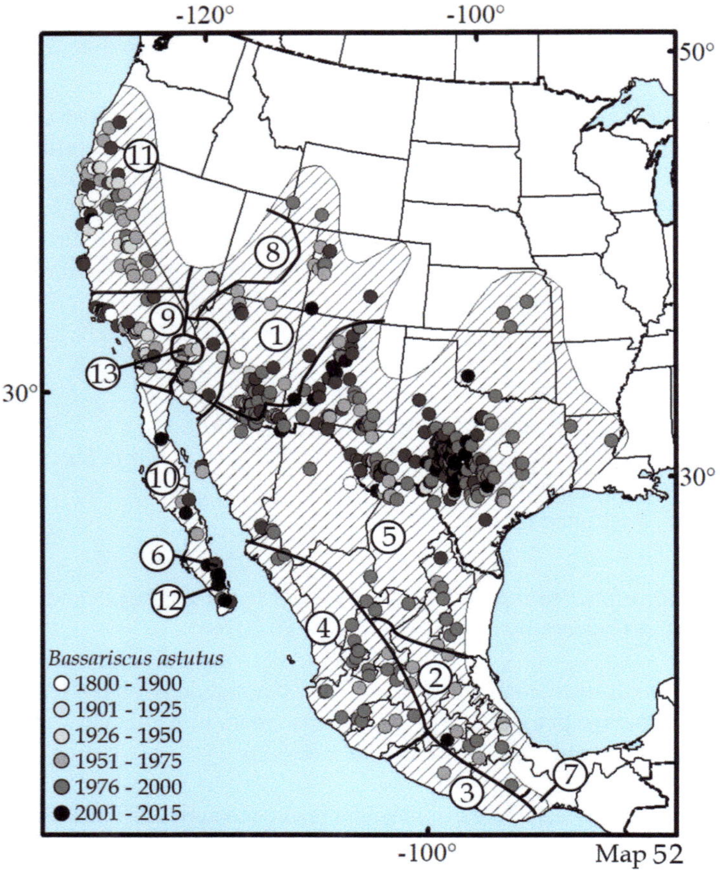

Map 52 Distribution map of *Bassariscus astutus*

Additional Literature: Nelson and Goldman (1932), Grinnell et al. (1937a, b), Burt (1960), Hall (1981), Poglayen-Neuwall and Toweill (1988, 1989), Wozencraft (1989, 2005).

Bassariscus sumichrasti (Saussure, 1860)

Cacomixtle, cacomixtle tropical

1. *B. s. campechensis* (Nelson and Goldman, 1909). From Campeche east through Central America, including the whole Yucatán Peninsula.
2. *B. s. latrans* (Davis and Lukens, 1958). Restricted to Guerrero.
3. *B. s. oaxacensis* (Goodwin, 1956). Southwestern Oaxaca and northeastern Chiapas.
4. *B. s. sumichrasti* (Saussure, 1860). Veracruz and Oaxaca.
5. *B. s. variabilis* (Peters, 1874). Form Chiapas south through Nicaragua.

Conservation: *Bassariscus sumichrasti* is listed as subject to Special Protection in the Official Mexican Standard (DOF 2019) and as Least Concern by the IUCN (Pino et al. 2020).

Characteristics: Cacomixtle is small sized; total length 790.0–1003.0 mm, condylobasal length 83.0–84.7 mm. **Dorsal pelage grayish brown;** fur of underparts pale yellow; **claws not retractile; ears sharp, ears wide with a white border; first and second upper incisors bicuspid; upper carnassial triangular**.

Comments: *Bassariscus sumichrasti* ranges from Guerrero and southern Veracruz south through Central America (Map 53). It can be found in sympatry, or nearly so, with *Potos flavus, Procyon lotor,* and *Nasua narica. B. sumichrasti* can be distinguished from all other species of Procyonidae by its dorsal pelage grayish to grayish brown; tail very long, similar to the head-and-body length and clearly marked with alternating black and white rings; rhinarium not extended or mobile. From *B. astutus*, by having dorsal pelage brownish, claws not retractile, and ears sharp.

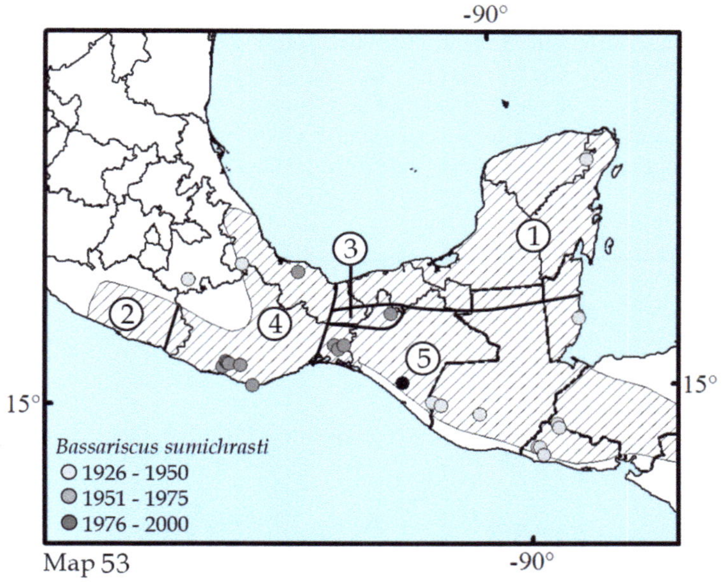

Map 53 Distribution map of *Bassariscus sumichrasti*

Additional Literature: Nelson and Goldman (1932), Burt (1960), Hall (1981), Poglayen-Neuwall (1989), Emmons and Freer (1990), Nowak (2005), Wozencraft (2005).

Genus ***Potos*** Geoffroy Saint-Hilaire and Cuvier, 1795

Kinkajou has the pelage **thick, soft, and woolly;** upperparts from tawny olive, yellowish tawny, clay color, to wood brown, with a black mid-dorsal stripe in some individuals; fur of underparts from deep fulvous yellow, buff to brownish yellow; muzzle dark brown to blackish; **head rounded and muzzle pointed; tail long and prehensile; ears short**; claws short and sharply pointed; digits partially webbed; limbs short; **two ventral cutaneous glands; skull short, rounded, and highly arched; rostrum short and broad**; braincase large, sharply constricted in the postorbital region; **postorbital processes well developed from frontals, slightly from the zygomatic process**; sagittal crest usually developed on the forward half of the braincase; palate flat, short, and extending only a short distance posterior to the plane of the maxillary tuberosities; auditory bullae markedly flattened; **mandible heavy, ramus high, straight, and slightly produced backward; teeth large; a short diastema between the incisors and canines**; incisors large, upper third larger; **canines large, flattened, emarginate lingually, deeply grooved labially, nearly straight**; premolars large; first upper molars twice the size of the second molars, subquadrate, and flat with a raised outer rim.

Potos flavus (Schreber, 1774)

Kinkajou, martucha

1. *P. f. chiriquensis* Allen, 1904b. From the Tehuantepec Isthmus south through Chiapas, including the Yucatán Peninsula.
2. *P. f. nocturnus* Wied-Neuwied, 1826. From Veracruz and Michoacán south through the Tehuantepec isthmus.

Conservation: *Potos flavus* is listed as subject to Special Protection in the Norma Oficial Mexicana (DOF 2019) and as Least Concern by the IUCN (Helgen et al. 2016a, b, c).

Characteristics: Kinkajou is medium sized; total length 845.0–1300.0 mm; condylobasal length 79.1 a 95.1 mm. **Dorsal pelage homogeneous from intense gold to pale and abdomen more ochre; tail prehensile of the same color as the back, without rings, prehensile, almost as long as the head-and-body length;** ears rounded; three upper and lower premolars; parallel molariform line; jaw very wide in the articular process region; the following conditions not met at the same time in a given specimen; first molars almost as wide as long; no diastema between the premolars; rostrum wider at the premolar level than at the canine level.

Comments: *Potos flavus* ranges throughout the Neotropics, from the east and south of the Sierra Madre, from southeastern Tamaulipas, southeastern San Luis Potosí, and Michoacán south through the lowlands to South America (Map 54). It can be found in sympatry, or nearly so, with *Bassariscus sumichrasti, Procyon lotor,* and *Nasua narica* and can be distinguished from all the species of Procyonidae by its homogeneous dorsal coloration varying from intense to pale gold, with the abdomen more ochre; tail prehensile, almost as long as the head-and-body length and with the same color as the back, without dark-colored rings; skull short, rounded, highly arched, and rostrum short and broad.

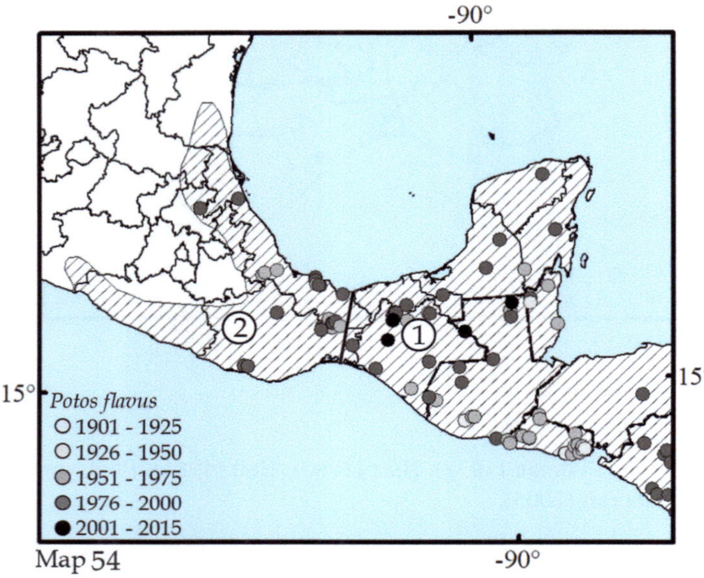

Map 54 Distribution map of *Potos flavus*

Additional Literature: Hollister (1915), Pocock (1921, 1928), Leopold (1959), Grzimek (1975), Hernández-Camacho (1977), Husson (1978), Hall (1981), Nowak and Paradiso (1983), Ford and Hoffmann (1988), Wozencraft (2005).

Genus *Nasua* Storr, 1780

Coati has the pelage coloration from pale brown to reddish to almost black, which can be overlaid with whitish, yellow, or silver shades; **neck and shoulders yellow to white; muzzle, chin, and throat whitish; eyes masked with pale umber to brown; thin whitish streaks extending from the muzzle between and over the eyes, and with whitish to yellowish post-auricular patches**; **ears short, rounded, and white at the tip**; fur of underparts from yellowish to dark brown; feet blackish; tail longer than the body, no prehensile, ring pattern often yellow and brown, sometimes scarcely visible; head long with the rostral portion elongated; **snout upturned, mobile**; forefeet claws long, slightly curved; hindfeet claws strong, sharp, and markedly curved; **skull long and narrow; rostrum elongated and compressed laterally; interorbital constriction slight; sagittal crest well developed;** palate flat or hollow posteriorly, with distinct lateral grooves, and extending far beyond the plane of the last molars; palatine bones deeply notched on each side, with long processes extending laterally to the maxillary tuberosities; vomer usually attached to the palatal bones; foramen oval, small; auditory bullae small, smoothly rounded, rising at a sharp angle from the tube of the external auditory meatus; teeth medium sized; **crowns relatively high, with sharp and strong cusps**; last upper premolars longer than wide; **first upper molars subquadrate, length and breadth about equal to or longer than wide**.

Nasua narica (Linnaeus, 1766)
White-Nosed Coati, coatí

1. *N. n. molaris* Merriam, 1902. From Arizona and southern Texas south through Veracruz and Oaxaca.
2. *N. n. narica* (Linnaeus, 1766). From Veracruz and Oaxaca south through Chiapas.
3. *N. n. nelsoni* Merriam, 1901. Restricted to Cozumel Island, Quintana Roo.
4. *N. n. yucatanica* Allen, 1904. Yucatán Peninsula.

Conservation: *Nasua narica nelsoni* is listed as Threatened in the Norma Oficial Mexicana (DOF 2019) and *N. narica* is listed as Least Concern by the IUCN (Cuarón et al. 2016h).

Characteristics: White-nosed coati is medium sized; total length 850.0–1340.0 mm, condylobasal length 100.8–128.9 mm. **Dorsal pelage from reddish orange, brown to gray; males with grizzly cream hair; fur of underparts paler than in the back; white spots above the eyes and the snout lower part; tail very long, greater than 50% of the head-and-body length, with or without rings colored** in paler shades and commonly displaced with the tail in vertical position; **rhinarium extended and mobile;** four upper and lower premolars; arched molariform line; jaws in normal proportion throughout its extension; first molars almost as wide as long; diastema between the premolars and rostrum breadth at the premolar height less than the breadth at the height of canines.

Comments: *Nasua nelsoni* is currently regarded as a subspecies of *N. narica* (Decker and Wozencraft 1991). *Nasua narica* ranges from Arizona and southern New Mexico south through to Central America, except for the Baja California Peninsula and the Altiplano Mexicano (Map 55). It can be found in sympatry, or nearly so, with *Bassariscus sumichrasti, Potos flavus, Procyon lotor,* and *P. pygmaeus* and can be distinguished from these species by having the dorsal pelage from reddish orange brown to gray, with spots above the eyes; lower part of the snout white, and rhinarium extended and mobile.

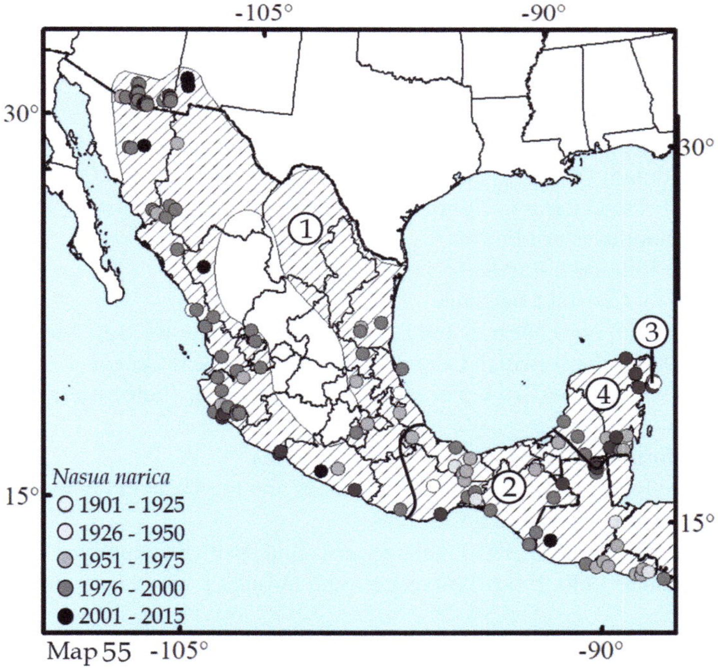

Map 55 Distribution map of *Nasua narica*

Additional Literature: Hollister (1915), Pocock (1921), Hershkovitz (1951), Eisenberg (1989), Decker and Wozencraft (1991), Gompper (1995), Gompper and Decker (1998), Wozencraft (2005).

Genus *Procyon* Storr, 1780

Dorsal pelage speckled, from grayish to brown; **rostrum with a marked black mask;** central rostrum dark, including the rhinarium; snout with whitish sides; fur of underparts yellowish; **tail short but ringed from dark to whitish shades;** four upper and lower premolars; arched molariform line; if the following condition is not met, first molars almost as wide as long and diastema between the premolars and rostrum breadth at the premolar level less than the breadth at the canine level. The keys were elaborated based on the review of specimens and the following sources: Goldman (1950), Hall (1981), Villa-Meza et al. (2011), and Álvarez-Castañeda et al. (2017a).

1. Total length less than 670.0 mm; diastema between the second and third upper premolars; skull length less than 100.0 mm; upper molar length less than 36.0 mm (Isla Cozumel, México)............................*Procyon pygmaeus* (p. 179)

1a. Total length generally greater than 670.0 mm; no diastema between the first and third upper premolars; skull length greater than 100.0 mm; upper molar length greater than 36.0 mm (southern Canada south through Central America, including the Tres Marías islands, México)..*Procyon lotor* (p. 178)

Procyon lotor (Linnaeus, 1758)

Raccoon, mapache

1. *P. l. auspicatus* Nelson, 1930. Restricted to Key Vaca, Florida.
2. *P. l. elucus* Bangs, 1898. Florida Peninsula.
3. *P. l. excelsus* Nelson and Goldman, 1930. From eastern Washington and Oregon west to western Montana, Idaho, and northern Nevada.
4. *P. l. fuscipes* Mearns, 1914. Texas, Coahuila, Nuevo León, and Tamaulipas.
5. *P. l. grinnelli* Nelson and Goldman, 1930. Southern Baja California Peninsula.
6. *P. l. hernandezii* Wagler, 1831. From Zacatecas to Nayarit south through the Tehuantepec isthmus.
7. *P. l. hirtus* Nelson and Goldman, 1930. From Alberta to western Ontario south through southeastern New Mexico, northern Texas, and Arkansas.
8. *P. l. incautus* Nelson, 1930. Restricted to Torch Key, Florida.
9. *P. l. inesperatus* Nelson, 1930. Restricted to Matecumbe Key, Florida.
10. *P. l. insularis* Merriam, 1898. Restricted to the Tres Marías Islands, Nayarit.
11. *P. l. litoreus* Nelson and Goldman, 1930. Eastern Georgia and southeastern South Carolina.
12. *P. l. lotor* (Linnaeus, 1758). From Ontario to Illinois and North Carolina northeast to New Brunswick.
13. *P. l. marinus* Nelson, 1930. Southwestern Florida.
14. *P. l. maritimus* Dozier, 1948. Rehoboth Bay and Saxis Island, Maryland.
15. *P. l. megalodous* Lowery, 1943. Coastal Louisiana.
16. *P. l. mexicanus* Baird, 1958. From New Mexico and Eastern Arizona south through Nayarit and Zacatecas.
17. *P. l. pacificus* Merriam, 1899. Southern British Columbia, Washington, and Oregon.
18. *P. l. pallidus* Merriam, 1900. Utah, western Colorado, southern Nevada, southeastern California, Arizona, and northwestern Sonora.
19. *P. l. psora,* Gray, 1842. California and northwestern Baja California.
20. *P. l. shufeldti* Nelson and Goldman, 1931. Lowlands of Chiapas and the Yucatán Peninsula.
21. *P. l. solutus* Nelson and Goldman, 1931. South Carolina.
22. *P. l. vancouverensis* Nelson and Goldman, 1930. Vancouver Island, British Columbia.
23. *P. l. varius* Nelson and Goldman, 1930. From Tennessee south through Louisiana to Georgia.

Conservation: *Procyon lotor insularis* is listed as Endangered in the Norma Oficial Mexicana (DOF 2019) and *P. lotor* is listed as Least Concern by the IUCN (Timm et al. 2016i).

Characteristics: Raccoon is medium sized; **total length 634.0–950.0 mm in males and 603.0–909.0 mm in females; condylobasal length 94.3–125.8 mm in males and 89.4–115.9 mm in females. Digits paler than limbs; without a diastema between the first and third upper premolars**.

Comments: *Procyon insularis* was previously elevated to full species (Nelson and Goldman 1931), but is currently regarded as a subspecies of *P. lotor* (Helgen and Wilson 2005). *Procyon lotor* ranges from throughout southern Canada south through Central America, including the Tres Marías Islands, México (Map 56). It can be found in sympatry, or nearly so, with *Bassariscus astutus, B. sumichrasti, Nasua narica,* and *Potos flavus*. *P. lotor* can be distinguished from these species by having a marked black mask on the rostrum, tail short but ringed from dark to whitish shades; body chubby. From *P. pygmaeus*, by its largesize, with a total length greater than 670.0 mm.

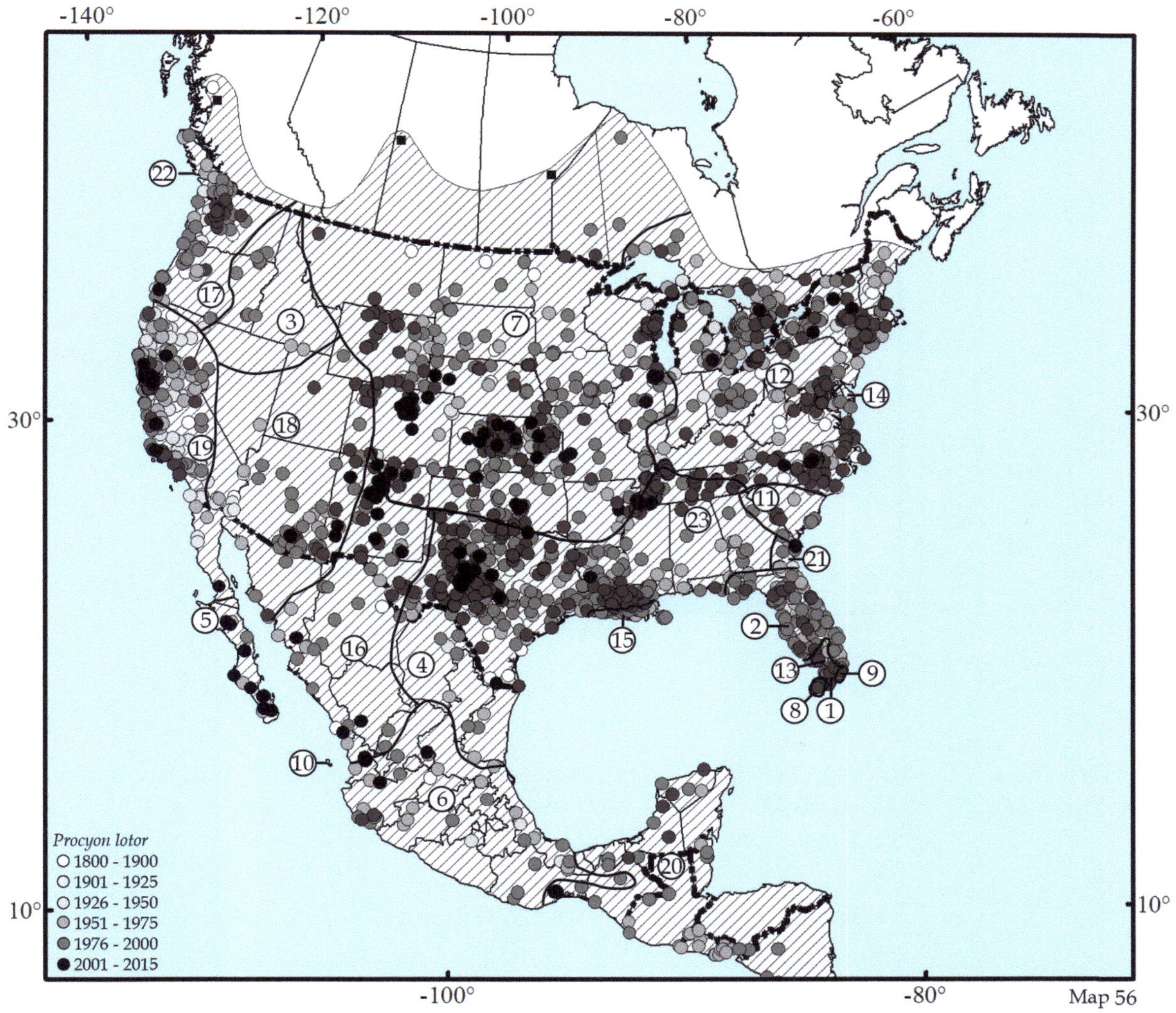

Map 56 Distribution map of *Procyon lotor*

Additional Literature: Merriam (e), Pocock (1921), Goldman (1950), Koopman et al. (1957), Paradiso (1969), Lotze and Anderson (1979), Helgen and Wilson (2003, 2005).

Procyon pygmaeus Merriam, 1901

Cozumel Raccoon, mapache de Cozumel

Monotypic.

Conservation: *Procyon pygmaeus* is listed as Endangered in the Norma Oficial Mexicana (DOF 2019) and as Critically Endangered (C2a (ii); D: small population size projected continuing decline with less than 50 mature individuals in one subpopulation) by the IUCN (Cuarón et al. 2016a).

Characteristics: Cozumel raccoon is the smallest known, **total length on average 667.0 mm in males and 665.0 mm in females; condylobasal length on average 93.7 mm in males and 91.9 mm in females. Digits darker than limbs; with a diastema between the second and third upper premolars**.

Comments: *Procyon pygmaeus* is restricted to Cozumel Island off the coast of the Yucatán Peninsula, México (Map 57). It can be found in sympatry, or nearly so, with *Nasua narica*, and can be distinguished from it by its marked black mask on the rostrum; tail short but ringed from dark to whitish shades, and body chubby. From *P. lotor*, by its smaller size, with total length less than 670.0 mm.

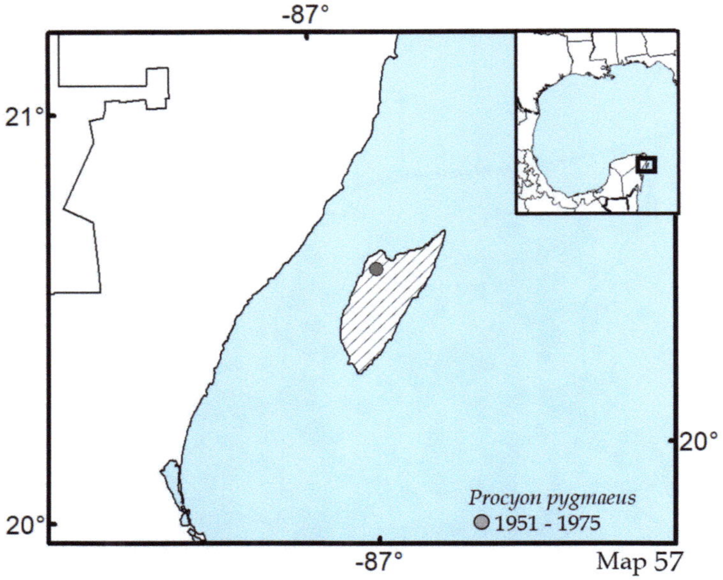

Map 57 Distribution map of *Procyon pygmaeus*

Additional Literature: Merriam (1901e), Hollister (1915), Pocock (1921), Goldman (1950), Hall (1981), Lazell (1981), Helgen and Wilson (2003, 2005), Villa-Meza et al. (2011), McFadden and Meiri (2013).

Family Ursidae Fisher de Waldheim, 1817

The family Ursidae includes bears, characterized by their **big size (1100.0–2600.0 cm in total length); tail short; foot posture plantigrade; claws long, curved, and nonretractile; mastoid and occipital processes equal in size; carnassial poorly developed; molars with flat broad crowns; last upper molars very large and elongated antero-posteriorly**. All species in North America belong to the subfamily Ursinae.

Subfamily Ursinae Fisher de Waldheim, 1817

All species within this subfamily are of the genus *Ursus* and have a complex cuspidation on the molars; masseteric fossa on the lower jaw not divided into two fossae by a bony septum; alisphenoid canal present.

Genus *Ursus* Linnaeus, 1758

General coloration varying from white to black, big body size; tail short; foot posture plantigrade; claws long, curved, and nonretractile. The keys were elaborated based on the review of specimens and the following Hall (1981).

1. Body always entirely white; first and second upper molar length less than palatal width; second upper molars proportionally larger than the first upper molars (Fig. 33; circumpolar distribution) .... *Ursus maritimus* (p. 184)

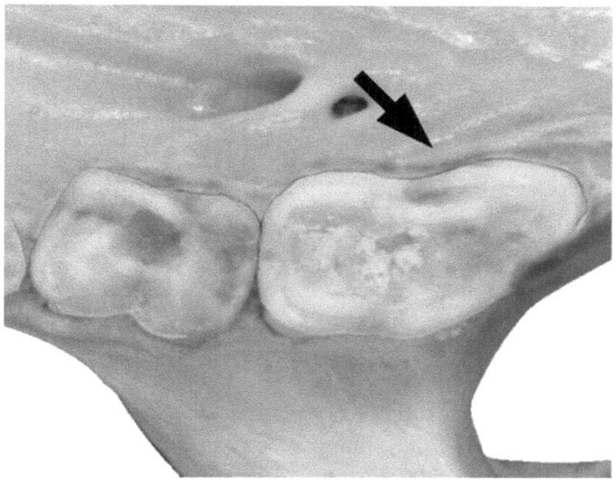

Fig. 33 Second upper molars proportionally larger than the first upper molars like in *Ursus maritimus*

1a. Body never entirely white; however, it can have paler colorations; first and second upper molar length greater than palatal width; second upper molars not proportionally larger than the first upper molars (Fig. 34) 2

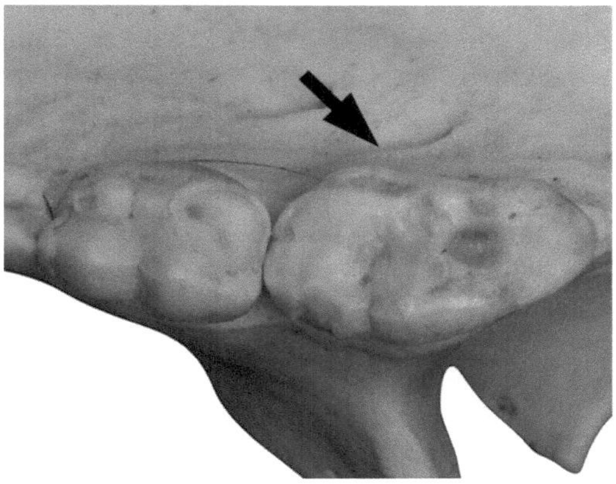

Fig. 34 Second upper molars are not proportionally larger than the first upper molars

2. Upperpart of the hips and back at the same height; claws of forefeet slightly longer than those of the hind feet. Second upper molar length less than 29.5 mm (from Alaska south through California, United States, to Zacatecas and Tamaulipas, México, and east through to the Atlantic coast, United States) .... *Ursus americanus* (p. 182)

2a. Upperpart of the back over the hips; claws of forefeet twice as long as those of the hind feet. Second upper molar length greater than 29.5 mm (from Alaska south through California, United States, and Chihuahua, México, and east through to Colorado and New Mexico, United States) .... *Ursus arctos* (p. 183)

Ursus americanus Pallas, 1780

American Black Bear, oso negro

1. *U. a. altifrontalis* Elliot, 1903. From coast of British Columbia to northern California.
2. *U. a. amblyceps* Baird, 1859. From Colorado and southwestern Kansas south through northern Chihuahua and southwestern Texas.
3. *U. a. americanus* Pallas, 1780. From Alaska to the Labrador Peninsula south through Texas and Arkansas, to Georgia.
4. *U. a. californiensis* Miller, 1900. Central and western California.
5. *U. a. carlottae* Osgood, 1901. Restricted to Queen Charlotte Island, British Columbia.
6. *U. a. cinnamomum* Audubon and Bachman, 1854. From western Washington to Montana south through Wyoming and Utah.
7. *U. a. emmonsii* Dall, 1895. Southwestern Alaska and northwestern British Columbia.
8. *U. a. eremicus* Merriam, 1904. Coahuila, Nuevo León, Tamaulipas, Zacatecas, and San Luis Potosí.
9. *U. a. floridanus* Merriam, 1896. Florida Peninsula to southern Mississippi, Alabama, and Georgia.
10. *U. a. hamiltoni* Cameron, 1957. Prince Edward Island, Newfoundland.
11. *U. a. kermodei* Hornaday, 1905. Coastal British Columbia.
12. *U. a. luteolus* Griffith, 1821. Southeastern Texas, southern Louisiana, and Mississippi.
13. *U. a. machetes* Elliot, 1903. Chihuahua, Durango, Sinaloa, and western Zacatecas.
14. *U. a. perniger* Allen, 1910. Restricted to the Kenai Peninsula, Alaska.
15. *U. a. pugnax* Swarth, 1911. Restricted to the Dall, Kupreanof, Mitkof, Prince of Wales, San Alberto, Wrangell, and Alaska Islands.
16. *U. a. vancouveri* Hall, 1928. Restricted to Vancouver Island, British Columbia.

Conservation: *Ursus americanus eremicus* is listed as Special Protection in the Norma Oficial Mexicana (DOF 2019) and *U. americanus* as Least Concern by the IUCN (Garshelis et al. 2016). It has been extirpated from large portions of its historic ranges intentionally and by habitat loss.

Characteristics: Black bear is medium sized, being the smallest member of the family; **total length 1300.0–1840.0 mm in males and 1220.0–1640.0 mm in females; condylobasal length 254.0–300.0 mm in males and 222.0–260.0 mm in females. Dorsal pelage black to very dark brown;** snout pale brown; **upperpart of the hips and back at same level; claws of forefeet slightly longer but not double in size, than those of hind feet;** skull triangular when tracing a profile line between the palate and the rostrum; nasal conducts stand out to the front through the suture between the nasals and the maxilla; **first and second upper molar length greater than palatal width; second upper molar length less than 29.5 mm**.

Comments: *Ursus americanus* ranges through a large area of Canada and the United States south through Sinaloa, Durango, Zacatecas, Coahuila, Nuevo León, and Tamaulipas, México, except for eastern Oregon, central and eastern Nevada, southeastern California, southwestern Arizona, western Utah, and Idaho (Map 58). It can be found in sympatry, or nearly so, with *U. arctos*, from which it can be distinguished by having the upper part of the hips and back at the same height; forefoot claws slightly longer than hindfoot claws, and second upper molar length less than 29.5 mm.

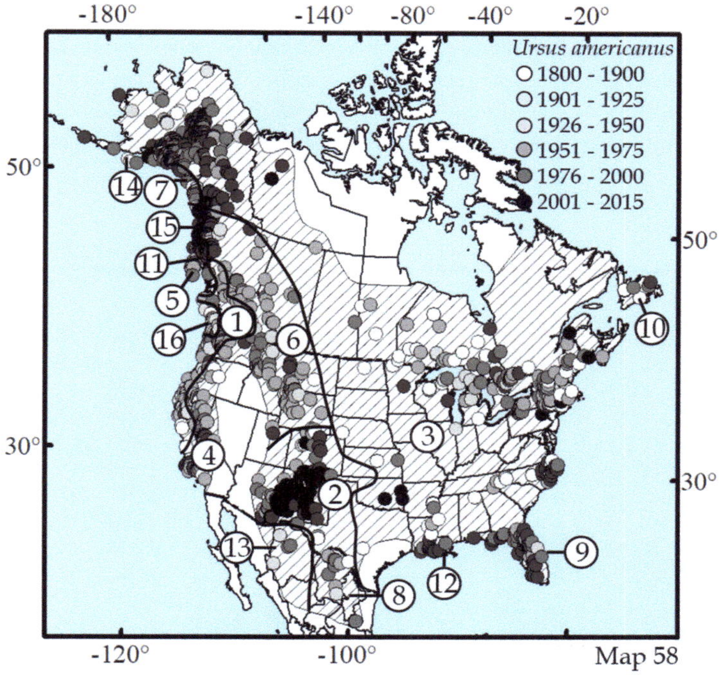

Map 58 Distribution map of *Ursus americanus*

Additional Literature: McTaggart Cowan (1938), Hall (1981), Nowak (1991), Larivière (2001), Kennedy et al. (2002), Williamson (2002), Pelton (2003), Faries et al. (2013), Scheick and McCown (2014), Puckett et al. (2015).

Ursus arctos Linnaeus, 1758

Brown Bear, oso pardo

1. *U. a. alascensis* Merriam, 1896. Throughout Alaska, except for the Iliamna Lake, southwest into the Alaska Peninsula.
2. *U. a. californicus* Merriam, 1896. From northern Baja California north through the border between California and Oregon.
3. *U. a. dalli* Merriam, 1896. Most northeastern British Columbia.
4. *U. a. gyas* Merriam, 1902. From Iliamna Lake southwest into the Alaska Peninsula.
5. *U. a. horribilis* Ord, 1815. From the Yukon and Mackenzie south through Sinaloa, Durango, and Zacatecas.
6. *U. a. middendorfii* Merriam, 1896. Kodiak Islands, Alaska.
7. *U. a. sitkensis* Merriam, 1896. Northwestern British Columbia.
8. *U. a. stikeenensis* Merriam, 1914. From the border between California and Oregon north through the border between British Columbia and Alaska.

Conservation: *Ursus arctos horribilis* is listed as Extinct in the Norma Oficial Mexicana (DOF 2019). *U. arctos* is listed as Threatened by the United States Endangered Species Act (FWS 2022) and Special Concern and Extirpated from Québec in the Canadian Endangered Species Protection Act, and as Least Concern by the IUCN (McLellan et al. 2017). *U. arctos* has been extirpated intentionally from large portions of their historic ranges and by habitat lost, including all México.

Characteristics: Grizzle bear is large sized; average total length 1640.0 mm in males and 1510.0 mm in females; condylobasal length 284.0–424.0 mm in males and 285.0–366.0 mm in females. **Dorsal pelage from brown to darker brown but never black;** snout of the same color than all the head; **back higher than any other part of body**; **claws in forefoot double the size of those in hindfoot;** skull no in triangular shape when tracing a profile line between the palate and rostrum; nasals do not stand out frontward over the suture between the nasal and maxilla; first and second upper molars length greater than palatal width.

Comments: *Ursus arctos* was the most widely distributed bear in North America. Its ranges from northern Canada south through northern México, but has been extirpated from great area of its range (Map 59). It can be found in sympatry, or nearly so, with *U. americanus* and *U. maritimus. U. arctos* can be distinguished from *U. americanus* by having the back upperpart over the upper parts of the hips; forefoot claws double in length as hindfoot claws; second upper molar length greater than 29.5 mm. From *U. maritimus*, by having a brown or grayish coloration but never entirely white; first and second upper molars greater than the palatal width.

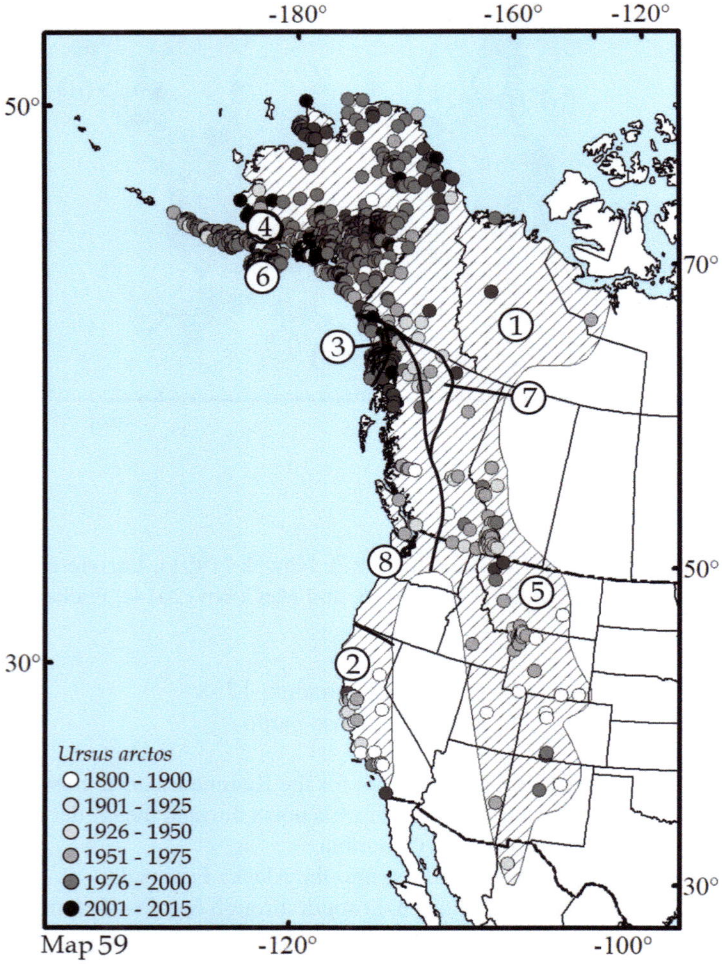

Map 59 Distribution map of *Ursus arctos*

Additional Literature: Merriam (1918), Guilday (1968), Honacki et al. (1982), Hall (1984), Brown (1985), Pasitschniak-Arts (1993), Mattson (1998), Larivière (2001), Mattson and Merrill (2002), Schwartz et al. (2003a, b), Mowat et al. (2013).

Ursus maritimus Phipps, 1774

Polar Bear, oso polar

Monotypic.

Conservation: *Ursus maritimus* is listed as Threatened by the United States Endangered Species Act (FWS 2022), as Special Concern by the Canadian Endangered Species Protection Act, and as Vulnerable (A3c: population reduction in the last 10 years 30% or less projected to be met in future data obtained by decline in area of occupancy) by the IUCN (Wiig et al. 2015).

Characteristics: Polar bear is the largest member of the family; total length 2300.0–2600.0 mm in males and 1900.0–2100.0 mm in females; condylobasal length 350.0–409.0 mm in males and 308.0–380.0 mm in females. **Pelage white or buffy**; hips upper-part and back at the same level; forefoot claws twice the size of hindfoot claws; skull triangular when tracing a profile line between the palate and the rostrum; nasal conducts stand out to the front through the suture between the nasals and the maxilla; **first and second upper molar length less than the palatal width**.

Comments: *Ursus maritimus* ranges throughout the northern coast and seas of Canada and Alaska and the western coast of Alaska (Map 60). It can be found in sympatry, or nearly so, with *U. arctos,* from which it can be differentiated by always being entirely white; first and second upper molar length less than the palatal width.

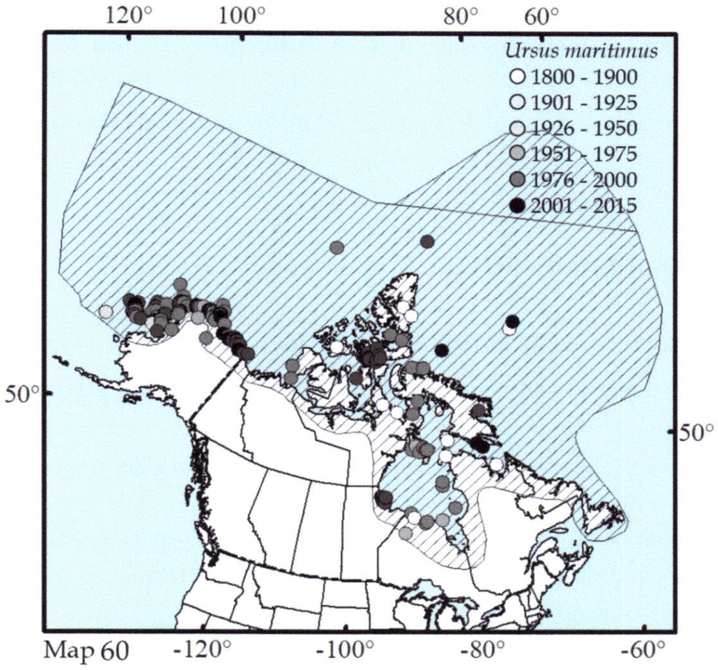

Map 60 Distribution map of *Ursus maritimus*

Additional Literature: Harington (1966), Manning (1971), Wilson (1976), DeMaster and Stirling (1981), Amstrup et al. (1986), Larivière (2001), Amstrup (2003).

Suborder Feliformia Kretzoi, 1945

The species within this suborder are generally digitigrade; claws retractile; **bulbourethral glands and vesicula seminalis always present;** baculum usually small; **rostrum short;** jaws relatively small with a few teeth; **carnassial teeth specialized** for eating meat (except for viverrids); **auditory bullae double-chambered, composed of two bones joined by a septum**.

Family Felidae Fisher de Waldheim, 1817

The family Felidae includes all cats, lions, cougars, jaguars, and panthers, characterized by a small to medium size; body slim; ears sharp and excellent sight; **five digits in the forefeet and four in the hind feet; claws retractable within a protective sheath when not in use**; tail short or long but not bushy; digitigrade; **skull short, rounded dorsally and rostrum very short; carnassial very well developed**; paroccipital and mastoid processes similar in size; no aliesphenoid canal; last molars small and rounded. Cosmopolitan distribution in the wild, except for Australia, Antarctica, Greenland, Madagascar, New Guinea, and New Zealand. Two subfamilies range in North America: Felinae and Pantherinae. The differences between these two families refer to ossification characteristics in the throat, allowing Pantherinae to roar while Felinae cannot. The keys were elaborated based on the review of specimens and the following sources: Lawlor (1979), Hall (1981), and Álvarez-Castañeda et al. (2017a).

1a. Dorsal pelage spotted or black; skull length greater than 237.0 mm (from Arizona, New Mexico, Louisiana, and Texas south through South America) .... subfamily Pantherinae; *Panthera onca* (p. 197)

1. Dorsal pelage uniform or spotted; skull length less than 237.0 mm (throughout México, the United States, and the Great Plains in Canada) .... subfamily Felinae (p. 186)

Subfamily Felinae Fischer von Waldheim, 1817

The cats of this subfamily are characterized **by having a bony hyoid;** for this reason, they are able to purr but not to roar.

1. Tail short, less than 25% of the head-and-body length; hairs on the head sides extended, three upper molariforms in each jaw .... *Lynx* (p. 192)

1a. Tail long, greater than 25% of the head-and-body length; hairs on the head sides not extended; four upper molariforms in each jaw 2

2. Total length greater than 1,400.0 mm; skull length greater than 158.0 mm; posterior border of the temporal bones with a frontal process (Fig. 35; from central Canada south through South America) .... *Puma concolor* (p. 195)

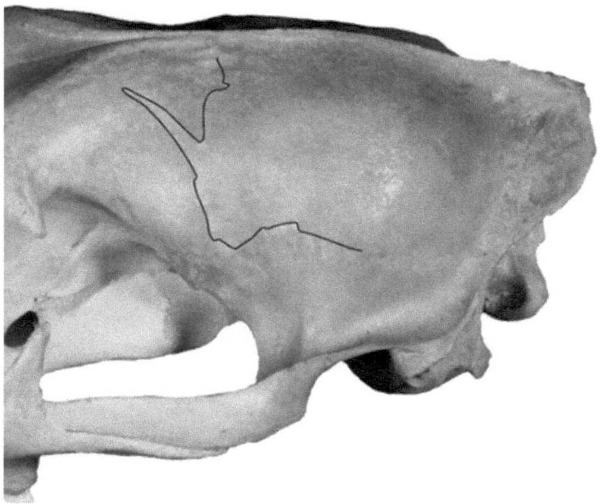

Fig. 35 Posterior border of the temporal bones with a frontal process like in *Puma concolor*

2a. Total length less than 1,400.0 mm, skull length less than 158.0 mm; posterior border of temporal bones without a frontal process (Fig. 36) 4

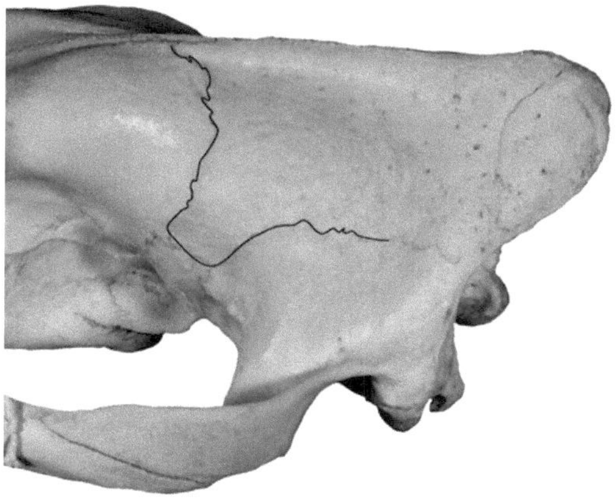

Fig. 36 Posterior border of the temporal bones without a frontal process

4. Dorsal pelage homogeneous; skull slightly arched in lateral view; frontal border of nasals extending beyond the anterior palatine foramen so that the nasals cover the palatine foramina in dorsal view (Fig. 37; from Sinaloa and Veracruz south through South America) .... *Herpailurus yagouaroundi* (p. 188)

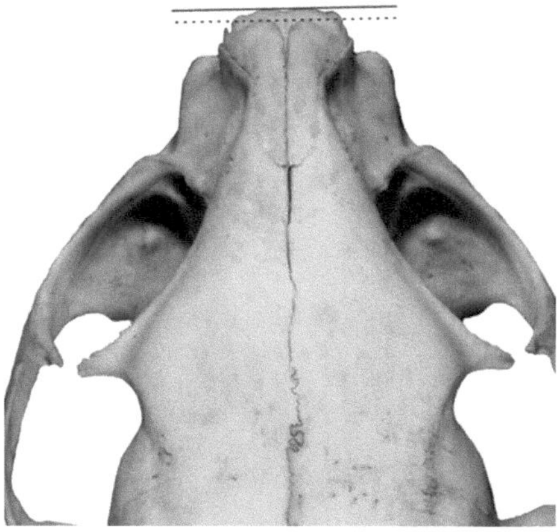

Fig. 37 The frontal border of nasals extending beyond the anterior palatine foramen so that the nasals cover the palatine foramina in dorsal view like in *Herpailurus yagouaroundi*

4a. Dorsal pelage homogeneous or with spots or stripes; skull strongly arched in lateral view; frontal border of nasals behind the anterior border of the palatine foramen so that the palatine foramina can be seen in dorsal view (Fig. 38) 5

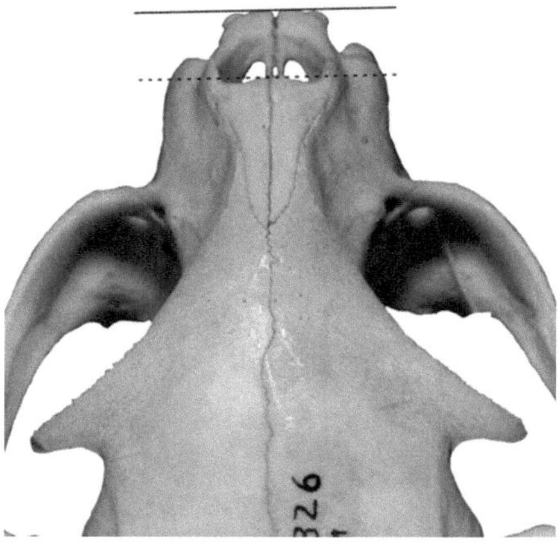

Fig. 38 The frontal border of nasals behind the anterior border of the palatine foramen so that the palatine foramina can be seen in dorsal view

5. Dorsal pelage homogeneous or with stripes but never spots; supraorbital process wide and long; in dorsal view, skull with a high part at the level of the supraocular processes .... *Felis sylvestris* (p. 188)

5a. Dorsal pelage with spots or stripes but never homogeneous; supraorbital process narrow and short; in dorsal view, skull without a high part at the level of the supraocular processes .... *Leopardus* (p. 189)

Felis sylvestris Schreber, 1777
Domestic Cat, gato doméstico

F. s. lybica Lataste, 1885. Introduced.

Conservation: *Felis sylvestris* is listed as introduced species and Least Concern by the IUCN (Yamaguchi et al. 2015).

Characteristics: Species subject to a domestication process that has introduced a great variation in its morphology; therefore, it is not possible to define a general morphology that can be used to define the species.

Comments: Housecat was domesticated from *Felis sylvestris lybica* (Driscoll et al. 2007, Macdonald et al. 2010) and has been considered a separate species, *F. catus*, or a subspecies, *F. s. catus*: however, that classification was considered arbitrary and without genetic support (Macdonald et al. 2010; Driscoll et al. (2007). *F. sylvestris* has been recorded as wildlife only from insolated areas of North America. *F. sylvestris* can be distinguished from *Leopardus* species by having a color pattern never with spots or stripes and a supraorbital process wide and long. From *Herpailurus,* by its larger size, limbs proportionally larger; head size in proportion to body size; skull strongly arched in lateral view.

Additional Literature: Gentry et al. (2004), Driscoll et al. (2007), Macdonald et al. (2010), Yamaguchi et al. (2015).

Genus ***Herpailurus*** Severtzow, 1858

Jaguarundi has a uniform unspotted pelage; it can be found in four main color patterns: blackish, brownish-gray, reddish-yellow, or chestnut; **body elongated; limbs short; ears small and rounded; head more elongated than rounded, proportionally smaller in relation to the body;** skull long; rostrum short; temporal crests almost always separated; sagittal crest rarely complete; narrowest part of the interorbital region situated forward, just behind the nasomaxillary line; inner chamber of auditory bullae never markedly inflated anteriorly, not projecting far beyond the low partition. *Herpailurus yagouaroundi* was previously considered *Felis,* but Groves (1982) and Newman et al. (1985) proposed its assignment as *Herpailurus* . It has been suggested that *Herpailurus* be a synonym with *Puma* (Johnson et al. 2006; Eizirik et al. 2008); nevertheless, *Herpailurus* is still in use (Wozencraft 2005; Segura et al. 2013).

Herpailurus yagouaroundi (È. Geoffroy Saint-Hilaire, 1803)
Jaguarundi, Yaguarundi

1. *H. y. cacomitli* (Berlandier, 1859). From southern Texas south through central Veracruz.
2. *H. y. fossata* (Mearns, 1901). From central Veracruz and Oaxaca south through Chiapas, including the Yucatán Peninsula.
3. *H. y. tolteca* (Thomas, 1898). From southern Arizona south through central Oaxaca.

Conservation: *Herpailurus yagouaroundi* is listed as Threatened in the Norma Oficial Mexicana (DOF 2019) and by the United States Endangered Species Act (FWS 2022), and as Least Concern by the IUCN (Caso et al. 2015).

Characteristics: Jaguarundi is small sized; total length 888.0–1,372.0 mm, condylobasal length 81.2–109.2 mm. **Dorsal pelage homogeneous, grizzly, found in two main color shades: iron gray or reddish-brown**; no spots; fur of underparts similar to dorsal fur; jaw upper part around the lips a little darker than at the back; ears rounded and small, without a white spot; body elongated; **limbs proportionally short** and tail long; skull slightly arched in side view; **front nasal border extending beyond the anterior palatine foramen in dorsal view so that nasals cover the palatine foramina**.

Comments: *Herpailurus yagouaroundi* ranges from Sinaloa and Veracruz across the tropical lowlands and coastal areas of México south through South America, including the Yucatán Peninsula (Map 61). It can be found in sympatry, or nearly so, with *Leopardus pardalis, L. weidii, Puma concolor,* and *Panthera onca. H. yagouaroundi* can be distinguished from the other genera of Felidae by its dorsal coloration homogeneous, grizzly in the coloration shades mentioned above, without spots, and with limbs proportionally shorter.

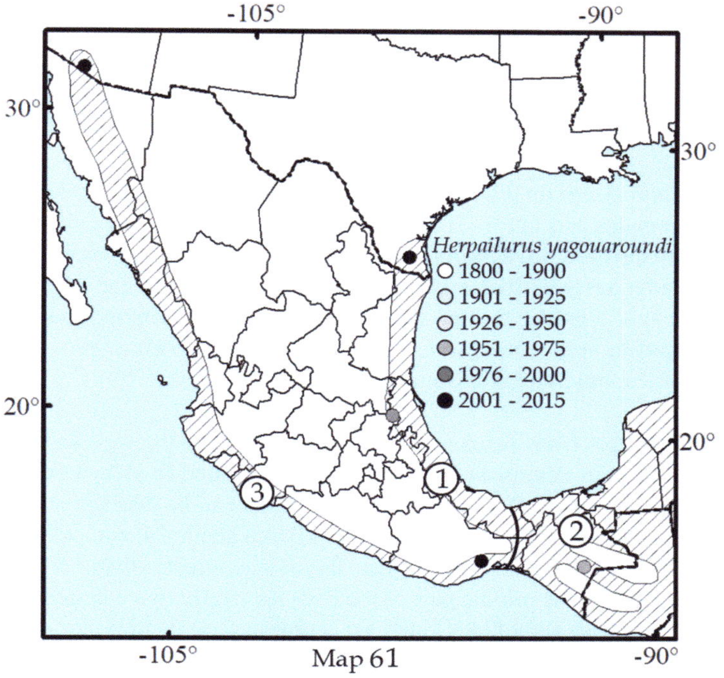

Map 61 Distribution map of *Herpailurus yagouaroundi*

Additional Literature: Guggisberg (1975), Hall (1981), Groves (1982), Eisenberg (1989), Nowak (1992), de Oliveira (1998a), Wozencraft (2005), Johnson et al. (2006), Eizirik et al. (2008), Agnarsson et al. (2010), Segura et al. (2013), Hunter (2015).

Genus *Leopardus* Gray, 1842

Dorsal pelage pale to reddish brown, with **darker brown spots surrounded by discontinuous lines of darker color; spots at the middle part of the body extended in different length; width ratios**; fur of underparts white to pale gray**;** upperpart of the jaw in homogeneous color; area around the lips white; **skull strongly arched in side view; frontal border of the nasals behind the anterior palatine foramen in dorsal view**; skull length less than 120.0 mm and upper carnassial greater than 12.7 mm. *Leopardus* was previously considered *Felis* but Groves (1982) and Newman et al. (1985) proposed its assignment as *Leopardus.* The keys were elaborated based on the review of specimens and the following sources: Goldman (1943); Hall (1981), and Álvarez-Castañeda et al. (2017a).

1. Head-and-body length greater than 650.0 mm; dorsal pelage sand brown to pale yellow, with blackish brown spots; hind limb length greater than 145.0 mm; weight greater than 5.0 kg; skull length greater than 120.0 mm; upper carnassial length greater than 12.7 mm (from Arizona and Texas, United States, south through South America).. *Leopardus pardalis* (p. 190)

1a. Head-and-body length less than 700.0 mm; dorsum grayish pale brown to tawny-colored, with blackish brown spots; hind limb length less than 145.0 mm; weight less than 5.0 kg; skull length less than 120.0 mm; upper carnassial length less than 12.7 mm (from southern Sonora, México and Texas, United States, south through South America).. *Leopardus wiedii* (p. 191)

Leopardus pardalis (Linnaeus, 1758)

Ocelot, ocelote

1. *L. p. albescens* (Pucheran, 1855). From Texas south through Tamaulipas, including Coahuila and Nuevo León.
2. *L. p. nelsoni* (Goldman, 1925). From central Sinaloa south through central Oaxaca.
3. *L. p. pardalis* (Linnaeus, 1758). From Veracruz and northern Oaxaca south through Chiapas, including the Yucatán Peninsula.
4. *L. p. sonoriensis* (Goldman, 1925). From Arizona south through central Sinaloa.

Conservation: *Leopardus pardalis* is listed as Endangered in the Norma Oficial Mexicana (DOF 2019) and as Least Concern by the IUCN (Paviolo et al. 2015).

Characteristics: Ocelot is medium sized; total length 1,057.0–1,230.0 mm in males and 1,050.0–1,189.0 mm in females; condylobasal 122.0–132.0 mm in males and 111.3–114.7 mm in females. Background color cream to tawny-yellow, reddish to gray; **coat with solid or open-centered dark spots at the middle of the body, generally two times larger than wide, in the neck as streaks; fur of underparts whitish with dark spots occasionally;** ear back black with a central white spot; ears rounded; tail proportionally short, similar to hind limbs in length; **head proportionally small relative to body size; frontal paws larger than hind paws;** very similar to *L. wiedii* but larger; postorbital process relatively shorter and strong; sagittal crest present in the supraoccipital process in some specimens.

Comments: *Leopardus pardalis* ranges from Texas to the east and Arizona to the west south through the tropical lowlands to South America, including the Yucatán Peninsula (Map 62). It can be found in sympatry, or nearly so, with *Herpailurus yagouaroundi, L. weidii, Puma concolor,* and *Panthera onca. L. pardalis* can be distinguished from the other species of the family Felidae by its smaller size, with a total length less than 1400.0 mm and the dorsal pelage with a pattern of darker spots or stripes over the yellowish coloration. From *L. wiedii*, by its larger size, greater than 650.0 mm in total length; coat with solid or open-centered dark spots; spot at the middle part of the body generally twice larger than wide; neck with two longitudinal streaks of similar pattern; tail with eight black rings, on average.

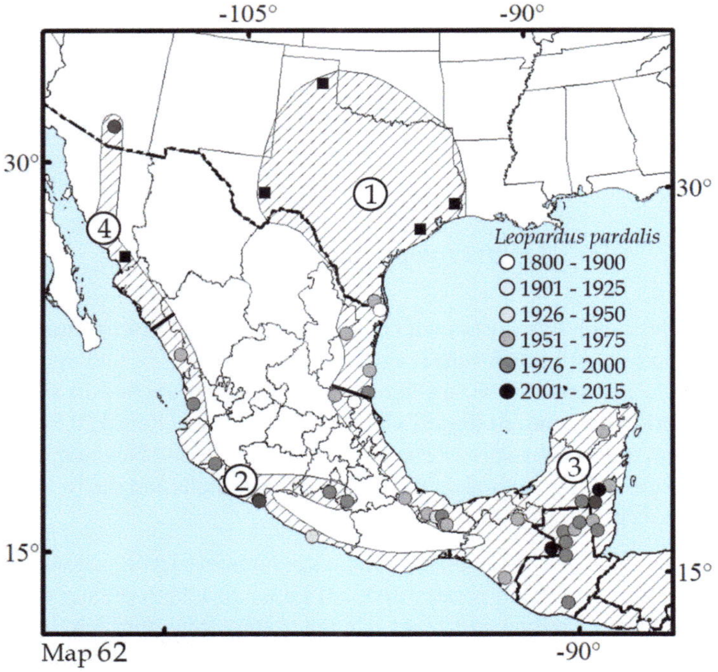

Map 62 Distribution map of *Leopardus pardalis*

Additional Literature: Allen (1919a), Goldman (1943), Brown (1989), Redford and Eisenberg (1992), Nowell and Jackson (1996), Murray and Gardner (1997), Sunquist and Sunquist (2002), González et al. (2003), Johnson et al. (2006), Eizirik et al. (2008), Hunter (2015).

Leopardus wiedii (Schinz, 1821)

Margay, tigrillo

1. *L. w. cooperi* (Goldman, 1943). Texas, Nuevo León, and northern Tamaulipas.
2. *L. w. glauculus* (Thomas, 1903). From southern Sonora south through central Oaxaca.
3. *L. w. oaxacensis* (Nelson and Goldman, 1931). From central Tamaulipas and Oaxaca south through Tehuantepec isthmus.
4. *L. w. salvinia* Pocock, 1941. Pacific side of Chiapas.
5. *L. w. yucatanicus* (Nelson and Goldman, 1931). Gulf side of Chiapas and the Yucatán Peninsula.

Conservation: *Leopardus wiedii* is listed as Endangered in the Norma Oficial Mexicana (DOF 2019) and as Near Threatened by the IUCN (de Oliveira et al. 2015).

Characteristics: Margay is small sized; total length 862.0–1,300.0 mm in males and 805.0–1,029.0 m in females; condylobasal length 81.1–94.0 mm. Background color grayish to cinnamon; coat with dark brown or black **open or closed spots, generally extended in the middle part of the body in a proportion three times larger than wide, and with streaks; fur of underparts white;** tail long, longer than hind limb length, about 70% of the head-and-body length, marked with twelve dark rings and black at the tip; **head proportionally smaller relative to body size;** skull very similar to *L. pardalis* but smaller; postorbital process relatively longer and slender; sagittal crest absent.

Comments: *Leopardus wiedii* ranges from southern Texas to the east and southern Chihuahua and Sonora to the west south through the tropical lowlands of México to South America, including the Yucatán Peninsula (Map 63). It can be found in sympatry, or nearly so, with *Herpailurus yagouaroundi, L. pardalis, Puma concolor,* and *Panthera onca. L. wiedii* can be distinguished from *Panthera* and *Puma* by its smaller size, with a total length less than 1400.0 mm; dorsal pelage with a pattern of darker spots or stripes over a yellowish coloration. From *L. pardalis,* by its smaller size, less than 650.0 mm in total length; coat with solid or open-centered dark spots; spots at the middle part of the body generally extended in a proportion three times larger than wide; neck with two or more longitudinal streaks with different patterns; tail with ten or more black rings.

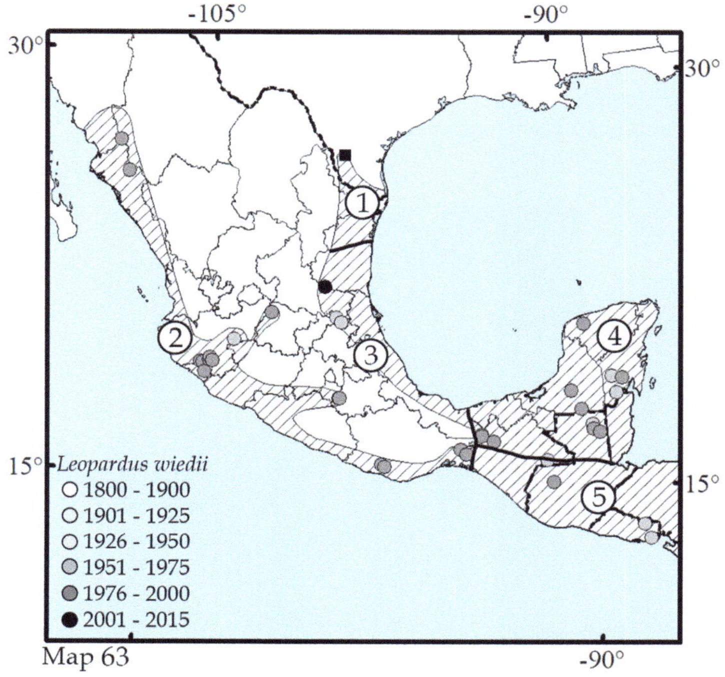

Map 63 Distribution map of *Leopardus wiedii*

Additional Literature: Allen (1919a), Pocock (1941), Goldman (1943), Redford and Eisenberg (1992), Oliveira (1998b), Sunquist and Sunquist (2002), Johnson et al. (2006), Eizirik et al. (2008), Hunter (2015).

Genus *Lynx* Kerr, 1792

Dorsal pelage speckled, spotted, and striped in gray and brown shades; fur of underparts white with spots; **tail very short; ears triangular, ending with a lock of black hair from small to large; cheeks fringed with short or large hair;** skull not arched in side view; anterior border of the nasals at the same level or extending beyond the anterior border of the palatine foramen in dorsal view; **two upper premolars**. *Lynx* has been considered a full genus (Mattyushkin 1979; García-Perea 1992), previously regarded as a junior synonym of *Felis* (Hall 1981) or a subgenus of *Felis* (Hemmer 1978; McKenna and Bell 1997). The keys were elaborated based on the review of specimens and the following sources: Hall (1981).

1. Ears with a black hair tuft less than 35.0 mm; tail only black dorsally at the tip; tail long, greater than the hindfoot length; anterior condyloid foramen confluent with the posterior foramen lacerum (Fig. 39; from central Canada south through Oaxaca, México)............ *Lynx rufus* (p. 194)

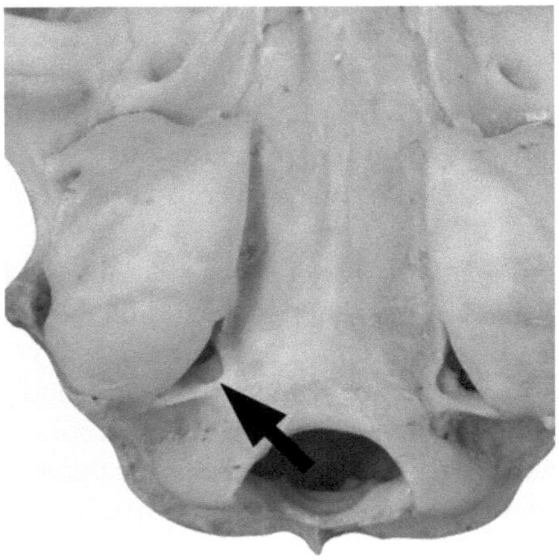

Fig. 39 The anterior condyloid foramen confluent with the posterior foramen lacerum like in *Lynx rufus*

1a. Ears with a long black hair tuft greater than 35.0 mm; tail completely black at the tip; tail short, less than the hindfoot length; anterior condyloid foramen separated from the posterior foramen lacerum (Fig. 40; from Alaska, United States, and Canada south through Oregon to New Hampshire, United States)............ *Lynx canadensis* (p. 193)

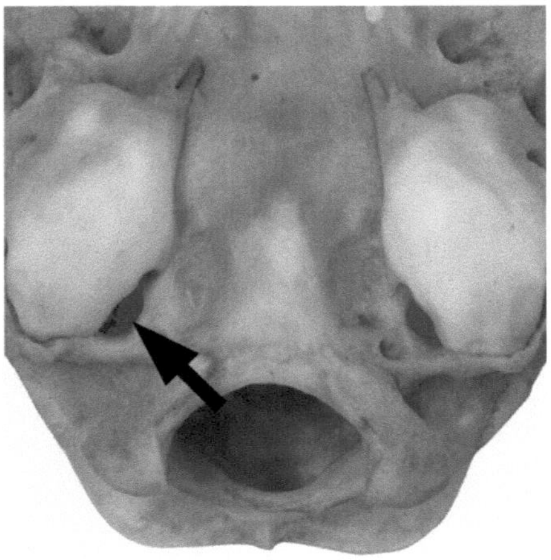

Fig. 40 The anterior condyloid foramen separated from the posterior foramen lacerum like in *Lynx canadensis*

Lynx canadensis Kerr, 1792

Canada Lynx, lince canadiense

1. *L. c. canadensis* Kerr, 1792. From Oregon to Maryland north through all Canada and Alaska, including Utah and Colorado.
2. *L. c. subsolanus* Bangs, 1897. Restricted to Newfoundland.

Conservation: *Lynx canadensis* is listed as Threatened by the United States Endangered Species Act (FWS 2022) and as Least Concern by the IUCN.

Characteristics: Canada lynx is medium sized; total length 902.0–1,570.0 mm in males and 820.0–962.0 mm in females; condylobasal length 121.0–138.0 mm in males and 112.0–128.0 mm in females. Dorsal pelage yellow gray to grizzled grayish brown; spots absent on the back; fur of underparts whitish with spots, especially on the inner limbs; **ears triangular, ending in a long black hair tuft; cheeks with a ruff of large hair; tail black at the tip; hind limbs longer than forelimbs; paws covered with long and dense hair, more than 75.0 mm wide**; anterior condiloid foramen separated from the posterior foramen lacerum.

Comments: *Lynx canadensis* was recognized as a distinct species from *L. lynx* (Mattyushkin 1979; Kurten and Anderson 1980; García-Perea 1992). *L. canadensis* ranges from central United States (Oregon, Nebraska, Iowa, Indiana, Ohio, and West Virginia) north throughout Canada (Map 64). It can be found in sympatry, or nearly so, with *L. rufus* and *Puma concolor. L. canadensis* can be distinguished from *Puma* by its short tail, less than 25% of the head-and-body length, and the hair on the head sides extended. From *L. rufus*, by its triangular ears ending in a long black hair tuft; tail black at the tip; hind limbs longer than forelimbs and paws covered with long and dense hair.

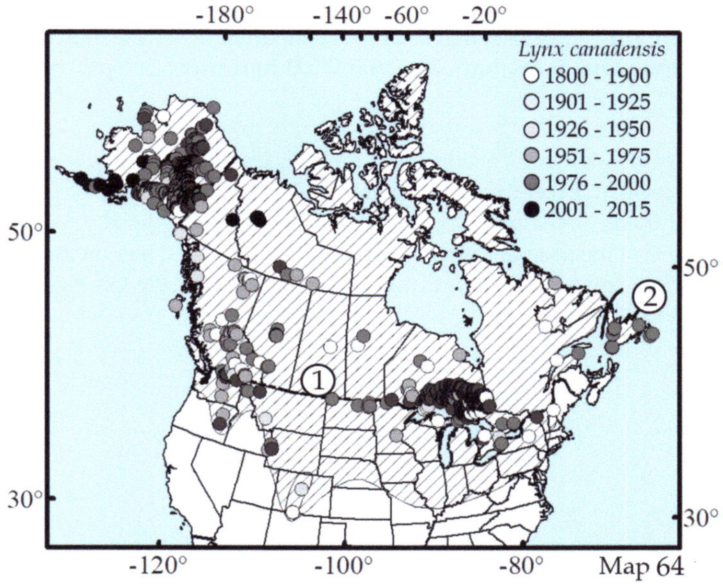

Map 64 Distribution map of *Lynx canadensis*

Additional Literature: Hemmer (1978), Mattyushkin (1979), Kurtén and Anderson (1980), Hall (1981), Tumlison (1987), Lariviére and Walton (1997), García-Perea (1992), McKenna and Bell (1997), Buskirk et al. (2000), Sunquist and Sunquist (2002), Poole (2003), Schwartz et al. (2003a, b), Johnson et al. (2006), Hunter (2015), Kitchner et al. (2017), Lavoie et al. (2019).

Lynx rufus (Schreber, 1777)

Bobcat, gato montés

1. *L. r. baileyi* Merriam, 1890. From central Nevada to central Kansas south through Durango.
2. *L. r. californicus* Mearns, 1897. From northern-central California to northwestern Baja California.
3. *L. r. escuinapae* Allen, 1903. From central Sonora, Sinaloa, east through southern Tamaulipas south through Puebla.
4. *L. r. fasciatus* Rafinesque, 1817. From Southwestern British Columbia to northwestern California.
5. *L. r. floridanus* Rafinesque, 1817. From Louisiana, Arkansas, and southeastern Missouri east through southwestern Virginia and south through Florida.
6. *L. r. gigas* Bangs, 1897. From southeastern Ontario, northeastern Vermont, and New Hampshire east through Nova Scotia.
7. *L. r. oaxacensis* Goodwin, 1963. Restricted to Oaxaca.
8. *L. r. pallescens* Merriam, 1899. From central British Columbia east through Saskatchewan and south through central Nevada and central Kansas.
9. *L. r. peninsularis* Thomas, 1898. Baja California Peninsula.
10. *L. r. rufus* (Schreber, 1777). From southwestern Saskatchewan south through central Oklahoma and east through Atlantic Coast between Virginia and New Hampshire, excluding Minnesota and Wisconsin.
11. *L. r. superiorensis* Peterson and Downing, 1952. Southeastern Manitoba, Ontario, Minnesota, Wisconsin, and northwestern Minnesota.
12. *L. r. texensis* Allen, 1895. Texas, Coahuila, Nuevo León, and Tamaulipas.

Conservation: *Lynx rufus* is listed as Least Concern by the IUCN (Kelly et al. 2016).

Characteristics: Bobcat is medium sized; total length 787.0–1,252.0 mm in males and 710.0–1,219.0 mm in females; condylobasal length 109.0–125.0 mm in males and 103.0–117.0 mm in females. Dorsal pelage gray to reddish brown; spot present or absent; fur of underparts whitish to pale brown, with spots especially on the inner part of the limbs; **ears triangular, ending in a small black hair tuft; cheeks fringed by whitish short hair; tail black at the tip, fur of underparts white; paws not covered with long and dense hair, less than 75.0 mm wide;** anterior condiloid foramen confluent with the posterior foramen lacerum.

Comments: *Lynx rufus* ranges from southern Canada south through Oaxaca, México, including all the United States and central México to the Eje Neovolcánico, except for the coastal plains of México (Map 65). It can be found in sympatry, or nearly so, with *L. canadensis* and *Puma concolor. L. rufus* can be differentiated from *Herpailurus, Leopardus, Panthera,* and *Puma* by its short tail, less than 25% of the head-and-body length, and the extended hair on the head sides. From *L. canadensis,* by having triangular ears ending in a small black hair tuft; only the dorsal tip of the tail black; hind and forelimbs of similar size and naked paws.

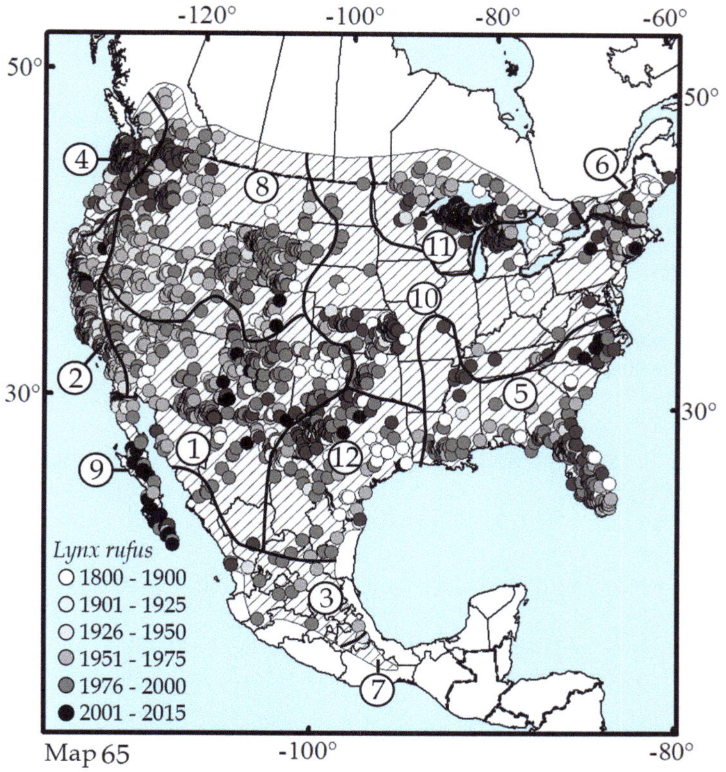

Map 65 Distribution map of *Lynx rufus*

Additional Literature: Hemmer (1978), Mattyushkin (1979), Hall (1981), Lariviére and Walton (1997), McKenna and Bell (1997), García-Perea (1992), Sunquist and Sunquist (2002), Johnson et al. (2006), Millions and Swanson (2007), Eizirik et al. (2008), Hass (2009), Riley et al. (2010), Hunter (2015).

Genus *Puma* Jardine, 1834

Puma has the dorsal **coloration grizzled gray or dark brown to shades from buff to ferruginous; with a more intense middorsal line from the top of the head to the tail base**; shoulders and flanks paler; fur of underparts dull whitish; sides of muzzle black; chin, medial muzzle, and throat white; **ears black externally**; body slender; ears small and rounded; **tail long, cylindrical, and about one-third of total length;** skull short and broad; sagittal crest convex in outline; lambdoid crest with a deep lateral concavity; **spicule-like bregmatic parietal processes extending anteromedially over the frontals, approaching or reaching the temporal ridge and fusing with the frontals in old specimens; nasals expanded distally**; palate wide; grooves on the canines obsolescent or absent. The genus of this species was previously considered *Felis* but Groves (1982) and Newman et al. (1985) proposed its assignment as *Puma.*

Puma concolor (Linnaeus, 1771)

Cougar, puma

1. *P. c. azteca* (Merriam, 1901). From Arizona and New Mexico south through Jalisco and Morelos and northwest through southwestern San Luis Potosí.
2. *P. c. browni* (Merriam, 1903). Northern Baja California, eastern California, western Arizona, and southern Nevada.
3. *P. c. californica* (May, 1896). From southwestern Oregon south through northwestern Baja California, including western Nevada.
4. *P. c. coryi* (Bangs, 1899). From Louisiana and Arkansas east through South Carolina, including the entire Florida Peninsula.
5. *P. c. couguar* (Kerr, 1792). From central Ontario, eastern Illinois, and eastern Missouri east through the Atlantic coast from South Virginia to New Brunswick.
6. *P. c. hippolestes* (Merriam, 1897). From southern Idaho and northern Nevada west through western Wisconsin, Iowa, and Kansas.
7. *P. c. improcera* (Phillips, 1912). From central Baja California Sur south through the Baja California Peninsula.
8. *P. c. kaibabensis* (Nelson and Goldman, 1931). Nevada, Western Utah, and northern Arizona.
9. *P. c. mayensis* (Nelson and Goldman, 1929). From Veracruz and Michoacán south through Chiapas, including the Yucatán Peninsula.
10. *P. c. missoulensis* (Goldman, 1943). From southern Yukon south through eastern Oregon, Idaho, Montana, North Dakota, and southwestern Ontario.
11. *P. c. oregonensis* (Rafinesque, 1832). Southwestern British Columbia, Washington, and Oregon.
12. *P. c. schorgeri* (Jackson, 1955). From western Ontario, Wisconsin, Iowa, and Kansas east through Illinois, Michigan, and central Ontario.
13. *P. c. stanleyana* (Goldman, 1938). From western New Mexico, Oklahoma, and Texas south through eastern Chihuahua, Coahuila, Nuevo León, and Tamaulipas.
14. *P. c. vancouverensis* (Nelson and Goldman, 1932). Restricted to Vancouver Island, British Columbia.

Conservation: *Puma concolor coryi* and *P. c. couguar* are listed as Endangered by the United States Endangered Species Act (FWS 2022) and as Least Concern by the IUCN (Nielsen et al. 2015).

Characteristics: Cougar or puma is large sized; total length 2,200.0–2,300.0 mm in males and 2,000.0–2,100.0 mm in females; condylobasal length 152.0–159.0 mm in males and 140.0–156.0 mm in females. **Dorsal pelage uniform in pale to grayish brown shades;** color, texture, and hair length vary along its geographic range; fur of underparts whitish; area around the lips white and medium area of the rostrum darker; tail long, black at the tip; **temporal bones with small processes in the upper part of the skull inserted in frontals, which can be seen in dorsal or lateral views; sagittal crest convex in outline; lamboid crest with a deep lateral concavity**.

Comments: *Puma concolor* ranges from southern Yukon, Mackenzie, central Saskatchewan, and southern Manitoba, Ontario, and Québec, Canada, south through South America, including all México and the United States (Map 66). It can be found in sympatry, or nearly so, with *Herpailurus yagouaroundi, Leopardus pardalis, L. weidii, Lynx rufus,* and *Panthera onca. P. concolor* can be distinguished from *P. onca* by having the dorsal pelage uniform in coloration; posterior border of temporal bones with a process on the frontals. From the other genus of the family Felidae, by its total and skull lengths larger than 1400.0 and 158.0 mm, respectively.

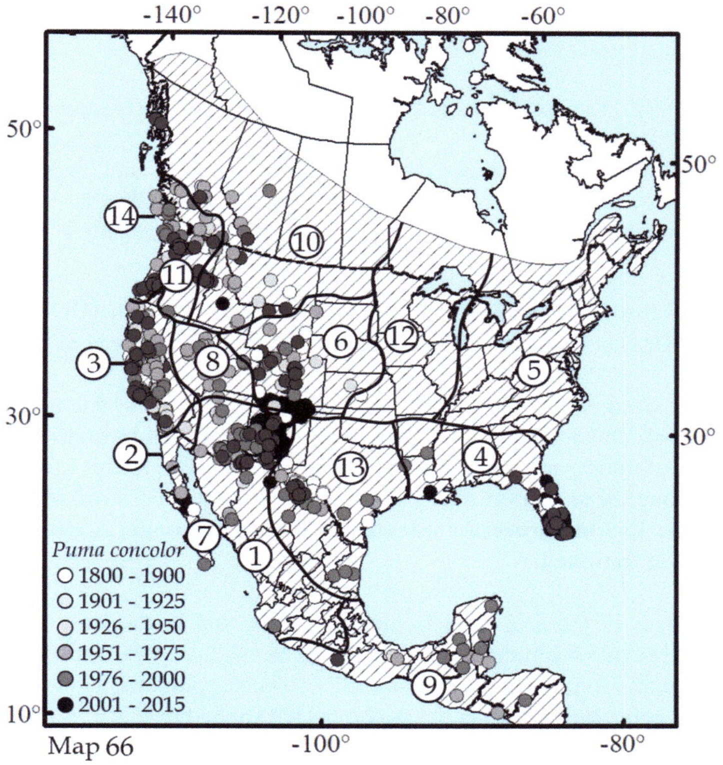

Map 66 Distribution map of *Puma concolor*

Additional Literature: Young and Goldman (1946), Currier (1983), Culver et al. (2000), Sunquist and Sunquist (2002), Laundré and Clark (2003), Johnson et al. (2006), Eizirik et al. (2008), Hass (2009), Hunter (2015).

Subfamily Pantherinae Pocock, 1917

The cats of this subfamily are characterized **in not having a bony hyoid, for that reason they are able to roar.**

Genus ***Panthera*** Oken, 1816

Jaguars have two main color patterns. **The first varies from pale ochraceous buff to pale golden tawny, heavily spotted with black irregularly circular, crescentic, or broken spots that tend to form rosettes or endorsing smaller spots. The second pattern is a dark coloration from deep brown to blackish until dark spots can be seen on the back and flanks**; size large; tail large and tapering; ears small, short, and rounded; pelage short and rather bristly; skull short and broad; **sagittal crest well developed; lambdoid crest with a high process in the central back of the skull**; spicule-like bregmatic processes of parietals not extending anteromedially over the frontals; nasals expanded distally; palate wide; **canine base width greater than molar width; braincase relatively small relative to the skull size**. The genus was previously considered *Felis* but Groves (1982) and Newman et al. (1985) proposed assigning it as *Panthera.*

Panthera onca (Linnaeus, 1758)

Jaguar, jaguar

1. *P. o. arizonensis* (Goldman, 1932). From Arizona and New Mexico south through southern Sonora.
2. *P. o. centralis* (Mearns, 1901). Eastern Chiapas.
3. *P. o. goldmani* (Mearns, 1901). From Veracruz south through northern Chiapas and the Yucatán Peninsula.
4. *P. o. hernandesii* (Gray, 1858). From southern Sonora south through southern Chiapas.
5. *P. o. veraecrucis* (Nelson and Goldman, 1933). From Louisiana, Texas, and Tamaulipas.

Conservation: *Panthera onca* is listed as Endangered in the Norma Oficial Mexicana (DOF 2019) and by the United States Endangered Species Act (FWS 2022), and as Near Threatened by the IUCN (Quigley et al. 2017).

Characteristics: Jaguar is large sized; total length 1,727.0–2,419.0 mm, condylobasal length 199.5–237.5 mm. Dorsal pelage yellowish pale brown to reddish brown **with darker brown spots surrounded by stripes in a darker color or blackish shades but not continuous**; fur of underparts white but conserving the spotted pattern; upper jaw of a uniform coloration; area around the lips white; tail long; head proportionally large. There is a melanic form; in some cases, the spotted pattern can be observed; **temporal bones lacking processes inserted in the frontal bone; sagittal crest more developed in the back of the skull; lambdoid crest flattened**.

Comments: *Panthera onca* ranges in the east from Louisiana, Texas, and in the west from Arizona and New Mexico through the tropical lowlands of México south through South America, including the Yucatán Peninsula (Map 67). It can be found in sympatry, or nearly so, with *Herpailurus yagouaroundi, Leopardus pardalis, L. weidii,* and *Puma concolor. P. onca* can be distinguished from *Puma concolor* by having the dorsal pelage spotted or black; if black, darker brown spots can be seen surrounded by stripes in blackish shades; posterior border of the temporal bones with a process on the frontal and sagittal crest more developed in the back of the skull. From the other genus of the family Felidae, by its larger total and skull lengths, 1400.0 and 158.0 mm, respectively.

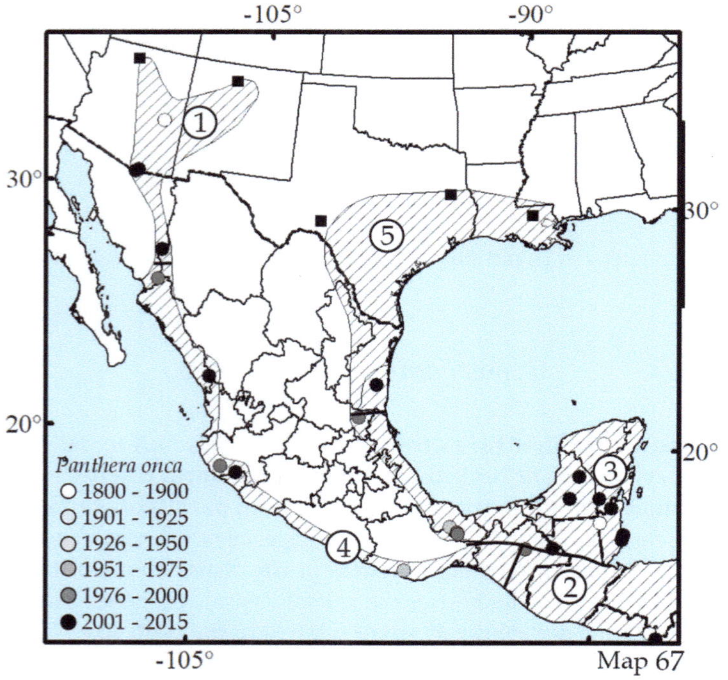

Map 67 Distribution map of *Panthera onca*

Additional Literature: Hall (1981), Seymour (1989), Swank and Teer (1989), Sunquist and Sunquist (2002), Johnson et al. (2006), Eizirik et al. (2008), Hunter (2015).

Order Chiroptera

Order Chiroptera Blumenbach, 1779

Chiroptera include all bats. It is the second most diverse order of mammals, containing approximately 20% of all species, and the only that possess true flight. To achieve this unique locomotory attribute, the front limbs are modified into wings with a naked and double-layered membrane stretched over greatly elongated phalanges and metacarpals; this membrane may extend to the hind limbs and enclose the tail in many species. Other attributes accommodating flight include thumb free of the wing membrane and usually clawed; knee joint directed outward and posteriorly; ears well developed, often of large size; very small eyes in most species; a reduced ulna of limited functionality; a large radius; an usually keeled sternum and greatly developed pectoral musculature; a well-developed clavicle; a dorsally directed glenoid fossa of the scapula; generally tubular limb bones; thoracic and cervical vertebrae that lack neural spines; a dome-shaped skull with an usually inflated braincase and a frontal region both inflated and concave. The distribution is nearly cosmopolitan. Bats are divided into two suborders: Megachiroptera (often diurnal bats limited to the Old World) and Microchiroptera (strictly nocturnal bats with worldwide distribution). Eight families occur in North America.

Suborder Microchiroptera Dobson, 1875

Use of echolocation; tragus present; eyes comparatively smaller than in Macrochiroptera; second finger closely bound to the third; humerus with large trochlea usually articulating with the scapula. All New World bats are in this suborder. Eight families occur in North America: Emballonuridae, Molossidae, Mormoopidae, Natalidae, Noctilionidae, Phyllostomidae, Thyropteridae, and Vespertilionidae. The keys were elaborated based on the review of specimens and the following sources: Miller (1907), Lawlor (1979), Hall (1981), Álvarez et al. (1994), and Álvarez-Castañeda et al. (2015, 2017a).

1. Noseleaf present (Fig. 1); with or without tail...family Phyllostomidae (p. 246)

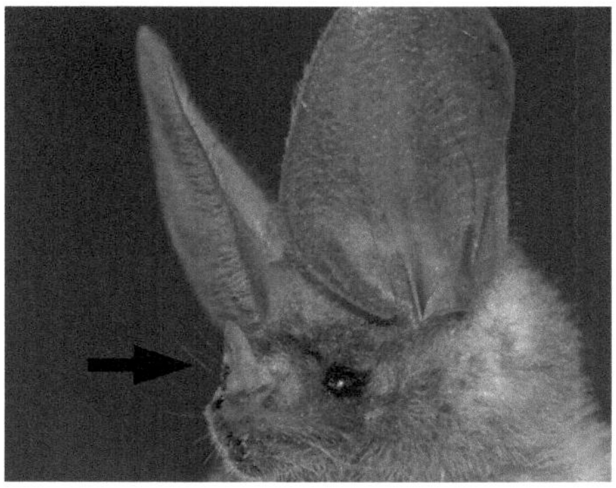

Fig. 1 Noseleaf present

S. T. Álvarez-Castañeda, *Mammals of North America - Volume 1*, https://doi.org/10.1007/978-3-031-41661-3_5

1a. Noseleaf absent; tail present 2

2. Extreme distal part of the tail projecting above the middle of the uropatagium, never reaching the posterior border (Fig. 2).. 3

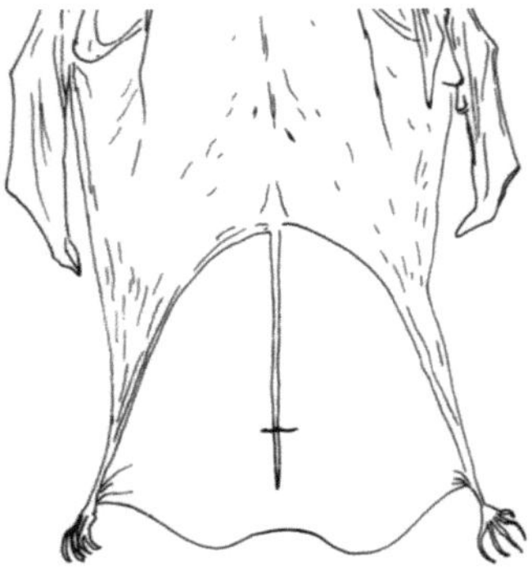

Fig. 2 Extreme distal part of the tail projecting above the middle of the uropatagium

2a. Uropatagium completely encloses the tail; tail sometimes extending past the posterior border 5

3. Second wing digit without phalanges, only with the metacarpal.. family Emballonuridae (p. 201)

3a. Second wing digit with phalanges 4

4. Hind feet with very large, laterally compressed claws; foot length in dry measurement greater than 70% of the tibia length.. family Noctilionidae (p. 244)

4a. Hind feet with claws not clearly flat or sharp; foot length less than 70% of the tibia length .... family Mormoopidae (p. 234)

5. Adhesive disks in the wrist and in the ankle (from Veracruz and coast of Chiapas south through South America) .... family Thyropteridae; *Thyroptera tricolor* (p. 322)

5a. No adhesive disks.. 6

6. Ears forming a visor over the eyes (Fig. 3); tail extending at least 7.0 mm distally from the uropatagial border.. family Molossidae (p. 212)

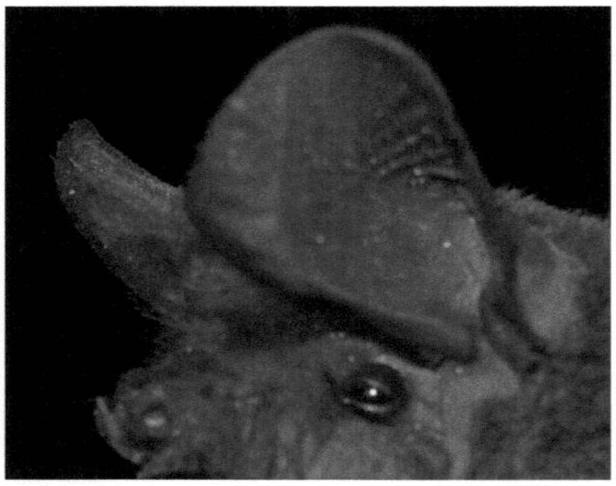

Fig. 3 Ears forming a visor over the eyes

6a. Ears not forming a visor and widely separated in the dorsal medium part; distal end of the tail posterior to the border of the uropatagium less than 7.0 mm 7

7. Ears funnel shaped (Fig. 4); tibia length greater than 47% of the forearm; braincase spherical .... family Natalidae (p. 241)

Fig. 4 Ears with funnel shaped

7a. Ears not funnel shaped; tibia length less than 47% of the forearm; nostrils open laterally or forward, behind a horseshoe-shaped fold (Fig. 5); braincase not spherical .... family Vespertilionidae (p. 323)

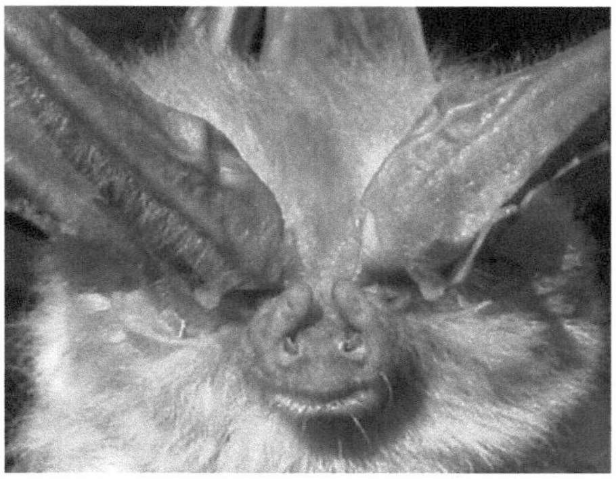

Fig. 5 The nostrils open laterally or forward, behind a horseshoe-shaped fold

Family Emballonuridae Gervais, 1855

The family Emballonuridae includes sac-winged bats characterized by a small to medium size; **glandular sac in the plagiopatagium, close to the elbow (in most genera) or at the tail base; tail emerging from the uropatagium dorsal surface; postorbital process prominent;** ears moderately large and round, frequently joined at the base without a ventral extension under the eyes; tragus small; third digit with two phalanges. Range pantropical. Belonging to superfamily Emballonuroidea. All species found in North America are in the subfamily Emballonuridae. The keys were elaborated based on the review of specimens and the following sources: Lawlor (1979), Hall (1981), Álvarez et al. (1994), and Álvarez-Castañeda et al. (2017a).

1. Hair white; broad postorbital process almost confluent with the supraorbital crest (Fig. 6; from Nayarit and Veracruz, México, through South America).. subgenus *Diclidurus; Diclidurus albus* (p. 206)

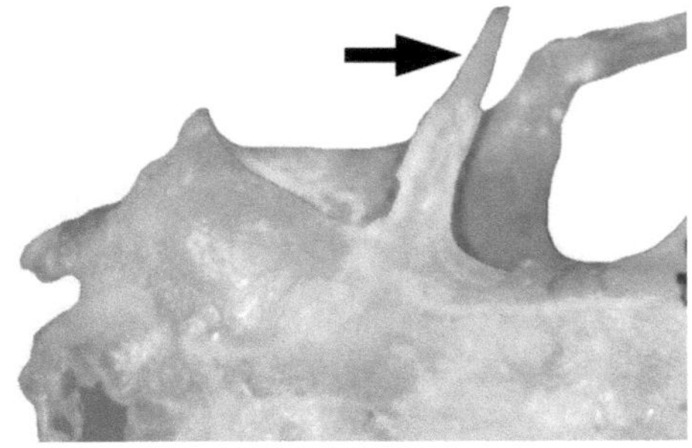

Fig. 6 Postorbital process broad almost confluent with the supraorbital crest like in *Diclidurus albus*

1a. Hair not white; postorbital process thin 2

2. No wing sac in propatagium 3

2a. Wing sac in propatagium (Fig. 7) 4

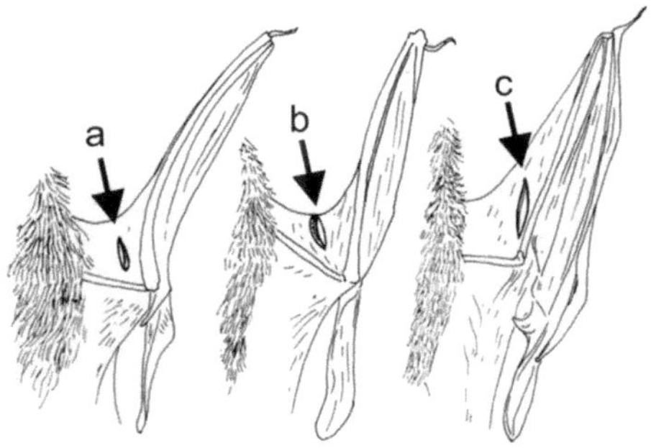

Fig. 7 Wing sac in propatagium

3. Forearm length less than 41.0 mm, with five to seven locks of pale-colored hair; basisphenoid cavity divided by a septum (Fig. 8; from Veracruz, Oaxaca and southern Yucatán Peninsula, México, south through South America) .... *Rhynchonycteris naso* (p. 209)

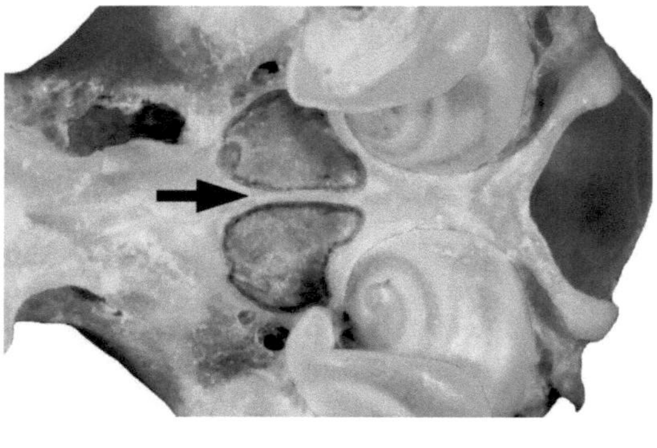

Fig. 8 The basisphenoid cavity is divided by a septum

3a. Forearm length greater than 41.0 mm, without locks of pale-colored hair; basisphenoid cavity not divided by septum (Fig. 9; from south Veracruz, México, south through South America) .... *Centronycteris centralis* (p. 205)

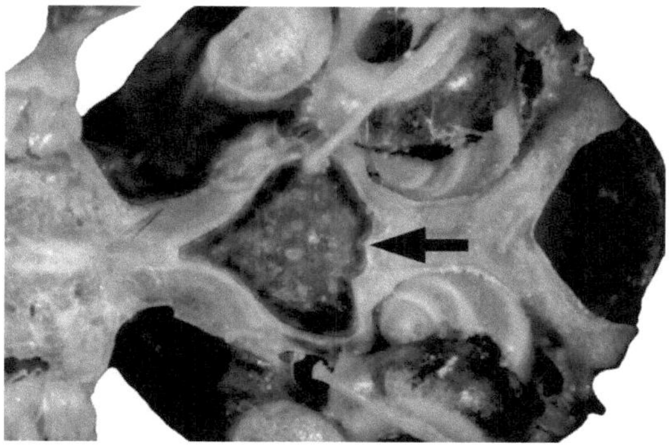

Fig. 9 The basisphenoid cavity is not divided by septum

4. Two contrasting parallel white stripes on the back from the shoulders to the hips; wing sac in the propatagium close to the forearm, opening toward the forearm (Fig. 7c); nasal and frontal bones straight with an angle greater than 160° .... *Saccopteryx* (p. 210)

4a. No dorsal stripes; wing sac in the propatagium separated from the forearm, opening toward the tip of the wing or body (Fig. 7b); nasal and frontal bones with an angle less than 160° 5

5. Forearm length less than 41.0 mm; wing sac in the central part of the propatagium, opening toward the body (Fig. 7a); anterior part of the face inflated (Fig. 10); basisphenoid cavity often divided by a septum .... *Balantiopteryx* (p. 204)

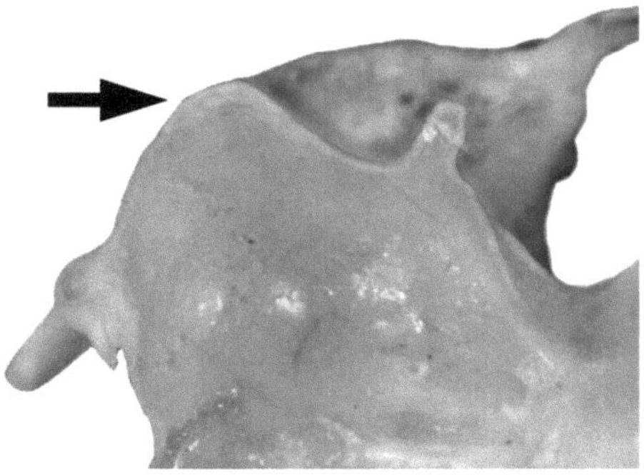

Fig. 10 The anterior part of the face inflated

5a. Forearm length greater than 41.0 mm; wing sac close to the anterior border of the propatagium, opening toward the tip of the wing (Fig. 7); anterior part of the face not inflated (Fig. 11); basisphenoid cavity not divided .... *Peropteryx* (p. 207)

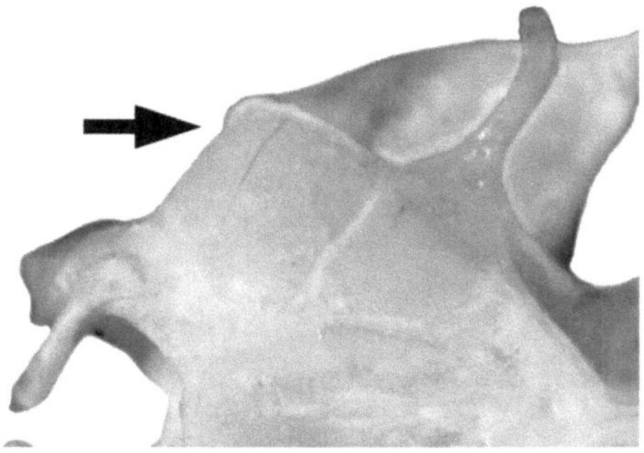

Fig. 11 The anterior part of the face is not inflated

Genus *Balantiopteryx* Peters, 1867

Dorsal color matte brown or pale gray; wing sac in males in the center of the propatagium, flask shaped, and with an opening directed proximally; ears slightly rounded distally, outer margin straight or slightly convex; premaxillary bones separated by the maxilla; basisphenoid cavity divided or not by a septum; **anterior region of the rostrum with two semi-spherical protuberances**; first upper premolars approximately one-quarter smaller than the second premolars. The keys were elaborated based on the review of specimens and the following sources: Hall (1981), Arroyo-Cabrales and Jones (1988a, b), Álvarez et al. (1994), and Álvarez-Castañeda et al. (2017a).

1. Forearm length greater than 38.2 mm; dorsal color grayish; wing membrane with a white border; interpterygoid fossa narrow, V-shaped (from southern Sonora and Veracruz, including the southern Baja California Peninsula, México, south through Central America, except for the Yucatán Peninsula) .... *Balantiopteryx plicata* (p. 205)

1a. Forearm length less than 38.9 mm; overall color brown; wing membrane without a white border; interpterygoid fossa broadly U-shaped (from southern Veracruz and northern Oaxaca, México, south through Guatemala)............................ .... *Balantiopteryx io* (p. 204)

Balantiopteryx io Thomas, 1904

Thomas's sac-winged bat, murciélago pardo de sacos en las alas

Monotypic

Conservation: *Balantiopteryx io* is listed as Least Concern by the IUCN (Lim 2015).

Characteristics: Thomas's sac-winged bat is small sized; forearm length 36.0–38.8 mm, skull length 12.4–12.9 mm. **Dorsal color dark brown**; fur of underparts grayish brown; i**nterpterygoid fossa broad and U-shaped; basisphenoid pits large, longer than broad, not divided by a medium septum**.

Comments: *Balantiopteryx io* occurs in the most humid Neotropical regions of México, including southern Veracruz, northeastern Oaxaca, Tabasco, and Chiapas, south through Guatemala (Map 1). Throughout its distribution, *B. io* can be found in sympatry, or nearly so, with *B. plicata*, from which it can be distinguished by its dorsal pelage brownish instead of grayish; absence of a white wing border; maxillary toothrow length less than 5.0 mm; forearm length less than 39.0 mm; skull length usually less than 13.0 mm; interpterygoid fossa broad, U-shaped; basisphenoid pits large, longer than broad, not divided by a medial septum.

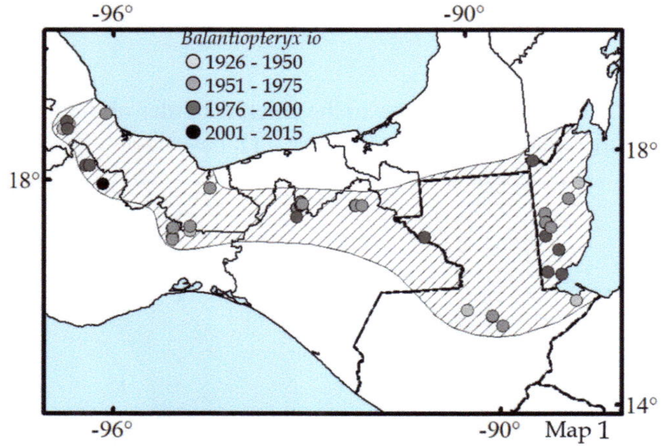

Map 1 Distribution map of *Balantiopteryx io*

Additional Literature: Sanborn (1936, 1937), Jones (1966), Villa-R. (1967), Kirkpatrick et al. (1975), Arroyo-Cabrales and Knox Jones (1988a, b), Jones and Hood (1993), Simmons (2005).

Balantiopteryx plicata Peters, 1867

Gray sac-winged bat, murciélago gris de sacos en las alas

1. *B. p. pallida* Burt, 1948. Southern Sonora, northern Sinaloa, and southern Baja California Sur.
2. *B. p. plicata* Peters, 1867. From central Sinaloa and Tamaulipas south through Central America.

Conservation: *Balantiopteryx plicata* is listed as Least Concern by the IUCN (Lim et al. 2016a).

Characteristics: Gray sac-winged bat is also small but larger than *B. io*; forearm length 38.3–45.0 mm, skull length 11.5–14.8 mm. **Dorsal pelage pale gray** giving the impression of being dull; **wing membrane with a white border;** fur of underparts pale; forearm in pinkish shades; **interpterygoid fossa narrow, V-shaped; basisphenoid pits smaller, divided by a medial septum**.

Comments: *Balantiopteryx plicata* ranges through the Neotropical Region of México, from southern Sonora and central Veracruz south through Costa Rica (Map 2). The record from Cueva Sabinas, near Valles, San Luis Potosí (Villa 1967) cannot be confirmed; for this reason, it has not been included. This species is found in sympatry with *B. io* only in the most tropical humid area of México. *P. plicata* differs from *B. io* by its grayish rather than brownish dorsal color; not white on the wing edge, and maxillary toothrow length equal to or greater than 5.0 mm.

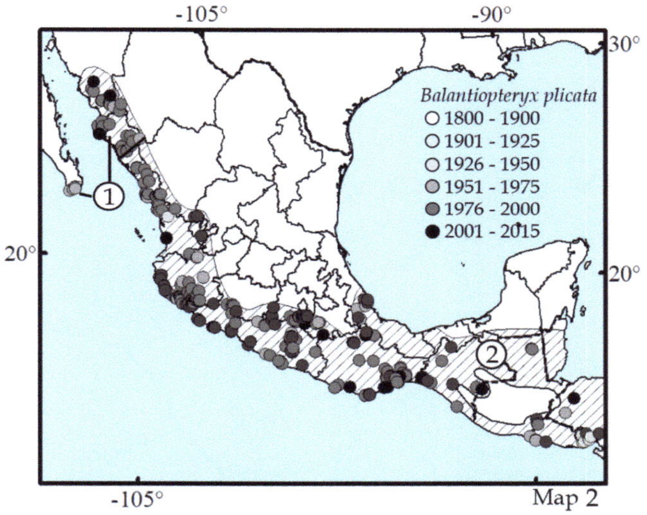

Map 2 Distribution map of *Balantiopteryx plicata*

Additional Literature: Miller (1907), Sanborn (1937), Martínez and Villa (1938), Álvarez (1968), Phillips and Jones (1969), Jones et al. (1972), Lópes-Forment (1981), Arroyo-Cabrales and Jones (1988a), Jones and Hood (1993), Simmons (2005).

Genus *Centronycteris* Gray, 1838

Small size; pelage long and soft; **without a wing sac**; angle between the rostrum and the frontal bone wide; lower orbit edge so little expanded that the edge of the toothrow can be seen from above; **first upper premolars with two distinct cusps, anterior and posterior; basisphenoid cavity divided by a median septum** (Sanborn 1937).

Centronycteris centralis Thomas, 1912

Thomas's shaggy bat, murciélago lanudo

Monotypic.

Conservation: *Centronycteris centralis* is listed as subject to Special Protection in the Norma Oficial Mexicana (DOF 2019) and as Least Concern by the IUCN (Arroyo-Cabrales et al. 2015c).

Characteristics: Thomas's shaggy bat is the smallest member of the family; forearm length 42.0–46.0 mm, skull length 11.2–12.6 mm. Dorsal pelage speckled yellowish or grayish brown with pale, poorly contrasting parallel bands on the back; fur of underparts pale yellowish to grayish; no wing or tail sac; **nose very sharp giving the impression of having a small proboscis;** premaxillary bone separated from the maxilla; basisphenoid cavity not divided by a septum, and the first premolars approximately one-third of the size of the second premolars.

Comments: *Centronycteris centralis* was previously considered a subspecies of *C. maximiliani* (Simmons and Handley 1998). *C. centralis* ranges in the most humid Neotropical region of México, including southern areas of Veracruz, Tabasco, Campeche, Quintana Roo, and northern Chiapas, south through South America (Map 3). It differs from the other species by the absence of a sac in the propatagium.

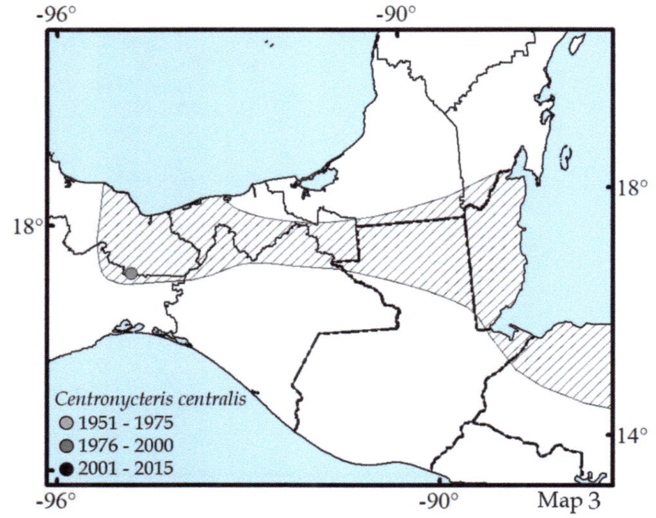

Map 3 Distribution map of *Centronycteris centralis*

Additional Literature: Álvarez-Castañeda and Álvarez (1991), Simmons and Handley (1998), Olguín-Monroy et al. (2008), Santos–Moreno (2000).

Genus *Diclidurus* Wied-Neuwied, 1820

Medium size; **white color; wing sac absent but with a very little sac close to the tip of the tail in the uropatagium**; nose simple; tail short, perforating the uropatagium; **postorbital processes wide, almost confluent with the supraorbital crests;** braincase large and flattened anteriorly, **abruptly descending to the rostrum;** rostrum broad, bearing lateral ridges.

Subgenus *Diclidurus* Wied-Neuwied, 1820

Same characteristics as the genus.

Diclidurus albus Wied-Neuwied, 1820
Northern ghost bat, murciélago blanco

D. a. virgo Thomas, 1903. Only subspecies in North America.

Conservation: *Diclidurus albus* is listed as Least Concern by the IUCN (Lim et al. 2016b).

Characteristics: Northern ghost bat is medium sized; forearm length 63.0–69.0 mm, skull length 17.0–19.6 mm. **Dorsal pelage and fur of underparts white; no wing sac but with a very little sac close to the tip of the tail;** premaxillary bone separated from the maxilla; **postorbital process a little wide**, and the first premolars approximately less than one-fourth of the size of the second premolars.

Comments: *Diclidurus* was previously in the subfamily Diclidurinae, but Simmons (2005) allocated the genus to the subfamily Emballonuridae. Ojasti and Linares (1971) treated North American specimens as the separate species, *D. virgo*; Jones and Hood (1993) disagreed, treating *D. virgo* as a junior synonym of *D. albus*. The current treatment (e.g., Hood and Gardner 2008) follows Thomas (1903) in assigning *D. virgo* as a subspecies of *D. albus*. The range of *D. albus* is associated with palms from Nayarit and Veracruz, excluding the Yucatán Peninsula, México, south through South America (Map 4). It can be distinguished from all other species from México to the north by being the only white bat of North America; it lacks a wing sac but has a very small sac close to the tip of the tail.

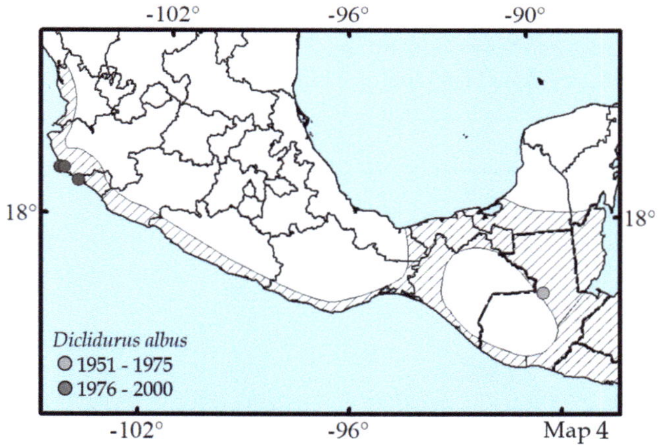

Map 4 Distribution map of *Diclidurus albus*

Additional Literature: Goodwin (1969), Hall (1981), Ceballos and Medellín (1988).

Genus *Peropteryx* Peters, 1867

Dorsal pelage dark reddish brown to dark brown; fur of underparts dark brown; hair long; snout almost bare and very elongated; **wing sac in the anterior propatagium border, opening directed distally; wing membrane extending to the ankle; ears dark brown or black;** skull with an angle between the nasal and frontal bone, approximately 115°; premaxillary bones separated by the maxilla; basisphenoid cavity not divided by a septum; the first upper premolars approximately one-quarter of the size of the second premolars. North American *Peropteryx* are in nominate subgenus. The keys were elaborated based on the review of specimens and the following sources: Lawlor (1979), Hall (1981), Álvarez et al. (1994), Yee (2000), and Álvarez-Castañeda et al. (2017a).

1. Forearm length greater than 45.0 mm; ears black; skull length greater than 16.0 mm (from southern Veracruz, Oaxaca, and Tabasco, México, south through South America) ....*Peropteryx kappleri* (p. 208)

1a. Forearm length less than 48.2 mm; ears dark brown, not black; skull length less than 16.0 mm (from Oaxaca and Tabasco, México, south through South America) .... *Peropteryx macrotis* (p. 209)

Subgenus *Peropteryx* Peters, 1867

Wings black; wing sac near the upper edge of the propatagium and opening outward; skull, in lateral view, with a very sharp angle between the expanded rostrum and the braincase; basisphenoid pit not divided.

Peropteryx kappleri Peters, 1867

Greater dog-like bat, murciélago perro mayor

P. k. kappleri Peters, 1867. Only subspecies in North America.

Conservation: *Peropteryx kappleri* is listed as subject to Special Protection in the Norma Oficial Mexicana (DOF 2019) and as Least Concern by the IUCN (Davalos et al. 2018).

Characteristics: Greater dog-like bat is small sized; **forearm length 45.0–52.0** mm, skull length 16.0–17.8 mm. Dorsal pelage dark reddish brown to dark brown; fur of underparts dark brown; fur long; **ears black;** skull similar to *P. macrotis* but larger.

Comments: *Peropteryx kappleri* is distributed in the humid Neotropical region of México from southern Veracruz, Tabasco, and Chiapas south through South America (Map 5). It differs from *P. macrotis* and *P. kappleri* by its larger size, forearm greater than 45.0 mm; skull length greater than 17.8 mm, and black ears.

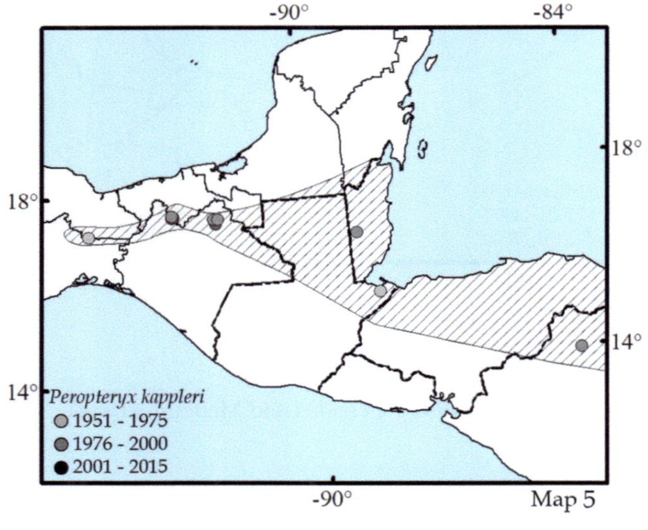

Map 5 Distribution map of *Peropteryx kappleri*

Additional Literature: Villa-R. (1967), Goodwin (1969), Koopman (1993), Jones and Hood (1993), Simmons and Voss (1998), Yee (2000).

Peropteryx macrotis (Wagner, 1843)

Lesser dog-like bat, murciélago perro menor

Monotypic.

Conservation: *Peropteryx macrotis* is listed as Least Concern by the IUCN (Barquez et al. 2015).

Characteristics: Lesser dog-like bat is small sized; **forearm length 43.0–48.0 mm, skull length 12.0–15.0 mm.** Dorsal pelage reddish brown to dark brown; fur of underparts grayish brown; snout almost naked and very elongated; **ears dark brown but not black;** skull similar to *P. kappleri* but smaller.

Comments: *Peropteryx macrotis* is monotypic (Simmons 2005). It is distributed in the Neotropical region of México from the Pacific coast from Oaxaca and southern Tabasco to northern Chiapas and the Yucatán Peninsula south through South America (Map 6). *P. macrotis* can be differentiated from *P. kappleri,* by its smaller size; forearm length less than 45.0 mm; skull length less than 17.8 mm, and brown ears.

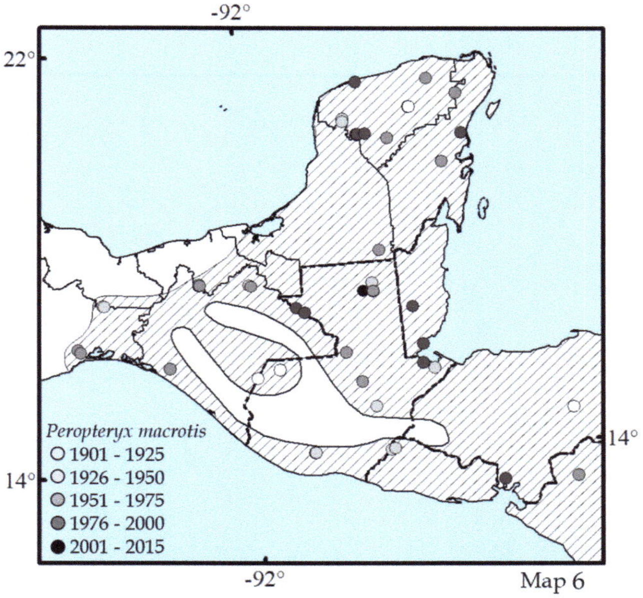

Map 6 Distribution map of *Peropteryx macrotis*

Additional Literature: Sanborn (1937), Benedict (1957), Goodwin and Greenhall (1961), Villa-R. (1967), Willig (1983), Arroyo-Cabrales and Álvarez (1990), Jones and Hood (1993), Koopman (1993), Simmons and Voss (1998), Yee (2000).

Genus *Rhynchonycteris* Peters, 1867

Small size; **without wing sacs, with a series of five to seven locks of pale hairs on the forearm**; uropatagium haired to the border of the tail; **muzzle greatly elongated;** the first upper premolars large, triangular, possessing a cingulum provided with anterior and posterior smaller cusps; **angle between the rostrum and forehead near 170°** (Sanborn 1937).

Rhynchonycteris naso (Wied-Neuwied, 1820)

Proboscis bat, murciélago narigón

Monotypic.

Conservation: *Rhynchonycteris naso* is listed as subject to Special Protection in the Norma Oficial Mexicana (DOF 2019) and as Least Concern by the IUCN (Lim and Miller 2016).

Characteristics: Proboscis bat is the smallest species in the family; forearm length 35.0–41.0 mm, skull length 11.2–12.6 mm. Dorsal pelage and fur of underparts yellowish to grayish brown; no wing or tail sac; ears shaped as a sickle; wings extending to the ankles; **five to seven locks of pale-colored hair on the forearms**; premaxillary bone separated from the maxilla; basisphenoid cavity divided by a septum; first upper premolars with distinct anterior and posterior cusps.

Comments: *Rhynchonycteris naso* ranges in the Neotropical region of México, from Veracruz, Tabasco, Oaxaca, and the southern of the Yucatán Peninsula south through South America (Map 7). *R. naso* can be distinguished by its smaller size, forearm length less than 40.0 mm; skull length less than 12.6 mm; no wing sac; a very sharp nose that mimic a small proboscis, and five to seven locks of pale-colored hair on the forearms.

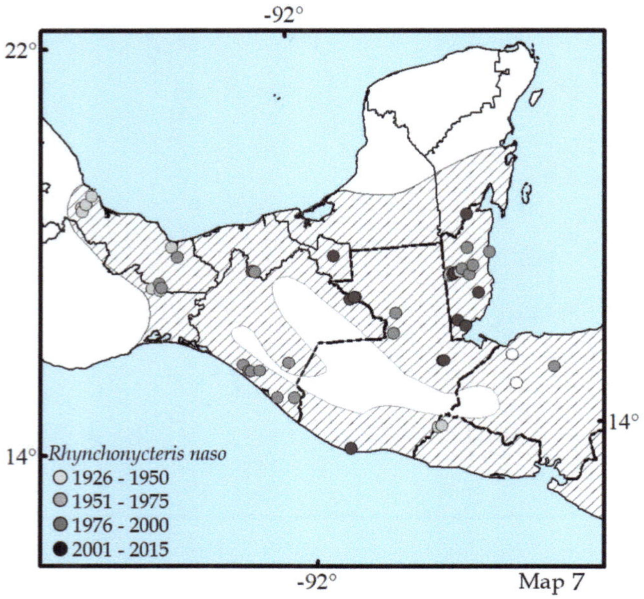

Map 7 Distribution map of *Rhynchonycteris naso*

Additional Literature: Miller (1907), Sanborn (1937), Husson (1962), Brown et al. (1971), Koopman (1982), Plumpton and Jones (1992), Jones and Hood (1993), Simmons (2005).

Genus *Saccopteryx* Illiger, 1811

Dorsal pelage brownish or blackish with two well-defined parallel, pale dorsal stripes; fur of underparts brownish or blackish; a **wing sac well developed in males, parallel to the forearm near the elbow and opening on the upper surface of the antebrachial membrane;** premaxillary bones separated from the maxilla; **angle between the nasal** bones **and the frontal bone narrow**; large basisphenoid cavity divided by a septum; first upper premolars approximately one-quarter of the size of the second premolars; **postorbital processes long and relatively wide**. The keys were elaborated based on the review of specimens and the following sources: Lawlor (1979), Hall (1981), Álvarez et al. (1994), Yancey et al. (1998a, b), and Álvarez-Castañeda et al. (2017a).

1. Forearm length greater than 43.0 mm and the head-and-body length greater than 45.0 mm; membranes blackish; dorsal pelage blackish, with two dorsal and parallel whitish stripes from the shoulders to the hips; maxillary toothrow length greater than 5.8 mm (from Jalisco and Veracruz, México, south through South America) .... *Saccopteryx bilineata* (p. 211)

1a. Forearm length less than 43.0 mm and the head-and-body length less than 45.0 mm; membranes brown; dorsal pelage tawny or chocolate brown, with two contrasting dorsal and parallel non-whitish stripes; maxillary length less than 5.8 mm (from Chiapas, México, south through South America).. *Saccopteryx leptura* (p. 211)

Saccopteryx bilineata (Temminck, 1838)

Greater sac-winged bat, murciélago rayado mayor

1. *S. b. bilineata* (Temminck, 1838). From Veracruz south through Central America.
2. *S. b. villai* Álvarez and González–Ruíz, 2000. From Jalisco south through Central America

Conservation: *Saccopteryx bilineata* is listed as Least Concern by the IUCN (Solari 2015d).

Characteristics: Greater sac-winged bat is small sized; forearm length 44.0–48.0 mm, skull length 15.7–15.9 mm. **Membranes and pelage blackish, with two contrasting parallel stripes on the dorsum;** skull similar to *S. leptura* but larger; skull length greater than 15.0 mm, and maxillary toothrow length greater than 5.8 mm.

Comments: *Saccopteryx bilineata* ranges in the Neotropical region of México, from Jalisco and Veracruz southward in the lowlands, including the Yucatán Peninsula, south through South America (Map 8). *S. bilineata* can be differentiated from other species of the genus *Saccopteryx* by having the membranes and pelage blackish, contrasting with the brownish color of *S. lepturus*, and larger skull and toothrow.

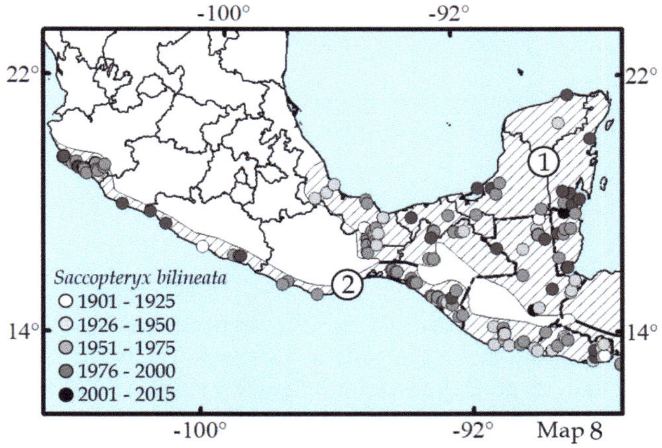

Map 8 Distribution map of *Saccopteryx bilineata*

Additional Literature: Sanborn (1937), Jones et al. (1973), Hall (1981), Linares (1986), Eisenberg (1989), Jones and Hood (1993), Koopman (1993), Simmons and Voss (1998), Yancey et al. (1998a), Álvarez and González–Ruíz (2000), Lim et al. (2004).

Saccopteryx leptura (Schreber, 1774)

Lesser sac-winged bat, murciélago rayado menor

Monotypic.

Conservation: *Saccopteryx leptura* is listed as subject to Special Protection in the Norma Oficial Mexicana (DOF 2019) and as Least Concern by the IUCN (Solari 2015e).

Characteristics: Lesser sac-winged bat is small sized; forearm length 37.0–43.0 mm, skull length 13.1–14.4 mm. **Membranes and pelage tawny to chocolate brown, with two parallel dorsal stripes well defined but minimally contrasting**. Skull similar to *S. bilineata* but smaller; skull length less than 15.0 mm and maxillary toothrow length less than 5.8 mm.

Comments: *Saccopteryx leptura* is distributed only in the humid areas along the coast of Chiapas, México, south through South America (Map 9). It can be differentiated from *S. bilineata* by having the membranes and pelage tawny to chocolate brown rather than blackish, less distinct dorsal stripes, and smaller skull and maxillary toothrow.

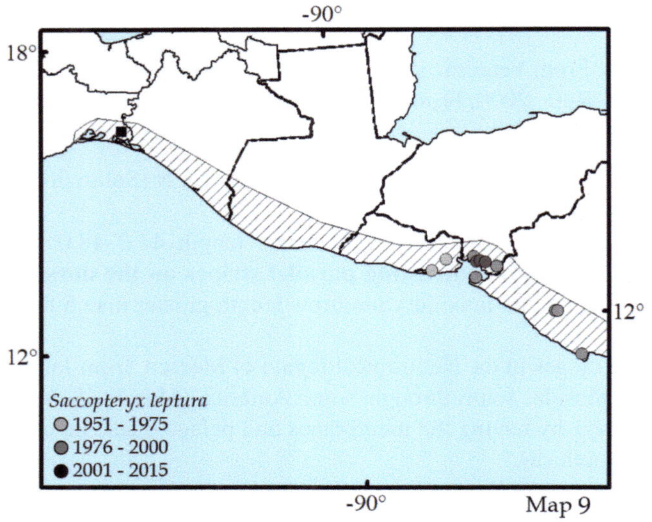

Map 9 Distribution map of *Saccopteryx leptura*

Additional Literature: Sanborn (1937), Bradbury and Vehrencamp (1976), Hall (1981), Linares (1986), Eisenberg (1989), Jones and Hood (1993), Koopman (1993), Yancey et al. (1998a, b), Lim et al. (2004).

Family Molossidae Gervais, 1856

Medium to large sized; **wings long and narrow, adapted for fast flight in open spaces; tail extending well beyond the margin of the relatively short uropatagium;** pinna generally large, sharp or rounded, frequently bound to the forehead and generally projecting backward more than vertically overhead; **tragus very small or absent**; **first phalanx of the third finger bent toward the metacarpal when the wing is folded; thumb relatively big; first and fifth digits of the feet with a fringe of stiff bristles**. Distributed in the New World from southwestern Canada south through the West Indies and Central America south through South America as far south as northern Chile and Argentina. Belonging to the superfamily Vespertilionoidea. The North American species are members of the subfamily Molossinae.

Subfamily Molossinae Gervais, 1856

Medium to large sized; wings long and narrow; **tail extending well beyond the margin of the relatively short uropatagium with more than two caudal vertebrae;** ears generally large, sharp or rounded, frequently bound to the forehead and generally projecting backward more than vertically overhead; tragus very small or absent; first phalanx of the third finger bent toward the metacarpal when the wing is folded; thumb relatively big; first and fifth digits of the feet with a fringe of stiff bristles; **seventh cervical and first thoracic vertebrae not fused**. The keys were elaborated based on the review of specimens and the following sources: Miller (1907), Lawlor (1979), Hall (1981), Álvarez et al. (1994), and Álvarez-Castañeda et al. (2017a).

1. Upper lip without perpendicular grooves; premaxillary bones in contact anteriorly, sometimes showing a notch but not extending posteriorly of the incisor roots 2

1a. Upper lip with perpendicular grooves; premaxillary bones separated anteriorly by a medial emargination extending posterior to the incisor roots (Fig. 12) 5

Fig. 12 Premaxillary bones separated anteriorly by a medial emargination extending posterior to the incisor roots

2. Ears tapering toward the base; antitragus rounded with a low neckline notch on the base; hair long on the hips with bristles extending from the fur; breadth of upper incisors (from extreme interior to exterior) of the same size or greater than tooth length .... *Molossus* (p. 222)

2a. Ears not tapering toward the base; if the antitragus is not rounded, hair on the hips short; breadth of the upper incisors much less than tooth length 3

3. Antitragus rounded with a low neckline on the base; palatine dome-shaped (from Nayarit, Estado de México, and southern Veracruz, México, south through South America) .... *Promops centralis* (p. 232)

3a. Antitragus not rounded and without a low neckline on the base; palatine plain or slightly arched 4

4. Uropatagium thick and rigid; hair on the dorsum of the middle part of the femur; rostrum notably flat; rostral length equal to breadth at the level of the lacrimal (from Jalisco, México, south through Central America) .... *Cynomops mexicanus* (p. 214)

4a. Uropatagium not thick or rigid; no hair on the dorsum of the middle part of the femur; rostrum subcylindrical (Fig. 13); rostral length less than breadth at the level of the lacrimal .... *Eumops* (p. 215)

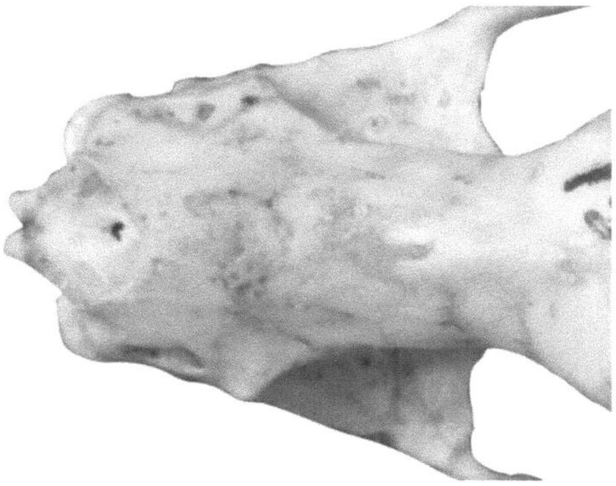

Fig. 13 The rostrum as subcylindrical as in *Eumops*

5. Ears not extending considerably beyond the tip of the nose when laid over the face; ears not conjoined at the base; second phalange of the fourth finger greater than 5.0 mm; anterior breadth of the rostrum distinctly greater than the interorbital breadth (from southern Oregon to North Carolina, United States, through Central America, except for the Yucatán Peninsula) .... *Tadarida brasiliensis* (p. 233)

5a. Ears extending considerably beyond the tip of the nose when laid over the rostrum; ears conjoined at the base; second phalange of fourth finger less than 5.0 mm; anterior breadth of the rostrum greater than interorbital breadth... *Nyctinomops* (p. 228)

Genus *Cynomops* Thomas, 1920

Small to medium sized, terminal part of the tail extending distally from the uropatagium; **uropatagium thick and rigid, hair on the middle dorsal femur**; ears short, orbicular, extending a little more than halfway from the eyes to the nostrils and not connected to each other; keel low but distinct; **rostrum notably flat; rostral length equal to its breadth at the level of the lacrimal;** nasal branches of the premaxillaries joined anteriorly; two pairs of lower incisors (Goodwin and Greenhall 1961).

Cynomops mexicanus (Jones and Genoways, 1967)

Mexican dog-faced bat, murciélago cara de perro mexicano

Monotypic.

Conservation: *Cynomops mexicanus* is listed as subject to Special Protection in the Norma Oficial Mexicana (DOF 2019) and as Least Concern by the IUCN (Rodriguez and Miller 2015).

Characteristics: Mexican Dog-faced bat is medium sized; forearm length 34.0–38.0 mm, skull length 17.0–20.6 mm. Dorsal pelage dark brown to blackish; fur of underparts grayish brown; antitragus nearly square, ears not joined on the head; upper lip without grooves; **uropatagium thick and rigid; hair present on the middle dorsal femur; rostrum notably flat; rostral length equal to its breadth at the level of the lacrimal**; no emargination in the palate between the upper incisors.

Comments: *Cynomops* was previously considered a subgenus of *Molossops*, but Barques et al. (1993), Simmons and Voss (1998), Solari et al. (1999), and Peters et al. (2002) included the species of *Molossops* with ranges in North America within *Cynomops*. *C. mexicanus*, described originally as a subspecies of *C. greenhalli* (Jones and Genoways 1967; see also Koopman 1994), was elevated to a full species by Simmons and Voss (1998) and Peters et al. (2002). *C. mexicanus* ranges from Jalisco south to Central America along the Pacific coastal plains (Map 10). It can be found in sympatry, or nearly so, with species of *Eumops*, *Molossus, Nyctinomops*, and *Tadarida*. This species can be distinguished from all the other sympatric species by having a thick and rigid uropatagium and hair on the middle dorsal femur; a notably flat rostrum, and rostral length equal to its breadth at the level of the lacrimal.

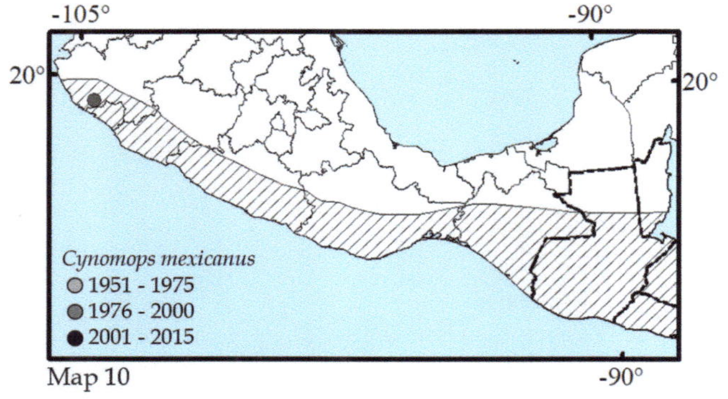

Map 10 Distribution map of *Cynomops mexicanus*

Additional Literature: Barques et al. (1993), Koopman (1994) Simmons and Voss (1998), Solari et al. (1999), Peters et al. (2002), Simmons (2005).

Genus *Eumops* Miller, 1906

Medium to large sized; dorsal pelage brown, reddish brown, dark brown, grayish, or black; fur of underparts pale; **large cartilaginous ears extending beyond the nose, resembling a cap;** antitragus varying in shape; **snout wide and flat**; tail extending beyond the uropatagium; skull without a palate emargination between the upper incisors; **skull robust, with the face practically at the same level as the braincase;** the first premolars very small; **rostral length greater than the lacrimal width.** The keys were elaborated based on the review of specimens and the following sources: Sanborn (1932), Eger (1977); Lawlor (1979), Hall (1981), Álvarez et al. (1994), Kiser (1995), and Álvarez-Castañeda et al. (2017a).

1. Forearm length greater than 52.0 mm; skull length greater than 22.0 mm 2

1a. Forearm length less than 52.0 mm; skull length less than 22.0 mm 6

2. Forearm length greater than 73.0 mm; skull length greater than 30.0 mm (from California, Nevada, Arizona, New Mexico, and Texas, United States, south through Jalisco, Puebla, Hidalgo, and Querétaro, México) .... *Eumops perotis* (p. 220)

2a. Forearm length less than 73.0 mm; skull length less than 30.0 mm 3

3. Forearm length averaging greater than 64.0 mm; toothrow length greater than 11.0 mm (from Arizona, United States, south through Central America) .... *Eumops underwoodi* (p. 221)

3a. Forearm length averaging less than 64.0 mm; toothrow length less than 11.0 mm 4

4. Tragus small and sharp, measuring approximately 3.5 mm from the low neckline; lacrimal crest developed (Fig. 14; from Oaxaca and southern Veracruz south through South America) .... *Eumops auripendulus* (p. 216)

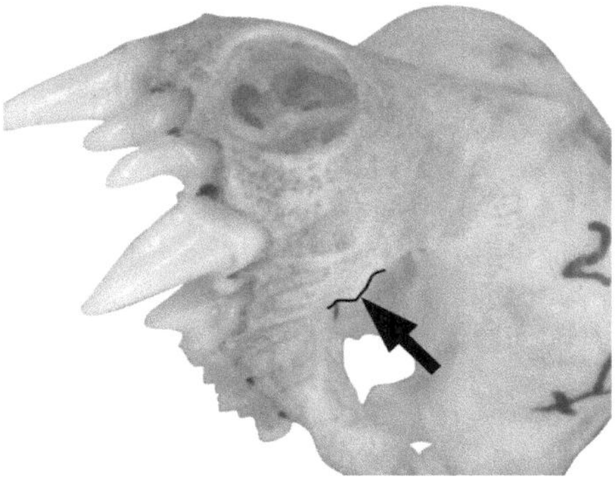

Fig. 14 Lacrimal crest developed like in *Eumops auripendulus*

4a. Tragus large, square, length approximately 4.5 mm from the notch; no lacrimal crest 5

5. Forearm length greater than 61.0 mm; skull length greater than 25.0 mm; condylobasal pits proportionally shorter (Restricted to the southern Florida Peninsula, United States) .... *Eumops floridanus* (p. 217)

5a. Forearm length less than 61.0 mm; skull length less than 25.0 mm; condylobasal pits proportionally larger (from Jalisco and Veracruz, México, south through Central America) .... *Eumops ferox* (p. 217)

6. Pelage brown; hairs short and dark at the base; basisphenoid pits large and deep (Fig. 15); third commissure of the third upper molars of the same length as the second commissure (Chiapas and Quintana Roo, México, south through South America) .... *Eumops hansae* (p. 218)

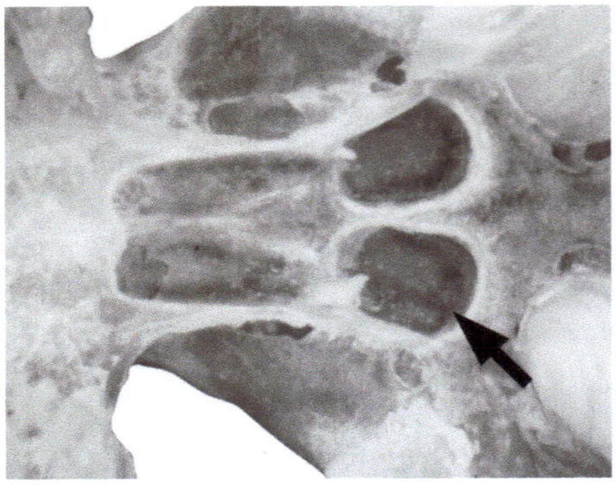

Fig. 15 Basisphenoid pits are as large and deep as in *Eumops hansae*

6a. Pelage grayish brown, hairs long and paler at the base; basisphenoid pits small and shallow; third commissure of the third upper molars not of the same length as the second commissure (from Veracruz and Chiapas, México, south through Central America, including Yucatán peninsula)..*Eumops nanus* (p. 219)

Eumops auripendulus (Shaw, 1800)

Black bonneted bat, murciélago negro de gorra

E. a. auripendulus (Shaw, 1800). Only subspecies in North America.

Conservation: *Eumops auripendulus* is listed as Least Concern by the IUCN (Barquez et al. 2015a).

Characteristics: Black bonneted bat is medium sized; forearm length 57.0–63.0 mm, skull length 23.0–25.2 mm. Dorsal pelage dark brown to blackish; fur of underparts dark grayish brown; **tragus small and sharp, measuring approximately 3.5 mm from the low neckline; with a notorious lacrimal crest**.

Comments: *Eumops auripendulus* ranges from the Tehuantepec isthmus south through South America, except for the northwestern Yucatán Peninsula (Map 11). The other species of *Eumops* that can be found in sympatry, or nearly so, are *E. ferox* and *E. underwoodi*. *E. auripendulus* is the third-largest species of the genus *Eumops*. It has as a distinctive characteristic, a small and sharp tragus, measuring approximately 3.5 mm from the low neckline, and a notorious lacrimal crest. From *E. ferox,* it differs by having a darker color; tragus more pointed; shorter skull and less robust; shallower basisphenoid pits; zygomatic arches subparallel; mastoid width and postorbital constriction proportionally narrower.

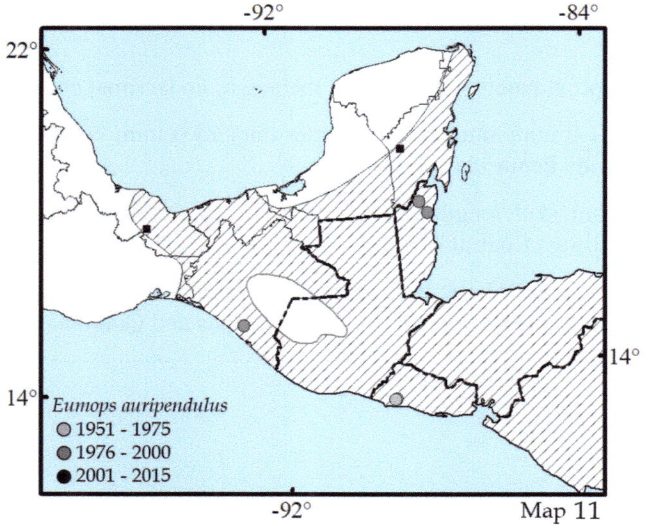

Map 11 Distribution map of *Eumops auripendulus*

Additional Literature: Allen (1900), Goodwin (1946), Husson (1962), Eger (1974, 1977), Freeman (1981), Baker and Genoways (1978), Eisenberg (1989), Álvarez-Castañeda and Álvarez (1991), Barquez et al. (1993), Koopman (1994), Kiser (1995), López-González (1998), Best et al. (2002), Naranjo-Piñera et al. (2005), Simmons (2005)

Eumops ferox (Gundlach, 1861)

Central American bonneted bat, murciélago de gorra centroamericano

Monotypic.

Conservation: *Eumops ferox* is listed as Least Concern by the IUCN (Solari 2019d).

Characteristics: Central America bonneted bat is medium sized; forearm length 59.0–61.0 mm, skull length 21.0–25.0 mm. Dorsal pelage grayish brown to grayish with long hairs that stand out on the hips; fur of underparts pale grayish brown; **tragus big and square with an approximate length of 4.5 mm**; basisphenoid pit proportionally larger, relative to condylobasal length.

Comments: *Eumops ferox* is considered for México because *E. glaucinus* does not range in this country (Mcdonough et al. 2008). *E. ferox* ranges from Veracruz to the east and Jalisco to the west south through Central America, including the Yucatán Peninsula (Map 12). Other *Eumops* species that can be found in sympatry, or nearly so, with *E. ferox* are *E. auripendulus, E. underwoodi,* and probably *E. perotis* in its northern range. *E. ferox* can be differentiated by its big and square tragus with an approximate length of 4.5 mm. From *E. auripendulus,* by having a paler coloration; tragus less pointed; larger skull and more robust; deeper basisphenoid pits; zygomatic arches converging anteriorly, and mastoid width and postorbital constriction proportionally wider.

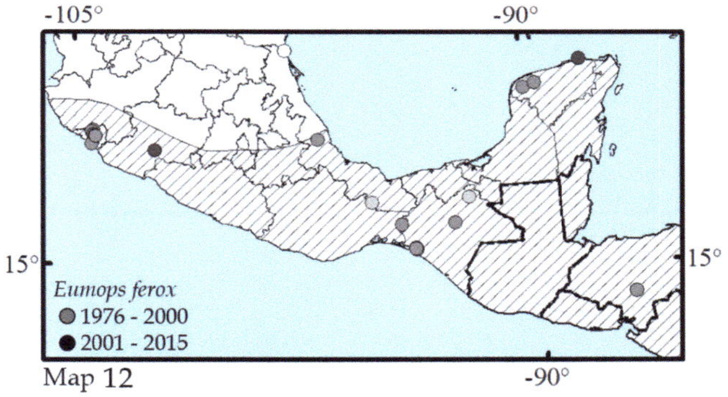

Map 12 Distribution map of *Eumops ferox*

Additional Literature: Miller (1907), Eger (1977), Freeman (1981), Hall (1981), Best et al. (1997), Baker et al. (2002), Villalpando and Álvarez (2000), Timm and Genoways (2004).

Eumops floridanus (Allen, 1932)

Florida bonneted bat, murciélago de gorra de Florida

Monotypic.

Conservation: *Eumops floridanus* is restricted to a few areas in the Florida Peninsula; it is listed as Endangered by the United States Endangered Species Act (FWS 2022) and as Vulnerable (C2a (i): small population size in decline, less than 1000 of mature individuals in each subpopulation) by the IUCN (Solari 2016d). However, it is considered one of the most critically endangered mammalian species in North America (Timm and Genoways 2004).

Characteristics: Florida bonneted bat is medium sized; forearm length 60.8–66.0 mm, skull length 25.2–27.2 mm. Dorsal pelage black to brownish gray or cinnamon brown; hair short and glossy sharply bicolored, white at the base; fur of underparts pale; **tragus big and square with an approximate length of 4.5 mm; basisphenoid pits proportionally shorter relative to the condylobasal length; palate proportionally narrower**.

Comments: *Eumops floridanus* was described as a full species from a Pleistocene deposit (Allen 1932) and was previously considered a subspecies of *E. glaucinus,* based on the Pleistocene fossil and modern southern Florida populations (Koopman 1971; Map 13). Morphological analyses and its bacular shape support its assignment as a valid species (Timm and Genoways 2004). However, it is not recognized by Simmons (2005). *E. floridanus* has a range restricted to the southern Florida Peninsula and is considered one of the most restricted distributions of any bat species in the New World. No other species of *Eumops* may be found in sympatry; however, it can be differentiated from *E. ferox* by its large size.

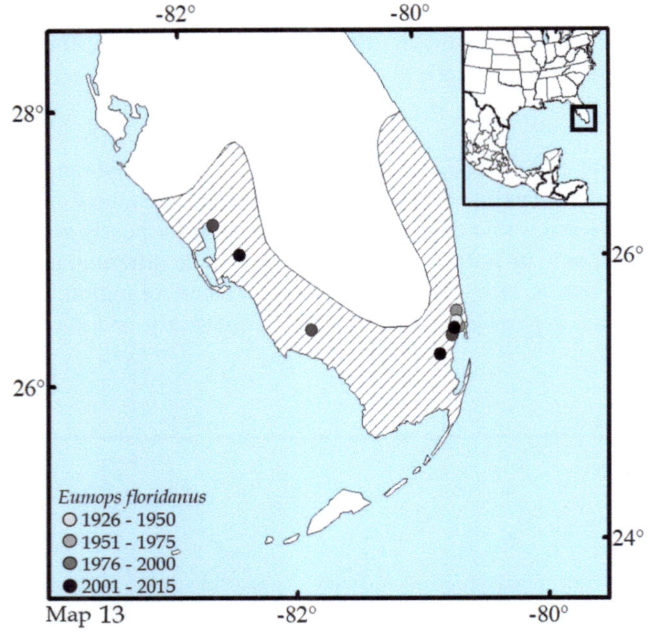

Map 13 Distribution map of *Eumops floridanus*

Additional Literature: Miller (1907), Eger (1977), Freeman (1981), Hall (1981), Barquez et al. (1993), López-González (1998), Best et al. (1997), Vannatta et al. (2020).

Eumops hansae Sanborn, 1932

Sanborn's bonneted bat, murciélago de gorra pigmeo

Monotypic.

Conservation: *Eumops hansae* is listed as Least Concern by the IUCN (Pineda and Rodriguez 2015).

Characteristics: Sanborn's bonneted bat is small sized; forearm length 37.0–42.0 mm, skull length 18.4–21.7 mm. Dorsal pelage dark brown or blackish; fur of underparts dark grayish; antitragus in the shape of a half-circle, wide at the base; **basisphenoid pits big and deep; third commissure of the third upper molars of the same length as the second commissure**.

Comments: In North America, *Eumops hansae* is only known from three areas in tropical forest lowlands of México's southern border with Guatemala and Belize (Map 14). It can be found in sympatry, or nearly so, with *E. auripendulus, E. nanus,* and *E. underwoodi,* from which it can be differentiated by its small size; hair short and brown; basisphenoid pits big and deep; third and second commissure of third upper molars of the same size.

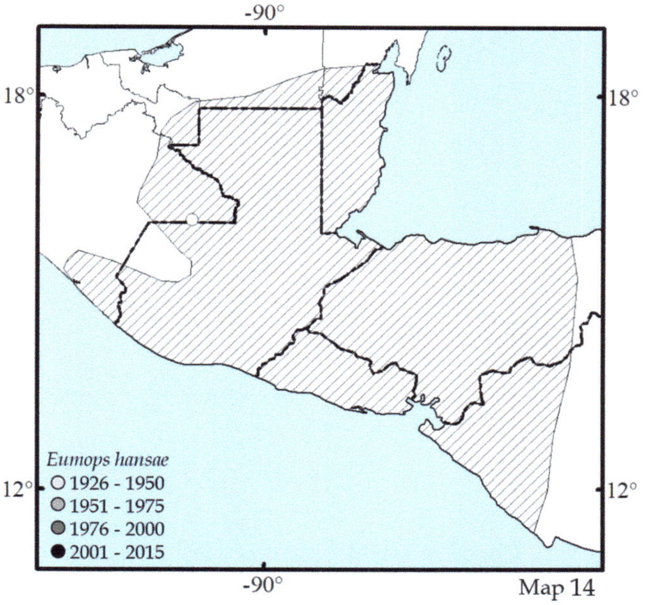

Map 14 Distribution map of *Eumops hansae*

Additional Literature: Sanborn (1932), Gardner et al. (1970), Eger (1977), Freeman (1981), Hall (1981), Eisenberg (1989), Álvarez and Álvarez-Castañeda (1990), Álvarez-Castañeda and Álvarez (1991), Kiser (1995), Best et al. (2001).

Eumops nanus (Miller, 1900)

Eastern lesser bonneted bat, murciélago de gorra enano

Monotypic.

Conservation: *Eumops nanus* is listed as Least Concern by the IUCN (Solari 2017a).

Characteristics: Eastern lesser bonneted bat is small sized; forearm length 39.0–48.0 mm, skull length 16.4–20.1 mm. **Dorsal pelage brown to pale grayish brown, no long hairs on the hips; fur of underparts pale grayish**; antitragus in the shape of a half-circle, wide at the base; **basisphenoid pits small and shallow; third commissure of the third upper molars not of the same length as the second commissure**.

Comments: *Eumops nanus* was previously considered *E. bonariensis,* but the latter is not distributed in México (Eger 2008). *E. nanus* ranges from Veracruz and Chiapas south through Central America, including the Yucatán peninsula (Map 15). It can be found in sympatry, or nearly so, with *E. auripendulus, E. hansae,* and *E. underwoodi,* from which it can be differentiated by its small size; short fur and brown; basisphenoid pits small and shallow; third and second commissure of third upper molars of the same size.

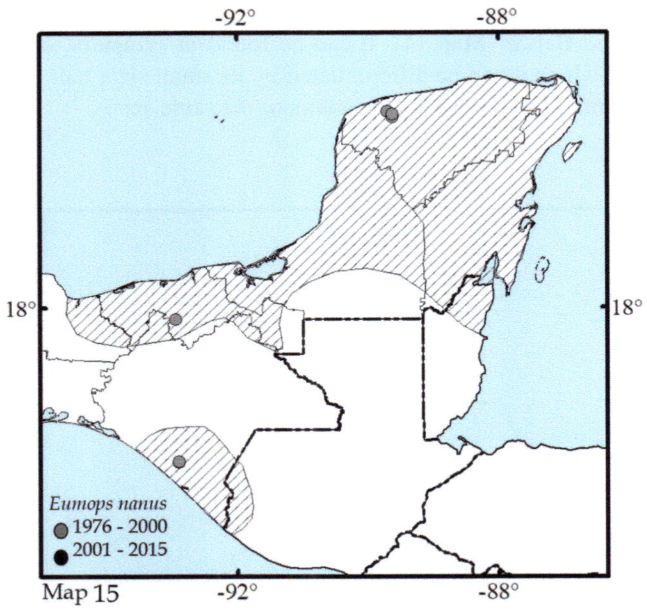

Map 15 Distribution map of *Eumops nanus*

Additional Literature: Sanborn (1932), Goodwin (1946), Handley (1955), Eger (1977), Dolan and Honeycutt (1978), Redford and Eisenberg (1992), Koopman (1994), Hunt et al. (2003), Simmons (2005).

Eumops perotis (Schinz, 1821)

Greater bonneted bat, murciélago de gorra mayor del oeste

E. p. californicus (Merriam, 1890). Only subspecies in North America.

Conservation: *Eumops perotis* is listed as Least Concern by the IUCN (Barquez and Diaz 2015).

Characteristics: Greater bonneted bat is the largest in size; **forearm length 72.0–83.0 mm, skull length 30.3–32.9 mm.** Dorsal pelage brown, white at the base; fur of underparts pale; antitragus in the shape of a half-circle, wide at the base; snout wide and flat; skull largest, see measurements.

Comments: *Eumops perotis* ranges are associated with the mountainous regions from central California to Texas south through Jalisco, Zacatecas, Nayarit, Hidalgo, and Querétaro (Map 16). It can be found in sympatry, or nearly so, with *E. underwoodi* and *E. ferox* at its southern range. *E. perotis* is the largest species of *Eumops* in North America; it differs from the other species by its forearm and skull length greater than 72.0 and 30.0 mm, respectively.

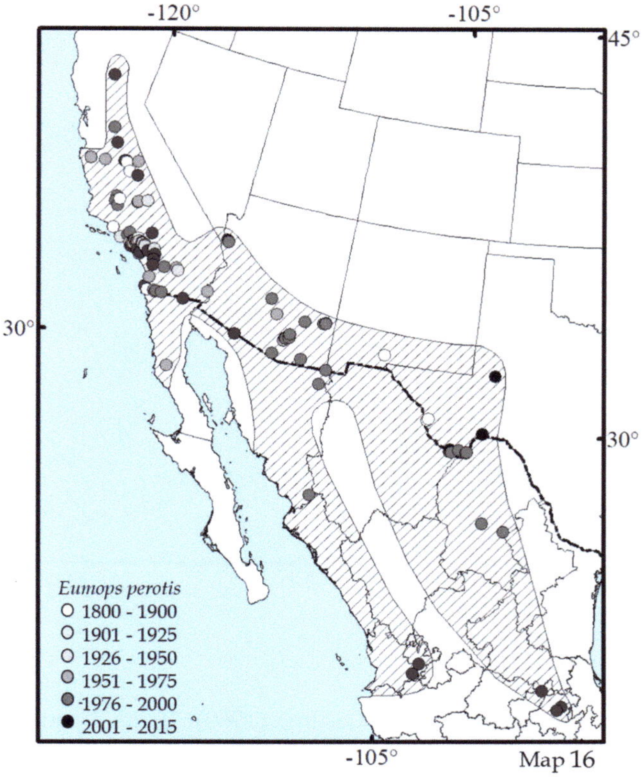

Map 16 Distribution map of *Eumops perotis*

Additional Literature: Miller (1907), Grinnell (1918), Benson (1947), Barbour and Davis (1969), Koopman (1971, 1978), Anderson (1972), Eger (1977), Carter and Dolan (1978), Hoffmeister (1986), Redford and Eisenberg (1992), Kiser (1995), Best et al. (1996b), Simmons (2005).

Eumops underwoodi Goodwin, 1940

Underwood's bonneted bat, murciélago de gorra del Pacífico

1. *E. u. sonoriensis* Benson, 1947. Restricted to the northern Sonora and southern Arizona.
2. *E. u. underwoodi* Goodwin, 1940. From northern Sinaloa southward to Central America.

Conservation: *Eumops underwoodi* is listed as Least Concern by the IUCN (Miller et al. 2016a).

Characteristics: Underwood's bonneted bat is medium sized; **forearm length 66.0–71.0 mm, skull length 28.8–29.9 mm.** Dorsal pelage grayish to reddish brown; **some longer hairs stand out from the hips**; antitragus in the shape of a half-circle, wide at the base; without a notorious lacrimal crest.

Comments: *Eumops underwoodi* ranges from southern Arizona and Sonora south through Central America, including the southern Yucatán Peninsula; in addition, it can be found in the Balsas Basin from Michoacán to Morelos (Map 17). It can be found in sympatry, or nearly so, with *E. auripendulus, E. hansae, E. nanus,* and *Nyctinomops macrotis*. *E. underwoodi* differs from the other species of *Eumops* by being the second-largest in size, with forearm length 66.0–71.0 mm, and skull length 28.8–29.9 mm. From *E. perotis*, the most similar species, by its slightly smaller size (forearm length less than 73.0 mm and skull less than 30.0 mm). From *Nyctinomops macrotis*, by the smooth upper lips.

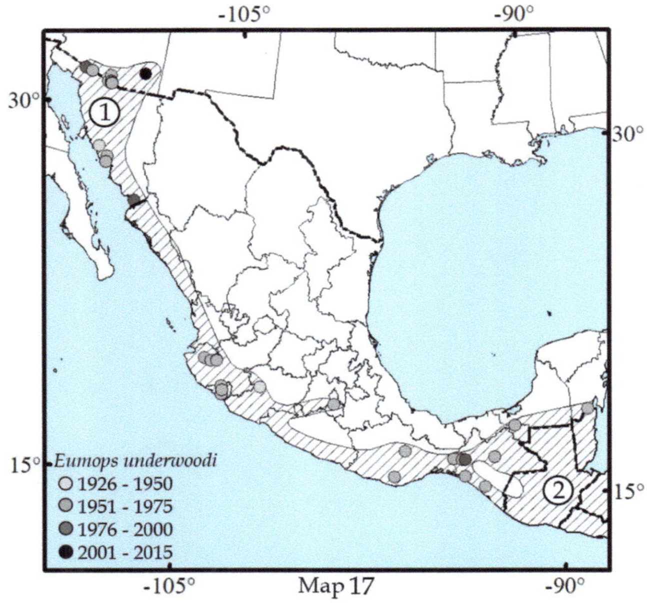

Map 17 Distribution map of *Eumops underwoodi*

Additional Literature: Benson (1947), Cockrum and Gardner (1960), Barbour and Davis (1969), Álvarez and Aviña (1964), Goodwin (1969), Freeman (1981), Watkins et al. (1972), Eger (1977), Dolan and Carter (1979), Milner et al. (1990), Kiser (1995), Simmons (2005).

Genus *Molossus* E. Geoffroy, 1805

Small to medium sized; **dorsal pelage generally dull black** but occasionally orange or chocolate brown; ears **medium size, attached at the base;** fur of underparts grayish or brownish; **tragus small and rounded, as wide as high, notched at the base, suborbicular in outline, and noticeably thickened; hair long on the hips; ears shorter than face and narrow**; skull with upper incisor breadth same or greater than tooth length; no second upper premolar or second lower incisor; hypocone very small and larger protocone occupying entire inner side of tooth; **sagittal and lambdoid crest s conspicuously developed, especially in males;** auditory bullae much better developed than usual in the family (inner margin not emarginate); **basisphenoid cavities distinct;** no palatal emargination on the anterior palate between upper incisors. The keys were elaborated based on the review of specimens and the following sources: Lawlor (1979), Hall (1981), Álvarez et al. (1994), Gónzalez-Ruiz et al. (2011), and Álvarez-Castañeda et al. (2017a).

1. Forearm length greater than 44.0 mm; skull length greater than 19.0 mm; breadth through the third upper molars greater than 8.2 mm 2

1a. Forearm length less than 41.0 mm; skull length less than 18.8 mm; breadth through the third upper molars less than 8.2 mm 4

2. Dorsal pelage usually dull black; dorsal hair unicolor; forearm length generally greater than 47.0 mm (from southern Sonora and Nuevo León, México, south through Central America)............ *Molossus nigricans* (p. 226)

2a. Dorsal pelage usually chocolate brown, never black; dorsal hair noticeably bicolored (basal half of hair white, distal half chocolate brown); length of forearm generally less than 49.0 mm............ 3

3. Dorsal pelage dark chocolate brown; length of skull less than 21.0 mm; toothrow length 12.0 mm in males and less than 11.5 mm in females (Yucatán Peninsula, México)............ *Molossus alvarezi* (p. 223)

3a. Dorsal pelage from medium to dark brown; length of skull greater than 21.0 mm; toothrow length 11.8 mm in males and greater than 11.4 mm in females (from Sinaloa, México, south through South America) .... *Molossus sinaloae* (p. 227)

4. Dorsal fur long (greater than 4.0 mm), noticeably bicolored (basal half white, distal half; from Sinaloa and San Luis Potosí, México, south through Central America).......... *Molossus molossus* (p. 225)

4a. Dorsal fur short (less than 3.5 mm), unicolor or undistinguishably bicolored 5

5. Sagittal and lambdoidal crests noticeably developed; width across third upper molars usually less than 7.8 mm (from Chiapas, México, south through South America) .... *Molossus coibensis* (p. 224)

5a. Sagittal and lambdoid crests poorly developed; width across the third upper molars usually greater than 7.8 mm (from Sinaloa south through Chiapas and south Quintana Roo, México) .... *Molossus aztecus* (p. 224)

Molossus alvarezi González-Ruiz et al., 2011

Álvarez's mastiff bat, murciélago mastín de Yucatán

Monotypic.

Conservation: *Molossus alvarezi* is listed as Data Deficient by the IUCN (Solari 2016h).

Characteristics: Álvarez's mastiff bat is medium sized; forearm length from 44.0 to 47.8 mm, skull length 19.1–20.1 mm. **Dorsal pelage dark chocolate brown**; hairs banded, with the basal half white, clearly contrasting with the darker distal half; **lambdoid crest is poorly developed, with their distal edge narrower than the braincase**.

Comments: *Molossus alvarezi* is known only from the northern Yucatán Peninsula (Map 18). It can be found in sympatry, or nearly so, with *M. molossus, M. nigricans,* and *M. sinaloae*. *M. alvarezi* can be distinguished from *M. molossus* by its larger size. From *M. nigricans*, by its smaller size (see measurements), and from *M. sinaloae,* by its skull length less than 21.0 mm and lambdoid crest s less developed, with their distal edge narrower than the braincase (González-Ruiz et al. 2011).

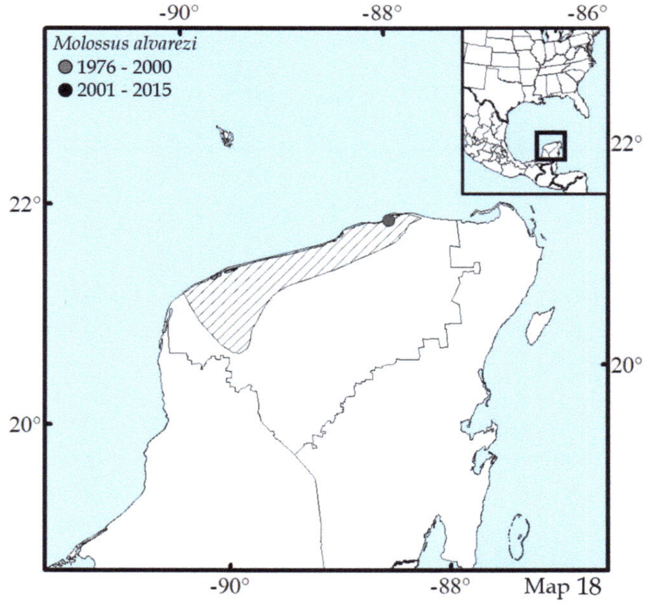

Map 18 Distribution map of *Molossus alvarezi*

Additional Literature: Goodwin (1959), Freeman (1981), Dolan (1989).

Molossus aztecus de Saussure, 1860

Aztec mastiff bat, murciélago mastín azteca

Monotypic.

Conservation: No population of *Molossus aztecus* is listed as Least Concern by the IUCN (Solari 2019s).

Characteristics: Aztec mastiff bat is small sized; forearm length from 36.6 to 38.4 mm, skull length 17.2–19.0 mm. Dorsal pelage pale to dark brown; ears shorter than the face and narrow; **sagittal and lambdoid crest s poorly developed; width across the upper third molars usually greater than 7.8 mm**.

Comments: *Molossus aztecus* was previously considered a subspecies of *M. molossus* (Dolan 1989). The range is poorly understood, as Dolan (1989), in his revision of the genus, listed only two confirmed localities in México (one in Jalisco and the type locality at Amecameca, "at the foot of Popocatepetl," México). He did not confirm the identification of any other specimen from México. Specimens from other localities, such as the Gulf side and southern México, treated as *M. molossus aztecus*, are best considered *M. molossus* until further studies confirm Dolan's restricted range (Map 19). It can be found in sympatry, or nearly so, with *M. molossus* and *M. sinaloae. M. aztecus* differs from *M. molossus* by having longer forearms but a shorter skull and a narrower rostrum, hair with the basal half white, and more pincerlike incisors. While both species are difficult to differentiate when alive, multivariate morphological analyses are unambiguous in their distinction (Dolan 1989). From *M. sinaloae, M. aztecus* differs by its smaller size (forearm and skull length less than 44.0 and 19.0 mm, respectively).

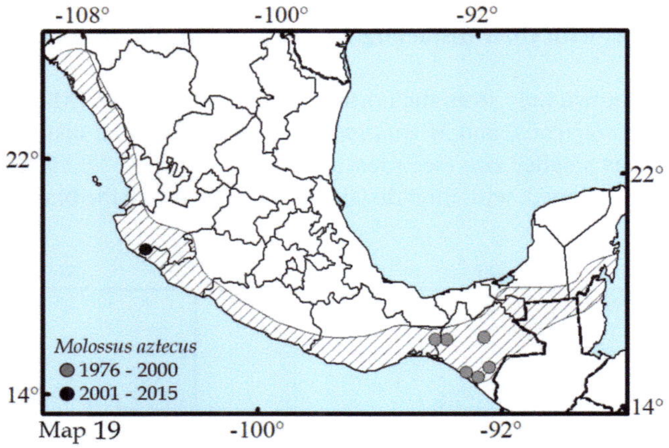

Map 19 Distribution map of *Molossus aztecus*

Additional Literature: Goodwin (1959), Freeman (1981), Koopman (1993, 1994), Álvarez-Castañeda and Álvarez (1991), Koopman (1993), López-González and Presley (2001), Naranjo Piñera et al. (2005), Simmons (2005).

Molossus coibensis Allen, 1904

Coiban mastiff bat, murciélago mastín de Coiban

Monotypic.

Conservation: *Molossus coibensis* is listed as Least Concern by the IUCN (Sampiao et al. 2017).

Characteristics: Coiban mastiff bat is small sized; forearm length from 32.6 to 36.8 mm, skull length 15.6–18.4 mm. Dorsal pelage pale to dark brown, **with a white or cream band at the base of hairs; sagittal and lambdoid crest noticeably developed; width across the third upper molars usually less than 7.8 mm; upper incisors spatulate in shape**.

Comments: *Molossus coibensis* was previously considered a subspecies of *M. molossus* (Dolan 1989). It has only been recorded from moist tropical regions along the coast of Chiapas along the border with Guatemala south to South America (Map 20). It can be found in sympatry, or nearly so, with *M. molossus, M. nigricans,* and *M. sinaloae*. *M. coibensis* is very small in relation to *M. nigricans* and *M. sinaloae* (forearm and skull length less than 36 and 18.0 mm, respectively). It can be distinguished from *M. molossus* and *M. sinaloae* by its dorsal hair less than 3.0 mm in length with a shorter pale band at the base, and shorter forearms; upper incisors spatulate, sagittal and lambdoid crests less developed in *M. coibensis* than in either *M. aztecus* or *M. molossus*, pectoral region of *M. coibensis* more robust than in *M. molossus*, and fur primarily deep black in color but grading into chocolate brown as the pelage wears.

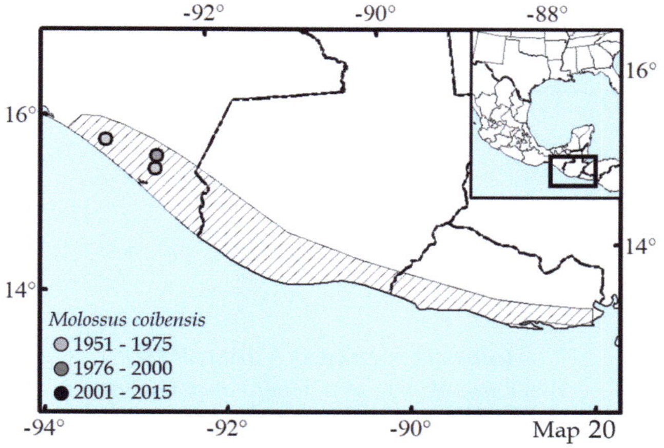

Map 20 Distribution map of *Molossus coibensis*

Additional Literature: Allen (1904d), Freeman (1981), Koopman (1993, 1994), Simmons and Voss (1998), Naranjo-Piñera et al. (2005), Simmons (2005).

Molossus molossus (Pallas, 1766)

Pallas' mastiff bat, murciélago mastín peludo

Monotypic.

Conservation: *Molossus molossus* is listed as Least Concern by the IUCN (Barquez et al. 2015k).

Characteristics: Pallas´ mastiff bat is small sized; forearm length from 37.8 to 40.5 mm, skull length 15.8–18.4 mm. Dorsal pelage from pale grayish brown to dark brown, with long hairs on the hips; fur of underparts grayish brown; short forearm; hairs white at basal first one-quarter or only at the insertion point; skull narrow; with moderately developed sagittal crest; incisors more spatulate than pincerlike.

Comments: *Molossus aztecus* and *M. coibensis* were previously considered a subspecies of *M. molossus* (Dolan 1989). The range of *M. molossus* is not fully understood (see above comments for *M. aztecus*), although its range is generally mapped along the eastern and western coasts of México and south through Central America (Map 21). Dolan (1989), in his review of the genus *Molossus*, tentatively assigned specimens from Colima, Chiapas, Coahuila, and Tamaulipas to *M. molossus* but without supporting data. The only known area where *M. molossus* and *M. aztecus* can be found in sympatry, or nearly so, is in Jalisco. A more complete review of museum specimens is required to define the ranges of each of the three current species originally grouped together under *M. molossus*. For comparisons between *M. aztecus* and *M. coibensis*, see those accounts.

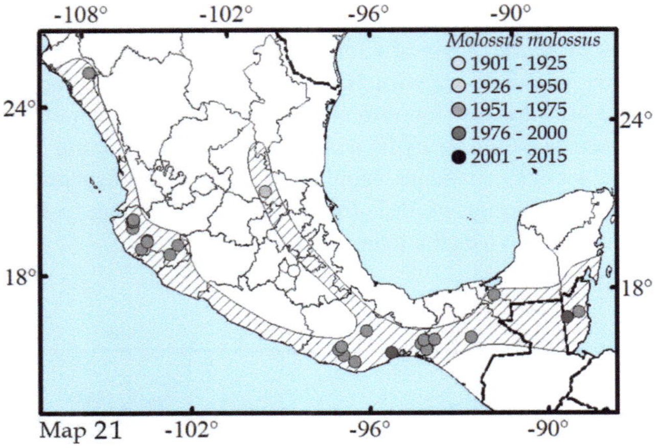

Map 21 Distribution map of *Molossus molossus*

Additional Literature: Goodwin (1959), Freeman (1981), Simmons and Voss (1998), Naranjo-Piñera et al. (2005), Simmons (2005).

Molossus nigricans Miller, 1902

Black mastiff bat, murciélago mastín negro

Monotypic.

Conservation: *Molossus nigricans* (as *M. rufus*) is listed as Least Concern by the IUCN (Barquez et al. 2015l).

Characteristics: Black mastiff bat is medium sized; **forearm length 47.0–54.4 mm, skull length 21.1–24.5 mm. Dorsal pelage generally dull black**, occasionally orange or chocolate brown; fine bristles on the back; fur of underparts dark grayish brown; width of basisphenoid cavity less than 1.0 mm, and sagittal crest unusually well developed.

Comments: *Molossus nigricans* was previously regarded as *M. ater* and included the specimens from Oaxaca that were considered *M. pretiosus macdougallii.* Dolan (1989) documented that *M. pretiosus* does not extend its range into México and treated *M. nigricans* as *M. rufus* (Dolan 1989). Loureiro et al. (2020) established the current taxonomy. *M. nigricans* ranges in the mountains and coastal plains along both coasts of México from southern Sonora and Tamaulipas east through San Luis Potosí and Querétaro and south through Central America (Map 22). It can be in sympatry, or nearly so, with *M. alvarezi, M. aztecus, M. coibensis, M. molossus,* and *M. sinaloae*. *M. nigricans* can be distinguished from all other species of *Molossus*, except for *M. sinaloae*, by its dull black pelage color and larger size, forearm and skull length greater than 47.0 and 21.0 mm, respectively. It differs from *M. sinaloae* by the width of basisphenoid cavity less than 1.0 mm, a broader and stouter skull, shorter hair (2.0–2.5 mm), a usually well-developed sagittal crest, particularly in males, and a short and square muzzle.

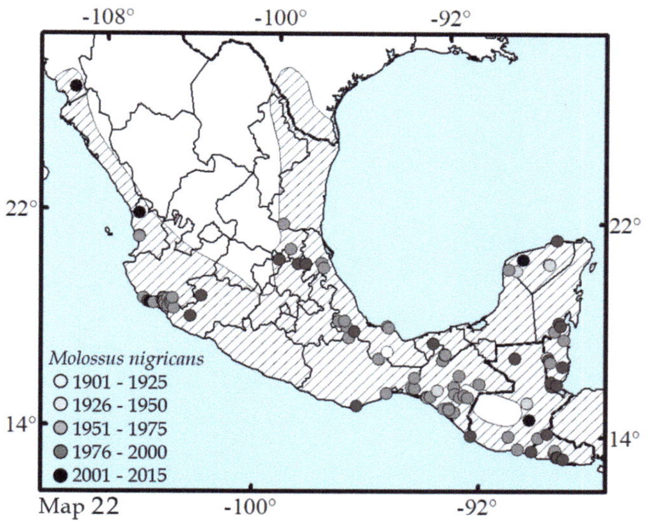

Map 22 Distribution map of *Molossus nigricans*

Additional Literature: Goodwin (1959, 1969), Jones et al. (1962), Álvarez (1963), Watkins et al. (1972), Freeman (1981), Dolan (1989), Naranjo-Piñera et al. (2005), Simmons (2005), García–Estrada et al. (2006), Loureiro et al. (2020).

Molossus sinaloae Allen, 1906

Sinaloan mastiff bat, murciélago mastín de Sinaloa

Monotypic.

Conservation: *Molossus sinaloae* is listed as Least Concern by the IUCN (Miller et al. 2016i).

Characteristics: Sinaloan mastiff bat is medium sized; forearm length from 46.8 to 50.8 mm, skull length 21.1–23.1 mm. Dorsal pelage reddish brown but never dark and blackish brown; fur of underparts brownish; **basal half of hair white, clearly contrasting with the distal half; lambdoid crests well developed with their distal edge wider than the braincase; width of** the **basisphenoid cavity larger than 1.0 mm**.

Comments: *Molossus sinaloae* ranges from Sinaloa south along the Pacific Coast through South America; specimens from the Yucatán Peninsula, previously regarded as this species, were reassigned as *M. alvarezi* by González-Ruiz et al. (2011; Map 23). It can be found in sympatry, or nearly so, with *M. aztecus, M. coibensis, M. molossus,* and *M. nigricans*. This species can be distinguished from all other species in the genus, except for *M. nigricans,* by its larger size, forearm, and skull length greater than 47.0 and 21.0 mm, respectively. From *M. nigricans,* by having the cavity floor of the basisphenoid greater than 1.0 mm, a narrower skull, and lighter bicolored long hair (Dolan 1989), and from *M. alvarezi,* by its skull length 21.0 mm, well-developed lambdoid crests in their distal edges, and a broader braincase (González-Ruiz et al. 2011).

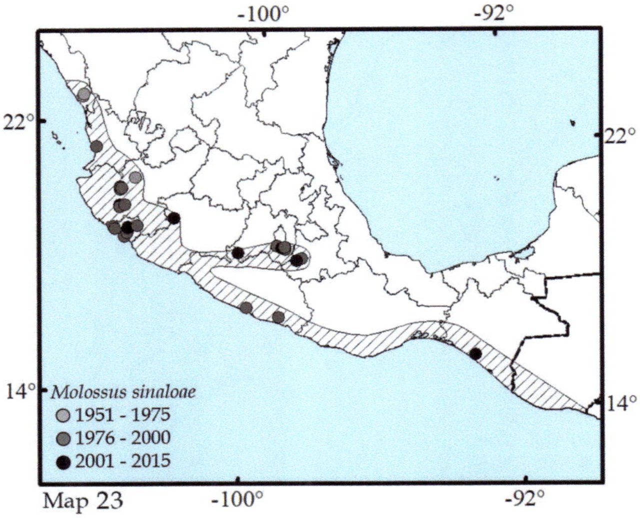

Map 23 Distribution map of *Molossus sinaloae*

Additional Literature: Goodwin (1959), Goodwin and Greenhall (1961), Álvarez and Aviña (1964), Watkins et al. (1972), Jones et al. (1971), Freeman (1981), Dolan (1989), Jennings et al. (2002), Navarro et al. (1990), Álvarez-Castañeda (1996), Van Staaden and Jones (1997), Simmons and Voss (1998), Álvarez and Ocaña (1999), Jennings et al. (2002), Naranjo-Piñera et al. (2005), Simmons (2005).

Genus *Nyctinomops* Miller, 1902

Dorsal pelage dark brown, reddish brown, dull brown, or grayish; **antitragus small and rectangular; ears long, joined at the base and extending beyond rostrum; upper lip grooves present**; **skull with emargination in palate between the upper incisors, anterior breadth of face slightly greater than interorbital breadth**; posterior margin of palate at same height as line tangential to posterior margin of last molars. All the species within this genus was previously placed within the genus *Tadarida* (Freeman 1981). The keys were elaborated based on the review of specimens and the following sources: Lawlor (1979), Hall (1981), Freeman (1981), Kumirai and Jones (1990), Álvarez et al. (1994), and Álvarez-Castañeda et al. (2017a).

1. Forearm length greater than 55.0 mm; skull length greater than 21.5 mm (California, Wyoming, Colorado, and Texas, United States, south to Oaxaca, México) .... *Nyctinomops macrotis* (p. 231)

1a. Forearm length less than 55.0 mm; skull length less than 21.5 mm 2

2. Hair on the hips longer remainder of body; dorsal pelage color grayish; first upper molars without hypocone keel; distal border of palatine posterior to third molars (from California, Arizona, New Mexico, and Texas, United States, south through Baja California Peninsula, Michoacán, and Estado de México) .... *Nyctinomops femorosaccus* (p. 229)

2a. Hair length uniform over entire body; dorsal pelage color brown; first upper molars with hypocone keel; distal border of the palatine at the same level as the posterior border of the third molars 3

3. Forearm length greater than 45.0 mm; skull length greater than 20.0 mm (Sonora and Tamaulipas southward to Chiapas, México) .... *Nyctinomops aurispinosus* (p. 229)

3a. Forearm length less than 45.0 mm; skull length greater than 20.0 mm (from Guerrero, Morelos, and Tamaulipas, México, south through Central America) .... *Nyctinomops laticaudatus* (p. 230)

Nyctinomops aurispinosus (Peale, 1848)
Peale's free-tailed bat, murciélago de cola libre

Monotypic.

Conservation: *Nyctinomops aurispinosus* is listed as Least Concern by the IUCN (Solari 2019e).

Characteristics: Peale's free-tailed bat is medium sized; forearm length 47.8–51.4 mm, skull length 20.5–21.5 mm. Dorsal pelage reddish brown to dark brown; fur of underparts pale; **ears rectangular; tragus small and quadrate; antitragus elongated**; **posterior margin of the palate terminating at the same level as the third upper molar;** sagittal crest poorly to moderately developed.

Comments: *Nyctinomops aurispinosus* is distributed along both coastal plains from the Tehuantepec isthmus north through Tamaulipas and Sonora (Map 24). It can be found in sympatry, or nearly so, with *N. femorosaccus, N. laticaudatus*, and *N. macrotis. N. aurispinosus* can be distinguished from the other species by its size, smaller than *N. macrotis* but larger than *N. laticaudatus* (see measurements). From *N. femorosaccus,* by a larger thumb, naked membranes, and posterior margin of the palate that terminates at the same level as the third upper molar.

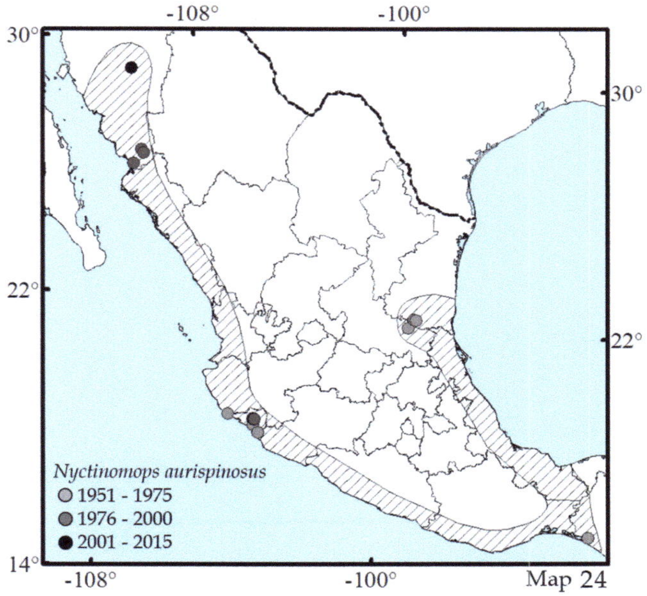

Map 24 Distribution map of *Nyctinomops aurispinosus*

Additional Literature: Sanborn (1941), Carter and Davis (1961), Gardner (1963), Baker et al. (1967), Baker and Jones (1972), Jones et al. (1972), Jones and Arroyo-Cabrales (1990), Kumirai and Jones.

Nyctinomops femorosaccus (Merriam, 1889)
Pocketed free-tailed bat, murciélago de cola libre de bolsa

Monotypic.

Conservation: *Nyctinomops femorosaccus* is listed as Least Concern by the IUCN (Arroyo-Cabrales and Álvarez-Castañeda 2015b).

Characteristics: Pocketed free-tailed bat is medium sized; forearm length 46.0–50.0 mm, skull length 19.3–19.8 mm. Dorsal pelage gray to dull brown; fur of underparts pale; **thumbs short; few and thin hairs on the membranes, especially on the uropatagium; ears more rugose, larger and more square in shape; tragus small and rounded; antitragus elongated and narrow; braincase less inflated; posterior margin of the palate terminating posterior to the third upper molar; cheek-teeth small; postero-internal margin of the first and second lower molars squared;** sagittal crest poorly to moderately developed.

Comments: *Nyctinomops femorosaccus* ranges are associated with the mountainous regions of western México, including the Baja California Peninsula, and Arizona, New Mexico, and Texas in the southwestern United States (Map 25). It can be found in sympatry, or nearly so, with *N. aurispinosus, N. laticaudatus,* and *N. macrotis*. *N. femorosaccus* can be distinguished from its congeners by having shorter thumbs; few and thin hairs on the membranes, especially on the uropatagium; more rugose ears; posterior margin of the palate terminating posterior to the third upper molar; smaller cheek-teeth; postero-internal margin of the first and second lower molars squared in shape.

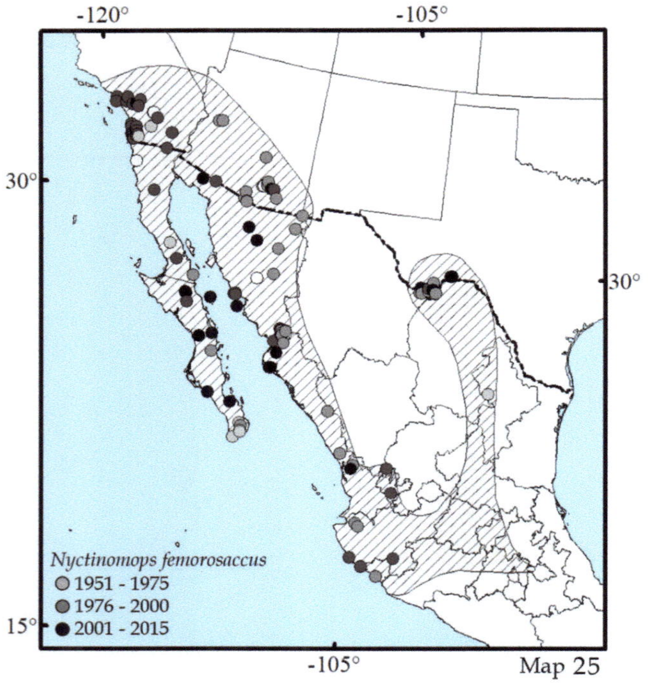

Map 25 Distribution map of *Nyctinomops femorosaccus*

Additional Literature: Miller (1902), Benson (1940), Gardner (1963), Barbour and Davis (1969), Jones et al. (1972), Jones et al. (1988), Hoffmeister (1986), Kumirai and Jones.

Nyctinomops laticaudatus (Geoffroy Saint-Hilaire, 1805)

Broad-eared free-tailed bat, murciélago de cola libre de orejas grandes

1. *N. l. ferruginea* (Goodwin, 1954). From the Tehuantepec isthmus north to southern Tamaulipas, Puebla, and Oaxaca.
2. *N. l. yucatanicus* Miller, 1902. From the Tehuantepec isthmus south through Chiapas, including the Yucatán Peninsula.

Conservation: *Nyctinomops laticaudatus* is listed as Least Concern by the IUCN (Barquez et al. 2015m).

Characteristics: Broad-eared free-tailed bat is medium sized; forearm length 41.0–45.0 mm, skull length 16.7–18.5 mm. Dorsal pelage reddish brown to dark brown; hairs whitish at the base; fur of underparts pale, hairs whitish at the base and yellowish at the tip; **wing membranes entirely naked and semitransparent; ears rectangular in shape but wider at the base than at the tip; tragus small and subquadrate, with the lateral side inclined; antitragus elongated but narrow; posterior margin of the palate terminating slightly posterior to the third upper molar, not as far as in *N. femorosaccus*;** sagittal crest poorly to moderately developed.

Comments: *Nyctinomops laticaudatus* ranges in the Neotropical region from Veracruz, México (except for the wetlands of southern Veracruz and Tabasco) south through Central America (Map 26). It can be found in sympatry, or nearly so, with *N. macrotis, N. aurispinosus,* and *N. femorosaccus*. From the first two, *N. laticaudatus* can be distinguished by its smaller size and from the latter by the greater size of the thumbs and the naked, semitransparent membranes.

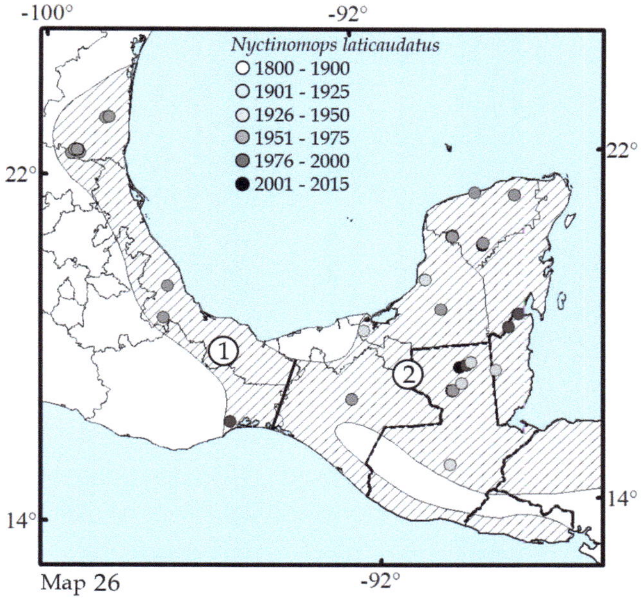

Map 26 Distribution map of *Nyctinomops laticaudatus*

Additional Literature: Shamel (1931), Goodwin and Greenhall (1961), Villa-R. (1960), Jones and Álvarez (1962), Jones and Arroyo-Cabrales (1990), Kumirai and Jones Milner et al. (1990), Sánchez-Hernández and Romero (1995), López-González (1998), Ávila-Flores et al. (2002), Riechers–Pérez (2004), Naranjo Piñera et al. (2005), Simmons (2005).

Nyctinomops macrotis (Gray, 1839)

Big free-tailed bat, murciélago de cola libre mayor

Monotypic.

Conservation: *Nyctinomops macrotis* is listed as Least Concern by the IUCN (Barquez et al. 2015n).

Characteristics: Big free-tailed bat is the largest in size; **forearm length 58.0–64.0 mm, skull length 22.2–24.0 mm.** Dorsal pelage dark brown; fur of underparts pale; **ears very large and subrectangular, wider at the tip and narrower at the base; tragus and antitragus medium sized and shaped as a circle;** basisphenoid pits deep; skull very large.

Comments: *Nyctinomops macrotis* ranges in desert areas with cliffs or mountains of temperate North America, including the states of California, Utah, Oklahoma, Colorado, Arizona, Nevada, New Mexico, and Texas in the United States south to the Eje Neovolcánico Trasversal in central México, except for the central and southern Baja California Peninsula (Map 27). *N. macrotis* is the largest species of the genus in North America and can be distinguished from all other species of *Nyctinomops* by its forearm and skull lengths greater than 58.0 and 22.0 mm, respectively.

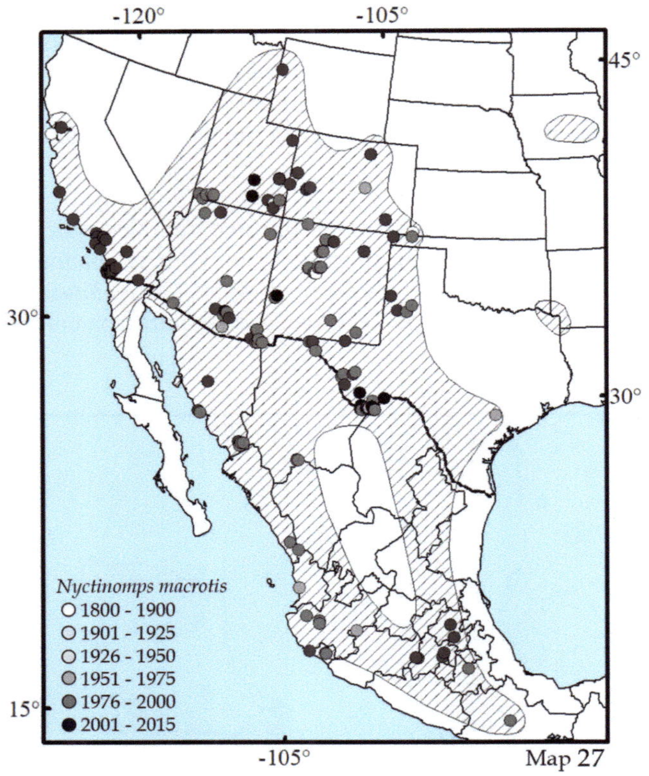

Map 27 Distribution map of *Nyctinomops macrotis*

Additional Literature: Barbour and Davis (1969), Hall (1981), Hoffmeister (1986), Milner et al. (1990), Timm and Genoways (2003), Riechers–Pérez (2004), Naranjo-Piñera et al. (2005), Simmons (2005).

Genus *Promops* Gervais, 1856

Medium sized; ears short and rounded, extending barely to the nostrils when laid forward; **antitragus rounded with a low neckline at the base**; no long hairs on the hips; pad of muzzle small but distinct and without processes; skull **short, robust, broad and seem rounded when viewed from above**; sagittal crest pronounced; rostrum markedly short and deep; **palate strongly dome-shaped;** tips of upper incisors diverge.

Promops centralis Thomas, 1915

Big crested mastiff bat, gran murciélago mastín crestado

P. c. centralis Thomas, 1915. Only subspecies in North America.

Conservation: *Promops centralis* is listed as Least Concern by the IUCN (Solari 2019u).

Characteristics: Big crested mastiff bat is medium sized; forearm length 51.0–57.0, skull length 21.0–21.8 mm. Dorsal pelage grayish brown, dark brown, or blackish; no long hairs on the hips; fur of underparts dark grayish brown; **circular antitragus with a distinct notch at the base**; tail length greater than half of the head-and-body length; **palatine bones dome-shaped;** no emargination on the palate between the upper incisors.

Comments: *Promops centralis* ranges in the tropical dry zone of the Pacific side of México from Jalisco southward, including the Balsas Basin and the Yucatán Peninsula (Map 28). This species can be distinguished from all other molossids by its triangular ears with the tip flat and narrow toward the base and not protruding from the crown, and a distinctly dome-shaped palatine.

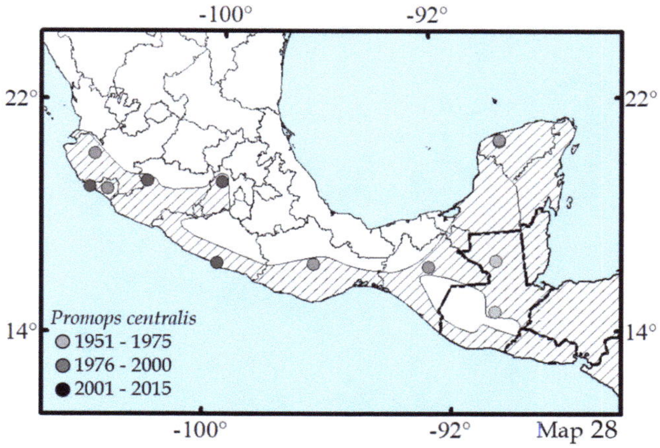

Map 28 Distribution map of *Promops centralis*

Additional Literature: Álvarez and Aviña (1964), Goodwin (1969), Ojasti and Linares (1971), Watkins et al. (1972), Birney et al. (1974), Genoways and Williams (1979), Freeman (1981), Hall (1981), Sánchez-Cordero et al. (1993), Van Staaden and Jones (1997), Vidal-López and Martínez-Coronel (2000), Simmons (2005).

Genus *Tadarida* Rafinesque, 1814

Medium sized; **ears not conjoined at the base, short, and not extending considerably beyond the tip of the nose when laid over the face; grooves in the upper lip;** second phalange of the fourth finger greater than 5.0 mm; anterior breadth of the rostrum distinctly greater than the interorbital breadth; single emargination in the upper incisors.

Tadarida brasiliensis (Geoffroy Saint-Hilaire, 1824)

Brazilean free-tailed bat, murciélago de cola libre mexicano

1. *T. b. cynocephala* (Le Conte, 1831). From Texas and Arkansas east through South Carolina and Florida, including a population in Ohio.
2. *T. b. intermedia* Shamel, 1931. From southern Chiapas south through Costa Rica.
3. *T. b. mexicana* (Saussure, 1860). From southern Oregon to Nebraska south through Chiapas and Campeche.

Conservation: *Tadarida brasiliensis* is listed as Least Concern by the IUCN (Barquez et al. 2015o).

Characteristics: Brazilian free-tailed bat is small sized; forearm length 41.0–45.0 mm, skull length 14.6–18.4 mm. Dorsal pelage gray, grayish to dark brown; fur of underparts pale; antitragus small and wide, triangle shaped; **ears large but not extending over the muzzle and not attached at the base**; grooves in the upper lip; single emargination in the upper incisors; **breadth of the anterior face much greater than the interorbital region**.

Comments: *Tadarida brasiliensis* is extremely widely distributed, ranging from approximately 43° North latitude southward through South America, excluding the Yucatán Peninsula, México (Map 29). It can be found in sympatry, or nearly so, with all other species, from which can be distinguished by having ears that do not surpass the snout nor are attached at the base; cranially, the anterior breadth of the rostrum is distinctly greater than the interorbital breadth.

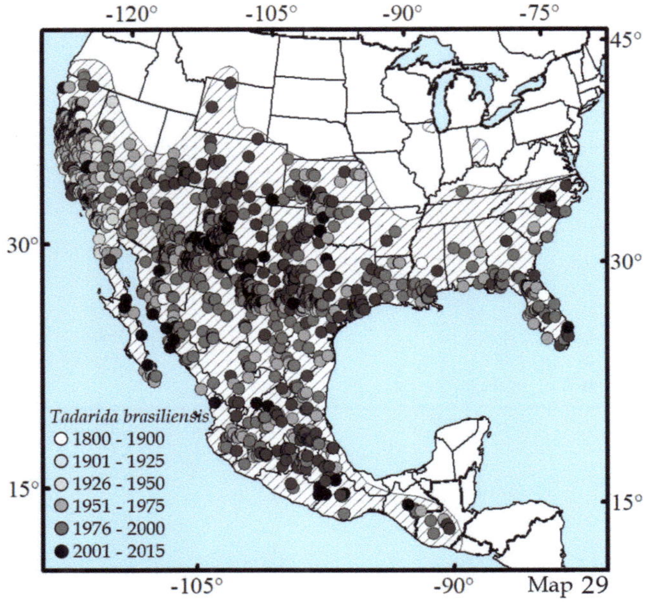

Map 29 Distribution map of *Tadarida brasiliensis*

Additional Literature: Shamel (1931), Villa-R. and Cockrum (1962), Jones et al. (1965), Davis (1966a, b), Barbour and Davis (1969), Goodwin (1969), Lowery (1974), Hall (1981), Wilkins (1989), Álvarez-Castañeda and Álvarez (1991), Koopman (1993), Riechers-Pérez (2004), Naranjo Piñera et al. (2005), Simmons (2005).

Family Mormoopidae Saussure, 1860

The family Mormoopidae includes "mustached bats," characterized by the unique presence of a filiform development on the lips and a stripe of stiff hair on the muzzle. The family contains two genera of small to medium sized bats with a rudimentary noseleaf muzzle; conspicuous skin folds on the chin; **tragus bordered with secondary skin folds; ears large, frequently joined at the base with low ventral extensions below the eyes; tail short, protruding from the dorsal surface of the large uropatagium;** third digit with three phalanges. Distributed from México through South America. It belongs to the superfamily Noctilionoidea and without subfamily divisions. The keys were elaborated based on the review of specimens and the following sources: Lawlor (1979), Hall (1981), Álvarez et al. (1994), and Álvarez-Castañeda et al. (2017a).

1. Folds under the lower lip, with one continuing from the ears to beneath the mouth commissure; frontal bone rising strongly above the plane of the rostrum (Fig. 16; almost at a 90° angle); the second lower premolars not noticeably smaller than the first lower premolars (from Arizona and Texas, United States, south through South America) .... *Mormoops megalophylla* (p. 235)

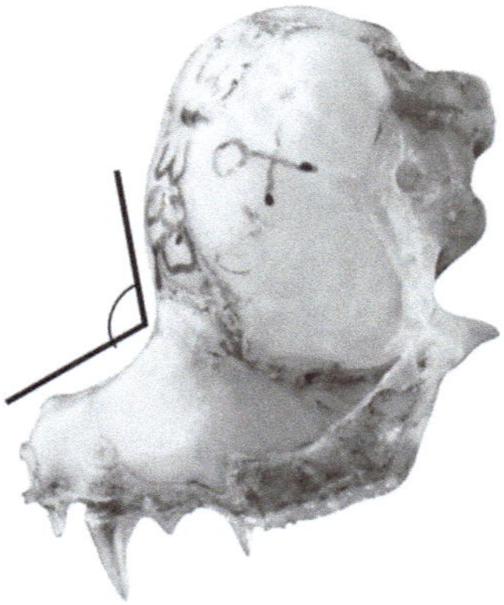

Fig. 16 The frontal bone rising strongly above the plane of the rostrum like in *Mormoops megalophylla*

1a. No folds beneath the lower lip or close to the ears; frontal bone not rising strongly above the plane of the rostrum (Fig. 17); second lower premolars noticeably smaller than the first lower premolars........................*Pteronotus* (p. 236)

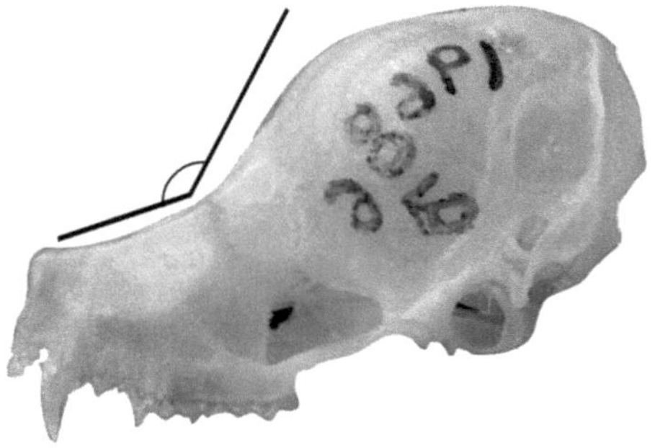

Fig. 17 The frontal bone not rising strongly above the plane of the rostrum like in *Pteronotus*

Genus *Mormoops* Leach, 1821

Medium sized; ears short and rounded, strongly connected dorsally on the rostrum by two conspicuous ridges; lower edge of the pinna confluent with the lower lip; ornamentation of the lower lip complex, with many intricate folds and scallops, forming a shield-like plate with many wartlike tubercles in front of the narrow central tubercle; a secondary transverse ridge below the lower lip further complicates its structure; a labio-nasal plate complex with the nostrils surrounded by separate pads and separated by a long, wartlike ridge; margin above and between the nostrils with several long, irregularly shaped tubercles; tragus complex, secondary fold prominent; pubescence restricted to several long hairs and numerous short bristles on the cranial edge; rostrum strongly upturned; forehead rising abruptly, almost at a right angle relative to the rostrum; zygomatic arches angled as a result of a rostro-braincase flexion; infraorbital foramen at the maxillary root of the zygomatic arch; incisive foramina large; basioccipital region relatively broad between the bullae and the basisphenoid with a cuplike depression. *Mormoops megalophylla* was considered within the genus *Aello* by Hall (1981). The keys were elaborated based on the review of specimens and the following sources: Smith (1972), Hall (1981), Álvarez et al. (1994), and Álvarez-Castañeda et al. (2017a).

Mormoops megalophylla Peters, 1864

Ghost-faced bat, murciélago con cara arrugada

M. m. megalophylla Peters, 1864. Only subspecies in North America.

Conservation: *Mormoops megalophylla* is listed as Least Concern by the IUCN (Davalos et al. 2019).

Characteristics: Ghost-faced bat is medium sized; forearm length 52.0–57.0 mm, skull length 13.6–14.9 mm. Dorsal pelage pale yellow or orange to dark brown; fur of underparts paler than the dorsum; many folds on the rostrum and both lips; muzzle seemingly absent; tail free at the middle of the uropatagium; **frontal bone rising strongly above the rostrum level; the second lower premolars not noticeably smaller than the first lower premolars.**

Comments: *Mormoops megalophylla* ranges throughout México and south Texas and Arizona, except for the central and northern Baja California Peninsula, northwest of Sonora, and the Altiplanicie Mexicana (Map 30). It can be distinguished from other species in sympatry, or nearly so, by its dermal folds under the lower lip and one that continues from the ears to below the corner of the mouth; the frontal bone rises strongly on the face plane (almost at an angle of 90°), and the second lower premolars not noticeably smaller than the first lower premolars.

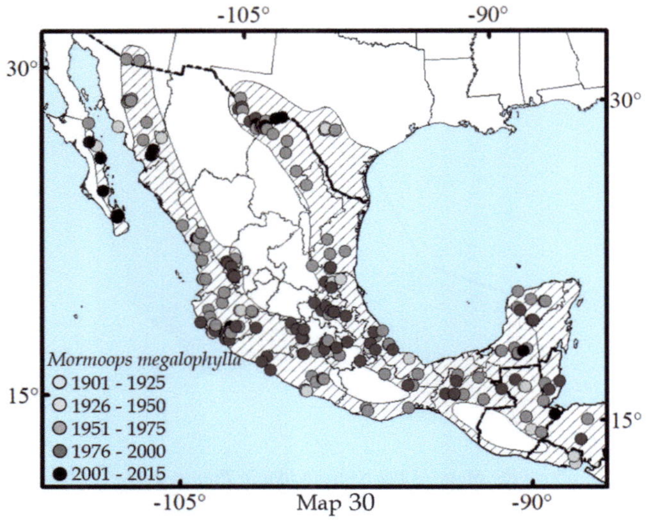

Map 30 Distribution map of *Mormoops megalophylla*

Additional Literature: Cockrum (1961), Davis and Carter (1962), Easterla (1968a), Smith (1972), Álvarez-Castañeda and Álvarez (1991), Rezsutek and Cameron (1993), Simmons (2005).

Genus *Pteronotus* Gray, 1838

Medium to large sized; dorsal pelage and fur of underparts orange to dark brown; no folds on the rostrum; **wing membranes fused or not dorsally at the midline**; ears pointed, twice as long as wide; **tail free at the middle uropatagium**; skull with the **rostrum more or less at the same level as the frontal bone**; second lower premolars noticeably smaller than the first lower premolars; basioccipital breadth between the auditory bullae of the same or less breadth than the auditory bullae. *Pteronotus* has three subgenera, *Chilonycteris* (*P. personatus*), *Phyllodia* (*P. mesoamericanus* and *P. mexicanus*), and *Pteronotus* (*P. davyi* and *P. gymnonotus*). The keys were elaborated based on the review of specimens and the following sources: Smith (1972), Hall (1981), Herd (1983), Álvarez et al. (1994), and Álvarez-Castañeda et al. (2017a).

1. Wing membranes fused dorsally at the midline, giving a naked-like appearance; mastoid breadth greater than the braincase breadth........subgenus *Pteronotus* 2

1a. Wing membranes not fused dorsally at the midline; mastoid breadth equal to or less than the braincase........3

2. Forearm length greater than 50.0 mm; maxillary tooth length greater than 7.1 mm (from Veracruz, Tabasco, and Chiapas, México, south through South America)........*Pteronotus gymnonotus* (p. 241)

2a. Forearm length less than 50.0 mm; maxillary tooth length less than 7.1 mm (from Sonora and Tamaulipas, México, south through South America)........*Pteronotus davyi* (p. 240)

3. Forearm length less than 48.0 mm; condylobasal length less than 17.0 mm; basioccipital bone not constricted between the auditory bullae, wider than the pterygoid fossa (from Sonora and southern Tamaulipas, México, south through South America)........subgenus *Chilonycteris; Pteronotus personatus* (p. 237)

3a. Forearm length greater than 48.0 mm; condylobasal length greater than 17.0 mm; basioccipital bone constricted between the auditory bullae, almost as wide as the pterygoid fossa........subgenus *Phyllodia* 4

4. Ventral fur with "frosted" appearance; body larger on average (from Tamaulipas and Oaxaca, México, south through Central America)........*Pteronotus mesoamericanus* (p. 238)

4a. Ventral fur pale drab gray; smaller body on average (from Sonora and Chihuahua south to Oaxaca, México)........*Pteronotus mexicanus* (p. 239)

Subgenus *Chilonycteris* Gray, 1839

Wing membranes attached high on the sides of the body, leaving a narrow hair-covered band on the back; nasal plate simple to moderately complex, margin above each nostril smooth and irregular; tragus relatively simple, spatulate, with a moderately well developed and shelf-like secondary fold or with a prominent secondary fold covered with many long and short hairs; antero-medial edge of the ears with two to four, occasionally as many as six, tooth-like serrations; dorsal pelage distinctly tricolored; skull profile relatively flat, although the rostrum may be slightly upturned; rostral breadth almost always less than, or equal to, the maxillary toothrow length; diastema usually present between the outer incisors and the canine; lower incisors reduced, trilobed and short; tympanic ring large, almost completely covering the auditory bullae; basioccipital broad between the bullae, with two large, extremely shallow, oval shaped pits.

Pteronotus personatus (Wagner, 1843)

Wagner's mustached bat, murciélago bigotudo menor

P. p. psilotis (Dobson, 1878). Only subspecies in North America.

Conservation: *Pteronotus personatus* is listed as Least Concern by the IUCN (Davalos et al. 2016).

Characteristics: Wagner's mustached bat is small sized; **forearm length 42.0–48.0 mm, skull length 13.4–15.2 mm.** Dorsal pelage and fur of underparts brown to dark brown; no folds on the rostrum; **wing membranes not fused dorsally at the midline**; upper incisors reduced, with a small diastema between the outer incisors and the canines; inner incisor pair clearly forked, narrow and rounded at the base.

Comments: *Pteronotus personatus* was proposed to have ranges only in South America; for this reason, the species with ranges in North America should be assigned to *P. psilotis* (Dávalos 2006). However, this change has not been accepted, so I am still using *P. personatus* until further detailed analyses are published. *P. personatus* ranges in the warm tropical regions of both coasts from Sonora and Tamaulipas south through Central America (Map 31). It can be found in sympatry, or nearly so, with *P. davyi*, *P. gymnonotus*, *P. mesoamericanus*, and *P. mexicanus*. *P. personatus* can be distinguished from the rest of the species by having no folds below the lower lip or close to the ears; wing membranes not joined together in the middle back; forearm and condylobasal lengths less than 48.0 and 17.0 mm, respectively.

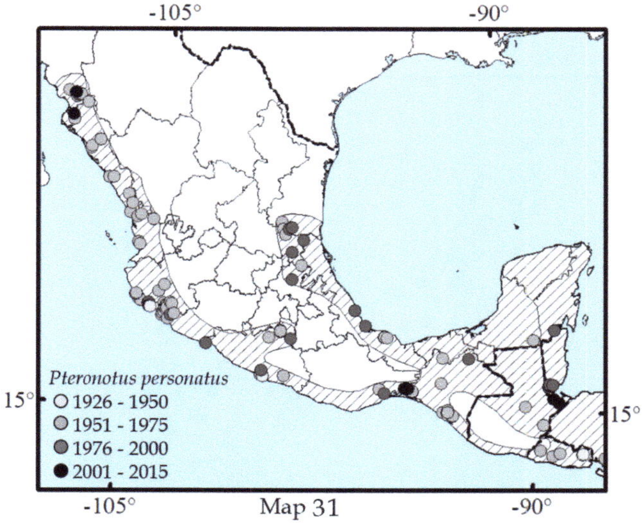

Map 31 Distribution map of *Pteronotus personatus*

Additional Literature: Villa (1967), Smith (1972), Jones et al. (1973, 1988), Medellín et al. (1997), De la Torre and Medellín (2010).

Subgenus *Phyllodia* Gray, 1843

Wing membranes attached low on the sides of the body; back covered by a wide band of short stiff hairs; labio-nasal plate simple; margin above each nostril with several irregular wartlike tubercles; emargination between the nostrils deep; tragus simple, lanceolate, and secondary fold small; antero-medial edge of the ears smooth and lanceolate portion broad; ears connected by two low, inconspicuous ridges that fuse on the muzzle top forming a prominent rostral tubercle; profile relatively flat, rostrum not elevated; forehead sloping gradually onto a rather long shape; braincase broad; upper incisors robust and peglike, inner pair distinctly bifurcate and broad, rounded heel; lower incisors also heavily constructed, inner pair trilobed, outer pair bilobed; tympanic rings smaller relative to those of other subgenera, covering approximately one-third of the auditory bullae; **basioccipital narrowly constricted between the auditory bullae**; with two narrow and deep longitudinal furrows.

Pteronotus mesoamericanus Smith, 1972
Common mustached bat, murciélago bigotudo mayor

Monotypic.

Conservation: *Pteronotus mesoamericanus* is listed as Least Concern by the IUCN (Solari 2016e).

Characteristics: Common mustached bat is medium sized; **forearm length 55.5–61.9 mm, skull length 19.0–20.7 mm**. Dorsal pelage dark brown; pectoral and abdominal regions with the hair basal portion nearly Prout's brown, giving an overall dark brown appearance to the pelage; **underparts pale cinnamon-buff at the tip, giving a "frosted" appearance;** head and nape of the same color; throat and chin regions usually somewhat paler; no folds on the rostrum; wing membrane large (see measurements), not joined on the back; upper incisors big, wedge shaped; inner incisor pair clearly forked, wide at the base.

Comments: *Pteronotus mesoamericanus* was recognized as a distinct species (Clare et al. 2013; Pavan and Marroig 2016, 2017), previously regarded as a subspecies of *P. parnellii* (Smith 1972). *P. mesoamericanus* ranges from Tamaulipas and Oaxaca south through Central América, including the southern Mexican highlands (Map 32). It can be found in sympatry, or nearly so, with *P. davyi, P. gymnonotus*, *P. mexicanus,* and *P. personatus. P. mesoamericanus* can be distinguished from *P. davyi, P. gymnonotus*, and *P. personatus* by not having folds below the lower lip or close to the ears; wing membranes not fused dorsally at the midline, and forearm and condylobasal lengths greater than 48.0 and 17.0 mm, respectively. From *P. mexicanus*, by its larger size, a darker coloration; fur of underparts with "frosted" appearance.

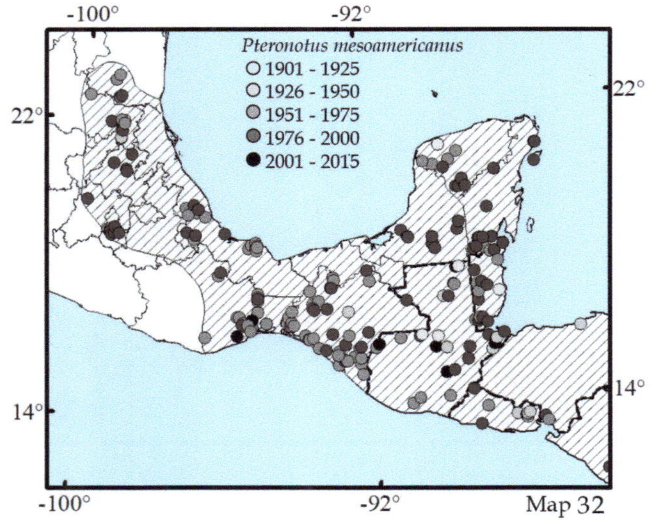

Map 32 Distribution map of *Pteronotus mesoamericanus*

Additional Literature: Hall and Kelson (1959), Cockrum and Bradshaw (1963), Walker (1975), Herd (1983), Álvarez-Castañeda and Álvarez (1991), Simmons (2005).

Pteronotus mexicanus (Miller, 1902)

Mexican mustached bat, murciélago bigotudo mexicano

Monotypic

Conservation: *Pteronotus mexicanus* (as *Pteronotus parnellii*) is listed as Least Concern by the IUCN (Solari 2016f).

Characteristics: Common mustached bat is medium sized; **forearm length 55.0–60.0, skull length 18.5–20.2 mm**. Dorsal pelage cinnamon-buff, brownish at the base and tip, seen nearly monocolored, primarily from a weak expression of a whitish central zone on the back; **pectoral and abdominal regions pale drab gray with a buffy tinge**, slightly paler on the neck and chin; head and nape of the same color; no folds on the rostrum; wing membranes not joined on the back; large-sized, see measurements; big upper incisors, wedge shaped; internal incisor pair clearly forked, wide at the base.

Comments: *Pteronotus mexicanus* was recognized as a distinct species (Pavan and Marroig 2016, 2017), previously as a subspecies of *P. parnellii* (Smith 1972). *P. mexicanus* ranges from Sonora south through Oaxaca (Map 33). It can be found in sympatry, or nearly so, with *P. davyi, P. mesoamericanus*, and *P. personatus*. *P. mexicanus* can be distinguished from *P. davyi* and *P. personatus* by not having folds below the lower lip or close to the ears; wing membranes not fused dorsally at the midline forearm and condylobasal lengths greater than 48.0 and 17.0 mm, respectively. From *P. mesoamericanus*, by its smaller size and paler coloration; fur of underparts without a frosted appearance.

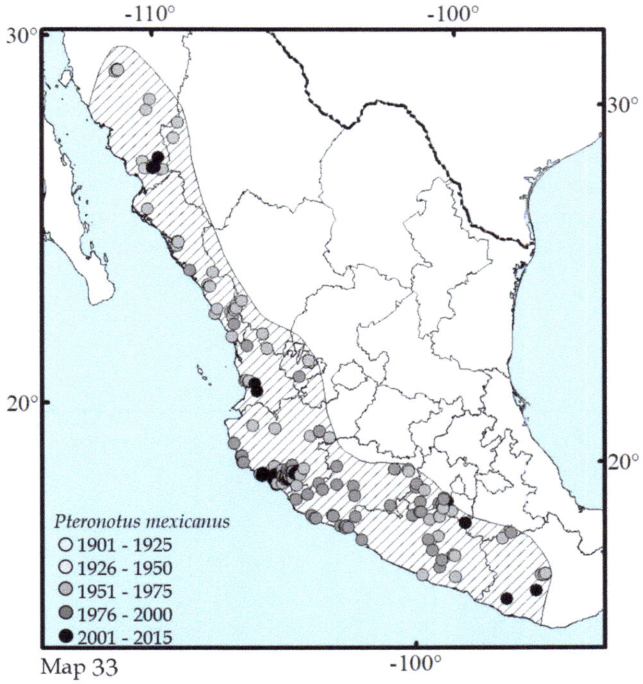

Map 33 Distribution map of *Pteronotus mexicanus*

Additional Literature: Hall and Kelson (1959), Cockrum and Bradshaw (1963), Walker (1975), Herd (1983), Álvarez-Castañeda and Álvarez (1991), Simmons (2005).

Subgenus *Pteronotus* Gray, 1838

Wing membranes fused dorsally at the midline, resulting in a naked-backed appearance; labio-nasal plate moderately complex in structure, margin above each nostril ornamented by a series of irregularly shaped, wart-like tubercles, and lateral spikes present on either side of the nostrils; tragus moderately complex, spatulate, with a prominent secondary fold and with moderately long hairs ornamenting the distal borders; antero-medial edge of the pinna smooth; ears not noticeably connected; pelage monocolored or only slightly bicolored dorsally; skull profile relatively flat, but rostrum short and noticeably upturned; forehead relatively abruptly elevated onto the elongated braincase; rostrum extremely broad relative to length, maxillary toothrow relatively short; upper incisors reduced in size, inner pair bifurcate and with a narrow rounded heel directed toward the canine; inner and outer incisors compressed together; diastema between the outer incisors and the canine; lower incisors reduced and trilobed, with a short rounded heel; tympanic ring large, nearly covering the auditory bullae; basioccipital between the auditory bullae extremely broad and cup shaped, having two shallow and oval shaped pits.

Pteronotus davyi Gray, 1838

Davy's naked-backed bat, murciélago de espalda desnuda menor

P. d. fulvus (Thomas, 1892). Only subspecies in North America.

Conservation: *Pteronotus davyi* is listed as Least Concern by the IUCN (Solari and Davalos 2019).

Characteristics: Davy's naked-backed bat is small sized; **forearm length 43.0–49.0 mm, skull length 13.9–15.6 mm.** Dorsal pelage and fur of underparts orange to dark brown; no folds on the rostrum; **wing membranes fused dorsally at the midline;** skull morphologically similar but smaller than *P. gymnonotus*.

Comments: *Pteronotus davyi* was proposed to range only in South America and the species present in North America should be assigned to *P. fulvus* (Davalos 2006); however, this change has not been accepted, so I still use *P. davyi* until further detailed analyses are published. It ranges in the warm tropical regions of both coasts from Sonora, Nuevo León, and Tamaulipas south through South America (Map 34). It can be found in sympatry, or nearly so, with *P. gymnonotus, P. mesoamericanus*, *P. mexicanus,* and *P. personatus. P. davyi* can be distinguished from the rest of the species by having no folds below the lower lip or close to the ears; wing membranes fused dorsally at the midline; forearm length and toothrow less than 50.0 and 7.1 mm, respectively.

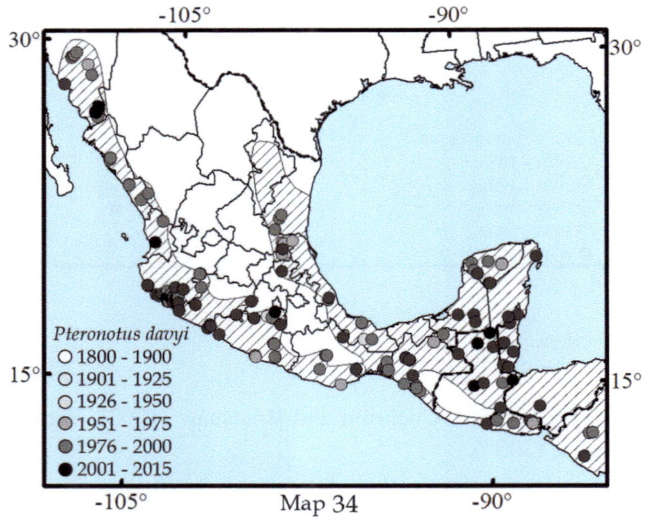

Map 34 Distribution map of *Pteronotus davyi*

Additional Literature: Goodwin and Greenhall (1961), Villa-R. et al. (1967), Álvarez (1968), Jiménez-Guzmán (1968), Smith (1972), Bateman and Vaughan (1974), Koopman (1978, 1982), Herd (1983), Adams (1989), Álvarez-Castañeda and Álvarez (1991), Simmons (2005), Guevara-Chumacero et al. (2010).

Pteronotus gymnonotus (Natterer, 1843)

Big naked-backed bat, murciélago de espalda desnuda mayor

Monotypic.

Conservation: *Pteronotus gymnonotus* is listed as Threatened in the Norma Oficial Mexicana (DOF 2019) and as Least Concern by the IUCN (Solari 2019n).

Characteristics: Big naked-backed bat is medium sized; **forearm length 50.0–55.0 mm, skull length 15.4–17.2 mm.** Dorsal pelage and fur of underparts orange to brown; **wing membranes fused dorsally at the midline;** skull morphologically similar but larger than *P. davyi*.

Comments: *Pteronotus gymnonotus* ranges from Veracruz, Tabasco, and Chiapas south through South America (Map 35). It can be found in sympatry, or nearly so, with *P. davyi, P. mesoamericanus,* and *P. personatus. P. gymnonotus* can be distinguished from the rest of the species by not having folds under the lower lip or close to the ears; wing membranes fused dorsally at the midline; forearm length and toothrow greater than 50.0 and 7.1 mm, respectively.

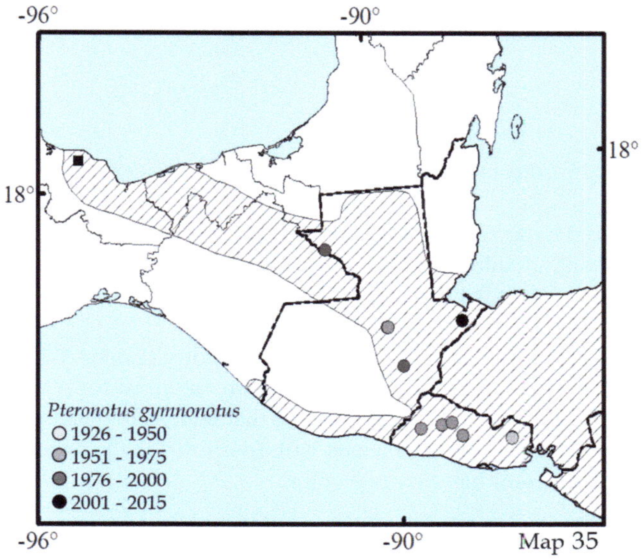

Map 35 Distribution map of *Pteronotus gymnonotus*

Additional Literature: Smith (1972), Hall (1981), Álvarez-Castañeda and Álvarez (1991), Koopman (1993), Simmons (2005), Pavan and Marroig (2016, 2017), Pavan and Tavares (2020).

Family Natalidae Gray, 1866

The family Natalidae includes "funnel-eared bats," characterized by ears funnel shaped; small sized; muzzle smooth; **pinna large, funnel shaped,** separated from each other, no ventral extensions under the eyes; tragus **small, triangular**; third digit with two phalanges; thumbs relatively small; tail long, fully enclosed in a large membrane. It is distributed in the tropics and subtropics of North and South America, including the Antilles. This family belongs to the superfamily Vespertilionoidea.

Genus *Natalus* Gray, 1838

Dorsal pelage and fur of underparts yellowish to reddish; fur long and silky; **ears funnel shaped, no folds along the upper external border (outer base slightly in front of the inner base); tragus thickened, distorted;** limbs and tail very long, each exceeding the head-and-body length, greater than 80% of the head-and-body length; muzzle long with lips and chin simple; nostrils simple, directed outward and slightly downward; skull with a **spherical braincase that rises abruptly from the rostrum**; greatest braincase breadth about equal to its depth, including the auditory bullae; rostrum about three-fourths as long as the braincase; six upper and lower molariform teeth (Miller 1907). The keys were elaborated based on the review of specimens and the following Álvarez-Castañeda et al. (2017a).

1. Lesser presence of hair close to the base of the digits in the limbs; no folds in the upper external border of the ears (from Sonora, Tamaulipas, and south Baja California Sur, México, south through Central America).. ....*Natalus mexicanus* (p. 243)

1a. Greater presence of hair close to the base of the digits of the limbs; two to four folds on the upper external border of the ears (from Sonora south through Guerrero, probably in Veracruz, México) ....*Natalus lanatus* (p. 242)

Natalus lanatus Tejedor, 2005

Haired funnel-eared bat, murciélago orejas de embudo peludo

Monotypic

Conservation: *Natalus lanatus* (as *Natalus mexicanus*) is listed as Least Concern by the IUCN (Solari 2019k).

Characteristics: Haired funnel-eared bat is small sized; forearm length 35.4–38.6 mm, skull length 15.0–16.4 mm. Dorsal pelage and fur of underparts yellowish to reddish; **locks of hair close to the base of the digits in the limbs**; no skull characteristic has been recorded to differentiate it from *N. mexicanus*.

Comments: *Natalus lanatus* has raised several questions about its validity (López-Wilchis et al. 2012). It is known only from Sonora to Guerrero and probably in Veracruz. The two species recognized for *Natalus*, *N. lanatus* and *N. mexicanus* (Map 36), can be found in sympatry, or nearly so. *N. lanatus* can be distinguished by having a greater amount of hair, emphasizing the presence of a hair tuft at the base of the digits; ears with four folds in the upper outer edge; hair bicolored, darker at the base than at the tip, while *N. mexicanus* has only one color.

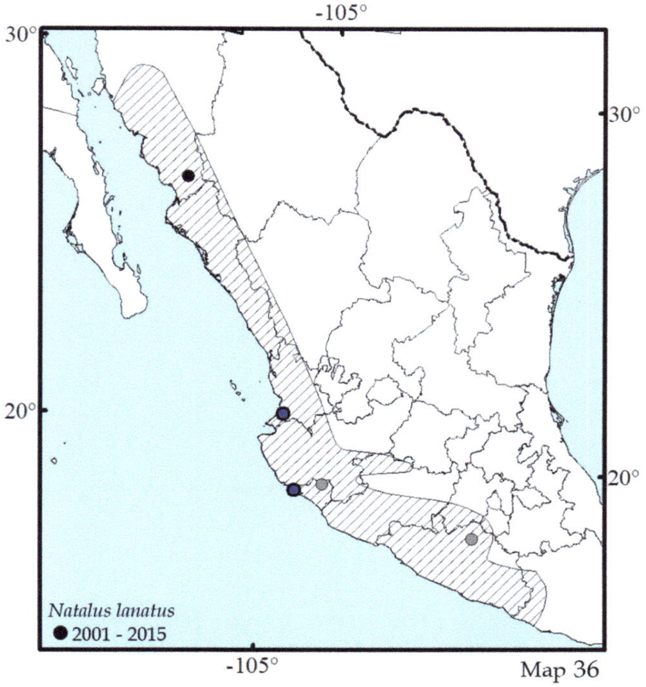

Map 36 Distribution map of *Natalus lanatus*

Additional Literature: Tejedor (2005, 2006, 2011), Simmons (2005), López-González and García-Mendoza (2006), Torres-Morales et al. (2010).

Natalus mexicanus Miller, 1902

Mexican funnel-eared bat, murciélago orejas de embudo mexicano

Monotypic

Conservation: *Natalus mexicanus* is listed as Least Concern by the IUCN (Solari 2019k).

Characteristics: Mexican funnel-eared bat is small sized; forearm length 35.3–40.0 mm, skull length 15.2–16.5 mm. Dorsal pelage and fur of underparts yellowish to reddish; **no locks of hair close to the digits at the base of the limbs**; no skull characteristic has been recorded to differentiate it from *N. lanatus*.

Comments: *Natalus mexicanus* is used instead of *N. stramineus* (Tejedor 2006, 2011). *N. mexicanus* is distributed in the warm tropical regions of both coasts from Sonora and Tamaulipas south through Central America, including the southern Baja California Peninsula. The two species recognized for *Natalus*, *N. lanatus* and *N. mexicanus* (Map 37), can be found in sympatry, or nearly so. *N. mexicanus* can be distinguished by having a smaller amount of hair, no hair tuft near at the base of the digits and without folds in the upper outer edge of the ears.

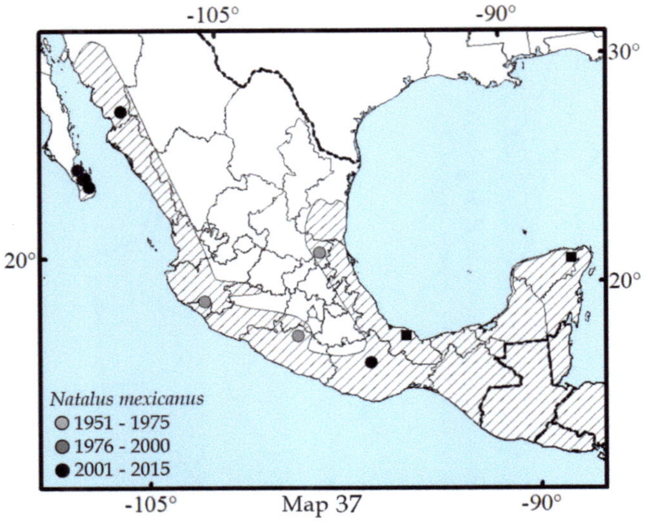

Map 37 Distribution map of *Natalus mexicanus*

Additional Literature: Miller (1902), Baker and Greer (1962), Jones et al. (1965), Álvarez (1968), Matson and Patten (1975), Hall (1981), Arroyo-Cabrales et al. (1997), Simmons (2005), López-Wilchis et al. (2012, 2020), Santos-Moreno et al. (2016).

Family Noctilionidae Gray, 1821

The family Noctilionidae includes "bulldog bats" because their face resembles this dog breed. These species are mainly insectivorous and considered fishing bats; **medium sized (forearm 55.0–90.0 mm); full lips, folded, with a medium cleft and forming different pouches on the cheeks; pinna large, pointed, and separated without a ventral extension under the eyes; tragus small; third digit with two phalanges**. Range restricted to tropical America. It belongs to the superfamily Noctilionoidea.

Genus ***Noctilio*** Linnaeus, 1766

Medium or large sized; dorsal coloration pale gray to orange yellow with a paler stripe on the back; **hind feet large with nails very well developed, long and flat; hind limb length and claws almost equivalent to the head-and-body length; head and lips large; ears projecting frontward, twice longer than wide**; skull strongly built with a **well-defined mastoid region in shape of a "shelf"; sagittal crest greatly enlarged**; teeth large and well developed. *Noctilio* has two subgenera, each monotypic: *Dirias* (*N. albiventris*) and *Noctilio* (*N. leporinus*). The keys were elaborated based on the review of specimens and the following sources: Davis (1973), Hall (1981), Álvarez et al. (1994), and Álvarez-Castañeda et al. (2017a).

1. Limb length less than 23.0 mm and forearm less than 73.0 mm; skull length less than 25.0 mm (from Chiapas, México, south through South America) .... subgenus *Dirias; Noctilio albiventris* (p. 245)

1a. Limb length greater than 23.0 mm and forearm greater than 73.0 mm; skull length greater than 25.0 mm (from Sinaloa and south Veracruz, México, south through South America)............ subgenus *Noctilio*; *Noctilio leporinus* (p. 245)

Subgenus ***Dirias*** Miller, 1904

Limbs and feet less elongated; length of tibia and foot together barely exceeding 40% of the total length; first and second upper molars very slightly emarginated posteriorly, so the spaces between teeth are scarcely noticeable; **lypocone relatively larger than in *Noctilio* and connected by a high**, conspicuous commissure, extending from the protocone to the metacone.

Noctilio albiventris Desmarest, 1818

Lesser bulldog bat, murciélago pescador menor

N. a. minor Osgood, 1910. Only subspecies in North America.

Conservation: *Noctilio albiventris* is listed as subject to Special Protection in the Norma Oficial Mexicana (DOF 2019) and as Least Concern by the IUCN (Barquez et al. 2015h).

Characteristics: Lesser bulldog bat is medium sized; **forearm length 58.4 mm, skull length 18.1–21.6 mm.** Dorsal pelage pale gray to orange yellow; skull similar to *N. leporinus* but smaller, see measurements.

Comments: *Noctilio albiventris* has been only recorded in the coast of Chiapas (Map 38). It can be distinguished from other species in sympatry, or nearly so, by its larger limbs, with very well-developed long and flat nails, limb length almost equivalent to the head-and-body length; large thumbs; ears projecting forward, twice longer than wide. *N. albiventris* is very small relative to *N. leporinus*, with smaller forearm length (73.0 mm) and skull length (25.0 mm).

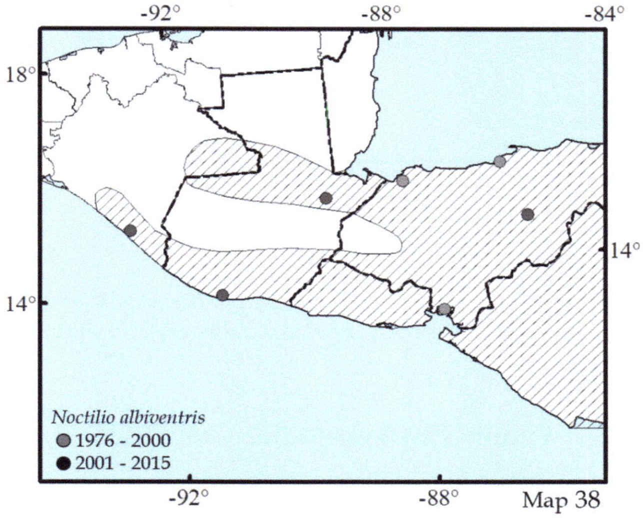

Map 38 Distribution map of *Noctilio albiventris*

Additional Literature: Miller (1907), Davis (1973, 1976), Hood and Pitocchelli (1983), Álvarez-Castañeda and Álvarez (1991), Vargas-Contreras et al. (2008).

Subgenus *Noctilio* Linnaeus, 1758

Tibia and foot length together about 60% of the total length; first and second upper molars subequal; four cusps, three styles, and commissures well developed; **lypocone relatively smaller than in *Dirias*.**

Noctilio leporinus (Linnaeus, 1758)

Great bulldog bat, murciélago pescador mayor

N. l. mastivus (Vahl, 1797). Only subspecies in North America.

Conservation: *Noctilio leporinus* is listed as Least Concern by the IUCN (Barquez et al. 2015i).

Characteristics: Great bulldog bat is large sized; **forearm length 81.6–88.1 mm, skull length 25.3–26.4 mm.** Dorsal pelage pale gray to orange yellow, with a paler stripe on the back; skull similar to *N. albiventris* but larger, see measurements.

Comments: *Noctilio leporinus* ranges in the warm tropical regions of both coastal plains near the sea, from Sonora and Tamaulipas south through South America (Map 39). It can be distinguished from other species in sympatry, or nearly so, by its larger hind feet; very nails well developed, long and flat; limb length almost equivalent to the head-and-body length; large thumbs; ears projecting forward twice longer than wide. *N. leporinus* is larger than *N. albiventris* with forearm length greater than 73.0 mm and skull length greater than 25.0 mm.

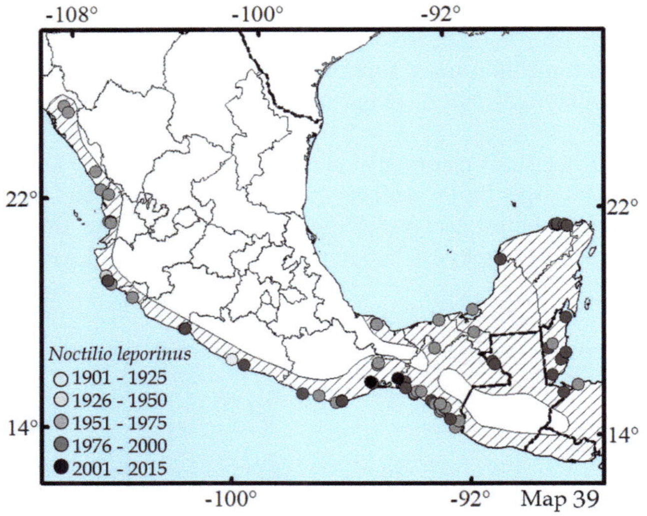

Map 39 Distribution map of *Noctilio leporinus*

Additional Literature: Miller (1907), Allen (1937), Watkins et al. (1972), Davis (1973, 1976), Jones et al. (1973), Hood and Pitocchelli (1983), Hood and Jones (1984), Álvarez-Castañeda and Álvarez (1991).

Family Phyllostomidae Gray, 1825

The family Phyllostomidae includes "leaf- or spear-nosed bats," characterized by having a filiform appendage resembling a spear on their nose; variable size, from small to relatively large (forearm length 25.0–110.0 mm); **noseleaf on the muzzle erect and conspicuous (rudimentary in some genera)**; pinna from small to large, generally separated, without a ventral extension under the eyes; tragus small; third digit with three phalanges. Restricted to tropical America. It belongs to the superfamily Noctilionoidea, with nine subfamilies in North America: Carolliinae, Desmodontinae, Glossophaginae, Glyphonycterinae, Lonchorhininae, Macrotinae, Micronycterinae, Phyllostominae, and Stenodermatinae. The keys were elaborated based on the review of specimens and the following sources: Lawlor (1979), Hall (1981), Álvarez et al. (1994), and Álvarez-Castañeda et al. (2017a).

1. Thumbs well developed and functional for walking, with a bearing on the ventral side; upper incisors shaped as a knife (Fig. 18); teeth 26 or less .... subfamily Desmodontinae (p. 252)

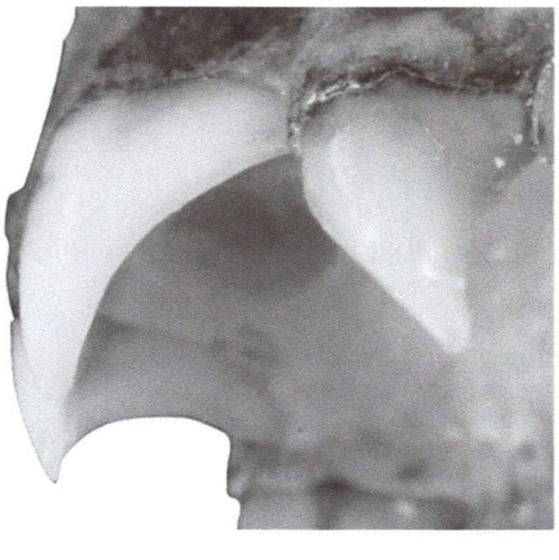

Fig. 18 Upper incisors as shaped as a knife

1a. Thumbs not developed and not functional for walking, without a bearing on the ventral side; upper incisors small and cylindrical; teeth more than 26 2

2. Noseleaf as long (greater than 25.0 mm) as the ears; dorsal border of the rostrum strongly convex with a deep depression between the eye sockets (Fig. 19; southern Veracruz, Oaxaca, Campeche, Quintana Roo, and Chiapas) .... subfamily Lonchorhininae; *Lonchorhina aurita* (p. 274)

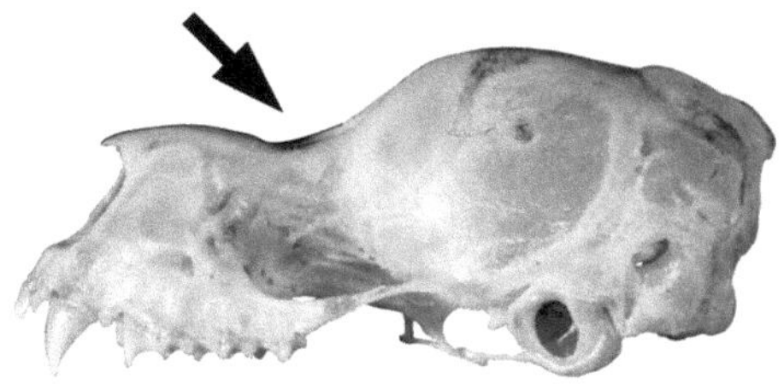

Fig. 19 The dorsal border of the rostrum is strongly convex with a deep depression between the eye sockets *Lonchorhina aurita*

2a. Noseleaf (less than 25.0 mm) much shorter than the ears; dorsal border of the rostrum not convex; no depression between the eye sockets 3

3. Ears twice the head size and tail tip free from the uropatagium; total length less than 110.0 mm; auditory bullae large, its maximum diameter larger than the distance between the auditory bullae .... subfamily Macrotinae (p. 275)

3a. Ears not twice the head size; in case of having a similar proportion, tail tip not free from the uropatagium; auditory bullae small, its maximum diameter equal to the distance between the auditory bullae 4

4. Rostrum thin and extended; distance between the eyes and the tragus less than the distance between the eyes and the lower lip; tongue long and hairy distally; muzzle cylindrical; teeth without a W shaped crown, and proportionally small; many species with a diastema between the molariforms (Fig. 20) .... subfamily Glossophaginae (p. 256)

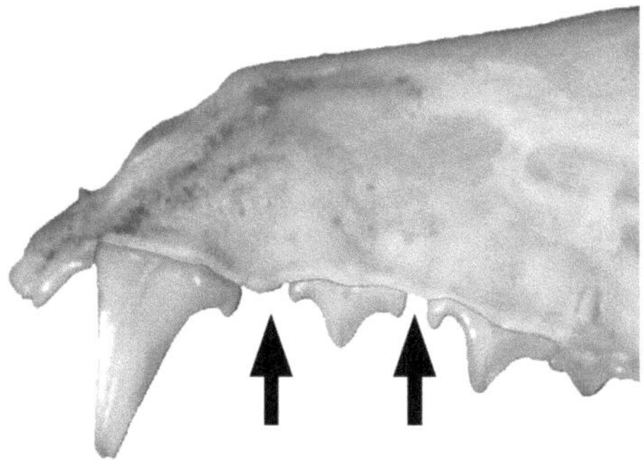

Fig. 20 With a diastema between the molariforms present in Glossophaginae

4a. Rostrum short and wide; distance between the eyes and the tragus larger than the distance between the eyes and the lower lip; muzzle not cylindrical; teeth with a crown W shaped or not but proportional to the muzzle size; molariforms without a diastema 5

5. Lower lip with a round central and large wart surrounded by smaller warts; tail short; each dorsal hair with three color bands; molars with a cutting crest but not W shaped .... subfamily Carolliinae (p. 248)

5a. No large, round, and central wart on the lower lip, although it can have several warts; tail present or absent; each hair in homogeneous color or with two bands, one basal and another distal; cutting molars W shaped or not 6

6. Tail absent; noseleaf spear shaped; molars with the crown not W shaped .... subfamily Stenodermatinae (p. 293)

6a. Tail present; if absent, the uropatagium extends to the level of the calcar; forearm length greater than 75.0 mm; noseleaf extended; molar crown W shaped 7

7. Ears not joined by a band between them; one or two lower incisors; in case of two, the second or third lower premolars of the same size .... subfamily Phyllostominae (p. 281)

7a. Ears joined by a high band; two lower incisors besides having the second and third lower premolars of different size 9

8. Fifth metacarpal shorter and third metacarpal longer; second upper incisors bifid; four upper premolars straight .... subfamily Micronycterinae (p. 277)

8a. Fourth metacarpal shorter and third or fifth metacarpal longest; second upper incisors small and unicuspid; four upper premolars slightly recurved .... subfamily Glyphonycterinae (p. 271)

Subfamily Carolliinae Miller, 1924

The subfamily Carolliinae includes short-tailed bats characterized by the trochiter impinging on the scapula; upper molars so modified that the W shaped pattern is markedly altered or almost obliterated; upper molars lacking commissures tilled styles; lower molars with cusps strictly lateral; crowns of upper and lower molars trenchant.

Genus *Carollia* Gray, 1838

Medium sized; dorsal pelage grayish or yellowish brown to dark orange; fur of underparts paler than in the dorsum; **row of small warts on the chin U-shaped, with a large central wart**; **hair with three bands; band color changing in relation to the species, mainly on the basal band;** upper and lower molars with a cutting crown; zygomatic arches incomplete; auditory bullae small. The keys were elaborated based on the review of specimens and the following sources: Pine (1972), Hall (1981), Álvarez et al. (1994), and Álvarez-Castañeda et al. (2017a).

1. Forearm length greater than 41.0 mm; lower jaw V-shaped (Fig. 21); in dorsal view, second lower incisors covered by the canine cingulum; lingual border of the upper molar series straight (from Veracruz, México, south through South America) .... *Carollia perspicillata* (p. 250)

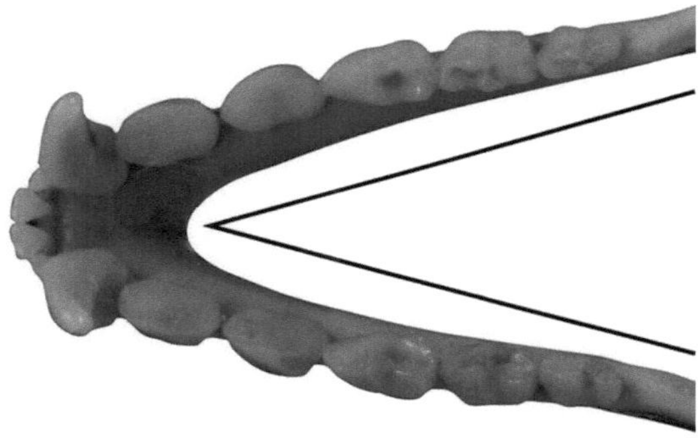

Fig. 21 Lower jaw V-shaped like in *Carollia perspicillata*

1a. Forearm length less than 41.0 mm; lower jaw with trend to be U-shaped (Fig. 22); in dorsal view, second lower incisors not covered by the canine cingulum; lingual border of the upper molariform series convex and with a step 2

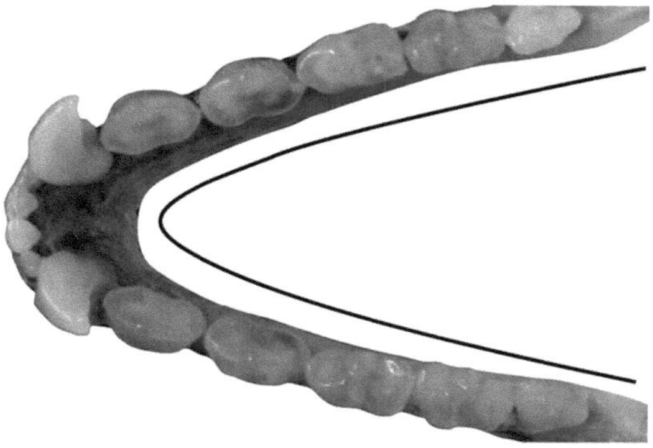

Fig. 22 Lower jaw with trend to be U-shaped

2. Forearms with hair; nape area covered by hair with rings of a well-defined color, so the dark basal wide band contrasts strongly with the hair tip; rostrum more robust and extended; braincase less spherical; palate long; mental symphysis ventral tip located above the ventral border of the mandible branch (Fig. 23; from Veracruz, México, south through Central America) .... *Carollia sowelli* (p. 250)

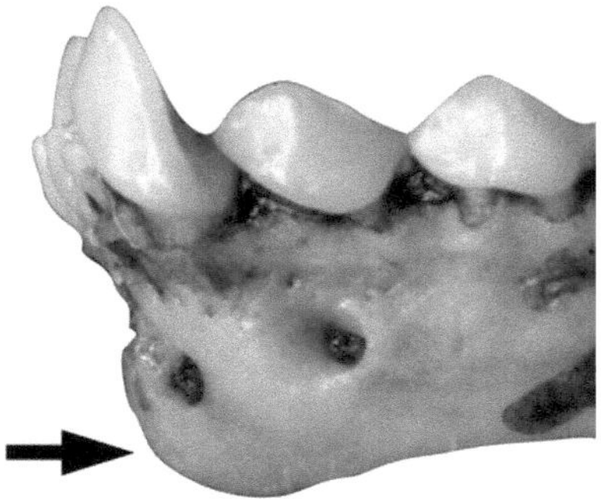

Fig. 23 Mental symphysis ventral tip located above the ventral border of the mandible branch like in *Carollia sowelli*

2a. Forearm with scarce hair; nape area covered by hair with indistinct rings, the basal band not markedly contrasting with the distal part; rostrum less robust and short; braincase more spherical; palate short; mental symphysis ventral tip located under the mandible branch ventral border (Fig. 24; from Jalisco, México, south through Central America) .... *Carollia subrufa* (p. 251)

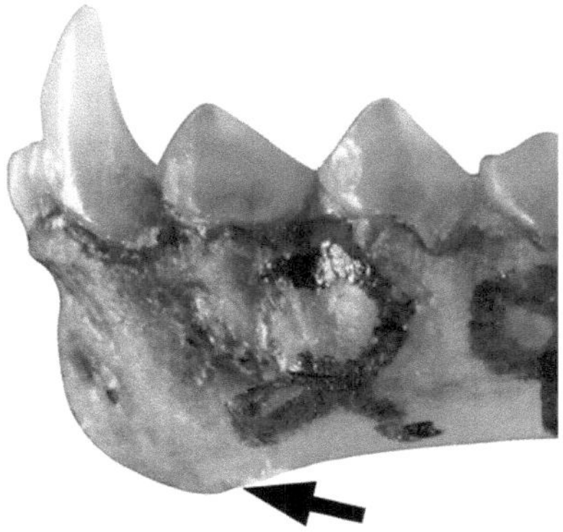

Fig. 24 Mental symphysis ventral tip located under the mandible branch ventral border like in *Carollia subrufa*

Carollia perspicillata (Linnaeus, 1758)
Seba's short-tailed bat, murciélago colicorta

C. p. azteca Saussure, 1860. Only subspecies in North America.

Conservation: *Carollia perspicillata* is listed as Least Concern by the IUCN (Barquez et al. 2015c).

Characteristics: Seba's short-tailed bat is medium sized; forearm length 41.0–45.0 mm, skull length 21.4–25.2 mm. Dorsal pelage grayish or yellowish brown to dark orange; fur of underparts paler than the dorsal part; **hair basal band brown; jaw in V-shape; second incisor covered by the canine cingulum**.

Comments: *Carollia perspicillata* ranges from central Veracruz south through Central America, including the Yucatán Peninsula (Map 40). It can be found in sympatry, or nearly so, with *C. subrufa* and *C. sowelli* in the tropical region of México. They are distinguished from other genera by having a row of small warts on the U-shaped chin, with a large central one, and hair with three bands. *C. perspicillata* is the largest species, with forearm length greater than 41.0 mm; jaw V-shaped; in dorsal view, the second lower incisors are covered by the canine cingulum and the lingual edge of the upper molar series is straight.

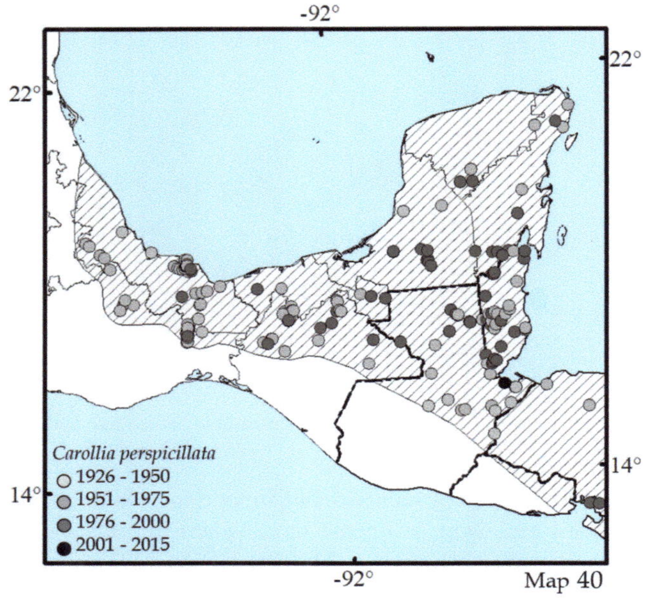

Map 40 Distribution map of *Carollia perspicillata*

Additional Literature: Miller (1907), Goodwin and Greenhall (1961), Pine (1972), Barbee and Fugler (1977), McLellan (1984), Owen et al. (1984), Fleming (1988), Álvarez-Castañeda and Álvarez (1991), Cloutier and Thomas (1992), Hoffmann and Baker (2003), Hoffmann et al. (2008), Gonçalves Da Silva et al. (2008).

Carollia sowelli Baker et al., 2002
Sowell's short-tailed bat, murciélago de colicorta pardo

Monotypic.

Conservation: *Carollia sowelli* is listed as Least Concern by the IUCN (Miller et al. 2015a).

Characteristics: Sowell's short-tailed bat is medium sized; forearm length 37.0–41.0 mm, skull length 21.0–24.1 mm. Dorsal pelage grayish brown; fur of underparts paler than in the dorsum; hair with three bands, darker at the base and extending to approximately one-half of the hair length; **hair in the nape area with rings of well-defined color, so that the dark basal wide band contrasts strongly with the hair tip**; **forearm with hair**; jaw U-shapes; second lower incisors not covered by the canine cingulum; braincase less spherical; palate longer; **mental symphysis ventral tip located under the ventral border of the mandible branch**.

Comments: *Carollia brevicauda* has been restricted to South America, so the specimens known as *C. brevicauda* for North America were assigned to a new species, *C. sowelli* (Baker et al. 2002). *C. sowelli* ranges from Veracruz south through Central America (Map 41). It can be found in sympatry, or nearly so, with *C. perspicillata* and *C. subrufa* in the tropical region of Oaxaca and Chiapas. This species ranges in the tropical coastal plain in Veracruz to the south, and in northern Chiapas and the Yucatán Peninsula. *C. sowelli* differs from *C. perspicillata* by its smaller size, with forearm length less than 41.0 mm; U-shaped jaw; in dorsal view, the second lower incisors are not covered by the canine cingulum and the lingual border of the upper molar series is curved. From *C. subrufa,* by having hair on the forearms; back neck hair with three bands of well-defined coloration, dark and broad at the base, strongly contrasting with the hair tip; face more robust and elongated, with the cranial box less globose; palate long and ventral chin symphysis tip below the ventral edge of the mandibular branch.

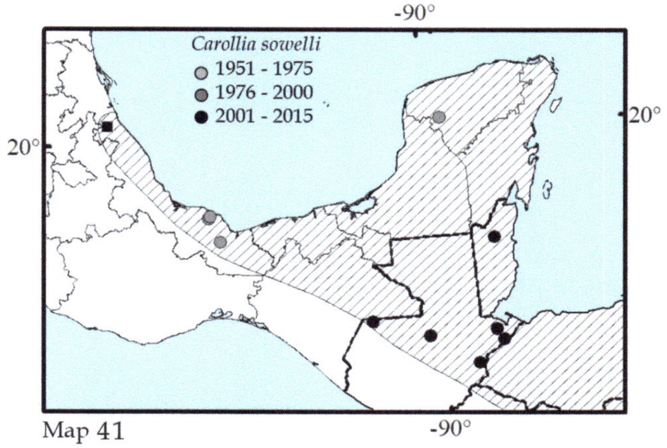

Map 41 Distribution map of *Carollia sowelli*

Additional Literature: Pine (1972), Swanepoel and Genoways (1979), McLellan (1984), Álvarez-Castañeda and Álvarez (1991), Cloutier and Thomas (1992), Medellín et al. (2000), Hoffman and Baker (2003), Simmons (2005), Gonçalves Da Silva et al. (2008).

Carollia subrufa (Hahn, 1905)

Gray short-tailed bat, murciélago colicorta gris

Monotypic.

Conservation: *Carollia subrufa* is listed as Least Concern by the IUCN (Miller et al. 2015a, b, c, d).

Characteristics: Sowell's short-tailed bat is medium sized; forearm length 37.0–41.0 mm, skull length 21.0–24.1 mm. Dorsal pelage grayish brown; fur of underparts paler than in the dorsum; hair with three bands, darker at the base and extending approximately to one-half of the hair length; **hair in the nape area with rings of well-defined color, so that the dark and wide band at the base contrasts strongly with the hair tip**; **forearms with hair**; jaw U-shaped; second lower incisors not covered by the canine cingulum; braincase less spherical; palate long; **ventral mental symphysis tip located under the mandible branch ventral border**.

Comments: *Carollia subrufa* ranges from Jalisco south through Central America (Map 42). It can be found in sympatry, or nearly so, with *C. perspicillata* and *C. sowelli* in the tropical region of Oaxaca and Chiapas. *C. subrufa* differs from *C. perspicillata* by its smaller size, with forearm length smaller than 41.0 mm; jaw U-shaped; in dorsal view, second lower incisors not covered by the canine cingulum and the lingual border, upper molar series curved. From *C. sowelli,* by having scarce hair in the forearms; the hair on the neck lacking well-defined rings and, therefore, the basal band does not contrast strongly with the distal band. Also, *C. subrufa* has a less robust and shorter face; a more globular cranial cavity; a shorter palate, and ventral chin symphysis tip located above the ventral border of the mandibular branch.

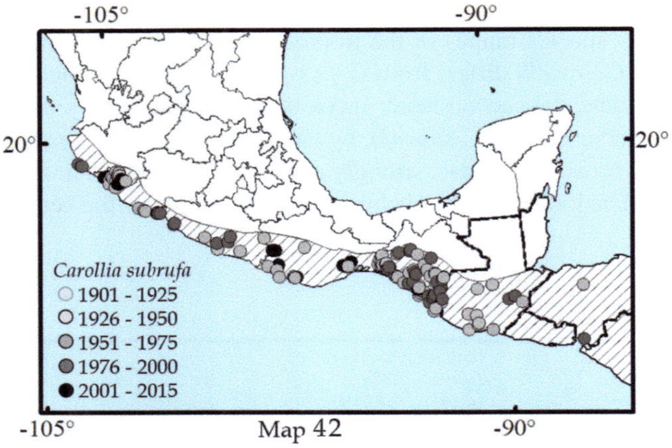

Map 42 Distribution map of *Carollia subrufa*

Additional Literature: Hahn (1905), Goodwin and Greenhall (1961), Pine (1972), Swanepoel and Genoways (1979), McLellan (1984), Álvarez-Castañeda and Álvarez (1991), Cloutier and Thomas (1992), Emmons and Feer (1997), Villalpando and Álvarez (2000), Baker et al. (2002), Hoffmann and Baker (2003), Ortega et al. (2008).

Subfamily Desmodontinae Wagner, 1840

The subfamily Desmodontinae includes vampire bats characterized by being Medium sized; **nostrils surrounded by dermal outgrowths forming a rudimentary noseleaf**; **tuberosities of the humerus of nearly equal size** and both distinctly exceeding the head; fibula large, extending to the head of the tibia; **stomach slender, with a caecum-like structure;** all along, bones of the limbs and wings deeply grooved for accommodating muscles, which is especially noticeable in the tibia, fibula, and femur; **teeth highly specialized for cutting, all trace of the crushing surface absent and cheek-teeth so reduced that the length of the entire upper row is less than 1.7 times the length of the canines along the alveolus.** The keys were elaborated based on the review of specimens and the following sources: Miller (1913), Lawlor (1979), Hall (1981), Álvarez et al. (1994), and Álvarez-Castañeda et al. (2017a).

1. Limbs dorsally hairy; teeth 26; outer lower incisors seven-lobed; four lower teeth behind the canine (from southern Texas, United States, Nuevo León and southern Tamaulipas, México, south through South America)........................ *Diphylla ecaudata* (p. 255)

1a. Bare limbs; teeth less than 26; outer lower incisor with two lobes maximum; three lower teeth behind the canine........................ 2

2. Wing tips white; thumb length about one-eighth of the third finger length; three upper molariforms on each side; inner lower incisors with a large median lobe and a minute outer lobe; with a preorbital process (from Tamaulipas, México, south through South America)........................ *Diaemus youngii* (p. 254)

2a. Wing tip of the same color as the body; thumb length about one-fifth of the third finger length; two upper molariforms on each side; inner lower incisors deeply two-lobed; without a preorbital process (from Sonora, Chihuahua, Nuevo León, and Tamaulipas, México, south through South America)........................ *Desmodus rotundus* (p. 253)

Genus *Desmodus* Wied-Neuwied, 1826

Ears relatively small, separate, and pointed; **thumbs long, with a distinct basal pad, and its length about one-fifth of the third finger length**; calcar reduced to a small nubbin; tail absent; **20 teeth**; palate deeply concave transversely; upper incisors greatly enlarged, with a long, curved, and sharp edge; **lower incisors minute and bilobed**; canines large, sharp-edged posteriorly; cheek-teeth minute, **two upper molariforms on each side**; rostrum extremely reduced, little more than a support for the large incisors and canines; palate deeply concave laterally, slightly concave longitudinally; braincase large and broad posteriorly, markedly narrower anteriorly; auditory bullae covering more than one-half of the cochlear surface.

Desmodus rotundus (Geoffroy Saint–Hilaire, 1810)

Common vampire bat, vampiro común

D. r. murinus Wagner, 1840. Only subspecies in North America.

Conservation: *Desmodus rotundus* is listed as Least Concern by the IUCN (Barquez et al. 2015e).

Characteristics: Common vampire bat is medium sized; forearm length 53.0–65.0 mm, skull length 25.0–26.5 mm. Dorsal pelage grayish brown, occasionally paler; fur of underparts grayish brown; ears triangular and greater than wide at the base, extending beyond the crown in lateral view; small nasal leaves; uropatagium absent or very small and with little hair; **calcar very small;** tail absent; thumbs well developed, useful for walking, **with two pads**; **total number of teeth 20**; **lower incisors with two lobes and relatively small**; coronoid process at the same level than the jaw condyle.

Comments: *Desmodus rotundus* ranges in all tropical parts of México, including both coastal plains south through South America, the lowlands of central México, and the Yucatán Peninsula (Map 43). It can be distinguished from other species in sympatry, or nearly so, such as *Diaemus youngii* and *Diphylla ecaudata*, by having 26 teeth; two pads in the thumbs and a very small calcar. From *Diaemus youngii,* by having wing tips and body of the same color; two upper molariforms and without a preorbital process. From *Diphylla ecaudata,* by having hairless limbs.

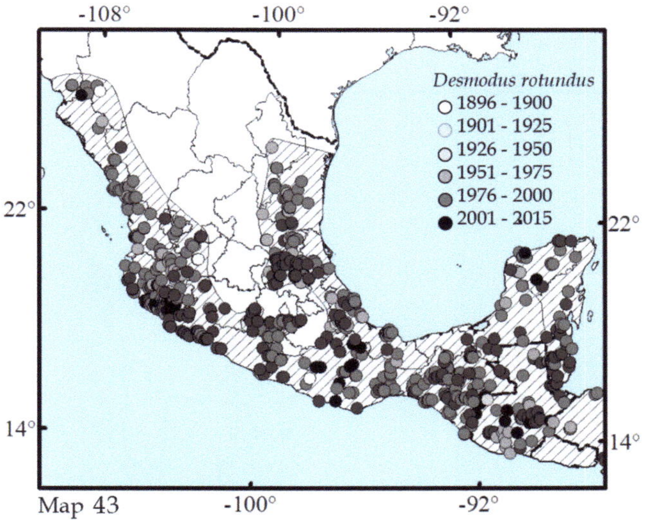

Map 43 Distribution map of *Desmodus rotundus*

Additional Literature: Miller (1907), Goodwin and Greenhall (1961), Husson (1962), Villa-R. (1966), Anderson (1972), Walker (1975), Schmidt (1978), Greenhall et al. (1983), Naranjo Piñera et al. (2005), Simmons (2005), López–González and García-Mendoza (2006), Torres-Morales et al. (2010), Elizalde-Arellano et al. (2010).

Genus *Diaemus* Miller, 1906

Ears relatively small and pointed; **thumbs short, about one-eighth of the third finger length and with two pads on the underside; metacarpal short**; **wings white at the tip**; **hairs on the feet extending ca. 5 mm beyond the tips of the toes**; insertion of the wing membrane on the back halfway between the lateral line and the dorsal midline; **20 teeth;** upper incisors greatly enlarged, with a long, curved, sharp edge; cutting edges of the **lower incisors entire or notched, not bilobed, and their tips slightly curved inward**; second upper molars very small and usually present; **rostrum larger than the braincase**; auditory bullae inflated; preorbital process present; **coronoid process higher than the mandibular condyle**.

Diaemus youngii (Jentink, 1893)

White-winged vampire bat, vampiro de alas blancas

D. y. cypselinus Thomas, 1928. Only subspecies in North America.

Conservation: *Diaemus youngii* is listed as Least Concern by the IUCN (Barquez et al. 2015f).

Characteristics: White-winged vampire bat is medium sized; forearm length 54.8–56.4 mm, skull length 25.4–26.7 mm. Dorsal pelage brown to reddish brown; **wing whitish at the tip**; fur of underparts grayish brown; ears triangular and longer than wide at the base, extending beyond the crown in lateral view; small nasal leaves; **tragus hairy on the anterior surface**; uropatagium absent and with little hair; **calcar absent;** tail absent; thumbs very well developed, useful for walking, **with one pad**; **total number of teeth 22**; **upper incisors blunt, with no lobes**; coronoid process higher than the jaw condyle.

Comments: *Diaemus* was previously considered a junior synonym of *Desmodus* (Handley 1976; Anderson 1977), nowadays is a full genus (Hall 1981; Greenhall and Schutt 1996; Simmons 2005). *D. youngii* ranges in all the tropical regions of the Gulf side of México, from Tamaulipas through South America, including the lowlands of Chiapas. It has not been recorded in the Yucatán Peninsula (Map 44). It lives in sympatry, or nearly so, with *Desmodus* and *Diphylla*, from which it can be distinguished by its wings white at the tip; a total number of 22 teeth; one pad in the thumb, and with no calcar. *D. youngii* is a rare species throughout its ranges in North America

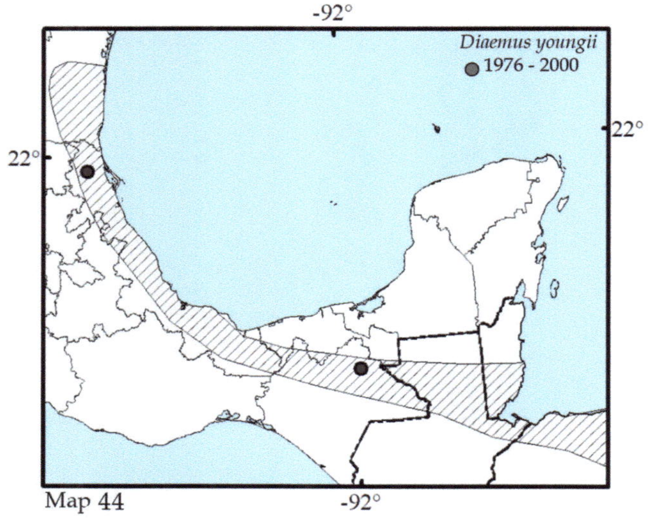

Map 44 Distribution map of *Diaemus youngii*

Additional Literature: Miller (1907), de la Torre (1956), Goodwin and Greenhall (1961), Villa-R. (1967), Schmidt (1978), Naranjo Piñera et al. (2005), Sánchez-Hernández et al. (2005), Sosa et al. (2005).

Genus *Diphylla* Spix, 1823

Ears shorter and rounded; **thumbs short and lacking a basal pad;** calcar well formed but small; **uropatagium absent and with many hairs that continue in the limbs**; **26 teeth;** upper incisors greatly enlarged, with a long, curved, and sharp edge; two pairs of upper incisors, the outer pair minute; **second upper molars present but reduced**; lower incisors relatively large; **the inner pair with four lobes and the outer pair with seven lobes**; interorbital region high and broad than other Desmodontinae.

Diphylla ecaudata Spix, 1823

Hairy-legged vampire bat, vampiro de patas peludas

Monotypic.

Conservation: *Diphylla ecaudata* is listed as Least Concern by the IUCN (Sampaio et al. 2016a).

Characteristics: Hairy-legged vampire bat is medium sized; forearm length 49.0–56.0 mm, skull length 22.0–23.4 mm. Dorsal pelage grayish brown; fur of underparts gray; ears short, rounded, wider than long and not extending over the crown in lateral view; small nasal leaves; **uropatagium absent and with many hairs that continue in the limbs; calcar approximately 3.0 mm**; tail absent; thumbs very well developed and useful for walking**, with no pads**; **total number of teeth 26; lower incisors with four lobes and relatively large**; coronoid process higher than the jaw condyle.

Comments: *Diphylla ecaudata* ranges in all the tropical parts of the Gulf Coastal Plains from southern Texas south through South America, including the lowlands of Chiapas and the Yucatán Peninsula (Map 45). It can be distinguished from other species in sympatry, or nearly so, with *Desmodus rotundus* and *Diaemus youngii* by its dorsally hairy limbs; 26 teeth; no pads in the thumb and a calcar of proximally 3.0 mm.

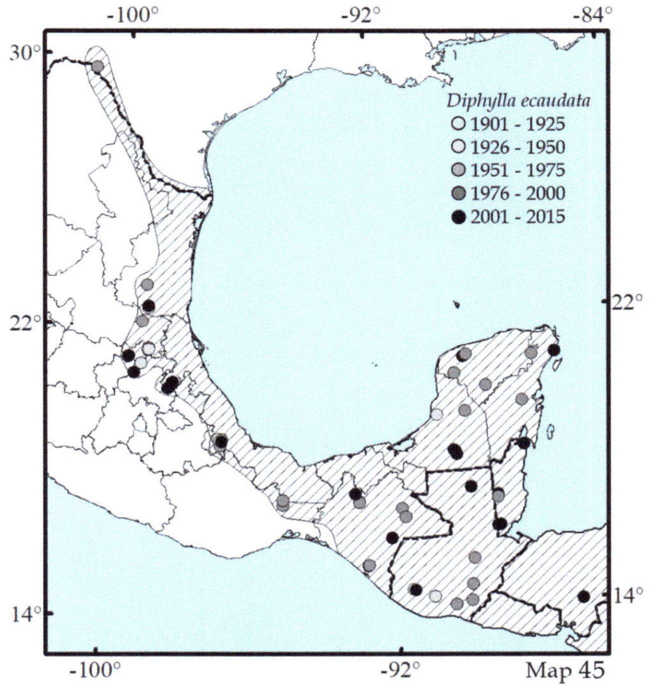

Map 45 Distribution map of *Diphylla ecaudata*

Additional Literature: Miller (1907), Ruschi (1951), Villa-R. (1967), Reddell (1968), McNab (1969), Jones and Carter (1976), Bhatnagar (1978), Schmidt (1978), Greenhall et al. (1984), Medellín et al. (2000), Naranjo Piñera et al. (2005), Simmons (2005), Elizalde-Arellano et al. (2007, 2008), Hoffmann et al. (2008), Pech-Canché et al. (2010).

Subfamily Glossophaginae Bonaparte, 1845

The subfamily Glossophaginae includes long-tongued bats characterized by being small sized; **rostrum elongated at different levels; noseleaf well developed; tongue long and highly extensible, its surface armed with conspicuous bristle-like papillae; upper molars never large, with styles reduced and closely approximated to the paracone and metacone, related to the obsolescence of the commissures, nearly obliterating the W-pattern**; lower molars with five typical cusps but reduced in height, especially paraconid; **commissures rudimentary; upper molars elongated**. The keys were elaborated based on the review of specimens and the following sources: Lawlor (1979), Hall (1981), Webster and Jones (1984a, b), Álvarez et al. (1994), and Álvarez-Castañeda et al. (2017a).

1. Five or six upper molariforms and six lower molariforms 2

1a. Four upper molariforms and five lower molariforms 7

2. Tail absent; six upper molariforms (from Sinaloa and Tamaulipas, México, south through South America) .... tribe Choeronycterini; *Anoura geoffroyi* (p. 257)

2a. Tail present; five upper molariforms 3

3. Lower incisors well developed .... tribe Glossophagini; *Glossophaga* (p. 263)

3a. Lower incisors absent (sometimes extremely small) .... tribe Choeronycterini 4

4. Hair with three inconspicuous color bands, mainly in the nape; pterygoid processes convex internally (Fig. 25; from Jalisco and Veracruz, México, south through Central America) .... *Hylonycteris underwoodi* (p. 260)

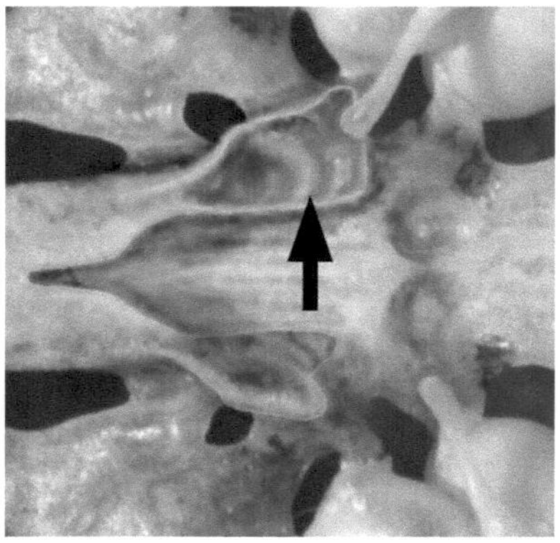

Fig. 25 The pterygoid processes convex internally like in *Hylonycteris underwoodi*

4a. Hair without three inconspicuous color bands; pterygoid processes strongly concave internally (Fig. 26) 5

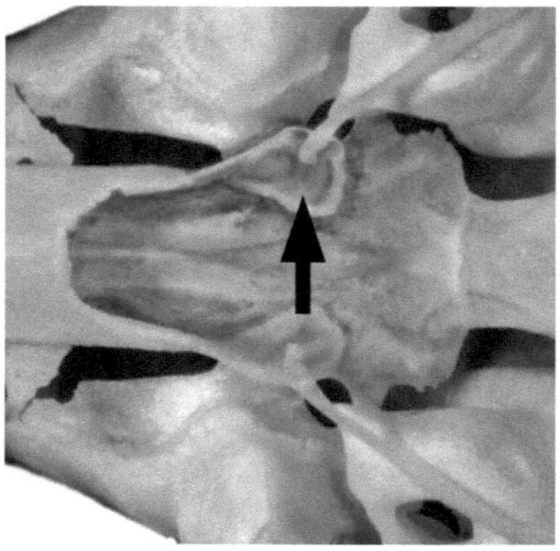

Fig. 26 The pterygoid processes strongly concave internally

5. Forearm length less than 38.0 mm; rostrum in normal proportion for the subfamily; cusp of the lower premolars very similar in size between them (from Sinaloa and southern Veracruz, México, south through South America, except for the Yucatán Peninsula) .... *Choeroniscus godmani* (p. 258)

5a. Forearm length greater than 38.0 mm; rostrum notably long; cusp of middle lower premolars larger than the cusp of the other lower premolars 6

6. Rostral length less than one-half of the total skull length, which is less than 31.5 mm (from California, Arizona, New Mexico, and Texas, United States, south through Central America, except for the Yucatán Peninsula, México) .... *Choeronycteris mexicana* (p. 259)

6a. Rostral length greater than one-half of skull length, which is greater than 31.5 mm (Jalisco, Colima, Michoacán, Guerrero, and lowlands of Morelos, México) .... *Musonycteris harrisoni* (p. 262)

7. Forearm length less than 50.0 mm; skull length less than 22.0 mm (from Chiapas, México, south through South America) .... tribe Choeronycterini; *Lichonycteris obscura* (p. 261)

7a. Forearm length greater than 50.0 mm; skull length greater than 22.0 mm .... tribe Glossophagini; *Leptonycteris* (p. 269)

Tribe Choeronycterini

Genus *Anoura* Gray, 1838

No external tail; noseleaf small; **rostrum elongated and jaw larger than maxilla;** upper incisors small, styliform, inner incisors smaller than outer incisors; **outer upper incisors noticeably elliptical in** cross-section**, anterior surface with a triangular outline**; first upper premolars much smaller than the others, separated from the canine by a space about equal to its base length and from the second premolar; **six upper molariforms**; **zygomatic arches incomplete or imperfectly ossified**; basisphenoid cavities obsolete.

Anoura geoffroyi Gray, 1838

Geoffroy's tailless bat, murciélago sin cola gris

(*see addenda, valid name *Anoura peruana lasiopyga*)

A. g. lasiopyga (Peters, 1868). Only subspecies in North America.

Conservation: *Anoura geoffroyi* is listed as Least Concern by the IUCN (Solari 2016a).

Characteristics: Geoffroy's tailless bat is medium sized; forearm length 40.0–45.0 mm, skull length 24.5–27.0 mm. Dorsal pelage dark brown; fur of underparts grayish brown; noseleaf small; **face elongated and jaw larger than the maxilla**; **no tail and uropatagium reduced and hairy**; **six upper molariforms**; three upper and lower molars; lower incisors present; zygomatic arches incomplete; rostrum shorter than the braincase.

Comments: *Anoura geoffroyi* ranges in both coasts of tropical México, in the Pacific side from Sinaloa and in the Gulf of México from Tamaulipas south through South America (Map 46). It can be found in sympatry, or nearly so, with *Choeroniscus, Choeronycteris, Glossophaga, Hylonycteris, Lichonycteris, Leptonycteris*, and *Musonycteris*. It differs from other genera by being Medium sized; no tail; reduced uropatagium with hair, and six upper molariforms.

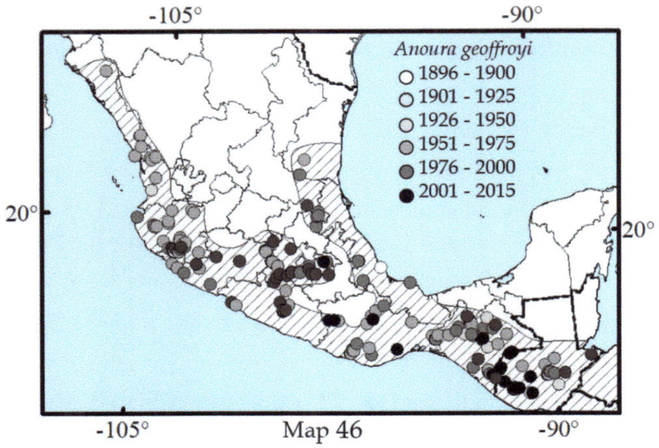

Map 46 Distribution map of *Anoura geoffroyi*

Additional Literature: Sanborn (1933), Handley (1960), Goodwin and Greenhall (1961), Hall (1981), Álvarez-Castañeda and Álvarez (1991), Emmons and Feer (1997), Reid (1997, 2009), Ortega and Alarcón-D. (2008), Mantilla-Meluk and Baker (2010), Segura-Trujillo and Navarro-Pérez (2010).

Genus *Choeroniscus* Thomas, 1928

Long-tongued bats are small sized; **forearm length 31.0–35.0 mm; noseleaf small and triangular**; ears small and rounded; **rostrum long but short within the family**; tail well developed, about half as long as the femur, extending less than halfway to the edge of the very wide uropatagium; calcar present but weak; upper incisors very small, in pairs, in contact with each other, separated from the canine by a distinct space and in the middle by a space nearly twice as great; upper molars distinctly spaced; three molars almost identical in shape, the third slightly smaller than either of the others; **rostrum small, less than one-half of the skull length**; zygomatic incomplete.

Choeroniscus godmani (Thomas, 1903)

Godman's long-tongued bat, murciélago de lengua larga

Monotypic.

Conservation: *Choeroniscus godmani* is listed as Least Concern by the IUCN (Tavares and Molinari 2015).

Characteristics: Godman's Long-tailed bat is small sized; **forearm length 31.0–35.0**, skull length 19.2–20.0 mm. Dorsal pelage grayish brown; each hair with two color bands; fur of underparts pale; noseleaf small and wide; **rostrum long but short within the family;** jaw greater than the maxilla; tail on the middle uropatagium; forearm base with hairs; uropatagium is connected at foot base; five upper molariforms; three upper and inferior molars; lower incisors absent; incomplete zygomatic arch; **rostrum small, less than half of skull length**; very small upper incisors.

Comments: *Choeroniscus godmani* ranges in the tropical region of México in both coasts, from Sinaloa and Veracruz south through South America, except for the Yucatán Peninsula (Map 47). It can be found in sympatry, or nearly so, with *Anoura, Choeronycteris, Glossophaga, Hylonycteris, Lichonycteris, Leptonycteris*, and *Musonycteris.* It differs from other genera by having the jaw longer than the maxilla; elongated face and absence of inferior incisors.

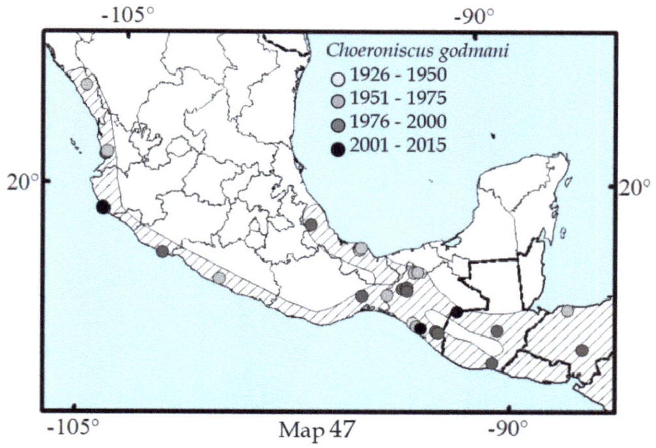

Map 47 Distribution map of *Choeroniscus godmani*

Additional Literature: Villa-R. (1967), Álvarez and González-Quintero (1970), Patton and Gardner (1971), Gardner (1977), Álvarez and Sánchez-Casas (1997a), Villalpando and Álvarez (2000), Solmsen and Schilemann (2007).

Genus *Choeronycteris* Tschudi, 1844

Small sized; noseleaf small; **ears small and rounded**; face and jaw very long, jaw larger than the maxilla; **tail well developed, about half as long as the femur, extending less than halfway to the edge of the very wide uropatagium**; calcar present and distinct but weak; upper incisors very small, in pairs, the teeth of each pair in contact with each other, separated from the canine by a distinct space and in the middle by a space nearly twice as great; upper molars distinctly spaced; three molars almost identical in shape, the third molar slightly smaller than either of the others; rostrum very greatly elongated; zygomatic incomplete; **pterygoids strongly concave on the inner side, hamulars in contact with the auditory bullae**.

Choeronycteris mexicana Tschudi, 1844

Mexican long-tongued bat, murciélago trompudo mexicano

Monotypic.

Conservation: *Choeronycteris mexicana* is listed as Threatened in the Norma Oficial Mexicana (DOF 2019) and as Near Threatened (A2c: population reduction (over 10 years in equal to or less than 30%) may not have ceased or may not be understood or may not be reversible, based on decline in area of occupancy, extent of occurrence and/or habitat quality) by the IUCN (Solari 2018b).

Characteristics: Mexican long-tongued bat is small sized; forearm length 43.0–49.0 mm, skull length 29.3–31.1 mm. Dorsal pelage grayish brown; fur of underparts pale; noseleaf small; **ears small and rounded**; face very long and jaw larger than the maxilla; tail occupies one-third of the uropatagium; forearm base with few hairs; uropatagium connected to the ankles; five upper molariforms; three upper and lower molars; no lower incisors; incomplete zygomatic arch; rostrum larger than the braincase; pterygoid process strongly concave toward inside.

Comments: *Musonycteris* was previously considered within the genus *Choeronycteris* (Handley 1966). Later, it was considered again as a full genus (Phillips and Jones 1971; Webster et al. 1982). *Choeronycteris mexicana* ranges from California, Nevada, Arizona, New Mexico, and Texas south through México except for the coastal plain of the Gulf of México and the Yucatán Peninsula (Map 48). It can be found in sympatry, or nearly so, with *Anoura, Choeroniscus, Glossophaga, Hylonycteris, Lichonycteris, Leptonycteris*, and *Musonycteris*. It differs from other genera by having small rounded ears; forearm length more than 38.0 mm and without lower incisors.

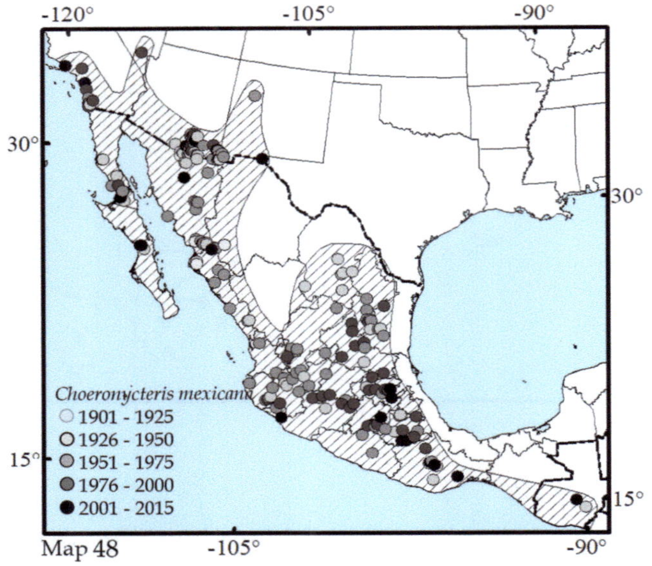

Map 48 Distribution map of *Choeronycteris mexicana*

Additional Literature: Miller (1907), Baker (1956) Schaldach and McLaughlin (1960), Phillips (1971), Phillips et al. (1977), Arroyo-Cabrales et al. (1987), Álvarez-Castañeda and Álvarez (1991), Simmons (2005), Álvarez-Castañeda et al. (2015, 2017a).

Genus *Hylonycteris* Thomas, 1903

Small sized; **each dorsal hair with three inconspicuous color bands**; noseleaf small; jaw very extended and notoriously greater than the maxilla; **uropatagium connected to the ankles**; five upper molariforms; lower incisors absent; upper molars distinctly spaced, with the parastyle and metastyle absent, and mesostyle greatly reduced; three molars almost identical in shape, but the third slightly smaller than either of the others; zygomatic arches incomplete; rostrum shorter than the braincase; **pterygoid processes not particularly modified, their inner surface being convex and the hamular processes not coming in contact with the auditory bullae**.

Hylonycteris underwoodi Thomas, 1903

Underwood's long-tongued bat, murciélago lengüilargo moreno

1. *H. u. minor* Phillips and Jones, 1971. From Southern Nayarit south through Coastal Oaxaca.
2. *H. u. underwoodi* Thomas, 1903. From Central Veracruz and Puebla south through Chiapas, including the Yucatán Peninsula.

Conservation: *Hylonycteris underwoodi* is listed as Least Concern by the IUCN (Miller et al. 2016b).

Characteristics: Undewood's long-tongued bat is small sized; forearm length 31.0–34.0 mm, skull length 20.2–23.0 mm. Dorsal pelage dark brown to blackish; **each hair with three inconspicuous color bands**; fur of underparts paler; noseleaf small; jaw very extended and notoriously greater than the maxilla; tail at the middle uropatagium; **uropatagium connected to the ankles**; five upper molariforms; three upper and lower molars; lower incisors absent; zygomatic arches incomplete; rostrum shorter than the braincase; **pterygoid processes convex internally**.

Comments: *Hylonycteris underwoodi* ranges in both tropical coasts of México, from Jalisco and Veracruz south through Central America, except Yucatán Peninsula (Map 49). It can be found in sympatry, or nearly so, with *Anoura, Choeroniscus, Glossophaga, Lichonycteris,* and *Leptonycteris*. It differs from other genera by the presence of three bands of notorious color in nape hairs; dorsal coloration dark brown to blackish and its small size, with forearm and skull lengths 34.0 and 23.0 mm, respectively. The diagnostic characteristic is having the pterygoid processes internally convex. The most similar species is *Choeroniscus godmani*, from which it differs externally because the hair has only two bands of coloration.

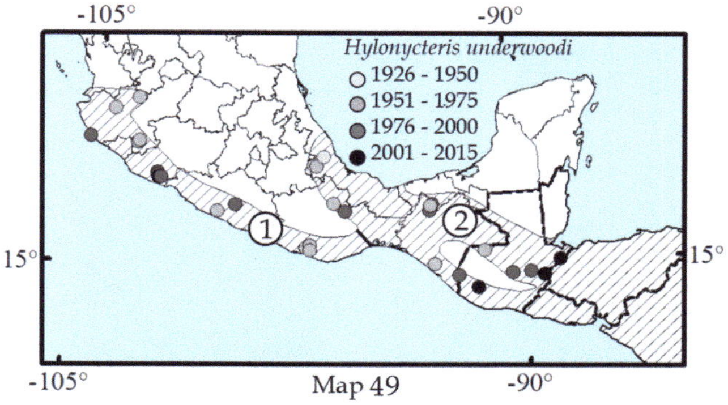

Map 49 Distribution map of *Hylonycteris underwoodi*

Additional Literature: Thomas (1903), Miller (1907), Davis and Carter (1962), Jones (1964), Goodwin (1969), Phillips (1971), Phillips and Jones (1971), Jones and Homan (1974); Álvarez-Castañeda and Álvarez (1991), Almazán-Catalán et al. (2009), Segura-Trujillo and Navarro-Pérez (2010).

Genus *Lichonycteris* Thomas, 1895

Small sized; **dorsal fur with three very clear color bands;** tail distinct, about half as long as the femur, extending scarcely to the middle uropatagium; **uropatagium connected to the foot near the digits; upper incisors evenly and widely spaced between the canines, their crowns narrow though scarcely trenchant, longer than high, with the crown of the inner teeth distinctly emarginate on the cutting edge, so that it appears bilobed in frontal view**; five upper molariforms; two upper and lower molars; lower incisors; incomplete zygomatic arch; rostrum shorter than the braincase; upper incisors with spaces between them.

Lichonycteris obscura Thomas, 1895

Dark long-tongued bat, murciélago lengüilargo oscuro

Monotypic.

Conservation: *Lichonycteris obscura* is listed as Least Concern by the IUCN (Solari 2018f).

Characteristics: Dark long-tongued bat is small sized; forearm length 31.0–35.0 mm, skull length 17.8–19.9 mm. Dorsal pelage brown to blackish; **each hair with three very clear color bands;** fur of underparts grayish brown; noseleaf small; face long and jaw larger than the maxilla; tail at the middle uropatagium; forearm base hairy; **uropatagium connected to the foot near the digits**; five upper molariforms; two upper and lower molars; lower incisors; incomplete zygomatic arch; rostrum shorter than the braincase; upper incisors with spaces between them.

Comments: *Lichonycteris obscura* ranges in the lowlands of northern Chiapas south through South America (Map 50). It can be found in sympatry, or nearly so, with *Anoura, Choeroniscus, Glossophaga,* and *Hylonycteris*. *L. obscura* differs from other genera by the presence of three very pale bands in each hair, which is a very distinctive trait, especially in the nape.

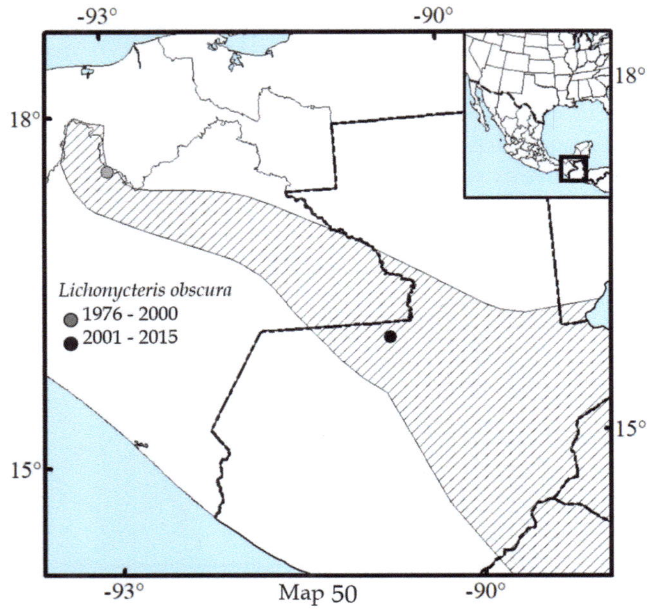

Map 50 Distribution map of *Lichonycteris obscura*

Additional Literature: Álvarez-Castañeda and Álvarez (1991), Medellín et al. (2000), Naranjo-Piñera et al. (2005), Zamora-Gutierrez and Ortega (2020).

Genus *Musonycteris* Schaldach and McLaughlin, 1960

Small sized; **snouts very large**; noseleaf small; ears small and rounded; tail well developed; calcar present; without lower incisors; **molariforms narrow and delicate in structure, with reduced lingual elements, widely separated; space between the second and third premolars approximately equal to the length of the second premolar; nasals elevated dorsally**; **rostrum long, representing more than 50% of the skull length**; **hamular processes of the pterygoids enlarged, extending onto the ventral surface of the auditory bullae to the level of the anterior margin of the meatus**; basioccipital and basisphenoid produced into a high and thin median ventral crest.

Musonycteris harrisoni Schaldach and McLaughlin, 1960

Banana bat, murciélago hocicudo del Balsas

Monotypic.

Conservation: *Musonycteris harrisoni* is listed as Endangered in the Norma Oficial Mexicana (DOF 2019) and as Vulnerable (C1: small population size with an estimated continuing decline of at least 10% within 10 years) by the IUCN (Arroyo-Cabrales and Ospina-Garces 2015).

Characteristics: Banana bat is small sized; forearm length 41.6–42.9 mm, skull length 32.0–35.0 mm. Dorsal pelage pale to medium brown with paler areas in the middle back and shoulders; fur of underparts paler than in the dorsum; jaw larger than the maxilla; tail occupying one-third of the uropatagium; forearm base with few hairs; uropatagium connected to the ankles; **rostrum larger than the head**; five upper molariforms; three upper and lower molars; lower incisors absent; zygomatic arches incomplete; **rostrum greater than the braincase**.

Comments: *Musonycteris harrisoni* was previously considered within the genus *Choeronycteris* (Handley 1966). Later, it was considered again as a full genus (Phillips and Jones 1971, Webster et al. 1982). *M. harrisoni* ranges only in Jalisco, Colima, Michoacán, southern Morelos, and Guerrero, associated with desert vegetation of the Balsas Basin (Map 51). It can be distinguished from other species in sympatry, or nearly so, as those of the genera *Anoura, Choeronycteris, Glossophaga*, and *Leptonycteris*, by its face length greater than one-half of the skull length, which is greater than 31.5 mm.

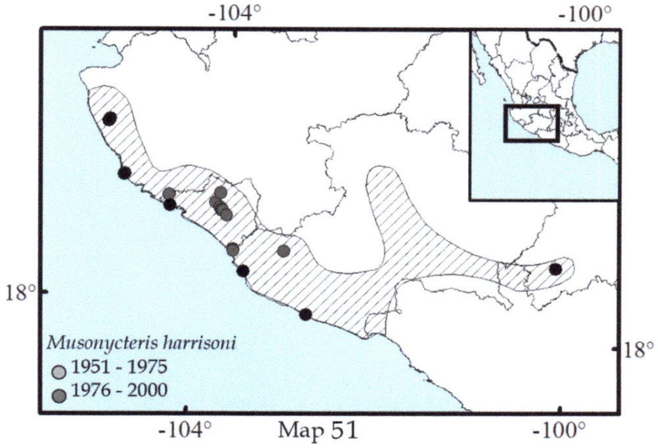

Map 51 Distribution map of *Musonycteris harrisoni*

Additional Literature: Schaldach and McLaughlin (1960), Phillips (1971), Kennedy et al. (1984), Álvarez-Castañeda and López-Forment (1995), Álvarez and Sánchez-Casas (1997a), Medellín et al. (1997), Tellez and Ortega (1999), Stoner et al. (2002), Baker et al. (2003c), Wang et al. (2003), Ortega et al. (2009).

Tribe Glossopagini

Genus *Glossophaga* Geoffroy, 1818

Small sized; dorsal and ventral pelage in different brown shades; fur of underparts paler than in the dorsum; noseleaf small, merged to the upper lip; rostrum long and jaw of the same size as the maxilla; tail one-third of the uropatagium length and not as long as the tibia, sometimes difficult to see; **upper incisors well developed, forming a continuous row almost from canine to canine; lower incisors of different shape among species;** three upper and lower molars; crown of inner teeth longer than high, cutting edge straight to slightly oblique; **braincase large and elongated but low, rounded, and smooth; rostrum somewhat shorter than the braincase, low and weak**; zygomatic arches complete; lacrimal inflation small but distinct; basisphenoid cavities distinct but shallow; auditory bullae small, covering less than one-half of the surface of the cochleae, with their greatest diameter slightly greater than the basioccipital width. The keys were elaborated based on the review of specimens and the following sources: Miller (1913), Goodwin (1942a, b), Lawlor (1979), Hall (1981), Álvarez et al. (1994), and Álvarez-Castañeda et al. (2017a).

1. Lower incisors big, usually in contact between themselves forming a continuous arch between the canines (Fig. 27); upper incisors very procumbent (Fig. 28; from Sonora and Tamaulipas, México, south through Central America)......... .... *Glossophaga mutica* (p. 268)

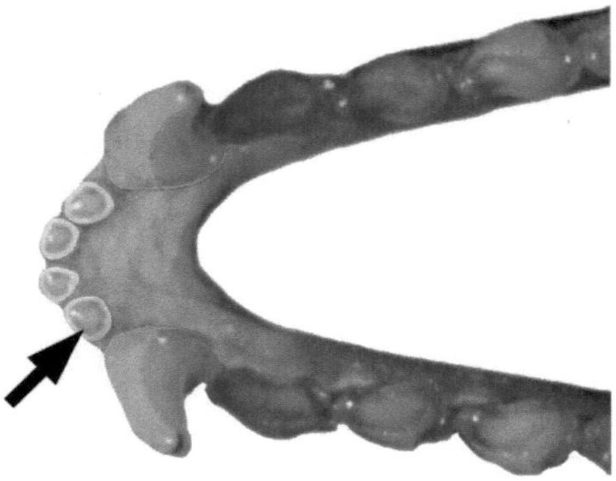

Fig. 27 Lower incisors are big, usually in contact between themselves forming a continuous arch between the canines like in *Glossophaga mutica*

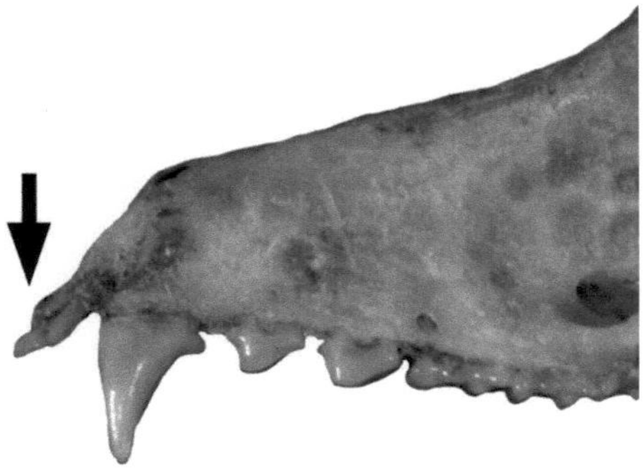

Fig. 28 The upper incisors are very procumbent like in *Glossophaga mutica*

1a. Lower incisors separated between themselves by distinguishable spaces, generally reduced in size (Fig. 29); upper incisors not clearly procumbent (only in *G. morenoi*) 3

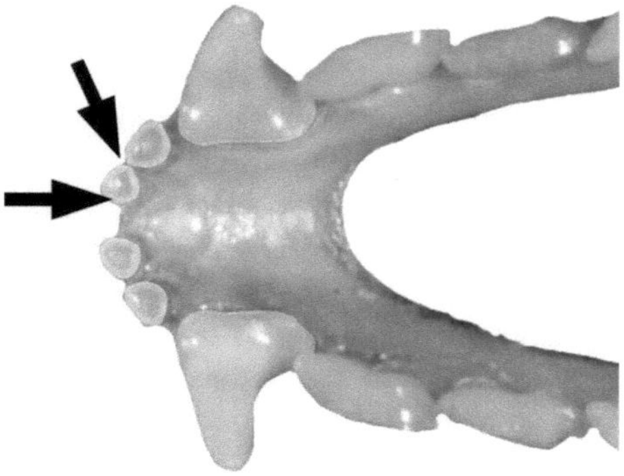

Fig. 29 Lower incisors separated between themselves by distinguishable spaces, generally reduced in size

2. Upper incisors slightly procumbent, inner incisors larger than the outer incisors; anterior premaxillary border long, beyond the canines; lower incisors grouped at each side with a medium diastema (Fig. 30; from Jalisco and Colima south through Chiapas, México) .... *Glossophaga morenoi* (p. 267)

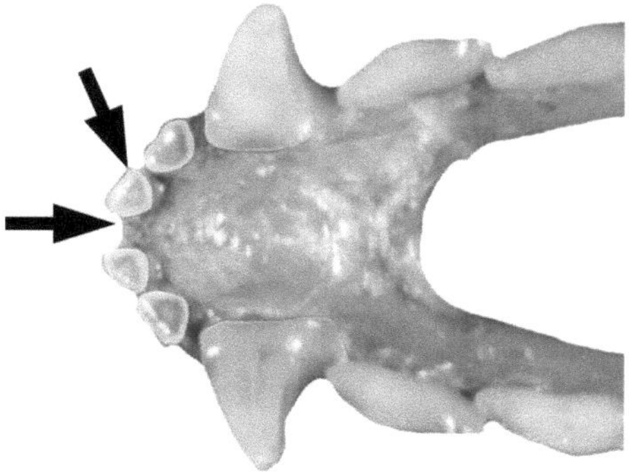

Fig. 30 Lower incisors grouped at each side with a medium diastema like in *Glossophaga morenoi*

2a. Upper incisors not procumbent, inner pair subequal to or smaller than the outer incisors; anterior premaxillary border not long, almost in line with the canines; lower incisors separated between them..3

3. First molar with the anterior cusps separated; lower incisors small, nail shaped, and with a diastema between each of them (Fig. 30); sphenoid process subterminal, flat, pterygoid wings absent (from Sinaloa, México, south through South America, except for the Yucatán Peninsula) ....*Glossophaga commissarisi* (p. 265)

3a. First molar with anterior cusps very close together; lower incisors big, separated by a diastema (Fig. 29); sphenoid process not flat; with pterygoid wings (from Guerrero, Estado de México, and central Veracruz, México, south through Central America).. *Glossophaga leachii* (p. 266)

Glossophaga commissarisi Gardner, 1962

Commissaris's long-tongued bats, murciélago lengüilargo pequeño

1. *G. c. commissarisi* Gardner, 1962. From Veracruz south through Central America.
2. *G. c. hespera* Webster and Jones, 1982. Sinaloa, Nayarit, and Jalisco.

Conservation: *Glossophaga commissarisi* is listed as Least Concern by the IUCN (Miller et al. 2016c).

Characteristics: Commissaris's long-tongued bats are small sized; forearm length 31.3–35.0 mm, skull length 19.1–21.0 mm. Dorsal pelage brown to dark brown; fur of underparts paler than in the dorsum; **small lower incisors nail shaped with a diastema between each of them**.

Comments: *Glossophaga commissarisi* is distributed in the tropical region of México from Sinaloa south through South America, except for the Yucatán Peninsula (Map 52). It can be found in sympatry, or nearly so, with *Anoura, Choeroniscus, Choeronycteris, Leptonycteris, Musonycteris,* and with their congeners *G. morenoi, G. leachii,* and *G. soricina* throughout its ranges in the western coast of México. *G. commissarisi* differs from its congeners by having smaller incisors clove shaped with a diastema between each of them. The species morphologically most similar is *G. leachii.* From other genera of the subfamily Glossophaginae, by its forearm length less than 40.0 mm and its hair with one or two color bands, never three.

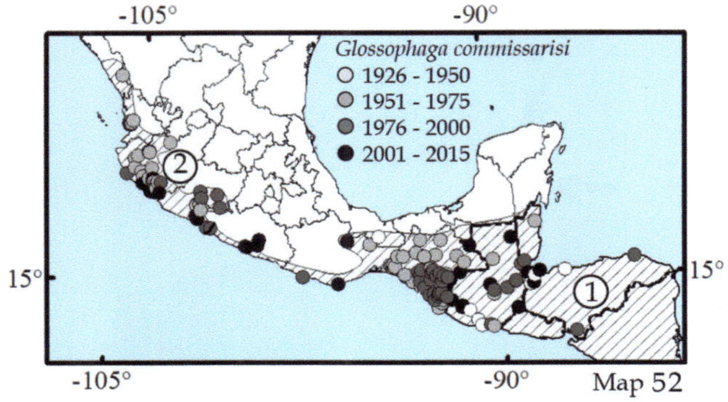

Map 52 Distribution map of *Glossophaga commissarisi*

Additional Literature: Sánchez-Hernández and Gaviño de la Torre (1988), Álvarez-Castañeda and Álvarez (1991), Webster (1993), Webster and Jones (1982a, 1984a, 1993), Naranjo Piñera et al. (2005), Álvarez-Castañeda et al. (2015, 2017a).

Glossophaga leachii (Gray, 1844)

Gray's long-tongued bats, murciélago lengüilargo

Monotypic.

Conservation: *Glossophaga leachii* is listed as Least Concern by the IUCN (Miller et al. 2016d).

Characteristics: Gray's long-tongued bats are small sized; forearm length 35.4–39.0 mm, skull length 20.1–21.3 mm. Dorsal pelage reddish to grayish brown; fur of underparts paler than in the dorsum; **lower incisors relatively large, separated by a diastema among them**.

Comments: *Glossophaga leachii* was elevated to full species (Webster and Jones 1980), previously considered within *G. soricina*. *G. leachii* ranges in the tropical region of México, from Guerrero, Estado de México, and central Veracruz south through Central America; it is also found in parts of central México (Map 53). It can be found in sympatry, or nearly so, with *Anoura, Choeroniscus, Choeronycteris, Leptonycteris* and their congeners *G. commissarisi*, *G. morenoi*, and *G. mutica*, with the first two in the tropical western coast of México and *G. soricina* in the rest of its distribution. *G. leachii* differs from its congeners because the lower incisors are relatively large, separated by a diastema. From other genera of the subfamily Glossophaginae, by the presence of larger lower incisors; forearm length less than 40.0 mm and hair with one or two color bands, never three.

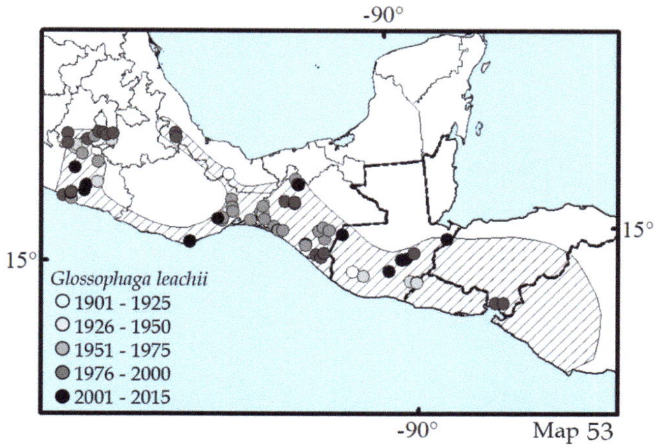

Map 53 Distribution map of *Glossophaga leachii*

Additional Literature: Miller (1900), Jones and Carter (1976), Webster and Jones (1984a), Álvarez-Castañeda and Álvarez (1991), Webster (1993), Segura-Trujillo and Navarro-Pérez (2010), Álvarez-Castañeda et al. (2015, 2017a).

Glossophaga morenoi Martínez and Villa-R., 1938

Western long-tongued bats, murciélago lengüilargo del oeste

1. *G. m. mexicana* Webster and Jones, 1980. Central southern Oaxaca southward.
2. *G. m. morenoi* Martínez and Villa-R., 1938. From central northern Oaxaca northward.

Conservation: *Glossophaga morenoi* is listed as Least Concern by the IUCN (Arroyo-Cabrales et al. 2015b).

Characteristics: Western long-tongued bats are small sized; forearm length 32.0–36.5 mm, skull length 21.1–22.7 mm. Dorsal pelage pale brown to brown; fur of underparts paler than in the dorsum; **lower incisors grouped at each side of a medium diastema**.

Comments: *Glossophaga morenoi* was previously a synonym of *G. mexicana* (Webster and Jones 1980), however Gardner (1986) considered it a junior synonym of *G. morenoi*. *G. mexicana brevirostris* (Webster and Jones 1980) was described from Yautepec, near Tequesquitengo, Morelos, which is the type locality of *G. morenoi* without morphological characteristics that can be used for distinguishing it; for this reason, *G. mexicana brevirostris* is considered a junior synonym of *G. m. morenoi*. *G. morenoi* ranges from Colima and Jalisco south through Chiapas, México (Map 54). It can be found in sympatry, or nearly so, with *Anoura, Choeroniscus, Choeronycteris, Leptonycteris,* and *Musonycteris,* as well as with *G. commissarisi*, *G. leachii*, and *G. soricina* throughout the western Mexican coast, and with *G. leachii* and *G. soricine* in central México. It differs from its congeners by having the lower incisors grouped in pairs on each side of a central diastema. From other genera of the subfamily Glossophaginae, by the presence of larger lower incisors; forearm length less than 40.0 mm and hair with one or two color band, never three.

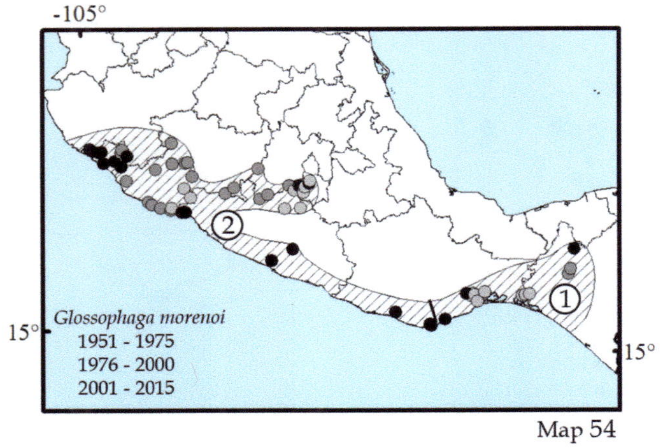

Map 54 Distribution map of *Glossophaga morenoi*

Additional Literature: Webster and Jones (1984a, b, 1985), Webster (1993), Sánchez-Casas and Álvarez (2000), Sheeler-Gordon and Smith (2001), Barragán et al. (2010), Álvarez-Castañeda et al. (2015, 2017a).

Glossophaga mutica Merriam, 1898

Tres Marías Island long-tongued bat, murciélago lengüilargo de Islas Tres Marías

1. *G. m. handleyi* Webster and Jones, 1980. From southern Sonora and Tamaulipas south through Central America, including the Yucatán Peninsula
2. *G. m. mutica* Merriam, 1898. Restricted to the Tres Marías Islands, Nayarit.

Conservation: *Glossophaga mutica* (as *Glossophaga soricina*) is listed as Least Concern by the IUCN (Barquez et al. 2015g).

Characteristics: Tres Marías Island long-tongued bat is small sized; forearm length 33.0–38.4 mm, skull length 20.6–22.8 mm. Dorsal pelage reddish to grayish brown; fur of underparts paler than in the dorsum; hair white at the base; lower incisors big, usually in contact between them, forming a continuous arch between the canines.

Comments: *Glossophaga mutica* was elevated to full species (Webster and Jones 1980; Calahorra-Oliart et al. 2021), previously regarded as a subspecies of *G. soricina. G. soricina* is restricted to South America (Calahorra-Oliart et al. 2021). The revision of the *G. soricina* complex does not define subspecies, but shows morphological differences between the populations of the Tres Marías Islands and the mainland (Calahorra-Oliart et al. 2021), so I am keeping them as two different subspecies. *G. mutica* ranges in the tropical region of México from Sonora and Tamaulipas south through Central America, including the Yucatán Peninsula and the Tres Marías Islands (Map 55). It can be found in sympatry, or nearly so, with *Anoura, Choeroniscus*, *Choeronycteris, Leptonycteris,* and their congeners *G. commissarisi*, *G. morenoi,* and *G. leachii* in the tropical west coast of México, and with *G. morenoi* and *G. leachii* in central México. In the rest of its range, it is the only species of *Glossophaga* present. *G. mutica* differs from its congeners by its lower incisors being larger and usually in contact with each other, forming a continuous arch between the canines; the pelage tends to be grayish brown, mainly with white areas at the base. From other genera of the subfamily Glossophaginae, by its larger lower incisors; forearm length less than 40.0 mm and hair with one or two color bands, never three.

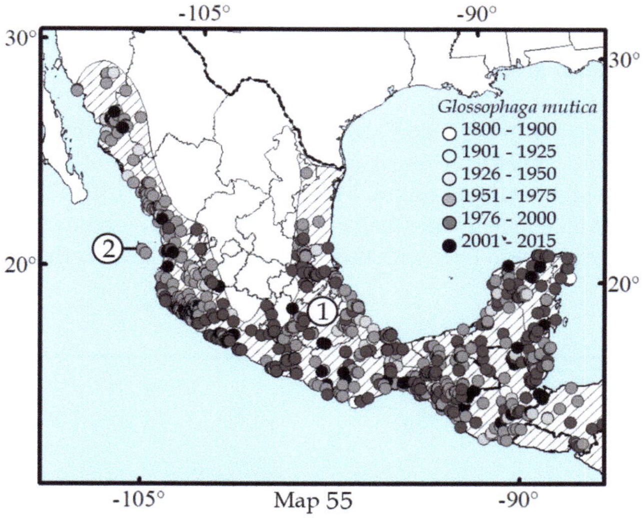

Map 55 Distribution map of *Glossophaga mutica*

Additional Literature: Nelson (1899), Miller (1900), Webster (1980, 1984a, 1993), Hall (1981), Álvarez et al. (1991), Álvarez-Castañeda and Álvarez (1991), Sánchez-Casas and Álvarez (2000), Hoffmann and Baker (2003), Akins et al. (2007), Gardner (2008), Torres-Morales et al. (2010), Álvarez-Castañeda et al. (2015, 2017a).

Genus *Leptonycteris* Lydekker, 1891

Medium sized; dorsal pelage grayish brown, reddish brown to gray; each hair with two color bands, proximal band white; fur of underparts pale; noseleaf small; rostrum long and jaw of the same size as the maxilla; tail absent; **hair in the uropatagium present or absent, in relation to the species**; uropatagium very narrow and connecting with the ankles; calcar small but distinct; upper incisors rather large, forming an almost continuous line between the canines or separated by a median space into two pairs, crowns projecting forward; crown of inner incisors about as long as high, cutting edge straight, nearly horizontal; crown of outer incisors higher than the crown of the inner incisors, its length about one-half of its height; lower incisors well developed, with low, flat, rounded crowns; two upper and lower molars; zygomata slender but complete; rostrum shorter than the braincase. The keys were elaborated based on the review of specimens and the following sources: Hall (1981), Hensley and Wilkins (1988), Álvarez et al. (1994), and Álvarez-Castañeda et al. (2017a).

1. Pelage long and soft; uropatagium hairy and hair extending beyond its posterior border; third digit length greater than 105.0 mm; skull breadth greater than 42% of the total length, wider portion in the mastoid region (from Texas, United States, south through to western Chiapas, México) .... *Leptonycteris nivalis* (p. 269)

1a. Pelage short and rough; uropatagium scarcely hairy and with few hairs extending to its posterior border; third digit length less than 105.0 mm; skull breadth less than 42% of the total length, wider portion in the zygomatic region (from Arizona, New Mexico, and Texas, United States south through Central America) .... *Leptonycteris yerbabuenae* (p. 270)

Leptonycteris nivalis (de Saussure, 1860)

Mexican long-nosed bat, murciélago magueyero mayor

Monotypic.

Conservation: *Leptonycteris nivalis* is listed as Threatened in the Norma Oficial Mexicana (DOF 2019), as Endangered by the United States Endangered Species Act (FWS 2022), and as Critically Endangered (A2c: population reduction in the last 10 years 80% or less and may not be reversible by decline in area of occupancy) by the IUCN (Medellín 2016a).

Characteristics: Mexican long-nosed bat is medium sized; **forearm length 56.0–61.0 mm, skull length 26.4–28.3 mm.** Dorsal pelage grayish brown to gray; fur of underparts paler than in the dorsum; **uropatagium hair and some as bangs; forearm base with hair**; **skull breadth greater than 42%** of the **total length**.

Comments: *Leptonycteris nivalis* ranges in the desert and tropical dry regions from southern Texas south through eastern Chiapas (Map 56). It can be found in sympatry, or nearly so, with *Anoura, Choeroniscus, Choeronycteris,* and their congeners *L. yerbabuenae* in the Mexican highlands and central México. This species differs from other genera by the presence of lower incisors; forearm length and skull length greater than 53.0 and 25.0 mm, respectively. From *L. yerbabuenae*, by having the uropatagium with hair as a fringe; forearm base with hair, and skull wide, greater than 42% of the total length.

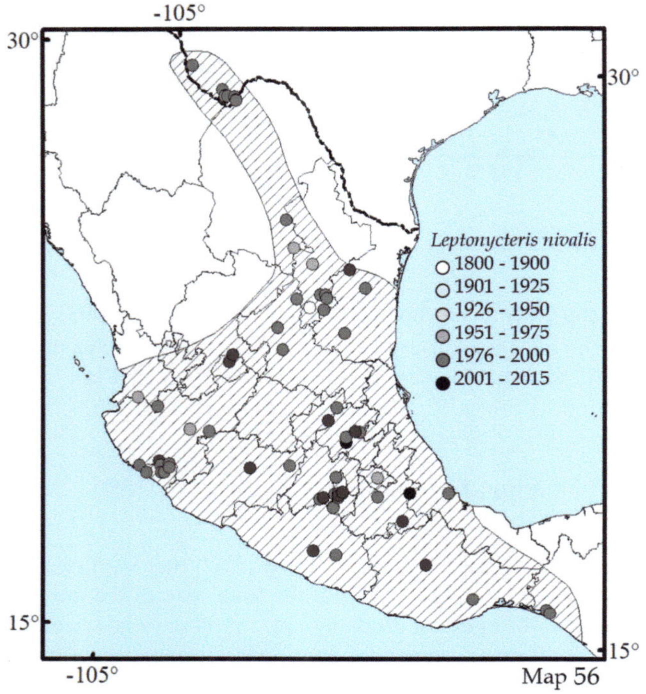

Map 56 Distribution map of *Leptonycteris nivalis*

Additional Literature: Hoffmeister (1957), Davis and Carter (1962), Barbour and Davis (1969), Phillips (1971), Watkins et al. (1972), Davis (1974), Greenbaum and Phillips (1974), Hensley and Wilkins (1988), Álvarez-Castañeda and Álvarez (1991), Moreno–Valdez et al. (2004), Segura-Trujillo and Navarro-Pérez (2010).

Leptonycteris yerbabuenae Martínez and Villa, 1940

Lesser long-nosed bat, murciélago magueyero menor

Monotypic.

Conservation: *Leptonycteris yerbabuenae* is listed as Threatened in the Norma Oficial Mexicana (DOF 2019), as Endangered by the United States Endangered Species Act (FWS 2022), and as Near Threatened by the IUCN (Medellín 2016b).

Characteristics: Lesser long-nosed bat is medium sized; **forearm length 53.0–57.0 mm, skull length 25.4–26.9 mm**. Dorsal pelage reddish to grayish brown; fur of underparts paler than in the dorsum; **uropatagium with few hairs and very few as bangs**; **forearm base bare**; **skull breadth less than 42%** of the **total length.**

Comments: *Leptonycteris yerbabuenae* was elevated to full species (Simmons 2005); it was previously regarded as a subspecies of *L. curasoe* based on wing morphometry (Arita and Humphrey 1988). *L. yerbabuenae* is distributed in the desert and tropical dry regions of México on both coasts south through Central America, including the Baja California Peninsula, Arizona, New Mexico, and Texas (Map 57). It can be found in sympatry, or nearly so, with species of *Anoura, Choeroniscus, Choeronycteris, Glossophaga, Musonycteris,* and their congeners *L. nivalis* only in the Altiplanicie Mexicana and central México; in the rest of its range, it is the only species of this genus present. *L. yerbabuenae* differs from other genera by the presence of lower incisors; forearm and skull lengths greater than 53.0 and 25.0 mm, respectively. From *L. nivalis,* by having the uropatagium with very few fringes; forearm base bare; uropatagium attached to the ankles and skull width less than 42% of the total length.

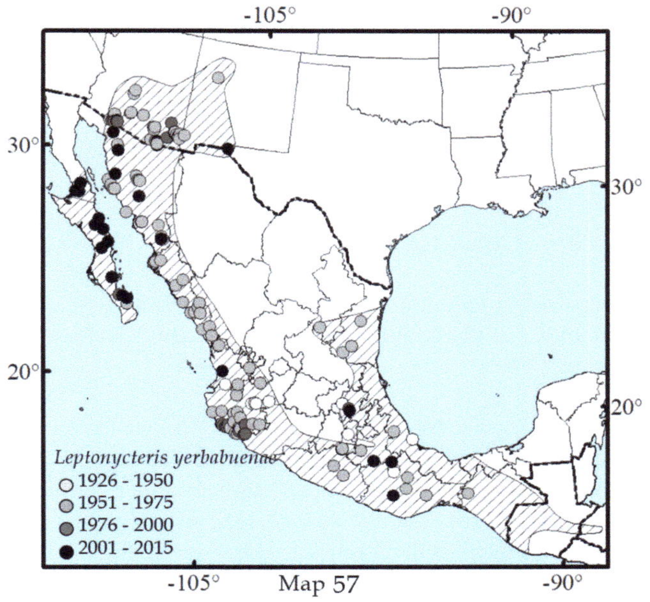

Map 57 Distribution map of *Leptonycteris yerbabuenae*

Additional Literature: Hoffmeister (1957), Davis and Carter (1962), Phillips (1971), Hoffmeister (1986), Arita and Humphrey (1988), Hensley and Wilkins (1988), Álvarez-Castañeda and Álvarez (1991), Arita (1999), Baker et al. (2000), Cole and Wilson (2006), Álvarez–Castañeda et al. (2008a, c).

Subfamily Glyphonycterinae Baker et al., 2016

The subfamily Glyphonycterinae is characterized by the dorsal pelage dark grayish brown; fur of underparts yellowish to pale gray; **ears relatively large and pointed; chin with a smooth pad; fourth metacarpal shorter and third or fifth metacarpal longest;** calcar shorter than the hind foot; **third upper premolars larger than the fourth upper premolars**; **second upper incisors small and unicuspid; four upper premolars slightly recurved; braincase expanded**. The keys were elaborated based on the review of specimens and the following sources: Andersen (1906a), Hall (1981), Álvarez et al. (1994), and Álvarez-Castañeda et al. (2017a).

1. Fifth metacarpal longest; third upper premolars larger than the four upper premolars (from southern Sinaloa and Veracruz, México, south through South America)....... *Glyphonycteris sylvestris* (p. 272)

1a. Third metacarpal longest; third upper premolars smaller than the four upper premolars (from Chiapas, México, and Central America)....... *Trinycteris nicefori* (p. 273)

Genus *Glyphonycteris* Thomas, 1896

Small sized; **dorsal pelage with three bands, blackish at the base, middle whitish, and blackish at the tip**; fur of underparts dark gray at the base and pale gray at the tip; without a skin band between the ears across the forehead; calcar of the same size as or greater than the limbs; upper incisors markedly unequal, completely filling the space between the canines; middle upper incisors with the anterior face flat; outer incisors scarcely extending beyond the cingulum of the inner incisors; upper premolars with the main cusp situated distinctly in front of the middle crown; teeth with the inner border well developed, secondary cusp separated from the main cusp by a deep longitudinal groove; lower incisors trifid, with a division indicated nearly to the base of the crown by grooves along the anterior face; antorbital region distinctly inflated; basisphenoid cavities deep and distinct, with a sharp median dividing ridge.

Glyphonycteris sylvestris Thomas, 1896

Tricolored big-eared bat, murciélago tricolor

Monotypic.

Conservation: *Glyphonycteris sylvestris* is listed as Least Concern by the IUCN (Solari 2018d).

Characteristics: Tricolored big-eared bat is medium sized; forearm length 37.0–43.0 mm, skull length 20.0–21.4 mm. **Dorsal pelage with three bands, blackish at the base, middle whitish, and blackish at the tip**. Other characteristics as the same as in the genus.

Comments: *Glyphonycteris* was previously considered a subgenus of *Micronycteris* (Sanborn 1949), nowadays is regarded as a full genus (Simmons 1996). *G. sylvestris* ranges from southern Sinaloa and Veracruz south through South America (Map 58). It can be distinguished from other species in sympatry, or nearly so, such as those of the genera *Lampronycteris, Micronycteris,* and *Trinycteris,* by its dorsal pelage with three bands, blackish at the base, middle whitish, and blackish at the tip, and without a skin band between the ears across the forehead.

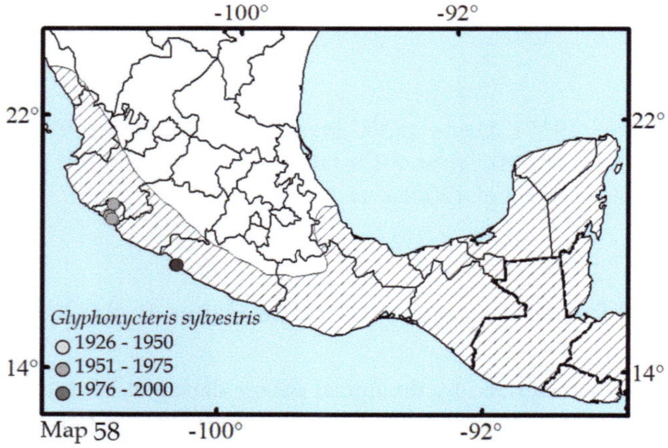

Map 58 Distribution map of *Glyphonycteris sylvestris*

Additional Literature: Simmons and Voss (1998), Sánchez–Hernández et al. (2002), Wang et al. (2003), Naranjo Piñera et al. (2005), Arroyo–Cabrales et al. (2008).

Genus *Trinycteris* Sanborn, 1949

Small sized; **without a skin band between the ears across the forehead**; **noseleaf narrow and tip pointed**; **fourth metacarpal shortest and third longest**; calcar of the same size as or greater than the limbs; upper incisors very unequal, not filling the space between the canines; inner tooth large, crown at least twice as high as wide, slightly projecting and somewhat oblique, so that the middle pair of teeth is strongly in contact distally; **outer incisors small**; lower incisors small, faintly trifid; two last lower premolars approximately of the same size; **forehead rising rather abruptly, at an angle of 45° with respect to the rostrum; auditory bullae small, well developed, covering about half of the cochleae, and their diameter equal to the distance between them**.

Trinycteris nicefori (Sanborn, 1949)

Niceforo's big-eared bat, murciélago orejón raro

Monotypic.

Conservation: *Trinycteris nicefori* is listed as Least Concern by the IUCN (Tavares and Burneo 2015).

Characteristics: Niceforo's big-eared bat is small sized; forearm length 35.0–38.6 mm, skull length 19.8–20.5 mm. Dorsal pelage grayish to dark brown; sometimes with gray stripes slightly notorious on the back; dark mask on the face; fur of underparts yellowish gray, contrasting with the dorsum; **no bands between the ears**; cushions on the chin V-shaped; calcar smaller than the limbs; ears big and pointed; **fourth metacarpal shortest**; braincase rising at an angle of 45° with respect to the rostrum; auditory bullae small, their diameter equivalent to the distance between them; the last two lower premolars approximately of the same size; **second upper incisors small and unicuspid**.

Comments: *Trinycteris* was considered a subgenus within the genus *Micronycteris* (Sanborn 1949); nowadays, it is regarded as a full genus (Simmons 1996; Wetterer et al. 2000). In North America, it is only known from Chiapas, México (Map 59). It can be distinguished from other species in sympatry, or nearly so, such as those of the genus *Micronycteris*, by the absence of a high band between the ears and third and fourth upper premolars of different size, and from *Lampronycteris*, by the fourth metacarpal being the shortest one and the second upper incisors being small and unicuspid.

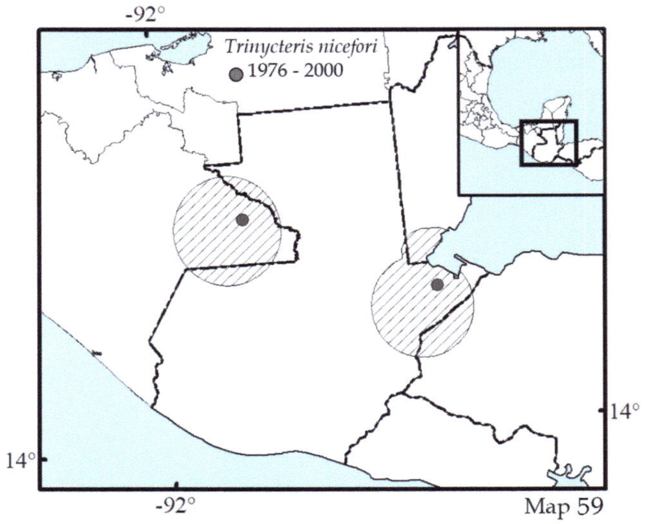

Map 59 Distribution map of *Trinycteris nicefori*

Additional Literature: Medellín et al. (1985), Álvarez-Castañeda and Álvarez (1991), Simmons and Voss (1998), Porter et al. (2007).

Subfamily Lonchorhininae Gray, 1866

This subfamily is represented in North America by only *Lonchorhina aurita*. It is characterized by the dorsal pelage and fur of underparts reddish brown to dark brown; **noseleaf very large, of the same size as the ears, and these at least twice the head length**; limbs and tail very long; tail tip extending beyond the uropatagium; **rostrum with a notoriously concave process over the nasals**; braincase supraoccipital region slightly over the level of the braincase roof; basisphenoid cavities large and deep; second lower premolars smaller than the rest; forearm length 49.0–54.0 mm, skull length 20.3–21.2 mm.

Genus ***Lonchorhina*** Tomes, 1863

Medium sized; ears large and separate; **noseleaf very large, horizontal portion with a thickened ridge in front of the nasal openings**; tail much longer than the femur, produced to the edge of the wide uropatagium; lower incisor crowns wide, width about equal to height; outer upper incisors with a distinctly hooked point close to the inner tooth and a flatfish heel extending outward and backward; second lower premolars smaller than the others; **distinct concavity at the rostrum between the orbits; nasals with a strongly convex dorsal profile, anterior end of bones projecting over the nares as a conspicuous, slightly downward-curved point**; middle of the braincase low, not rising conspicuously above the occiput; basisphenoid cavities very large and distinct, so carried forward that the anterior edge overhangs slightly; auditory bullae small.

Lonchorhina aurita Tomes, 1863

Common sword-nosed bat, murciélago nariz de espada

L. a. aurita Tomes, 1863. Only subspecies in North America.

Conservation: *Lonchorhina aurita* is listed as Threatened in the Norma Oficial Mexicana (DOF 2019) and as Least Concern by the IUCN (Solari 2015b).

Characteristics: Common sword-nosed bat is medium sized; forearm length 45.0–50.0 mm, skull length 20.0–21.5 mm. Dorsal pelage reddish brown to dark brown to almost black; fur of underparts paler than in the dorsum. Other characteristics same as in the genus.

Comments: *Lonchorhina aurita* ranges in the most humid and warmest region of México, from southern Veracruz, Oaxaca, Campeche, Tabasco, Quintana Roo, and Chiapas (Map 60). It can be found in sympatry, or nearly so, with. It differs from the rest of the species by having the noseleaf of the same size as the ears, which are at least twice the head length; skull with the dorsal profile of the nasals strongly convex, anterior end of the bones projecting over the nares as a conspicuous, slightly downward-curved point.

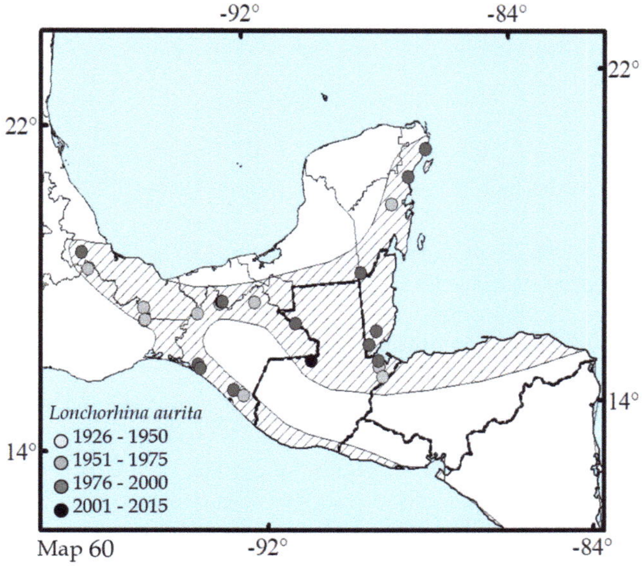

Map 60 Distribution map of *Lonchorhina aurita*

Additional Literature: Goodwin (1946), Goodwin and Greenhall (1961), Hernández-Camacho and Cadena (1978), Baker (1979), Lassieur and Wilson (1989), Álvarez-Castañeda and Álvarez (1991), Baker et al. (2000, 2003).

Subfamily Macrotinae Van Den Bussche, 1992

The subfamily Macrotinae includes big-eared bats with only one genus, *Macrotus*, and its characteristics are the same as for the genus.

Genus ***Macrotus*** Gray, 1843

Medium sized; dorsal pelage yellowish, reddish brown to grayish brown; hair long and white at the base; fur of underparts pale; **ears long and rounded, twice the head size and with a noseleaf;** tragus lanceolate**;** lower lip with a triangular pad bearing a longitudinal groove; tail tip free from the uropatagium; **middle upper incisors chisel shaped and big**; outer incisors less developed; rostrum moderately long and distinctly flattened above; braincase forming only a very slight angle with the rostrum; auditory bullae conspicuously enlarged, covering almost the entire cochleae, their greatest diameter distinctly greater than the space between them; basisphenoid cavity less distinct. The keys were elaborated based on the review of specimens and the following sources: Anderson and Nelson (1965), Hall (1981), Álvarez et al. (1994), and Álvarez-Castañeda et al. (2017a).

1. Pelage in yellow shades; interorbital breadth less than 3.8 mm; breadth through the external border of the upper canines less than 3.6 mm, sum of these two measures less than 7.3 mm (from California, southern Nevada, and Arizona, Unites States south through northern Sinaloa and Baja California Peninsula, México) .... *Macrotus californicus* (p. 276)

1a. Pelage in grayer shades; interorbital breadth greater than 3.8 mm; breadth throughout the external border of the upper canines greater than 3.6 mm, sum of these two measures greater than 7.3 mm (from Sinaloa and Querétaro, México, south through Guatemala) .... *Macrotus waterhousii* (p. 277)

Macrotus californicus Baird, 1858

California big-eared bat, murciélago orejón norteño

Monotypic.

Conservation: *Macrotus californicus* is listed as Least Concern by the IUCN (Solari 2018h).

Characteristics: California big-eared bat is medium sized; forearm length 44.6–57.5 mm, skull length 20.0–25.8 mm. **Dorsal pelage yellowish but tending to be pale yellow**; fur of underparts paler than in the dorsum; both species of *Macrotus* are very similar, but in *M. californicus* the interorbital width and the **width across the upper canines outer edge tend to be smaller than 3.8 and 3.6 mm, respectively; sum of these two measures always less than 7.3 mm**.

Comments: *Macrotus waterhousii* was reinstated as a distinct species (Davis and Baker 1974; Simmons 2005); it was previously considered a subspecies of *M. californicus* (Anderson 1969; Hall 1981). *M. californicus* ranges from southern California, southern Nevada, and virtually all of Arizona south through the Baja California Peninsula, Sonora, and northern Sinaloa. An isolated population has been recorded in San Luis Potosí (not shown in the map). The two species of *Macrotus* (Map 61) can be found in sympatry, or nearly so, with in central and southern Sinaloa. *M. californicus* can be distinguished from *M. waterhousii* by its pale yellowish dorsal coloration; the interorbital width and the width across the upper canine outer edge tend to be less than 3.8 and 3.6 mm, respectively.

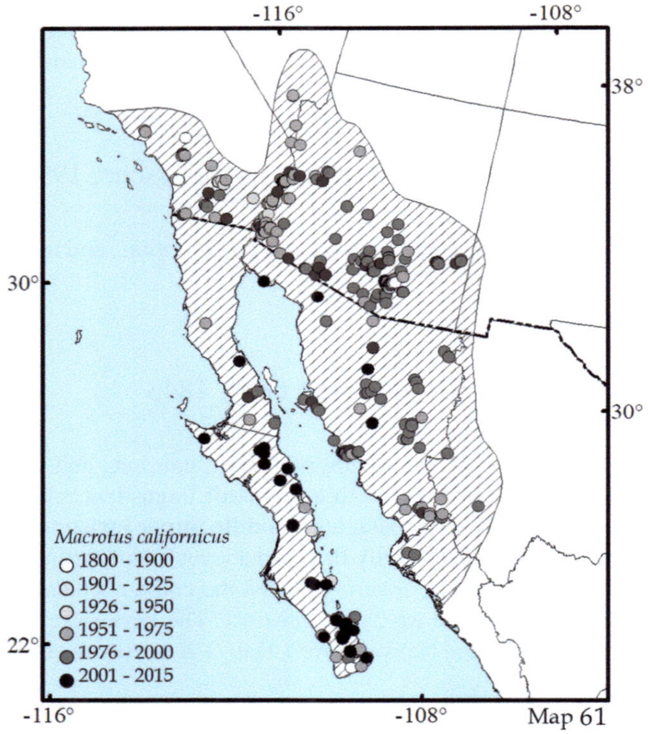

Map 61 Distribution map of *Macrotus californicus*

Additional Literature: Anderson and Nelson (1965), Álvarez-Castañeda and Patton (1999), Simmons and Conway (2001), Álvarez-Castañeda et al. (2008c), Frick et al. (2008), Torres-Morales et al. (2008).

Macrotus waterhousii Gray, 1843

Waterhous's big-eared bat, murciélago orejón sureño

1. *M. w. bulleri* Allen, 1890. From central Sinaloa and Chihuahua south through Jalisco, Guanajuato, and Querétaro.
2. *M. w. mexicanus* Saussure, 1860. From Jalisco, Guanajuato, and Querétaro south through Chiapas.

Conservation: *Macrotus waterhousii* is listed as Least Concern by the IUCN (Solari 2018i).

Characteristics: Waterhous's big-eared bats are medium sized; forearm length 49.0–54.0 mm, skull length 22.1–26.8 mm. **Dorsal pelage from reddish brown to grayish brown**; fur of underparts paler than in the dorsum; both species of *Macrotus* are very similar, but in *M. waterhousii* the interorbital width and the **width across the upper canine outer edge tend to be larger than 3.8 and 3.6 mm, respectively; the sum of these two measures is always greater than 7.3 mm**.

Comments: *Macrotus waterhousii* was reinstated as a distinct species (Davis and Baker 1974; Simmons 2005); it was previously regarded as a subspecies of *M. californicus* by Anderson (1969) and Hall (1981). *M. waterhousii* ranges from Sinaloa and Nuevo León south through Oaxaca, including central México. The two species of *Macrotus* can be found in sympatric distribution in central and southern Sinaloa (Map 62). *M. waterhousii* can be distinguished from *M. californicus* by its pale yellowish dorsal coloration and the interorbital width and the width across the upper canine outer edge smaller than 3.8 and 3.6 mm, respectively.

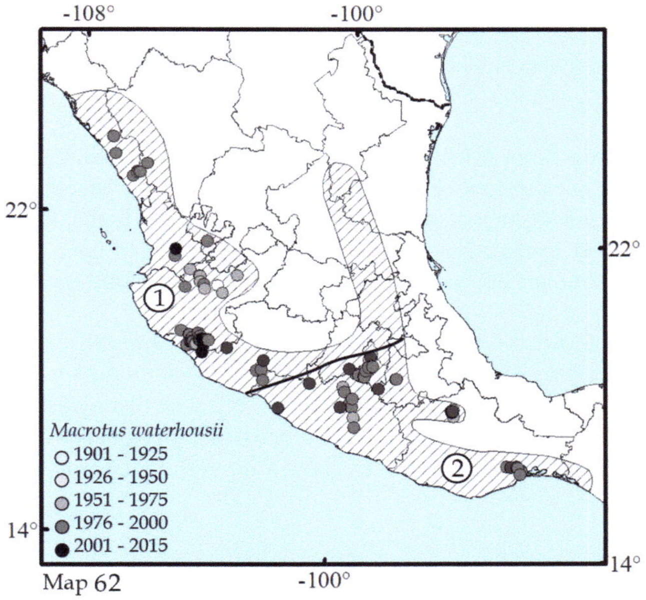

Map 62 Distribution map of *Macrotus waterhousii*

Additional Literature: Anderson and Nelson (1965), Álvarez-Castañeda and Patton (1999), Baker et al. (2000), Simmons and Conway (2001), McCarthy and Pérez (2006).

Subfamily Micronycterinae Van Den Bussche, 1992

The subfamily Micronycterinae includes big-eared bats characterized by its small size; generally yellowish color; ears large; tail extending to the middle of the interfemoral membrane; skull slender, lightly constructed; rostrum narrow and shorter than the braincase; **frontal region rising from the axis of the rostrum at an angle of approx. 45° (varying among species);** auditory bullae small, greatest diameter of one approximating the interbullar space. The keys were elaborated based on the review of specimens and the following sources: Andersen (1906a), Hall (1981), Álvarez et al. (1994), and Álvarez-Castañeda et al. (2017a).

1. Ears connected by a high band with a middle notch; third upper premolars of the same size as the second premolars..*Micronycteris* (p. 279)

1a. Ears not connected by a band; third upper premolars slightly smaller than the fourth upper premolars (from Chiapas and Oaxaca, México, south through South America) ....*Lampronycteris brachyotis* (p. 278).

Genus *Lampronycteris* Sanborn, 1949

Small sized; **without a skin band between the ears across the forehead**; **fifth metacarpal shortest and third longest**; calcar shorter than the foot, with claws; upper incisors very unequal, not filling the space between the canines; inner tooth large, crown at least twice as high as wide, slightly projecting and somewhat oblique, so that the middle pair of teeth is strongly in contact distally; **outer incisors small**; lower incisors small, faintly trifid; **second upper premolars longer and narrower**; forehead rising rather abruptly with respect to the rostrum; auditory bullae small, well developed, covering about one-half of the cochleae, their diameter equal to the distance between them.

Lampronycteris brachyotis (Dobson, 1879)

Yellow- throated big-eared bat, murciélago orejón de garganta amarilla

Monotypic.

Conservation: *Lampronycteris brachyotis* is listed as Threatened in the Norma Oficial Mexicana (DOF 2019) and as Least Concern by the IUCN (Solari 2018e).

Characteristics: Yellow-throated big-eared bat is small sized; forearm length 40.0–42.0 mm, skull length 21.0–21.9 mm. Dorsal pelage orange to dark brown; neck intense orange; fur of underparts yellowish contrasting with the dorsum; **no bands between the ears**; cushions on the chin V-shaped; calcar of the same size as the limbs; **fifth metacarpal shortest**; braincase rising at an angle of 45° with respect to the rostrum; auditory bullae small and their diameter equivalent to the distance between them; the two last lower premolars approximately of the same size; **second incisors bifid**

Comments: *Lampronycteris* was considered a subgenus within the genus *Micronycteris* (Sanborn 1949), nowadays it is regarded as a full genus (Simmons 1996; Wetterer et al. 2000). *Lampronycteris brachyotis* ranges in the most tropical and wettest regions of México, including Chiapas and Oaxaca (Map 63). It can be distinguished from other species in sympatry, or nearly so, such as those of the genus *Micronycteris*, by the absence of a high band between the ears and the third and fourth upper premolars of different size, and from *Trinycteris,* by having the fifth metacarpal as the shortest one and the second upper incisors forked.

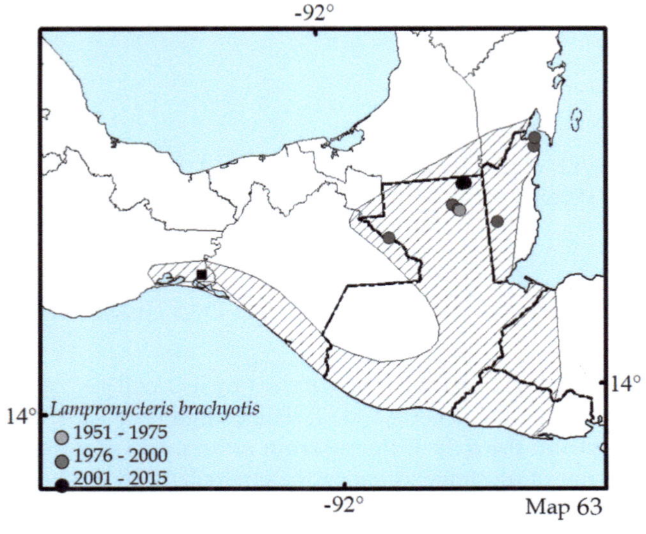

Map 63 Distribution map of *Lampronycteris brachyotis*

Additional Literature: Miller (1900), Andersen (1906a), Goodwin and Greenhall (1961), Davis et al. (1964), Goodwin (1969), Carter et al. (1981), Álvarez-Castañeda and Álvarez (1991), Simmons and Voss (1998), Briones-Salas and Sánchez-Cordero (2004), García–Estrada et al. (2006), Porter et al. (2007).

Genus *Micronycteris* Gray, 1866

Small sized; dorsal pelage reddish, pale brown to grayish brown; fur of underparts pale or white; **skin band between the ears across the forehead high and notched**; cushions in the cheeks V-shaped; calcar of the same size or greater than the limb; upper incisors very unequal, completely filling the space between the canines; inner incisors larger, crown at least twice higher than wide, slightly projecting and somewhat oblique; middle incisors strongly in contact distally; outer incisors scarcely extending beyond the canine cingulum, its crown flat, with a slight cusp on the inner side and roundish in outline; lower incisors small, simple, forming a continuous slightly convex row between the canines; two last lower premolars approximately of the same size; forehead rising rather abruptly at an angle of 45° with respect to the rostrum**; auditory bullae small, well developed, covering about one-half of the cochleae, their diameter equal to the distance between them**. Two subgenera (*Micronycteris* and *Schizonycteris*) are present in North America. The keys were elaborated based on the review of specimens and the following sources: Andersen (1906a), Hall (1981), Álvarez et al. (1994), and Álvarez-Castañeda et al. (2017a).

1. Interauricular membrane low with a shallow notch at the midline; fur of underparts brown; profile between the face and the braincase curved (Fig. 31; from Jalisco and Tamaulipas, México, south through South America)............................ ....subgenus *Micronycteris*; *Micronycteris microtis* (p. 279)

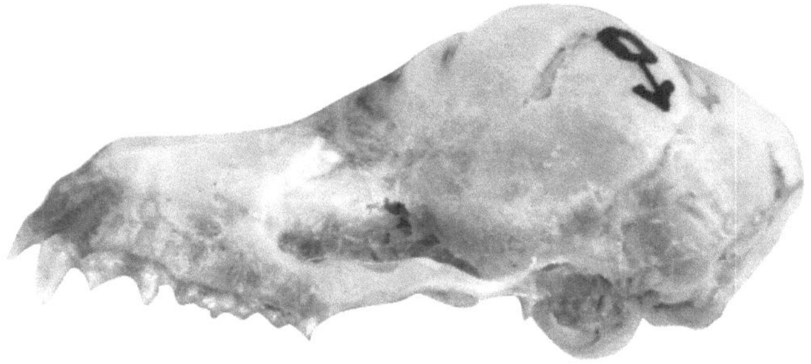

Fig. 31 The profile between the face and the braincase curved

1a. Interauricular membrane moderate to high, with a moderate to deep midline notch dividing the membrane into two triangular flaps; fur of underparts white, pale gray, or buff, paler than the dorsal fur; profile between the face and braincase almost straight (from Quintana Roo, Oaxaca, and Chiapas, México, south through South America) .... subgenus *Schizonycteris*; *Micronycteris schmidtorum* (p. 280)

Subgenus *Micronycteris* Gray, 1866

Small sized; low interauricular membrane with a shallow notch at the midline; ventral fur dark; calcar longer than the foot; second phalanx of fourth wing digit shorter than the first phalanx; mastoid breadth less than the zygomatic breadth.

Micronycteris microtis Miller, 1898

Common little big-eared bats, murciélago orejón pequeño

M. m. mexicana Miller, 1898. The only subspecies in North America.

Conservation: *Micronycteris microtis* is listed as Least Concern by the IUCN (Solari and Camacho 2019).

Characteristics: Common little big-eared bat is small sized; forearm length 32.0–37.0 mm, skull length 17.7–20.2 mm. Dorsal pelage reddish to grayish brown; fur of underparts pale brown, not contrasting with the dorsal fur; a **band between the ears with a small notch**; profile between the face and the braincase curved.

Comments: *Micronycteris microtis* was reinstated as a distinct species (Simmons 1996; Wetterer et al. 2000); it was previously considered a subspecies of *M. megalotis* (Sanborn 1949). *M. microtis* ranges in the tropical regions of México from Jalisco and Tamaulipas south through South America and the Yucatán Peninsula (Map 64). It can be found in sympatry, or nearly so, with all the species of the subfamily Micronycterinae in the central and southern areas of its distribution. It can be differentiated from *Lampronycteris* and *Trinycteris* by the presence of a high band between the ears; third and fourth upper premolars of the same size. From *M. schmidtorum,* by a deep cleft in the band between the ears; fur of underparts brown, and its skull profile.

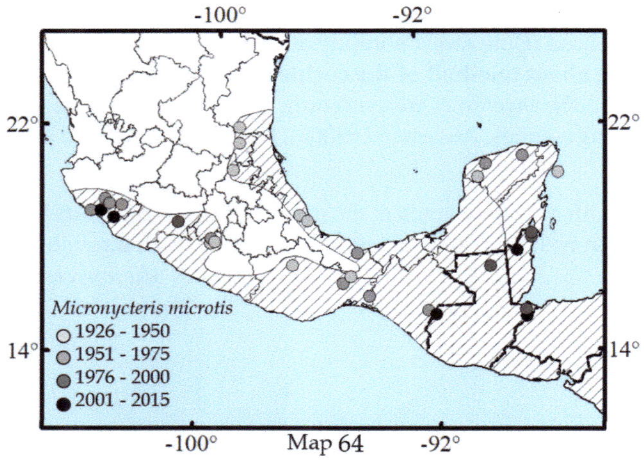

Map 64 Distribution map of *Micronycteris microtis*

Additional Literature: Andersen (1906a), Álvarez (1963), Goodwin (1969), Medellín et al. (1985), Genoways and Williams (1986), Ramírez-Pulido et al. (1986), Alonso-Mejía and Medellín (1991), Simmons and Voss (1998), Fonseca et al. (2007), Porter et al. (2007).

Subgenus *Schizonycteris* Porter et al., (2007)

Moderate to high interauricular membrane with a moderate to deep midline notch dividing the membrane into two triangular flaps; ventral fur white, pale gray, or buff, paler than in the dorsum; lower incisors not hypsodont, crown height not more than twice the crown width.

Micronycteris schmidtorum Sanborn, 1935

Schmidt's little big-eared bats, murciélago orejón centroamericano

Monotypic.

Conservation: *Micronycteris schmidtorum* is listed as Threatened in the Norma Oficial Mexicana (DOF 2019) and as Least Concern by the IUCN (Sampaio et al. 2016b).

Characteristics: Schmidt's little big-eared bat is small sized; forearm length 34.0–38.0 mm, skull length 18.0–19.8 mm. Dorsal pelage pale brown; fur of underparts white, contrasting strongly with the dorsal fur; **band between the ears with a deep cleft**; **profile between the face and the braincase almost straight**.

Comments: *Micronycteris schmidtorum* ranges in Quintana Roo, Oaxaca, and Chiapas (Map 65). It can be found in sympatry, or nearly so, with all the species of the subfamily Micronycterinae. *M. schmidtorum* can be differentiated from *Lampronycteris* and *Trinycteris* by the presence of a high band between the ears; third and fourth upper premolars of the same size. From *M. microtis,* by a shallow cleft in the band between the ears, underparts pale gray, and its skull profile.

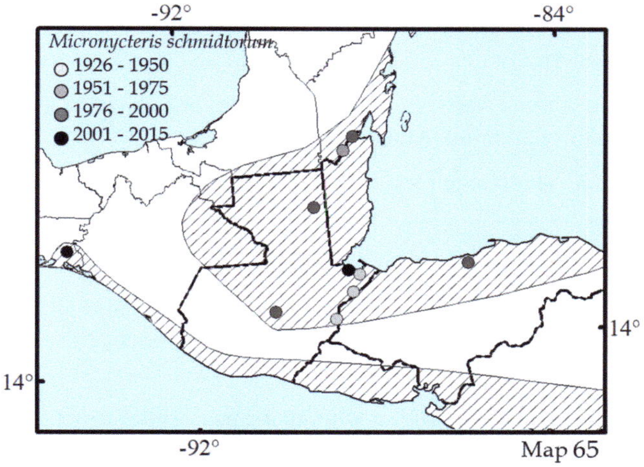

Map 65 Distribution map of *Micronycteris schmidtorum*

Additional Literature: Sanborn (1949), Medellín et al. (1985), Genoways and Williams (1986), Álvarez-Castañeda and Álvarez (1991), Simmons and Voss (1998), Wetterer et al. (2000), Fonseca et al. (2007), Porter et al. (2007).

Subfamily Phyllostominae Gray, 1825

The subfamily Phyllostominae is characterized by its medium to large size; noseleaf present; humerus with a definite secondary impingement (no articulation) with the scapula; lower lip without platelike outgrowths; teeth essentially normal, although in some genera, the mesostyle of the first and second upper molars and the paraconid and metaconid of the first and second lower molars are reduced; epitrochlear large, having a slightly developed spiny process. The keys were elaborated based on the review of specimens and the following sources: Goodwin (1942a, b), Lawlor (1979), Hall (1981), Álvarez et al. (1994), and Álvarez-Castañeda et al. (2017a).

1. One pair of lower incisors 2

1a. Two pairs of lower incisors 4

2. Three lower premolars 3

2a. Two lower premolars........ 9

3. Forearm length greater than 70.0 mm; second lower premolars displaced lingually to the series of teeth (from southern Veracruz and southern Oaxaca, México, south through South America) .... tribe Vampyrini *Chrotopterus auritus* (p. 290)

3a. Forearm length less than 70.0 mm; second lower premolars displaced from the series of teeth toward the lip........ 8

4. Two lower premolars (from southern Oaxaca and southern Veracruz, México, south through South America, excluding Yucatán Peninsula)........ tribe Phyllostomini *Phyllostomus discolor* (p. 287)

4a. Three lower premolars 5

5. Forearm length greater than 95.0 mm; distance between the internal borders of the second upper molars (palatine) less than the diameter of the same molars (from southern Veracruz and Oaxaca, México, south through South America) .... tribe Vampyrini *Vampyrum spectrum* (p. 292)

5a. Forearm length less than 95.0 mm, distance between the internal borders of the second upper molars greater than the diameter of the same molars 6

6. Second premolars in normal position, separating the first and third premolars (from Veracruz and Chiapas, México, south through South America) .... *Phylloderma stenops* (p. 286)

6a. Second lower premolars displaced lingually, the first and second premolars almost in contact 7

7. Lower part of the tail membrane with warts; rostral length less than the braincase breadth (from Tabasco and Chiapas, México, south through South America) .... *Macrophyllum macrophyllum* (p. 285)

7a. Lower part of the tail membrane without warts; rostral length approximately same as the braincase breadth (from southern Veracruz and Oaxaca south through South America) .... *Trachops cirrhosus* (p. 289)

8. Forearm length less than 40.0 mm; skull length less than 22.0 mm (from central Veracruz, México, south through South America) .... *Tonatia bakeri* (p. 288)

8a. Forearm length greater than 40.0 mm; skull length greater than 22.0 mm .... *Lophostoma* (p. 283)

9. Wing membrane connected to the limb side close to the base of the external digit; a white stripe in the middle dorsal part; auditory bullae large (from southern Campeche and eastern Tabasco, México, south through South America) .... *Gardnerycteris keenani* (p. 282)

9a. Wing membrane connected at the level of the ankles; without a dorsal white stripe; auditory bullae small (from southern Veracruz, Oaxaca, and the Yucatán Peninsula, México, south through South America) .... tribe Vampyrini *Mimon cozumelae* (p. 291)

Tribe Phyllostomini

Genus *Gardnerycteris* Hurtado and Pacheco, 2014

Medium sized; dorsal stripe absent, faint, or very marked; muzzle giving the appearance of bare skin; noseleaf narrow and lanceolate, its length at least twice its width, with inter-branch vibrissa and long hairs around the limbus; ears fleshy, lanceolate, with the tips parallel, without projections; gular gland always present; thumbs always covered with hair and the proximal one-third of forearms always bare; chin with a medium deep groove; tail longer than the limbs; upper canines with accessory cusps; lower incisors bilobed or trilobed; nasal bones short; braincase not expanded anteriorly and posterior border rounded or angular; rostral length one-quarter to one-third of the skull length; zygomatic arches laterally flattened; postorbital constriction narrower than the space between the canines; accessory foramina behind the incisors; auditory bullae high; basisphenoid cavities deep with a mid-high septum.

Gardnerycteris keenani (Handley, 1960)

Striped hairy-nosed bat, murciélago dorado nariz peluda

Monotypic.

Conservation: *Gardnerycteris keenani* (as *Gardnerycteris crenulatum*) is listed as Threatened in the Norma Oficial Mexicana (DOF 2019) and as Least Concern by the IUCN (Solari 2019e).

Characteristics: Striped hairy-nosed bat is medium sized; forearm length 46.0–55.0 mm, skull length 20.9–23.3 mm. **Dorsal and ventral pelage blackish with a yellowish stripe in the middle**; **ears pale on the base and dark at the tip**; long tail embedded within the uropatagium; **noseleaf large and wide, "serrated," and not merged with the upper lip; small warts at both sides of a V-shaped cushion**; calcar greater than hind limb length; tail long, one-third of the uropatagium length; **auditory bullae large,** expanded and slim; rostrum wide; basisphenoid cavities wide but not deep, separated by a middle septum; small projection in the anterior and superior parts of the rostrum barely developed; one pair of lower incisors and two lower premolars.

Comments: The species review of *Mimon* shows that the genus needs to be split in two. *M. crenulatum* is considered within the genus *Gardnerycteris* (Hurtado and Pacheco 2014). The review of *Gardnerycteris* shows that the subspecies *G. crenulatum keenani* should be considered a full species (Hurtado and D'Elias 2018). *G. keenani* ranges in North America only in southwestern Campeche and eastern Tabasco (Map 66). It can be distinguished from other species in sympatry, or nearly so, by its large, wide noseleaf, "serrated" and not merged with the upper lip, with small warts at both sides of a V-shaped cushion; calcar longer than the hind limb length and tail long, one-third of the uropatagium length; skull expanded and slim; rostrum wide; basisphenoid cavities wide but not deep, separated by a middle septum. From *M. cozumelae,* by its dorsal and ventral pelage blackish with a yellowish stripe in the middle and larger auditory bullae.

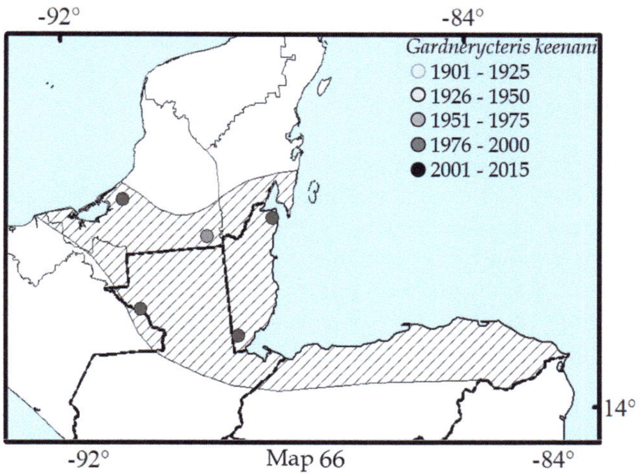

Map 66 Distribution map of *Gardnerycteris keenani*

Additional Literature: Handley (1960), Jones and Carter (1976), Koopman (1978), Naranjo Piñera et al. (2005).

Genus ***Lophostoma*** d'Orbigny, 1836

Medium sized; dorsal pelage gray to grayish brown; fur of underparts paler than in the dorsum; **ears big and rounded, with folds in the outer side, perpendicular to the ear main axis;** tragus tip pointed an relatively very small in relation to the ear size; **noseleaf lower part merged with the upper lip; warts small at both sides of a groove on the chin;** wings connected to the ankles; tibias very large and tail small; uropatagium large with the calcar well developed: skull robust; braincase rising at an angle of 45° with respect to the rostrum; auditory bullae small; basisphenoid cavities absent; **sagittal crest in the frontal bone to the supraoccipital bone**; upper canines markedly large; second lower premolars markedly reduced. The keys were elaborated based on the review of specimens and the following sources: Goodwin (1942a, b), Hall (1981), Álvarez et al. (1994), and Álvarez-Castañeda et al. (2017a).

1. Forearm less than 40.0 mm; skull less than 22.0 mm; mastoid breadth less than the zygomatic breadth (from southern Veracruz, northern Chiapas, and southern Yucatán Peninsula, México, south through Honduras) .... *Lophostoma brasiliense* (p. 261)

1a. Forearm greater than 40.0 mm; skull length greater than 22.0 mm; mastoid breadth greater than the zygomatic breadth (from southern Veracruz, Tabasco, Campeche, northern Chiapas, and southern Yucatán Peninsula, México, south through Central America) .... *Lophostoma evote* (p. 261)

Lophostoma brasiliense Peters, 1867

Pygmy round-eared bat, murciélago orejas redondas brasileño

(*see addenda, valid name *Lophostoma nicaraguae*)

Monotypic.

Conservation: *Lophostoma brasiliense* is listed as Threatened in the Norma Oficial Mexicana (DOF 2019), and as Least Concern by the IUCN (Sampaio et al. 2016c).

Characteristics: Pygmy round-eared bat is medium sized; **forearm length 32.0–36.0 mm, skull length 26.8–29.2 mm. Dorsal pelage brown; fur of underparts and neck hair of the same color as the dorsal fur but paler; second lower premolars less than one-third in size of the first premolars**.

Comments: *Lophostoma brasiliense* was originally considered within the genus *Tonatia,* but it was later demonstrated not to be monophyletic, so *T. brasiliense* was designated to the genus *Lophostoma* (Lee et al. 2002). *L. brasiliense* ranges in the wettest areas of southern México, from southern Veracruz, northern Chiapas, and the southern Yucatán Peninsula south through Honduras (Map 67). It can be distinguished from other species in sympatry, or nearly so, by its lower noseleaf part merged with the upper lip; small warts at both sides of a groove on the chin and a sagittal crest in the frontal bone. From *L. evote,* by its larger size and the sagittal crest well developed in the frontal bone.

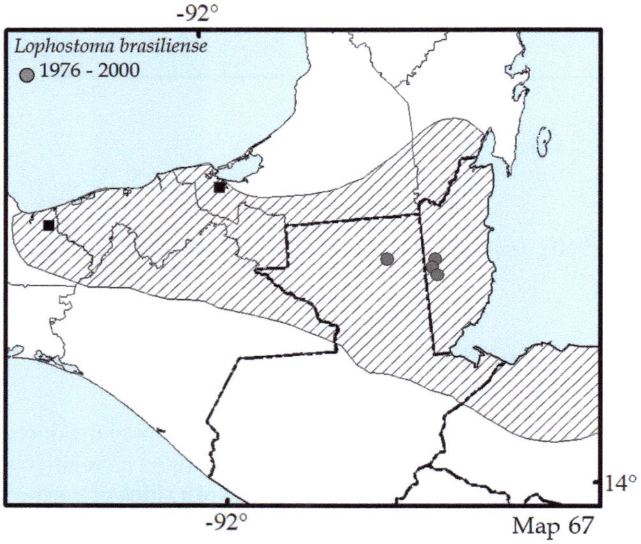

Map 67 Distribution map of *Lophostoma brasiliense*

Additional Literature: Goodwin (1942a), Davis and Carter (1978), Medellín and Arita (1989), Álvarez-Castañeda and Álvarez (1991), Medellín et al. (2000), Naranjo Piñera et al. (2005).

Lophostoma evote (Davis and Carter, 1978)

Davis's round-eared bat, murciélago orejas redondas mesoamericano

Monotypic.

Conservation: *Lophostoma evote* is listed as Threatened in the Norma Oficial Mexicana (DOF 2019) and as Least Concern by the IUCN (Solari 2018g).

Characteristics: Davis's round-eared bat is medium sized; **forearm length 47.0–54.0 mm, skull length 18.3–21.4 mm. Dorsal pelage gray to grayish brown; fur of underparts grayish brown;** neck pale grayish; **second lower premolars more than one-third in size of the first premolars**.

Comments: *Lophostoma evote* was originally considered within the genus *Tonatia,* but it was later demonstrated not to be monophyletic, so *T. evote* was designated to the genus *Lophostoma* (Lee et al 2002). *L. evote* was originally considered *Tonatia evote* (Davis and Carter 1978). *T. silvicola* is considered a junior synonym of *L. evote* (Medellín and Arita 1989). *L. evote* ranges from southern Veracruz, Tabasco, Campeche, northern Chiapas, and the southern Yucatán Peninsula south through Central America (Map 68). It can be distinguished from other species in sympatry, or nearly so, such as those of the same genus, by its noseleaf lower part merged with the upper lip; small warts at both sides of a groove on the chin and a sagittal crest in the frontal bone. From *L. brasiliense*, by its smaller size and having the sagittal crest in the frontal bone less developed.

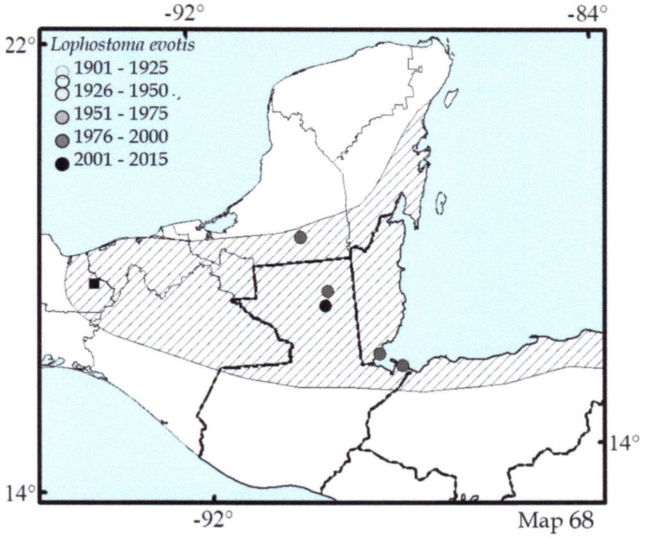

Map 68 Distribution map of *Lophostoma evote*

Additional Literature: Sánchez-H. et al. (1986), Álvarez-Castañeda and Álvarez (1991), Porter et al. (2003), Baker et al. (2004), Hoffmann et al. (2008).

Genus *Macrophyllum* Gray, 1838

Medium sized; ears big, separate, and pointed; **noseleaf large, greater than one-half of the ear length and not merged to the upper lip**; limbs notoriously long; **warts in the lower part of the tail membrane;** tail longer than the femur and continued to the outer uropatagium edge; **tibia, uropatagium, feet, and claws elongated;** middle upper incisors projecting more conspicuously and with crowns; lower incisors relatively wider; middle lower premolars very small; **first upper molars not much larger than the outer incisors**; braincase rising at an angle of 45° with respect to the rostrum; **rostrum very short, less than the braincase breadth**; auditory bullae small, not covering half of the cochleae; basisphenoid cavities obsolete.

Macrophyllum macrophyllum (Schinz, 1821)

Long-legged bat, murciélago de patas largas

Monotypic.

Conservation: *Macrophyllum macrophyllum* is listed as Threatened in the Norma Oficial Mexicana (DOF 2019) and as Least Concern by the IUCN (Rodriguez and Pineda 2015).

Characteristics: Long-legged bat is medium sized; forearm length 34.0–37.0 mm, skull length 16.3–17.5 mm. Dorsal pelage dark grayish brown; ears big and pointed; noseleaf large, greater than one-half of the ear length and not merged to the upper lip; limbs notoriously long; **warts in the lower part of the tail membrane**; braincase rising at an angle of 45° with respect to the rostrum; **rostrum very short, less than the braincase breadth**; auditory bullae small; basisphenoid cavities slightly notorious; second lower premolar smaller than the other lower premolars.

Comments: *Macrophyllum macrophyllum* ranges from southern Tabasco and northern Chiapas south through South America (Map 69). It can be distinguished from other species in sympatry, or nearly so, such as those of the same genus, by the presence of warts in the lowest part of the tail membrane, and a very short rostrum, less than the braincase breadth.

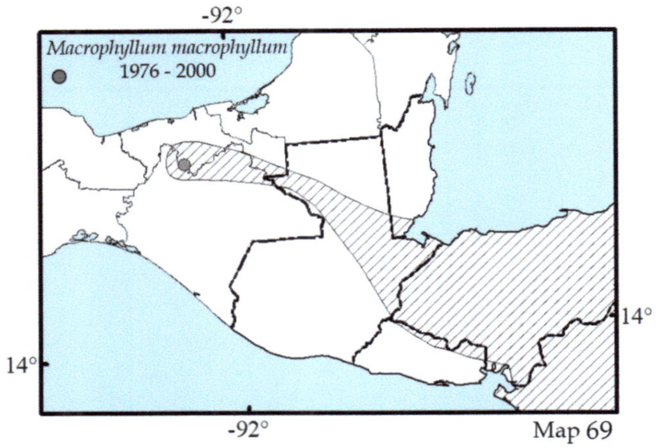

Map 69 Distribution map of *Macrophyllum macrophyllum*

Additional Literature: Miller (1907), Goldman (1920), Goodwin (1946), Felten (1956), Harrison (1975), Álvarez-Castañeda and Álvarez (1991), Naranjo Piñera et al. (2005), Hernández–Mijangos et al. (2008).

Genus *Phylloderma* Peters, 1865

Medium sized; **face skin pinkish**; ears big and rounded; **noseleaf base merged within the upper lip; pelage and wings frequently with white spots**; wings white at the tip; tail one-half of the uropatagium length; upper incisors unequal, completely filling the space between the canines; lower incisors small, simple, forming a continuous slightly convex row between the canines; second lower premolars smaller than other genera; skull robust and wide; nasals rising abruptly behind the nares; rostrum deep and short; sagittal crest well developed; zygomatic arches thick; basisphenoid cavities slightly notorious; symphysis menti long.

Phylloderma stenops Peters, 1865

Pale-faced bat, Murciélago cara pálida

P. s. septentrionalis Goodwin, 1940. Only subspecies in North America.

Conservation: *Phylloderma stenops* is listed as Least Concern by the IUCN (Solari 2015c) and as under Special Protection by the Norma Oficial Mexicana (DOF 2019).

Characteristics: Pale-faced bat is medium sized; forearm length 69.0–83.0 mm, skull length 33.0–35.5 mm. Dorsal pelage reddish brown to brown; **face skin pinkish**; fur of underparts pale gray; ears big and rounded; **noseleaf base merged with the upper lip; pelage and wings frequently with white spots**; wings white at the tip; robust and wide; sagittal crest well developed; zygomatic arches thick; basisphenoid cavities slightly notorious; internal incisors forked; second lower premolars small.

Comments: *Phylloderma stenops* ranges from Veracruz and Chiapas, México, south through South America (Map 70). It can be distinguished from other species in sympatry, or nearly so, such as those of the same genus, by its pinkish facial skin; noseleaf base merged with the upper lip; pelage and wings frequently with white spots.

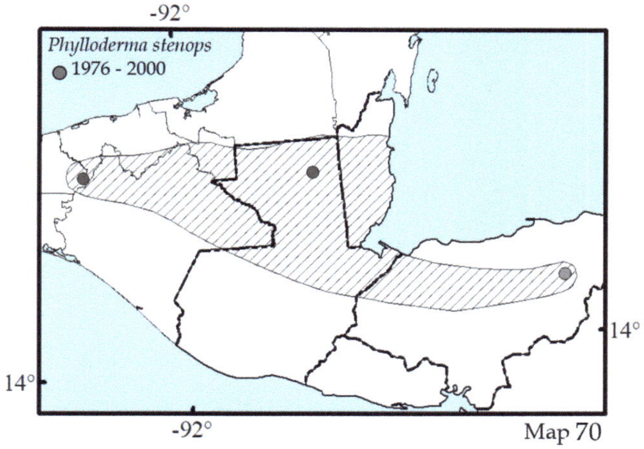

Map 70 Distribution map of *Phylloderma stenops*

Additional Literature: Miller (1907), Jones and Carter (1976), Baker et al. (1988), Álvarez-Castañeda and Álvarez (1991), Van Den Bussche and Baker (1993), Emmons (1997), Medellín et al. (2000), Naranjo Piñera et al. (2005).

Genus *Phyllostomus* Lacépède, 1799

Medium sized; ears triangular, moderately large and widely separate; noseleaf well developed, though simple, merged to the upper lip; **groove on the chin with big warts as beans around it**; calcar as big as the limbs; upper incisors unequal, completely filling the space between the canines; lower incisors small, simple, forming a continuous slightly convex row between the canines; **two pairs of lower incisors and lower premolars;** rostrum heavy, robust, and broad; sagittal crest high and well developed; zygomatic arches heavy, with the anterior and posterior portions slightly expanded; basisphenoid cavities evident, although rather shallow; auditory bullae small and flat, covering about one-half of the surface of the cochleae.

Phyllostomus discolor Wagner, 1843

Pale spear-nosed bat, murciélago pálido de nariz de lanza

P. d. verrucossus (Elliot, 1905). Only subspecies in North America.

Conservation: *Phyllostomus discolor* is listed as Least Concern by the IUCN (Barquez et al. 2015j)

Characteristics: Pale spear-nosed bat is medium sized; forearm length 60.0–68.0 mm, skull length 28.0–31.9 mm. Dorsal pelage brown to grayish brown; rostrum skin dark brown; fur of underparts notoriously paler than in the dorsum; ears triangular and long; noseleaf merged to the upper lip; **groove on the chin with big warts as beans around it**; calcar as big as the limbs; robust and wide; sagittal crest well developed; zygomatic arches thick; basisphenoid cavities slightly notorious; **two pairs of lower incisors and lower premolars**.

Comments: *Phyllostomus discolor* ranges from southern Veracruz and Oaxaca south through South America, excluding the Yucatán Peninsula (Map 71). It can be distinguished from other species in sympatry, or nearly so, such as those of the same genus, by its groove on the chin with big warts as beans around it and two pairs of lower incisors and lower premolars.

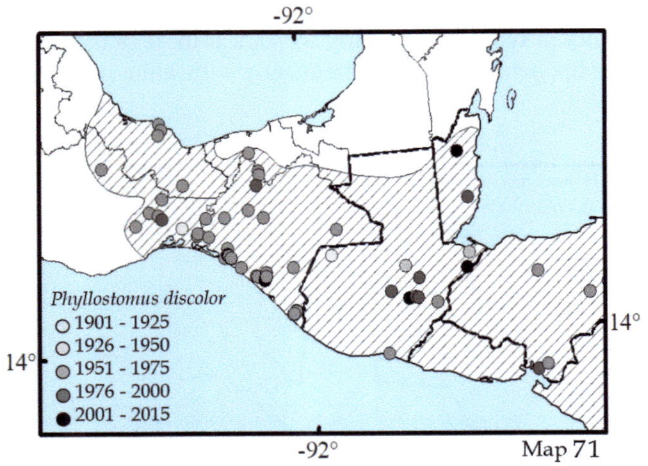

Map 71 Distribution map of *Phyllostomus discolor*

Additional Literature: Dobson (1878), Miller (1907), Sanborn (1936), Felten (1956), Goodwin and Greenhall (1961), Davis and Carter (1962), Handley (1980), Baker et al. (1989a), Álvarez-Castañeda and Álvarez (1991), Kwiecinski (2006), Gonçalves da Silva et al. (2008), Medellín et al. (2008).

Genus *Tonatia* Gray, 1827

Medium sized; noseleaf well developed; **horseshoe of noseleaf fused to the upper lip**; **pale-colored stripes on the head barely perceptible; ears big and rounded,** separate or joined across the forehead; forearm hairy; tail shorter than the femur, extending to about the middle of the broad uropatagium; upper incisors unequal, completely filling the space between the canines; lower incisors small, simple, forming a continuous, slightly convex row between the canines; **second lower premolars markedly reduced**; zygomatic arches with an abrupt expansion near the anterior base; forehead rising rather abruptly, at an angle of 45° with respect to the rostrum; basisphenoid cavities absent; auditory bullae small, covering less than one-half of the surface of the cochleae; sagittal crest well developed, not divided anteriorly but continued forward as a stripe traceable to the posterior extremity of the nasals.

Tonatia bakeri Williams, Willig, and Reid, 1995

Stripe-headed round-eared bat, murciélago de orejas redondas

Monotypic.

Conservation: *Tonatia bakeri* (as *Tonatia saurophila*) is listed as Threatened in the Norma Oficial Mexicana (DOF 2019) and as Least Concern by the IUCN (Solari 2018o).

Characteristics: Stripe-headed round-eared bat is medium sized; forearm length 56.0–61.0 mm, skull length 26.3–30.6 mm. Dorsal pelage and fur of underparts grayish brown to dark brown with grizzly shades; hairs on the shoulders white at the base and at the tip; hairs on the top of the head white at the tip, forming a pale-colored stripe between the ears; **ears big and rounded**; **noseleaf horseshoe fused to the upper lip**; **pale-colored stripes on the head barely perceptible;** forearms hairy; tail long, embedded within the uropatagium; skull robust, braincase rising at an angle of 45° with respect to the rostrum; auditory bullae small; basisphenoid cavities absent; sagittal crest well defined; upper canines big; **second lower premolars markedly reduced**; lambdoid region inflated and extended.

Comments: *Tonatia bakeri* was previously considered a subspecies of *T. saurophila* (Williams et al. 1995); because *T. saurophila* only ranges in Jamaica, British West Indies, I have used *T. bakeri* for North America (Basantes et al. 2020) it ranges from eastern Chiapas south through South America (Map 72). It can be distinguished from other species in sympatry, or nearly so, such as those of the same genus by its big and rounded ears; noseleaf horseshoe fused to the upper lip with barely perceptible pale-colored stripes on the head, and the second lower premolars markedly reduced.

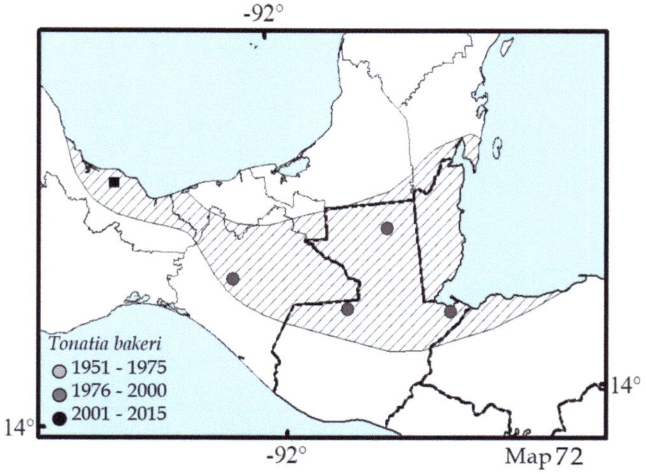

Map 72 Distribution map of *Tonatia bakeri*

Additional Literature: Miller (1907), Davis and Carter (1978), Koopman (1978), Genoways and Williams (1984), Medellín and Arita (1989), Morales–Malacara and Juste (2002), Lee et al. (2002), Escobedo-Morales et al. (2006).

Genus *Trachops* Gray, 1847

Medium sized; **ears big and rounded**; **noseleaf horseshoe fused to the upper lip**; **head with pale-colored stripes barely perceptible;** forearms hairy; tail long, embedded within the uropatagium and much shorter than the femur; **second lower premolars markedly reduced**; skull elongated and rounded; interorbital region smoothly rounded and somewhat depressed; posterior zygomatic expansion angular-emarginate in the front; anterior expansion barely indicated; auditory bullae small and flat, covering about one-half of the surface of the cochleae, with height fully equal to width; sagittal crest high and broadly expanded, shelf-like paroccipital processes; zygomata heavy, slightly expanded both anteriorly and posteriorly; basisphenoid cavities evident, although rather shallow.

Trachops cirrhosus (Spix, 1823)

Fringe-lipped bat, murciélago de labio verrugoso

(*see addenda, valid name *Trachops coffini*)

T. c. coffini Goldman, 1925. Only subspecies in North America.

Conservation: *Trachops cirrhosus* is listed as Threatened in the Norma Oficial Mexicana (DOF 2019) and as Least Concern by the IUCN (Miller et al. 2015a, b, c, d).

Characteristics: Fringe-lipped bat is medium sized; forearm length 57.0–65.0 mm, skull length 27.7–29.2 mm. Dorsal pelage orange brown to grayish brown; fur of underparts paler than the dorsum; ears big and rounded; **warts extend as a cushion on the lips and chin**; tail short, embedded within the uropatagium; calcar big; **skull extended in the interorbital region as a sand clock**; **rostral length approximately equal to the braincase breadth;** zygomatic arches robust; outer incisors strongly reduced compared with the inner incisors.

Comments: *Trachops cirrhosus* ranges from southern Veracruz and Oaxaca south through South America (Map 73). It can be distinguished from other species in sympatry, or nearly so, such as those of the same genus, by its warts extending as a cushion on the lips and chin; interorbital region shaped as a sand clock, and rostral length approximately equal to the braincase breadth.

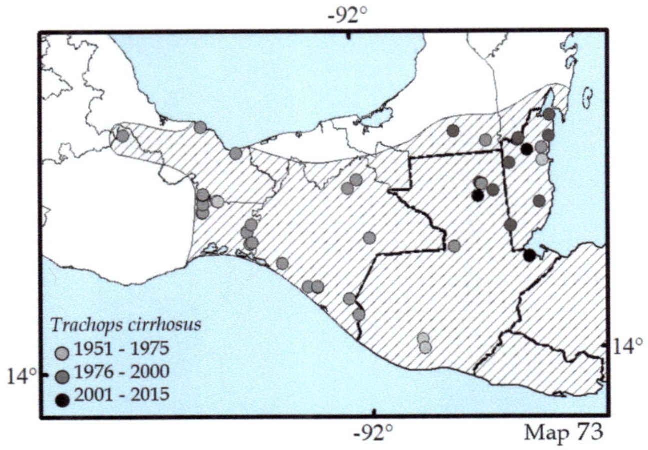

Map 73 Distribution map of *Trachops cirrhosus*

Additional Literature: Miller (1907), Goodwin and Greenhall (1961), Swanepoel and Genoways (1979), Willig (1983), Álvarez-Castañeda and Álvarez (1991), Koopman (1993), Medellín et al. (2000), Cramer et al. (2001), Naranjo Piñera et al. (2005).

Tribe Vampyrini

Genus *Chrotopterus* Peters, 1865

Medium sized; **ears very big and rounded; wings white at the tip; lips and chin nearly smooth;** hair **unusually long and soft; outer incisors very small, almost hidden between the canines and the inner incisors; second lower premolars displaced lingually to the teeth series;** skull robust and wide; rostrum and interorbital region subcylindrical; sagittal crest well developed; zygomatic arches thick; face and interorbital region subcylindrical; with basisphenoid cavities barely perceptible; auditory bullae small, covering less than one-half of the cochlea, diameter nearly twice the height at the inner edge; paroccipital expansion small but distinct.

Chrotopterus auritus (Peters, 1856)

Woolly false vampire bat, falso vampiro lanudo

C. a. auritus (Peters, 1856). Only subspecies in North America.

Conservation: *Chrotopterus auritus* is listed as Threatened in the Norma Oficial Mexicana (DOF 2019) and as Least Concern by the IUCN (Barquez et al. 2015d).

Characteristics: Woolly false vampire bat is medium sized; **forearm length 77.2–83.5 mm**, skull length 35.3–38.5 mm. Dorsal pelage grayish brown to dark grayish; fur of underparts silver gray to whitish; hair long; **ears very big and rounded; wings white at the tip;** skull robust and wide; sagittal crest well developed; zygomatic arches thick; face and interorbital region subcylindrical; with basisphenoid cavities barely perceptible; **second lower premolars displaced lingually from the teeth series**.

Comments: *Chrotopterus auritus* ranges from southern Veracruz and Oaxaca south through South America, including the Yucatán Peninsula (Map 74). It can be found in sympatry, or nearly so, with all the species of the subfamily Phyllostominae. It can be differentiated from all the other genera by having the following combination of characters: a larger forearm length, from 77.2 to 83.5 mm; wings white at the tip; ears very big and rounded; second lower premolars displaced lingually to the teeth series.

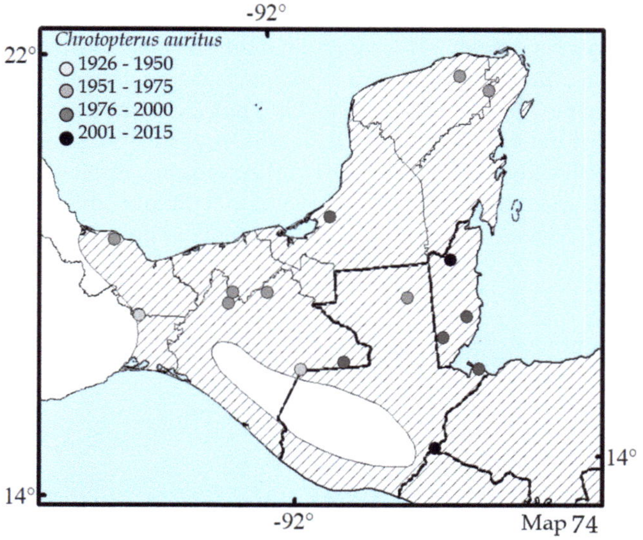

Map 74 Distribution map of *Chrotopterus auritus*

Additional Literature: Miller (1907), Villa-R. (1967), McNab (1969), Swanepoel and Genoways (1979), Medellín (1988, 1989), Álvarez-Castañeda and Álvarez (1991), Medellín et al. (2000), Briones-Salas and Sánchez-Cordero (2004).

Genus *Mimon* Gray, 1847

Medium sized; **dorsal and ventral pelage brilliant old gold; no dorsal white stripe**; **noseleaf large and wide, "serrated," not merged with the upper lip; small warts at both sides of a V-shaped cushion**; calcar greater than hind limb length; tail long, one-third of the uropatagium length; **inner upper incisors scarcely oblique and separated from the canines by a distinct space**; lower incisors small, simple, forming a continuous, slightly convex row between the canines; one pair of lower incisors and two lower premolars; **skull slender, with a rather broadly arched rostrum; zygomatic without an anterior or posterior expansion;** rostrum wide; **basisphenoid cavities broad and shallow, separated by a prominent median ridge;** small projection in the anterior upper part of the rostrum barely developed; auditory bullae small, covering about one-half of the cochlear surface but unusually narrow.

Mimon cozumelae Goldman, 1914

Cozumelan golden bat, murciélago dorado de Cozumel

Monotypic.

Conservation: *Mimon cozumelae* is listed as Threatened in the Norma Oficial Mexicana (DOF 2019) and as Least Concern by the IUCN (Arroyo-Cabrales et al. 2015a).

Characteristics: Cozumelan golden bat is medium sized; forearm length 53.0–61.0 mm, skull length 25.3–27.4 mm. **Dorsal and ventral pelage brilliant old gold; no dorsal white stripe**; fur of underparts paler than in the dorsum; **noseleaf large and wide, "serrated," not merged with the upper lip; small warts at both sides of a V-shaped cushion**; calcar greater than hind limb length; tail long, one-third of the uropatagium length; **auditory bullae small;** expanded and slim; rostrum wide; basisphenoid cavities wide but not deep, separated by a middle septum; small projection in the anterior upper part of the rostrum barely developed; one pair of lower incisors and two lower premolars.

Comments: *Mimon cozumelae* was elevated to full species (Burton and Engstrom 2001; Simmons 2005; Williams and Genoways 2008); it was previously a subspecies of *M. bennettii* (Gregorin et al. 2008). *M. cozumelae* ranges from southern Veracruz, Oaxaca, Campeche, Chiapas, Tabasco, and the Yucatán Peninsula, including Cozumel Island, south through South America (Map 75). It can be distinguished from other species in sympatry, or nearly so, such as those of the same genus, by its large, wide, and "serrated" noseleaf not merged with the upper lip, with small warts at both sides of a V-shaped cushion; calcar greater than the hind limb length and long tail, one-third of the uropatagium length; skull expanded and slim; rostrum wide; basisphenoid cavities wide but not deep, separated by a middle septum; auditory bullae small with a small anterior projection. From *M. crenulatum,* by its dorsal and ventral pelage brilliant old gold and small auditory bullae.

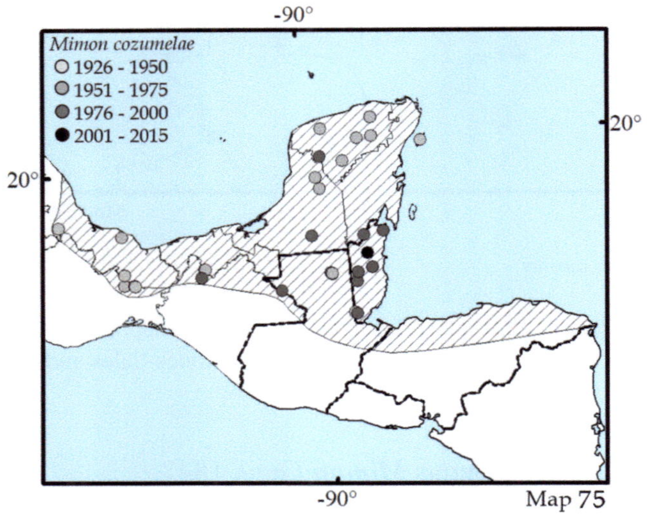

Map 75 Distribution map of *Mimon cozumelae*

Additional Literature: Schaldach (1964), Gardner et al. (1970), Álvarez-Castañeda and Álvarez (1991), Koopman (1993), Ortega and Arita (1997).

Genus *Vampyrum* Rafinesque, 1815

Largest bat in North America; ears moderately large, widely separated; noseleaf well developed, although simple; noseleaf and ears in paler color; **muzzle markedly elongated**; tail absent; **wingspan approximately 1000 mm**; calcar on average 37.0 mm; uropatagium very wide; **middle upper incisors low and wide, their cutting edges strongly oblique but entire, almost in contact with the canines laterally**; outer incisors pushed slightly forward out of the tooth row, barely rising to the level of the cingulum of the canines; upper molars with the outer cusps very large, markedly higher than the protocone, due to the unusual elongation of the outer cusps rather than the suppression of the protocone**; rostrum subcylindrical;** braincase breadth less than one-third of the greatest length; **sagittal crest well developed, especially in the occipital region**; paroccipital expansions distinct, although not very large, strongly concave beneath; auditory bullae very small, covering less than one-half of the cochlear surface, height at the inner edge less than the diameter; **symphysis menti very long, jaws strongly compressed above it**.

Vampyrum spectrum (Linnaeus, 1758)

Spectral bat, murciélago espectral

Monotypic.

Conservation: *Vampyrum spectrum* is listed as Endangered in the Norma Oficial Mexicana (DOF 2019) and as Near Threatened by the IUCN (Solari 2018p).

Characteristics: Spectral bat is the largest bat in North America; **forearm length 100.0–107.0 mm, skull length 49.2–52.2 mm.** Dorsal pelage orange to dark brown; slightly notorious stripe from the shoulders to the hips; fur of underparts grayish brown; noseleaf and ears in paler color; **wingspan approximately 1000 mm**; calcar on average 37.0 mm; skull robust with the sagittal crest well developed, projecting backward at the foramen magnum level; upper inner incisors well developed; **distance between the second upper molar (palatine) internal borders less than the diameter of the same molars**.

Comments: *Vampyrum spectrum* ranges in Chiapas, Oaxaca, and Veracruz, México south through South America (Map 76). It can be distinguished from other species in sympatry, or nearly so, such as those of the same genus, by its largest size, making it the largest bat in North America, forearm length greater than 95.0 mm and skull length greater than 49.2 mm; distance between second upper molar (palatine) internal borders less than transverse diameter of the same molars.

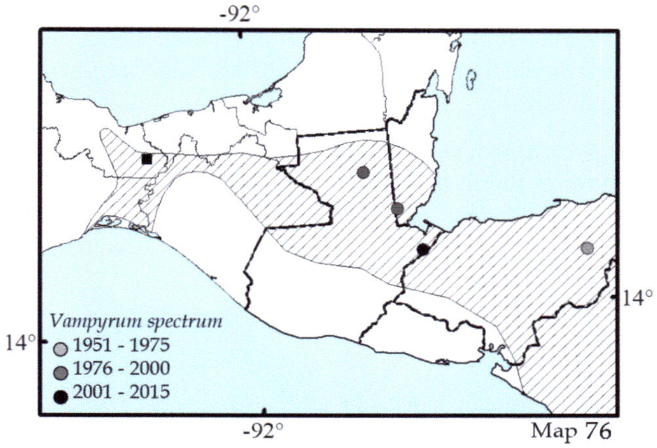

Map 76 Distribution map of *Vampyrum spectrum*

Additional Literature: Miller (1907), Goodwin and Greenhall (1961), Gardner (1977), Baker and Genoways (1978), Swanepoel and Genoways (1979), Navarro and Wilson (1982), Santos-Moreno and Alfaro Espinoza (2009), Santos-Moreno et al. (2010).

Subfamily Stenodermatinae Gervais, 1856

The subfamily Stenodermatinae is characterized by its medium to large size; noseleaf usually present but sometimes markedly reduced or absent; trochiter impinging on the scapula; upper molars with cusps and commissures so reduced that the W shaped pattern is obliterated; upper molars lacking styles; crowns of upper and lower molars markedly flattened and with well-developed cusps arising from the crushing surface. The keys were elaborated based on the review of specimens and the following sources: Lawlor (1979), Hall (1981), Álvarez et al. (1994), and Álvarez-Castañeda et al. (2017a).

1. Uropatagium absent and limbs very hairy; molar crowns with well-developed cusps on both margins of one central groove (Fig. 32).. tribe Sturnirini *Sturnira* (p. 319)

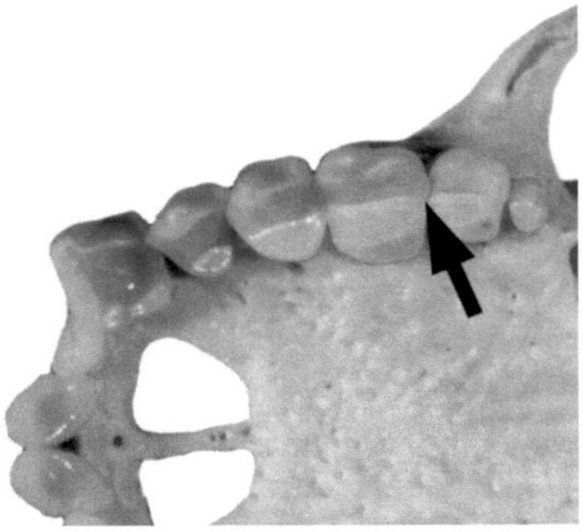

Fig. 32 Molar crowns with well-developed cusps on both margins of one central groove like in *Sturnira*

1a. Uropatagium present, sometimes markedly reduced, and limbs not very hairy; molar crown corrugated, not on both margins of one central groove .... tribe Stenodermatini 2

2. Rostrum bare and with skin folds; rostrum less than one-third of the braincase length; jaw in half-circle shape (Fig. 33); upper canines with a basal groove in the front side (from Sinaloa and southern Tamaulipas, México, south through South America) .... *Centurio senex* (p. 301)

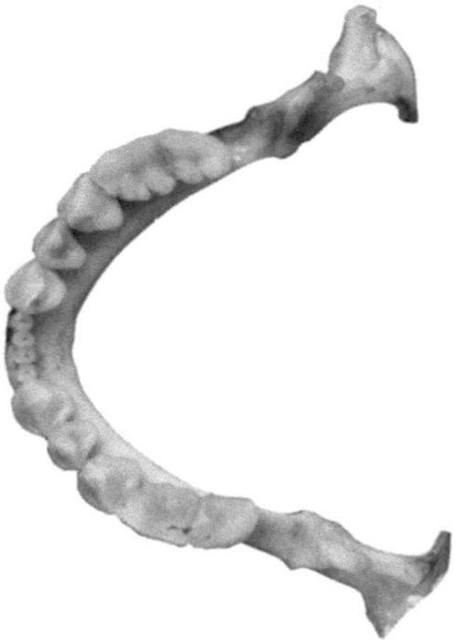

Fig. 33 The jaw is in half-circle shape like in *Centurio senex*

2a. Rostrum covered with hair; rostrum greater than one-third of the braincase length; jaw extended; upper canines without a basal groove in the front side 3

3. Nasal cavities extending close to the interorbital region (Fig. 34), giving the impression of no nasals .... *Chiroderma* (p. 302)

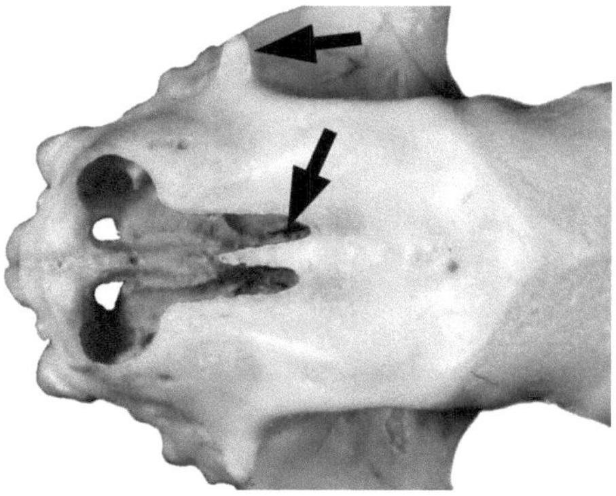

Fig. 34 Nasal cavities extend close to the interorbital region like in *Chiroderma*

3a. Nasal cavities normal, not extending backward, near the interorbital region 4

4. Each shoulder with a white patch and a second patch just behind the posterior insertion of the ears; dactylopatagium minus transparent (only known from Cozumel Island, México) .... *Phyllops falcatus* (p. 312)

4a. White patch on each shoulder and behind the ears absents; dactylopatagium minus not transparent 5

5. Upper inner incisors slightly larger than the outer incisors (but not twice in size) and not markedly different in shape 6

5a. Upper inner incisors at least twice the size of the outer incisors and differing in shape 9

6. Rostral length approximately three-fourths of the braincase length, and its height at the level of the second premolars greater than one-half of the braincase height (Fig. 35) .... *Uroderma* (p. 314)

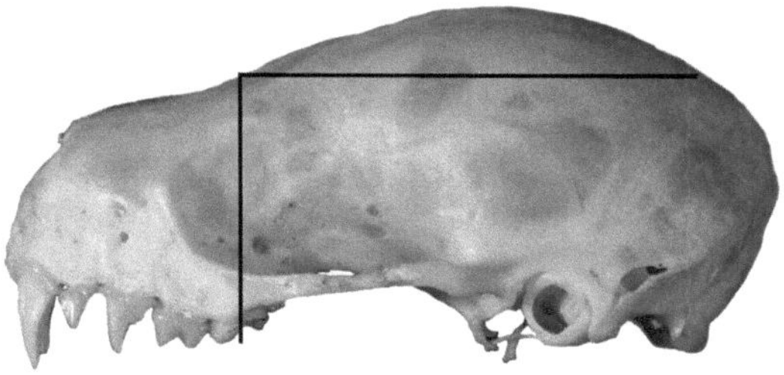

Fig. 35 Rostral length approximately three-fourths of the braincase length, and its height at the level of the second premolars greater than one-half of the braincase height

6a. Rostral length slightly greater than one-half of the braincase length, and its height less than one-half of the braincase length; rostrum height at the level of the second premolars less than one-half of the braincase height (Fig. 36) 7

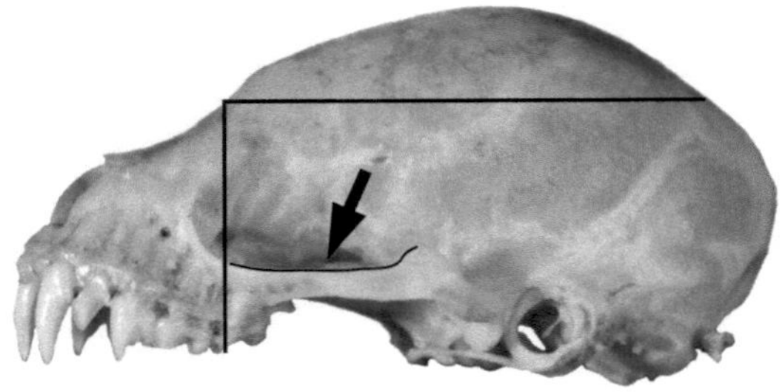

Fig. 36 Rostrum height at the level of the second premolars less than one-half of the braincase height

7. Facial bands in suede color; inner upper incisors unicuspid; third lower molars well developed, making the maxillary bone look bulky (Fig. 37; from Jalisco and Tamaulipas, México, south through South America)... *Enchisthenes hartii* (p. 311)

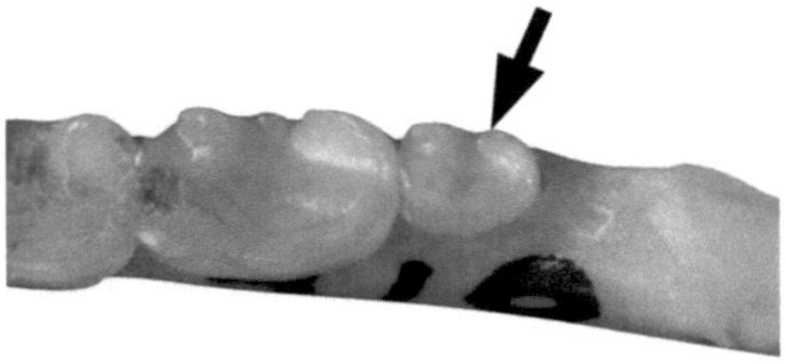

Fig. 37 The third lower molars well developed, making the maxillary bone look bulky like in *Enchisthenes hartii*

7a. Facial bands whitish; inner upper incisors forked; third lower molars absent, when present, they are so small that do not make the bone bulky (Fig. 38) 8

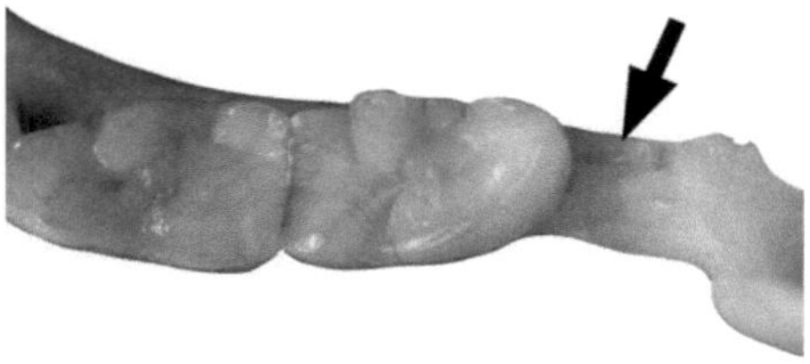

Fig. 38 The third lower molars are absent, when present, they are so small that do not make the bone bulky

8. Forearm length less than 48.0 mm; skull length less than 24.0 mm ....*Dermanura* (p. 306)

8a. Forearm length greater than 48.0 mm; skull length greater than 24.0 mm...*Artibeus* (p. 297)

9. Forearm length less than 35.0 mm; first lower molars without inner and posterior cusps (Fig. 39); occlusal figure similar to that of the preceding tooth (from Oaxaca, Veracruz, and Campeche, México, south through South America)............. ....*Vampyressa thyone* (p. 318)

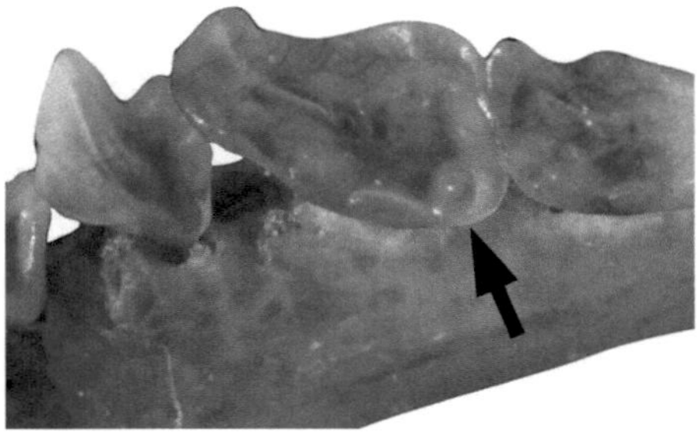

Fig. 39 The first lower molars without inner and posterior cusps like in *Vampyressa thyone*

9a. Forearm length greater than 35.0 mm; first lower molars with inner and posterior cusps well developed (Fig. 40); occlusal figure markedly different from that of the preceding tooth 10

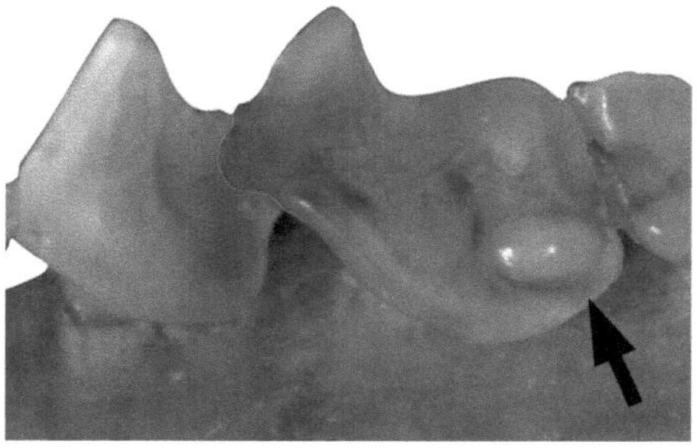

Fig. 40 The first lower molars with inner and posterior cusps well developed

10. Forearm less than 45.0 mm; three upper molars at each side, second upper molars with a large metacone (from Veracruz and Oaxaca, México, south through South America, except for the Yucatán Peninsula) .... *Platyrrhinus helleri* (p. 313)

10a. Forearm greater than 45.0 mm; two upper molars at each side, the second upper molars without a metacone (from southern Veracruz and Oaxaca, México, south through South America) .... *Vampyrodes major* (p. 318)

Tribe Stenodermatini

Genus *Artibeus* Leach, 1821

Medium to large sized; dorsal and ventral pelage brown in different shades to grayish brown; shoulder spots absent; **facial stripes well marked but not on the back**; ears and noseleaf small; uropatagium with a barely noticeable U-shape **and with hair**; limbs hairy; tail absent; skull wide; rostrum almost as long as wide between the lacrimal bones and well distinguished from the braincase; palate broad, approximately one-half of its length to the glenoid pit; inner upper incisors bilobed and larger than the outer upper incisors; molars large, with a grinding surface. The keys were elaborated based on the review of specimens and the following sources: Hall (1981), Álvarez et al. (1994), and Álvarez-Castañeda et al. (2017a).

1. Forearm length usually greater than 64.0 mm; skull length usually greater than 30.0 mm; preorbital and postorbital processes well developed (Fig. 41; from Sinaloa, Durango, and Tamaulipas, México, south through South America) .... *Artibeus lituratus* (p. 300)

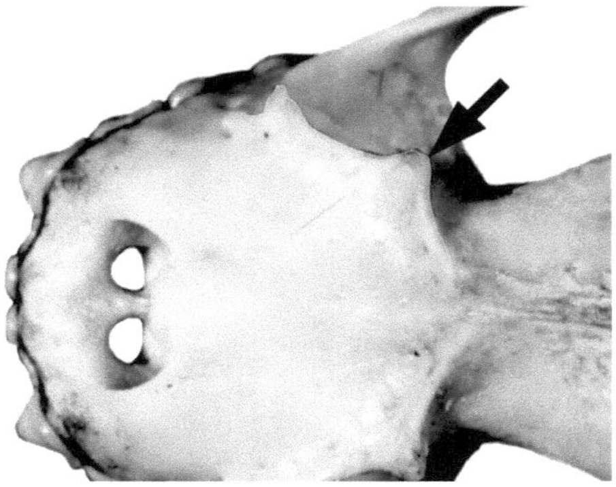

Fig. 41 Preorbital and postorbital processes are as well developed as in *Artibeus lituratus*

1a. Forearm length usually less than 64.0 mm; skull length usually less than 30.0 mm; preorbital and postorbital processes barely developed or absent (Fig. 42) 2

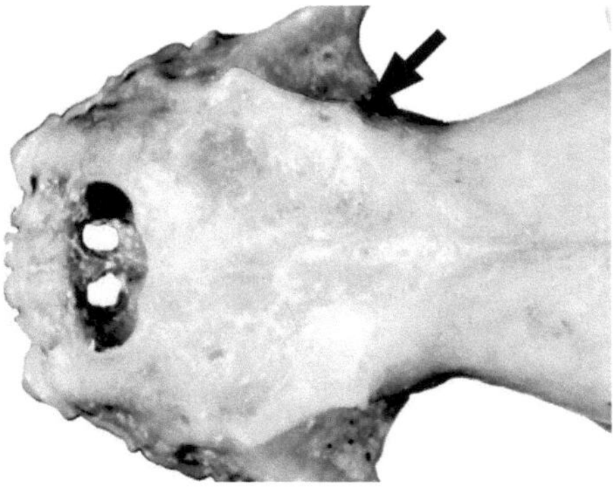

Fig. 42 Preorbital and postorbital processes are barely developed or absent

2. Uropatagium hairy, with hairs projecting backward from the membrane border as bangs; dorsal color silver gray (from Sonora south through Guerrero, México) .... *Artibeus hirsutus* (p. 398)

2a. Uropatagium bare and without bangs; dorsal color brown (from Tamaulipas, and Sinaloa, México, south through South America) .... *Artibeus jamaicensis* (p. 300)

Artibeus hirsutus Andersen, 1906

Hairy fruit-eating bat, murciélago frutero de patas peludas

Monotypic.

Conservation: *Artibeus hirsutus* is listed as Least Concern by the IUCN (Arroyo-Cabrales and Álvarez-Castañeda 2015a).

Characteristics: Hairy fruit-eating bat is large sized; forearm length 52.0–59.7 mm, skull length 25.6–27.3 mm. Dorsal pelage with silver gray shades; fur of underparts paler than in the dorsum; **uropatagium with hairs on the border; tibias with hairs**; preorbital and postorbital processes not well developed; sagittal crest present.

Comments: *Artibeus hirsutus* is an endemic species of México, ranging from Sonora to Guerrero (Map 77). It can be found in sympatry, or nearly so, with all the species of the subfamily Stenodermatinae. The genus *Artibeus* differs from other genera by its large size (forearm length greater than 52.0 mm); facial stripes well marked but absent on the back; uropatagium barely developed, U-shaped. From other species, by being the smallest species of *Artibeus* ranging in North America, with hairs on the borders of the uropatagium and tibias.

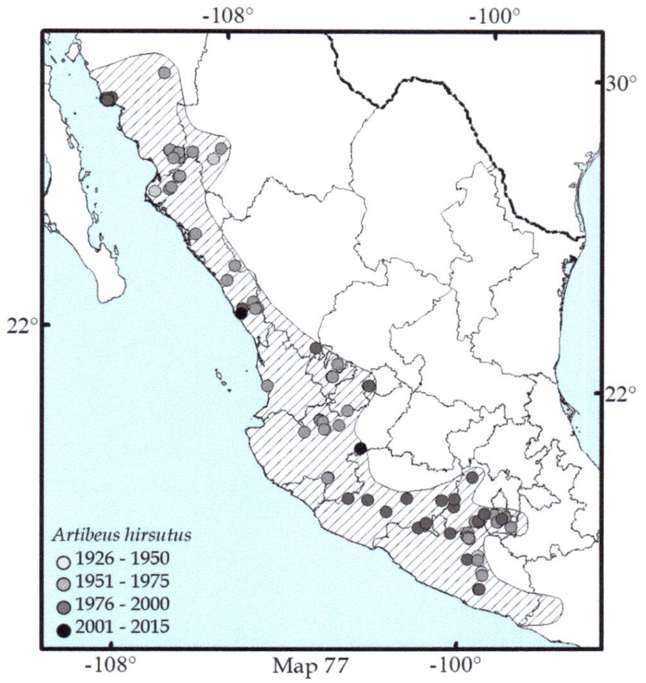

Map 77 Distribution map of *Artibeus hirsutus*

Additional Literature: Andersen (1908), Anderson (1960), Burt and Stirton (1961), Davis and Carter (1964), Webster and Jones (1983), Álvarez and Álvarez-Castañeda (1996), Guerrero et al. (2004), Álvarez-Castañeda et al. (2008c), Segura-Trujillo and Navarro-Pérez (2010).

Artibeus jamaicensis Leach, 1821

Jamaican fruit-eating bat, murciélago frutero común

1. *A. j. paulus* Davis, 1970. From southern coast of Tehuantepec isthmus to Chiapas.
2. *A. j. richardsoni* Allen, 1908. Central and northern Chiapas.
3. *A. j. triomylus* Handley, 1966. From Sinaloa south through Tehuantepec isthmus.
4. *A. j. yucatanicus* Allen, 1904. From southern Tamaulipas south through the Yucatán Peninsula.

Conservation: *Artibeus jamaicensis* is listed as Least Concern by the IUCN (Miller et al. 2016e).

Characteristics: Jamaica fruit-eating bat is large sized; forearm length 55.0–64.0 mm, skull length 26.2–31.6 mm. Dorsal pelage gray to grayish brown; fur of underparts paler than in the dorsum; uropatagium border hairless; limbs bare; preorbital and postorbital processes not well developed; sagittal crest present; third lower molars in the two mandibular branches.

Comments: *Artibeus triomylus* is considered a junior synonym of *A. jamaicensis* (Larsen et al. 2007, 2010); it was previously regarded as full species (Guerrero et al. 2004). This species is probably the most common bat in tropical North America; it can be found throughout the lowlands from Sinaloa and Tamaulipas south through South America (Map 78). It can be found in sympatry, or nearly so, with all the species of the subfamily Stenodermatinae. The genus *Artibeus* differs from other genera by its large size (forearm length greater than 52.0 mm), the facial stripes are well marked but absent on the back and the uropatagium is barely developed, U-shaped. From the other species, by not having the uropatagium and tibias with hair on the borders and the preorbital and postorbital processes not well developed.

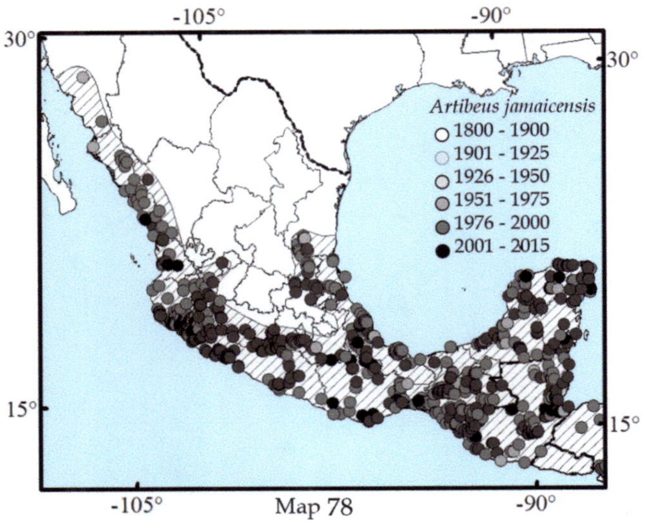

Map 78 Distribution map of *Artibeus jamaicensis*

Additional Literature: Dalquest et al. (1952), Davis and Carter (1964), Handley (1966), Jones and Carter (1976), Jones (1978), Álvarez-Castañeda and Álvarez (1991), Handley (1991), Álvarez-Castañeda (1996), Ortega and Castro-Arellano (2001).

Artibeus lituratus (Olfers, 1818)

Great fruit-eating bat, murciélago frutero gigante

1. *A. l. koopmani* Wilson, 1991. Restricted to Islas Marías, Nayarit.
2. *A. l. palmarum* Allen and Chapman, 1897. From Sinaloa and Veracruz southward to all tropical México.

Conservation: *Artibeus lituratus* is listed as Least Concern by the IUCN (Barquez et al. 2015b).

Characteristics: Great fruit-eating bat is large sized; **forearm length 69.0–78.0 mm, skull length 29.7–34.0 mm.** Dorsal pelage usually brown in different shades; fur of underparts grayish brown; uropatagium border hairless; limbs bare; **preorbital and postorbital processes well developed**; sagittal crest well marked.

Comments: *Artibeus intermedius* is considered junior synonym of *A. lituratus* (Redondo et al. 2008, Marchán–Rivadeneira et al. 2012), previously a full species (Davis 1984). *A. lituratus* range in all warm lowlands from Sinaloa and Tamaulipas south through South America (Map 79). It can be found in sympatry, or nearly so, with all the species of the subfamily Stenodermatinae. The genus *Artibeus* differs from the other genera in its large size (forearm length greater than 52.0 mm); facial stripes well marked but absent on the back, and the uropatagium little developed in U-shape. *A. lituratus* differs from other *Artibeus* species in being the largest species in North America; with preorbital and postorbital processes well developed.

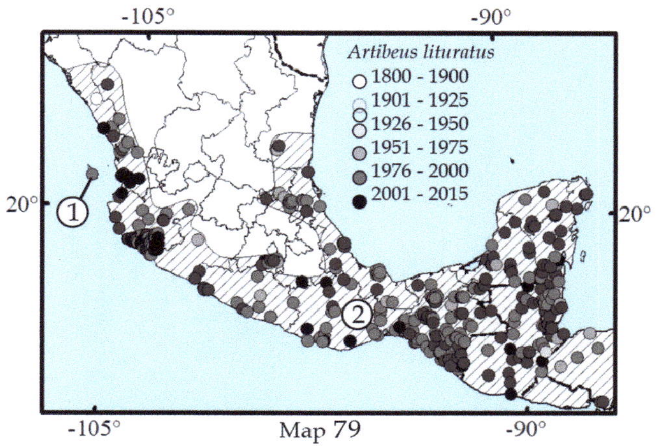

Map 79 Distribution map of *Artibeus lituratus*

Additional Literature: Miller (1907), Andersen (1908), Goodwin and Greenhall (1961), Álvarez-Castañeda and Álvarez (1991), Álvarez-Castañeda (1996), Guerrero et al. (2004), Simmons (2005).

Genus *Centurio* Gray, 1842

Medium sized; ears **with a conspicuous inner lappet; no true noseleaf**, but the entire face and neck complicated by a mass of naked dermal outgrowths; **rostrum very short and bare**; uropatagium hairy; tail absent; **25 or more white stripes between the fourth and fifth wing fingers; shoulders with a white spot;** uropatagium moderately wide; tail absent; calcar well developed; **upper canines with a cavity on their base; rostrum and palate extremely short; palate length approximately one-half of its breadth; skull rounded;** auditory bullae very small, covering much less than one-half of the cochlear surface; **jaw in half-circle shape**.

Centurio senex Gray, 1842

Wrinkle-faced bat, murciélago cara de viejo

C. s. senex Gray, 1842. Only subspecies in North America.

Conservation: *Centurio senex* is listed as Least Concern by the IUCN (Miller et al. 2016f).

Characteristics: Wrinkle-faced bat is medium sized; forearm length 41.0–45.0 mm, skull length 17.3–18.9 mm. Dorsal pelage yellowish brown to grayish brown; without facial or dorsal stripes; fur of underparts paler than in the dorsum; ears yellowish; **noseleaf absent;** uropatagium hairy and well developed; tail absent; **rostrum bare and very short**; **more than 25 white stripes between the fourth and fifth wing fingers; shoulders with a white spot**; **rostrum and palate extremely short; palate length approximately one-half of its breadth; upper canines with a cavity on their base; skull rounded; jaw in half-circle shape**.

Comments: *Centurio senex* ranges from Sinaloa and southern Tamaulipas south through South America (Map 80). It can be distinguished from other species in sympatry, or nearly so, such as those of the same genus, by its very short and naked rostrum; they have more than 25 white stripes on the wings, between the fourth and fifth fingers and a white spot on the shoulders; the rostrum and palate are extremely short; the upper canines have a cavity on their base, and the jaw is shaped as a half-circle.

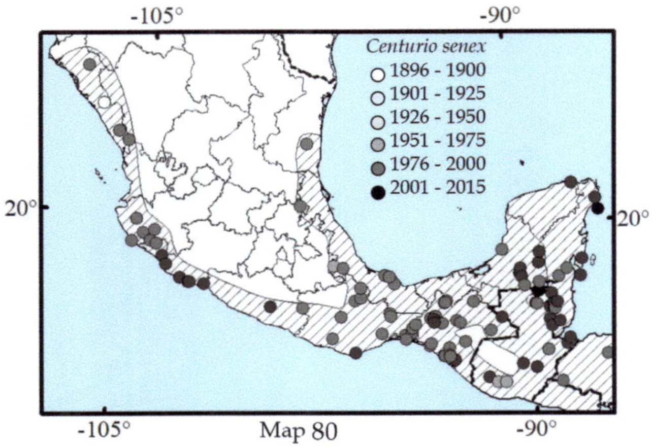

Map 80 Distribution map of *Centurio senex*

Additional Literature: Miller (1907), Goodwin and Greenhall (1961), Paradiso (1967), Jones et al. (1971), Snow et al. (1980), Medellín et al (2000), Villalpando and Álvarez (2000), Arroyo-Cabrales et al. (2008).

Genus *Chiroderma* Peters, 1860

Medium sized; dorsal and ventral pelage dense buff to dark brown or dark gray; individual hairs with three well-defined bands, the middle band always wider and paler than the basal and terminal bands; shoulder spots absent; **four facial stripes present in most individuals; the dorsal stripe from the interscapular region to the base of the uropatagium conspicuous or faint** (may be absent in some *C. villosum*); ears and noseleaf small, without borders in paler shades; **uropatagium well developed and with little hair but not fringed**; tail absent; **nasals markedly reduced giving the appearance of being absent, their place occupied by an emargination extending back from the nares to between the orbits; rostrum not clearly differentiated from the braincase**; inner upper incisor teeth different from the external upper incisors, which are less than one-half the size of the inner upper incisors; **lacrimal processes well developed**. The keys were elaborated based on the review of specimens and the following sources: Hall (1981), Álvarez et al. (1994), Álvarez-Castañeda et al. (2017a), and Garbino et al. (2020).

1. Facials and dorsal stripes slightly marked or absent; noseleaf spear with a notched tip and uniformly brown with the lateral margins slightly pale; nasal notch long, extending back into the interorbital region, close to the postorbital processes; rostrum relatively short; lacrimal process less developed (Fig. 43; from Veracruz, and Oaxaca, México, south through South America) .... *Chiroderma villosum* (p. 305)

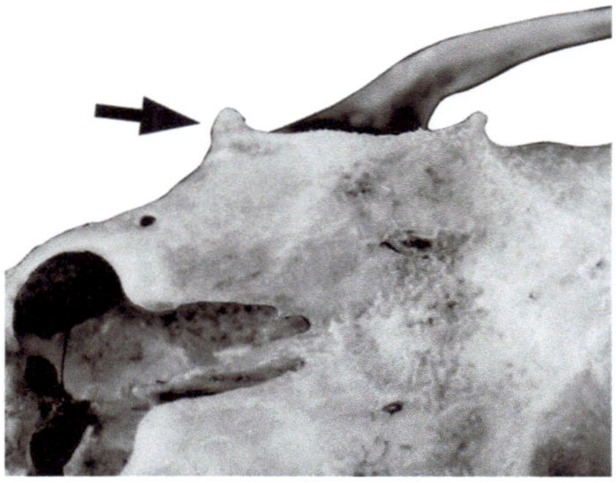

Fig. 43 Lacrimal process less developed like in *Chiroderma villosum*

1a. Facial and dorsal stripes well marked; noseleaf spear with a simple tip and lateral margins pale; nasal notch short, ending near the level of the postorbital constriction; relatively long rostrum; lacrimal process well developed (Fig. 34)....2

2. Skull length on average greater than 25.0 mm; nasal notch extending posterior to the interorbital region; height of the lower canine crown relatively sorter and wider; first lower premolar proportional larger (Fig. 44; from Veracruz and northwestern Oaxaca, México, south through South America) .... *Chiroderma salvini* (p. 303)

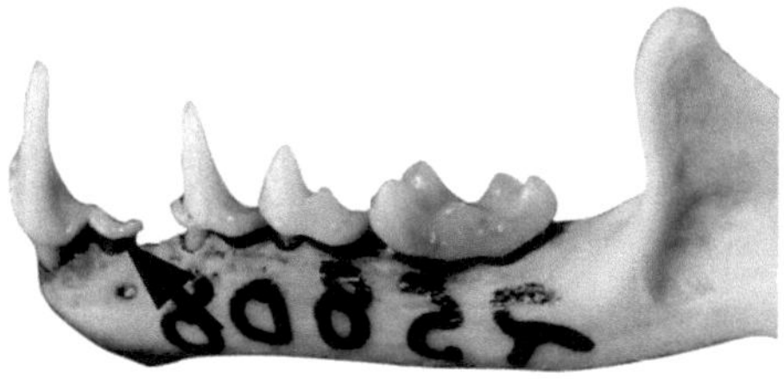

Fig. 44 The first lower premolar proportional as larger as in *Chiroderma salvini*

2a. Skull length on average less than 25.0 mm; nasal notch reaching only the interorbital region anterior margin; height of the lower canine crown relatively larger and narrow; first lower premolar proportional smaller (Fig. 45; from Sinaloa and Durango south through Oaxaca, México) .... *Chiroderma scopaeum* (p. 304)

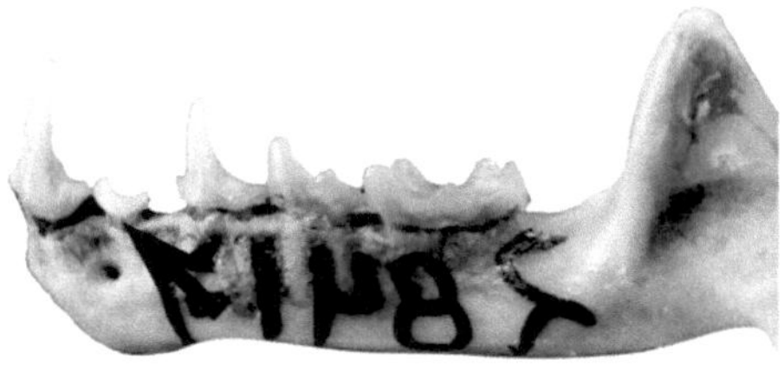

Fig. 45 The first lower premolar proportional as smaller as in *Chiroderma scopaeum*

Chiroderma salvini Dobson, 1878

Salvin's big-eyed bat, murciélago de grandes ojos pardo

Monotypic.

Conservation: *Chiroderma salvini* is listed as Least Concern by the IUCN (Aguirre et al. 2015).

Characteristics: Salvin's big-eyed bat is medium sized; forearm length 45.0–53.0 mm, skull length 25.2–28.3 mm. Dorsal pelage dark brown; fur of underparts pale grayish brown; no spots on the shoulders; **facial and dorsal stripes well marked**; noseleaf spear with a simple tip and lateral margins pale; lacrimal **processes well developed; rostrum relatively long; nasal notch short, extending posterior to the interorbital region**.

Comments: *Chiroderma scopaeum* was elevated to full species (Garbino et al. 2020); it was previously regarded as a subspecies of *C. salvini* (Handley 1966). *C. salvini* ranges from Veracruz and northwestern Oaxaca south through South America, except for the Yucatán Peninsula (Map 81). This species can be differentiated from other sympatric genera by its well-marked facial and dorsal stripes; uropatagium developed and with little hair but not fringed; nasals are much reduced, giving the appearance of being absent. With the skull in hand, the differentiation is very easy, but externally it can be confused with *Platyrrhinus*, *Uroderma*, and *Vampyrodes*, as all these genera have stripes on the face and a white dorsal stripe. *C. salvini* can be differentiated from *Platyrrhinus helleri* by its larger size; forearm more than 45.0 mm, and the uropatagium developed with little hair and not fringed with white hair. From *Uroderma convexum*, *U. davisi*, and *U. magnirostrum* in the uropatagium developed and with little hair. From *Vampyrodes caraccioli* by having the uropatagium developed with little hair and not fringed with white hair. From *C. scopaeum*, by having the skull length on average greater than 25.0 mm; the nasal noch extends posterior to the interorbital region; lower canine crown approximately of the same height as the coronoid process. From *C. villosum*, by having the facial and dorsal stripes well developed; longer forearms on average; basal and apical bands of the dorsal hair of the same color; noseleaf spear with a simple tip and lateral margins pale; a relatively longer rostrum and a shorter nasal notch; postorbital processes less pointed.

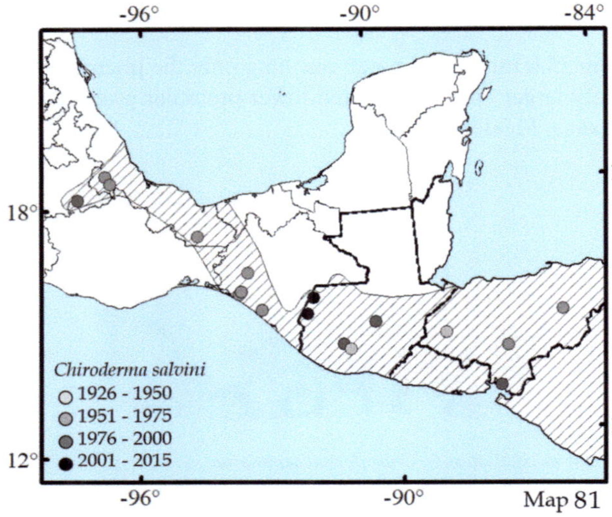

Map 81 Distribution map of *Chiroderma salvini*

Additional Literature: Goodwin (1958), Álvarez-Castañeda and Álvarez (1991), González-Ruíz et al. (2000), Muñiz-Martínez et al. (2003), Segura-Trujillo and Navarro-Pérez (2010).

Chiroderma scopaeum Handley, 1966

Dwarf big-eyed bat, murciélago de grandes ojos enano

Monotypic.

Conservation: *Chiroderma scopaeum* (as *Chiroderma salvini*) is listed as Least Concern by the IUCN (Aguirre et al. 2015).

Characteristics: Dwarf big-eyed bat is medium sized; forearm length 43.5–46.9 mm, skull length 23.6–25.4 mm. Dorsal pelage pale brown to dark brown; no spots on the shoulders; **facial and dorsal stripes well marked, a dorsal stripe extending from the interscapular region to the posterior end of the body**; noseleaf spear with a simple tip and brown with lateral margins whitish; fur of underparts pale grayish brown; **lacrimal processes well developed**; braincase rounded and less massive than in *C. salvini*; postorbital constriction relatively wide; **rostrum relatively long; nasal notch short, reaching only its interorbital anterior margin**.

Comments: *Chiroderma scopaeum* was elevated to full species (Garbino et al. 2020); it was previously a subspecies of *C. salvini* (Handley 1966). *C. scopaeum* ranges from Sinaloa and Durango south through Oaxaca (Map 82). This species can be differentiated from other sympatric genera by its well-marked facial and dorsal stripes; uropatagium developed and with little hair but not fringed; nasals markedly reduced, giving the appearance of being absent. With the skull in hand, the differentiation is very easy, but externally it can be confused with *Uroderma* and *Vampyrodes*, as all these genera have facial stripes and a white dorsal stripe. *C. scopaeum* can be differentiated from *Uroderma magnirostrum* by the uropatagium developed and with little hair. From *Vampyrodes caraccioli,* by the uropatagium developed with little hair and not fringed with white hair. From *C. salvini,* by having on average a smaller size and paler dorsal pelage, skull length on average less than 25.0 mm; a relatively broader postorbital constriction; the nasal notch reaches only its interorbital anterior margin; the lower canine crown height relative shorter than the coronoid process. From *C. villosum,* because the facial and dorsal stripes are well developed; longer forearms on average; basal and apical bands of the dorsal hair of the same color; noseleaf spear with a simple tip and lateral margins pale; relatively longer rostrum and a shorter nasal notch; postorbital processes less pointed.

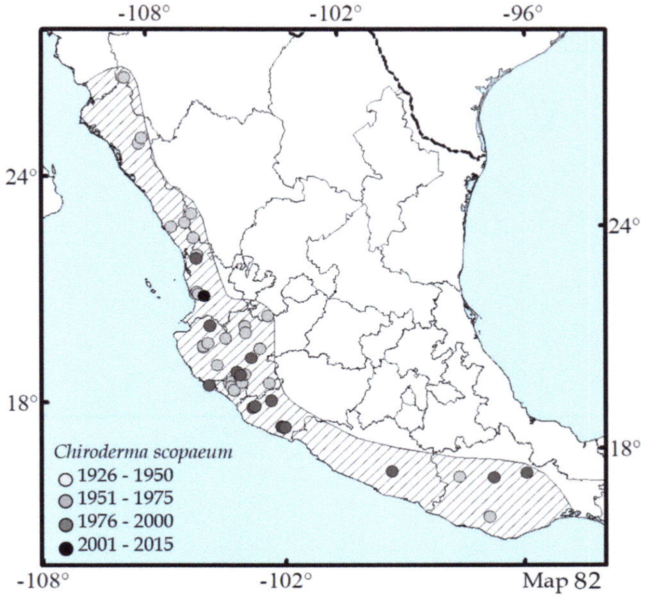

Map 82 Distribution map of *Chiroderma scopaeum*

Additional Literature: Goodwin (1958), Álvarez-Castañeda and Álvarez (1991), González-Ruíz et al. (2000), Muñiz-Martínez et al. (2003), Segura-Trujillo and Navarro-Pérez (2010).

Chiroderma villosum Peters, 1860

Hairy big-eyed bat, murciélago de grandes ojos güero

C. v. jesupi Allen, 1900. Only subspecies in North America.

Conservation: *Chiroderma villosum* is listed as Least Concern by the IUCN (Solari 2015a).

Characteristics: Hairy big-eyed bat is medium sized; forearm length 42.0–50.0 mm, skull length 24.0–27.2 mm. Dorsal pelage gray to grayish brown; **facial and dorsal stripes barely noticeable or absent**; fur of underparts pale grayish brown; **noseleaf tip notched in about three-quarters and nearly uniformly pale brown, with the central noseleaf rib of a more pinkish shade and the lateral margin slightly paler**; lacrimal **processes barely developed**; braincase relatively deep and globose; orbits relatively large; sagittal crest sometimes present; nasal notch long and extending back into the interorbital region close to the postorbital processes.

Comments: *Chiroderma villosum* ranges from Veracruz and Oaxaca south through South America, including the Yucatán Peninsula (Map 83). This species can be differentiated from other sympatric genera by its facial and dorsal stripes barely noticeable or absent; uropatagium developed and with little hair but not fringed; nasals are markedly reduced, giving the appearance of being absent. With the skull in hand, the differentiation is very easy, but externally it can be confused in the field with *Platyrrhinus*, *Uroderma,* and *Vampyrodes*, as all these genera have facial stripes and white dorsal stripes, which are less notorious on *C. villosum*. *C. villosum* can be differentiated from *Platyrrhinus helleri* by its larger size; forearm length more than 41.0 mm and the uropatagium developed with little hair and not fringed with white hair. From *Uroderma convexum, U. davisi,* and *U. magnirostrum*, by the uropatagium developed and with little hair. From *Vampyrodes caraccioli,* by the uropatagium developed with little hair and not fringed with white hair, in many cases because the dorsal stripe is not clear. *C. villosum* can be confused with *Dermanura,* but in *Dermanura* the facial stripes are well marked. *C. villosum* can be differentiated from *C. salvini* and *C. scopaeum* by having the facial and dorsal stripes slightly marked or absent; forearm shorter on average; apical bands of dorsal hair lighter than at the tip; noseleaf spear with a notched tip and uniformly brown; relatively shorter rostrum and longer nasal notch, ending near the postorbital constriction level; postorbital processes pointed.

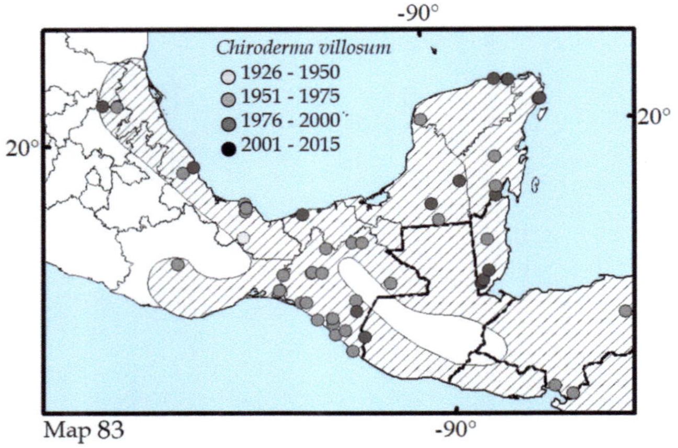

Map 83 Distribution map of *Chiroderma villosum*

Additional Literature: Goodwin (1958), Handley (1960), Álvarez-Castañeda and Álvarez (1991), Medellín et al. (2000), Naranjo Piñera et al. (2005), Mejenes-López et al. (2010), Garbino et al. (2020).

Genus *Dermanura* Gervais, 1856

Medium sized; dorsal and ventral pelage pale-colored to dark brown or grayish brown; **facial stripes well marked but not on the back**; ears and noseleaf small; **uropatagium barely developed, U-shaped,** and with or without hairs in the border; limbs hairy or hairless; tail absent; skull wide; rostrum almost as long as wide between the lacrimal bones, clearly distinguished from the braincase; palate broad, approximately one-half of its length to the glenoid fossa; inner upper incisors bilobed and larger than the outer upper incisors; molars large and with a grinding surface. The keys were elaborated based on the review of specimens and the following sources: Lawlor (1979), Hall (1981), Álvarez et al. (1994), and Álvarez-Castañeda et al. (2017a).

1. Uropatagium less than 7.0 mm broad and with the posterior border clearly hairy (Fig. 46)..2

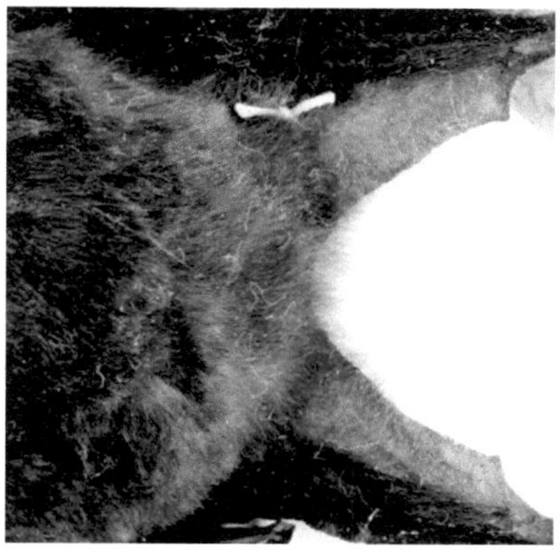

Fig. 46 Uropatagium broad and with the posterior border clearly hairy

1a. Uropatagium greater than 7.0 mm broad and with the posterior border almost bare (Fig. 47) 3

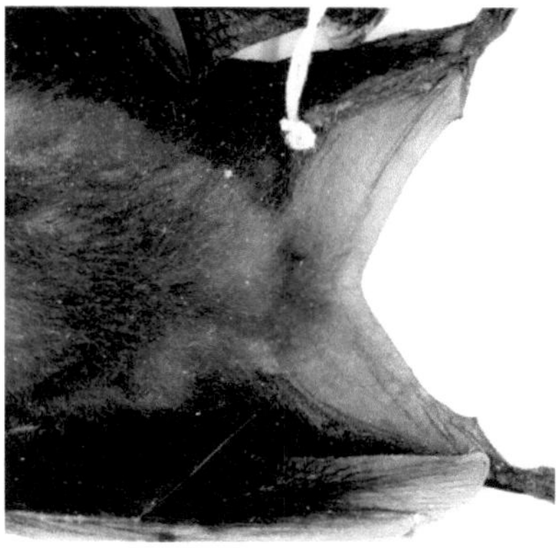

Fig. 47 Uropatagium greater than 7.0 mm broad and with the posterior border almost bare

2. Forearm length less than 42.0 mm; uropatagium with scarce hair; skull usually less than 21.0 mm; maxillary toothrow length usually less than 7.0 mm (from Sinaloa, Veracruz, and Nuevo León, México, south through Central America) ....*Dermanura tolteca* (p. 309)

2a. Forearm length greater than 42.0 mm; uropatagium with abundant hair; skull usually greater than 21.0 mm; maxillary toothrow length usually greater than 7.0 mm (from Sinaloa and Tamaulipas, México, south Central America).............. .... *Dermanura azteca* (p. 308)

3. Uropatagium breadth at the level of the upper calcar greater than the palm breadth of the hind limbs (characteristics measured live); two upper and three lower molars; first upper molars with the talon cusp always thin; yugal without a process in the dorsal side (Fig. 36; from southern Veracruz, Tabasco, and the southern Yucatán Peninsula, México, south through South America)..*Dermanura watsoni* (p. 310)

3a. Uropatagium breadth at the level of the calcar same as the palm breadth of the hind limbs (characteristics measured live); two lower and two upper molars; first upper molars with a wide or thin talon cusp; yugal showing a process on the dorsal side (from Sinaloa and Veracruz, México, south through South America).. ....*Dermanura phaeotis* (p. 308)

Dermanura azteca (Andersen, 1906)

Aztec fruit-eating bat, murciélago frutero menor de patas peludas

1. *D. a. azteca* (Andersen, 1906). From Southern Sinaloa, and Nuevo León and Tamaulipas south through Oaxaca.
2. *D. a. minor* (Davis, 1969). Restricted to Chiapas.

Conservation: *Dermanura azteca* is listed as Least Concern by the IUCN (Solari 2016b).

Characteristics: Aztec Fruit-eating bat is medium sized; **forearm length 42.0–49.0 mm, skull length 21.0–23.8 mm**. Dorsal pelage blackish; fur of underparts paler than in the dorsum; ears and noseleaf small; **uropatagium very narrow, less than 7.0 mm, with abundant hair, covering it almost completely**; limbs hairy; skull and toothrow length greater than 21.0 and 7.0 mm, respectively; ranging usually above 1500 m.

Comments: *Dermanura* was reinstated to the generic rank by Van Den Bussche et al. (1998); it was previously considered a subgenus of *Artibeus* (Simmons 1993). *D. azteca* ranges in the midlands and highlands in the Sierras Madres, from Sinaloa and Tamaulipas south to Central America (Map 84). Within the subfamily Stenodermatinae, *D. azteca* is likely the species with the least sympatric distribution with other species because of its highland habitat. The genus *Dermanura* differs from the other genera by its medium size (forearm length less than 49.0 mm); facial stripes well marked but absent on the back, and the uropatagium slightly developed, U-shaped. *D. azteca* differs from other *Dermanura* species by being the largest (forearm and skull length greater than 41.0 and 21.0 mm, respectively); uropatagium hairy and very narrow, less than 7.0 mm.

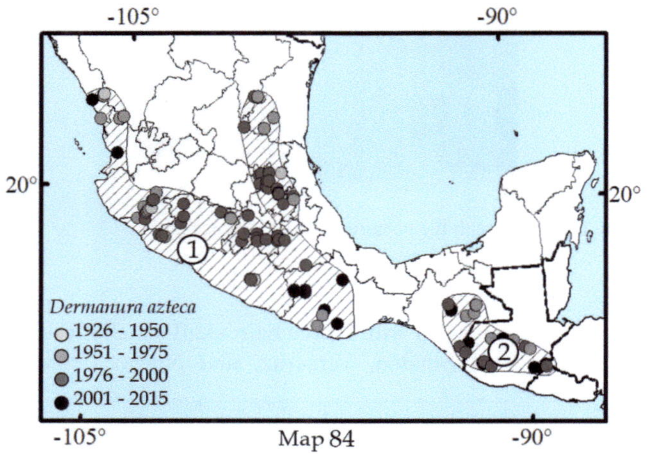

Map 84 Distribution map of *Dermanura azteca*

Additional Literature: Andersen (1906b, 1908), Davis (1958, 1969), Jones and Carter (1976), Webster and Jones (1982c), Álvarez-Castañeda and Álvarez (1991), Álvarez-Castañeda (1996), Peterson et al. (2004), Guerrero et al. (2008), Hoofer et al. (2008), Solari et al. (2009).

Dermanura phaeotis Miller, 1902

Pygmy fruit-eating bat, murciélago frutero pigmeo

1. *D. p. nana* (Andersen, 1906). From Sinaloa south through Oaxaca.
2. *D. p. palatina* (Davis, 1970). Southern coastal of Oaxaca and Chiapas.
3. *D. p. phaeotis* Miller, 1902. From Veracruz and Oaxaca east to northern Chiapas, including the Yucatán Peninsula.

Conservation: *Dermanura phaeotis* is listed as Least Concern by the IUCN (Miller et al. 2015a, b, c, d).

Characteristics: Pygmy fruit-eating bat is medium sized; forearm length 35.0–40.0 mm, skull length 17.5–20.5 mm. Dorsal pelage pale brown to grayish brown; fur of underparts paler than in the dorsum; with facial stripes well marked but not on the back; ears and noseleaf small; **uropatagium wide, more than 7.0 mm and with scarce hair**; limbs with scarce hair; tail absent; two upper and lower molars; **first upper molars with a wide or thin talon cusp**.

Comments: *Dermanura* was reinstated to the generic rank by Van Den Bussche et al. (1998); it was previously considered a subgenus of *Artibeus* (Simmons 1993). *Dermanura phaeotis* ranges from Sinaloa and southern Veracruz south through South America (Map 85). It can be found in sympatry, or nearly so, with all the species of the subfamily Stenodermatinae. The genus *Dermanura* differs from the other genera by its medium size (forearm length less than 49.0 mm); facial stripes well marked but absent on the back and the uropatagium barely developed, U-shaped. *D. phaeotis* differs from other *Dermanura* species by having a wide uropatagium (more than 7.0 mm) with scarce hair.

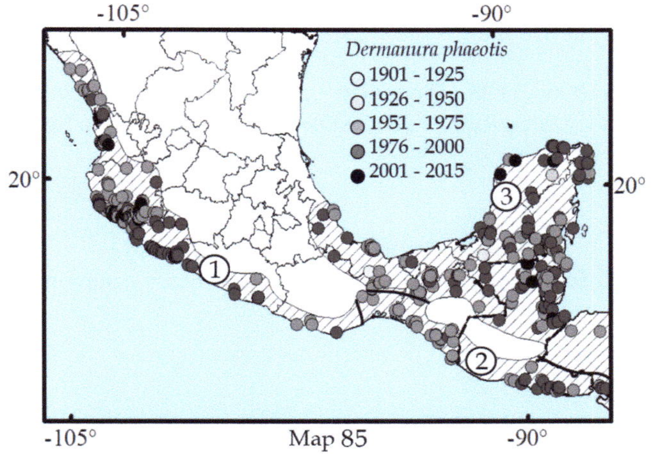

Map 85 Distribution map of *Dermanura phaeotis*

Additional Literature: Andersen (1906b, 1908), Goodwin (1934), Jones and Lawlor (1965), Jones (1966), Álvarez (1968), Davis (1969, 1970), Timm (1985), Álvarez-Castañeda and Álvarez (1991), Hoofer et al. (2008), Solari et al. (2009), Larsen et al. (2010).

Dermanura tolteca (Saussure, 1860)

Toltec fruit-eating bat, murciélago frutero menor de patas desnudas

1. *D. t. hespera* (Davis, 1969). From central Sinaloa south through Chiapas.
2. *D. t. tolteca* (Saussure, 1860) From Nuevo León south through northern Chiapas, including the Yucatán Peninsula.

Conservation: *Dermanura tolteca* is listed as Least Concern by the IUCN (Rodriguez and Cajas 2015).

Characteristics: Toltec fruit-eating bat is medium sized; forearm length 39.0–42.0 mm, skull length 19.0–21.0 mm. Dorsal pelage dark grayish brown; fur of underparts grayish pale brown; ears and noseleaf small; **uropatagium very narrow, less than 7.0 mm, covered with hair**; limbs hairy; skull and toothrow length less than 21.0 and 7.0 mm, respectively.

Comments: *Dermanura* was reinstated to the generic rank by Van Den Bussche et al. (1998) it was previously considered a subgenus of *Artibeus* (Simmons 1993). *Dermanura tolteca* ranges from central Sinaloa, Veracruz, and Nuevo León south through Central America, excluding the Yucatán Peninsula (Map 86). It can be found in sympatry, or nearly so, with all the species of the subfamily Stenodermatinae. The genus *Dermanura* differs from the other genera by its medium size (forearm length less than 49.0 mm); facial stripes well marked but absent on the back and the uropatagium barely developed, U-shaped. *D. tolteca* differs from other *Dermanura* species by having a narrow uropatagium (less than 7.0 mm).

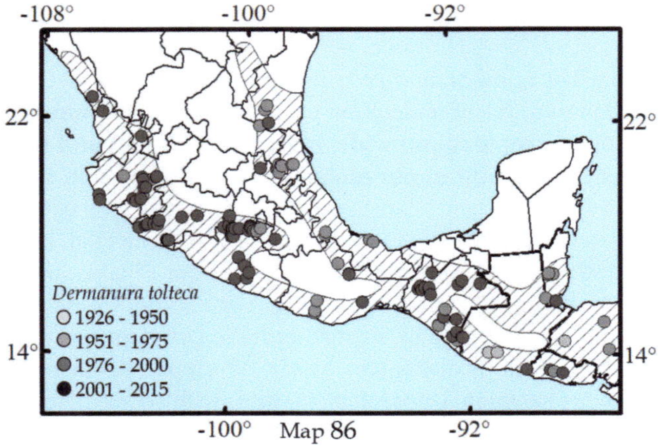

Map 86 Distribution map of *Dermanura tolteca*

Additional Literature: Andersen (1906b, 1908), Davis (1969), Webster and Jones (1982d); Álvarez-Castañeda and Álvarez (1991), Álvarez-Castañeda (1996), González-Ruíz et al. (2000), Hoofer et al. (2008), Solari et al. (2009), Williams-Guillén and Perfecto (2010).

Dermanura watsoni (Thomas, 1901)

Thomas's fruit-eating bat, murciélago frutero menor centroamericano

Monotypic.

Conservation: *Dermanura watsoni* is listed as subject to Special Protection in the Norma Oficial Mexicana (DOF 2019) and as Least Concern by the IUCN (Solari 2016c).

Characteristics: Thomas's lesser fruit-eating bat is medium sized; forearm length 35.0–41.0 mm, skull length 18.7–21.2 mm. Dorsum grayish brown; fur of underparts paler than in the dorsum; ears with a paler border; **uropatagium wide, more than 7.0 mm, naked**; limbs with very little hair; two upper and three lower molars; **first upper molars with the talon cusp always thin**.

Comments: *Dermanura* was reinstated to the generic rank by Van Den Bussche et al. (1998); it was previously considered a subgenus of *Artibeus* (Simmons 1993). *Dermanura watsoni* ranges from southern Veracruz, Tabasco, Campeche, and northern Chiapas south through South America (Map 87). It can be found in sympatry, or nearly so, with only those species of the subfamily Stenodermatinae ranging in tropical areas. The genus *Dermanura* differs from the other genera by its medium size (forearm length less than 49.0 mm); facial stripes well marked but absent on the back, and the uropatagium barely developed, U-shaped. *D. watsoni* differs from other *Dermanura* species by having a wide naked uropatagium (more than 7.0 mm) and the first upper molars with the talon cusp always thin.

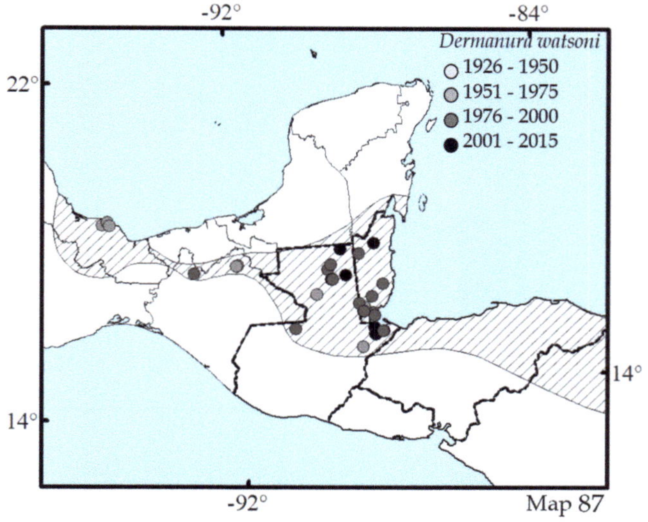

Map 87 Distribution map of *Dermanura watsoni*

Additional Literature: Andersen (1906b, 1908), Davis (1958, 1969), Álvarez-Castañeda and Álvarez (1991), Gonçalves da Silva et al. (2008), Hoofer et al. (2008), Solari et al. (2009).

Genus *Enchisthenes* K. Andersen, 1906

Small sized; **dorsal pelage dark chocolate brown, same coloration in the chin and neck**; **facial stripe faint in suede color but distinguishable**; short uropatagium with little hair but showing some on the border; tail absent; **inner upper incisors with a simple shape; third upper and lower molars well developed,** wide in appearance; molars big and with a grinding surface; rostrum almost as long as the breadth between the lacrimal bones and well differentiated from the braincase.

Enchisthenes hartii (Thomas, 1892)

Velvety fruit-eating bat, murciélago frutero negro

Monotypic.

Conservation: *Enchisthenes hartii* is listed as subject to Special Protection in the Norma Oficial Mexicana (DOF 2019) and as Least Concern by the IUCN (Solari 2018c).

Characteristics: Velvety fruit-eating bat is small sized; forearm length 38.0–42.0 mm, skull length 20.1–22.0 mm. **Dorsal pelage dark chocolate brown, of the same coloration in the chin and neck**; fur of underparts paler than in the dorsum; no spots on the shoulders; **facial stripe faint in suede color but distinguishable**; ears and noseleaf small, in blackish shades; short uropatagium with little hair but showing some on the border; limbs with little hair; tail absent; skull wide in appearance; rostrum almost as long as the breadth between the lacrimal bones and well differentiated from the braincase; palate wide, approximately one-half of its length to the glenoid pit; inner upper incisors not bilobed and larger than the outer upper incisors; molars big and with a grinding surface; **third upper and lower molars well developed**.

Comments: *Enchisthenes* was considered cospecific with *Artibeus* (Goodwin 1969) and reinstated to the generic rank (Hall 1981; Van Den Bussche et al. 1998). *E. hartii* ranges from Jalisco and Tamaulipas south through South America, except for northern Yucatán Peninsula (Map 88). It can be found in sympatry, or nearly so, with species of the subfamily Stenodermatinae ranging in tropical areas. The most similar species are *Dermanura phaeotis* and *D. watsoni*. *E. hartii* can be differentiated from both species by its dorsal pelage dark chocolate brown of the same color in the chin and neck; facial stripe faint in suede color but distinguished, and the third upper and lower molars well developed.

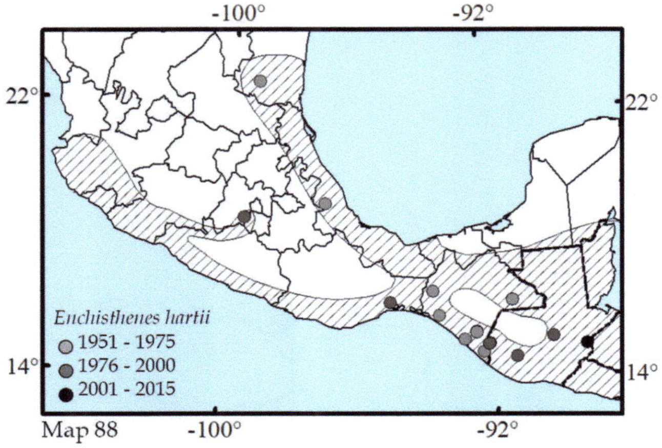

Map 88 Distribution map of *Enchisthenes hartii*

Additional Literature: Miller (1907), Andersen (1908), Goodwin (1940), Swanepoel and Genoways (1979), Álvarez-Castañeda and Álvarez (1991), Van Den Bussche et al. (1993), Arroyo-Cabrales and Owen (1996, 1997), Baker et al. (2000), González-Ruíz et al. (2000), Naranjo Piñera et al. (2005), Almazán-Catalán et al. (2009).

Genus *Phyllops* Petes, 1865

Skull similar to *Dermanura* but broader; rostrum flattened; supraorbital ridges thickened and rounded; interpterygoid space extending anteriorly to the level of the first upper molars; inner upper incisors with the crown higher than long and without a distinct secondary cusp; first and second upper molars with the hypocone much lower than the protocone; first lower molars with a well-developed metaconid connected with a ridge on the inner side of the protoconid.

Phyllops falcatus (Gray, 1839)

Cuban fig-eating bat, murciélago higüero de Cuba

Monotypic.

Conservation: *Phyllops falcatus* is listed as Least Concern by the IUCN (Solari et al. 2019). New record for continental America.

Characteristics: Cuban fig-eating bat is medium sized; forearm length 38.6–48.8 mm, skull length 19.0–20.7 mm. Dorsal pelage grayish brown with hairs mostly tricolored, darker at the base and at the tip, and with a pale yellow median band; **with a small white patch on each shoulder and another behind the posterior ear insertion;** fur of underparts overall paler than in the dorsal pelage; **dactylopatagium minus transparent.**

Comments: *Phyllops falcatus* has ranges restricted to the Great Antilles plus one recent record from Cozumel Island, México (Map 89; Rivas-Camo et al. 2020). Five specimens have been recorded from Cozumel Island, one of them a lactating female (Rivas-Camo et al. 2020), so these could belong to a resident population. Rivas-Camo et al. (2020) mention that the possible presence of *P. falcatus* in Cozumel could have been carried by a hurricane that passed by Cuba and continued its path to the Yucatán Peninsula. *P. falcatus* can be differentiated from all the other sympatric species by having a white patch on each shoulder and just behind the posterior ear insertion, and a transparent dactylopatagium minus.

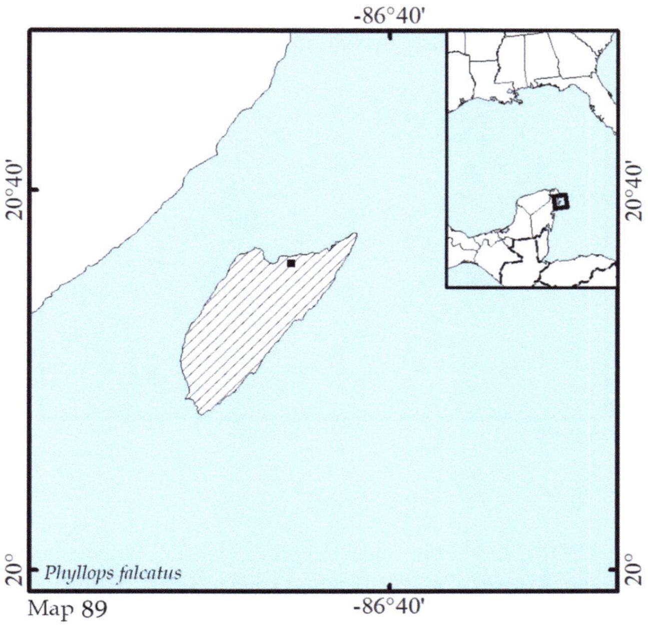

Map 89 Distribution map of *Phyllops falcatus*

Additional Literature: Miller (1907), Hall (1981), Da Cunha-Tavares and Mancina (2008).

Genus *Platyrrhinus* Saussure, 1860

Small sized; **white facial stripes marked and one dorsal stripe strongly marked from the crown to the hips**; **ears and noseleaf brown with the border marked in creamy or whitish shades**; tail absent; **uropatagium narrow, with a conspicuous fringe, with hair 2.0–6.0 mm long;** inner upper incisors different from the outer upper incisors, which are bilobed, approximately one-half of the size of the inner ones; rostrum not clearly differentiated from the braincase.

Platyrrhinus helleri (Peters, 1866)

Heller's broad-nosed bat, murciélago listado nariz ancha

Monotypic.

Conservation: *Platyrrhinus helleri* is listed as Least Concern by the IUCN (Arroyo-Cabrales and Reid 2016).

Characteristics: Heller's broad-nosed bat is small sized; **forearm length 37.0–41.0 mm, skull length 21.7–22.9 mm**. Dorsal pelage and fur of underparts paler to dark or reddish brown; no spots on the shoulders; **facial stripes marked and one dorsal white stripe strongly marked from the crown to the hips**; **ears and noseleaf brown with the border marked in creamy or whitish shades**; uropatagium barely developed and with white hair bangs; tail absent; rostrum not clearly differentiated from the braincase; inner upper incisors different from the outer upper incisors, which are bilobed and approximately one-half of the size of the inner ones.

Comments: *Platyrrhinus helleri* ranges from central Veracruz and Oaxaca south through South America, except for the Yucatán Peninsula (Map 90). This species can be differentiated from all the other sympatric species by having white facial stripes, one dorsal strongly marked from the crown to the hips and by its smaller size (forearm length and skull length less than 41.0 and 22.9 mm, respectively); ears and noseleaf brown with the border marked in creamy or whitish shades. From *Chiroderma* and *Uroderma,* by having a fringed white hair on the uropatagium border.

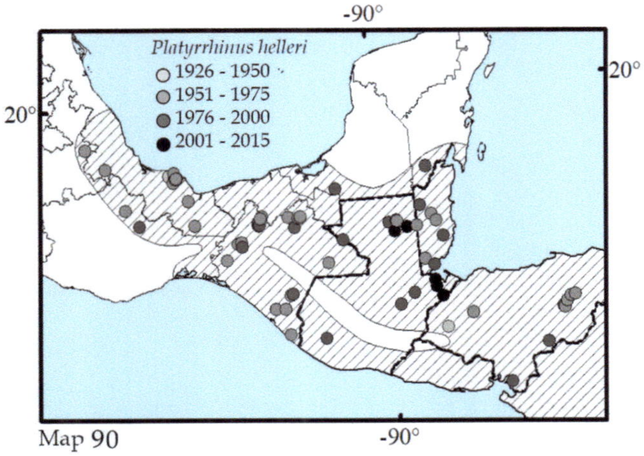

Map 90 Distribution map of *Platyrrhinus helleri*

Additional Literature: Goodwin (1946), Goodwin and Greenhall (1961), Tamsitt and Valdivieso (1963), Gardner and Carter (1972), Jones and Carter (1976), Álvarez-Castañeda and Álvarez (1991); Ferrell and Wilson (1991), Velazco (2005), Gonçalves da Silva et al. (2008), Velazco et al. (2010).

Genus *Uroderma* Peters, 1866

Medium sized; dorsal and ventral pelage brown to dark gray; **facial and dorsal stripes very clear to slightly marked or absent, according to the species**; ears and noseleaf brown with the borders marked in yellow or whitish shades; **uropatagium slightly developed, U-shaped, and bare;** tail absent; **inner upper incisors different from the external outer upper incisors, which are bilobed; rostrum variable according to the species;** rostral length much greater than the lacrimal breadth and about equal to the distance across the first molars; depth through the lacrimal region nearly equal to the lacrimal breadth and more than one-half of the braincase depth. The keys were elaborated based on the review of specimens and the following sources: Davis (1968), Hall (1981), Álvarez et al. (1994), and Álvarez-Castañeda et al. (2017a).

1. Ear border not whitish; facial bands very thin or absent; dorsal profile of skull straight (Fig. 35); mesoetmoid (bone separating the nasal conducts) rhomboidal in frontal view (Fig. 48; from Michoacán, México, south through South America) .... *Uroderma magnirostrum* (p. 317)

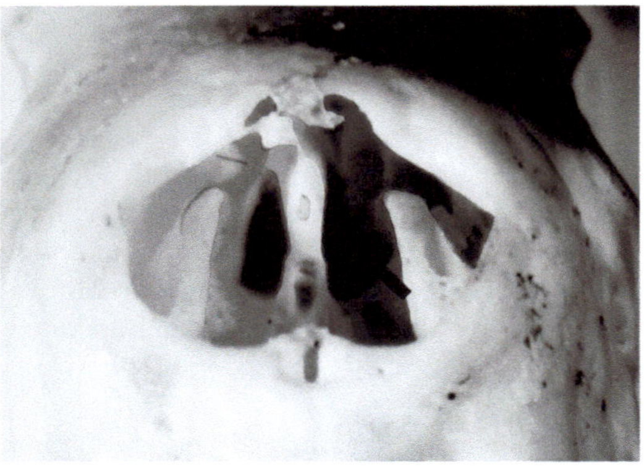

Fig. 48 The mesoetmoid bone rhomboidal in frontal view like in *Uroderma magnirostrum*

1a. Ear border whitish in dry specimens, brilliant yellow in live ones; facial bands well marked; dorsal profile of skull rising notably in the frontal region; mesoetmoid (bone separating the nasal conducts) in the shape of an extended rectangle in frontal view 2

2. Restricted to the Pacific coast of Chiapas and Oaxaca, México .... *Uroderma davisi* (p. 316)

2a. Ranging through the Gulf side from southern Veracruz south through South America .... *Uroderma convexum* (p. 315)

Uroderma convexum Lyon, 1902

Northern tent-making bat, murciélago acampador del norte

U. c. molaris Davis, 1968. Veracruz, Campeche, Tabasco, Chiapas, and Quintana Roo.

Conservation: *Uroderma convexum* (as *Uroderma bilobatum*) is listed as Least Concern by the IUCN (Solari 2019r).

Characteristics: Northern tent-making bat is medium sized; forearm length 40.0–44.0 mm, skull length 21.0–24.5 mm. Dorsal pelage grayish brown to dark gray; fur of underparts grayish brown; **facial stripes well marked; one dorsal white stripe; ears and noseleaf brown with the borders marked in yellow or whitish shades; rostrum not clearly differentiated from the braincase, not cylindrical or low; dorsal skull profile rising notably in the frontal region; mesoetmoid in the shape of an extended rectangle in frontal view.** Note: *U. convexum* and *U. davisi* can be differentiated only by genetic analyses.

Comments: *Uroderma convexum* was previously regarded as part of *U. bilobatum.* The karyotype and a molecular review showed that *U. bilobatum* had to be separated in three different species, *U. convexum, U. davisi,* and *U. bilobatum* (Greenbaum 1981; Baker 1981; Hoffmann et al. 2003; Mantilla-Meluk 2014). *U. convexum* ranges from central Veracruz south through South America, excluding the Yucatán Peninsula (Map 91). It can be differentiated from the species of *Chiroderma* by the uropatagium nearly naked. From *Platyrrhinus helleri,* by its larger size; forearm more than 41.0 mm and the uropatagium without white fringed hair. From *Vampyrodes,* by the absence of white fringed hair. From *U. convexum*, by having allopatric distribution with respect to other species of *Uroderma*.

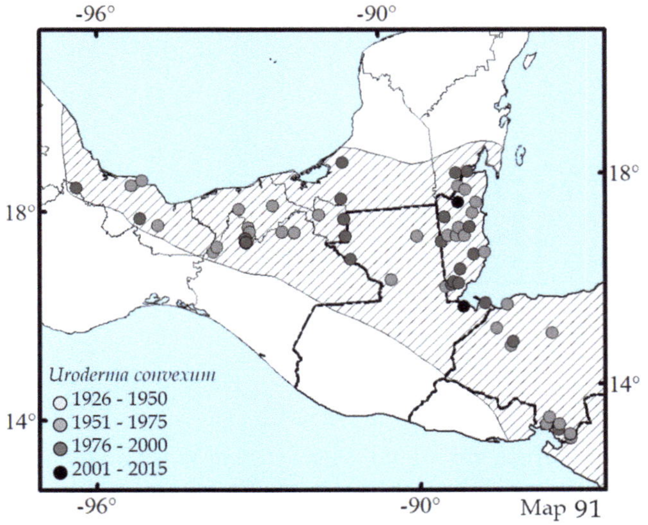

Map 91 Distribution map of *Uroderma convexum*

Additional Literature: Miller (1907), Andersen (1908), Davis (1968), Baker and McDaniel (1972), Baker et al. (1972), Swanepoel and Genoways (1979), Baker and Clark (1987), Álvarez-Castañeda and Álvarez (1991), Naranjo Piñera et al. (2005), Simmons (2005).

Uroderma davisi Baker and McDaniel, 1972

Soconusco tent-making bat, murciélago acampador del Soconusco

Monotypic.

Conservation: *Uroderma davisi* (as *Uroderma bilobatum*) is listed as Least Concern by the IUCN (Solari 2019r).

Characteristics: Soconusco tent-making bat is medium sized; forearm length 40.0–44.0 mm, skull length 21.0–24.5 mm. Dorsal pelage grayish brown to dark gray; fur of underparts grayish brown; **facial stripes well marked; one dorsal white stripe; ears and noseleaf brown with the borders marked in yellow or whitish shades**; **rostrum not clearly differentiated from the braincase, not cylindrical or low; dorsal skull profile rising notably in the frontal region; mesoetmoid in the shape of an extended rectangle in frontal view.** Note: *U. convexum* and *U. davisi* can be differentiated only by genetic analyses.

Comments: *Uroderma davisi* was previously regarded as part of *U. bilobatum.* The karyotype and a molecular review showed that *U. bilobatum* had to be separated in three different species, *U. convexum, U. davisi,* and *U. bilobatum* (Greenbaum 1981; Baker 1981; Hoffmann et al. 2003; Mantilla-Meluk 2014). *U. davisi* ranges from Chiapas south through El Salvador (Map 92). It can be differentiated from the species of *Chiroderma* in the uropatagium nearly naked. From *Platyrrhinus helleri,* by its larger size; forearm more than 41.0 mm and the uropatagium without white fringed hair. From *Vampyrodes,* by the absence of white fringed hair. *U. davisi* and *U. magnirostrum* can be found in sympatric distribution in the coasts of Chiapas, and *U. davisi* can be differentiated from *U. magnirostrum* by having the ears and noseleaf brown with the borders marked in yellow or whitish shades; facial stripes well marked; one dorsal white stripe; the mesoetmoid is in the shape of an extended rectangle in frontal view.

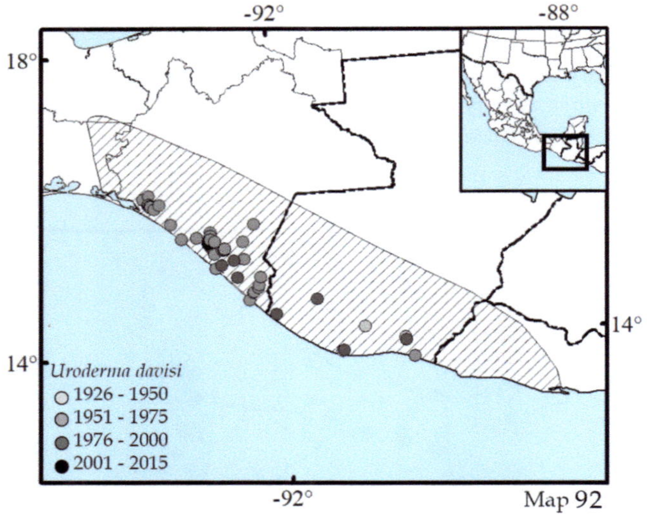

Map 92 Distribution map of *Uroderma davisi*

Additional Literature: Miller (1907), Andersen (1908), Davis (1968), Baker and McDaniel (1972), Baker et al. (1972), Swanepoel and Genoways (1979), Baker and Clark (1987), Álvarez-Castañeda and Álvarez (1991), Hoffmann et al. (2003), Simmons (2005).

Uroderma magnirostrum Davis, 1968

Brown tent-making bat, murciélago acampador pálido

Monotypic.

Conservation: *Uroderma magnirostrum* is listed as Least Concern by the IUCN (Solari 2015f).

Characteristics: Brown tent-making bat is medium sized; forearm length 41.0–45.0 mm, skull length 21.9–24.9 mm. Dorsal color brown to grayish brown; fur of underparts paler than in the dorsum; **facial stripes slightly marked or absent; one dorsal white stripe slightly distinguishable, sometimes only few** paler **hairs are seen**; **ears and noseleaf all brown with margins paler only on the base**; **rostrum not clearly differentiated from the braincase, cylindrical and high; skull straight in dorsal view; mesoetmoid rhomboidal in frontal view**.

Comments: *Uroderma magnirostrum* ranges from Michoacán south through South America (Map 93). It can be found in sympatry, or nearly so, with. It can be differentiated from the species of *Chiroderma* in the uropatagium nearly naked. From *Platyrrhinus helleri,* in its larger size; forearm more than 41.0 mm, and the uropatagium without white fringed hair. From *Vampyrodes,* by the absence of white fringed hair. Both species of *Uroderma* can be found in sympatric distribution in the coasts of Oaxaca and Chiapas, and can be differentiated from *U. davisi* by having the facial stripes slightly marked or absent; the dorsal white stripe barely distinguishable, sometimes with only few paler hairs; mesoetmoid is rhomboidal in frontal view.

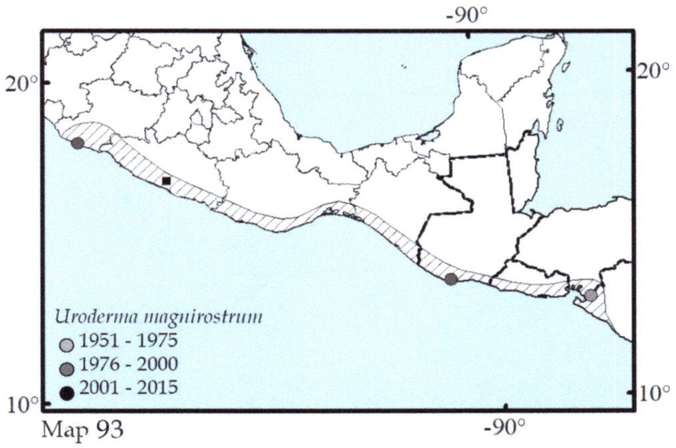

Map 93 Distribution map of *Uroderma magnirostrum*

Additional Literature: Miller (1907), Andersen (1908), Davis (1968), Álvarez-Castañeda and Álvarez (1991), Naranjo Piñera et al. (2005).

Genus *Vampyressa* Thomas, 1900

Small sized; **dorsal pelage pale brown**; facial stripes not distinguishable and back stripe absent; **ears and tragus yellowish**; noseleaf brown; tail absent; **inner upper incisors different from the outer upper incisors, with the latter bilobed and approximately one-half of the size of the former**; first lower molars without the posterior inner cusp; rostrum clearly differentiated from the braincase.

Vampyressa thyone Thomas, 1909

Northern little yellow-eared bat, murciélago cara pintada menor

(*see addenda, split in two species *Vampyressa villai*)

Monotypic.

Conservation: *Vampyressa thyone* is listed as Least Concern by the IUCN (Tavares et al. 2015).

Characteristics: Northern little yellow-eared bat is small sized; **forearm length 29.0–34.0 mm, skull length 18.5–19.0 mm. Dorsal pelage pale brown**; fur of underparts pale grayish brown; no spots on the shoulders, facial stripes not distinguishable and absent on the dorsum; **ears and tragus yellowish**; noseleaf brown; uropatagium slightly developed and with little hair at the center; tail absent; rostrum clearly differentiated from the braincase; **inner upper incisors different from the outer upper incisors, latter bilobed and approximately one-half of the size of the former**; first lower molars without the posterior inner cusp.

Comments: *Vampyressa thyone* was previously considered *V. pusilla thyone* (Lim et al. 2003). *Vampyressa thyone* ranges from Oaxaca, Campeche, northern Chiapas, and southern Quintana Roo south through South America (Map 94). It can be differentiated from all the other sympatric species by its smaller size (forearm and skull length less than 34.0 and 19.0 mm, respectively); from the subfamily Stenodermatinae, by its dorsal pelage pale brown and ears and tragus yellowish; inner upper incisors different from the outer upper incisors and the latter bilobed and approximately one-half of the size of the former.

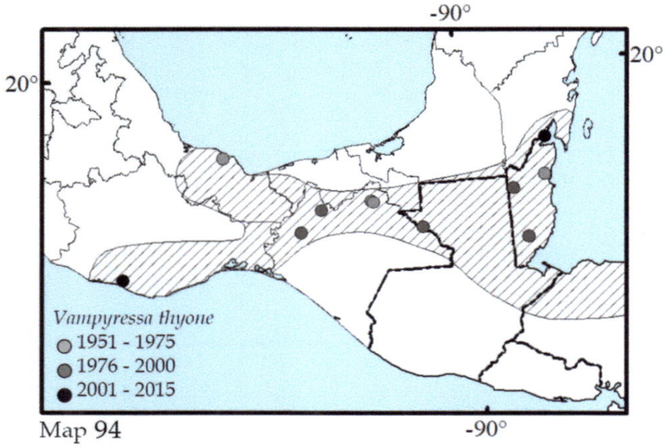

Map 94 Distribution map of *Vampyressa thyone*

Additional Literature: Miller (1907), Handley (1966), Baker et al. (1973), Davis (1975), Lewis and Wilson (1987), Ramírez-Pulido et al. (1983), Álvarez-Castañeda and Álvarez (1991), Naranjo-Piñera et al. (2005).

Genus ***Vampyrodes*** Thomas, 1900

Medium sized; facial stripes well defined; **white stripe from the crown to the hips**; uropatagium barely developed and with white hair bangs; tail absent; inner upper incisors different from the outer upper incisors, the latter bilobed and approximately one-half of the size of the former; **two upper premolars and two upper molars**; last upper molars well developed; rostrum not clearly differentiated from the braincase.

Vampyrodes major Allen, 1908

Great stripe-faced bat, murciélago cara rayada mayor

Monotypic.

Conservation: *Vampyrodes major* is listed as Least Concern by the IUCN (Solari 2016g) and as under Special Protection by the Norma Oficial Mexicana (DOF 2019).

Characteristics: Great stripe-faced bat is medium sized; **forearm length 52.0–54.5 mm, skull length 27.4–28.3 mm**. Dorsal pelage pale yellow to reddish brown; fur of underparts grayish brown; no spots on the shoulders; facial stripes well defined; a marked white stripe from the crown to the hips; ears and noseleaf brown with borders marked in creamy or whitish shades; uropatagium barely developed and with white hair bangs; tail absent; rostrum not clearly differentiated from the braincase; inner upper incisors different from the outer upper incisors, the latter bilobed and approximately one-half of the size of the former; **two upper premolars and two molars**; last upper molars well developed.

Comments: *Vampyrodes major* was previously considered a subspecies of *V. caraccioli* and a different species (Velazco and Simmons 2011). This species ranges from central Veracruz and Oaxaca south through South America, excluding the Yucatán Peninsula (Map 95). *V. major* can be differentiated from *Platyrrhinus helleri* by its larger size; forearm length greater than 41.0 mm. From the species of *Chiroderma* and *Uroderma*, by having a short uropatagium with white fringed hair.

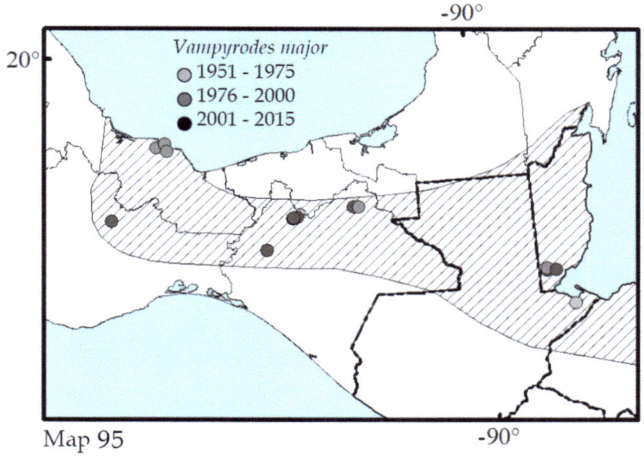

Map 95 Distribution map of *Vampyrodes major*

Additional Literature: Miller (1907), Jones and Carter (1976), Lewis and Wilson (1987), Willis et al. (1990), Medellín et al. (2000), Naranjo Piñera et al. (2005).

Tribe Sturnirini

Genus *Sturnira* Gray, 1842

Medium sized; dorsal and ventral pelage orange brown, brown to grayish brown; **males may show yellow, orange, or reddish spots on the shoulders;** no facial stripes; fur of underparts paler than in the dorsum; ears and noseleaf small; without an uropatagium and hairy; limbs hairy; tail absent; **molars with a groove in the middle, with lateral crests**; sagittal crests slightly developed; rostrum greater than one-half of the braincase; **lower incisors bilobed or trilobed according to the species**. The keys were elaborated based on the review of specimens and the following sources: Hall (1981), Álvarez et al. (1994), and Álvarez-Castañeda et al. (2017a).

1. Dorsal pelage in brown shades; lower incisors trilobed (Fig. 49; the lower incisors should be observed under a magnifying glass; from Sonora and Tamaulipas, México, south through Central America) .... *Sturnira parvidens* (p. 321)

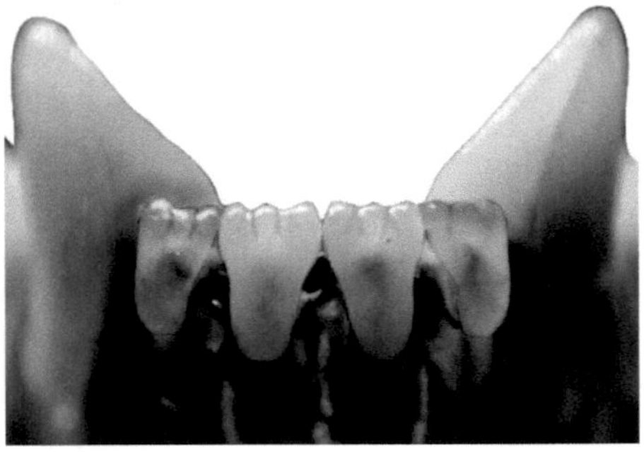

Fig. 49 Lower incisors trilobed like in *Sturnira parvidens*

1a. Dorsal pelage in grayish shades; lower incisors bilobed (Fig. 50; from Sinaloa and Tamaulipas, México, south through South America) .... *Sturnira hondurensis* (p. 320)

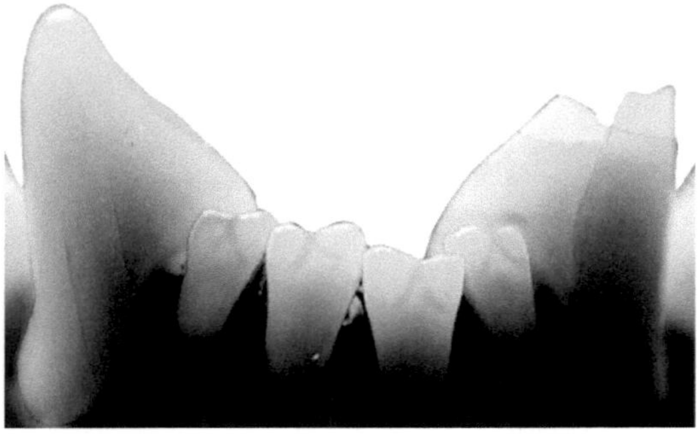

Fig. 50 Lower incisors bilobed like in *Sturnira hondurensis*

Subgenus *Sturnira* Gray, 1842

Four lower incisors normally, subequal in size and all functional.

Sturnira hondurensis Goodwin, 1940

Highland yellow-shouldered bat, murciélago de charreteras mayor

1. *S. h. hondurensis* Goodwin, 1940. From Tamaulipas and Michoacán southern to Chiapas, including the Yucatán Peninsula.
2. *S. h. occidentalis* Jones and Phillips, 1964. From southern Sinaloa south through Colima.

Conservation: *Sturnira hondurensis* is listed as Least Concern by the IUCN (Solari 2017c).

Characteristics: Highland yellow-shouldered bat is medium sized; forearm length 41.0–45.0 mm, skull length 22.8–24.1 mm. **Dorsal pelage normally grayish brown**; fur of underparts pale grayish brown; some individuals orange brown; males may show orange or dark reddish spots on the shoulders; no facial stripes; **lower incisors bilobed**.

Comments: The genus *Sturnira* was previously included in its own subfamily, Sturnirinae; it is currently considered part of the subfamily Phyllostominae. I followed Gardner (2008) in considering that *S. ludovici* is a subspecies of *S. oporaphilum* restricted to South America, and that the species ranging in México is *S. hondurensis* (Goodwin 1940). *S. hondurensis* ranges from Sinaloa and Tamaulipas south through South America (Map 96). It can be found in sympatry, or nearly so, with many species of the subfamily Stenodermatinae; it can be differentiated from all of them by not having the uropatagium and with the limbs very hairy relative to the skull; molar crowns with well-developed cusps on both margins of the central groove. *S. hondurensis* can be differentiated from *S. parvidens* by its general dorsal pelage normally grayish brown and the lower incisors bilobed.

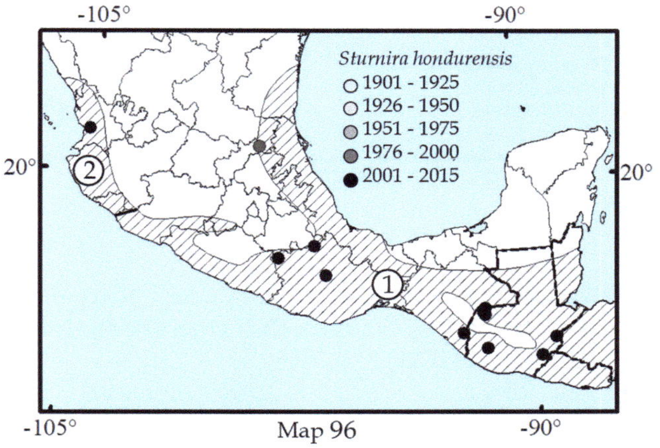

Map 96 Distribution map of *Sturnira hondurensis*

Additional Literature: Miller (1907), Goodwin (1969), Davis (1980); Hall (1981), Álvarez-Castañeda and Álvarez (1991), Sánchez-Hernández et al. (2002), Naranjo Piñera et al. (2005), Velazco and Patterson (2014), Hernández-Canchola et al. (2021b).

Sturnira parvidens Goldman, 1917

Little yellow-shouldered bat, murciélago de charreteras menor

Monotypic.

Conservation: *Sturnira parvidens* is listed as Least Concern by the IUCN (Solari 2019q).

Characteristics: Little yellow-shouldered bat is medium sized; forearm length 37.0–45.0 mm, skull length 21.5–23.7 mm. **Dorsal pelage normally orange brown**, sometimes grayish or with orange shades; males may show yellow, orange, or reddish spots on the shoulders; fur of underparts paler than in the dorsum; **lower incisors trilobed**.

Comments: The genus *Sturnira* was previously included in its own subfamily, Sturnirinae; currently, it is considered part of the subfamily Phyllostominae. I used *S. parvidens* instead of *S. lilium* (Velazco and Patterson 2013). *S. parvidens* ranges from Sonora and Tamaulipas south through Central America, including the Yucatán Peninsula (Map 97). It can be found in sympatry, or nearly so, with many species of the subfamily Stenodermatinae; it can be differentiated from all of them by not having the uropatagium and with very hairy limbs relative to the skull; molar crowns with well-developed cusps on both margins of the central groove. *S. parvidens* can be differentiated from *S. hondurensis* by a general dorsal pelage normally orange brown and the lower incisors trilobed.

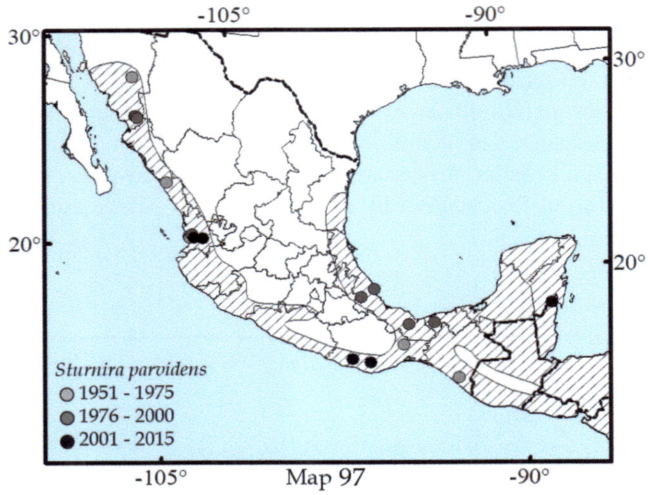

Map 97 Distribution map of *Sturnira parvidens*

Additional Literature: Miller (1907), Goodwin (1969), Jones and Carter (1976), Davis (1980), Willig (1983), Gannon et al. (1989), Álvarez-Castañeda and Álvarez (1991), González-Ruíz et al. (2000), Naranjo Piñera et al. (2005), Gardner (2008), Hernández-Canchola and León-Paniagua (2021).

Family Thyropteridae Miller, 1907

Bats known as "disk-winged bats." The family has only one genus, *Thyroptera,* characterized by its small size; **pinna large, funnel shaped, separated, with no ventral extension under the eyes; tragus small; prominent adhesive disks in the wrists and ankles; third digit with three phalanges; thumbs markedly reduced but with well-developed claws;** third and fourth limb digits with syndactyly (including the claws). They are distributed in the American tropics and belong to the superfamily Nataloidea. No subfamilies are known in this family.

Only one species has been found in North America (in the most humid areas of Tabasco and Chiapas). *Thyroptera tricolor* is characterized by its dark brown or reddish brown dorsal color; fur of underparts yellowish or whitish with an extended snout; ears pale brown; tail extending approximately 5.0 mm out from the uropatagium; **suction cups on the thumbs and soles**; calcar long, with two "borders"; rostrum well differentiated from the braincase and from one-third of it; interorbital region constrained; auditory bullae small; braincase inflated; forearm length 34.0–38.0 mm, skull length 14.4–15.1 mm.

Thyroptera tricolor Spix, 1823

Small sized; **pinna large funnel shaped, separated, with no ventral extension under the eyes; tragus small; prominent adhesive disks in the wrists and ankles; third digit with three phalanges; thumbs markedly reduced but with well-developed claws;** third and fourth limb digits with syndactyly.

Thyroptera tricolor Spix, 1823

Disk-winged bat, murciélago con discos en las alas

T. t. albiventer (Tomas, 1856). From Veracruz, southern Campeche, and Chiapas south through South America.

Conservation: *Thyroptera tricolor* is listed as subject to Special Protection in the Norma Oficial Mexicana (DOF 2019) and as Least Concern by the IUCN (Tavares and Mantilla 2015).

Characteristics: Disk-winged bat is medium sized; **forearm length 32.0–38.5 mm, skull length 14.4–15.1 mm**. Dorsal pelage reddish brown to darker; fur of underparts white. Other characteristics same as in the genus.

Comments: *Thyroptera tricolor* ranges from southernmost Veracruz and the coast of Chiapas south through South America (Map 98). It can be distinguished from other species in sympatry, or nearly so, such as those of the same genus, by its adhesive disks on the thumbs and soles, and muzzle with a small wart-like excrescence above the nostrils.

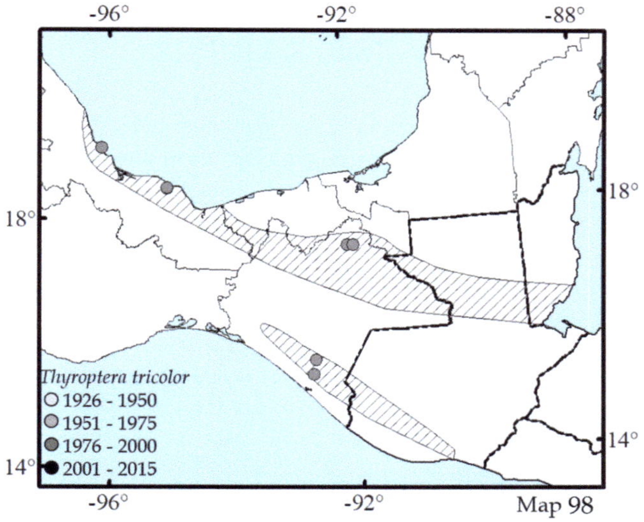

Map 98 Distribution map of *Thyroptera tricolor*

Additional Literature: Wilson and Findley (1977), Álvarez-Castañeda and Álvarez (1991), Koopman (1993), Pine (1993), Naranjo-Piñera et al. (2005), Simmons (2005).

Family Vespertilionidae Gray, 1821

The family Vespertilionidae comprises the most diversified bats. Small to medium sized (forearm length from 24.0 to 90.0 mm); face smooth; pinna variable, from small to large, pointed or rounded, usually separated, with no ventral extensions under the eyes; **tragus present and prominent; tail long, enclosed in a relatively large uropatagium;** third digit with two ossified phalanges, the third phalanx cartilaginous; no postorbital process; cheek-teeth tritubercular. Distribution practically cosmopolitan (all America, Eurasia, Africa, and Australia). Belonging to the superfamily Vespertilionoidea. Two subfamilies in North America: Myotiinae and Vespertilioninae. Some authors consider the tribe Antrozoini at family level (Antrozoidae). I conserve as subfamily (following Roehrs et al. 2010). The keys were elaborated based on the review of specimens and the following sources: Miller (1897, 1907), Hall (1981), Álvarez et al. (1994), and Álvarez-Castañeda et al. (2017a).

1. Teeth 34 or more, if having 34 teeth, ears small and less than the head length; inner upper incisors smaller than the outer upper incisors, bifid in occlusal view......................subfamily Myotiinae; *Myotis* (p. 324)

1a. Teeth 34 or less, if having 34 teeth, ears large and greater than the head length; inner upper incisors of the same size as the outer upper incisors, not clearly bifid in occlusal view...................... subfamily Vespertilioninae (p. 355)

Subfamily Myotinae Tate, 1942

The subfamily Myotiinae is characterized by being Small to medium sized; tail within the uropatagium, in some species extending slightly beyond the uropatagium margin; ears in general proportionally small and less than the head length; nostrils simple, sometimes tubularly elongated but never bordered by special outgrowths; three lower incisors in all known genera; **inner upper incisors smaller than the outer upper incisors and bifid in occlusal view.**

Genus ***Myotis*** Kaup, 1829

Small sized; dorsal and ventral pelage varying in coloration according to the species; **six upper and lower molariforms, the first two small sized, with two species in which some specimens can have one less molariform**; rostrum well differentiated from the braincase; teeth proportionally big with cutting cusps W shaped; an emargination between the canines including the nasals and palate; upper incisors well developed and outer incisors larger than the inner incisors; canines well developed with a cingulum. The mouse-eared bats have four different morphotypes in North America: myotis, leuconoe, selysius, and pyzonyx; for this reason, no external morphological characteristics can be used to distinguish the genera. I use morphotypes instead of subgenera.

Myotis morphotype. Medium to large sized; ears long, directed to the front of the rostrum; wings broad; limbs long but with small feet; calcar small to medium sized, keeled or unkeeled; uropatagium reduced (compared to other ecomorphotypes); skull larger, with a low braincase but with a well-developed sagittal crest; skull thick; rostrum sharp; mandible elongated with a high coronoid process; lower incisors generally reduced and with lobes; premolars reduced; carnassial extended; molars with reduced accessory cusps.

Leuconoe morphotype. Medium to large sized; ears medium sized; wings short; limbs elongated with short tibia and large feet (greater than 50% of the tibia), and a long calcar keeled or unkeeled; tail relatively short; uropatagium less broad than in the selysius morphotype, and characterized by having hair; skull large to medium sized, with the braincase moderately flattened; rostrum conical; sagittal crest developed; mandible elongated with a coronoid process not highly developed; teeth elongated, with a lesser reduction in the lower premolars; carnassial elongated but less developed than in the myotis morphotype; molars with the protonocular developed.

Selysius morphotype. Small to medium sized; ears small; wings long; tail and limbs long; feet small with the calcar large and keeled; skull small with a bulging braincase; rostrum short; teeth small; sagittal crest poorly developed or absent; lower incisors very small, premolars developed and more bunodonts than in the other morphotypes; carnassial barely developed; jaw small with a low coronoid process.

Pyzonyx morphotype. Large sized; ears long, directed to rostrum; wings long and broad; tail long; limbs and feet elongated; calcar keeled; toenails very long and flat laterally; skull large and strong; rostrum long; sagittal crest well developed; jaw and teeth larger. The keys were elaborated based on the review of specimens and the following sources: Miller and Allen (1928), LaVal (1973a, b), Hall (1981), Álvarez et al. (1994), and Álvarez-Castañeda et al. (2017a) (Fig. 51).

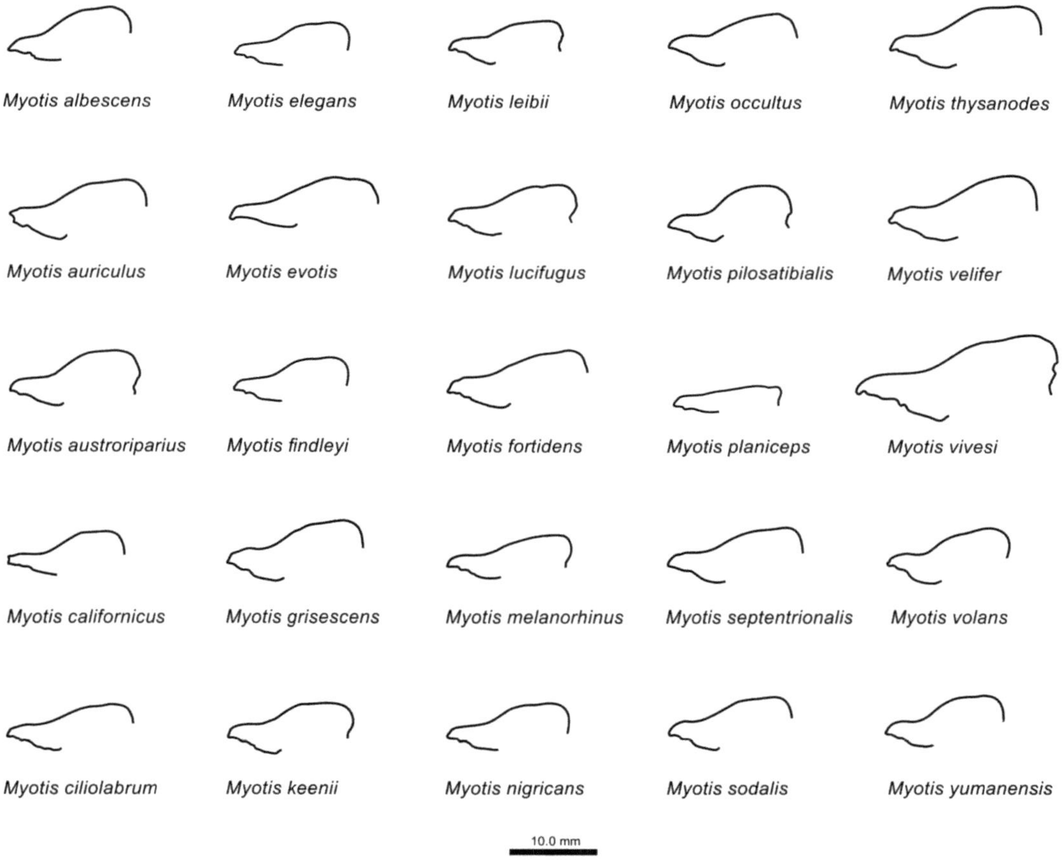

Fig. 51 Side view of the shape of the skulls of the main *Myotis* species, in which the profile and the proportional in size between them can be observed

1. Forearm length greater than 56.0 mm; toenails very long and flat laterally; skull length greater than 19.0 mm (eastern and western coast Baja California Peninsula and coast of Sonora, México) .... *Myotis vivesi* (p. 352)

1a. Forearm length less than 56.0 mm; toenails proportional to hind feet size; skull length less than 19.0 mm 2

2. Dorsal hair white at the tip. Contrasting strongly with its base; calcar unkeeled (from southern Veracruz, Tabasco, and Chiapas, México, south through South America) .... *Myotis albescens* (p. 329)

2a. Dorsal hair pale at the tip but never white; calcar keeled 3

3. Forearm length less than 28.0 mm; braincase flat; angle between the rostrum and the frontal bone approximately 180° (Coahuila, Zacatecas, and Nuevo León, México) .... *Myotis planiceps* (p. 346)

3a. Forearm length greater than 28.0 mm; braincase not flat; angle between the rostrum and the frontal bone less than 170° 4

4. Teeth 34–36 5

4a. Teeth 38 6

5. Dorsal pelage yellowish; calcar keeled (Fig. 52); always two upper and two lower premolars (from southern Sonora and Veracruz, south through Chiapas, México) .... *Myotis fortidens* (p. 338)

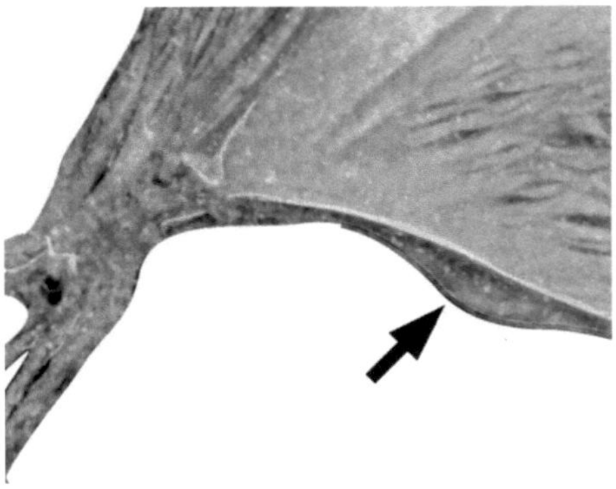

Fig. 52 The calcar keeled presents in *Myotis fortidens*

5a. Dorsal pelage dark brown to blackish; calcar unkeeled; two or three upper and lower premolars (from Colorado and Arizona, United States, south through Central México)..*Myotis occultus* (part, p. 344)

6. Specimens Medium to large sized; ears longer, directed to the front of the face; wings broad; uropatagium reduced and narrow; larger skull; braincase generally wider than high; rostrum sharp; jaw branch elongated; coronoid process high..........7

6a. Specimens medium to small sized; ears medium to small, not directed to the front of the face; wings narrow; uropatagium large and broad; skull large to small; braincase higher than wide; rostrum slightly sharp to short; jaw branch wide; coronoid process not high..........12

7. Total skull length greater than 16.3 mm; sagittal crest well developed..........8

7a. Total skull length less than 16.3 mm; sagittal crest not well developed or absent..........9

8. Uropatagium with distinguishable hair bangs at the border; breadth through the molars greater than 6.2 mm (from British Columbia, Canada and South Dakota, United States, south through Chiapas, México)..........*Myotis thysanodes* (p. 349)

8a. Uropatagium without hair bangs at the border; breadth through the molars less than 6.2 mm (from Arizona and New Mexico, United States, south through Jalisco, Guanajuato, and Tamaulipas, México)..........*Myotis auriculus* (part, p. 330)

9. Ear length less than 16.0 mm, not extending in front of the nose when folded over the face; posterior braincase higher than at the middle (from Colorado and Arizona, United States, south through Central México)..........*Myotis occultus* (part, p. 344)

9a. Ear length greater than 16.0 mm, extending more than 2.0 mm in front of the nose when folded over the face; posterior braincase at the same level as the middle..........10

10. Ear length greater than 18.0 mm; toothrow length equal to or greater than 6.0 mm; posterior braincase at the same level as the middle (from British Columbia, Alberta, and Saskatchewan, Canada east through North Dakota and New Mexico, United States, and south through the Baja California Peninsula, México)..........*Myotis evotis* (p. 335)

10a. Ear length less than 18.0 mm; toothrow length equal to or less than 6.3 mm; posterior braincase lower than the middle..........11

11. Ears and patagium black; shoulders with darker spots; uropatagium with distinguishable hair bangs at the border; cingulum of canines wider than 0.74 mm (British Columbia, Canada and Washington, United States)..........*Myotis keenii* (p. 340)

11a. Ears and patagium brown; shoulders of the same color as the back; uropatagium with non-distinguishable hair bangs at the border; cingulum of canines narrower, greater or less than 0.74 mm (from Saskatchewan, Canada, east through Newfoundland and south through Mississippi to South Carolina, United States).......... *Myotis septentrionalis* (p. 347)

12. Maxillary toothrow equal to or greater than 6.0 mm (from California, Arizona, New Mexico, and Kansas, United States, south through Central America, except for the Yucatán Peninsula, México)..........*Myotis velifer* (p. 350)

12a. Maxillary toothrow equal to or less than 6.0 mm 13

13. Limb length (dry measurement) less than 50% of the tibia length 14

13a. Limb length (dry measurement) greater than 50% of the tibia length 16

14. Black mask usually absent; forehead rising abruptly from the rostrum; anterior braincase higher than the posterior part (Fig. 53); no sagittal crest (from Alaska, British Columbia, Canada and Wyoming, United States, south through Chiapas, México) .... *Myotis californicus* (p. 332)

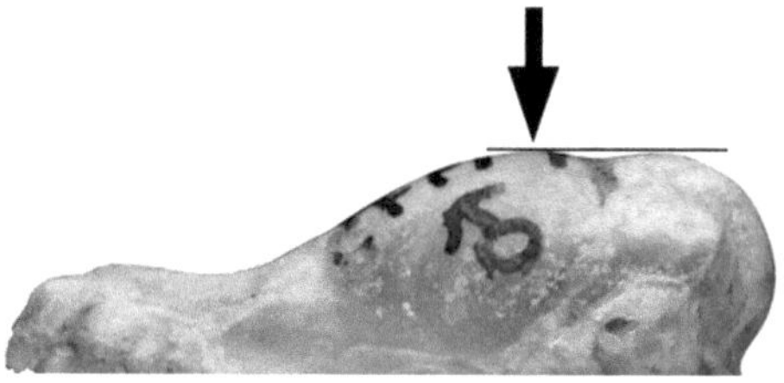

Fig. 53 The anterior braincase is higher than the posterior part

14a. Black mask present; forehead not rising abruptly from the rostrum; posterior braincase higher than at the middle (Fig. 54); generally with a sagittal crest 15

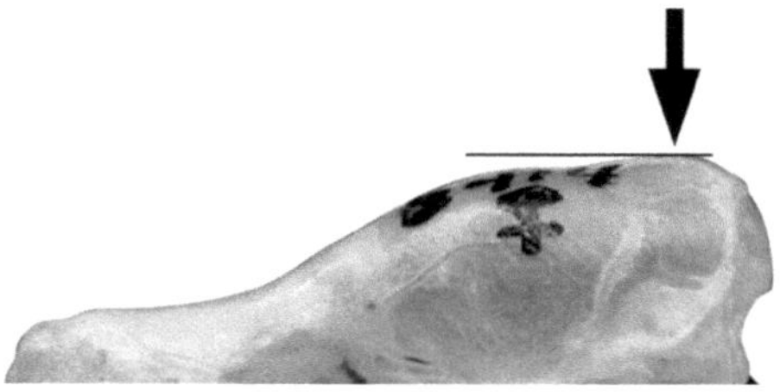

Fig. 54 The posterior braincase higher than at the middle

15. Dorsal coloration dark yellow to brown, not contrasting with the black face mask, ears, and wings (from Ontario and Québec, Canada, and New York, United States, south through North Carolina, Tennessee, Georgia, and Missouri, United States) .... *Myotis leibii* (p. 341)

15a. Dorsal coloration pale brown, contrasting with the face mask, ears, and wings (Alberta, Saskatchewan, Idaho, Montana, the Dakotas, Wyoming, Nebraska, Colorado, and Altiplanicie Mexicana) .... *Myotis ciliolabrum* (p. 335)

16. Sagittal crest absent 17

16a. Sagittal crest present, may be very small 22

17. Braincase supraoccipital region above the level of the braincase roof (Fig. 55) 18

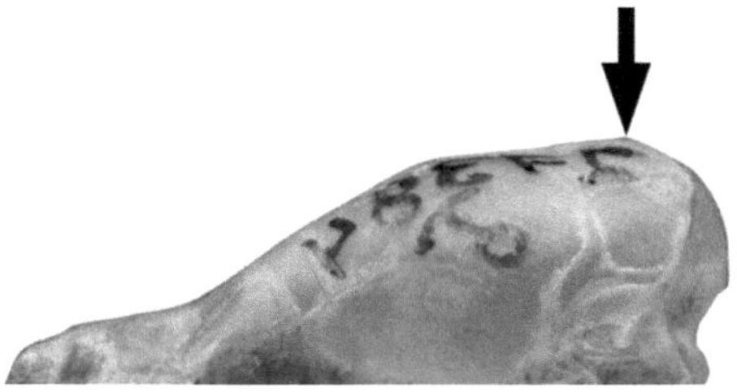

Fig. 55 Braincase supraoccipital region above the level of the braincase roof

17a. Braincase supraoccipital region and roof at the same level 20

18. Coloration pale brown; ranging in tropical México (southern Tamaulipas, México, south through Central America) .... *Myotis elegans* (p. 335)

18a. Coloration generally chestnut brown to dark brown, with pale brown specimens occasionally found in northern México 19

19. Hair fine, fluffy, and glossy; calcar keeled; hind foot length 9.0 mm on average (from Arkansas, Missouri, and Illinois east through New Hampshire and New Jersey, and south through Florida, United States) .... *Myotis sodalis* (part; p. 348)

19a. Hair rough, shaggy, and not glossy; calcar unkeeled; hind foot on average 10.0 mm .... *Myotis lucifugus* (p. 342)

20. Tibia length greater than 41% of the forearm length; fur of underparts grayish or whitish; external border of the pterygoid extending laterally in a laminar shape (Fig. 56); braincase spherical (from British Columbia, Canada, south through central México) .... *Myotis yumanensis* (p. 354)

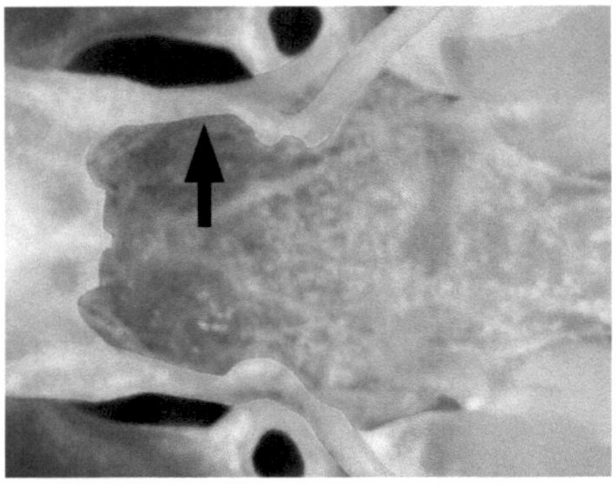

Fig. 56 The external border of the pterygoid extending laterally in a laminar shape

20a. Tibia length less than 41% of the forearm length; fur of underparts brown; external pterigoid border straight (Fig. 57); braincase not spherical 21

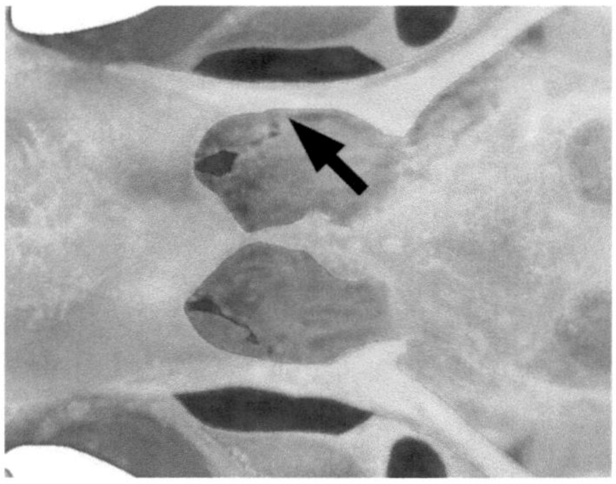

Fig. 57 The external border of the pterygoid straight

21. Tail length less than 97% of the forearm length; rostrum breadth at the canine level less than 4.7 mm; interorbital breadth less than 3.2 mm (endemic from Islas Marías, México) .... *Myotis findleyi* (p. 337)

21a. Tail length greater than 97% of the forearm length; rostrum breadth at the canine level greater than 4.7 mm; interorbital breadth greater than 3.2 mm (from Nayarit and Tamaulipas, México, south through South America) .... *Myotis nigricans* (p. 343)

22. Dorsal hair without bands, generally in gray shades, with brown specimens occasionally found; uropatagium attached to the ankles (from Missouri and Louisiana east through Virginia and Florida, United States).. ....*Myotis grisescens* (p. 339)

22a. Dorsal hair with bands, never in gray shades; uropatagium attached to the thumb base.........23

23. Ears large, extending greater than 6.0 mm in front of the nose when folded over the face; total cranial length greater than 15.5 mm (from Arizona and New Mexico, United States, south through Jalisco, Michoacán, and Tamaulipas, México) ....*Myotis auriculus* (part, p. 330)

23a. Ears small, extending less than 6.0 mm in front of the nose when folded over the face; total cranial length less than 15.5 mm 24

24. Wings with hair from the elbows to the knees; calcar keeled; supraoccipital region of the skull inflated (Fig. 58; from British Columbia and Alberta, Canada and North Dakota, United States south through Oaxaca, México)........ ....*Myotis volans* (p. 353)

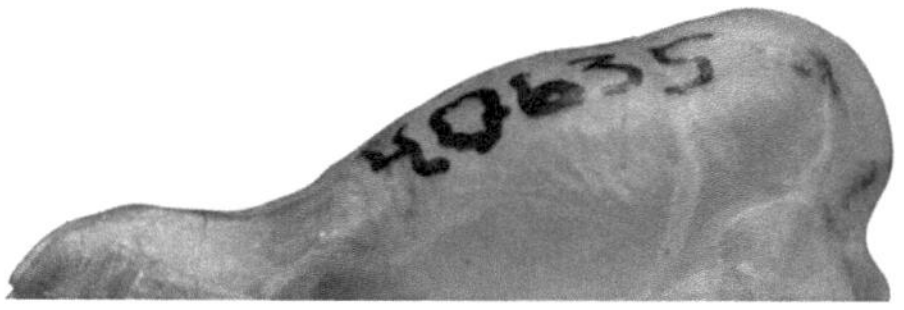

Fig. 58 The supraoccipital region of the skull inflated

24a. Wings without hair from the elbows to the knees; calcar with a small keel or unkeeled; braincase inflated throughout but not particularly at the supraoccipital region.........25

25. Calcar with a small keel (from Missouri, Texas, and Oklahoma east through North Carolina and south through Florida, United States, except for the Appalachian Mountains)........ *Myotis austroriparius* (p. 331)

25a. Calcar unkeeled.........26

26. Supraoccipital region above the level of the braincase roof (from Arkansas, Oklahoma, Missouri, and Iowa east through New Hampshire and south through North Carolina and northern Florida, United States)........ ....*Myotis sodalis* (part, p. 348)

26a. Supraoccipital region and braincase roof at the same level (from Oaxaca and Tamaulipas, México, south through Central America) .... *Myotis pilosatibialis* (p. 345)

Myotis albescens (Geoffroy Saint-Hilaire, 1806)

Silver-tipped myotis, murciélago ratón plateado

Monotypic.

Conservation: *Myotis albescens* is listed as subject to Special Protection in the Norma Oficial Mexicana (DOF 2019) and as Least Concern by the IUCN (Barquez et al. 2016c).

Characteristics: Silver-tipped myotis is small sized; forearm length 33.0–38.5 mm, skull length 13.7–15.2 mm. **Dorsal pelage blackish with white or cream hair at the tip**; fur of underparts gray; no face mask, rostrum skin and ears blackish; ears spear shaped and tragus narrow and pointed; uropatagium bare from the knee to the tail but with thin hair bangs on the border; membranes blackish; limbs relatively long; calcar unkeeled; braincase not rising abruptly from the rostrum; supraoccipital region of the braincase at a slightly lower level than the braincase roof. Morphotype leuconoe.

Comments: *Myotis albescens* ranges from southern Veracruz, Tabasco, and the northern-central Chiapas south through South America (Map 99). It can be distinguished from other species in sympatry, or nearly so, such as *M. elegans*, *M. fortidens*, *M. nigricans,* and *M. pilosatibialis* by having the dorsal pelage blackish with white or cream hair at the tip and fur of underparts gray.

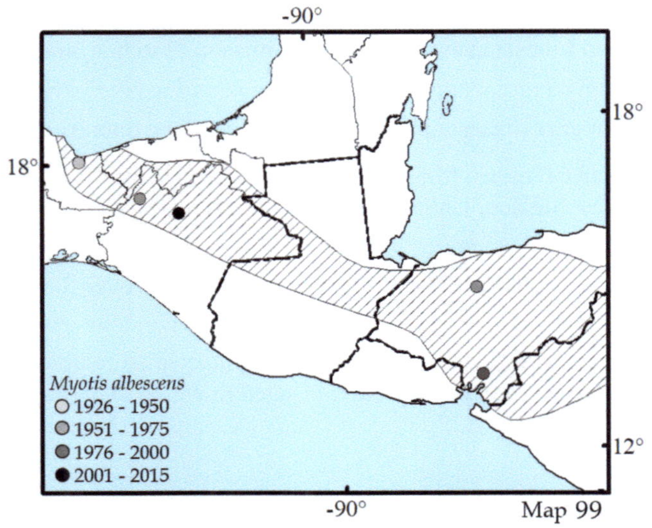

Map 99 Distribution map of *Myotis albescens*

Additional Literature: Miller and Allen (1928), LaVal (1973a), Carter and Dolan (1978), Álvarez-Castañeda and Álvarez (1991), López-González et al. (2001), Naranjo-Piñera et al. (2005), Stadelmann et al. (2007), Braun et al. (2009).

Myotis auriculus Baker and Stains, 1955

Southwestern myotis, murciélago ratón suroccidental

1. *M. a. apache* Hoffmeister and Krutzsch, 1955. From Arizona, New Mexico, and Coahuila south through Jalisco.
2. *M. a. auriculus* Baker and Stains, 1955. Nuevo León, Tamaulipas, and Veracruz.

Conservation: *Myotis auriculus* is listed as Least Concern by the IUCN (Arroyo-Cabrales et al. 2017).

Characteristics: Southwestern myotis is small sized; forearm length 36.0–40.0 mm, skull length 15.7–17.0 mm. Dorsal pelage yellowish brown; fur of underparts cream; no face mask, **rostrum skin and ear base pinkish; ears large, their length close to one-half of the forearm length; if the ears are folded over the rostrum, they go 2.0 mm beyond the nose;** tragus large and narrow; uropatagium bare from the knees to the tail; membranes brown; third metacarpal larger than the fourth metacarpal; calcar unkeeled; supraoccipital region and braincase roof almost at the same level; sagittal crest present. Morphotype myotis.

Comments: *Myotis auriculus* was considered a subspecies of *M. evotis* (Hall and Kelson 1959). This species ranges in both Sierras Madres of México from Arizona and New Mexico, south through Jalisco, Guanajuato, Tamaulipas, and Veracruz (Map 100). It can be distinguished from other species in sympatry, or nearly so, such as *M. californicus*, *M. ciliolabrum*, *M. evotis*, *M. lucifugus*, *M. occultus*, *M. thysanodes*, *M. velifer*, *M. volans*, and *M. yumanensis*, by having the rostrum skin and ear base pinkish; ears large and rounded, their length close to one-half of the forearm length and, if folded over the rostrum, they go 2.0 mm beyond the nose; tragus narrow and long, deep black in coloration.

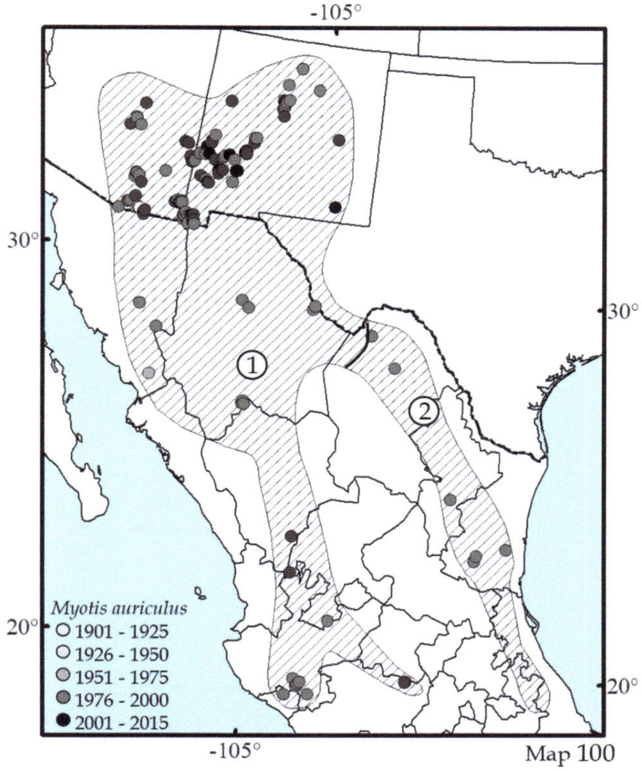

Map 100 Distribution map of *Myotis auriculus*

Additional Literature: Miller and Allen (1928), Baker and Stains (1955), Hoffmeister and Krutzsch (1955), Barbour and Davis (1969), Genoways and Jones (1969), Findley (1960, 1972), Fitch and Shump (1979), Williams and Findley (1979), Warner (1982), Villalpando and Álvarez (2000), Stadelmann et al. (2007), Arroyo-Cabrales et al. (2008).

Myotis austroriparius (Rhoads, 1897)

Southeastern myotis, murciélago ratón del sureste

Monotypic.

Conservation: *Myotis austroriparius* is listed as Least Concern by the IUCN (Arroyo-Cabrales and Álvarez-Castañeda 2017f).

Characteristics: Southeastern myotis is small sized; forearm length 33.0–39.0 mm, skull length 13.5–15.2 mm. **Dorsal pelage gray to bright orange brown, woolly in appearance; with no contrast between the hair base and the tip**; fur of underparts paler and with a marked contrast between the base and the tip; no face mask; muzzle, ears, tragus and membranes brownish; uropatagium with no hair; no prominent keel on the calcar; braincase highly inflated and rising abruptly from the rostrum; supraoccipital region and braincase roof almost at the same level; sagittal crest usually present. Morphotype leuconoe.

Comments: *Myotis austroriparius* ranges in the southeastern United States from Texas, Oklahoma, Arkansas, eastern Missouri, Illinois, and Indiana, western Kentucky, and Tennessee eastward to North Carolina except for the southern Florida Peninsula and the Appalachian Mountains (Map 101). The diagnostic characteristic for identifying the species is the small and particular baculum shape. *M. austroriparius* can be differentiated from the sympatric species *M. grisescens*, *M. keenii*, *M. leibii*, and *M. sordalis* by having the dorsal pelage varying from gray to bright orange brown with woolly appearance; hair not contrasting between the base and the tip.

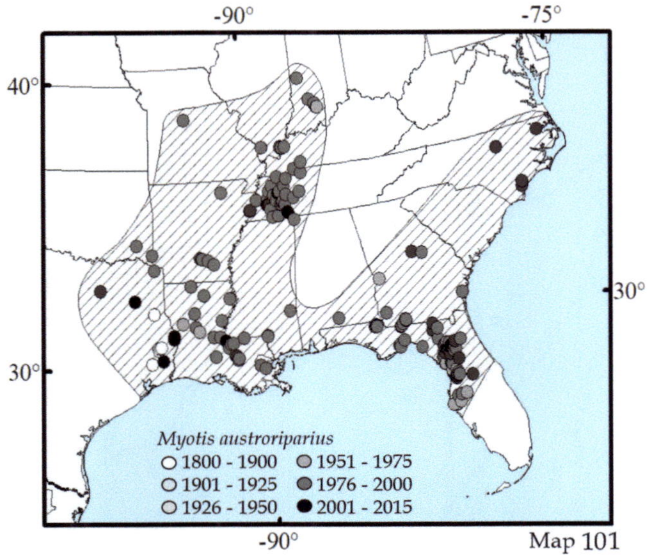

Map 101 Distribution map of *Myotis austroriparius*

Additional Literature: Miller and Allen (1928), Davis and Rippy (1968), Barbour and Davis (1969), LaVal (1970), Lowery (1974), Hall (1981), Jones and Manning (1989), Stadelmann et al. (2007), Simmons (2005).

Myotis californicus (Audubon and Bachman, 1842)

California myotis, murciélago ratón de California

1. *M. c. californicus* (Audubon and Bachman, 1842). From central New Mexico, Arizona, southwestern Texas, and Sonora south through Nuevo León to Sinaloa.
2. *M. c. caurinus* Miller, 1897. Coastal area from British Columbia to San Francisco Bay in California.
3. *M. c. mexicanus* (Saussure, 1860). From Nuevo León to Sinaloa south through Chiapas.
4. *M. c. stephensi* Dalquest, 1946. From Nevada, Utah, and Colorado south through central New Mexico, Arizona, and Sonora.

Conservation: *Myotis californicus* is listed as Least Concern by the IUCN (Arroyo-Cabrales and Perez 2017).

Characteristics: California myotis is small sized; forearm length 31.0–37.0 mm, skull length 12.6–14.2 mm. Dorsal pelage yellowish; fur of underparts cream or yellowish; **muzzle, chin, rostrum, ears, and tragus blackish forming a black face mask usually absent; uropatagium membrane black**; ears spear shaped and tragus narrow; uropatagium with little hair from the knees to the tail; **tail tip not sticking out of the uropatagium; hind foot proportionally very small**; hair dark at the base contrasting with a paler shade at the tip; ears pale yellow appearing to be pointed and small; if they fold on the face, they do not go more than 2.0 mm over the nose; calcar keeled; braincase rising abruptly from the rostrum, which is long; anterior braincase higher than the posterior one; no sagittal crest. Morphotype selysius.

Comments: *Myotis californicus* ranges in all the western areas, from Alaska, western Montana, and Wyoming, Canada, south through Oaxaca and Chiapas, México, including western Colorado, New Mexico, and southwestern Texas mainly throughout México except for northeastern Tamaulipas, the Yucatán Peninsula, and from the coastal plains of Jalisco southward (Map 102). It can be distinguished from other species in sympatry, or nearly so, such as *M. auriculus*, *M. ciliolabrum*, *M. evotis*, *M. lucifugus*, *M. occultus*, *M. thysanodes*, *M. velifer*, *M. volans*, and *M. yumanensis* by having the rostrum, ears, and membranes black; the tail tip does not stick out of the uropatagium; hind foot proportionally very small; the braincase rises abruptly from the rostrum and the anterior braincase is higher than the supraoccipital region.

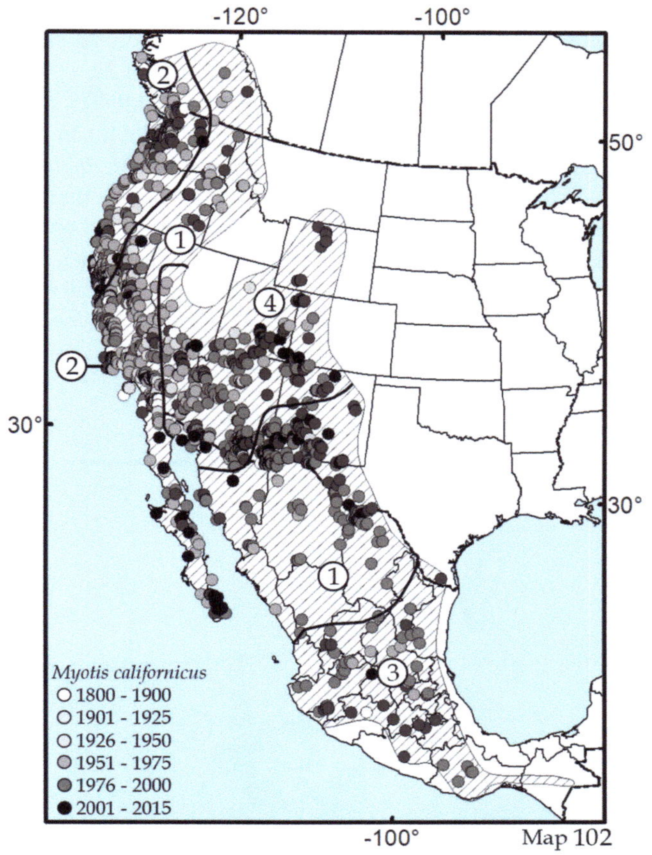

Map 102 Distribution map of *Myotis californicus*

Additional Literature: Miller and Allen (1928), Findley (1972), Bogan (1974, 1978), van Zyll de Jong (1984), Simpson (1993), Yancey (1997), Bogan (1999), Villalpando and Álvarez (2000), Rodríguez and Ammerman (2004), Arroyo-Cabrales et al. (2005), Simmons (2005), Naranjo Piñera (2007), Stadelmann et al. (2007), Álvarez-Castañeda et al. (2008a, b, c), Frick et al. (2008).

Myotis ciliolabrum (Merriam, 1886)

Western small-footed myotis, murciélago ratón occidental

1. *M. c. ciliolabrum* (Merriam, 1886). From western North America, mainly in the Great Plains, in Alberta, Saskatchewan, Idaho, Montana, the Dakotas, Wyoming, Nebraska, and Colorado
2. *M. c. melanorhinus* (Merriam, 1890). From Alberta and Saskatchewan into the northern Altiplanicie Mexicana, and among Montana, Wyoming, east Texas, Colorado, Washington, and California.

Conservation: *Myotis ciliolabrum* is listed as Least Concern by the IUCN (Arroyo-Cabrales and Álvarez-Castañeda 2017g).

Characteristics: Western small-footed myotis is small sized; forearm length 23.4–33.0 mm, skull length 13.0–14.5 mm. Dorsal pelage pale brown yellowish; **fur of underparts nearly white but blackish at the base; snout, chin, rostrum, ears, and tragus blackish, forming a black face mask; uropatagium membrane blackish with hair**; **tail tip sticking out about 1.5–2.5 mm of the uropatagium;** limbs proportionately small; calcar keeled, ending in a small lobe; braincase appearing flat and rising gradually from the rostrum; braincase supraoccipital region above the braincase roof level; sagittal crest poorly developed or absent. Morphotype selysius.

Comments: *Myotis ciliolabrum* was previously considered a synonym of *M. subulatus subulatus*. However, *M. subulatus* was not considered a valid species (Glass and Baker 1965, 1968) or different from *M. leibii* and *M. melanorhinus* (van Zyll de Jong 1984). Nowadays *M. melanorhinus* is considered a valid subspecies of *M. ciliolabrum* (Ammerman et al. 2016). *M. ciliolabrum* ranges in western North America, from the Great Plains, in Alberta, Saskatchewan, Idaho, Montana, the Dakotas, Wyoming, Nebraska, Colorado, and Altiplanicie Mexicana (Map 103). It can be distinguished from other species in sympatry, or nearly so, such as *M. auriculus*, *M. californicus*, *M. evotis*, *M. lucifugus, M. occultus, M. thysanodes*, *M. velifer*, *M. volans*, and *M. yumanensis*, by its small size; with a black face mask that includes the muzzle, chin, rostrum, ears and tragus, contrasting with the paler hair elsewhere; hind feet short, less than one-half of the tibia; distinctly keeled calcar and black wing membranes. From *M. californicus,* by having a black face mask that contrast with the general coloration, and by its paler coloration; the skull has a less steeply sloping frontal area, which is flatter and broader. From *M. leibii,* by a flatter cranium, and the dorsal pelage with a distinct golden sheen. From *M. melanorhinus,* by having the fur of underparts nearly white but blackish at the base.

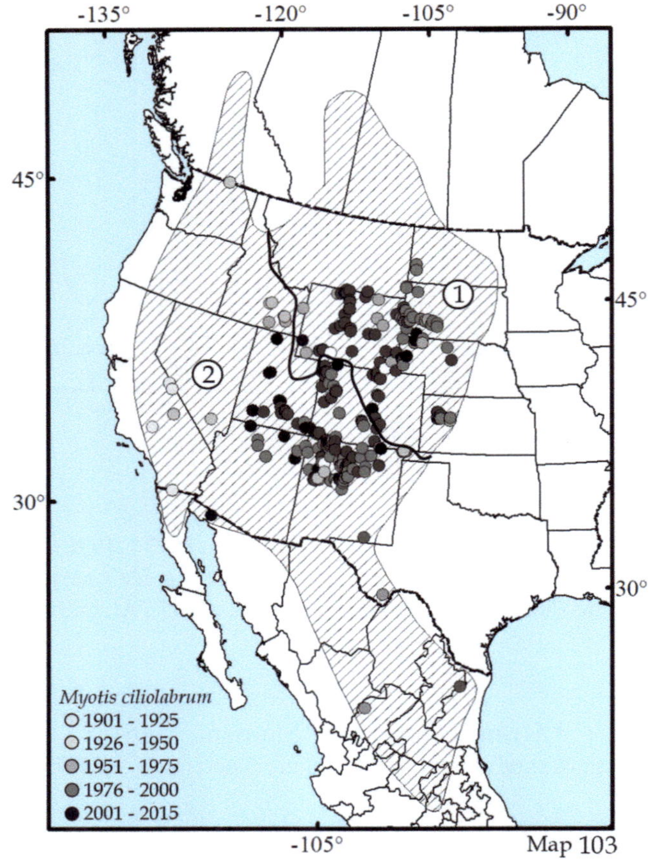

Map 103 Distribution map of *Myotis ciliolabrum*

Additional Literature: Miller and Allen (1928), Findley (1972), Bogan (1974), Hoffmeister (1986), Verts and Carraway (1998), Holloway and Barclay (2001), Rodriguez and Ammerman (2004), Stadelmann et al. (2007).

Myotis elegans Hall, 1962

Elegant myotis, murciélago ratón elegante

Monotypic.

Conservation: *Myotis elegans* is listed as Least Concern by the IUCN (Miller and Rodriguez 2016a).

Characteristics: Elegant myotis is small sized; forearm length 31.9–34.1 mm, skull length 12.5–13.4 mm. Dorsal pelage orange to reddish brown; fur of underparts pale orange or yellowish; rostrum skin pinkish with a mustache on the upper lip; no face mask; ears and tragus grayish brown, narrow and pointed; uropatagium bare from the knees to the tail; hair brown at the base, contrasting with a lighter shade at the tip; membranes blackish; calcar keeled; sagittal crest absent; braincase not inflated but flattened; braincase supraoccipital region above the braincase roof level. Morphotype selysius.

Comments: *Myotis elegans* ranges from Tamaulipas south through Central America, excluding the Yucatán Peninsula, highlands, and southeastern Chiapas (Map 104). It can be distinguished from other species in sympatry, or nearly so, such as *M. fortidens*, *M. nigricans*, *M. pilosatibialis*, and *M. velifer*, by its small size (forearm and skull length less than 34.1 and 13.4 mm, respectively); foot greater than 50% of the tibia length and a pale brown coloration.

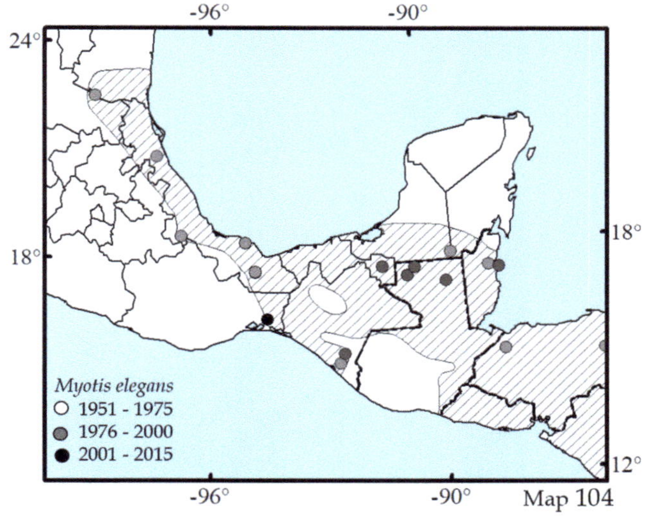

Map 104 Distribution map of *Myotis elegans*

Additional Literature: LaVal (1973a), Álvarez-Castañeda and Álvarez (1991), Naranjo-Piñera et al. (2005), Simmons (2005), Stadelmann et al. (2007).

Myotis evotis (Allen, 1864)

Long-eared myotis, murciélago ratón de oeste

1. *M. e. evotis* (Allen, 1864). From British Columbia, Alberta, and Saskatchewan south through the tip of the Baja California Peninsula, central Arizona, and New Mexico.
2. *M. e. pacificus* Dalquest, 1943. Coastal, from British Columbia south through northern California, including northern Idaho and western Montana.
3. *M. e. milleri* Elliot, 1903. Restricted to northwestern Baja California.

Conservation: *Myotis evotis* is listed as special protection in México (NOM 2010) and as Least Concern by the IUCN (Arroyo-Cabrales and Álvarez-Castañeda 2017h).

Characteristics: Long-eared myotis is small sized; forearm length 35.5–41.0 mm, skull length 15.0–16.5 mm. Dorsal pelage pale to medium brown; fur of underparts paler than in the dorsum; **ears large, more than 18.0 mm; muzzle, chin, rostrum, ears, and tragus blackish forming a black face mask; if the ears fold over the rostrum, they go more than 5.0 mm beyond the nose;** uropatagium with few dispersed hairs; calcar with a small keel; braincase not rising abruptly on the rostrum; supraoccipital region and braincase roof at the same level; sagittal crest slightly developed but present. Morphotype myotis.

Comments: *Myotis milleri* is a subspecies of *M. evotis* (Manning 1993; Simmons 2005), previously considered a full species (Findley 1972; Reducker et al. 1983). *M. evotis* ranges from the Great Plains in New Mexico, Colorado, eastern Kansas, Nebraska, the Dakotas, Saskatchewan, Alberta, and British Columbia westward, and south through the northwestern of the Baja California Peninsula (Map 105). It can be distinguished from other species in sympatry, or nearly so, such as *M. auriculus*, *M. californicus*, *M. ciliolabrum*, *M. lucifugus*, *M. occultus*, *M. sordalis*, *M. thysanodes*, *M. velifer*, *M. volans*, and *M. yumanensis*, by its large ears (more than 18.0 mm) and notoriously blackish; if they fold over the rostrum, they go more than 5.0 mm beyond the nose; braincase top at the same level as the supraoccipital region; sagittal crest scarcely developed but present.

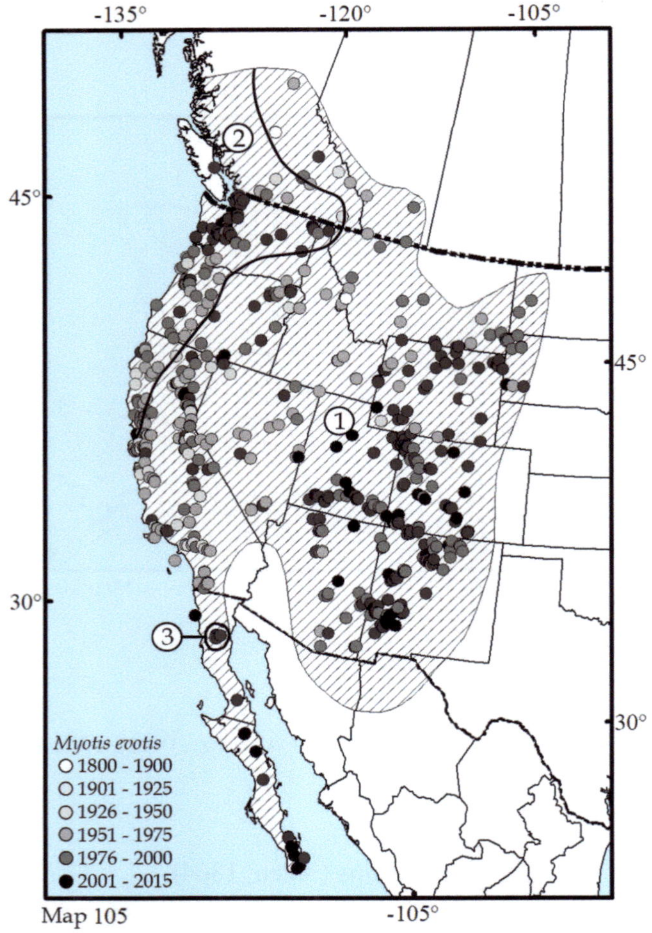

Map 105 Distribution map of *Myotis evotis*

Additional Literature: Miller and Allen (1928), Grinnell (1933), Genoways and Jones (1969), Jones and Choate (1978), Koopman (1984), van Zyll de Jong (1985), Manning and Jones (1989), Álvarez-Castañeda and Bogan (1997), Bogan (1999), Álvarez-Castañeda et al. (2008c), Stadelmann et al. (2007).

Myotis findleyi Bogan, 1978

Findley's myotis, murciélago ratón de las Islas Marías

Monotypic.

Conservation: *Myotis findleyi* is listed as Endangered (B1ab (i, iii): extent of occurrence less than 5000 km^2 in five or less populations and continuing decline in extent of occurrence and quality of habitat) by the IUCN (Arroyo-Cabrales and Ospina-Garces 2016a).

Characteristics: Findley's myotis is small sized; forearm length 29.5–33.2, skull length 11.5–12.2 mm. Dorsal pelage dark brown to blackish; hair light brown at the tip, creating a contrast; fur of underparts blackish at the base and brown at the tip; no face mask, plagiopatagium brown; limbs proportionally large; calcar with a keel barely developed; braincase spherical and inflated, rising abruptly from the rostrum; sagittal crest present but barely developed. Morphotype selysius.

Comments: The individuals from Islas Tres Marías, Nayarit (Map 106) were originally considered *Myotis californicus mexicanus* (Hall 1981); but it was elevated to a full species as *M. findleyi* (Bogan 1978). *M. findleyi* is the only species of *Myotis* present on the Tres Marías Archipelago, and differs from *M. californiacus* by having the limb length (dry measurement) greater than 50% of the tibia length. Nelson (1899) recorded the specimens as *M. nigricans*.

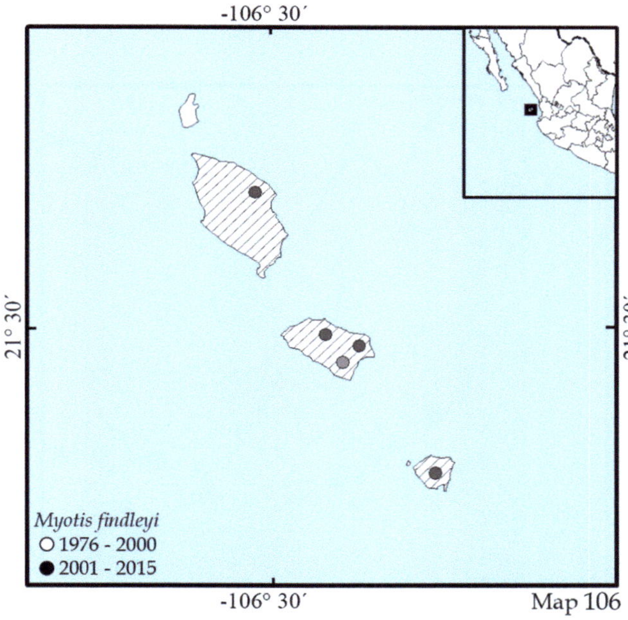

Map 106 Distribution map of *Myotis findleyi*

Additional Literature: Miller and Allen (1928), Villa-R. (1967), LaVal (1973a), Hall (1981), Wilson (1991), Koopman (1993), Simmons (2005).

Myotis fortidens Miller and Allen, 1928

Cinnamon myotis, murciélago ratón canela

1. *M. f. fortidens* Miller and Allen, 1928. Coastal areas from Sinaloa and Veracruz south through Chiapas and Campeche.
2. *M. f. sonoriensis* Findley and Jones, 1967. Sonora and northern Sinaloa.

Conservation: *Myotis fortidens* is listed as Least Concern by the IUCN (Perez et al. 2017).

Characteristics: Cinnamon myotis is small sized; forearm length 35.5–40.0 mm, skull length 14.8–15.5 mm. Dorsal pelage brown with blackish shades at the base; fur of underparts paler than in the dorsum; no face mask; snout, ears, and tragus brown, similar to the dorsal coloration; ears spear shaped and tragus narrow; uropatagium naked from the knees to the tail tip; calcar unkeeled; sagittal crest well developed; braincase not rising abruptly; supraoccipital region and braincase roof approximately at the same level. Morphotype leuconoe.

Comments: *Myotis fortidens* ranges in both coastal plains of México, from Sonora and Veracruz south through Chiapas (Map 107). This species does not have any distinctive characteristics that can be used to differentiate it from the other sympatric species of *Myotis* such as *M. albescens*, *M. californicus*, *M. nigricans*, *M. velifer*, *M. volans* and *M. yumanensis*. *M. fortidens* can be identified by the absence of the diagnostic characteristics of other species. Its characteristics are no face mask; snout, ears, and tragus brown, very similar to the dorsal coloration, with a very low contrast; ears spear shaped; tragus long and narrow; uropatagium naked from the knees to the tail tip; calcar unkeeled.

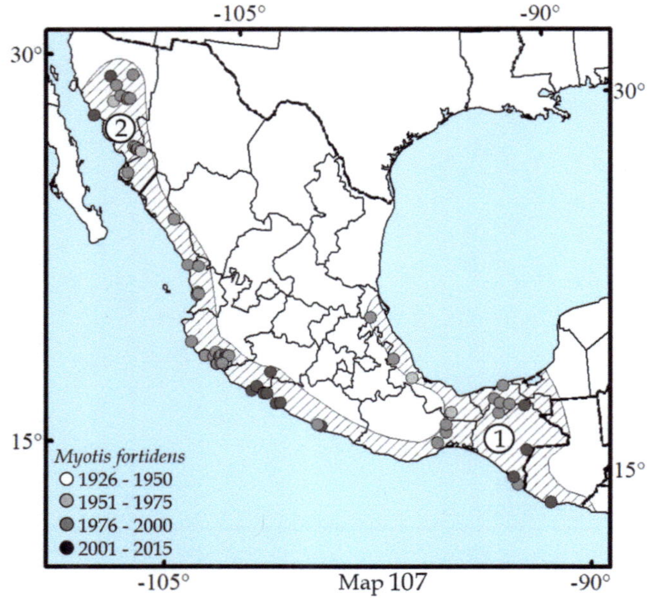

Map 107 Distribution map of *Myotis fortidens*

Additional Literature: Findley and Jones (1967), LaVal (1973a), Álvarez-Castañeda and Álvarez (1991), Bogan (1999), Villalpando and Álvarez (2000), Hoofer and Van den Bussche (2003), Simmons (2005), Sánchez-Hernández et al. (2010), Roehrs et al. (2010).

Myotis grisescens Howell, 1909
Gray myotis, murciélago ratón gris

Monotypic.

Conservation: *Myotis grisescens* is listed as Endangered by the United States Endangered Species Act (FWS 2022). However, it is listed as Vulnerable (A4bce: population reduction in the past 10 years equal to or greater than 30% including both the past and the future, and maybe not reversible, based on data obtained by an index of abundance, a decline in the area of occupancy, and the effects of introduced taxa) by the IUCN (Solari 2018j).

Characteristics: Gray myotis is small sized; forearm length 41.0–45.0 mm, skull length 14.8–15.5 mm. **Dorsal pelage uniformly monochromatic gray from the base to the tip;** fur of underparts paler and grizzle with pale gray; however, specimens sometimes cinnamon brown dorsally with fur of underparts pale buff; no face mask; snout, ears, and tragus brown; ears spear shaped; wing membrane attached to the ankles instead of the toe base; calcar unkeeled; braincase not rising abruptly; braincase supraoccipital region above the braincase roof level; sagittal crest present. Morphotype leuconoe.

Comments: *Myotis grisescens* ranges in southeastern United States, in Kansas, Oklahoma, Arkansas, Louisiana, Missouri, Illinois, Indiana, Kentucky, Tennessee, Virginia, the Carolinas, Georgia, Alabama, Mississippi, and eastern Florida (Map 108). It can be distinguished from other species in sympatry, or nearly so, such as *M. australoriparius*, *M. keenii*, *M. leibii*, *M. lucifugus,* and *M. sordalis*, by its monochromic pelage, in which the fur is uniformly gray from the base to the tip, fur of underparts paler and grizzled with pale gray. A less frequent coloration is dorsally cinnamon brown with fur of underparts pale buff; the uropatagium is attached to the ankles.

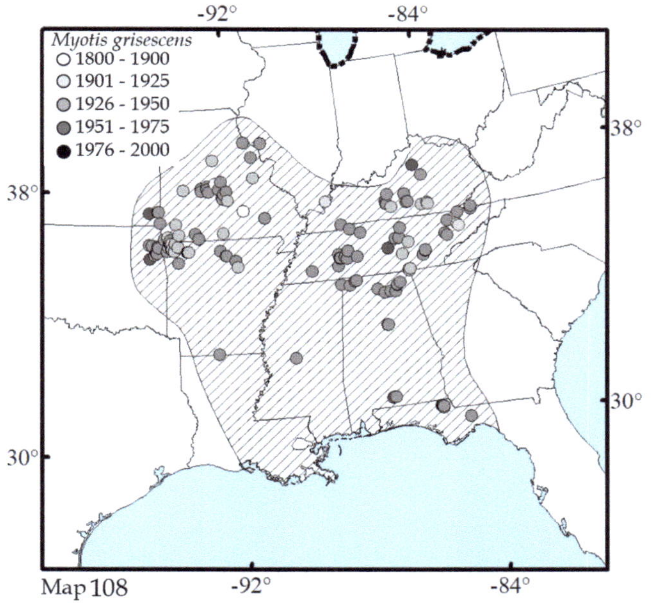

Map 108 Distribution map of *Myotis grisescens*

Additional Literature: Howell (1909), Miller and Allen (1928), LaVal (1967), Barbour and Davis (1969), Findley (1972), Schwartz and Schwartz (1981), Jones and Birney (1988), Decher and Choate (1995), Stadelmann et al. (2007).

Myotis keenii (Merriam, 1895)

Keen's myotis, murciélago ratón de la Columbia Británica

Monotypic.

Conservation: *Myotis keenii* is listed as Least Concern by the IUCN (Arroyo-Cabrales and Álvarez-Castañeda 2017i).

Characteristics: Keen's myotis is medium sized; forearm length 35.9–37.0 mm, skull length 14.7–15.0 mm. **Dorsal pelage brown with dark shoulder spots**; fur of underparts paler than in the dorsum; no face mask, however, rostrum and ears very dark brown; ears large (18.0–19.0 mm), if they fold over the rostrum, they go more than 2.0 mm beyond the nose; patagium of similar color as the dorsum; **border free from the uropatagium with distinguishable hair bangs**; hind foot less than 60% of the tibia length; third, fourth, and fifth metacarpals are of about equal in length; calcar with a small keel; **canine cingulum wider than 0.74 mm**; braincase not rising abruptly on the rostrum; braincase supraoccipital region above braincase roof level; sagittal crest little developed but present. Morphotype myotis.

Comments: *Myotis keenii* previously considered *M. septentrionalis* as a subspecies (Zyll de Jong 1979). Its ranges are restricted to the western British Columbia and Washington (Map 109). It can be distinguished from other species in sympatry, or nearly so, such as *M. leibii*, *M. lucifugus*, *M. septentrionalis*, and *M. sordalis*, by having the dorsal pelage brown with dark shoulder spots, no face mask but a very dark brown rostrum; large ears (18.0–19.0 mm) that, when folded over the rostrum, go more than 2.0 mm beyond the nose.

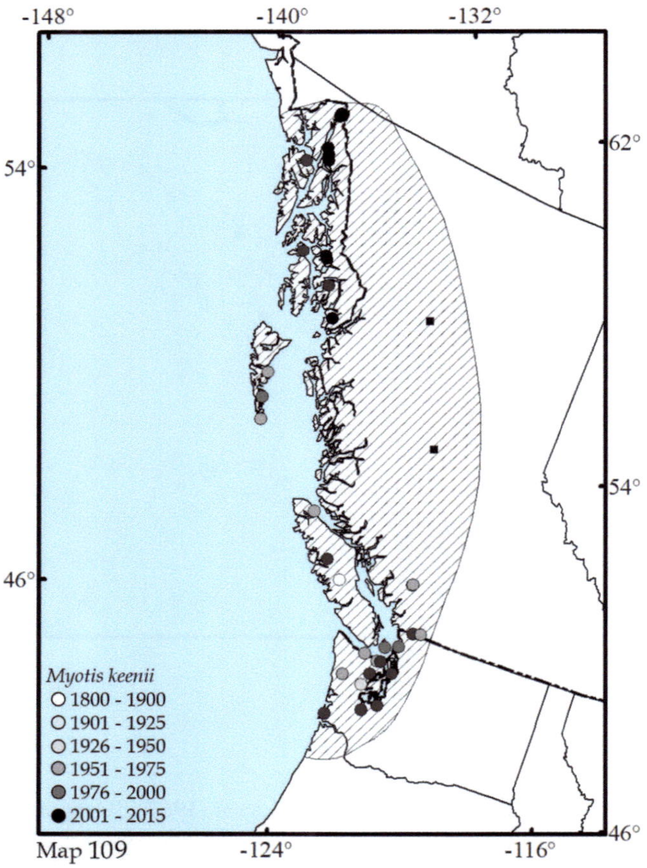

Map 109 Distribution map of *Myotis keenii*

Additional Literature: Miller and Allen (1928), Jones (1964), Barbour and Davis (1969), Genoways and Jones (1969), Findley (1960, 1972), Fitch and Shump (1979), Williams and Findley (1979), Stadelmann et al. (2007).

Myotis leibii (Audubon and Bachman, 1842)

Eastern small-footed myotis, murciélago ratón oriental

Monotypic.

Conservation: *Myotis leibii* is listed as Endangered (A4bce: population reduction in the last 10 years 50% or less include past, future, and may not be reversible with bases on the index of abundance, area od occupancy, and introduced taxa) by the IUCN (Solari 2018k).

Characteristics: Eastern small-footed myotis is small sized; forearm length 31.4–33.0 mm, skull length 13.0–14.0 mm. **Dorsal pelage dark yellow brown, black at the base; fur of underparts cinereous; muzzle, chin, rostrum, ears, and tragus blackish, forming a black face mask; uropatagium membrane black, with hair; tail tip sticking out about 1.5–2.5 mm of the uropatagium;** limbs proportionately small; calcar keeled ending in a small lobe; braincase appearing flat and rising gradually from the rostrum; braincase supraoccipital region above the braincase roof level; sagittal crest barely developed or absent. Morphotype selysius.

Comments: *Myotis leibii* was previously considered a subspecies of *M. subulatus* (van Zyll de Jong 1984). However, *M. subulatus* was not considered a valid species (Glass and Baker 1965, 1968) or different from *M. ciliolabrum* (van Zyll de Jong 1984). *M. leibii* ranges from Ontario, Québec, and New England south through Georgia and Alabama, and westward into Oklahoma (Map 110). It can be distinguished from other species in sympatry, or nearly so, such as *M. keenii*, *M. lucifugus*, *M. septentrionalis*, and *M. sordalis,* by its small size; black face mask that includes the muzzle, chin, rostrum, ears, and tragus, with a low contrast with the paler hair elsewhere; black wing membranes; small hind feet, less than one-half of the tibia length; calcar distinctly keeled. From *M. californicus,* by having a black face mask contrasting with the general paler coloration; skull less steeply sloping in the frontal area, flatter and broader. From *M. leibii,* by having a flatter cranium, and the dorsal pelage has a distinct golden sheen.

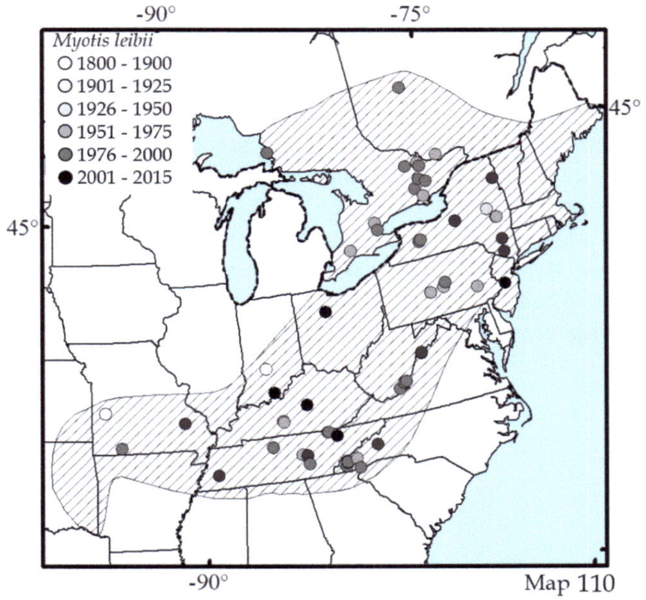

Map 110 Distribution map of *Myotis leibii*

Additional Literature: Miller and Allen (1928), Allen (1939), Bogan (1974), Godin (1977), McDaniel et al. (1982), van Zyll de Jong (1985), Koopman (1993), Best and Jennings (1997), Stadelmann et al. (2007).

Myotis lucifugus (Le Conte, 1831)

Little brown bat, pequeño murciélago ratón norteamericano

1. *M. l. alascensis* Miller, 1897. Coast from British Columbia to California, and northern Idaho and western of Montana.
2. *M. l. carissima* Thomas, 1904. From the Dakotas, Kansas, Colorado, and northern New Mexico west through Washington, Oregon, California, and central British Columbia.
3. *M. l. lucifugus* (Le Conte, 1831). From Alaska east through central Canada to Newfoundland south through Florida, except for the central Great Plains and southwestern United States.
4. *M. l. pernox* Hollister, 1911. Restricted to central western Alberta.
5. *M. l. relictus* Harris, 1974. Restricted to Sierra Nevada, California.

Conservation: *Myotis lucifugus* is listed as Endangered (A3be: population reduction in the last 10 years 50% or less projected to be met in future data obtain by index of abundance and introduced taxa) by the IUCN (Solari 2021).

Characteristics: Little brown bat is small sized; forearm length 36.0–40.0 mm, skull length 15.7–17.0 mm. Dorsal pelage brown; fur of underparts paler than in the dorsum; no face mask but rostrum and ears darker; ears spear shaped; tragus long and narrow; uropatagium dark brown; calcar unkeeled; braincase supraoccipital region above the braincase roof level; sagittal crest present.

Comments: *Myotis occultus, M. l. alascensis, M. l. carissima, M. l. pernox,* and *M. l. relictus* were elevated to full species (Morales and Carstens 2018); however, the latter four are still considered subspecies of *M. lucifugus* (Hoffmeister 1986; Francis and Simmons 2022). *M. lucifugus* ranges from Alaska east through central Canada to Newfoundland south through eastern Florida, except for the central Great Plains and southwestern United States (Map 111). This species does not have any diagnostic characteristics that can be used to distinguish it from the other sympatric species of *Myotis*. Its identification is based on the absence of the characteristics of the other species. However, some individuals can have a reduced number of upper and lower premolars. The species with sympatric distribution, or nearly so, are *M. austroriparius, M. californicus, M. ciliolabrum, M. evotis, M. grisescens, M. keenii, M. leibii, M. occultus, M. septentrionalis, M. sordalis, M. thysanodes, M. velifer, M. volans,* and *M. yumanensis*. *M. lucifugus* differs from the other species just mentioned based on the following traits. From *M. californicus, M. ciliolabrum,* and *M. leibii* by having larger hind feet, greater than one-half of the tibia length. From *M. evotis, M. keenii, M. occultus, M. thysanodes,* and *M. velifer,* by its small size (forearm and skull length less than 40.0 and 17.0 mm, respectively). From *M. septentrionalis*, by not having large ears extending more than 2.0 mm over the nose when folded over the face, and a long sharply pointed tragus. From *M. sordalis,* by not having a keeled calcar and hairs in the hind feet beyond the toes. From *M. volans*, by not having a shorter tibia and less hair on the wing undersurface. From *M. yumanensis,* by having larger dorsal hairs, darker at the tip; pelage glossy; ears darker; tragus smaller with the border not semicircular; tail relatively shorter; braincase rising at a less abrupt angle from the rostrum; a greater mastoid breadth.

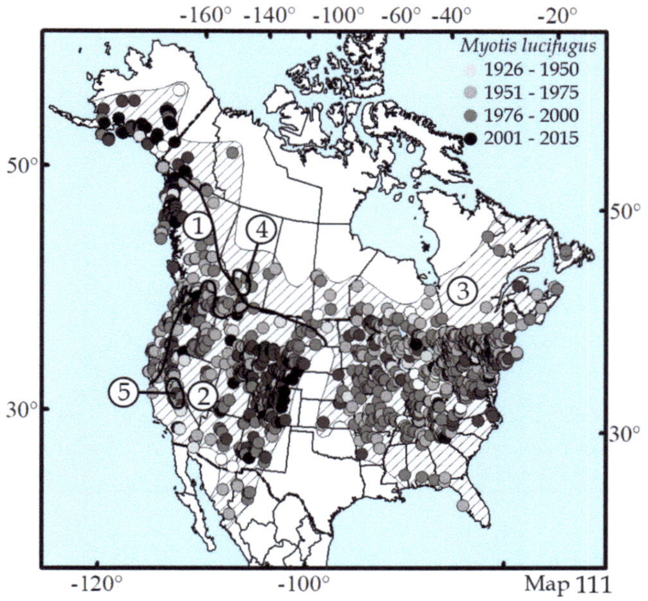

Map 111 Distribution map of *Myotis lucifugus*

Additional Literature: Miller and Allen (1928), Findley and Jones (1967), Davis and Rippy (1968), Álvarez and Ramírez-Pulido (1972), Harris (1974), Williams and Findley (1979), Fenton and Barclay (1980), Bogan (1999), Piaggio et al. (2002), Stadelmann et al. (2007).

Myotis nigricans (Schinz, 1821)

Black myotis, murciélago ratón negro

1. *M. n. carteri* LaVal, 1973. Coast of Nayarit and Jalisco.
2. *M. n. nigricans* (Schinz, 1821). From southern Tamaulipas south through Oaxaca and Chiapas.

Conservation: *Myotis nigricans carteri* is listed as Especial Protection in the Norma Oficial Mexicana (DOF 2019). However, *M. nigricans* is considered Least Concern by the IUCN (Solari 2019h).

Characteristics: Black myotis is small sized; forearm length 31.0–39.6 mm, skull length 12.8–14.5 mm. **Dorsal pelage dark brown to blackish**, occasionally reddish brown, blackish at the base; fur of underparts cream, yellowish, or brown at the tip **but blackish at the base; hair 6.0–8.0 mm in length**; no face mask; membranes, rostrum, and ears blackish; ears spear shaped and tragus narrow; calcar unkeeled; sagittal crest absent; braincase not rising very abruptly and slightly inflated; braincase supraoccipital region below the roof level. Morphotype selysius.

Comments: *Myotis carteri* was listed as a subspecies of *M. nigricans* (Corbet and Hill 1980; Simmons 2005); it was previously considered a full species (Ceballos 2014). *M. nigricans* ranges from Tamaulipas and Nayarit south through South America (Map 112). It can be distinguished from other species in sympatry, or nearly so, such as *M. californicus*, *M. elegans*, *M. fortidens*, *M. pilosatibialis*, *M. thysanodes*, and *M. velifer*, by the combination of the following characteristics: forearm 2.0–5.0 mm longer than the third metacarpal; hair tip normally contrasting slightly with the base; dorsal uropatagium with fur but not above the knees.

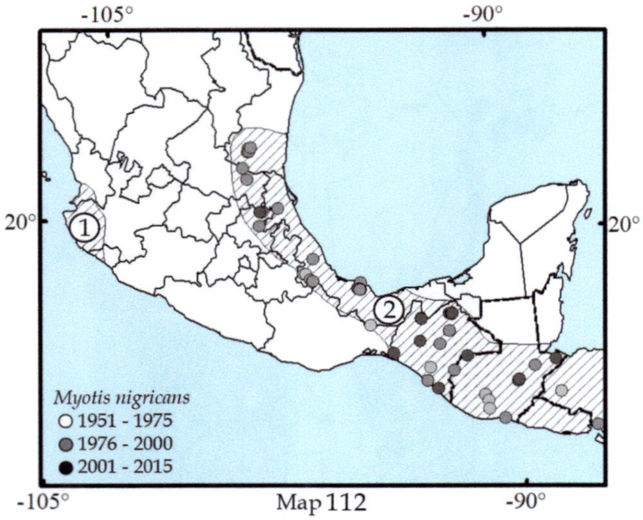

Map 112 Distribution map of *Myotis nigricans*

Additional Literature: LaVal (1973a), Miller and Allen (1928), Wilson and LaVal (1974), Bogan (1978), López-González (2001), Ruedi and Mayer (2001), Villalpando and Álvarez (2000), Stadelmann et al. (2007).

Myotis occultus Hollister, 1909

Arizona myotis, murciélago ratón de Arizona

Monotypic.

Conservation: *Myotis occultus* is listed as Least Concern by the IUCN (Solari 2018l).

Characteristics: Arizona myotis is small sized; forearm length 33.0–41.0 mm, skull length 14.0–16.0 mm. Dorsal pelage brilliant yellowish brown, dark at the base and pale buff at the tip; fur of underparts yellowish to pale gray; no face mask; snout and ears large, dark brown or blackish; limbs relatively large; calcar unkeeled; braincase not rising abruptly; braincase supraoccipital region and roof almost at the same level; without a sagittal crest. Morphotype leuconoe.

Comments: *Myotis occultus* was previously considered a subspecies of *M. lucifugus* (Findley and Jones 1967; Hoffmeister 1986). It ranges from Utah and Colorado to the Eje Neovolcánico. It is found mainly in the central and western to the Altiplanicie Mexicana (Map 113). *M. occultus* does not have any distinctive characteristics that can be used for differentiating it from the other species of *Myotis*. However, some individuals can have a reduced number of upper and lower premolars. The species with sympatric distribution, or nearly so, are *M. auriculus*, *M. californicus*, *M. evotis*, *M. lucifugus, M. thysanodes*, *M. velifer*, *M. volans*, and *M. yumanensis.* The characteristics of *M. occultus* include absence of a face mask; snout, ears, and tragus brown, very similar to the dorsal coloration, with very low contrast; ears spear shaped; tragus long and narrow; uropatagium naked from the knees to the tail tip; calcar unkeeled.

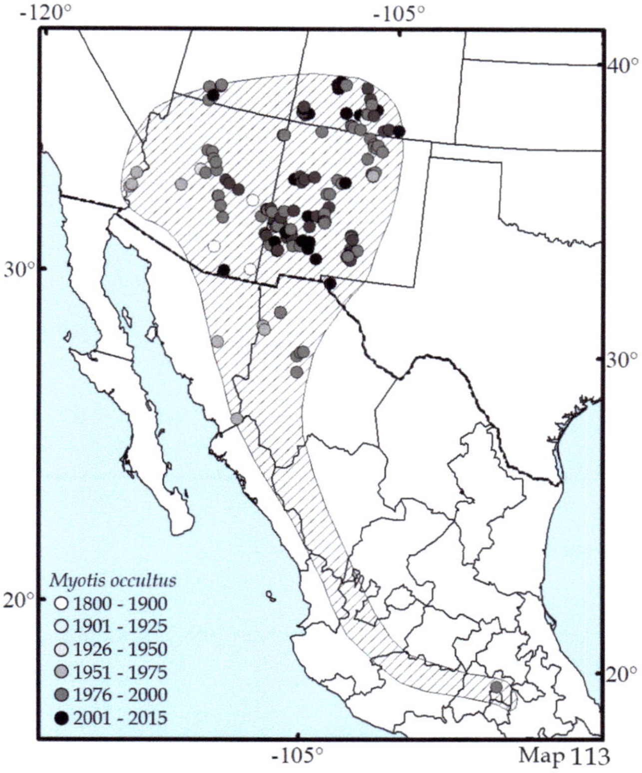

Map 113 Distribution map of *Myotis occultus*

Additional Literature: Miller and Allen (1928), Barbour and Davis (1970), Fenton and Barclay (1980), Piaggio et al. (2002), Stadelmann et al. (2007).

Myotis pilosatibialis LaVal, 1973

Hairy-legged myotis, murciélago ratón patas peludas

Monotypic.

Conservation: *Myotis pilosatibialis* (as *Myotis keaysi*) is listed as Least Concern by the IUCN (Barquez and Diaz 2016b).

Characteristics: Hairy-legged myotis is small sized; forearm length 32.0–34.0 mm, skull length 12.6–14.7 mm. Dorsal pelage grayish brown to reddish dark brown; fur of underparts paler than in the dorsum; no face mask; rostrum skin and ears base pinkish brown; ears spear shaped and tragus narrow grayish brown; **uropatagium with hair in its first half, the second half with less hair tufts**; membranes blackish brown; hair in the uropatagium extending along the tibia to the limbs; calcar unkeeled; braincase rising abruptly from the rostrum, spherical; supraoccipital region and braincase roof at the same level; sagittal crest present. Morphotype selysius.

Comments: *Myotis pilosatibialis* was elevated to full species (Mantilla-Meluk and Muñoz-Garay 2014); it was previously considered a subspecies of *M. keaysi* (LaVal 1973a). *M. pilosatibialis* ranges from Tamaulipas and Oaxaca south through Central America, including the Yucatán Peninsula (Map 114). It differs from the other sympatric species of *Myotis*, such as *M. albescens, M. elegans, M. fortidens, M. nigricans,* and *M. thysanodes*, by having the uropatagium with hair in its first half, the second half with less hair tufts. *M. pilosatibialis* differs from the other sympatric species of *Myotis* by having hair in the first uropatagium half; the second with less hair arranged in tuftson transverse lines.

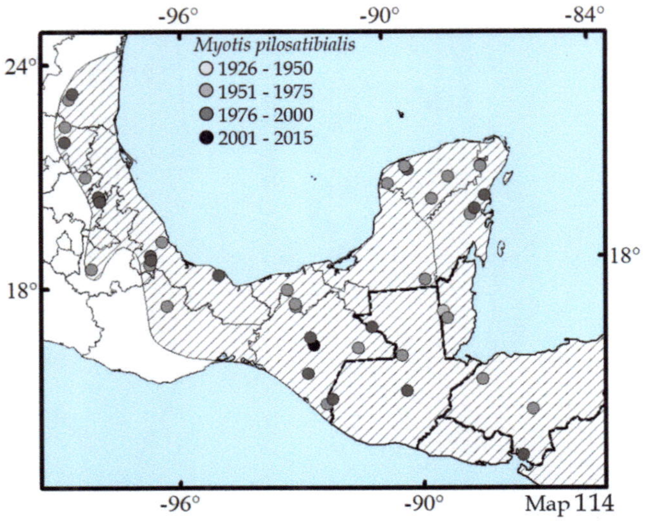

Map 114 Distribution map of *Myotis pilosatibialis*

Additional Literature: Hall (1981), Álvarez-Castañeda and Álvarez (1991), Koopman (1993), Hoofer and Van den Bussche (2003), Hernández-Meza et al. (2005), Roehrs et al. (2010).

Myotis planiceps Baker, 1955

Flat-headed myotis, murciélago ratón de cabeza plana

Monotypic.

Conservation: *Myotis planiceps* is listed as Endangered (D: less than 250 mature individuals) by the IUCN (Arroyo-Cabrales and Ospina-Garces 2016b) and as Endangered by the Norma Oficial Mexicana (DOF 2019).

Characteristics: Flat-headed myotis is small sized; forearm length 25.6–27.5 mm, skull length 13.5–14.5 mm. Dorsal pelage brown, blackish at the base and cinnamon brown at the tip; fur of underparts yellowish; hairs silky long, 8.0–9.5 mm in length; no face mask; calcar unkeeled; **skull flatter than in any other species of *Myotis*, where the braincase is barely differentiated in height from the rostrum; molars relatively small relative to the palatal surface.** Morphotype leuconoe.

Comments: *Myotis planiceps* ranges in a very small area of Coahuila, Durango, Nuevo León, and Zacatecas (Map 115). It can be distinguished from other species in sympatry, or nearly so, such as *M. auriculus*, *M. californicus*, *M. ciliolabrum*, *M. thysanodes*, and *M. velifer*, by having the head flat, with the braincase approximately at the same level as the rostrum.

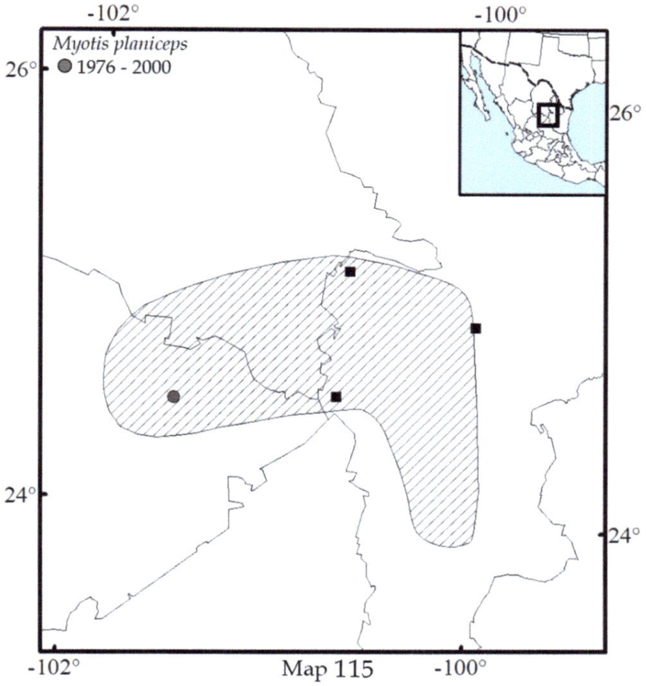

Map 115 Distribution map of *Myotis planiceps*

Additional Literature: Baker (1955), Jimenez (1968), Matson and Patten (1975), Matson (1975), Arroyo-Cabrales et al. (2005), Simmons (2005).

Myotis septentrionalis (Trouessart, 1897)

Northern myotis, murciélago ratón grande del norte

Monotypic.

Conservation: *Myotis septentrionalis* is listed as Threatened by the United States Species Act (FWS 2022). However, it is listed as Near Threatened (A3bce: Population reduction projected in the future by index of abundance, area of occupancy, and effects of introduced taxa) by the IUCN (Solari 2018m) and the Canadian Endangered Species Protection Act.

Characteristics: Northern myotis is large sized; forearm length 34.0–38.0 mm, skull length 14.6–15.4 mm. Dorsal pelage pale brown without a spot in the shoulders; fur of underparts paler than in the dorsum; **no face mask; rostrum and ears large (14.0–19.0 mm), blackish to dark brown; if they fold over the rostrum, they go more than 2.0 mm beyond the nose;** uropatagium with no or very few hairs dispersed, similar in color to the dorsal coloration; hind feet less than 60% of the tibia length; third, fourth, and fifth metacarpals about equal in length; calcar unkeeled; **canine cingulum narrower, less or more than 0.74 mm**; braincase not rising abruptly on the rostrum; supraoccipital region and braincase roof almost at the same level; sagittal crest barely developed or absent. Morphotype myotis.

Comments: *Myotis septentrionalis* was previously considered a subspecies of *Myotis keenii* (Zyll de Jong 1979). *M. septentrionalis* ranges from the Great Plains of Canada and the United States, from Saskatchewan east through New Newfoundland and south through Mississippi to South Carolina; in addition, it inhabits the western coast of British Columbia and Alberta (Map 116, not in the map). It can be distinguished from other species in sympatry, or nearly so, such as *M. keenii, M. leibii, M. lucifugus*, and *M. sordalis*, by its larger (14.0–19.0 mm) and dark brown blackish ears, which go more than 2.0 mm beyond the nose if they fold over the rostrum; uropatagium with no or very few dispersed hairs; third, fourth, and fifth metacarpals are about equal in length. From *M. lucifugus,* by its longer ears and tragus; relatively longer tail; however, *M. lucifugus* has a glossier pelage. From *M. evotis,* by its darker pelage and paler membranes. From *M. keenii,* by the absence of dark shoulder spots and the border free from the uropatagium with distinguishable hair bangs.

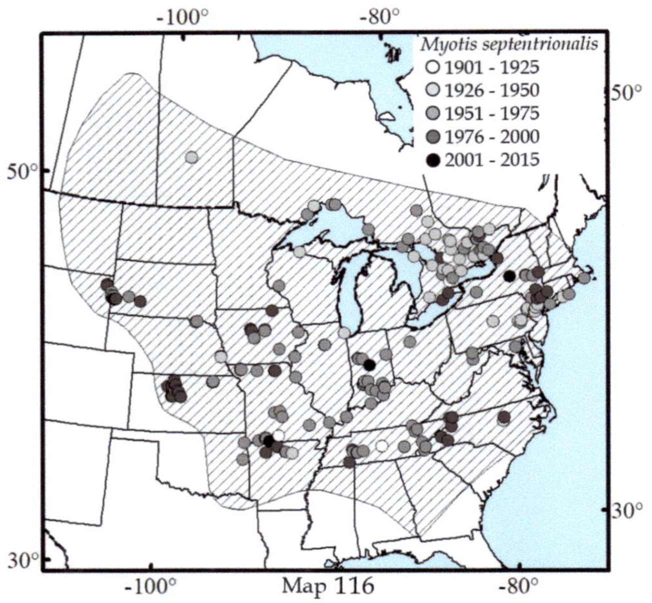

Map 116 Distribution map of *Myotis septentrionalis*

Additional Literature: Miller and Allen (1928), Findley (1972), Caire et al. (1979), Williams and Findley (1979), Zyll de Jong (1985), Koopman (1993), Cáceres and Barclay (2000), Simmons (2005), Stadelmann et al. (2007).

Myotis sodalis Miller and Allen, 1928

Indiana myotis, murciélago ratón de Indiana

Monotypic.

Conservation: *Myotis sodalis* is listed as Endangered by the United States Species Act (FWS 2022). However, it is listed as Near Threatened (without details) by the IUCN (Arroyo-Cabrales and Ospina-Garces 2016c).

Characteristics: Indiana myotis is small sized; forearm length 36.0–40.6 mm, skull length 14.2–15.0 mm. Dorsal pelage pale brown to black, most commonly chestnut gray; fur of underparts with slaty hair at the base and grayish-white at the tip; no face mask; muzzle, ears and membranes brown; tragus small, less than one-half of the ear length; **tail projecting 1.0–2.0 mm out of the uropatagium; uropatagium with some hair at the knee level; calcar long with a low keel**; braincase supraoccipital region above the braincase roof level; sagittal crest very low, sometimes absent in young individuals.

Comments: *Myotis sodalis* ranges from Arkansas, Oklahoma, Missouri, and Iowa eastward to New Hampshire and North Carolina, plus one record from northern Florida, not including the coastal plains in the Atlantic Coast (Map 117). It can be distinguished from other species in sympatry, or nearly so, such as *M. grisescens, M. keenii, M. leibii, M. lucifugus,* and *M. septentrionalis*, by its tail projecting out of the uropatagium (1.0–2.0 mm); some hair at the knee level; calcar long with a low keel, and related to cavernous limestone areas.

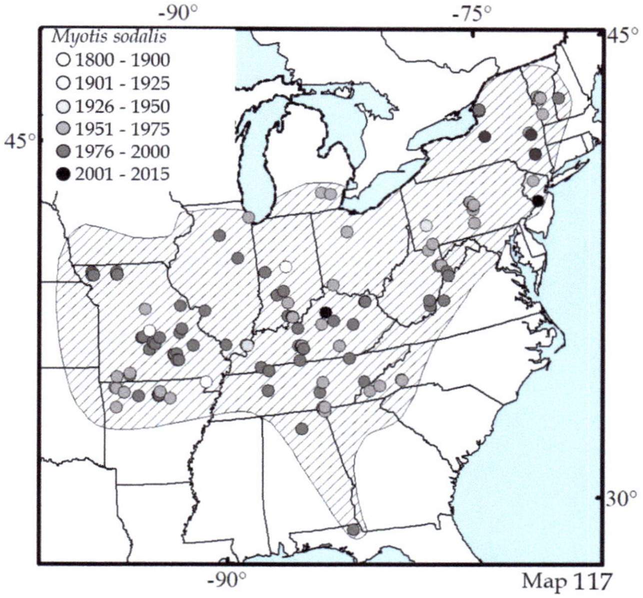

Map 117 Distribution map of *Myotis sodalis*

Additional Literature: Miller and Allen (1928), Lyon (1936), Glass and Ward (1959), Hall (1962), Barbour and Davis (1969), Findley (1972), LaVal and LaVal (1980), Thomson (1982), Cáceres and Barclay (2000), Stadelmann et al. (2007).

Myotis thysanodes Miller, 1897

Fringed myotis, murciélago ratón peludo

1. *M. t. aztecus* Miller and Allen, 1928. From Hidalgo and Estado de México, south through Chiapas.
2. *M. t. pahasapensis* Jones and Genoways, 1967. Restricted to western Wyoming, eastern South Dakota, and Nebraska.
3. *M. t. thysanodes* Miller, 1897. From Hidalgo and Estado de México northward to Montana, British Columbia, and Washington.

Conservation: *Myotis thysanodes* is listed as Least Concern by the IUCN (Arroyo-Cabrales and de Grammont 2017b).

Characteristics: Fringed myotis is the largest mainland *Myotis;* **forearm length 40.0–44.0 mm**, skull length 16.2–17.2 mm. Dorsal pelage yellowish to cream; fur of underparts whitish to cream; no face mask; rostrum skin and ears brown, similar in color to the dorsal region; ears go 3.0 mm beyond the nose when folded forward in fresh specimens; **uropatagium with little hair from the knees to the tail but with notorious bangs from the calcar to the tail tip**; hair blackish at the base, contrasting with the pale color at the tip; membranes blackish; calcar not distinctly keeled, robust and wide; braincase supraoccipital region and roof almost at the same level; sagittal crest well developed. Morphotype myotis.

Comments: *Myotis thysanodes* ranges in western North America from southern British Columbia south through the highlands of Chiapas, excluding all the coastal states of México and the Baja California Peninsula; in the United States, from New Mexico, Colorado, and Idaho westward, and one population in the area of South Dakota, Nebraska, and Wyoming (Map 118). It can be distinguished from other species in sympatry, or nearly so, such as *M. auriculus*, *M. californicus*, *M. ciliolabrum*, *M. evotis*, *M. lucifugus*, *M. occultus*, *M. velifer*, *M. volans*, and *M. yumanensis*, by the presence of a well-developed fringe of hair on the uropatagium posterior edge (backlighting view). The species most similar to *M. thysanodes* and with which it could be misidentified is *M. evotis*; however, *M. thysanodes* has shorter ears on average (16.0–19.0 mm).

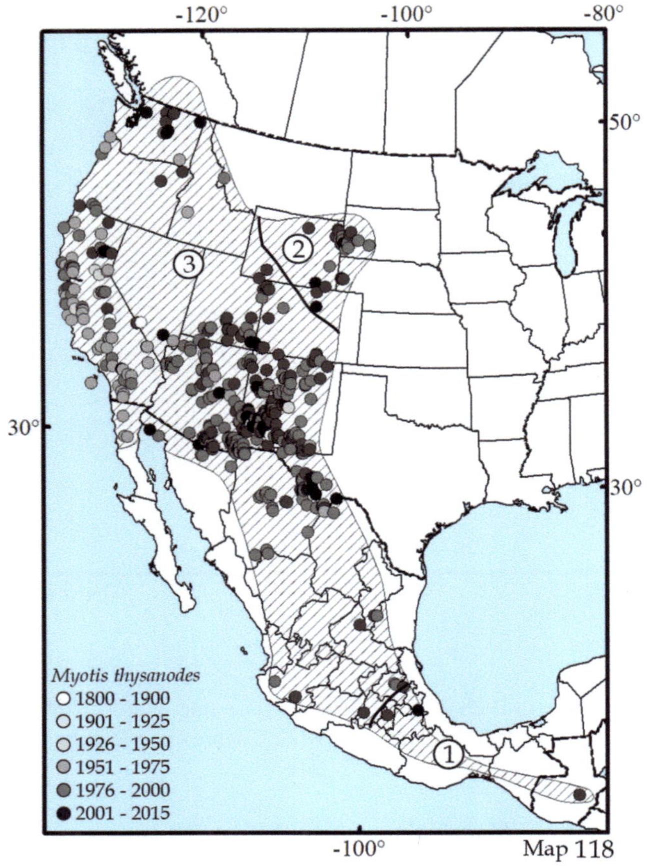

Map 118 Distribution map of *Myotis thysanodes*

Additional Literature: Miller and Allen (1928), Hall and Kelson (1959), Williams (1968), Jones and Choate (1978), Williams and Findley (1979), Boyce (1980), O'Farrell and Studier (1980), Manning and Jones (1988b), Bogan (1999), Ruedi and Mayer (2001), Stadelmann et al. (2007), Álvarez-Castañeda et al. (2008c)

Myotis velifer (Allen, 1890)

Cave myotis, murciélago orejas de ratón de las cuevas

1. *M. v. brevis* Vaughan, 1954. Northern Sonora, California, and Arizona.
2. *M. v. grandis* Hayward, 1970. Southern Kansas, Oklahoma, and Northern Texas.
3. *M. v. incautus* (Allen, 1896). From Chihuahua, New Mexico, and Texas south through Durango, Zacatecas, and Tamaulipas.
4. *M. v. peninsularis* Miller, 1898. Southern of Baja California Sur.
5. *M. v. velifer* (Allen, 1890). From Durango, Zacatecas, and Tamaulipas south through Chiapas.

Conservation: *Myotis peninsularis* is listed as Endangered (B1ab (i, iii): extent of occurrence less than 5000 km^2 in five or less populations and continuing decline in extent of occurrence and quality of habitat) by the IUCN (Arroyo-Cabrales and Ospina-Garces 2016d). However, it is now considered a subspecies of *M. velifer*, so it should be considered Least Concern by the IUCN.

Characteristics: Cave myotis is large sized; forearm length 41.0–47.0 mm, skull length 14.2–17.6 mm. Dorsal pelage grayish brown, **a small area with little hair between the shoulders**; fur of underparts cream or yellowish; no face mask; rostrum and ears dark grayish brown, tragus narrow and pointed; uropatagium bare from the knees to the tail; membranes brown; ears small, and if they bend over the rostrum, they do not go more than 2.0 mm beyond the nose; calcar well developed, ending in one lobe but unkeeled; rostrum robust and wide; braincase supraoccipital region above the braincase roof level; sagittal crest well developed. Morphotype leuconoe.

Comments: *Myotis peninsularis* is a subspecies of *M. velifer* (Nájera-Cortazar and Álvarez-Castañeda 2015); it was previously considered a full species (Álvarez-Castañeda and Bogan 1998). *M. velifer* ranges from California to Texas, Oklahoma, and southern Kansas south through all México, except for the coastal plains of the Gulf side, the Yucatán Peninsula, and the coasts of Oaxaca and Chiapas (Map 119). It can be distinguished from other species in sympatry, or nearly so, such as *M. auriculus*, *M. californicus*, *M. ciliolabrum*, *M. evotis*, *M. lucifugus*, *M. occultus*, *M. thysanodes*, *M. volans*, and *M. yumanensis*, by the presence of a small area with little hair between the shoulders. The most similar species is *M. thysanodes*, from which it can be differentiated by the absence of a fringe of hair on the uropatagium posterior edge.

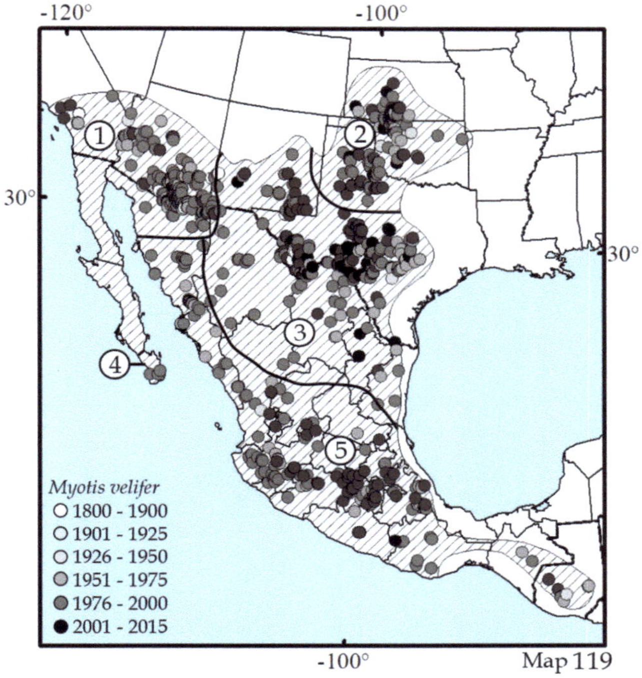

Map 119 Distribution map of *Myotis velifer*

Additional Literature: LaVal (1973a), Miller and Allen (1928), Glass and Ward (1959), Jones et al. (1971), Williams and Findley (1979), Fitch et al. (1981), Bogan (1999), Villalpando and Álvarez (2000), Hoofer et al. (2003), Stadelmann et al. (2007), Álvarez-Castañeda et al. (2008c).

Myotis vivesi Menegaux, 1901

Fish-eating myotis, murciélago ratón pescador

Monotypic.

Conservation: *Myotis vivesi* is listed as Endangered in the Norma Oficial Mexicana (DOF 2019) and as Vulnerable (A2ae: population reduction in the past 10 years equal to or greater than 30% and maybe not reversible based on direct observation and the effects of introduced taxa) by the IUCN (Arroyo-Cabrales and Ospina-Garces 2016e).

Characteristics: Fish-eating myotis is the largest *Myotis* in North America; **forearm length 59.0–62.5 mm, skull length 21.0–22.0 mm.** Dorsal pelage medium brown to chestnut-colored; fur of underparts whitish; **limbs, toes, and finger claws markedly extended and developed; limbs almost one-half of the head-and-body length;** no face mask; rostrum skin and ears brown; **ears large, tragus narrow and pointed**; calcar unkeeled; **big size, larger than 20.0 mm**; braincase not rising abruptly; braincase supraoccipital region above the braincase roof level. Morphotype pyzonyx.

Comments: *Myotis vivesi* is a sea fisher bat that can be found in the seashore of the Gulf of California off the Baja California Peninsula and southern Sonora, and at the Baja California Sur western coast (Map 120). It can be distinguished from other species in sympatry, or nearly so, such as *M. californicus, M. evotis, M. thysanodes, M. velifer, M. volans,* and *M. yumanensis*, by its large size; forearm length greater than 56.0 mm; toe claws very long, flat laterally, and nearly of the same length as the tibia.

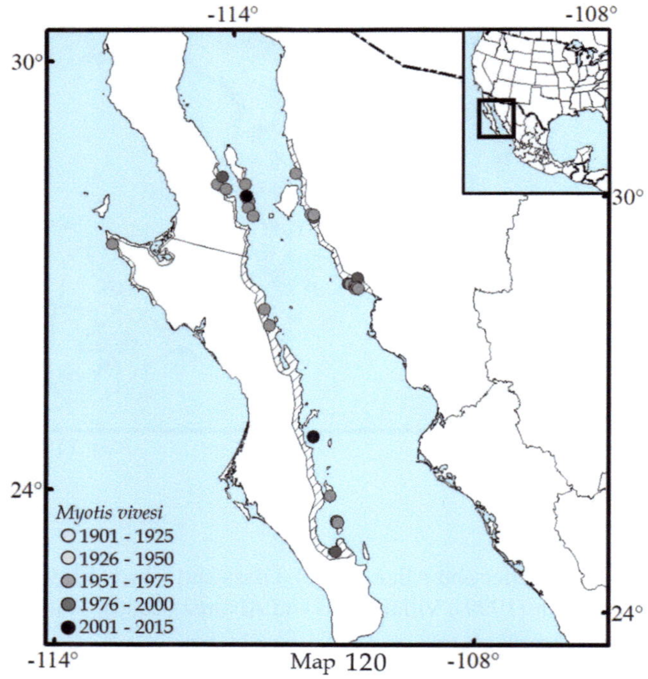

Map 120 Distribution map of *Myotis vivesi*

Additional Literature: Miller and Allen (1928), Cockrum and Bradshaw (1963), Baker and Patton (1967), Fish et al. (1991), Blood and Clark (1998), Stadelmann et al. (2004, 2007), Álvarez-Castañeda et al. (2008c), Frick et al. (2008).

Myotis volans (Allen, 1866)

Long-legged myotis, murciélago ratón de patas largas

1. *M. v. amotus* Miller, 1914. Trans-Mexican Volcanic Belt from Colima to Veracruz.
2. *M. v. interior* Miller, 1914. From northwestern Washington to North Dakota south through California, Arizona, and Durango
3. *M. v. longicrus* (True, 1886). British Columbia, coastal Oregon, and central California, including Alberta
4. *M. v. volans* (Allen, 1866). Baja California Peninsula.

Conservation: *Myotis volans* is listed as Least Concern by the IUCN (Solari 2019i).

Characteristics: Long-legged myotis is small sized; forearm length 35.2–41.2 mm, skull length 12.2–15.0 mm. Dorsal pelage dark to reddish brown; fur of underparts paler than in the dorsum; no face mask; snout and ears brown; snout with hair up to the nose; **ears small and rounded**, and if folded on the rostrum, they do not go more than 2.0 mm beyond the nose; **inner patagium with little hair in an imaginary line from the elbows to the knees**; calcar keeled, skull delicate; rostrum short; braincase rising abruptly from the rostrum; braincase supraoccipital region above the braincase roof level. Morphotype leuconoe.

Comments: *Myotis volans* ranges mainly in western North America, from British Columbia and Alberta south through the highlands of Oaxaca, and from North Dakota to Texas and westward, including the Baja California Peninsula (Map 121). It can be distinguished from other species in sympatry, or nearly so, such as *M. auriculus*, *M. californicus*, *M. ciliolabrum*, *M. evotis, M. lucifugus, M. occultus, M. velifer, M. thysanodes,* and *M. yumanensis*, by having the dorsal pelage generally dark brown to blackish; if other coloration is present, the hair is blackish at the tip and always long, 6.0–8.0 mm.

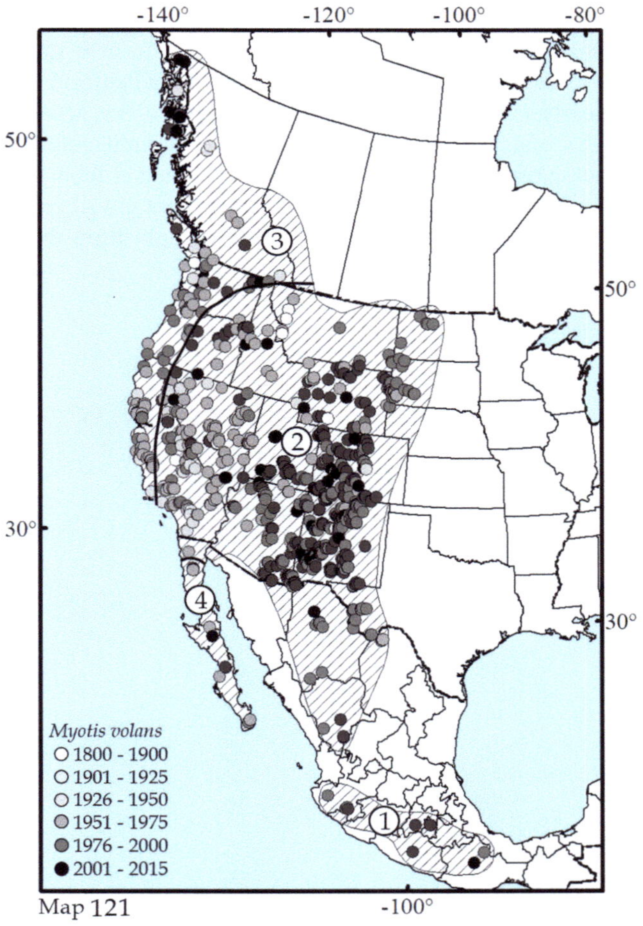

Map 121 Distribution map of *Myotis volans*

Additional Literature: Miller and Allen (1928), Baker and Phillips (1965), Findley (1972), Williams and Findley (1979), Warner and Czaplewski (1984), Stadelmann et al. (2007), Álvarez-Castañeda et al. (2008c).

Myotis yumanensis (Allen, 1864)

Yuma myotis, murciélago ratón pequeño

1. *M. y. lambi* Benson, 1947. Restricted to central Baja California.
2. *M. y. lutosus* Miller and Allen, 1928. From Sinaloa, southern Durango, and Zacatecas south through Jalisco, Michoacán, and Estado de México.
3. *M. y. oxalis* Dalquest, 1947. Central Valley of California.
4. *M. y. saturatus* Miller, 1897. From coastal Washington south through southern California.
5. *M. y. sociabilis* Grinnell, 1914. From southeastern British Columbia to northern California and Nevada east through Montana and Idaho.
6. *M. y. yumanensis* (Allen, 1864). From Nevada, Utah, Colorado, and Oklahoma south through Sonora, Baja California Peninsula, and Durango.

Conservation: *Myotis yumanensis* is listed as Least Concern by the IUCN (Solari 2019j).

Characteristics: Yuma myotis is small sized; forearm length 32.0–38.0 mm, skull length 13.0–14.2 mm. Dorsal pelage speckled yellowish to brown; it varies according to the subspecies; fur of underparts pale yellowish; no face mask, membranes pale brown; ears small, if folded on the rostrum, they do not go more than 2.0 mm beyond the nose; **limbs relatively large and robust, greater than 41% of the forearm length**; tail extending out of the uropatagium; **hind feet proportionally large and with hair**; calcar straight, long, and unkeeled; **braincase rising notoriously from the rostrum; supraoccipital region and braincase roof almost at the same level, almost spherical**; sagittal crest absent. Morphotype leuconoe.

Comments: *Myotis yumanensis* ranges mainly in western North America, from British Columbia south through central México, and from Wyoming to Texas westward, including the Baja California Peninsula (Map 122). It can be distinguished from other species in sympatry, or nearly so, such as *M. auriculus*, *M. californicus*, *M. ciliolabrum*, *M. evotis*, *M. lucifugus*, *M. occultus*, *M. velifer*, *M. thysanodes*, and *M. volans*, by its small size and limbs relatively large and robust, greater than 41% of the forearm length; the braincase rises notoriously from the rostrum and is almost spherical in shape. From *M. lucifugus*, by having shorter dorsal hairs with paler tips; the pelage is duller and not glossy; paler ears; larger tragus with a semicircular border; tail relatively longer; the braincase rises at an abrupt angle from the rostrum, and mastoid breadth is smaller.

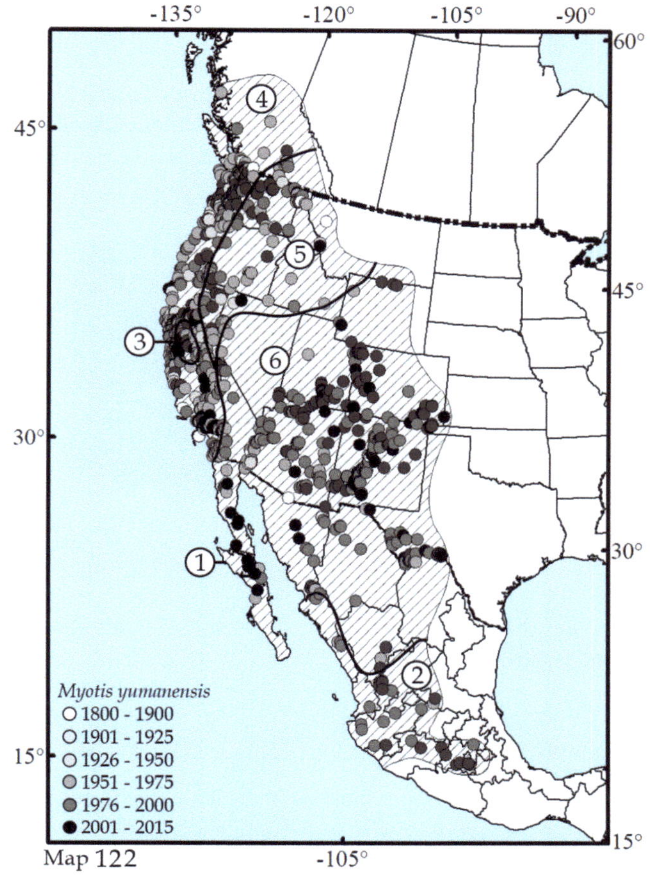

Map 122 Distribution map of *Myotis yumanensis*

Additional Literature: Miller and Allen (1928), Dalquest (1947), Glass and Baker (1968), Harris (1974), Herd and Fenton (1983), Álvarez et al. (1994), Ruedi and Mayer (2001), Villalpando and Álvarez (2000), Stadelmann et al. (2007), Weller et al. (2007), Álvarez-Castañeda et al. (2008c), Frick et al. (2008), Braun et al. (2015).

Subfamily Vespertilioninae Gray, 1821

The subfamily Vespertilioninae is characterized by its medium to large size; tail within the uropatagium, in a few species extending slightly beyond the uropatagium margin; ears ranging from proportionally small to large; sternum slender, its entire length considerably more than twice greater than the presternum width; median lobe much smaller than the body of the presternum; six ribs connected with the sternum; seventh cervical vertebra not fused with the first dorsal vertebra; scapula with the coracoid curved outward; nostrils simple, sometimes tubularly elongated but never bordered by special outgrowths; three lower incisors in all known genera. The species considered within the tribe follow (Roehrs et al. 2010). The keys were elaborated based on the review of specimens and the following sources: Miller (1907), Hall and Dalquest (1950), Lawlor (1979), Hall (1981), Álvarez et al. (1994), and Álvarez-Castañeda et al. (2017a).

1. Nostrils open frontward behind a horseshoe-shaped fold or have a "piglike" shape; ears and tragus large; if folded frontward, they go 3.0 mm over the nose .... Tribe Antrozoini 2

1a. Nostrils open laterally or vertically; ears and tragus wide; if folded frontward, only in a few species they go less than 3.0 mm over the nose 3

2. Hair paler basally than at the tip; two lower incisors at each side; midfrontal area slightly convex (from British Columbia, Canada, Washington east through Kansas, United States, south through central México).......... .... *Antrozous pallidus* (p. 357)

2a. Hair basally darker than at the tip; three lower incisors at each side; frontal area flat (from Veracruz, México, south through Central America) .... *Bauerus dubiaquercus* (p. 358)

3. Tip of dorsal pelage silver-white; interfemoral membrane furred above on the basal half; rostrum wide with two concavities (Fig. 59; from southeastern Alaska, British Columbia, and southern Canada south through Texas, including all the United States) .... tribe Eptesicini *Lasionycteris noctivagans* (p. 343)

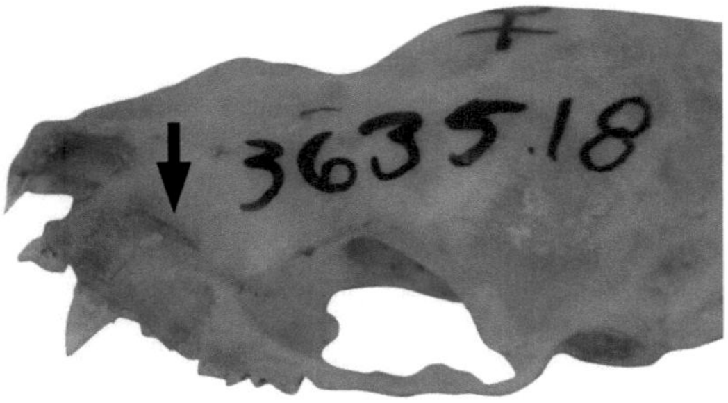

Fig. 59 Rostrum wide with two concavities like in *Lasionycteris noctivagans*

3a. Tip of dorsal pelage similar or paler than the medium part; interfemoral membrane, if is furred above, all completed; rostrum narrow without two concavities 4

4. Two upper premolars at each side 5

4a. One upper premolar at each side 10

5. Uropatagium hairless or with scarce hair; two upper incisors at each side; metacarpal of the third, fourth, and fifth toes approximately equal in length 6

5a. Uropatagium with abundant hair at the base, color speckled in ash, yellow, or reddish-cinnamon shades; one upper incisor at each side; metacarpal of the third, fourth, and fifth toes progressively shorter .... tribe Lasiurini *Lasiurus* (p. 344)

6. Ears large, twice the head length; auditory bullae large .... tribe Plecotini 7

6a. Ears short, less than twice the head length; auditory bullae small .... tribe Perimyotini 9

7. Two white dorsal spots; two lower premolars; lower canines small and forked (from Montana and California, United States, south through Durango, México) .... *Euderma maculatum* (p. 390)

7a. No white dorsal spots; three lower premolars; lower canines well developed and unicuspid 8

8. Nostrils not extending backward; rhinarium gland between the nostrils and the eyes barely noticeable; calcar keeled; basal interauricular lobe projecting in leaf shape; braincase breadth greater than one-half of the skull length (from California and Utah, United States, south through central México) .... *Idionycteris phyllotis* (p. 391)

8a. Nostrils extending backward; glandular bulkiness clearly visible between the nostrils and the eyes; calcar unkeeled; no interauricular basal lobe; braincase breadth less than one-half of the skull length .... *Corynorhinus* (p. 386)

9. Limb length less than one-half of the tibia length; dorsal skull outline (side view) more or less straight; palatine extending much more to the back of the last molar (from Washington east through Oklahoma, United States and south through Guerrero, Estado de México, Morelos, and Veracruz, México) .... *Parastrellus hesperus* (p. 383)

9a. Limb length greater than one-half of the tibia length; dorsal outline of the skull concave; palatine extending slightly to the back of the last molar (from South Dakota and New Mexico east through New Hampshire and south through Central America) .... *Perimyotis subflavus* (p. 385)

10. One upper incisor at each side 11

10a. Two upper incisors at each side .... tribe Eptesicini *Eptesicus* (p. 368)

11. Third, fourth, and fifth metacarpals (wing) similar; skull height much less than one-half of its length 12

11a. Third, fourth, and fifth metacarpals (wing) progressively shorter; skull short and high; braincase height, including the auditory bullae, approximately one-half of the cranial length .... tribe Lasiurini *Lasiurus* (p. 372)

12. Dorsal pelage yellowish or brown; third lower incisors much smaller than the other lower incisors ....tribe Antrozoini *Rhogeessa* (p. 359)

12a. Dorsal pelage blackish; three lower incisors of the same size (from Nebraska and Texas east through New Hampshire and south through Veracruz, México) ....tribe Nycticeinin *Nycticeius humeralis* (p. 382)

Tribe Antrozoini

Genus *Antrozous* Allen 1862

Medium sized; **nose similar to that of a pig** instead of the typical noseleaf, snout blunt, piglike; ears large, extending well beyond the muzzle when laid forward, separate; **midfrontal area slightly convex**; braincase high; rostrum long, more than one-half of the braincase length; basisphenoid cavities absent; tympanic ring large, covering most of the cochlea, **with two lower incisors**.

Antrozous pallidus (LeConte, 1856)

Pallid bat, murciélago pálido

1. *A. p. bunkeri* Hibbard, 1934. Restricted to central western Oklahoma.
2. *A. p. minor* Miller, 1902. Extreme southern Nevada, southeastern California, southwestern Arizona, northwestern Sonora, and the Baja California Peninsula.
3. *A. p. pacificus* Merriam, 1897. From Central Oregon along the coast south through Los Angeles Valley, California.
4. *A. p. packardi* Martin and Schmidly, 1982. From western Sonora south through Jalisco.
5. *A. p. pallidus* (LeConte, 1856). From southern British Columbia, Oregon, southern Idaho, Wyoming, and western Oklahoma south through Guanajuato, Querétaro, and San Luis Potosí, including southwestern California and northwestern Baja California.

Conservation: *Antrozous pallidus* is listed as Least Concern by the IUCN (Arroyo-Cabrales and de Grammont 2017a).

Characteristics: Pallid bat is large sized; forearm length 50.0–57.0 mm, skull length 18.6–23.6 mm. **Dorsal pelage pale yellowish; fur of underparts whitish**; ears large, greater than 25.0 mm; eyes large, pointed, and frontward; snout in rhombus shape; **nose similar to that of a pig**; rostrum well differentiated from the braincase and greater than one-half of the braincase length; basisphenoids absent; braincase high; **two lower incisors at each side**; **third lower molars talonoid, length approximately one-half of the trigonid;** midfrontal area slightly convex.

Comments: *Antrozous pallidus* was part of the subfamily Nyctophilinae and was previously considered a full family, Antrozoidae (Simmons 1998). Genetic studies are still considered it a subfamily of Vespertilionidae (Koopman 1993, 1994; Hoofer and Den Bussche 2001; Simmons 2005). *Bauerus* was a subgenus of *Antrozous,* genetic and morphometric analyses showed it should be considered at the genus level (Engstrom and Wilson 1981; Engstrom et al. 1987b). *A. p. cantwelli* and *A. p. obscurus* were considered junior synonyms of *A. p. pallidus* (Martin and Schmidly 1982). *A. pallidus* ranges from southern British Columbia south through the Eje Neovolcánico and east through from Wyoming and Texas (Map 123). It can be distinguished from other species in sympatry, or nearly so, such as those of the same genus, by its nostrils that open frontward behind a horseshoe-shaped fold, its larger ears and eyes; without a noseleaf, and ears not joined basally across the midline. Although all *Antrozous* species have allopatric distribution, it can be distinguished from *B. dubiaquercus* by having a less pronounced sagittal crest; relatively larger auditory bullae; third upper molars with the talonid approximately one-half of the length of the trigonid, and two lower incisors on each side.

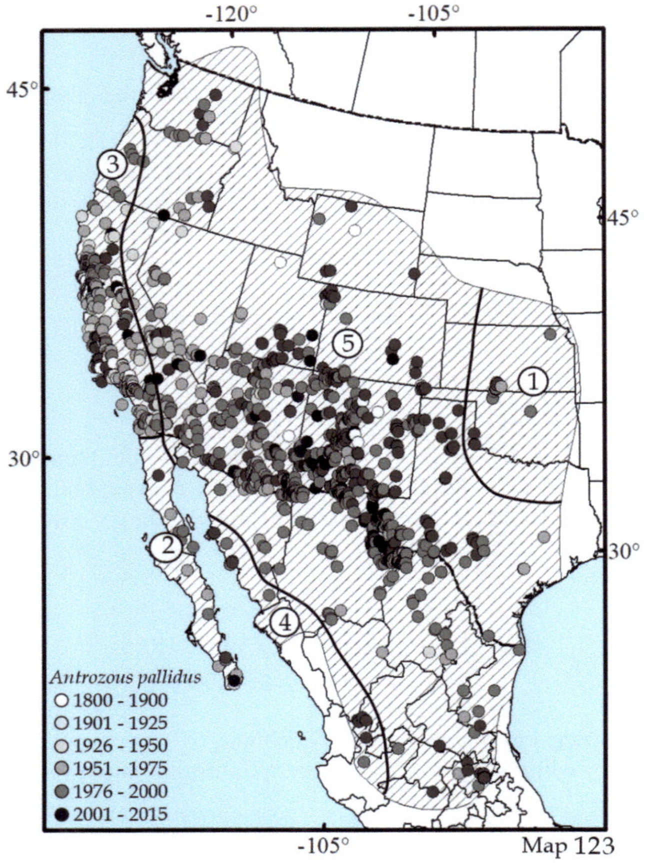

Map 123 Distribution map of *Antrozous pallidus*

Additional Literature: Ingles (1965), Hoffmeister (1971), Armstrong (1972), Banfield (1974), Hall (1981), Martin and Schmidly (1982), Hermanson and O'Shea (1983), McKenna and Bell (1997), Koopman (1993, 1994), Bogan (1999), Weyandt and Van den Bussche (2007), Arroyo-Cabrales et al. (2008), Frick et al. (2009).

Genus *Bauerus* Van Gelder, 1959

Medium sized; pelage darker at the base than at the tip; **nostrils open frontward, behind a horseshoe-shaped fold**; ears large, extending well beyond the muzzle when laid forward, separate; **midfrontal area flat**; braincase high; rostrum long, more than one-half longer than the braincase; basisphenoid cavities absent; tympanic ring large, covering most of the cochlea; **with three lower incisors** (Engstrom and Wilson 1981; Hall 1981).

Bauerus dubiaquercus (Van Gelder, 1959)

Van Gelder's bat, murciélago pálido tropical

1. *B. d. dubiaquercus* (Van Gelder, 1959). Restricted to María Magdalena Island, Nayarit (not in the map).
2. *B. d. meyeri* Pine, 1966. From southern Veracruz south through Central America.

Conservation: *Bauerus dubiaquercus* is listed as Near Threatened by the IUCN (Solari 2018a).

Characteristics: Van Gelder's bat is large sized; forearm length 49.0–56.0 mm, skull length 19.9–21.5 mm. **Dorsal pelage dark brown**; fur of underparts yellowish, dark brown at the base; ears and tragus large; if folded frontward, they go over the nose; **nostrils open frontward, behind a horseshoe-shaped fold;** uropatagium large and with little hair; membranes blackish; rostrum well differentiated from the braincase and greater than one-half of the braincase length; basisphenoid cavities absent, braincase high; auditory bullae relatively small; **third lower molars with the talonoid approximately equal in length to the trigonid; three lower incisors at each side, but some specimens from the Pacific coast with only two**; frontal area flat.

Comments: *Bauerus dubiaquercus* was part of the subfamily Nyctophilinae and was previously considered a full family, Antrozoidae (Simmons 1998). Genetic studies still consider it a subfamily of Vespertilionidae (Koopman 1993, 1994; Hoofer and Den Bussche 2001; Simmons 2005). *Bauerus* was a subgenus of *Antrozous,* but genetic and morphometric analyses showed they should be considered at a genus level (Engstrom and Wilson 1981; Engstrom et al. 1987b). *B. dubiaquercus* ranges from southern Veracruz and southern Quintana Roo south through Central America (Map 124). One population is restricted to the Tres Marías Islands in Nayarit (not in map). It can be distinguished from other species in sympatry, or nearly so, such as those of the same genus, by its nostrils that open frontward behind a horseshoe-shaped fold, large ears, and dark brown coloration. Although all *Bauerus* species have allopatric distribution, it can be distinguished from *Antrozous pallidus* by having a more pronounced sagittal crest; relatively small auditory bullae; third upper molars with the talonid approximately equal in length to the trigonid and the third lower incisors spicule-like.

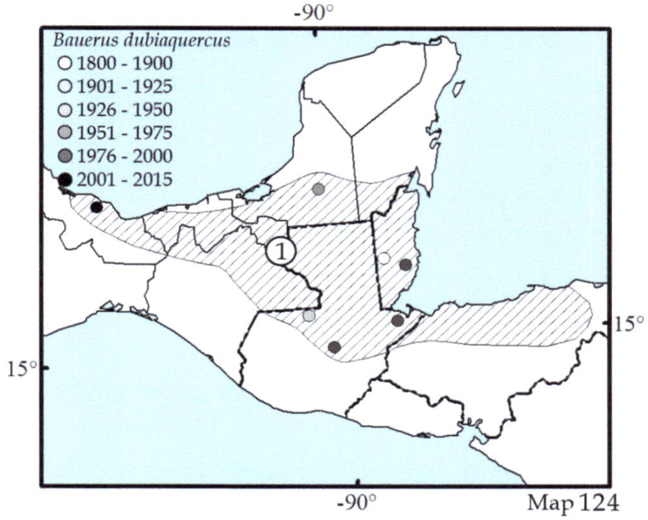

Map 124 Distribution map of *Bauerus dubiaquercus*

Additional Literature: Van Gelder (1959a), Pine et al. (1971), Martin and Schmidly (1982), Simmons (1998), Girón et al. (2010), Roehrs et al. (2010).

Genus ***Rhogeessa*** Allen, 1866

Small sized; dorsal and ventral coloration yellow brownish; dorsal hairs with three color bands and ventral hairs with two; dorsal uropatagium bare; skull short and wide relative to other species of the same subfamily; rostrum well differentiated from the braincase; teeth proportionally small, with cutting cusps in W-shape; four upper molariforms; one well developed upper incisor in each mandibular branch; canines large and with a cingulum; braincase not very high; two upper incisors at each side; three upper molars of the same size; digital crest well developed; **third lower incisors unicuspid, of smaller size than the others, and almost hidden by the canine cingulum.** *Rhogeessa* has two subgenera, *Baeodon* (*R. alleni* and *R. gracilis*) and *Rhogeessa,* with two groups of species: *tumida* (*R. aeneus, R. bickhami, R. genowaysi,* and *R. tumida*) and *parvula* (*R. parvula* and *R. mira*). The keys were elaborated based on the review of specimens and the following sources: LaVal (1973a, b), Hall (1981), Álvarez et al. (1994), Baird et al. (2015), and Álvarez-Castañeda et al. (2017a).

1. Three lower incisors of the same size .... subgenus *Rhogeessa* 2

1a. Third lower incisors spicule shaped, less than one-twentieth as large as the first or second incisors, nearly concealed beneath the cingulum of the canine .... subgenus *Baeodon* 3

2. Metacarpal of the third digit averaging 2.2 times greater than the first-phalange length; skull length greater than 14.5 mm; third lower incisors unicuspid and almost hidden by the canine cingulum (from Jalisco and San Luis Potosí south through Oaxaca, México) .... *Rhogeessa alleni* (p. 361)

2a. Metacarpal of the third digit averaging 2.2 times less than the first-phalange length; skull length less than 14.5 mm; third lower incisors frequently bicuspid (from Zacatecas and Nayarit south through Oaxaca, México) .... *Rhogeessa gracilis* (p. 362)

3. Forearm length on average less than 26.0 mm; skull length less than 11.6 mm; cingulum lingual of the upper canines without a cuspid (endemic of Balsas Basin, México) .... *Rhogeessa mira* (p. 365)

3a. Forearm length on average greater than 26.0 mm; skull length greater than 11.6; cingulum lingual of the upper canines usually with a cusp 4

4. Uropatagium hairy only at the base; third lower incisors usually greater than one-fourth of the size of the second lower incisors (from Sonora south through Chiapas, México) .... *Rhogeessa parvula* (p. 366)

4a. Uropatagium very hairy from the base to the middle tibia; third lower incisors usually less than one-fourth of the size of the second lower incisors (the *tumida* group includes cryptic species that can only be clearly differentiated with genetic and distribution data) 5

5. Pelage and ears paler; angle more elevated between the rostrum and the braincase; rostrum shorter; upper incisors larger and thinner; third upper molars smaller 6

5a. Pelage and ears darker; angle less elevated between the rostrum and the braincase; rostrum larger; upper incisors not so large and thin; third upper molars in proportion to the rest 7

6. Cryptic species can only be differentiated by genetic analysis (from Tamaulipas, México, south through Central America) .... *Rhogeessa tumida* (p. 367)

6a. Cryptic species can only be differentiated by genetic analysis (restricted to the Yucatán Peninsula, México) .... *Rhogeessa aeneus* (p. 365)

7. Hypocones of the second and third upper molars three-fourths of the second upper molar length and one-third in breadth (endemic of the coast of Chiapas, México) .... *Rhogeessa genowaysi* (p. 364)

7a. Hypocones of the second and third upper molars similar in size; second and third upper molars of the same breadth (coastal plain of Chiapas, México, and Guatemala) .... *Rhogeessa bickhami* (p. 364)

Subgenus *Baeodon* Miller, 1897

Small sized; dorsal pelage yellowish; ears triangular; rostrum, ears, and membranes blackish; calcar unkeeled; upper incisors simple, unicuspid, nearly one-half as high as the canines; three lower incisors**, the third functionless, spicule shaped, less than one-twentieth as large as the first or second incisors, nearly concealed beneath the cingulum of the canines**; third upper molars with the posterior portion slightly more reduced than in *Rhogeessa*.

Rhogeessa alleni Thomas, 1892

Yellow bat, murciélago amarillo grande

Monotypic.

Conservation: *Rhogeessa alleni* (as *Baeodon alleni*) is listed as Least Concern by the IUCN (Solari 2019a).

Characteristics: Yellow bat is small sized; forearm length 30.8–34.4 mm, skull length 14.7–15.4 mm. Dorsal pelage with three hair bands, basal grayish, middle brown, and tip brown; fur of underparts with two bands, basal grayish and tip light ocher; dorsal uropatagium bare; **third digit metacarpal averaging 2.2 times greater than the first-phalange length;** skull large-sized.

Comments: *Baeodon* is a subgenus of *Rhogeessa.* However, Hoofer and Van Den Bussche (2003) stated that *Baeodon* should be considered a genus, using *Baeodon alleni.* I continue using *R. alleni* (Baird et al. 2008). *R. alleni* ranges from Jalisco and San Luis Potosí south through Oaxaca by the midlands (Map 125). It can be found in sympatry, or nearly so, with *R. gracilis, R. mira,* and *R. parvula. R. alleni* differs from the rest of the species of *Rhogeessa* by its third metacarpal finger 2.2 times longer than the first phalanx length, on average. Specimens of *Rhogeessa* are distinguished from other species mainly by their small size, forearm, and skull lengths less than 35.0 and 15.5 mm, respectively, and their coloration in pale yellowish to yellowish brown shades.

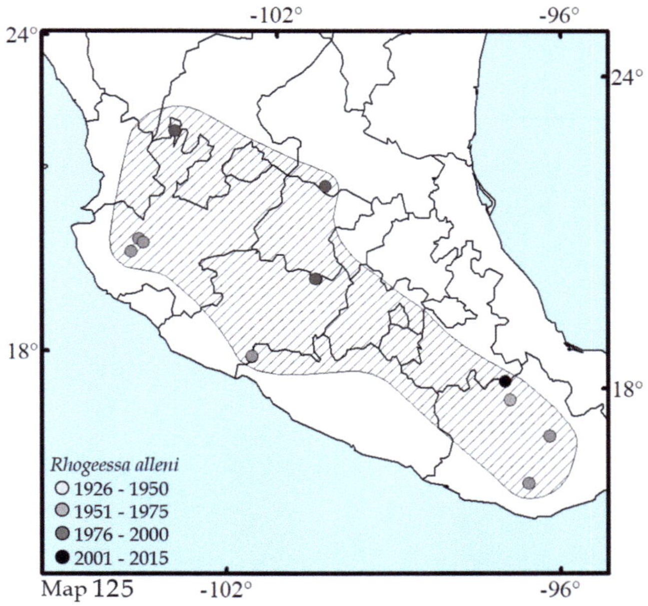

Map 125 Distribution map of *Rhogeessa alleni*

Additional Literature: Goodwin (1958), LaVal (1973b), Villalpando and Álvarez (2000), Roehrs et al. (2010).

Rhogeessa gracilis Miller, 1897

Slender yellow bat, murciélago amarillo delgado

Monotypic.

Conservation: *Rhogeessa gracilis* is listed as Least Concern by the IUCN (Solari 2019b).

Characteristics: Slender yellow bat is small sized; forearm length 30.2–33.4 mm, skull length 12.9–13.8 mm. **Dorsal pelage with three hair bands, dark grayish brown at the base, brown at the middle, and light ocher brown at the tip;** fur of underparts with two color bands, dark grayish brown at the base and paler brown at the tip; dorsal uropatagium with little hair scattered almost to the knees.

Comments: *Baeodon* is a subgenus of *Rhogeessa.* However, Hoofer and Van Den Bussche (2003) stated that *Baeodon* should be considered a genus. *R. gracilis* was previously assigned to the subgenus *Rhogeessa,* but genetic analysis showed it belongs within *Baeodon* (Baird et al. 2008). *R. gracilis* ranges from Nayarit and Zacatecas south through Oaxaca through the midlands of the Sierra Madre del Sur but not on the coastal plain and tropical region (Map 126). It can be found in sympatry, or nearly so, with *R. alleni, R. mira,* and *R. parvula. R. gracilis* differs from other species of *Rhogeessa* by its dorsal coloration formed by hair with three bands, dark grayish brown at the base, brownish at the middle, and pale brownish at the tip. These three bands are not observed in the other species of *Rhogeessa*. Specimens of *Rhogeessa* are distinguished from other species mainly by their small size, forearm, and skull lengths less than 35.0 and 15.5 mm, respectively, and their coloration in pale yellowish to yellowish brown shades.

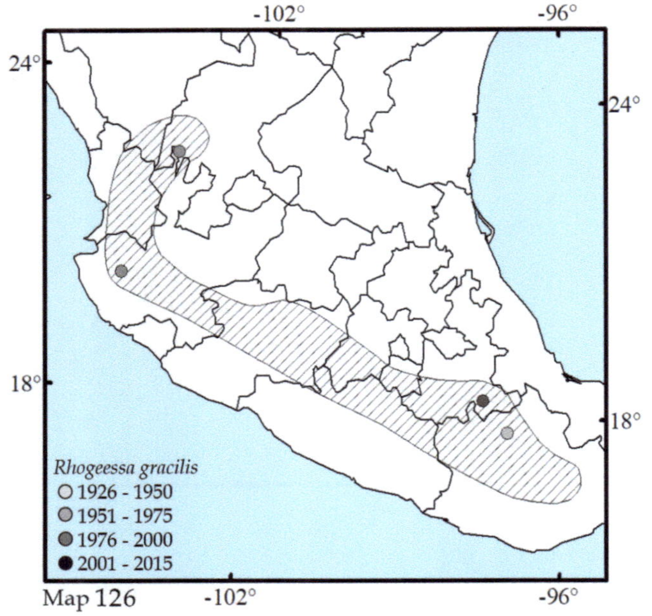

Map 126 Distribution map of *Rhogeessa gracilis*

Additional Literature: Goodwin (1958), Jones (1977).

Subgenus *Rhogeessa* Allen, 1866

Small sized; dorsal pelage yellowish; ears triangular; rostrum, ears, and membranes blackish; calcar unkeeled; upper incisors simple, unicuspid, nearly one-half as high as the canines; three lower incisors, **outer lobe of the three lower incisors obsolete, inner lobe practically absent also in the outer lower incisors, which are thus much reduced in size compared with the other two incisors, although with a cutting edge formed by the middle lobe**; upper premolars without a cusp on the inner side.

Rhogeessa aeneus Goodwin, 1958

Yucatán yellow bat, murciélago amarillo de Yucatán

Monotypic.

Conservation: *Rhogeessa aeneus* is listed as Least Concern by the IUCN (Solari 2019o).

Characteristics: Yucatán yellow bat is small sized; forearm length 25.1–32.3 mm, skull length 11.7–12.2 mm. Dorsal pelage yellowish brown, yellow at the base and brown at the tip; fur of underparts pale yellow or cream; rostrum and ears dark brown or blackish; tragus long and narrow; membranes blackish; ears with notorious glands. Note: *Rhogeessa tumida* and *R. aeneus* can only be separated by karyotypes.

Comments: *Rhogeessa aeneus* was elevated to full species (Audet et al. 1993); it was previously included within *R. tumida*. *R. aeneus* is the only species of *Rhogeessa* inhabiting the Yucatán Peninsula (Map 127). It can be found in sympatry, or nearly so, with *R. tumida*, from which it can only be differentiated by genetic methods. Specimens of *Rhogeessa* are distinguished from other species mainly by their small size, forearm, and skull lengths less than 35.0 and 15.5 mm, respectively, and their coloration in yellowish pale buff to yellowish brown shades.

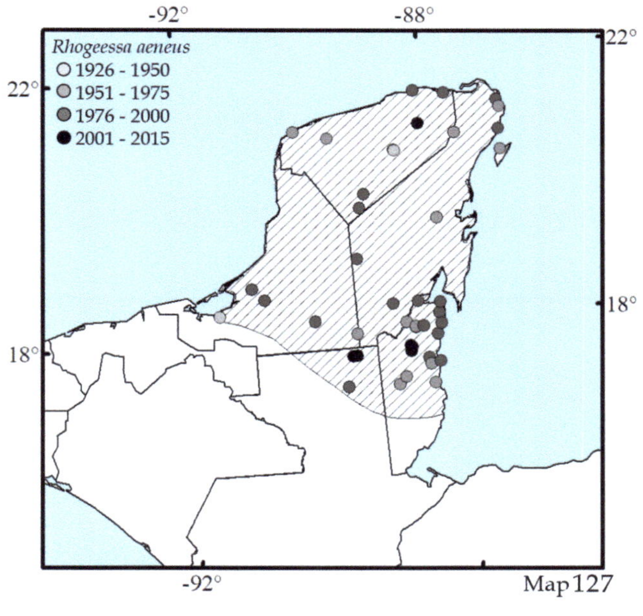

Map 127 Distribution map of *Rhogeessa aeneus*

Additional Literature: Goodwin (1958), LaVal (1973b), Koopman (1993), Genoways and Baker (1996), Rydell et al. (2002), Simmons (2005) Roots and Baker (2007), Baird et al. (2008, 2009), Baird (2010).

Rhogeessa bickhami Baird et al., 2012

Brickhami's yellow bat, murciélago amarillo de la costa de Chiapas

Monotypic.

Conservation: *Rhogeessa bickhami* is listed as Least Concern by the IUCN (Solari 2017b).

Characteristics: Brickhami's yellow bat is small sized, skull length 11.2–12.9 mm, no records for forearm length. Dorsal pelage dark, yellowish gray to yellowish at the base and dark brown to blackish at the tip; fur of underparts pale brown; no hair bangs in the uropatagium; ears short and dark. NOTE: *Rhogeessa genowaysi* and *R. bickhami* can only be separated by karyotypes.

Comments: *Rhogeessa bickhami* is known from the coast of Chiapas, México, and Guatemala (Map 128). It can be found in sympatry, or nearly so, with *R. genowaysi* and *R. tumida*, from which it can only be differentiated by karyotype analyses and not by its morphology. Specimens of *Rhogeessa* are distinguished from other species mainly by their small size, forearm, and skull lengths less than 35.0 and 15.5 mm, respectively, and their coloration in yellowish pale buff to yellowish brown shades.

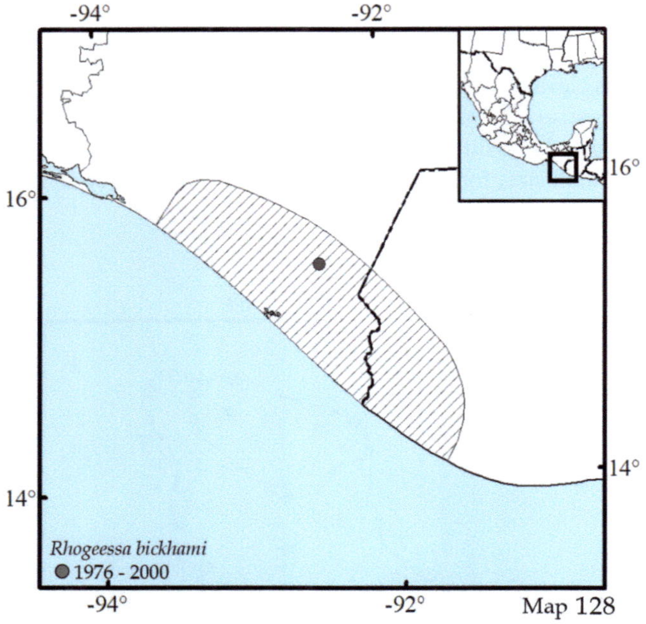

Map 128 Distribution map of *Rhogeessa bickhami*

Additional Literature: Baird et al. (2012).

Rhogeessa genowaysi Baker, 1984

Genoways's yellow bat, murciélago amarillo de Chiapas

Monotypic.

Conservation: *Rhogeessa genowaysi* is listed as Threatened by the Norma Oficial Mexicana (DOF 2019) and as Critically Endangered (B1ab (iii); extent of occurrence less than 100 km^2 in only one population and continuing decline in quality of habitat) by the IUCN (Arroyo-Cabrales 2016).

Characteristics: Genoways's yellow bat is small sized; forearm length 27.8–30.5 mm, skull length 12.0–13.1 mm. Dorsal pelage yellowish brown, yellow at the base and brown at the tip; fur of underparts pale yellow or cream; rostrum and ears dark brown or blackish; tragus long and narrow; tail extending throughout the uropatagium; membranes blackish; ears with notorious glands. NOTE: *Rhogeessa genowaysi* and *R. bickhami* can only be separated by karyotypes.

Comments: *Rhogeessa genowaysi* is only known from the coast of Chiapas, México (Map 129). It can be found in sympatry, or nearly so, with *R. bickhami* and *R. tumida*, from which it can only be differentiated by karyotype analyses and not by its morphology. Specimens of *Rhogeessa* are distinguished from other species mainly by their small size, forearm, and skull lengths less than 35.0 and 15.5 mm, respectively, and their coloration in yellowish light to yellowish brown shades.

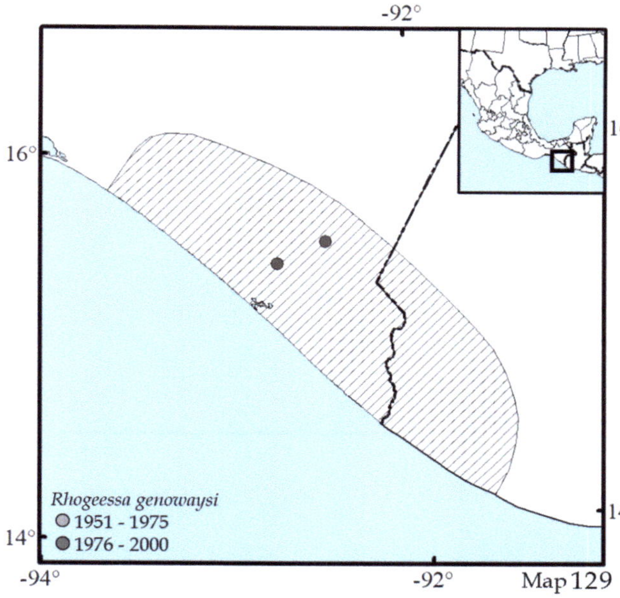

Map 129 Distribution map of *Rhogeessa genowaysi*

Additional Literature: Baker and Patton (1967), Baker (1984b), Baker et al. (1985), Genoways and Baker (1996), Roots and Baker (1998), Naranjo Piñera et al. (2005), Baird et al. (2008, 2012).

Rhogeessa mira LaVal, 1973

Least yellow bat, murciélago amarillo pigmeo

Monotypic.

Conservation: *Rhogeessa mira* is listed as Special Protection in the Norma Oficial Mexicana (DOF 2019) and as Vulnerable (B1ab (iii): Extent of occurrence less than 20,000 km^2 severely fragmented and under continuing decline in area, extent and/or quality of habitat) by the IUCN (Arroyo-Cabrales and Ospina-Garces 2016f).

Characteristics: Least yellow bat is the **smallest bat in North America; forearm length 24.5–25.9 mm, skull length 11.0–11.6 mm.** Dorsal pelage formed by two hair bands, yellow brown to yellowish at the tip and more yellowish at the base but not contrasting with the tip; fur of underparts with two bands similar to those on the back; dorsal uropatagium with little hair grizzly to the knees and a little further; skull smallest, see measurements.

Comments: *Rhogeessa mira* is only known from the Balsas Basin in Michoacán and the limit with Guerrero (Map 130). It can be found in sympatry, or nearly so, with *R. alleni, R. gracilis,* and *R. parvula. R. mira* is distinguished by its small size, being one of the smallest mammals worldwide; forearm length less than 30.0 mm and skull less than 11.6 mm. Specimens of *Rhogeessa* are distinguished from other species mainly by their small size, forearm, and skull lengths less than 35.0 and 15.5 mm, respectively, and their coloration in yellowish pale to yellowish brown shades.

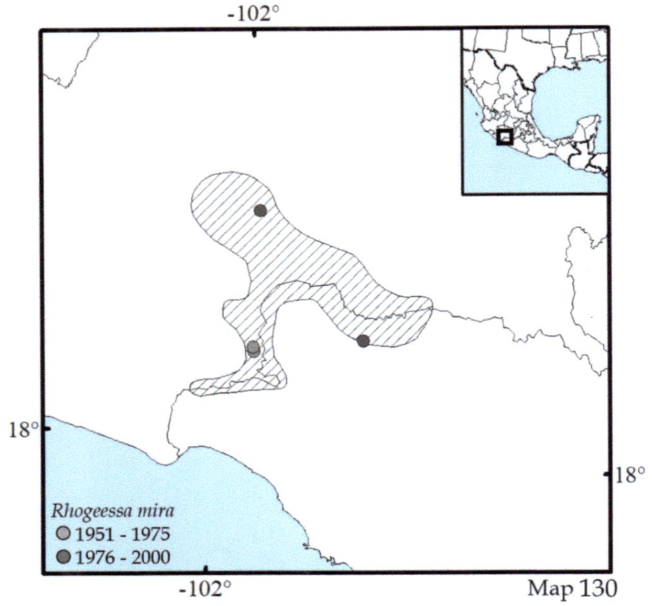

Map 130 Distribution map of *Rhogeessa mira*

Additional Literature: LaVal (1973b), Baker et al. (1985), Polaco and Muñiz-Martínez (1987), Genoways and Baker (1996), Villalpando-R. and Arroyo-Cabrales (1996), Arroyo-Cabrales and Polaco (1997), Roehrs et al. (2010).

Rhogeessa parvula Allen, 1866

Little yellow bat, murciélago amarillo pequeño

1. *R. p. major* Goodwin, 1958. From Sonora south through Chiapas.
2. *R. p. parvula* Allen, 1866. Restricted to the Tres Marías Islands, Nayarit.

Conservation: *Rhogeessa parvula* is listed as Least Concern by the IUCN (Solari 2019p).

Characteristics: Little yellow bat is small sized; forearm length 25.8–32.8 mm, skull length 11.7–12.0 mm. Dorsal hair with two color bands, chestnut brown or intense suede at the tip and pale gray to pale suede at the base but not contrasting with the tip; fur of underparts with two color bands, suede to ocher at the tip and slightly paler at the base; dorsal uropatagium with little hair grizzly to the middle between the knees and the limbs; **third lower incisors usually greater than one-fourth of the size of the second lower incisors**.

Comments: *Rhogeessa parvula* ranges from Sonora south through Chiapas (Map 131). It can be found in sympatry, or nearly so, with *R. alleni, R. bickhami, R. genowaysi, R. gracilis,* and *R. mira. R. parvula* is distinguished from *R. bickhami* and *R. genowaysi* because the uropatagium has hair only at the base, contrasting with a hairy uropatagium in the middle portion. From *R. mira,* by its ~30% larger size. From *R. alleni*, by its smaller size in all the measurements. From *R. gracilis,* by its dorsal hairs only with two bands, contrasting with the three bands in *R. gracilis*. Specimens of *Rhogeessa* are distinguished from other species mainly by their small size, forearm, and skull lengths less than 35.0 and 15.5 mm, respectively, and their coloration in yellowish pale to yellowish brown shades.

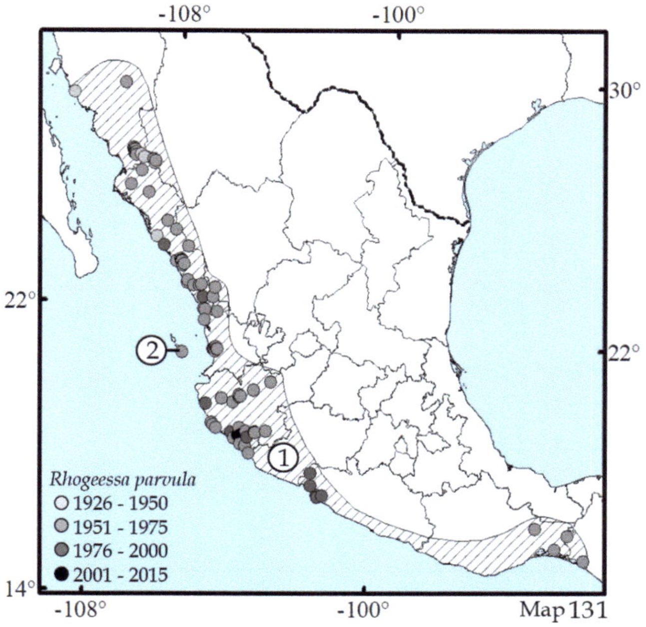

Map 131 Distribution map of *Rhogeessa parvula*

Additional Literature: Goodwin (1958), Álvarez and Aviña (1965), Jones et al. (1972), LaVal (1973b), Baker et al. (1985), Wilson (1991), Bogan (1999), Villalpando and Álvarez (2000), Hoofer and Van den Bussche (2003), Roots and Baker (2007), Baird et al. (2008).

Rhogeessa tumida Allen, 1866

Black-winged little yellow bat, murciélago amarillo de alas negras

Monotypic.

Conservation: *Rhogeessa tumida* is listed as Least Concern by the IUCN (Miller et al. 2016g).

Characteristics: Black-winged little yellow bat is small sized; forearm length 25.0–32.0 mm, skull length 11.2–12.9 mm. Dorsal pelage yellowish brown, yellow at the base and brown at the tip; fur of underparts pale yellow or cream; rostrum and ears dark brown or blackish; tragus large and narrow; membranes blackish; ears with notorious glands [a brilliant area swollen on the ear external surface above the head crown]. Note: *Rhogeessa tumida* and *R. aeneus* can only be separated by karyotypes.

Comments: *Rhogeessa tumida* was elevated to full species (LaVal 1973b); it was previously regarded as a subspecies of *R. parvula* (Hall and Kelson 1959). It ranges from Tamaulipas south through Central America, except for the Yucatán Peninsula (Map 132). It can be found in sympatry, or nearly so, with *R. aeneus, R. bickhami,* and *R. genowaysi*, from which it can only be differentiated by karyotype analyses and not by its morphology.

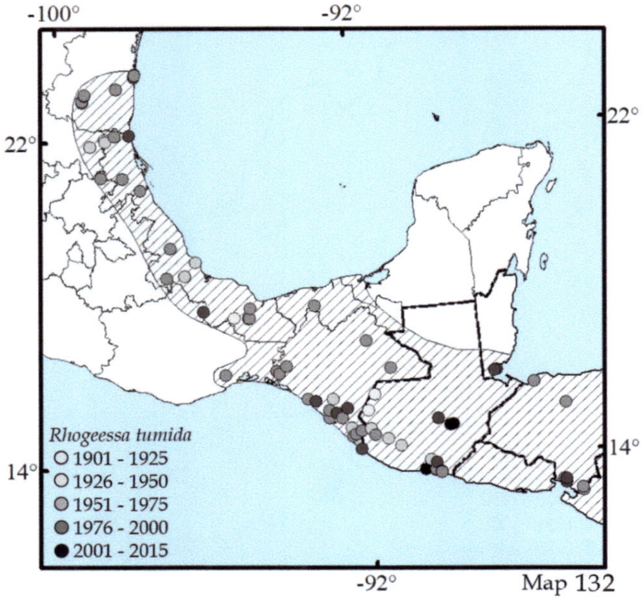

Map 132 Distribution map of *Rhogeessa tumida*

Additional Literature: Goodwin (1958), Brown et al. (1971), Jones et al. (1971), Carter and Jones (1978), Baker (1984b), Vonhof (2000), Baird et al. (2008).

Tribe Eptesicini

Genus *Eptesicus* Rafinesque, 1820

Medium sized; dorsal and ventral pelage with different brown shades; ears triangular with the tragus curved frontward; rostrum, ears, membranes, and hair base in different color according to the species; tail extending to the uropatagium border; rostrum well differentiated from the braincase; teeth proportionally large, with cutting cusps W shaped; **four upper molariforms; two upper incisors on each maxillary branch, well developed, the inner upper incisors larger than the outer upper incisors, the latter separated from the canines by a distance equivalent to their diameter; lower incisors subequal and imbricate; canines well developed and with a cingulum.** All species in North America are included within the subgenus *Eptesicus*. The keys were elaborated based on the review of specimens and the following sources: Miller (1897), Hall (1981), Álvarez et al. (1994), and Álvarez-Castañeda et al. (2017a).

1. Second upper molars greater than 2.4 mm in length by 1.8 mm in breadth; average skull length greater than 17.2 mm (from southern Canada south through South America, except for Tabasco and the Yucatán Peninsula)........................... .... *Eptesicus fuscus* (*see addenda, p. 370)

1a. Second upper molars less than 2.4 mm in length by 1.8 mm in breadth; average skull length less than 17.2 mm..........2

2. Forearm length on average greater than 42.5 mm; dorsal hair length greater than 8.0 mm; second upper molars greater than 1.8 × 1.6 mm (from Veracruz, México, south through South America).................... *Eptesicus brasiliensis* (p. 369)

2a. Forearm length on average less than 42.5 mm; dorsal hair length less than 8.0 mm; second upper molars less than 1.8 × 1.6 mm (from Jalisco and Tamaulipas, México, south through South America).......... *Eptesicus furinalis* (p. 370)

Subgenus *Eptesicus* Rafinesque, 1820

Large size; upper surface of forearm, tail, and tibia not thickly sprinkled with pointed, horny excrescences; braincase relatively narrow anteriorly and long; rostrum high; upper toothrows less concave on the inner side and less convergent anteriorly.

Eptesicus brasiliensis (Desmarest, 1819)

Brazilian brown bat, murciélago pardo de Brasil

(*see addenda, valid name *Neoeptesicus brasiliensis*)

E. b. brasiliensis (Desmarest, 1819). The only subspecies in North America.

Conservation: *Eptesicus brasiliensis* is listed as Least Concern by the IUCN (Barquez et al. 2016a).

Characteristics: Brazilian brown bat is medium sized; forearm length 39.0–48.0 mm, skull length 15.9–18.1 mm. Dorsal pelage dark orange brown to blackish; fur of underparts orange or yellowish brown; rostrum skin pinkish; membranes blackish; hair blackish at the base; uropatagium naked; **second upper molars greater than 1.8 × 1.6 mm**.

Comments: *Eptesicus andinus* was previously considered a subspecies of *E. brasiliensis but* is currently regarded as a full species (Simmons and Voss 1998). *E. brasiliensis* ranges in the most humid Neotropical region of México from Veracruz south through South America (Map 133). It can be found in sympatry, or nearly so, with *E. furinalis* and *E. fuscus*. *E. brasiliensis* is the smallest species within *Eptesicus* in North America and its measurements on average are smaller than those of *E. furinalis* and *E. fuscus*. The most similar species is *E. furinalis*, from which it differs by its hair length greater than 8.0 mm.

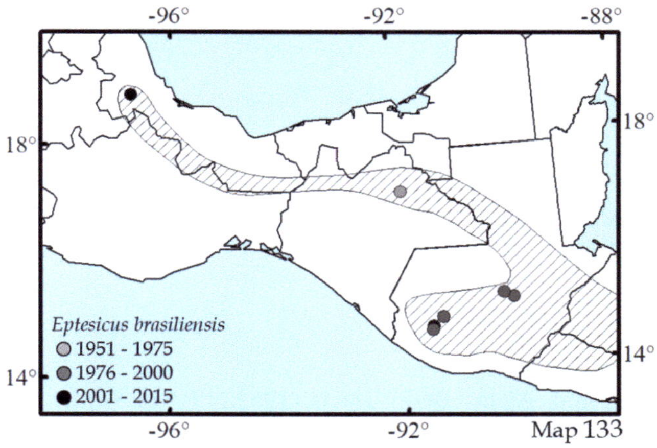

Map 133 Distribution map of *Eptesicus brasiliensis*

Additional Literature: Davis (1965, 1966a, b), Shump and Shump (1982a), Davis and Gardner (2008).

Eptesicus furinalis (d'Orbigny and Gervais, 1847)

Argentine brown bat, murciélago pardo tropical

(*see addenda, valid name *Neoeptesicus furinalis*)

E. f. gaumeri (Allen, 1897). Only subspecies in North America.

Conservation: *Eptesicus furinalis* is listed as Least Concern by the IUCN (Barquez et al. 2016b).

Characteristics: Argentine brown bat is medium sized; forearm length 37.0–41.0 mm, skull length 15.0–17.1 mm. Dorsal pelage dark brown to blackish; fur of underparts orange gray; ears dark brown; rostrum skin pinkish; membranes blackish; hair blackish at the base; uropatagium bare; **second upper molars less than 1.8 × 1.6 mm**.

Comments: *Eptesicus furinalis* ranges from Jalisco and Tamaulipas, México, south through South America (Map 134). It can be found in sympatry, or nearly so, with *E. brasiliensis* and *E. fuscus*. Some specimens of *E. furinalis* may have forearm length greater than 49.0 mm, similar to *E. fuscus*. The difference is that *E. furinalis* can be found in tropical forests and *E. fuscus* is associated with temperate forests. It differs from *E. brasiliensis* by its hair length, which is shorter, less than 8.0 mm, and its smaller-than-average size. The genus *Eptesicus* differs from other genera by having four upper molariforms, two upper incisors well developed, and the outer upper incisors larger than the inner ones, separated from the canines by a distance equivalent to the incisor diameter.

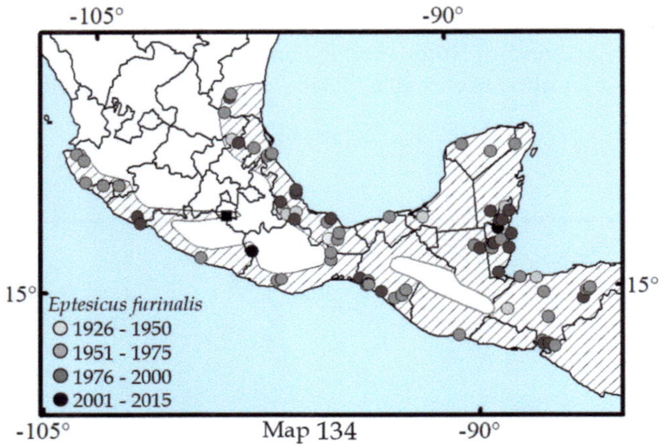

Map 134 Distribution map of *Eptesicus furinalis*

Additional Literature: Dobson (1885), Allen (1897), Davis (1965, 1966a), Williams (1978a), Álvarez-Castañeda and Álvarez (1991); Mies et al. (1996), Villalpando and Álvarez (2000), Davis and Gardner (2008).

Eptesicus fuscus (Palisot de Beauvois, 1796)

Big brown bat, murciélago pardo grande

(*see addenda, valid name *Eptesicus miradorensis*)

1. *E. f. bernardinus* Rhoads, 1902. From British Columbia to California.
2. *E. f. fuscus* (Palisot de Beauvois, 1796). From the Great Plains eastward.
3. *E. f. miradorensis* (Allen, 1866). From Tamaulipas, San Luis Potosí, Durango, and Sinaloa southward.
4. *E. f. osceola* Rhoads, 1902. Florida Peninsula.
5. *E. f. pallidus* Young, 1908. From the Great Plains westward, including northern México, except for the coastal states of the United States.
6. *E. f. peninsulae* (Thomas, 1898). Southern Baja California Peninsula

Conservation: *Eptesicus fuscus* is listed as Least Concern by the IUCN (Miller et al. 2016h).

Characteristics: Big brown bat is medium sized; forearm length 39.0–54.0 mm, skull length 15.1–23.0 mm* NOTE: the specimens distributed in México tend to be in the upper side of the length range, greater than 17.0 mm. Dorsal pelage from yellowish, reddish brown to brown; fur of underparts orange or brown; rostrum, ears, and membranes blackish; hair blackish at the base; **uropatagium dorsally naked**; **second upper molars greater than 2.4 mm in length by 1.8 mm in breadth**.

Comments: *Eptesicus fuscus* ranges from southern Canada south through South America, except for Tabasco and the Yucatán Peninsula (Map 135). It can be found in sympatry, or nearly so, with the three species of *Eptesicus*. *E. fuscus* is the largest species of *Eptesicus* in North America; forearm greater than 49.0 mm, but some specimens of *E. furinalis* with forearm length greater than 49.0 mm. The difference is that *E. fuscus* is associated with temperate forests and *E. furinalis* with tropical forests. *Eptesicus* differs from other genera by having four upper molariforms, two upper incisors well developed and the outer upper incisors larger than the inner incisors, and separated from the canines by a distance equivalent to the incisor diameter "See adeendum".

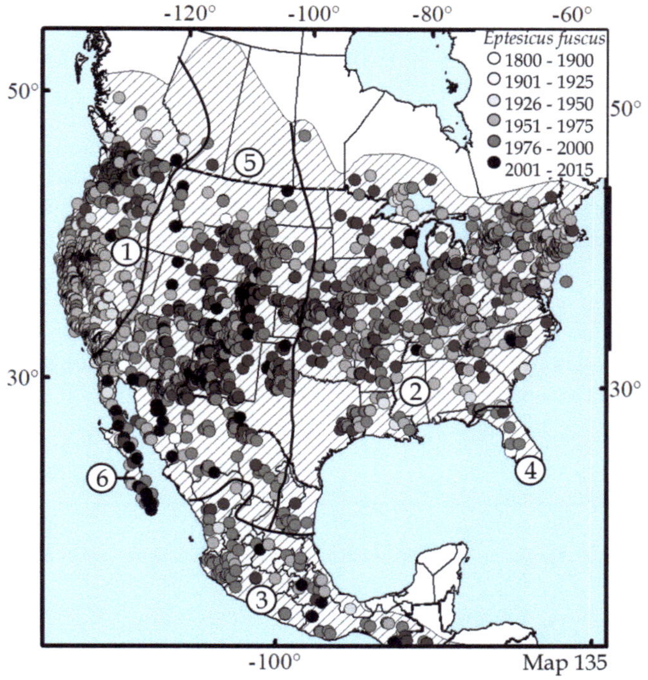

Map 135 Distribution map of *Eptesicus fuscus*

Additional Literature: Manville and Young (1965), Davis (1965, 1966a, b), van Zyll de Jong (1985), Kurta and Baker Álvarez-Castañeda and Álvarez (1991), Bogan (1999), Villalpando and Álvarez (2000), Álvarez-Castañeda et al. (2008c), Frick et al. (2008).

Genus *Lasionycteris* Peters, 1866

Medium sized; **ears short, nearly as broad as long**; tragus short, straight, and bluntly rounded, its width more than one-half the length of the anterior margin; **uropatagium furred above the basal half;** teeth structure as in *Myotis*, but inner incisors strongly bicuspid and outer incisors simple; **third upper molars absent**; first and second upper molars with the hypocone distinctly indicated; **third upper molars with the crown area more than one-half of the first upper molars**, metacone nearly as large as the paracone, and three well developed commissures; **skull shape flattened, with a broad rostrum; braincase depth, including the auditory bullae, about three-fourths of the mastoid breadth**; sagittal crest obsolete; interorbital region wide and flattish; upper surface of rostrum with a distinct concavity on each side between the lacrimal region and the nares.

Lasionycteris noctivagans (Le Conte, 1831)

Silver-haired bat, murciélago plateado

Monotypic.

Conservation: *Lasionycteris noctivagans* is listed as subject to Special Protection in the Norma Oficial Mexicana (DOF 2019) and as Least Concern by the IUCN (Solari 2019f).

Characteristics: The silver-haired bat is small sized; forearm length 38.0–45.0 mm, skull length 15.7–16.9 mm. **Pelage grizzly black with silver**; rostrum skin blackish; **ears small and rounded, blackish with paler spots at the base; uropatagium base covered with abundant hair**; a square emargination between the canines that includes the nasals and the palate; skull short and wide; contact surface of the third upper molars less than one-third of the first upper molars.

Comments: *Lasionycteris noctivagans* ranges from southeastern Alaska, British Columbia, and southern Canada south through Texas, including all the United States. It is absent from southeastern and southwestern United States (Map 136). It can be distinguished from other species in sympatry, or nearly so, by its silver color pelage; forearm length greater than 38.0 mm and rostrum wide and with two concavities.

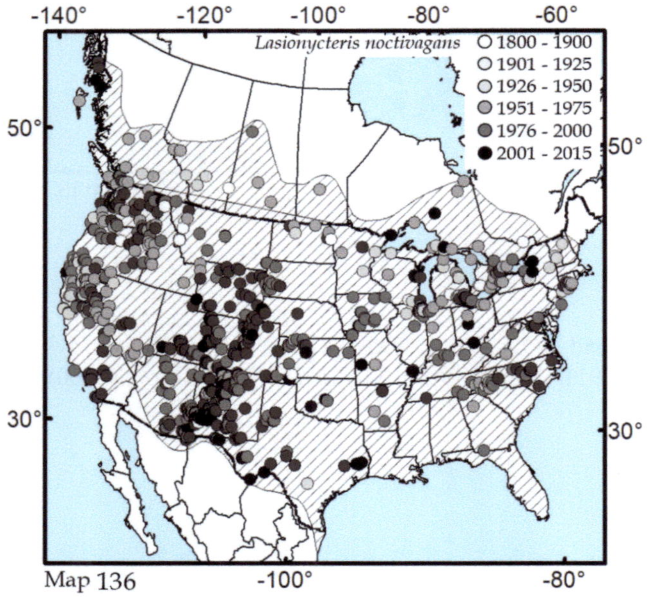

Map 136 Distribution map of *Lasionycteris noctivagans*

Additional Literature: Merriam (1884), Miller (1907), Grinnell (1918), Hoffmeister (1970), Jackson (1961), Williams and Findley (1979), Kunz .

Genus Lasiurus

This genus includes all species within the genus *Lasiurus* (sensu *latos*). Medium sized; dorsal coloration very variable among species; **interfemoral membrane large, with most of its upper surface furred; ears short and rounded; third, fourth, and fifth fingers progressively shortened; skull short and broad; nares and palatal emargination wide and shallow; third upper molars much reduced,** with their crown area less than one-third the crown area of the first upper molars. The species considered within the tribe are the following (Tate 1942; Koopman 1994; Hoofer and Van Den Bussche 2003; Roehrs et al. 2010).

There is a controversy regarding the acceptance of only a single genus, *Lasiurus* (Ziegler et al. 2016, Novaes et al. 2018), or three genera, *Aeorestes, Dasypterus*, and *Lasiurus* (Baird et al. 2015, 2017). Authors have interpreted molecular data in different ways. Based on the arguments exposed and on the works by Roehrs et al. (2010) and Hoofer and Van Den Bussche (2003), which address the whole family, I considered only *Lasiurus* for the present work, with three subgenera: *Aeorestes, Dasypterus*, and *Lasiurus* (Ziegler et al. 2016, Novaes et al. 2018). The keys were elaborated based on the review of specimens and the following sources: Miller (1897), Hall (1981), Álvarez et al. (1994), and Álvarez-Castañeda et al. (2017a).

1. Yellowish or reddish color; one upper premolar at each side .... subgenus (*Dasypterus*) 2

1a. Ash or reddish color; one or two upper premolars at each side; if two, the first one very little 4

2. Total length greater than 119.0 mm; upper toothrow greater than 6.0 mm (from New Jersey, United States, and Sinaloa, México, south through Central America).. *Lasiurus intermedius* (p. 377)

2a. Total length less than 119.0 mm; upper toothrow less than 6.0 mm.. 3

3. Pelage intense yellow especially on the uropatagium; cryptic species can only be differentiated by genetic analysis (from California and Texas, United States, south through Michoacán and Guanajuato, México) ....
....*Lasiurus xanthinus* (p. 378)

3a. Pelage pale yellow, especially on the uropatagium; the cryptic species can only be differentiated by genetic analysis (from Texas, United States, and Guerrero, México, south through South America) ....*Lasiurus ega* (p. 375)

4. Pelage ash-colored; total length greater than 120.0 mm (from southern Canada south through Guatemala) ....
....subgenus (*Aeorestes*) *Lasiurus cinereus* (p. 374)

4a. Pelage reddish or brownish; total length less than 120.0 mm.. subgenus (*Lasiurus*) 5

5. Upper parts not brick red to rusty red; lacrimal ridge not developed (Fig. 60; from New York south through Texas).. *Lasiurus seminolus* (p. 381)

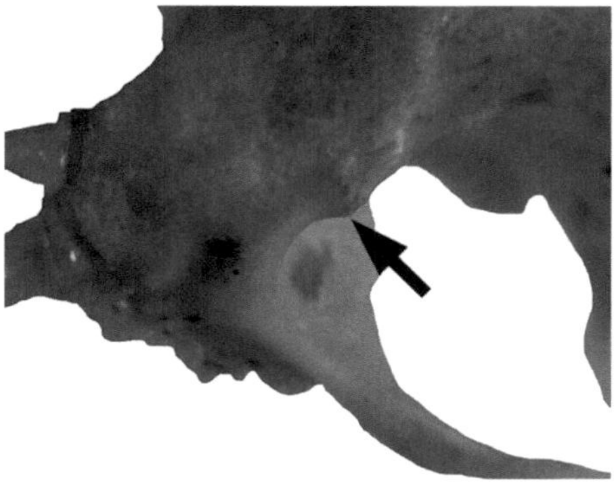

Fig. 60 Lacrimal ridge not developed like in *Lasiurus seminolus*

5a. Upper parts brick red to rusty red, frequently washed with white; lacrimal ridge present (Fig. 61)................................6

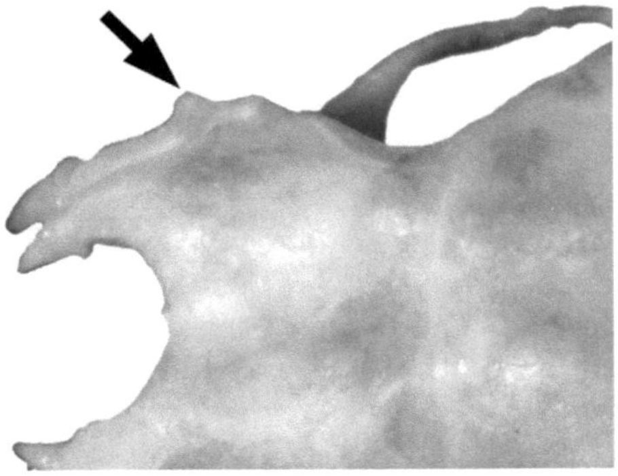

Fig. 61 Lacrimal ridge present

6. Dorsal coloration red to brownish red, with a mix of ash-colored hairs, general appearance opaque; upperpart of the uropatagium grayish red, opaque; tending to be larger sized; rostrum large (from Alberta and New Brunswick, Canada, south through Coahuila and Tamaulipas, México) .... *Lasiurus borealis* (p. 379)

6a. Dorsal coloration intense red to rusty red, with cinnamon shades in some specimens, with a few showing a mix of ash-colored shades; upperpart of the uropatagium intense oxide red with cinnamon shades; tending to be small in size; rostrum small (from Saskatchewan west through British Columbia, Canada, and south through South America) ....*Lasiurus frantzii* (p. 380)

Subgenus *Aeorestes* Fitzinger, 1870

Medium sized; dorsal and ventral pelage grizzly with gray; ears small and rounded; **uropatagium covered with abundant hair, with the same pattern as in the back, that overhangs the uropatagium border;** skull short and wide; one emargination between the canines including the nasals and the palate; contact surface of third upper molars less than one third of the first molar; one or two upper premolars; braincase high and rounded, rostrum sloping away rapidly in front so that its upper surface is nearly in line with that of braincase; lacrimal crest varying in size.

Lasiurus cinereus (Palisot de Beauvois, 1796)
Hoary bat, murciélago canoso

L. c. cinereus (Palisot de Beauvois, 1796). Only subspecies in North America.

Conservation: *Lasiurus cinereus semotus* is listed as Endangered by the United States Endangered Species Act (FWS 2022) and *L. cinereus* as Least Concern by the IUCN (Gonzalez et al. 2016a).

Characteristics: Hoary bat is medium sized; forearm length **46.0–57.0 mm, skull length 17.0–18.5 mm**. **Dorsal pelage and fur of underparts grizzly with gray, white, brown**; rostrum and throat with yellowish lines; **white locks of hair at thumb base**; ears small and rounded; rostrum skin and ears blackish; **with lacrimal crest;** short and wide; one emargination between the canines including the nasals and the palate; braincase high and rounded; rostrum sloping away rapidly in the front, so that its upper surface is nearly in line with the braincase.

Comments: The genus *Lasiurus* was considered *Nycteris* (Hall 1981). A molecular review of the Lasiurine showed that the genus *Lasiurus* should be separated into three different genera, using the generic name *Aeorestes* for hoary bats (Baird et al. 2015), but the change was not accepted (Ziegler et al. 2016, Novaes et al. 2018). *L. cinereus* ranges from southern Canada south through Guatemala (Map 137). It can be found in sympatry, or nearly so, with all species of *Lasiurus* in North America. *L. cinereus* is characterized by the uropatagium covered entirely by hair and with an ash color mixed with different shades; the skull has an emargination between the canines that includes the nasals and the palate. *L. cinereus* and *L. seminolus* share a similar coloration pattern, differentiated by the forearm length greater than 46.0 mm and a tuft of white hair at the thumb base.

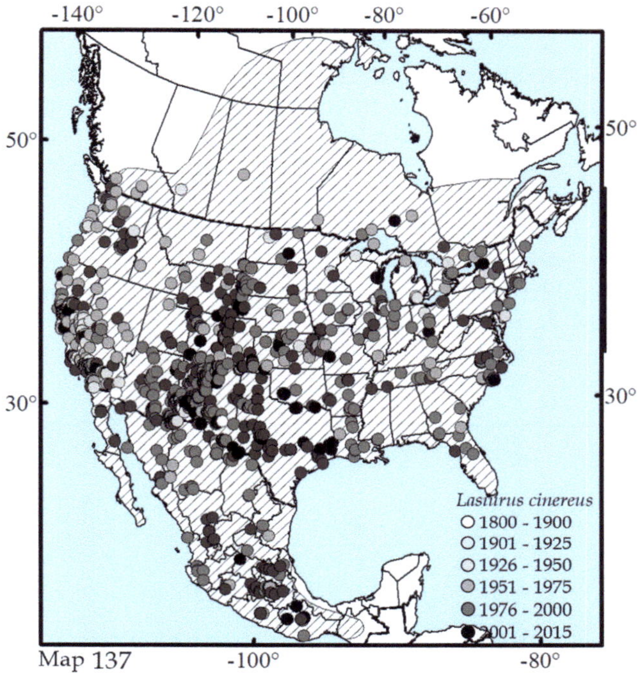

Map 137 Distribution map of *Lasiurus cinereus*

Additional Literature: Miller (1897), Hall and Jones (1961), Jackson (1961), Barbour and Davis (1969), Watkins et al. (1972), Hall (1981), Shump and Shump (1982a, b), Álvarez-Castañeda and Álvarez (1991), Bogan (1999), Villalpando and Álvarez (2000), Álvarez-Castañeda et al. (2008c), Baird et al. (2021).

Subgenus ***Dasypterus*** Reuter, 1871

Medium sized; dorsal and ventral pelage yellowish; ears small and rounded; **uropatagium very large and covered with abundant hair, with the same pattern as in the back that overhangs the border;** skull short and wide; one emargination between the canines including the nasals and the palate; braincase high and rounded, rostrum sloping away rapidly in the front, so it is nearly in line with the braincase; lacrimal crest varying in size.

Lasiurus ega (Gervais, 1855)

Southern yellow bat, murciélago amarillo del sur

L. e. panamensis (Thomas, 1901). The only subspecies in North America.

Conservation: *Lasiurus ega* is listed as Least Concern by the IUCN (Barquez and Diaz 2016a).

Characteristics: Southern yellow bat is medium sized; forearm length 48.0–58.0 mm, skull length 14.7–16.2 mm. Dorsal pelage dull yellow or grayish; fur of underparts pale, dull yellow, or cream; rostrum hair and ears pinkish; ears small and rounded but higher than wide.

Comments: The genus *Lasiurus* was considered *Nycteris* (Hall 1981). A molecular review of the Lasiurine showed that the subgenus *Dasypterus* should be considered a full genus (Baird et al. 2015), but the change was not accepted (Ziegler et al. 2016, Novaes et al. 2018). *L. xanthinus* was previously considered a subspecies of *L. ega* (Baker et al. 1988). *L. ega* ranges from Texas and Guerrero south through South America (Map 138). It can be found in sympatry, or nearly so, with *L. cinereus, L. intermedius,* and *L. borealis,* and is allopatric with respect to *L. xanthinus*. *L. ega* is characterized by having the uropatagium entirely covered by hair; the skull has an emargination between the canines that includes the nasals and the palate; it is differentiated from *L. intermedius* by its smaller size, total length 119.0 mm, and its smaller toothrow, 6.0 mm. *L. ega* is cryptic with *L. xanthinus*, both can be differentiated only by molecular methods and their distribution area. It differs from all other species of *Lasiurus* by its yellowish color.

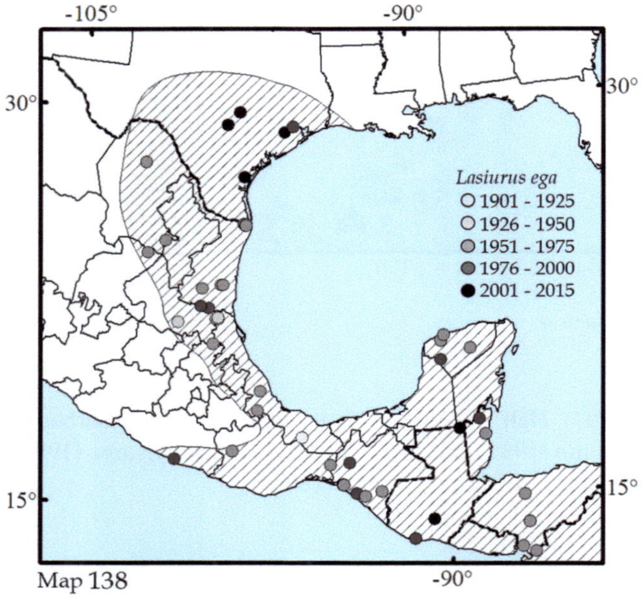

Map 138 Distribution map of *Lasiurus ega*

Additional Literature: Dobson (1878), Allen (1906), Handley (1960), Hall and Jones (1961), Husson (1962), Shump and Shump (1982a), Álvarez-Castañeda and Álvarez (1991), Kurta and Lehr Hoofer et al. (2006), Roehrs et al. (2010), Baird et al. (2021).

Lasiurus intermedius Allen, 1862

Northern yellow bat, murciélago amarillo del norte

L. i. intermedius Allen, 1862. The only subspecies in North America.

Conservation: *Lasiurus intermedius* is listed as Least Concern by the IUCN (Miller and Rodriguez 2016b).

Characteristics: Northern yellow bat is medium sized; forearm length 43.0–47.0 mm, skull length 17.8–18.5 mm. Dorsal pelage dull or grayish yellow; fur of underparts pale or cream yellow; rostrum skin and ears pinkish; ears small and rounded but higher than wide.

Comments: The genus *Lasiurus* was considered *Nycteris* (Hall 1981). A molecular review of the Lasiurine showed that the subgenus *Dasypterus* should be considered a full genus (Baird et al. 2015), but the change was not accepted (Ziegler et al. 2016, Novaes et al. 2018). *L. intermedius* ranges from New Jersey and Sinaloa south through Central America, including the Florida and Yucatán Peninsulas (Map 139). It can be found in sympatry, or nearly so, with *L. intermedius, L. xanthinus,* and *L. ega* in North America. *L. intermedius* is characterized by the uropatagium entirely covered by hair; the skull has an emargination between the canines that includes the nasals and the palate. It is differentiated from *L. xanthinus* and *L. ega* by its larger size, total length 119.0 mm and toothrow length 6.0 mm. It is distinguished from all other species of *Lasiurus* by its yellowish color.

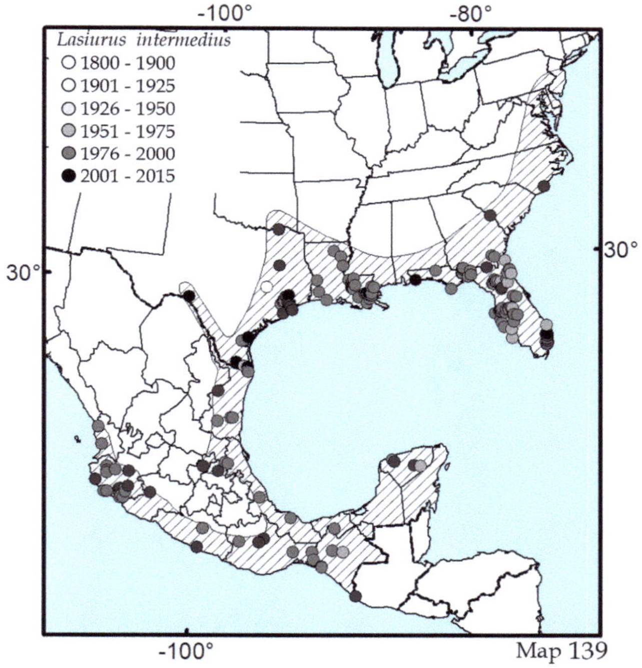

Map 139 Distribution map of *Lasiurus intermedius*

Additional Literature: Handley (1959, 1960), Hall and Jones (1961), Lowery (1974), Webster et al. (1980), Shump and Shump (1982a), Álvarez-Castañeda and Álvarez (1991), González-Ruíz et al. (2000), Mejenes-López et al. (2010), Baird et al. (2015, 2021).

Lasiurus xanthinus (Thomas, 1897)

Western yellow bat, murciélago amarillo del oeste

Monotypic.

Conservation: *Lasiurus xanthinus* is listed as Least Concern by the IUCN (Arroyo-Cabrales and Álvarez-Castañeda 2017e).

Characteristics: Western yellow bat is medium sized; forearm length 42.7–52.0, skull length 14.2–15.5 mm. Dorsal pelage dull yellow; fur of underparts pale yellow or cream; rostrum skin and ears pinkish; ears small and rounded but higher than wide.

Comments: The genus *Lasiurus* was considered *Nycteris* (Hall 1981). A molecular review of the Lasiurine showed that the subgenus *Dasypterus* should be considered a full genus (Baird et al. 2015), but the change was not accepted (Ziegler et al. 2016, Novaes et al. 2018). *L. xanthinus* was previously considered a subspecies of *L. ega* (Baker et al. 1988). It ranges from Texas, New Mexico, Arizona, and California south through Michoacán and Guanajuato (Map 140). It can be found in sympatry, or nearly so, with *L. cinereus*, *L. intermedius,* and *L. frantzii,* and is allopatric with respect to *L. ega*. It is characterized by having the uropatagium entirely covered by hair. The skull has an emargination between the canines that includes the nasals and the palate. *L. xanthinus* differs from *L. intermedius* by its smaller size, total length 119.0 mm, and smaller toothrow, 6.0 mm. It is cryptic species with *L. ega*, from which it is differentiated only by molecular methods and their distribution area, and from all other species of *Lasiurus*, by its yellowish color.

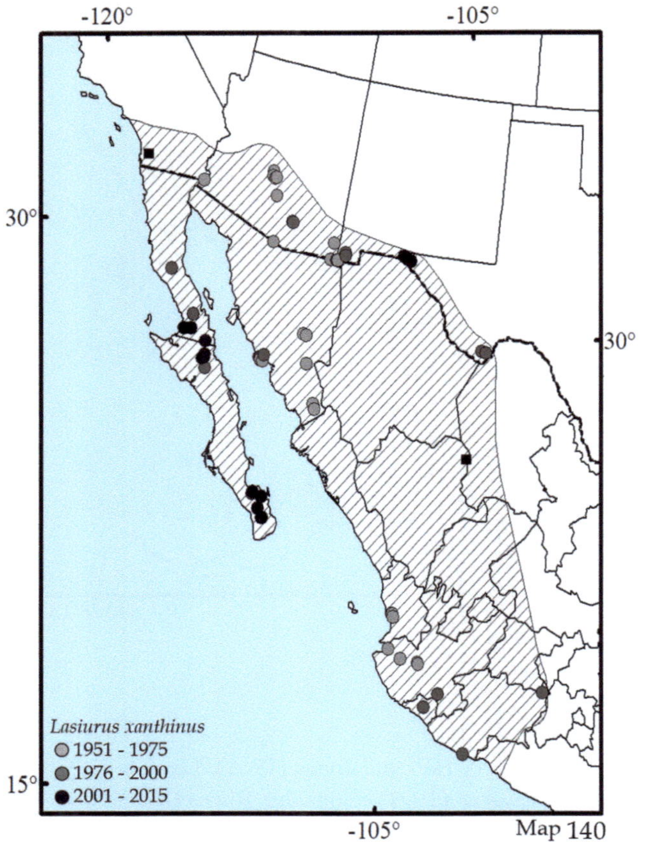

Map 140 Distribution map of *Lasiurus xanthinus*

Additional Literature: Dobson (1878), Allen (1906), Handley (1960), Hall and Jones (1961), Husson (1962), Kurta and Lehr (1995), Bogan (1999), Villalpando and Álvarez (2000), Álvarez-Castañeda et al. (2008c), Frick et al. (2008), Baird et al. (2021).

Subgenus *Lasiurus* Gray, 1831

Medium sized; dorsal and ventral pelage reddish; ears small and rounded; **uropatagium very large, covered with abundant hair, with the same patten as the back, that overhangs the border**; skull short and wide; one emargination between the canines including the nasals and the palate; anterior upper premolars minute, occupying the inner angle between the canines and the large premolars; contact surface of the third upper molars less than one-third of the first molars; one or two premolars; braincase high and rounded, rostrum sloping away rapidly in the front, so that its upper surface is nearly in line with the braincase; lacrimal crest varying in size.

Lasiurus borealis (Müller, 1776)

Eastern red bat, murciélago rojo del este

Monotypic.

Conservation: *Lasiurus borealis* is listed as Least Concern by the IUCN (Arroyo-Cabrales et al. 2016).

Characteristics: Eastern red bat is medium sized; forearm length 37.1–43.6 mm, skull length 11.5–13.5 mm. **Dorsal coloration red to brownish red, with a mix of ash-colored hairs, with a general opaque appearance;** fur of underparts paler than in the dorsum; **uropatagium upperpart grayish red, opaque**; hair on the neck not more brilliant; rostrum skin and ears reddish; **rostrum longer;** lacrimal crest.

Comments: The genus *Lasiurus* was considered *Nycteris* (Hall 1981). A molecular review of the Lasiurine showed that the genus *Lasiurus* should be separated into three different genera using the generic names *Aeorestes* for hoary bats and *Dasypterus* for yellow bats (Baird et al. 2015), but the change was not accepted (Ziegler et al. 2016, Novaes et al. 2018). *L. frantzii* was previously part of *L. borealis,* so the subspecies of *L. borealis* known as *L. b. teliotis,* and *L. b. frantzii* are considered under *L. frantzii* (Baker et al. 1988). *L. borealis* ranges from the Great Plains and southern parts of Alberta, Saskatchewan, Manitoba, Ontario, Québec to eastern New Brunswick, Canada south through Tamaulipas, Nuevo León, and Coahuila, and east through the Atlantic coast of the United States (Map 141). It can be found in sympatry, or nearly so, with *L. cinereus, L. seminolus*, and *L. ega but* is allopatric with respect to *L. frantzii.* Species of *Lasiurus* are characterized by having the uropatagium entirely covered by hair. The skull has an emargination between the canines that includes the nasals and the palate. *L. frantzii* can be differentiated from *L. borealis* by its dorsal coloration red to brownish red with a mix of ash-colored hairs and a general opaque appearance; uropatagium upperpart opaque, grayish red, tending to be larger in size; rostrum larger. From the other species of *Lasiurus*, by its reddish pelage.

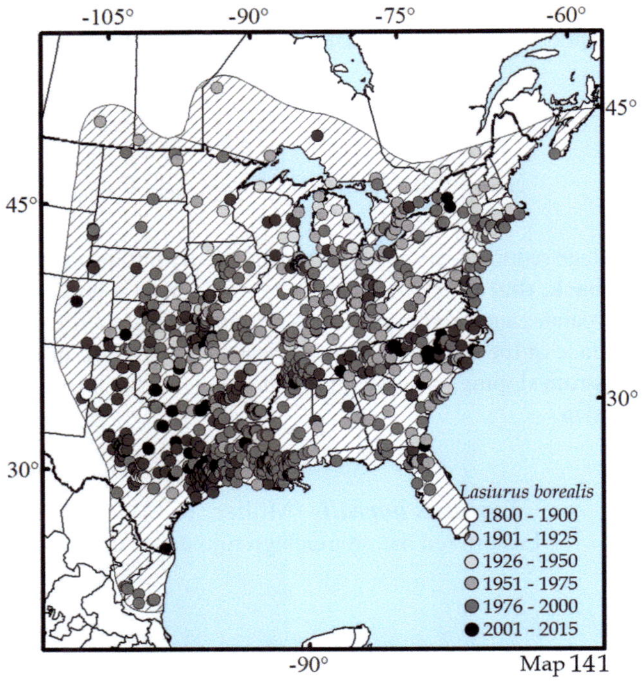

Map 141 Distribution map of *Lasiurus borealis*

Additional Literature: Barbour and Davis (1969), Hall and Jones (1961), Shump and Shump (1982a), Roehrs et al. (2010).

Lasiurus frantzii (Peters, 1871)

Western red bat, murciélago rojo del oeste

1. *L. f. frantzii* (Peters, 1871). From the Tehuantepec Isthmus southward.
2. *L. f. teliotis* (Allen, 1891). From British Columbia to the Tehuantepec Isthmus.

Conservation: *Lasiurus frantzii* (as *Lasiurus blossevillii*) is listed as Least Concern by the IUCN (Gonzalez et al. 2016b).

Characteristics: Western red bat is medium sized; forearm length 38.0–42.0 mm, skull length 11.8–13.0 mm. **Dorsal coloration intense red to rusty red, some specimens with cinnamon shades, few with a mix of ash shades;** fur of underparts paler than in the dorsum; **uropatagium upperpart intense rusty red with cinnamon shades**; hair on the neck more brilliant; rostrum skin and ears pinkish brown; **rostrum short**; lacrimal crest.

Comments: The genus *Lasiurus* was considered *Nycteris* (Hall 1981). A molecular review of the Lasiurine showed that the genus *Lasiurus* should be separated into three different genera using the generic names *Aeorestes* for hoary bats and *Dasypterus* for yellow bats (Baird et al. 2015), but the change was not accepted (Ziegler et al. 2016, Novaes et al. 2018). *L. frantzii* was previously regarded as part of *L. borealis,* so the subspecies of *L. borealis* known as *L. b. teliotis* and *L. b. frantzii* are considered under *L. frantzii* (Baker et al. 1988). *L. frantzii* ranges from the Great Plains of Alberta, and Saskatchewan west through British Columbia and south through South America, except for the Rocky Mountains and Sierra Nevada (Map 142). It can be found in sympatry, or nearly so, with *L. cinereus, L. ega, L. intermedius,* and *L. xanthinus* but is allopathic with *L. borealis*. *L. frantzii* is characterized by having the uropatagium entirely covered with hair to the tail; the skull has an emargination between the canines that includes the nasals and the palate. It can be differentiated from *L. borealis* by its dorsal coloration intense red to rusty red, with cinnamon shades in some specimens, with a mix of ash-colored hairs; uropatagium upperpart with an intense rusty with cinnamon shades tending to be smaller size; smaller rostrum. It can be differentiated from the other species of *Lasiurus* by its reddish pelage.

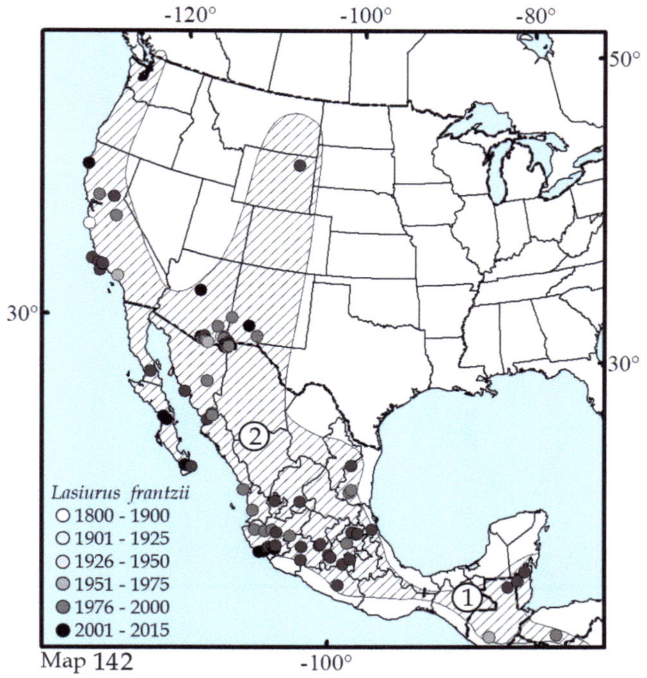

Map 142 Distribution map of *Lasiurus frantzii*

Additional Literature: Hall and Jones (1961), Barbour and Davis (1969), Shump and Shump (1982a), Bogan (1999), Villalpando and Álvarez (2000), Álvarez-Castañeda et al. (2008c), Frick et al. (2008).

Lasiurus seminolus (Rhoads, 1895)

Seminole bat, murciélago canoso de Florida

Monotypic.

Conservation: *Lasiurus seminolus* is listed as Least Concern by the IUCN (Solari 2019g).

Characteristics: Seminole bat is medium sized; forearm length on average 40.7 mm, skull length on average 11.2 mm. **Dorsal pelage mahogany brown grizzly with pale-colored gray; throat and neck whitish**; fur of underparts paler than in the dorsum; ears small and rounded; ears pinkish brown; lacrimal crestshort and wide; one emargination between the canines including the nasals and the palate; braincase high and rounded, rostrum sloping away rapidly in front so that its upper surface is nearly in line with that of braincase.

Comments: The genus *Lasiurus* was considered *Nycteris* (Hall 1981). A molecular review of the Lasiurine showed that genus *Lasiurus* should be separated into three different genera, using the generic name *Aeorestes* for hoary bats (Baird et al. 2015), but the change was not accepted (Ziegler et al. 2016, Novaes et al. 2018). *L. seminolus* ranges in southeastern United States from Texas to New York, including the entire Florida Peninsula. No record exists for México (Map 143). It can be found in sympatry, or nearly so, with *L. cinereus, L. intermedius,* and *L. borealis*. This species is characterized by having the uropatagium entirely covered by hair and with ash hairs mixed with different shades. The skull has an emargination between the canines that includes the nasals and the palate. *L. seminolus* and *A. cinereus* share the same coloration pattern and *L. seminolus* is differentiated by having the forearm length less than 46.0 mm and not having a tuft of white hair at the thumb base.

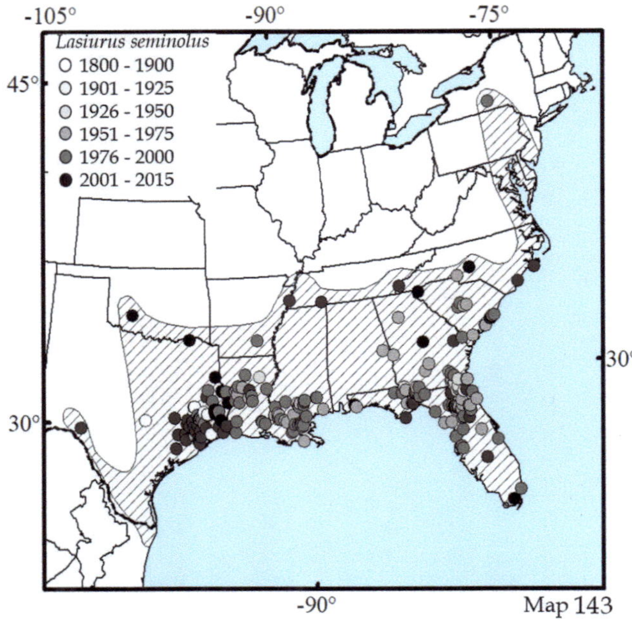

Map 143 Distribution map of *Lasiurus seminolus*

Additional Literature: Golley (1962), Lowery (1974), Sealander (1979), Lee et al. (1982), Shump and Shump (1982a), Schmidly (1983), Kennedy et al. (1984), Wilkins (1987).

Tribe Nycticeiini

Genus *Nycticeius* Rafinesque, 1819

Medium sized; dorsal pelage blackish brown; **ears triangular with the tragus curved frontward**; rostrum, ears, and membranes blackish; hair blackish at the base; calcar unkeeled; upper incisors simple, unicuspid, nearly one-half as high as the canines; **three lower incisors, the outer incisors larger than either of the others**; upper premolars without a cusp on the inner side; **third lower molars with the crown area more than one-half that of first and second lower molars; the mesostyle, metacone, and three commissures of the third lower molars well developed**.

Nycticeius humeralis (Rafinesque, 1818)

Evening bat, murciélago del anochecer

1. *N. h. humeralis* (Rafinesque, 1818). From Texas northward to Nebraska and Iowa, and eastward to the Atlantic coast.
2. *N. h. mexicanus* Davis, 1944. From Texas south through Veracruz.
3. *N. h. subtropicalis* Schwartz, 1951. Restricted to the southern Florida Peninsula.

Conservation: *Nycticeius humeralis* is listed as Least Concern by the IUCN (Solari 2019l).

Characteristics: Evening bat is medium sized; forearm length 33.5–39.0 mm, skull length 14.0–14.7 mm. Dorsal pelage blackish brown; fur of underparts brown; **ears triangular with the tragus curved frontward**; rostrum, ears, and membranes blackish; hair blackish at the base; tail extending to the uropatagium tip; calcar unkeeled; skull short, robust, wide, and rostrum well differentiated from the braincase; teeth proportionally smaller; with an emargination between the canines including the nasals and the palate, four upper molariforms in each mandibular branch well developed; canines large and with a cingulum; braincase not very high; one upper incisor at each side; **three lower incisors**.

Comments: *Nycticeius humeralis* ranges from the Great Plains eastward from Nebraska and Texas east through New Hampshire and south through Veracruz, México (Map 144). It can be distinguished from other species in sympatry, or nearly so, such as those of *Corynorhinus, Eptesicus, Lasiurus,* and *Rhogeessa*, by having triangular ears with the tragus curved forward, in addition to having three lower incisors in each mandibular branch.

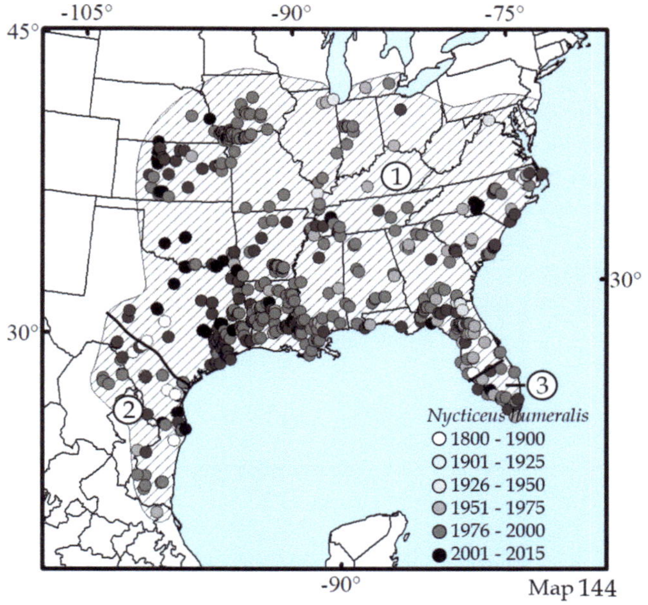

Map 144 Distribution map of *Nycticeius humeralis*

Additional Literature: Allen (1893), Miller (1897, 1907), Davis (1944b), Barbour and Davis (1969), Watkins (1969, 1972).

Tribe Perimyotini

Genus *Parastrellus* Hoofer et al., 2006

Small sized; **face mask and ears black, contrasting with the rest of the body**; ears usually short and broad; tragus less acutely pointed; calcar unkeeled; **limb length less than one-half of the tibia length**; third upper and lower premolars absent; outer upper incisors not as large as the inner upper incisors but extending considerably beyond the cingulum; rostrum well differentiated from the braincase; an emargination between the canines, including the nasals and the palate; supraoccipital region above the braincase roof level; **dorsal skull outline more or less straight; palatine extending beyond the back of the last molars**.

Parastrellus hesperus (Allen, 1864)

American pipistrelle, pequeño murciélago del oeste

1. *P. h. hesperus* (Allen, 1864). From Oregon, Colorado, Sonora, and Sinaloa westward.
2. *P. h. maximus* (Hatfield, 1936). From New Mexico and Chihuahua westward.

Conservation: *Parastrellus hesperus* (as *Pipistrellus hesperus*) is listed as Least Concern by the IUCN (Solari 2019m).

Characteristics: Western pipistrelle is small sized; forearm length 26.0–32.0 mm, skull length 10.0–11.2 mm. Dorsal pelage from yellowish brown to medium gray; fur of underparts paler than in the dorsum; **black face mask and ears contrasting with the rest of the body**; calcar unkeeled; **limb length less than one-half of the tibia length;** rostrum well differentiated from the braincase; teeth proportionally big, with cutting cusps W shaped; an emargination between the canines including the nasals and the palate; five upper molariforms; upper incisors well developed and outer incisors bigger than the inner incisors; canines well developed and with a cingulum; supraoccipital region above the level of the braincase roof; **dorsal skull outline more or less straight; palatine extending beyond the back of the last molars**.

Comments: *Parastrellus hesperus* was originally considered within the genus *Pipistrellus* (Miller 1897). In the present work, I use *Parastrellus hesperus* following Hoofer et al. (2006). *P. hesperus* ranges from Washington, Utah, and Colorado south through Michoacán, Estado de México, Morelos, Hidalgo, and northern Veracruz (Map 145). It can be found in sympatry, or nearly so, with many species of *Myotis*, with which it can be confused. The main differences are its small size (forearm length less than 32.0 mm); presence of a black face mask, low number of teeth (total = 34) and inner upper incisors of the same size as the outer incisors, besides being bifid. From *Perimyotis subflavus*, because its limb length is less than one-half of the tibia length, and the palatine extends much more to the back of the last molars. From the other species of the subfamily Vespertilioninae, by having only two premolars on each side; scarce or no hair in the uropatagium, and by its small size.

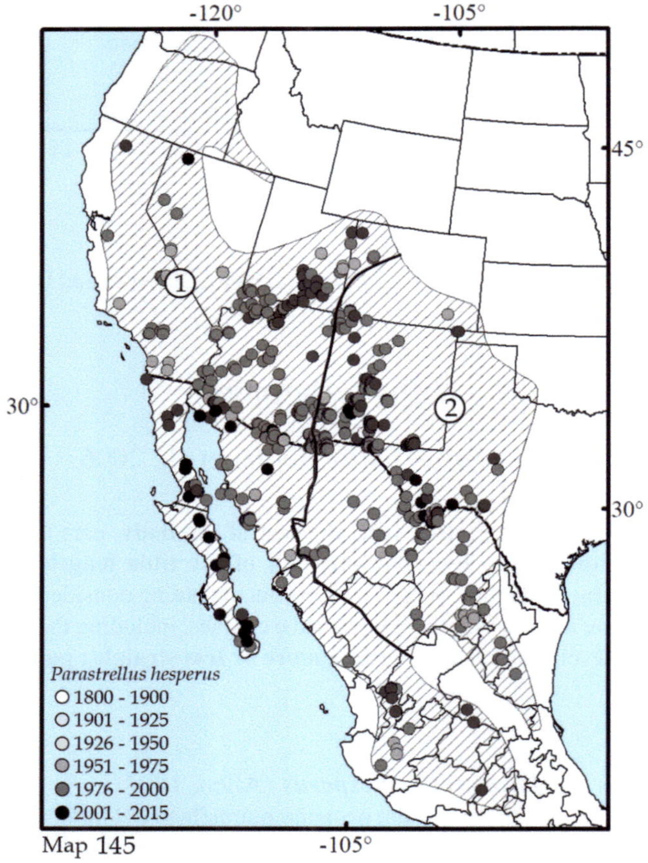

Map 145 Distribution map of *Parastrellus hesperus*

Additional Literature: Hall and Dalquest (1950), Bogan (1999), Álvarez-Castañeda et al. (2008c), Frick et al. (2008).

Genus *Perimyotis* Menu, 1984

Small sized; without a strong black mask in the rostrum; ears usually short and broad; tragus less acutely pointed; membranes blackish; **limb length greater than one-half of the tibia length;** third upper and lower premolars absent; outer upper incisors not as large as the inner incisors but extending considerably beyond the cingulum; rostrum well differentiated from the braincase; an emargination between the canines including the nasals and the palate; braincase roof above the supraoccipital level; **palatine extending slightly to the back of the last molar**.

Perimyotis subflavus (Cuvier, 1832)

Eastern pipistrelle, pequeño murciélago del este

1. *P. s. clarus* (Baker, 1954). Restricted to northern Coahuila.
2. *P. s. floridanus* (Davis, 1957). From southeastern Georgia and the Florida Peninsula.
3. *P. s. subflavus* (Cuvier, 1832). From Veracruz northward to Wyoming, eastward to Nova Scotia and to all the Atlantic coast. Include southeastern Canada.
4. *P. s. veraecrucis* (Ward, 1891). From Veracruz south through Guatemala, except for the northern of the Yucatán Peninsula.

Conservation: *Perimyotis subflavus* is listed as under Special Protection by the United States Endangered Species Act (FWS 2022), and the Norma Oficial Mexicana (DOF 2019), and as Vulnerable (A3bce: population reduction in the past 10 years equal to or greater than 30% projected to be met in the future based on data obtained by an index of abundance, area of occupancy, and the effects of introduced taxa) by the IUCN (Solari 2018n).

Characteristics: Eastern pipistrelle is small sized; forearm length 31.0–35.0 mm, skull length 12.4–13.1 mm. Dorsal pelage dark grizzly brown with longer yellow hairs; fur of underparts brown; tragus wide and with a round tip; no strong black mask in the rostrum; uropatagium with yellow hairs scattered from the knees to the tail; membranes blackish; **limb length greater than one-half of the tibia length**; rostrum well differentiated from the braincase; teeth proportionally big, with cutting cusps W shaped; an emargination between the canines including the nasals and the palate; five upper molariforms; upper incisors well developed, outer incisors larger than the internal incisors; canines well developed with a cingulum; braincase supraoccipital region at a lower level than the braincase roof; **palatine extending very little to the back of the last molars**.

Comments: *Perimyotis subflavus* was originally considered in the subgenus *Perimyotis* of the genus *Pipistrellus* (Miller 1897). The present work uses *Perimyotis subflavus* (Menu 1984; Hoofer and Van den Bussche 2003). *P. subflavus* ranges from the Great Plains to the eastern from South Dakota, Ontario, and New Mexico east through New Hampshire and south through Central America, including the southern Yucatán Peninsula (Map 146). It can be found in sympatry, or nearly so, with many species of *Myotis*, with which it could be confused. The main differences are its small size (forearm length less than 35.0 mm); low number of teeth (total = 34) and inner upper incisors of the same size as the outer incisors and not clearly bifid. From *Parastrellus hesperus*, because its limb length is greater than one-half of the tibia length; in the skull, the palatine extends slightly to the back of the last molar. From the other species of the subfamily Vespertilioninae, by having only two premolars on each side and scarce or no hair in the uropatagium, in addition to its small size.

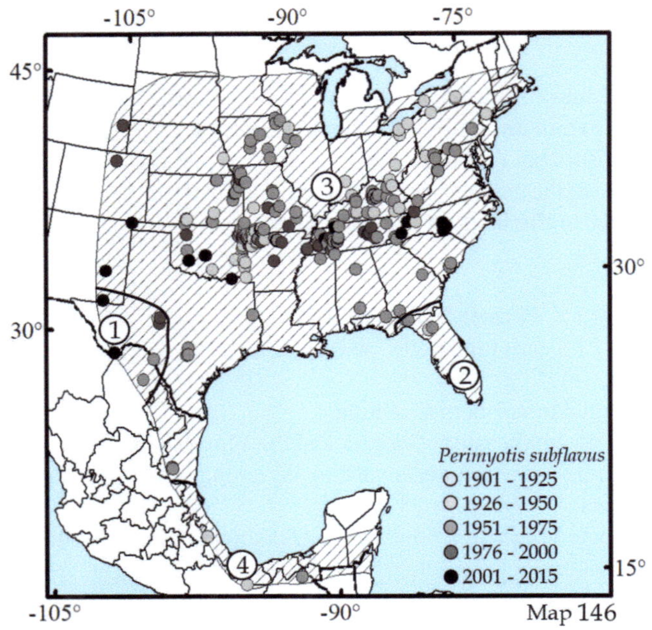

Map 146 Distribution map of *Perimyotis subflavus*

Additional Literature: Nason (1948), Hall and Dalquest (1950), Davis (1959), Barbour and Davis (1969), Laval and LaVal (1980), Fujita and Kunz (1984), Bogan (1999), Hoofer et al. (2006).

Tribe Plecotin

Genus *Corynorhinus* Allen, 1865

Medium sized; dorsal and ventral pelage reddish brown to grayish brown; **ears and tragus very large, much longer than the head, joined across the forehead**; **ears more than 75% of the forearm length**; **nostrils opening upward, their orifice continues backward by a slit-like extension**; **glandular masses on the muzzle rising high above the nostrils**; tail tip beyond the uropatagium; **uropatagium with stripes of hair, which vary in number according to the species**; membranes pale brown or grayish brown; skull **small; sagittal crest in the frontal bone and very low or absent in the supraoccipital bone**; middle braincase higher, rostrum proportionally short; **lacrimal region rounded**; auditory bullae large, rounded, and their greatest diameter about three times the distance between them. The keys were elaborated based on the review of specimens and the following sources: Handley (1959), Hall (1981), Álvarez et al. (1994), and Álvarez-Castañeda et al. (2017a).

1. Ventral hairs white or whitish at the tip, in strong contrast with the blackish shade at the base; median postpalatal process triangular, with a broad base; first upper incisors bifid; anterointernal cusp usually present on the cingulum of the fourth upper premolars (New Jersey, Indiana, and Ohio south through Florida and west through Oklahoma and Texas, United States)........ *Corynorhinus rafinesquii* (p. 388)

1a. Ventral hairs brownish or buff at the tip, often not strongly contrasting with the slate gray or brownish shade at the base; median postpalatal process styliform, with a narrow base; first upper incisors either bifid or not; anterointernal cusp present on the cingulum of the fourth upper premolars........2

2. Dorsal hairs dark sooty-brown, with a slight contrast between the tip and the base; tragus length less than one-third of the ear length, less than 13.0 mm; usually less than eight stripes of hair on the uropatagium; skull length usually less than 15.5 mm (from Chihuahua and Coahuila south through Michoacán and Puebla, México)........ *Corynorhinus mexicanus* (p. 387)

2a. Dorsal hairs brown, with contrasting shades between the tip and the base; tragus length more than one-third of the ear length, greater than 13.0 mm; usually eight stripes of hair in the uropatagium; skull length usually greater than 15.5 mm (from British Columbia, Canada and South Dakota, United States, south through Puebla and Oaxaca, México and east through North Carolina and Pennsylvania, United States) ....*Corynorhinus townsendii* (p. 389)

Corynorhinus mexicanus Allen, 1916

Mexican big-eared, murciélago orejudo mexicano

(*see addenda, split in two species *Corynorhinus leonpaniaguae*)

Monotypic.

Conservation: *Corynorhinus mexicanus* is listed as Near Threatened by the IUCN (Solari 2019c).

Characteristics: Mexican big-eared bat is medium sized; forearm length 39.0–45.0 mm, skull length 14.5–15.9 mm. **Dorsal pelage dark sooty-brown and opaque; fur of underparts paler than in the dorsum, with hair paler gray at the tip and blackish at the base;** tragus small, grayish, much less than 13.0 mm, less than one-third of the ear length; **usually less than eight stripes of hair**; membranes brownish gray; supraoccipital region lower than the braincase roof; median postpalatal process styliform and with a narrow base.

Comments: *Corynorhinus* was elevated to full genus (Tumlison and Douglas 1992); it was previously regarded as a subgenus of *Plecotus,* known as *Plecotus mexicanus*. *C. mexicanus* was considered a subspecies of *C. townsendii* (Hall and Kelson 1959) but is a full species (Handley 1966). *C. mexicanus* ranges from northward to Chihuahua and Nuevo León south through Michoacán and Veracruz (Map 147). It can be distinguished from other species in sympatry, or nearly so, by its enlarged nasal opening, with a glandular bulge clearly visible between the nasal opening and the eyes, in addition to having very large ears and tragus, greater than the head length. Although all *Corynorhinus* species are allopatric, *C. mexicanus* can be distinguished from *C. townsendii* by having less than eight transverse hair stripes in the uropatagium that are clearly distinguished "See adeendum".

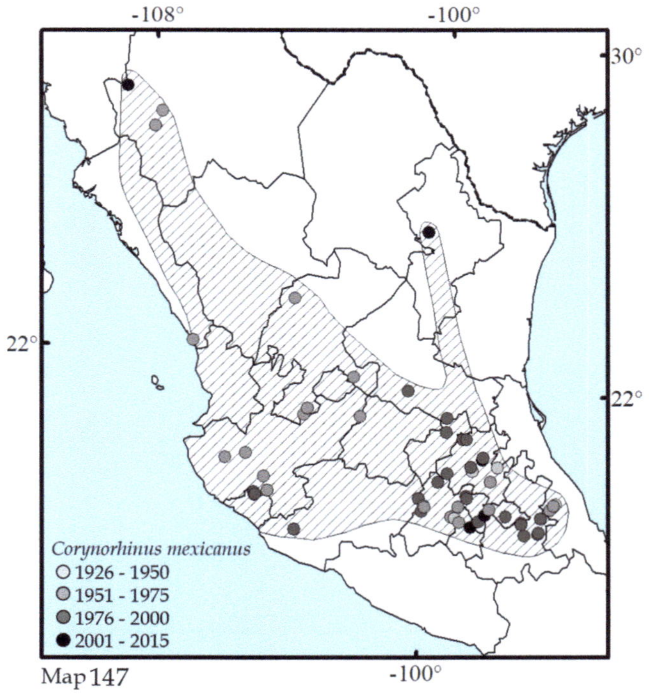

Map 147 Distribution map of *Corynorhinus mexicanus*

Additional Literature: Allen (1916), Handley (1959), Baumgardner et al. (1977), Jones (1977), Morales-Malacara and López-W. (1990), Tumlison (1991, 1992), Villalpando and Álvarez (2000), Hoofer and Van den Bussche (2003), Piaggio and Perkins (2005), Álvarez-Castañeda et al. (2008c), Roehrs et al. (2010).

Corynorhinus rafinesquii (Lesson, 1827)

Rafinesque's big-eared bat, murciélago orejudo norteamericano

1. *C. r. rafinesquii* (Lesson, 1827). From eastern Oklahoma, Texas, Louisiana, Arkansas, Mississippi, Alabama, Georgia, Florida, North and South Carolina.
2. *C. r. macrotis* (Le Conte, 1831). From northeastern Arkansas, northern Mississippi, Alabama, and Georgia, and western North Carolina north through Indiana and Ohio.

Conservation: *Corynorhinus rafinesquii* is listed as Least Concern by the IUCN (Arroyo-Cabrales and Álvarez-Castañeda 2017b).

Characteristics: Rafinesque's big-ear bat is medium sized; forearm length 38.8–43.5 mm, skull length 13.2–15.1 mm. **Dorsal pelage brown, paler at the tip than at the base; fur of underparts with hairs white or whitish at the tip, strongly contrasting with the blackish shade at the base;** tragus brown, small, less than 20.0 mm, less than one-third of the ear length; uropatagium **usually with more than 12 stripes of hair**; membranes brown; supraoccipital region lower than the braincase roof; **median postpalatal process triangular, with a broad base**.

Comments: *Corynorhinus* was previously considered a subgenus of *Plecotus* but is currently regarded as a genus (Tumlison and Douglas 1992). *C. rafinesquii* is a different species from *C. townsendii* (Bogdanowicz et al. 1998). *C. rafinesquii* ranges in the southeastern United States, from Texas and Oklahoma eastward and from Illinois to New Jersey south through the Florida Peninsula (Map 148). It can be distinguished from other species in sympatry, or nearly so, by its enlarged nasal opening, with a glandular bulge clearly visible between the nasal opening and the eyes, in addition to having very large ears and tragus, greater than the head length. Although all *Corynorhinus* species are allopatric, *C. rafinesquii* can be distinguished from *C. townsendii* by the dorsal pelage brown, paler at the tip than at the base; fur of underparts with hair white or whitish at the tip and blackish at the base, usually with more than 12 stripes of hair in the uropatagium.

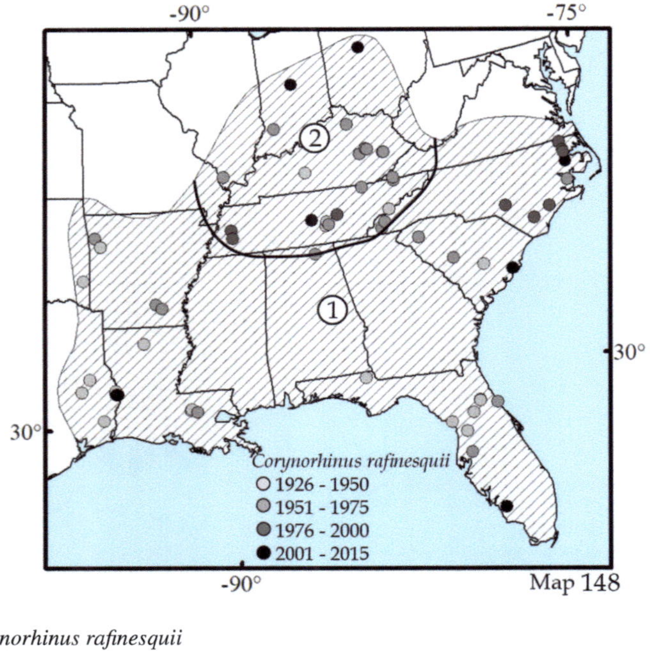

Map 148 Distribution map of *Corynorhinus rafinesquii*

Additional Literature: Handley (1959), Baker and Ward (1967), Barbour and Davis (1974), Lowery (1974), Jones and Suttkus (1975), Jones (1977), Piaggio and Perkins (2005).

Corynorhinus townsendii (Cooper, 1837)

Townsend's big-eared bat, murciélago orejudo pálido

1. *C. t. australis* Handley, 1955. From Chihuahua and Coahuila south through Oaxaca.
2. *C. t. ingens* Handley, 1955. From Iowa to Missouri.
3. *C. t. pallescens* Miller, 1897. From British Columbia to northern México.
4. *C. t. townsendii* (Cooper, 1837). From British Columbia to California, by the coastal plain.
5. C. *t. virginianus* Handley, 1955. From Kentucky to Virginia.

Conservation: *Corynorhinus townsendii ingens* is listed as Endangered by the United States Endangered Species Act (FWS 2022) and *C. townsendii* as Least Concern by the IUCN (Arroyo-Cabrales and Álvarez-Castañeda 2017c).

Characteristics: Townsend's big-eared bat is medium sized; forearm length 39.0–47.6 mm, skull length 15.2–17.2 mm. **Pelage brownish, although grayish at the base; fur of underparts paler brown at the tip and similar to the dorsal pelage at the base, with low contrast between the tip and the base; tragus brown, very large, much greater than 20.0 mm, greater than one-third of the ear length; uropatagium usually with eight stripes of hair**; membranes pale brown; **supraoccipital region and braincase roof almost at the same level**; median postpalatal process styliform with a narrow base.

Comments: *Corynorhinus* was previously considered a subgenus of *Plecotus but* is now considered a genus (Tumlison and Douglas 1992), known as *Plecotus townsendii. C. townsendii* ranges in British Columbia, Canada, and South Dakota, United States, south through Puebla and Oaxaca, México, and east through North Carolina and Pennsylvania, United States, including the Baja California Peninsula; with two isolated populations in the Ozark Plateau, Arkansas, and the Appalachian Mountains (Map 149). It can be distinguished from other species in sympatry, or nearly so, by its enlarged nasal opening with a glandular bulge clearly visible between the nasal opening and the eyes, in addition to having very large ears and tragus, greater than the head length. Although all *Corynorhinus* species are allopatric, *C. townsendii* can be distinguished from *C. mexicanus* by having eight transverse hair stripes in the uropatagium that are clearly distinguishable; all the species with the first upper incisors not bifid, except for *C. t. inges*. From *C. rafinesquii*, by having the dorsal pelage brown with the base grayish; fur of underparts light brown at the tip, usually with less than eight stripes of hair in the uropatagium.

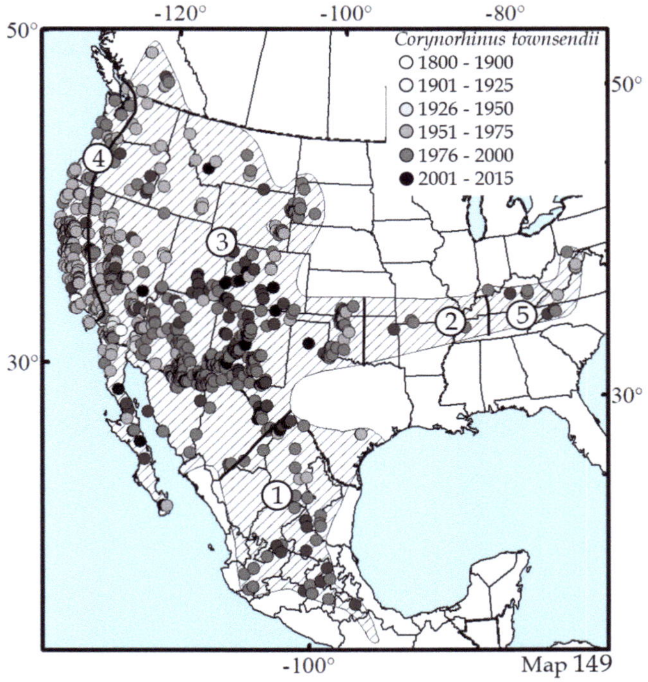

Map 149 Distribution map of *Corynorhinus townsendii*

Additional Literature: Handley (1959), Villa (1966), Barbour and Davis (1969), Hoffmeister (1970), Humphrey and Kunz (1976), Jones (1977), Kunz and Martin Tumlison (1991), Bogan (1999), Piaggio and Perkins (2005), Álvarez-Castañeda et al. (2008c).

Genus *Euderma* Allen, 1891

Medium sized; ears very large, as long as three-fourths of the forearm; forehead with a low band membrane; tragus well developed; nostrils simple; **general coloration nearly black, with conspicuous white marks on the back;** upper incisors alike in form, slender, with well-developed cingula, outer incisors somewhat larger than the inner incisors; lower incisors trifid, middle lobe larger; first upper premolar minute; **skull large, rounded, though braincase rather low and rostrum excessively weak**; zygomatic expanded abruptly at the middle; auditory bullae very large, elongated, **their greatest diameter about equal to the toothrow length except incisors**, inner edge strongly emarginated anteriorly.

Euderma maculatum (Allen, 1891)

Spotted bat, murciélago orejudo manchado

Monotypic.

Conservation: *Euderma maculatum* is listed as subject to Special Protection in the Norma Oficial Mexicana (DOF 2019) and as Least Concern by the IUCN (Arroyo-Cabrales and Álvarez-Castañeda 2017c).

Characteristics: Spotted bat is medium sized; forearm length 48.0–51.0 mm, skull length 18.4–19.0 mm. **Dorsal pelage blackish with white spots on the back; fur of underparts white with a white collar; ears intensely pink, very big, greater than 35.0 mm, almost of the same size as the forearms**; patagium brown; extending particularly in the braincase with the middle higher; zygomatic arches stronger and with an upper process in the middle part; auditory bullae very large, its greater axis almost equal in length as the maxillary toothrow.

Comments: *Euderma maculatum* ranges from southern Montana and California south through Durango, with presence in the Rocky Mountains (Map 150). It can be distinguished from other species in sympatry, or nearly so, such as those of the genera *Eptesicus, Idionycteris, Lasiurus,* and *Corynorhinus*, by its black coloration with three white spots on the back and large ears, almost the size of the forearms, intensely pink.

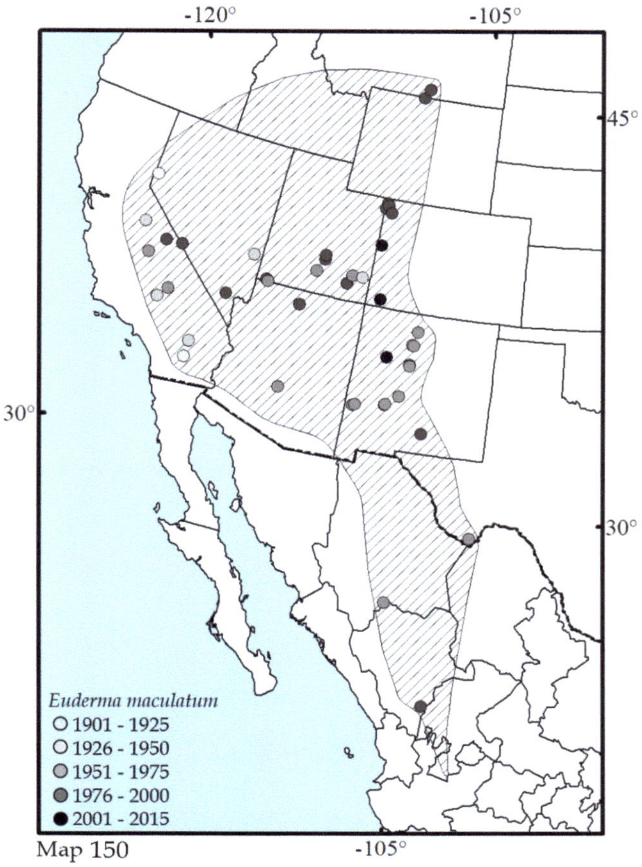

Map 150 Distribution map of *Euderma maculatum*

Additional Literature: Handley (1959), Easterla (1965, 1970, 1971), Gardner (1965), Watkins (1997), López–González (2003).

Genus *Idionycteris* Anthony, 1923

Medium sized; **ears and tragus very large, much longer than the head, with an accessory basal lobe of the auricle developed, projecting from the ear lappet**; **ears greater than 75% of the forearm length**; tail tip beyond the uropatagium border; **uropatagium with hair stripes varying in number according to the species**; **calcar having a well-developed keel; sagittal crest small in the frontal bone and very low or absent in the supraoccipital bone**; **braincase breadth greater than one-half of the skull length**; **lacrimal region rounded**; auditory bullae large, rounded, their greatest diameter equal to about three times the distance between them.

Idionycteris phyllotis (Allen, 1916)

Allen's big-eared bat, murciélago orejudo negro

1. *I. p. phyllotis* (Allen, 1916). Only subspecies in North America.

Conservation: *Idionycteris phyllotis* is listed as Least Concern by the IUCN (Arroyo-Cabrales and Álvarez-Castañeda 2017d).

Characteristics: Allen's big-eared bat is medium sized; forearm length 41.8–49.0 mm, skull length 16.6–17.4 mm. Dorsal pelage brownish yellow; fur of underparts cream; **ears enormous, greater than 35.0 mm, with skin flaps on the ear base projecting forward; nostrils not extending backward; rhinarium gland between the nostrils and the eyes barely noticeable,** middle part of braincase higher, rostrum proportionally short; lacrimal region rounded.

Comments: *Idionycteris* was previously considered a subgenus of *Plecotus* (Handley 1959), known as *Plecotus phyllotis*, and *Idionycteris phyllotis* within the genus *Euderma* (Frost and Timm 1992). It is considered a full genus following Bogdanowicz et al. (1998). *I. phyllotis* ranges in the desert areas from southern Utah and California south through Jalisco and Ciudad de México, México (Map 151). It can be distinguished from other species in sympatry, or nearly so, such as those of the genera *Corynorhinus, Eptesicus, Euderma, Lasiurus,* and *Rhogeessa,* by its very large ears, greater than 35.0 mm, with flaps of skin on the ear base that are projected forward; nostrils not elongated backward and rhinarium gland between the nostrils and the eyes barely noticeable.

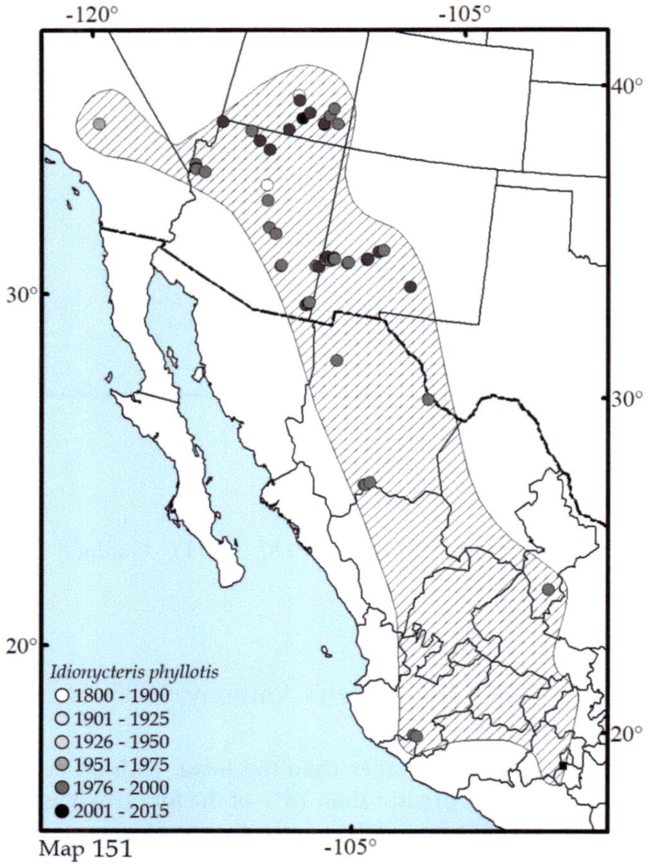

Map 151 Distribution map of *Idionycteris phyllotis*

Additional Literature: Anthony (1923), Genoways and Jones (1967), Anderson (1972), Williams and Findley (1979), Czaplewski (1983), Villalpando and Álvarez (2000).

Order Cingulata

Order Cingulata Illiger, 1811

The order Cingulata includes all armadillos, characterized by **the body covered by keratin shields overlaid in bony plates; cheek-teeth homodont; skull elongated and flattened; rostrum elongated;** extra zygapophysis in the posterior part of the thoracic and lumbar vertebrae. Distribution restricted to the American Continent. Two families occur in North America: Dasypodidae and Chlamyphoridae, classification following Gibb et al. (2016). The keys were elaborated based on the review of specimens and the following sources: Lawlor (1979), Hall (1981), and Álvarez-Castañeda et al. (2017a).

1. Body covered by a shell formed with nine ossified dermal scutes; tail also with dermal plates; mastoid breadth less than 40% of the skull length (from Colorado to South Carolina south through Central America) .... family Dasypodidae; subgenus *Dasypus*; *Dasypus novemcinctus* (p. 383)

1a. Body covered by a hard shell formed by more than nine ossified dermal scutes; tail without dermal plates; mastoid breadth greater than 40% of the skull length (from Chiapas south through Central America) .... family Chlamyphoridae; subfamily Tolypeutinae, *Cabassous centralis* (p. 395)

Family Dasypodidae Gray, 1821

The family Dasypodidae includes armadillos, characterized by teeth small, subcylindrical, allegedly lacking enamel; first two teeth laterally compressed and last half or less (in occlusal area) as large as the others; toothrow short, terminating anterior to the root of the zygoma; snout long; **pterygoids meeting below the posterior nares;** ears close together on the occiput; **7–11 movable bands in the carapace between the anterior and posterior bucklers;** pattern of sculpture on the scutes V shaped; tail bands arranged in rings, third digit not unusually enlarged; young polyembryonic.

Subfamily Dasypodinae Gray, 1821

Genus ***Dasypus*** Linnaeus, 1758

Dorsal pelage yellowish brown to dark brown; **with 7–11 movable ossified dermal scutes in the shell between the anterior and posterior shields**; hairs arising from pits occurring at the intersection of the furrow surrounding large plaques with radial furrows that separate secondary (peripheral) plaques, which in turn surround each large plaque; snout long; **ears close together on the head back; tail scutes arranged in rings; third digit not unusually enlarged; teeth small, peg-like, single rooted, subcylindrical, canines and incisors absent; without enamel, first two teeth laterally compressed and last half or less (in occlusal area) as large as the others;** toothrow short, terminating anterior to the root of the zygoma; pterygoids meeting below the posterior nares.

S. T. Álvarez-Castañeda, *Mammals of North America - Volume 1*, https://doi.org/10.1007/978-3-031-41661-3_6

Subgenus *Dasypus* Linnaeus, 1758

Medium sized with **7–11 movable ossified dermal scutes in the shell between the anterior and posterior shields; females produce four genetically identical offspring.**

Dasypus novemcinctus Linnaeus, 1758
Nine-banded armadillo, armadillo de nueve bandas

D. n. mexicanus Peters, 1864. Only subspecies in North America.

Conservation: *Dasypus novemcinctus* is listed as Least Concern by the IUCN (Loughry et al. 2014).

Characteristics: Nine-banded armadillo is medium sized; total length 615.0–800.0 mm; skull length 85.5–100.0 mm. Dorsal pelage ocher to dark brown; **body covered by nine dermal scutes; tail long with dermal plates, approximately two-thirds of the body size;** two large central nails; skull length between the anterior palatine border and the first unicuspid tooth alveolus, which is of the same length as the toothrow; no unicuspid teeth behind the posterior palatine border.

Comments: *Dasypus novemcinctus davisi* (Russell 1953) is considered a junior synonym of *D. n. mexicanus* (Gardner 2005b). *D. novemcinctus* ranges from central United States, Colorado to South Carolina south through Central America, including the Florida Peninsula (Map 1). It can be distinguished from other species in sympatry, or nearly so, such as by its body covered with nine ossified dermal scutes; long tail with dermal scutes, approximately two-thirds of the body size.

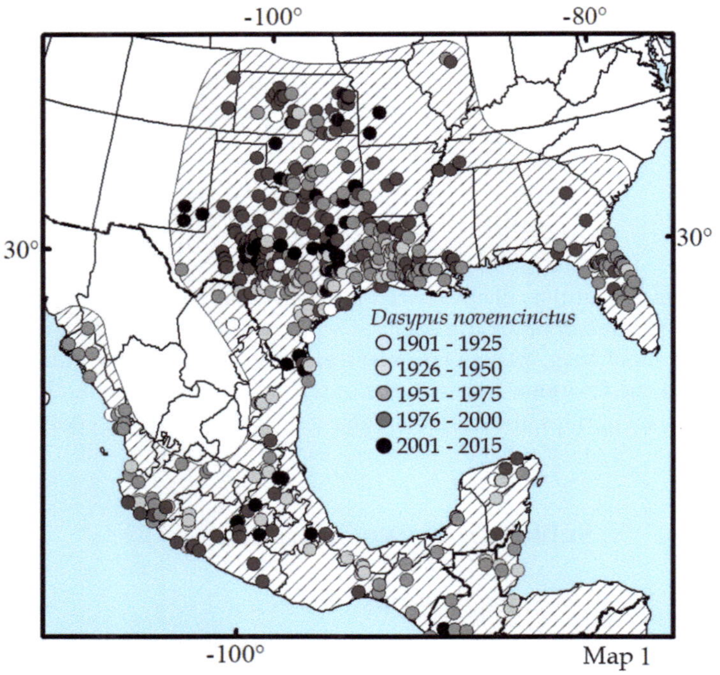

Map 1 Distribution map of *Dasypus novemcinctus*

Additional Literature: Kalmbach (1943), Hall (1955), Humphrey (1974), Wetzel and Mondolfi (1979), Hall (1981), McBee and Baker (1982), Wetzel (1985a, b), Gibb et al. (2016), Feijó et al. (2018, 2019).

Family Chlamyphoridae Pocock, 1924

The family Chlamyphoridae includes naked-tailed armadillos characterized by having **between 11 and 13 ossified dermal scutes on the back; three to four bands on the neck; scales on the tail not segmentally arranged;** forefoot with five digits, each bearing a claw; third digit and its claw much enlarged.

Subfamily Tolypeutinae Gray, 1865

Species within the subfamily are dark brown; heavily armored with a thick, leathery shell usually segmented into three bands; armor covering the tail, head, feet, and back; **tail very stout and immobile; third middle toes of the hind feet bearing a thick claw.**

Genus ***Cabassous*** McMurtrie, 1831

Dorsal pelage from yellowish brown to dark brown; **with 11–21 movable ossified dermal scutes in the carapace; scales on the tail not segmentally arranged or absent**; forefoot with five digits, each bearing a claw; claws enlarged, flattened, and ventrally concave; third digit and its claw markedly enlarged; incisors and canines absent; rostrum short.

Cabassous centralis (Miller, 1899)

Northern naked-tailed armadillo, armadillo de cola desnuda

Monotypic.

Conservation: *Cabassous centralis* is listed as Endangered in the Norma Oficial Mexicana (DOF 2019) and as Data Deficient by the IUCN (Tirira et al. 2014).

Characteristics: Northern naked-tailed armadillo is medium sized; total length 500.0–573.0 mm; skull length 78.0–80.0 mm. Dorsal pelage dark grayish brown with the shell border yellowish; **body covered by a shell formed by 10–13 dermal ossified dermal scutes but none on the tail, which is short, approximately one-half of the body size**; central nail very large, one-half of the paw size; anterior palatal border and first unicuspid tooth alveolus length one-half of the toothrow length; three unicuspid teeth behind the posterior palatine border.

Comments: *Cabassous centralis* has been recorded from Chiapas only and ranges south through Central America (Map 2). It can be distinguished from other species in sympatry, or nearly so, by its body covered by a shell formed by 10–13 dermal ossified dermal scutes but none on the tail, and tail short, approximately one-half of the body size.

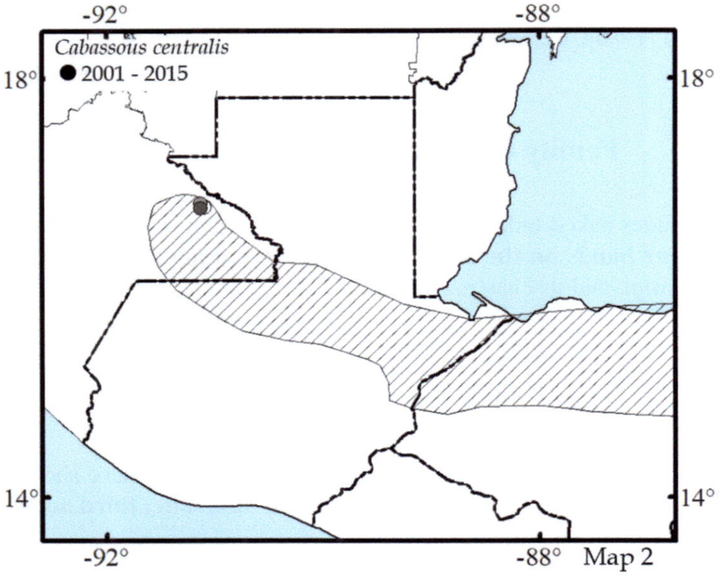

Map 2 Distribution map of *Cabassous centralis*

Additional Literature: Allen (1904b), Wetzel (1980), Hall (1981), Cuarón et al. (1989), Gardner (2005b), Hayssen et al. (2013), Gibb et al. (2016), Juárez-López et al. (2017).

Order Eulipotyphla

Order Eulipotyphla Waddell et al., 1999

Eulipotyphla includes shrews and moles characterized by their small size; feet plantigrade and pentadactyl; pollex and hallux not opposable; digits with claws; **snout usually long and pointed;** skull shape primitive; braincase low; **orbital and temporal fossae confluent; tympanic bone usually annular, either entirely free or enclosed in an entotympanic bullae, or incorporated into the auditory bullae; basisphenoid and petrosal both taking part in the enclosure of the tympanic cavity; vertical portion of the palatine bone excluded from the orbit by the maxilla; angular process of the mandible typically not inflected;** incisors may be reduced or enlarged; canines can be caniniform, incisiform, or premolariform; premolars acuminate, the posteriormost frequently sectorial; molars lophodont or bunodont; lower molars usually with five sharp tubercles and upper molars tritubercular or quadritubercular; distal ends of the tibia and fibula usually fused; glenoid fossa for the humerus widely separated from the sternum. Distribution nearly cosmopolitan. Two families occur in North America: Soricidae and Talpidae. The keys were elaborated based on the review of specimens and the following sources: Jackson (1938), Lawlor (1979), Hall (1981), and Álvarez-Castañeda et al. (2017a).

1. Zygomatic arches absent..........family Soricidae (p. 397)

1a. Zygomatic arches present..........family Talpidae (p. 498)

Family Soricidae Fischer, 1814

The family Soricidae includes all shrews, characterized by having a **condyloid process with two condyles in the lower jaw**; small sized (60.0–330.0 mm); **upper incisors large, hooked, with a cusp at the proximal teeth base.** Cosmopolitan distribution. Only one subfamily ranges in North America, Soricinae, with three tribes: Blarinini, Notiosoricini, and Soricini. The keys were elaborated based on the review of specimens and the following sources: Jackson (1928), Hall (1981), and Álvarez-Castañeda et al. (2017a).

1. Tail long, equal to or greater than 40% of the head-and-body length; first upper incisors with a small internal tine (between the two; Fig. 1); five unicuspid teeth in the upper toothrow (Fig. 2); all teeth pigmented.......... ....tribe Soricini; *Sorex* (p. 402)

S. T. Álvarez-Castañeda, *Mammals of North America - Volume 1*, https://doi.org/10.1007/978-3-031-41661-3_7

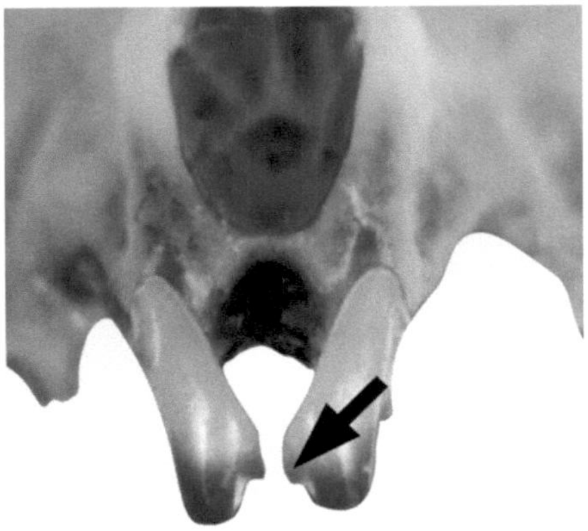

Fig. 1 First upper incisors with a small internal tine

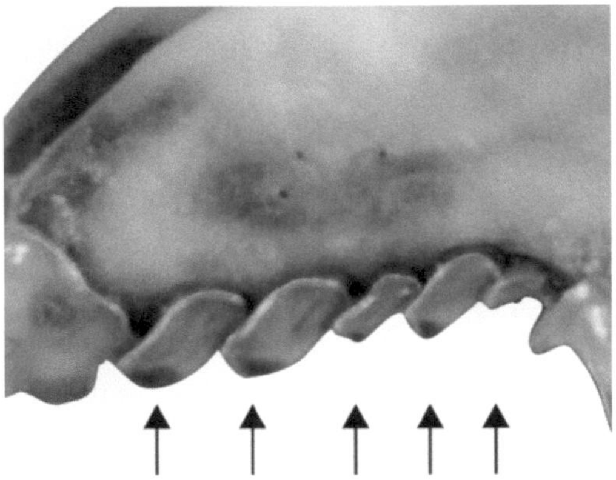

Fig. 2 Five unicuspid teeth in the upper toothrow

1a. Tail short, generally less than 40% of the head-and-body length, if the tail greater than 40% of the head-and-body length, ears within the pelage; first upper incisors without internal ridges (Fig. 3); three or four unicuspid teeth in the upper toothrow; not all teeth pigmented 2

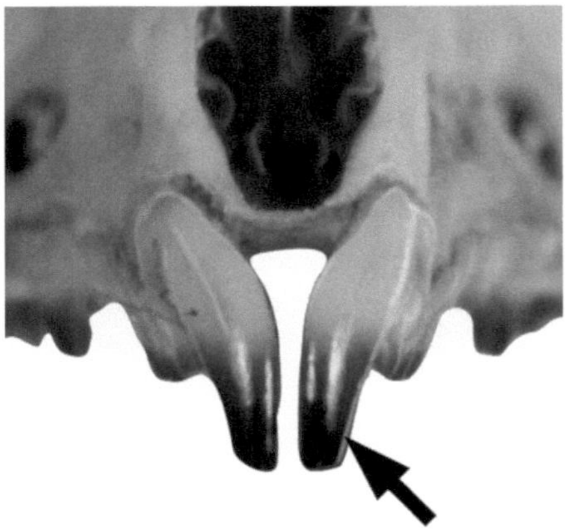

Fig. 3 First upper incisors without internal ridges

2. Ears inconspicuous, concealed by the pelage: upper toothrow with three or four unicuspid teeth; first three upper unicuspid teeth with secondary cusps; all teeth pigmented .... tribe Blarinini 3

2a. Ears conspicuous, not concealed by the pelage; upper toothrow with three unicuspid teeth; first three upper unicuspid teeth without secondary cusps; not all teeth pigmented.. tribe Notiosoricini 4

3. Tail short, less than 20% of the total length; upper toothrow with five unicuspid teeth without secondary cusps; total teeth 32..*Blarina* (p. 400)

3a. Tail medium sized, greater than 20% of the total length; upper toothrow with four unicuspid teeth with secondary cusps; total teeth 30... *Cryptotis* (p. 404)

4. Without scent glands on the flanks; total length greater than 105.0 mm; head-and-body length greater than 81.0 mm; tail length greater than 37.0 mm and slightly bicolored; upper toothrow with four unicuspid teeth; all teeth unpigmented (Fig. 4; endemic to the Pacific Coastal Plains southern Sinaloa south through Oaxaca, México) ....
....*Megasorex gigas* (p. 426)

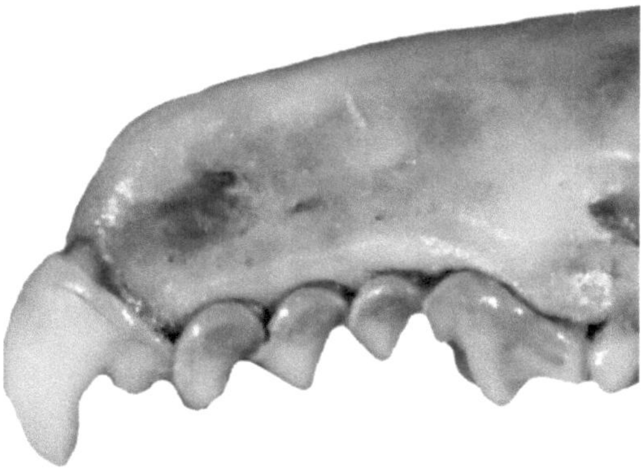

Fig. 4 All teeth unpigmented

4a. With scent glands on the flanks; total length less than 105.0 mm; head-and-body length less than 74.0 mm; tail length less than 37.0 mm and unicolored, same as the dorsal pelage; upper toothrow with three unicuspid teeth; teeth and molars with a distinct pale pigment (Fig. 5) ....*Notiosorex* (p. 427)

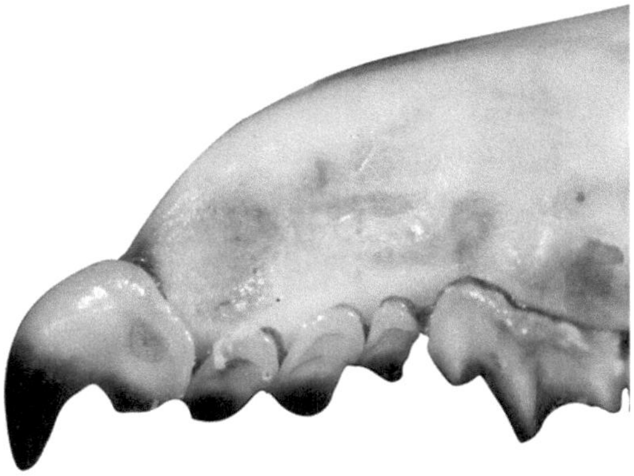

Fig. 5 Teeth and molars with a distinct pale pigment

Genus *Blarina* (Gray, 1838)

Dorsal pelage silver to black, some species with hair brown at the tip; limbs short and body robust; ears inconspicuous and concealed by the pelage; eyes small; rostrum pointed and somewhat proboscis-like; **tail short, less than 20% of the total length,** faintly to indistinctly bicolored, haired, and slightly flattened dorsoventrally; **total teeth 32; five upper teeth unicuspid, first and second teeth largest and subequal, third and fourth teeth much smaller and subequal, fifth teeth very small; lobe in the base of the middle incisors elongated anteroposteriorly; braincase not arched anteroposteriorly, the highest point at the lambdoid suture**. The keys were elaborated based on the review of specimens and the following sources: Merriam (1895b), Hall (1981), Benedict et al. (2006), and Álvarez-Castañeda et al. (2017a).

1. Total length greater than 105.0 mm; condylobasal length greater than 20.0 mm; first lower incisors less procumbent and set at an angle equal to or less than 17° from the horizontal ramus of the dentary (from Ontario, Canada south through Kansas and east through Massachusetts to North Carolina, United States) .... *Blarina brevicauda* (p. 400)

1a. Total length less than 105.0 mm; condylobasal length less than 20.0 mm; first lower incisors more procumbent and set at an angle equal to or greater than 17° from the horizontal ramus of the dentary 2

2. Head-and-body length greater than 81.0 mm; condylobasal length greater than 20.0 mm; skull wide, greater than 11.5 mm (from Nebraska east through Missouri and south through Oklahoma, Arkansas, and Louisiana, United States) .... *Blarina hylophaga* (p. 402)

2a. Head-and-body length less than 81.0 mm; condylobasal length less than 20.0 mm; skull wide, less than 11.5 mm 3

3. Dorsal pelage dull plumbeous washed with brownish shades; hind feet less than 13.0 mm (from western Texas east through Virginia and south through Florida, not including the southern Florida Peninsula, United States) .... *Blarina carolinensis* (p. 401)

3a. Dorsal pelage slate black; hind feet greater than 13.0 mm (restricted to southwestern coast of the Florida Peninsula, in the vicinity of Royal Palm, United States) .... *Blarina shermani* (p. 403)

Tribe Blarinini

Blarina brevicauda (Say, 1823)

Northern short-tailed shrew, musaraña de cola corta del norte

1. *B. b. aloga* Bangs, 1902. Restricted to Martha's Vineyard Island, Massachusetts.
2. *B. b. brevicauda* (Say, 1823). From North Dakota and Nebraska east through Wisconsin and Illinois.
3. *B. b. cumberlandensis* Webster et al., 2011. Restricted to the Cumberland Plateau central Tennessee and western-central Kentucky.
4. *B. b. delmarvensis* Webster et al., 2011. Southern Delaware and eastern shore of Maryland and Virginia.
5. *B. b. jerryrchoatei* Webster et al., 2011. Kansas River valleys, Kansas.
6. *B. b. knoxjonesi* Webster, 1996. South of Pamlico River, eastern North Carolina.
7. *B. b. talpoides* (Gapper, 1830). From eastern Ontario to western Québec.

Conservation: *Blarina brevicauda* is listed as Least Concern by the IUCN (Cassola 2016b).

Characteristics: Northern short-tailed shrew is medium sized for shrews; **total length 106.0–141.0 mm; condylobasal 23.1–25.2 mm**. Dorsal pelage darker slate, with considerable variation in size and coloration; fur of the underparts ashy-plumbeous, paler in summer; **pelage glossy**; **skull largest within *Blarina***, massive, and angular; **braincase markedly angular for *Blarina*, with the highest point at the lambdoid suture; jaw ramus angular, bent rather abruptly upward opposite to the last molar**.

Comments: *Blarina carolinensis*, *B. hylophaga*, and *B. shermani* were recognized as distinct species (George et al. 1981; Jones et al. 1992; Hutterer 2005; Benedict et al. 2006), previously regarded as subspecies of *B. brevicauda* (Merriam 1895b; Hamilton 1955; Hall 1981). *B. b. angusta*, *B. b. churchi*, *B. b. hooperi*, *B. b. kirtlandi*, *B. b. manitobensis*, *B. b. pallida*, and *B. b. telmalestes* are considered junior synonyms of *B. berlandieri*; and *B. b. compacta* of *B. b. aloga* (Woodman 2018). *B. brevicauda* ranges from Saskatchewan east through Newfoundland, south through Missouri and northern Georgia, and east through the Atlantic coast (Map 1). All the species of *Blarina* have allopatric distribution; however, *B. brevicauda* can be distinguished from the other species of *Blarina* by its karyotype and greater size, total length greater than 105.0 mm, and condylobasal length greater than 20.0 mm, including *B. carolinensis* and *B. hylophaga*. From *B. carolinensis*, by having unicuspid teeth linear and the fifth tooth visible when viewed from the labial side.

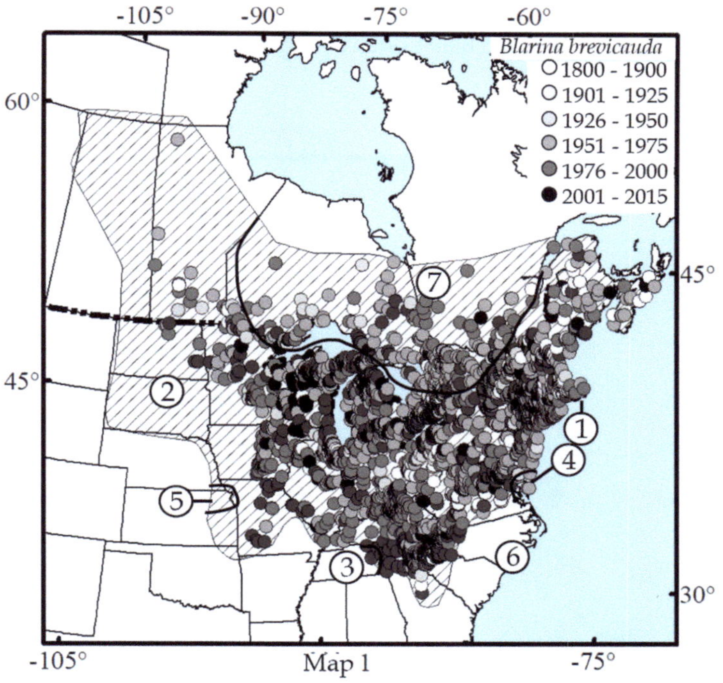

Map 1 Distribution map of *Blarina brevicauda*

Additional Literature: Merriam (1895b), Hamilton (1955), Easterla (1968d), George et al. (1986).

Blarina carolinensis (Bachman, 1837)

Southern short-tailed shrew, musaraña de cola corta del sur

1. *B. c. carolinensis* (Bachman, 1837). From southern Missouri east through Virginia and central Florida Peninsula.
2. *B. c. minima* Lowery, 1943. Restricted to southeastern Texas and Louisiana north through Mississippi Basin to southern Illinois.
3. *B. c. peninsulae* Merriam, 1895. Restricted to southern half of Florida Peninsula.
4. *B. c. plumbea* Davis, 1941. Only known from Aransas National Wildlife Refuge, Texas.

Conservation: *Blarina carolinensis* is listed as Least Concern by the IUCN (Cassola 2016c).

Characteristics: Southern short-tailed shrew is medium sized for shrews; total length 72.0–105.0 mm, condylobasal 18.3–19.6 mm. Dorsal pelage dull plumbeous washed with brownish shades, mainly in summer; fur of the underparts paler than in the dorsum; rostrum relatively pointed and furred; vibrissae long and white; ears and eyes conspicuous; tail faintly bicolored, haired, and slightly flattened dorsoventrally; **hind feet proportionally smaller**; **first lower incisors more procumbent and set at an angle equal to or greater than 17° from the horizontal ramus of the dentary**.

Comments: *Blarina carolinensis* was recognized as a distinct species (Jones et al. 1992; Hutterer 1993, 2005; Benedict et al. 2006), previously regarded as a subspecies of *B. brevicauda* (Merriam 1895b). *B. carolinensis* includes the previous subspecies of *B. brevicauda*: *B. b. carolinensis*, *B. b. minima*, *B. b. peninsularis*, and *B. b. shermani* (McCay 2001). *B. peninsulae* was elevated to full species (Hutterer 2005; Woodman 2018), distinct from *B. carolinensis*, based on the distinct karyotype and morphology (George et al. 1982; Genoways and Choate 1998). However, a morphological study of *Blarina* in the Florida Peninsula did not find differences at species level and proposed it as a subspecies of *B. carolinensis* (Benedict et al. 2006). The same study proposed that *B. b. shermani* should be considered a valid species (Benedict et al. 2006). *B. carolinensis* ranges from eastern Texas and Arkansas east through Virginia and North Carolina and south through the Florida Peninsula and all the states in the Gulf of México coast, except for the Appalachian Mountains (Map 2). All the species of *Blarina* have allopatric distribution; however, *B. carolinensis* can be distinguished from the other species of *Blarina* by its karyotype and smaller size; head-and-body length less than 81.0 mm; condylobasal length less than 20.0 mm; skull wide, less than 11.5 mm and by its karyotype, and from *B. brevicauda*, by having the unicuspid teeth more nearly linear, with the fifth tooth generally hidden when viewed from the labial side.

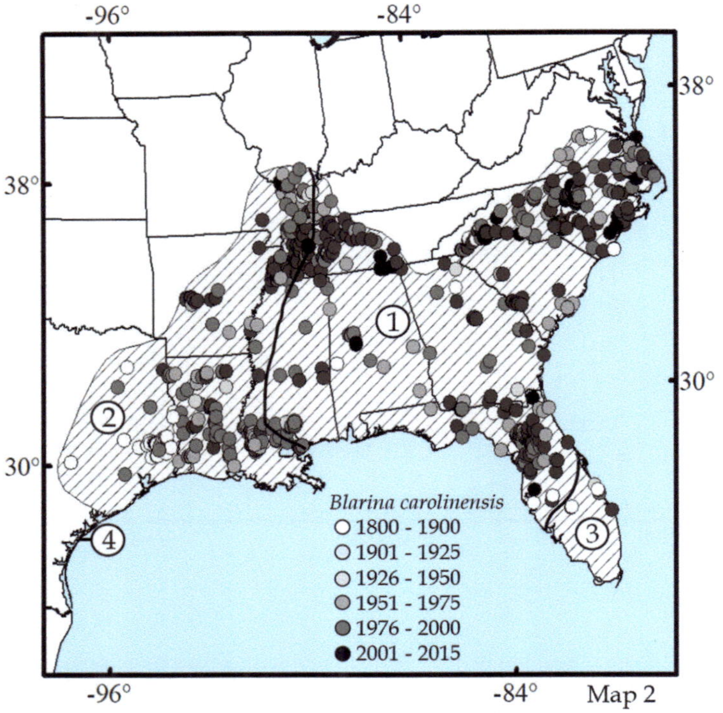

Map 2 Distribution map of *Blarina carolinensis*

Additional Literature: Genoways and Choate (1972, 1998), Tate et al. (1980), George et al. (1982), McCay (2001).

Blarina hylophaga Elliot, 1899

Elliot's short-tailed shrew, musaraña de cola corta de las planicies

Monotypic.

Conservation: *Blarina hylophaga* is listed as Least Concern by the IUCN (Cassola 2016d).

Characteristics: Elliot's short-tailed shrew is medium sized for shrews; total length 92.0–120.0 mm; condylobasal length 18.4–24.0 mm. Dorsal pelage pale plumbeous to blackish plumbeous; fur of the underparts ashy-plumbeous; skull largest, massive, and angular; braincase with the highest point near the lambdoid suture; jaw ramus less angular; **upper premolars and molars larger in relation to the other species**.

Comments: *Blarina hylophaga* was recognized as a distinct species (George et al. 1981; Jones et al. 1992; Hutterer 2005), previously regarded as a subspecies of *B. brevicauda* (Hall 1981). It ranges from southern Nebraska east through Missouri and Illinois, and south through Oklahoma, Arkansas, and Louisiana (Map 3). All the species of *Blarina* have allopatric distribution; however, *B. hylophaga* can be distinguished from *B. brevicauda* by its smaller size, total length less than 105.0 mm; condylobasal length less than 20.0 mm; first lower incisors with an angle equal to or greater than 17° from the horizontal ramus of the dentary. From *B. carolinensis*, by its larger size, the head-and-body length greater than 81.0 mm; condylobasal length greater than 20.0 mm; and skull width greater than 11.5 mm.

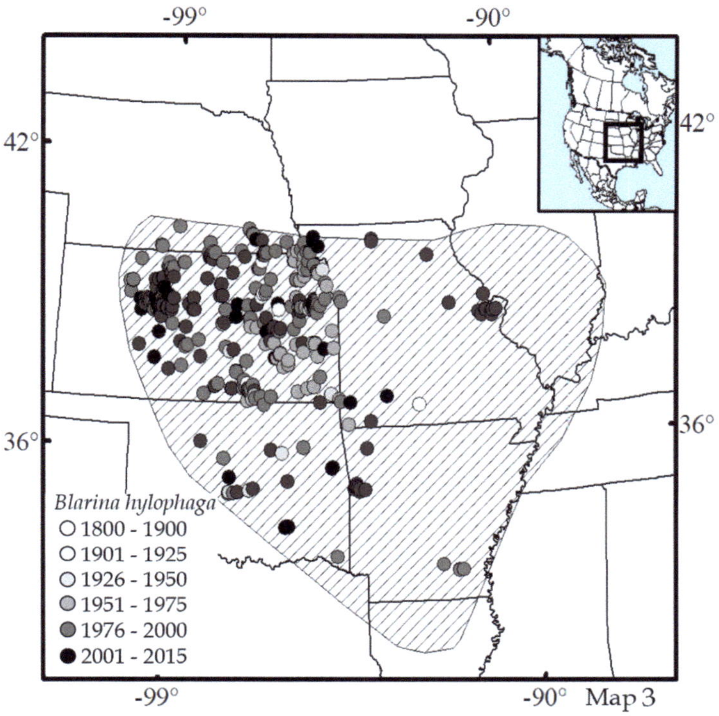

Map 3 Distribution map of *Blarina hylophaga*

Additional Literature: George et al. (1986).

Blarina shermani Hamilton, 1955

Sherman's short-tailed shrew, musaraña de cola corta de Florida

Monotypic.

Conservation: *Blarina shermani* (as *B. carolinensis*) is listed as Least Concern by the IUCN (Cassola 2016c). However, the subspecies *B. b. shermani* may be extinct, because it is only known from the type locality (Layne 1992).

Characteristics: Sherman's short-tailed shrew is the largest in the Florida Peninsula; total length 94.2–98.1 mm; condylobasal length 19.2–21.3 mm. Dorsal pelage slate black without traces of brown; fur of the underparts ashy-plumbeous; **hind feet proportionally larger**; skull similar to *B. carolinensis* but somewhat larger and heavier; pterygoids broader and more massive than in the other species; upper molariform teeth clearly larger, heavier, and less emarginate posteriorly.

Comments: *Blarina shermani* was recognized as a distinct species (Benedict et al. 2006), previously regarded as a subspecies of *B. carolinensis*. However, genetic analyses need to be done to confirm the specific status. *B. shermani* is restricted to southwestern coast of the Florida Peninsula, in the vicinity of Royal Palm (Map 4). All the species of *Blarina* have allopatric distribution; however, *B. shermani* can be distinguished from *B. carolinensis* by its darker coloration; larger body, and skull size.

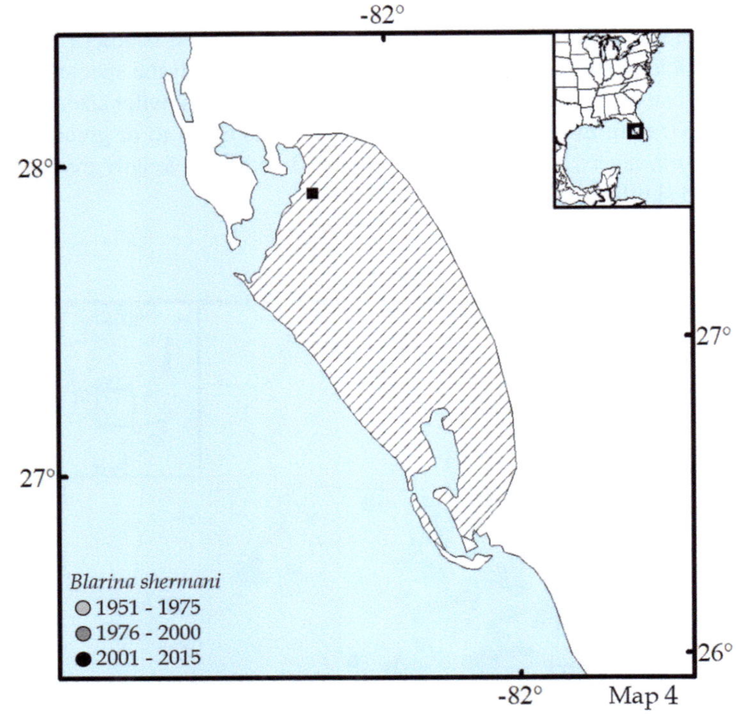

Map 4 Distribution map of *Blarina shermani*

Additional Literature: Hamilton (1955).

Genus *Cryptotis* Pomel, 1848

Dorsal pelage paler to darker brown; tail length less than 50% of the head-and-body length; **ears very small, covered by the pelage; tips of the teeth reddish (30 teeth); four upper unicuspid teeth visible; upper unicuspid teeth with a secondary peak.** The keys were elaborated based on the review of specimens and the following sources: Choate (1970), Hall (1981), and Álvarez-Castañeda et al. (2017a).

1. Total length greater than 123.0 mm; tail length greater than 40.0 mm; condylobasal length greater than 22.0 mm; palatine length greater than 9.4 mm; distance from the first upper unicuspid teeth to the last upper molars greater than 8.4 mm (endemic to northern Oaxaca, México) .... *Cryptotis magnus* (p. 414)

1a. Total length less than 123.0 mm; tail length less than 40.0 mm; condylobasal length usually less than 21.0 mm; palatine length less than 9.4 mm; distance from the first upper unicuspid teeth to the last upper molars length than 8.4 mm 2

2. Tail slightly bicolored; fourth upper unicuspid tooth completely visible in lateral view 3

2a. Tail not bicolored; fourth upper unicuspid tooth not completely visible in lateral view (Fig. 6), one can be hidden or absent 6

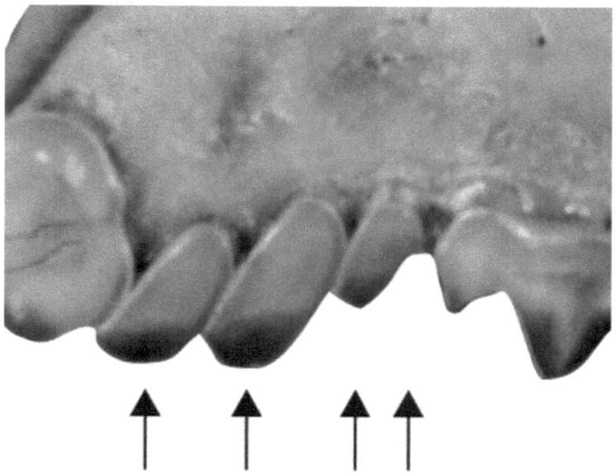

Fig. 6 Fourth upper unicuspid tooth not completely visible in lateral view

3. Claw length on the middle forefoot digit greater than 2.5 mm (greater than 2.5% of the total length); cranial breadth greater than 9.7 mm; zygomatic processes bulbous (Fig. 7; endemic to southern Oaxaca City, México) .... *Cryptotis peregrinus* (p. 421)

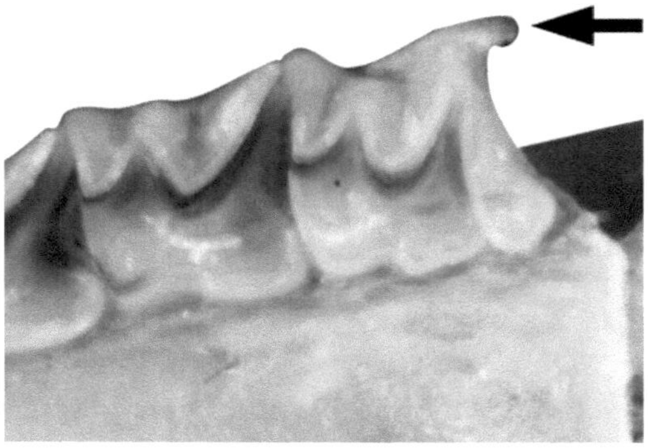

Fig. 7 Zygomatic processes bulbous

3a. Claw length on the middle forefoot digit less than 2.0 mm (less than 2.6% of the total length); cranial breadth less than 9.7 mm; zygomatic processes sharply pointed (Fig. 8) 4

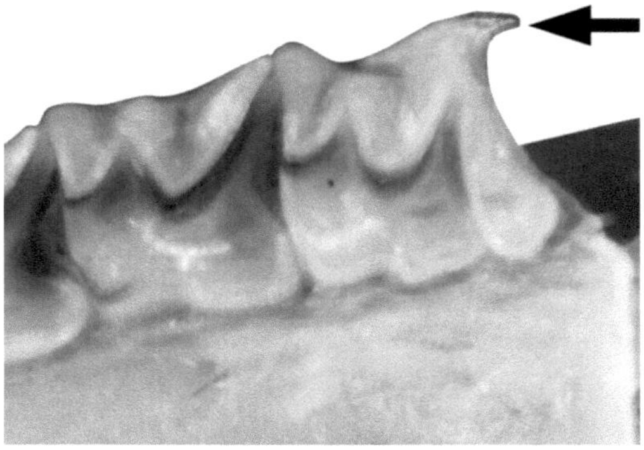

Fig. 8 Zygomatic processes sharply pointed

4. Dorsal pelage paler to medium brown; fur of the underparts white distally; claw length on the middle forefoot digit 1.0–1.8 mm (1.4–2.4% of the total length); palatine length less than 7.5 mm; distance from the first upper unicuspid teeth to the last upper molars less than 7.0 mm 16

4a. Dorsal pelage grayish; fur of the underparts slightly paler gray; claw length on the middle forefoot digit 1.5–1.6 mm and 1.2–1.6% of the total length; palatine length greater than 7.5 mm; distance from the first upper unicuspid teeth to the last upper molars equal to or greater than 7.0 mm 5

5. Dorsal pelage grayish; head-and-body length greater than 76.0 mm; skull length greater than 20.0 mm (endemic to Selva Lacandona, Chiapas, México and El Petén, Guatemala) .... *Cryptotis lacandonensis* (p. 413)

5a. Dorsal pelage paler grayish; head-and-body length less than 76.0 mm; skull length less than 20.0 mm (Campeche, Chiapas, Quintana Roo and Yucatán, México) .... *Cryptotis mayensis* (p. 415)

6. Tail length usually less than 24.0 mm; palatine length equal to or less than 7.2 mm; cranial breadth equal to or less than 8.8 mm; interorbital breadth equal to or less than 4.4 mm; first lower incisor with three denticles and very shallow interdenticular spaces (Fig. 9) 16

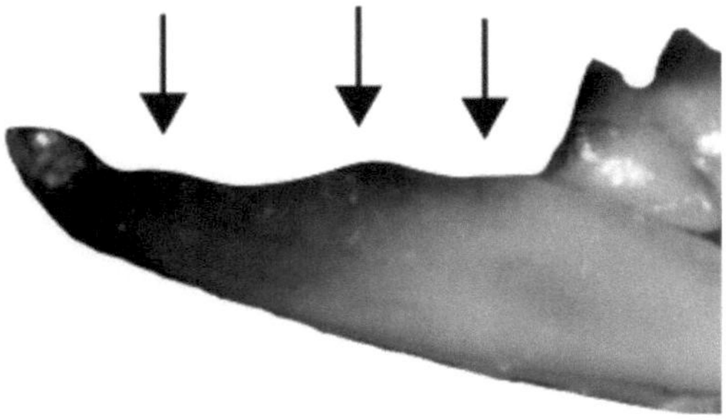

Fig. 9 First lower incisor with three denticles and very shallow interdenticular spaces

6a. Tail length usually greater than 26.0 mm; palatine length greater than 7.2 mm; cranial breadth greater than 8.8 mm; interorbital breadth greater than 4.4 mm; first lower incisors with two denticles and interdenticular spaces usually deep (Fig. 10) 7

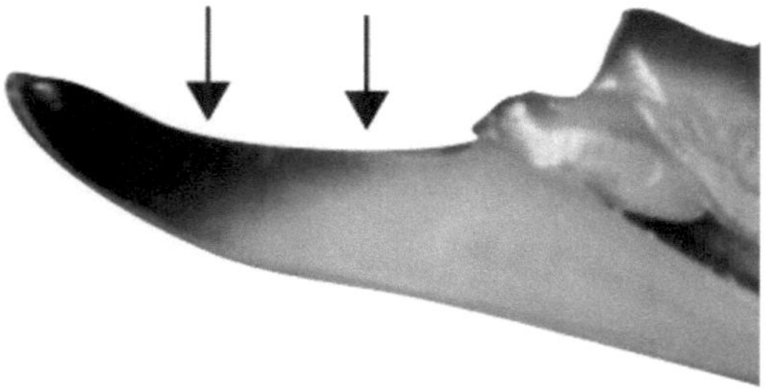

Fig. 10 First lower incisors with two denticles and interdenticular spaces usually deep

7. Condylobasal length greater than 21.0 mm; cranial breadth greater than 11.0 mm; distance from the first upper unicuspid teeth to the third upper molars greater than 8.0 mm (endemic to southeastern Chiapas, México, south through Central America) .... *Cryptotis goodwini* (*see addenda, p. 411)

7a. Condylobasal length less than 21.0 mm; cranial breadth usually less than 11.0 mm; distance from the first upper unicuspid teeth to the last upper molars less than 8.0 mm 8

8. Palatine length usually less than 8.1 mm and maxillary breadth less than 6.4 mm 9

8a. Palatine length usually greater than 8.1 mm and maxillary breadth greater than 6.4 mm 11

9. Dorsal pelage dark silvery gray with a white stripe in the proximal half and dark reddish brown at the tip; fur of the underparts dark silvery gray proximally, blond at the tip (restricted to the highlands of eastern Chiapas, México, Guatemala, and Belize) .... *Cryptotis tropicalis* (p. 425)

9a. Dorsal pelage dark brown, no pale-colored division along its length; fur of the underparts dark silvery gray at the base and dark brown or black distally 10

10. Fur of the underparts dark brown at the tip; zygomatic processes bulbous and flaring, usually projecting at the level of the posterior edge of the third upper molars (Fig. 7; restricted to the highlands of Sierra Madre Oriental, from Hidalgo south through Chiapas, México) .... *Cryptotis mexicanus* (p. 417)

10a. Fur of the underparts white at the tip; zygomatic processes projecting at the level of the second upper molars (from northern Sierra Madre Oriental in Tamaulipas south through Querétaro, Hidalgo, and Veracruz, México) .... *Cryptotis obscurus* (p. 419)

11. Pelage on the middle dorsum gray to dark gray; three or four unicuspid teeth in the upper toothrow 12

11a. Pelage on the middle dorsum brown to dark brown; four unicuspid teeth in the upper toothrow (zygomatic processes extending posteriorly and ventrolaterally below the occlusal surface of the teeth) 14

12. Dorsal pelage very dark gray with silvery shades distally, giving a frosted appearance; three or four upper unicuspid teeth (restricted to southern Sierra Madre del Sur in Oaxaca, México) .... *Cryptotis phillipsii* (p. 422)

12a. Dorsal pelage uniformly in blackish or brown shades but without a frosted appearance; four upper unicuspid teeth always present 13

13. Dorsal pelage blackish gray and venter pale gray; condylobasal length less than 20.0 mm; palatine length 7.5–8.2 mm; zygomatic processes extended, small, similar in length to the third upper molars; mandible length less than 8.5 mm; distance from the first unicuspid teeth to the last lower molars less than 5.6 mm (restricted to the central highlands of Chiapas, México, south through Central America) .... *Cryptotis merriami* (p. 416)

13a. Dorsal and ventral pelage uniformly dark brownish gray; condylobasal length equal to or greater than 20.0 mm; palatine length 8.5–9.5 mm; zygomatic processes large and elliptic, larger than the third upper molars; mandible length greater than 8.5 mm; distance from the first unicuspid teeth to the last lower molars greater than 5.6 mm (restricted to Chiapas, México) .... *Cryptotis griseoventris* (p. 412)

14. Dorsal pelage medium brown and underparts white at the tip; zygomatic processes pointed (endemic to the Eje Neovolcánico Transversal from Colima to Puebla, México) .... *Cryptotis alticola* (p. 408)

14a. Dorsal pelage dark brown and underparts with or without a pale coloration at the tip; zygomatic processes elliptic 15

15. Fur of the underparts only slightly paler brown than in the dorsum; claw length on the middle forefoot digit less than 2.8 mm and less than 2.8% of the total length (endemic to the San Martín Tuxtla Volcano in Veracruz, México) .... *Cryptotis nelsoni* (p. 418)

15a. Fur of the underparts much paler than in the dorsum; claw length on the middle forefoot digit usually greater than 2.8–3.8 mm and greater than 2.9% of the total length (endemic to the Sierra Madre del Sur, from Guerrero through Oaxaca, México) .... *Cryptotis goldmani* (p. 410)

16. Occurs in central Texas to the north (South Dakota south through Texas and east through the Atlantic coast, including the Florida Peninsula, United States) .... *Cryptotis parvus* (p. 420)

16a. Occurs in southern Texas to the south 17

17. Occurs in the Valle de México only .... *Cryptotis soricinus* (p. 424)

17a. Occurs in other parts of México and southeastern Texas 18

18. Occurs from southeastern Texas south through Coahuila, Nuevo León, and Tamaulipas southwest through Michoacán, México .... *Cryptotis berlandieri* (p. 409)

18a. Occurs in southeastern San Luis Potosí south through Oaxaca and central Chiapas, México .... *Cryptotis pueblensis* (p. 423)

Cryptotis alticola (Merriam, 1895)

Merriam's small-eared shrew, musaraña tropical de montaña

Monotypic.

Conservation: *Cryptotis alticola* is listed as subject to Special Protection in the Norma Oficial Mexicana (DOF 2019) and as Data Deficient by the IUCN (Woodman et al. 2019).

Characteristics: Merriam's small-eared shrew is small but large sized relative to the shrews of other genera; total length 95.0–117.0 mm; condylobasal length 19.4–20.7 mm. Dorsal pelage medium brown; fur of the underparts paler with whitish shades at the tip; tail pelage same as the body pelage; **hind feet larger than in other sympatric species**; four unicuspid teeth present in the upper toothrow, last unicuspid teeth partially or totally covered in lateral view; first lower incisors with one deep valley between two crests; zygomatic process extending posteriorly and ventrolaterally to below the teeth occlusal surface; **braincase narrow and less sharply angular laterally; teeth unicuspid with thick and blunt crowns**. Species group *mexicanus*.

Comments: *Cryptotis alticola* was reinstated as a distinct species (Woodman and Timm 1999), previously regarded as a subspecies of *C. goldmani* (Choate 1970). It is endemic to the Eje Neovolcánico, from Colima and Jalisco east through Puebla (Map 5). It can be found in sympatry, or nearly so, with *C. goldmani*, *C. parvus*, and *C. obscurus*. It can be distinguished from *C. parvus* and *C. obscurus* by its generally larger size; condylobasal length greater than 19.3 mm; claw length on the middle forefoot digit greater than 3.0 mm (2.9% of the total length). From *C. goldmani* and *C. obscurus*, by having the dorsal pelage medium brown and the zygomatic process pointed.

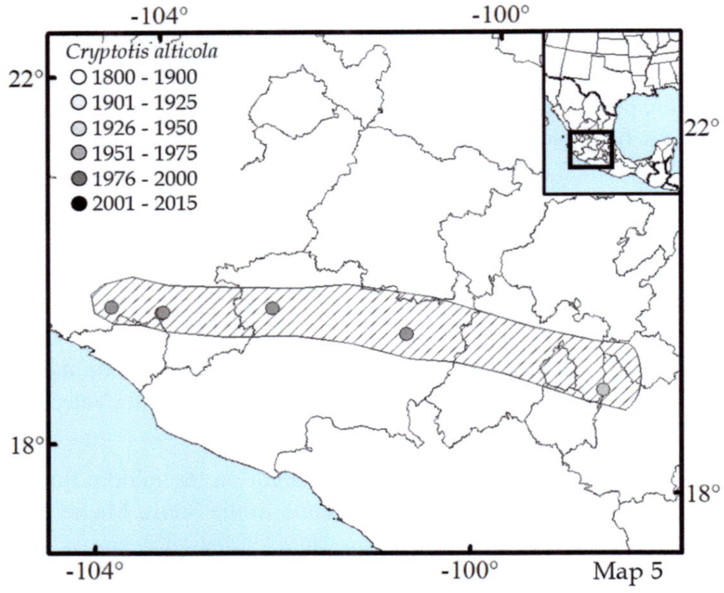

Map 5 Distribution map of *Cryptotis alticola*

Additional Literature: Álvarez and Sánchez-Casas (1997b), Hutterer (2005), Woodman (2005), Carraway (2007), Guevara et al. (2015).

Cryptotis berlandieri (Baird, 1858)

Northeastern México least shrew, musaraña tropical del noreste de México

Monotypic.

Conservation: *Cryptotis berlandieri* (as *Cryptotis parvus*) is not listed by the IUCN (Woodman et al. 2016c).

Characteristics: Northeastern México least shrew is small within the species group *parvus* but large sized relative to the shrews of other genera; total length 67.0–80.0 mm; condylobasal length 15.2–16.9 mm. Dorsal pelage dark to medium brown; fur of the underparts paler than in the dorsum; claw length on the middle digit of the manus (1.0–1.8 mm) among the shortest of the short-clawed *Cryptotis* (1.4–2.4% of the total length); four upper unicuspid teeth ranging from not visible, partially visible, to completely visible in lateral view; first lower incisors with three denticles and very shallow interdenticular spaces. Species group *parvus*.

Comments: *Cryptotis berlandieri* was reinstated as a distinct species (Woodman 2018), previously regarded as a subspecies of *C. parvus* (Davis 1941). It ranges from southeastern Texas south Coahuila, Nuevo León, and Tamaulipas southwest through Michoacán (Map 6). It can be distinguished from other species in sympatry, or nearly so, such as *C. alticola* by its smaller size, tail length less than 24.0 mm; palatine length less than 7.2 mm; cranial breadth less than 8.8 mm; and interorbital breadth less than 4.4 mm, and by having the first lower incisors with three denticles and with very shallow interdenticular spaces. From *C. pueblensis* and *C. soricina*, by its paler to medium brown dorsal pelage; underpart hairs medium silvery gray proximally, white at the tip; breadth across the second upper molars 4.1–4.9 mm, and condylobasal length usually 15.2–16.5 mm.

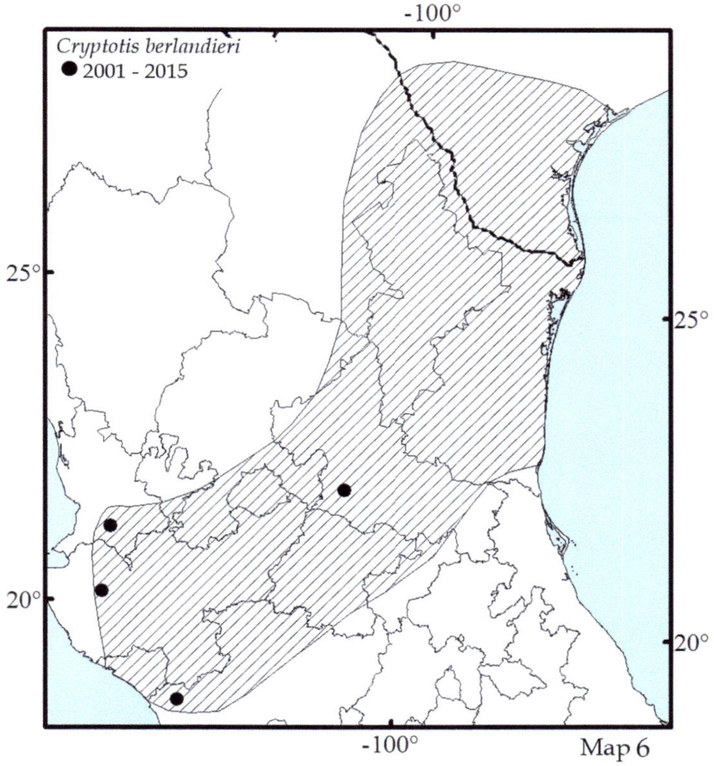

Map 6 Distribution map of *Cryptotis berlandieri*

Additional Literature: Baird (1858), Schaldach (1966), Choate (1970), Woodman and Timm (1999), Woodman (2005), Carraway (2007), Guevara et al. (2015).

Cryptotis goldmani (Merriam, 1895)

Goldman's small-eared shrew, musaraña tropical de la Sierra Madre del Sur

1. *C. g. goldmani* (Merriam, 1895). Restricted to Guerrero.
2. *C. g. machetes* (Merriam, 1895). Restricted to Oaxaca.

Conservation: *Cryptotis goldmani* is listed as subject to Special Protection in the Norma Oficial Mexicana (DOF 2019) and as Least Concern by the IUCN (Cassola 2016j).

Characteristics: Goldman's small-eared shrew is small but large sized relative to the shrews of other genera; total length 93.0–110.0 mm; condylobasal length 18.7–20.7 mm. Dorsal pelage dark gray, pale-colored at the tip; fur of the underparts dark gray, paler at the tip (variation among subspecies); **head and shoulders more plumbeous**; tail slightly bicolored; claw length on the middle forefoot digit 2.7–3.8 mm (2.5–3.6% of the total length); four unicuspid teeth in the upper toothrow, last unicuspid teeth partially or totally covered in lateral view; first lower incisors with one deep valley between two crests; zygomatic process extending posteriorly and ventrolaterally to below the teeth occlusal surface; **braincase relatively flat, only slightly elevated above the rostrum**; teeth unicuspid, narrower at the base. Species group *mexicanus*.

Comments: *Cryptotis goldmani* was recognized as a distinct species (Choate 1970), previously regarded as a subspecies of *C. mexicanus* (Merriam 1895b). *C. griseoventris* was assigned as a subspecies of *C. goldmani* (Jackson 1933; Choate 1970) but is recognized as a distinct species (Woodman and Timm 1999). *C. goldmani* ranges in the highlands of Sierra Madre del Sur from Guerrero through Oaxaca (Map 7). It can be found in sympatry, or nearly so, with *C. magnus*, *C. mexicanus*, *C. parvus*, *C. peregina*, and *C. phillipsii*. It can be distinguished from *C. magnus* by its total length less than 123.0 mm; tail less than 40.0 mm; condylobasal length less than 21.0 mm, and distance from the first upper unicuspid teeth to the third upper molars less than 8.4 mm. From *C. mexicanus*, by having the dorsal and ventral hair paler at the tip. From *C. parvus*, by having the first lower incisors with two crests and deep valleys; condylobasal length greater than 18.7 mm. From *C. peregrinus*, by its narrow skull and zygomatic plate less than 1.3 mm; fourth upper unicuspid teeth partially visible; zygomatic processes not bulbous and extending posteriorly and ventrolaterally. From *C. phillipsii*, by the first lower incisors lacking a deep valley besides a "bump" at the base of the interdenticular spaces.

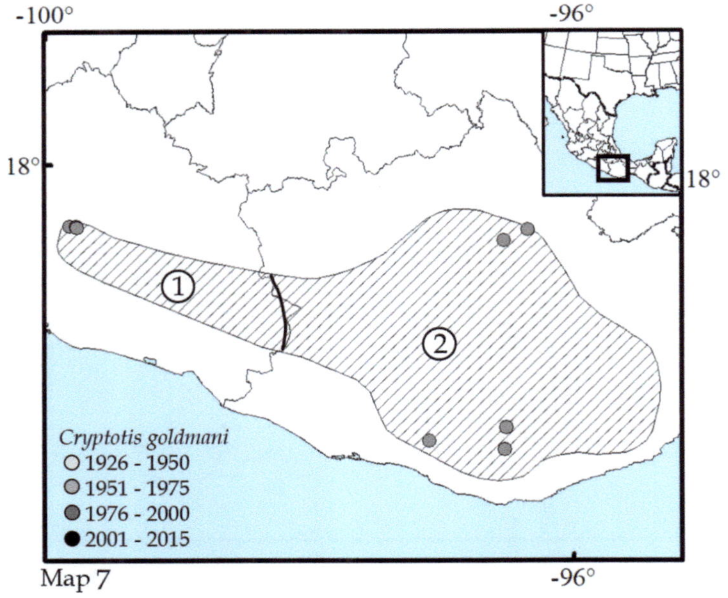

Map 7 Distribution map of *Cryptotis goldmani*

Additional Literature: Goodwin (1969), Choate and Fleharty (1974), Álvarez et al. (1987), Carraway (2007), Guevara et al. (2015).

Cryptotis goodwini Jackson, 1933

Goodwin's small-eared shrew, musaraña tropical de Chiapas

(*see addenda, split in two species *Cryptotis woodmani*)

Monotypic.

Conservation: *Cryptotis goodwini* is listed as Least Concern by the IUCN (Cuarón and de Grammont 2017).

Characteristics: Goodwin's small-eared shrew is small but large sized relative to the shrews of other genera; total length 103.0–128.0 mm; condylobasal length 21.0–22.8 mm. Dorsal pelage dark brown; fur of the underparts paler brown with silvery shades; tail slightly bicolored; claw on the middle forefoot digit approximately 3.7 mm (3.4% of the total length); four upper unicuspid teeth visible but not the last unicuspid teeth in lateral view; first lower incisors with two crests; first lower incisors with two denticles. Species group *mexicanus*.

Comments: *Cryptotis goodwini* is restricted to southern highlands above 1200 m in Chiapas south through Central America (Map 8). It can be found in sympatry, or nearly so, with *C. griseoventris*, *C. parvus*, and *C. tropicalis*. This species can be distinguished from all other species of *Cryptotis* by its condylobasal length greater than 21.0 mm. From *C. griseoventris*, by having a deep interdenticular space between the first lower incisors; coronoid process height between 4.7 and 4.8 mm. From *C. parvus*, by having the first lower incisors with two crests; total length greater than 101.0 and condylobasal length greater than 21.0 mm. From *C. tropicalis*, by its condylobasal length greater than 21.0 mm; distance from the first upper unicuspid teeth to the third upper molars greater than 8.0 mm and dorsal pelage dark brown "See adeendum".

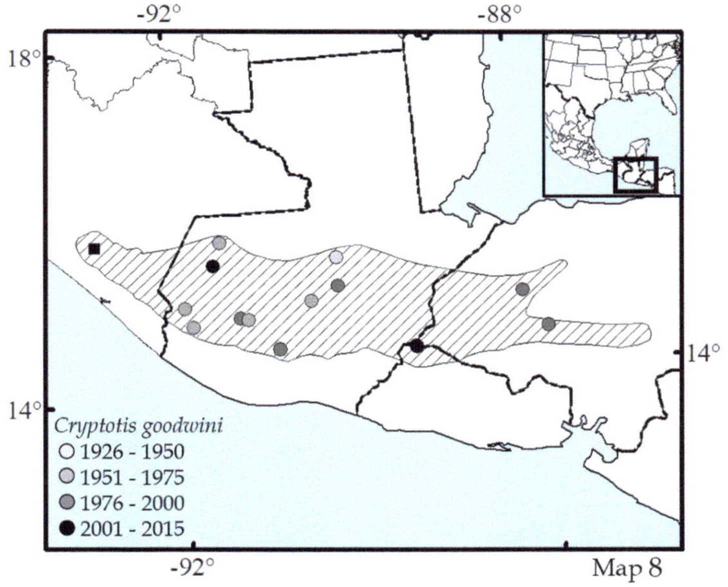

Map 8 Distribution map of *Cryptotis goodwini*

Additional Literature: Jackson (1933), Choate and Fleharty (1974), Woodman (2005, 2010), Carraway (2007), Woodman and Stephens (2009), Woodman et al. (2012), Guevara et al. (2015).

Cryptotis griseoventris Jackson, 1933

Guatemalan broad-clawed shrew, musaraña tropical de Guatemala

Monotypic.

Conservation: *Cryptotis griseoventris* is listed as Endangered (B1ab (i, iii): extent of occurrence less than 5000 km^2, severely fragmented or known to exist at no more than five locations, and under continuing decline in area, extent of occurrence and quality of habitat) by the IUCN (Roach and Naylor 2017).

Characteristics: Guatemalan broad-clawed shrew is small but larger sized relative to the shrews of other genera; total length 100.0–110.0 mm; condylobasal length 19.4–20.7 mm. **Dorsal pelage dark grayish brown; fur of the underparts dark grayish brown**; tail slightly bicolored; three or four upper unicuspid teeth, if four, the last unicuspid teeth partially or totally covered in lateral view; first lower incisors with one deep valley between two crests. Species group *mexicanus*.

Comments: *Cryptotis griseoventris* was reinstated as a distinct species (Woodman and Timm 1999); it was previously regarded as a subspecies of *C. goldmani goldmani* (Choate 1970). Its ranges are restricted to the highlands above 2000 m in central Chiapas (Map 9). It can be found in sympatry, or nearly so, with *C. goodwini*, *C. mexicanus*, *C. parvus*, and *C. tropicalis*. *C. griseoventris* can be distinguished from all other species of *Cryptotis* by its dark brown grayish fur of the underparts and first lower incisors with a moderately deep interdenticular space between them. From *C. goodwini*, by having the coronoid process height from 4.1 to 4.3 mm. From *C. mexicanus*, by having the fur of the underparts and the dorsum dark grayish brown, while it is not grayish in *C. mexicanus*. From *C. parvus*, by having the first lower incisors with two crests; total length greater than 100.0 mm; condylobasal length greater than 19.5 mm. From *C. parvus* and *C. tropicalis*, by having the zygomatic process from medium large to elliptic.

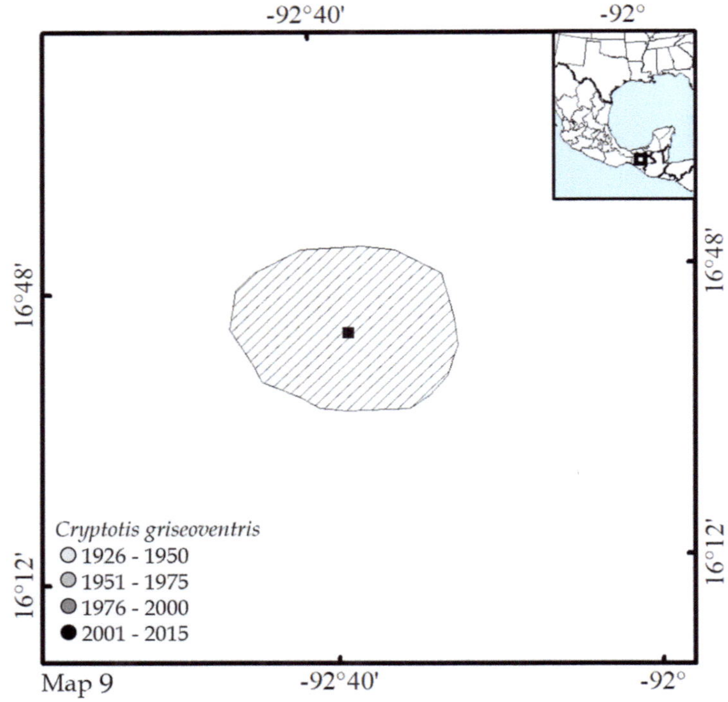

Map 9 Distribution map of *Cryptotis griseoventris*

Additional Literature: Jackson (1933), Hutterer (2005), Woodman (2005, 2010), Carraway (2007), Woodman and Stephens (2009), Sánchez-Cordero et al. (2014), Guevara et al. (2015).

Cryptotis lacandonensis Guevara et al., 2014

Lacandona small-eared shrew, musaraña tropical de la Lacandona

Monotypic.

Conservation: *Cryptotis lacandonensis* has been recently described; for this reason, it has not been included in any risk list. However, it should to be listed, given its seemingly restricted distribution.

Characteristics: Lacandona small-eared shrew is small but large sized relative to the shrews of other genera; total length 78.0–81.0 mm; **condylobasal length 20.1–20.5 mm**. Dorsal pelage deep vinaceous-gray to dark plumbago gray; fur of the underparts paler than in the dorsum; **tail on average 42.5% of the head-and-body length, of the same color as in the ventral part**; four upper unicuspid teeth visible in lateral view. Species group *nigrescens*.

Comments: *Cryptotis lacandonensis* is only known from two localities in the Selva Lacandona, Chiapas, and the other in El Petén, Guatemala (Map 10). It can be distinguished from other sympatric species, such as *C. tropicalis* and probably *C. mayensis*, by having the average tail length 42.5% of the head-and-body length and of the same color as the underparts.

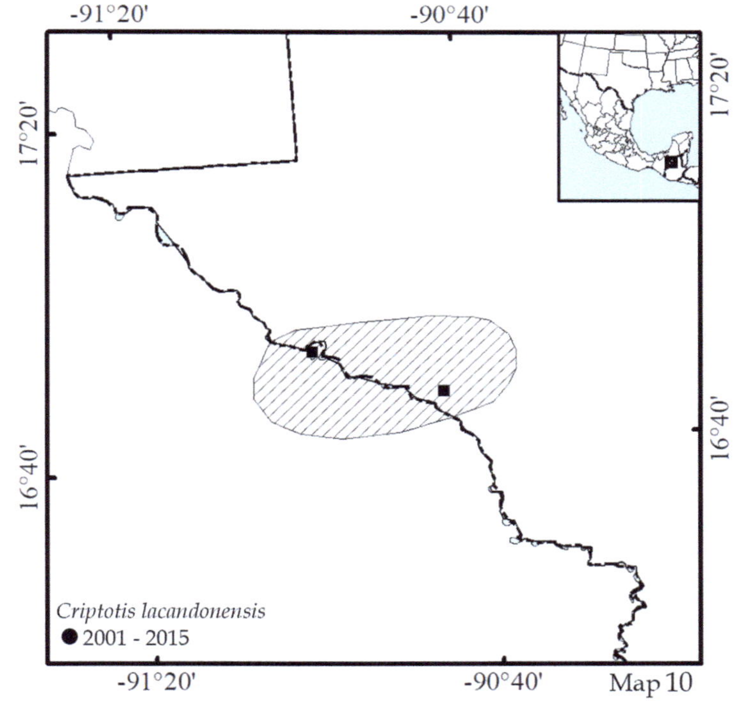

Map 10 Distribution map of *Cryptotis lacandonensis*

Additional Literature: Guevara et al. (2014a, 2015), Pérez et al. (2019), Lorenzo et al. (2019b).

Cryptotis magnus (Merriam, 1895)

Big small-eared shrew, musaraña tropical gigante

Monotypic.

Conservation: *Cryptotis magnus* is listed as subject to Special Protection (as *C. magnus*) in the Norma Oficial Mexicana (DOF 2019) and as Vulnerable (B1ab (iii): extent of occurrence less than 20,000 km^2, severely fragmented, and under continuing decline in area, extent, and quality of the habitat) by the IUCN (Cuarón and de Grammont 2018a).

Characteristics: Big small-eared shrew is small but large sized relative to the shrews of other genera; **total length 123.0–141.0 mm; condylobasal length 21.6–23.6 mm. Dorsal pelage uniformly dark brown**; fur of the underparts slightly paler than in the dorsum; **first lower incisors with three cusps and a deep interdenticular space between them**; four upper unicuspid teeth visible, but the last unicuspid teeth not visible in lateral view; zygomatic process bulbous and extending posteriorly and ventrolaterally to below the teeth occlusal surface; **braincase profile strongly convex, with the highest point near the junction of the posterior and middle thirds; upper premolars large and broad, with the antero-internal angle broadly rounded off**.

Comments: *Cryptotis magnus* is only known from northern Oaxaca, from 1500 to 2800 m (Map 11). It can be found in sympatry, or nearly so, with *C. goldmani*, *C. mexicanus*, *C. parvus*, and *C. peregrinus*. *C. magnus* can be distinguished from all other species of *Cryptotis* by its total length greater than 123.0 mm; condylobasal length greater than 21.6 mm; dorsal pelage and the tail uniformly dark brown; first lower incisors with three cusps and a marked interdenticular space between them.

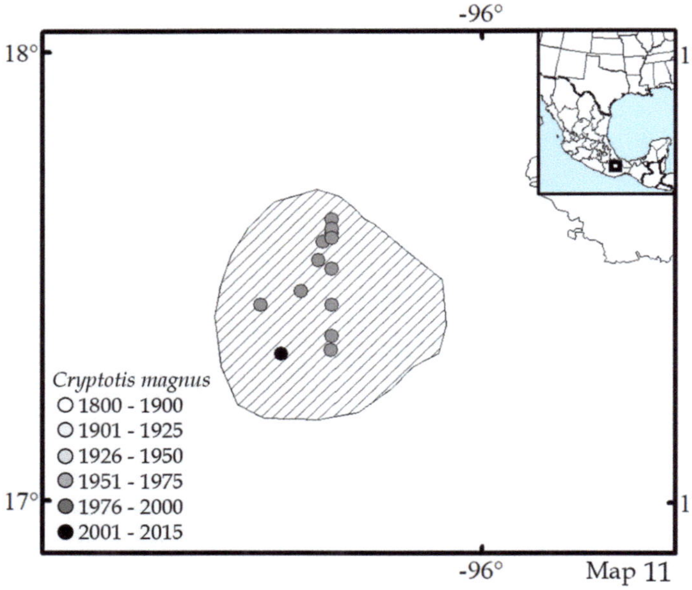

Map 11 Distribution map of *Cryptotis magnus*

Additional Literature: Merriam (1895b), Choate (1970), Choate and Fleharty (1974), Robertson and Rickart (1976), Carraway (2007), Guevara et al. (2015).

Cryptotis mayensis (Merriam, 1901)

Yucatán small-eared shrew, musaraña tropical Maya

Monotypic.

Conservation: *Cryptotis mayensis* is listed as subject to Special Protection in the Norma Oficial Mexicana (DOF 2019) and as Least Concern by the IUCN (Cuarón et al. 2016b).

Characteristics: Yucatán small-eared shrew is small but large sized relative to the shrews of other genera; total length 64.0–102.0 mm; condylobasal length 18.9–20.5 mm. Dorsal pelage dark gray, with fine "pepper and salt" appearance; fur of the underparts plumbeous, overlaid with ash brown, strongest on the throat and breast; nose darker; tail with the same color as underparts; braincase lower, practically continuing plane of the rostrum; four upper unicuspid tooth visible inside view; first lower incisors with two denticles and deep interdenticular space; larger first and second unicuspid teeth and swollen, with inner tubercles obsolete; larger unicuspid teeth and broader. Species group *nigrescens*.

Comments: *Cryptotis mayensis* was recognized as a distinct species (Woodman and Timm 1999), previously was junior synonym of *C. nigrescens* (Choate 1970) and its ranges is restricted to the Yucatán Peninsula in México, Guatemala, and Belize (Map 12). It is also known from owl pellets collected in Guerrero (Choate 1970; Ramírez-Pulido and Sánchez-Hernández 1972). The record is completely out of the range and habitat, the most probable explanation is that it could be a different species (Monroy-Gamboa 2021). Guerrero's record does not is consider within the range of *C. mayensis*. *C. mayensis* cannot be found in sympatry with other *Cryptotis* species within the Yucatán Peninsula, but it could be in sympatry with *C. lacandonesis* in Chiapas. It can be differentiated from it by having the first lower incisors with two denticles with deep interdenticular space and dorsal pelage dark gray.

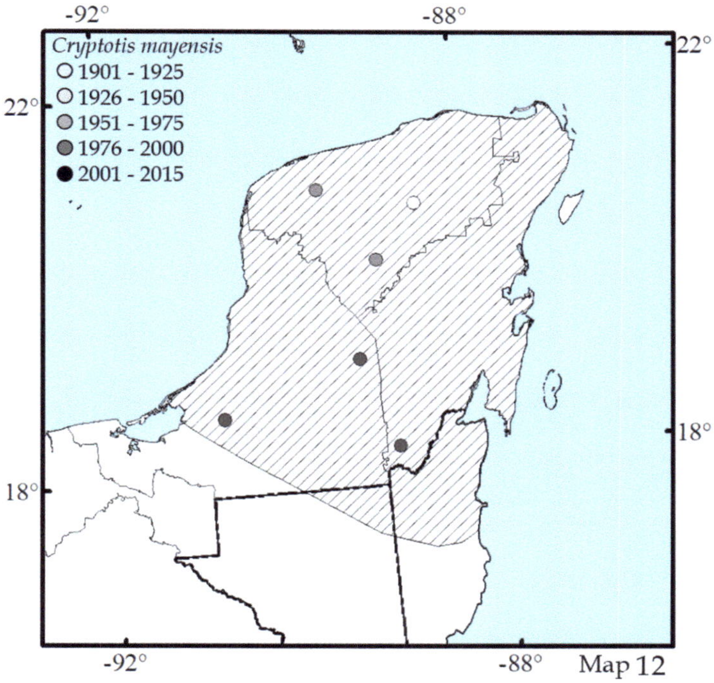

Map 12 Distribution map of *Cryptotis mayensis*

Additional Literature: Merriam (1901e), Woodman (2005), Carraway (2007), Guevara et al. (2015).

Cryptotis merriami Choate, 1970

Merriam's small-eared shrew, musaraña tropical de tierras altas

Monotypic.

Conservation: *Cryptotis merriami* is listed Least Concern by the IUCN (Woodman et al. 2016b).

Characteristics: Merriam's small-eared shrew is small but large sized in relation to the shrews of other genera; total length 87.0–107.0 mm; condylobasal length 18.8–19.8 mm. Dorsal pelage blackish gray; fur of the underparts paler gray; tail bicolored following the body pattern; three or four upper molars unicuspid; **zygomatic process extending dorsolaterally at the alveoli level of the third and second upper molars.** Species group *nigrescens*.

Comments: *Cryptotis merriami* was reinstated as a distinct species (Woodman and Timm 1999), previously regarded as a subspecies of *C. nigrescens* (Choate 1970). This species can be considered as being in a third subgenus (George 1988) and its ranges extend from Chiapas south through Central America. However, Woodman (2018) does not include Chiapas within its range (Map 13). It can be found in sympatry, or nearly so, with *C. tropicalis* and *C. parvus*. It can be distinguished from these species by its zygomatic process extending dorsolaterally at the alveoli level of the third and second upper molars. From *C. tropicalis*, by its fur of the underparts paler. From *C. parvus*, by having two denticles with a deep interdenticular space between the first lower incisors.

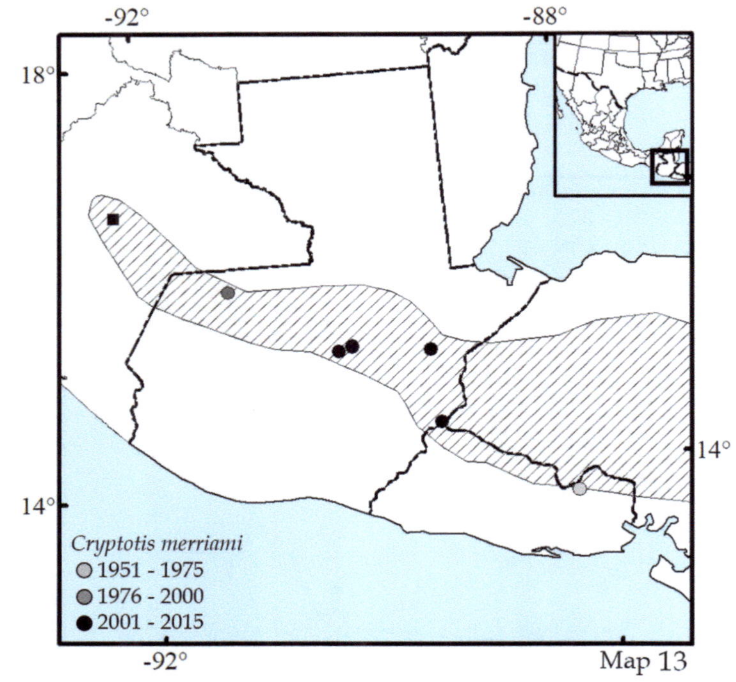

Map 13 Distribution map of *Cryptotis merriami*

Additional Literature: Carraway (2007), Guevara et al. (2015).

Cryptotis mexicanus (Coues, 1877)

Mexican small-eared shrew, musaraña tropical mexicana

Monotypic.

Conservation: *Cryptotis mexicanus* is listed as Least Concern by the IUCN (Cassola 2016k).

Characteristics: Mexican small-eared shrew is small but large sized relative to the shrews of other genera; total length 83.0–112.0 mm; condylobasal length 17.7–19.9 mm. Dorsal pelage dark brown, **darkest on the back and rump**; fur of the underparts darker at the tip; tail slightly bicolored; **feet and tail blackish**; four upper unicuspid teeth, the last unicuspid tooth partially or totally covered in lateral view; first lower incisors with one deep valley between two crests; **zygomatic processes bulbous and projecting at the posterior edge of the third upper molars; zygomatic plate 1.7–2.3 mm; braincase inflated, rising abruptly above the rostrum plane;** cusplet on the inner side strongly developed and chestnut brown at the tip; all the teeth chestnut brown at the tip. Species group *mexicanus*.

Comments: *Cryptotis nelsoni*, *C. obscurus*, and *C. peregrinus* were recognized as a distinct species (Woodman and Timm 1999), previously regarded as subspecies of *C. mexicanus* (Merriam 1895b; Choate 1970). *C. mexicanus* ranges in the highlands of the Sierra Madre Oriental from Hidalgo south through Chiapas, including Puebla, Veracruz, and Oaxaca (Map 14). It can be found in sympatry, or nearly so, with *C. goldmani*, *C. griseoventris*, *C. magnus*, *C. merriami*, *C. obscurus*, *C. parvus*, and *C. peregrinus*. *C. mexicanus* can be distinguished from *C. goldmani* by its dorsal and ventral hair darker or of the same color at the tip. From *C. griseoventris*, by having the dorsal pelage in brown colors and the fur of the underparts paler, while in *C. griseoventris* the hair is dark grayish brown dorsally and in the underparts. From *C. magnus*, by its total length less than 123.0 mm; tail less than 40.0 mm; condylobasal length less than 21.0 mm and distance from the first upper unicuspid teeth to the third upper molars less than 8.4 mm. From *C. obscurus*, by not having the dorsal pelage with a grizzly appearance and the zygomatic processes not projecting ventrally at the level of the second upper molars. From *C. merriami*, by the zygomatic process not extending dorsolaterally at the level of the alveoli of the third and second upper molars. From *C. parvus*, by having two crests and deep valleys in the first lower incisors; zygomatic processes bulbous and projecting at the third upper molar posterior edge. From *C. peregrinus*, by its narrow skull and zygomatic plate less than 1.3 mm; zygomatic processes not bulbous and extending posteriorly and ventrolaterally.

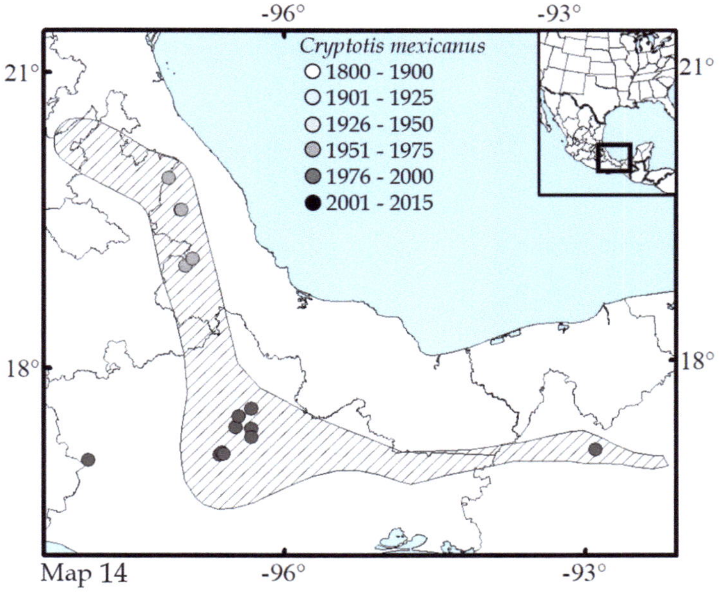

Map 14 Distribution map of *Cryptotis mexicanus*

Additional Literature: Choate (1973), Choate and Fleharty (1974), Woodman (2005), Carraway (2007), Guevara and Cervantes (2013), Guevara et al. (2015), Guevara (2017).

Cryptotis nelsoni (Merriam, 1895)

Nelson's small-eared shrew, musaraña tropical de Catemaco

Monotypic.

Conservation: *Cryptotis nelsoni* is listed as subject to Special Protection in the Norma Oficial Mexicana (DOF 2019) and as Endangered (B1ab (i, iii): extent of occurrence less than 5000 km^2, severely fragmented or known from a single location, and continuing decline in extent of occurrence, area, and quality of habitat) by the IUCN (Matson et al. 2018).

Characteristics: Nelson's small-eared shrew is small but large sized relative to the shrews of other genera; total length 100.0–110.0 mm; condylobasal length 19.2–20.2 mm. Dorsal pelage dark brown, including the tail; four unicuspid teeth in the upper toothrow; the last molar partially or totally covered in lateral view; first lower incisors with one deep valley between two crests; zygomatic process elliptic, extending posteriorly and ventrolaterally to the teeth occlusal surface; braincase large, **flat, and not abruptly elevated above the rostrum plane; interpterygoid fossa broad; unicuspid teeth with the inner cusplet nearly obsolete**. Species group *mexicanus*.

Comments: *Cryptotis nelsoni* was reinstated as a distinct species by Woodman and Timm (1999); it was previously regarded as a junior synonym of *C. mexicanus* (Choate 1970). This species is known only from the western slope of the San Martín Tuxtla volcano, Veracruz (Map 15), and has allopatric distribution. It differs from nearly all species by its uniformly dark brown color on the back and the tail; braincase flatter and not abruptly elevated above the rostrum plane; interpterygoid fossa broader and unicuspid teeth with the inner cusplet nearly obsolete.

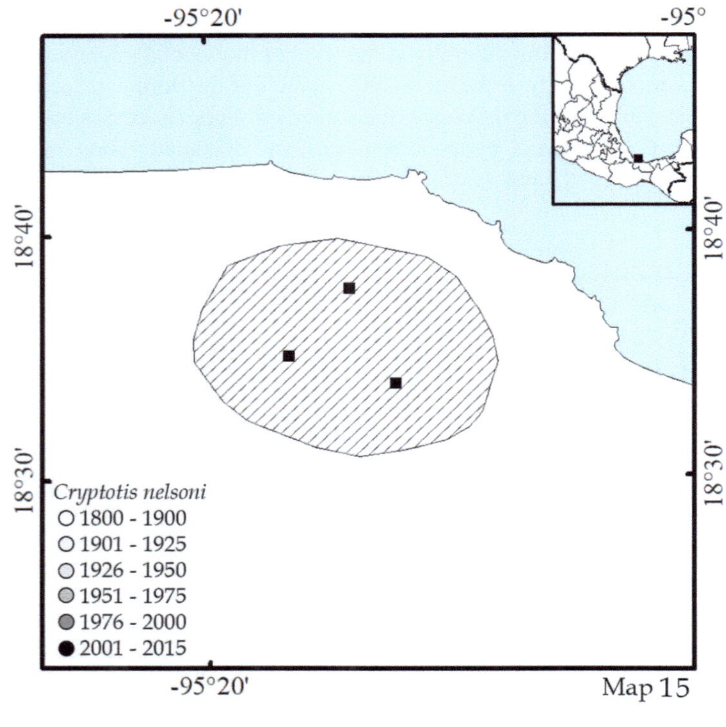

Map 15 Distribution map of *Cryptotis nelsoni*

Additional Literature: Merriam (1895b), Choate (1973), Woodman (2005), Carraway (2007), Guevara and Cervantes (2010), Guevara et al. (2015), Guevara and Sánchez-Cordero (2018).

Cryptotis obscurus (Merriam, 1895)

Grizzled Mexican small-eared shrew, musaraña tropical de Hidalgo

Monotypic.

Conservation: *Cryptotis obscurus* is listed as subject to Special Protection in the Norma Oficial Mexicana (DOF 2019) and as Least Concern by the IUCN (Roach and Naylor 2016a).

Characteristics: Grizzled Mexican small-eared shrew is small but large sized relative to the shrews of other genera; total length 89.0–99.0 mm; condylobasal length 17.2–19.6 mm. **Dorsal pelage with two bands, dark gray at the base with silver shades and brown at the tip; fur of the underparts with two bands, dark gray with silver shades at the base and silvery at the tip;** tail slightly bicolored; four upper unicuspid teeth, the last unicuspid tooth partially or totally covered in lateral view; first lower incisors with one deep valley between two crests; **zygomatic processes projecting ventrally at the level of the second upper molars**. Species group *mexicanus*.

Comments: *Cryptotis obscurus* was reinstated as a distinct species by Woodman and Timm (1999); it was previously regarded as a junior synonym of *C. mexicanus* (Choate 1970). It ranges in the highlands, from 1040 to 2500 m in Hidalgo, Querétaro, San Luis Potosí, Tamaulipas, and Veracruz (Hutterer 2005; Map 16). It can be found in sympatry, or nearly so, with *C. alticola*, *C. mexicanus*, and *C. parvus*. *C. obscurus* can be distinguished from these species by its dorsal pelage with a grizzly appearance and the zygomatic processes projecting ventrally at the level of the second upper molars.

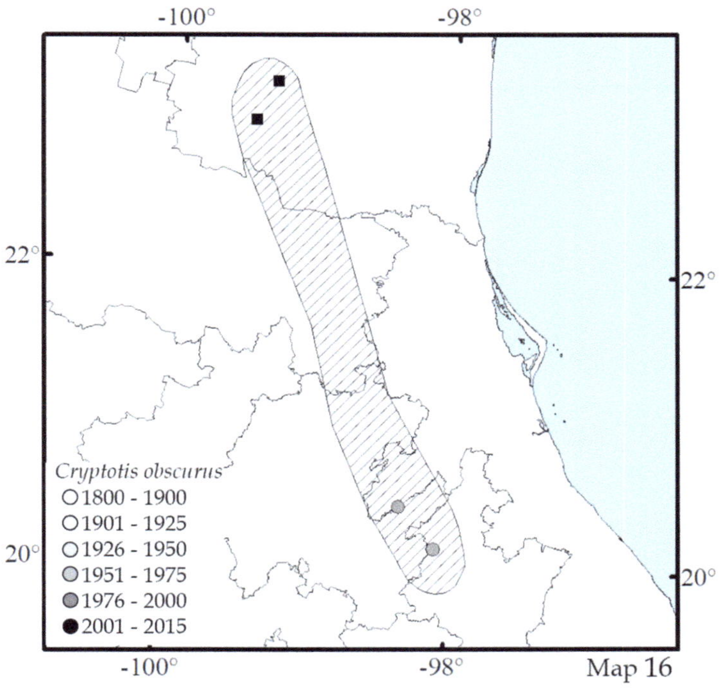

Map 16 Distribution map of *Cryptotis obscurus*

Additional Literature: Merriam (1895b), Choate (1973), Woodman and Timm (1999), Woodman (2005), Carraway (2007), Guevara et al. (2015).

Cryptotis parvus (Say, 1823)

North American least shrew, musaraña tropical de Norteamérica

1. *C. p. classon* Bole and Moulthrop, 1942. Restricted to Ohio.
2. *C. p. floridanus* (Merriam, 1895). Restricted to the Florida Peninsula.
3. *C. p. harlani* (Duvernoy, 1842) Restricted to Illinois, Michigan, and Indiana.
4. *C. p. parvus* (Say, 1823). From South Dakota south through Texas, and east through the Atlantic coast from Massachusetts south through the Florida Peninsula.

Conservation: *Cryptotis parvus* is listed as Least Concern by the IUCN (Woodman et al. 2016c).

Characteristics: North American least shrew is small but large sized relative to the shrews of other genera; total length 67.0–103.0 mm; condylobasal length less than 18.0 mm. Dorsal pelage dark to pale brown; fur of the underparts paler than in the dorsum; claw length on the middle forefoot digit less than 2.0 mm; four upper unicuspid teeth visible in lateral view [not visible in some populations]; **first lower incisors with three denticles with a very shallow interdenticular space between them**. Species group *parvus*.

Comments: *Cryptotis parvus berlandieri*, *C. p. pueblensis*, *C. p. soricinus*, and *C. p. tropicalis* were elevated to full species (Handley and Choate 1970; Choate 1970; Woodman 2018). *C. parvus* ranges in all eastern United States from South Dakota south through Texas, and east through the Atlantic coast, including the Florida Peninsula (Map 17). It can be found in sympatry, or nearly so, with *C. berlandieri*, from which it can be distinguished only using genetic data.

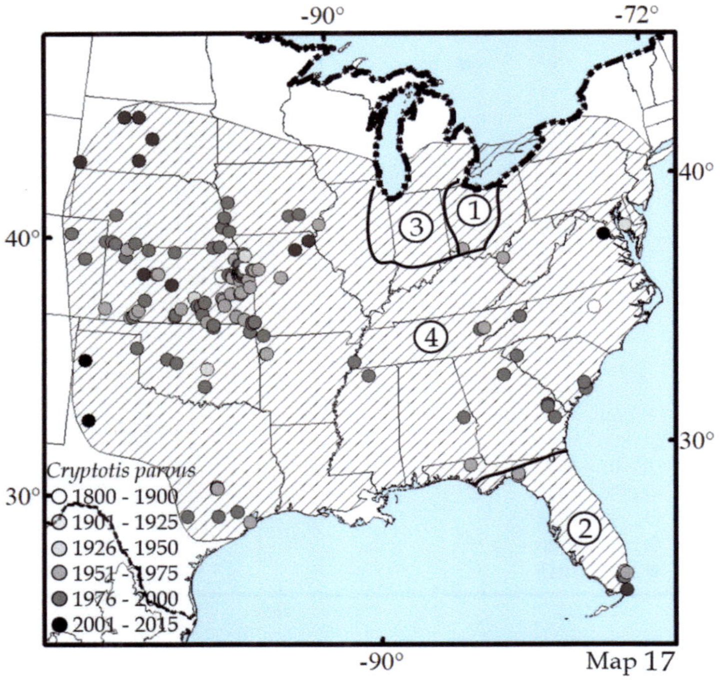

Map 17 Distribution map of *Cryptotis parvus*

Additional Literature: Merriam (1895b), Goodwin (1932), Hall and Kelson (1959), Davis (1966b), Packard and Judd (1968), Choate and Fleharty (1974), Whitaker (1974), Bowles (1975), Hoffmeister (1977), Hall (1981), Woodman (2005), Carraway (2007), Guevara et al. (2015).

Cryptotis peregrinus (Merriam, 1895)

Oaxacan small-eared shrew, musaraña tropical de Oaxaca

Monotypic.

Conservation: *Cryptotis peregrinus* is listed as subject to Special Protection in the Norma Oficial Mexicana (DOF 2019) and as Data Deficient by the IUCN (Álvarez-Castañeda et al. 2016x).

Characteristics: Oaxacan small-eared shrew is small but large sized relative to the shrews of other genera; total length 99.7–102.5 mm; condylobasal length 18.4–19.9 mm. Dorsal pelage dark brown; fur of the underparts slightly paler; tail slightly bicolored; **forefoot claw length on the middle digit from 2.5% to 3.1% of the total length**; four upper unicuspid teeth visible in lateral view; **skull very broad; zygomatic processes bulbous and extending posteriorly and ventrolaterally to below the tooth occlusal surface; inner cusplet of the unicuspid teeth nearly obsolete and without chestnut brown tips; second upper molars with the postero-internal lobe larger than the antero-internal lobe**. Species group *mexicanus*.

Comments: *Cryptotis peregrinus* was recognized as a distinct species (Woodman and Timm 1999), previously regarded as a subspecies of *C. mexicanus* (Merriam 1895b). It is currently known in only from two localities: Sierra de Cuatro Venados and Sierra Yucuyacua, over 2860 m in Oaxaca (Map 18). It can be found in sympatry, or nearly so, with *C. goldmani*, *C. magnus*, *C. mexicanus*, *C. parvus*, and *C. phillipsii*. *C. peregrinus* can be distinguished from all these species by having the posterior inner cusplet of unicuspid teeth obsolete; second upper molars with the postero-internal lobe larger than the antero-internal lobe; four unicuspid teeth completely visible in lateral view; skull very broad; zygomatic plate narrow (1.3–1.7 mm); zygomatic processes bulbous and extending posteriorly and ventrolaterally to below the occlusal surface of the teeth.

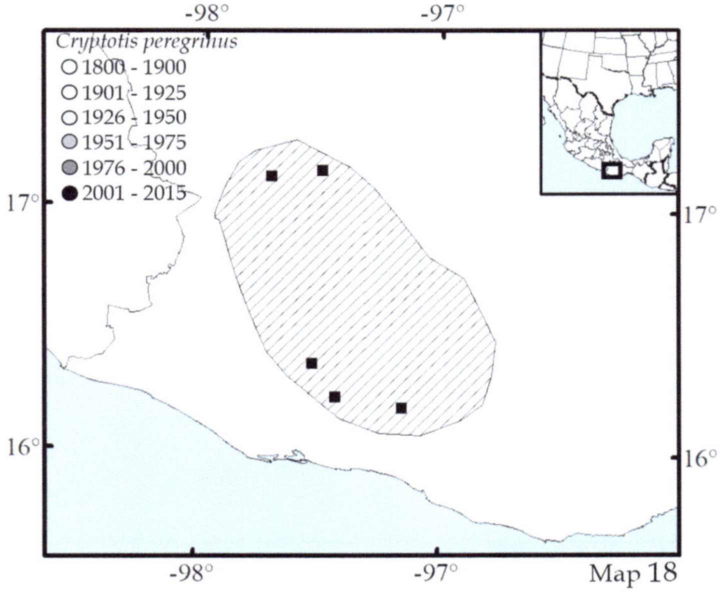

Map 18 Distribution map of *Cryptotis peregrinus*

Additional Literature: Schaldach (1966), Choate (1969, 1973), Woodman (2005), Carraway (2007), Lavariega et al. (2012), Guevara et al. (2015).

Cryptotis phillipsii (Schaldach, 1966)

Phillips' small-eared shrew, musaraña tropical de la sierra de Oaxaca

Monotypic.

Conservation: *Cryptotis phillipsii* is listed as Vulnerable (B1ab (iii): extent of occurrence less than 20,000 km^2 severely fragmented and continuing decline in area, extent, and quality of habitat) by the IUCN (Álvarez-Castañeda et al. 2018o).

Characteristics: Phillips' small-eared shrew is small but larger sized relative to the shrews of other genera; total length 104.0–108.0 mm; condylobasal length 17.9–20.4 mm. **Dorsal pelage very dark gray, silver at the tip, giving a frosted appearance**; fur of the underparts paler than in the dorsum but with an increase in presence of silver shades; tail uniformly very dark gray; zygomatic plate 1.9–2.2 mm; three or four upper unicuspid teeth, when four are present, the last unicuspid tooth is not visible in lateral view; first lower incisors with one deep valley between two crests **besides a "bump" on the base of the interdenticular spaces**. Species group *mexicanus*.

Comments: *Cryptotis phillipsii* was recognized as a distinct species (Woodman and Timm 2000), previously was junior synonym of *C. mexicanus peregrinus* (Schaldach 1966). Its ranges in the Sierra de Miahuatlán, south Oaxaca, has been recorded between elevations of 1060 and 2600 m (Map 19). It can be found in sympatry, or nearly so, with *C. goldmani*, *C. parvus*, and *C. peregrinus*. *C. phillipsii* can be distinguished from all these species by having the first lower incisors with two crests and deep valleys besides a "bump" at the base of the interdental spaces. From *C. goldmani* and *C. peregrinus*, by its claw length of middle digit of the forefoot 1.9–2.0 mm.

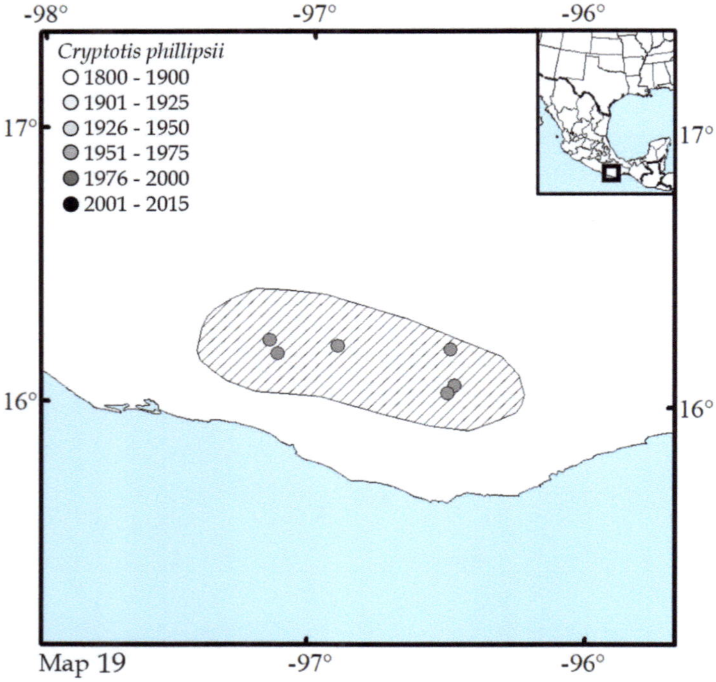

Map 19 Distribution map of *Cryptotis phillipsii*

Additional Literature: Choate (1969), Woodman (2005), Carraway (2007), Guevara et al. (2015).

Cryptotis pueblensis Jackson, 1933

Southern México least shrew, musaraña tropical del sur de México

Monotypic.

Conservation: *Cryptotis pueblensis* (as *Cryptotis parvus*) is not listed by the IUCN (Woodman et al. 2016c).

Characteristics: Southern México least shrew is small but large sized relative to the shrews of other genera; total length 85.0–103.0 mm; condylobasal length 16.4–17.6 mm. Dorsal pelage medium-dark brown; fur of the underparts paler than in the dorsum; tail slightly bicolored; claw length on the middle forefoot digit 1.0–2.0 mm and 1.6–2.2% of the total length; four upper unicuspid teeth visible in lateral view [not visible in some populations]; first lower incisors with three denticles with very shallow interdenticular spaces. Species group *parvus*.

Comments: *Cryptotis pueblensis* was recognized as a distinct species (Woodman 2018), previously regarded as a subspecies of *C. parvus* (Choate 1970). *C. pueblensis* ranges from the southeastern corner of San Luis Potosí south through Oaxaca and central Chiapas (Map 20). It can be found in sympatry, or nearly so, with *C. berlandieri*, *C. goldmani*, *C. goodwini*, *C. griseoventris*, *C. merriami*, *C. mexicanus*, *C. peregrinus*, *C. phillipsii*, *C. soricina*, and *C. tropicalis*. *C. pueblensis* (except those species of group *parvus*) can be distinguished from these species by its smaller size and having the first lower incisors with three denticles with very shallow interdenticular spaces. From *C. griseoventris*, by having the fur of underparts paler, not dark, and not having moderately deep interdenticular spaces between the first lower incisors. From *C. mayensis*, by having a deep interdenticular spaces between the first lower incisors; dorsal pelage not dark gray; condylobasal length usually 16.4–17.6 mm; palatilar length 6.7–7.3 mm, and length from the first upper unicuspid teeth to the third upper molars 5.7–6.8 mm. From *C. merriami*, by its zygomatic process not extending dorsolaterally at the level of the alveoli of the third and second upper molars. From *C. mexicanus*, by having the zygomatic processes not bulbous and not projecting at the posterior edge of the third upper molars. From *C. peregrinus*, by its narrow skull and zygomatic plate less than 1.3 mm; zygomatic processes not bulbous and extend posteriorly and ventrolaterally. From *C. phillipsii*, by not having deep valleys in the first lower incisors besides a "bump" at the base of the interdenticular spaces. From *C. tropicalis*, by the condylobasal length 16.4–17.6 mm; palatilar length 6.7–7.3; distance from the first upper unicuspid teeth to the third upper molars 5.7–6.8 mm; dorsal pelage dark brown; underparts pale to medium brown. From *C. berlandieri* and *C. soricina*, by a combination of condylobasal length usually 16.4–17.6 mm and length of the third lower molars plus width of the first lower molars plus width of the second lower molars plus width of the third lower molars usually 4.3–5.1 mm.

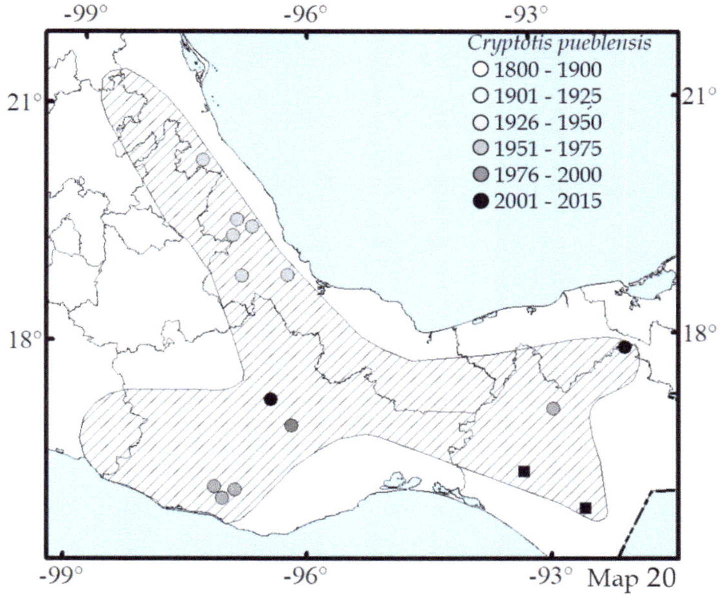

Map 20 Distribution map of *Cryptotis pueblensis*

Additional Literature: Jackson (1933), Schaldach (1966), Choate (1969, 1973), Woodman and Timm (1999), Woodman (2005, 2018), Carraway (2007), Lavariega et al. (2012), Guevara et al. (2015).

Cryptotis soricinus (Merriam, 1895)

Valle de México least shrew, musaraña tropical del Valle de México

Monotypic.

Conservation: *Cryptotis soricinus* (as *C. parvus soricinus*) is listed as subject to Special Protection in the Norma Oficial Mexicana (DOF 2019) and is not listed by the IUCN (Woodman et al. 2016c).

Characteristics: Valle de México least shrew is small but large sized relative to the shrews of other genera; total length 75.0–90.0 mm; condylobasal length 16.1–17.7 mm. Dorsal pelage dark brown; underparts paler than in the dorsum; tail slightly bicolored; claw length on the middle forefoot digit 2.0 mm, 2.2% of the total length; four upper unicuspid teeth visible in lateral view [not visible in some populations]; first lower incisors with three denticles with very shallow interdenticular spaces. Species group *parvus*.

Comments: *Cryptotis soricinus* was described as a full species (Merriam 1895b; Woodman 2018); it was previously regarded as a subspecies of *C. parvus* (Choate 1970). *C. soricinus* is restricted to the Valley of México, including Estado de México, Morelos, and Ciudad de México (Map 21). It can be found in sympatry, or nearly so, with *C. alticola*, *C. berlandieri*, and *C. pueblensis*. *C. soricinus* can be distinguished from these species by its smaller size; tail length less than 24.0 mm; palatine length less than 7.2 mm; cranial breadth less than 8.8 mm, and interorbital breadth less than 4.4 mm, and by having the first lower incisors with three denticles with very shallow interdenticular spaces. From *C. berlandieri*, by having a pale brown venter. From *C. pueblensis*, by having the breadth across the second upper molars 3.8–4.1 mm, and from both, by a combination of the condylobasal length 16.1–17.7 mm and length of the third lower molars plus width of the first lower molars plus width of the second lower molars plus width of the third lower molars usually 3.6–4.8 mm.

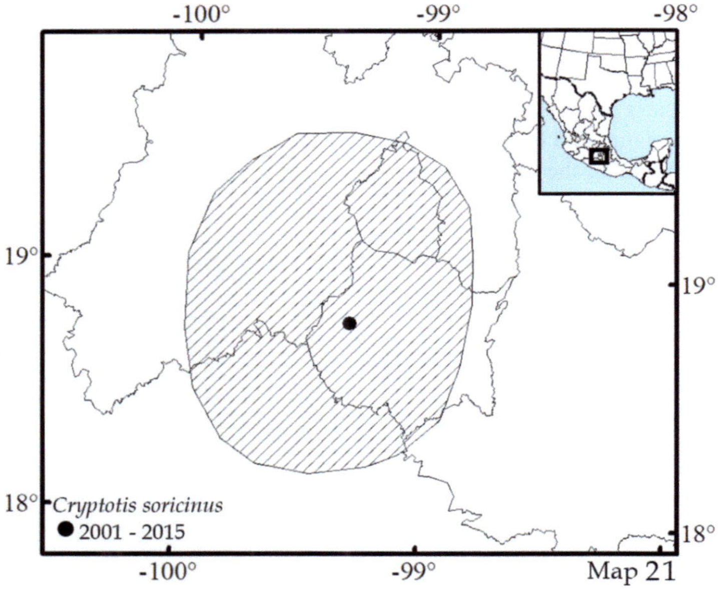

Map 21 Distribution map of *Cryptotis soricinus*

Additional Literature: Schaldach (1966), Woodman and Timm (1999), Woodman (2005), Carraway (2007), Guevara et al. (2015).

Cryptotis tropicalis (Merriam, 1895)

Tropical small-eared shrew, musaraña de tropical

Monotypic.

Conservation: *Cryptotis tropicalis* (as *C. parvus*) is listed as subject to Special Protection in the Norma Oficial Mexicana (DOF 2019) and as Data Deficient by the IUCN (Woodman 2019).

Characteristics: Tropical small-eared shrew is small but large sized relative to the shrews of other genera; total length 92.1–94.5 mm; condylobasal length 18.7–18.9 mm. **Dorsal pelage with changes in hair length, which has three color stripes: silver gray with silvery shades at the base, whitish in the middle, and reddish brown at the tip**; fur of the underparts with two stripes: silver gray at the base and pale yellowish brown at the tip; tail slightly bicolored; three or four upper unicuspid teeth; braincase essentially on the rostrum plane, with only a shallow sulcus; **second unicuspid teeth with a prominent inner cusplet and projecting well inward; third unicuspid teeth without an inner cusplet**. Species group *parvus*.

Comments: *Cryptotis tropicalis* was reinstated as a distinct species (Carraway 2007), previously regarded as a subspecies of *C. parvus* (Handley and Choate 1970). Its ranges in the eastern highlands of Chiapas, Guatemala, and Belize (Map 22), and can be found in sympatry, or nearly so, with *C. goodwini*, *C. merriami*, and *C. parvus*. *C. tropicalis* can be distinguished from *C. goodwini* and *C. parvus* by having the dorsal pelage with three color stripes: silver gray with silvery shades at the base, whitish in the middle, and reddish brown at the tip. From *C. goodwini*, by its condylobasal length less than 21.0 mm and distance from the first upper unicuspid teeth to the third upper molars less than 8.0 mm. From *C. parvus*, by having the first lower incisors with two crests; total length greater than 100.0 mm and condylobasal length greater than 21.0 mm. From *C. merriami*, by having a darker ventral pelage; breadth across the second upper molars 5.5–5.8 mm; mandible length 8.3–8.8 mm.

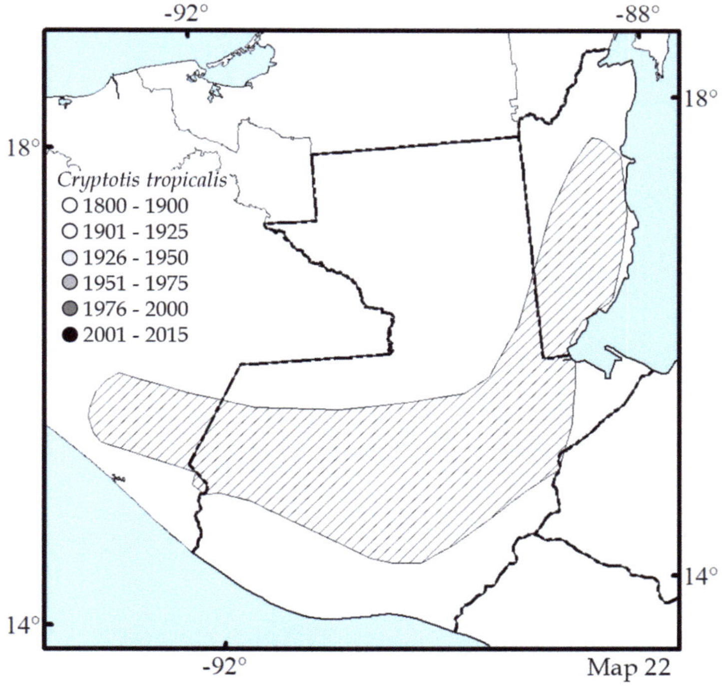

Map 22 Distribution map of *Cryptotis tropicalis*

Additional Literature: Merriam (1895b), Choate (1970), Hutterer (2005), Guevara et al. (2015).

Tribe Notiosoricini

Genus *Megasorex* Hibbard, 1950

Tail short and ears large; gray coloration in different shades; scent gland on the flanks; **three upper unicuspid teeth forming a uniform series; third teeth more than half as large as the second teeth and never minute; unicuspid teeth narrow-based and without an internal cusplet**; first incisors and unicuspid teeth slightly pigmented in labial view; **posterior borders of the upper fourth premolars and the first and second molars nearly fused when viewed from the occlusal surface.**

Megasorex gigas (Merriam, 1897)

Mexican shrew, musaraña gigante

Monotypic.

Conservation: *Megasorex gigas* is listed as Threatened in the Norma Oficial Mexicana (DOF 2019) and as Least Concern by the IUCN (Woodman et al. 2016a).

Characteristics: Mexican shrew is large sized; total length 81.0–87.0 mm; condylobasal length 22.8–23.8 mm. Dorsal pelage pale gray with brown; underparts with fur whitish at the tip; tail bicolored, **large size, greater than 105.0 mm**; **all teeth unpigmented in labial view;** four unicuspid teeth in the upper toothrow.

Comments: *Megasorex gigas* was described as a species of *Notiosorex* but was assigned to *Megasorex* (Hibbard 1950). It ranges from Sinaloa south through Oaxaca (Map 23). *M. gigas* lives in sympatry with *Notiosorex evotis* and can be distinguished from it and other sympatric species of shrews by having four unicuspid teeth in the upper toothrow; not pigmented in labial view; large size, greater than 105.0 mm. From *Notiosorex*, by having the lower glenoid furrow fused more laterally to the alisphenoid.

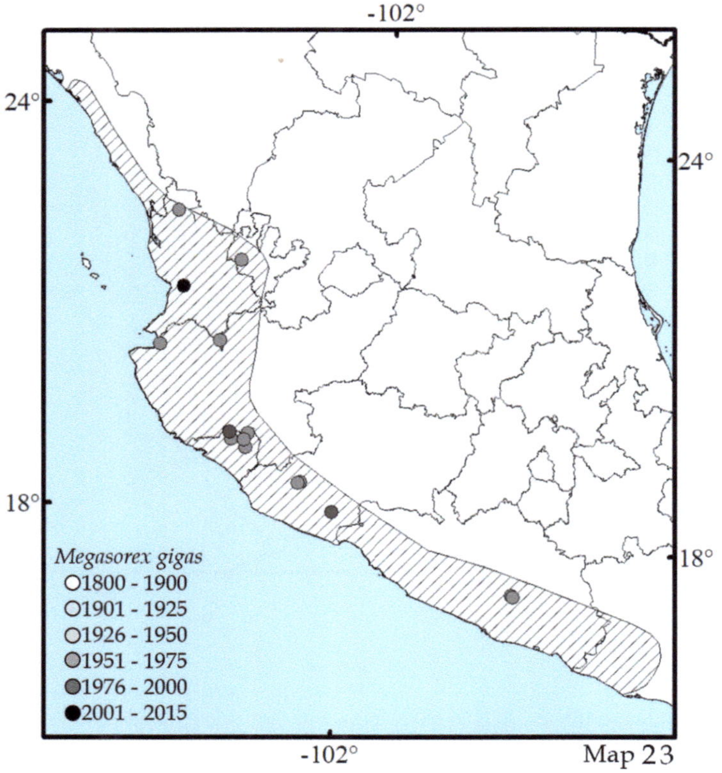

Map 23 Distribution map of *Megasorex gigas*

Additional Literature: Repenning (1967), Armstrong and Jones (1972a), Hall (1981), George (1986), Carraway (2007), Guevara et al. (2015), Camargo et al. (2021).

Genus *Notiosorex* (Coues, 1877)

Dorsal pelage gray in different shades; tail short and ears large; scent glands on the flanks; **three upper unicuspid teeth forming a uniform series, third teeth more than half as large as the second and never minute; unicuspid teeth narrow-based and without an internal cusplet**; anterior teeth pale orange at the tip and molars only white; first incisors and unicuspid teeth slightly pigmented in labial view; **posterior borders of the upper fourth premolars and the first and second molars "excavated" (deeply concave) when viewed from the occlusal surface.** The keys were elaborated based on the review of specimens and the following sources: Merriam (1985b), Hall (1981), Álvarez-Castañeda et al. (2017a), and Camargo and Álvarez-Castañeda (2019).

1. Pelage on the dorsum and abdomen with silvery shades at the tip; glenoid fossa roof not extending laterally in dorsal view of the skull (endemic of eastern Tamaulipas highlands, México) ....*Notiosorex villai* (p. 432)

1a. Pelage on the dorsum whitish at the base and dark grayish brown at the tip, on the venter white on the distal half; glenoid fossa roof extending laterally in dorsal view of the skull 2

2. Zygomatic plate breadth greater than 2.1 mm; combination at the level of the coronoid process usually equal to or greater than 4.5 mm; cranial breadth usually equal to or greater than 8.2 mm (endemic, from Sinaloa south through Michoacán, México).. *Notiosorex evotis* (p. 429)

2a. Zygomatic plate breadth less than 2.1 mm; combination at the level of the coronoid process equal to or less than 4.4 mm; cranial breadth usually equal to or less than 8.5 mm..3

3. Occurs in western Colorado River and the Gulf of California .... *Notiosorex tataticuli* (p. 430)

3a. Occurs in eastern Colorado River and the Gulf of California..4

4. Claw length on the middle forefoot digit less than 1.2 mm and less than 1.55% of the total length; total length usually less than 86.0 mm (restricted to Arizona, United States, and northern Sonora, México)..
....*Notiosorex cockrumi* (p. 428)

4a. Claw length on the middle forefoot digit greater than 1.2 mm and greater than 1.55% of the total length; total length usually greater than 84.0 mm, most individuals greater than 88.0 mm (from western Colorado River east through Arkansas, United States, and south through Ciudad de México)....................................... *Notiosorex crawfordi* (p. 428)

Notiosorex cockrumi Baker et al., 2003

Cockrum's gray shrew, musaraña del desierto de Arizona

Monotypic.

Conservation: *Notiosorex cockrumi* is listed as Least Concern by the IUCN (Timm et al. 2017b).

Characteristics: Cockrum's gray shrew is small but medium sized relative to other shrews; total length 78.0–86.0 mm; condylobasal length 15.7–16.5 mm. Dorsal pelage dark to medium gray; fur of the underparts bright at the tip; tail not bicolored, dorsally of the same color as on the back; **claw length on the middle forefoot digit less than 1.2 mm, less than 1.55% of the total length**; in lateral view the most posterior part of the lower incisors at the same level or behind the back of the first cusp [paraconid] of the first lower molar; breadth of the zygomatic plate ~10% to 13% of the condylobasal length.

Comments: *Notiosorex cockrumi* was described using genetic data (Baker et al. 2003b) and is not morphologically different from *N. crawfordi* (Carraway and Timm 2000). This species ranges from southern Arizona south through central Sonora (Map 24). It can be found in sympatry, or nearly so, with *N. crawfordi*, from which it can be distinguished by having the claw length on the middle forefoot digit less than 1.2 mm and less than 1.55% of the total length; the only clear way of distinguishing it is using genetic data. From *N. evotis*, by its small paroccipital process that does not extend to the ventral edge of the exoccipital in lateral view; smaller size; zygomatic plate breadth less than 2.1 mm; cranial breadth usually equal to or less than 8.5 mm. From *N. villai*, by having a relatively large paroccipital process separated and not extending to the ventral edge of the exoccipital in lateral view.

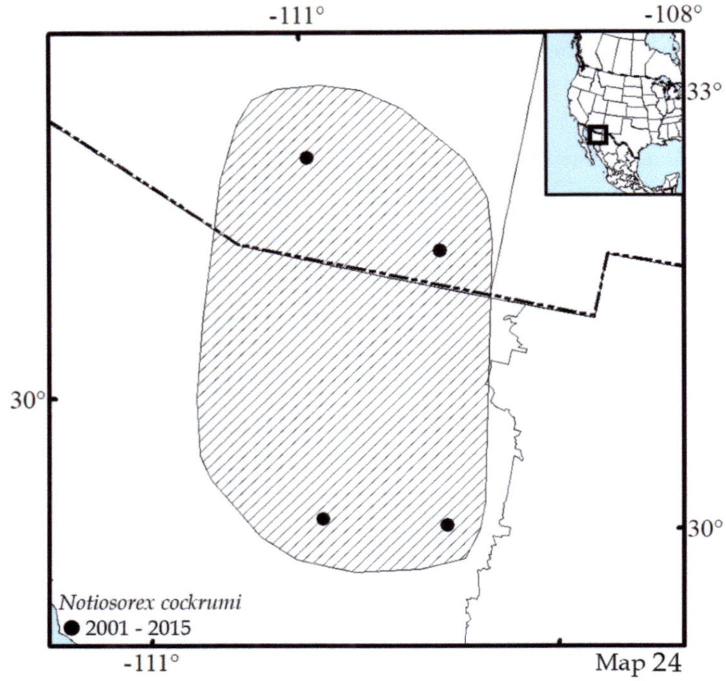

Map 24 Distribution map of *Notiosorex cockrumi*

Additional Literature: Baker and Hsu (1970), Carraway (2007), Baker et al. (2007b), Guevara et al. (2015), Camargo and Álvarez-Castañeda (2020), Camargo et al. (2021).

Notiosorex crawfordi (Coues, 1877)

Crawford's gray shrew, musaraña del desierto

(*see addenda, split in two species *Notiosorex carrawayae*)

Monotypic.

Conservation: *Notiosorex crawfordi* is listed as Threatened in the Norma Oficial Mexicana (DOF 2019) and as Least Concern by the IUCN (Timm et al. 2016k). Own experience in fieldwork shows that this is a very common species that does not need to be included in any risk list.

Characteristics: Crawford's gray shrew is small but medium sized relative to other shrews; total length 84.0–100.0 mm; condylobasal length 14.9–17.2 mm. Dorsal pelage dark to medium gray; fur of the underparts paler at the tip; tail not bicolored, same color as dorsal; in lateral view the most posterior part of the lower incisors at same level or behind the first cusp [paraconid] of the first lower molar; breadth of the zygomatic plate ~10% to 13% of the condylobasal length.

Comments: *Notiosorex crawfordi* and *N. cockrumi* are sympatric and not morphologically different (Carraway and Timm 2000), and only differ using genetic data (Baker et al. 2003b; Camargo and Álvarez-Castañeda 2020). *N. crawfordi* ranges from Utah and Colorado east through Oklahoma and Texas and south through Ciudad de México (Carraway 2007; Map 25). *N. crawfordi* can be found in sympatry, or nearly so, with *N. cockrumi*, from which it can be distinguished by having the claw length on the middle forefoot digit greater than 1.2 mm and greater than 1.55% of the total length. From *N. evotis*, by its small paroccipital process that does not extend to the ventral edge of the exoccipital in lateral view; smaller size; zygomatic plate breadth less than 2.1 mm; cranial breadth usually equal to or less than 8.5 mm. From *N villai*, by a relatively large and separated paroccipital process that does not extend to the ventral edge of the exoccipital in lateral view.

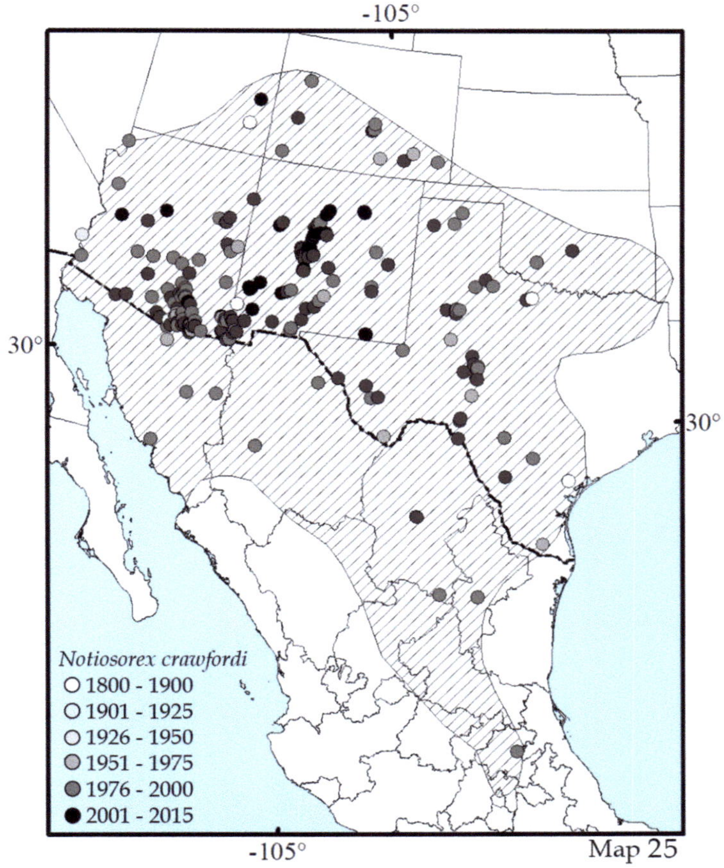

Map 25 Distribution map of *Notiosorex crawfordi*

Additional Literature: Armstrong and Jones (1972b), Guevara et al. (2015), Camargo et al. (2021).

Notiosorex evotis (Coues, 1877)

Large-eared gray shrew, musaraña del desierto del Pacífico

Monotypic.

Conservation: *Notiosorex evotis* is listed as Threatened in the Norma Oficial Mexicana (DOF 2019) and as Least Concern by the IUCN (Castro-Arellano et al. 2016b). Own experience in fieldwork shows that this is a very common species and does not need to be included in any risk list.

Characteristics: Large-eared gray shrew is small but medium sized relative to other shrews; total length 84.0–88.0 mm; condylobasal length 16.4–18.4 mm. Dorsal pelage dark to medium gray; fur of the underparts paler at the tip; tail not bicolored, of the same color as in the dorsum; in lateral view, the most posterior part of the lower incisors at the level of or behind the first cusp [paraconid] of the first lower molars; **breadth of the zygomatic plate greater than 2.1 mm, ~13% to 15% of the condylobasal length; low-set paroccipital process extending in an oblique angle from the skull to the ventral edge of the exoccipital in lateral view**.

Comments: *Notiosorex evotis* was recognized as a distinct species (Jones et al. 1962; Camargo et al. 2022), previously regarded as a subspecies of *N. crawfordi* (Merriam 1895b). It ranges from Sinaloa south through northern Michoacán at the limit with Jalisco (Carraway and Timm 2000; Map 26) and has allopatric distribution with respect to other species of *Notiosorex*; however, it can be distinguished from *N. cockrumi* and *N. crawfordi* by its larger size, the zygomatic plate breadth greater than 2.1 mm, and cranial breadth usually greater than 8.5 mm. From *N. villai*, by having the roof of the glenoid fossa extending laterally from the cranium in dorsal view. From all other shrews, by having three unicuspid teeth in the upper toothrow with a pale pigment and a scent gland on the flanks; low-set paroccipital process extending at an oblique angle from the skull to the ventral edge of the exoccipital in lateral view.

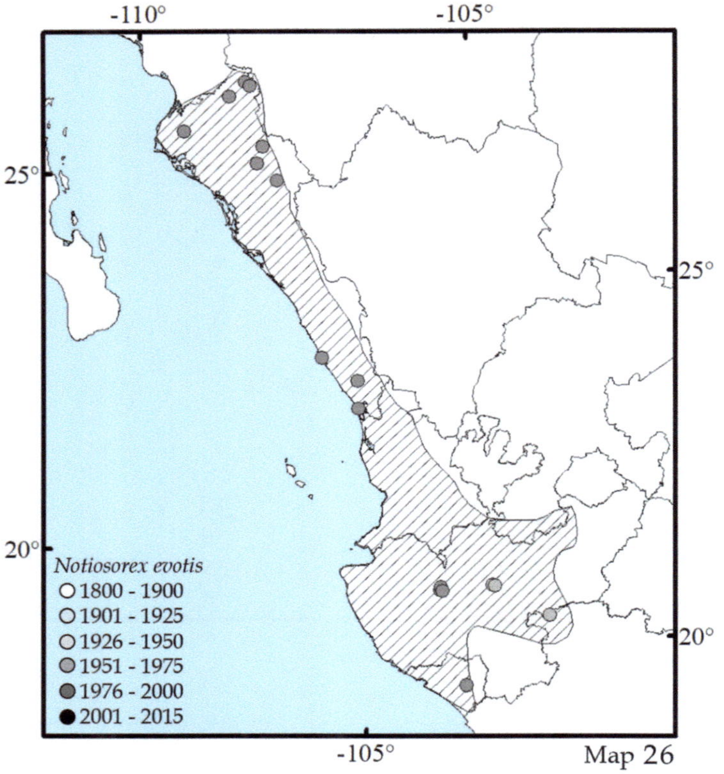

Map 26 Distribution map of *Notiosorex evotis*

Additional Literature: Carraway (2007), Guevara et al. (2015), Camargo and Álvarez-Castañeda (2020).

Notiosorex tataticuli Camargo and Álvarez-Castañeda, 2020

Ticul's gray shrew, musaraña de las Californias

1. *N. t. arroyoi* Camargo and Álvarez-Castañeda, 2020. Restricted to San Martín Island, Baja California.
2. *N. t. ocanai* Camargo and Álvarez-Castañeda, 2020. From southern-central California and southwestern Nevada south through Baja California.
3. *N. t. tataticuli* Camargo and Álvarez-Castañeda, 2020. Restricted to the Baja California Sur.

Conservation: *Notiosorex tataticuli* (as *N. crawfordi*) is listed as Threatened in the Norma Oficial Mexicana (DOF 2019) and as Least Concern by the IUCN (Timm et al. 2016k). Own experience in fieldwork shows that this is a very common species that does not need to be included in any risk list.

Characteristics: Ticul's gray shrew is small but medium sized relative to other shrews; total length 85.0–101.0 mm; condylobasal length 15.5–17.9 mm. Dorsal pelage pale gray to dark gray; fur of the underparts paler; tail not bicolored, of the same color as in the dorsum; in lateral view, the most posterior part of the lower incisors at the same level or behind the back of the first cusp [paraconid] of the first lower molars; breadth of the zygomatic plate ~10% to 13% of the condylobasal length.

Comments: *Notiosorex tataticuli* was previously included in the ranges of *N. crawfordi* (Baker et al. 2003b; Carraway 2007), but a genetic analysis shows that the populations west to the Colorado River were different from those to the east (Baker et al. 2003b; Camargo and Álvarez-Castañeda 2020). *N. tataticuli* is known from west through the Colorado River in Nevada and California south through the Baja California Peninsula (Map 27). It has allopatric distribution with respect to other species of *Notiosorex*; however, it can be distinguished from other shrews by having three unicuspid teeth in the upper toothrow with a light pigment and a scent gland on the flanks.

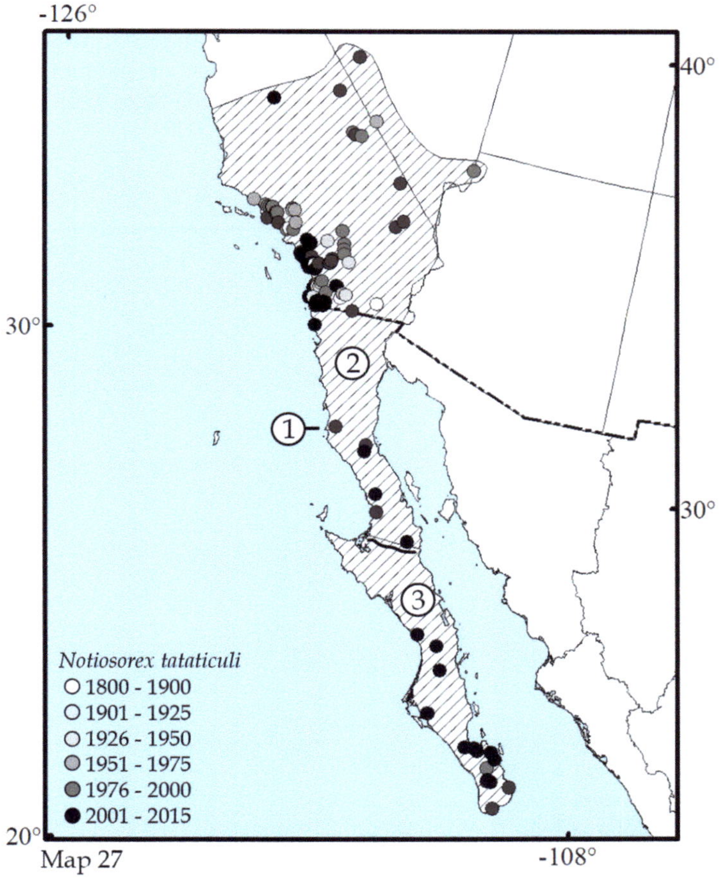

Map 27 Distribution map of *Notiosorex tataticuli*

Additional Literature: Hall (1981), Manning (2014).

Notiosorex villai Carraway and Timm, 2000

Villa's gray shrew, musaraña de Tamaulipas

Monotypic.

Conservation: *Notiosorex villai* is listed as Threatened in the Norma Oficial Mexicana (DOF 2019) and as Vulnerable (B1ab (iii): extent of occurrence less than 20,000 km^2 severely fragmented and under continuing decline in area, extent and quality of habitat) by the IUCN (Timm and Matson 2018).

Characteristics: Villa's gray shrew is small but medium sized relative to other shrews; total length 88.0–90.0 mm; condylobasal length 16.9–17.2 mm. Dorsal pelage medium to dark gray with **silvery shades**; fur of the underparts paler at the tip; a very dark gray broad stripe on the back; tail not bicolored, of the same color as the back; in lateral view, the most posterior part of the lower incisors at the level of or behind the first cusp [paraconid] of the first lower molars; **narrow zygomatic plate, ~8% of the condylobasal length; roof of the glenoid fossa not extending laterally from the cranium when the skull is viewed dorsally; small paroccipital process lying against and extending to the ventral edge of the exoccipital in lateral view**.

Comments: *Notiosorex villai* is known only from southwestern Tamaulipas in all the western dry area of the Sierra Madre Oriental (Carraway and Timm 2000; Map 28). It has allopatric distribution with respect to other species of *Notiosorex*; however, it can be distinguished by its small paroccipital process lying against and extending to the ventral exoccipital edge in lateral view. From other shrews, by having three unicuspid teeth in the upper toothrow with a light pigment and a scent gland on the flanks.

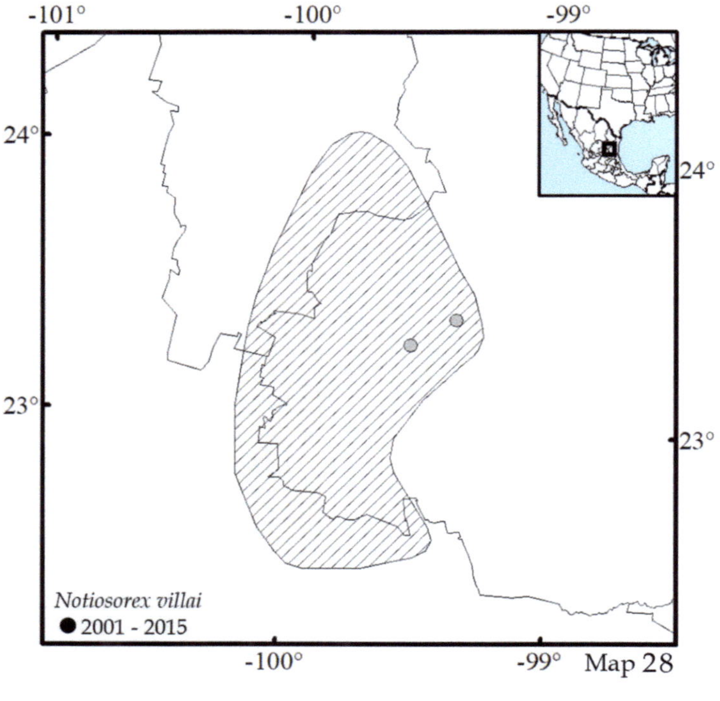

Map 28 Distribution map of *Notiosorex villai*

Additional Literature: Carraway (2007), Guevara et al. (2015).

Tribe Soricini

Genus *Sorex* Linnaeus, 1758

Dorsal pelage brown in different shades; **feet relatively long and narrow; tail length greater than 50% of the head-and-body length; five upper unicuspid teeth**. The assignment of the subgenera *Sorex* and *Otiosorex* follows Findley (1955b), Diersing and Hoffmeister (1977), Junge and Hoffman (1981), and George (1988). *S. cristobalensis*, *S. mediopua*, *S. ventralis*, and *S. veraecrucis* are included in the *saussurei* species group; given that *S. saussurei* is in the subgenus *Sorex*, all the species have been included in this subgenus, although specific analyses need to be made. Species within the subgenus *Otiosorex*: *S. albibarbis*, *S. bendirii*, *S. cinereus*, *S. dispar*, *S. emarginatus*, *S. fumeus*, *S. haydeni*, *S. hoyi*, *S. ixtlanensis*, *S. jacksoni*, *S. longirostris*, *S. lyelli*, *S. milleri*, *S. monticolus*, *S. nanus*, *S. navigator*, *S. obscurus*, *S. oreopolus*, *S. orizabae*, *S. ornatus*, *S. pacificus*, *S. palustris*, *S. preblei*, *S. pribilofensis*, *S. rohweri*, *S. sonomae*, *S. tenellus*, *S. ugyunak*, and *S. vagrans*. Species within the subgenus *Sorex*: *S. arcticus*, *S. maritimensis*, *S. minutissimus*, and *S. tundrensis*. *Incertae sedis* are: *S. arizonae*, *S. altoensis*, *S. chiapensis*, *S. cristobalensis*, *S. macrodon*, *S. mediopua*, *S. merriami*, *S. mutabilis*, *S. salvini*, *S. saussurei*, *S. sclateri*, *S. stizodon*, *S. trowbridgii*, *S. veraecrucis*, and *S. ventralis*. The definition of the subgenera *Otiosorex* and *Sorex* was based on morphological characteristics. However, phylogenetic analyses (George 1988) showed that the morphological characteristics are not in accordance with the lineage within the genus *Sorex*; for this reason, the species of these two subgenera are combined in the key. In it, the subgenus of each species is given between parentheses before the species name. Note: Several authors mention that the genus *Sorex* should be reviewed, particularly in the tropical region of North America. The current nomenclature is used herein, although it is highly likely that it will change in the near future. The keys were elaborated based on the review of specimens and the following sources: Jackson (1928), Hall (1981), Carraway (1990), Álvarez-Castañeda et al. (2017a), and Woodman (2018).

1. Third unicuspid teeth equal to or larger than the fourth; upper teeth weakly pigmented; ridges extending from the apices of the unicuspid teeth medially toward the cingulum but incomplete and not ending in internal cusplets; postmandibular foramen usually well developed 2

1a. Third unicuspid teeth smaller than the fourth; upper teeth usually pigmented; ridges extending from the cingulum apices uninterrupted by an anteroposterior groove; postmandibular foramen absent in most species 21

2. Occurs south of 30° N in México 3

2a. Occurs north of 30°N in northwestern Baja California, United States, and Canada 9

3. Total length equal to or greater than 120.0 mm; tail length greater than 50.0 mm; forefoot length greater than 13.0 mm; condylobasal length greater than 18.0 mm; cranial breadth greater than 8.3 mm; maxillary breadth equal to or greater than 5.5 mm; distance from the first upper unicuspid teeth to the third upper molars equal to or greater than 7.0 mm (endemic to Chiapas, México) .... *Sorex* (*incertae sedis*) *sclateri* (p. 492)

3a. Total length less than 120.0 mm; tail length less than 50.0 mm; forefoot length equal to or less than 13.0 mm; condylobasal length less than 18.0 mm; cranial breadth equal to or less than 8.3 mm; maxillary breadth equal to or less than 5.5 mm; distance from the first upper unicuspid teeth to the third upper molars equal to or less than 7.0 mm 4

4. Occurs from Eje Neovolcánico Transversal northward 5

4a. Occurs from Eje Neovolcánico Transversal southward 7

5. Maxillary breadth less than 4.5 mm; breadth across the second upper molars less than 4.0 mm; distance from the first upper unicuspid teeth to the third upper molars less than 4.0 mm (endemic to Coahuila and Nuevo León, México) .... *Sorex* (*Otiosorex*) *milleri* (part; p. 455)

5a. Maxillary breadth greater than 4.5 mm; breadth across second upper molars greater than 4.0 mm; distance from the first upper unicuspid teeth to the third upper molars greater than 4.0 mm 6

6. Dorsal hairs with a medial wide blond band and a narrow dark brown band at the tip, giving the pelage an overall medium brown appearance; fur of the underparts white at the tip; interdenticular space between the first lower incisors deep (from southeastern Arizona, southwestern New Mexico, Sonora, and Chihuahua, México) .... *Sorex* (*incertae sedis*) *arizonae* (part; p. 481)

6a. Dorsal hairs with a medial narrow blond band and a narrow dark brown band at the tip, giving the pelage an overall medium-dark brown appearance; fur of the underparts blond at the tip; interdenticular space between the first lower incisors shallow (restricted to Durango, Zacatecas, and northern Jalisco, México) .... *Sorex* (*Otiosorex*) *emarginatus* (p. 446)

7. Only known from San Cristóbal de las Casas, Chiapas, México .... *Sorex* (*incertae sedis*) *stizodon* (p. 493)

7a. Occurs in the Eje Neovolcánico 8

8. Fur with a band of hair pale reddish white medially and dark brown at the tip; hips and rump with guard hairs uniformly dark, except for scattered white hairs on the flanks; venter hair pale red at the tip; interorbital region markedly elongated (Eje Neovolcánico, from eastern Jalisco east through Puebla, México) .... *Sorex* (*Otiosorex*) *oreopolus* (p. 462)

8a. Pelage on the dorsum blond medially and dark brown at the tip; hips and rump with brown guard hairs extending 1.0–1.5 mm beyond the dorsal pelage; venter hair with the proximal two-thirds silvery gray and the distal third white; interorbital constriction broad 9

9. Total length less than 76.0 mm; condylobasal length less than 13.6 (restricted to central Alaska, United States) .... *Sorex* (*Sorex*) *minutissimus* (p. 479)

9. Total length greater than 76.0 mm; condylobasal length greater than 13.6 mm 10

10. Posterior border of the infraorbital foramen posterior to the plane of the space between the first and second upper molars (from the highlands of the Appalachian Mountains in Maine south through North Carolina and New Brunswick, United States) .... *Sorex* (*Otiosorex*) *dispar* (p. 444)

10a. Posterior border of the infraorbital foramen even with, or anterior to, the plane of the space between the first and second upper molars 11

11. Restricted to southeastern Arizona and southwestern New Mexico, United States, and Sonora and Chihuahua, México .... *Sorex* (incertae sedis) *arizonae* (part; p. 481)

11a. Occurs throughout Canada and the United States 12

12. Tail length equal to or less than 30% of the total head-and-body length; condylobasal length on average greater than 17.0 mm; ranges in Canada, Alaska, and northeastern United States 13

12a. Tail length greater than 30% of the total head-and-body length; condylobasal length less than 17.0 mm; ranges throughout Canada and the United States 15

13. No stripe of pigment on the inside at the center of the lower edge of the first lower incisors; if two branch of the mandibular are attached, this characteristic cannot be seen (from Yukon and the Northwest Territories west through Québec, Canada, and south through South Dakota, Minnesota, Michigan, and Wisconsin, United States) .... *Sorex* (*Sorex*) *arcticus* (p. 477)

13a. Stripe of pigment on the inside at the center of the lower edge of the first lower incisors; if two branch of the mandibular are attached, this characteristic cannot be seen 14

14. More than 25% of the labial side of the first lower incisors pigmented; height of the coronoid process equal to or less than 4.2 mm; mandible length equal to or less than 7.9 mm; mandibular branch narrow, molars wide; height of the coronoid process from the first lower molars to the ventral edge of the mandible less than 5.4 mm (from Yukon and the Northwest Territories, Canada, west through Alaska, United States) .... *Sorex* (*Sorex*) *tundrensis* (p. 480)

14a. Less than 25% of the labial side of the first lower incisors pigmented; height of the coronoid process equal to or greater than 4.2 mm; mandible length equal to or greater than 7.9 mm; mandibular branch wide, molars narrow; height from the coronoid process of the first lower molars to the ventral edge of the mandible greater than 5.4 mm (restricted to Nova Scotia and southern New Brunswick, United States) .... *Sorex* (*Sorex*) *maritimensis* (p. 478)

15. Maxillary breadth greater than 4.6 mm; fourth unicuspid teeth wider than large; lingual face of the unicuspid teeth often lacking pigment; skull flattened to some degree (from Indiana, Tennessee, and Georgia northeast through New Brunswick, United States, and southern Québec and Ontario, Canada) .... *Sorex* (*Otiosorex*) *fumeus* (p. 448)

15a. Maxillary breadth less than 4.6 mm; fourth unicuspid teeth more or less similar in length and width; lingual face of the unicuspid teeth often pigmented; skull inflated 16

16. Lower toothrow length less than 4.1 mm (from Washington and western Wyoming west through California and south through northern Utah and Nevada, United States) .... *Sorex* (*Otiosorex*) *preblei* (p. 469)

16a. Lower toothrow length greater than 4.1 mm........ 17

17. Only occurs in tundra areas of north Alaska and Canada........ *Sorex* (*Otiosorex*) *ugyunak* (p. 474)

17a. Occurs south of 60°........ 18

18. Combined length of the two first molars equal to or greater than 2.4 mm (from Arizona and New Mexico north through Washington and Montana, and from California east through Nebraska and the Dakotas, United States) .... *Sorex* (*incertae sedis*) *merriami* (part; p. 488)

18a. Combined length of the two first molars equal to or less than 2.4 mm 19

19. Small sized; maxillary toothrow length less than 5.2 mm (from Alberta and Manitoba south through Kansas and Missouri, in association with the Great Plains, United States)........ *Sorex* (*Otiosorex*) *haydeni* (p. 449)

19a. Larger sized; maxillary toothrow length greater than 5.2 mm........ 20

20. Only occurs in Sierra Nevada, California, United States........ *Sorex* (*Otiosorex*) *lyelli* (p. 454)

20a. Occurs from Alaska east through Newfoundland and south through Washington and Utah, in all the northern states of the United States and the Appalachian Mountains south through northern Georgia, Tennessee, and western South Carolina, United States .... *Sorex* (*Otiosorex*) *cinereus* (p. 443)

21. In lingual view of the mandible, the most posterior margin of the incisors at the same level or posterior to the paraconid of the first lower molars; third unicuspid teeth disk shaped, markedly compressed anteroposteriorly; fifth unicuspid teeth minute; first and second unicuspid teeth with a distinct internal ridge with an internal cusp; accessory medial tine on the anterior surface of the first upper incisors relatively large and long (Fig. 11)........ 22

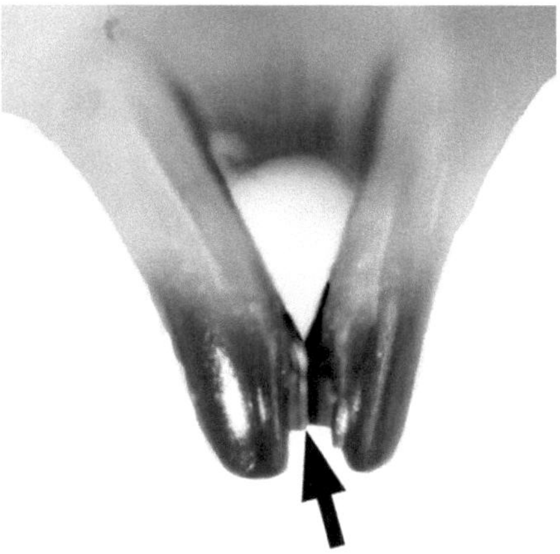

Fig. 11 Accessory medial tine on the anterior surface of the first upper incisors relatively large and long

21a. In lingual view of the mandible, the most posterior margin of the incisors anterior to the paraconid of the first lower molars; third unicuspid teeth similar to the others; fifth unicuspid teeth can be small but not minute; first and second unicuspid teeth without a distinct internal ridge with an internal cusp; accessory medial tine on the anterior surface of the first upper incisors if present, shorter and not as above (Fig. 60 in chapter "Order Chiroptera Blumenbach, 1779") 23

22. Pelage coloration somewhat more reddish brown; condylobasal length on average less than 14.3 mm. (from Mackenzie, eastern Manitoba, North and South Dakota east through the Labrador Peninsula and New England, excluding the northern tundra zones and including southern Indiana, Kentucky, western Tennessee, and northern Georgia, United States)........ *Sorex* (*Otiosorex*) *hoyi* (p. 450)

22a. Pelage coloration somewhat less reddish brown; condylobasal length on average greater than 14.3 mm (from Manitoba and North and South Dakota west through Alaska and British Columbia, excluding the northern tundra zones, with disjunct populations in the southern Rockies, United States) .... *Sorex* (*Otiosorex*) *eximius* (p. 447)

23. Pelage tricolor; ranges restricted to the islands in the Bering Sea, Alaska 24

23a. Pelage bicolored or monocolor; no restricted ranges to the islands in the Bering Sea, Alaska 25

24. Main axis of the first lower incisors at an angle of 8° or greater horizontally from the mandible branch (restricted to Saint Paul Island, Bering Sea, Alaska, United States) .... *Sorex* (*Otiosorex*) *pribilofensis* (p. 470)

24a. Main axis of the first lower incisors at an angle of 8° or less horizontally from the mandible branch (restricted to Saint Lawrence Island, Bering Sea, Alaska, United States) .... *Sorex* (*Otiosorex*) *jacksoni* (p. 452)

25. Occurs in México 26

25a. Occurs in northwestern Baja California, United States, and Canada 42

26. Condylobasal length on average less than 17.9 mm; distance from the first upper unicuspid teeth to the third upper molars less than 6.5 mm 27

26a. Condylobasal length on average greater than 17.9 mm; distance from the first upper unicuspid teeth to the third upper molars greater than 6.5 mm 33

27. First upper unicuspid teeth with a medial crest usually pigmented above; pigment always separated 28

27a. First upper unicuspid teeth with a medial crest included within the tooth pigment; pigment in a long stripe at the anteromedial edge 29

28. Belly coloration pale; skull not flattened; first lower incisors with pigment in two sections (from Michoacán east through western Veracruz, México) .... *Sorex* (*Otiosorex*) *orizabae* (p. 463)

28a. Belly coloration lighter but not pale; skull flattened; first lower incisors with pigment in one section (from western California, United States, south through Baja California Peninsula, México) .... *Sorex* (*Otiosorex*) *ornatus* (part; p. 464)

29. Dorsal pelage blond or pale red; hips and rump with brown to dark brown guard hairs greater than 0.75 mm beyond the dorsal pelage; maxillary breadth greater than 4.95 mm; first lower incisors with two denticles and shallow interdenticular spaces 30

29a. Dorsal pelage pale brown; hips and rump with dark brown guard hairs extending 0.33 mm beyond the dorsal pelage; maxillary breadth equal to or less than 4.95 mm; first lower incisors with three denticles and deep interdenticular spaces 32

30. Pelage on the dorsum pale red medially, dark brown at the tip; hips and rump with dark brown guard hairs extending 0.75 mm beyond the dorsal pelage; hair on the venter proximal two-thirds dark gray and the distal third blond; tail distinctly bicolored (endemic to the Eje Neovolcánico and highlands of Veracruz, Puebla, and Oaxaca, México) .... *Sorex* (*incertae sedis*) *ventralis* (p. 497)

30a. Pelage on the dorsum blond medially, dark brown at the tip; hips and rump with brown guard hairs extending 1.0–1.5 mm beyond the dorsal pelage; hair on the venter with the proximal two-thirds silvery gray medially and the distal third white; tail uniform in color or slightly bicolored 31

31. Hips and rump with dark brown guard hairs extending less than 1.5 mm beyond the dorsal pelage, not pale at the tip; first lower incisors with a shallow interdenticular space (from Jalisco east through Puebla, México) .... *Sorex* (*incertae sedis*) *saussurei* (p. 491)

31a. Hips and rump with dark brown guard hairs extending greater than 1.5 mm beyond the dorsal pelage, pale at the tip; first lower incisors with a deep interdenticular space (restricted to the highlands of Oaxaca, México) .... *Sorex* (*incertae sedis*) *salvini* (p. 490)

32. Maxillary breadth less than 4.5 mm; breadth across the second upper molars less than 4.0 mm; distance from the first upper unicuspid teeth to the third upper molars less than 4.0 mm; distance from the lower canines to the third lower molars less than 4.5 mm; height of the coronoid process less than 3.5 mm (endemic to Coahuila and Nuevo León, México) .... *Sorex* (*Otiosorex*) *milleri* (part; p. 455)

32a. Maxillary breadth greater than 4.5 mm; breadth across the second upper molars greater than 4.0 mm; distance from the first upper unicuspid teeth to the third upper molars greater than 4.0 mm; distance from the lower canines to the third lower molars greater than 4.5 mm; height of the coronoid process greater than 3.5 mm..57

33. Tail slightly bicolored; first upper incisors with a medial tine above the pigment (from the highlands of Chiapas, México, and western Guatemala)..*Sorex* (*incertae sedis*) *chiapensis* (p. 484)

33a. Tail uniform in color, same as the dorsal pelage; first upper incisors with or without a medial tine above the pigment34

34. Condylobasal length greater than 19.0 mm; cranial breadth greater than 9.6 mm; first lower incisors with three denticles and long stripe of pigment not present at the anteromedial edge 35

34a. Condylobasal length less than 19.0 mm; cranial breadth usually less than 9.6 mm; first lower incisors with two or three denticles and with or without a long stripe of pigment at the anteromedial edge 36

35. Upperparts with fur very dark brown; fur of the underparts only slightly paler than on the back; hips and rump with dark brown guard hairs extending less than 2.0 mm beyond the dorsal pelage; first upper incisors with a medial tine above the pigment (endemic to the boundaries of Veracruz, Oaxaca, and Puebla, México)... .... *Sorex* (*incertae sedis*) *macrodon* (p. 486)

35a. Upperparts with fur reddish blond and dark brown at the tip; fur of the underparts paler reddish brown; hips and rump with pale reddish blond and dark brown guard hairs extending greater than 1.8 mm beyond the dorsal pelage; first upper incisors with a medial tine within the pigmented band (from highland of Guerrero and Oaxaca, México) ....*Sorex* (*incertae sedis*) *mutabilis* (p. 489)

36. Cranial breadth greater than 9.0 mm...37

36a. Cranial breadth less than 9.0 mm..40

37. First upper incisors with a medial tine at the interface of the pigmented and unpigmented areas; first lower incisors with deep interdenticular spaces (endemic to Jalisco, Michoacán, and Estado de México, México) ....*Sorex* (*incertae sedis*) *mediopua* (p. 487)

37a. First upper incisors with a medial tine above the pigment 38

38. Tail length less than 50.0 mm; first lower incisors with deep or shallow interdenticular spaces and pigmented in one or two sections...39

38a. Tail length greater than 50.0 mm; first lower incisors with deep interdenticular spaces and pigmented in one section 41

39. Hips and rump with brown guard hairs extending greater than 1.5 mm beyond the dorsal pelage; hind foot length less than 14.0 mm; skull breadth less than 8.6 mm; first lower incisors with deep interdenticular spaces and pigmented in two sections (restricted to the highlands of central and Pacific coast of Chiapas, México, and western Guatemala)........ .... *Sorex* (*incertae sedis*) *cristobalensis* (p. 485)

39a. Hips and rump with brown guard hairs extending less than 1.5 mm beyond the dorsal pelage; hind foot length equal to or greater than 14.0 mm; skull breadth equal to or greater than 8.6 mm; first lower incisors with shallow interdenticular spaces and pigmented in one section 41

40. Fur of the upperparts reddish blond medially and dark brown at the tip; hips and rump with pale reddish blond and dark brown guard hairs extending greater than 1.9 mm beyond the dorsal pelage; first lower incisors with three denticles (endemic to Guerrero and Oaxaca, México)...*Sorex* (*Otiosorex*) *ixtlanensis* (p. 451)

40a. Fur of the upperparts dark gray at the base, medially pale brown, and white at the tip, giving the pelage an overall medium-dark reddish brown appearance; hips and rump with paler brown guard hairs extending less than 1.9 mm beyond the dorsal pelage; first lower incisors with two denticles...41

41. Tail length greater than 50.0 mm (from the Tehuantepec Isthmus south through Chiapas, México) ....*Sorex* (*incertae sedis*) *veraecrucis* (p. 496)

41a. Tail length less than 50.0 mm (from the Tehuantepec Isthmus north through Coahuila and Durango, México)..*Sorex* (incertae sedis) *altoensis* (p. 482)

42. Dorsal pelage grayish, never brownish; hind foot length greater than 18.0 mm; condylobasal length greater than 19.0 mm 43

42a. Dorsal pelage always brownish; hind foot length less than 18.0 mm; condylobasal length less than 19.0 mm............46

43. Hind feet slightly fimbriated; rostrum long and markedly curved downward; rostrum vertical depth at the level of the third unicuspid teeth equal to approximately one-half of the distance from the anterior border of the infraorbital foramen to the posterior border of the first incisors (from southern British Columbia south through northwestern California, United States).. *Sorex (Otiosorex) bendirii* (p. 441)

43a. Hind feet distinctly fimbriated; rostrum short and slightly curved downward; rostrum vertical depth at the level of the third unicuspid teeth less than one-half of the distance from the anterior border of the infraorbital foramen to the posterior border of the first incisors 44

44. Smaller average size, upperparts with paler pelage; unicuspid teeth much narrower (from Alaska, Yukon, and Mackenzie south through California, Arizona, and New Mexico, including Vancouver Island) ....*Sorex (Otiosorex) navigator* (p. 459)

44a. Larger average size, upperparts with darker pelage; unicuspid teeth much broader (Great Plains of Canada and United States)..45

45. Fur of the underparts white, in strong contrast with the back pelage; chin not paler than the belly; tail sharply bicolored, darker dorsally, contrasting with the white fur of the underparts (from Mackenzie and Alberta east through Ontario and the Great Lakes. In the western south through mid-California, Nevada, Utah, and New Mexico, with one population in the White Mountains, Arizona, United States) .... *Sorex (Otiosorex) palustris* (p. 458)

45a. Fur of underparts pale gray; chin paler than the belly; tail bicolored, blackish dorsally and near the tip, whitish ventrally from one-half to two-thirds of its length (restricted to New England of Saint Lawrence River and Great Lakes, United States)..*Sorex (Otiosorex) albibarbis* (p. 440)

46. Usually no postmandibular foramen (Fig. 12) or, if present, small and often only on one ramus of the mandible; unicuspid teeth with a ridge, usually pigmented on the lingual face running from the apex to the cingulum, sometimes ending in a pigmented cusplet 47

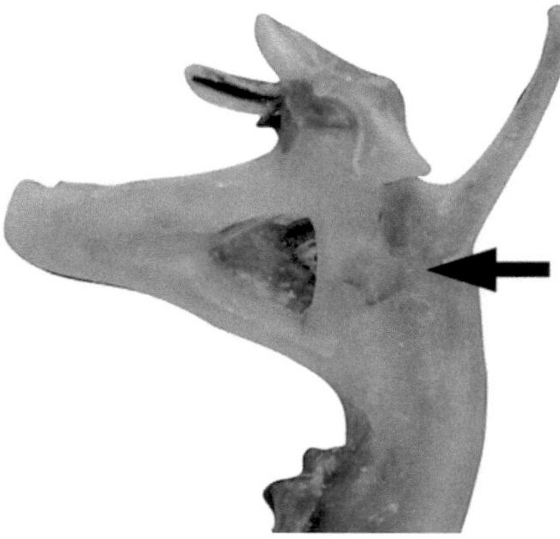

Fig. 12 Usually no postmandibular foramen

46a. Usually well-developed postmandibular foramen (Fig. 13); unicuspid teeth without a pigmented ridge on the lingual face 57

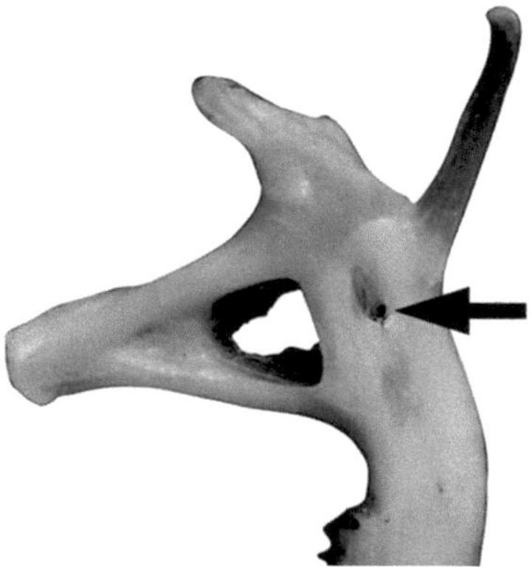

Fig. 13 Usually well-developed postmandibular foramen

47. Lower toothrow length less than 4.1 mm 48

47a. Lower toothrow length greater than 4.1 mm 49

48. Pigment of the first lower incisors in one or two continuous spots (from Montana east through South Dakota and south through Arizona and New Mexico, United States) .... *Sorex* (*Otiosorex*) *nanus* (p. 458)

48a. Pigment of the first lower incisors in three spots (from Illinois and Oklahoma south through Louisiana east through the Atlantic coast of Maryland to Florida, United States) .... *Sorex* (*Otiosorex*) *longirostris* (p. 453)

49. Combined length of the two first molars equal to or less than 2.4 mm 50

49a. Combined length of the two first molars equal to or greater than 2.4 mm 51

50. Pigment of the first lower incisors in one continuous spot (from eastern-central California and western-central Nevada, United States) .... *Sorex* (*Otiosorex*) *tenellus* (p. 473)

50a. Pigment of the first lower incisors not in one continuous spot but in two or three (restricted to western Washington and southwestern British Columbia, United States) .... *Sorex* (*Otiosorex*) *rohweri* (p. 471)

51. Foramen magnum plate relatively dorsal, with greater area into the supraoccipital and less in the basioccipital; skull flattened (western California, United States, south through Baja California Peninsula, México) .... *Sorex* (*Otiosorex*) *ornatus* (part; p. 464)

51a. Foramen magnum plate relatively ventral, with less area into the supraoccipital and greater in the basioccipital; skull not flattened 52

52. Medial tine begins mainly dorsally to the pigmented area on the first upper incisors 53

52a. Medial tine contained entirely within the pigmented area of the first upper incisors 55

53. Occurs in western Canada and the United States but not in western California and Oregon (from southern British Columbia and southwestern Alberta south through the mountains of the central California and east through western Wyoming, United States) .... *Sorex* (*Otiosorex*) *vagrans* (p. 475)

53. Occurs in western California and Oregon, south of the Columbia River 54

54. Lower canine length equal to or smaller than the premolar length (from southern Oregonto San Francisco Bay, California, United States) .... *Sorex* (*Otiosorex*) *sonomae* (p. 472)

54a. Lower canine length smaller than the premolar length (from Langara, Alaska south through northwestern California, United States) .... *Sorex* (*Otiosorex*) *pacificus* (p. 466)

55. Dorsal pelage darker; fur of the underparts brownish suffusion; longer unicuspid toothrow and broader rostrum at the level of the first unicuspid tooth (from California, Nevada, Utah, and Colorado south through Durango, México and from Montana to New Mexico, United States) .... *Sorex* (*Otiosorex*) *monticolus* (part; p. 425)

55a. This key is applicable only for specimens from New Mexico; dorsal pelage pale; fur of the underparts gray; unicuspid toothrow short and rostrum narrow at the level of the first unicuspid teeth...56

56. Total and tail lengths usually less than 110.0 mm and 44.0 mm, respectively; protoconid of the first lower molars comparatively lower (from highlands of southeastern Arizona, southwestern New Mexico, Sonora, and Chihuahua, México) ....*Sorex* (incertae sedis) *arizonae* (part; p. 481)

56a. Total and tail lengths usually greater than 110.0 mm and 44.0 mm, respectively; protoconid of the first lower molars comparatively higher (from Alaska and western Canada south through Utah and New Mexico, United States) ....*Sorex* (*Otiosorex*) *obscurus* (p. 460)

57. Third upper unicuspid teeth usually distinctly smaller than the fourth, sometimes subequal (from southern British Columbia south through north of Los Angeles, California, and east through western Nevada, United States)................ .... *Sorex* (*incertae sedis*) *trowbridgii* (p. 494)

57. Fourth upper unicuspid teeth usually distinctly smaller than the third unicuspid teeth, sometimes subequal 58

58. Medial tine on the anterior face of the first upper incisors; palate less broad; from Gila River southward (from highlands of southeastern Arizona, southwestern New Mexico, United States, Sonora, and Chihuahua, México) ....*Sorex* (incertae sedis) *arizonae* (part; p. 481)

58. No medial tine on the anterior face of the first upper incisors; palate broad; from Gila River northward (from Arizona and New Mexico north through Washington and Montana and from California east through Nebraska and the Dakotas, United States).. *Sorex* (*incertae sedis*) *merriami* (part; p. 488)

Subgenus *Otiosorex* De kay, 1842

Third unicuspid teeth smaller than the fourth; upper unicuspid teeth usually with pigmented ridges extending from the apices of the cingulum, uninterrupted by an anteroposterior groove; postmandibular foramen absent in most species (Findley 1955b; Diersing and Hoffmeister 1977; Junge and Hoffman 1981).

Sorex albibarbis (Cope, 1862)

White-bearded water shrew, musaraña de agua de barba blanca

Monotypic.

Conservation: *Sorex albibarbis* (as *S. palustris*) is listed as Least Concern by the IUCN (Cassola 2016do).

Characteristics: White-bearded water shrew is small sized but large among *Sorex* species in North America; total length 147.2–157.8 mm; condylobasal length 18.4–19.2 mm. Dorsal pelage blackish slate with very sparingly mixed hair paler at the tip; fur of the underparts pale gray with a dusky coloration, varying geographically; **chin whitish or grayish; tail bicolored, blackish dorsally and near the tip, whitish ventrally from one-half to two-thirds of the tail length**; feet large; hind feet very long, broad for swimming; toes slightly webbed; skull similar to *S. palustris*, but anterior unicuspid teeth narrower and molars less deeply excavated posteriorly.

Comments: *Sorex albibarbis albibarbis*, *S. a. gloveralleni*, *S. a. labradorensis*, *S. a. punctulatus*, and *S. a. turneri* were considered subspecies of *S. palustris*, but genetic analyses showed that they can be considered a different species, *S. albibarbis* (Mycroft et al. 2011; Woodman 2018). *S. albibarbis* ranges in New England south of the Saint Lawrence River and the Great Lakes (Map 29). It can be found in sympatry, or nearly so, with *S. cinereus*, *S. hoyi*, and *S. fumeus*. *S. albibarbis* can be distinguished from all these species by having the chin whitish or grayish; tail bicolored, blackish dorsally and near the tip, whitish ventrally from one-half to two-thirds of the tail length. Except for *S. palustris*, by having the inner side of the unicuspid teeth without a secondary cusp; four upper incisors well developed; braincase broad; feet conspicuously fringed with bristle-like hairs.

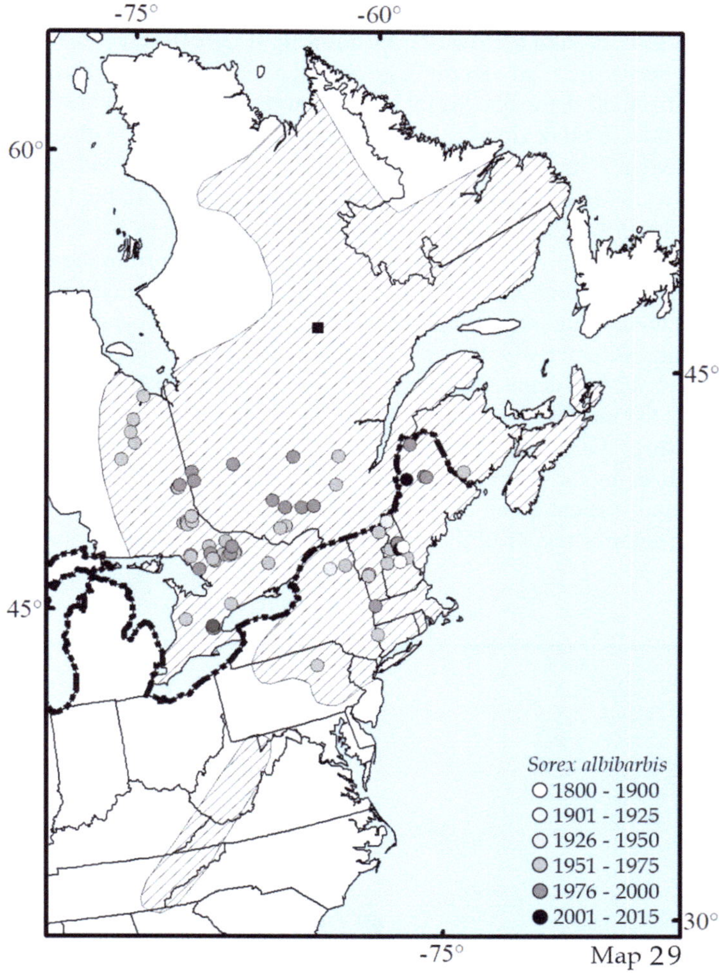

Map 29 Distribution map of *Sorex albibarbis*

Additional Literature: Merriam Allen (1915a, b), Jackson (1925, 1928), Findley (1955a), Junge and Hoffman (1981), van Zyll de Jong (1983a), Beneski and Stinson (1987), Whitaker and Hamilton (1998).

Sorex bendirii (Merriam, 1884)

Marsh shrew, musaraña de pantano

1. *S. b. albiventer* Merriam, 1895. Restricted to the Olympic Peninsula, Washington.
2. *S. b. bendirii* (Merriam, 1884). From southwestern British Columbia south through San Francisco Bay, California.
3. *S. b. palmeri* Merriam, 1895. From the Columbia River south through northwestern California.

Conservation: *Sorex bendirii* is listed as Least Concern by the IUCN (Cassola 2016dg).

Characteristics: Marsh shrew is small sized **but larger within shrews; total length 147.0–174.0 mm; condylobasal length 20.8–23.8 mm**. **Dorsal pelage dull sooty plumbeous to sooty brown**; fur of the underparts faintly paler; on the Olympic Peninsula of Washington, it may be whitish; tail long and dusky throughout; feet large, with the third and fourth **hind foot toes webbed at the base;** skull large and robust; **fifth unicuspid teeth larger than in other shrews, sometimes double or bifid under unworn conditions**; **lower first and second molars with a large internal cusp which rises posteriorly to form a distinct secondary cusp**.

Comments: *Sorex bendirii* was considered the only species of the subgenus *Atophyrax* (Merriam 1884). It ranges from southwestern British Columbia south through northwestern California (Map 30). It can be found in sympatry, or nearly so, with *S. cinereus*, *S. eximus*, *S. longirostris*, *S. navigator*, *S. pacificus*, *S. rohweri*, *S. sonomae*, *S. trowbridgii*, and *S. vagrans*. *S. bendirii* can be distinguished from all these species by being a large, velvety, black shrew; hind foot toes webbed at the base; fifth unicuspid teeth larger than in other shrews; first and second lower molars with a distinct large internal secondary cusp. From *S. albibarbis*, by having a straighter rostrum; tail distinctly bicolored and paler ventrally. From *S. rohweri*, *S. trowbridgii*, and *S. vagrans*, by having the dorsal pelage grayish, never brownish; hind foot length greater than 18.0 mm and condylobasal length greater than 19.0 mm. From *S. cinereus* and *S. sonomae*, by having the third unicuspid teeth smaller than the fourth; upper unicuspid teeth usually pigmented on the ridges extending from the apices of the cingulum and uninterrupted by an anteroposterior groove. From *S. longirostris*, by having the third unicuspid teeth equal to or larger than the fourth; upper teeth weakly pigmented; ridges extending from the apices of the unicuspid teeth medially toward the cingulum but incomplete, and postmandibular foramen usually well developed. From *S. navigator*, by having the hind feet not fimbriated; underparts never silver gray; vertical depth of the rostrum at the third unicuspid teeth greater than one-half of the distance from the anterior border of the infraorbital foramen to the posterior border of the first incisors. From *S. pacificus*, by having the pelage of the upperparts grayish, never brownish; hind feet greater than 18.0 mm and condylobasal length greater than 19.0 mm. From *Sorex palustris*, by having the rostrum proportionately shorter; anterior premaxillary barely shallower dorsoventrally than the middle, and vertical depth of the rostrum at the level of the third unicuspid teeth about equal to or one-half of the distance from the anterior border of the infraorbital foramen to the posterior border of the first incisors.

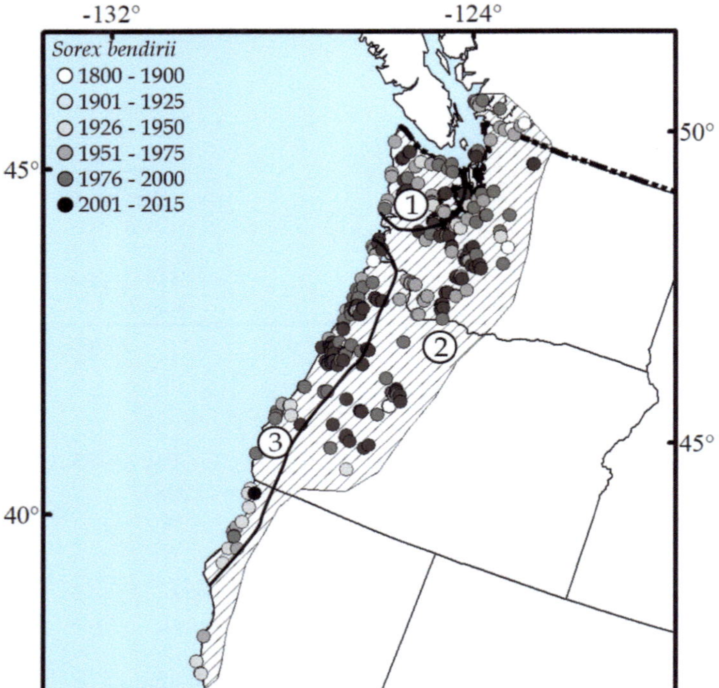

Map 30 Distribution map of *Sorex bendirii*

Additional Literature: Dalquest (1948), Findley (1955a), Hall and Kelson (1959), Cowan and Guiguet (1965), Pattie (1973), van Zyll de Jong (1983a), Hoffmann (1987), George (1988), Carraway (1990), Galindo-Leal and Zuleta (1997), Demboski and Cook (2001).

Sorex cinereus Kerr, 1792
Masked shrew, musaraña canosa

1. *S. c. cinereus* Kerr, 1792. From western Alaska to Washington and east through Québec, eastern Newfoundland, New Jersey and south through Iowa.
2. *S. c. fontinalis* Hollister, 1911. Restricted to southeastern Pennsylvania, New Jersey, Delaware, and Maryland.
3. *S. c. hollisteri* Jackson, 1925. Restricted to western Alaska.
4. *S. c. lesueurii* (Duvernoy, 1842). Illinois, Indiana, and Michigan.
5. *S. c. miscix* Bangs, 1899. Western Newfoundland and Québec.
6. *S. c. nigriculus* Green, 1932. Restricted to Tuckahoe River, southern New Jersey.
7. *S. c. streatori* Merriam, 1895. From the southeastern coast of Alaska south through by the coastal plains of Washington.

Conservation: *Sorex cinereus* is listed as Least Concern by the IUCN (Cassola 2016dh).

Characteristics: Masked shrew is small sized; total length 71.0–111.0 mm; condylobasal length 14.6–16.9 mm. *Sorex cinereus* has a strong variation between its winter and summer pelage. In winter, dorsal pelage grayish fuscous to brown, extending throughout the flanks. In summer, pelage more brownish than in winter, from blackish to mummy brown, usually extending throughout the flanks, **sometimes with a distinct buffy brown lateral ribbon**; fur of the underparts smoke gray to pale smoke gray or pale grayish olive to deep olive-buff; snout long and pointed; tail essentially bicolored throughout the year, brownish gray dorsally and buffy ventrally; teeth tips chestnut brown, including the inner ridges of the upper unicuspid teeth; third unicuspid teeth smaller than the fourth, fifth unicuspid teeth small; lower first and second molars with an internal secondary cusp not distinctly large; posterior border of the infraorbital foramen even within, or anterior to, the plane of the space between the first and second upper molars.

Comments: *Sorex cinereus* was previously a complex of several species; many of those are distinct species, such as *Sorex haydeni* (George 1988), *S. jacksoni* (van Zyll de Jong 1982, 1999a, b), *S. rohweri* (Rausch et al. 2007), and *Sorex ugyunak* (van Zyll de Jong 1991b). *S. c. acadicus* and *S. c. ohionensis* are considered junior synonyms of *S. c. cinereus* (Woodman 2018). The taxonomic status of *S. c. fontinalis* is still questionable; for this reason, I still consider it a subspecies of *S. cinereus* (Kirkland 1977; Kirkland and Levengood 1987; Hutterer 2005; Woodman 2018). *S. cinereus* ranges from Alaska east through Newfoundland and south through Washington, western Montana, and Utah, in all the northern states of the United States and the Appalachian Mountains south through northern Georgia, Tennessee, and western South Carolina (Map 31). It can be found in sympatry, or nearly so, with *S. albibarbis*, *S. arcticus*, *S. bendirii*, *S. dispar*, *S. eximus*, *S. fumeus*, *S. haydeni*, *S. hoyi*, *S. longirostris*, *S. maritimensis*, *S. merriami*, *S. obscurus*, *S. nanus*, *S. navigator*, *S. pacificus*, *S. palustris*, *S. preblei*, *S. rohweri*, *S. sonomae*, *S. trowbridgii*, *S. tundrensis*, *S. ugyunak*, and *S. vagrans*. *S. cinereus* can be distinguished from *S. albibarbis*, *S. navigator*, and *S. palustris* by having the hind feet not fimbriated; vertical depth of the rostrum at the level of the third unicuspid teeth greater than one-half of the distance from the anterior border of the infraorbital foramen to the posterior border of the first incisors. From *S. arcticus*, *S. maritimensis*, and *S. tundrensis*, by having the tail length greater than 30% of the total head-and-body length; condylobasal length less than 17.0 mm; third unicuspid teeth smaller than the fourth, and upper unicuspid teeth usually with pigmentation on the ridges, which extent from the apices to the cingulum. From *S. bendirii*, *S. nanus*, *S. obscurus*, *S. pacificus*, *S. rohweri*, *S. trowbridgii*, and *S. vagrans*, by having the third unicuspid teeth equal to or larger than the fourth; upper teeth weakly pigmented, ridges extending from the apices of the unicuspid teeth medially toward the cingulum but incomplete and not ending in internal cusplets. From *S. dispar*, by having the posterior border of the infraorbital foramen even with or anterior to the plane of the space between the first and second upper molars. From *S. fumeus*, by having the maxillary breadth less than 4.6 mm; fourth unicuspid teeth more or less similar in length and width, and unicuspid teeth lingual face often pigmented; skull inflated. From *S. haydeni*, by its total length greater than 99.0 mm and maxillary toothrow length greater than 5.2 mm. From *S. hoyi*, by having the third unicuspid teeth similar to the others; the fifth unicuspid teeth can be small but not minute; first and second unicuspid teeth without a distinct internal ridge with an internal cusp; accessory medial tine on the anterior surface of the first upper incisors and in the mandible in lingual view; most posterior margin of the incisors anterior to the paraconid of the first lower molars. From *S. longirostris*, by having all teeth more pigmented; skull greater, longer, and rostrum narrower. From *S. lyelli*, by having the skull not noticeably flatter; larger palatal length; larger and wider rostrum, and narrow interorbital breadth. From *S. merriami*, by having the combined length of the two first molars equal to or greater than 2.4 mm. From *S. preblei*, by its larger size with the rostrum relatively long; lower toothrow length, greater than 4.1 mm; condylobasal length greater than 14.6 mm and lower toothrow length greater than 4.1 mm. From *S. sonomae*, by having the zygomatic processes of the maxillary finely pointed. From *S. ugyunak*, mainly by its distribution area, as *S. ugyunak* is restricted to the tundra.

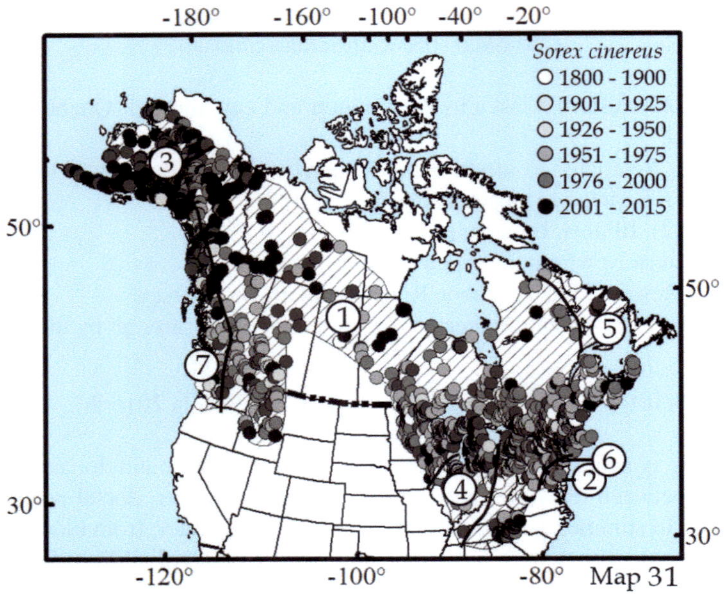

Map 31 Distribution map of *Sorex cinereus*

Additional Literature: Jackson (1925), Diersing and Hoffmeister (1981), Junge and Hoffman (1981), van Zyll De Jong (1982, 1983c), Stewart et al. (1989), van Zyll De Jong and Kirkland (1989), Stewart et al. (1993), Stewart and Baker (1994b, 1997), Demboski and Cook (2003), Whitaker (2004).

Sorex dispar Batchelder, 1911

Long-tailed shrew, musaraña de cola larga

Monotypic.

Conservation: *Sorex dispar* is listed as Least Concern by the IUCN (Whittaker et al. 2016).

Characteristics: Long-tailed shrew is small sized; total length 103.0–139.0 mm; condylobasal length 17.3–18.5 mm. Dorsal pelage mouse-gray; fur of the underparts lighter than in the dorsum; anterior edge of the zygomatic plate posterior to the plane separating the first and second upper molars; rostrum unusually long and narrow; third and fourth upper unicuspid teeth usually equal in size, if different, the third unicuspid teeth slightly smaller; unicuspid teeth widely spaced and relatively narrow; unicuspid teeth with the lingual face of the ridge often unpigmented.

Comments: *Sorex gaspensis* was recognized as a distinct species (French and Kirkland 1983); it was previously considered a subspecies of *S. dispar* (Rhymer et al. 2004); however, the only significant difference between the two taxa is size (Kirkland and Van Deusen 1979; French and Kirkland 1983, Hutterer 2005), and the genetic analyses show that they are very similar. Therefore, I still consider it as a subspecies of *S. dispar*. Mitochondrial and nuclear genetic markers among *S. d. blitchi*, *S. dispar*, and *S. d. gaspensis* have found the lack of the genetic structure at the subspecies level; for this reason, *S. d. blitchi* and *S. d. gaspensis* are considered junior synonyms of *S. dispar* (Schwartz 1956; Kirkland and Van Deusen 1979; Scott and van Zyll de Jong 1989; Woodman 2018). *S. dispar* ranges in the Appalachian Mountains, from Nova Scotia and southeastern New Brunswick, including Cape Breton Island (Map 32). It can be found in sympatry, or nearly so, with *S. cinereus*, *S. fumeus*, *S. hoyi*, *S. maritimensis*, and *S. palustris*. *S. dispar* can be distinguished from *S. cinereus* by having the posterior border of the infraorbital foramen posterior to the plane of the space between the first and second upper molars. From *S. fumeus*, by having the posterior border of the infraorbital foramen posterior to the plane of the space between the first and second upper molars. From *S. hoyi*, by having the third unicuspid teeth similar to the others; the fifth unicuspid teeth can be small but not minute; the first and second unicuspid teeth without a distinct internal ridge with an internal cusp; an accessory medial tine on the anterior surface of the first upper incisors and in mandible lingual view; most posterior margin of the incisors anterior to the paraconid of the first lower molars. From *S. maritimensis*, by having the third unicuspid teeth smaller than the fourth; upper unicuspid teeth usually with pigmentation on the ridges extending from the apices of the cingulum, uninterrupted by an anteroposterior groove. From *S. palustris*, by having the hind feet not fimbriated; vertical depth of the rostrum at the level of the third unicuspid teeth greater than one-half of the distance from the anterior border of the infraorbital foramen to the posterior border of the first incisors.

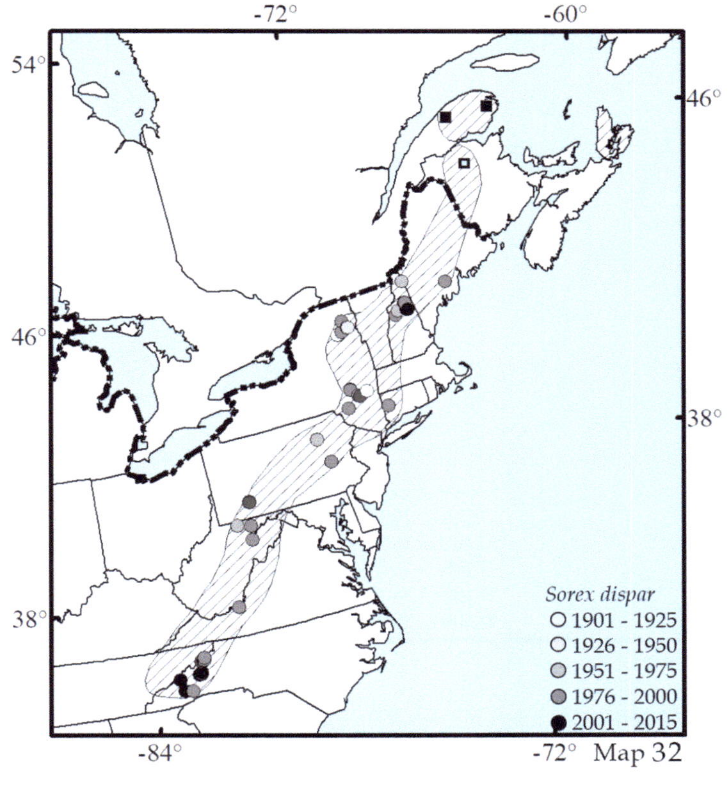

Map 32 Distribution map of *Sorex dispar*

Additional Literature: Peterson and Symansky (1963), Kirkland (1981, 1999), Scott (1987), George (1988), Shafer and Stewart (2007), Diersing (2019).

Sorex emarginatus Jackson, 1925

Zacatecas shrew, musaraña de Zacatecas

Monotypic.

Conservation: *Sorex emarginatus* is listed as Least Concern by the IUCN (Matson et al. 2017a).

Characteristics: Zacatecas shrew is small sized; total length 100.0–113.0 mm; condylobasal length 18.4–19.9 mm. Dorsal pelage dark brown; hair with a pale brown band in the middle and dark at the tip**; feet relatively long and narrow; tail length greater than 50% of the head-and-body length** and bicolored; first lower incisors with two crests and shallow valleys between them.

Comments: *Sorex emarginatus* was elevated to full species (Carraway 2007); it was previously a subspecies of *S. oreopolus* (Findley 1955a). Its ranges are restricted to southern Durango, Zacatecas, and northern Jalisco (Map 33). It can be found in sympatry, or nearly so, with *S. mediopua*, *S. saussurei*, *S. ventralis*, and *S. veraecrucis*. *S. emarginatus* can be distinguished from *S. mediopua*, *S. saussurei*, *S. ventralis*, and *S. veraecrucis* by having feet relatively long and narrow; tail length greater than 50% of the head-and-body length and bicolored; third unicuspid teeth smaller than the fourth; upper unicuspid teeth usually with pigmentation on the ridges extending from the apices of the cingulum and uninterrupted by an anteroposterior groove.

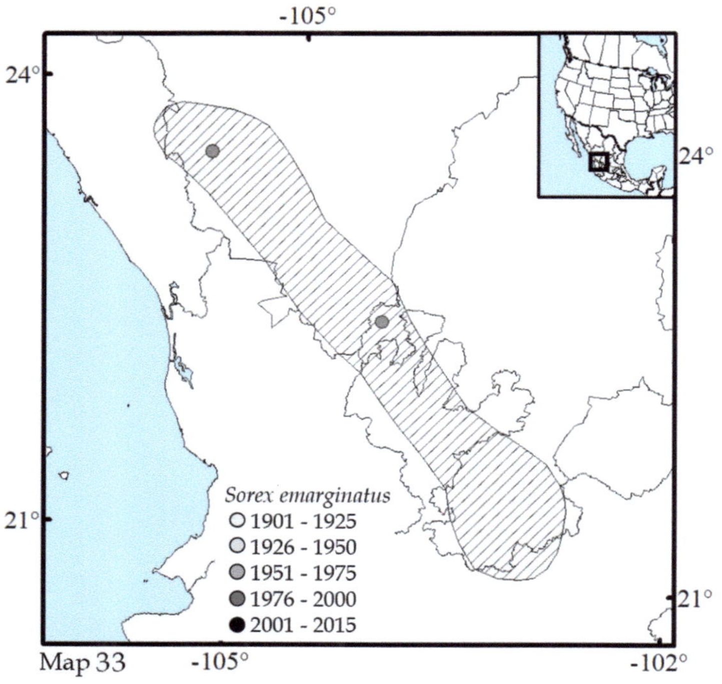

Map 33 Distribution map of *Sorex emarginatus*

Additional Literature: Jackson (1925, 1928), Diersing and Hoffmann (1977), Álvarez and Polaco (1984), Matson and Baker (1986), George (1988), Hutterer (2005), Carraway (2007).

Sorex eximius Osgood, 1901

Western pygmy shrew, musaraña pigmea del oeste

1. *S. e. eximius* Osgood, 1901. From Alaska and Yukon.
2. *S. e. montanus* (Brown, 1966). From Colorado and Wyoming.

Conservation: *Sorex eximius* (as *S. hoyi*) is listed as Least Concern by the IUCN (Cassola 2016dk).

Characteristics: Western pygmy shrew is **the smallest shrew in western North America**; **total length 87.0–105.0 mm; condylobasal length 15.0–16.2 mm.** Dorsal pelage sepia or slightly paler; fur of the underparts smoke gray tinged with hazelnut or wine-buff, with buffy shades on the neck and chest, and occasionally in the belly; **tail length shorter than the head-and-body length** (without considering the head), sepia to mummy dorsally and pinkish to hazelnut ventrally; skull large, rostrum relatively long; anterior and posterior borders of the skull gently rounded, not angular or truncate; dentition heavy; **third unicuspid teeth disk shaped, markedly compressed anteroposteriorly; fifth unicuspid teeth minute;** braincase high, long, and very narrow; interorbital constriction relatively broad, and rostrum medium sized; **lower jaw relatively large, heavy, with the styliform angular process very long**.

Comments: *Sorex eximius* was recognized as a distinct species (Stewart et al. 2003; Hope et al. 2020), previously regarded as a subspecies of *S. hoyi* (Jackson 1925). *S. hoyi* was the only species of the subgenus *Microsorex* (Baird 1857), now included in the subgenus *Otiosorex* (George 1980). *S. eximius* ranges from Manitoba, North and South Dakota west through Alaska and British Columbia, excluding the northern tundra zones, with disjunct populations in the southern Rockies (Map 34). It can be found in sympatry, or nearly so, with *S. arcticus*, *S. bairdi*, *S. bendirii*, *S. cinereus*, *S. haydeni*, *S. merriami*, *S. nanus*, *S. navigator*, *S. obscurus*, *S. pacificus*, *S. palustris*, *S. preblei*, *S. rohweri*, *S. trowbridgii*, *S. tundrensis*, *S. ugyunak*, and *S. vagrans*. *S. eximius* can be distinguished from all these species by its smaller size (total length less than 105.0 mm and condylobasal length less than 16.2 mm); tail length shorter than body length; third unicuspid teeth disk shaped, markedly compressed anteroposteriorly; fifth unicuspid teeth minute; lower jaw relatively small, heavy, with the styliform angular process very long. From *S. hoyi*, by having the pelage coloration somewhat less reddish brown; larger size; condylobasal length on average greater than 14.3 mm; longer palate and heavier dentition; skull somewhat more depressed, through both the braincase and the rostrum.

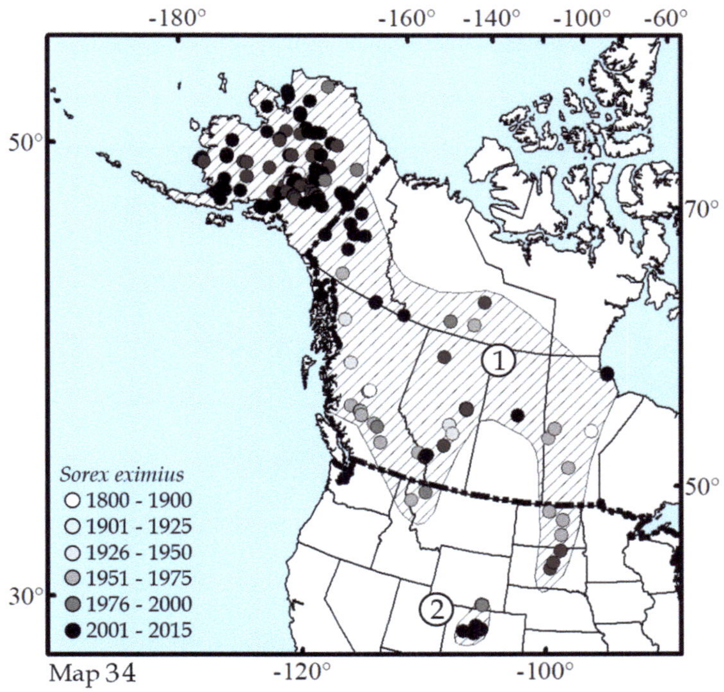

Map 34 Distribution map of *Sorex eximius*

Additional Literature: Jackson (1928), Long (1974, 1999a), Diersing (1980), Junge and Hoffman (1981), Baker (1983), George (1988), Feldhamer (1993), Long (1999a), Hutterer (2005), Shafer and Stewart (2007), Woodman (2018).

Sorex fumeus Miller, 1895

Smoky shrew, musaraña de las Smoky Mountains

1. *S. f. fumeus* Miller, 1895. From southern Ontario, Québec, New York, Vermont, and New Hampshire, south through North Carolina and Georgia.
2. *S. f. umbrosus* Jackson, 1917. From northern New York, Vermont, and New Hampshire northeastern through New Brunswick and Nova Scotia.

Conservation: *Sorex fumeus* is listed as Least Concern by the IUCN (Cassola 2016di).

Characteristics: Smoky shrew is small but medium sized within the genus; total length 101.0–122.0 mm; condylobasal length 17.4–18.5 mm. Dorsal pelage dorsally dull brown or slaty to deep gray or almost blackish; fur of the underparts slightly paler and somewhat silvery; tail bicolored, fuscus dorsally and paler and washed with pale yellow ventrally; old specimens may have whitish-tipped hairs; skull large; palate broad; third unicuspid teeth larger than the fourth; each unicuspid teeth wider than long; medial pigmented ridge extending from the apices of the unicuspid teeth to nearly one-half of the distance to the cingulum, often absent; small postmandibular foramen may be present on at least one mandible.

Comments: *Sorex fumeus* ranges from Indiana, Tennessee, Georgia, and North Carolina northeastern through New Brunswick and southern Québec and Ontario (Map 35). It can be found in sympatry, or nearly so, with *S. albibarbis*, *S. arcticus*, *S. cinereus*, *S. dispar*, *S. hoyi*, *S. longirostris*, *S. maritimensis*, and *S. palustris*. *S. fumeus* can be distinguished from *S. albibarbis*, *S. navigator*, and *S. palustris* by having the hind feet not fimbriated; vertical depth of the rostrum at the level of the third unicuspid teeth greater than one-half of the distance from the anterior border of the infraorbital foramen to the posterior border of the first incisors. From *S. arcticus* and *S. maritimensis*, by having the tail length greater than 30% of the total head-and-body length and condylobasal length less than 17.0 mm. From *S. cinereus*, by having the maxillary breadth greater than 4.6 mm; unicuspid teeth wider than long and unicuspid teeth lingual face often non-pigmented and skull flattened to some extent. From *S. dispar*, by having the posterior border of the infraorbital foramen even with or anterior to the plane of the space between the first and second upper molars. From *S. hoyi*, by having the third unicuspid teeth similar to the others; the fifth unicuspid teeth can be small but not minute; first and second unicuspid teeth without a distinct internal ridge with an internal cusp; an accessory medial tine on the anterior surface of the first upper incisors, in lingual view. From *S. longirostris*, by having the third unicuspid teeth equal to or greater than the fourth; upper teeth weakly pigmented; ridges extending from the apices of the unicuspid teeth medially toward the cingulum but incomplete, and postmandibular foramen usually well developed.

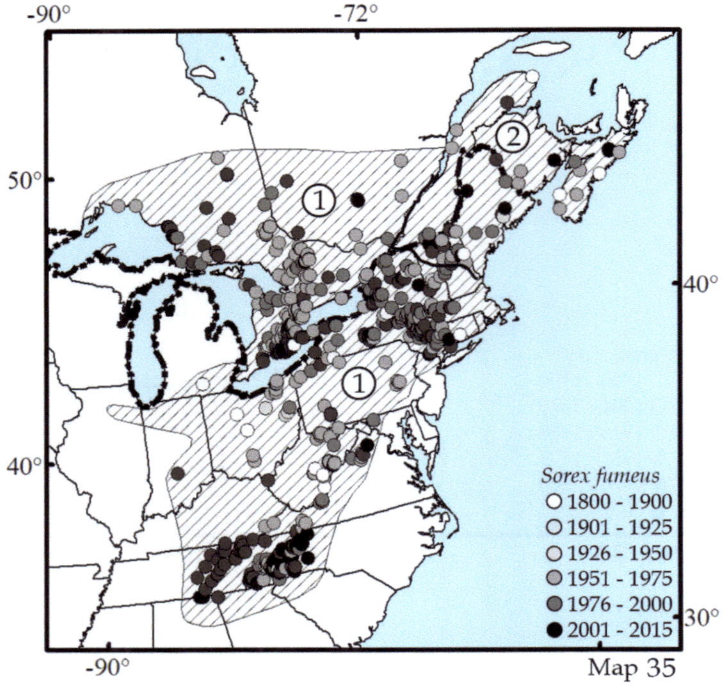

Map 35 Distribution map of *Sorex fumeus*

Additional Literature: Jackson (1928), Hamilton (1940), Green (1943), Wharton (1968), Junge and Hoffman (1981), van Zyll De Jong (1983a), Owen (1984), George (1988), Jannett and Oehlenschlager (1994), Whitaker (1999a).

Sorex haydeni Baird, 1857

Prairie shrew, musaraña de las praderas

Monotypic.

Conservation: *Sorex haydeni* is listed as Least Concern by the IUCN (Cassola 2016dj).

Characteristics: Prairie shrew is small sized; total length 83.3–99.8 mm; condylobasal length 14.1–15.9 mm. *Sorex haydeni* shows a strong variation between the winter and summer pelages. In winter, dorsal pelage brown or drab and sometimes fuscous posteriorly and not extending onto the flanks. In summer, brown to olive brown; fur of the underparts pale olive-gray to pale smoke gray and paler; flanks paler, contrasting with the dorsal and ventral coloration; tail dorsally cinnamon brown and ventrally chestnut brown or pinkish buff; palate relatively shorter and rostrum broader; dentition densely pigmented.

Comments: *Sorex haydeni* was formerly considered a subspecies of *S. cinereus* (Jackson 1925; Brunet et al. 2002), but it is currently regarded as a distinct species (George 1988; Demboski and Cook 2003, Baker et al. 2003a; Hutterer 2005). Its ranges from Saskatchewan and Manitoba south through Kansas and Missouri (Map 36). It can be found in sympatry, or nearly so, with *S. arcticus*, *S. cinereus*, *S. eximus*, *S. hoyi*, *S. longirostris*, *S. merriami*, *S. nanus*, *S. palustris*, and *S. preblei*. *S. haydeni* can be distinguished from *S. arcticus* by having the tail length greater than 30% of the total head-and-body length, and condylobasal length less than 17.0 mm. From *S. cinereus*, by having total length less than 99.0 mm and maxillary toothrow length less than 5.2 mm. From *S. longirostris* and *S. nanus*, by having the third unicuspid teeth equal to or larger than the fourth; upper teeth weakly pigmented; ridges extending from the apices of the unicuspid teeth medially toward the cingulum but incomplete, and the postmandibular foramen usually well developed. From *S. lyelli*, by having the skull slightly smaller and less flattened, with the interorbital region narrower; rostrum relatively wider and with a shorter upper toothrow. From *S. merriami*, by having the combined length of the two first molars equal to or greater than 2.4 mm. From *S. palustris*, by having the hind feet not fimbriated; vertical depth of the rostrum at the third unicuspid teeth greater than one-half of the distance from the anterior border of the infraorbital foramen to the posterior border of the first incisors. From *S. preblei*, by its larger size with the rostrum relatively long; lower toothrow length greater than 4.1 mm; condylobasal length greater than 14.6 mm and lower toothrow length greater than 4.1 mm.

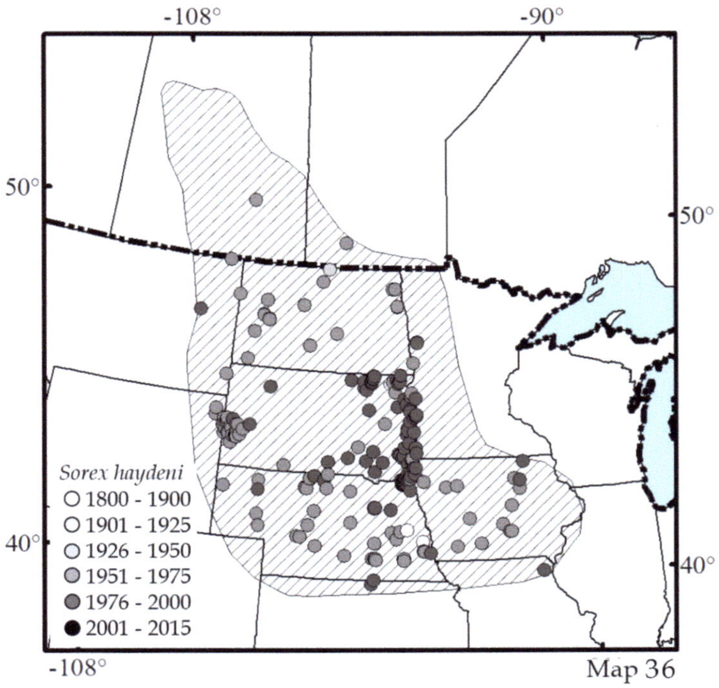

Map 36 Distribution map of *Sorex haydeni*

Additional Literature: Merriam (1895c), Jackson (1928), van Zyll de Jong (1982), Junge and Hoffman (1981), Frey and Moore (1990), Stewart et al. (1993).

Sorex Baird, 1857

American pygmy shrew, musaraña pigmea

1. *S. h. hoyi* Baird, 1857. From northwestern Manitoba east through Newfoundland, Québec, and New Brunswick.
2. *S. h. thompsoni* Baird, 1857. From southern Wisconsin, northern Illinois, and eastern Michigan, northeast through New Brunswick.
3. *S. h. winnemana* (Preble, 1910). From northeastern Georgia and eastern Tennessee northeast through southeastern Pennsylvania.

Conservation: *Sorex hoyi* is listed as Least Concern by the IUCN (Cassola 2016dk).

Characteristics: American pygmy shrew is the smallest **shrew in eastern North America**; **total length 78.0–88.0 mm; condylobasal length 14.3–15.0 mm.** Dorsal pelage brown and olive brown; fur of the underparts smoke gray, sometimes slightly tinged warm buff, with buffy shades on the neck and chest, and occasionally in the belly; **tail length shorter than the head-and-body length** (without considering the head), indistinctly bicolored, mummy brown dorsally and drab ventrally, darkening toward the tip; skull relatively long, narrow, markedly depressed; anterior and posterior borders of the skull gently rounded, not angular or truncate; dentition moderate; **third unicuspid teeth disk shaped, markedly compressed anteroposteriorly; fifth unicuspid teeth minute;** braincase low, long, and very narrow; interorbital constriction relatively broad and rostrum medium sized; **lower jaw relatively large, heavy, with the styliform angular process very long**.

Comments: *Sorex hoyi* was considered the only species of the subgenus *Microsorex* (Baird 1857); the current treatment has assigned it in the subgenus *Otiosorex* (George 1980). *S. extimus* was recognized as a distinct species (Stewart et al. 2003; Hope et al. 2020), previously regarded as a subspecies of *S. hoyi* (Jackson 1925). *S. h. alnorum* is a junior synonym of *S. h. hoyi* (Diersing 1980; Woodman 2018). *S. hoyi* ranges from eastern Manitoba and North and South Dakota east through the Labrador Peninsula and New England, excluding the northern tundra zones and including southern Indiana, Kentucky, and western Tennessee (Map 37). It can be found in sympatry, or nearly so, with *S. albibarbis*, *S. arcticus*, *S. cinereus*, *S. dispar*, *S. eximus*, *S. haydeni*, *S. fumeus*, *S. longirostris*, *S. maritimensis*, and *S. palustris*. *S. hoyi* can be distinguished from all these species by its smaller size, total length less than 88.0 mm and condylobasal length less than 15.0 mm; tail length shorter than the body length; third unicuspid teeth disk shaped, markedly compressed anteroposteriorly; fifth unicuspid teeth minute; lower jaw relatively small, heavy, with the styliform angular process very long. From *S. eximus*, by having the pelage coloration somewhat more reddish brown; condylobasal length on average less than 14.3; shorter palate and light dentition; skull somewhat less depressed both through the braincase and the rostrum.

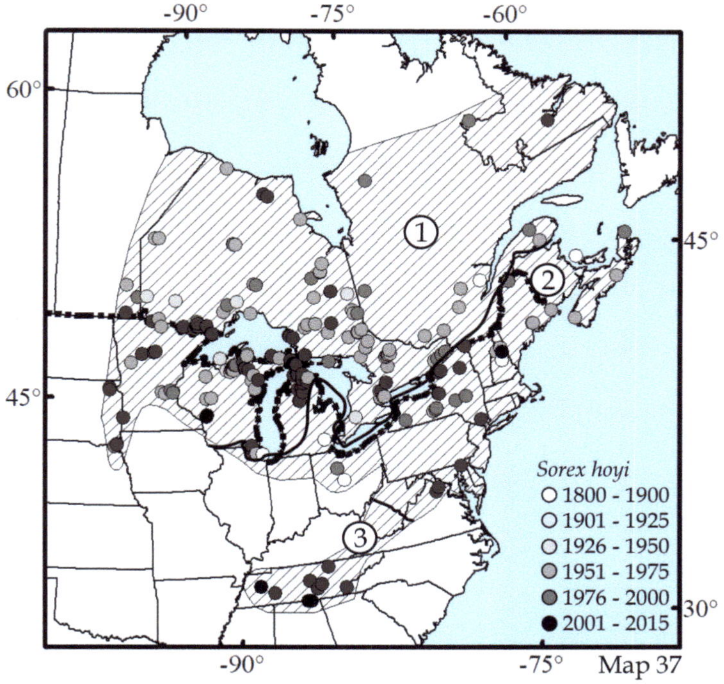

Map 37 Distribution map of *Sorex hoyi*

Additional Literature: Jackson (1928), Long (1974, 1999a), Junge and Hoffman (1981), Baker (1983), Whitaker and French (1984), Mengak (1987), George (1988), Feldhamer (1993), Long (1999a), Hutterer (2005), Shafer and Stewart (2007), Czech et al. (2017).

Sorex ixtlanensis Carraway, 2007

Ixtlán shrew, musaraña de Ixtlán

Monotypic.

Conservation: *Sorex ixtlanensis* is listed as Data Deficient by the IUCN (Matson et al. 2019).

Characteristics: Ixtlán shrew is small sized; total length 117.0–135.0 mm; condylobasal length 18.2–19.7 mm. Dorsal pelage with two color bands, one pale reddish medially and dark brown at the tip; fur of the underparts pale reddish brown; guard hair on the hips projecting backward as "thorns" and of uniform color; 32 teeth pigmented.

Comments: *Sorex ixtlanensis* was formerly considered part of *S. veraepacis mutabilis* (Carraway 2007). It is difficult to determine the separate distribution of this newly elevated species (Carraway 2007). This species is only known from the highlands of Oaxaca and Guerrero (Map 38), but its exact distribution is not known; there is ongoing confusion with *S. veraepacis mutabilis*. *S. ixtlanensis* is sympatric, or nearly so, with *S. macrodon*, *S. mutabilis*, *S. ventralis*, and *S. veraecrucis*. It can be distinguished from *S. macrodon* by having the condylobasal length on average less than 19.5 mm; cranial breadth usually less than 9.6 mm; first lower incisors with two or three denticles and a long stripe of pigment present or absent at the anteromedial edge. From *S. mutabilis*, by having the condylobasal length less than 17.9 mm and the distance from the first upper unicuspid teeth to the third upper molars less than 6.5 mm. From *S. ventralis*, by having the condylobasal length less than 17.9 mm and the distance from the distance from the first upper unicuspid teeth to the third upper molars less than 6.5 mm. From *S. veraecrucis*, by having the upperparts reddish blond medially, dark brown at the tip; hips and rump with pale reddish blond and dark brown guard hairs extending greater than 1.9 mm beyond the dorsal pelage; fur of the underparts pale reddish brown at the tip and first lower incisors with three denticles.

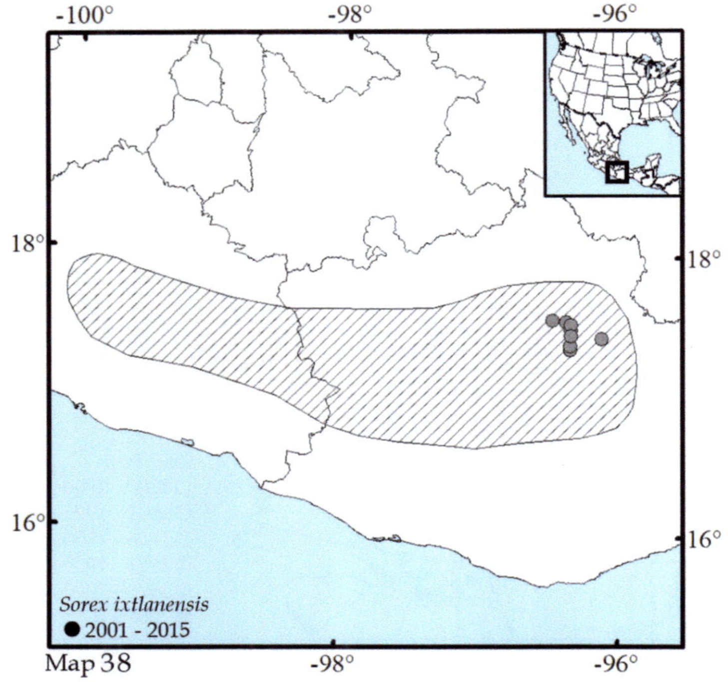

Map 38 Distribution map of *Sorex ixtlanensis*

Additional Literature: Guevara et al. (2015), Woodman (2018).

Sorex jacksoni Hall and Gilmore, 1932

Saint Lawrence Island shrew, musaraña de la isla Saint Lawrence

Monotypic.

Conservation: *Sorex jacksoni* is listed as Least Concern by the IUCN (Naylor and Roach 2016).

Characteristics: Saint Lawrence Island shrew is small sized; total length 83.0–100.2 mm; condylobasal length 14.3–16.0 mm. Dorsal pelage very dark brown to black; fur of the underparts grayish brown; pelage tricolored; flanks paler and contrasting strongly with the dark dorsum; tail indistinctly bicolored, brown to brownish-black dorsally and paler ventrally; postmandibular canals well developed; pigmented ridges lacking on unicuspid teeth.

Comments: *Sorex jacksoni* was reinstated as a distinct species (van Zyll de Jong 1982, 1999a, b); it was previously regarded as a subspecies of *S. cinereus* (Hoffmann and Peterson 1967; Rausch and Rausch 1995) and is endemic to Saint Lawrence Island, in the Bering Sea (Map 39), Alaska, with no other sympatric species of *Sorex*.

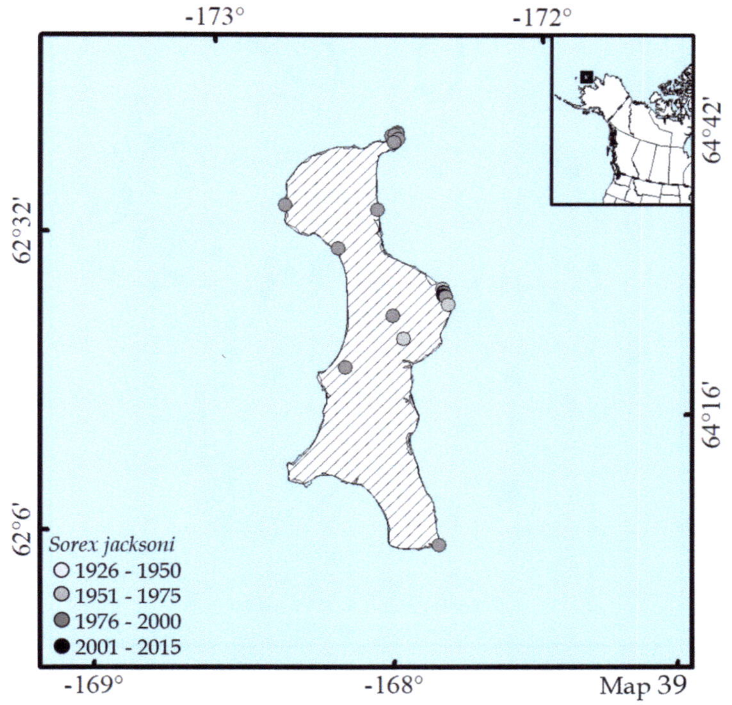

Map 39 Distribution map of *Sorex jacksoni*

Additional Literature: Hall and Gilmore (1932), van Zyll de Jong (1983c, 1991b), Jones et al. (1986), Hutterer (1993, 2005), Baker et al. (2003a), Demboski and Cook (2003).

Sorex longirostris Bachman, 1837

Southeastern shrew, musaraña del suroeste

1. *S. l. eionis* Davis, 1957. Restricted to central-western coast of Florida.
2. *S. l. fisheri* Merriam, 1895. Restricted to a small area in Virginia and North Carolina.
3. *S. l. longirostris* Bachman, 1837. From southwestern Missouri, northwestern Illinois east through southern Pennsylvania, south through Louisiana and the Florida Peninsula.

Conservation: *Sorex longirostris* is listed as Least Concern by the IUCN (Cassola 2016dl).

Characteristics: Southeastern shrew is small sized; total length 83.5–87.0 mm; condylobasal length 14.0–16.5 mm. Dorsal pelage chestnut brown; fur of the underparts ashy tinged with drab, markedly contrasting with the dorsal coloration; tail short, ventrally dark and pale brownish dorsally; ears large and conspicuous; skull flat; one of the smallest species in North America; **third unicuspid teeth clearly smaller than or nearly equal to the fourth;** first and second unicuspid teeth subequal; third unicuspid teeth about half the size of the second unicuspid teeth and clearly smaller than the fourth; fifth unicuspid teeth relatively large.

Comments: Junge and Hoffman (1981) suggested that *Sorex longirostris fisheri* may be a full species, but it is still considered a subspecies of *S. longirostris* (Jones et al. 1992; Hutterer 1993, 2005). *S. longirostris* ranges in southeastern United States, from southern Maryland and West Virginia south through central Florida, west through northwestern Illinois, eastern Oklahoma, and eastern Louisiana (Map 40). It can be found in sympatry, or nearly so, with *S. cinereus*, *S. fumeus*, *S. haydeni*, and *S. hoyi*. *S. longirostris* can be distinguished from *S. cinereus*, *S. fumeus*, and *S. haydeni* by having the tail short; third unicuspid teeth smaller than the fourth teeth; upper unicuspid teeth usually with pigmentation on the ridges extending from the apices of the cingulum and uninterrupted by an anteroposterior groove. From *S. cinereus*, by having all teeth more lightly pigmented; skull smaller, shorter, and rostrum broader. From *S. hoyi*, by having the third unicuspid teeth similar to the others; the fifth unicuspid teeth can be small but not minute; first and second unicuspid teeth without a distinct internal ridge with an internal cusp; an accessory medial tine on the anterior surface of the first upper incisors and in the mandible in lingual view; most posterior margin of the incisors anterior to the paraconid of the first lower molars. From *S. nanus*, by its higher cranium; constriction broader and more swollen; palate broader and arched, anterior rostrum broader, shorter, and less attenuated.

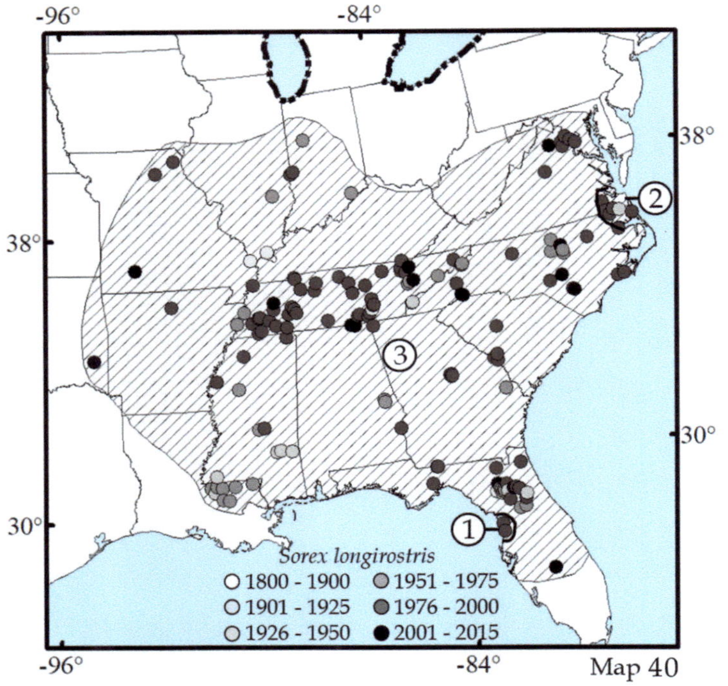

Map 40 Distribution map of *Sorex longirostris*

Additional Literature: Merriam (1895c), Jackson (1928), French (1980), George (1988), Taylor and Wilkinson (1988), Jones et al. (1991), Woodman (2018).

Sorex lyelli Merriam, 1902

Mount Lyell shrew, musaraña del Monte Lyell

Monotypic.

Conservation: *Sorex lyelli* is listed as Least Concern by the IUCN (NatureServe 2017).

Characteristics: Mount Lyell shrew is small sized; total length 99.5–103.0 mm; condylobasal length 15.2–15.4 mm. Dorsal pelage brown to drab, almost olive brown in some specimens; fur of the underparts pale olive-gray to pale smoke, with buff shades; flanks paler than the dorsum; tail dorsally olive brown to buff brown, darker at the tip, ventrally pale, dirty; skull flatter through the braincase; interorbital region wide.

Comments: *Sorex lyelli* is restricted to a small area of the eastern-central Sierra Nevada, California, including the areas in and around Yosemite National Park (Map 41). It can be found in sympatry, or nearly so, with *S. merriami*, *S. obscurus*, *S. palustris*, *S. tenellus*, and *S. vagrans*. *S. lyelli* can be distinguished from *S. cinereus* by having the skull noticeably flatter; rostrum shorter and broader; smaller palatal length, and wider interorbital breadth. From *S. merriami*, by having the combined length of the two first molars equal to or greater than 2.4 mm. From *S. obscurus*, *S. tenellus*, and *S. vagrans*, by having the third unicuspid teeth equal to or larger than the fourth; upper teeth weakly pigmented; ridges extending from the apices of the unicuspid teeth medially toward the cingulum but incomplete and not ending in internal cusplets. From *S. palustris*, by having the hind feet not fimbriated, vertical depth of the rostrum at the level of the third unicuspid teeth greater than one-half of the distance from the anterior border of the infraorbital foramen to the posterior border of the first incisors.

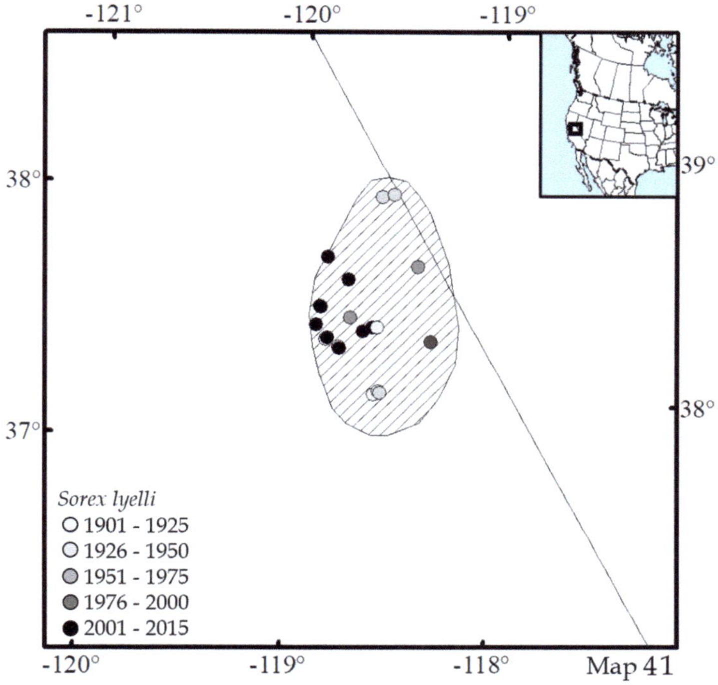

Map 41 Distribution map of *Sorex lyelli*

Additional Literature: Jackson (1928), Grinnell (1933), Williams (1984), van Zyll de Jong (1991a, b), Hutterer (2005), Woodman (2018).

Sorex milleri Jackson, 1947

Carmen mountain shrew, musaraña de Sierra del Carmen

Monotypic.

Conservation: *Sorex milleri* is listed as Special Protection in the Norma Oficial Mexicana (DOF 2019) and as Vulnerable (B1ab (iii): extent of occurrence less than 20,000 km^2 severely fragmented and under continuing decline in area, extent, and quality of habitat) by the IUCN (Matson and de Grammont 2018).

Characteristics: Carmen Mountain shrew is small sized; total length 93.0–107.0 mm; condylobasal length 14.7–16.0 mm. Dorsal pelage pale brown; fur of the underparts whitish and bicolored at the tip; in lateral view, most posterior part of the first lower incisors in front of the first cusp [paraconid] of the first lower molars; fourth unicuspid teeth larger than the third unicuspid teeth; 32 teeth, pigmented; first lower incisors with two deep valleys between three crests.

Comments: *Sorex milleri* was considered a subspecies of *S. cinerus* (van Zyll de Jong and Kirkland 1989), but this assignment was not accepted (Hutterer 2005). It is restricted to the Sierra Madre Oriental in Coahuila and Nuevo León (Map 42) and does not have sympatric species.

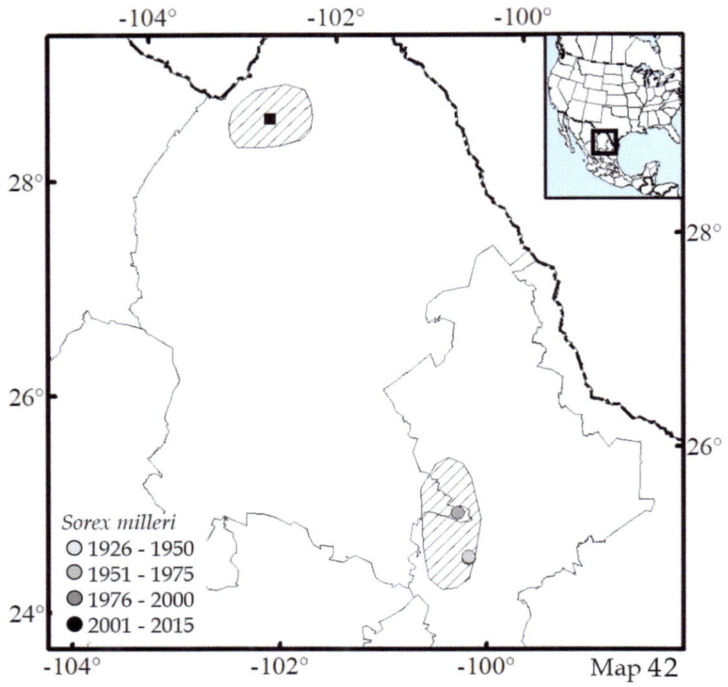

Map 42 Distribution map of *Sorex milleri*

Additional Literature: Jackson (1947), Carraway (2007), Hope et al. (2012), Guevara et al. (2015).

Sorex monticolus Merriam, 1890

Montane shrew, musaraña oscura

1. *S. m. monticolus* Merriam, 1890. From eastern Arizona and western New Mexico south through by the highlands to southern Durango.
2. *S. m. neomexicanus* Bailey, 1913. From Capitan and Sacramento mountains in southern-central New Mexico.

Conservation: *Sorex monticola* is listed as Threatened in the Norma Oficial Mexicana (DOF 2019) and as Least Concern by the IUCN (Matson et al. 2016b).

Characteristics: Montane shrew is small sized; total length 86.0–128.0 mm; condylobasal length 16.2–16.7 mm. Dorsal pelage pale sepia brown, flanks usually barely paler than the back; fur of the underparts pale gray pinkish buff; hair on the hips projecting backward as "thorns" and bicolored; **tail bicolored, dorsally brownish and ventrally pinkish buff**; rostrum and teeth relatively heavy; third unicuspid teeth smaller than the fourth; first lower incisors with deep valleys between them; **rostrum proportionally shorter than other species of *Sorex*.**

Comments: *Sorex monticolus* is recognized as a distinct species, different from *S. vagrans* (Carraway 1990; Jones et al. 1992; Hutterer 1993, 2005; Alexander 1996; Smith and Belk 1996). Several populations of the *vagrans* complex were considered full species, such as *S. bairdi* (Hennings and Hoffmann 1977), *S. neomexicanus* (Alexander 1996), *S. orizabae* (Carraway 2007), *S. pacificus* (Carraway 1990), and *S. sonomae* (Carraway 1990). *S. m. parvidens* was previously considered a subspecies of *S. ornatus* by Woodman (2012). *S. m. alascensis*, *S. m. obscurus*, S. *m. shumaginensis*, and *S. m. soperi* are subspecies of *S. obscurus* (Alexander 1996; Woodman 2012). However, Woodman (2018) classified *S. neomexicanus* at the subspecies level. *S. monticolus* is widespread, ranging from California, Nevada, Utah, and Colorado south through Durango and from Montana to New Mexico (although it is patchily distributed from the highlands; Map 43). It can be found in sympatry, or nearly so, with *S. arizonae*, *S. merriami*, *S. obscurus*, and *S. vagrans*. *S. monticolus* can be distinguished from *S. arizonae* and *S. merriami* by usually lacking the postmandibular foramen or, if present, it is small and often only on one mandible; unicuspid teeth with a ridge, usually pigmented on the lingual face, running from the apex to the cingulum, sometimes ending in a pigmented cusplet. From *S. merriami*, by having the third unicuspid teeth smaller than the fourth; upper unicuspid teeth usually with pigmentation on the ridges extending from the apices of the cingulum and uninterrupted by an anteroposterior groove. From *S. obscurus*, by having the palate narrower on average; rostrum, smaller, narrower, and more attenuated; dentition weaker, especially unicuspid teeth. From *S. palustris*, by having the hind feet not fimbriated; vertical depth of the rostrum at the level of the third unicuspid teeth greater than one-half of the distance from the anterior border of the infraorbital foramen to the posterior border of the first incisors.

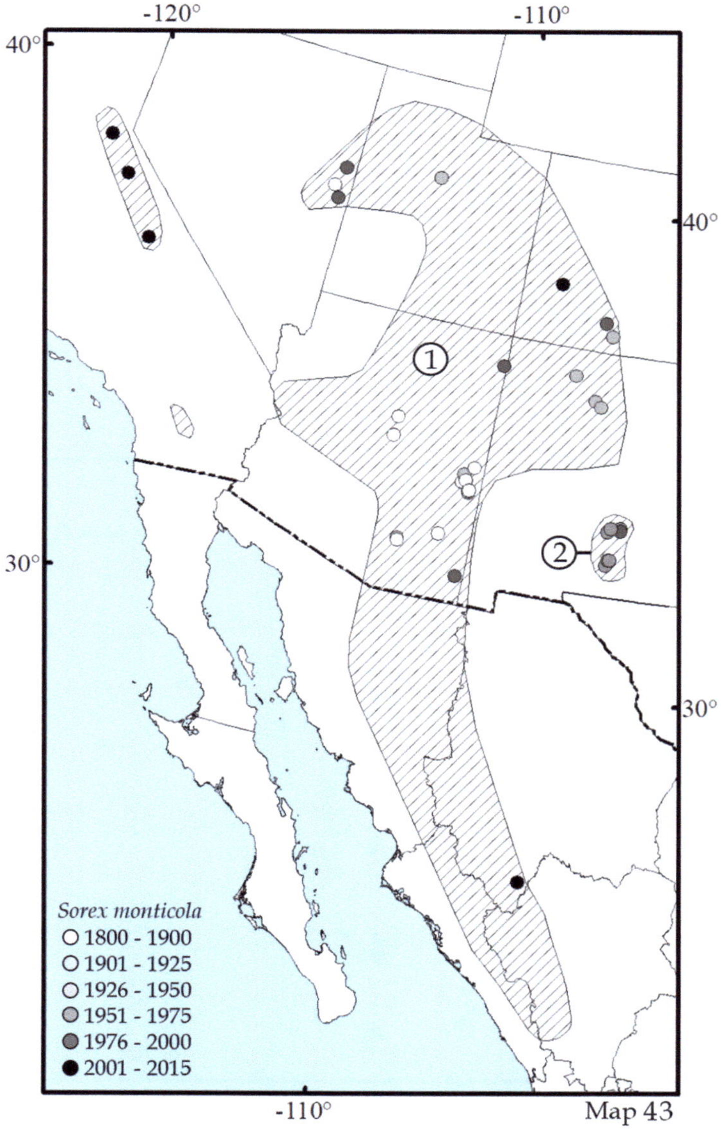

Map 43 Distribution map of *Sorex monticolus*

Additional Literature: Jackson (1928), Hawes (1975), Hennings and Hoffmann (1977), Junge and Hoffman (1981), van Zyll de Jong (1982, 1983c), George (1988), George and Smith (1991), Stewart and Baker (1994b), Demboski and Cook (2001), Carraway (2007), Shafer and Stewart (2007), Guevara et al. (2015), Woodman (2018).

Sorex nanus Merriam, 1895

Dwarf shrew, musaraña enana

Monotypic.

Conservation: *Sorex nanus* is listed as Least Concern by the IUCN (Cassola 2016dn).

Characteristics: Dwarf shrew is small sized; total length 91.0–105.0 mm; **condylobasal length 13.8–14.8 mm**. Dorsal pelage sepia brown, with the back darker; underparts and feet with hairs grayish ash; tail bicolored, dorsally of similar color to the back, darker at the tip, ventrally whitish; skull very flat, small, and slender; a pigmented ridge on unicuspid teeth, extending from the apices to the cingulum, palate usually well developed, narrow; **accessory cusp on the first upper incisors well developed and heavily** pigmented, lacking a postmandibular canal.

Comments: *Sorex nanus* is considered a distinct species from *S. tenellus* (George 1988; Jones et al. 1992; Hutterer 2005). It ranges from central Montana east through South Dakota and south through central Arizona and New Mexico (Map 44). It can be found in sympatry, or nearly so, with *S. cinereus*, *S. eximus*, *S. haydeni*, *S. merriami*, *S. monticolus*, *S. obscurus*, *S. palustris*, and *S. preblei*. *S. nanus* can be distinguished from these species by its smaller size; condylobasal length less than 15.2 mm and accessory cusp on the first upper incisors well developed and heavily pigmented. From *S. cinereus*, *S. haydeni*, *S. merriami*, and *S. preblei*, by having the third unicuspid teeth smaller than the fourth; upper unicuspid teeth usually with pigmentation on the ridges extending from the apices of the cingulum and uninterrupted by an anteroposterior groove. In addition, from *S. preblei*, by having the dentary below the lower first molars less deep than the height of the lower first molar at the metaconid and a higher (less than 2.9 mm) coronoid process. From *S. monticolus* and *S. obscurus*, by usually having a well-developed postmandibular foramen and an unpigmented ridge on the unicuspid teeth lingual face. From *S. palustris*, by having the hind feet not fimbriated; vertical depth of the rostrum at the level of the third unicuspid teeth greater than one-half of the distance from the anterior border of the infraorbital foramen to the posterior border of the first incisors.

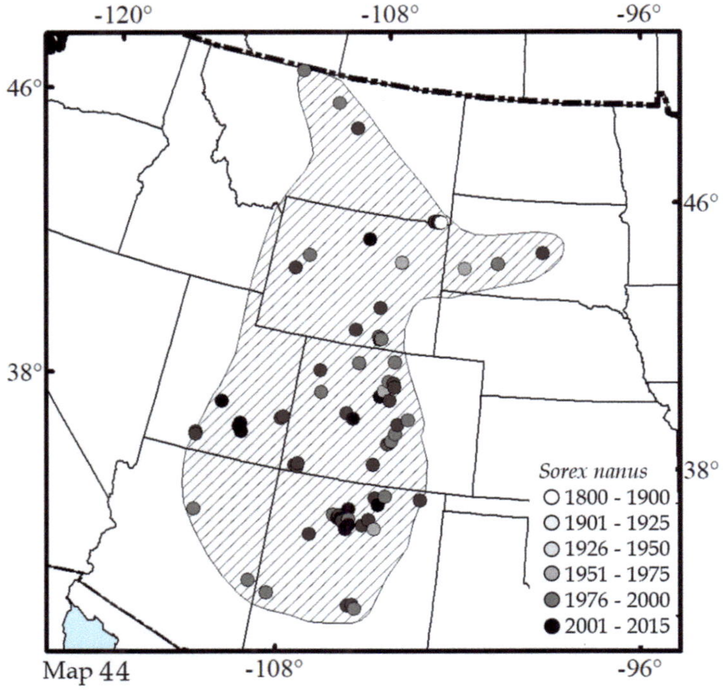

Map 44 Distribution map of *Sorex nanus*

Additional Literature: Jackson (1928), Findley (1955b), Spencer (1966), Hoffmann and Peterson (1967), Repenning (1967), Diersing and Hoffmeister (1977), Hoffmann and Owen (1980), MacCracken et al. (1985), Berna (1990), George (1990), Backlund (1995), Woodman (2018).

Sorex navigator (Baird, 1858)

Western water shrew, musaraña de agua occidental

1. *S. n. alaskanus* Merriam, 1900. Restricted to the flats around Gustavus and Bartlett Cove, Alaska.
2. *S. n. brooksi* Anderson, 1934. Restricted to Vancouver Island.
3. *S. n. navigator* (Baird, 1858). From Alaska, Yukon, and Mackenzie south through California, Arizona, and New Mexico.

Conservation: *Sorex navigator* (as *S. palustris*) is listed as Least Concern by the IUCN (Cassola 2016do).

Characteristics: Western water shrew is small **but large sized among the *Sorex*; total length 127.0–163.3 mm**; condylobasal length 18.5–19.3 mm. Dorsal pelage plumbeous, hoary; **fur of the underparts silver gray to dull white, chest and anal region pale gray whitish**; tail bicolored, dusky dorsally and whitish ventrally; feet large; hind feet very long and broad, suitable for swimming; toes slightly webbed; skull similar to *S. palustris* but braincase flatter; palate and interpterygoid fossa narrower.

Comments: *Sorex navigator* was recognized as a distinct species (Mycroft et al. 2011), previously with *S. n. brooksi* were considered a subspecies of *S. palustris* (Merriam 1985; Anderson 1934; Hutterer 2005). *S. alaskanus* was described as a subspecies of *S. navigator* (Merriam 1900), later as a subspecies of *S. palustris* (Elliot 1901), as a distinct species (Jackson 1928; Hutterer 1993, 2005), and as a subspecies of *S. navigator* (Nagorsen et al. 2017, Woodman 2018). *S. navigator* ranges from Alaska, Yukon, and Mackenzie south through California, Arizona, and New Mexico, including Vancouver Island (Map 45). It can be found in sympatry, or nearly so, with *S. bendirii*, *S. cinereus*, *S. eximus*, *S. obscurus*, *S. palustris*, *S. rohweri*, *S. trowbridgii*, and *S. vagrans*. *S. navigator* can be distinguished from these species by having the venter silver gray and, except for *S. palustris*, the inner side of the unicuspid teeth without a secondary cusp; four upper incisors well developed; braincase broad and feet conspicuously fringed with bristle-like hairs. From *S. hoyi* and *S. palustris*, by having sagittal and lambdoidal crests greatly developed, and a well-developed inframaxillary ridge extending along the unicuspid row. From *S. palustris*, by its smaller size; upperparts with a paler coloration; unicuspid teeth markedly narrower; ratio of the first upper unicuspid teeth length/distance from the first to the fifth upper unicuspid teeth greater than 0.25 mm, and ratio of the coronoid process height/dentary length greater than 0.53 mm.

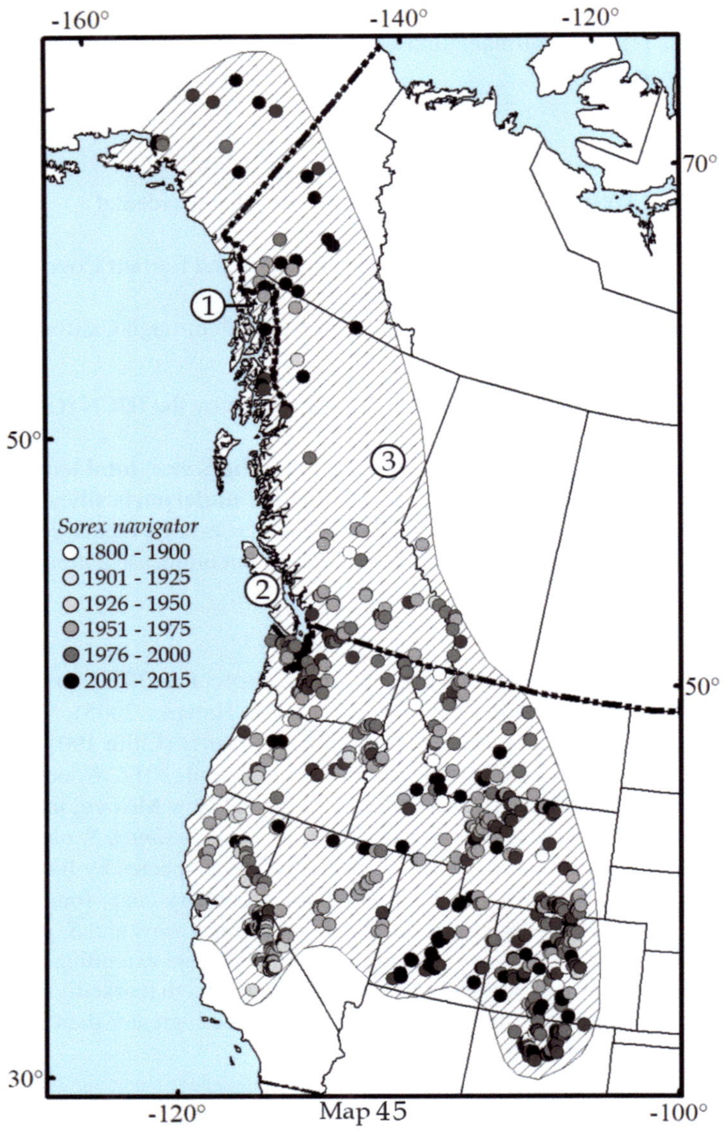

Map 45 Distribution map of *Sorex navigator*

Additional Literature: Jackson (1928), Junge and Hoffman (1981), Jones et al. (1982), George (1988), Harris (1999), Shafer and Stewart (2007), Nagorsen et al. (2017), Woodman (2018).

Sorex obscurus Merriam, 1895

Darker shrew, musaraña oscura

1. *S. o. alascensis* Merriam, 1895. From southwestern Alaska south through northwestern British Columbia.
2. *S. o. obscurus* Merriam, 1895. From western Alaska east through Manitoba, and south through Utah and New Mexico.
3. S. *o. shumaginensis* Merriam, 1900. Western Alaska.
4. *S. o. soperi* Anderson and Rand, 1945. Eastern Saskatchewan and western Manitoba.

Conservation: *Sorex obscurus* (as *S. monticolus*) is listed as Threatened in the Norma Oficial Mexicana (DOF 2019) and as Least Concern by the IUCN (Matson et al. 2016b).

Characteristics: Montane shrew is small sized; total length 108.0–120.0 mm; condylobasal length 16.4–17.2 mm. Dorsal pelage brown; fur of the underparts ashy gray, whitish at the tip; hair covering the hips projecting backward as "thorns" and bicolored; hind feet small; **tail short, bicolored, brownish dorsally and whitish ventrally, near-dark all round at the tip**; third unicuspid teeth smaller than the fourth; first lower incisors with two deep valleys between three crests; **rostrum proportionally shortest within *Sorex***.

Comments: *Sorex obscurus* was recognized as a distinct species (Alexander 1996; Woodman 2012) including the subspecies *S. o. alascensis*, *S. o. obscurus*, S. *o. shumaginensis*, and *S. o. soperi;* it was previously regarded as a subspecies of *S. monticolus* (Carraway 1990; Alexander 1996), and *S. vagrans* (Hall 1981). *S. obscurus* is widespread, ranging from Alaska and western Canada south through Utah and New Mexico (Map 46). It can be found in sympatry, or nearly so, with *S. arcticus*, *S. cinereus*, *S. eximus*, *S. merriami*, *S. monticolus*, *S. palustris*, *S. preblei*, *S. tundrensis*, and *S. vagrans*. *S. obscurus* can be distinguished from *S. arcticus* and *S. tundrensis*, by having the third unicuspid teeth smaller than the fourth; upper unicuspid teeth usually with pigmentation on the ridges extending from the apices of the cingulum, uninterrupted by an anteroposterior groove. From *S. merriami*, by usually lacking the postmandibular foramen or, if present, it is small and often only on one mandible; unicuspid teeth with a ridge, usually pigmented, on the lingual face, running from the apex to the cingulum, sometimes ending in a pigmented cusplet. From *S. cinereus*, *S. merriami*, and *S. preblei*, by having the third unicuspid teeth smaller than the fourth; upper unicuspid teeth usually with pigmentation on the ridges extending from the apices of cingulum and uninterrupted by an anteroposterior groove. From *S. hoyi*, by having the third unicuspid tooth similar to the others; the fifth unicuspid teeth can be small but not minute; first and second unicuspid teeth without a distinct internal ridge with an internal cusp, accessory medial tine on the anterior surface of the first upper incisors and in the mandible in lingual view; most posterior margin of the incisors anterior to the paraconid of the first lower molars. From *S. monticolus*, by having the palate wider on average; rostrum larger, wider, and less attenuate; dentition stronger, especially in unicuspid teeth. From *S. navigator* and *S. palustris*, by having the hind feet not fimbriated; vertical depth of the rostrum at the level of the third unicuspid teeth greater than one-half of the distance from the anterior border of the infraorbital foramen to the posterior border of the first incisors. From *S. vagrans*, which can be found in a narrow distribution band, by having a medial tine beginning above the main pigmented area on the first upper incisors; smaller size; tail shorter and smaller teeth robust and harder.

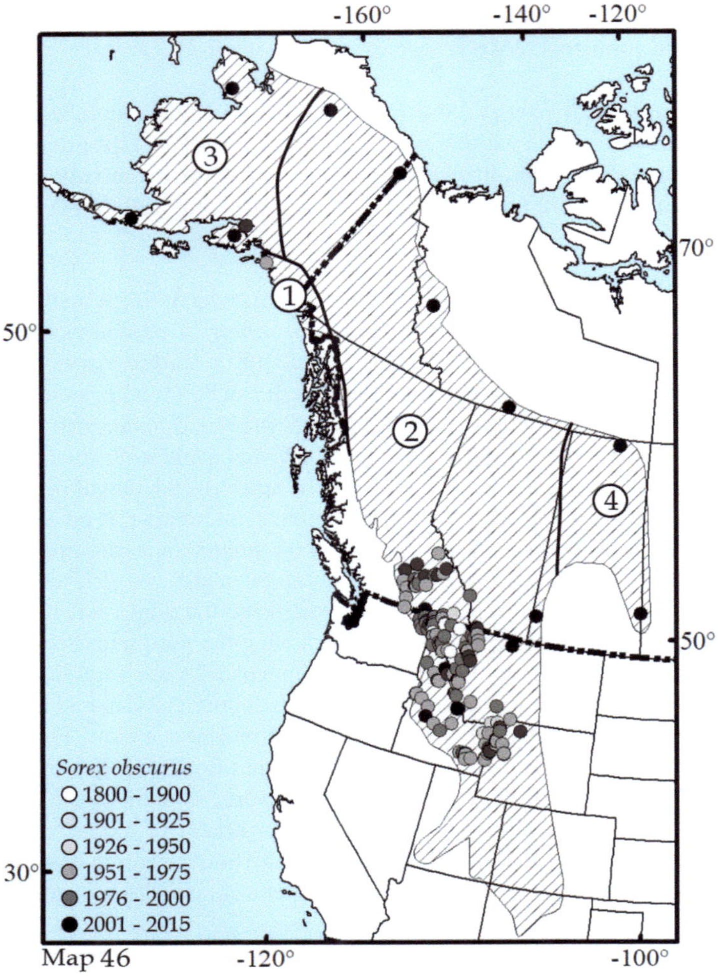

Map 46 Distribution map of *Sorex obscurus*

Additional Literature: Jackson (1928), Hawes (1975), Hennings and Hoffmann (1977), Junge and Hoffman (1981), van Zyll de Jong (1982, 1983c), Carraway (2007), George (1988), George and Smith (1991), Jones et al. (1992), Hutterer (1993, 2005), Stewart and Baker (1994b), Smith and Belk (1996), Shafer and Stewart (2007), Guevara et al. (2015).

Sorex oreopolus Merriam, 1892

Mexican long-tailed shrew, musaraña mexicana

Monotypic.

Conservation: *Sorex oreopolus* is listed as Least Concern by the IUCN (Álvarez-Castañeda et al. 2017g).

Characteristics: Mexican long-tailed shrew is small sized; total length 90.0–117.0 mm; condylobasal length 16.8–18.5 mm. Dorsal fur with two bands, pale reddish and white at the base and dark brown at the tip; fur of the underparts dark gray at the base and light red at the tip; **tail bicolored all round at the tip, soiled whitish ventrally;** fourth upper unicuspid teeth larger than the third upper unicuspid teeth; 32 teeth pigmented; **interorbital region especially elongated**.

Comments: *Sorex oreopolus* was listed with three different subspecies (Hall 1981), *S. o. oreopolus*, *S. o. emarginatus*, and *S. o. ventralis* (Findley 1955a), but all three were recognized as distinct species; for this reason, *S. oreopolus* is monotypic (Carraway 2007). *S. oreopolus* ranges in the highlands of the Eje Neovolcánico from Colima and eastern Jalisco east through western Puebla (Map 47). It can be found in sympatry, or nearly so, with *S. mediopua*, *S. orizabae*, *S. saussurei*, *S. ventralis*, and *S. veraecrucis*. *S. oreopolus* can be distinguished from all these species by having the tail bicolored all round at the tip and soiled whitish ventrally; interorbital region especially elongated. From *S. mediopua*, *S. orizabae*, *S. ventralis*, and *S. veraecrucis*, by having the third unicuspid teeth smaller than the fourth; upper unicuspid teeth usually with pigmentation on the ridges extending from the apices of cingulum, uninterrupted by an anteroposterior groove. From *S. emarginatus*, *S. saussurei*, and *S. ventralis*, by having the first upper unicuspid teeth with a medial crest usually above the pigment, and the pigment always separated from the body pigment.

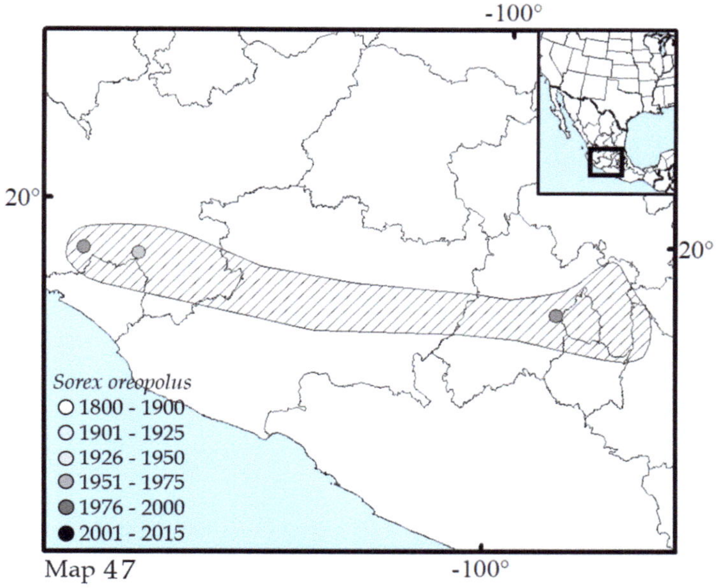

Map 47 Distribution map of *Sorex oreopolus*

Additional Literature: Diersing and Hoffmeister (1977), Hennings and Hoffmann (1977), Junge and Hoffman (1981), Ruiz-Soberanes and Gómez-Álvarez (2010), Guevara et al. (2015), Montiel-Parra et al. (2019).

Sorex orizabae Merriam, 1895

Orizaba long-tailed shrew, musaraña de Orizaba

Monotypic.

Conservation: *Sorex orizabae* is listed as Least Concern by the IUCN (Álvarez-Castañeda et al. 2017h).

Characteristics: Orizaba long-tailed shrew is the smallest **shrew in México**; total length 92.0–108.0 mm; condylobasal length 16.2–17.2 mm. Dorsal pelage with two bands, one with whitish hair in the middle and dark brown at the tip giving the appearance of speckled; fur of the underparts bicolored, dark gray at the base and pale whitish at the tip; **the only species with a pale-colored belly in México**; fourth unicuspid teeth larger than the third unicuspid teeth; first lower incisors with two deep valleys between three crests.

Comments: *Sorex orizabae* was reinstated as a distinct species (Carraway 2007), previously regarded as a subspecies of *S. vagrans* (Hennings and Hoffmann 1977) and later of *S. oreopolus* (Junge and Hoffman 1981). *S. orizabae* is known only from the highlands of Michoacán, Ciudad de México, Estado de México, Morelos, Puebla, Tlaxcala, eastern Hidalgo, and western Veracruz (Map 48). It can be found in sympatry, or nearly so, with *S. macrodon*, *S. mediopua*, *S. oreopolus*, *S. saussurei*, and *S. ventralis*. *S. orizabae* can be distinguished from these species by its small size and with a pale-colored belly. From *S. macrodon* and *S. mediopua*, by having the condylobasal length on average less than 17.9 mm and the length from the first upper unicuspid teeth to the third upper molars less than 6.5 mm. From *S. oreopolus*, *S. saussurei*, and *S. ventralis*, by having the third unicuspid teeth smaller than the fourth; upper unicuspid teeth usually with pigmentation on the ridges extending from the apices of cingulum and uninterrupted by an anteroposterior groove.

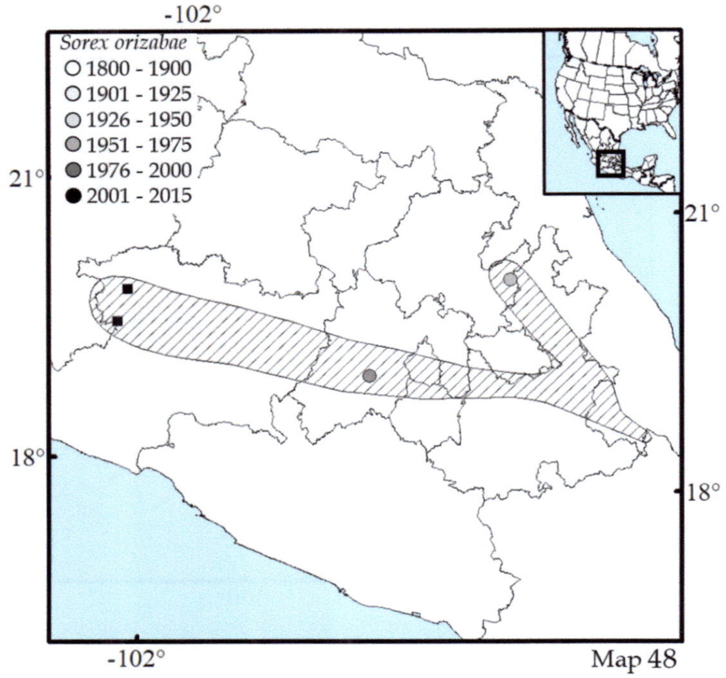

Map 48 Distribution map of *Sorex orizabae*

Additional Literature: Findley (1955a), Guevara et al. (2015), Woodman (2018).

Sorex ornatus Merriam, 1895

Ornate shrew, musaraña ornamentada

1. *S. o. californicus* Merriam, 1895. Restricted to San Pablo Bay, Suisun Bay, and San Joaquin River to central Santa Barbara, California.
2. *S. o. juncensis* Nelson and Goldman, 1909. Restricted to San Quintin Valley, Baja California.
3. *S. o. lagunae* Nelson and Goldman, 1909. Restricted to Sierra la Laguna, Baja California Sur.
4. *S. o. ornatus* Merriam, 1895. From Monterey and Sierra Nevada in California south to northwestern Baja California.
5. *S. o. parvidens* Jackson, 1921. San Bernardino and San Gabriel Mountains, California.
6. *S. o. relictus* Grinnell, 1932. Restricted to Buena Vista Lake, California.
7. *S. o. salarius* von Bloeker, 1939. Restricted to Monterey County, California.
8. *S. o. salicornicus* von Bloeker, 1932. Coastal area of Los Angeles County, California.

Conservation: *Sorex ornatus* and *S. juncensis* are listed as Threatened by the Norma Oficial Mexicana (DOF 2019), as Endangered (relictus) by the Species Act of the United States, and as Least Concern by the IUCN (Álvarez-Castañeda et al. 2016ag).

Characteristics: Ornate shrew is small sized; total length 80.0–110.0 mm; condylobasal length 15.6–17.1 mm. Dorsal and ventral ash gray to dark brown, **darker on the rump strongly contrasting with the shoulders, and anterior back clear gray, mainly in summer**; fur of the underparts whitish; hair on the hips projecting backward as small "thorns"; tail indistinctly bicolored, paler ventrally on the proximal half; first lower incisors with two deep valleys between three crests; upper third unicuspid teeth smaller than the fourth; medial tine of the upper first incisor teeth contained entirely within the pigmented area; upper toothrow not crowded, with a distinct triangular space between the posteromedial edge of the fifth unicuspid teeth and the anteromedial edge of the premolars; **foramen magnum relatively dorsal, positioned more into the supraoccipital and less into the basioccipital**.

Comments: *Sorex juncensis* was previously considered a junior synonym of *S. ornatus* (Camargo and Álvarez-Castañeda 2018), and *S. o. sinuosus* as a subspecies of *S. ornatus* (Maldonado et al. 2001, 2004), nowadays within *S. vagrans* (Woodman 2018). *S. ornatus* ranges from the northern San Francisco Bay and southern Buck Lake south through by the coastal region of California south through San Pedro Mártir, Baja California, with a discontinuous population in Sierra de la Laguna, Baja California Sur (Map 49). It can be found in sympatry, or nearly so, with *S. palustris*, *S. trowbridgii*, and *S. vagrans*. *S. ornatus* can be distinguished from species of the subgenus *Otisorex* by the following combination of the characters: upper third unicuspid teeth smaller than the fourth and condylobasal length from 14.7 to 17.1 mm; medial tine of the upper first incisor teeth contained entirely within the pigmented area, and upper toothrow not crowded, with a distinct triangular space between the posteromedial edge of the fifth unicuspid teeth and the anteromedial edge of the premolars, in addition, by having a darker coloration on the rump, strongly contrasting with the shoulders, and anterior back light gray, mainly in summer. From *S. palustris*, by having the hind feet not fimbriated; vertical depth of the rostrum at the level of the third unicuspid teeth greater than one-half of the distance from the anterior border of the infraorbital foramen to the posterior border of the first incisors. From *S. trowbridgii* and *S. vagrans*, by usually having no postmandibular foramen or, if present, it is small and often only on one mandible; unicuspid teeth with a ridge, usually pigmented, on the lingual face, running from the apex to the cingulum, sometimes ending in a pigmented cusplet, and skull flattened.

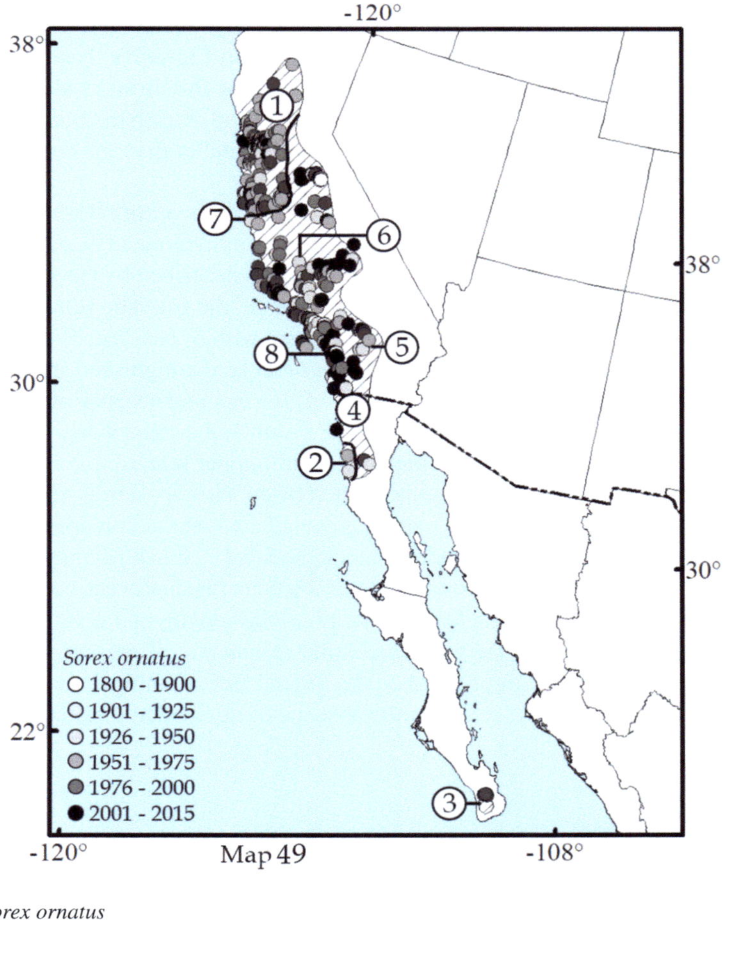

Map 49 Distribution map of *Sorex ornatus*

Additional Literature: Jackson (1928), Findley (1955a), Hoffmann and Peterson (1967), Repenning (1967), Williams (1979), Walker (1980), Brown and Rudd (1981), Junge and Hoffman (1981), Owen and Hoffmann (1983), George (1988), Carraway (2007), Woodman (2012, 2018), Guevara et al. (2015).

Sorex pacificus Coues, 1877

Pacific shrew, musaraña del Pacífico

1. *S. p. bairdi* Merriam, 1895. Restricted to the extreme northwestern Oregon.
2. *S. p. calvertensis* Cowan, 1941. Restricted to the Calvert Islands, British Columbia.
3. *S. p. cascadensis* Carraway, 1990. Restricted to Cascade Range of Oregon.
4. *S. p. elassodon* Osgood, 1901. Different Islands of northwestern British Columbia such as Graham, Queen Charlotte, Langara, in Alaska to Admiralty, Baranof, Dall, Duke, Forester, Kuiu, Kupreanof, Mitkof, Prince of Wales, and Zarembo.
5. *S. p. insularis* Cowan, 1941. Restricted to the Bardswell group Islands, British Columbia.
6. *S. p. isolatus* Jackson, 1922. Vancouver Island, British Columbia.
7. *S. p. longicauda* Merriam, 1895. Northwestern coast of British Columbia.
8. *S. p. malitiosus* Jackson, 1919. Only known to Warren and Coronation Island, Alaska.
9. *S. p. pacificus* Coues, 1877. From Mendocino County, California, to southwestern Oregon.
10. *S. p. permiliensis* Jackson, 1918. Restricted to northern Oregon.
11. *S. p. prevostensis* Osgood, 1901. Restricted to the Queen Charlotte Islands, British Columbia.
12. *S. p. setosus* Elliot, 1899. Southwestern British Columbia and Washington.

Conservation: *Sorex pacificus* is listed as Least Concern by the IUCN (Hammerson 2016).

Characteristics: Pacific shrew is small but larger sized among long-tailed *Sorex* in North America; total length 149.5–153.6 mm; condylobasal length 21.4–23.0 mm. Dorsal pelage in summer **uniform cinnamon rufous** and in winter darker throughout with the back black tipped; fur of the underparts similar to the dorsum; tail length about equal to the body length without the head; hind feet large; **ears conspicuous; skull large and massive**; braincase broad and flattened; squamosal bone strongly developed, forming a horizontal shelf; **medial tines in the incisors absent**; first and second unicuspid teeth subequal; third unicuspid teeth about one-half of the size of the second; fourth teeth larger than the third but not of the size of the second; fifth unicuspid teeth variable in size, usually slightly smaller than the third.

Comments: *Sorex pacificus* formerly was recognized as a subspecies of *S. vagrans* (Carraway 1985). *S. p. sonomae* was elevated to full species (Carraway 1990). *S. yaquinae* was regarded as synonymous of *S. p. pacificus* (Carraway 1990). The new subspecies *S. p. cascadensis* from the Cascade Range of Oregon was described by Carraway (1990). *S. pacificus* ranges from Langara, Alaska, south through western Oregon to the Coast Range, the Cascade Range and the coastal area of northwestern California (Map 50). It can be found in sympatry, or nearly so, with *S. bendirii*, *S. cinereus*, *S. eximus*, *S. obscurus*, *S. palustris*, *S. sonomae*, *S. trowbridgii*, and *S. vagrans*. *S. pacificus* can be distinguished from these species by having ears conspicuous; skull large and massive, suggesting the morphology of *Blarina*, and absence of the medial tines. From *S. bendirii*, by having upper parts always brownish; hind foot less than 18.0 mm and condylobasal length less than 19.0 mm. From *S. cinereus*, by having the third unicuspid teeth smaller than the fourth; upper unicuspid teeth usually with pigmentation on the ridges extending from the apices of cingulum and uninterrupted by an anteroposterior groove. From *S. obscurus*, by having the medial tine very small and beginning above the main pigmented area on the first upper incisors. From *S. palustris*, by having the hind feet not fimbriated; vertical depth of the rostrum at the level of the third unicuspid teeth greater than one-half of the distance from the anterior border of the infraorbital foramen to the posterior border of the first incisors. From *S. sonomae*, by having the lower canine length smaller than the premolar length, and maxillary zygomatic processes finely pointed. From *S. trowbridgii*, by usually lacking the postmandibular foramen or, if present, it is small and often only on one mandible; unicuspid teeth with a ridge, usually pigmented on the lingual face, running from the apex to the cingulum, sometimes ending in a pigmented cusplet. From *S. vagrans*, by the absence of the medial tines in the first upper incisors.

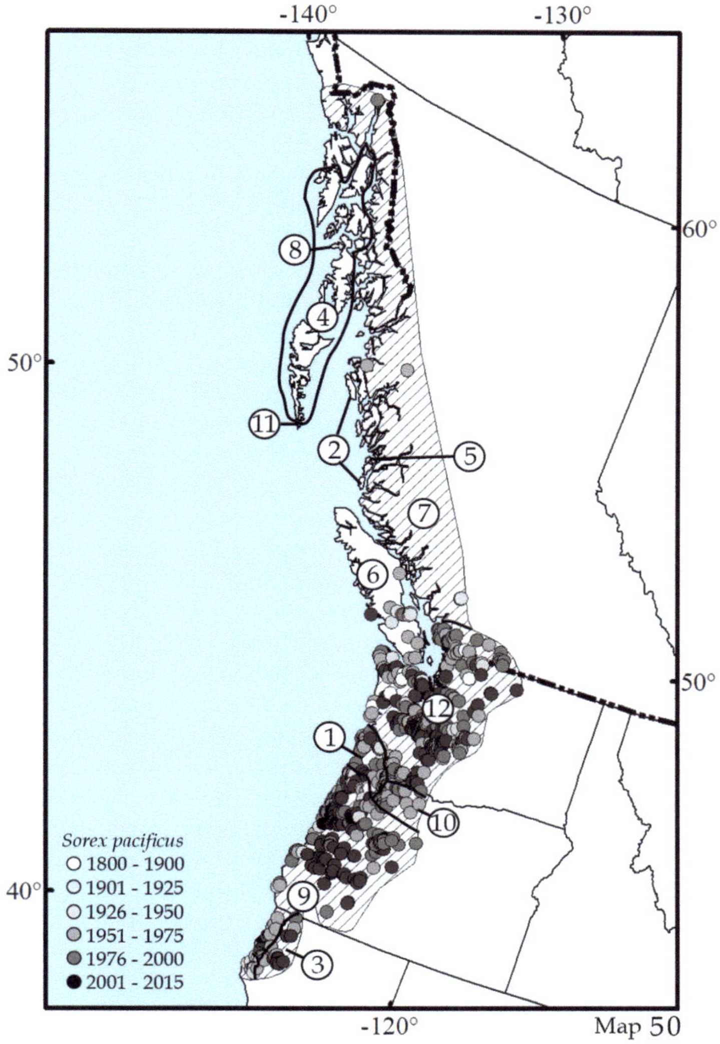

Map 50 Distribution map of *Sorex pacificus*

Additional Literature: Jackson (1928), Bailey (1936), Findley (1955a, b), Johnson and Ostenson (1959), Hoffman (1971), Hennings and Hoffmann (1977), Junge and Hoffman (1981), Carraway (1988), George (1988), Verts and Carraway (1998), Demboski and Cook (2001) Hutterer (2005), Woodman (2018).

Sorex palustris Richardson, 1828

Water shrew, musaraña de agua Americana

1. *S. p. hydrobadistes* Jackson, 1926. Northeastern South Dakota, central Minnesota, northern Wisconsin, and Michigan.
2. *S. p. palustris* Richardson, 1828. From Mckenzie and Alberta east through central Ontario and northern Wisconsin.

Conservation: *Sorex palustris* is listed as Least Concern by the IUCN (Cassola 2016do).

Characteristics: Water shrew is small but large sized among the *Sorex* in North America; total length 150.0–159.4 mm; condylobasal length 19.1–21.7 mm. Dorsal pelage dusky, with little hoary shades; fur of the underparts dull white, chest and anal region pale gray whitish; **tail sharply bicolored, blackish dorsally and all round near the tip, white ventrally, white band narrower than the black band**; feet large with the outer side dark and the inner side whitish; hind feet very long, broad, suitable for swimming; toes slightly webbed; braincase elevated above the rostrum plane; palatal and postpalatal notches rather broad; molars large and heavy; unicuspid teeth only slightly imbricated, first and second unicuspid teeth very broad and subequal, third much smaller and fourth larger than the third.

Comments: *Sorex alaskanus* was previously considered a subspecies of *S. palustris* (Junge and Hoffman 1981; Jarrell and MacDonald 1989; Jones et al. 1992; Harris 1999); it was elevated to full species (Hall 1981; Beneski and Stinson 1987; George 1988; Hutterer 1993, 2005; Carraway 1995; Baker et al. 2003a). *S. albibarbis albibarbis*, *S. a. gloveralleni*, *S. a. labradorensis*, *S. a. punctulatus*, and *S. a. turneri* are subspecies of *S. albibarbis* (Mycroft et al. 2011; Woodman 2018), and *S. p. navigator* and *S. p. brooksi* are full species, as *S. navigator* (Mycroft et al. 2011). *S. p. hydrobadistes*, *S. p. labradorensis*, *S. p. punctulatus*, and *S. p. turneri* need to be directly assessed before being assigned to either *S. palustris* or *S. albibarbis* (Mycroft et al. 2011) but are maintained within *S. palustris*. *S. palustris* ranges from Mackenzie and Alberta east through Ontario and the Great Lakes. In the west, south through mid-California, Nevada, Utah, and New Mexico, with a population in the White Mountains, Arizona (Map 51). It can be found in sympatry, or nearly so, with *S. arcticus*, *S. cinereus*, *S. eximus*, *S. dispar*, *S. fumeus*, *S. haydeni*, *S. hoyi*, *S. lyelli*, *S. maritimensis*, *S. merriami*, *S. nanus*, *S. navigator*, *S. obscurus*, *S. ornatus*, *S. pacificus*, *S. preblei*, *S. sonomae*, *S. tenellus*, *S. trowbridgii*, *S. tundrensis*, and *S. vagrans*. *S. palustris* can be distinguished from these sympatric species, except for *S. albibarbis* and *S. navigator*, by having the inner side of the unicuspid teeth without a secondary cusp, four upper incisors well developed; braincase broad, and feet conspicuously fringed with bristle-like hairs. From *S. albibarbis*, by having the fur of the underparts white and the tail sharply bicolored. From *S. navigator*, by its larger size, upperparts darker and unicuspid teeth broader; ratio of the first upper unicuspid length/distance from the first to the fifth upper unicuspid teeth less than 0.25 mm; ratio of the height of the coronoid process/dentary length less than 0.53 mm. From *S. arcticus*, *S. maritimensis*, and *S. tundrensis*, by having the third unicuspid teeth smaller than the fourth; upper unicuspid teeth usually with pigmentation on the ridges extending from the apices of cingulum and uninterrupted by an anteroposterior groove. From *S. hoyi*, by having the third unicuspid teeth similar to the others; the fifth unicuspid teeth can be small but not minute; first and second unicuspid teeth without a distinct internal ridge with an internal cusp; an accessory medial tine on the anterior surface of the first upper incisors and in the mandible in lingual view, and the most posterior margin of the incisors anterior to the paraconid of the first lower molars. From *S. sonomae*, by having the maxillary zygomatic processes finely pointed.

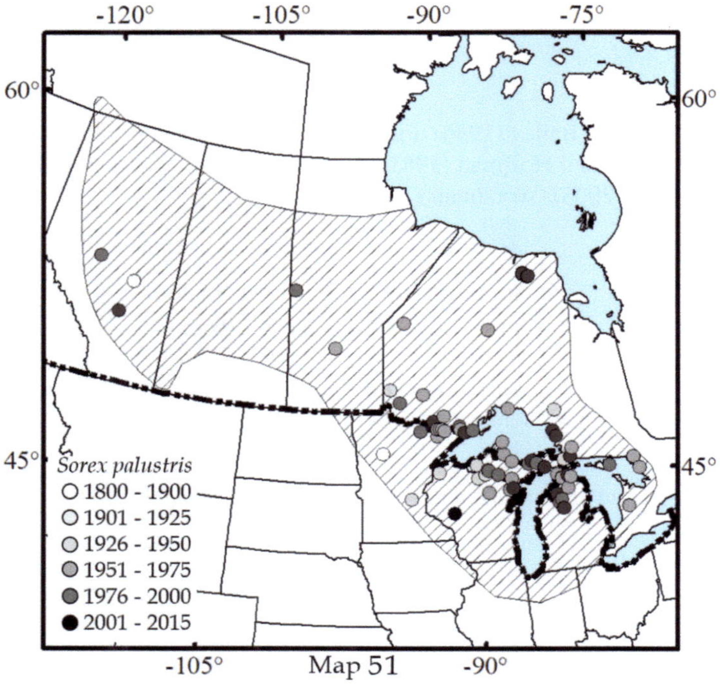

Map 51 Distribution map of *Sorex palustris*

Additional Literature: Jackson (1928), Dalquest (1948), Johnson (1951), Findley (1955a), Hall (1981), van Zyll de Jong (1983a), Hoffmeister (1986), Beneski and Stinson (1987), George (1988), Shafer and Stewart (2007), Nagorsen et al. (2017), Woodman (2018).

Sorex preblei Jackson, 1922

Preble's shrew, musaraña del noroeste

Monotypic.

Conservation: *Sorex preblei* is listed as Least Concern by the IUCN (Cassola 2019).

Characteristics: Preble's shrew is very small sized; **total length 77.0–95.0 mm; condylobasal length 13.8–15.1 mm**. Dorsal pelage grayish; **fur of the underparts silvery; snout long and pointed**; eyes small; ears somewhat conspicuous; tail bicolored, olive brown above and hazel ventrally, darkened toward the tip; **medial tine on the edge of the first upper incisors long, acutely pointed, and within the pigmented area**.

Comments: *Sorex preblei* ranges from Washington and western Wyoming west through northwestern California and south through northern Utah and Nevada (Map 52). It can be found in sympatry, or nearly so, with *S. cinereus*, *S. eximus*, *S. haydeni*, *S. merriami*, *S. nanus*, *S. obscurus*, *S. palustris*, and *S. vagrans*. *S. preblei* can be distinguished from *S. cinereus*, *S. haydeni*, and *S. merriami* by its smaller size with the rostrum relatively short; lower toothrow length less than 4.1 mm and condylobasal length less than 14.6 mm. From *S. haydeni*, by having the maxillary breadth equal to or less than 4.2 mm. From *S. merriami*, by having a medial tine on the edge of the first upper incisors; maxillary breadth equal to or less than 4.2 mm; back grayish and underparts silvery-colored. From *S. obscurus*, *S. nanus*, and *S. vagrans*, by having the third unicuspid teeth equal to or larger than the fourth; upper teeth weakly pigmented; ridges extending from the apices of the unicuspid medially toward the cingulum but incomplete and not ending in internal cusplets. From *S. nanus*, by having the dentary below the first lower molars much deeper than the height of the lower first molars at the metaconid and a higher (greater than 2.9 mm) coronoid process. From *S. palustris*, by having the hind feet not fimbriated; vertical depth of the rostrum at the level of the third unicuspid teeth greater than one-half of the distance from the anterior border of the infraorbital foramen to the posterior border of the first incisors.

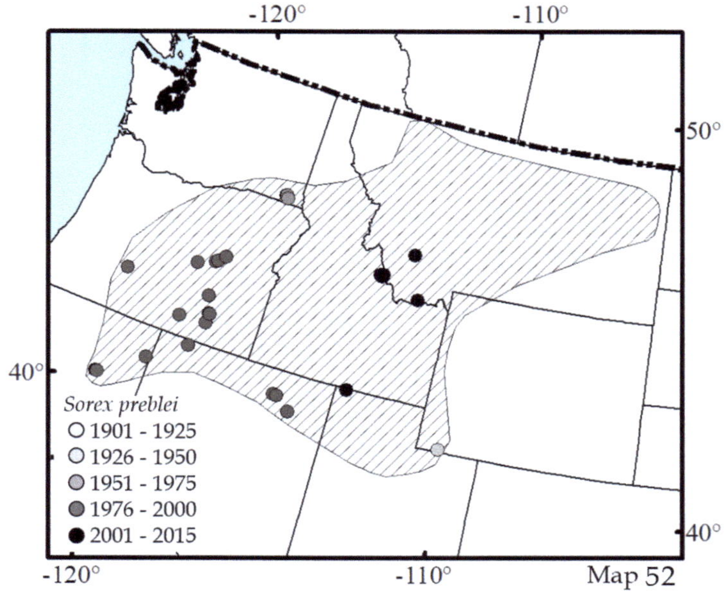

Map 52 Distribution map of *Sorex preblei*

Additional Literature: Jackson (1922, 1928), Findley (1955a), Hoffmann and Pattie (1968), Hoffmann et al. (1969), Hoffmann and Fisher (1978), Junge and Hoffman (1981), Larrison and Johnson (1981), Verts and Carraway (1984), George (1988), Ports and George (1990), Cornely et al. (1992), Long and Hoffmann (1992), Demboski and Cook (2003), Shohfi et al. (2006).

Sorex pribilofensis Merriam, 1895

Pribilof island shrew, musaraña de la isla Pribilof

Monotypic.

Conservation: *Sorex pribilofensis* is listed as Endangered (B1ab (iii) + 2ab (iii): extent of occurrence less than 5000 km^2 in five or less populations and under continuing decline in area, extent and quality of habitat + area of occupancy less than 500 km^2, severely fragmented and under continuing decline in area, extent and quality of habitat) by the IUCN (Matson 2018).

Characteristics: Pribilof Island shrew is small sized; total length 92.0–103.0 mm; condylobasal length 15.5–16.0 mm. Dorsal pelage with the back chocolate brown and the flanks dull ochereous buff; **fur of the underparts whitish; chin, neck, and feet white**; tail sharply bicolored, dorsally brown and ventrally white; skull short and thick; interorbital constriction greatly swollen; palatopterygoids very broad.

Comments: *Sorex pribilofensis* was formerly regarded as *S. cinereus*; it was later recognized as a distinct species by Demboski and Cook (2003). *S. hydrodromus* was considered the name of the species, with ranges in Saint Paul Island. However, the type specimen of *S. hydrodromus* is morphologically different from the samples of shrews from Pribilofs Island (van Zyll de Jong 1982). Consequently, the correct name should be *S. pribilofensis* (Rausch and Rausch 1997; Jones et al. 1997; Baker et al. 2003a; Hutterer 2005). *S. pribilofensis* is restricted to Saint Paul Island, Alaska, being the only species of *Sorex* in the islands (Map 53).

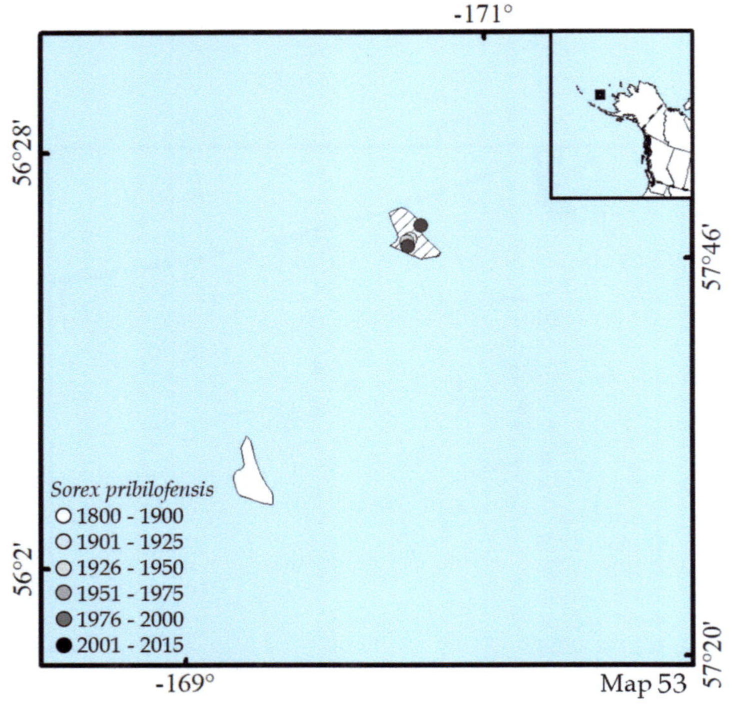

Map 53 Distribution map of *Sorex pribilofensis*

Additional Literature: Jackson (1928), van Zyll de Jong (1982), Byrd and Norvell (1993), Esteva et al. (2010), Amori et al. (2011).

Sorex rohweri Rausch et al., 2007

Olympic shrew, musaraña de la península Olimpia

Monotypic.

Conservation: *Sorex rohweri* is listed as Least Concern by the IUCN (Cassola 2017b).

Characteristics: Olympic shrew is small sized; total length 80.0–110.0 mm; condylobasal length 15.0–16.5 mm. Dorsal pelage grayish brown to dark brown, extending throughout the flanks; fur of the underparts paler than in the dorsum; snout long and pointed; tail bicolored, dorsally brownish and ventrally buffy; teeth tips chestnut brown, including the inner ridges of the upper unicuspid teeth; third unicuspid teeth smaller than the fourth, fifth unicuspid teeth small; first and second lower molars with an internal secondary cusp not distinctly large.

Comments: *Sorex rohweri* was previously misidentified as *S. cinereus* but is now recognized as a distinct species based on multiple morphological characters (Rausch et al. 2007). This species has a wide range on the Olympic Peninsula and the surrounding regions in western Washington and also ranges to the north through British Columbia (Map 54). It can be found in sympatry, or nearly so, with *S. bendirii*, *S. cinereus*, *S. eximus*, *S. obscurus*, *S. navigator*, *S. trowbridgii*, and *S. vagrans*. *S. rohweri* can be distinguished from *S. bendirii*, by having the dorsal pelage always brownish; hind foot length less than 18.0 mm and condylobasal length less than 19.0 mm. From *S. cinereus*, by having the third unicuspid teeth smaller than the fourth; upper unicuspid teeth usually with pigmentation on the ridges extending from the apices of cingulum and uninterrupted by an anteroposterior groove. From *S. obscurus* and *S. trowbridgii*, by having the combined length of the two first molars equal to or less than 2.4 mm. From *S. navigator*, by having the hind feet not fimbriated; vertical depth of the rostrum at the level of the third unicuspid teeth greater than one-half of the distance from the anterior border of the infra-orbital foramen to the posterior border of the first incisors. From *S. vagrans*, by having the combined length of the two first molars equal to or less than 2.4 mm.

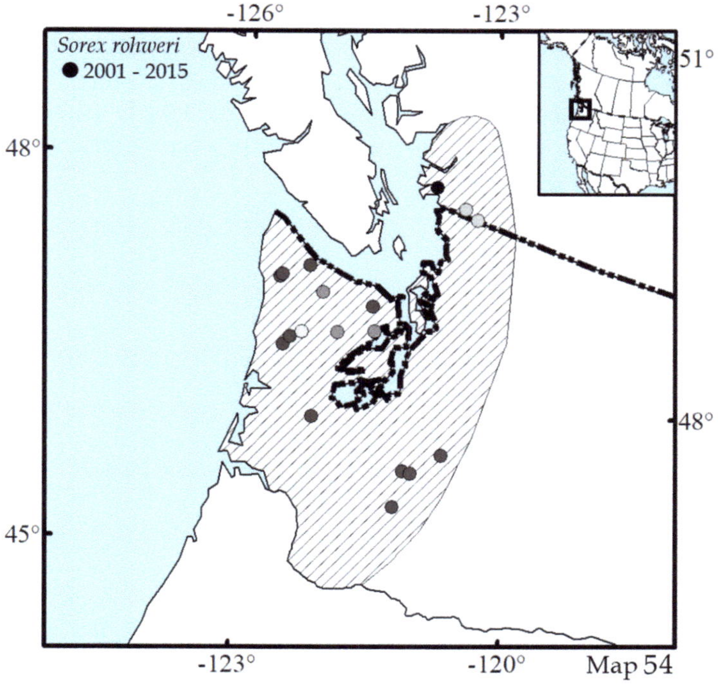

Map 54 Distribution map of *Sorex rohweri*

Additional Literature: George (1988), Dannelid (1989).

Sorex sonomae Jackson, 1921

Fog shrew, musaraña de la niebla

1. *S. s. sonomae* Jackson, 1921. Coastal areas of Oregon and northwestern California.
2. *S. s. tenelliodus* Carraway, 1990. Oregon south through west slope of the Cascade Mountains to Hilt, Siskiyou County, California.

Conservation: *Sorex sonomae* is listed as Least Concern by the IUCN (Cassola 2017c).

Characteristics: Fog shrew is small; total length 105.0–180.0 mm; condylobasal length 21.0–21.7 mm. Dorsal pelage paler dark brown to medium pale brown, neutral black at the base; fur of the underparts dark brown, neutral very dark gray at the base; tail with a similar coloration to the back; skull large and robust; **maxillary zygomatic processes rounded**; **coronoid process broad, heavy, rugose, and deep**; first upper incisors about the same length of the entire pigmented region; unicuspid teeth large, robust, straight-sided, and tightly appressed from near the alveolus to the crown; first and second upper unicuspid teeth almost square; fifth upper unicuspid teeth usually rectangular and usually with a gap between the fifth upper unicuspid teeth and the fourth premolars; however, the fourth upper premolars never overlay the fifth unicuspid teeth.

Comments: *Sorex sonomae* was described as *S. pacificus sonomae* (Jackson 1921), later as a subspecies of *S. vagrans sonomae* (Findley 1955a), and then recognized as a distinct species (Carraway 1990). *S. sonomae* ranges from south Oregon to south through San Francisco Bay, California (Map 55). It can be found in sympatry, or nearly so, with *S. bendirii*, *S. cinereus*, *S. obscurus*, *S. pacificus*, *S. palustris*, *S. trowbridgii*, and *S. vagrans*. *S. sonomae* can be distinguished from these species, by having the maxillary zygomatic processes rounded (not finely pointed as in the other species). From *S. bendirii*, by having the dorsal pelage always brownish, hind foot length less than 18.0 mm, and condylobasal length less than 19.0 mm. From *S. cinereus*, by having the third unicuspid teeth smaller than the fourth, upper unicuspid teeth usually with pigmentation on the ridges extending from the apices of the cingulum and uninterrupted by an anteroposterior groove. From *S. obscurus*, by having the medial tine beginning above the main pigmented area on the first upper incisors. From *S. pacificus*, by having the lower canine length equal to or smaller than the premolar length. From *S. palustris*, by having the hind feet not fimbriated; vertical depth of the rostrum at the level of the third unicuspid teeth greater than one-half of the distance from the anterior border of the infraorbital foramen to the posterior border of the first incisors. From *S. trowbridgii*, by usually lacking the postmandibular foramen or, if present, it is small and often only on one mandible; unicuspid teeth with a ridge, usually pigmented, on the lingual face, running from the apex to the cingulum, sometimes ending in a pigmented cusplet. From *S. vagrans*, by having the maxillary zygomatic processes rounded and the coronoid process broad, heavy, rugose, and deep.

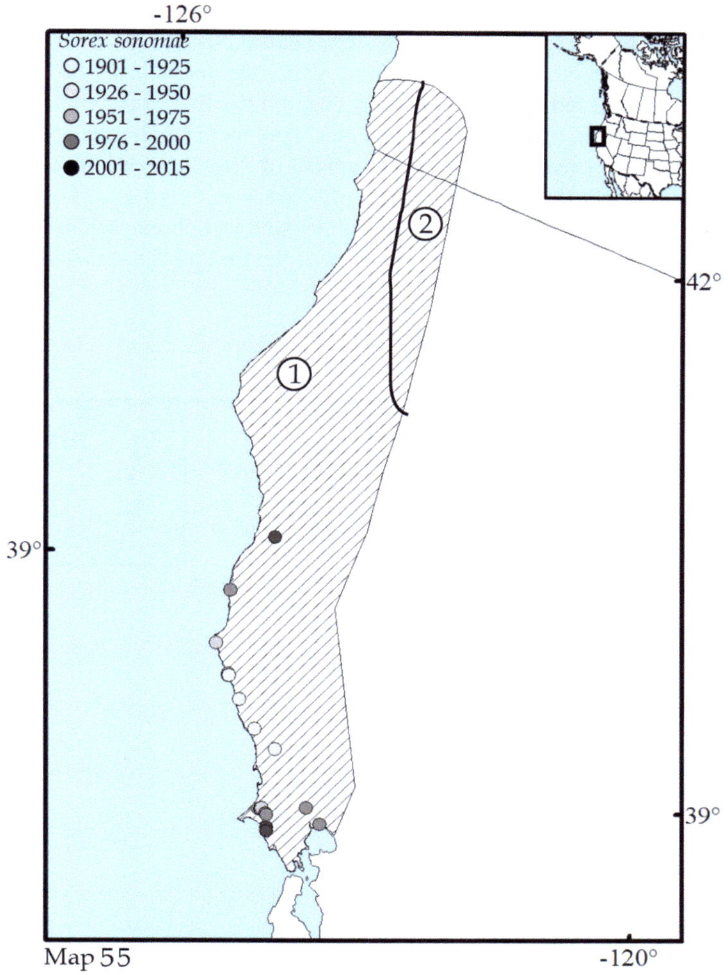

Map 55 Distribution map of *Sorex sonomae*

Additional Literature: Jackson (1928), Bailey (1936), Johnson and Ostenson (1959), Hennings and Hoffmann (1977), Carraway (1985, 1988), Verts and Carraway (1998), Hutterer (2005), Esteva et al. (2010), Woodman (2018).

Sorex tenellus Merriam, 1895

Inyo shrew, musaraña de las montañas Inyo

Monotypic.

Conservation: *Sorex tenellus* is listed as Least Concern by the IUCN (Cassola 2016dp).

Characteristics: Inyo shrew is small sized; total length 85.0–103.0 mm; condylobasal length 14.5–15.3 mm. **Dorsal pelage pale ash gray;** fur of the underparts white; feet white; tail bicolored, dark dorsally and white ventrally; skull very flat, small, and slender; palate narrow; pigmented ridge on unicuspid teeth extending from the apices to the cingulum, palate usually well developed and narrow; **accessory cusp on the first upper incisors developed and heavily pigmented**; lacking a postmandibular canal.

Comments: *Sorex tenellus* was previously considered a different species from *S. nanus* (George 1988; Jones et al. 1992; Hutterer 2005). It ranges in eastern-central California and western-central Nevada (Map 56). It can be found in sympatry, or nearly so, with *S. lyelli*, *S. merriami S. palustris*, and *S. vagrans. It can be distinguished* from all these species by its small size; condylobasal length less than 15.2 mm and an accessory cusp on the first upper incisors well developed and heavily pigmented. From *S. lyelli* and *S. merriami*, by having the third unicuspid teeth smaller than the fourth; upper unicuspid teeth usually with pigmentation on the ridges extending from the apices of cingulum and uninterrupted by an anteroposterior groove. From *S. ornatus*, by having a skull smaller and markedly narrower. From *S. palustris*, by having the hind feet not fimbriated; vertical depth of the rostrum at the level of the third unicuspid teeth greater than one-half of the distance from the anterior border of the infraorbital foramen to the posterior border of the first incisors. From *S. vagrans*, by having the combined length of the two first molars equal to or less than 2.4 mm.

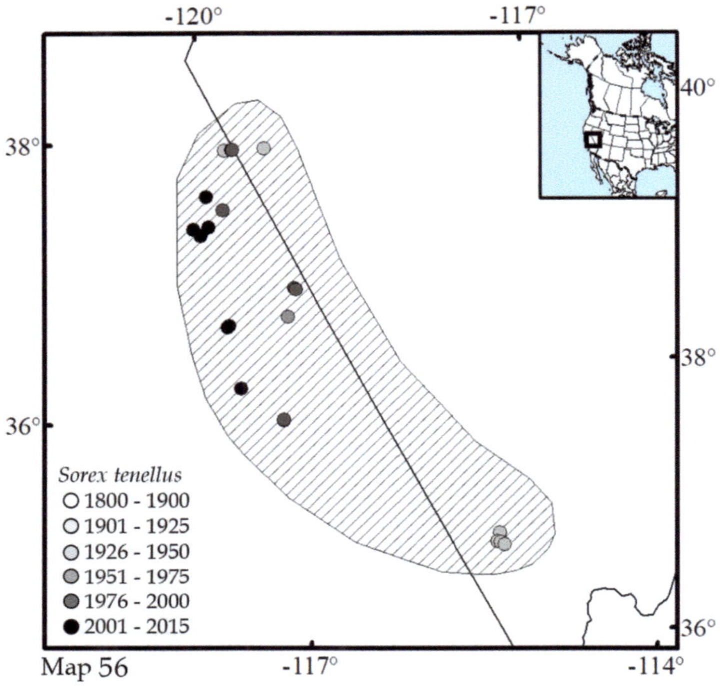

Map 56 Distribution map of *Sorex tenellus*

Additional Literature: Jackson (1928), Findley (1955a), Ingles (1965), Spencer (1966), Hoffmann and Peterson (1967), Repenning (1967), Diersing and Hoffmeister (1977), Hoffmann and Owen (1980), Williams (1984), Rickart et al. (2004), Shohfi et al. (2006), Esteva et al. (2010).

Sorex ugyunak Anderson and Rand, 1945

Barren ground shrew, musaraña del norte

Monotypic.

Conservation: *Sorex ugyunak* is listed as Least Concern by the IUCN (Cassola 2016dr).

Characteristics: Barren ground shrew is small sized; total length 74.0–103.0 mm; condylobasal length 15.8–17.5 mm. **Dorsal pelage in summer tricolored, medium brown on the back, pale brown or grayish brown on the flanks, and pale-colored on the underparts**; postmandibular canals well developed; with no pigmented ridges on unicuspid teeth.

Comments: *Sorex ugyunak* was recognized as a distinct species (van Zyll de Jong 1991a, b; Jones et al. 1992; Hutterer 1993, 2005), previously regarded as a subspecies of *S. cinereus* (Anderson and Rand 1945). *S. ugyunak* ranges in the tundra from the west of Hudson Bay west through northern Alaska (Map 57). It can be found in sympatry, or nearly so, with *S. cinereus*, *S. hoyi*, and *S. tundrensis*. *S. ugyunak* can be distinguished from *S. cinereus* mainly by its distribution area, as *S. ugyunak* is restricted to the tundra. From *S. hoyi*, by having the third unicuspid teeth similar to the others; the fifth unicuspid teeth can be small but not minute; first and second unicuspid teeth without a distinct internal ridge with an internal cusp; an accessory medial tine on the anterior surface of the first upper incisors and in the mandible in lingual view; most posterior margin of the incisors anterior to the paraconid of the first lower molars. From *S. tundrensis*, by having the third unicuspid teeth smaller than the fourth; upper unicuspid teeth usually with pigmentation on the ridges extending from the apices of cingulum and uninterrupted by an anteroposterior groove.

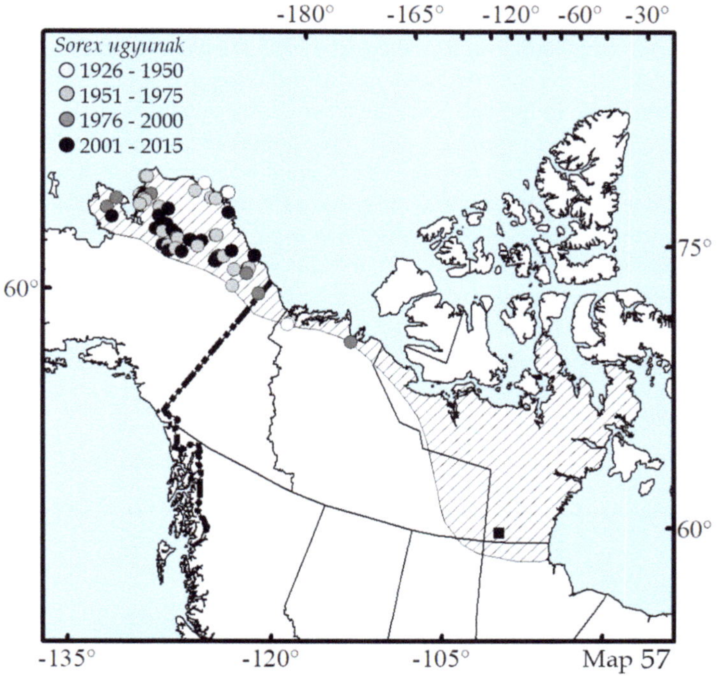

Map 57 Distribution map of *Sorex ugyunak*

Additional Literature: van Zyll de Jong (1982, 1983a, 1999b), Jones et al. (1986), Esteva et al. (2010).

Sorex vagrans Baird, 1857

Vagrant shrew, musaraña vagabunda

1. *S. v. halicoetes* Grinnell, 1913. Restricted to southern and eastern San Francisco Bay, California.
2. *S. v. paludivagus* von Bloeker, 1939. Restricted to Monterey Bay area, California.
3. *S. v. sinuosus* Grinnell, 1913. Restricted to Grizzly Island, California.
4. *S. v. vagrans* Baird, 1857. From southern British Columbia and western Montana south through highlands of California and Nevada, including northwestern Utah and western Wyoming.
5. *S. v. vancouverensis* Merriam, 1895. Vancouver, Bowen, Saltspring, Saturna, and South Pender Islands, southwestern British Columbia.

Conservation: *Sorex vagrans* is listed as Least Concern by the IUCN (Matson et al. 2016a).

Characteristics: Vagrant shrew is one of the smallest shrews; total length 80.0–94.0 mm; condylobasal length 14.9–15.5 mm. Dorsal pelage dark brown, varying to almost russet; fur of the underparts ashy; tail dorsally dusky and ventrally pale; skull with not particular characteristics.

Comments: *Sorex vagrans* was a complex of several species, many of which are recognized as distinct species, as *S. bairdi* (Hennings and Hoffmann 1977), *S. orizabae* (Carraway 2007), *S. monticolus* (Carraway 1990), *S. pacificus* (Carraway 1990), and *S. sonomae* (Carraway 1990). *S. v. sinuosus* was considered a subspecies of *S. ornatus* (Maldonado et al. 2001, 2004); nowadays, it is within *S. vagrans* (Woodman 2018). *S. vagrans* ranges from southern British Columbia south through Monterey, California, and east through Idaho, western Montana, western Wyoming, and northern Utah (Map 58). It can be found in sympatry, or nearly so, with *S. arizonae*, *S. bendirii*, *S. cinereus*, *S. hoyi*, *S. lyelli*, *S. merriami*, *S. navigator*, *S. obscurus*, *S. ornatus*, *S. pacificus*, *S. palustris*, *S. preblei*, *S. rohweri*, *S. sonomae*, *S. tenellus*, and *S. trowbridgii*. *S. vagrans* can be distinguished from *S. arizonae* by having larger medial tines on the first upper incisors; palate breadth measured across the fourth upper premolars greater than the breadth measured across the third upper molars; palate wide, especially as reflected by the breadth across the unicuspid fourth upper premolars; fifth unicuspid teeth slightly smaller. From *S. bendirii*, by having the dorsal pelage brownish, never grayish; hind foot length less than 18.0 mm and condylobasal length less than 19.0 mm. From *S. cinereus*, *S. eximus*, *S. lyelli*, *S. merriami*, and *S. preblei*, by having the third unicuspid teeth smaller than the fourth; upper unicuspid teeth usually with pigmentation on the ridges extending from the apices of cingulum and uninterrupted by an anteroposterior groove. From *S. hoyi*, by having the third unicuspid teeth similar to the others; the fifth unicuspid teeth can be small but not minute; first and second unicuspid teeth without a distinct internal ridge with an internal cusp; accessory medial tine on the anterior surface of the first upper incisors and in the mandible in lingual view; the most posterior margin of the incisors anterior to the paraconid of the first lower molars. From *S. obscurus*, by being smaller; tail shorter; larger teeth, paler, and upper first incisors with a pigmentation extending no further than the dorsal limits of the medial accessory cuspules. From *S. longirostris*, by having unpigmented or lightly pigmented lines that often separate the cuspules. From *S. navigator* and *S. palustris*, by the hind feet not extended; vertical depth of the rostrum at the level of the third unicuspid teeth greater than one-half of the distance from the anterior border of the infraorbital foramen to the posterior border of the first incisors. From *S. ornatus*, by usually having a well-developed postmandibular foramen; non-pigmented ridge on the unicuspid teeth lingual face and skull not flattened. From *S. pacificus*, by having medial tines in the first upper incisors. From *S. rohweri* and *S. tenellus*, by having the combined length of the two first molars equal to or greater than 2.4 mm. From *S. sonomae*, by having finely pointed maxillary zygomatic processes. From *S. trowbridgii*, interorbital breadth less than 3.5 mm; tail monocolor and a medial tine beginning above the main pigmented area on the first upper incisors.

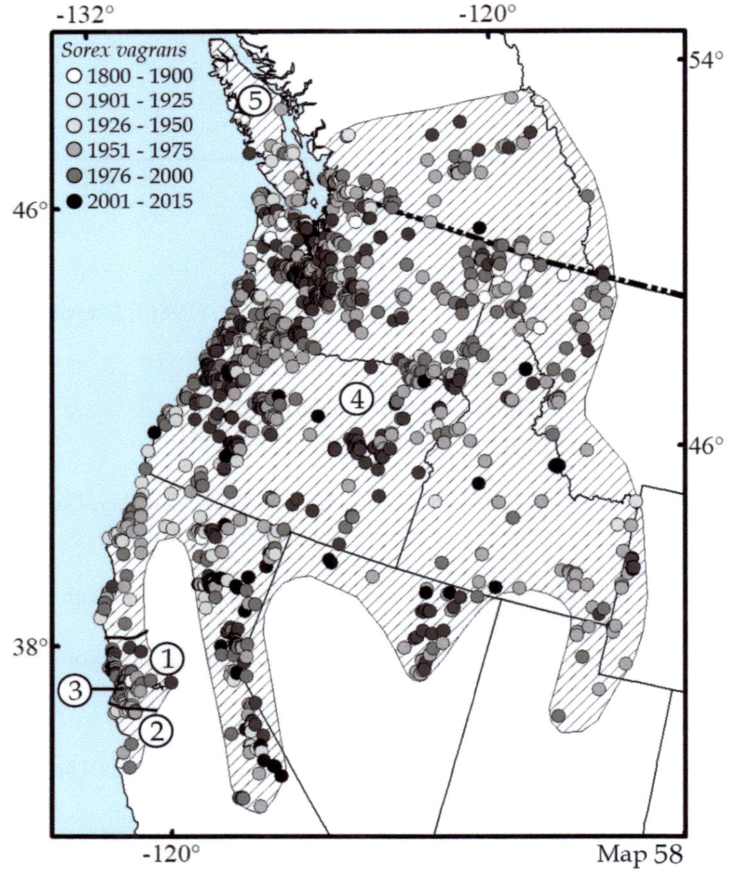

Map 58 Distribution map of *Sorex vagrans*

Additional Literature: Jackson (1928), Findley (1955b), Repenning (1967), Brown (1974), Junge and Hoffman (1981), van Zyll de Jong (1982, 1983a), Carraway (1987), George (1988), Smith (1988), George and Smith (1991), Woodward (1994), Nagorsen (1996), Demboski and Cook (2001), Gillihan and Foresman (2004), Alexander (1996), Carraway (1990, 2007).

Subgenus ***Sorex*** Linnaeus, 1758

Third unicuspid teeth larger than the fourth; upper teeth weakly pigmented; ridges extending from the apices of the unicuspid teeth medially toward the cingulum but incomplete, not ending in internal cusplets; postmandibular foramen usually well developed.

Sorex arcticus Kerr, 1792

Arctic shrew, musaraña del Ártico

1. *S. a. arcticus* Kerr, 1792. From western Yukon east through the Labrador Peninsula, and south through the border with the United States.
2. *S. a. laricorum* Jackson, 1925. From Northern Dakota to Wisconsin and eastern Michigan.

Conservation: *Sorex arcticus* is listed as Least Concern by the IUCN (Cassola 2016df).

Characteristics: Arctic shrew is small sized; total length 100.0–124.0 mm; condylobasal length 19.1–20.0 mm. Dorsal pelage very dark brown to black; fur of the underparts grayish brown; **pelage tricolored**; flanks paler and contrasting strongly with the dark dorsum; tail indistinctly bicolored, brown to brownish-black dorsally and paler ventrally; postmandibular canals well developed; no pigmented ridges on unicuspid teeth.

Comments: *Sorex arcticus maritimensis* (Smith 1939) and *S. a. tundrensis* (Bell and Hall 1956) were formerly subspecies but later elevated to full species (Youngman 1975; Volobouev and van Zyll de Jong 1988). *Sorex arcticus* ranges from Yukon and the Northwest Territories west through Québec and south through South Dakota, Minnesota, Michigan, and Wisconsin (Map 59). It can be found in sympatry, or nearly so, with *S. cinereus*, *S. eximus*, *S. fumeus*, *S. haydeni*, *S. hoyi*, *S. obscurus*, and *S. palustris*. *S. arcticus* can be distinguished from all these species by having well-developed postmandibular canals; no pigmented ridges on unicuspid teeth, and pelage tricolored. From *S. cinereus*, by having the tail length equal to or less than 30% of the total head-and-body length; condylobasal length on average greater than 17.0 mm. From *S. fumeus* and *S. haydeni*, by having the tail length less than 30% of the total head-and-body length, and condylobasal length greater than 17.0 mm. From *S. hoyi*, by having the third unicuspid teeth similar to the others, the fifth unicuspid teeth can be small but not minute; first and second unicuspid teeth without a distinct internal ridge with an internal cusp; an accessory medial tine on the anterior surface of the first upper incisors and in the mandible in lingual view; most posterior margin of the incisors anterior to the paraconid of the first lower molars. From *S. obscurus*, by having the third unicuspid teeth smaller than the fourth; upper unicuspid teeth usually with pigmentation on the ridges extending from the apices of cingulum and uninterrupted by an anteroposterior groove. From *S. palustris*, by having the hind feet not fimbriated; vertical depth of the rostrum at the level of the third unicuspid teeth greater than one-half of the distance from the anterior border of the infraorbital foramen to the posterior border of the first incisors.

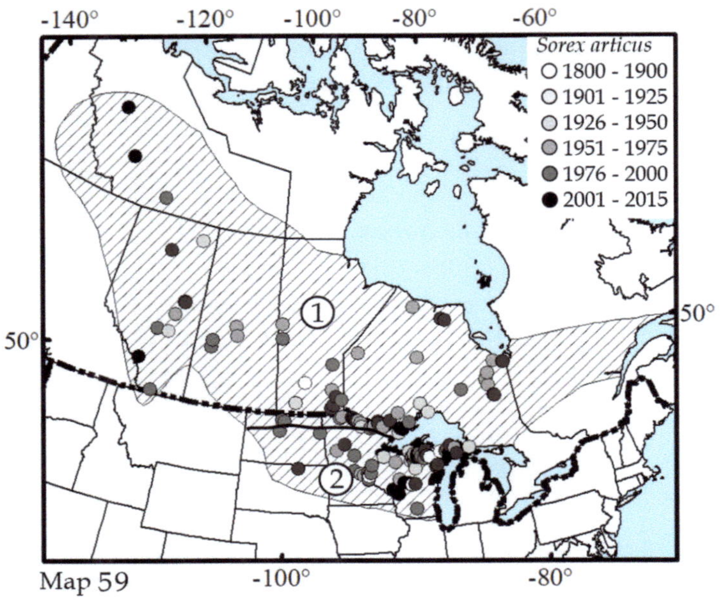

Map 59 Distribution map of *Sorex arcticus*

Additional Literature: Jackson (1925, 1928), Findley (1955a), Sokolov and Orlov (1980), Junge et al. (1983), Ivanitskaya et al. (1986), Okhotina (1983), van Zyll de Jong (1983c), Hoffmann (1985), George (1988), Hutterer (1993, 2005), Kirkland and Schmidt (1996), Stewart et al. (2002).

Sorex maritimensis Smith, 1939

Maritime shrew, musaraña de Nueva Escocia

Monotypic.

Conservation: *Sorex maritimensis* is listed as Least Concern by the IUCN (Reid 2016b).

Characteristics: Maritime shrew is small sized; total length 101.0–125.0 mm; condylobasal length 17.8–18.8 mm. Dorsal pelage tricolored: dark brown to black on the back, paler brown on the flanks, and grayish brown on the underparts; tail indistinctly bicolored, brown to brownish-black dorsally and paler ventrally; postmandibular canals well developed; no pigmented ridges on unicuspid teeth; height from the coronoid process of the first lower molars to the ventral edge of the mandible greater than 5.4 mm.

Comments: *Sorex maritimensis* was formerly included in *S. arcticus*, later recognized as a distinct species (Stewart et al. 2002; Baker et al. 2003a; Hutterer 2005). This species ranges in Nova Scotia and New Brunswick (Stewart et al. 2002; Map 60). It can be found in sympatry, or nearly so, with *S. cinereus*, *S. dispar*, *S. fumeus*, *S. hoyi*, and *S. palustris*. *S. maritimensis* can be distinguished from all these species by having the tail length equal to or less than 30% of the total head-and-body length. From *S. cinereus* and *S. fumeus*, by having the condylobasal length on average greater than 17.0 mm. From *S. dispar*, by having the posterior border of the infraorbital foramen even with, or anterior to, the plane of the space between the first and second upper molars. From *S. hoyi*, by having the third unicuspid teeth similar to the others, the fifth unicuspid teeth can be small but not minute; first and second unicuspid teeth without distinct internal ridge with internal cusp; an accessory medial tine on the anterior surface of the first upper incisors, and in the mandible in lingual view; most posterior margin of the incisors anterior to the paraconid of the first lower molars. From *S. palustris*, by having the hind feet not fimbriated; vertical depth of the rostrum at the level of the third unicuspid teeth greater than one-half of the distance from the anterior border of the infraorbital foramen to the posterior border of the first incisors.

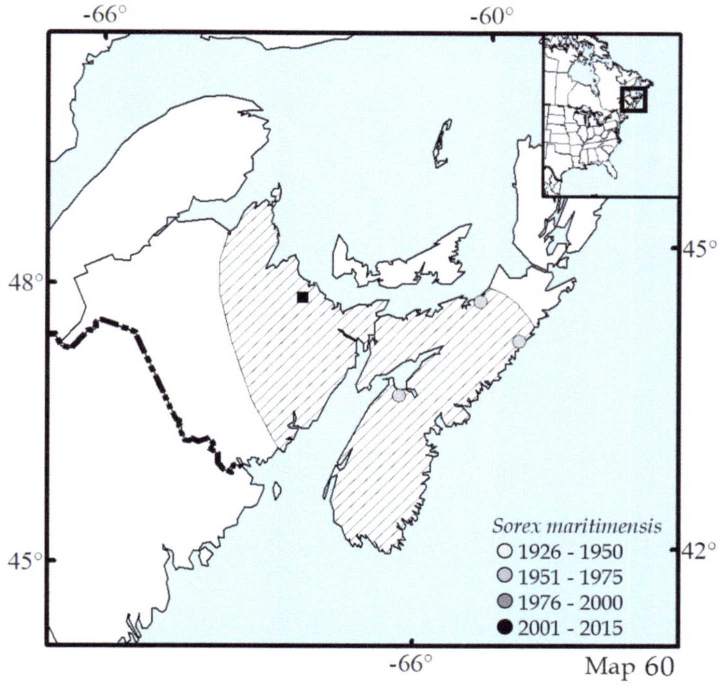

Map 60 Distribution map of *Sorex maritimensis*

Additional Literature: Findley (1955a), van Zyll de Jong (1983c), Volobouev and van Zyll de Jong (1988), George (1988).

Sorex minutissimus Zimmermann, 1780

Lesser shrew, Musaraña menor

Monotypic.

Conservation: *Sorex minutissimus* (as *S. yukonicus*) is listed as Least Concern by the IUCN (Henttonen et al. 2016).

Characteristics: Lesser shrew is small sized; **total length 68.0–76.0 mm; condylobasal length 13.1–13.6 mm.** Dorsal pelage nearest fuscous and flanks near gray; fur of the underparts pale smoke gray; tail bicolored, dorsally of the same color as the back, ventrally whiter than the underparts; breath of the second upper molars slightly greater than (3.09 mm); postmandibular canal on the jaw well developed.

Comments: *Sorex yukonicus* was first described as a full species (Dokuchaev 1997) but was later considered a subspecies of *S. minutissimus* (Hope et al. 2010). It ranges from northern Scandinavia through European Russia and Siberia to the Pacific coast and the Hokkaido and Sakhalin Islands; in America, it is restricted to central Alaska (Corbet 1978; Hope et al. 2010; Map 61). *S. yukonicus* can be found in sympatry, or nearly so, with *S. cinereus*, *S. hoyi*, *S. palustris*, *S. obscurus*, and *S. tundrensis.* It can be distinguished from *S. cinereus*, *S. hoyi*, *S. palustris*, and *S. obscurus*, by having a postmandibular canal on the jaw well developed and by its smaller size. From *S. tundraensis*, by having total length less than 80.0 mm.

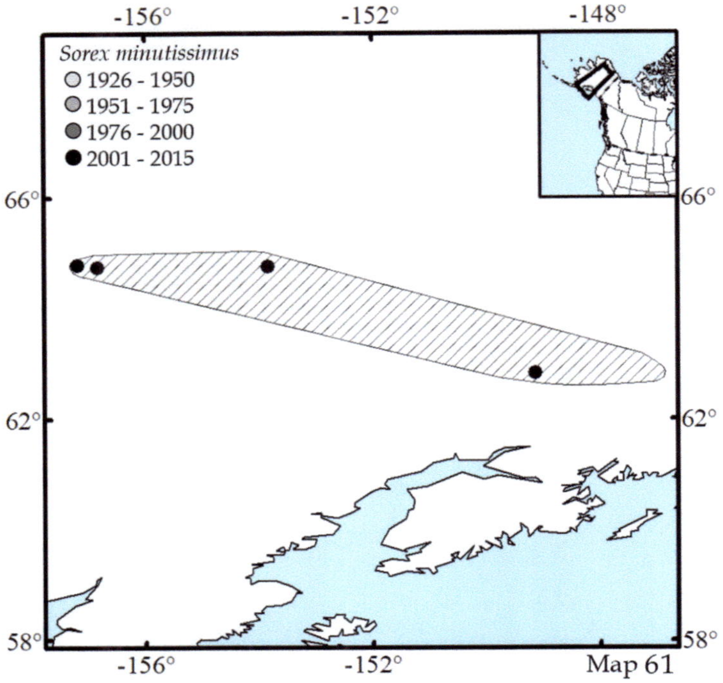

Map 61 Distribution map of *Sorex minutissimus*

Additional Literature: Sulkava (1999), Dokuchaev (1994).

Sorex tundrensis Merriam, 1900

Tundra shrew, musaraña de la tundra

Monotypic.

Conservation: *Sorex tundrensis* is listed as Least Concern by the IUCN (Tsytsulina et al. 2016).

Characteristics: Tundra shrew is small sized; total length 83.0–120.0 mm; condylobasal length 17.0–18.5 mm. **Dorsal pelage in summer tricolored: dark brown on the back, pale brown or brown gray on the flanks, and pale grayish on the underparts**; postmandibular canals well developed; lacking pigmented ridges on unicuspid teeth.

Comments: *Sorex tundrensis* was formerly included in *S. arcticus* and later recognized as a distinct species (Youngman 1975). It ranges in the tundra and forest zones from Yukon and Northwest Territories west through Alaska (Map 62). It can be found in sympatry, or nearly so, with *S. cinereus*, *S. eximus*, *S. navigator*, *S. obscurus*, *S. palustris*, and *S. ugyunak*. *S. tundrensis* can be distinguished from *S. hoyi*, *S. navigator*, *S. obscurus*, and *S. palustris*, by having the third unicuspid teeth larger than the fourth; upper teeth weakly pigmented; ridges extending from the apices of the unicuspid medially toward the cingulum but incomplete and not ending in internal cusplets; postmandibular foramen usually well developed. From *S. hoyi*, by having the third unicuspid teeth similar to the others; the fifth unicuspid teeth can be small but not minute; first and second unicuspid teeth without a distinct internal ridge with an internal cusp; an accessory medial tine on the anterior surface of the first upper incisors and in the mandible in lingual view; most posterior margin of the incisors anterior to the paraconid of the first lower molars. From *S. cinereus* and *S. ugyunak*, by having the tail length equal to or less than 30% of the total head-and-body length and condylobasal length on average greater than 17.0 mm. From *S. palustris*, by having the hind feet not fimbriated, vertical depth of the rostrum at the level of the third unicuspid teeth greater than one-half of the distance from the anterior border of the infraorbital foramen to the posterior border of the first incisors.

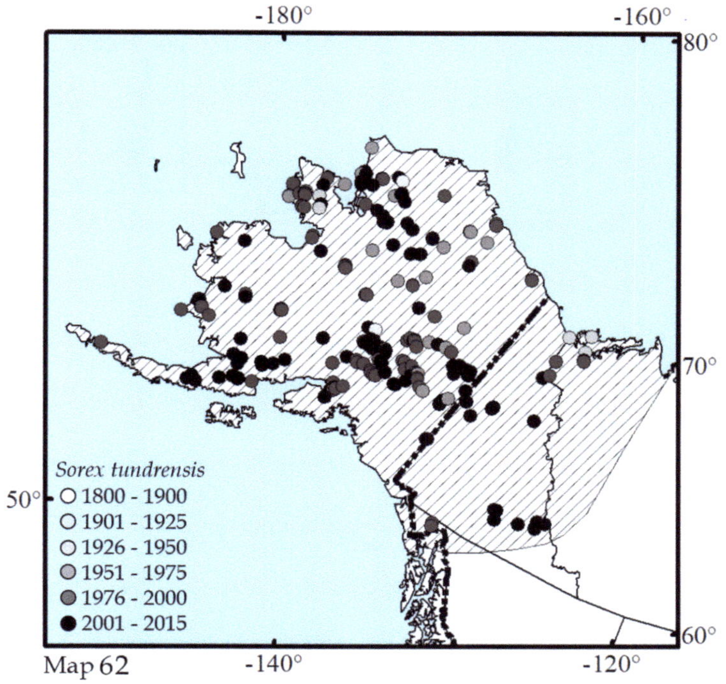

Map 62 Distribution map of *Sorex tundrensis*

Additional Literature: Merriam (1900), Findley (1955a), Junge et al. (1983), Ivanitskaya et al. (1986), Okhotina (1983), Hoffmann (1987), George (1988), van Zyll de Jong (1991b, 1999a), Fumagalli et al. (1999).

Incertae sedis

Sorex arizonae Diersing and Hoffmeister, 1977

Arizona shrew, musaraña de Arizona

Monotypic.

Conservation: *Sorex arizonae* is listed as Endangered in the Norma Oficial Mexicana (DOF 2019) and as Least Concern by the IUCN (Woodman et al. 2016d).

Characteristics: Arizona shrew is small sized; total length 92.0–109.0 mm; condylobasal length 15.5–16.4 mm. Dorsal pelage medium brown; hair with a pale brown band in the middle and dark at the tip; fur of the underparts bicolor, whitish at the tip; **medial tines on the first upper incisors**, **postmandibular foramen well defined**, lack of the pigmentation on the lingual face of the upper unicuspid teeth; third unicuspid teeth larger than or subequal to the fourth; **palate narrow, palate breadth measured across the fourth upper premolars less than the breadth measured across the third upper molars**; first lower incisors with one deep valley between two crests.

Comments: *Sorex arizonae* can be considered in a third subgenus (George 1988). Its ranges are restricted to southeastern Arizona, southwestern New Mexico, and the highlands of Sonora and Chihuahua (Map 63). It can be found in sympatry, or nearly so, with *S. merriami* and *S. monticolus*, from which it can be distinguished by having a well-defined postmandibular foramen; lack of the pigmentation on the lingual face of the upper unicuspid teeth; medial tines on the first upper incisors; third unicuspid teeth larger than or subequal to the fourth; narrow palate, with the palate breadth measured across the fourth upper premolars less than the breadth measured across the third upper molars; tail length and total length greater; fifth unicuspid teeth slightly larger; third unicuspid teeth equal to or slightly larger than the fourth unicuspid teeth*;* upper molars with cusps not W-shaped as in *S. merriami but* with the metacone-parastyle length noticeably greater than the paracone-parastyle length.

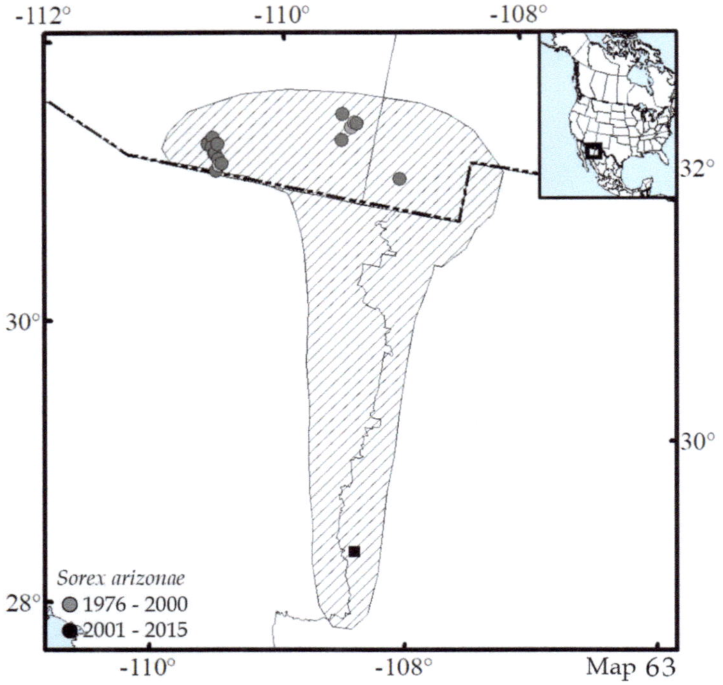

Map 63 Distribution map of *Sorex arizonae*

Additional Literature: Diersing and Hoffmeister (1977), Caire et al. (1978), Conway and Schmitt (1978), Hoffmeister (1986), Simons and Van Pelt (1999), Lee and Hoffmeister (2003), Simons and Hoffmeister (2003), Carraway (2007), Guevara et al. (2015), Maldonado et al. (2015), Woodman (2018).

Sorex altoensis Carraway, 2007

Mexican Highlands shrew, musaraña de las tierras altas de México

Monotypic.

Conservation: *Sorex altoensis* (as *veraecrucis*) is listed as subject to Special Protection in the Norma Oficial Mexicana (DOF 2019) and as Least Concern by the IUCN (Matson et al. 2017b).

Characteristics: Mexican highlands shrew is small sized; total length 98.0–128.0 mm; condylobasal length 17.5–18.6 mm. **Dorsal pelage with a medium silvery gray band proximally, a white or blond band medially, and dark brown at the tip, giving an overall medium brown appearance**; fur of the underparts medium silvery gray proximally, blond at the tip; guard hair on the hips projecting backward 1.0–1.5 mm as "thorns"; **tail hair evenly colored medium silvery gray proximally, blond at the tip**; fourth unicuspid teeth larger than the third; first lower incisors with deep valleys between the crests.

Comments: *Sorex altoensis* was recognized as a distinct species (Woodman 2018) following the genetic data of Esteva et al. (2010), previously regarded as a subspecies of *S. veraecrucis* (Carraway 2007) or *S. salvini* (Ramírez-Pulido et al. 2014). Its ranges from Oaxaca north through, along the Gulf slope, to Coahuila, Nuevo León, and Tamaulipas, and on the Pacific slope, to western Durango (Map 64). It can be found in sympatry, or nearly so, with *S. emarginatus*, *S. ixtlanensis*, *S. macrodon*, *S. mediopua*, *S. milleri*, *S. oreopolus*, *S. orizabae*, *S. monticolus*, *S. saussurei*, *S. ventralis*, and *S. veraecrucis*. *S. altoensis* can be distinguished from *S. emarginatus*, *S. oreopolus*, and *S. saussurei* by having the third unicuspid teeth larger than the fourth; upper teeth weakly pigmented; ridges extending from the apices of the unicuspid teeth medially toward the cingulum but incomplete and not ending in internal cusplets. From *S. ixtlanensis*, by having the pelage of the upperparts dark gray at the base, medially paler brown, and dark brown reddish brown appearance at the tip; fur of the underparts pale reddish brown; hips and rump with pale reddish blond and dark brown guard hairs extending 1.7–2.0 mm beyond the dorsal pelage; first upper incisors with a medial tine within a pigment band and first lower incisors with two denticles. From *S. macrodon*, by having the condylobasal length less than 18.6 mm; cranial breadth usually less than 8.3 mm; first lower incisors with two or three denticles and the long stripe of the pigment present or absent at the anteromedial edge. From *S. mediopua*, by having the first upper incisors with a medial tine above the pigment. From *S. milleri*, *S. oreopolus*, *S. orizabae*, and *S. monticolus*, by its tail evenly colored similar to the back. From *S. milleri* and *S. oreopolus*, by the incisors with shallow interdenticular spaces; hips and rump with dark brown guard hairs extending 1.0–1.5 mm beyond the dorsal pelage. From *S. ventralis*, by having the condylobasal length greater than 17.9 mm; length from the first upper unicuspid teeth to the third upper molars greater than 6.5 mm. From *S. veraecrucis*, only using genetic data.

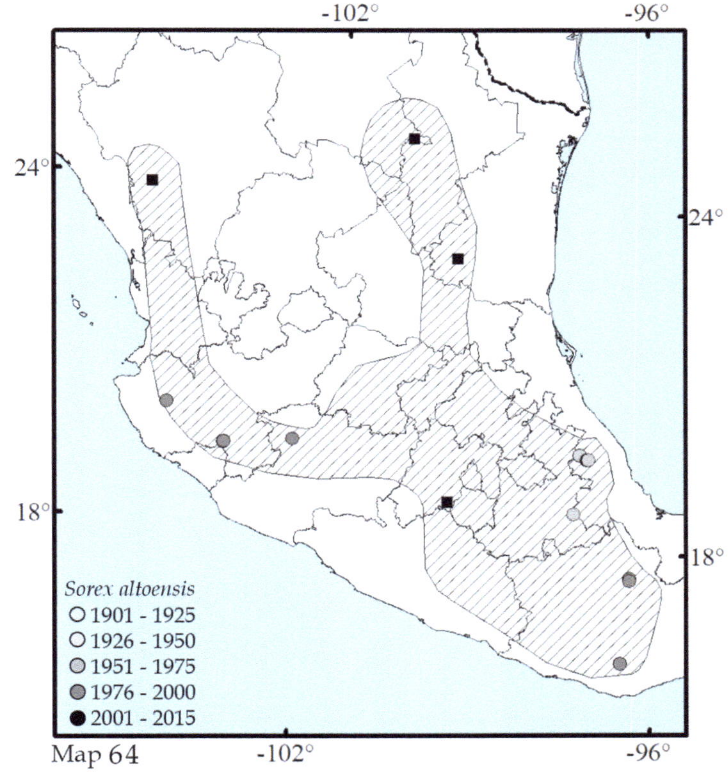

Map 64 Distribution map of *Sorex altoensis*

Additional Literature: Jackson (1925), Lavariega et al. (2012), Esteva et al. (2010).

Sorex chiapensis Jackson, 1925

Chiapas shrew, musaraña de Chiapas

Monotypic.

Conservation: *Sorex chiapensis* (as *Sorex veraepacis chiapensis*) is listed as Threatened in the Norma Oficial Mexicana (DOF 2019) and as Least Concern by the IUCN (Matson et al. 2017e).

Characteristics: Chiapas shrew is small sized; total length 100.0–135.0 mm; condylobasal length 19.4–19.8 mm. Dorsal pelage dark brown to dark reddish brown; fur of the underparts paler than in the dorsum; tail long and bicolored; fourth unicuspid teeth larger than the third; 32 teeth pigmented; first lower incisors with two deep valleys between three crests.

Comments: *Sorex chiapensis* was recognized as a distinct species (Matson and Ordoñez-Garza 2017; Woodman 2018), previously regarded as a subspecies of *S. veraepacis* (Carraway 2007; Woodman et al. 2012). *S. chiapensis* is restricted to the highlands of Chiapas and western Guatemala (Map 65). It can be found in sympatry, or nearly so, with *S. cristobalensis* and *S. stizodon*. *S. chiapensis* can be distinguished from *S. cristobalensis*, by having the hips and rump with darker brown, non-tipped guard hairs extending 1.7 mm beyond the dorsal pelage; tail darker dorsally and paler ventrally; first upper incisors with pigment in one section; interorbital and cranial breadths greater than 3.9 and 8.5 mm, respectively. From *S. stizodon*, by having the first upper unicuspid teeth with a medial crest included within the teeth pigment; pigment in a long stripe at the anteromedial edge; second unicuspid teeth not much larger than the first; teeth pigmented.

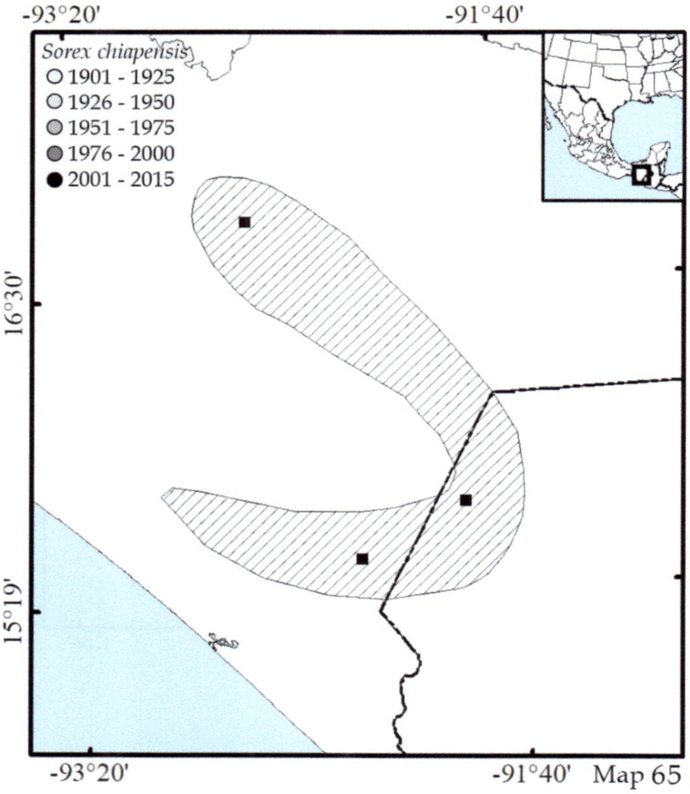

Map 65 Distribution map of *Sorex chiapensis*

Additional Literature: Jackson (1925).

Sorex cristobalensis Jackson, 1925

Cristobal shrew, musaraña de Cristóbal

Monotypic.

Conservation: *Sorex cristobalensis* (as *Sorex veraecrusis cristobalensis*) is listed as Least Concern by the IUCN (Matson et al. 2017d).

Characteristics: Cristóbal shrew is small sized; total length 100.0–127.0 mm; condylobasal length 17.9–18.4 mm. Dorsal pelage dull clove brown; fur of the underparts slightly paler reddish brown at the tip; guard hairs on the hips paler brown, projecting backward; tail evenly colored; skull large; rostrum wide and long; fourth unicuspid teeth larger than the third; 32 teeth pigmented; first lower incisors with deep valleys and pigmented in two sections.

Comments: *Sorex cristobalensis* was recognized as a distinct species (Matson and Ordóñez-Garza 2017) and was originally described as a subspecies of *S. saussurei* (Jackson 1925), later as a subspecies of *S. veraecrusis* (Carraway 2007) and of *S. salvini* (Ramírez-Pulido et al. 2014). Woodman (2018), following Esteva et al. (2010), considered *S. cristobalensis* as a subspecies of *S. salvini*. *S. cristobalensis* is restricted to the highlands of central and the Pacific coast of Chiapas and western Guatemala (Map 66). It can be found in sympatry, or nearly so, with *S. chiapensis* and *S. stizodon*. *S. cristobalensis* can be distinguished from *S. chiapensis* by having the hips and rump with pale brown, pale-tipped guard hairs extending 1.7 mm beyond the dorsal pelage, tail evenly colored, first upper incisors with pigment in two sections; interorbital and cranial breadth less than 3.9 and 8.5 mm, respectively. From *S. sclateri*, by having the first upper incisors with a deep interdenticular space, and by its smaller size, with the condylobasal length less than 18.4 mm. From *S. stizodon*, by having the hips and rump with pale brown, clear-tipped guard hairs extending 1.7 mm beyond the dorsal pelage; first upper incisors with a deep interdenticular space with pigment in two sections and pigment in a long stripe at the anteromedial edge.

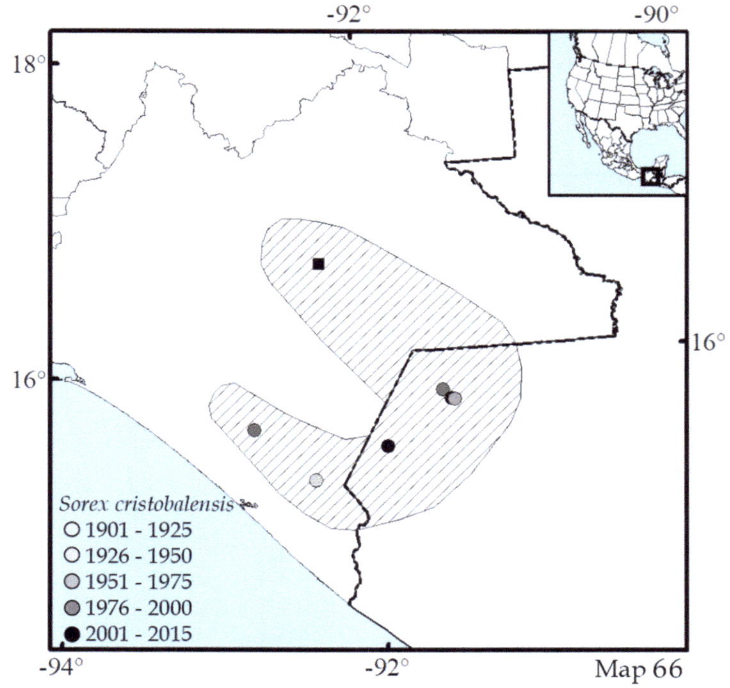

Map 66 Distribution map of *Sorex cristobalensis*

Additional Literature: Hall (1981).

Sorex macrodon Merriam, 1895

Large-teethed shrew, musaraña de dientes grandes

Monotypic.

Conservation: *Sorex macrodon* is listed as Vulnerable (B1ab (iii): extent of occurrence less than 20,000 km^2, severely fragmented and under continuing decline in area, extent, and quality of habitat) by the IUCN (de Grammont and Matson 2018) and listed as Threatened in the Norma Oficial Mexicana (DOF 2019).

Characteristics: Large-teethed shrew is small sized; total length 126.0–131.0 mm; condylobasal length 19.1–19.9 mm. **Dorsal pelage a mixed of sepia and black, black usually predominating, mainly on the posterior half of the back**; fur of the underparts seal brown; guard hairs on the hips projecting backward as "thorns"; tail not bicolored but blackish dorsally and paler ventrally; feet blackish; skull large and heavy with a large braincase; **molariform teeth very large and massive**; first lower incisors with two deep valleys between three crests; palate and interpterygoid fossa broad.

Comments: *Sorex macrodon* is known from Oaxaca, Puebla, and Veracruz (Map 67). It can be found in sympatry, or nearly so, with *S. ixtlanensis*, *S. mutabilis*, *S. orizabae*, *S. ventralis*, and *S. veraecrucis*. It can be distinguished from all these species by having the dorsal pelage color mixed of sepia and black, with black usually predominating, mainly on the posterior half of the back, and molariform teeth very large and massive. In addition, from *S. ixtlanensis*, by having the condylobasal length greater than 19.5 mm on average; cranial breadth greater than 9.6 mm; first lower incisors with three denticles and no long stripe of the pigment at the anteromedial edge; first upper incisors with a medial tine above the pigment. From *S. orizabae*, by having the condylobasal length on average greater than 17.9 mm; length from the first upper unicuspid teeth to the third upper molars greater than 6.5 mm. From *S. ventralis*, by having the third unicuspid teeth smaller than the fourth; upper unicuspid teeth usually with pigmentation on the ridges extending from the apices of cingulum and uninterrupted by an anteroposterior groove. From *S. veraecrucis*, by having the condylobasal length greater than 19.0 mm; cranial breadth greater than 9.6 mm; first lower incisors with three denticles and no long stripe of the pigment at the anteromedial edge, and first upper incisors with a medial tine above the pigment. From *S. mutabilis*, by having the upperparts very dark brown; fur of the underparts only slightly paler than the back, hips and rump with dark brown guard hairs extending less than 2.0 mm beyond the dorsal pelage, and first upper incisors with a medial tine above the pigment.

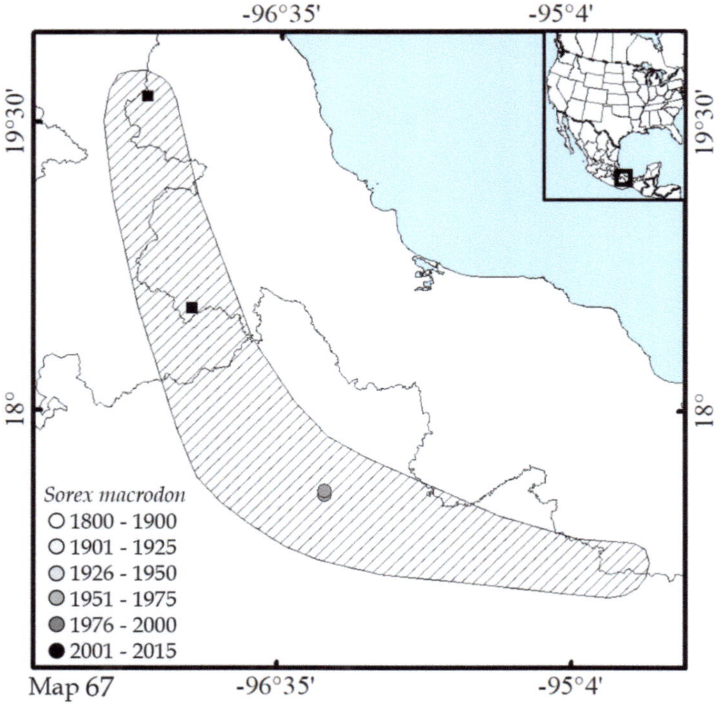

Map 67 Distribution map of *Sorex macrodon*

Additional Literature: Jackson (1928), Junge and Hoffman (1981), Carraway (2007), Guevara et al. (2015), Woodman (2018).

Sorex mediopua Carraway, 2007

Jalisco shrew, musaraña de Jalisco

Monotypic.

Conservation: *Sorex mediopua* is listed as Least Concern by the IUCN (Matson et al. 2017b).

Characteristics: Jalisco shrew is small sized; total length 102.0–120.0 mm; condylobasal length without data. Dorsal pelage with two color bands, silver gray medially and dark brown at the tip; fur of the underparts silver gray at the base, lighter at the tip, and evenly colored; fourth unicuspid teeth larger than the third.

Comments: *Sorex mediopua* was previously considered part of *S. saussurei saussurei* (Carraway 2007). The ranges of *S. mediopua* and *S. saussurei* are not well known. S. *mediopua* is known from Jalisco, Michoacán, and Estado de México (Map 68). It can be found in sympatry, or nearly so, with *S. emarginatus*, *S. oreopolus*, *S. orizabae*, *S. saussurei*, and *S. veraecrucis*. *S. mediopua* can be distinguished from *S. emarginatus*, *S. oreopolus*, and *S. saussurei*, by having the third unicuspid teeth larger than the fourth; upper teeth weakly pigmented; ridges extending from the apices of the unicuspid teeth medially toward the cingulum but incomplete and not ending in internal cusplets. From *S. orizabae*, by having the condylobasal length greater than 17.9 mm and distance from the first upper unicuspid teeth to the third upper molars greater than 6.5 mm. From *S. veraecrucis*, by having the first upper incisors with a medial tine at the interface of the unpigmented and pigmented areas, and first lower incisors with deep interdenticular spaces.

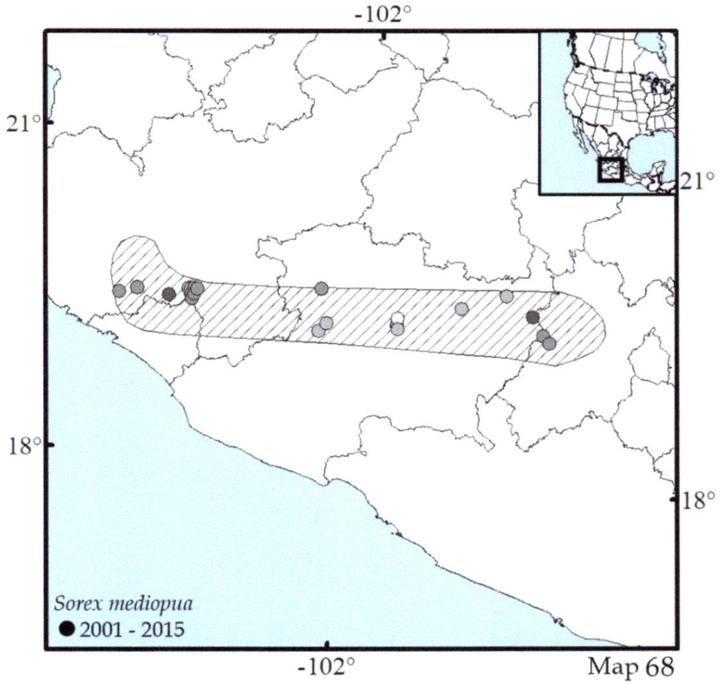

Map 68 Distribution map of *Sorex mediopua*

Additional Literature: Guevara et al. (2015), Woodman (2018).

Sorex merriami Dobson, 1890

Merriam's shrew, musaraña del Great Basin

Monotypic.

Conservation: *Sorex merriami* is listed as Least Concern by the IUCN (Cassola 2016dm).

Characteristics: Merriam's shrew is small sized; total length 85.0–107.0 mm; condylobasal length 15.1–16.9 mm. **Dorsal pelage ash gray or drab with a buffy tinge; hair of the underparts and flanks pure white**; feet whitish; **tail whitish dorsally and buffy white ventrally**; skull short, broad, and broaden; braincase not elevated above the plane of the rostrum; rostrum and palate remarkably broad and short; unicuspid series short, crowded, nearly vertical but slightly imbricating; **second unicuspid teeth largest; first, third and fourth unicuspid teeth subequal, fifth smaller**.

Comments: *Sorex merriami* ranges from the extreme northern Washington and Montana south through Arizona and New Mexico, east through the Dakotas and south to New Mexico, except for all the Snake River Basin (Map 69). It can be found in sympatry, or nearly so, with *S. arizonae*, *S. cinereus*, *S. eximus*, *S. haydeni*, *S. lyelli*, *S. monticolus*, *S. nanus*, *S. obscurus*, *S. palustris*, *S. preblei*, *S. tenellus*, *S. trowbridgii*, and *S. vagrans*. *S. merriami* can be distinguished from all these species by having the zygomatic process extending dorsolaterally at the alveoli of the level of the third and second upper molars. From *S. arizonae*, by not having medial tines on the first upper incisors; palate wide, especially as reflected by its breadth across the unicuspid fourth upper premolars; tail length and total length smaller; fifth unicuspid teeth slightly smaller; third unicuspid teeth equal to or slightly larger than the fourth unicuspid teeth, and upper molars with W-shaped cusps. From *S. cinereus*, *S. haydeni*, and *S. lyelli*, by having the combined length of the two first molars equal to or less than 2.4 mm. From *S. monticolus*, *S. obscurus*, *S. nanus*, and *S. vagrans*, by having the third unicuspid teeth equal to or larger than the fourth; upper teeth weakly pigmented; ridges extending from the apices of the unicuspid medially toward the cingulum but incomplete and not ending in internal cusplets. From *S. monticolus* and *S. obscurus*, by having usually a well-developed postmandibular foramen and no pigmented ridge on the lingual face of the unicuspid teeth. From *S. palustris*, by having the hind feet not fimbriated; vertical depth of the rostrum at the level of the third unicuspid teeth greater than one-half of the distance from the anterior border of the infraorbital foramen to the posterior border of the first incisors. From *S. preblei*, by its larger size with the rostrum relatively long; lower toothrow length greater than 4.1 mm; condylobasal length greater than 14.6 mm and lower toothrow length greater than 4.1 mm. From *S. tenellus* and *S. trowbridgii*, by having the third unicuspid teeth equal to or larger than the fourth; upper teeth weakly pigmented, ridges extending from the apices of the unicuspid teeth medially toward the cingulum but incomplete and not ending in internal cusplets.

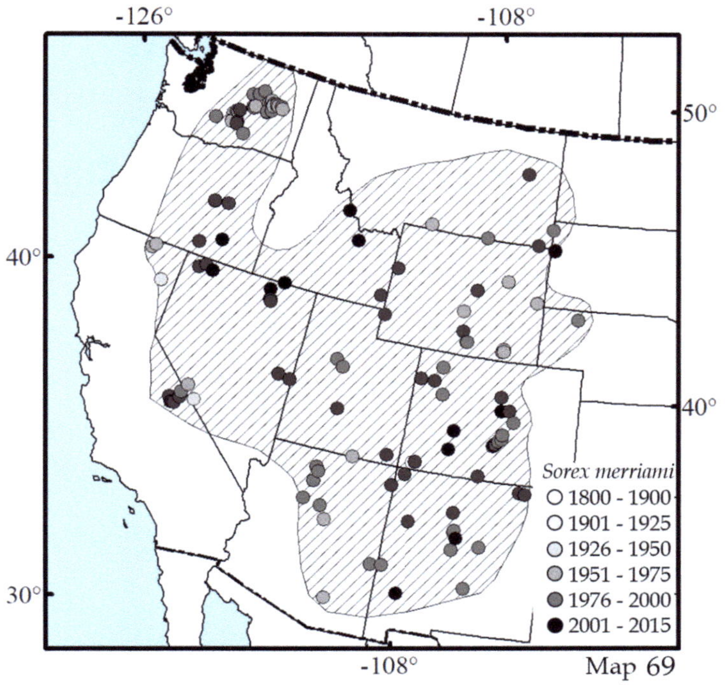

Map 69 Distribution map of *Sorex merriami*

Additional Literature: Jackson (1928), Benson and Bond (1939), Johnson and Clanton (1954), Armstrong and Jones (1971b), Diersing and Hoffmeister (1977), George (1988), Verts and Carraway (1998), Benedict et al. (1999), Hutterer (2005), Woodman (2018).

Sorex mutabilis Merriam, 1898

Mutable shrew, musaraña de Oaxaca

Monotypic.

Conservation: *Sorex mutabilis* (as *Sorex veraepacis*) is listed as Threatened in the Norma Oficial Mexicana (DOF 2019) and as Least Concern by the IUCN (Matson et al. 2017e).

Characteristics: Mutable shrew is small sized; total length 95.0–120.0 mm; condylobasal length 19.0–19.4 mm. Dorsal pelage dark brown to dark reddish brown; fur of the underparts paler than in the dorsum; tail long and bicolored; fourth unicuspid teeth larger than the third; 32 teeth pigmented; first lower incisors with two deep valleys between three crests.

Comments: *Sorex mutabilis* was elevated to full species (Esteva et al. 2010; Matson and Ordoñez-Garza 2017; Woodman 2018); it was previously regarded as a subspecies of *S. veraepacis* (Merriam 1989; Carraway 2007; Woodman et al. 2012). It ranges from central Guerrero south through the highlands of Oaxaca (Map 70). It can be found in sympatry, or nearly so, with *S. ixtlanensis*, *S. macrodon*, and *S. ventralis*. *S. mutabilis* can be distinguished from all the species just mentioned based on the following traits. From *S. ixtlanensis*, by having the condylobasal length greater than 19.0 mm; cranial breadth greater than 9.6 mm; first lower incisors with three denticles and no long stripe of the pigment at the anteromedial edge, and first upper incisors with a medial tine above the pigment. From *S. macrodon*, by having the upperparts reddish blond medially and dark brown at the tip; fur of the underparts pale reddish brown; hips and rump with pale reddish blond and dark brown guard hairs extending greater than 1.8 mm beyond the dorsal pelage and first upper incisors with a medial tine within a pigment band. From *S. ventralis*, by having the condylobasal length on average greater than 17.9 mm and distance from the first upper unicuspid teeth to the third upper molars greater than 6.5 mm.

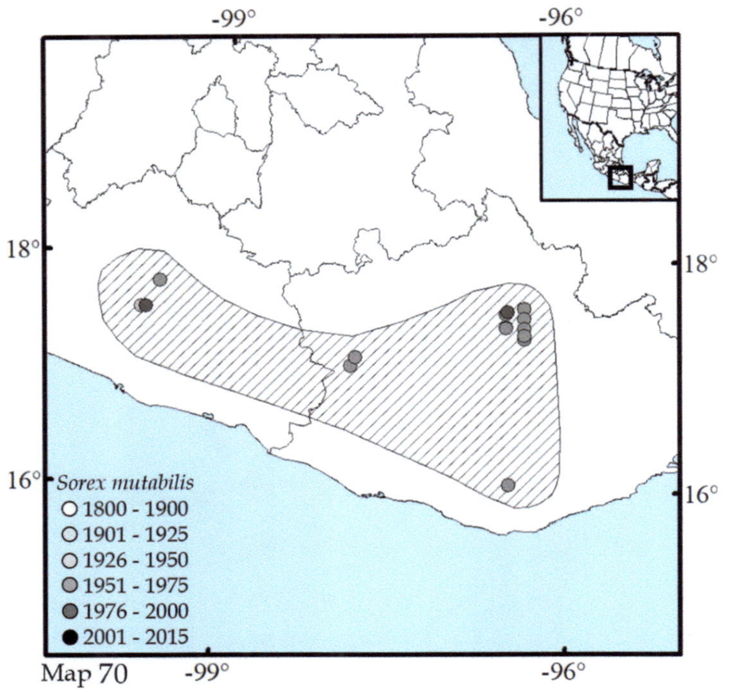

Map 70 Distribution map of *Sorex mutabilis*

Additional Literature: Guevara et al. (2015).

Sorex salvini Merriam, 1897

Salvini's shrew, musaraña de la Sierra Madre del Sur

1. *S. s. oaxacae* Jackson, 1925. From Mountains near Ozolotepec, Oaxaca.

Conservation: *Sorex salvini* (as *Sorex veraecrucis*) is listed as Least Concern by the IUCN (Matson et al. 2017d).

Characteristics: Salvini's shrew is small sized; total length 101.0–108.0 mm; condylobasal length 17.8–18.1 mm. Dorsal pelage with three stripes silver gray at the base, pale brown in the middle, and dark brown at the tip; **belly whitish silvery gray and white at the tip; neck and chest drab gray**; **hips and rump with brown guard hairs extending 1.9 mm beyond the dorsal pelage and lighter at the tip**; tail dark and relatively longer; in lateral view, first lower incisors most posterior part in front of the first cusp [paraconid] of the first lower molars; **interorbital constriction broad;** first lower incisors with a deep interdenticular space.

Comments: *Sorex salvini* was recognized as a distinct species (Esteva et al. 2010; Woodman 2018), previously regarded as a subspecies of *S. saussurei* (Jackson 1925) and *S. veraecrucis* (Carraway 2007). *S. s. oaxacae* ranges are restricted to the mountains near Ozolotepec, Oaxaca (Map 71); no other species of *Sorex* are found in sympatry or near its range.

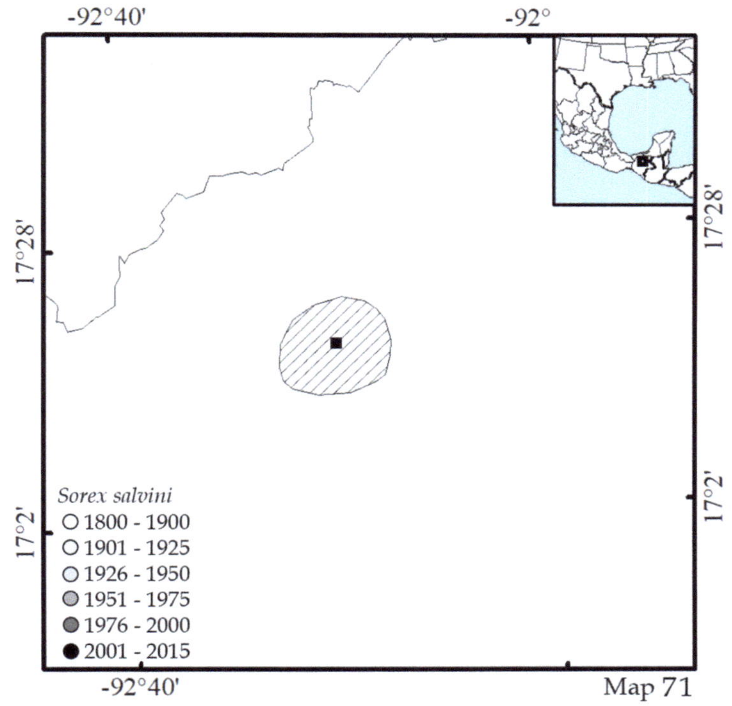

Map 71 Distribution map of *Sorex salvini*

Additional Literature: Merriam (1897), Hall (1981), George (1988), Guevara et al. (2015).

Sorex saussurei Merriam, 1892

Saussure's shrew, musaraña común de montaña

Monotypic.

Conservation: *Sorex saussurei* is listed as Least Concern by the IUCN (Cuarón et al. 2017).

Characteristics: Saussure's shrew is small sized; total length 100.0–110.0 mm; condylobasal length 17.8–18.6 mm. Dorsal pelage with three stripes silver gray at the base, pale brown in the middle, and dark brown at the tip; **belly whitish silvery gray; neck and chest drab gray**; **hips and rump with brown guard hairs extending 1.0–1.5 mm beyond the dorsal pelage**; tail dark and relatively longer; in lateral view, first lower incisors most posterior part in front of the first cusp [paraconid] of the first lower molars; **interorbital constriction broad**; first lower incisors with a shallow interdenticular space.

Comments: *Sorex saussurei* had five subspecies in México (Hall 1981), *S. s. cristobalensis*, *S. s. oaxacae*, *S. s. veraecrucis* (Jackson 1925), and *S. s. salvini* (Jackson 1928); it is now considered monotypic (Carraway 2007). *S. mediopua* was previously part of *S. s. saussurei*. The ranges of *S. mediopua* and *S. saussurei* are not very well differentiated (Carraway 2007). Given this confusion, it is difficult to assess this species. However, the ranges of *S. saussurei* were restricted to the highlands of the Eje Neovolcánico from Jalisco east through Puebla (Guevara et al. 2015; Map 72). It can be found in sympatry, or nearly so, with *S. emarginatus*, *S. mediopua*, *S. oreopolus*, *S. orizabae*, *S. mutabilis*, and *S. ventralis*. *S. saussurei* can be distinguished from all the other species just mentioned based on the following traits. From *S. emarginatus*, by having the fourth upper unicuspid teeth greater than the third. From *S. oreopolus*, by having the first upper unicuspid teeth with a medial crest included within the teeth pigment, and pigment in a long stripe at the anteromedial edge. From *S. orizabae*, by having the first upper unicuspid teeth with a medial crest included within the teeth pigment; pigment in a long stripe at the anteromedial edge. From *S. ventralis* and *S. mutabilis*, by having the dorsal coloration blond medially, dark brown at the tip; hips and rump with brown guard hairs extending 1.0–1.5 mm beyond the dorsal pelage; fur of the underparts with the proximal two-thirds silvery gray and the distal third white, and the tail evenly colored or slightly bicolored. From *S. mutabilis*, maxillary breadth greater than 4.95 mm; first lower incisors with two denticles and shallow interdenticular spaces; first lower incisors with pigment in the first section.

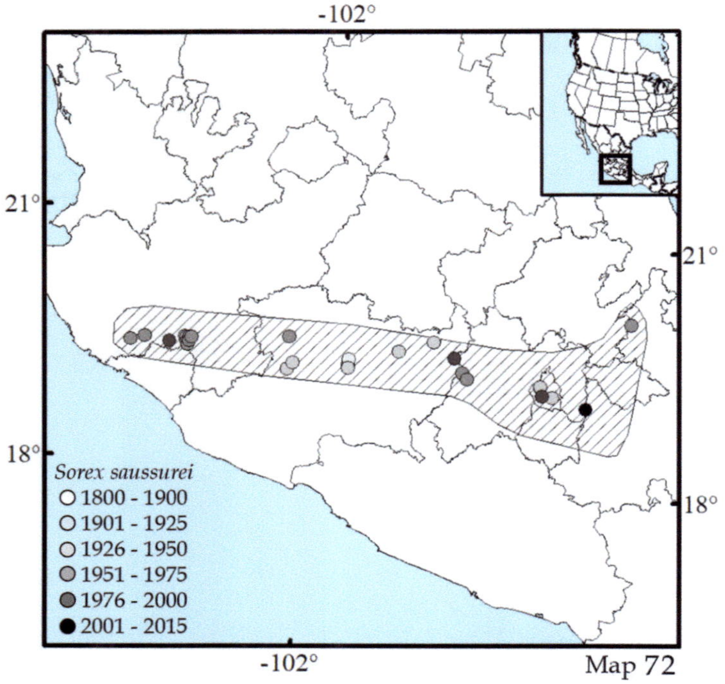

Map 72 Distribution map of *Sorex saussurei*

Additional Literature: Merriam (1892a, b, 1897), Junge and Hoffman (1981), George (1988), Woodman (2018).

Sorex sclateri Merriam, 1897

Sclater's shrew, musaraña selvática

Monotypic.

Conservation: *Sorex sclateri* is listed as Threatened in the Norma Oficial Mexicana (DOF 2019) and as **Critically Endangered** (B1ab (iii); extent of occurrence less than 100 km^2 in only one population and under continuing decline in area, extent and quality of habitat) by the IUCN (Cuarón and de Grammont 2018b).

Characteristics: Sclater's shrew is small sized; total length 120.0–126.0 mm; condylobasal length 18.4–19.9 mm. Dorsal pelage, including the tail, blackish brown; fur of the underparts slightly paler; long hair on the hips extending over; uniform brown hair; in lateral view, first lower incisors most posterior part in front of the first cusp [paraconid] of the first lower molar; 32 teeth pigmented; first lower incisors with three crests and shallow valleys between them.

Comments: *Sorex sclateri* is known only from the type locality in Tumbalá and San Antonio Buenavista, Chiapas (Map 73; Carraway 2007), and does not have sympatric species. However, it can be distinguished from *S. cristobalensis* by having the first upper incisors with a shallow and large interdenticular space and condylobasal length greater than 18.4 mm.

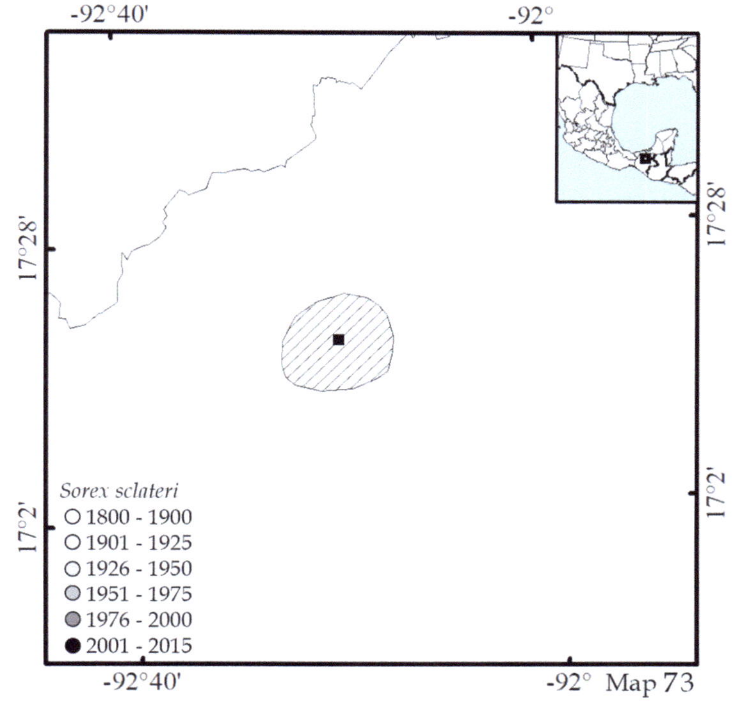

Map 73 Distribution map of *Sorex sclateri*

Additional Literature: Jackson (1928), Carraway (2007), Guevara et al. (2015), Woodman (2018).

Sorex stizodon Merriam, 1895

San Cristobal shrew, musaraña de San Cristóbal de Las Casas

Monotypic.

Conservation: *Sorex stizodon* is listed as Threatened in the Norma Oficial Mexicana (DOF 2019) and as **Critically Endangered** (B1ab (iii); extent of occurrence less than 100 km^2 in only one population and under continuing decline in area, extent, and quality of habitat) by the IUCN (Cuarón et al. 2018).

Characteristics: San Cristóbal shrew is small sized; total length 107.0 mm; condylobasal length 17.5 mm. Body hair mixed sepia brown and dusky; guard hairs on the hips projecting backward as "thorns"; tail indistinctly bicolored, dusky dorsally and pale ventrally, skull shorter and broader, with the braincase inflated and a shorter rostrum; **second unicuspid teeth much larger than the first and larger than in any other species of *Sorex***; first lower incisors with three cusps and very shallow valleys between them; teeth barely pigmented; third and fourth unicuspid teeth approximately equal in size; first upper molars wider than long.

Comments: *Sorex stizodon* is only known from the type locality at Reserva Ecológica Huitepec in Chiapas (Map 74) and can be found in sympatric, or nearly so, with *S. chiapensis* and *S. cristobalensis*. *S. stizodon* can be distinguished from these species by having the first upper unicuspid teeth with a median crest usually above the pigment; second unicuspid teeth much larger than the first and pigment scarce, always separated from the body pigment. From *S. cristobalensis*, by having the hips and rump with brown not pale-tipped guard hairs extending beyond the dorsal pelage.

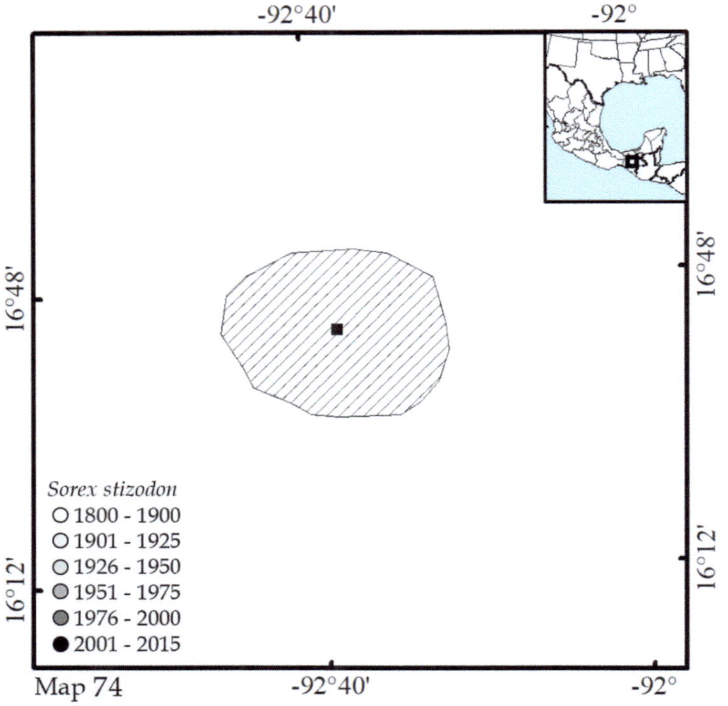

Map 74 Distribution map of *Sorex stizodon*

Additional Literature: Naranjo and Espinoza-Medinilla (2001), Carraway (2007), Guevara et al. (2014b, 2015), Woodman (2018).

Sorex trowbridgii Baird, 1857

Trowbridge's shrew, musaraña de la costa oeste

1. *S. t. destructioni* Scheffer and Dalquest, 1942. Destruction Island, Washington.
2. *S. t. humboldtensis* Jackson, 1922. Northwestern coast of California.
3. *S. t. mariposae* Grinnell, 1913. From southern Oregon south through the highlands to Sequoia National Park, including the eastern slope of the Sierra Nevada mountains, California.
4. *S. t. montereyensis* Merriam, 1895. From Mendocino County south through the Santa Barbara area, California.
5. *S. t. trowbridgii* Baird, 1857. From southwestern British Columbia to northwestern California.

Conservation: *Sorex trowbridgii* is listed as Least Concern by the IUCN (Cassola 2016dq).

Characteristics: Trowbridge's shrew is medium sized; total length 119.5–121.5 mm; condylobasal length 16.8–17.5 mm. Dorsal pelage blackish slate or sooty plumbeous; fur of the underparts dull plumbeous; ears almost hidden by the pelage; tail sharply bicolored, blackish dorsally and whitish ventrally; feet flesh-colored; braincase globular; palate narrow; first and second unicuspid teeth smaller and narrow; **medial tine small, high on the anterior face of the first upper incisors**.

Comments: *Sorex trowbridgii* can be considered in a third subgenus (George 1988). It ranges in coastal southwestern British Columbia south of Burrard Inlet south through Coast Ranges to Santa Barbara County and one branch south throughout Sierra Nevada, California, and western Nevada (Map 75). It can be found in sympatry, or nearly so, with *S. bendirii*, *S. cinereus*, *S. merriami*, *S. navigator*, *S. obscurus*, *S. ornatus*, *S. pacificus*, *S. palustris*, *S. rohweri*, *S. sonomae*, and *S. vagrans*. *S. trowbridgii* can be distinguished from all these species by its small medial tine high on the anterior face of the first upper incisors; tail sharply bicolored and upperparts dark almost concolor with the upperparts. From *S. bendirii*, by having the dorsal pelage always brownish; hind foot length less than 18.0 mm and condylobasal length less than 19.0 mm. From *S. cinereus* and *S. merriami*, by having the third unicuspid teeth smaller than the fourth upper unicuspid teeth, usually with pigmentation on the ridges extending from the apices of cingulum and uninterrupted by an anteroposterior groove. From *S. obscurus*, by usually having a well-developed postmandibular foramen and no pigmented ridge on the unicuspid teeth lingual face. From *S. navigator* and *S. palustris*, by having the hind feet not fimbriated; vertical depth of the rostrum at the level of the third unicuspid teeth greater than one-half of the distance from the anterior border of the infraorbital foramen to the posterior border of the first incisors. From *S. ornatus* and *S. pacificus*, by usually having a well-developed postmandibular foramen; no pigmented ridge on the unicuspid teeth lingual face. In addition, from *S. ornatus*, by having the skull not flattened. From *S. rohweri*, by having the combined length of the two first molars greater than 2.4 mm. From *S. sonomae*, by having the maxillary zygomatic processes finely pointed and usually a well-developed postmandibular foramen; no pigmented ridge on the unicuspid teeth lingual face. From *S. vagrans*, by having a medial tine contained entirely within the pigmented area of the first upper incisors; interorbital breadth equal to or greater than 3.5 mm and the tail bicolored.

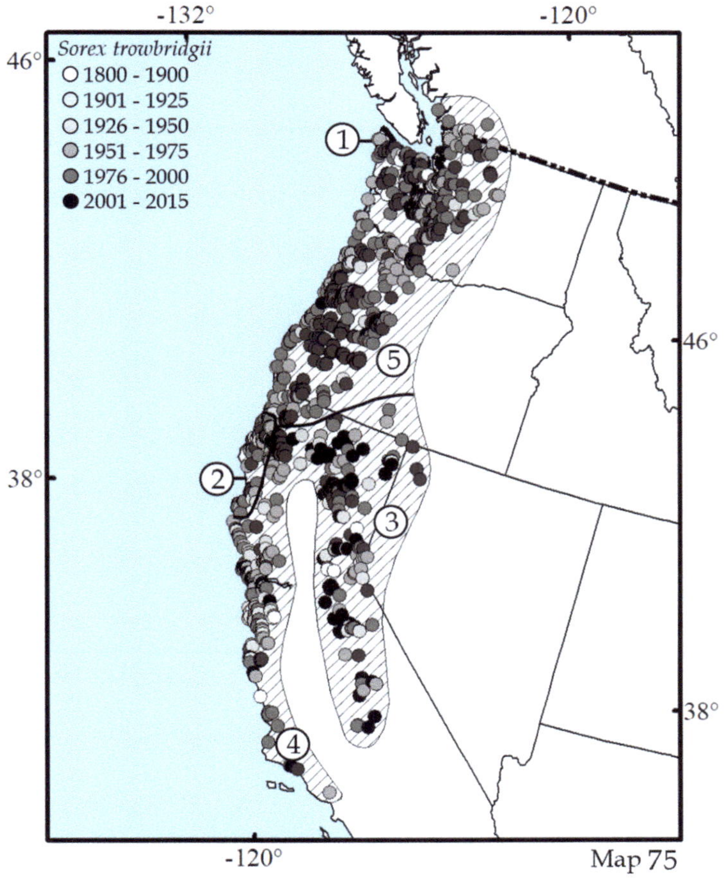

Map 75 Distribution map of *Sorex trowbridgii*

Additional Literature: Jackson (1928), Repenning (1967), Junge and Hoffmann (1979), van Zyll de Jong (1983a), Carraway (1987), George (1989), Verts and Carraway (1998), Zima et al. (1998), Woodman (2018).

Sorex veraecrucis Jackson, 1925

Veracruz shrew, musaraña de Veracruz

Monotypic.

Conservation: *Sorex veraecrucis* is listed as subject to Special Protection in the Norma Oficial Mexicana (DOF 2019) and as Least Concern by the IUCN (Matson et al. 2017d).

Characteristics: Veracruz shrew is small sized; total length 100.0–126.0 mm; condylobasal length 17.8–18.4 mm. Dorsal pelage pale brown; fur of the underparts paler at the tip; guard hairs on the hips projecting backward as "thorns"; tail with uniform color, slightly bicolored in some specimens; fourth unicuspid teeth larger than the third; first lower incisors with deep valleys between crests.

Comments: *Sorex veraecrucis* was recognized as a distinct species (Carraway 2007), previously regarded as a subspecies of *S. saussurei* (Jackson 1925). *Sorex altoensis* was formerly a subspecies of *S. veraecrucis* (Carraway 2007) but is now regarded as a valid species (Woodman 2018) following the genetic data of Esteva et al. (2010). *S. veraecrucis* is restricted to the highlands of Oaxaca south through Chiapas (Map 76) and can be found in sympatry, or nearly so, with *S. macrodon*, *S. orizabae*, and *S. ventralis*. *S. veraecrucis* can be distinguished from *S. macrodon* by having the condylobasal length less than 19.0 mm; cranial breadth usually less than 9.6 mm; first lower incisors with two or three denticles and the long stripe of the pigment present or absent at the anteromedial edge. From *S. ventralis*, by having the condylobasal length greater than 17.9 mm; distance from the first upper unicuspid teeth to the third upper molars greater than 6.5 mm.

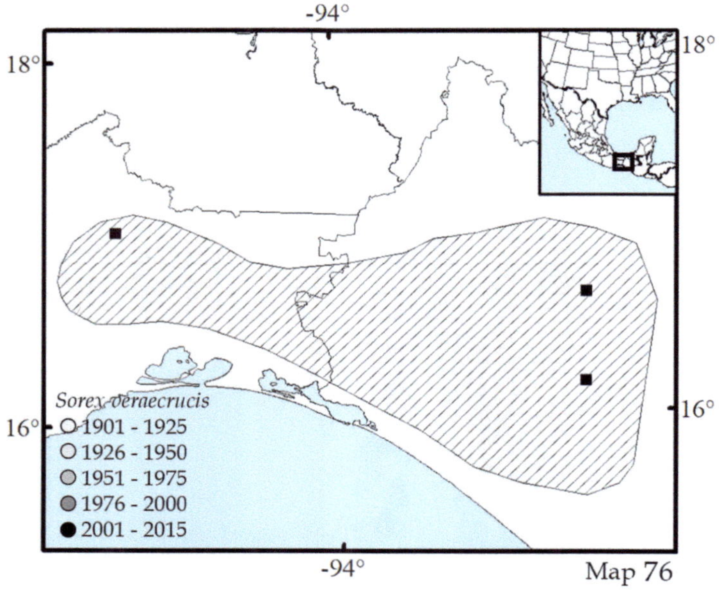

Map 76 Distribution map of *Sorex veraecrucis*

Additional Literature: Lavariega et al. (2012).

Sorex ventralis Merriam, 1895

Chestnut-bellied shrew, musaraña de vientre castaño

Monotypic.

Conservation: *Sorex ventralis* is listed as Least Concern by the IUCN (Matson et al. 2017c).

Characteristics: Chestnut-bellied shrew is small sized; total length 97.0–104.0 mm; condylobasal length 16.6–18.2 mm. Dorsal pelage with two bands, reddish in the middle and dark brown at the tip; fur of the underparts dull chestnut brown; cover hair on the hips projecting backward as small "thorns" of the same color as the body; **hips and rump with dark brown guard hairs extending 0.75 mm beyond the dorsal pelage**; **tail bicolored, dusky dorsally and soiled whitish ventrally, with a demarcation line usually distinct**; fourth unicuspid teeth larger than the third; first lower incisors with two crests and very shallow valleys between them.

Comments: *Sorex ventralis* is referred to as an unnamed subgenus by George (1988) and ranges in the highlands of central México, in the central area of the Eje Neovolcánico and the highlands of Veracruz, Puebla, and Oaxaca (Map 77). It can be found in sympatry, or nearly so, with *S. emarginatus*, *S. ixtlanensis*, *S. macrodon*, *S. mutabilis*, *S. oreopolus*, *S. orizabae*, *S. saussurei*, and *S. veraecrucis*. *S. ventralis* can be distinguished from all the other species just mentioned based on the following traits. From *S. emarginatus*, *S. oreopolus*, and *S. saussurei*, by having the tail bicolored, dusky dorsally and soiled whitish ventrally with a demarcation line usually distinct. From *S. emarginatus*, by not having the third upper unicuspid teeth greater than the fourth and the first lower incisors with a long stripe of the pigment on the anteromedial edge. From *S. oreopolus*, by having the first upper unicuspid teeth with a median crest included within the teeth pigment and pigment in a long stripe at the anteromedial edge. From *S. saussurei*, by having the dorsal coloration pale red medially and dark brown at the tip, hips and rump with dark brown guard hairs extending 0.75 mm beyond the dorsal pelage, fur of the underparts in the proximal two-thirds dark gray and on the distal third blond, and the tail is distinctly bicolored. From *S. ixtlanensis*, *S. macrodon*, *S. veraecrucis*, and *S. mutabilis*, by having the condylobasal length less than 17.9 mm and the distance from the first upper unicuspid teeth to the third upper molars less than 6.5 mm.

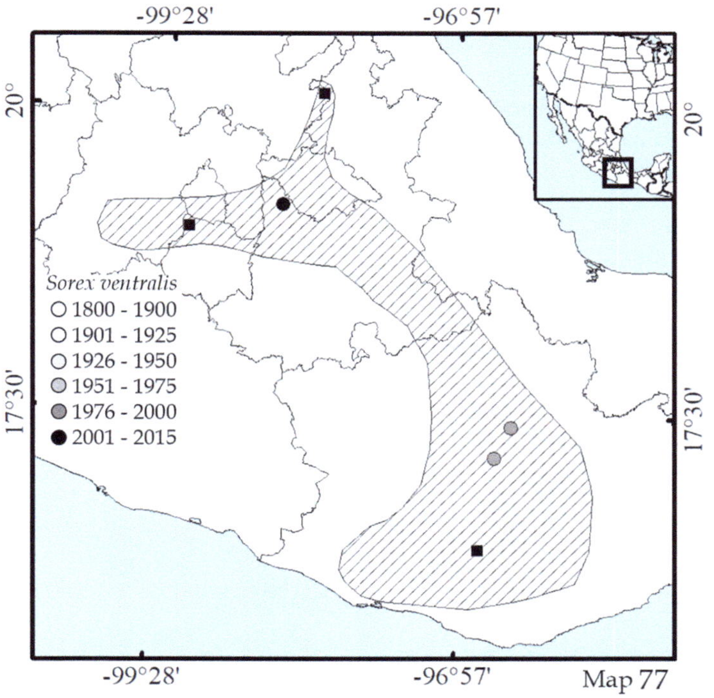

Map 77 Distribution map of *Sorex ventralis*

Additional Literature: Hall (1981), Junge and Hoffman (1981), George (1988), Carraway (2007), Guevara et al. (2015).

Family Talpidae Fisher, 1814

The family Talpidae includes moles characterized by subterranean life; small sized (2.4–20.0 cm). Distribution restricted to Euroasia, Africa, and North America. Two subfamilies range in North America: Scalopinae and Talpinae. The keys were elaborated based on the review of specimens and the following sources: Jackson (1915) and Hall (1981).

1. Tail greater than 25% of the total length; palm of forefeet less broad than long and without a circular fringe of fleshy processes in the anterior end of the snout; two lower premolars (from southern British Columbia, Canada, south through Monterey Bay, California, including the Sierra Nevada Mountains, United States) ....subfamily Talpinae; *Neurotrichus gibbsii* (p. 512)

1a. Tail less than 25% of the total length; palm of the forefeet broader than long, if these conditions are present, circular fringe of the fleshy processes in the anterior end of the snout; more than two lower premolars.. .... Subfamily Scalopinae (p. 498)

Subfamily Scalopinae Gill, 1875

The species of Scalopinae can be differentiated from those of Talpinae based on the mtDNA phylogeny. Two tribes range in North America: Condylurini and Scalopini. The keys were elaborated based on the review of specimens and the following sources: Jackson (1915), and Hall (1981).

1. Anterior end of the snout with a circular fringe of the fleshy processes; nostrils from circular to oval; nasal opening of the skull obliquely upward (from eastern Manitoba to North Dakota east through the Atlantic coast and south through Georgia and northern Florida, United States) .... tribe Condylurini; *Condylura cristata* (p. 500)

1a. Anterior end of the snout without a circular fringe of the fleshy processes; nostrils not circular to oval; nasal opening of the skull anteriorly .... tribe Scalopini 2

2. Tail densely haired; nostrils lateral; interior basal projection of the upper molars broad and lobed, with a trilobed basal shelf (Fig. 14); auditory bullae incomplete (Fig. 15; from southern Ontario east through New Brunswick south through Ohio, Connecticut, the Appalachian Mountains to western North Carolina, United States).. .... *Parascalops breweri* (p. 501)

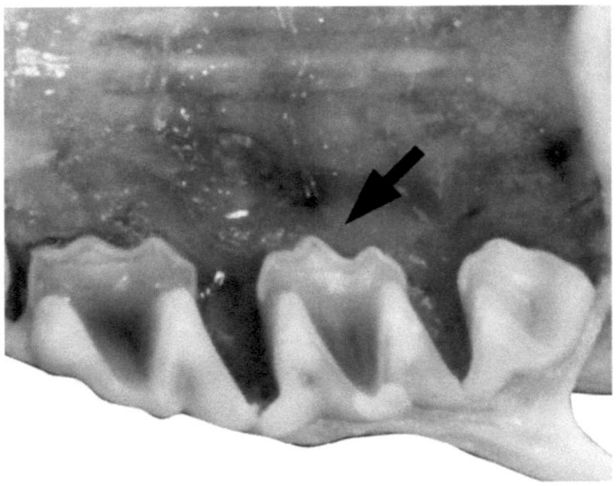

Fig. 14 Interior basal projection of the upper molars broad and lobed, with a trilobed basal shelf

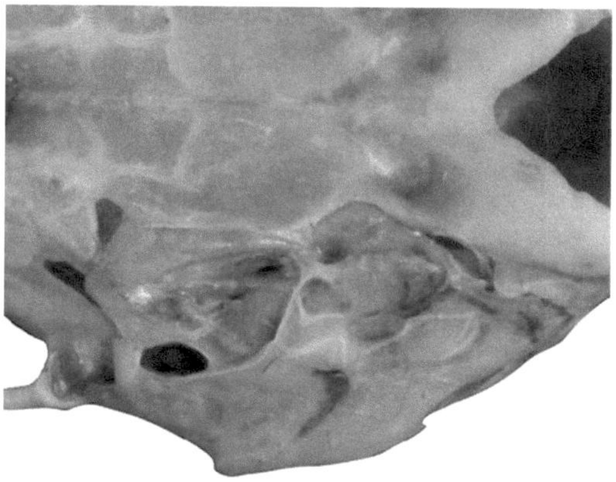

Fig. 15 Auditory bullae incomplete

2a. Tail naked or with few hairs; nostrils superior; interior basal projection of the upper molars narrow and simple (Fig. 16); auditory bullae complete (Fig. 17) 3

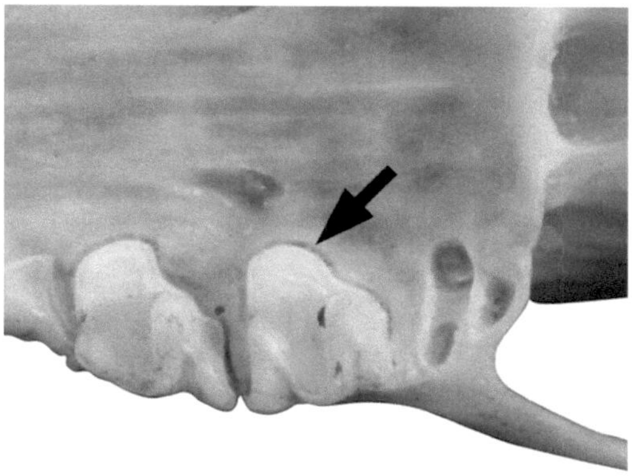

Fig. 16 Interior basal projection of the upper molars narrows and simple

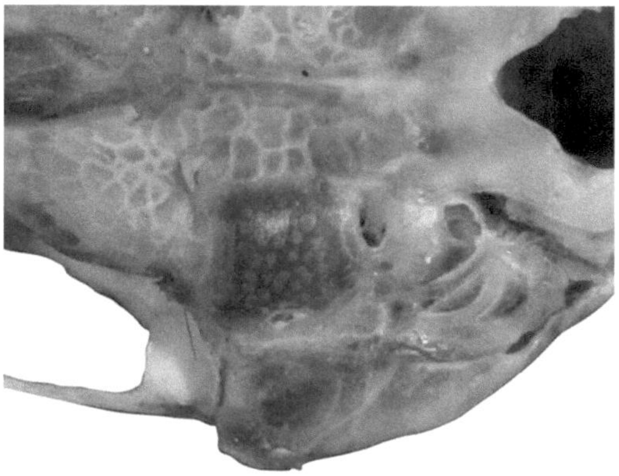

Fig. 17 Auditory bullae complete

3. Tail thin and naked; fingers webbed; two lower incisors, eight lower teeth in total (from southern Ontario, southern South Dakota east through Massachusetts, south through the Florida Peninsula, United States, and Tamaulipas, México. Patchily distributed from southwestern Texas south through Coahuila and Tamaulipas) .... *Scalopus aquaticus* (p. 502)

3a. Tail fleshy and with little hair; without webbed digits; three lower incisors, 10 or 11 lower teeth in total .... *Scapanus* (p. 504)

Tribe Condylurini

Genus *Condylura* Illiger, 1811

Dorsal pelage blackish brown to nearly black; **nose with a ring of 22 fleshy appendages; tail as long as the head-and-body length**; tail scaled and annulated; forefeet palms as broad as long; premaxilla extending much beyond the nasals anteriorly; first upper incisors large, curved, and projecting anteriorly.

Condylura cristata (Linnaeus, 1758)

Star-nosed mole, topo nariz de estrella

1. *C. c. cristata* (Linnaeus, 1758). From Manitoba and western North Dakota east through the Labrador Peninsula and south through North Virginia.
2. *C. c. nigra* Smith, 1940. Restricted to Nova Scotia.
3. *C. c. parva* Paradiso, 1959. From north Florida and central North Carolina north through North Virginia.

Conservation: *Condylura cristata* is listed as Least Concern by the IUCN (Cassola 2016h).

Characteristics: Star-nosed mole is medium sized within moles, with a broad latitudinal variation in size, with northern specimens larger than the southern; total length 161.0–206.0 mm; skull length 32.1–35.2 mm. Dorsal pelage blackish brown to nearly black; **nose with a ring of 22 fleshy appendages; tail approximately as long as the body**; forefoot width less than its length; first upper incisors large, curved, and projecting anteriorly; four upper and lower premolars; **first and second upper molars trilobed; anterior nasal opening of the skull obliquely upward**.

Comments: *Condylura cristata parva* was previously considered a subspecies of *C. c. nigra*, with no discussion (Peterson and Yates 1980). *C. cristata* ranges from eastern Manitoba, Minnesota, Wisconsin, northern Indiana, and Ohio east through to the Atlantic coast of Labrador and Nova Scotia south through Georgia and northern Florida, including the Appalachian Mountains (Map 78). This species differs from the other sympatric species, such as *Scalopus aquaticus* and *Parascalopus breweri*, by having the anterior end of the snout with 22 circular fringes of the fleshy processes; the nostrils vary from circular to oval; tail approximately as long as the body; four upper and lower molars; first and second upper molars with the interior basal projection trilobed.

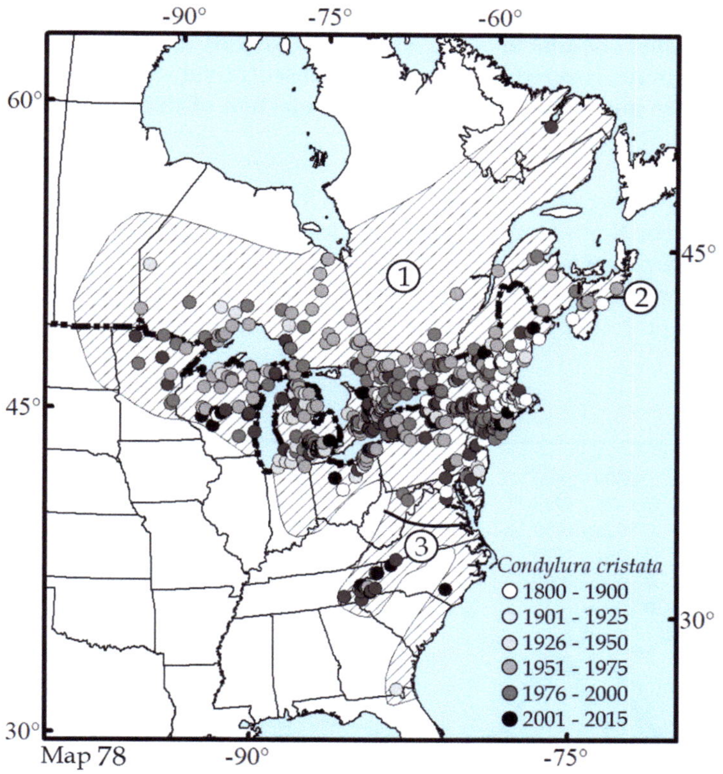

Map 78 Distribution map of *Condylura cristata*

Additional Literature: Hall and Kelson (1959), Yates (1978), Hutterer (2005).

Tribe Scalopini

Genus *Parascalops* True, 1894

Dorsal pelage blackish; **tail densely covered with long coarse hairs** and less than 25% of the head-and-body length; **nostrils lateral;** tail short, less than the head-and-body length, constricted at the base and annulated; snout shorter than on *Scalopus* and *Scapanus*, with a median longitudinal groove on its anterior half; forefeet palms as long as broad and toes not webbed; **auditory bullae incomplete; upper first and second molars trilobed with an internal basal projection broad and lobed**.

Parascalops breweri (Bachman, 1842)

Hairy-tailed mole, topo de cola peluda

Monotypic.

Conservation: *Parascalops breweri* is listed as Least Concern by the IUCN (Cassola 2016cg).

Characteristics: Hairy-tailed mole is small within moles; total length 139.0–153.0 mm, skull length 31.0–33.8 mm. Dorsal pelage blackish; **tail short, less than one-fourth of the total length, slightly constricted at the base and densely covered with long coarse hairs; eyes minute, covered by hair; pinnae absent;** forefoot width equal to or greater than its length; toes not webbed; **auditory bullae incomplete**; **interior basal projection of the upper molars broad and lobed with a trilobed basal shelf.**

Comments: *Parascalops breweri* ranges from southern Ontario to New Brunswick south through Ohio, Connecticut, the Appalachian Mountains to western North Carolina (Map 79). It can be distinguished from other species in sympatry, or nearly so, such as *Condylura cristata* and *Scalopus aquaticus*, by having the tail short, less than one-fourth of the total length, slightly constricted at the base, and densely covered with long hairs; eyes minute, covered by hair; pinnae absent; auditory bullae incomplete; interior basal projection of the upper molars broad and lobed with a trilobed basal shelf.

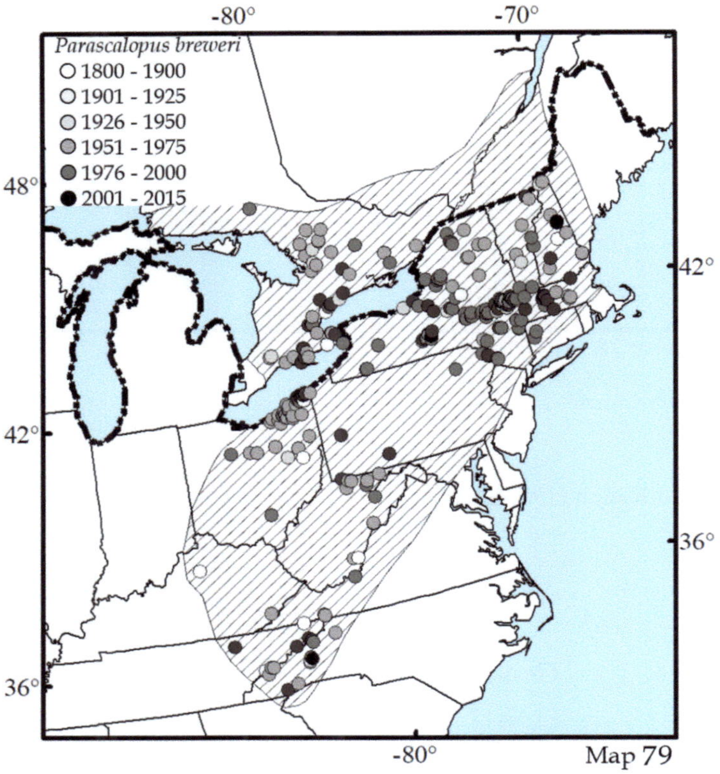

Map 79 Distribution map of *Parascalops breweri*

Additional Literature: True (1896), Jackson (1915), Hall and Kelson (1959), Peterson (1966), Yates and Schmidly (1975), Hallett (1978, 1999), Hutterer (2005).

Genus *Scalopus* Saint-Hilaire, 1803

Tail length less than one-fourth of the head-and-body length, indistinctly annulated and nearly naked; eyes minute, concealed in hair; nose elongated into a distinct snout, naked anterior to the nostrils; **forefeet palms broader than long and toes webbed; hind feet without tubercles and toes webbed**; frontal region flat and rostrum short; **interparietal bone short and narrow;** anterior nares of the skull directed forward; auditory bullae complete; **external pterygoid region markedly inflated posteriorly and slightly inflated anteriorly;** first upper incisors long and broad; first and second molars subequal.

Scalopus aquaticus (Linnaeus, 1758)

Eastern mole, topo de la costa este

1. *S. a. aereus* Bangs, 1896. Southwestern Kansas, northeastern Texas, Oklahoma, Arkansas, and Louisiana.
2. *S. a. alleni* Baker, 1951. Central-eastern Texas.
3. *S. a. anastasae* Bangs, 1898. Anastasia Island, Florida.
4. *S. a. aquaticus* (Linnaeus, 1758). From eastern Tennessee and western North Carolina northeast through Massachusetts.
5. *S. a. australis* Chapman, 1893. From southeastern Georgia south through the central and eastern of Florida Peninsula.
6. *S. a. bassi* Howell, 1939. Restricted to Sarasota County, Florida.
7. *S. a. caryi* Jackson, 1914. Central western Nebraska, northwestern Kansas, and northeastern Colorado.
8. *S. a. howelli* Jackson, 1914. Mississippi, Alabama, northwestern Florida, and central northern Georgia.
9. *S. a. inflatus* Jackson, 1914. Southeastern Texas.
10. *S. a. machrinoides* Jackson, 1914. Southern Minnesota, Iowa, Missouri, eastern Nebraska, and Kansas.
11. *S. a. machrinus* (Rafinesque, 1832). From Tennessee north through Michigan, southern Ontario, and Ohio.
12. *S. a. montanus* Baker, 1951. Restricted to the Sierra del Carmen, Coahuila.
13. *S. a. nanus* Davis, 1942. Eastern Texas.
14. *S. a. parvus* Rhoads, 1894. Central-western Florida Peninsula.
15. *S. a. porteri* Schwartz, 1952. Restricted to southwestern Florida Peninsula.
16. *S. a. texanus* Allen, 1891. Restricted to Presidio County, Texas.

Conservation: *Scalopus aquaticus* is listed as Endangered in the Norma Oficial Mexicana (DOF 2019) and as Least Concern by the IUCN (Matson et al. 2016c).

Characteristics: Eastern mole is medium sized among moles; total length 129.0–148.0 mm in females, 142.0–164.0 mm in males; skull length 30.9–32.9 mm in females; 31.0–33.8 mm in males. Dorsal pelage silver to blackish; nose distinct snout, naked anterior to the nostrils; **tail less than one-fourth of the total length, thin and naked**; **digits of the forefeet and hind feet webbed;** forefoot width equal to or greater than its length; intraorbital foramen very small; interparietal bone large but not rectangular; nares directed forward; **two lower incisors**; eight lower teeth in total and increasing in size posteriorly.

Comments: *Scalopus aquaticus* ranges from southern Ontario and southern South Dakota east through Massachusetts and south through the Florida Peninsula (except the southeast) and northern Tamaulipas, being patchily distributed from south-western Texas south through Coahuila and Tamaulipas (Map 80). It can be distinguished from other species in sympatry, or nearly so, such as *Condylura cristata* and *Parascalopus breweri*, by having the tail less than one-fourth of the total length, thin and naked; webbed digits in forefeet and hind feet; two lower incisors. From *C. cristata*, by not having a ring of 22 fleshy appendages in the nose. From *P. breweri*, by not having the tail slightly constricted at the base and densely covered with long coarse hairs, and eyes minute and covered with hair.

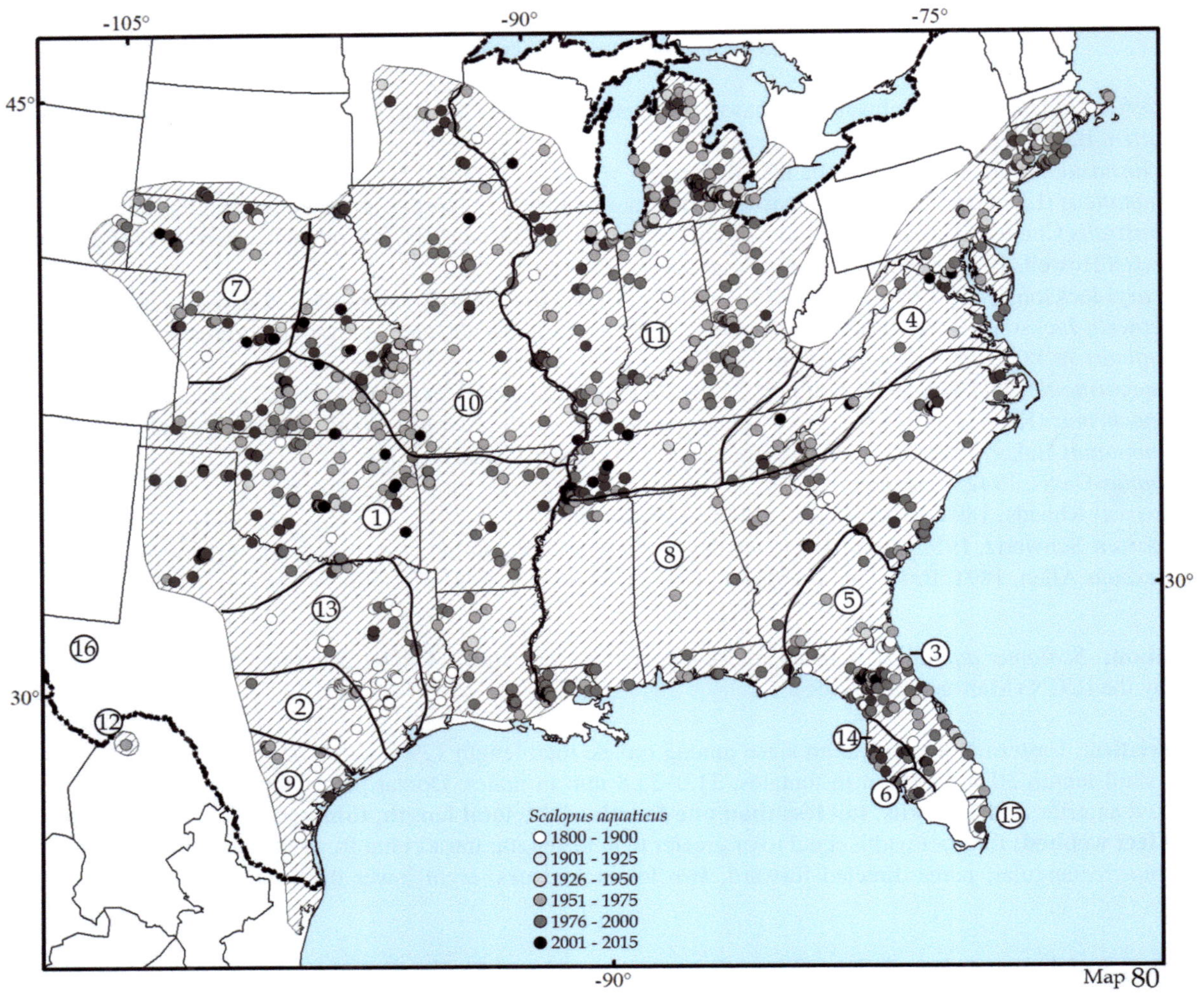

Map 80 Distribution map of *Scalopus aquaticus*

Additional Literature: Jackson (1914, 1915), Baker (1951), Yates and Schmidly (1975, 1977, 1978), Gureev (1979), Yates (1999d), Hutterer (2005), Guevara et al. (2015).

Genus *Scapanus* Pomel, 1848

Dorsal pelage brown in different shades; **tail length less than one-fifth of the head-and-body length, indistinctly annulated, tapered toward the tip and bearing coarse hairs**; eyes minute, concealed in hair; nose with a short snout and less truncated than in *Scalopus*, apical portion naked; forefeet palms as broad as long, toes not webbed; **hind feet bearing one to three (usually two) distinct tubercles and not webbed;** skull flattened; braincase relatively broad; interparietal bone large, somewhat rectangular; auditory bullae complete and depressed; first upper incisors long and broad; first and second molars subequal. The keys were elaborated based on the review of specimens and the following sources: Jackson (1915), Hall (1981), and Álvarez-Castañeda and Cortés-Calva (2021).

1. Dorsal pelage usually brown to gray; all unicuspid teeth with different spaces between them and usually crowded (Fig. 18); rostrum short and broad 2

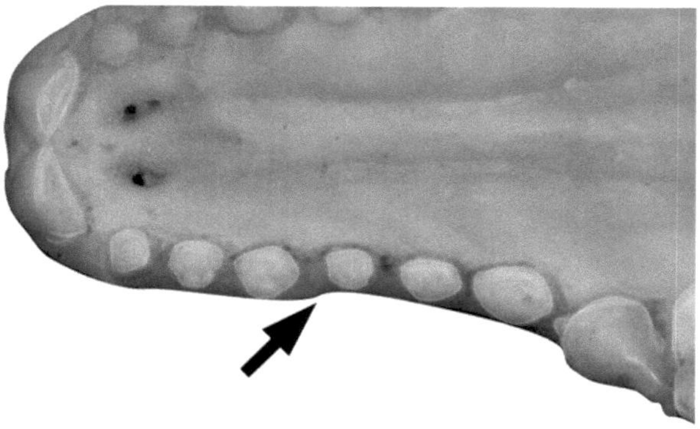

Fig. 18 All unicuspid teeth with different spaces between them and usually crowded

1a. Dorsal pelage almost black; all unicuspid teeth with similar spaces between them and not crowded (Fig. 19); rostrum long and narrow 4

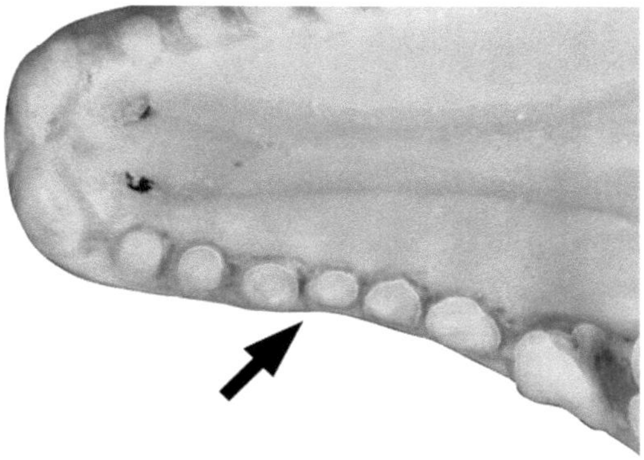

Fig. 19 All unicuspid teeth with similar spaces between them and not crowded

2. Skull total length less than 32.5 mm; less than seven unicuspid teeth posterior to the incisors in the maxilla and mandible (endemic to Sierra San Pedro Mártir, Baja California, México)........ *Scapanus anthonyi* (p. 506)

2a. Skull total length greater than 32.5 mm; seven unicuspid teeth posterior to the incisors in maxilla and mandible 3

3. Total length greater than 161.0 mm; skull length greater than 34.0 mm in males and greater than 33.4 mm in females; ratio of the mastoid breadth to the greatest skull length less than 49%, including the population of Alameda Island, California (from southern Sierra Nevada and Transverse Range in Pacific north through southern Oregon and Western Nevada, United States)........ *Scapanus latimanus* (p. 507)

3a. Total length less than 161.0 mm; skull length less than 34.0 mm in males and less than 33.4 mm in females; ratio of the mastoid breadth to the greatest skull length greater than 49%, not including the population of Alameda Island, California (from southern Sierra Nevada and Transverse Range in Pacific south through northern Baja California, including the eastern slope of Sierra Nevada, United States)........ *Scapanus occultus* (p. 509)

4. Total length on average greater than 200.0 mm; sublacrimal–maxillary ridge well developed (Fig. 20); skull length greater than 40.0 mm (from southern British Columbia south through the Humboldt Bay area, California, United States) .... *Scapanus townsendii* (p. 511)

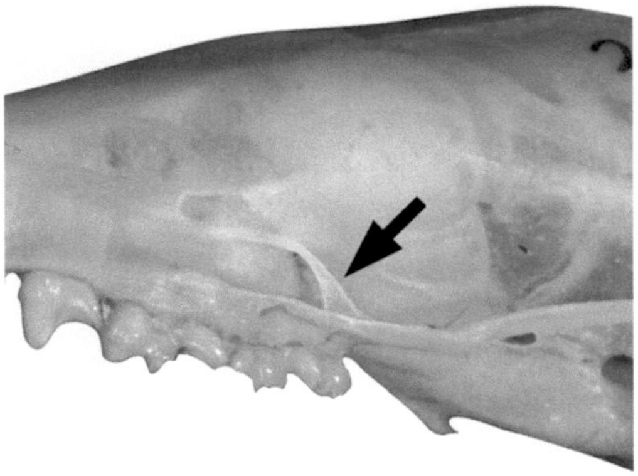

Fig. 20 Sublacrimal–maxillary ridge well developed

4a. Total length on average less than 200.0 mm; sublacrimal–maxillary ridge barely developed (Fig. 21); skull length less than 40.0 mm (from southern British Columbia south through north California, and east through western Idaho, United States)........ *Scapanus orarius* (p. 510)

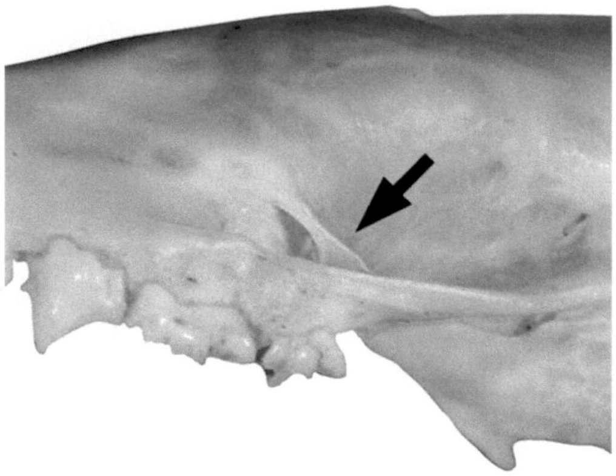

Fig. 21 Sublacrimal–maxillary ridge barely developed

Scapanus anthonyi Allen, 1893

Anthony's Mexican mole, topo de San Pedro Mártir

Monotypic.

Conservation: *Scapanus anthonyi* (as *S. latimanus anthonyi*) is listed as Endangered in the Norma Oficial Mexicana (DOF 2019) and as Least Concern (as *S. latimanus*) by the IUCN (Matson et al. 2016d).

Characteristics: Anthony's Mexican mole **is the smallest member of the genus; total length 141.0–146.0 mm, skull length 30.3–31.9 mm**. Dorsal pelage pale brown to reddish black; fur of the underparts paler than in the dorsum; color varying according to the season; intraorbital foramen small; interparietal bone large with a rectangular shape; three lower incisors; **two or three premolars but never four**.

Comments: *Scapanus anthonyi* was recognized as a distinct species (Allen 1893; Jackson 1915; Huey 1936; Yates and Salazar-Bravo 2005; Álvarez-Castañeda and Cortés-Calva 2021), previously regarded as a subspecies of *S. latimanus* (Palmer 1937; Hutterer 2005). This species is Restricted to the Sierra de San Pedro Mártir, Baja California (Map 81), and has allopatric distribution with respect to the other species of *Scapanus*.

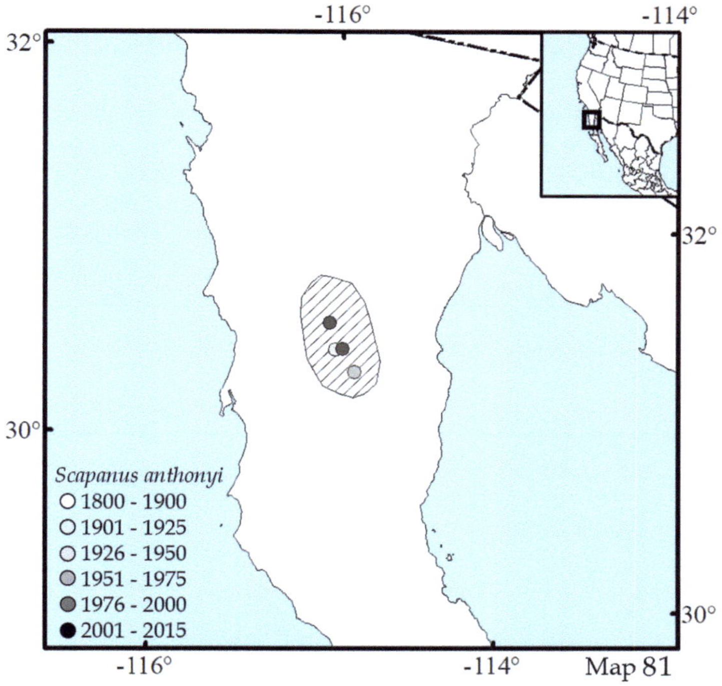

Map 81 Distribution map of *Scapanus anthonyi*

Additional Literature: Verts and Carraway (2001a), Hutterer (2005), Guevara et al. (2015).

Scapanus latimanus (Bachman, 1842)

Broad-footed mole, topo de patas anchas

1. *S. l. campi* Grinnell and Storer, 1916. Restricted to western slope of Sierra Nevada, California.
2. *S. l. caurinus* Palmer, 1937. From the border between Oregon and California in western California south through San Francisco Bay.
3. *S. l. dilatus* True, 1894. Southern Oregon, western Nevada, and northwestern California.
4. *S. l. insularis* Palmer, 1937. Restricted to San Angel Island, California.
5. *S. l. latimanus* (Bachman, 1842). From the San Francisco Bay to Santa Barbara area, California.
6. *S. l. minusculus* Bangs, 1899. Restricted to El Dorado County, California.
7. *S. l. monoensis* Grinnell, 1918. Restricted to Mono Lake area, western central California, and eastern central Nevada.
8. *S. l. parvus* Palmer, 1937. Restricted to Alameda Island, California.
9. *S. l. sericatus* Jackson, 1914. Restricted to eastern slope of Sierra Nevada.

Conservation: *Scapanus latimanus* is listed as Threatened in the Norma Oficial Mexicana (DOF 2019) and as Least Concern by the IUCN (Matson et al. 2016d).

Characteristics: Broad-footed mole is larger in size among moles; total length 170.0–175.0 mm, skull length 34.1–37.8 mm. Dorsal pelage pale brown; fur of the underparts paler than in the dorsum; color varying according to the season; intraorbital foramen small; interparietal bone large with a rectangular shape; four premolars.

Comments: *Scapanus latimanus anthonyi* was considered a subspecies of *S. latimanus* (Palmer 1937; Hutterer 2005) but was later recognized as a distinct species (Allen 1893; Jackson 1915; Huey 1936; Yates and Salazar-Bravo 2005; Álvarez-Castañeda and Cortés-Calva 2021). *S. l. grinnelli* and *S. l. monoensis* were subspecies of *S. latimanus*, now considered subspecies of *S. occultus* (Álvarez-Castañeda and Cortés-Calva 2021). *S. latimanus* ranges from southern-central Oregon and western Nevada south through the southern Sierra Nevada, including the eastern slope and Transverse Range in the Pacific, California (Map 82). It can be distinguished from other species in sympatry, or nearly so, such as *S. orarius* and *S. townsendii*, by having the dorsal pelage in brown to gray shades; all unicuspid teeth with different spaces between them and usually crowded; rostrum short and broad. From *S. occultus*, by its larger size (see measurements). Some specimens of *S. occultus* have one less molar (these are genetically different; Álvarez-Castañeda and Cortés-Calva 2021). From *N. gibbsii*, by having 44 teeth instead of 36; forefoot as wide as long, greater than 130.0 mm, and tail length less than 25% of the total length.

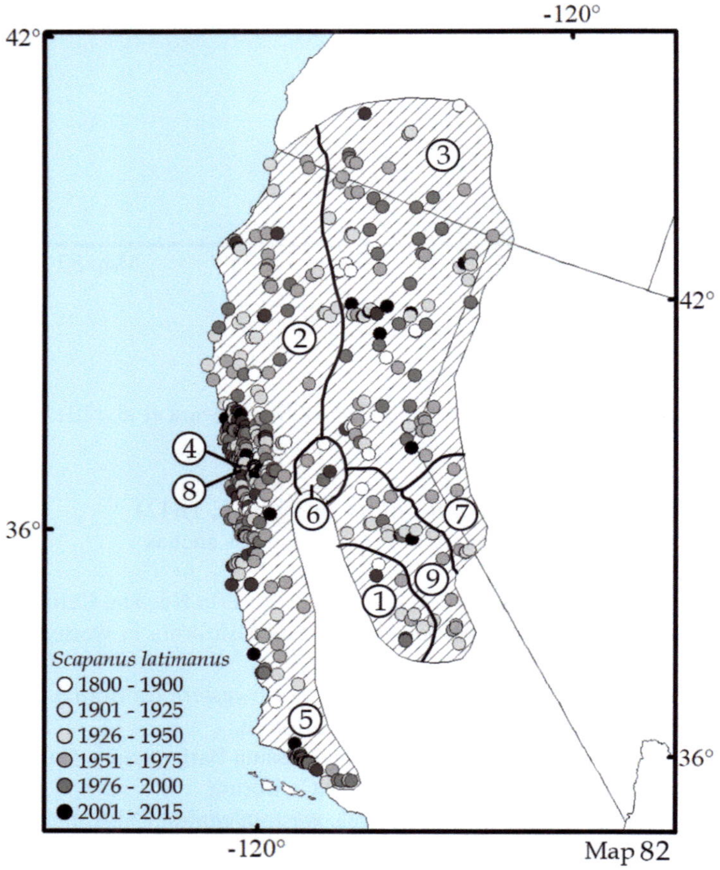

Map 82 Distribution map of *Scapanus latimanus*

Additional Literature: Carraway et al. (1993), Yates (1999b), Verts and Carraway (2001a), Guevara et al. (2015).

Scapanus occultus Grinnell and Swarth, 1912

West Coast mole, topo de la costa oeste

1. *S. o. grinnelli* Jackson, 1914. Restricted to eastern Sierra Nevada.
2. *S. o. occultus* Grinnell and Swarth, 1912. From central Sierra Nevada and Santa Barbara area in California south through northern Baja California.

Conservation: *Scapanus occultus* (as *S. latimanus*) is listed as Least Concern by the IUCN (Matson et al. 2016d).

Characteristics: Broad-footed mole is medium sized within moles; total length 132.0–160.0 mm, skull length 30.9–34.1 mm. Dorsal pelage pale brown; fur of the underparts paler than in the dorsum; color varying according to the season; intraorbital foramen small; interparietal bone large with a rectangular shape; four or three premolars; rostrum relatively long and narrow.

Comments: *Scapanus latimanus grinnelli*, *S. l. monoensis*, and *S. l. occultus* were previously considered subspecies of *S. latimanus* (Palmer 1937; Hutterer 2005); they are now subspecies of *S. occultus* (Álvarez-Castañeda and Cortés-Calva 2021). *S. occultus* ranges from northern Baja California north through southern Sierra Nevada, including the eastern slope and the Transverse Range in the Pacific, California (Map 83). It can be distinguished from other species in sympatry, or nearly so, such as *S. latimanus*, by its smaller size and some specimens have three premolars (these are genetically different; Álvarez-Castañeda and Cortés-Calva 2021).

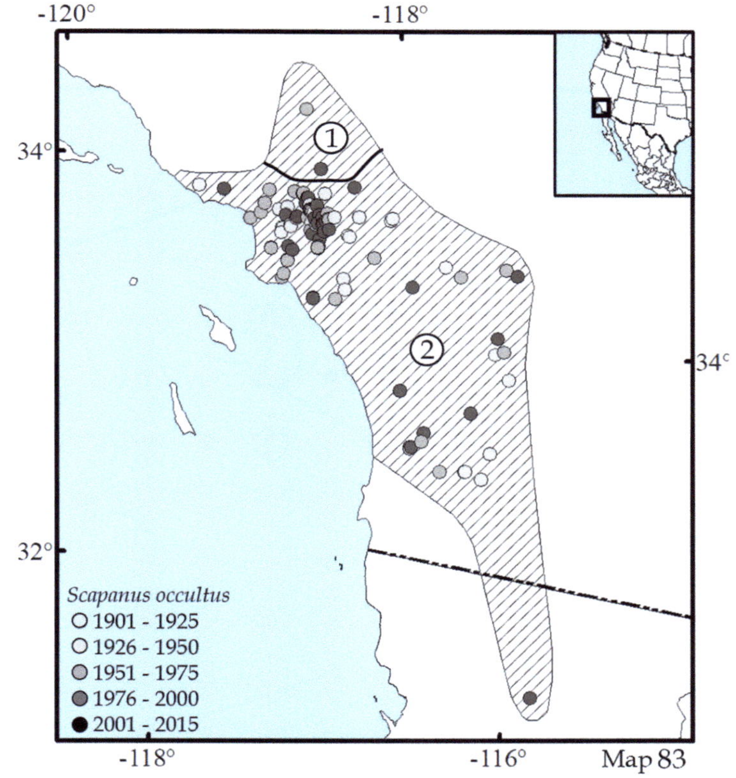

Map 83 Distribution map of *Scapanus occultus*

Additional Literature: Jackson (1915), Verts and Carraway (2001a), Yates and Salazar-Bravo (2005).

Scapanus orarius True, 1896

Pacific mole, topo de la costa

1. *S. o. orarius* True, 1896. From southern British Columbia south through north California.
2. *S. o. schefferi* Jackson, 1915. Southeastern Washington, northern Oregon, and western Idaho.
3. *S. o. yakimensis* Dalquest and Scheffer, 1944. Restricted to central Washington.

Conservation: *Scapanus orarius* is listed as Least Concern by the IUCN (Cassola 2016cz).

Characteristics: Coast mole is medium sized within moles; total length 133.0–190.0 mm, skull length 32.0–35.1 mm. Dorsal pelage velvet-like dark gray; limbs muscular, short, and close to the body; snout with its apical portion naked to the nasals anterior portion; neck short; tail scantily haired and short; auditory meatus concealed within the hair; eyes small; rostrum relatively narrow and long; sublacrimal-maxillary ridge not much developed; mandible relatively weak; unicuspid teeth evenly spaced and uncrowded; jaw horizontal ramus narrow and weak; teeth small, specially the first incisors.

Comments *Scapanus orarius* ranges from southern British Columbia south through Humboldt Bay in California and east through western Idaho (Map 84). It can be distinguished from other species in sympatry, or nearly so, such as *Neurotrichus gibbsii*, *S. latimanus*, and *S. townsendii.* From *N. gibbsii*, by having 44 teeth instead of 36; forefoot as wide as long and greater than 130 mm; tail less than 25% of the total length. From *S. latimanus*, by having the dorsal pelage blackish; all the unicuspid teeth with similar spaces between them, not crowded, and rostrum long and narrow. From *S. townsendii*, by its smaller size; with total length on average less than 200.0 mm; skull length less than 40.0 mm; upper incisors small, nearly two times the first unicuspid teeth width, and sublacrimal–maxillary ridge barely developed.

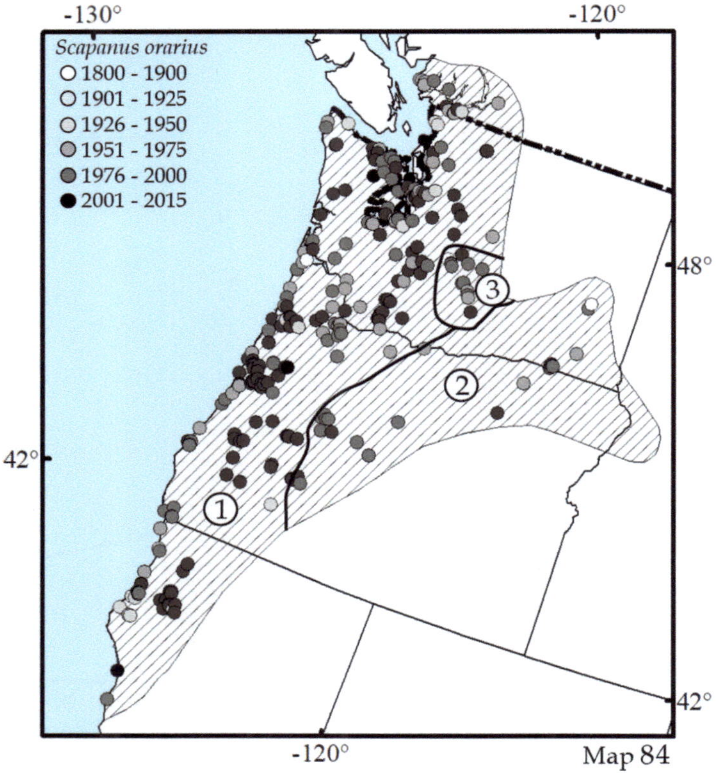

Map 84 Distribution map of *Scapanus orarius*

Additional Literature: Jackson (1915), Yates (1978), Hartman and Yates (1985), Carraway et al. (1993).

Scapanus townsendii (Bachman, 1839)

Snow mole, topo de la nieve

1. *S. t. olympicus* Johnson and Yates, 1980. Restricted to above 1000 m of the Olympic Mountains, Washington.
2. *S. t. townsendii* (Bachman, 1839). From southern British Columbia south through Humboldt Bay area, California.

Conservation: *Scapanus townsendii* is listed as Least Concern by the IUCN (Cassola 2016da).

Characteristics: Snow mole is large sized within moles; total length 190.0–237.0 mm, skull length 39.4–44.6 mm. Dorsal pelage short and black velvet-like; shoulders heavy; limbs muscular, short, and close to the body; snout long; neck short; tail short, almost naked; without pinnae; eyes minute; skull compressed dorsoventrally; rostrum long and narrow; interparietal bone large and broad.

Comments: *Scapanus townsendii* ranges from southern British Columbia by the coastal plains to southern Humboldt Bay, California (Map 85). It can be distinguished from other species in sympatry, or nearly so, such as *Neurotrichus gibbsii*, *S. latimanus*, and *S. orarius*. From *N. gibbsii*, by having 44 teeth instead of 36; forefoot as wide as long and greater than 130 mm; tail less than 25% of the total length. From *S. latimanus*, by having the dorsal pelage blackish; all the unicuspid teeth with similar spaces between them and not crowded; rostrum long and narrow. From *S. orarius*, by its larger size, with total length on average greater than 200.0 mm, skull length greater than 40.0 mm; upper incisors large nearly three times the first unicuspid teeth width and the sublacrimal–maxillary ridge well developed.

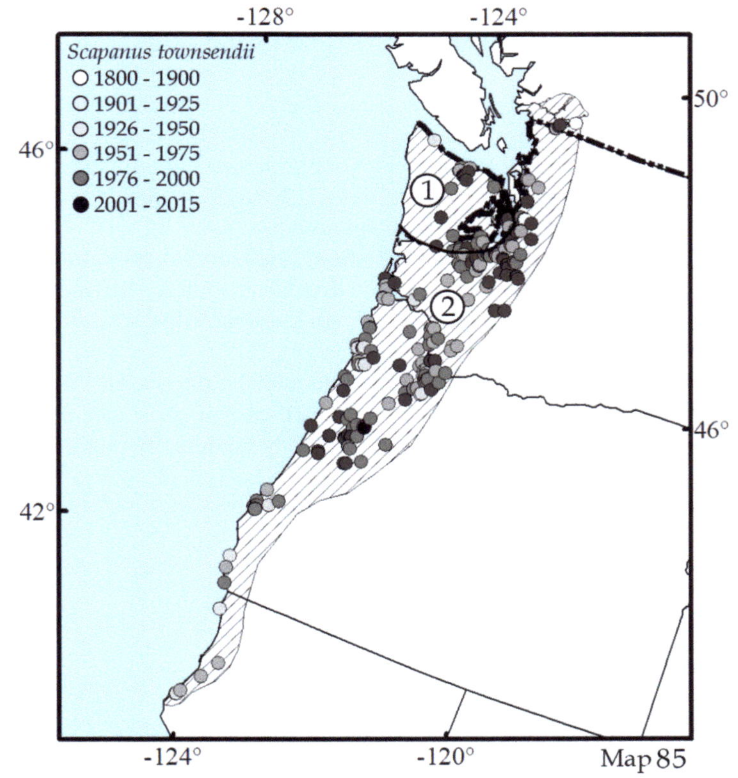

Map 85 Distribution map of *Scapanus townsendii*

Additional Literature: Jackson (1915), Johnson and Yates (1980), Yates and Greenbaum (1982), Yates and Pedersen (1982), Carraway et al. (1993), Verts and Carraway (1998), Yates (1999c).

Subfamily Talpinae Fisher, 1814

The species of Talpinae can be differentiated from those of the Scalopinae based on the mtDNA phylogeny. One tribe ranges in North America: tribe Neurotrichini.

Tribe Neurotrichini

Genus *Neurotrichus* Günther, 1880

Dorsal pelage from dark mouse-gray to blackish mouse-gray; **tail length nearly one-half of the head-and-body length;** tail constricted at the base with scales, annulated, and with coarse hairs; **forefeet palms longer than wide, toes not webbed; hind foot soles with six tubercles;** braincase broad; interorbital constricted, scarcely developed; zygomata short; auditory bullae incomplete; upper molars with a bicuspid internal edge; first and second molars subequal and third molars smaller.

Neurotrichus gibbsii (Baird, 1858)

Shrew mole, topo musaraña

1. *N. g. gibbsii* (Baird, 1858). From southern British Columbia to northern California, including Destruction Island, and excluding western Washington.
2. *N. g. hyacinthinus* Bangs, 1897. From Humboldt Bay south through Monterey Bay, California.
3. *N. g. minor* Dalquest and Burgner, 1941. Western coast of Washington.

Conservation: *Neurotrichus gibbsii* is listed as Least Concern by the IUCN (Cassola 2016cc).

Characteristics: Shrew mole is small within moles, varying widely in size latitudinally, with northern specimens larger than the southern; total length 100.0–126.0 mm; skull length 21.1–22.3 mm. Dorsal pelage nearly black to sooty blue black; **forefeet only moderately modified for digging**; **eyes inconspicuous;** without webbed digits**; six tubercles on the hind foot sole; nose elongated and flattened dorsoventrally; tail about one-third of the total length, wide, sparsely haired, and with transverse annular rows of scales**; forefoot less wide than long; auditory bullae incomplete; braincase broad and skull scarcely constricted interorbitally; upper molars with a bicuspid internal edge.

Comments: *Neurotrichus gibbsii* ranges from southern British Columbia by the coastal plains to Monterey Bay, California, including Sierra Nevada Mountains (Map 86), and differs from the other sympatric species of *Scapanus* by having inconspicuous eyes; nose elongated and flattened dorsoventrally; forefeet only moderately modified for digging and six tubercles on the hind foot sole; tail about one-third of the total length, wide, sparsely haired, and with transverse annular rows of the scales.

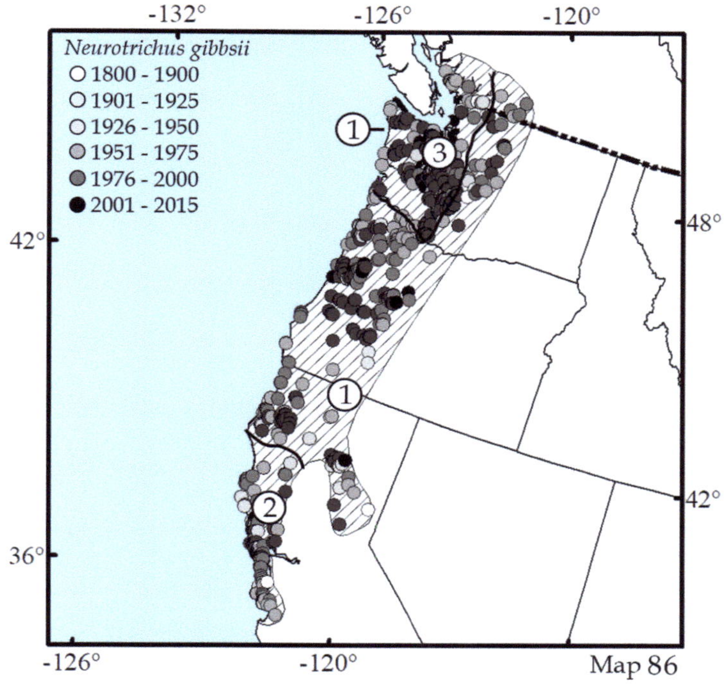

Map 86 Distribution map of *Neurotrichus gibbsii*

Additional Literature: Jackson (1915), Dalquest and Orcutt (1942), Reed (1951), Yates and Schmidly (1975), Hall (1981), Lewis (1983), Carraway and Verts (1991b), Yates (1999a), Hutterer (2005).

Order Lagomorpha

Order Lagomorpha Brandt, 1855

Lagomorpha include hares, rabbits, and pikas that are characterized by **two pairs of upper incisors, the second smaller, peg-shaped, located directly behind the first pair,** contrasting with rodents that have only one pair of the incisors; **tail indistinct or very small**; ears small to large; and **maxilla fenestrated or with orifices**. Distribution is nearly cosmopolitan, except for Antarctica, introduced in Australia. Two families occur in North America: Ochotonidae and Leporidae. The keys were elaborated based on the review of specimens and the following sources: Howell (1924), Hall (1951, 1981), and Álvarez-Castañeda et al. (2017a).

1. Ears long, larger than wide; forelegs shorter than the hind legs; hindfeet greater than 40.0 mm; six molariforms in the maxilla; maxilla fenestrated (Fig. 1); nasals wider posteriorly (Fig. 2)....................................Family Leporidae (p. 517)

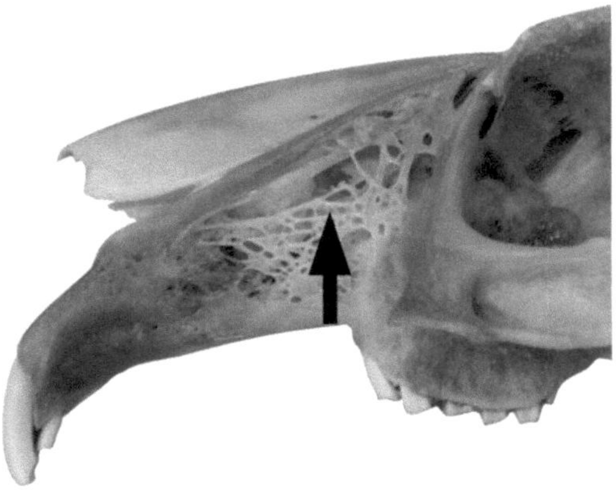

Fig. 1 Maxilla fenestrated

S. T. Álvarez-Castañeda, *Mammals of North America - Volume 1*, https://doi.org/10.1007/978-3-031-41661-3_8

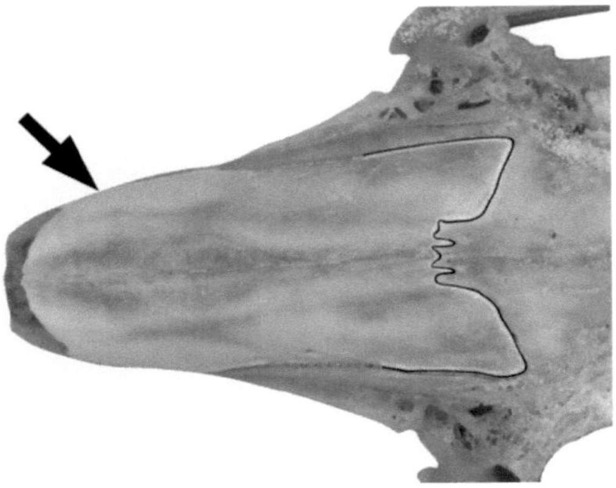

Fig. 2 Nasals wider posteriorly

1a. Ears short, similar in width and length; forelegs scarcely shorter than the hind legs; hindfeet less than 40.0 mm; five molariforms in the maxilla; maxilla not fenestrated (Fig. 3); nasals wider anteriorly (Fig. 4) .... Family Ochotonidae (p. 554)

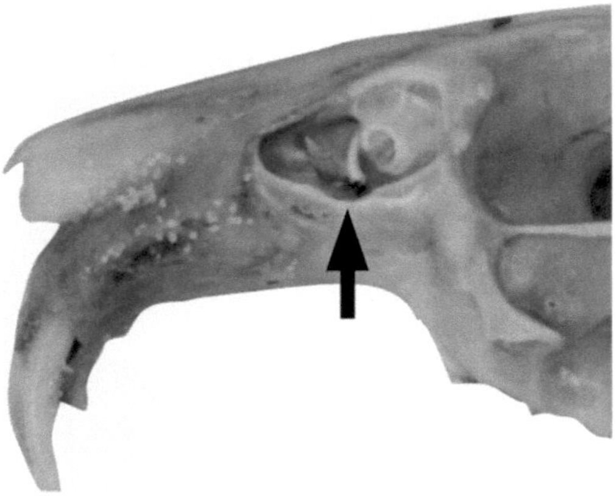

Fig. 3 Maxilla not fenestrated

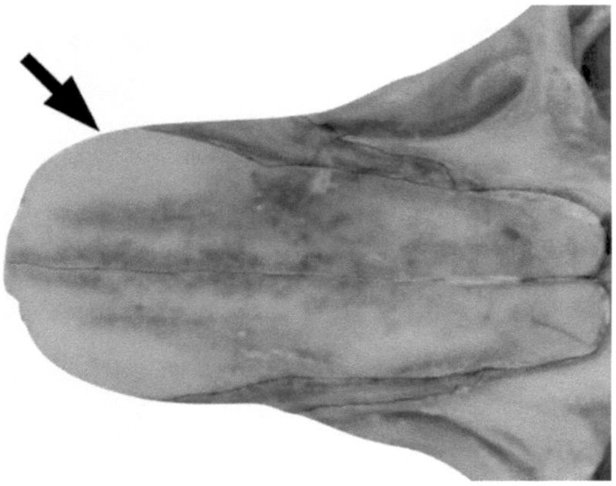

Fig. 4 Nasals wider anteriorly

Family Leporidae Fischer de Waldheim, 1817

The family Leporidae includes rabbits and hares characterized by **ears longer than wide; tail small; two pairs of the incisors, one behind the other; maxilla fenestrated; supraorbital process fan shaped; and nasals wider posteriorly**. Distribution is virtually cosmopolitan, except for Antarctica and some oceanic islands. The keys were elaborated based on the review of specimens and the following sources: Howell (1924), Hall (1951, 1981), and Álvarez-Castañeda et al. (2017a).

1. Ears larger than the head length, if similar size, with distribution north of 45° N latitude; hindfoot length greater than 105.0 mm; dorsal pelage tending to be speckled in gray shades; interparietal bone absent (Fig. 5)...........*Lepus* (p. 518)

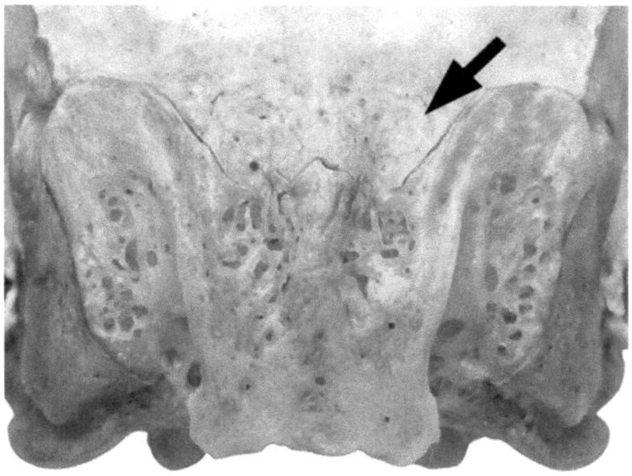

Fig. 5 The interparietal bone absent

1a. Ear length equal to or less than the head length; hindfoot length less than 105.0 mm; dorsal pelage tending to be brown or gray to blackish gray; interparietal bone present (Fig. 6)..2

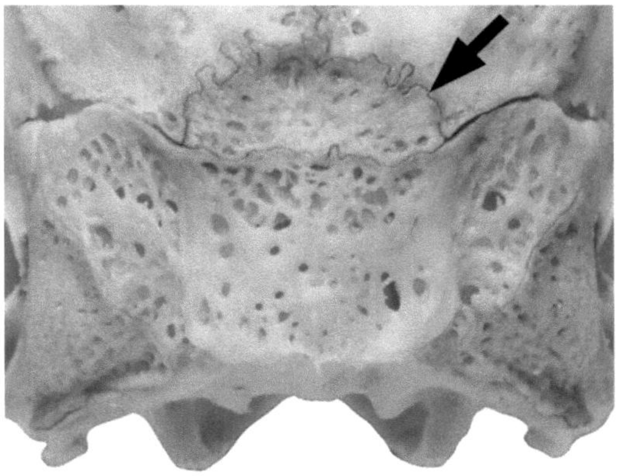

Fig. 6 The interparietal bone present

2. Ear length less than 40.0 mm, ears rounded; hindfoot length less than 65.0 mm; fur grayish yellowish brown to dark cinnamon; anterior extension of the supraorbital process absent or present only as a little tip (Fig. 7); auditory bullae larger than the foramen magnum (endemic of the Eje Volcánico Transversal, from Pico de Orizaba west through the Nevado de Toluca, México) .... *Romerolagus diazi* (p. 533)

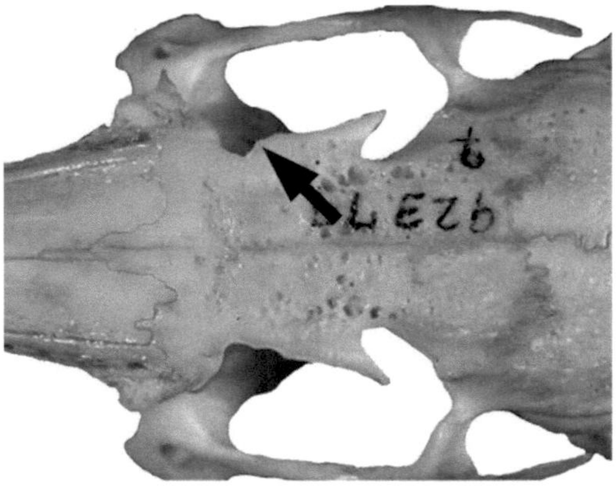

Fig. 7 Anterior extension of the supraorbital process absent or present only as a little tip

2a. Ear length greater than 40.0 mm, length twice the breadth; hindfoot length equal to or greater than 65.0 mm; pelage pale or dark brown, speckled but not grayish; anterior extension of the supraorbital process present (Fig. 8), if absent, auditory bullae less than the *foramen magnum*..3

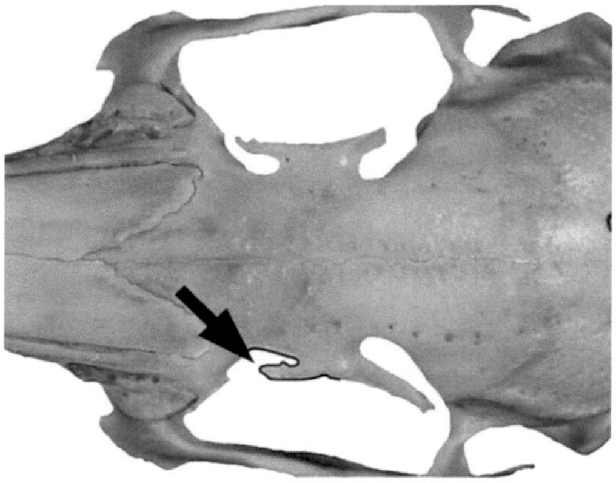

Fig. 8 Anterior extension of the supraorbital process present

3. Ears in general of the same color as dorsum; bridge across the palate between the molars wide... *Oryctolagus cuniculus* (p. xxx)

3a. Ears in a different color than dorsum; bridge across the palate between the molars narrow................*Sylvilagus* (p. 534)

Genus *Lepus* Linnaeus, 1758

Dorsal pelage tends to be speckled in gray shades. However, some specimens are white or black, according to the species and season; rostrum rounded; **ears very large,** greater than the **head length; hind legs very large;** tail small and haired; **hind feet greater than 105.0 mm; no interparietal bone**; posterior supraorbital process large; anterior supraorbital process varying in size or absent; auditory bullae varying in size according to the species but in general relatively larger relative to *Sylvilagus*. *Lepus* has four subgenera in North America: *Lepus* (*L. arcticus* and *L. othus*), *Macrotolagus* (*L. alleni*), *Poecitolagus* (*L. americanus*), and *Proeulagus* (*L. altamirae, L californicus, L. callotis, L. flavigularis,* and *L. townsendii*). The keys were elaborated based on the review of specimens and the following sources: Nelson (1909), Howell (1924), Hall (1951, 1981), and Álvarez-Castañeda et al. (2017a).

1. Ears small, length less than 90.0 mm, similar to the head length; distributed in cold areas..2

1a. Ears large, length greater than 90.0 mm, at least 1.5 times the head length; distributed in desert and cold areas 4

2. Hindfoot length less than 19% of the total length; incisors procumbent (Fig. 9; Canada and Greenland north of 60° N in the tundra, Northwest Territories, Nunavut, the Arctic islands, Newfoundland, Québec, and Manitoba, Canada) ....*Lepus* (*Lepus*) *arcticus* (p. 521)

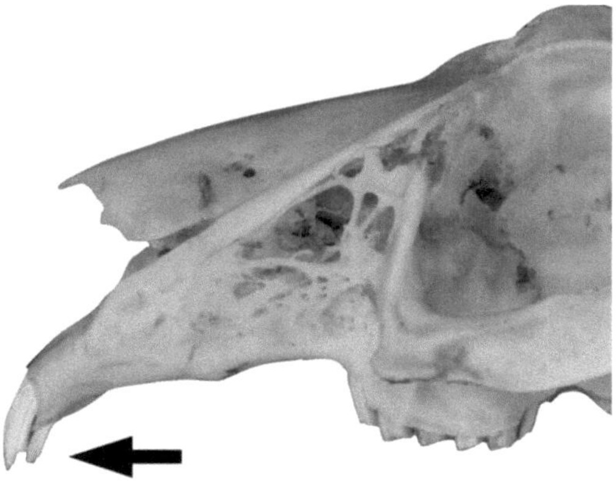

Fig. 9 Incisors procumbent like in *Lepus arcticus*

2a. Hindfoot length greater than 19% of the total length; incisors not procumbent (Fig. 10, restricted to Alaska)..............3

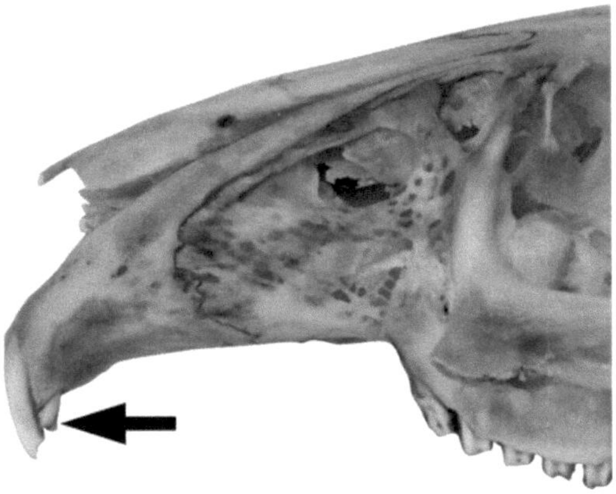

Fig. 10 Incisors not procumbent

3. Ears small, length less than 77.0 mm; basilar length greater than 67.0 mm; incisors proportionally small (from Alaska east through the Labrador Peninsula and south through California, Utah, New Mexico, South Dakota, Minnesota, Wisconsin; east through New York and the Appalachian Mountains, United States) ....*Lepus* (*Poecitolagus*) *americanus* (p. 525)

3a. Ears large, length greater than 77.0 mm; basilar length less than 67.0 mm; incisors as in other species (restricted to western and northwestern Alaska, United States)..*Lepus* (*Lepus*) *othus* (p. 522)

4. Tail all white all year round; ear length less than 17% of the total length (from southern regions of Canada, British Columbia east through Ontario; in the United States, from west through California, Nevada, New Mexico, Utah, Colorado, Missouri, and Illinois, United States).. *Lepus* (*Proeulagus*) *townsendii* (p. 532)

4a. Dorsal tail not white all year round; ear length greater than 17% of the total length...5

5. Ears externally gray or white at the tip (not black)...6

5a. Ears externally black at the tip...8

6. Ear length from the low neckline greater than 130.0 mm (when dry) and 137.0 mm (when fresh); skull length greater than 105.0 mm and toothrow greater than 18.0 mm (from southern-central Arizona, United States, south through Pacific Coastal Plains to northern Nayarit, southwestern Chihuahua, and Tiburón Island, México) .... *Lepus* (*Macrotolagus*) *alleni* (p. 523)

6a. Ear length from the low neckline less than 130.0 mm (when dry) and 137.0 mm (when fresh); skull length less than 105.0 mm and toothrow length less than 18.0 mm...7

7. Ears ocher; auditory bullae very small (Fig. 11; endemic to the Tehuantepec Isthmus, Oaxaca, México) .... *Lepus* (*Proeulagus*) *flavigularis* (p. 531)

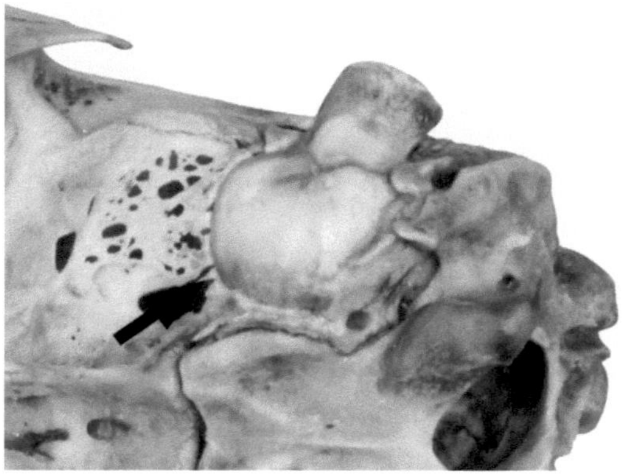

Fig. 11 The auditory bullae very small like in *Lepus flavigularis*

7a. Ears dark ocher, grayish, white, and black; auditory bullae large (Fig. 12; from New Mexico south through northern Oaxaca) .... *Lepus* (*Proeulagus*) *callotis* (p. 530)

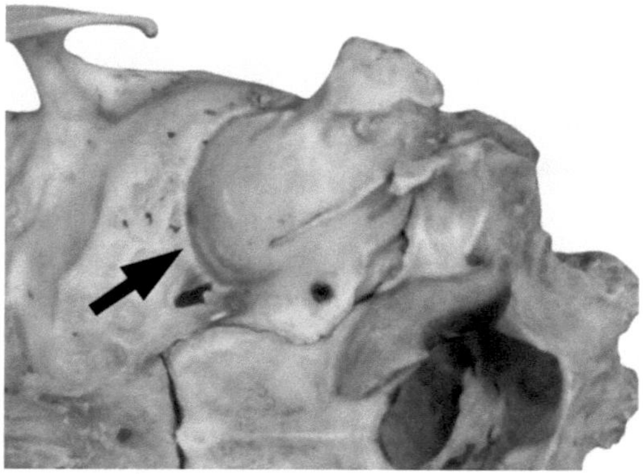

Fig. 12 The auditory bullae large

8. Ears proportionally short, less than 23% of the total length; does not range in desert areas; distributed in Ontario around the Great Lakes and south of the Canadian Shield, New York, Pennsylvania, Massachusetts, and Connecticut, United States .... *Lepus* (*Eulagos*) *europaeus* (p. 521)

8a. Ears proportionally large, greater than 23% of the total length; ranges in desert areas, from Washington east through South Dakota and south through Guanajuato and Querétaro, including the Baja California Peninsula 9

9. Black spot on the nape; black spot at the tip of the ears large and sides paler; flanks with a coloration similar to that of the upper parts, not colored as the center of the back (widely distributed from Washington east through South Dakota, United States, and south through San Luis Potosí, Querétaro, and Zacatecas, including the Baja California Peninsula, México) ....*Lepus* (*Proeulagus*) *californicus* (p. 527)

9a. Having black lines on the nape; black spot in the tip of the ears is very small and sides in paler coloration (range Restricted to southeastern Tamaulipas and southeastern San Luis Potosí, México)... .... *Lepus* (*Proeulagus*) *altamirae* (p. 526)

Subgenus *Eulagos* Gray, 1867

Large sized; ears moderately large; skull length nearly 95.0 mm; skull broad and arched; postorbital process wide, and triangular; its posterior process not touching the side of the cranium; rostrum long and broad; nasals and rostrum wide and heavy.

Lepus europaeus Pallas, 1778

European Hare, liebre europea

Conservation: *Lepus europaeus* is listed as Least Concern by the IUCN (Hacklander and Schai-Braun 2019).

Characteristics: European hare is large sized; total length 550.0–650.0 mm, basilar length 90.0–103.0 mm. Dorsal pelage grizzled yellow-brown; shoulders, legs, neck, and throat rufous; tail dorsally black and ventrally; ears tips black. Does not turn completely white in the winter.

Comments: *Lepus europaeus* was introduced in Canada, and the United States to the Great Lakes Area. It can be found in sympatry, or nearly so, with *L. americanus*, from which can be distinguished by having the ears large with a length greater than 90.0 mm and at least 1.5 times the head length; it is not white in winter.

Additional Literature: Honacki et al. (1982), Whitaker and Hamilton (1998), Hoffmann and Smith (2005), Hackländer and Schai-Braun (2018), and Bock (2020).

Subgenus *Lepus* Linnaeus, 1758

Large sized; ears moderately large; skull length nearly 100.0 mm; **skull broad and strongly arched; postorbital process wide, triangular, and heavy; its posterior process not touching the side of the cranium; rostrum short and broad**; nasals and rostrum wide and heavy; spines of the lumbar vertebrae well marked.

Lepus arcticus Ross, 1819

Arctic Hare, liebre del Ártico

1. *L. a. andersoni* Nelson, 1934. Northern Mackenzie and Keewatin.
2. *L. a. arcticus* Ross, 1819. Bylot Island, Canada.
3. *L. a. bangsii* Rhoads, 1896. Northeastern Newfoundland.
4. *L. a. banksicola* Manning and Macpherson, 1958. Banks Island, Canada.
5. *L. a. hubbardi* Handley, 1952. Prince Patrick Island, Canada.
6. *L. a. labradorius* Miller, 1899. Northern Labrador Peninsula, Canada.
7. *L. a. monstrabilis* Nelson, 1934. Ellesmere and Devon Islands, Canada.

Conservation: *Lepus arcticus* is listed as Least Concern by the IUCN (Smith and Johnston 2019a).

Characteristics: Arctic hare is large sized; total length 550.0–635.0 mm, basilar length 72.0–80.0 mm. Dorsal pelage gray in summer to white in winter, hair white all the way to the base; **ears always black at the tip with a longitudinal black strip**; braincase relatively broad; **frontal anterior region with a pronounced depression**; **incisors procumbent**.

Comments: *Lepus arcticus* was recognized as a distinct species (Flux and Angermann 1990; Hoffmann and Smith 2005); it was previously regarded as a subspecies of *L. timidus* (Honacki et al. 1982; Dixon et al 1983) or of *L. othus* (Chapman et al. 1983; Dixon et al. 1983). *L. arcticus* ranges in association with the tree line in Canada at the Northwest Territories, Nunavut, the Arctic islands, Newfoundland, Québec, and Manitoba (Map 1). It can be found in sympatry, or nearly so, with *L. americanus*, from which can be distinguished by having the incisors procumbent and relatively larger and hindfeet proportionally smaller.

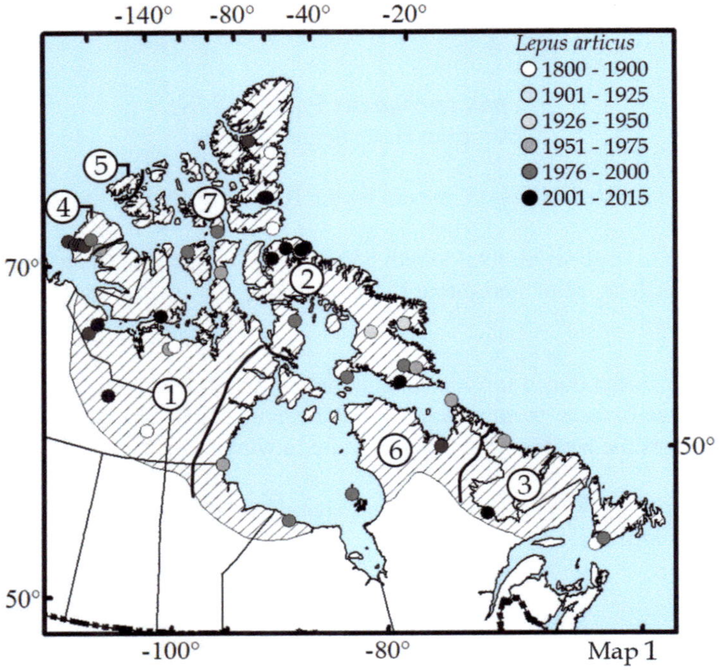

Map 1 Distribution map of *Lepus arcticus*

Additional Literature: Nelson (1909), Howell (1936), Hall (1951b), Honacki et al. (1982), Best and Henry (1994a), and Schai-Braum and Hackländer (2018).

Lepus othus Merriam, 1900

Alaskan Hare, liebre de Alaska

1. *L. o. othus* Merriam, 1900. Western and northern coast of Alaska.
2. *L. o. poadromus* Merriam, 1900. Southwestern Alaska.

Conservation: *Lepus othus* is listed as Least Concern by the IUCN (Smith and Johnston 2019b).

Characteristics: Alaskan hare is large sized; total length 565.0–690.0 mm, basilar length 73.0–83.3 mm. Dorsal pelage brownish in summer to white in winter; nose, face side, and head top darker as the back cinnamon-buff, white in winter, hair not white all the way to the base; ears black at the tip in a narrow longitudinal line; **claws stout**; frontal bones broader at the interorbital region; supraoccipital shield narrow; **upper incisors strongly recurved**; upper molars heavier.

Comments: *Lepus arcticus* and *L. timidus* have been conspecific of *L. othus* (Chapman et al. 1983; Dixon et al. 1983) but are now considered distinct species (Flux and Angermann 1990; Hoffmann and Smith 2005). *L. othus* only ranges in the Arctic tundra of western and northwestern Alaska (Map 2). It has allopatric distribution with respect to the other species of *Lepus* and can be distinguished from *L. americanus* by its larger ears, first upper incisors describing an arc of a circle, and tail always white.

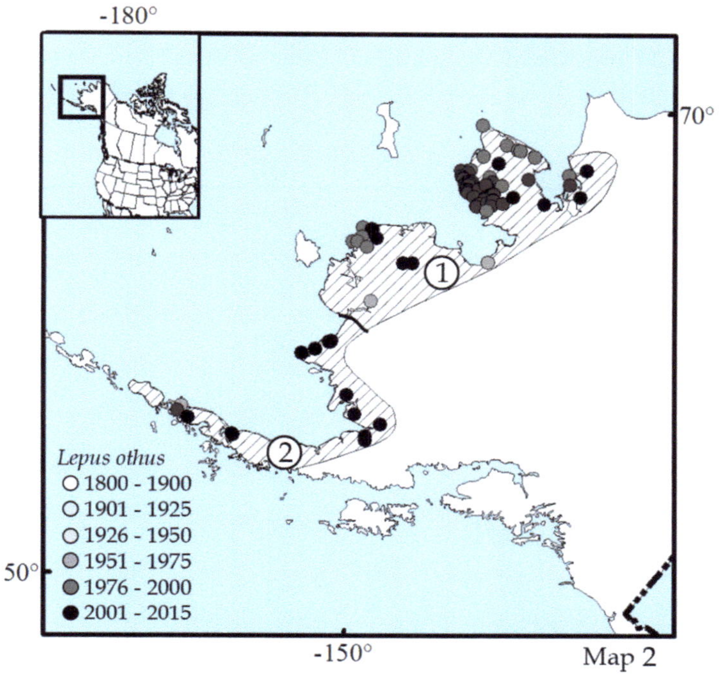

Map 2 Distribution map of *Lepus othus*

Additional Literature: Nelson (1909), Howell (1936), Best and Henry (1994b), Beever (2018).

Subgenus *Macrolagus* Mearns, 1896

Large sized; **ears longer** than the **head or hindfeet; skull narrow and elongated, less arched**; **postorbital process meeting the side of the cranium and enclosing a large foramen; nasals large and narrow;** neural spines slightly larger; radius slender, about equal in length than the humerus.

Lepus alleni Mearns, 1890
Antelope Jackrabbit, liebre antílope

1. *L. a. alleni* Mearns, 1890. From central Sonora to Arizona.
2. *L. a. palitans* Bang, 1900. From southern Sonora to Nayarit.
3. *L. a. tiburonensis* Townsend, 1912. Restricted to Tiburón Island, Sonora.

Conservation: *Lepus alleni tiburonensis* is listed as Special protection in México (DOF 2019) and *L. alleni* is listed as Least Concern by the IUCN (Lorenzo and Brown 2019).

Characteristics: Antelope jackrabbit is the **largest member of the genus; total length 553.0–670.0 mm, basilar length 81.5 mm**. Dorsal pelage speckled with black; flanks, hips, and rump pale gray; fur of the underparts, chin, and throat white; neck pale orange; nape plumbeous; **ears nearly naked with white hairs at the tip and borders**; tail plumbeous black dorsally and pale gray ventrally; skull large; supraorbital frontal process slightly arched; nasals, premaxillary teeth, and molars massive; rostrum large and heavy; frontal region wide; groove in the anterior supraorbital process small or vestigial.

Comments: *Lepus altamirae* was considered a subspecies of *L. alleni* (Hall 1951b), later of *L. californicus* (Hall 1981), and it is currently recognized as a distinct species (Vargas et al. 2019). *L. alleni* ranges from southern Arizona south through northern Nayarit, southwestern Chihuahua, and Tiburón Island (Map 3). It can be found in sympatry, or nearly so, with *L. californicus* and *L. callotis*, from which it can be distinguished by its pale gray flanks, larger size, and larger ears.

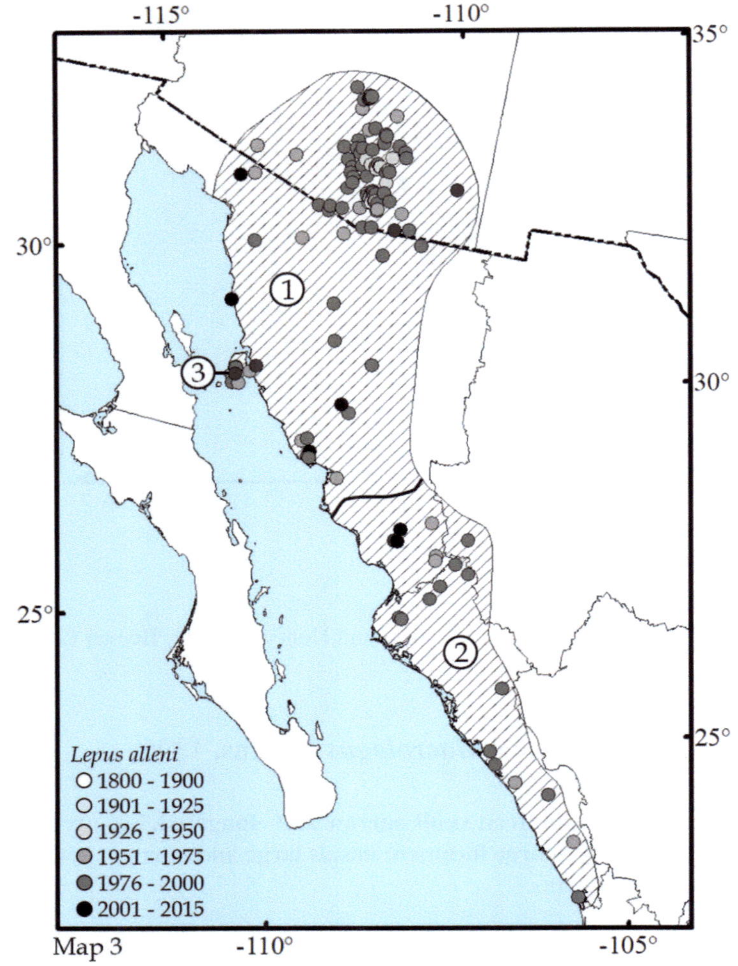

Map 3 Distribution map of *Lepus alleni*

Additional Literature: Nelson (1909), Honacki et al. (1982), Hoffmeister (1986), Best and Henry (1993), Brown et al. (2014, 2018a, b), and Lorenzo et al. (2015).

Subgenus *Poecitolagus*

Size small; **ears shorter** than the **head or hind feet**; ears blackish at the tip, posteriorly of the same color as the rest of the ears; **foot soles densely furred;** skull with a relatively broad braincase; **anterior region of the frontal bones with a pronounced depression; incisors proportionally smaller.**

Lepus americanus Erxleben, 1777

Snowshoe Hare, liebre de las nieves

1. *L. a. americanus* Erxleben, 1777. From Alberta to Newfoundland.
2. *L. a. bairdii* Hayden, 1869. From southern Alberta to northern New Mexico and Nevada.
3. *L. a. cascadensis* Nelson, 1907. Cascade mountain range in British Columbia and Washington.
4. *L. a. columbiensis* Rhoad, 1895. Southwestern Alberta and southeastern British Columbia.
5. *L. a. dalli* Merriam, 1900. From central British Columbia north through Alaska, Yukon, and Mackenzie.
6. *L. a. klamathensis* Merriam, 1899. Central Oregon and northeastern California.
7. *L. a. oregonus* Orr, 1934. Western Oregon.
8. *L. a. pallidus* Cowan, 1938. Central western British Columbia.
9. *L. a. phaeonotus* Allen, 1899. Southern Manitoba, Ontario, Minnesota, Wisconsin, and northwestern Michigan.
10. *L. a. pineus* Dalquest, 1942. Western Washington and northern Idaho.
11. *L. a. seclusus* Baker and Hankins, 1950. Northcentral Wyoming.
12. *L. a. struthopus* Bangs, 1898. Labrador Peninsula.
13. *L. a. tahoensis* Orr, 1933. Western central California.
14. *L. a. virginianus* Harla, 1825. From southern Ontario and Québec south through Pennsylvania, and the Appalachian Mountains.
15. *L. a. washingtonii* Baird, 1855. Western coast of Washington and Oregon.

Conservation: *Lepus americanus* is listed as Least Concern by the IUCN (Mills and Smith 2019).

Characteristics: Snowshoe hare is small sized; total length 413.0–518.0 mm, basilar length 62.5–71.2 mm. Dorsal pelage in summer grizzled rusty to grayish brown, with a middle dorsal line blackish; flanks buffy and belly white; face and limbs cinnamon brown; ears brownish, black at the tip and with white or creamy borders. Fur almost entirely white in winter, although the underfur remains gray; eyelids blackish; **ears blackish at the tip, posteriorly of the same color as the rest of the ears; foot soles densely furred**; braincase relatively broad; **anterior region of the frontal bones with a pronounced depression; incisors proportionally smaller**.

Comments: *Lepus americanus* is widespread in Canada and northern United States, from Alaska east through the Labrador Peninsula and south through California, Utah, New Mexico, South Dakota, Minnesota, Wisconsin, and east through New York and the Appalachian Mountains (Map 4). It can be found in sympatry, or nearly so, with *L. arcticus, L. californicus, L. europaeus, L. othus,* and *L. townsendii. L. americanus* can be distinguished from *L. articus* by not having procumbent incisors and with relatively smaller and proportionally larger hindfeet. From *L. californicus* and *L. townsendii,* by having the ears very small, similar in length to the head. From *L. europaeus* can be found in sympatry, or nearly so, in the Great Lakes Area. *L. americanus* can be distinguished by having the ears small with a length of less than 90.0 mm, similar in length to the head length and white in winter. From *L. othus,* by its shorter ears, first upper incisors inscribing a wide arc of a circle, and the tail is not always white.

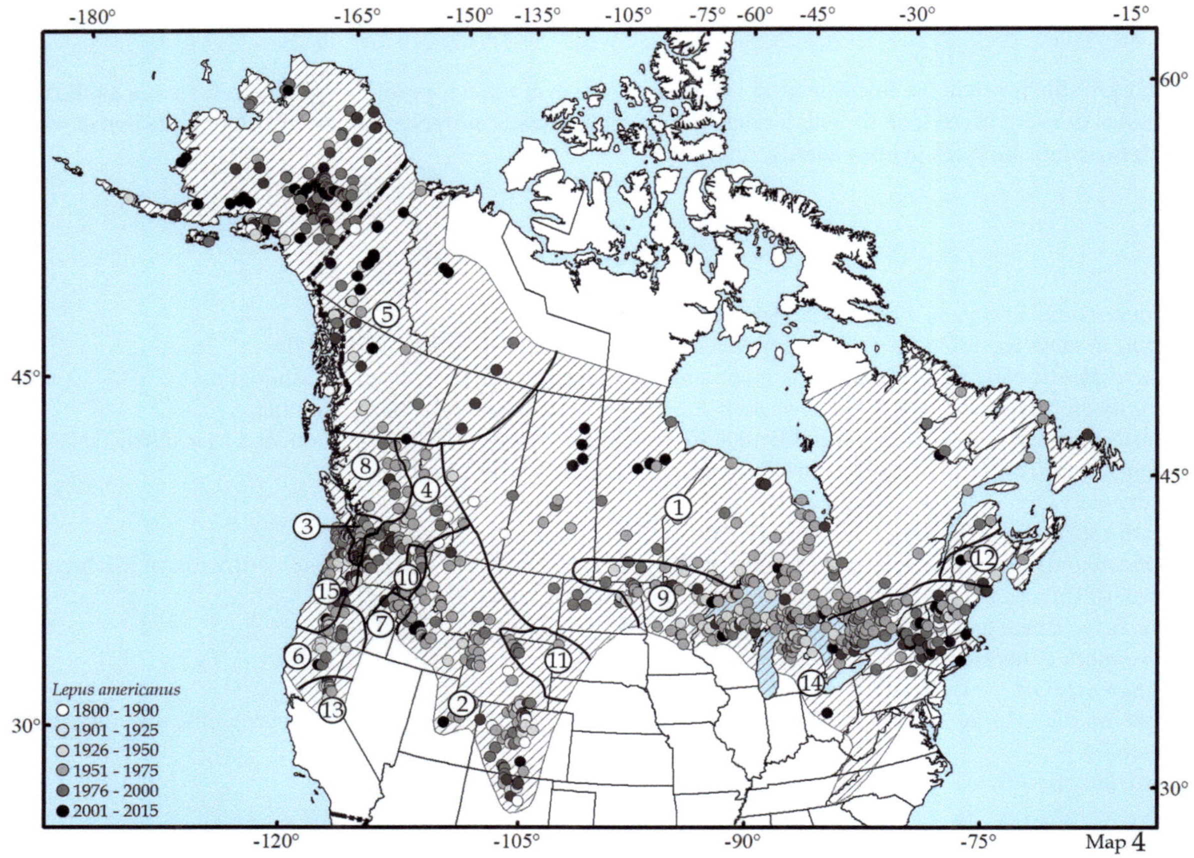

Map 4 Distribution map of *Lepus americanus*

Additional Literature: Nelson (1909), Hall (1951b), Honacki et al. (1982), Best and Hill (1994), Melo-Ferreira et al. (2014), Krebs and Murray (2018).

Subgenus ***Proeulagus*** Lyon, 1903

Medium sized; skull length about 75.0 mm; **postorbital processes rather slender and free; nasals and rostrum narrow;** spines of the anterior dorsal vertebrae relatively shorter and lumbar vertebrae low and triangular.

Lepus altamirae Nelson, 1904

Tamaulipas Jackrabbit, liebre de Tamaulipas

Monotypic

Conservation: *Lepus altamirae* is listed within *L. californicus* as Least Concern by the IUCN (Brown et al. 2019). The species needs an evaluation, which will probably categorize it as Threatened.

Characteristics: Tamaulipas jackrabbit is medium sized; **total length 465.0–630.0 mm, and basilar length 67.3–76.9 mm.** Dorsal pelage speckled brownish to grizzle brown, **with back blackish and flanks very pale, contrasting with the back**; fur of the underparts white or pale orange; chest brownish orange; **ears externally black at the tip; tail plumbeous black dorsally and white ventrally; nape with two dark lines instead of a central spot**; supraorbital process of the frontal bones slightly arched; nasals and premaxillary bones and molars appearing not massive; well developed groove in the anterior region of the frontal process; braincase long and narrow.

Comments: *Lepus altamirae* was recognized as a distinct species (Vargas et al. 2019), previously regarded as a subspecies of *L. californicus* (Hall 1981) and *L. alleni* (Hall 1951). The percentage of genetic differences from other species is very small, so it is probably not a valid species. *L. altamirae* is restricted to southeastern Tamaulipas and southwestern San Luis Potosí (Silva-Caballero and Rosas-Rosas 2022; Map 5). It can be found in sympatry, or nearly so, with *L. californicus*, from which it can be distinguished by having black lines on the nape, a black spot at the ear tip very small, and flanks of paler coloration.

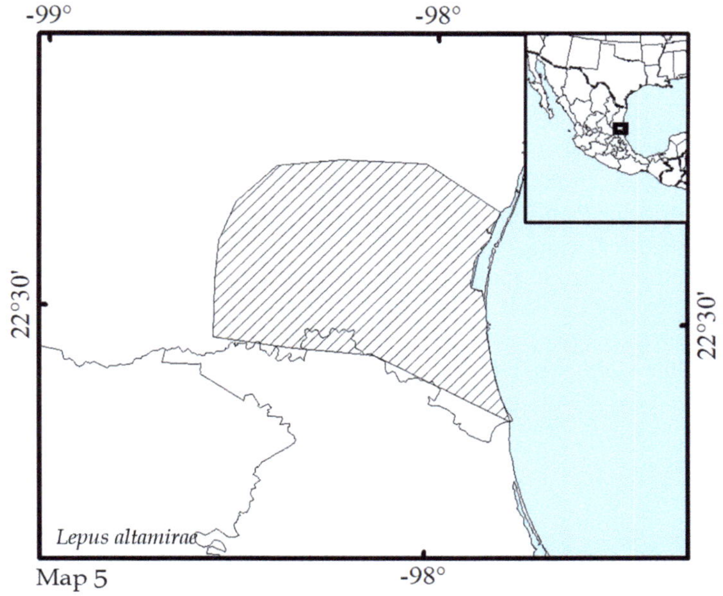

Map 5 Distribution map of *Lepus altamirae*

Additional Literature: Nelson (1904, 1909), Goldman (1951).

Lepus californicus Gray, 1837
Black-Tailed Jackrabbit, liebre de cola negra

1. *L. c. asellus* Miller, 1899. From southern Nuevo León to Zacatecas south through Querétaro.
2. *L. c. bennettii,* Gray, 1843. Costal California, from Los Angeles Valley south through northwestern Baja California.
3. *L. c. californicus,* Gray, 1837. Costal California, from Los Angeles Valley north through Oregon.
4. *L. c. curti* Hall, 1951. Island off Matamoros, Tamaulipas.
5. *L. c. deserticola* Mearns, 1896. From Idaho and southwestern Montana south through northwestern Sonora and northeastern Baja California.
6. *L. c. eremicus* Allen, 1894. Sonora and Southern Arizona.
7. *L. c. festinus* Nelson, 1904. Hidalgo.
8. *L. c. insularis* Bryant, 1891. Espíritu Santo Island, Baja California Sur.
9. *L. c. magdalenae* Nelson, 1907. Magdalena and Margarita islands and western Baja California Sur.
10. *L. c. martirensis* Stowell, 1895. Central Baja California Peninsula.
11. *L. c. melanotis* Mearns, 1890. From Texas and northwestern New Mexico north through South Dakota and east through Arkansas and Missouri.
12. *L. c. merriami* Mearns, 1896. Texas, western Coahuila, Nuevo León, and Tamaulipas.
13. *L. c. richardsonii* Bachman, 1839. San Joaquin Valley, California.
14. *L. c. sheldoni* Burt, 1933. Carmen Island, Baja California Sur.
15. *L. c. texianus* Waterhouse, 1848. From Zacatecas north through Idaho and Colorado.
16. *L. c. wallawalla* Merriam, 1904. Northeastern California, northwestern Nevada, Oregon, and southern Washington.
17. *L. c. xanti* Thomas, 1898. Southern and western Baja California Peninsula.

Conservation: *Lepus californicus insularis, L. c. magdalenae,* and *L. c. sheldoni* are listed as Endangered in México (DOF 2019), and *L. c. insularis* is listed as Vulnerable (D1: very small population size, with less than 1000 mature individuals) by the IUCN (Brown et al. 2019).

Characteristics: Black-tailed jackrabbit is medium sized; total length 465.0–630.0 mm, basilar length 67.3–76.9 mm. Dorsal pelage speckled brownish to grizzle brown, **back blackish and flanks paler**; fur of the underparts white or pale orange; chest brownish orange; **ears externally black at the tip; tail plumbeous black dorsally and white ventrally; nape with a central dark spot**; color and size varying in different geographic areas; supraorbital process of the frontal bones slightly arched; nasals, premaxillary bones, and molars not massive; anterior region of the frontal process with a well developed groove; braincase long and narrow.

Comments: *Lepus insularis* is a junior synonym of *L. californicus* (Álvarez-Castañeda and Lorenzo 2017), previously considered a different species (Dixon et al. 1983). *L. californicus* is the most frequent hare in the North American desert and is easily observable in these areas, from Washington east through South Dakota and south through Guanajuato and Queretaro, including the Baja California Peninsula (Map 6). It can be found in sympatry, or nearly so, with *L. alleni, L. americanus, L. callotis,* and *L. townsendii. L. californicus* can be distinguished from *L. alleni* by its smaller size and ears and with the flanks with a similar color pattern to the upper parts. From *L. americanus,* by having larger ears, more than the head length. From *L. callotis,* by having the flanks with the same pattern as the upper parts; ears black at the tip and with a more general brown coloration. From *L. townsendii,* by having a black dorsal stripe on the tail, from the latter, by a terminal black patch outside the ears and the white area reduced in the flanks. No skull characteristics are diagnostic to differentiate these species.

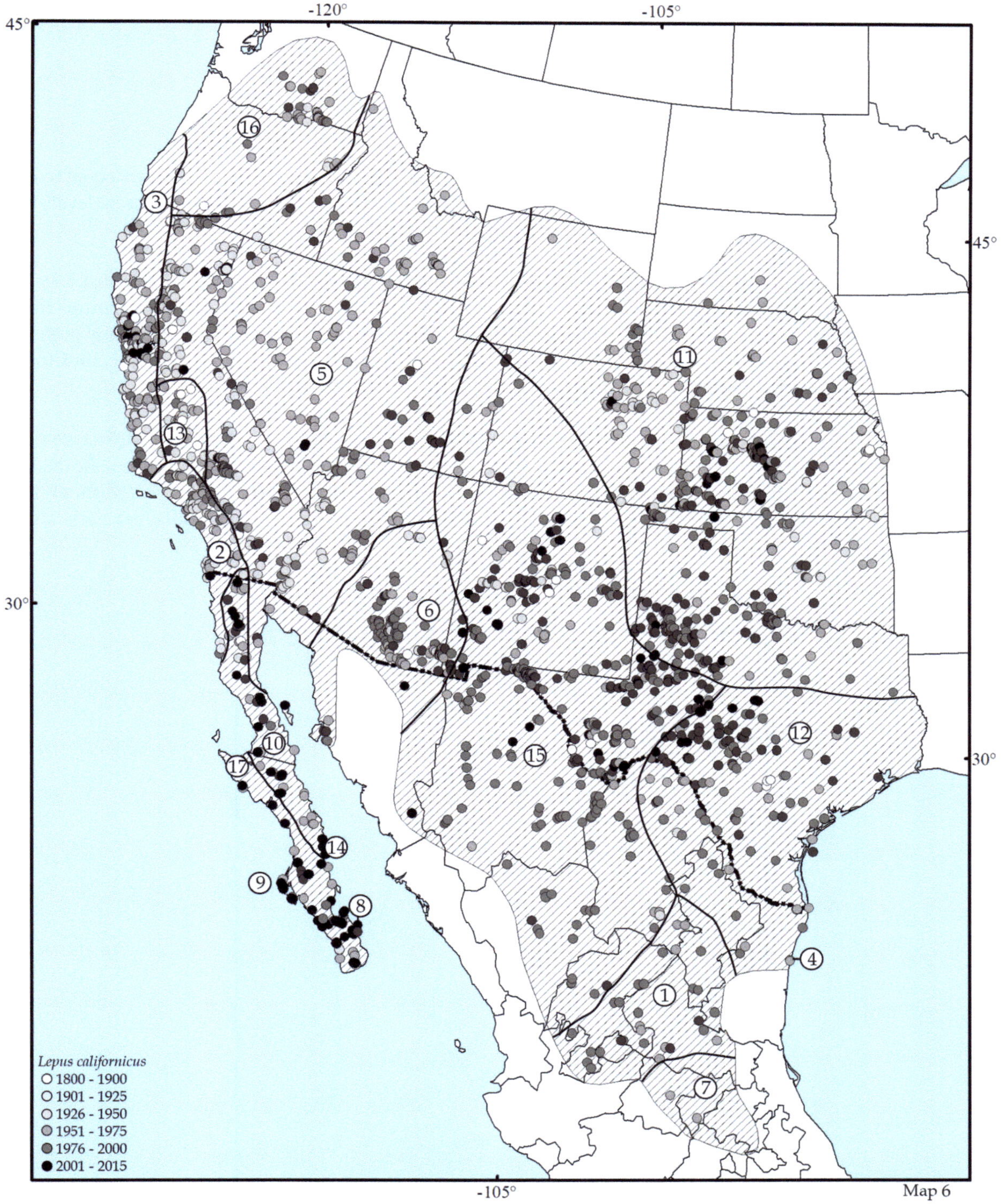

Map 6 Distribution map of *Lepus californicus*

Additional Literature: Nelson (1909), Howell (1936), Hall (1951b), Thomas and Best (1994), Best (1996), Cervantes et al. (1996), Beever et al. (2018a), Lorenzo et al. (2018b).

Lepus callotis Wagler, 1830

White-Sided Jackrabbit, liebre de costados blancos

1. *L. c. callotis* Wagler, 1830. From central Durango south through Oaxaca.
2. *L. c. gaillardi* Mearns, 1896. Central Durango and Chihuahua.

Conservation: *Lepus callotis* is listed as Vulnerable (A2bcde: population reduction in the last 10 years equal to or greater than 30% which may not be reversible based on an index of abundance, decline in area of occupancy, actual levels of exploitation, and effects of introduced taxa) by the IUCN (Brown and Smith 2019).

Characteristics: White-sided jackrabbit is medium sized; total length 432.0–598.0 mm, basilar length 64.3–73.8 mm. Dorsal pelage speckled with yellowish to ochraceous cinnamon and blackish shades; **ventral sides, rump, limbs, and thighs white or light gray, contrasting with the dorsal color**; shoulders and neck yellowish; ears creamy at the tip, may have blackish shades throughout except at the tip; tail plumbeous black dorsally and pale gray ventrally; **auditory bullae large,** its length less than that the molariform toothrow.

Comments: *Lepus callotis* ranges in the central highlands and throughout central México, from New Mexico south through northern Oaxaca (Map 7). It can be found in sympatry, or nearly so, with *L. alleni* and *L. californicus*. It can be distinguished from *L. alleni* by its smaller size and ears and flanks with a similar pattern to the upper parts. From *L. californicus*, by its pale gray flanks, ears are white at the tip and with a more general yellowish coloration. There are no skull characteristics allowing their separation.

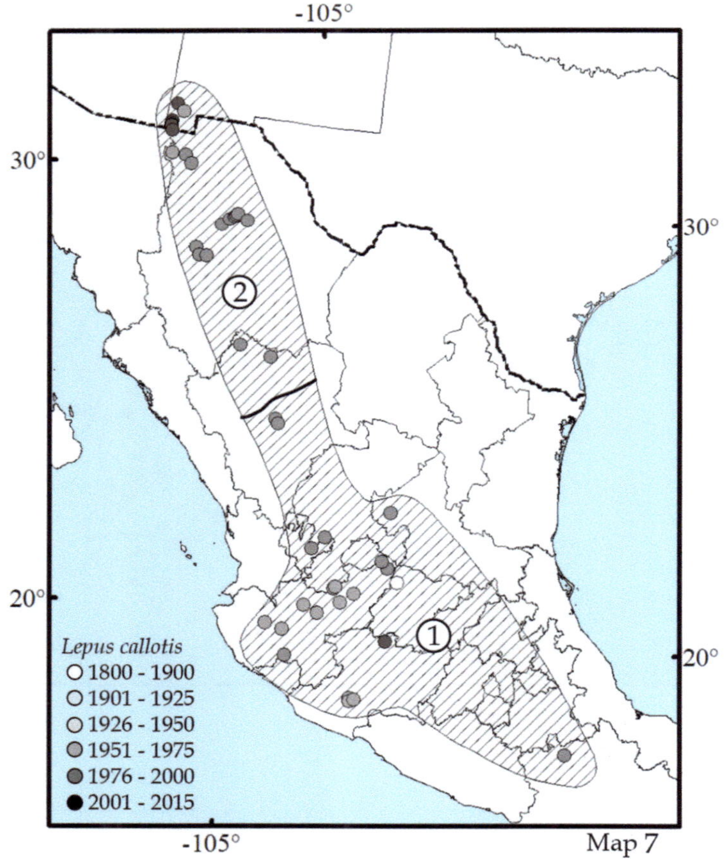

Map 7 Distribution map of *Lepus callotis*

Additional Literature: Nelson (1909), Howell (1936), Anderson and Gaunt (1962), Hoffmann and Smith (2005), and Brown et al. (2018c, d).

Lepus flavigularis Wagner, 1844

Tehuantepec Jackrabbit, liebre de Tehuantepec

Monotypic

Conservation: *Lepus flavigularis* is listed as Endangered in México (DOF 2019) and as Endangered (A2b + 3c; B1ab (i, ii, iii, v) + 2ab (i, ii, iii, v); C1; D: population reduction in the last 10 years equal to or greater than 50% and may not be reversible based on an index of abundance + population reduction equal to or greater than 50% projected within the next 10 years based on the decline in area of occupancy, extent of occurrence and quality of habitat + extent of occurrence less than 5000 km^2 severely fragmented and under continuing decline in extent of occurrence, area of occupancy, area, extent, and quality of habitat, and number of subpopulations + area of occupancy less than 500 km^2 and an observed, estimated, or projected continuing decline of at least 20% in 5 years + number of mature individuals less than 250) by the IUCN (Cervantes et al. 2008).

Characteristics: Tehuantepec jackrabbit is medium sized; total length 565.0–610.0 mm, basilar length 70.0–77.8 mm. Dorsal pelage speckled in bright ochraceous buff with black shades; underparts and flanks with white hairs; hips and limbs in grayish shades; **chest and neck yellowish**; ears pale yellow at the tip; tail black dorsally and white ventrally; **auditory bullae proportionally very small**; supraorbital process notch in the anterior region absent.

Comments: No congeneric species of *Lepus* are sympatric with *L. flavigularis*. It is found only in the eastern grasslands of Oaxaca and the extreme west of Chiapas (Map 8). It is now considered extirpated from Chiapas, leaving its main colonies around Laguna Inferior and Superior, Oaxaca.

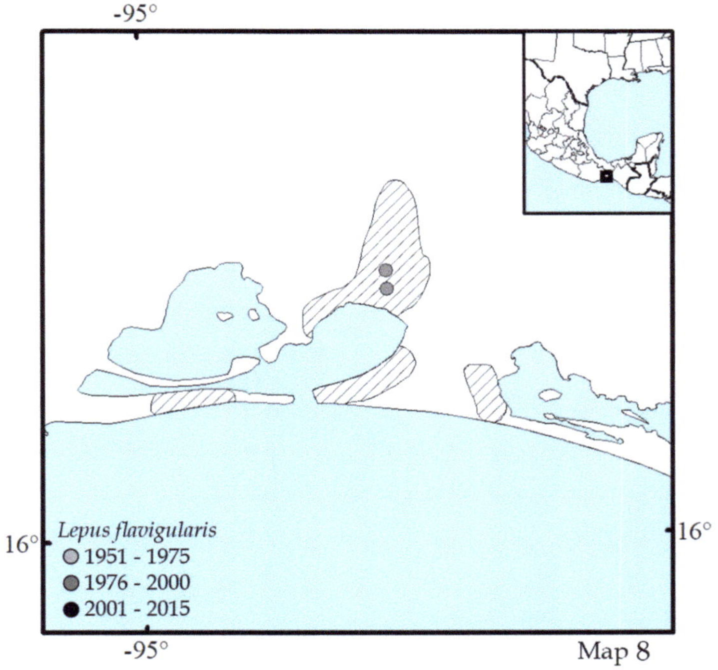

Map 8 Distribution map of *Lepus flavigularis*

Additional Literature: Nelson (1909), Howell (1936), Cervantes (1993), and Lorenzo et al. (2008, 2011, 2014, 2015, 2018f).

Lepus townsendii Bachman, 1839

White-Tailed Jackrabbit, liebre de cola blanca

1. *L. t. campanius* Hollister, 1915. From Alberta, Montana, Wyoming, central Colorado, and northern New Mexico east through Ontario, Minnesota, Illinois, and Arkansas.
2. *L. t. townsendii* Bachman, 1839. From Idaho, western Wyoming, and central Colorado west through Washington, Oregon, and California.

Conservation: *Lepus townsendii* is listed as Least Concern by the IUCN (Brown and Smith 2019).

Characteristics: White-tailed jackrabbit is medium sized; total length 565.0–618.0 mm, basilar length 67.6–74.0 mm. Dorsal pelage in summer yellowish to grayish brown; fur of the underparts white to pale gray; neck darker; in winter, white with the possibility of a buffy tinge in the ears, face, back, and feet; populations with a southern distribution not changing to a winter coloration; **ears always black at the tips, and a large part of the ears pale gray; tail always completely white**; incisors procumbent.

Comments: *Lepus townsendii* ranges in the Great Plains of the United States and Canada, and east of the Cascade Mountains, from British Columbia east through Ontario, and from the western Appalachian Mountains west through California, Nevada, New Mexico, Utah, Colorado, Missouri, and Illinois (Map 9). It can be found in sympatry, or nearly so, with *L. americanus* and *L. californicus. L. townsendii* can be distinguished from *L. americanus* by having larger ears, more than the head length. From *L. californicus,* by having the tail color white dorsally; skull relatively short, broad, and arched, with larger molariform teeth.

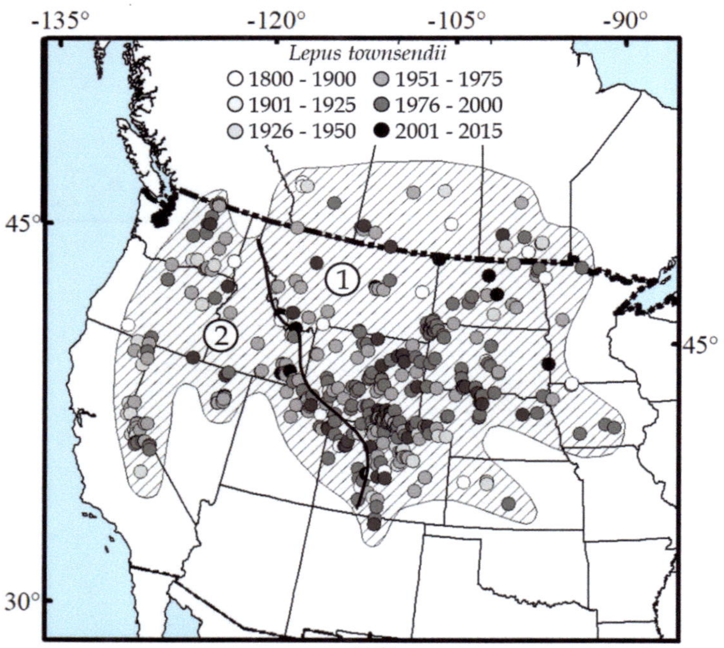

Map 9 Distribution map of *Lepus townsendi*

Additional Literature: Nelson (1909), Howell (1936), Lim (1987), Beever et al. (2018b).

Romerolagus Merriam, 1896

Dorsal pelage grizzled buffy brown or dull cinnamon brown; **ears rounded with a length less than 40.0 mm**; underparts dingy gray. **Anterior projection of supraorbital process absent or present only as a little tip**; jugal projecting posteriorly past squamosal root of zygomatic arch more than halfway to external auditory meatus; **auditory bullae larger than the foramen magnum**.

Romerolagus diazi (Ferrari-Pérez, 1893)

Teporingo, zacatuche

Monotypic

Conservation: *Romerolagus diazi* is listed as Endangered in México (DOF 2019) and as Endangered (B1ab (i, ii, iii, iv, v) + 2ab (i, ii, iii, iv, v): extent of occurrence less than 5000 km^2 in five or less populations and under continuing decline in extent of occurrence, area of occupancy, area, extent and quality of habitat, number of subpopulations, and number of mature individuals + area of occupancy less than 500 km^2 in five or less populations and under continuing decline in extent of occurrence, area of occupancy, area, extent, and quality of habitat, number of subpopulations, and number of mature individuals) by the IUCN (Velázquez and Guerrero 2019).

Characteristics: Teporingo is small sized; total length 234.0–311.0 mm, basilar length 47.1–47.2 mm. **Dorsal and lateral pelage antimony yellow interspersed with black, resulting in a grizzled appearance**; fur of the underparts pale gray; **hair black at the base and tip, yellow in the middle;** eye-socket region and cheeks pale yellowish; **ears very small relative to the rest of the rabbits, practically round**; forelegs scarcely shorter than the hind legs; hindfeet less than 55.0 mm; **tail vestigial**; interparietal bone present; **supraorbital anterior process absent or very small; auditory bullae length greater than the foramen magnum**.

Comments: *Romerolagus diazi* is associated with the alpine scrublands from the Popocatépetl, Iztaccíhuatl, and Chichinautzin volcanoes west through the Nevado de Toluca Volcano (Xinantécatl; Map 10). It can be found in sympatry, or nearly so, with *Sylvilagus cunicularius* and *S. floridanus*. It can be distinguished from these species by its small and rounded ears, less than 40.0 mm in length; hindfoot length less than 65.0 mm; hair grayish with yellowish brown to dark cinnamon shades, and auditory bullae larger than the foramen magnum.

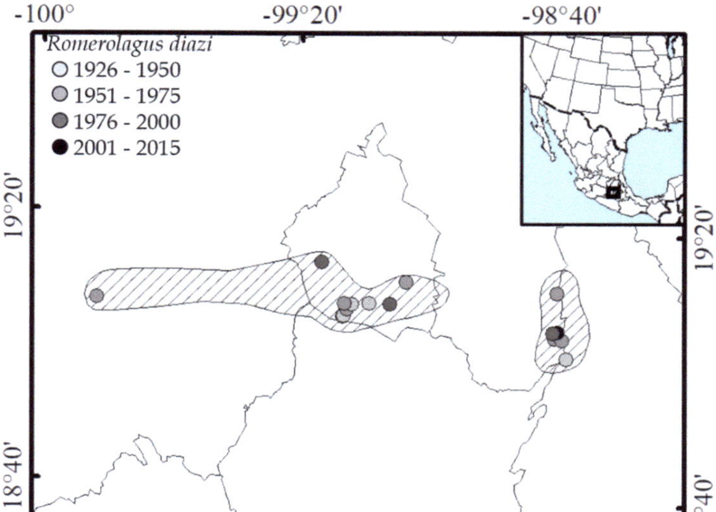

Map 10 Distribution map of *Romerolagus diazi*

Additional Literature: Merriam (1896), Nelson (1909), Howell (1936), Cervantes et al. (1990), Velázquez et al. (1993), Velazquez (1994), Rizo-Aguilar et al. (2016), Vázquez and Gopar-Merino (2018).

Genus *Sylvilagus* Gray, 1867

Dorsal pelage tending to be grizzled in gray or brown, no white specimens (except albino); rostrum rounded; **ears shorter than the head; forelegs and hind legs large but less than 105.0 mm;** tail small, white or very pale ventrally, cotton appearance; **with the interparietal bone**; posterior supraorbital process large, anterior supraorbital process varying in size or absent; auditory bullae varying in size according to the species but generally relatively smaller relative to *Lepus*. The keys were elaborated based on the review of specimens and the following sources: Nelson (1909), Howell (1924), Hall (1981), Álvarez-Castañeda et al. (2017a), and Cano-Sánchez et al. (2022).

1. Total length less than 310.0 mm; anterior extension of the supraorbital process more than half the length of the posterior extension (Fig. 13); first upper molariforms with one reentrant angle on the anterior face; reentrant angle of the second upper molariforms not crenated (Fig. 14; from Great Basin in Montana, Wyoming, Utah, Washington, Nevada, northeastern California, and eastern Oregon, United States) .... *Sylvilagus idahoensis* (p. 547)

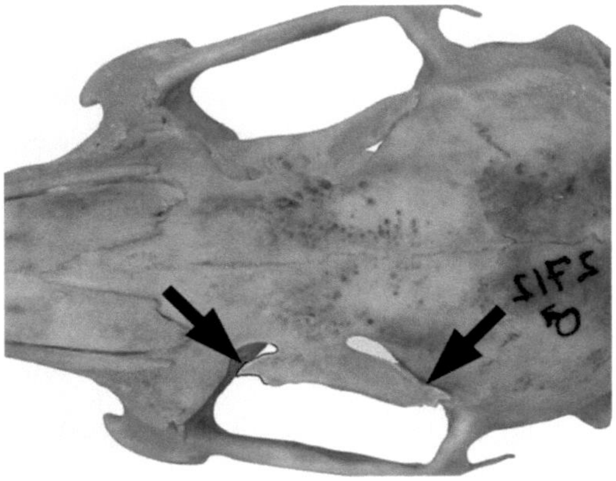

Fig. 13 The anterior extension of the supraorbital process more than half the length of the posterior extension like in *Sylvilagus idahoensis*

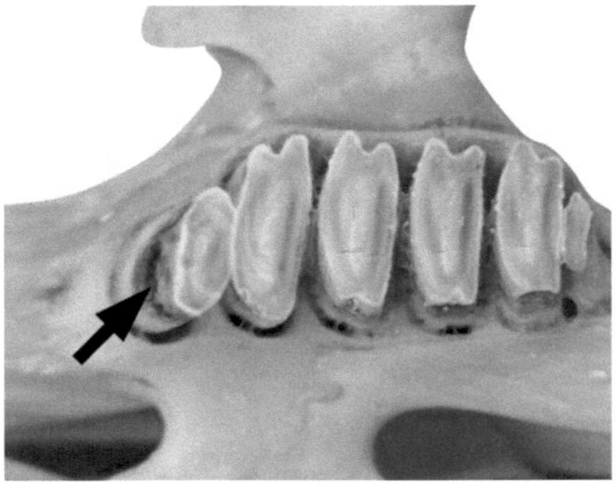

Fig. 14 Reentrant angle of the second upper molariforms not crenated present in *Sylvilagus idahoensis*

1a. Total length greater than 310.0 mm; anterior extension of the supraorbital process less than half the length of the posterior extension or entirely absent (Fig. 15); first upper molariforms with more than one (usually three) reentrance angle on the anterior face; reentrance angle of the second upper molariforms crenated (Fig. 16) 2

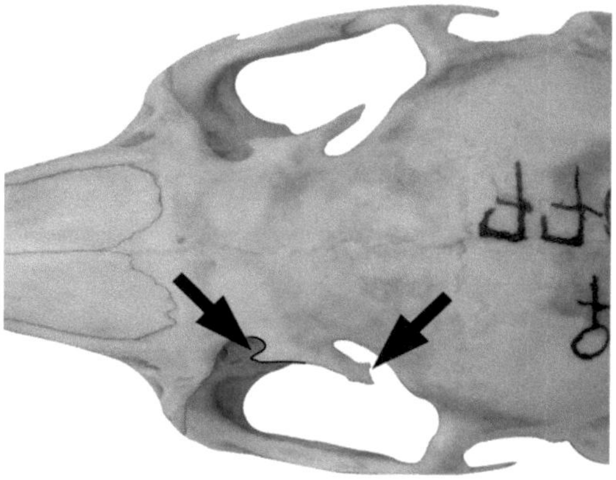

Fig. 15 The anterior extension of the supraorbital process is less than half the length of the posterior extension or entirely absent

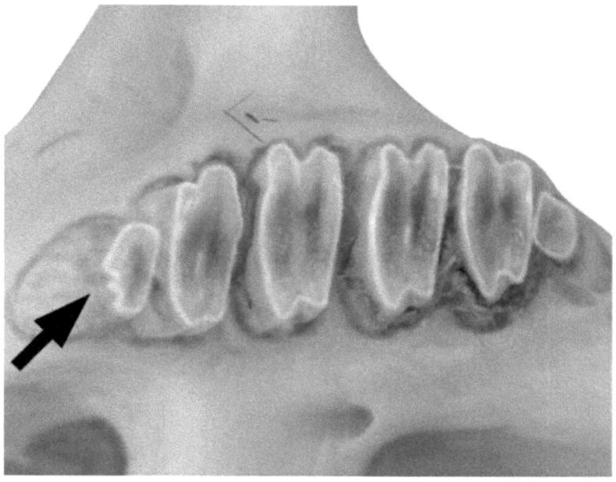

Fig. 16 First upper molariforms with more than one (usually three) reentrance angle on the anterior face; reentrance angle of the second upper molariforms crenated

2. Anterorbital extension of the supraorbital process absent, or a great part fused to the braincase (Fig. 7) 3

2a. Anterorbital extension of the supraorbital process present and posterior extension free, with a space with the braincase (Fig. 8) 6

3. A black spot between the ears; ear length greater than 13% of the total length 4

3a. Without a black spot between the ears; ear length less than 13% of the total length 5

4. Restricted to the region of New England in the United States (from Maine, Vermont, New Hampshire, Connecticut, eastern New York, Massachusetts, and Rhode Island) .... *Sylvilagus transitionalis* (p. 553)

4a. Restricted to the highlands of the Appalachian Mountains in the United States (from Pennsylvania, Maryland, Virginia, West Virginia, Tennessee, Kentucky, North Carolina, South Carolina, Georgia, and Alabama) .... *Sylvilagus obscurus* (p. 550)

4. Total length greater than 450.0 mm; ear length greater than 61.0 mm; tail white; basilar length greater than 63.0 mm (from Oklahoma and Texas east through North Carolina and Georgia, and from the Gulf of México coast north through Arkansas, Missouri, and Kentucky, United States) .... *Sylvilagus aquaticus* (p. 537)

4a. Total length less than 450.0 mm; ear length less than 61.0 mm; tail very small and gray; basilar length less than 63.0 mm (from Florida and Georgia along the coast north through Virginia, United States) .... *Sylvilagus palustris* (p. 551)

6. Ventral pelage white to white with yellowish shades; auditory bullae large; anteroposterior bullae length greater than 16% of the skull length (from Montana and North Dakota south through Puebla and Tlaxcala, and from California east through South Dakota, Nebraska, Oklahoma, and Texas, including the Baja California Peninsula, México) .... *Sylvilagus audubonii* (p. 538)

6a. Ventral pelage not white, if white, with gray hair interspersed and an orange spot on the neck; auditory bullae small, less than 16% of the skull length 7

7. Anterior supraorbital process joined to the skull (Fig. 7) 8

7a. Anterior supraorbital process free (Fig. 17) 9

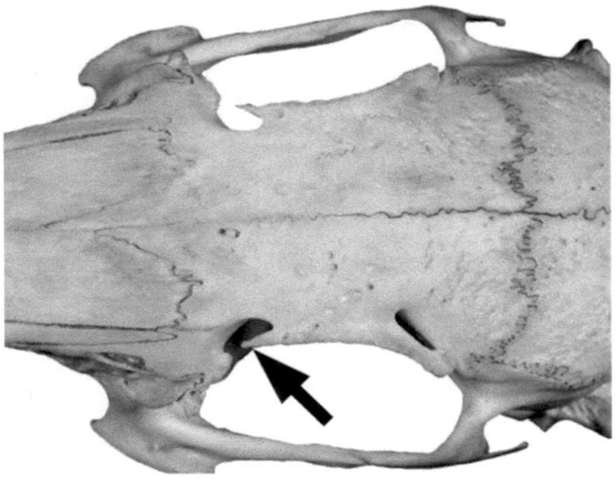

Fig. 17 The anterior supraorbital process is free

8. Tail length less than 30.0 mm; ear length (when dry) from the low neckline less than 50.0 mm; dorsal pelage dark yellowish brown; tail dark yellowish dorsally, not white (from Tamaulipas south through Chiapas, México) .... *Sylvilagus gabbi* (p. 545)

8a. Tail length greater than 30.0 mm; ear length (in dry) from the low neckline greater than 50.0 mm; dorsal pelage pale; tail white dorsally (endemic to the Tres Marías Islands, Nayarit, México) .... *Sylvilagus graysoni* (p. 546)

9. Dorsal pelage in red shades; tail yellowish ventrally, not white (endemic of Sierra Madre of Guerrero, México) .... *Sylvilagus insonus* (p. 548)

9a. Dorsal pelage in brown shades; tail white ventrally 10

10. Hindfoot length less than 81.0 mm; ventral hair grayish at the base (from the Columbia River south through Baja California Peninsula, México) .... *Sylvilagus bachmani* (p. 539)

10a. Hindfoot length greater than 81.0 mm; ventral hair white or yellowish at the base 1

11. Total length greater than 476.0 mm; interorbital breadth usually greater than 19.3 mm (from Sinaloa south through Oaxaca, México) .... *Sylvilagus cunicularius* (p. 543)

11a. Total length less than 476.0 mm; interorbital breadth less than 19.3 mm 12

12. Restricted to the Great Basin and Rocky Mountains and plateaus; diameter of the external auditory meatus greater than the crown length of the last three molariforms (from British Columbia, Alberta, and Saskatchewan south through California, Arizona, and New Mexico, and from California east through North Dakota, South Dakota, and Nebraska, United States) .... *Sylvilagus nuttallii* (p. 549)

12a. Rest of the United States and México; diameter of the external auditory meatus less than the crown length of the last three molariforms 13

13. Two basisphenoid foramina; no ear process (restricted to the Guadalupe Mountains in Texas and New Mexico, Chicos, and Davis Mountains in Texas, United States, and to Sierra La Madera in Coahuila, México) .... *Sylvilagus robustus* (p. 552)

13a. One basisphenoid foramen; ear process starts on the external base of the auditory meatus 14

14. Restricted to the highlands of the Manzano Mountains of New Mexico, United States .... *Sylvilagus cognatus* (p. 541)

14a. Occurs throughout México, except for the Baja California Peninsula, western-central Pacific Coast, and Sonora Desert where *Sylvilagus robustus* has been recorded from British Columbia, Saskatchewan, Ontario, and Québec, Canada, south through South America. In the western United States, in Washington and Oregon and from Montana south through Texas east through the Atlantic coast, United States .... *Sylvilagus floridanus* (*see addenda, p. 544)

Sylvilagus aquaticus (Bachman, 1837)

Swamp Rabbit, conejo de los pantanos

1. *S. a. aquaticus* (Bachman, 1837). From Texas, Oklahoma, and southwestern Kansas east through Georgia, including western Tennessee, Kentucky, and southwestern Indiana, Illinois, and Missouri.
2. *S. a. littoralis* Nelson, 1909. Coast of Texas, Louisiana, and Mississippi.

Conservation: *Sylvilagus aquaticus* is listed as Least Concern by the IUCN (Lanier and Nielsen 2019).

Characteristics: Swamp rabbit is **the largest member of the genus; total length 452.0–552.0 mm, basilar length 60.9–72.3 mm**. Pelage of the dorsum, rump, tail dorsally, and hind legs are chestnut brown to dark rusty red; fur of the underparts in the central abdomen white with the rest of the belly buff to pale brown; eye ring prominent cinnamon-colored; ears medium sized; supraorbital process joined to the skull along most of its extent, some specimens with a foramen between the skull and the supraorbital process; rostrum width at the level of the maxillary molariforms greater than the interorbital width; **auditory bullae smaller than the foramen magnum**.

Comments: *Sylvilagus aquaticus* ranges from Oklahoma and Texas east through North Carolina and Georgia, and from the Gulf of México coast north through Arkansas, Missouri, and Kentucky (Map 11). It can be found in sympatry, or nearly so, with *S. floridanus* and *S. palustris*. *S. aquaticus* can be distinguished from all other species of *Sylvilagus* in North America by being the largest sized (greater than 450.0 mm). From *S. palustris*, having the auditory bullae smaller than the foramen magnum, and from *S. floridanus,* by being larger.

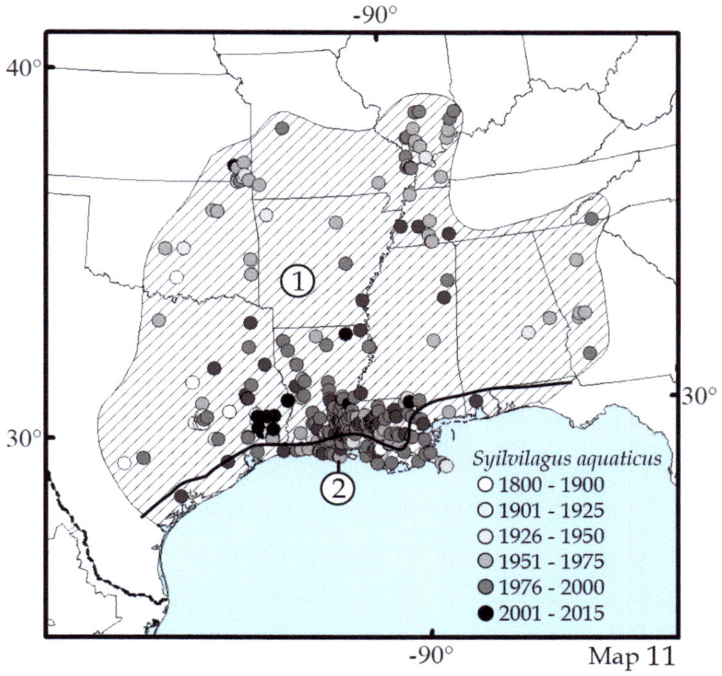

Map 11 Distribution map of *Sylvilagus aquaticus*

Additional Literature: Nelson (1909), Hall and Kelson (1959), Golley (1962), Lowery (1974), Chapman and Feldhamer (1981), Robinson et al. (1983, 1984), Chapman and Ceballos (1990), Nielsen and Berkman (2018a).

Sylvilagus audubonii (Baird, 1858)

Desert Cottontail, conejo del desierto

1. *S. a. arizonae* (Allen, 1877). From western Utah, central Nevada, and California south through southern Sonora.
2. *S. a. audubonii* (Baird, 1858). Central California.
3. *S. a. baileyi* (Merriam, 1897). From northern Utah, Colorado, and Kansas north through Montana and North Dakota.
4. *S. a. cedrophilus* Nelson, 1907. From central Utah and Colorado south through central Arizona and New Mexico.
5. *S. a. confinis* (Allen, 1898). Mainly all the Baja California Peninsula.
6. *S. a. goldmani* (Nelson, 1904). Southern Sonora and northern Sinaloa.
7. *S. a. minor* (Mearns, 1896). From southwestern Arizona and New Mexico south through Coahuila and central Durango.
8. *S. a. neomexicanus* Nelson, 1907. From southwestern Kansas and western New Mexico east through southern Texas.
9. *S. a. parvulus* (Allen, 1904). From Central Texas, Coahuila, Zacatecas, and Durango south through Puebla and Tlaxcala.
10. *S. a. sanctidiegi* (Miller, 1899). From Los Angeles Valley in California south through northwestern Baja California.
11. *S. a. vallicola* Nelson, 1907. Central-western California.

Conservation: *Sylvilagus audubonii* is listed as Least Concern by the IUCN (Smith and Brown 2019a).

Characteristics: Desert cottontail is medium sized; total length 350.0–420.0 mm, basilar length 51.9–54.5 mm. Dorsal pelage speckled yellowish gray with black; fur of the underparts white to white with yellowish shades; nape pale orange, cream spot around the eyes; **ears with sparse hairs on the inner surface, black at the tip, long pointed and larger than the hind foot length**; hind and forelimbs pale orange and hind limb upperparts white; supraorbital anterior process present; **auditory bullae length equal to or greater than the toothrow length**.

Comments: *Sylvilagus audubonii* ranges from Montana and North Dakota south through Puebla and Tlaxcala, and from California east through South Dakota, Nebraska, Oklahoma, and Texas, including the Baja California Peninsula. *S. audubonii* (Map 12) can be found in sympatry, or nearly so, with *S. bachmani, S. cognatus, S. cunicularious, S. idahoensis, S. floridanus,* and *S. nuttallii. S. audubonii* can be distinguished from these species by having the ventral region mainly completely white to white with yellowish shades and large auditory bullae, greater than 16% of the skull length. From *S. bachmani,* by having ears with sparse hairs on the inner surface, black at the tip, long, pointed, and larger than the hind foot length. From *S. cognatus* and *S. floridanus,* by not having a small white spot between the ears; underparts white and ears long and pointed. From *S. cunicularius,* by its smaller size, less than 450.0 mm in total length. From *S. idahoensis,* by its larger size, greater than 300.0 mm in total length. From *S. nuttallii,* by having ears with sparse hairs on the inner surface, black at the tip, long, pointed, and larger than the hind foot length; auditory bullae length equal to or greater than the toothrow length.

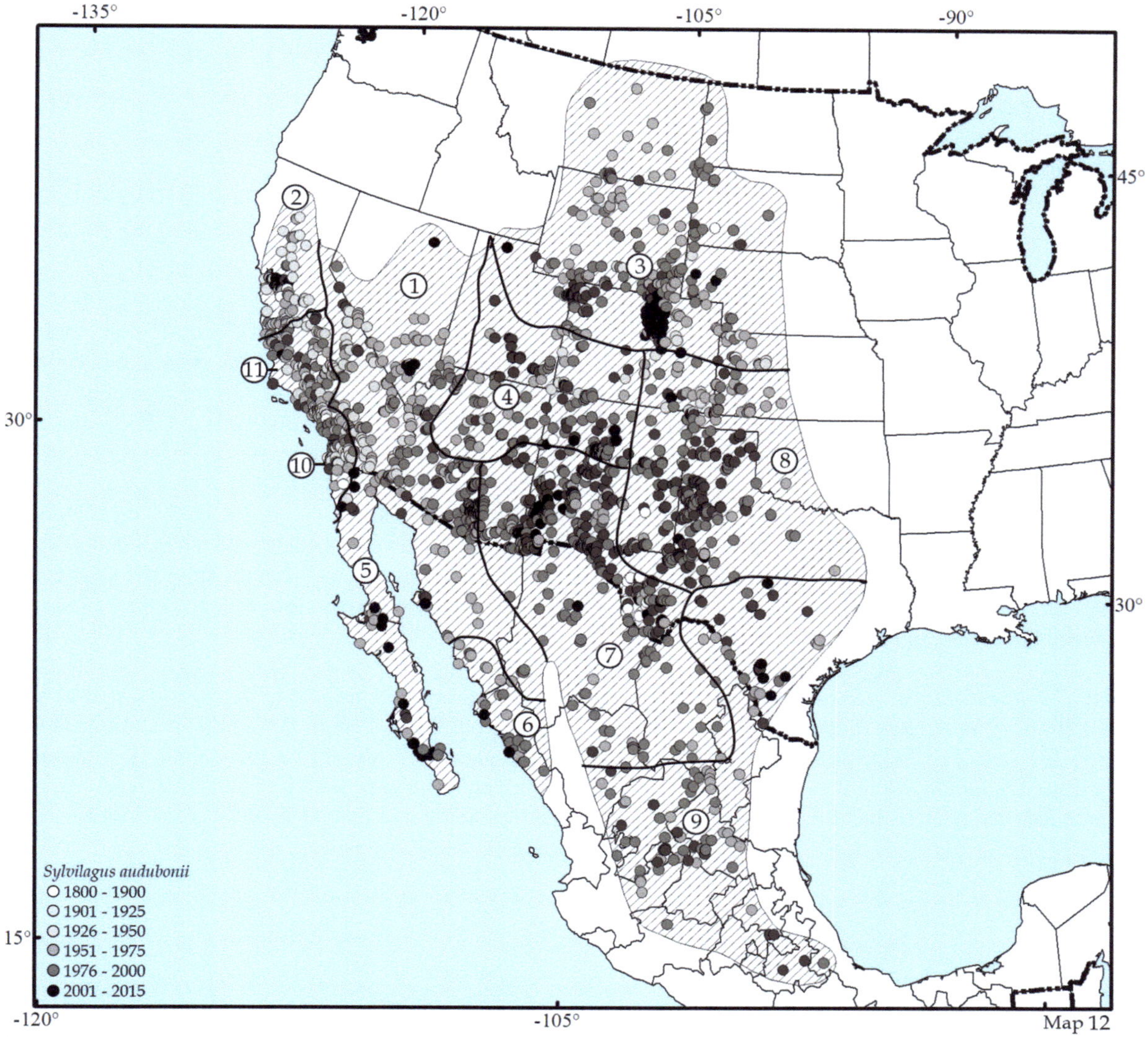

Map 12 Distribution map of *Sylvilagus audubonii*

Additional Literature: Nelson (1909), Orr (1940), Hall (1951b), Chapman and Willner (1978), Chapman and Ceballos (1990), Lorenzo et al. (2013), Laundré (2018).

Sylvilagus bachmani (Waterhouse, 1839)

Brush Rabbit, conejo de los arbustos

1. *S. b. bachmani* (Waterhouse, 1839). Coast of California, from San Luis Obispo to San Francisco Bay.
2. *S. b. cerrosensis* (Allen, 1898). Cedros Island, Baja California.
3. S. *b. cinerascens* (Allen, 1890). Northwestern Baja California and Southern California.
4. *S. b. exiguus* Nelson, 1907. Central Baja California Peninsula.
5. *S. b. howelli* Huey, 1927. Central and northern Baja California.
6. *S. b. macrorhinus* Orr, 1935. From San Francisco Bay south through Monterey Bay, California.

7. *S. b. mansuetus* Nelson, 1907. San José Island, Baja California Sur.
8. *S. b. mariposae* Grinnell and Storer, 1916. Western slope of Sierra Nevada, California.
9. *S. b. peninsularis* (Allen, 1898). Southern Baja California Sur.
10. *S. b. riparius* Orr, 1935. Restricted to the Stanislaus County, California.
11. *S. b. rosaphagus* Huey, 1940. From Santo Tomas north through Ensenada, Baja California.
12. *S. b. tehamae* Orr, 1935. From southern Oregon south through central California.
13. *S. b. ubericolor* (Miller, 1899). From San Francisco Bay, California north through the Columbia River.
14. *S. b. virgulti* Dice, 1926. Restricted to the slopes southwestern of the Central Valley.

Conservation: *Sylvilagus bachmani cerrosensis* and *S. b. mansuetus* are listed as Endangered in México (DOF 2019), *S. b. riparius* by the United States Endangered Species Act (FWS 2022), and *S. b. mansuetus* as Critically Endangered (B1ab (ii, iii, v): extent of occurrence less than 100 km^2 in only one population and under continuing decline in area of occupancy, area, extent, and quality of habitat, and number of mature individuals) by the IUCN (Lorenzo and Álvarez-Castañeda 2011). *S. b. peninsularis* is possibly extinct from the Baja California Peninsula (Lorenzo et al. 2011, 2013), and *S. bachmani* is listed as Least Concen by the IUCN (Kelly et al. 2019).

Characteristics: Brush rabbit is small sized; total length 300.0–375.0 mm, basilar length 49.0–53.7 mm. Dorsal pelage dark brown or yellowish gray; fur of the underparts grayish white; sides grayish yellow; nape with a small pale orange brown spot; spot around the eyes cream; **ears naked inside, not black at the tip, rounded and shorter than the hind foot length**; fore and hind limbs pale orange with hind foot upperparts white; **tail small and brown above**; supraorbital anterior process absent; auditory bullae length less than the toothrow length.

Comments: *Sylvilagus mansuetus* was a full species and is now regarded as a subspecies of *S. bachmani* (Álvarez-Castañeda and Lorenzo 2016). *S. bachmani* ranges throughout the Pacific coast from Oregon south through Baja California Peninsula, México (Map 13). It can be found in sympatry, or nearly so, with *S. audubonii*, except for the islands. *S. bachmani* can be distinguished from *S. audubonii* in that the ventral region is not completely white or with few yellowish hairs; ears not black at the tip, rounded, and shorter than the hind foot length, and auditory bullae less than 16% of the skull length.

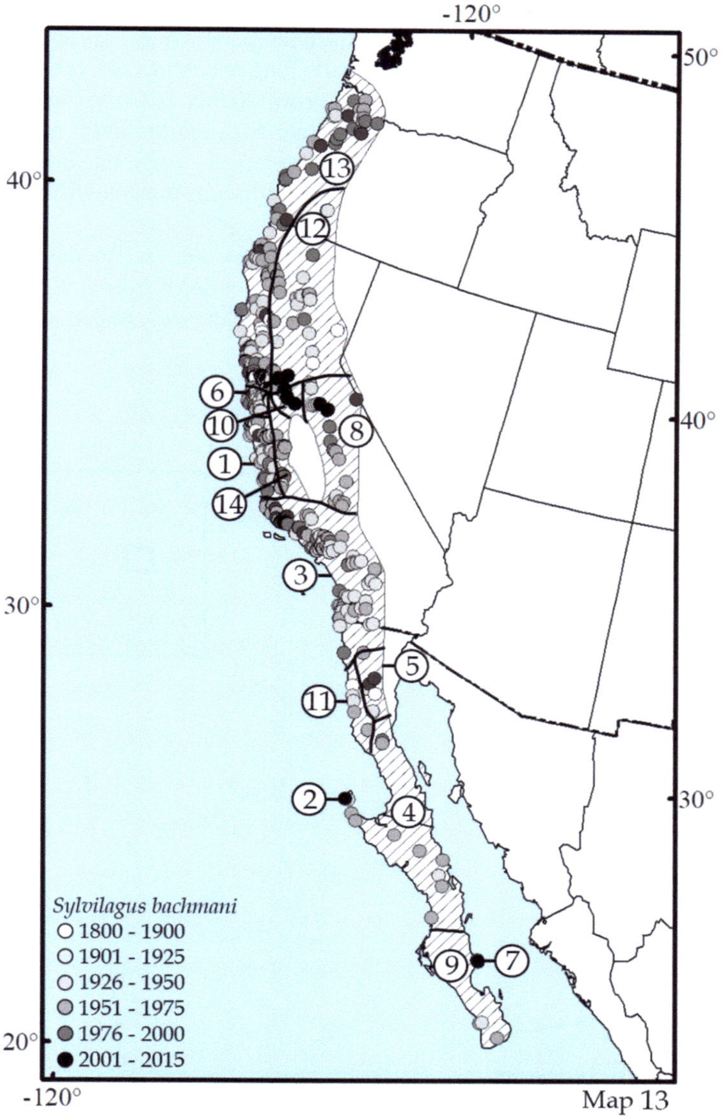

Map 13 Distribution map of *Sylvilagus bachmani*

Additional Literature: Nelson (1909), Chapman (1974), Cortés-Calva et al. (2013), Lorenzo et al. (2013, 2018a), Kelly (2018).

Sylvilagus cognatus Nelson, 1907

Manzano Mountain Cottontail, conejo de las montañas Manzano

Monotypic

Conservation: *Sylvilagus cognatus* is listed as Endangered (B1ab (iii): extent of occurrence less than 5,000 km^2 in five or less populations and under continuing decline in area, extent, and quality of habitat) by the IUCN (Smith and Ruedas 2016).

Characteristics: Manzano mountain cottontail is medium sized; total length 375.0–463.0 mm, basilar length is 52.0–59.7 mm. Dorsal pelage speckled in pale gray buff; fur of the underparts white; hips and limbs in grayish shades; flanks paler than the back; nape and limbs orange; hind feet proportionally long and white; tail brown dorsally and white ventrally; **small white spot between the ears**; ears slightly more buffy and brown than *S. robustus*; supraorbital anterior process present; postorbital process thin, tapering, and touching the skull; suture between the nasal and frontal bones with a smooth appearance; auditory bullae length less than the toothrow length but relatively large; one basisphenoid foramen; ear process starts on the auditory meatus external base. Note: *S. robustus* and *S. floridanus* must be differentiated genetically.

Comments: *Sylvilagus cognatus* was considered a subspecies of *S. floridanus* (Ruedas 1998) and is restricted to the Manzano Mountains of New Mexico (Map 14). This species can be differentiated from *S. floridanus* and *S. robustus* only by genetic analyses. From *S. audubonii,* by having a small white spot between the ears; underparts not snow white; ears not long and pointed.

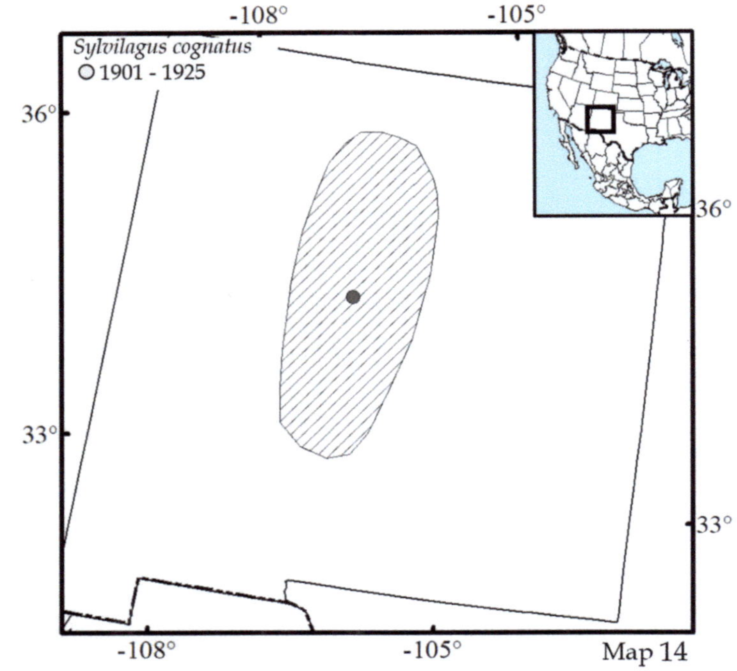

Map 14 Distribution map of *Sylvilagus cognatus*

Additional Literature: Orr (1940), Ruedas (2018a), Smith and Ruedas (2016).

Sylvilagus cunicularius (Waterhouse, 1848)

Mexican Cottontail, conejo mexicano

1. *S. c. cunicularius* (Waterhouse, 1848). From Central Michoacán south through Oaxaca and east through Tlaxcala and Veracruz.
2. *S. c. insolitus* (Allen, 1890). From Sinaloa south through central Michoacán.
3. *S. c. pacificus* (Nelson, 1904). Coastal Michoacán, Guerrero, and Oaxaca.

Conservation: *Sylvilagus cunicularius* is listed as Least Concern by the IUCN (Lorenzo and Lanier 2019a).

Characteristics: Mexican cottontail is large sized; total length 485.0–515.0 mm, basilar length 60.0–68.8 mm. Dorsal pelage pale yellowish brown and head more yellowish with ears long and blackish; fur of the underparts white, scarcely interspersed with blackish; limbs and nape in reddish rust brown shades; supraorbital anterior process present; **posterior projection of the supraorbital varying from those that have the tip free to those attached to the braincase**; auditory bullae process length less than the toothrow length.

Comments: *Sylvilagus cunicularius* ranges from Sinaloa south through Oaxaca along the coast and in central México (Map 15). It can be found in sympatry, or nearly so, with *S. audubonii, S. gabbi*, and *S. floridanus*. It can be distinguished by its large size. From *S. audubonii,* by its larger size; fur of the underparts yellowish, never white; small auditory bullae, less than 16% of the skull length. From *S. gabbi,* by having the supraorbital anterior part free.

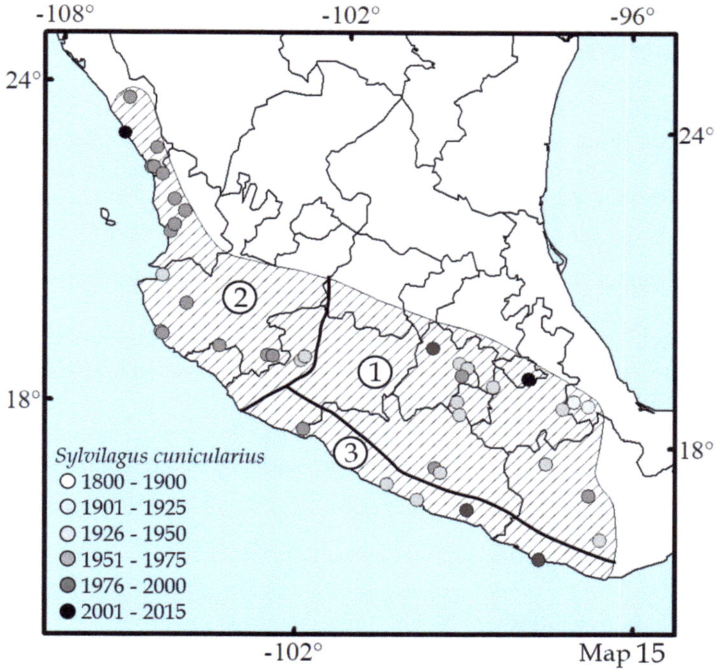

Map 15 Distribution map of *Sylvilagus cunicularius*

Additional Literature: Nelson (1909), Cervantes et al. (1992), Lorenzo et al. (2018c).

Sylvilagus floridanus (Allen, 1890)

Eastern Cottontail, conejo del este

(*see addenda, *Sylvilagus yucatanicus* full species)

1. *S. f. alacer* Bangs, 1896. From southern Kansas and Missouri south through Texas, Louisiana, and Mississippi.
2. *S. f. ammophilus* Howell, 1939. Central-eastern coast of the Florida Peninsula.
3. *S. f. aztecus* (Allen, 1890). Southern Oaxaca.
4. *S. f. chapmani* (Allen, 1899). From central Texas south through Tamaulipas, Coahuila, and Nuevo León.
5. *S. f. chiapensis* Nelson, 1904. Chiapas.
6. *S. f. cognathus* Nelson, 1907. Colorado and New Mexico.
7. *S. f. connectens* (Nelson, 1904). From southern Tamaulipas south through central Oaxaca, including western San Luis Potosí, Hidalgo, and Puebla.
8. *S. f. floridanus* (Allen, 1890). Florida Peninsula, except for the eastern coast.
9. *S. f. hesperius* Hoffmeister and Lee, 1963. Restricted to Arizona.
10. *S. f. hitchensi* Mearns, 1911. Restricted to Smith Island, Virginia.
11. *S. f. holzneri* (Mearns, 1896). From central Arizona and New Mexico south through Nayarit, Jalisco, and Zacatecas.
12. *S. f. llanensis* Blair, 1938. From southern Kansas and Colorado south through central Texas.
13. *S. f. macrocorpus* Diersing and Wilson, 1980. Nayarit.
14. *S. f. mallurus* (Thomas, 1898). From Massachusetts along the coast south through north of Florida and Alabama, including Tennessee.
15. *S. f. mearnsii* Allen, 1894. From central Kansas to Minnesota east through Pennsylvania to Vermont.
16. *S. f. orizabae* (Merriam, 1893). Altiplano Mexicano from Coahuila and Nuevo León south through southern Puebla.
17. *S. f. paulsoni* Schwartz, 1956. Most of southeastern Florida.
18. *S. f. restrictus* Nelson, 1907. Coastal area from Colima to Michoacán.
19. *S. f. russatus* (Allen, 1904). Southern Veracruz.
20. *S. f. similis* Nelson, 1907. From Kansas and Colorado north through southwestern Saskatchewan and Manitoba.
21. *S. f. subcinctus* (Miller, 1899). Northern Michoacán, Jalisco, and western Guanajuato.
22. *S. f. yucatanicus* (Miller, 1899). Northwestern of Yucatán and Campeche.

Conservation: *Sylvilagus floridanus* is listed as Least Concern by the IUCN (Nielsen and Lanier 2019).

Characteristics: Eastern cottontail is medium sized; total length 375.0–463.0 mm, basilar length 52.0 to 59.7 mm. Dorsal pelage speckled in yellowish shades with black; fur of the underparts white; hips and limbs in grayish shades; sides paler than the back; nape and limbs orange; hind feet proportionally long and white; tail brown dorsally and white ventrally; **a small white spot between the ears**; supraorbital anterior process present; postorbital process thin, tapering and touching the skull; suture between the nasal and frontal bones smooth in appearance; auditory bullae length less than the toothrow length but relatively large; one basisphenoid foramen; ear process starts on the auditory meatus external base. Note: *S. robustus* and *S. cognatus* must be differentiated genetically.

Comments: *Sylvilagus robustus* and *S. cognatus* were considered subspecies of *S. floridanus*. *S. floridanus* has the widest distribution of any other species of *Sylvilagus* in North America. It ranges from southern Canada, British Columbia, Saskatchewan, Ontario, and Québec south through South America. In western United States, it can be found in Washington and Oregon and from Montana south through Texas east through the Atlantic coast. In México, it is absent in the Mexican highlands and the Baja California Peninsula (Map 16). It can be found in sympatry, or nearly so, with *S. aquaticus, S. audubonii, S. cunicularius, S. gabbi, S. obscurus,* and *S. transitionalis*. *S. floridanus* can be distinguished from *S. aquaticus* by being larger (greater than 450.0 mm). From *S. audubonii,* by having a small white spot between the ears. From *S. cunicularius,* by its smaller size. From *S. gabbi,* by its ears larger than the head length and the supraorbital anterior process free. From *S. obscurus,* by the absence of a black spot between the ears; ears anterior edge without black hairs, and supraorbital process developed. From *S. transitionalis,* by having a white spot on the forehead, absence of a black spot between the ears, and ears anterior edge without black hairs. Supraorbital process developed and the postorbital process narrow, touching the skull. From *S. robustus* and *S. cognatus,* it must be differentiated genetically "See adeendum".

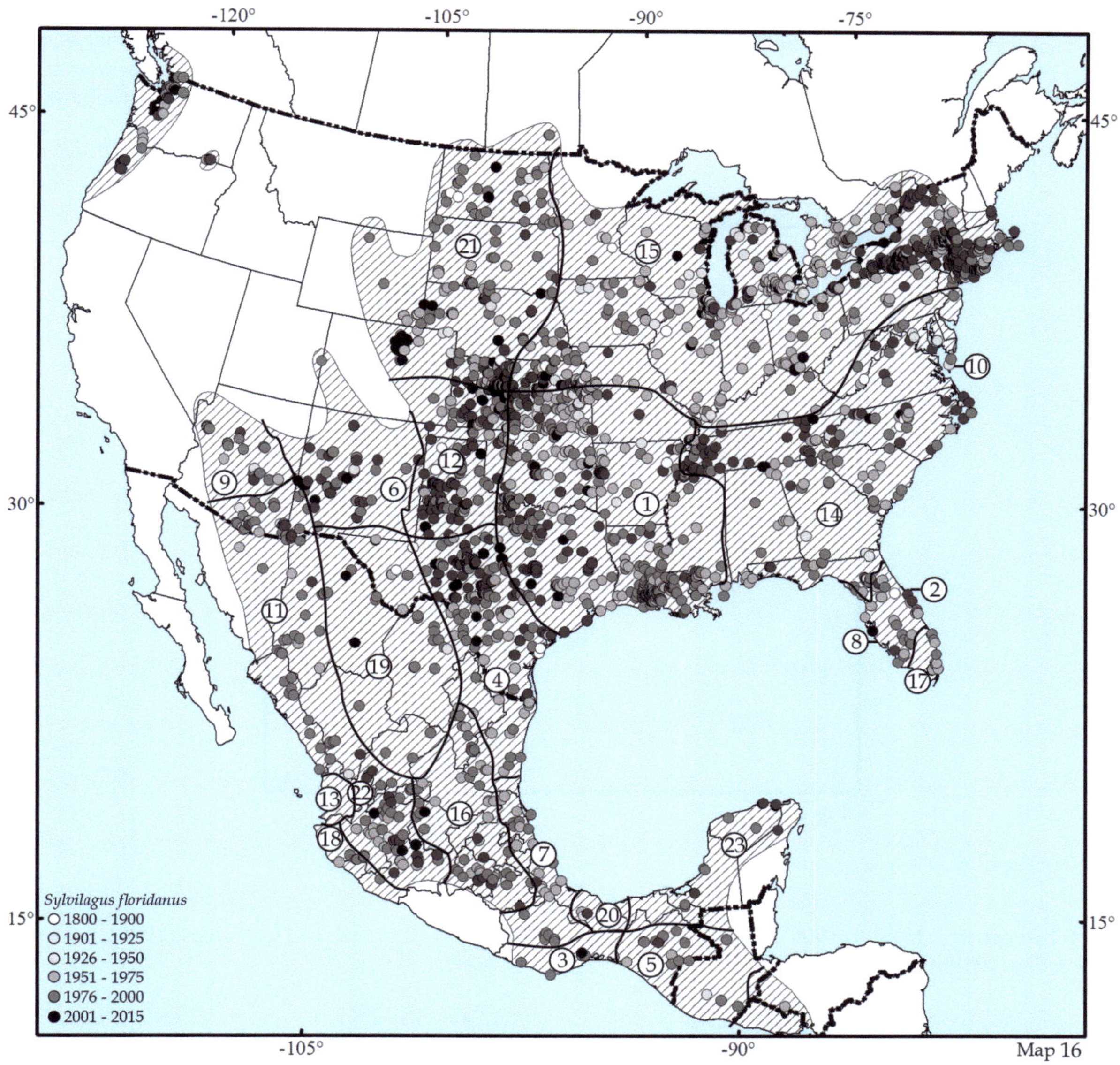

Map 16 Distribution map of *Sylvilagus floridanus*

Additional Literature: Nelson (1909), Chapman et al. (1980), Ruedas (1998), Cervantes et al. (1999b), Nielsen and Berkman (2018b).

Sylvilagus gabbi (Allen, 1877)

Tropical Cottontail, conejo tropical

S. g. truei (Allen, 1890). From southern Tamaulipas south through southern Campeche and Chiapas.

Conservation: *Sylvilagus gabbi* is listed as Least Concern by the IUCN (Ruedas and Smith 2019a)

Characteristics: Tropical cottontail is small sized; total length 380.0–420.0 mm, basilar length 59.1–63.3 mm. Dorsal pelage speckled in yellowish brown shades with black; fur of the underparts white interspersed with gray; nape and throat orange; limbs orange brown; ears very small, less than the head length; tail very small, brown dorsally, and dark yellow ventrally; ears shorter than the hind feet; supraorbital anterior process absent; auditory bullae length less than the toothrow length.

Comments: *Sylvilagus gabbi* was considered under *S. brasiliensis* but was elevated to full species (Ruedas and Salazar-Bravo 2007). *S. gabbi* ranges throughout the tropical region of México, from Tamaulipas south through Chiapas, except for the Yucatán Peninsula (Map 17). It can be found in sympatry, or nearly so, with *S. cunicularius* and *S. floridanus. S. gabbi* can be distinguished from *S. cunicularius* by having the interior part of the supraorbital process joined to the skull. From *S. floridanus,* by having very small ears, less than the head length; supraorbital process joined to the skull.

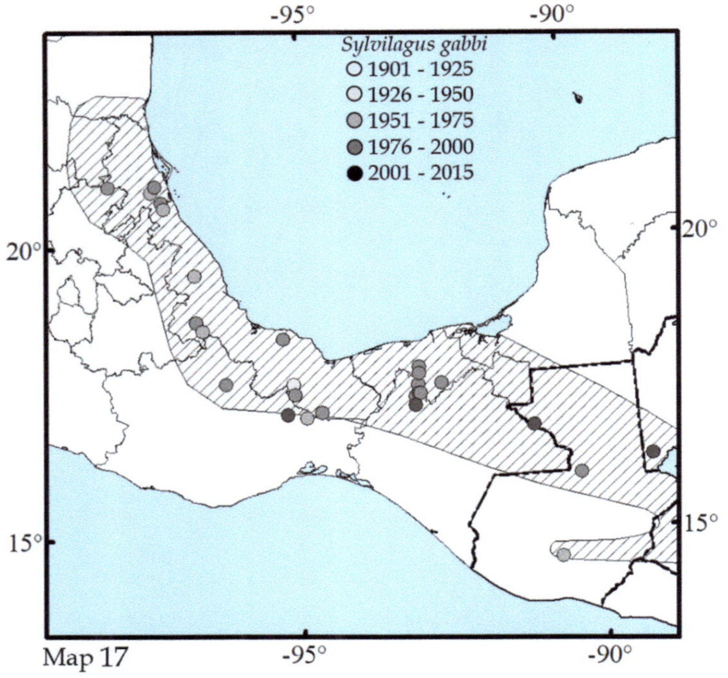

Map 17 Distribution map of *Sylvilagus gabbi*

Additional Literature: Nelson (1909), Yates et al. (1979), Diersing (1981), Lorenzo and Cervantes (1995), Cervantes et al. (1999a), Ruedas (2018b).

Sylvilagus graysoni (Allen, 1877)

Tres Marías Islands Cottontail, conejo de las Islas Marías

1. *S. g. badistes* Diersing and Wilson, 1980. San Juanito Island, Nayarit.
2. *S. g. graysoni* (Allen, 1877). María Madre Island, Nayarit.

Conservation: *Sylvilagus graysoni* is listed as Endangered in México (DOF 2019) and as Endangered (B1ab (iii) + 2ab (iii): extent of occurrence less than 5000 km^2 in five or less populations and under continuing decline in area, extent, and quality of habitat + area of occupancy less than 500 km^2 and under continuing decline in area, extent and quality of habitat) by the IUCN (Lorenzo and Lanier 2019b).

Characteristics: Tres Marias Islands cottontail is medium sized; total length 436.9–466.1 mm, condylobasal length 78.2–79.9 mm. Dorsal pelage reddish, more intense on the hips and nape; fur of the underparts white and throat brown; sides pale reddish; anterior and supraorbital posterior process absent; auditory bullae length less than the toothrow.

Comments: *Sylvilagus graysoni* can be found at the María Madre, María Magdalena, and María Cleofas Islands (Map 18). *S. g. badistes* is only found at San Juanito Island. *S. graysoni* is the only species of Leporidae on the islands, so they cannot be confused with other species.

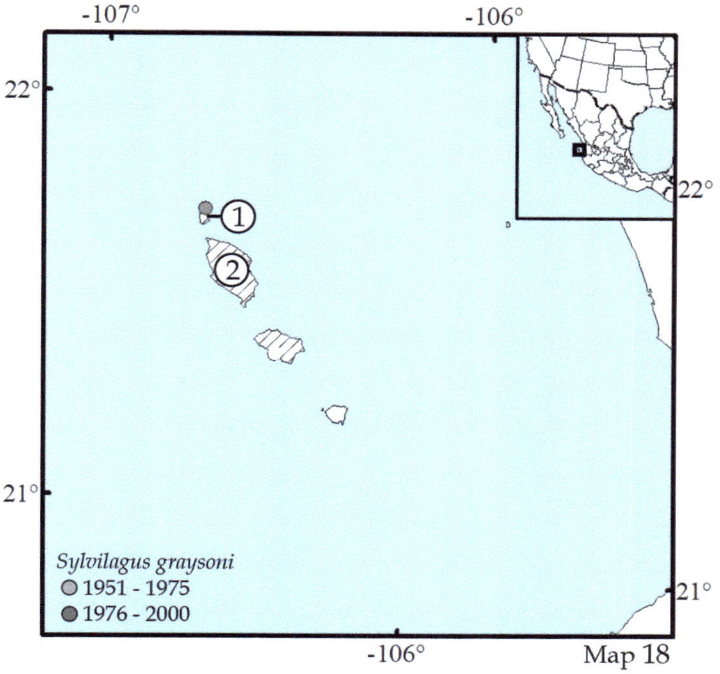

Map 18 Distribution map of *Sylvilagus graysoni*

Additional Literature: Nelson (1909), Diersing and Wilson (1980), Chapman and Ceballos (1990), Chapman and Flux (1990), Wilson (1991), Lorenzo et al. (1993, 2018d), Cervantes (1997).

Sylvilagus idahoensis (Merriam, 1891)

Pygmy Rabbit, conejo pigmeo

Monotypic

Conservation: *Sylvilagus idahoensis* (as *Brachylagus idahoensis*) is listed as Least Concern by the IUCN (Rachlow et al. 2016). Populations of the state of Washington of the counties of Douglas, Grant, Lincoln, Adams, and Benton are considered Endangered by the United States Endangered Species Act (FWS 2022).

Characteristics: Pygmy rabbit is the smallest member of the genus; **total length 252.0–310.0 mm, basilar length 36.9–41.0 mm**. **Dorsal and lateral pelage buffy gray; nape, chest, and forelimbs cinnamon buff; rostrum short and pointed**; **ears very small, rounded, and densely haired inside and outside, with the edge buff**; **hind feet covered with thick dense hair**; interparietal bone present; **larger supraorbital process relative to other *Sylvilagus*; auditory bullae and braincase relative larger**; anterior palatine foramina broad posteriorly; molariforms relative smaller; **first upper molars anterior surface with a single reentrant angle**.

Comments: *Brahylagus* was previously considered a genus (Hoffmann and Smith 2005) or a subgenus within *Sylvilagus* (Grinnell et al. 1930), but it is currently considered *Sylvilagus idahoensis* (Cano-Sánchez et al. 2022). *S. idahoensis* ranges at the Great Basin and some adjacent intermountain areas in Montana, Wyoming, Utah, Washington, Nevada, northeastern California, and eastern Oregon (Map 19). It can be found in sympatry, or nearly so, with *S. audubonii* and *S. nuttallii*. *S. idahoensis* can be distinguished from all the other species by its very small size, less than 300 mm; ears rounded and densely haired inside and outside, with the edge buff; hind legs very short and hind feet covered with thick dense hair; larger supraorbital process relative to *Sylvilagus*; auditory bullae and braincase relatively larger; first upper molars anterior surface with a single reentrant angle.

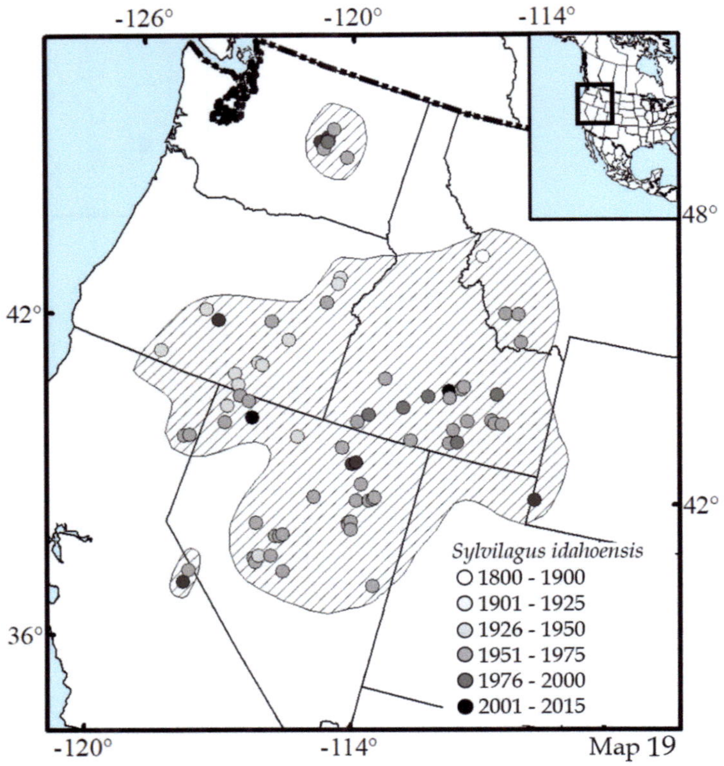

Map 19 Distribution map of *Sylvilagus idahoensis*

Additional Literature: Miller (1900), Nelson (1909), Davis (1939), Green and Flinders (1980), Hoffmann and Smith (2005), Rachlow et al. (2018).

Sylvilagus insonus (Nelson, 1904)

Omiltemi Cottontail, conejo de Omilteme

Monotypic

Conservation: *Sylvilagus insonus* is listed as Endangered in México (DOF 2019) and Data Deficient by the IUCN (Lorenzo et al. 2019a).

Characteristics: Omiltemi cottontail is medium sized; total length 430.0–440.0 mm, basilar length 57.0–59.2 mm. Dorsal pelage reddish rust with black, sides gray interspersed with black; fur of the underparts white, except for the brown spot on the neck; **tail yellowish (not white) ventrally**; nape with more reddish shades; ears speckled with blackish brown; blackish parts toward the anterior border and at the tips; supraorbital anterior process present; auditory bullae length less than the toothrow length.

Comments: *Sylvilagus insonus* ranges only at Omiltemi region of Sierra Madre del Sur (Map 20), an area with strong ravines, steep slopes, and many social issues, which makes the region difficult for conducting field work. The other sympatric species that can be found in the area is *S. cunicularius*, from which it differs by its dorsal rufous black coloration; hindfeet white; ears longer; tail longer and bicolored (Diersing 1981; Cervantes and Lorenzo 1997).

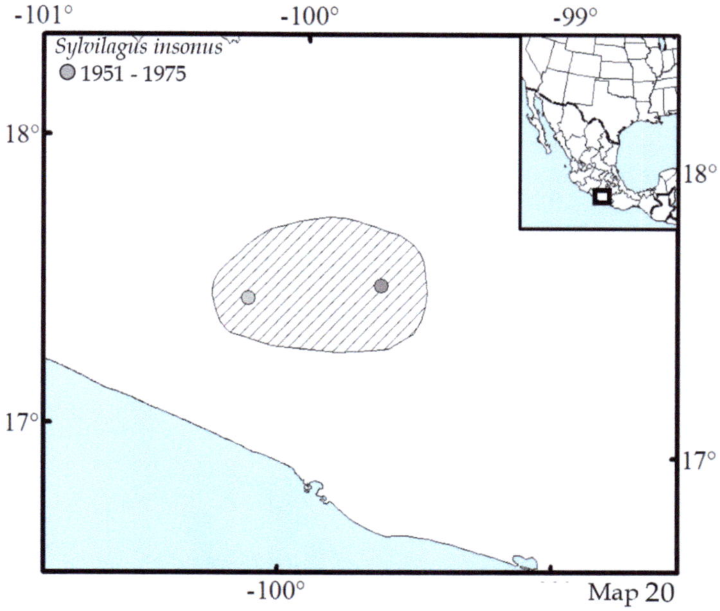

Map 20 Distribution map of *Sylvilagus insonus*

Additional Literature: Nelson (1909), Hershkovitz (1950), Lorenzo et al. (2018e).

Sylvilagus nuttallii (Bachman, 1837)

Mountain Cottontail, conejo de las montañas

1. *S. n. grangeri* (Allen, 1895). From southern Alberta and Saskatchewan south through northern Arizona, Nevada, and California.
2. *S. n. nuttallii* (Bachman, 1837). From southern British Columbia south through central California.
3. *S. n. pinetis* (Allen, 1894). Colorado, northern New Mexico, and northeastern Arizona.

Conservation: *Sylvilagus nuttallii* is listed as Least Concern by the IUCN (Smith and Brown 2019b).

Characteristics: Mountain cottontail is small sized; total length 335.0–375.0 mm, basilar length 48.2–52.3 mm. Dorsal pelage, rump, dorsal tail, and hind legs chestnut brown to dark rusty red; fur of the underparts in the central abdomen white with the rest of the belly buff to pale brown; eye ring prominent cinnamon-colored; **ears relatively short, rounded at the tip and with the inner surface noticeably haired and lined with white hair**; tail large grizzled and dark above; hind feet densely covered with long hair; supraorbital process small and with an abruptly pointed anterior projection; postorbital process long and slender; supraoccipital posterior shield pointed or rounded; auditory bullae medium sized; **diameter of the external auditory meatus greater** than the **crown length of the last three molariforms**.

Comments: *Sylvilagus nuttallii* ranges include the intermountain areas of western United States and southern Canada, from British Columbia, Alberta, and Saskatchewan south through California, Arizona, and New Mexico, and from California east through North Dakota, South Dakota, and Nebraska (Map 21). It can be found in sympatry, or nearly so, with *S. audubonii, S. idahoensis,* and *S. floridanus*, from which it can be distinguished by its smallest size, less than 370.0 mm. From *S. audubonii*, by having the underparts not white; ears relatively short, rounded at the tip, inner surface noticeably haired and lined with white hair; diameter of the external auditory meatus greater than the crown length of the last three molariforms. From *S. idahoensis*, by its larger size, greater than 300.0 mm in total length.

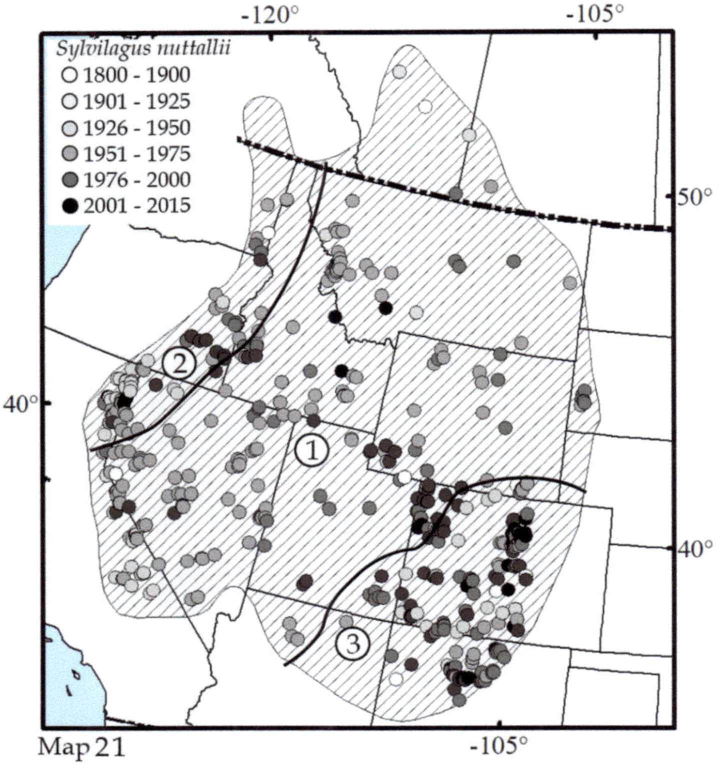

Map 21 Distribution map of *Sylvilagus nuttallii*

Additional Literature: Nelson (1909), Hoffmeister and Lee (1963), Chapman (1975b), Beever and French (2018).

Sylvilagus obscurus Chapman et al., 1992

Appalachian Cottontail, conejo de las montañas Apalaches

Monotypic

Conservation: *Sylvilagus obscurus* is listed as Near Threatened by the IUCN (Barry and Lanier 2019).

Characteristics: Appalachian cottontail is medium sized; total length 390.0–430.0 mm, basilar length 53.0–57.0 mm. Pelages and skull characteristics very similar to those of *S. floridanus but* **auditory bullae larger.**

Comments: *Sylvilagus obscurus* was separated from *S. transitionalis* by morphometric analysis and karyotype (Ruedas 1986; Chapman et al. 1992); molecular data showed a limited variation (Litvaitis et al. 1997). It was previously considered a subspecies of *S. transitionalis. S. obscurus* ranges in the highlands in the eastern United States, Pennsylvania, Maryland, Virginia, West Virginia, Tennessee, Kentucky, North Carolina, South Carolina, Georgia, and Alabama (Map 22). It can be found in sympatry, or nearly so, with *S. floridanus,* from which it can be distinguished by the presence of a black spot between the ears and never a white spot on the forehead, ear anterior edge with black hair and a supraorbital process obsolete or short. Postorbital process thin and rarely touching the skull. *S. obscurus* can be distinguished from *S. transitionalis* by its different cytotype (2n = 46).

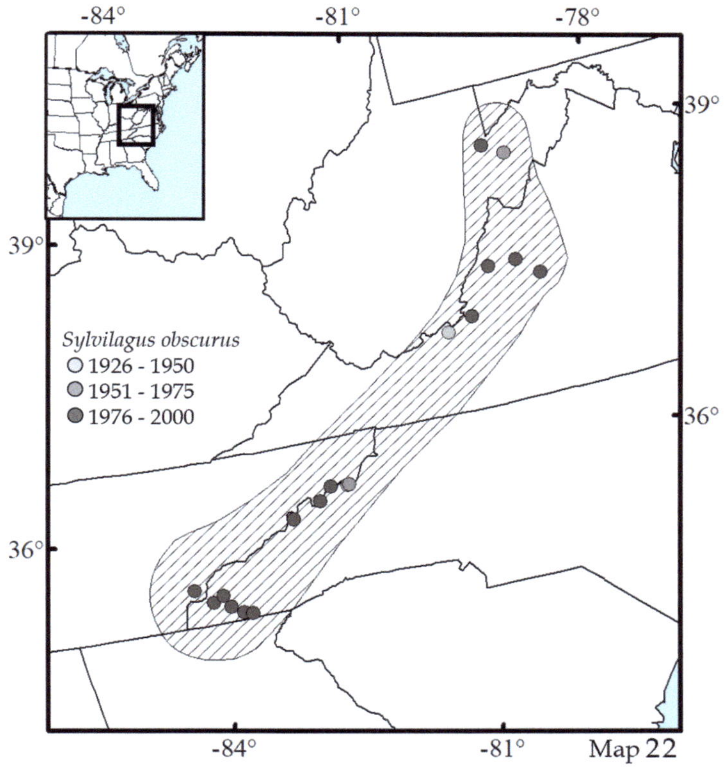

Map 22 Distribution map of *Sylvilagus obscurus*

Additional Literature: Ruedas et al. (1989), Barry (2018), Ruedas et al. (2017), Edelman (2019).

Sylvilagus palustris Bachman, 1837

Marsh Rabbit, conejo de los humedales

1. *S. p. hefneri* Lazell, 1984. Restricted to Key West, Florida.
2. *S. p. paludicola* (Miller and Bangs, 1894). Mainly throughout the Florida Peninsula.
3. *S. p. palustris* (Bachman, 1837). Southern Alabama and northern Florida north through southern Virginia and Delaware.

Conservation: *Sylvilagus palustris hefneri* is listed as Endangered by the United States Endangered Species Act (FWS 2022) and as Least Concern by the IUCN (Faulhaber and Smith 2008)

Characteristics: Marsh rabbit is medium sized; total length 430.0–442.0 mm, basilar length 61.7–66.5 mm. Dorsal, rump, dorsal tail, and hind leg pelage chestnut-brown to dark brown; fur of the underparts reddish brown; **ears short and broad**; feet short and dark reddish to buffy; **tail short and grayish ventrally**; supraorbital process joined to the skull along most of its extent.

Comments: *Sylvilagus palustris* ranges from Florida and Georgia along the coast north through Virginia (Map 23). It can be found in sympatry, or nearly so, with *S. aquaticus*, from which it can be distinguished by being smallest and having the auditory bullae equal to or larger than the foramen magnum.

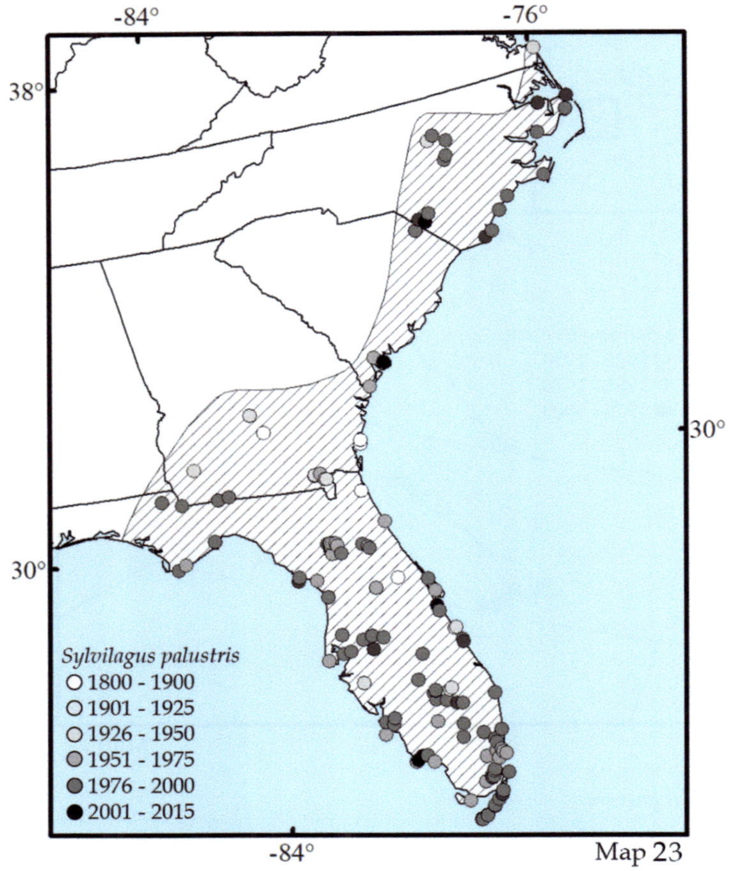

Map 23 Distribution map of *Sylvilagus palustris*

Additional Literature: Nelson (1909), Chapman and Willner (1981), Robinson et al. (1983, 1984), McCleery et al. (2018).

Sylvilagus robustus (Bailey, 1905)

Robust Cottontail, conejo del suroeste

Monotypic

Conservation: *Sylvilagus robustus* is listed as Endangered (B2ab (iii) c (iv): area of occupancy less than 500 km^2 in five or less populations and under continuing decline in area, extent, and quality of habitat and extreme fluctuations in the number of mature individuals) by the IUCN (Ruedas and Smith 2019b).

Characteristics: Robust cottontail is medium sized; total length 380.0–400.0 mm, basilar length 57.0–59.7 mm. Dorsal pelage speckled in yellowish shades with black; hips and limbs in grayish shades; flanks paler than the back; fur of the underparts white; nape and limbs orange; hind feet proportionally large and white; tail brown dorsally and white ventrally; supraorbital anterior process present; auditory bullae length less than the toothrow; **two basisphenoid foramina; no ear process**.

Comments: *Sylvilagus robustus* was considered a subspecies of *S. floridanus* (Ruedas 1998). No other species of *Sylvilagus* are present in its range. Its ranges are restricted to the Guadalupe Mountains in Texas and New Mexico, Chicos, and Davis Mountains in Texas, and Sierra La Madera, Coahuila (Map 24). This species can be differentiated from *S. cognatus* and *S. floridanus* only by genetic analyses.

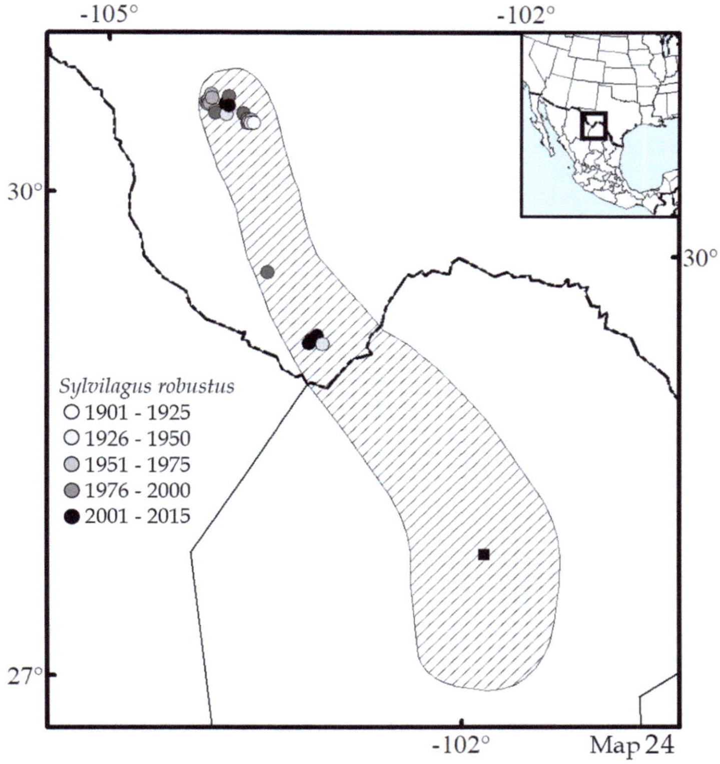

Map 24 Distribution map of *Sylvilagus robustus*

Additional Literature: Ruedas and Dowler (2018).

Sylvilagus transitionalis (Bangs, 1895)

New England Cottontail, conejo de Nueva Inglaterra

Monotypic

Conservation: *Sylvilagus transitionalis* is listed as Vulnerable (A2ace: population reduction in the last 10 years equal to or greater than 30% and may not be reversible based on direct observation, decline in area of occupancy, extent of occurrence and quality of habitat, and effects of introduced taxa) by the IUCN (Litvaitis and Lanier 2019).

Characteristics: New England cottontail is medium sized; total length 386.0–415.0 mm, basilar length 52.9–56.5 mm. Dorsal color speckled pinkish buff to ochraceous buff; back overline distinctive black wash; fur of the underparts white to white with yellowish shades; ears short and rounded, with the anterior edge covered with distinct black hair; with a black spot between the ears; supraorbital process obsolete or short; postorbital process thin, tapering and rarely touching the skull; suture between the nasal and frontal bones jagged and irregular; **auditory bullae small.**

Comments: *Sylvilagus obscurus* may be considered a subspecies of *S. transitionalis* (Ruedas 1986; Litvaitis et al. 1997). *S. transitionalis* is widespread across all of New England, Maine, Vermont, New Hampshire, Connecticut, eastern New York, Massachusetts, and Rhode Island (Map 25). It can be found in sympatry, or nearly so, with *S. floridanus,* from which it can be distinguished by the presence of a black spot between the ears and never a white spot in the forehead, ears anterior edge with black hair and supraorbital process obsolete or short, postorbital process thin and rarely touching the skull, and by its different cytotype (2n = 46).

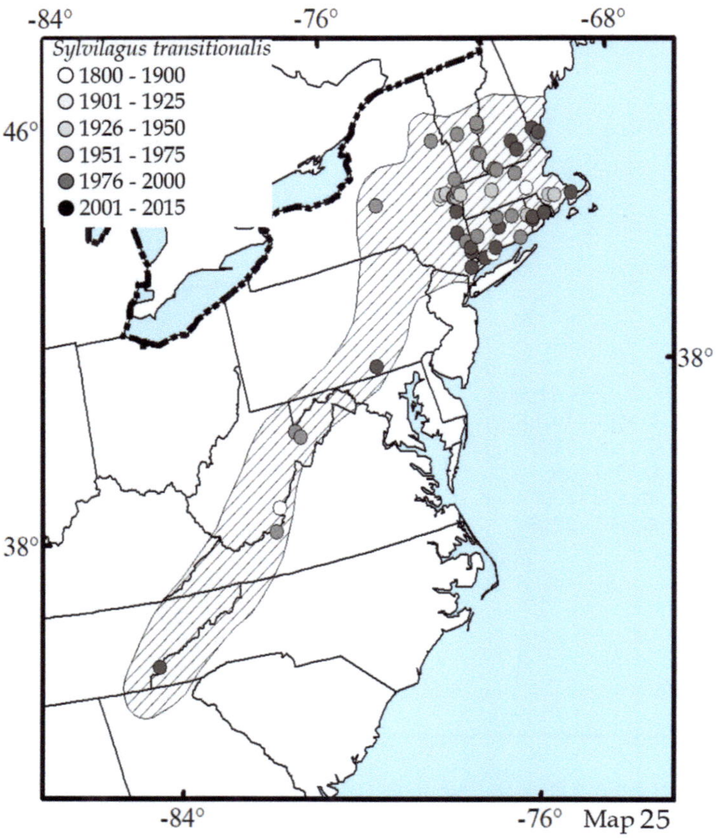

Map 25 Distribution map of *Sylvilagus transitionalis*

Additional Literature: Nelson (1909), Chapman (1975a), Ruedas et al. (1989), Chapman et al. (1992), Litvaitis et al. (2018).

Family Ochotonidae Thomas, 1897

The family Ochotonidae includes pikas, characterized by **fore and hindlimbs of similar size; ears rounded and short; tail very small, indistinct;** two pairs of incisors, one behind the other; **nasals wider anteriorly** than **posteriorly**; rostrum slender; **maxilla with one large fenestra; no postorbital process in the frontal bones; jugal with a large projection backward**. Distributed in the northern hemisphere.

Genus ***Ochotona*** Link, 1795

Dorsal coloration grayish or drab; ears short and rounded, similar in width and length; forelegs slightly shorter than the hind legs; hind feet less than 40.0 mm; tailless; with two pairs of incisors, one behind the other; five molariforms in the maxilla; nasals wider anteriorly than posteriorly; rostrum slender; **maxilla with one large fenestra;** no postorbital process in the frontal bones; **jugal with a large projection backward**. Only one subgenus in North America: *Pika* (*O. collaris* and *O. princeps*). The keys were elaborated based on the review of specimens and the following sources: Howell (1924) and Hall (1951, 1981).

1. With a distinct grayish spot between the neck and the back; neck buff contrasting with the dorsal coloration; auditory bullae generally greater than 27% of the skull length (restricted to central and southern Alaska, northwestern British Columbia, and the western Northwest Territory, Canada)... *Ochotona collaris* (p. 513)

1a. Without a grayish spot between the neck and the back, and neck buff; auditory bullae generally less than 25% of the skull length (restricted to the mountainous areas of western United States and southwestern Canada) .... *Ochotona princeps* (p. 514)

Subgenus *Pika* Lacépede, 1799

Braincase and whole skull very flat, interorbital region rather broad and flat, not arched; no foramina in the anterior area of the frontal bones; **opening in the maxilla roundly triangular and single; incisor foramina constricted into two unequal portions.**

Ochotona collaris (Nelson, 1893)

Collared Pika, pica de collar

Monotypic

Conservation: *Ochotona collaris* is listed as Least Concern by the IUCN (Lanier and Hik 2016).

Characteristics: Collared pika is very small sized; total length 175.0–195.0 mm, skull length 42.0–44.2 mm. Dorsal and lateral coloration drab grayish brown; **buff collar from the ears to the neck contrasting with the dorsal coloration; fur of the underparts creamy white;** ears with a white margin**; auditory bullae relatively larger, generally greater than 27% of the skull length**.

Comments: *Ochotona collaris* and *O. princeps* have been considered conspecific (Broabooks 1965; Youngman 1975) or conspecific with *O. alpina* (Corbet 1978). Weston (1981) and Hoffmann and Smith (2005) indicated that *O. alpina, O. collaris,* and *O. princeps* are distinct species. *O. collaris* ranges in the mountains of central and southern Alaska, Yukon Territory, northwestern British Columbia, and western Northwest Territory (Map 26). It has allopatric distribution with respect to other species of *Ochotona;* however, it can be distinguished from *O. princeps* by the presence of a gray patch on the nape, the shoulders forming an indistinct collar, and a creamy-buff patch on the head side; fur of the underparts white; skull broader; auditory bullae generally greater than 27% of the total skull length; and nasals shorter and interpterygoid fossa broader and more spatulated.

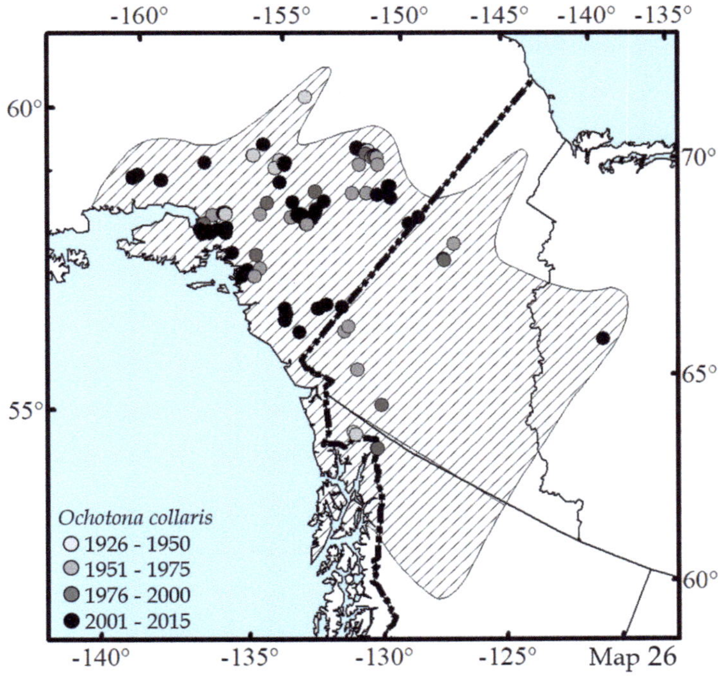

Map 26 Distribution map of *Ochotona collaris*

Additional Literature: Howell (1924), Banfield (1961), Banfield (1961), Broabooks (1965), Youngman (1975), MacDonald and Jones (1987), Smith et al. (1990), Lanier (2018).

Ochotona princeps (Richardson, 1828)

American Pika, pica americana

1. *O. p. princeps* (Richarson, 1828). Patchily distributed from central British Columbia and Alberta Rocky Mountains south through Idaho and Montana. Isolated populations in Montana, Wyoming, Nevada, Idaho, Utah, and Colorado.
2. *O. p. fenisex* Osgood, 1863. Patchily distributed along the Columbia River, from central British Columbia to southern Oregon on the Coast Mountains, and Cascade Range.
3. *O. p. saxatilis* Bangs, 1899. Patchily distributed in the southern Rocky Mountains south of the Colorado River in Utah, Colorado, and New Mexico.
4. *O. p. schisticeps* (Merriam, 1889). Patchily distributed in California and Nevada. Isolated populations in Nevada, eastern Oregon, and southwestern Utah.
5. *O. p. uinta* Hollister, 1912. Patchily distributed in the Uinta Mountains and the Wasatch Range, central Utah.

Conservation: *Ochotona princeps* is listed as Least Concern by the IUCN (Smith and Beever 2016).

Characteristics: American pika is very small sized; total length 160.0–215.0 mm, skull length 38.8–47.3 mm. Dorsal and lateral coloration grayish to cinnamon brown; **fur of the underparts whitish with a buffy wash; without a grayish collar spot;** ears with a white margin; **auditory bullae relatively smaller, generally less than 25% of the skull length**.

Comments: *Ochotona princeps* and *O. collaris* have been considered conspecific (Broabooks 1965; Youngman 1975). Weston (1981) and Hoffmann and Smith (2005) indicated that *O. collaris* and *O. princeps* are separated species. Hafner and Smith (2010) revised the subspecific taxonomy of the American Pika considering only five instead of the previous 36 (Hall 1981). *Ochotona princeps princeps* includes *O. p. clamosa, O. p. cuppes, O. p. figginsi, O. p. goldmani, O. p. howelli, O. p. lemhi, O. p. lutescens, O. p. nevadensis, O. p. obscura, O. p. saturate, O. p. saxatilis, O. p. ventorum,* and *O. p. wasatchensis. O. p. fenisex* includes *O. p. brooksi, O. p. brunnescens, O. p. fenisex, O. p. fumosa, O. p. littoralis,* and *O. p. septentrionalis. O. p. saxatilis* includes *O. p. incana, O. p. lasalensis,* and *O. p. nigrescens. O. p. schisticeps* includes *O. p. albata, O. p. cinnamomea, O. p. fuscipes, O. p. jewetti, O. p. muiri, O. p. schisticeps, O. p. sheltoni,* and *O. p. taylori. O. p. uinta* includes *O. p. barnesi, O. p. moorei, O. p. utahensis,* and *O. p. uinta. O. princeps* ranges from British Columbia and Alberta and is patchily distributed from south through central California, Nevada, Utah, and northern New Mexico (Map 27). It has allopatric distribution with respect to other species of *Ochotona*; however, it can be distinguished from *O. princeps* by the collar absence; fur of the underparts buffy wash; and auditory bullae generally less than 25% of the total skull length.

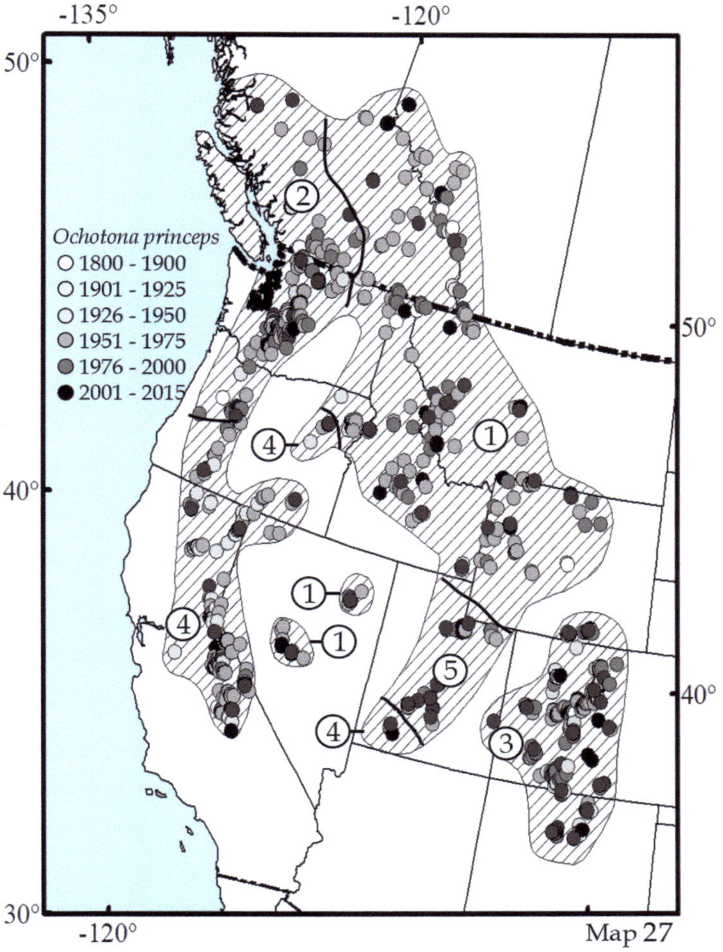

Map 27 Distribution map of *Ochotona princeps*

Additional Literature: Hall (1981), Smith and Weston (1990), Hafner and Sullivan (1995), Galbreath et al. (2009, In press), Smith et al. (2018).

Order Perissodactyla

Order Perissodactyla Owen, 1848

The order Perissodactyla includes horses, donkeys, and tapirs, characterized by unguligrade limbs; **third digit of each front and hindfoot larger than the others, in symmetry with the axis; digits usually one or four in the fore or hind feet; femur with a third trochanter; astragals with a pulley-like surface dorsally and flattened underneath.** Distribution nearly cosmopolitan. Two suborders, Ceratomorpha and Hippomorpha, each with one family, Tapiridae and Equidae (introduced), occur in North America.

1. Nose without proboscis; one functional digit per foot; mane present, bushy; postorbital bar present; molars with a very complex lophodont pattern ....suborder Ceratomorpha; family Equidae (p. 561)

1a. Nose modified into a moveable proboscis; three or four functional digits per foot; mane absent; postorbital bar absent; molars with a simple pattern of cusp and transverse ridges (from the Tehuantepec isthmus south through Central America, including the Yucatán Peninsula) .... suborder Hippomorpha; family Tapiridae; *Tapirus bairdii* (p. 559)

Suborder Ceratomorpha Woods, 1937

Family Tapiridae Gray, 1821

The family Tapiridae includes the tapir, characterized by **forefeet with four digits and hindfeet with three digits; nose modified into a moveable proboscis; nasal opening in the skull very large and hollow; body heavy with short limbs; skin smooth, with little hair.** Distributed in tropical areas of America and Asia. It belongs to the suborder Ceratomorpha.

Genus ***Tapirus*** Brünnch, 1772

Dorsal pelage dark grayish brown with white spots on the neck, cheeks, and borders of the ears; body stocky; neck long; **upper lip long and prehensile**; ears short; tail short; limbs short and stout; **four toes in the forefeet and three in the hind feet, third digit longest;** without a postorbital bar; anterior nasal openings enlarged, recessed; nasal bones small and arched; teeth brachydont, without cement; molars bilophodont, without extra cusps or crests.

S. T. Álvarez-Castañeda, *Mammals of North America - Volume 1*, https://doi.org/10.1007/978-3-031-41661-3_9

Subgenus *Tapirella* Palmer, 1903

***Tapirella bairdii* is characterized by its dark grayish brown coloration with white spots on theneck, cheeks, and borders of the ears; proboscis small; nasals markedly reduced;** nasal opening not in the distal part of the skull. **Total length 2020.0 mm, condylobasal length 395.0 mm.**

Tapirus bairdii (Gill, 1865)
Baird's tapir, tapir Americano

Monotypic

Conservation: *Tapir bairdii* is listed as Endangered in the Norma Oficial Mexicana (DOF 2019) and as Endangered (A2abcd + 3bcde: population reduction in the last 10 years equal to or greater than 50% and may not be reversible based on data obtained by direct observation, index of abundance, decline in area of occupancy, extent of occurrence and quality of habitat, and actual levels of exploitation + population reduction equal to or greater than 50% projected in the future based on the index of abundance, decline in area of occupancy, extent of occurrence and quality of habitat, actual levels of exploitation, and effects of introduced taxa) by the IUCN (García et al. 2016).

Characteristics: Baird's tapir is very large in size; total length 2,010.0–2120.0 mm, skull length 420.0 to 450.0 mm. Other characteristics same as in the genus.

Comments: The genus *Tapirella* is used instead of *Tapirus* (Ashley et al. 1996; Groves and Grubb 2011). However, the change is currently not accepted until more in-depth molecular and morphological studies are used to delimit the species within the clade. On the other hand, given the commercial use of the species, I maintain the name *Tapirus*. *Tapirus bairdii* ranges in tropical areas, from the Tehuantepec isthmus south through Central America, including the Yucatán Peninsula (Map 1). It lives in sympatry with *Mazama pandora* and *M. temama,* from which it can be distinguished by having the third digit of each fore and hind foot larger than the others, in symmetry with the axis**;** a small proboscis, and with a color pattern dark grayish brown with white spots on the neck, cheeks, and borders of the ears.

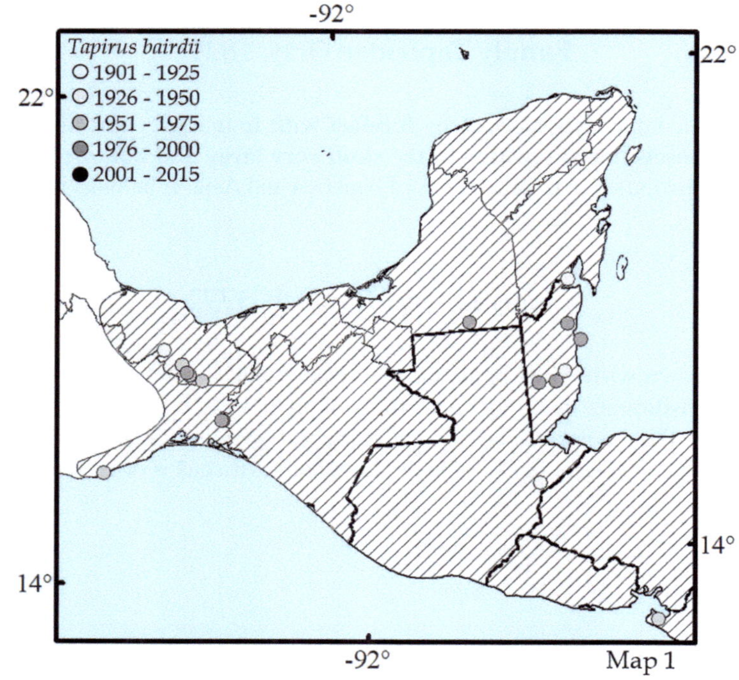

Map 1 Distribution map of *Tapirus bairdii*

Additional Literature: Hershkovitz (1954), Mendoza et al. (2013).

Suborder Hippomorpha Woods, 1937

Family ***Equidae*** Gray, 1821

The family Equidae includes horses and donkeys, characterized by one distinct digit in all the feet**; mane present and bushy; limbs long and slender,** with a single hoof on the third digit**; orbit enclosed by a postorbital bar; molars with a highly complex lophodont pattern.**

1 Ears long and thick; hoofs small and more upright; mane and tail thin; five vertebrates in the neck .... *Equus africanus* (p. 561)

1a. Ears small and wide; hoofs large; mane and tail wide; six vertebrates in the neck .... *Equus ferus* (p. 561)

Tribe Equini

Genus ***Equus*** Linnaeus, 1758

Dorsal coloration is in different colors or patterns; head directed downward; **limbs long and slender, with a single hoof on the third digit; molariforms very high crowned (hypsodont) and with a complex grinding surface of exposed enamel, dentine, and cement;** nasal bone long and narrow, with the anterior end free.

Equus africanus (von Heuglin and Fitzinger, 1866)

Donkey, burro silvestre

Conservation: *Equus africanus* is listed as an invasive species and as Critically Endangered by the IUCN in the wild populations (Moehlmann et al. 2015).

Comments: *Equus africanus* was considered *E. asinus* (Grubb, 2005), with current treatment as *E. africanus* (Gentry 2006). *E. africanus* has small feral populations in different parts of North America. The only native species of Artiodactyla in America is *Tapir bairdii* and can be distinguished by the absence of the proboscis; one functional digit per foot; mane present; postorbital bar present; and molars with a highly complex lophodont pattern.

Additional Literature: Grubb (2005), Gentry (2006), Moehlmann et al. (2015).

Equus ferus Boddaert, 1785

Feral horse, caballo silvestre

Conservation: *Equus ferus* is listed as an invasive species and Endangered by the IUCN in the wild populations (King et al. 2015).

Comments: *Equus ferus* has small feral populations in different parts of North America. The only native species of Artiodactyla in America is *Tapir bairdii* and can be distinguished by the absence of proboscis; one functional digit per foot; mane present; postorbital bar present; and molars with a very complex lophodont pattern.

Additional Literature: Groves (1986), Steiner and Ryder (2011), Grubb (2005), King et al. (2015).

Order Pilosa

Order Pilosa Flower, 1883

The order Pilosa includes anteaters, tamanduas, and sloths, characterized by an extra zygapophysis in the posterior part of the thoracic and lumbar vertebrae; **teeth absent; body not covered with horny scutes; pterygoids meeting at the midline, palate extending posteriorly to the auditory bullae**. Distribution restricted to the American continent. Only two families with one species each occur in North America: Cyclopedidae and Myrmecophagidae. The order Pilosa was previously included within the order Edentata (Simpson 1945) or the order Xenarthra (Wilson and Reeder 1993). The keys were elaborated based on the review of specimens and the following sources: Hall (1981) and Álvarez-Castañeda et al. (2017a).

1. Rostrum short and not tube-like (Fig. 1); total length less than 500.0 mm; skull length less than 70.0 mm (from southern Veracruz south through South America, including eastern Oaxaca).. Family Cyclopedidae; *Cyclopes dorsalis* (p. 564)

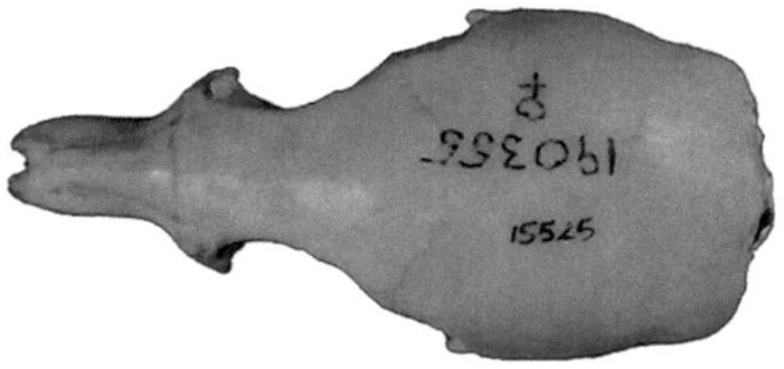

Fig. 1 Rostrum as short and not tube-like as in *Cyclopes dorsalis*

1a. Rostrum extended and tube-like (Fig. 2); total length greater than 500.0 mm; skull length greater than 70.0 mm (from Guerrero and Tamaulipas south through South America, including the Yucatán Peninsula)..Family Myrmecophagidae; *Tamandua mexicana* (p. 565)

S. T. Álvarez-Castañeda, *Mammals of North America - Volume 1*, https://doi.org/10.1007/978-3-031-41661-3_10

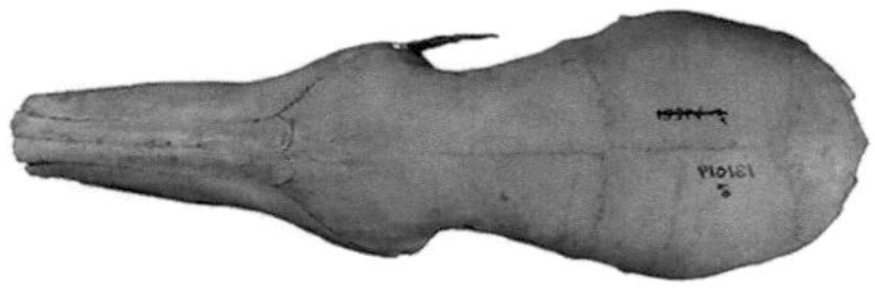

Fig. 2 Rostrum as extended and tube-like as in *Tamandua mexicana*

Family Cyclopedidae Opcock, 1924

The family Cyclopedidae includes the two-digit anteater or honey ant bear, characterized by their small size; no teeth, and long sticky tongue allowing them to collect insects; **two digits in the forefeet with strong claws; muzzle short and not tube shaped, and less than the braincase; tail long, prehensile, and hairy; zygomatic arches absent.** Distribution restricted to tropical America. No subfamilies are known.

Genus *Cyclopes* Gray, 1821

General pelage gray to yellow, darker dorsally; fur of the underparts with streaks and paler shades; pelage long and dense with a silver gloss; tail longer than the head-and-body length and prehensile, bare below; forefeet with two claws, hind feet with four claws; skull with a **proportionally shorter rostrum and a larger braincase; glenoid well separated from the porus acusticus; braincase strongly recurved; zygomatic arches absent.**

Cyclopes dorsalis (Gray, 1865)

Silky anteater, osito hormiguero dorado

C. d. mexicanus Hollister, 1914. Only subspecies in North America.

Conservation: *Cyclopes dorsalis* (as *Cyclopes didactylus*) is listed as Endangered in México (DOF 2019) and as Least Concern by the IUCN (Miranda et al. 2014).

Characteristics: Silky anteater is medium sized; **total length 398.0–422.0 mm, skull length on average 40.0–53.5 mm.** Dorsal pelage brilliant golden gray; fur of the underparts cream yellow, usually with a dark line on the abdomen; **muzzle short and not tube shaped; tail long prehensile and hairy; forefeet with two claws, hind feet with four claws**; **rostral length less than that the braincase length; zygomatic arches absent**.

Comments: *Cyclopes dorsalis* was described as a species (Gray 1865) and later considered a subspecies of *C. didactylus* (Trouessart 1905), but molecular and morphological analyses show that it should be reconsidered a full species (Miranda et al. 2018). Given that the revision by Miranda et al. (2018) is only at the species level, I continue using *mexicanus* as the subspecies distributed in México. *C. dorsalis* ranges from southern Veracruz and eastern Oaxaca south through South America, not including the Yucatán Peninsula (Map 1). *C. dorsalis* can be found in sympatry with *Tamandua mexicana*, from which it differs by having the rostrum short, not tube shaped, and by its smaller size, less than 500.0 mm and skull length less than 70.0 mm.

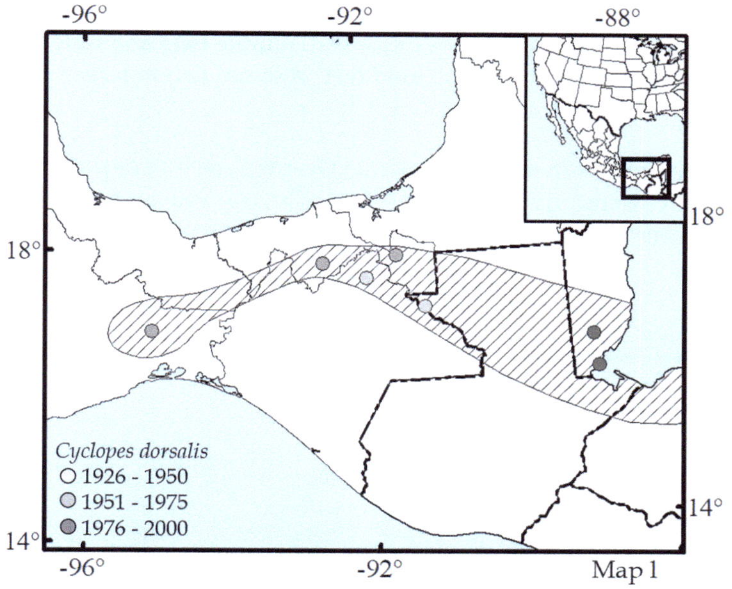

Map 1 Distribution map of *Cyclopes dorsalis*

Additional Literature: Wetzel (1982), Gaudin and Branham (1998), Gardner (2005c), Hayssen et al. (2011).

Family Myrmecophagidae Gray, 1825

The family Myrmecophagidae includes anteaters or tamandua, characterized by the lack of teeth; tongue long and sticky that allows them to collect insects; **hind limbs with strong claws and more than two digits; muzzle long and tube shaped, larger than that braincase; zygomatic arches present only in the anterior part**. Distribution restricted to tropical America. No subfamilies are known.

Genus ***Tamandua*** Gray, 1825

Dorsal pelage with a **large patch across the back, like a vest worn backward against a pale yellow background**; pelage short, dense, slightly rigid, and bright; **rostrum elongated with a tubular head**; teethless; tongue long, slender and sticky; eyes small; ears short and rounded; tail prehensile, nearly as long as the than the head-and-body length, furred dorsally across about one-third of its length but naked on the distal two-thirds, sometimes with irregular dark spots ; **five conspicuous digits on the hindfeet and four digits on the forefeet; middle forefoot digit with an extremely large claw, two medium and one small**; **four pairs of orbital foramina and posterior border of the infraorbital foramen distinctly crescent shaped; rostrum length larger than the braincase**; **zygomatic arches present only in the anterior part;** jugal attached to the maxilla.

Tamandua mexicana (Saussure, 1860)

Northern Tamandua, oso hormiguero

1. *T. m. hesperia* Davis, 1955. From Guerrero and the western lowlands of Oaxaca.
2. *T. m. mexicana* (Saussure, 1860). From central Veracruz and southern Oaxaca south through Central America.

Conservation: *Tamandua mexicana hesperia* and *T. m. mexicana* are listed as Endangered in México (DOF 2019) and as Least Concern by the IUCN (Ortega Reyes et al. 2014).

Characteristics: Northern Tamandua is medium sized; **total length 1,000.0 –1,300.0 mm; skull length 123.0–128.0 mm.** Dorsal pelage cream to gold with a spot resembling a blackish vest; **muzzle long and tube shaped; arms strong with one large claw, two medium and one smaller; five in the hind feet; rostrum length larger than the braincase; zygomatic arches present only in the anterior part.**

Comments: *Tamandua mexicana* ranges from Tamaulipas and Guerrero south through South America (Map 2), including the Yucatán Peninsula. It lives in sympatry with *Cyclopes dorsalis*, from which differs by having the rostrum tube-like, extended; the forefeet have only one extremely large claw; large sized, greater than 500.0 mm and skull length greater than 70.0 mm.

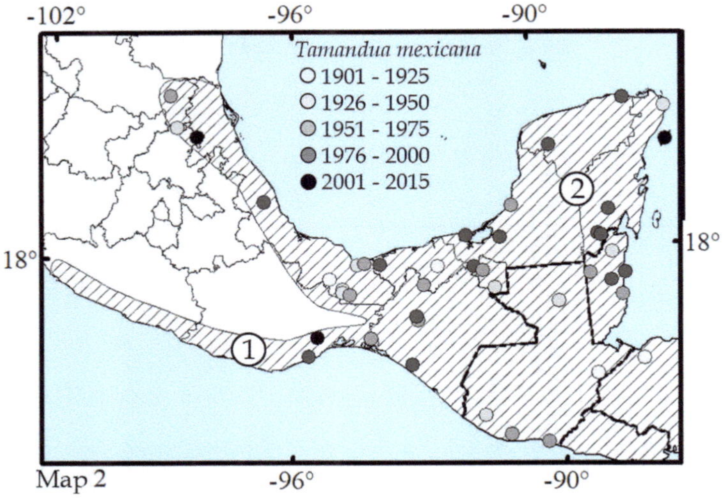

Map 2 Distribution map of *Tamandua mexicana*

Additional Literature: Allen (1904c), Wetzel (1975, 1985a), Eisenberg (1989), Gardner (2005c), Gardner and Naples (2007), Navarrete and Ortega (2011).

Order Primates

Order Primates Linnaeus, 1758

Primates include all monkeys; it is one of the mammal groups that do not have unique characteristics to help differentiate them from the rest of the mammals, but they have a combination of traits that allows their differentiation. Among those that stand out are fore and hind feet with five digits or plantigrades; **thumbs with nails, generally opposable** (at least in the hands); presence of a clavicle; vision stereoscopic and in color; **brain hemispheres well developed**; nails flat. Their distribution is tropical with few exceptions. One order, Haplorrhini, with two families, Atelidae and Cercopithecidae (introduced), ranges in North America.

Suborder Haplorhini Pocock, 1918

Face hairless except for particularly well-defined regions; nostrils complete, surrounded by naked skin without a lateral cleft; orbital face forward, completely separated from the temporal fossa by a bony shelf joined to the postorbital bar of the skull; lower incisors vertical; lacrimal bone within the orbit; auditory bullae reduced or absent; all digits bearing flattened nails, except for the Callitrichidae, which have a nail on the first digit of feet and pointed claws on the remainder digits. The keys were elaborated based on the review of specimens and the following sources: Hall (1981), and Álvarez-Castañeda et al. (2017a).

1. Nostrils well separated, directed laterally; auditory bullae not extending laterally as a bony tube .... suborder Strepsirrhini; family Atelidae (p. 567)

1a. Nostrils close together, directed forward or downward; auditory bullae extending laterally as a bony tube .... suborder Haplorrhini; family Cercopithecidae; *Macaca mulatta* (p. 572)

Family Atelidae Gray, 1825

The family Atelidae includes spider monkeys, howler monkeys, and wooly monkeys, characterized by having fingernails and being capable of climbing and scaling; **tail long, prehensile, haired and with a sensitive tactile area; pollex only slightly opposable or absent; hallux opposable; nostrils well separated, directed laterally**. Distribution restricted to tropical America. Two subfamilies range in North America: Atelinae and Alouattinae.

S. T. Álvarez-Castañeda, *Mammals of North America - Volume 1*, https://doi.org/10.1007/978-3-031-41661-3_11

1. Fore and hind legs are as long as the head-and-body length; in lateral view, the jaw is equal to one-fourth of the head height. Bread of the descending branch of the jaw less than the distance between the auditory meatus and the infraorbital channel (Fig. 1); less than 34% of the skull length behind the glenoid pits (from Jalisco and Tamaulipas south through Central America, including the Yucatán Peninsula)....................................subfamily Atelinae; *Ateles geoffroyi* (p. 571)

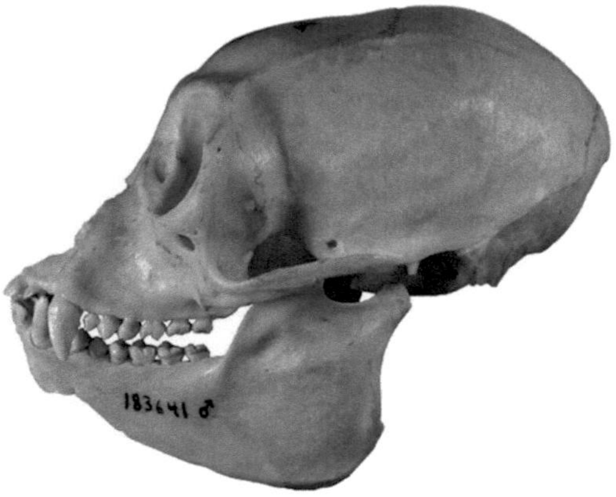

Fig. 1 Breadth of the descending branch of the jaw less than the distance between the auditory meatus and the infraorbital channel

1a. Fore and hind legs shorter than the head-and-body length; in lateral view, the jaw equals one-third of the head height. Breadth of the descending branch of the jaw is greater than the external distance between the auditory meatus and the infraorbital channel (Fig. 2); more than 34% of the skull length behind the glenoid pits ....subfamily Alouattinae; *Alouatta* (p. 568)

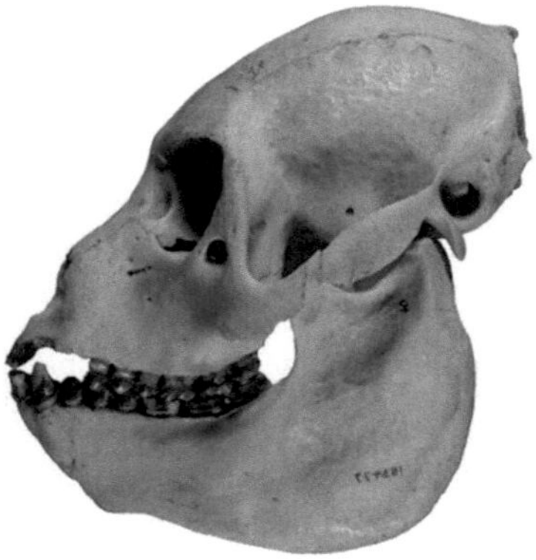

Fig. 2 Breadth of the descending branch of the jaw greater than the external distance between the auditory meatus and the infraorbital channel

Subfamily Alouattinae Trouessart, 1897

The characteristics of the species in the subfamily Alouattinae are the same as in the genus *Alouatta* because it is the only genus in the subfamily.

Genus ***Alouatta*** Lacépède, 1799

Dorsal and ventral pelage different according to the species; fore and hind legs less than the head-and-body length; in lateral view, the jaw equals one-third of the head height, although this is due to the effect of the beard; **upper canines practically three times the length of the incisors; braincase extended, approximately 1.5 times the length of the rostrum**. The keys were elaborated based on the review of specimens and the following sources: Hall (1981), and Álvarez-Castañeda et al. (2017a).

1. Brown pelage. Although it may have golden locks, interorbital breadth generally less than 12.0 mm; condylobasal length generally less than 103.0 mm (from southern Veracruz, Tabasco, Campeche, and central Chiapas south through Central America)........*Alouatta palliata* (p. 569)

1a. Black pelage. Interorbital breadth generally greater than 12.0 mm; condylobasal length usually greater than 103.0 mm (Restricted to the Yucatán Peninsula)........*Alouatta pigra* (p. 570)

Alouatta palliata (Gray, 1849)

Brown howler, mono aullador café

A. p. mexicana Merriam, 1902. Only subspecies in North America.

Conservation: *Alouatta palliata* is listed as Endangered in México (DOF 2019) and as Vulnerable (A4cd: population reduction in the last 10 years equal to or greater than 30% including the past and the future, and may not be reversible based on a decline in area of occupancy, extent of occurrence and quality of habitat, and actual or potential levels of exploitation) by the IUCN (Cortés-Ortíz et al. 2020a).

Characteristics: Brown howler is medium sized; total length 1104.0–1145.0 mm; occipitonasal length 94.4–100.9 mm. **Dorsal and ventral pelage dark brown, although it may have golden locks, especially on the abdomen**; fore and hind legs bare, in blackish color; skull similar to *A. pigra*.

Comments: *Alouatta pigra* is considered different from *A. palliata* (Cortés-Ortiz et al. 2003). *A. palliata* ranges from southern Veracruz, Tabasco, southern Campeche, and northern Chiapas south through Central America (Map 1). It lives in sympatry with *Ateles geoffroyi,* from which it can be differentiated by having the fore and hind legs less than the head-and-body length; breadth of the descending branch of the jaw greater than the distance between the auditory meatus and the infraorbital channel. Both species of *Alouatta* have allopatric distribution.

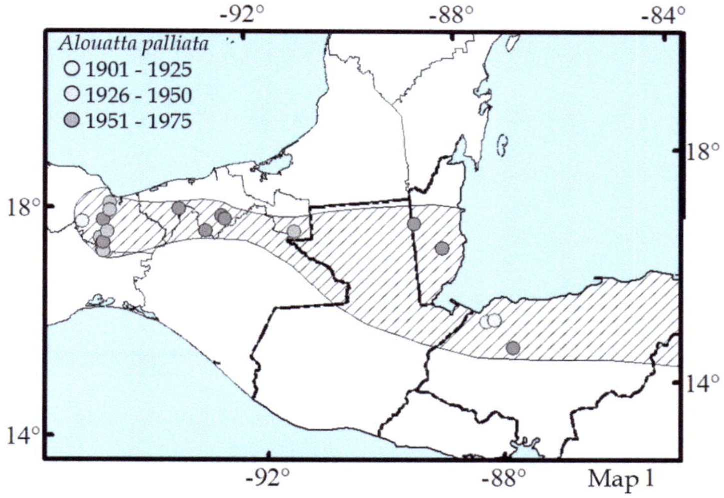

Map 1 Distribution map of *Alouatta palliata*

Additional Literature: Smith (1970), Hall (1981), Brandon-Jones (2006), Ortiz-Martínez et al. (2008), Cortés-Ortiz et al. (2007, 2015b).

Alouatta pigra Lawrence, 1933

Black howler, mono aullador negro

Monotypic.

Conservation: *Alouatta pigra* is listed as Endangered in México (DOF 2019), and *A. palliata* is listed as Vulnerable (A4cd: population reduction in the last 10 years equal to or greater than 30% including the past and the future, and may not be reversible based on the decline in the area of occupancy, extent of occurrence and quality of habitat, and potential levels of exploitation) by the IUCN (Cortés-Ortíz et al. 2020c).

Characteristics: Black howler is medium sized; total length 1100.0–1300.0 mm; occipitonasal length 106.5–121.0 mm. **Dorsal and ventral pelage completely black**; fore and hind feet monocolored; skull similar to *A. palliata.*

Comments: *Alouatta pigra* is considered different from *A. palliata* (Cortés-Ortiz et al. 2003, Groves 2005). Some authors considered *A. villosa* instead of *A. pigra* (Brandon-Jones 2006). *A. pigra* can be found in the Yucatán Peninsula (Map 2), living in sympatry with *Ateles geoffroyi*, from which it differs by having the fore and hind legs less than the head-and-body length; head length and breadth of the descending branch of the jaw greater than the distance between the auditory meatus and the infraorbital channel. Both species of *Alouatta* have allopatric distribution.

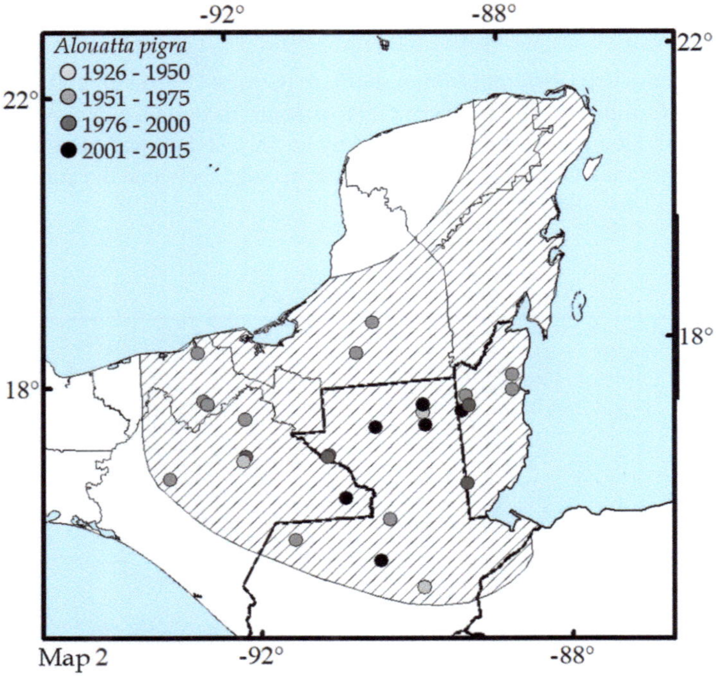

Map 2 Distribution map of *Alouatta pigra*

Additional Literature: Hall and Kelson (1959), Smith (1970), Hall (1981), Cortés-Ortiz et al. (2015).

Subfamily Atelinae Gray, 1825

Genus ***Ateles*** Geoffroy Saint-Hilaire, 1806

Color pattern brown to dark brown or blackish; **head small;** muzzle prominent; thumbs absent or poorly developed and appressed to the second digit; **limbs and tail exceptionally long relative to the body length; tail prehensile;** mandible not modified; **laryngeal pouch absent; ectopterygoid fossa absent;** braincase rounded and facial part of the skull short; inferior margin of the lower mandible approximately straight; molars small, each with four pronounced cusps.

Ateles geoffroyi Kuhl, 1820

Spider monkey, mono araña

1. *A. g. vellerosus* Gray, 1866. From Tamaulipas and Jalisco by the lowlands south through Chiapas.
2. *A. g. yucatanensis* Kellogg and Goldman, 1944. Yucatán Peninsula.

Conservation: *Ateles geoffroyi* is listed as Endangered in México (DOF 2019) and as Endangered (A4cd: population reduction in the last 10 years equal to or greater than 50% including the past and the future, and may not be reversible based on a decline in area of occupancy, extent of occurrence and quality of habitat, and potential levels of exploitation) by the IUCN (Cortes-Ortíz et al. 2020b).

Characteristics: Spider monkey is medium sized; total length 1129.0–1280 mm; occipitonasal length is 87.0–102.0 mm. Dorsal and ventral color variable, combining black, brown, and pale yellow regions; **fore and hind legs thin and longer than the head-and-body length**, and at least of a bicolored combination; **in lateral view, the jaw equals at least one-fourth of the head height;** head size proportionally small; upper canines less than double in length as the incisors; **braincase round, length greater than twice the rostrum length**.

Comments: *Ateles geoffroyi* ranges from Tamaulipas and Jalisco south through Central America, including the Yucatán Peninsula (Map 3). It lives in sympatry with *Alouatta palliata* and *A. pigra,* from which it differs by having the fore and hind legs as long as the head-and-body length; breadth of the descending branch of the jaw less than the distance between the auditory meatus and the infraorbital channel.

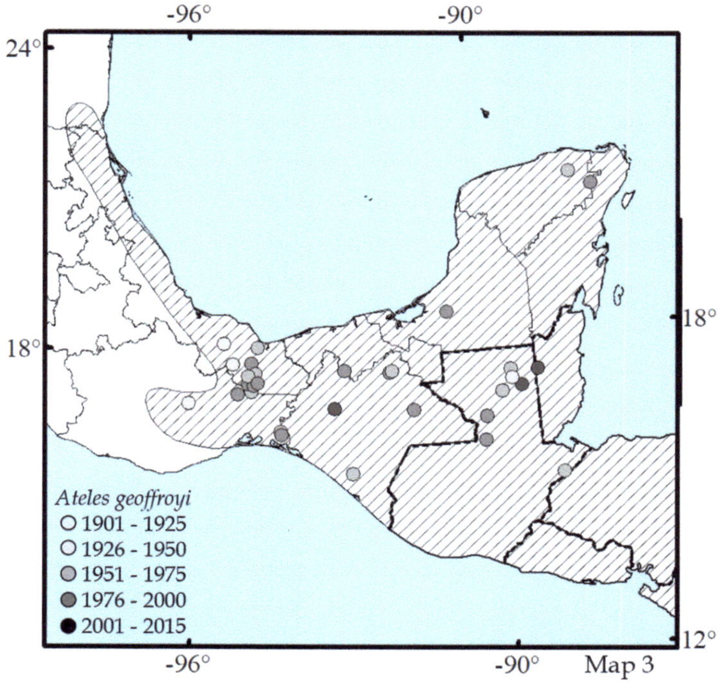

Map 3 Distribution map of *Ateles geoffroyi*

Additional Literature: Hall and Kelson (1959), Hall (1981), Collins and Dubach (2000), Collins (2008).

Family Cercopithecidae Linnaeus, 1758

The family Cercopithecidae includes macaques, characterized by **nostrils close together; cheek pouches present; tympanic forming a tubular external auditory meatus;** tail, when present, not prehensile. Family not native of North America.

Tribe Papionini

Genus *Macaca* Lacépède, 1799

Pelage varying from brown to black shades; individuals robust; arms and limbs about same in length; muzzles rounded with nostrils on the upper surface; cheeks with large pouches where they carry extra food; tail varying among species, from long to absent; body size differing according to the sex and species.

Macaca mulatta (Zimmermann, 1780)

Rhesus monkey, mono resus

Conservation: *Macaca mulatta* is listed as introduced species and Least Concern by the IUCN (Singh et al. 2020)

Characteristics: Rhesus monkey is medium sized; head-and-body length 47.0–53.0 cm. Dorsal pelage brown or gray, with a pink face, which is hairless.

Comments: Macaca mulatta has been recorded only from Silver Springs State Park, Florida, Morgan Island and the Sea Islands in South Carolina, and in one island in the Catemaco lagoon, Veracruz. *M. mulatta* differs from all the native species by having the nostrils close together, directed forward or downward; auditory bullae extending laterally as a bony tube.

Additional Literature: Lawlor (1979); Taub and Mehlman (1989); Singh et al. (2020).

Order Sirenia

Order Sirenia Illiger, 1811

Sirenia includes all manatees and sea cows that are characterized by **body fusiform; entirely aquatic; without neck, no visible separation of head from body; eyes small, eyelids imperfectly developed;** external auditory meatus small, ear absent; **pectoral appendages paddle shaped, digits not externally visible;** carpal elements vary between families; **pelvic appendages absent; tail flattened and laterally expanded, rounded or forked posteriorly; skin tough and rugose, naked or sparsely haired; bones of skull and postcranial skeleton massive and dense; nasal bones absent or vestigial;** six to eight cervical vertebrae; clavicles absent; sacrum absent; anterior caudal vertebrae having chevron bones; toothless or having incisors and molariform that are separated by wide diastema; tongue small, fixed, rugose; single pair of pectoral mammae; stomach of several chambers. Distribution is tropical in America and Asia. Two families occur in North America: Dugongidae and Trichechidae. The keys were elaborated based on the review of specimens and following Hall (1981).

4. Total length greater than nine meters; caudal fin regularly rounded; ranges in the Pacific Ocean (Restricted to the Commander Islands in the Bering Sea) .... family Dugongidae; *Hydrodamalis gigas* (extinct p. 575)

4a. Total length less than nine meters; caudal fin lobbed; ranges in the Atlantic Ocean (from North Carolina south through South America) .... family Trichechidae; *Trichechus manatus* (p. 573)

Family Trichechidae Linnaeus, 1758

The family Trichechidae includes manatees that are characterized by fusiform body; skin nearly hairless excepting spines on muzzle; **caudal fin regularly rounded, posterior margin not lobbed; upper deeply divided, each half capable of moving independently of other; nostrils valvelike; vestigial nails on pectoral appendages; rostrum short, formed above by union of premaxillae anterior to external nares;** six cervical vertebrae; five or six carpal elements rostrum only slightly convex ventrally; nasal bones present; **incisors rudimentary, concealed beneath homy plate and lost before maturity; molariform selenodont, rarely more than six upper and lower at any one time;** anterior molariform wear clown and fall out, replaced at posterior end of toothrow; two tuberculate, transverse ridges on molariform; bones heavy. No subfamilies are known.

S. T. Álvarez-Castañeda, *Mammals of North America - Volume 1*, https://doi.org/10.1007/978-3-031-41661-3_12

Genus *Trichechus* Linnaeus, 1758

Trichechus manatus Linnaeus, 1758
American manatee, manatí

Monotypic.

Conservation: *Trichechus manatus* is listed as Endangered in México (DOF 2019) and Vulnerable (C1: small population size decline 10% or less in 10 years) by the IUCN (Deutsch et al. 2008).

Characteristics: See the characteristics of the family.

Comments: *Trichechus manatus* ranges from Rhode Island south through Brazil in South America (Map 1). No other similar species can be found in sympatry. Can be differs from all other species by having caudal fin regularly rounded, posterior margin not lobbed; upper deeply divided, each half capable of moving independently of other; nostrils valvelike; vestigial nails on pectoral appendages; rostrum short, formed above by union of premaxillae anterior to external nares; incisors rudimentary, molariform selenodont, rarely more than six upper and lower at any one time.

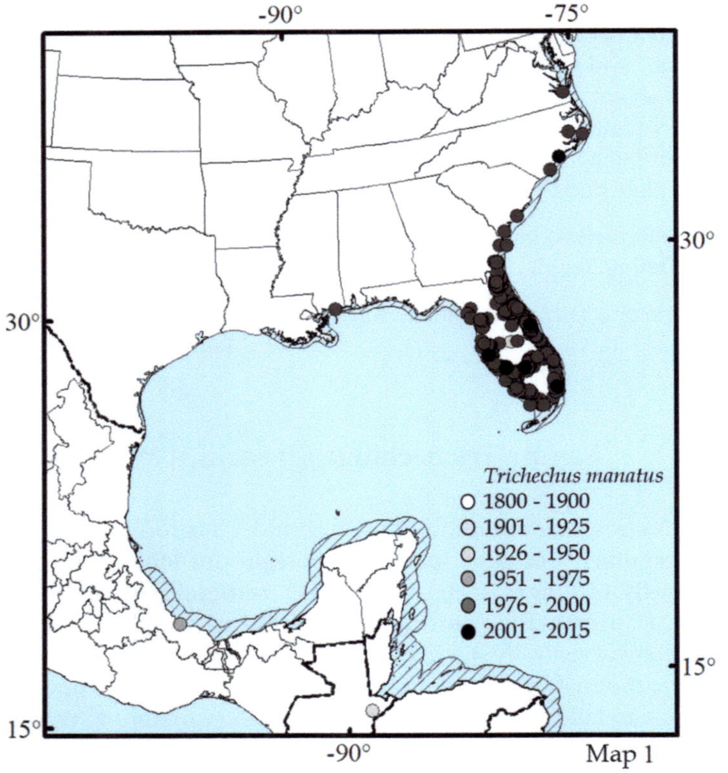

Map 1 Distribution map of *Trichechus manatus*

Additional Literature: Powell and Rathbun (1984), Lazcano-Barrero and Packard (1989), Husar (1978), Deutsch et al. (2003), Fertl et al. (2005), Vianna et al. (2006).

Family Dugongidae Gray, 1821

The family Dugongidae includes sea cows that are characterized by toothless; snout pointed downwards; small eyes located halfway between its nostrils and ears; eye with nictitating membrane; tongue small and remained in the back of the mouth; seven cervical, 17 thoracic, three lumbar, and 34 caudal vertebrae; without clavicles; scapula fan shaped, with the anterior border nearly straight; reduced phalanges and possibly did not have a manus at all.

Genus ***Hydrodamalis*** (Zimmermann, 1780)

Hydrodamalis gigas (Zimmermann, 1780)

Steller's sea cow, manati marino

Monotypic.

Conservation: *Hydrodamalis gigas* is listed as Extinct by the IUCN (Domning 2016).

Characteristics: Steller's sea cow is large sized; total length 8000–9000 mm, estimates of weight 24 tons. Hair on its body was sparse and its skin was brownish-black, with white patches on some individuals; back is smooth, but flanks rough with crater-like depressions most likely caused by parasites; tail fluke was forked; without dorsal fin; rostrum short; toothless.

Comments: *Hydrodamalis gigas* ranges only in the Bering Sea. Can be different from all other species by having **tail fluke was forked**; **without dorsal fin; rostrum short; toothless**.

Additional Literature: Forsten and Youngman (1982), Domning and Furusawa (1995); Domning et al. (2007).

Addenda

Page 75. *Mazama temama* (Kerr 1792). The subspecies of *Mazama temama temama* is restricted to México and Guatemala (Escobedo-Morales et al. 2023).

Page 257. *Anoura geoffroyi* Gray, 1838 was re-examinated and considering (Molinari et al. 2023) that the population present in México and central America as *A. g. lasiopyga* (Peters 1868) is in fact a population of *A. peruana* (von Tschudi 1844), for that reason need to be known as *A. peruana lasiopyga* (Peters 1868).

Page 284. *Lophostoma brasiliense* Peters, 1867 was considered to range from southern Veracruz, northern Chiapas, and the southern Yucatán Peninsula south through Honduras. However, was restricted to South America, including as a subspecies *L.nicaraguae* (Esquivel et al. 2022). The current species with range in México and northern Central America is the Mesoamerican round-eared bat, murciélago orejas redondas mesoamericano, *L. nicaraguae* (Goodwin 1942).

Page 289 *Trachops cirrhosus* (Spix 1823) was re-examinated and considering (Camacho et al. 2024) that its range is from South America throughout Costa Rica. The current species with range in México and northern Central America is the Central American fringe-lipped bat, murciélago de labio verrugoso de Centro America, *Trachops coffini* Goldman, 1925.

Page 318. *Vampyressa thyone* Thomas, 1909 from Oaxaca and Guerrero was considered as divergent species and describe as *Vampyressa villai* (Garbino et al. 2024) Villa's Little Yellow-eared Bat, Vampiresa de Villa is monotypic and is known only from lowlands of Oaxaca and Guerrero, México.

Vampyressa villai distinguish from *V. thyone* in having a smaller size in cranial dimensions and forearm length; a conspicuous fringe of white hairs along the upper lip and on the ventral margin of chin. In the skull with an oblique naso-premaxillary region in profile; absence of a sagittal crest; shorter hard palate, not extending to the level of the glenoid fossa, and fourth lower premolar protoconid narrower in lateral view (Gabino et al. 2024).

Page 369. *Eptesicus brasiliensis* (Desmarest 1819). This species is move to a new described genus as *Neoeptesicus brasiliensis* (Cláudio et al. 2023).

Page 370. *Eptesicus furinalis* (d'Orbigny and Gervais 1847). This species is move to a new described genus as *Neoeptesicus furinalis* (Cláudio et al. 2023).

Page 370. *Eptesicus fuscus miradorensis* (Allen 1866) was considered as full species and describe as *Eptesicus miradorensis* (Ramírez-Chaves et al. 2023). Tropical brown bat, murciélago pardo tropical is monotypic and is known only from southern Texas Tamaulipas, San Luis Potosí, Durango, and Sinaloa southward to Colombia and Venezuela. *Eptesicus miradorensis* distinguish from the other species of the genus by be the largest species of the genus with continental distribution in America (forearm: 48.54–51.13 mm); skull is robust (greatest length of the skull: 19.79–20.59 mm; condyle-canine length: 17.7–18.47 mm) and presents well developed sagittal and lambdoidal crests, with a triangular appearance in caudal view; dorsal fur is long (9.0–11.0 mm), lustrous, smooth, and shiny, with a brown color. Dorsal coloration is brown to golden yellow, dark to light, and bright, with two bands, the lower band being darker; underparts is lighter than the dorsal, with two bands, the lower band being darker (Ramírez-Chaves et al. 2023).

Page 387. *Corynorhinus mexicanus* (Allen 1916) from northern Sierra Madre Oriental was considered as divergent species and describe as *Corynorhinus leonpaniaguae* (López-Cuamatzi et al. 2024) León Paniagua's Big-eared Bat, Murciélago-mula de León-Paniagua is monotypic and is known only from Sierra Madre Oriental and mountains of the state of Coahuila, Nuevo León, and Tamaulipas, México.

S. T. Álvarez-Castañeda, *Mammals of North America - Volume 1*, https://doi.org/10.1007/978-3-031-41661-3

Corynorhinus leonpaniaguae distinguish from *C. mexicanus* in having a dorsal coloration brownish to grayish with dark bases fur and slightly lighter tips that appear to contrast between bands; have a double cuspid in the first upper incisor. *C. leonpaniaguae* is distinguished from *C. townsendii* in having the tragus length of less than 13 mm and a hairy uropatagium with nine or fewer interfemoral grooves.

Page 411. *Cryptotis goodwini* Jackson, 1933 from Chiapas was considered as divergent species and describe as *C. woodmani* (Guevara 2023). Woodman's Small-eared Shrew, musaraña tropical de El Triunfo is monotypic and is known only from the cloud forests in El Triunfo Biosphere Reserve, Chiapas, México.

Characteristics: is the larger *Cryptotis* within the *goodwini* species complex (110–127 mm). Dorsal pelage dark brown dorsum; fur of the underparts slightly paler sooty black or blackish mouse gray and venter; broad manus with broad and elongate foreclaws. The skull is large with broad interorbital region, wide unicuspids and second upper molars, vestigial entoconid of third upper molar; third upper molar small and relatively simple, with well-pigmented parastyle, paracrista, paracone, and protocone; upper molars moderately pigmented; coronoid process of mandible high; anterior border of the coronoid process of the mandible joining horizontal ramus at a relatively low obtuse angle; articular process tall and wide, with a moderately broad lower articular facet (Guevara 2023). Species group *mexicanus*.

Comments: *Cryptotis woodmani goodwini* is restricted to southern highlands above 1200 m in Chiapas México (as *C. goodwini* map 255). It can be found in sympatry, or nearly so, with *C. griseoventris, C. parvus,* and *C. tropicalis*. This species can be distinguished from all other species of *Cryptotis* by its condylobasal length greater than 21.0 mm, its remarkably larger external body and very dark fur, broad forefeet, and long and broad foreclaws. From *C. griseoventris,* by having a deep interdenticular space between the first lower incisors; coronoid process height between 4.7 and 4.8 mm. From *C. parvus,* by having the first lower incisors with two crests; total length greater than 101.0 and condylobasal length greater than 21.0 mm. From *C. tropicalis,* by its condylobasal length greater than 21.0 mm; distance from the first upper unicuspid teeth to the third upper molars greater than 8.0 mm and dorsal pelage dark brown.

Additional literature: Jackson (1933), Choate and Fleharty (1974), Woodman (2005, 2010), Carraway (2007), Woodman and Stephens (2009), Woodman et al. (2012), Guevara et al. (2015).

Page 428. *Notiosorex crawfordi* (Coues 1877) from Mexican Tableland was considered as divergent species and describe as *Notiosorex carrawayae* (Camargo et al. 2024). Carraway's Desert Shrew, musaraña desértica del Altiplano Mexicano is monotypic and is known from south of the Conchos River in Durango and Zacatecas south through Hidalgo and México City. There are additional records of *Notiosorex* in Hidalgo, México.

Page 544. *Sylvilagus floridanus yucatanicus* (Miller 1899) from northwestern of Yucatán and Campeche was considered as full species and describe as *Sylvilagus yucatanicus* (Ruedas et al. 2023). Yucatán cottontail, conejo de Yucatán is monotypic and is known only from the Yucatán peninsula including Campeche.

Glossary

Abdomen The part of the body (excepting the back) between the thorax (rib basket) and the pelvis.
Accessory cusp A small cusp usually situated peripheral to main biting or crushing surface of a tooth.
Aerial Pertaining to flying. Bats are the only truly aerial mammals.
Alisphenoid channel A passage in the sphenoid bone of the skull leading from just anterior to the foramen lacerum in the middle cranial fossa to the pterygopalatine fossa.
Alisphenoid A bony process of the sphenoid bone; there is one on each side, extending from the side of the body of the sphenoid and curving upward, laterally, and backward.
Allopatric Mode that occurs when biological populations become geographically isolated from each other.
Alveolar (plural, alveoli) Small cavity or pit, as a socket for a tooth. Alveolar length of a tooth-row therefore denotes the length of the row of the teeth, taken from the posteriormost place where the back tooth emerges from the bone to the anteriormost point where the front tooth in the row emerges from the bone.
Alveolus See alveolar.
Ambulatory Pertaining to a walking habit, as in bears and raccoons.
Angular process On a mandible, a third process at the back of the mandible and at the ramus level.
Annulation A circular or ringlike formation, as of the dermal scales on the tail of a mammal where one ring of scales that extends entirely around the tail is succeeded, posteriorly, by other rings.
Antebrachial membrane Wing membrane of bats between the arm and the anterior edge of the wing.
Anterior palatine foramen See incisive foramen.
Anterocone A cusp at the front of the tooth that may be divided into anterolabial and anterolingual conules in an upper teeth.
Anteroconid A cusp at the front of the tooth that may be divided into anterolabial and anterolingual conulids in a lower teeth.
Anterolabial cingulum A crest before the protoconid and protoflexid in a lower teeth.
Anteroloph A crest between the paracone and the anterolabial conule that may be connecting to a parastyle in an upper teeth.
Anterolophid A crest between the metaconid and the anterolabial conulid that may be connecting to a metastylid in a lower teeth.
Antitragus The antitragus is located just above the earlobe and points anteriorly. It is separated from the tragus by the intertragic notch.
Antler Antlers are a single structure composed of bone, cartilage, fibrous tissue, skin, nerves, and blood vessels, shed and regrow each year.
Arboreal Inhabiting or frequenting trees-contrasted with fossorial, aquatic, and cursorial.
Articular process On a mandible, the process ends in the articular condyle.
Articulate To join or connect two adjacent bones.
Auditory bulla (plural, auditory bullae) A hollow, bony prominence of rounded form (in most mammals formed by the tympanic bone) partly enclosing structures of the middle and inner ear.
Auditory canal Bony tubular passage between tympanic membrane and external auditory meatus.
Auditory meatus (external) Pathway running from the outer ear to the middle ear.
Auditory meatus (internal) Pathway within the petrous part of the temporal bone of the skull between the posterior cranial fossa and the inner ear.
Baculum Bone found in the penis of many placental mammals.

S. T. Álvarez-Castañeda, *Mammals of North America - Volume 1*, https://doi.org/10.1007/978-3-031-41661-3

Basal length Distance on skull from the anteriormost inferior border of the foramen magnum to a line connecting the anteriormost parts of the premaxillary bones.
Basihyal Helps them make their loud vocalizations.
Basilar length Distance on skull from the anteriormost inferior border of the foramen magnum to a line connecting the posteriormost margins of the alveoli of the first upper incisors.
Basioccipital Unpaired bone at base of occipital complex.
Basisphenoid Part of the sphenoid bone which formed part of the floor of the braincase and lay immediately above.
Bead A salient, rounded cordlike projecting ridge of bone, as in certain rodents where the superior border of the orbit is beaded.
Bone Hard supportive tissue cons1sting of cells distributed in a matrix of fibrous protein (collagen) and salt (chiefly calcium and phosphate).
Brachydont Type of dentition characterized by low-crowned teeth.
Braincase The part of the skull enclosing the brain.
Bristle Stiff hair.
Browtine First tine above the base on an antler.
Bunodont Molars with low cusps and rounded hills, which are effective crushing devices and often basically quadrate in shape.
Calcar In bats a process connected with the calcaneum (heel bone), helping to support the edge of the fold of skin that extends between the leg and tail.
Canal Perforation, or foramen, that tends to be elongated into a tube. Examples are alisphenoid canal, infraorbital canal.
Cancellous Having a spongy or porous structure.
Canine One of four basic kinds of mammalian teeth; anteriormost in maxilla (and counterpart of dentary); frequently elongate, unicuspid, and single-rooted; never more than one per quadrant.
Caniniform Having the general shape of a canine.
Carnassial pair Pair of large bladelike teeth (last upper premolar and first lower molar) that occlude with scissorlike (shearing) action, possessed by most modem carnivores.
Carnassial Blade-like teeth especially adapted for slicing and chopping.
Carnivore An animal that preys on other animals; an animal that eats the flesh of other animals; especially any mammal of the order carnivora.
Cartilage Relatively soft supportive tissue consisting of rounded cells in a matrix of polysaccharides and fibrous protein (collagen).
Cheekteeth Teeth behind the canines can be premolar and molars.
Cingulum A shelf at the margin of a tooth.
Claw Curved, pointed appendage found at the end of a toe or finger.
Condylar (articular) process On a mandible, the process ends in the articular condyle.
Condylobasal length Least distance on skull from a line connecting the posteriormost projections of the exoccipital condyles to connecting the anteriormost projections of the premaxillary bones.
Condyloid process See articular process.
Conspecific Belong to the same species.
Coronoid process The upward projecting process of the posterior part of the mandible, giving attachment on its outward side to the masseter muscle and on its inner side to the temporal muscle.
Cranium Collectively, bones that form upper part of skull (contains upper teeth and braincase); lower part of skull is the mandible.
Crepuscular Active at twilight, i.e., at dusk or at dawn.
Cursorial Pertaining to a running habit, as in artiodactyls and perissodactyls.
Cusp A pointed structure on a tooth.
Cuspidate Presence of cusps on a tooth.
Cusplet (or secondary cusp) A small cusp.
Dactylopatagium Membranous structure that assists an animal in flight (bats), is the portion found within the digits.
Deciduous dentition (or milk teeth) Juvenile teeth, those that appear first in lifetime of a mammal, consisting (if complete) of incisors, canines, and premolars; generally replaced by adult (permanent) dentition.
Decurve To curve downward.
Deflected Bent outward or laterally.

Dental formula (plural, formulae) A brief method for expressing the number and kind of teeth of mammals. The abbreviations i. (incisor), c. (canine), p. or pm. (premolar), and m. (molar) indicate the kinds in the permanent dentition, and the number in each jaw is written like a fraction, the figures above the horizontal line showing the number in the upper jaw, and those below, the number in the lower jaw. Uppercase is used for upper teeth and lowercase for lower. The dental formula of an adult coyote is i. 3/3 c. 1/1; p. 4/4; m. 2/3. [d.i., d.c. And d.p. (in place of i., c. And p.) Designate deciduous ("milk") teeth.]
Dentary Bone of lower jaw, forming half of mandible.
Dentine A calcareous material, harder and denser than bone, which composes the principal mass of a tooth.
Dentition The teeth, considered collectively, of any animal.
Depth of skull at anterior margin of basioccipital Measured from anterior end of ventral face of basioccipital, excluding median ridge, vertically to dorsal face of parietal excluding sagittal crest.
Depth of skull at posterior borders of first upper molar Measured from ventral face of palatine bones at posterior edge of upper molars to dorsal face of frontals in plane of postorbital processes of frontals.
Dermal plates Bony structure derived from intramembranous ossification forming components of the vertebrate skeleton including much of the skull, jaws, shoulder girdle and fin spines rays (lepidotrichia), and the shell of armadillos. Name to as a dermal bone; investing bone; membrane bone.
Deuterocone One of the cusps of a premolar tooth of a mammal corresponding in position (anteromedial) to the protocone of a true molar.
Dewclaws (or dewhoofs) Clawed or hoofed remnants of side toes in many carnivores and artiodactyls, located just above the main functional digits.
Diastema A vacant place or gap between teeth in a jaw.
Digital web Membranes between the fingers.
Digitigrade Walking on the toes with the heel and wrist permanently raised.
Digits Toes.
Dilambdodont Molars having a distinct ectoloph, shaped like two lambdas or a W shape.
Distribution range The geographic limit of a particular taxon's distribution is its range.
Diurnal Active by day-opposed to nocturnal.
Ectoloph A crest connecting the ectostyle to the mesocone in a upper teeth.
Ectolophid A crest connecting the ectostylid to the mesoconid in a lower teeth.
Ectostyle A stylid between the protocone and the hypocone, in the hypoflexe in an upper teeth.
Ectostylid A stylid between the protoconid and the hypoconid, in the hypoflexid in a lower teeth.
Enamel Hardest substance of the mammalian body and forming a thin layer that caps or partly covers a tooth.
Enamel plate A segment or portion of a tooth that is heavily invested with enamel (e.g., on pocket gopher teeth).
Endemic State of a species being found in a single defined geographic location.
Enteroloph A crest connecting the enterostyle to the mesocone in an upper teeth.
Enterostyle A style between the protocone and the hypocone, in the hypoflexus in an upper teeth.
Entoconid One of the main cusps, at the posterolingual side in a lower teeth.
Entolophid A crest attaching the entoconid to the hypoconid or median murid in a lower teeth.
Ethmoid Unpaired bone in the skull that separates the nasal cavity from the brain
External auditory meatus Round bony oriface or opening that is covered by eardrum.
Faeces (singular and plural). Intestinal excrement.
Feces (see faeces).
Femur (plural, femora). The proximal bone of the hindlimb.
Fenestrated Having a network of irregular perforations or holes.
Fenestration Opening; in current work applied to specialized openings in crania of lagomorphs and cervids.
Flange A laterally compressed or flattened portion of bone that increases the surface area.
Flipper A forelimb of an aquatic animal.
Foramen (pl. foramina) A hole, opening, or perforation through bone. An example is the foramen magnum.
Foramen lacerum A hole in the base of the skull located between the sphenoid bone, the apex of the petrous part of the temporal bone, and the basilar part of the occipital bone.
Foramen magnum The large opening in the back of a skull through which the spinal cord passes to become the medulla oblongata of the brain.
Foramen An open hole that is present in a bone, typically allow muscles, nerves, arteries, veins, or other structures to connect one part of the body with another.
Foramens Plural foramen.
Forearm The part of the forelimb between the elbow and wrist.

Forelimb Paired articulated appendages (limbs) attached on the cranial (anterior) end of a terrestrial tetrapod vertebrate's torso.

Formina Plural foramen.

Fossa (pl. fossae) A depression, usually forming a site of muscular attachment or bone articulation. Examples are glenoid fossa, mandibular fossa, masseteric fossa, temporal fossa.

Fossil Any preserved remains, impression, or trace of any once-living thing from a past geological age.

Fossorial Fitted for digging.

Frontal Pertaining to or designating the bone (paired) immediately in front of the parietal bone and behind the nasal.

Frontal appendages Bone growth (horns or antlers) arising from frontal bones.

Gestation period The period of carrying young in the uterus, as applied to placental mammals; the period of pregnancy.

Gland Synthesizes substances (such as hormones) for release into the bloodstream (endocrine gland) or into cavities inside the body or its outer surface (exocrine gland).

Glandular sac A sac which contains a gland.

Glenoid fossa A shallow, pyriform articular surface, which is located on the lateral angle of the scapula and articulates with the head of the humerus.

Guard-hairs The stiffer, longer hairs that grow up through the limber, shorter hairs (fur) of a mammal's pelage.

Habitat The kind of environment in which a species of organism is normally found.

Hallux Used to refer to the big toe of the hindfeet.

Hamular process of pterygoid A hook like process on the pterygoid bone.

Heterodont A dentition in which there are teeth of different forms (e.g., incisors and canines).

Heterothermic Having a variable body temperature.

Hibernation Torpidity especially in winter; the bodily temperature approximates that of the surroundings; the rate of respiration and the heart beat ordinarily are much slower than in an active mammal.

Hindlimb One of the paired articulated appendages (limbs) attached on the caudal (posterior) end of a terrestrial tetrapod.

Homodont Animal which possesses more than a single tooth morphology.

Hoof (pl. hoofs or hooves) Tip of a toe of an ungulate mammal, strengthened by a thick and horny keratin covering.

Horizontal ramus Lower jaw, the ramus baring the teeth, and anterior to the vertical ramus.

Horn A pointed, bony projection on the head of various animals, either the "true" horn, or other horn-like growths.

Hybrid An offspring resulting from cross-breeding.

Hyoid bone Helps them make their loud vocalizations.

Hypocone One of the main cusps, at the posterolingual side in an upper teeth.

Hypoconid One of the main cusps, at the posterolabial side in a lower teeth.

Hypsodont Dentition is characterized by high-crowned teeth and enamel that extends far past the gum line, which provides extra material for wear and tear.

Incisiform Having the general form of an incisor.

Incisive foramen (also referred to as anterior palatine foramen or palatine slit).

Incisive foramina The anterior palatine foramina (singular, foramen), of which there are two, in the bony roof of the anterior part of the cavity of the mouth at the juncture of the premaxillary bones and maxillary bones; transmit nasal branches of palatine arteries and nasopalatine ducts of jacobson.

Incisor Pertaining to or designating one of the teeth in front of the canine tooth; those in the upper jaw invariably are in the premaxillary bone.

Inflated Enlarged or expanded, not flattened or compressed.

Inflected Bent inward or medially. insertion – the site of attachment of a muscle (usually on a bone) on the more movable of the two elements, or bones, that are joined by the muscle.

Infraorbital canal A canal through the maxillary bone from the orbit to the face.

Infraorbital foramen Opening in maxilla from orbit onto the face (rostrum of cranium).

Inguinal Pertaining to or in the region of the groin.

Insectivorous Eating insects; preying or feeding on insects.

Interfemoral membrane In a bat the fold of skin stretching from hind legs to tail. The uropatagium.

Interorbital breadth Least distance across top of skull between orbits (eye sockets).

Interorbital constriction The least distance across the top of the skull between the orbits (eye sockets).

Interorbital region The region between the eye sockets; the region of the skull between the rostrum and the braincase.

Interparietal Pertaining to or designating the bone (rarely paired) immediately in front of the supraoccipital bone and between the two parietal bones.

Interpterygoidea fossa It is located in the mid-ventral part of the skull, posterior to the bony palate. Variations in it are important for the identification of some species.

Jugal Often called the malar or zygomatic. It is connected to the quadratojugal and maxilla, as well as other bones, which may vary by species.
Jugo-maxillary suture Juncture between jugal and maxillary bones at anterior margin of zygomatic arch.
Jugular foramen See posterior lacerate foramen.
Karyotype A preparation of the complete set of metaphase chromosomes in the cells of a species or in an individual organism, sorted by length, centromere location and other features.
Labial Pertaining to lips, for example labial side of tooth is that side nearer lips rather than tongue; lateral surface of a tooth.
Lacerum posterior Triangular hole in the base of skull, located between the sphenoid bone, the apex of the petrous part of the temporal bone, and the basilar part of the occipital bone.
Lachrymal Small and fragile bone of the facial skeleton, situated at the front part of the medial wall of the orbit.
Lacrimal See lachrymal.
Lacrimal pit Pit or opening in lacrimal bone containing tear duct.
Lambdoidal ridge Bony ridge formed at juncture of occiput and parietal bones.
Lamboidal crest See lambdoidal ridge
Lingual Pertaining to tongue, for example, lingual side of tooth is that side nearer tongue rather than lips, also medial surface of a tooth.
Litter The two or more young brought forth at one birth by a female mammal.
Loph A combining form used as the terminal part of certain words and denoting the ridges (or areas) composed of several cusps and styles on the occlusal face of a tooth, as protoloph.
Lophodont Teeth with differentiating patterns of ridges or lophs of enamel interconnecting the cusps on the crowns.
Lyre-shaped Shaped as the lyre, a musical instrument.
Malar See jugal.
Mamma The glandular organs for secreting milk.
Mammae (singular, mamma) The glandular organs for secreting milk.
Mandible tooth-row The row of teeth in one mandible bone; in most mammals all the premolars and molars on one side of the lower jaw.
Mandible Movable part of the jaw.
Mandibular fossa (or glenoid fossa) Concavity on ventral surface of zygomatic arm of squamosal with which the dentary articulates.
Marsupium The pouch is a fold of skin with a single opening that covers the teats in marsupials. Inside the pouch, the blind offspring attaches itself to one of the mother's teats and remains attached for as long as it takes to grow and develop to a juvenile stage.
Mask Coloration pattern in the rostrum of the species contrasting with the general color of the head.
Masseter muscle A muscle often consisting of several bands that originates on and adjacent to the zygomatic arch and inserts mostly in the masseteric fossa of the lower jaw.
Mastoid breadth Greatest distance across mastoid bones perpendiciJar to long axis of skull.
Mastoid bulla The part of bullar region covered by mastoid (mastoidal) bone.
Mastoid Designating or pertaining: to the mastoid bone (paired) or its process. This bone is bounded by the squamosal bone, the exoccipital bone, and the tympanic bone.
Maxilla Bone of the jaw formed from the fusion of two maxillary bones.
Maxillary breadth Width of skull from some designated place on the lateral face of the right maxillary bone (maxilla) to the corresponding place on the left maxillary bone; in shrews, across the ends of the zygomatic processes of the two maxillary bones.
Maxillary tooth-row The row of teeth in one maxillary bone; in most mammals all the premolars and molars on one side of the upper jaw.
Mesaxonic Having the axis of the foot formed by the middle digit (relating to the Perissodactyla).
Mesocone A conule in the median mure where the mesoloph is attached to it in an upper teeth.
Mesoconid A conulid in the median murid where the mesolophid is attached to it in a lower teeth.
Mesolophid A crest in front of the entoflexid, connected to the median murid.
Mesopterygoid fossa Shallow area posterior to internal nares and between pterygoid bones.
Mesostyle A style at the labial margin between the paracone and metacone in a upper teeth.
Metabolic water Water formed as an end product of combustion of foodstuffs in an animal's body.
Metacarpal Bone of the hand or forefoot between the wrist and fingers; when all the digits are present there are five more or less elongated metacarpal bones, one at the base of each digit.
Metacone One of the main cusps, at the posterolabial side in an upper teeth.

Metaconid One of the main cusps, at the anterolingual side in a lower teeth.
Metastylid A stylid in front of the metaconid in a lower teeth.
Milk teeth See deciduous dentition.
Miocene See geologic time.
Globular Globe-like, spherical.
Molar Teeth behind the premolar teeth, used primarily to grind food during chewing.
Molariform Molar-like teeth, either premolars or true molars.
Molt (moult) In a mammal, the act or process of shedding or casting off the hair, or outer layer of skin or horns; most mammals shed the hair once, twice, or three times annually. The cast off covering (obsolete). As a verb: to be shed (intransitive) or to shed (transitive).
Muzzle Projecting parts of the face (including the nose and mouth) of an animal.
Nail Keratinized projection at the tips of the digits; usually short, flat, and blunt.
Nares Openings, external and internal, of the nasal passage.
Nasal (paired) On the dorsal surface of the skull at its anterior end, form the bridge of the upper one third of the nose.
Nasal leaf Often large, lance-shaped nose, found in bats.
Nasal Paired bone of cranium situated on anterodorsal surface of skull.
Nocturnal Active by night-opposed to diurnal.
Nostril Either of the two orifices of the nose.
Occipital condyle Surface of articulation between cranium and first cervical vertebra (atlas); two such condyles in mammals, on either side of foramen magnum.
Occipital crest Bony ridge formed at occipital bones.
Occipital Cranial bone and the main bone of the back and lower part of the braincase.
Occipitonasal length Least distance between two vertical lines, one touching the posteriormost part of the skull above the foramen magnum (opening for the spinal cord) and the other touching the anteriormost part of the nasal bones or a nasal bone.
Occlusal Of or pertaining to the grinding or biting (occluding) surface of a tooth.
Opposable Capable of being placed opposite something else; said of the first toe of an opossum in the sense that it can be placed opposite each of the other toes on that same foot.
Orbit The cavity in the skull in which the eye and its appendages are situated; the eye socket.
Orbital ring Circular spot around the eye that generally contrasts with the rest of the rostrum coloration.
Orbitonasal length Distance on anterior part of skull from posterior margin of base of postorbital process of frontal bone to posteriormost part of anterior border of nasal bone on same side of skull.
Origin The site of attachment of a muscle (usually on a bone) on the less movable of the two elements, or bones, that are joined by the muscle.
Os clitoris Bone found in the clitoris of many placental mammals.
Os penis Bone found in the penis of many placental mammals.
Osseous Composed of, or resembling, bone; bony. Osseous tissue is bony tissue.
Overhairs The longer hairs of the pelage of a mammal that project above the fur (shorter hairs).
Palatal bridge Bony tooth-bearing plate (somewhat raised, part of paired maxillae) behind diastema, especially well developed in lagomorphs and rodents.
Palatal emargination Indentation at anterior margin of rostrum (in premaxillae), particularly in bats.
Palatal Bony roof of the mouth, made up of two palatine bones, two maxillary bones, and two premaxillary bones.
Palate The roof of the mouth, consisting of the structures that separate the mouth from the nasal cavity. The bony palate is composed of the following bones: premaxillae, maxillae, and palatines.
Palatine slit See incisive foramen.
Palmated Form like the palm of a hand; term applied to antlers in which at least some spaces between tines are filled with bony growth.
Paracone One of the main cusps, at the anterolabial side in an upper teeth.
Parastyle A style in front of the paracone in an upper teeth.
Paraxonic Having the axis of the foot between the third and fourth digits
Parietal Pertaining to or designating the parietal bone (paired) roofing the braincase. This bone is behind the frontal bone and in front of the occipital bones.
Paroccipital process Bony projection extending ventrally from, or located ventrally on, paroccriptal bone.
Patagium Membranous structure that assists an animal in gliding or flight and stretches between an animal's hind limbs is called the uropatagium or the interfemoral membrane.

Pectoral Pertaining to, or situated or occurring in or on, the chest.
Pelage Hair, fur.
Pencil Tuft of fur or hair, as a black pencil on the end of the tail of a mammal.
Penis bone Bone found in the penis of many placental mammals.
Pentadactyl Five toed.
Perforate Pierced by an opening or openings.
Petrous Part of the temporal bone is pyramid-shaped and is wedged in at the base of the skull between the sphenoid and occipital bones.
Phalanx (plural, phalanges) A bone, in a finger, distal to the metacarpus or a bone, in a toe, distal to the metatarsus.
Pinna (plural, pinnae) The projecting part of an ear.
Plagiopatagium Membranous structure that assists an animal in flight (bats), is the portion found between the last digit and the hindlimbs.
Plantar tubercles Glandular and swollen areas found on the mammals soles. The morphological relationships of such integumentary annexes is a taxonomic characteristic.
Plantigrade Walking with the toes and metatarsals flat on the ground.
Pollex Used to refer to the big toe of the forefeet.
Population A group of organisms of the same species who inhabit the same particular geographical area and are capable of interbreeding.
Postauricular Situated behind the auricle (pinna) of the ear, as a postauricular patch (ordinarily referring to a patch of fur differing in color from surrounding fur).
Postcanine teeth Collectively, teeth behind canines (premolars and molars).
Posterior lacerate foramen (or jugular foramen) Opening in basicranial region between tympanic bulla and basioccipital.
Posterior palatine foramen Small opening in hard palate near juncture of maxillae and palatines.
Posteroloph A crest at the back of the molar, connected to the hypocone in an upper teeth.
Posterostylid A crest on the posterolingual corner of the molar in a lower teeth.
Postorbital Situated behind the eye, as the postorbital process of the frontal bone or postorbital process of the jugal bone.
Postorbital bar Complete dorsoventral connection (bar) of bone posterior to orbit; results from fusion of postorbital process of frontal and postorbital process of jugal.
Postorbital process of frontal Bony projection of frontal bone posterior to orbit.
Postorbital process of jugal Bony projection of jugal bone posterior to orbit.
Pouch Fold of skin with a single opening that covers the teats in marsupials. Inside the pouch, the blind offspring attaches itself to one of the mother's teats and remains attached for as long as it takes to grow and develop to a juvenile stage.
Preglenoid crest Bony ridge or shelf on anterior part of glenoid (mandibular) fossa, especially well developed in camivores.
Prehensile Quality of an appendage or organ that has adapted for grasping or holding.
Premaxilla (or premaxillary) Paired bone in anterior of cranium; point of origin of upper incisors (when present).
Premaxillary Bone (paired) in the mammalian skull bearing the incisor teeth of the upper jaw; the premaxilla is situated in front of the maxilla.
Premolar Designating or pertaining to one of the teeth (a maximum of four on each side of upper jaw and lower jaw of mammals), in front of the true molars.
Prismatic Cheekteeth (especially of rodents) with well-developed triangles (or prisms) of enamel surrounding basins of dentine.
Proboscis Elongated nose or snout.
Process Small bony projection.
Procumbent Horizontally protruding condition of the incisors in some mammals, notably many marsupials, insectivores, and primates.
Propatagium Membranous structure that assists an animal in flight (bats), is the patagium present from the neck to the first digit.
Protocone One of the main cusps, at the anterolingual side in an upper teeth.
Protoconid One of the main cusps, at the anterolabial side in a lower teeth.
Protostyle A style in front of the protocone, in the protoflexus in an upper teeth.
Protostylid A stylid in front of the protoconid, in the protoflexid in a lower teeth.
Pterygoid Paired bone on ventral surface of cranium, posterior to palatine and anterior to alisphenoid; forms border of internal nares.
Pterygoid canal A passage in the sphenoid bone of the skull leading from just anterior to the foramen lacerum in the middle cranial fossa to the pterygopalatine fossa.

Quadrant One-fourth of the total complement of teeth; one side of upper or one side of lower jaw.
Re-entrant angle An infold of the enamel layer on the side, front, or back of a cheektooth, as in a molar of a muskrat or wood rat.
Rhinarium Furless skin surface surrounding the external openings of the nostrils in many mammals.
Rosette French diminutive of rose. For example markings like those of a jaguar.
Rostrum Part of the skull projecting in front of the orbits.
Rostrum breadth Least distance from lateral base of hamular process of lacrimal bone to corresponding point on opposite side of skull.
Ruminant Ungulate with specialized four-chambered digestive system; cud-chewing mammals.
Rut The breeding period, as in deer.
Sagittal crest The ridge of bone at the juncture of the two parietal bones resulting from the coalescence of the temporal ridges; in old individuals of many species of mammals the crest extends from the middle of the lambdoidal crest anteriorly onto the frontal bones and divides there into two temporal ridges, each of which extends anterolaterally on the posterior edge of the postorbital process of the frontal bone.
Saliva The fluid secreted by the glands discharging into the mouth.
Saltatorial Pertaining to hopping, as in kangaroo rats.
Scansorial Pertaining to mammals that climb by use of claws, such as squirrels and cats.
Scapula Bone that connects the humerus (upper arm bone) with the clavicle (collar bone).
Scutes Flat bony plates of dermal tissue covered by epidermis forming the outer shell of armadillos.
Secodont Blade-like teeth especially adapted for slicing and chopping, called carnassials.
Secondary cusp See cusplet.
Selenodont Cusp pattern of molars in which individual cusps are cresent shaped; highly developed in ungulates.
Selenodont Teeth with the major cusp elongated into a crescent-shaped ridge.
Semiprismatic Cheekteeth (especially of rodents) with partially closed (slight re-entrant angles) triangles (or prisms) of enamel surrounding dentine.
Septum A wall, dividing a cavity or structure into smaller ones.
Sesamoid A bone formed in a tendon.
Sexual dimorphism Conditions where the sexes of the same species exhibit different characteristics, particularly characteristics not directly involved in reproduction. Differences may include secondary sex characteristics, size, weight, color, markings, or behavioral or cognitive traits.
Skull Cranium plus mandible.
Snout Projecting parts of the face (including the nose and mouth) of an animal.
Sphenoid Unpaired bone of the neurocranium. It is situated in the middle of the braincase toward the front, in front of the basilar part of the occipital bone.
Squamosal Bones form the cheek series of the skull, bordered anteroventrally by the jugal and ventrally by the quadratojugal.
Subauricular spot A spot, patch of hair, distinctively colored immediately below the ear.
Subgenus Taxonomic rank directly below genus.
Supraorbital process of frontal The process of the frontal bone on the top rim of the orbit, as in a rabbit.
Supraorbital process Bony projection above orbit on frontal bone; especially well developed in lagomorphs.
Supraorbital shelf Small bony ridge on dorsal margin of orbit on frontal and parietal bones.
Suture Fairly rigid joint between two or more hard elements of an organism, with or without significant overlap of the elements.
Suture Point of contact (or juncture) and fusion between adjacent bones.
Sympatry Two related species or populations which exist in the same geographic area and thus frequently encounter one another.
Syndactylous A condition in which two or more digits are bound together in a common tube of skin. The underlying bones remain distinct.
Talon (id) A posterior "tail" or expansion on an upper (lower) cheektooth which produces a square outline to the tooth and expands the crushing surface.
Talonid Crushing heel toward the rear.
Tarsus The ankle.
Temporal muscle The muscle that originates on the posterodorsal and lateral portions of the braincase and inserts chiefly on the coronoid process of the lower jaw. It is especially large in carnivores.

Temporal ridge (paired) A curved, raised line on the side of the braincase marking the upper limit of attachment of the fascia of the temporal muscle. The temporal ridge is prominent on the parietal bone, frequently extends forward onto the frontal bone, and in some kinds of mammals extends backward onto the interparietal bone. When present, the sagittal crest is formed by the coalescence of the two temporal ridges.

Terrestrial Inhabiting the land, rather than the water, trees or air.

Tibia (plural, tibiae) The inner and usually the larger of the two bones of the hindlimb (leg) between the knee and the ankle.

Tooth-row Row of teeth in one mandible or maxilla bone; in most mammals all the premolars and molars on one side of the lower or upper jaw.

Torpid Having lost most of the power of exertion; dormant. A ground squirrel is torpid when it is hibernating.

Tragus Small pointed eminence of the outer ear.

Tricolor Having three colors. Said of hair on the back of mammals when hair has three bands each of a different color.

Trigonid Shearing ends toward the front of the jaw.

Trochanter Tubercle of the femur near its joint with the hip bone.

Trochater Tubercle of the femur near its joint with the hip bone.

Truncated Abruptly or sharply marked, having a square or broad end; sometimes appearing as cut off.

Tuberculo-sectorial A tritubercular tooth with sharp cutting edges (cusps).

Tympanic bulla See auditory bulla.

Tympanic bulla breadth From bottom of pit immediately posterior to external auditory meatus to medial face of bulla at right angle with longitudinal axis of skull.

Tympanic bulla depth Least distance from ventral face of basioccipital, excluding median ridge, to line touching ventral-most points of the two bullae.

Tympanic bulla length From posterior face to most anterior part of anterior border.

Tympanum Of, or pertaining to, bony ring, as in shrews, that does not form a complete bulla.

Type locality The place where a type specimen was obtained.

Ulna Long bone found in the forearm that stretches from the elbow to the smallest finger, and when in anatomical position, is found on the medial side of the forearm.

Underfur The short hair of a mammal; in temperate and boreal climates the underfur ordinarily is denser, made up of more hairs, than the longer and coarser overhair.

Underparts Underneath (ventral) side of a mammal (not the back or sides), as of a woods mouse with white underparts.

Ungulates Hoofed mammals, such as artiodactyls.

Unguligrade Walking on the nail or nails of the toes (hoof) with the heel/wrist and the digits permanently raised.

Unicuspid With single well-developed cusp.

Upper parts The top (dorsal) surface and all of the sides (not the belly, chest, or throat), as of a woods mouse with reddish-brown upper parts.

Uropatagium The interfemoral membrane of a bat; that is to say, the fold of skin that stretches from the hind legs to the tail.

Uterus Main female hormone-responsive, secondary sex organ of the reproductive system in most mammals.

Vibrissa Type of stiff, functional hair used by animals to sense their environment.

Vomer Unpaired facial bones of the skull located in the midsagittal line and articulates with the sphenoid, the ethmoid, the left and right palatine bones, and the left and right maxillary bones.

Whiskers Type of stiff, functional hair used by animals to sense their environment.

Zygomatic arch In the skull formed by the zygomatic process of the temporal bone and the temporal process of the zygomatic bone, the two being united by an oblique suture.

Zygomatic breadth Greatest distance across zygomatic arches of cranium at right angles to the long axis of skull.

Zygomatic notch A small notch in the zygomatic plate, a bony plate derived from the flattened front part of the zygomatic arch.

Zygomatic plate Platelike extension of maxilla in anterior part of zygoma.

References

Abramov, A. V. 1999. A taxonomic review of the genus *Mustela* (Mammalia, Carnivora). Zoosystematica Rossica 8:357–364.

Abramov, A.V. 2000a. The taxonomic status of the Japanese weasel, *Mustela itatsi* (Carnivora, Mustelidae). Zoologicheskii Zhurnal 79:80–88.

Abramov, A. 2000b. A taxonomic review of the genus *Mustela* (Mammalia, Carnivora). Zoosystematica Rossica 8:357–364.

Abramov, A.V. 2016. *Gulo gulo*. The IUCN Red List of Threatened Species 2016:e.T9561A45198537.

Abramson, N. I., and A. A. Lissovsky. 2012. Subfamily Arvicolinae. Pp. 127–141, *in* The mammals of Russia: A taxonomic and geographic reference (Pavlinov, I. Y., and A. A. Lissovsky, eds.). Moscow: KMK Scientific Press.

Abril, V. V., E. A. G. Carnelossi, S. González, and J. M. B Duarte. 2010. Elucidating the evolution of the red brocket deer *Mazama americana* complex (Artiodactyla, Cervidae). Cytogenetic and Genome Research 128:177–187.

Adams, D. R., and D. A. Sutton. 1968. A description of the baculum and os clitoris of *Eutamias townsendii ochrogenys*. Journal of Mammalogy 49:764–768.

Adams, J. K. 1989. *Pteronotus davyi*. Mammalian Species 346:1–5.

Adam, M. D., and J. P. Hayes. 1998. *Arborimus pomo*. Mammalian Species 593:1–5.

Adam, P. J. 2004. *Monachus tropicalis*. Mammalian Species 747:1–9.

Agnarsson, I., M. Kuntner, and L. J. May-Collado. 2010. Dogs, cats, and kin: A molecular species-level phylogeny of Carnivora. Molecular Phylogenetics and Evolution 54:726–745.

Aguirre, L., H. Mantilla, B. Miller, and L. Dávalos. 2015. *Chiroderma salvini*. The IUCN Red List of Threatened Species 2015:e.T4666A22037356.

Akins, J. B., M. L. Kennedy, G. D. Schnell, C. Sánchez-Hernández, M. de la L. Romero-Almaraz, M. C. Wooten, and T. L. Best. 2007. Flight speeds of three species of Neotropical bats: *Glossophaga soricina, Natalus stramineus,* and *Carollia subrufa*. Acta Chiropterologica 9:477–482.

Alexander, L. F. 1996. A morphometric analysis of geographic variation within *Sorex monticolus* (Insectivora: Soricidae). Miscellaneous Publications of the University of Kansas Natural History Museum 88:1–54.

Alexander, L. F., and B. J. Verts. 1992. *Clethrionomys californicus*. Mammalian Species 406:1–6.

Allen, G. M. 1916. Bats of the genus *Corynorhinus*. Bulletin of the Museum of Comparative Zoology 60:331–356.

Allen, G. M. 1919a. Notes on the synonymy and nomenclature of the smaller spotted cats of tropical America. Bulletin of the American Museum of Natural History 41:341–419.

Allen, G. M. 1919b. The American collared lemmings. Bulletin Museum Comparative Zoology 62:509–543.

Allen, G. M. 1930. The walrus in New England. Journal of Mammalogy 11:139–145.

Allen, G. M. 1942. Extintct and vanishing mammals of the western hemisphere. American Committee for International Wildlife Protection, Special Publication 11:1–61.

Allen, J. A. 1890. A review of some of the North American ground squirrels of the genus *Tamias*. Bulletin of the American Museum of Natural History 3:45–116.

Allen, J. A. 1893. On a collection of mammals from San Pedro Martir region of Lower California. Bulletin American Museum Natural History 5:181–202.

Allen, J. A. 1897. Description of a new vespertilionine bat from Yucatan. Bulletin American Museum of Natural History 9:231–232.

Allen, J. A. 1898. Descriptions of new mammals from western Mexico and Lower California. Bulletin of the American Museum of Natural History 10:143–158.

Allen, J. A. 1900. List of bats collected by Mr. H. H. Smith in the Santa Marta region of Colombia, with descriptions of new species. Bulletin of the American Museum of Natural History 13:87–94.

Allen, J. A. 1901. A preliminary study of the North American opossum of the Genus *Didelphis*. Bulletin of the American Museum of Natural History 14:149–188.

Allen, J. A. 1903. List of mammals collected by Mr. J. H. Batty in New Mexico and Durango, with descriptions of new species and subspecies. Bulletin of the American Museum of Natural History 19:587–612.

Allen, J. A. 1906. Mammals from the States of Sinaloa and Jalisco, Mexico, collected by J. H. Batty during 1904 and 1905. Bulletin of the American Museum of Natural History 22:191–262.

Allen, J. A. 1913. Ontogenetic and other variations in muskoxen, with a systematic review of the muskox group, recent and extinct. Memoryes of the American Museum of Natural History, new series 1:101–226.

S. T. Álvarez-Castañeda, *Mammals of North America - Volume 1*, https://doi.org/10.1007/978-3-031-41661-3

Allen, J. A. 1915a. Review of the south american Sciuridae. Bulletin of the American Museum of Natural History 34:147–309.
Allen, J. A. 1915b. Notes on American deer of the genus *Mazama*. Bulletin of the American Museum of Natural History 34:521–553.
Allen, J. A. 1919. Notes on the synonymy and nomenclature of the smaller spotted cats of tropical America. Bulletin of the American Museum of Natural History 1:341–419.
Allen, G. M. 1937. Sex dichromatism in Noctilio. Journal of Mammalogy18: 514.
Allen, J. A. 1939. Bats. Harvard University Press. Cambridge, U.S.A.
Allen, P. J. 2004. *Monachus tropicalis*. Mammalian Species 747:1–9.
Almazán-Catalán, J. A., A. Taboada-Salgado, C. Sánchez-Hernández, M. de la L. Romero-Almaraz, Q. Jiménez-Salmerón, and E. Guerrero-Ibarra. 2009. Registros de murciélagos para el estado de Guerrero, México. Acta Zoológica Mexicana (n. s.) 25:177–185.
Almendra, A. L., D. S. Rogers, and F. X. González-Cózatl. 2014. Molecular phylogenetics of the *Handleyomys chapmani* complex in Mesoamerica. Journal of Mammalogy 95:26–40.
Almendra, A. L., F. X. González-Cózatl, M. D. Engstrom, and D. S. Rogers. 2018. Evolutionary relationships and climatic niche evolution in the genus *Handleyomys* (Sigmodontinae: Oryzomyini). Molecular Phylogenetics and Evolution 128:12–25.
Alonso-Mejía, A., and R. A. Medellín. 1991. *Micronycteris megalotis*. Mammalian Species 376:1–6.
Alonso-Mejía, A., and R. A. Medellín. 1992. *Marmosa mexicana*. Mammalian Species 421:1–4.
Álvarez, J., M. R. Willig, J. K. Jones, Jr. and W. D. Webster. 1991. *Glossophaga soricina*. Mammalian Species 379:1–7.
Álvarez, T. 1960. Sinopsis de las especies mexicanas del género *Dipodomys*. Revista de la Sociedad Mexicana de Historia Natural 21:391–424.
Álvarez, T. 1961. Taxonomic status of some mice of the *Peromyscus boylii* group in eastern Mexico, with description of a new subspecies. University of Kansas Publications of the Museum of Natural History 14:111–120.
Álvarez, T. 1963. The recent mammals of Tamaulipas, Mexico. University of Kansas Publications, Museum of Natural History 14:363–473.
Álvarez, T. 1968. Notas sobre una coleccion de mamíferos de la región costera del Río Balsas entre Michoacán y Guerrero. Revista de la Sociedad Mexicana de Historia Natural 29:21–35.
Alvarez, T., and C. E. Aviña. 1965. Baedon [sic] alleni, Rhogeessa tumida major and R. p. parvula newly reported for Michoacan, with notes on the qualitative differentiation of the two Rhogeessas. Southwestern Naturalist 10:69–78.
Álvarez, T., and S. T. Álvarez-Castañeda. 1990. Cuatro nuevos registros de murciélagos (Chiroptera) del estado de Chiapas. Anales de la Escuela Nacional de Ciencias Biológicas 55:157–161.
Álvarez, T., and S. T. Álvarez-Castañeda. 1991. Análisis de la fauna de roedores del área de El Cedral, San Luis Potosí México. Anales del Instituto de Biología, Universidad Nacional Autónoma de México, Serie Zoología 62:169–180.
Álvarez, T., and S. T. Álvarez-Castañeda 1996. Aspectos biológicos y ecológicos de los murciélagos de Ixtapan del Oro, México. Pp. 169–182, *in* Contribution in Mammalogy: a memorial volume Honoring Dr. J. Knox Jones, Jr. (Genoways, H. H., and R. Baker, eds.). Special Publications, Museum Texas Tech University. Lubbock, U.S.A.
Álvarez, T., and C. E. Aviña. 1964. Nuevos registros en México de la familia Molossidae. Revista de la Sociedad Mexicana de Historia Natural 25:243–254.
Álvarez, T., and L. González-Quintero. 1970. Análisis polínico del contenido gástrico de murciélagos Glossophaginae de México. Anales de la Escuela Nacional de Ciencias Biológicas 18:137–165.
Álvarez, T., and N. González-Ruíz. 2000. Variación geográfica de *Saccopteryx bilineata* (Chiroptera: Emballonuridae) en México, con descripción de una nueva subespecie. Anales de la Escuela Nacional de Ciencias Biológicas 46:305–316.
Álvarez, T., and J. J. Hernández-Chávez. 1990. Cuatro nuevos registros del ratón de campo *Peromyscus* (Rodentia: Muridae) en el Estado de México, México. Anales de la Escuela Nacional de Ciencias Biológicas 33:163–173.
Álvarez, T., and J. J. Hernández-Chávez. 1993. Taxonomía del metorito *Microtus mexicanus* en el centro de México con la descripción de una nueva subespecie. Pp. 137–156, *in* Avances en el estudio de los mamíferos de México (Medellin, R. A., and G. Ceballos, eds.). Publicaciones Especiales. Asociación Mexicana de Mastozoología, Distrito Federal, México.
Álvarez, T., and A. Ocaña. 1999. Sinopsis de restos arqueozoológicos de vertebrados terrestres. Basada en informes del Laboratorio de Paleozoología del INAH. Instituto de Antropología e Historia, Colección Científica 386:1–108.
Álvarez, T., and O. Polaco. 1984. Estudio de los mamíferos capturados en La Michilía, sureste de Durango, México. Anales de la Escuela Nacional de Ciencias Biológicas 28:99–148.
Álvarez, T., and J. Ramírez-Pulido. 1968. Descripción de una nueva subspecie de *Spermophilus adocetus* (Rodentia, Sciuridae) de Michoacán, México y estado taxónomico de S. *a. arceliae* (Villa R., 1942). Revista de la Sociedad Mexicana de Historia Natural 29:181–189.
Álvarez, T., and J. Ramírez-Pulido. 1972. Notas acerca de murciélagos mexicanos. Anales de la Escuela Nacional de Ciencias Biológicas 19:167–178.
Álvarez, T., and N. Sánchez-Casas. 1997a. Notas sobre la alimentación de *Musonycteris* y *Choeroniscus* (Mammalia: Phyllostomidae) en México. Revista Mexicana de Mastozoología 2:113–115.
Álvarez, T., and N. Sánchez-Casas. 1997b. Contribución al conocimiento de los mamíferos, excepto Chiroptera y Rodentia, de Michoacán, México. Anales de la Escuela Nacional de Ciencas Biológicas 42:47–74.
Álvarez, T., S. T. Álvarez-Castañeda, and J. C. López-Vidal. 1994. Claves para los murciélagos de México. Publicación Especial, Centro de Investigaciones Biológicas de Baja California Sur y Escuela Nacional de Ciencias Biológicas, Instituto Politécnico Nacional. La Paz, México.
Álvarez, T., J. Arroyo-Cabrales, and M. González-Escamilla. 1987. Mamíferos (excepto Chiroptera) de la costa de Michoacán. Anales de la Escuela Nacional de Ciencias Biologicas 31:13–62.
Álvarez-Castañeda, S. T. 1994. Current status of the rice rat, *Oryzomys couesi peninsularis*. The Southwestern Naturalist 39:99–100.
Álvarez-Castañeda, S. T. 1996. Los mamíferos de Morelos. Centro de Investigaciones Biológicas del Noroeste, S. C. La Paz, México.
Álvarez-Castañeda, S. T. 1998. *Peromyscus pseudocrinitus*. Mammalian Species 601:1–3.
Álvarez-Castañeda, S. T. 2001. *Peromyscus sejugis*. Mammalian Species 658:1–3.
Álvarez-Castañeda, S. T. 2002. *Peromyscus hooperi*. Mammalian Species 709:1–3.
Álvarez-Castañeda, S. T. 2005a. *Peromyscus melanotis*. Mammalian Species 764:1–4.
Álvarez-Castañeda, S. T. 2005b. *Peromyscus winkelmanni*. Mammalian Species 765:1–3.
Álvarez-Castañeda, S. T. 2007. Systematics of the antelope ground squirrel (*Ammospermophilus*) from islands adjacent to the Baja California Peninsula. Journal of Mammalogy 88:1160–1169.

Álvarez-Castañeda, S. T. 2010. Phylogenetic structure of the *Thomomys bottae-umbrinus* complex in North America. Molecular Phylogenetics and Evolution 54:671–679.
Álvarez-Castañeda, S. T. 2016. *Peromyscus hooperi*. The IUCN Red List of Threatened Species 2016:e.T16667A22361329.
Álvarez-Castañeda, S. T. 2018a. *Megadontomys cryophilus*. The IUCN Red List of Threatened Species 2018a:e.T12940A22353517.
Álvarez-Castañeda, S. T. 2018b. *Neotoma nelsoni*. The IUCN Red List of Threatened Species 2018:e.T14592A22372306.
Álvarez-Castañeda, S. T. 2018c. *Neotoma palatina*. The IUCN Red List of Threatened Species 2018:e.T14593A22370476.
Álvarez-Castañeda, S. T. 2018d. *Peromyscus bullatus*. The IUCN Red List of Threatened Species 2018:e.T16653A22361454.
Álvarez-Castañeda, S. T. 2018e. *Peromyscus mekisturus*. The IUCN Red List of Threatened Species 2018:e.T16675A22362990.
Álvarez-Castañeda, S. T. 2018f. *Peromyscus sejugis*. The IUCN Red List of Threatened Species 2018:e.T16688A22364100.
Álvarez-Castañeda, S. T. 2018g. *Peromyscus stephani*. The IUCN Red List of Threatened Species 2018:e.T16692A22362173.
Álvarez-Castañeda, S. T. 2018h. *Peromyscus winkelmanni*. The IUCN Red List of Threatened Species 2018:e.T16695A22362540.
Álvarez-Castañeda, S. T. 2019a. *Peromyscus ochraventer*. The IUCN Red List of Threatened Species 2019:e.T16683A22363628.
Álvarez-Castañeda, S. T. 2019b. *Peromyscus slevini*. The IUCN Red List of Threatened Species 2019:e.T16690A143643850.
Álvarez-Castañeda, S. T. 2019c. *Neotoma phenax*. The IUCN Red List of Threatened Species 2019:e.T14594A22370414.
Álvarez-Castañeda, S. T., and T. Álvarez. 1991. Los murciélagos de Chiapas. Instituto Politécnico Nacional, Escuela Nacional de Ciencias Biológicas, Distrito Federal, México.
Álvarez-Castañeda, S. T., and M. Bogan. 1997. *Myotis milleri*. Mammalian Species 561:1–3.
Álvarez-Castañeda, S. T., and M. Bogan. 1998. *Myotis peninsularis*. Mammalian Species 573:1–2.
Álvarez-Castañeda, S. T., and I. Castro-Arellano. 2008a. *Nelsonia neotomodon*. The IUCN Red List of Threatened Species 2008:e.T14487A4438546.
Álvarez-Castañeda, S. T., and I. Castro-Arellano. 2008b. *Neotoma angustapalata*. The IUCN Red List of Threatened Species 2008:e.T14583A4446470.
Álvarez-Castañeda, S. T., and I. Castro-Arellano. 2016a. *Peromyscus eva*. The IUCN Red List of Threatened Species 2016:e.T16660A22360342.
Álvarez-Castañeda, S. T., and I. Castro-Arellano. 2016b. *Neotomodon alstoni*. The IUCN Red List of Threatened Species 2016:e.T14600A22345400.
Álvarez-Castañeda, S. T., and I. Castro-Arellano. 2019a. *Megadontomys thomasi*. The IUCN Red List of Threatened Species 2019:e.T12942A22353461.
Álvarez-Castañeda, S. T., and I. Castro-Arellano. 2019b. *Nelsonia goldmani*. The IUCN Red List of Threatened Species 2019:e.T14486A22338391.
Álvarez-Castañeda, S. T., and I. Castro-Arellano. 2019c. *Tylomys bullaris*. The IUCN Red List of Threatened Species 2019:e.T22570A22340265.
Álvarez-Castañeda, S. T., and I. Castro-Arellano. 2019d. *Tylomys tumbalensis*. The IUCN Red List of Threatened Species 2019:e.T22575A22340781.
Álvarez-Castañeda, S. T., y P. Cortés Calva. 1996. Anthropogenic extinction of the endemic deer mouse *Peromyscus maniculatus cineritius* on San Roque Island, Baja California Sur, México. Southwestern Naturalist 41:459–461.
Álvarez-Castañeda, S. T., and P. Cortés-Calva. 1999. Familia Muridae. Pp. 445–566, *in* Mamíferos del noroeste de México (Álvarez-Castañeda, S. T., and J. L. Patton, eds.). Centro de Investigaciones Biológicas del Noroeste, La Paz, México.
Álvarez-Castañeda, S. T., and P. Cortés-Calva. 2001. *Peromyscus sejugis*. Mammalian Species 658:1–2.
Álvarez-Castañeda, S. T., and P. Cortés-Calva. 2002a. *Peromyscus slevini*. Mammalian Species 705:1–2.
Álvarez-Castañeda, S. T., and P. Cortés-Calva. 2002b. Extirpation of bailey's pocket mouse, *Chaetodipus baileyi fornicatus* (Heteromyidae: Mammalia), from isla Montserrat, Baja California Sur, México. Western American Naturalist 62:496–497.
Álvarez-Castañeda, S. T., and P. Cortés-Calva. 2003a. *Peromyscus pembertoni*. Mammalian Species 734:1–2.
Álvarez-Castañeda, S. T., and P. Cortés-Calva. 2003b. *Peromyscus eva*. Mammalian Species 738:1–3.
Álvarez-Castañeda, S. T., and P. Cortés-Calva. 2011. Genetic evaluation of the Baja California rock squirrel *Otospermophilus atricapillus* (Rodentia: Sciuridae). Zootaxa 3138:35–51.
Álvarez-Castañeda, S. T., and N. González-Ruiz. 2009. *Peromyscus levipes*. Mammalian Species 824:1–6.
Álvarez-Castañeda, S. T., and N. González-Ruiz. 2008. Análisis preliminar de las relaciones filogenéticas entre los grupos de especies del género *Peromyscus*. Pp. 5–26, *in* Avances en el Estudio de los Mamíferos de México (Lorenzo, C., E. Espinoza, and J. Ortega, eds.). Publicaciones Especiales, Vol. II, Asociación Mexicana de Mastozoología, A. C., Distrito Federal, México.
Álvarez-Castañeda, S. T., and T. Lacher. 2016a. *Neotoma macrotis*. The IUCN Red List of Threatened Species 2016:e.T14597A22370553.
Álvarez-Castañeda, S. T., and T. Lacher. 2016b. *Peromyscus californicus*. The IUCN Red List of Threatened Species 2016:e.T16654A22361553.
Álvarez-Castañeda, S. T., and T. Lacher. 2017. *Chaetodipus arenarius*. The IUCN Red List of Threatened Species 2017:e.T92459293A123795361.
Álvarez-Castañeda, S. T. and T. Lacher. 2018. *Dipodomys gravipes*. The IUCN Red List of Threatened Species 2018:e.T6676A22227742.
Álvarez-Castañeda, S. T. and W. López-Forment. 1995. Datos sobre los mamíferos del área aledaña a Palpan, Morelos, México. Anales del Instituto de Biología, Universidad Nacional Autónoma de México, Serie Zoología 66:123–133.
Álvarez-Castañeda, S. T., and C. Lorenzo. 2016. Genetic evidence supports *Sylvilagus mansuetus* (Lagomorpha: Leporidae) as a subspecies of *S. bachmani*. Zootaxa. 4196:289–295.
Álvarez-Castañeda, S. T., y C. Lorenzo. 2017. Phylogeography and phylogeny of *Lepus californicus* (Lagomorpha: Leporidae) from Baja California Peninsula and adjacent islands. The Biological Journal of the Linnean Society 121:15–27.
Álvarez-Castañeda, S. T., and L. Méndez. 2003. *Oryzomys nelsoni*. Mammalian Species 735:1–2.
Álvarez-Castañeda, S. T., and L. Méndez. 2005. *Peromyscus madrensis*. Mammalian Species 774:1–3.
Álvarez-Castañeda, S. T., and R. W. Murphy. 2014. The endemic insular and peninsular species *Chaetodipus spinatus* (Mammalia, Heteromyidae) breaks patterns for Baja California. PlosOne 9:e116146.
Álvarez-Castañeda, S. T., and L. A. Nájera-Cortazar. 2020. Does islands populations change in size and form in relation to mainland counterparts? Journal of Mammalogy 101:373–385.
Álvarez-Castañeda, S. T., and A. Ortega-Rubio. 2003. Current status of rodents on islands in the Gulf of California. Biological Conservation 109:157–163.
Álvarez-Castañeda, S. T., and J. L. Patton. 1999. Mamíferos del Noroeste Mexicano I. Centro de Investigaciones Biológicas del Noroeste. La Paz, México 1:1–583.
Álvarez-Castañeda, S. T., and J. L. Patton. 2000. Mamíferos del Noroeste Mexicano II. Centro de Investigaciones Biológicas del Noroeste. La Paz, México 2:584–873.
Álvarez-Castañeda, S. T., and F. Reid. 2016. *Microtus mexicanus*. The IUCN Red List of Threatened Species 2016:e.T13443A115113184.

Álvarez-Castañeda, S. T., and E. Rios. 2010. A phylogenetic analysis of *Neotoma varia* (Rodentia: Cricetidae), a rediscovered, endemic, and threatened rodent from Datil Island, Sonora, Mexico. Zootaxa 2647:51–60.

Álvarez-Castañeda, S. T., and E. Rios. 2011. Revision of *Chaetodipus arenarius* (Rodentia: Heteromyidae). The Zoological Journal of the Linnean Society 160:213–228.

Álvarez-Castañeda, S. T., and E. Vázquez. 2018. *Heteromys spectabilis*. The IUCN Red List of Threatened Species 2018:e.T12077A22225458.

Álvarez-Castañeda, S. T., and E. Yensen. 1999. *Neotoma bryanti*. Mammalian Species 619:1–3.

Álvarez-Castañeda, S. T., P. Cortés-Calva, and C. Gómez-Machorro. 1998. *Peromyscus caniceps*. Mammalian Species 602:1–3.

Álvarez-Castañeda, S. T., P. Cortés-Calva, L. Méndez, and A. Ortega-Rubio. 2006. Sea of Cortes island development call for mitigation. BioSciences 56:825–829.

Álvarez-Castañeda, S. T., E. Rios, P. Cortés-Calva, N. González-Ruiz, and C. G. Suárez-Gracida. 2008c. Los mamíferos de las Reservas de El Valle de los Cirios y El Vizcaíno. Centro de Investigaciones Biológicas del Noroeste, S. C - Comisión Nacional para el Conocimiento y Uso de la Biodiversidad. Distrito Federal, México.

Álvarez-Castañeda, S. T., W. Z. Lidicker, and E. Rios. 2009. Revision of the *Dipodomys merriami* complex in the Baja California Peninsula, México. Journal of Mammalogy 90:992–1008.

Álvarez-Castañeda, S. T., G. Arnaud, P. Cortés-Calva, and L. Méndez. 2010a. Invasive migration of a mainland rodent to Santa Catalina Island and its effect on the endemic species *Peromyscus slevini*. Biological Invasions 12:437–439.

Álvarez-Castañeda, S. T., A. Gutiérrez, and M. De la Paz. 2010b. Rediscovery of the *Neotoma* population on Datil [Turner] Island, Sonora, México. Western North American Naturalist 70:437–440.

Álvarez-Castañeda, S. T., T. Álvarez, and N. González-Ruiz. 2015. Guía para la identificación de los mamíferos de México en campo y laboratorio/ Keys for identifying Mexican mammals in the field and in the laboratory. Centro de Investigaciones Biológicas del Noroeste, S. C. y Asociacion Mexicana de Mastozoología, A. C. Guadalajara, México.

Álvarez-Castañeda, S. T., I. Castro-Arellano, and T. Lacher. 2016a. *Neotamias merriami*. The IUCN Red List of Threatened Species 2016:e. T21358A22269203.

Álvarez-Castañeda, S. T., I. Castro-Arellano, and T. Lacher. 2016b. *Neotamias obscurus*. The IUCN Red List of Threatened Species 2016:e. T21359A22268421.

Álvarez-Castañeda, S. T., I. Castro-Arellano, and T. Lacher. 2016c. *Neotoma devia*. The IUCN Red List of Threatened Species 2016:e. T14586A22371569.

Álvarez-Castañeda, S. T., I. Castro-Arellano, and T. Lacher. 2016d. *Sigmodon fulviventer*. The IUCN Red List of Threatened Species 2016:e. T20212A22355460.

Álvarez-Castañeda, S. T., I. Castro-Arellano, and T. Lacher. 2016e. *Dipodomys deserti*. The IUCN Red List of Threatened Species 2016:e. T6686A22228301.

Álvarez-Castañeda, S. T., I. Castro-Arellano, and T. Lacher. 2016f. *Chaetodipus artus*. The IUCN Red List of Threatened Species 2016:e. T4327A115067951.

Álvarez-Castañeda, S. T., I. Castro-Arellano, and T. Lacher. 2016g. *Chaetodipus fallax*. The IUCN Red List of Threatened Species 2016:e. T4330A22226385.

Álvarez-Castañeda, S. T., I. Castro-Arellano, and T. Lacher. 2016h. *Chaetodipus formosus*. The IUCN Red List of Threatened Species 2016:e. T4331A22226290.

Álvarez-Castañeda, S. T., I. Castro-Arellano, and T. Lacher. 2016i. *Chaetodipus pernix*. The IUCN Red List of Threatened Species 2016:e. T4337A22225579.

Álvarez-Castañeda, S. T., I. Castro-Arellano, and T. Lacher. 2016j. *Chaetodipus rudinoris*. The IUCN Red List of Threatened Species 2016:e. T136837A22225520.

Álvarez-Castañeda, S. T., I. Castro-Arellano, and T. Lacher. 2016k. *Perognathus amplus*. The IUCN Red List of Threatened Species 2016:e. T16633A22224910.

Álvarez-Castañeda, S. T., I. Castro-Arellano, T. Lacher, and E. Vázquez. 2016l. *Sciurus oculatus*. The IUCN Red List of Threatened Species 2016:e.T20017A22246721.

Álvarez-Castañeda, S. T., I. Castro-Arellano, T. Lacher, and E. Vázquez. 2016m. *Notocitellus annulatus*. The IUCN Red List of Threatened Species 2016:e.T20479A22265951.

Álvarez-Castañeda, S. T., I. Castro-Arellano, T. Lacher, and E. Vázquez. 2016n. *Microtus californicus*. The IUCN Red List of Threatened Species 2016:e.T13427A22349460.

Álvarez-Castañeda, S. T., I. Castro-Arellano, T. Lacher, and E. Vázquez. 2016o. *Hodomys alleni*. The IUCN Red List of Threatened Species 2016:e.T10211A22379623.

Álvarez-Castañeda, S. T., I. Castro-Arellano, T. Lacher, and E. Vázquez. 2016p. *Peromyscus perfulvus*. The IUCN Red List of Threatened Species 2016:e.T16685A22364160.

Álvarez-Castañeda, S. T., I. Castro-Arellano, T. Lacher, and E. Vázquez. 2016q. *Peromyscus melanotis*. The IUCN Red List of Threatened Species 2016:e.T16678A115136970.

Álvarez-Castañeda, S. T., I. Castro-Arellano, T. Lacher, and E. Vázquez. 2016r. *Reithrodontomys chrysopsis*. The IUCN Red List of Threatened Species 2016:e.T19404A22385521.

Álvarez-Castañeda, S. T., I. Castro-Arellano, T. Lacher, and E. Vázquez. 2016s. *Reithrodontomys zacatecae*. The IUCN Red List of Threatened Species 2016:e.T19419A115151808.

Álvarez-Castañeda, S. T., I. Castro-Arellano, T. Lacher, and E. Vázquez. 2016t. *Sigmodon leucotis*. The IUCN Red List of Threatened Species 2016:e.T20215A22355853.

Álvarez-Castañeda, S. T., I. Castro-Arellano, T. Lacher, and E. Vázquez. 2016u. *Handleyomys melanotis*. The IUCN Red List of Threatened Species 2016:e.T15606A22328750.

Álvarez-Castañeda, S. T., I. Castro-Arellano, T. Lacher, and E. Vázquez. 2016v. *Cratogeomys goldmani*. The IUCN Red List of Threatened Species 2016:e.T136647A22216690.

Álvarez-Castañeda, S. T., I. Castro-Arellano, T. Lacher, and E. Vázquez. 2016w *Dipodomys nelsoni*. The IUCN Red List of Threatened Species 2016:e.T6690A22228791.

Álvarez-Castañeda, S. T., A. D. Cuarón, and P. C. de Grammont. 2016x. *Cryptotis peregrina*. The IUCN Red List of Threatened Species 2016:e.T136550A22284600.

Álvarez-Castañeda, S. T., T. Lacher, and E. Vázquez. 2016y. *Cratogeomys fumosus*. The IUCN Red List of Threatened Species 2016:e.T16026A22216828.

Álvarez-Castañeda, S. T., T. Lacher, and E. Vázquez. 2016z. *Neotamias durangae*. The IUCN Red List of Threatened Species 2016:e.T21357A22268753.

Álvarez-Castañeda, S. T., T. Lacher, and E. Vázquez. 2016aa. *Xerospermophilus perotensis*. The IUCN Red List of Threatened Species 2016:e.T20489A22264586.

Álvarez-Castañeda, S. T., T. Lacher, and E. Vázquez. 2016ab. *Neotamias bulleri*. The IUCN Red List of Threatened Species 2016:e.T21356A103309807.

Álvarez-Castañeda, S. T., T. Lacher., and E. Vázquez. 2016ac. *Dipodomys phillipsii*. The IUCN Red List of Threatened Species 2016:e.T92463503A22228991.

Álvarez-Castañeda, S. T., T. Lacher, and E. Vázquez. 2016ad. *Callospermophilus madrensis*. The IUCN Red List of Threatened Species 2016:e.T20485A22263330.

Álvarez-Castañeda, S.T., T. Lacher, and E. Vázquez. 2016ae. *Cratogeomys merriami. The IUCN Red List of Threatened Species* 2016:e.T16028A22217011.

Álvarez-Castañeda, S. T., I. Castro-Arellano, T. Lacher, E. Vázquez, and J. Arroyo-Cabrales. 2016af. *Sigmodon mascotensis*. The IUCN Red List of Threatened Species 2016:e.T20216A115157887.

Álvarez-Castañeda, S. T., J. Matson, N. Woodman, P. C. de Grammont, and G. A. Hammerson. 2016ag. *Sorex ornatus*. The IUCN Red List of Threatened Species 2016:e.T41408A115184758.

Álvarez-Castañeda, S. T., I. Castro-Arellano, and T. Lacher. 2016ah. *Sigmodon arizonae*. The IUCN Red List of Threatened Species 2016:e.T20211A22355540.

Álvarez-Castañeda, S. T., T. Álvarez, and N. González-Ruiz. 2017a. Keys for identifying Mexican Mammals. The Johns Hopkins University Press. Baltimore, U.S.A.

Álvarez-Castañeda, S. T., T. Lacher, and E. Vázquez. 2018a. *Cynomys mexicanus*. The IUCN Red List of Threatened Species 2018:e.T6089A22260873.

Álvarez-Castañeda, S. T., T. Lacher, and E. Vázquez. 2018b. *Habromys chinanteco*. The IUCN Red List of Threatened Species 2018:e.T9608A22376453.

Álvarez-Castañeda, S. T., T. Lacher, and E. Vázquez. 2018c. *Habromys schmidlyi*. The IUCN Red List of Threatened Species 2018:e.T136616A22376358.

Álvarez-Castañeda, S. T., T. Lacher, and E. Vázquez. 2018d. *Peromyscus caniceps*. The IUCN Red List of Threatened Species 2018:e.T16655A22361697.

Álvarez-Castañeda, S. T., T. Lacher, and E. Vázquez. 2018e. *Peromyscus guardia*. The IUCN Red List of Threatened Species 2018:e.T16664A22359973.

Álvarez-Castañeda, S. T., T. Lacher, and E. Vázquez. 2018f. *Peromyscus madrensis*. The IUCN Red List of Threatened Species 2018 e.T16671A22361002.

Álvarez-Castañeda, S. T., T. Lacher, and E. Vázquez. 2018g. *Peromyscus simulus*. The IUCN Red List of Threatened Species 2018:e.T16689A22362390.

Álvarez-Castañeda, S. T., T. Lacher, and E. Vázquez. 2018h. *Reithrodontomys hirsutus*. The IUCN Red List of Threatened Species 2018:e.T19409A22386368.

Álvarez-Castañeda, S. T., T. Lacher, I. Castro-Arellano, and E. Vázquez. 2018i. *Habromys lepturus*. The IUCN Red List of Threatened Species 2018:e.T9609A22376889.

Álvarez-Castañeda, S. T., T. Lacher, E. Vázquez, and J. Arroyo-Cabrales. 2018j. *Peromyscus zarhynchus*. The IUCN Red List of Threatened Species 2018:e.T16697A22362848.

Álvarez-Castañeda, S. T., T. Lacher, E. Vázquez, and J. Arroyo-Cabrales. 2018k. *Zygogeomys trichopus*. The IUCN Red List of Threatened Species 2018:e.T23323A22216603.

Álvarez-Castañeda, S. T., T. Lacher, E. Vázquez, J. Arroyo-Cabrales, and P. C. de Grammont. 2018l. *Megadontomys nelsoni*. The IUCN Red List of Threatened Species 2018:e.T12941A22353381.

Álvarez-Castañeda, S. T., T. Lacher, E. Vázquez, J. Arroyo-Cabrales, and P. C. de Grammont. 2018m. *Peromyscus melanocarpus*. The IUCN Red List of Threatened Species 2018:e.T16676A22363072.

Álvarez-Castañeda, S. T., E. Vázquez, I. Castro-Arellano, and T. Lacher. 2018n. *Habromys ixtlani*. The IUCN Red List of Threatened Species 2018:e.T136582A22376638.

Álvarez-Castañeda, S. T., A. D. Cuarón, and P. C. Grammont. 2018o. *Cryptotis phillipsii*. The IUCN Red List of Threatened Species 2018:e.T136639A22283893.

Álvarez-Castañeda, S. T., T. Lacher, and E. Vázquez. 2019a. *Microtus quasiater*. The IUCN Red List of Threatened Species 2019:e.T13453A22348115.

Álvarez-Castañeda, S. T., T. Lacher, and E. Vázquez. 2019b. *Peromyscus melanurus*. The IUCN Red List of Threatened Species 2019:e.T16679A22363320.

Álvarez-Castañeda, S. T., T. Lacher, and E. Vázquez. 2019c. *Peromyscus pseudocrinitus*. The IUCN Red List of Threatened Species 2019:e.T16687A143642485.

Álvarez-Castañeda, S. T., T. Lacher, E. Vázquez, and J. Arroyo-Cabrales. 2019d. *Peromyscus sagax*. The IUCN Red List of Threatened Species 2019:e.T136710A22362792.

Álvarez-Castañeda, S. T., I. Castro-Arellano, T. Lacher, E. Vázquez, and J. Arroyo-Cabrales. 2019e. *Reithrodontomys burti*. The IUCN Red List of Threatened Species 2019:e.T19403A22385577.

Álvarez-Castañeda, S. T., C. Lorenzo, C. A. Segura-Trujillo, and S. G. Pérez-Consuegra. 2019f. Two new species of *Peromyscus* from Chiapas, Mexico, and Guatemala. Pp. 543–558, *in* From field to laboratory: a memorial volume in honor of Robert J. Baker (Bradley, R. D., H. H. Genoways, D. J. Schmidly, and L. C. Bradley, eds.). Special Publications, Museum of Texas Tech University. Lubbock, U.S.A.

Álvarez-Castañeda, S. T., T. Lacher, E. Vázquez, and J. Arroyo-Cabrales. 2020. *Peromyscus polius*. The IUCN Red List of Threatened Species 2020:e.T16686A22364219.

Álvarez, T., J. Arroyo-Cabrales, and M. González-Escamilla. 1987. Mamíferos (excepto Chiroptera) de la costa de Michoacán. Anales de la Escuela Nacional de Ciencias Biológicas 31:13–62.

Álvarez-Castañeda, S. T., T. Álvarez, and N. González-Ruiz. 2017a. Keys for identifying Mexican Mammals. The Johns Hopkins University Press. Baltimore, U.S.A.

Álvarez-Castañeda, S. T., T. Álvarez, and N. González-Ruiz. 2017b. Keys for identifying Mexican Mammals. The Johns Hopkins University Press. Baltimore, U.S.A.

Álvarez-Castañeda, S. T., C. A. Segura-Trujillo, and N. González-Ruiz. 2021. *Peromyscus mekisturus*. Mammalian Species, in press.

Arnason, U., K. Bodin, A. Gullberg, C. Ledje, and S. Mouchaty. 1995. A molecular view of pinniped relationships with particular emphasis on the true seals. Journal of Molecualr Evolution 40:78–85.

American Society of Mammalogy. 2022. ASM's Mammal Diversity Database. https://www.mammaldiversity.org/ Accessed on 27 December 2022.

Amman, B. R., and R. D. Bradley. 2004. Molecular evolution in *Baiomys* (Rodentia: Sigmodontinae): evidence for a genetic subdivision in *B. musculus*. Journal of Mammalogy 85:162–166.

Ammerman, L. K., D. N. Lee, and R. S. Pfau. 2016. Patterns of Genetic Divergence among *Myotis californicus*, *M. ciliolabrum*, and *M. leibii* Based on Amplified Fragment Length Polymorphism. Acta Chiropterologica 18:337–347.

Amori, G., F. Chiozza, C. Rondinini, and L. Luiselli. 2011. Worldwide conservation hotspots for Soricomorpha focusing on endemic island taxa: an analysis at two taxonomic levels. Endangered Species Research 15:143–149.

Amstrup, S. C. 2003. Polar bear, *Ursus maritimus*. Pp. 587–610, *in* Wild Mammals of North America: Biology, Management, and Conservation (Feldhamer, G. A., B. C. Thomson, and J. A. Chapman, eds). John Hopkins University Press. Baltimore, USA.

Amstrup, S. C., I. Stirling, and J. W. Lentfer. 1986. Past and present status of polar bears in Alask. Wildlife Society Bulletin 14:241–254.

Andersen, J. M., Y. F. Wiersma, and G. Stenson. 2009. Movement patterns of hooded seals (*Cystophora cristata*) in the Northwest Atlantic Ocean during the post-mout and pre-breeding seasons. Journal of Northwest Atlantic Fisheries Science 42:1–11.

Andersen, K. 1906a. On the bats of the genera *Micronycteris* and *Glyphonycteris*. Annals and magazine of Natural History 18:50–65.

Andersen, K. 1906b. Brief diagnoses of a new genus and ten new forms of stenodermatous bats. Annals and magazine of Natural History 18:419–423.

Andersen, K. 1908. A monograph of the chiropteran genera *Uroderma, Enchisthenes,* and *Artibeus*. Proceedings of the Zoological Society of London 204–319.

Anderson, A. E., and O. C. Wallmo. 1984. *Odocoileus hemionus*. Mammalian Species 219:1–9.

Anderson, E. 1977. Pleistocene Mustelidae (Mammalia, Carnivora) from Fairbanks, Alaska. Bulletin of the Museum of Comparative Zoology 148:1–21.

Anderson, R. M. 1934. *Sorex palustris brooksi*, a new water shrew from Vancouver Island. The Canadian Field Naturalist 48:134.

Anderson, R. M. 1942. Canadian voles of the genus *Phenacomys* with description of two new Canadian subspecies. Canadian Field-Naturalist 56:56–60.

Anderson, R. M., and A. L. Rand. 1945. The Varying Lemming (Genus *Dicrostonyx*) in Canada. Journal of Mammalogy 26:301–306.

Anderson, S. 1954. Subspeciation in the meadow mouse, *Microtus montanus*, in Wyoming and Colorado. University of Kansas Publications, Museum of Natural History 7:489–506.

Anderson, S. 1959. Distribution, variation, and relationships of the montane vole, *Microtus montanus*. University of Kansas Publications, Museum of Natural History 9:415–511.

Anderson, S. 1960. The baculum of Microtine rodents. University of Kansas Publications, Museum of Natural History 12:181–216.

Anderson, S. 1962. Tree squirrels (*Sciurus colliaei* group) from western Mexico. American Museum Novitates 2093:1–13.

Anderson, S. 1964. The systematic status of *Perognathus artus* and *Perognathus goldmani* (Rodentia). American Museum Novitates 2184:1–27.

Anderson, S. 1966. Taxonomy of gophers, especially *Thomomys* in Chihuahua, Mexico. Systematic Zoology 15:189–198.

Anderson, S. 1969. *Macrotus waterhousii*. Mammalian Species 1:1–4.

Anderson, E. 1970. Quaternary evolution of the genus Martes (Carnivora, Mustelidae). Acta Zoologica Fennica 130:1–132.

Anderson, S. 1972. Mammals of Chihuahua: taxonomy and distribution. Bulletin of the American Museum of Natural History 148:149–410.

Anderson, S. and A. S. Gaunt. 1962. A classification of the white-sided jackrabbits of Mexico. American Museum Novitates 2088:1–16.

Anderson, S., and C. E. Nelson. 1965. A systematic revision of *Macrotus* (Chiroptera). American Museum Novitates 2212:1–39.

Andersen, J. M., Y. F. Wiersma, G. Stenson, M. O. Hammill, and A. Rosing-Asvid. 2009. Movement Patterns of Hooded Seals (Cystophora cristata) in the Northwest Atlantic Ocean during the Post-Moult and Pre-Breed Seasons. Journal of the Northwestern Atlantic Fisheries Sciences 42:1–11

Angerbjörn, A., and M. Tannerfeldt. 2014. *Vulpes lagopus*. The IUCN Red List of Threatened Species 2014:e.T899A57549321.

Anthony, H. E. 1923. Mammals from Mexico and South America. American Museum Novitates 54:1–10.

Anthony, H. E. 1932. A new genus of rodents from Yucatan. American Museum Novitates 586:1–3.

Appollonio, M. 1999. Dama dama. *in* The Atlas of European Mammals (Mitchell-Jones, A. J., G. Amori, W. Bogdanowicz, B. Kryštufek, P. J. H. Reijnders, F. Spitzenberger, M. Stübbe, J. B. M. Thissen, V. Vohralík, and J. Zima, eds). Academic Press, London, UK.

Arbogast, B. S. 1999. Mitochondrial DNA phylogeography of the New World flying squirrels (*Glaucomys*): implications for Pleistocene biogeography. Journal of Mammalogy 80:142–155.

Arbogast, B. S., R. A. Browne, and P. D. Weigl. 2001. Evolutionary genetics and Pleistocene biogeography of North American Tree squirrels (*Tamasciurus*). Journal of Mammalogy 82:302–319.

Arbogast, B. S., K. I. Schumacher, N. J. Kerhoulas, A. L. Bidlack, J. A. Cook, and G. J. Kenagy. 2017. Genetic data reveal a cryptic species of New World flying squirrel: *Glaucomys oregonensis*. Journal of Mammalogy 98:1027–1041.

Arcangeli, J., J. E. Light, and F. A. Cervantes 2018. Molecular and morphological evidence of the diversification in the gray mouse opossum, *Tlacuatzin canescens* (Didelphimorphia), with description of a new species. Journal of Mammalogy 99:138–158.
Arellano, E., D. S. Rogers, and F. A. Cervantes. 2003. Genic differentiation and phylogenetic relationships among tropical harvest mice (*Reithrodontomys*: Subgenus Aporodon). Journal of Mammalogy 84:129–143.
Arita, H. T. 1999. Southern long-nosed bat/*Leptonycteris curasoae*. Pp. 76–78, *in* The Smithsonian book of North American mammals (Wilson, D. E., and S. Ruff, eds.). Smithsonian Institution Press, Washington, U.S.A.
Arita, H. T., and S. R. Humphrey. 1988. Revision taxonómica de los murciélagos magueyeros del género *Leptonycteris* (Chiroptera: Phyllostomidae). Acta Zoológica Mexicana 29:1–60.
Arita, H. T., J. G. Robinson, and K. H. Redford. 1990. Rarity in neotropical forest mammals and its ecological correlates. Conservation Biology 4:181.
Armitage, K. B., D. W. Johns, and D. C. Andersen. 1979. Cannibalism among yellow-bellied marmots. Journal of Mammalogy 60:205–206.
Arizona Game and Fish Department. 2004a. *Sigmodon arizonae arizonae.* Arizona Game and Fish Department Heritage Management Data System. http://www.azgfd.com/w_c/edits/documents/Sigmaraz.D.pdf, accessed 15 December 2008.
Arizona Game and Fish Department. 2004b. *Sigmodon arizonae jacksoni.* Arizona Game and Fish Department Heritage Management Data System. http://www.azgfd.com/w_c/edits/documents/Sigmarja.D.pdf, accessed 15 December 2008.
Armstrong, D. M. 1971. Notes on variation in *Spermophilus tridecemlineatus* (Rodentia, Sciuridae) in Colorado and adjacent states, and description of a new subspecies. Journal of Mammalogy 52:528–536.
Armstrong, D. M. 1972. Distribution of mammals in Colorado. Museum of Natural History University of Kansas Monographs 3:1–415.
Armstrong, D. M., and J. K. Jones, Jr. 1971a. Mammals from the Mexican state of Sinaloa. I. Marsupialia, Insectivora, Edentata, Lagomorpha. Journal of Mammalogy 52:747–757.
Armstrong, D. M., and J. K. Jones, Jr. 1971b. *Sorex merriami*. Mammalian Species 2:1–2.
Armstrong, D. M., and J. K. Jones, Jr. 1972a. *Megasorex gigas*. Mammalian Species 16:1–2.
Armstrong, D. M., and J. K. Jones, Jr. 1972b. *Notiosorex crawfordi*. Mammalian Species 17:1–5.
Armstrong, D. M., J. P. Fitzgerald, and C. A. Meaney. 2011. Mammals of Colorado. University Press of Colorado. Colorado, U.S.A.
Árnason, U., A. Gullberg, S. Gretarsdotti, B. Ursing, and A. Janke. 2000. The mitochondrial genome of the sperm whale and a new molecular reference for estimating eutherian divergence dates. Journal of Molecular Evolution50:569–578
Arroyo-Cabrales, J. 2016. *Rhogeessa genowaysi*. The IUCN Red List of Threatened Species 2016:e.T19680A21989676.
Arroyo-Cabrales, J., and T. Álvarez. 1990. Restos óseos de murciélagos procedentes de las excavaciones en las grutas de Loltún. Colección científica: Serie prehistoria del Instituto Nacional de Antropología e Historia 194:1–103.
Arroyo-Cabrales, J., and S. T. Álvarez-Castañeda. 2015a. *Artibeus hirsutus*. The IUCN Red List of Threatened Species 2015:e.T2131A21996678.
Arroyo-Cabrales, J., and S. T. Álvarez-Castañeda. 2015b. *Nyctinomops femorosaccus*. The IUCN Red List of Threatened Species 2015:e.T14994A22010542.
Arroyo-Cabrales, J., and S. T. Álvarez-Castañeda. 2017a. *Corynorhinus rafinesquii*. The IUCN Red List of Threatened Species 2017:e.T17600A21976905.
Arroyo-Cabrales, J., and S. T. Álvarez-Castañeda. 2017b. *Corynorhinus townsendii*. The IUCN Red List of Threatened Species 2017:e.T17598A21976681.
Arroyo-Cabrales, J., and S. T. Álvarez-Castañeda. 2017c. *Euderma maculatum*. The IUCN Red List of Threatened Species 2017:e.T8166A22028573.
Arroyo-Cabrales, J., and S. T. Álvarez-Castañeda. 2017d. *Idionycteris phyllotis*. The IUCN Red List of Threatened Species 2017:e.T10790A21990019.
Arroyo-Cabrales, J., and S. T. Álvarez-Castañeda. 2017e. *Lasiurus xanthinus*. The IUCN Red List of Threatened Species 2017:e.T41532A22004260.
Arroyo-Cabrales, J., and S. T. Álvarez-Castañeda. 2017f. *Myotis austroriparius*. The IUCN Red List of Threatened Species 2017:e.T14147A22059907.
Arroyo-Cabrales, J., and S. T. Álvarez-Castañeda. 2017g. *Myotis ciliolabrum*. The IUCN Red List of Threatened Species 2017:e.T14153A22058110.
Arroyo-Cabrales, J., and S. T. Álvarez-Castañeda. 2017h. *Myotis evotis*. The IUCN Red List of Threatened Species 2017:e.T14157A22059133.
Arroyo-Cabrales, J., and S. T. Álvarez-Castañeda. 2017i. *Myotis keenii*. The IUCN Red List of Threatened Species 2017:e.T14171A22055579.
Arroyo-Cabrales, J., and P. C. de Grammont. 2017a. *Antrozous pallidus*. The IUCN Red List of Threatened Species 2017:e.T1790A22129152.
Arroyo-Cabrales, J., and P. C. de Grammont. 2017b. *Myotis thysanodes*. The IUCN Red List of Threatened Species 2017:e.T14206A22063246.
Arroyo-Cabrales, J., and J. K. Jones. 1988a. *Balantiopteryx plicata*. Mammalian Species 301:1–4.
Arroyo-Cabrales, J., and J. K. Jones. 1988b. *Balantiopteryx io* and *Balantiopteryx infusca*. Mammalian Species 313:1–3.
Arroyo-Cabrales, J., and S. Ospina-Garcés. 2015. *Musonycteris harrisoni*. The IUCN Red List of Threatened Species 2015:e.T14003A22099002.
Arroyo-Cabrales, J., and S. Ospina-Garcés. 2016a. *Myotis findleyi*. The IUCN Red List of Threatened Species 2016:e.T14159A22058800.
Arroyo-Cabrales, J., and S. Ospina-Garcés. 2016b. *Myotis planiceps*. The IUCN Red List of Threatened *Species* 2016:e.T14191A22066742.
Arroyo-Cabrales, J., and S. Ospina-Garcés. 2016c. *Myotis sodalis*. The IUCN Red List of Threatened Species 2016:e.T14136A22053184.
Arroyo-Cabrales, J., and S. Ospina-Garcés. 2016d. *Myotis peninsularis*. The IUCN Red List of Threatened Species 2016:e.T14189A22066405.
Arroyo-Cabrales, J., and S. Ospina-Garcés. 2016e. *Myotis vivesi*. The IUCN Red List of Threatened Species 2016:e.T14209A22069146.
Arroyo-Cabrales, J., and S. Ospina-Garcés. 2016f. *Rhogeessa mira*. The IUCN Red List of Threatened Species 2016:e.T19683A22007311.
Arroyo-Cabrales, J., and R. D. Owen. 1996. Intraspecific variation and phenetic affinities of *Dermanura hartii,* with reapplication of the specific name *Enchisthenes hartii.* Pp. 67–81, *in* Contributions in mammalogy: a memorial volume honoring Dr. J. Knox Jones, Jr. (Genoways, H. H., and R. J. Baker, eds.). The Museum, Texas Tech University, Lubbock, U.S.A.
Arroyo-Cabrales, J., and R. D. Owen. 1997. *Enchisthenes hartii*. Mammalian Species 546:1–4.
Arroyo-Cabrales, J., and S. Perez. 2017. *Myotis californicus*. The IUCN Red List of Threatened Species 2017:e.T14150A22061366.
Arroyo-Cabrales, J., and O. J. Polaco. 1997. *Rhogeessa mira*. Mammalian Species 550:1–2.
Arroyo-Cabrales, J., and F. Reid. 2016. *Platyrrhinus helleri*. The IUCN Red List of Threatened Species 2016:e.T88159886A88159952.
Arroyo-Cabrales, J., R. R. Hollander, and J. K. Jones, Jr. 1987. *Choeronycteris mexicana*. Mammalian Species 291:1–5.
Arroyo-Cabrales, J., R. A. van den Bussche, K. Haiduk-Sigler, R. K. Chesser, and R. J. Baker. 1997. Genic variation in island populations of *Natalus stramineus* (Chiroptera: Natalidae). Occasional Papers, Museum of Texas Tech University 1771: 1–9.

Arroyo-Cabrales, J., E. K. V. Kalko, R. K. Laval, J. E. Maldonado, R. A. Medellín, O. J. Polaco, and B. Rodríguez-Herrera. 2005. Rediscovery of the Mexican flat–headed bat *Myotis planiceps* (Vespertilionidae). Acta Chiropterologica 7:309–314.

Arroyo-Cabrales, J., O. J. Polaco, D. E. Wilson, and A. L. Gardner. 2008. Nuevos registros de murciélagos para el estado de Nayarit, México. Revista Mexicana de Mastozoología 12:141–162.

Arroyo-Cabrales, J., Miller, B., F. Reid, A. D. Cuarón, and P. C. de Grammont. 2015a. *Mimon cozumelae*. The IUCN Red List of Threatened Species 2015:e.T136561A21991024.

Arroyo-Cabrales, J., S. T. Álvarez-Castañeda, A. D. Cuarón, and P. C. de Grammont. 2015b. *Glossophaga morenoi*. The IUCN Red List of Threatened Species 2015:e.T9276A22108155.

Arroyo-Cabrales, J., B. Miller, F. Reid, A. D. Cuarón, and P. C. de Grammont. 2015c. *Centronycteris centralis*. The IUCN Red List of Threatened Species 2015:e.T136350A22023809.

Arroyo-Cabrales, J., Miller, B., F. Reid, A. D. Cuarón, and P. C. de Grammont. 2016. *Lasiurus borealis*. The IUCN Red List of Threatened Species 2016:e.T11347A22121017.

Arroyo-Cabrales, J., Miller, B., F. Reid, A. D. Cuarón, and P. C. de Grammont. 2017. *Myotis auriculus*. The IUCN Red List of Threatened Species 2017:e.T14145A22060698.

Ashley, M., J. Norman, and L. Stross. 1996. Phylogenetic analysis of the Perissodactyla family Tapiridae using mitochondrial cytochrome c oxidase (COII) sequences. Journal of Mammalian Evolution 3:315–326.

Astua de Moraes, D., D. Lew, L. P. Costa, and R. Pérez-Hernandez. 2016. Didelphis marsupialis. In: IUCN 2016. The IUCN Red List of Threatened Species. Version 2016. www.iucnredlist.org. Accessed February 24, 2021.

Audet, D., M. D. Engstrom, and M. B. Fenton. 1993. Morphology, Karyology, and echolocation calls of *Rhogeessa* (Chiroptera: Vespertilionidae) from the Yucatán Peninsula. Journal of Mammalogy 74:498–502.

Audet, A. M., C. B. Robbins, and S. Larivière. 2002. *Alopex lagopus*, Mammalian Species 713:1–10.

Aune, K., D. Jørgensen, and C. Gates. 2017. *Bison bison*. The IUCN Red List of Threatened Species 2017:e.T2815A123789863.

Aurioles-Gamboa, D., and F. J. Camacho-Ríos. 2007. Diet and feeding overlap of two otariids, *Zalophus californianus* and *Arctocephalus townsendi*: Implications to survive environmental uncertainty. Aquatic Mammals 33:315–326.

Aurioles-Gamboa, D., F. Elorriaga-Verplancken, and C. J. Hernández-Camacho. 2010. Guadalupe fur seal population status on the San Benito Islands, Mexico. Marine Mammal Science 26:402–408.

Austin, J., and N. Roach. 2019. *Podomys floridanus*. The IUCN Red List of Threatened Species 2019:e.T17830A22339074.

Astúa de Moraes, D., D. Lew, L. P. Costa, and R. Pérez-Hernandez. 2016. *Didelphis marsupialis*. In: IUCN 2016. The IUCN Red List of Threatened Species. Version 2016. www.iucnredlist.org. Accessed February 24, 2021.

Astúa, D., D. Lew, L. P. Costa, and R. Pérez-Hernandez. 2021. *Didelphis marsupialis* (amended version of 2016 assessment). The IUCN Red List of Threatened Species 2021:e.T40501A197310576.

Aurioles-Gamboa, D. 2015. *Arctocephalus townsendi*. The IUCN Red List of Threatened Species 2015:e.T2061A45224420.

Aurioles-Gamboa, D., and J. Hernández-Camacho. 2015. *Zalophus californianus*. The IUCN Red List of Threatened Species 2015:e.T41666A45230310.

Ávila-Flores, R., J. J. Flores-Martínez, and J. Ortega. 2002. *Nyctinomops laticaudatus*. Mammalian Species 697:1–6.

Ávila-Valle, Z. A., A. Castro-Campillo, L. León-Paniagua, I. H. Salgado-Ugalde, A. G. Navarro-Sigüenza, B. E. Hernández-Baños, and J. Ramírez-Pulido. 2012. Geographic variation and molecular evidence of the blackish deer mouse complex (*Peromyscus furvus*, Rodentia: Muridae). Mammalian Biology 77:166–177.

Avise J. C., M. H. Smith, R. K. Selander, T. E. Lawlor, and P. R. Ramsey. 1974. Biochemical polymorphism and systematics in the genus *Peromyscus*. V. Insular and mainland species of the subgenus *Haplomylomys*. Systematic Zoology 23:226–238.

Avise, C., J. Smith, and R. K. Selander. 1979. Biochemical polymorphism and systematics in the genus *Peromyscus* VII. Geographic differentiation in members of the *truei* and *maniculatus* species group. Journal of Mammalogy 60:177–192.

Backlund, D. C. 1995. New records for the Dwarf Shrew, Pygmy Shrew, and Least Shrew in South Dakota. Prairie Naturalist 27:63–64.

Bailey, B. 1929. Mammals of Sherburne County, Minnesota. Journal of Mammalogy 10:153–164.

Bailey, V. 1897. Revision of the American voles of the genus *Evotomys*. Proceedings of the Biological Society of Washington 11:113–138.

Bailey, V. 1900. Revision of American voles of the genus *Microtus*. North American Fauna 17:1–88.

Bailey, V. 1902a. Seven new mammals from western Texas. Proceedings of the Biological Society of Washington 15:117–120.

Bailey, V. 1902b. Synopsis of the North American species of *Sigmodon*. Proceedings of the Biological Society of Washington 15:101–116.

Bailey, V. 1905. Biological survey of Texas. North American Fauna, 25:1–222.

Bailey, V. 1915. Revision of the Pocket Gophers of the Genus *Thomomys*. North American Fauna 39:1–136.

Bailey, V. 1931. Mammals of New Mexico. North American Fauna 53:1–412.

Bailey, V. 1936. The mammals and life zones of Oregon. North American Fauna 55:1–416.

Baird, A. B., D. M. Hills, J. C. Patton, and J. W. Bickham. 2008. Evolutionary history of the genus *Rhogeessa* (Chiroptera: Vespertilionidae) as revaluated by mitochondrial DNA sequences. Journal of Mammalogy 89:744–754.

Baird, A. B., D. M. Hillis, J. C. Patton, and J. W. Bickham. 2009. Speciation by monobrachial centric fusions: A test of the model using nuclear DNA sequences from the bat genus *Rhogeessa*. Molecular Phylogenetics and Evolution 50:256–267.

Baird, A. B. 2010. Genetic identification of cryptic species: Anexample in *Rhogeessa*. Pp. 22–23, *in* Molecular approaches in natural resource conservation and management (DeWoody, J. A., J. W. Bickham, C. H. Michler, K. M. Nichols, O. E. Rhodes, Jr., and K. E. Woeste, eds.). Cambridge University Press. Cambridge, U.S.A.

Baird, A. B., M. R. Marchán-Rivadeneira, S. G. Pérez, and R. J. Baker. 2012. Morphological analysis and description of two new species of *Rhogeessa* (Chiroptera: Vespertilionidae) from the neotropics. Occasional Papers of the Museum, Texas Tech University 307:1–25.

Baird, A. B., J. K. Braun, M. A. Mares, J. C. Morales, J. C. Patton, C. Q. Tran, and J. W. Bickham. 2015. Molecular systematic revision of tree bats (Lasiurini): doubling the native mammals of the Hawaiian Islands. Journal of Mammalogy 96:1255–1274.

Baird, A. B., J. Braun, M. Engstrom, B. Lim, M. Mares, J. Patton, and J. Bickham. 2021. On the utility of taxonomy to reflect biodiversity: the example of *Lasiurini* (Chiroptera: Vespertilionidae). Therya 12:283.

Barquez, R., and M. Diaz. 2015. *Eumops perotis*. The IUCN Red List of Threatened Species 2015:e.T8247A97207171.

Barquez, R., and M. Diaz. 2016a. *Lasiurus ega*. The IUCN Red List of Threatened Species 2016:e.T11350A22119259.
Barquez, R., and M. Diaz. 2016b. *Myotis keaysi*. The IUCN Red List of Threatened Species 2016:e.T14170A22056048.
Barquez, R., B. Rodriguez, B. Miller, and M. Diaz. 2015a. *Eumops auripendulus*. The IUCN Red List of Threatened Species 2015:e.T8241A97206888.
Barquez, R., S. Perez, B. Miller, and M. Diaz. 2015b. *Artibeus lituratus*. The IUCN Red List of Threatened Species 2015:e.T2136A21995720.
Barquez, R., S. Perez, B. Miller, and M. Diaz. 2015c. *Carollia perspicillata*. The IUCN Red List of Threatened Species 2015:e.T3905A22133716.
Barquez, R., S. Perez, B. Miller, and M. Diaz. 2015d. *Chrotopterus auritus*. The IUCN Red List of Threatened Species 2015:e.T4811A22042605.
Barquez, R., S. Perez, B. Miller, and M. Diaz. 2015e. *Desmodus rotundus*. The IUCN Red List of Threatened Species 2015:e.T6510A21979045.
Barquez, R., S. Perez, B. Miller, and M. Diaz. 2015f. *Diaemus youngi*. The IUCN Red List of Threatened Species 2015:e.T6520A21982777.
Barquez, R., S. Perez, B. Miller, and M. Diaz. 2015g. *Glossophaga soricina*. The IUCN Red List of Threatened Species 2015:e.T9277A22107768.
Barquez, R., S. Perez, B. Miller, and M. Diaz. 2015h. *Noctilio albiventris*. The IUCN Red List of Threatened Species 2015:e.T14829A22019978.
Barquez, R., S. Perez, B. Miller, and M. Diaz. 2015i. *Noctilio leporinus*. The IUCN Red List of Threatened Species 2015:e.T14830A22019554.
Barquez, R., S. Perez, B. Miller, and M. Diaz. 2015j. *Phyllostomus discolor*. The IUCN Red List of Threatened Species 2015:e.T17216A22136476.
Barquez, R., B. Rodriguez, B. Miller, and M. Diaz. 2015k. *Molossus molossus*. The IUCN Red List of Threatened Species 2015:e.T13648A22106602.
Barquez, R., B. Rodriguez, B. Miller, and M. Diaz. 2015l. *Molossus rufus*. The IUCN Red List of Threatened Species 2015:e.T13644A22107969.
Barquez, R., B. Rodriguez, B. Miller, and M. Diaz. 2015m. *Nyctinomops laticaudatus*. The IUCN Red List of Threatened Species 2015:e.T14995A22011208.
Barquez, R., M. Diaz, and J. Arroyo-Cabrales. 2015n. *Nyctinomops macrotis* (errata version published in 2016). The IUCN Red List of Threatened Species 2015:e.T14996A97207443.
Barquez, R., M. Diaz, E. Gonzalez, A. Rodriguez, S. Incháustegui, and J. Arroyo-Cabrales. 2015o. *Tadarida brasiliensis*. The IUCN Red List of Threatened Species 2015:e.T21314A22121621.
Barquez, R., S. Perez, B. Miller, and M. Diaz. 2016a. *Eptesicus brasiliensis*. The IUCN Red List of Threatened Species 2016:e.T7916A22114459.
Barquez, R., S. Perez, B. Miller, and M. Diaz. 2016b. *Eptesicus furinalis*. The IUCN Red List of Threatened Species 2016:e.T7927A22118013.
Barquez, R., M. Diaz, R. Samudio Jr, and J. Arroyo-Cabrales. 2016c. *Myotis albescens*. The IUCN Red List of Threatened Species 2016:e.T14140A22049892.
Baker, R. H. 1951. Two new moles (genus *Scalopus)* from Mexico and Texas. University of Kansas Publications, Museum of Natural History 5:17–24.
Baker, R. H. 1954. The silky pocket mouse (*Perognathus flavus*) of Mexico. University of Kansas Publications, Museum of Natural History 7:339–347.
Baker, R. H. 1955. A new species of bat (genus *Myotis)* from Coahuila, Mexico. Proceedings of the Biological Society of Washington 68:165–166.
Baker, R. H. 1956. Mammals of Coahuila, México. University of Kansas Publications, Museum of Natural History 9:125–335.
Baker, R. H. 1969a. Cotton rat of the *Sigmodon fulviventer* group. Pp. 177–232, *in* Contribution in Mammalogy (Jones Jr., J. K., ed.). Miscellaneous Publication, Museum of Natural History, University of Kansas 51:1–428.
Baker, R. H. 1969b. Cotton rats of the *Sigmodon fulviventer* group. University of Kansas, Museum of Natural History, Miscellaneous Publications 51: 177–232.
Baker, R.H. 1983. Michigan mammals. Michigan State University Press. Lasing, U.S.A.
Baker, R. H. 1984a. Origin, classification, and distribution of the white-tailed deer. Pp. 1–18, *in* White-tailed deer: ecology and management (Halls, L. K. ed.), Stackpole Books. Harrisburg, U.S.A.
Baker, R. H., and B. P. Glas. 1951. The taxonomic status of the Pocket Gopher *Geomys bursarius* and *Geomys breviceps*. Proceedings of the Biological Society of Washington 64:55–58.
Baker, R. H., and J. K. Greer. 1962. Mammals of the Mexican state of Durango. Publications of the Museum, Michigan State University, Biological Series 2:25–154.
Baker, R. H., and C. J. Phillips. 1965. Mammals from EI Nevado de Colima, Mexico Journal of Mammalogy 46:691–693.
Baker, R. H., and K. A. Shump. 1977. *Sigmodon fulviventer*. Mammalian Species 94:1–4.
Baker, R. H., and K. A. Shump. 1978. *Sigmodon ochrognathus*. Mammalian Species 97:1–2.
Baker, R. H., R. G. Webb, and P. Dalby. 1967. Notes on reptiles and mammals from southern Zacatecas. The American Midland Naturalist 77:223–226.
Baker, R. H., and H. J. Stains. 1955. A new long-eared *Myotis* (*Myotis evotis*) from northeastern Mexico. University of Kansas Publications, Museum of Natural History 9:81–84.
Baker, R. J. 1979. Karyology. Pp. 107–155, *in* Biology of bats of the New World family Phyllostomatidae Part III (Baker, R. J., J. K. Jones, Jr., and D. C. Carter, eds.). Special Publications of the Museum, Texas Tech University 16:1–441.
Baker, R. J., and C. L. Clark. 1987. *Uroderma bilobatum*. Mammalian Species 279:1–4.
Baker, R. J., and H. H. Genoways. 1975. A new subspecies of *Geomys bursarius* (Mammalia: Geomyidae) from Texas and New Mexico. Occasional Papers, The Museum Texas Tech University 29:1–18.
Baker, R. J., and H. H. Genoways. 1978. Zoogeography of Antillean bats. Pp. 53–97, *in* Zoogeography in the Caribbean (Gill, F. B., ed.). Special Publication, Academy of Natural Sciences of Philadelphia 13:1–128.
Baker, R. J., and T. C. Hsu. 1970. Chromosomes of the desert shrew, *Notiosorex crawfordi* (Coues). The Southwestern Naturalist 14:448–449.
Baker, R. J., and J. K. Jones, Jr. 1972. *Tadarida aurispinosa* in Sonora, Mexico. The Southwestern Naturalist 17:308–309.
Baker, R. J., and V. R. McDaniel. 1972. A new subspecies of *Uroderma bilobatum* (Chiroptera, Phyllostomatidae) from Middle America. Occasional Papers, the Museum Texas Tech University 7:1–4.
Baker, R. J., and J. L. Patton. 1967. Karyotypes and karyotypic variation of North American vespertilionid bats. Journal of Mammalogy 41:270–286.
Baker, R. J., and C. Ward. 1967. Distribution of bats in southeastern Arkansas. Journal of Mammalogy 48:130–132.
Baker, R. J., and S. J. Williams. 1974. *Geomys tropicalis*. Mammalian Species 35:1–4.
Baker, R. J., H. H. Genoways, S. J. Williams, and J. W. Warner. 1973. Cytotypes and morphometries of two phyllostomatid bats, *Micronycteris hirsuta* and *Vampyressa pusilla*. Occasional Papers of the Museum, Texas Tech University 17: 1–10.

Baker, R. J., W. R. Atchley, and V. R. Mcdaniel. 1972. Karyology and morphometries of Peters' tent-making bat, Uroderma bilobatum Peters (Chiroptera, Phyllostomatidae). Systematic Zoology 21:414– 429.

Baker, R. J., J. W. Bickham, and M. J. Arnold. 1985. Chromosomal evolution in *Rhogeessa* (Chiroptera: Vespertilionidae): possible speciation by centric fusions. Evolution 39:233–243.

Baker, C. G. Dunn, and K. Nelson. 1988. Allozymic study of the relationships of Phylloderma and four species of Phyllostomus. Occasional Papers, Museum of Texas Tech University 125:1–14.

Baker, R. J., S. K. Davis, R. D. Bradley, M. J. Hamilton, and R. A. Van Den Busseh. 1989a. Ribosomal-DNA, mitochondrial-DNA, chromosomal, and allozymic studies on a contact zone in the pocket gopher, *Geomys*. Evolution 43:63–75.

Baker, R. J., C. S. Hood, and R. L. Honeycutt. 1989b. Phylogenetic relationships and classification of the higher categories of the New World bat family Phyllostomidae. Systematic Zoology 38:228–238.

Baker, R. J., C. A. Porter, J. C. Patton, and R. A. Van Den Bussche. 2000. Systematics of bats of the family Phyllostomidae based on RAG2 DNA sequences. Occasional Papers, The Museum of Texas Tech University 202:1–16.

Baker, R. J., S. Solari, and F. G. Hoffmann. 2002. A new Central American species from the *Carollia brevicauda* complex. Occasional Papers of the Museum, Texas Tech University 217:1–12.

Baker, R. J., M. B. O'Neill, and L. R. McAliley. 2003. A new species of desert shrew, Notiosorex, based on nuclear and mitochondrial sequence data. Occasional Papers of the Museum, Texas Tech University 222:1–12.

Baker, R. J., L. C. Bradley, R. D. Bradley, J. W. Dragoo, M. D. Engstrom, R. S. Hoffman, C. A. Jones, F. Reid, D. W. Rice, and C. Jones. 2003a. Revised checklist of North American mammals north of Mexico, 2003. Occasional Papers, Museum of Texas Tech University 229:1–23.

Baker, R. M. Fonseca, D. A. Parish, C. J. Phillips, and F. G. Hoffmann. 2004. New bat of the genus Lophostoma (Phyllostomidae: Phyllostominae) from northwestern Ecuador. Occasional Papers, Museum of Texas Tech University 232:1–16

Baker, R. J., M. M. Mcdonough, V. J. Swier, P. A. Larsen, J. P. Carrera, and L. K. Ammerman. 2009. New species of bonneted bat, genus *Eumops* (Chiroptera: Molossidae) from lowlands of western Ecuador and Peru. Acta Chiropterologica 11:1–13.

Bangs, O. 1896. A review of the squirrels of eastern North America. Proccedings of the Biological Society of Washington 10:145–167.

Banks, R. C. 1964. The mammals of Cerralvo Island, Baja California. Transactions of the San Diego Society of Natural History 13:397–404.

Banfield, A. W. F. 1961. Notes on mammals of the Kluane Game Santuary. Yukon Territory. Natural Museum Canada Bulletin 172:128–135.

Banfield, A. W. F. 1974. The mammals of Canada. University of Toronto Press. Toronto, Canada.

Banfield, A. W. F. 1987. The mammals of Canada. University of Toronto Press. Ontario, Canada.

Banfield, A. W. F. 1962a. A revision of the reindeer and Caribou, genus *Rangifer*. Bulletin Natural Museum Canada 177, Biological series 66:1–137.

Barbee, R. W., and C. M. Fugler. 1977. Varation in the three species of the chiropteran genus *Carollia* of north-western Amazonia. Elisha Mitchell Science Society Journal 93:101.

Barbour, R. W., and W. H. Davis. 1969. Bats of America. University Press of Kentucky. Lexington, U.S.A.

Barbour, R. W., and W. H. Davis. 1970. The status of *Myotis occultus*. Journal of Mammalogy 51:150–151.

Barbour, R. W., and W. H. Davis. 1974. Mammals of Kentucky. University Press of Kentucky. Lexington, U.S.A.

Barkalow, F. S., Jr., and M. Shorten. 1973. The world of the gray squirrel. J. B. Lippincott Company. Philadelphia, U.S.A.

Barnes, L. G., D. P. Domning, and C. E. Ray. 1985. Status of studies on fossil marine mammals. Marine Mammal Science 1:15–53.

Barquez, R. M., N. P. Giannini, and M. A. Mares. 1993. Guide to the bats of Argentina. Oklahoma Museum of Natural History, University of Oklahoma, Norman, U.S.A.

Barquez, R., B. Lim, B. Rodriguez, B. Miller, and M. Diaz. 2015. *Peropteryx macrotis*. The IUCN Red List of Threatened Species 2015:e. T16709A22101100.

Bateman, G. C., and T. A. Vaughan. 1974. Night activities of mormoopid bats. Journal of Mammalogy 55:45–65.

Barthelmes, E. L. 2016. Family Eretizontidae (New World Porcupines). Pp. 372–397, in Handbook of the mammals of the world 6. Lagomorpha and Rodentia I (Wilson, D. E., T. E. Lacher, Jr., and R. A. Mittermeier, eds.). Lynx editions. Barcelona, Spain.

Bartholomew, G. A., and R. A. Boolootlan. 1960. Numbers and population structure of pinnipeds on the California Chanel Islands. Journal of Mammalogy 41:366–375.

Barragán, F., C. Lorenzo, A. Morón, M. A. Briones-Salas, and S. López. 2010. Bat and rodent diversity in a fragmented landscape on the Isthmus of Tehuantepec, Oaxaca, México. Tropical Conservation Science 3:1–16.

Barry, R. 2018. Sylvilagus obscurus. Pp. 149–152, in Pikas, Rabbits and Hares of the World (Smith, A., C. H. Johnston, P. C. Alves, and K. Hackländer, eds.). The Johns Hopkins University Press/IUCN/SSC. Baltimore, U.S.A.

Barry, R., and H. C. Lanier. 2019. *Sylvilagus obscurus*. The IUCN Red List of Threatened Species 2019:e.T41301A45192437.

Bartels, M. A., and P. D. Thompson. 1993. *Spermophilus lateralis*. Mammalian Species 440:1–8.

Bartig, J. L., T. L. Best, and S. L. Burt. 1993. *Tamias bulleri*. Mammalian Species 438:1–4.

Batzli, G. O. 1999. Brown lemming/*Lemmus sibiricus*. Pp. 653–654, *in* The Smithsonian Book of North American Mammals (Wilson, D. E., and S. Ruff, eds.). Smithsonian Institution Press. Washington, U.S.A.

Batzli, G. O., and S. T. Sobaski. 1980. Distribution, abundance, and foraging patterns of ground squirrels near Atkasook, Alaska. Arctic and Alpine Research 12:501–510.

Basantes, M., Tinoco, N., Velazco, P., Hofmann, M., Rodriguez-Posada, M., Camacho, M. 2020. Systematics and Taxonomy of Tonatia saurophila Koopman & Williams, 1951 (Chiroptera, Phyllostomidae). ZooKeys 915:59—86.

Baumgardner, G. D., K. T. Wilkins, and D. J. Schmidly. 1977. Noteworthy additions to the bat fauna of the Mexican states of Tamaulipas (San Carlos Mountains) and Queretaro. Mammalia 41:237–238.

Baumgardner, G. D. 1991. *Dipodomys compactus*. Mammalian Species 369:1–4.

Baumgardner, G. D., and D. J. Schmidly. 1981. Systematics of the southern races of two species of kangaroo rats (*Dipodomys compactus* and *D. ordii*). Occasional Papers, The Museum, Texas Tech University 73:1–27.

Baumgartner, L. L. 1943. Pelage studies of fox squirrels (*Sciurus niger rufiventer*). The American Midland Naturalist 29:588–590.

Beauchamp-Martin, S. L., F. B. Stangl, Jr., D. J. Schmidly, R. D. Stevens, and R. D. Bradley. 2019. Systematic review of Botta's pocket gopher (*Thomomys bottae*) from Texas and southeastern New Mexico, with description of a new taxon. Pp. 515–542, *in* From field to laboratory: a memorial volume in honor of Robert J. Baker (Bradley, R. D., H. H. Genoways, D. J. Schmidly, and L. C. Bradley, eds.). Special Publications, Museum of Texas Tech University. Lubbock, U.S.A.

Bee, J. W., and E. R. Hall. 1956. Mammals of northern Alaska on the Arctic Slope. Miscelaneous Publications Museum Natural History, University of Kansas 8:1–309.

Beer, J. R. 1965. The interparietal in kangaroo rats. The Southwestern Naturalist 10:145–150.

Beever, E. A. 2018. *Lepus othus*. Pp. 204–205, *in* Pikas, Rabbits and Hares of the World (Smith, A., C. H. Johnston, P. C. Alves, and K. Hackländer, eds.). The Johns Hopkins University Press/IUCN/SSC. Baltimore, U.S.A.

Beever, E. A., and J. French. 2018. *Sylvilagus nuttallii*. Pp. 147–149, *in* Pikas, Rabbits and Hares of the World (Smith, A., C. H. Johnston, P. C. Alves, and K. Hackländer, eds.). The Johns Hopkins University Press/IUCN/SSC. Baltimore, U.S.A.

Beever, E. A., D. E. Brown, and C. Lorenzo. 2018a. *Lepus californicus*. Pp. 170–173, *in* Pikas, Rabbits and Hares of the World (Smith, A., C. H. Johnston, P. C. Alves, and K. Hackländer, eds.). The Johns Hopkins University Press/IUCN/SSC. Baltimore, U.S.A.

Beever, E. A., D. E. Brown, and J. Berger. 2018b. *Lepus townsendii*. Pp. 218–220, *in* Pikas, Rabbits and Hares of the World (Smith, A., C. H. Johnston, P. C. Alves, and K. Hackländer, eds.). The Johns Hopkins University Press/IUCN/SSC. Baltimore, U.S.A.

Bekoff, M. 1977. *Canis latrans*. Mammalian Species 79:1–9.

Belant, J., D. Biggins, D. Garelle, R. G. Griebel, and J. P. Hughes. 2015. *Mustela nigripes*. The IUCN Red List of Threatened Species 2015:e. T14020A45200314.

Belcher, R. L., and T. E. Lee, Jr. 2002. *Arctocephalus townsendi*. Mammalian Species 700:1–5.

Belk, M. C., and H. D. Smith. 1991. *Ammospermophilus leucurus*. Mammalian Species 368:1–8.

Bell, D. M., M. J. Hamilton, C. W. Edwards, L. E. Wiggins, R. Muñiz Martínez, R. E. Strauss, R. D. Bradley, and R. J. Baker. 2001. Patterns of karyotypic megaevolution in *Reithrodontomys*: evidence from a cytochrome-*b* phylogenetic hypothesis. Journal of Mammalogy 82:81–91.

Bello-Gutiérrez, J., R. Reyna-Hurtado, and W. Jorge. 2010. Central American red brocket deer *Mazama temama* (Kerr 1792). Pp. 166– 171, *in* Neotropical Cervidology: Biology and Medicine of Latin American Deer (Duarte, J. M. B., and S. González, eds.). Funep/IUCN. Jaboticabal, Brazil and Gland, Switzerland.

Bello, J., R. Reyna, and J. Schipper. 2016. *Mazama temama*. The IUCN Red List of Threatened Species 2016:e.T136290A22164644.

Benedict, F. A. 1957. Hair structure as a generic character in bats. University of California Publications in Zoology, 59:285–547.

Benedict, R. A., J. D. Druecker, and H. H. Genoways. 1999. New records and habitat information for *Sorex merriami* in Nebraska. Great Basin Naturalist 59:285–287.

Benedict, R. A., H. H. Genoways, and J. R. Choate. 2006. Taxonomy of the short-tailed shrews (genus *Blarina*) in Florida. Occasional Papers, Museum of Texas Tech University 251:1–19.

Benedict, B. D., A. A. Castellanos, and J. E. Light. 2019. Phylogeographic assessment of the Heermann's kangaroo rat (*Dipodomys heermanni*). Journal of Mammalogy 100:72–91.

Beneski, Jr., J. T., and D. W. Stinson. 1987. *Sorex palustris*. Mammalian Species 296:1–6.

Benson, D. L., and F. R. Gehlbach. 1979. Ecological and taxonomical notes on the rice rat (*Oryzomys couesi*) in Texas. Journal of Mammalogy 60:225–228.

Benson, S. B. 1933. Concealing coloration among some desert rodents of the southwestern United States. University of California Publications in Zoology 40:1–7.

Benson, S. B. 1940. New subspecies of the canyon mouse (*Peromyscus crinitus*) from Sonora, Mexico. Proceedings Biological Society of Washington 53:1–4.

Benson, S. B. 1947. Description of a mastiff bat (genus *Eumops*) from Sonora, Mexico. Proceedings of the Biological Society of Washington 60:133–134.

Benson, S. B., and R. M. Bond. 1939. Notes on *Sorex merriami* Dobson. Journal of Mammalogy 20:348–351.

Berna, H. J. 1990. Observations on the dwarf shrew (Sorex nanus) in northern Arizona. Great Basin Naturalist 50:161–165

Bertram, G. C. L., and C. K. R. Bertram. 1964. Manatees in the Guianas. Zoologica 49:115–120.

Berta, A., and M. Churchill. 2011. Pinniped taxonomy: review of currently recognized species and subspecies, and evidence used for their description. Mammal Review 42:207–234.

Best, T. L. 1978. Variation in kangaroo rats (genus *Dipodomys)* of the *heermanni* group in Baja California, Mexico. Journal of Mammalogy 59:160–175.

Best, T. L. 1981. Bacular variation in kangaroo rats (genus *Dipodomys*) of the *heermanni* group in Baja California, Mexico. The Southwestern Naturalist 25:529–534.

Best, T. L. 1982. Relationships between ecogeographic and morphologic variation of the agile kangaroo rat (*Dipodomys agilis*) in Baja California, Mexico. Bulletin of the Southern California Academy of Sciences 80:60–69.

Best, T. L. 1983a. Morphologic variation in the San Quintin kangaroo rat (*Dipodomys gravipes* Huey 1925). The American Midland Naturalist 109:409–413.

Best, T. L. 1983b. Intraspecific variation in the agile kangaroo rat *(Dipodomys agilis)*. Journal of Mammalogy 64:426–436.

Best, T. L. 1986. *Dipodomys elephantinus*. Mammalian Species 255:1–4.

Best, T. L. 1987. Sexual dimorphism and morphometric variation in the Texas kangaroo rat (*Dipodomys elator* Merriam 1894). The Southwestern Naturalist 32:53–59.

Best, T. L. 1988a. *Dipodomys nelsoni*. Mammalian Species 326:1–4.

Best, T. L. 1988b. *Dipodomys spectabilis*. Mammalian Species 311:1–10.

Best, T. L. 1991. *Dipodomys nitratoides*. Mammalian Species 381:1–7.

Best, T. L. 1992a. *Dipodomys margaritae*. Mammalian Species 400:1–3.

Best, T. L. 1992b. *Dipodomys venustus*. Mammalian Species 403:1–4.

Best, T. L. 1993a. *Tamias palmeri*. Mammalian Species 443:1–6.

Best, T. L. 1993b. *Tamias sonomae*. Mammalian Species 444:1–5.

Best, T. L. 1993c. *Perognathus inornatus*. Mammalian Species 450:1–5.

Best, T. L. 1993d. *Chaetodipus lineatus*. Mammalian Species 451:1–3.

Best, T. L. 1993e. *Tamias ruficaudus*. Mammalian Species 452:1–7.

Best, T. L. 1993f. Patterns of morphologic and morphometric variation in heteromyid rodents. Pp. 197–235, *in* Biology of the Heteromyidae (Genoways, H. H., and J. H. Brown, eds.). Special Publication, The American Society of Mammalogists 10:1–719.

Best, T. L. 1994a. *Perognathus alticolus*. Mammalian Species 463:1–4.

Best, T. L. 1994b. *Chaetodipus nelsoni*. Mammalian Species 484:1–6.

Best, T. L. 1995a. *Sciurus variegatoides*. Mammalian Species 500:1–6.

Best, T. L. 1995b. *Spermophilus mohavensis*. Mammalian Species 509:1–7.

Best, T. L. 1995c. *Sciurus oculatus*. Mammalian Species 498:1–3.

Best, T. L. 1995d. *Sciurus nayaritensis*. Mammalian Species 492:1–5.

Best, T. L. 1995e. *Sciurus deppei*. Mammalian Species 505:1–5.

Best, T. L. 1995f. *Sciurus colliaei*. Mammalian Species 497:1–4.

Best, T. L. 1995g. *Sciurus alleni*. Mammalian Species 501:1–4.

Best, T. L. 1995h. *Spermophilus annulatus*. Mammalian Species 508:1–4.

Best, T. L. 1995i. *Spermophilus adocetus*. Mammalian Species 504, 20:1–4.

Best, T. L. 1996. *Lepus californicus*. Mammalian Species 530:1–10.

Best, T. L., and H. T. Hill. 1994. Lepus arcticus. Mammalian Species. American Society of Mammalogists 457:1–9.

Best, T. L., and G. Ceballos. 1995. *Spermophilus perotensis*. Mammalian Species 507:1–3.

Best, T. L., and N. J. Granai. 1994a. *Tamias obscurus*. Mammalian Species 472:1–6.

Best, T. L., and N. J. Granai. 1994b. *Tamias merriami*. Mammalian Species 476:1–9.

Best, T. L., and T. H. Henry. 1993. *Lepus alleni*. Mammalian Species 424:1–8.

Best, T. L., and T. H. Henry. 1994a. *Lepus arcticus*. Mammalian Species 457:1–9.

Best, T. L., and T. H. Henry. 1994b. *Lepus othus*. Mammalian Species 458:1–5.

Best, T. L., and J. A. Lackey. 1985. *Dipodomys gravipes*. Mammalian Species 236:1–4.

Best, T. L., and J. A. Lackey. 1992a. *Chaetodipus artus*. Mammalian Species 418:1–3.

Best, T. L., and J. A. Lackey. 1992b. *Chaetodipus pernix*. Mammalian Species 420:1–3.

Best, T. L., and J. B. Jennings. 1997. *Myotis leibii*. Mammalian Species 547:1–6.

Best, T. L., and S. Riedel. 1995. *Sciurus arizonensis*. Mammalian Species 496:1–5.

Best, T. L., and G. D. Schnell. 1974. Bacular variation kangaroo rats (genus *Dipodomys*). The American Midland Naturalist 91:257–270.

Best, T. L., and M. P. Skupski. 1994a. *Perognathus flavus*. Mammalian Species 471:1–10.

Best, T. L., and M. P. Skupski. 1994b. *Perognathus merriami*. Mammalian Species 473:1–7.

Best, T. L. and H. H. Thomas. 1991a. *Dipodomys insularis*. Mammalian Species 374:1–3.

Best, T. L., and H. H. Thomas. 1991b. *Spermophilus madrensis*. Mammalian Species 378:1–2.

Best, T. L., R. M. Sullivan, J. A. Cook, and T. Yates. 1986. Chromosomal, genetic and morphologic variation in the agile kangaroo rat, *Dipodomys agilis* (Rodentia: Heteromyidae). Systematic Zoology 35:311–324.

Best, T. L., C. L. Lewis, K. Caesar, and A. S. Titus. 1990a. *Ammospermophilus interpres*. Mammalian Species 365:1–6.

Best, T. L., A. S. Titus, K. Caesar, and C. L. Lewis. 1990b. *Ammospermophilus harrisii*. Mammalian Species 366:1–7.

Best, T. L., A. S. Titus, C. L. Lewis, and K. Caesar. 1990c. *Ammospermophilus nelsoni*. Mammalian Species 367:1–7.

Best, T. L., J. L. Bartig, and S. L. Burt. 1992. *Tamias canipes*. Mammalian Species 411:1–5.

Best, T. L., S. L. Burt, and J. L. Bartig. 1993. *Tamias durangae*. Mammalian Species 437:1–4.

Best, T. L., S. L. Burt, and J. L. Bartig. 1994a. *Tamias quadrivittatus*. Mammalian Species 466:1–7.

Best, T. L., R. G. Clawson, and J. A. Clawson. 1994b. *Tamias speciosus*. Mammalian Species 478:1–9.

Best, T. L., R. G. Clawson, and J. A. Clawson. 1994c. *Tamias panamintinus*. Mammalian Species 468:1–7.

Best, T. L., R. K. Chesser, D. A. McCullough, and G. D. Baumgardner. 1996a. Genic and Morphometric Variation in Kangaroo Rats, Genus *Dipodomys*, from Coastal California. Journal of Mammalogy 77:785–800.

Best, T. L., W. M. Kiser, and P. W. Freeman. 1996b. *Eumops perotis*. Mammalian Species 534:1–8.

Best, T. L., W. M. Kiser, and J. C. Rainey. 1997. *Eumops glaucinus*. Mammalian Species 551:1–6.

Best, T. L., J. L. Hunt, L. A. McWilliams, and K. G. Smith. 2001. *Eumops hansae*. Mammalian Species 687:1–3.

Best, T. L., J. L. Hunt, L. A. Williams, and K. G. Smith. 2002. *Eumops auripendulus*. Mammalian Species 708:1–5.

Beever, E. A., D. E. Brown, and C. Lorenzo. 2018a. Lepus californicus. Pp. 170–173, in Pikas, Rabbits and Hares of the World (Smith, A., C. H. Johnston, P. C. Alves, and K. Hackländer, eds.). The Johns Hopkins University Press/IUCN/SSC. Baltimore, U.S.A.

Beever, E. A., D. E. Brown, and J. Berger. 2018b. Lepus townsendii. Pp. 218–220, in Pikas, Rabbits and Hares of the World (Smith, A., C. H. Johnston, P. C. Alves, and K. Hackländer, eds.). The Johns Hopkins University Press/IUCN/SSC. Baltimore, U.S.A.

Bhatnagar, K. P. 1978. Breech presentation in the hairy-legged vampire, *Diphylla ecaudata*. Journal of Mammalogy 59:864–866.

Bininda-Emonds, O. R. P., J. L. Gittleman, and A. Purvis. 1999. Building large trees by combining phylogenetic information: A complete phylogeny of the extant Carnivora (Mammalia). Biological Reviews 74:143–175.

Birkenholz, D. E. 1972. *Neofiber alleni*. Mammalian Species 15:1–4.

Birney, E. C., and H. H. Genoways. 1973. Chromosomes of *Spermophilus adocetus* (Mammalia; Sciuridae), with comments on the subgenetic affinities of the species. Experimentia 29:228–229.

Birney, E. C., and J. K. Jones, Jr. 1972. Woodrats (genus *Neotoma*) of Sinaloa, Mexico. Transactions of the Kansas, Academy of Science 74:197–211.

Birney, E. C. 1973. Systematics of three species of woodrats (genus *Neotoma*) in central North America. Miscellaneous Publications of the Museum of Natural History, University of Kansas 58:1–173.

Birney, E. C., J. B. Bowles., R. M. Timm, and S. L. Williams. 1974. Mammalian distributional records in Yucatán and Quintana Roo, with comments on reproduction, structure, and status of peninsular populations. Bell Museum Natural History, University of Minnesota, Occasional Papers 13:1–25.

Bisbal E. F. J. 1991. Distribución y taxonomía del venado Matacán (*Mazama* sp) en Venezuela. Acta Biológica Venezuelica 13:89–104.

Blair, W. F. 1940. Two cases of abnormal coloration in mammals. Journal of Mammalogy 21:461–462.

Blair, W. F. 1954. Mammals of the mesquite plains biotic district in Texas and Oklahoma, and speciation in the central grasslands. Texas Journal of Science 6:235–264.

Blanco, J. C. 1998. Mamíferos de España. Planeta. Madrid, España.

Blankenship, D. J., and G. L. Bradley. 1985. Electrophoretic comparison of two southern California chipmunks (*Tamias obscurus* and *Tamias merriami*). Bulletin of the Southern California Academy of Sciences 84:48–50.

Blair, F. W., A. P. Blair, P. Brodkorb, F. R. Cagle, and G. A. Moore. 1957. Vertebrates of the United States. McGraw-Hill. New York, U.S.A.

Bleich, V. C. 1977. *Dipodomys stephensi*. Mammalian Species 73:1–3.

Block, S. B., and E. G. Zimmerman. 1991. Allozymic variation and systematics of plains pocket gophers (*Geomys*) of south-central Texas. The Southwestern Naturalist 36:29–36.

Blois, J. and NatureServe (Clausen, M. K., G. Hammerson, and S. Cannings). 2008. *Arborimus pomo*. The IUCN Red List of Threatened Species 2008:e.T2018A9174950.

Blood, B. R., and M. K. Clark. 1998. *Myotis vivesi*. Mammalian Species 588:1–5.

Bobrinskii, N. A. 1965. Order Carnivora. Pp. 117–163, in Opre delitl'mlekopitayushchikh SSSR [Guide to the mammals of the USSR]. (N. A. Bobrinskii, B. A. Kuznetsov, and A. P. Kuzyakin, eds.). Proveshchenie, Moscow, Russia.

Bobrinskii, N. A., B. A. Kuznetsov, and A. P. Kuzyakin. 1944. Opredelitl' mlekopitayushchikh SSSR [Guide to the mammals of the U.S.S.R.]. Sovietskaya Nauka, Moscow, Russia.

Bock, A. 2020. *Lepus europaeus* (Lagomorpha: Leporidae), Mammalian Species 52:125–142.

Bogan, M. A. 1974. Identification of *Myotis californicus* and *M. leibii* in southwestern North America. Proceedings of the Biological Society of Washington 87:49–56.

Bogan, M. A. 1978. A new species of *Myotis* from the Islas Tres Marias, Nayarit, Mexico, with comments on variation in *Myotis nigricans*. Journal of Mammalogy 59:519–530.

Bogan, M. A. 1997. On the status of *Neotoma varia* from Isla Datil, Sonora. Pp. 81–88, *in* Life among the muses: papers in honor of James S. Findley (Yates, T. L., W. L. Gannon, and D. Wilson, eds.). Special Publication, the Museum of Southwestern Biology 3:1–290.

Bogan, M. A. 1999. Family Vespertilionidae. Pp. 139– 181, *in* Mamíferos del noroeste de México (Álvarez-Castañeda, S. T., and J. L. Patton, eds.). Centro de Investigaciones Biológicas del Noroeste, S. C., La Paz, México.

Boitani, L., M. Phillips, and Y. Jhala. 2018. *Canis lupus* (errata version published in 2020). The IUCN Red List of Threatened Species 2018:e. T3746A163508960.

Bogdanowicz, W., S. Kasper, and R. D. Owen. 1998. Phylogeny of Plecotine bats: reevaluation of morphological and chromosomal data. Journal of Mammalogy 79:78–90.

Bole, B. P., Jr., and P. N. Moulthrop. 1942. The Ohio Recent mammal collection in the Cleveland Museum of Natural History. Science Publications, Cleveland Museum Natural History 5:83–181.

Bonner, W. N. 1981. Grey seal *Halichoerus grypus* Fabricius. 1791. Pp. 111–144, *in* Handbook of marine mammals (Ridgway, S. H., and R. Harrison, eds). Academic Press. London, United Kingdom.

Boskovic, R., K. M. Kovacs, M. O. Hammill, and B. N. White. 1996. Geographic distribution of mitochondrial DNA haplotypes in grey seals (*Halichoerus grypus*). Canadian Journal of Zoology 74:1787–1796.

Boulware, J. T. 1943. Two new subspecies of kangaroo rats (genus *Dipodomys*) from southern California. University of California Publications in Zoology 46:391–396.

Boveng, P. 2016. *Phoca largha*. The IUCN Red List of Threatened Species 2016:e.T17023A45229806.

Bowen, D. 2016. *Halichoerus grypus*. The IUCN Red List of Threatened Species 2016:e.T9660A45226042.

Bowles, J. B. 1975. Distribution and biogeography of mammals of lowa. Special Publication Museum of Texas Tech University 9:1–184.

Bowyer, R. T., and D. M. Leslie. 1992. *Ovis dalli*. Mammalian Species 393:1–7.

Boyce, M. S. 1978. Climatic variability and body size variation in the muskrats (*Ondatra zibethicus*) of North America. Oecologia 36:1–19.

Boyce, M. S. 1980. First record of the fringe-tailed bat, *Myotis thysanodes,* from southeastern Wyoming. The Southwestern Naturalist 25:114–115.

Boyeskorov, G. 1999. New data on moose (*Alces*, Artiodactyla) Systematic. Säugetierkundliche Mitteilungen 44:3–13.

Bradbury, J. W., and S. L. Vehrencamp. 1976. Social organization and foraging in emballonurid bats: I. Field studies. Behavioral Ecology and Sociobiology 1:337–381.

Bradley, R. D. 2017. Genus *Reithrodontomys*. Pp. 367–383, *in* Handbook of the mammals of the world: rodents II (Wilson, D. E., T. E. Lacher, and R. A. Mittermeier, eds.). Lynx Edicions, Barcelona, Spain.

Bradley, R. D., and R. J. Baker. 1999. *Geomys knoxjonesi*. Pp. 486–487, *in* The Smithsonian Book of North American Mammals (Wilson, D. E., and S. Ruff, eds.). Smithsonian Institution Press. Washington, U.S.A.

Bradley, R. D., and M. R. Mauldin. 2016. Molecular data indicate a cryptic species in *Neotoma albigula* (Cricetidae: Neotominae) from northwestern Mexico. Journal of Mammalogy 97:187–199.

Bradley, R. D., and D. J. Schmidly. 1987. The glans penes and bacula in Latin American taxa of the *Peromyscus boylii* group. Journal of Mammalogy 68:595–616.

Bradley, R. D., D. J. Schmidly, and C. W. Kilpatrick. 1996. The relationships of *Peromyscus sagax* to the *P. boylii* and *P. truei* species groups in Mexico based on morphometric, karyotipic, and allozymic data. Pp. 95–106, *in* Contributions in Mammalogy: a memorial volume honoring Dr. J. K. Jones, Jr. (Genoways, H. H., and R. J. Baker, eds.). Museum Texas Tech University. Lubbock, U.S.A.

Bradley, R. D., D. J. Schmidley, and C. Jones. 1999. The northern rock mouse, *Peromyscus nasutus* (Mammalia: Rodentia), from Davis Mountain Texas. Occasional papers, Museum of Texas Tech University 190:1–3.

Bradley, R. D., I. Tiemann-Boege, C. W. Kilpatrick, and D. J. Schmidly. 2000. Taxonomic status of *Peromyscus boylii sacarensis*: interferences from DNA sequences of the mitochondrial cytochrome–*b* gene. Journal of Mammalogy 81:875–884.

Bradley, R. D., D. S. Carroll, M. L. Haynie, R. Muñiz-Martínez, M. J. Hamilton, and W. L. Kilpatrick. 2004a. A new species of *Peromyscus* from western Mexico. Journal of Mammalogy 85:1184–1193.

Bradley, R. D., F. Mendez-Harclerode, M. J. Hamilton, and G. Ceballos. 2004b. A new species of *Reithrodontomys* from Guerrero, Mexico. Occasional Papers of the Museum of Texas Tech University 231:1–12.

Bradley, R. D., N. D. Durish, D. S. Rogers, J. R. Miller, M. D. Engstrom, and C. W. Kilpatrick. 2007. Toward a molecular phylogeny for *Peromyscus*: evidence from mitochondrial cytochrome-*b* sequences. Journal of Mammalogy 88:1146–1159.

Bradley, R. D., L. K. Ammerman, R. J. Baker, L. C. Bradley, J. A. Cook, R. C. Dowler, C. Jones, D. J. Schmidly, F. B. Stangl, Jr., R. A. Van Den Bussche, and B. Würsig. 2014a. Revised checklist of North American mammals north of Mexico, 2014. Museum of Texas Tech University 327:1–27.

Bradley, R. D., N. Ordóñez-Garza, C. G. Sotero-Caio, H. M. Huynh, C. W. Kilpatrick, L. I. Iñiguez-Dávalos, and D. J. Schmidly. 2014b. Morphometric, karyotypic, and molecular evidence for a new species of *Peromyscus* (Cricetidae: Neotominae) from Nayarit, Mexico. Journal of Mammalogy 95:176–186.

Bradley, R. D., D. J. Schmidly, B. R. Amman, R. N. Platt II, K. M. Neumann, H. M. Huynh, R. Muñiz-Martínez, C. López-González, and N. Ordóñez-Garza. 2015. Molecular and morphometric data reveal multiple species in *Peromyscus pectoralis*. Journal of Mammalogy 96:446–459.

Bradley, R. D., N. Ordoñez-Garza, G. Ceballos, D. S. Rogers, and D. J. Schmidly. 2016a. A new species in the *Peromyscus boylii* species group (Cricetidae: Neotominae) from Michoacán, México. Journal of Mammalogy 97:154–165.

Bradley, R. D., M. Nuñez-Tabares, T. J. Soniat, S. Kerr, R. W. Raymond, and N. Ordóñez-Garza. 2016b. Molecular systematics and phylogeography of *Peromyscus nudipes* (Cricetidae: Neotominae). Pp. 201–214, *in* Contributions in Natural History: a memorial volume in honor of Clyde Jones (Manning, R. M., J. R. Goetze, and F. D. Yancey, II, eds.). Special Publications, Museum of Texas Tech University 65:1–273.

Bradley, R. D., N. Ordóñez-Garza, and L. C. Bradley. 2017. Oaxacan deermouse *Peromyscus oaxacensis*. Pp. 392, *in* Handbook of the Mammals of the World, Volume 7, Rodents II (Wilson, D. E., T. E. Lacher, Jr., and R. A. Mittermeier, eds.). Lynx Edicions. Barcelona, Spain.

Bradley, R. D., J. Q. Francis, R. N. Platt II , T. J. Soniat, D. Alvarez, and L. L. Lindsey. 2019. Mitochondrial DNA Sequence Data Indicate Evidence for Multiple Species within *Peromyscus maniculatus*. Special Publications of the Museum of Texas Tech University 70:1–59.

Bradley, R. D., N. Ordóñez-Garza, C. W. Thompson, E. A. Wright, G. Ceballos, C. W. Kilpatrick, and D. J. Schmidly. 2022a. Two new species of *Peromyscus* (Cricetidae: Neotominae) from the Transverse Volcanic Belt of Mexico. Journal of Mammalogy 103:255–274.

Bradley, R. D., C. W. Edwards, L. L. Lindsey, J. R. Bateman, M. N. B. Cajimat, M. L. Milazzo, C. F. Fulhorst, M. D. Matocq, and M. R. Mauldin. 2022b. Reevaluation of the phylogenetic relationships among Neotomini rodents (*Hodomys*, *Neotoma*, and *Xenomys*) and comments on the woodrat classification. Journal of Mammalogy 103:1221–1236.

Bradley, W. G., and L. Baker. 1965. Skull measurements of desert bighorn sheep from the Desert Game Range. Transactions of the Desert Bighorn Council 9:70–74.

Bradley, W. G., and L. R. Baker. 1967. Range of variation in Nelson bighorn sheep from the Desert Game Range and its taxonomic significance. Transactions of the Desert Bighorn Council 11:114–140.

Brandborg, S. M. 1955. Life history and management of the mountain goat in Idaho. Wildlife Bulletin of the Idaho Department Fish and Game 2:1–142.

Brandon-Jones, D. 2006. Apparent confirmation that *Alouatta villosa* (Gray, 1845) is a senior synonym of *A. pigra* Lawrence, 1933 as the species-group name for the black howler monkey of Belize, Guatemala, and Mexico. Primate Conservation 21:41–43.

Braun, J. K., and M. A. Mares. 1989. *Neotoma micropus*. Mammalian Species 330:1–9.

Braun, J. K., S. B. Gonzalez-Perez, G. M. Street, J. M. Mook, and N. J. Czaplewski. 1988. *Phenacomys ungava* (Rodentia: Cricetidae). Mammalian Species 45:18–29.

Braun, J. K., Q. D. Layman, and M. A. Mares. 2009. *Myotis albescens*. Mammalian Species 846:1–9.

Braun, J. K., A. A. Johnson, and M. A. Mares. 2011a. *Tamias umbrinus* (Rodentia: Sciuridae). Mammalian Species 43:216–227.

Braun, J. K., T. S. Eaton, and M. A. Mares. 2011b. *Marmota caligata* (Rodentia: Sciuridae). Mammalian Species 43:155–171.

Braun, J. K., B. Yang, S. B. González-Pérez, and M. A. Mares. 2015. *Myotis yumanensis* (Chiroptera: Vespertilionidae). Mammalian Species 47:1–14.

Braun, J. K., S. B. C. Coyner, and M. A. Mares. 2021. Modern extirpation of the Texas kangaroo rat, *Dipodomys elator*, in Oklahoma: changing land use and climate over a century of time as the road to eventual extinction. Therya 12:177–186.

Briggs, K. T., and V. G. Morejohn. 1976. Dentition, cranial morphology and evolution in elephant seals. Mammalia 40:199–222.

Briones-Salas, M., and V. Sánchez-Cordero. 2004. Mamíferos. Pp. 423–447, *in* Biodiversidad de Oaxaca (García-Mendoza, A. J., M. J. Ordoñez, and M. Briones-Salas eds.). Instituto de Biología, UNAM, Fondo Oaxaqueño para la Conservación de la Naturaleza, World Wildlife Fund. Distrito Federal, México.

Brinck, C., S. Erlinge, and M. Sandell. 1983. Anal sac secretion in mustelids: a comparison. Journal of Chemical Ecology 9:727–745.

Brito, D., D. Astúa, D. Lew, and N. de la Sancha. 2021. *Metachirus nudicaudatus* (amended version of 2015 assessment). The IUCN Red List of Threatened Species 2021:e.T40509A197311536.

Broadbooks, H. E. 1965. Ecology and distribution of the pikas of Washington and Alaska. The American Midland Naturalist 73:299–335.

Brook, S. M., J. Pluháček, R. Lorenzini, S. Lovari, M. Masseti, O. Pereladova, and S. Mattioli. 2018. *Cervus canadensis*. The IUCN Red List of Threatened Species 2018:e.T55997823A142396828.

Brown, R. J. 1974. A comparative study of the chromosomes of three species of shrews, Sorex bendirii, Sorex trowbridgii, and Sorex vagrans. Wasmann Journal of Biology 32:303–326.

Brown, D. E. 1984. Arizona's tree squirrels. Arizona Game and Fish Department. Phoenix, U.S.A.

Brown, D. E. 1985. The grizzly in the Southwest. University of Oklahoma Press, Norman,

Brown, D. E. 1989. The ocelot. Pp. 421–433, *in* Audubon wildlife report (Chandler, W. J., ed.). Harcourt Brace Jovanovich. San Diego, U.S.A.

Brown, D. E., and A. T. Smith. 2019. *Lepus callotis*. The IUCN Red List of Threatened Species 2019:e.T11792A45177499.

Brown, D. E., R. D. Babb, C. Lorenzo, and M. M. Altemus. 2014. Ecology of the Antelope Jackrabbit (*Lepus alleni*). The Southwestern Naturalist 59:577–589.

Brown, D. E., G. Beatty, J. E. Brown, and A. T. Smith. 2018a. History, status, and population trends of cottontail rabbits and jackrabbits in the western United States. Western Wildlife 5:16–42.

Brown, D. E., C. Lorenzo, and M. Altemus. 2018b. *Lepus alleni*. Pp. 159–163, *in* Pikas, Rabbits and Hares of the World (Smith, A., C. H. Johnston, P. C. Alves, and K. Hackländer, eds.). The Johns Hopkins University Press/IUCN/SSC. Baltimore, U.S.A.

Brown, D. E., C. Lorenzo, and M. B. Traphagen. 2018c. *Lepus callotis*. Pp. 173–176, *in* Pikas, Rabbits and Hares of the World (Smith, A., C. H. Johnston, P. C. Alves, and K. Hackländer, eds.). The Johns Hopkins University Press/IUCN/SSC. Baltimore, U.S.A.

Brown, D. E., C. Lorenzo, and S. T. Álvarez-Castañeda. 2019. *Lepus californicus*. The IUCN Red List of Threatened Species 2019:e. T41276A45186309.

Brown, D. E., M. B. Traphagen, C. Lorenzo, and M. Gomez-Sapiens. 2018d. Distribution, status and conservation needs of the white-sided jackrabbit, *Lepus callotis* (Lagomorpha). Revista Mexicana de Biodiversidad 89:313–324.

Brown, J. H., and G. A. Lieberman. 1973. Resource utilization and coexistence of seed-eating desert rodents in sand dune habitats. Ecology 54:788–797.

Brown, R. H., H. H. Genoways, and J. K. Jones, Jr. 1971. Bacula of some Neotropical bats. Mammalia 35:456–464.

Brown, R. J., and R. L. Rudd. 1981. Chromosomal comparisons within the *Sorex ornatus-S. vagrans* complex. Wassman Journal Biology 39:30–35.

Brunet, A. K., R. M. Zink, K. M. Kramer, R. C. Blackwell-Rago, S. L. Farrell, T. V. Line, and E. C. Birney. 2002. Evidence of introgression between masked shrews (*Sorex cinereus*), and prairie shrews (*Sorex haydeni*), in Minnesota. The American Midland Naturalist 147:116–122.

Brunner, S. 2004. Fur seals and sea lions (Otariidae): identification of species and taxonomic review. Systematics and Biodiversity 1:339–439.

Bryant, W. E. 1945. Phylogeny of Neartic Sciuridae. The American Midland Naturalist 33:257–390.

Bucher, J. E., and S. R. Hoffmann.1980. *Caluromys derbianus*. Mammalian Species 140:1–4.

Buck, C. L., and B. M. Barnes. 1999. Temperatures of hibernacula and changes in body composition of Arctic ground squirrels over winter. Journal of Mammalogy 80:1264–1276.

Buechner, H. K. 1960. The bighorn sheep in the United States, its past, present and future. Wildlife Monographs 4:1–174.

Bueler, L. E. 1973. Wild dogs of the world. Stein and Day, New York.

Burn, D. M., and A. M. Doroff. 2005. Decline in sea otter (*Enhydra lutris*) populations along the Alaska Peninsula, 1986–2001. Fishery Bulletin 103:270–279.

Burns, J. J. and F. H. Fay. 1972. Comparative morphology of the skull of the ribbon seal, Histriophoca fasciata, with remarks on systematics of Phocidae. Journal of Zoology 161:363–394.

Burns, J. C., J. R. Choate, and E. G. Zimmerman. 1985. Systematic relationships of Pocket Gophers (genus *Geomys*) on the central Great Plains. Journal of Mammalogy 66:102–118.

Burns, J. J. 1981a. Bearded seal *Erignathus barbatus* Erxleben, 1777. Pp. 145–170, *in* Handbook of marine mammals (Ridgway, S. H., and R. Harrison, eds.). Academic Press. New York, U.S.A.

Burns, J. J. 1981b. Ribbon seal *Phoca fasciata* Zimmerman, 1783. Pp. 89–109, *in* Handbook of marine mammals (Ridgway, S. H., and R. Harrison, eds). Academic Press. New York, U.S.A.

Burns, J. J., and F. H. Fay. 1972. Comparative morphology of the skull of the ribbon seal, *Histriophoca fasciata*, with remarks on systematics of Phocidae. Journal of Zoology 161:363–394.

Burt, M. S., and R. C. Dowler. 1999. Biochemical systematics of *Geomys breviceps* and two chromosomal races of *Geomys attwateri* in eastern Texas. Journal of Mammalogy 80:799–809.

Burt, S. L., and T. L. Best. 1994. *Tamias rufus*. Mammalian Species 460:1–6.

Burt, W. H. 1932. Descriptions of heretofore unknown mammals from islands in the Gulf of California, Mexico. Transactions of the San Diego Society of Natural History 7:161–182.

Burt, W. H. 1936. A study of the baculum in the genera *Perognathus* and *Dipodomys*. Journal of Mammalogy 17:145–156.

Burt, W. H. 1960. Bacula of North American mammals. Miscellaneous Publications of the Museum of Zoology, University of Michigan 113:1–76.

Burt, W. H., and F. S. Barkalow. 1942. A comparative study of the bacula of wood rats (subfamily Neotominae). Journal of Mammalogy 23:287–297.

Burt, W. H., and R. P. Grossenheider. 1964. A field guide to the mammals. The Riverside Press, Cambridge, U.S.A.

Burt, W. H., and E. T. Hooper. 1941. Notes on mammals from Sonora and Chihuahua, Mexico. Occasional Papers of the Museum of Zoology, University of Michigan 430:1–7.

Burt, W. H., and R. A. Stirton. 1961. The mammals of El Salvador. Miscelaneous Publication of Museum Zoology, University Michigan 117:1–69.

Burton, K. L., and M. D. Engstrom. 2001. Species diversity of bats (Mammalia: Chiroptera) in Iwokrama Forest, Guyana, and the Guianan subregion: implications for conservation. Biodiversity and Conservation 10:613–657.

Buskirk, S. W., L. F. Ruggiero, K. B. Aubry, K. B. Pearson, J. R. Squires, K. S. Mckelvey, G. M. Koehler, and C. J. Krebs. 2000. Comparative ecology of *Lynx* in North America. Pp. 397–418, *in* Ecology and Conservation of *Lynx* in the United States (Ruggiero, L. F., K. B. Aubry, S. W. Buskirk, G. M. Koehler, C. J. Krebs, K. S. McKelvey, and J. R. Squires, eds). University Press of Colorado. Boulder, U.S.A.

Byrd, G. V., and N. Norvell. 1993. Status of the Pribilof shrew based on summer distribution and habitat use. Northwestern Naturalist 74:49–54.

Cab-Sulub, L., and S. T. Álvarez-Castañeda. 2020. Analysis of the remaining habitat of an endemic species rediscovered. Mammalian Biology 100:307–314.

Cabrera, A. 1919. Genera *Mammalium*. Monotremata. Marsupialia. Museo Nacional de Ciencias Naturales. Madrid, Spain.

Cabrera, H., S. T. Álvarez-Castañeda, N. González-Ruíz, and J. P. Gallo-Reynoso. 2007. Distribution and natural history of Schmidly's Deermouse (*Peromyscus schmidlyi*). The Southwestern Naturalist 52:620–623.

Cáceres, M. C., and R. M. R. Barclay. 2000. *Myotis septentrionalis*. Mammalian Species 634:1–4.

Cáceres, N. C., and A. P. Carmignotto. 2006. *Caluromys lanatus*, Mammalian Species 803:1–6.

Caire, W. 1976. Phenetic relationships of pocket mice in the subgenus *Chaetodipus* (Rodentia: Heteromyidae). Journal of Mammalogy 57:375–378.

Caire, W. 1978. The distribution and zoogeography of the mammals of Sonora, Mexico. Ph.D. dissertation. The University of New Mexico. Albuquerque, U.S.A.

Caire, W., R. K. Laval, M. L. Laval, and R. Clawson. 1979. Notes on the ecology of *Myotis keenii* (Chiroptera, Vespertilionidae) in eastern Missouri. The American Midland Naturalist 102:404–407.

Caire, W., J. D. Taylor, B. P. Glass, and M. A. Mares. 1989. Mammals of Oklahoma. University of Oklahoma Press. Norman, U.S.A.

Caire, W., J. E. Vaughan, and V. E. Diersing. 1978. First record of *Sorex arizonae* (Insectivora: Soricidae) from Mexico. The Southwestern Naturalist 23:532–533.

Calahorra-Oliart, A., S. M. Ospina-Garcés, and L. León-Paniagua. 2021. Cryptic species in *Glossophaga soricina* (Chiroptera: Phyllostomidae): do morphological data support molecular evidence? Journal of Mammalogy 102:54–68.

Callahan, J. R. 1976. Systematics and biogeography of the *Eutamias obscurus* complex (Rodentia: Sciuridae). Ph.D. dissertation. The University of Arizona. Tucson, U.S.A.

Callahan, J. R. 1977. Diagnosis of *Eutamias obscurus* (Rodentia: Sciuridae). Journal of Mammalogy 58:188–201.

Callahan, J. R., and D. R. Davis. 1976. Desert chipmunks. The Southwestern Naturalist 21:127–130.

Caldwell, D. K. 1961. The harbor seal in South Carolina. Journal of Mammalogy 42:425.

Caldwell, D. K., and M. C. Caldwell. 1969. The harbor seal, *Phoca vitulina concolor*, in Florida. Journal of Mammalogy 50:379–380.

Camacho, M. A., P. A. Menendez-Guerrero, B. Horvath, D. Cadar, and J. Murienne. 2024. A polytypic species revisited: Phylogenetic and morphological variation, taxonomic status, and geographical distribution of *Trachops* (Chiroptera: Phyllostomidae). Journal of Mammalogy 103:1001–1021.

Camargo, I., and S. T. Álvarez-Castañeda. 2018. *Sorex ornatus juncensis* (Mammalia: Soricomorpha) rediscovering after 90 years without records, and taxonomical comments. Mammalia 83:193–197.

Camargo, I., and S. T. Álvarez-Castañeda. 2020. A new species and three subspecies of the desert shrew (*Notiosorex*) from the Baja California peninsula and California. Journal of Mammalogy 101:872–886.

Camargo, I., E. Rios, and S. T. Álvarez-Castañeda. 2016. Geographical extension of *Chaetodipus fallax* (Rodentia: Heteromyidae) in the Baja California peninsula. Mammalia 81:315–318.

Camargo, I, S. T. Álvarez-Castañeda, P. D. Polly, J. D. Stuhler, and J. E. Maldonado. 2021. Molecular phylogenetic and taxonomic status of the large-eared desert shrew *Notiosorex evotis* (Eulipotyphla: Soricidae), with the designation of a neotype. Journal of Mammalogy 103:1422–1330.

Camargo, I., P. D. Polly, S. T. Álvarez-Castañeda, and J. D. Stuhler. 2024. Paleophylogeography of *Notiosorex* desert shrews with description of a new species. Journal of Mammalogy 103:1067–1082.

Cameron, G. N., and S. R. Spencer. 1981. *Sigmodon hispidus* Mammalian Species 158:1–9.

Cameron, G. N., S. R. Spencer, B. D. Eshelman, L. R. Williams, and M. J. Gregory. 1988. Activity and burrow structure of Attwater's pocket gopher (*Geomys attwateri*). Journal of Mammalogy 69:667–677.

Cano-Sánchez, E., F. Rodriguez-Gomez, L. A. Ruedas, K. Oyama, L. Leon-Paniagua, A. Mastretta-Yanes, and A. Velazquez. 2022. Using Ultraconserved Elements to Unravel Lagomorph Phylogenetic Relationships. Journal of Mammalian Evolution 29:395–411.

Carleton, M. D. 1977. Interrelationships of populations of the *Peromyscus boylii* species group (Rodentia, Muridae) in western Mexico. Occasional Papers, Museum of Zoology, University of Michigan 675:1–47.

Carleton, M. D. 1979. Taxonomic status and relationships of *Peromyscus boylii* from El Salvador. Journal of Mammalogy 60:280–296.

Carleton, M. D. 1980. Phylogenetic relationships in neotomine-peromyscine-rodents (Muroidea) and a reappraisal of the dichotomy within New World Cricetinae. Miscellaneous Publications, Museum of Zoology, University of Michigan 157:93–108.

Carleton, M. D. 1989. Systematics and evolution in avance study *Peromyscus*. Pp. 7–141, *in* Advances in the study of *Peromyscus* (Kirkland Jr., G. L., and J. L. Lane, eds.). Texas Tech University Press. Texas, U.S.A.

Carleton, M. D., and J. Arroyo-Cabrales. 2009. Review of the *Oryzomys couesi* complex (Rodentia: Cricetidae: Sigmodontinae) in western Mexico. Pp. 93–127, *in* Systematic Mammalogy: contributions in honor of Guy G. Musser (Voss. R. S., and M. D. Carleton, eds.). Bulletin of the American Museum of Natural History 331:1–450.

Carleton, M. D., and G. G. Musser. 1984. Muroid rodents. Pp. 289–379, *in* Orders and families of recent mammals of the world (Anderson, S., and J. K. Jones, Jr., eds.) John Wiley and Sons. New York, U.S.A.

Carleton, M. D., and G. G. Musser. 1995. Systematic studies of oryzomyine rodents (Muridae: Sigmodontinae): definition and distribution of *Oligoryzomys vegetus* (Bangs, 1902). Proceedings of the Biological Society of Washington 108:338–369.

Carleton, M. D., and S. Olson. 1999. Amerigo Vespucci and the rat of Fernando de Noronha: A new genus and species of Rodentia (Muridae: Sigmodontinae) from a volcanic island off Brazil's continental shelf. American Museum Novitates 3256:59.

Carleton, M. D., D. E. Wilson, A. L. Gardner, and M. A. Bogan. 1982. Distribution and systematics of *Peromyscus* (Mammalia: Rodentia) of Nayarit, Mexico. Smithsonian Contributions in Zoology 352:1–46.

Carleton, M. D., R. D. Fisher, and A. L. Gardner. 1999. Identification and distribution of cotton rats, genus *Sigmodon* (Muridae: Sigmodontinae) of Nayarit, Mexico. Proceedings of the Biological Society of Washington 112:813–856.

Carleton, M. D., O. Sánchez, and G. Urbano-Vidales. 2002. A new species of *Habromys* (Muroidea: Neotominae) from Mexico, with generic review of species definitions and remarks on diversity patterns among Mesoamerica small mammals restricted to humid montane forest. Proceedings of the Biological Society of Washington 115:488–553.

Carraway, L. N. 1985. *Sorex pacificus*. Mammalian Species 231:1–5.

Carraway, L. N. 1987. Analysis of characters for distinguishing *Sorex trowbridgii* from sympatric *S. vagrans*. The Murrelet 68:29–30.

Carraway, L. N. 1988. Records of reproduction in *Sorex pacificus*. The Southwestern Naturalist 33:479–480.

Carraway, L. N. 1990. A morphologic and morphometric analysis of the "*Sorex vagrans* species complex" in the Pacific coast region. Special Publications of the Texas Tech University Museum 32:1–76.

Carraway, L. N. 2007. Shrews (Eulypotyphla: Soricidae) of Mexico. Monographs of the Western North American Naturalist 3:1–91.

Carraway, L. N., and P. K. Kennedy. 1993. Genetic variation in *Thomomys bulbivorus*, an endemic to the Willamette Valley, Oregon. Journal of Mammalogy 74:952–962.

Carraway, L. N., and B. J. Verts. 1985. *Microtus oregoni*. Mammalian Species 233:1–6.

Carraway, L. N., and B. J. Verts. 1991a. *Neotoma fuscipes*. Mammalian Species 386:1–10.

Carraway, L. N., and B. J. Verts. 1991b. *Neurotrichus gibbsii*. Mammalian Species 387:1–7.

Carraway, L. N., and B. J. Verts. 1994. *Sciurus griseus*. Mammalian Species 474:1–7.

Carraway, L. N., and R. M. Timm. 2000. Revision of the extant taxa of the genus *Notiosorex* (Mammalia: Insectivora: Soricidae). Proceeding Biological Society of Washington 113:302–318.

Carraway, L. N., L. F. Alexander, and B. J. Verts. 1993. *Scapanus townsendii*. Mammalian Species 434:1–7.

Carroll, L. E., and H. H. Genoways. 1980. *Lagurus curtatus*. Mammalian Species 124:1–6.

Carroll, D. S., L. L. Peppers, and R. D. Bradley. 2005. Molecular systematics and phylogeography of the *Sigmodon hispidus* species group. Pp. 87–100, *in* Contribuciones mastozoológicas en homenaje a Bernardo Villa (Sánchez-Cordero, V., and R. A. Medellín, eds.). Instituto de Biología e Instituto de Ecología, Universidad Nacional Autónoma de México y Comisión Nacional para el Conocimiento y Uso de la Biodiversidad. Distrito Federal, México.

Carr, S. M., and S. A. hicks. 1997. Are there two species of marten in North America? Genetic and evolutionary relationships within Martes. Pp. 15–20 in Martes: taxonomy, ecology, techniques, and management (Proulx, G., H. N. Bryant, and P. M. Woodard, eds.). Provincial Museum of Alberta, Edmonton, Alberta, Canada.

Carter, D. C., and W. B. Davis. 1961. *Tadarida aurispinosa* (Peale) (Chiroptera: Molossidae) in North America. Proceedings of the Biological Society of Washington 74:161–165.

Carter, D. C., and P. G. Dolan. 1978. Catalogue of type specimens of Neotropical bats in selected European museums. Special Publications, The Museum, Texas Tech University 15:1–136.

Carter, C. H., and H. H. Genoways. 1978. *Liomys salvini*. Mammalian Species 84:1–5.

Carter, D. C., and J. K. Jones, Jr. 1978. Bats from the Mexican state of Hidalgo. Occasional Papers, The Museum, Texas Tech University 54:1–12.

Carter, C. H., H. H. Genoways, R. S. Loregnard, and R. J. Baker. 1981. Observations on bats from Trinidad with a checklist of species occurring on the island. Occasional Papers Museum, Texas Tech University 72:1–27.

Carter, D. C., W. D. Webster, J. K. Jones, Jr., C. Jones, and R. D. Suttkus. 1985. *Dipodomys elator*. Mammalian Species 232:1–3.

Caso, A., T. de Oliveira, and S. V. Carvajal. 2015. *Herpailurus yagouaroundi*. The IUCN Red List of Threatened Species 2015:e.T9948A50653167.

Cassola, F. 2016a. *Arborimus albipes*. The IUCN Red List of Threatened Species 2016:e.T2017A22389204.

Cassola, F. 2016b. *Blarina brevicauda*. The IUCN Red List of Threatened Species 2016:e.T41451A115187102.

Cassola, F. 2016c. *Blarina carolinensis*. The IUCN Red List of Threatened Species 2016:e.T41452A115187223.

Cassola, F. 2016d. *Blarina hylophaga*. The IUCN Red List of Threatened Species 2016:e.T41453A115187348.

Cassola, F. 2016e. *Callospermophilus lateralis*. The IUCN Red List of Threatened Species 2016:e.T42468A22265474.

Cassola, F. 2016f. *Callospermophilus saturatus*. The IUCN Red List of Threatened Species 2016:e.T42562A22262657.

Cassola, F. 2016g. *Castor canadensis*. The IUCN Red List of Threatened Species 2016:e.T4003A22187946.

Cassola, F. 2016h. *Condylura cristata*. The IUCN Red List of Threatened Species 2016:e.T41458A115187740.

Cassola, F. 2016i. *Cratogeomys castanops*. The IUCN Red List of Threatened Species 2016:e.T16025A115131062.

Cassola, F. 2016j. *Cryptotis goldmani*. The IUCN Red List of Threatened Species 2016:e.T41371A22285527.

Cassola, F. 2016k. *Cryptotis mexicana*. The IUCN Red List of Threatened Species 2016:e.T41374A22286065.

Cassola, F. 2016l. *Cynomys gunnisoni*. The IUCN Red List of Threatened Species 2016:e.T42453A115189620.

Cassola, F. 2016m. *Cynomys leucurus*. The IUCN Red List of Threatened Species 2016:e.T42454A22261371.

Cassola, F. 2016n. *Cynomys ludovicianus*. The IUCN Red List of Threatened Species 2016:e.T6091A115080297.

Cassola, F. 2016o. *Dicrostonyx groenlandicus*. The IUCN Red List of Threatened Species 2016:e.T42618A115195764.

Cassola, F. 2016p. *Dicrostonyx hudsonius*. The IUCN Red List of Threatened Species 2016:e.T42619A115195917.

Cassola, F. 2016q. *Dicrostonyx nelsoni*. The IUCN Red List of Threatened Species 2016:e.T42620A22331765.

Cassola, F. 2016r. *Dicrostonyx nunatakensis*. The IUCN Red List of Threatened Species 2016:e.T6567A22332006.

Cassola, F. 2016s. *Dicrostonyx richardsoni*. The IUCN Red List of Threatened Species 2016:e.T42622A115196066.

Cassola, F. 2016t. *Dipodomys agilis*. The IUCN Red List of Threatened Species 2016:e.T6684A22228553.

Cassola, F. 2016u. *Dipodomys californicus*. The IUCN Red List of Threatened Species 2016:e.T42599A115193510.

Cassola, F. 2016v. *Dipodomys compactus*. The IUCN Red List of Threatened Species 2016:e.T6685A115083122.

Cassola, F. 2016w. *Dipodomys heermanni*. The IUCN Red List of Threatened Species 2016:e.T42600A22227931.

Cassola, F. 2016x. *Dipodomys microps*. The IUCN Red List of Threatened Species 2016:e.T42603A22227645.

Cassola, F. 2016y. *Dipodomys ordii*. The IUCN Red List of Threatened Species 2016:e.T6691A115083268.

Cassola, F. 2016z. *Dipodomys panamintinus*. The IUCN Red List of Threatened Species 2016:e.T42604A115193637.

Cassola, F. 2016aa. *Dipodomys venustus*. The IUCN Red List of Threatened Species 2016:e.T42605A22227166.

Cassola, F. 2016ab. *Geomys attwateri*. The IUCN Red List of Threatened Species 2016:e.T136380A22217970.

Cassola, F. 2016ac. *Geomys breviceps*. The IUCN Red List of Threatened Species 2016:e.T136840A115212715.

Cassola, F. 2016ad. *Geomys bursarius*. The IUCN Red List of Threatened Species 2016:e.T42588A115192675.

Cassola, F. 2016ae. *Geomys knoxjonesi*. The IUCN Red List of Threatened Species 2016:e.T136258A22218149.

Cassola, F. 2016af. *Geomys pinetis*. The IUCN Red List of Threatened Species 2016:e.T42589A115192878.

Cassola, F. 2016ag. *Geomys texensis*. The IUCN Red List of Threatened Species 2016:e.T9062A22217724.

Cassola, F. 2016ah. *Glaucomys sabrinus*. The IUCN Red List of Threatened Species 2016:e.T39553A22256914.

Cassola, F. 2016ai. *Glaucomys volans*. The IUCN Red List of Threatened Species 2016:e.T9240A115091392.

Cassola, F. 2016aj. *Heteromys desmarestianus*. The IUCN Red List of Threatened Species 2016:e.T47804700A115400334.

Cassola, F. 2016ak. *Ictidomys tridecemlineatus*. The IUCN Red List of Threatened Species 2016:e.T42564A22263122.

Cassola, F. 2016al. *Lemmiscus curtatus*. The IUCN Red List of Threatened Species 2016:e.T42624A115196202.

Cassola, F. 2016am. *Lemmus trimucronatus*. The IUCN Red List of Threatened Species 2016:e.T136712A115211700.

Cassola, F. 2016an. *Marmota broweri*. The IUCN Red List of Threatened Species 2016:e.T42455A22258026.

Cassola, F. 2016ao. *Marmota flaviventris*. The IUCN Red List of Threatened Species 2016:e.T42457A115189809.

Cassola, F. 2016ap. *Marmota monax*. The IUCN Red List of Threatened Species 2016:e.T42458A115189992.

Cassola, F. 2016aq. *Marmota olympus*. The IUCN Red List of Threatened Species 2016:e.T42459A22257452.

Cassola, F. 2016ar. *Microtus abbreviatus*. The IUCN Red List of Threatened Species 2016:e.T13425A22350031.

Cassola, F. 2016as. *Microtus canicaudus*. The IUCN Red List of Threatened Species 2016:e.T42625A22348218.

Cassola, F. 2016at. *Microtus chrotorrhinus*. The IUCN Red List of Threatened Species 2016:e.T42626A115196387.

Cassola, F. 2016au. *Microtus longicaudus*. The IUCN Red List of Threatened Species 2016:e.T42627A115196586.

Cassola, F. 2016av. *Microtus montanus*. The IUCN Red List of Threatened Species 2016:e.T42630A22346732.

Cassola, F. 2016aw. *Microtus ochrogaster*. The IUCN Red List of Threatened Species 2016:e.T42631A115196932.

Cassola, F. 2016ax. *Microtus oregoni*. The IUCN Red List of Threatened Species 2016:e.T42632A115197199.

Cassola, F. 2016ay. *Microtus pennsylvanicus*. The IUCN Red List of Threatened Species 2016:e.T13452A115114123.

Cassola, F. 2016az. *Microtus pinetorum*. The IUCN Red List of Threatened Species 2016:e.T42633A115197344.
Cassola, F. 2016ba. *Microtus richardsoni*. The IUCN Red List of Threatened Species 2016:e.T42634A115197660.
Cassola, F. 2016bb. *Microtus townsendii*. The IUCN Red List of Threatened Species 2016:e.T13487A115114983.
Cassola, F. 2016bc. *Microtus xanthognathus*. The IUCN Red List of Threatened Species 2016:e.T42628A22348489.
Cassola, F. 2016bd. *Myodes californicus*. The IUCN Red List of Threatened Species 2016:e.T42616A115195295.
Cassola, F. 2016be. *Myodes gapperi*. The IUCN Red List of Threatened Species 2016:e.T42617A115195411.
Cassola, F. 2016bf. *Napaeozapus insignis*. The IUCN Red List of Threatened Species 2016:e.T42612A115194392.
Cassola, F. 2016bg. *Neofiber alleni*. The IUCN Red List of Threatened Species 2016:e.T14520A22356567.
Cassola, F. 2016bh. *Neotamias alpinus*. The IUCN Red List of Threatened Species 2016:e.T42568A22266784.
Cassola, F. 2016bi. *Neotamias amoenus*. The IUCN Red List of Threatened Species 2016:e.T42569A115190467.
Cassola, F. 2016bj. *Neotamias canipes*. The IUCN Red List of Threatened Species 2016:e.T21364A22266973.
Cassola, F. 2016bk. *Neotamias cinereicollis*. The IUCN Red List of Threatened Species 2016:e.T42570A22267056.
Cassola, F. 2016bl. *Neotamias minimus*. The IUCN Red List of Threatened Species 2016:e.T42572A115190804.
Cassola, F. 2016bm. *Neotamias ochrogenys*. The IUCN Red List of Threatened Species 2016:e.T42573A22267475.
Cassola, F. 2016bn. *Neotamias panamintinus*. The IUCN Red List of Threatened Species 2016:e.T42574A22267550.
Cassola, F. 2016bo. *Neotamias quadrimaculatus*. The IUCN Red List of Threatened Species 2016:e.T42575A22267619.
Cassola, F. 2016bp. *Neotamias quadrivittatus*. The IUCN Red List of Threatened Species 2016:e.T42576A115191051.
Cassola, F. 2016bq. *Neotamias ruficaudus*. The IUCN Red List of Threatened Species 2016:e.T42577A22268024.
Cassola, F. 2016br. *Neotamias rufus*. The IUCN Red List of Threatened Species 2016:e.T42578A115191185.
Cassola, F. 2016bs. *Neotamias senex*. The IUCN Red List of Threatened Species 2016:e.T42579A22268269.
Cassola, F. 2016bt. *Neotamias siskiyou*. The IUCN Red List of Threatened Species 2016:e.T42580A22268201.
Cassola, F. 2016bu. *Neotamias sonomae*. The IUCN Red List of Threatened Species 2016:e.T42581A115191316.
Cassola, F. 2016bv. *Neotamias speciosus*. The IUCN Red List of Threatened Species 2016:e.T42582A115191427.
Cassola, F. 2016bw. *Neotamias townsendii*. The IUCN Red List of Threatened Species 2016:e.T42584A115191888.
Cassola, F. 2016bx. *Neotamias umbrinus*. The IUCN Red List of Threatened Species 2016:e.T42585A115192014.
Cassola, F. 2016by. *Neotoma cinerea*. The IUCN Red List of Threatened Species 2016:e.T42673A115200351.
Cassola, F. 2016bz. *Neotoma floridana*. The IUCN Red List of Threatened Species 2016:e.T42650A115199202.
Cassola, F. 2016ca. *Neotoma fuscipes*. The IUCN Red List of Threatened Species 2016:e.T14587A22371665.
Cassola, F. 2016cb. *Neotoma stephensi*. The IUCN Red List of Threatened Species 2016:e.T42651A115199398.
Cassola, F. 2016cc. *Neurotrichus gibbsii*. The IUCN Red List of Threatened Species 2016:e.T41468A115188045.
Cassola, F. 2016cd. *Ochrotomys nuttalli*. The IUCN Red List of Threatened Species 2016:e.T42674A115200634.
Cassola, F. 2016ce. *Ondatra zibethicus*. The IUCN Red List of Threatened Species 2016:e.T15324A22344525.
Cassola, F. 2016cf. *Oryzomys palustris*. The IUCN Red List of Threatened Species 2016:e.T42675A115200837.
Cassola, F. 2016cg. *Parascalops breweri*. The IUCN Red List of Threatened Species 2016:e.T41469A115188181.
Cassola, F. 2016ch. *Perognathus fasciatus*. The IUCN Red List of Threatened Species 2016:e.T42608A115194044.
Cassola, F. 2016ci. *Perognathus inornatus*. The IUCN Red List of Threatened Species 2016:e.T42609A22224670.
Cassola, F. 2016cj. *Perognathus longimembris*. The IUCN Red List of Threatened Species 2016:e.T16636A115135230.
Cassola, F. 2016ck. *Perognathus parvus*. The IUCN Red List of Threatened Species 2016:e.T42610A115194207.
Cassola, F. 2016cl. *Peromyscus attwateri*. The IUCN Red List of Threatened Species 2016:e.T42652A115199528.
Cassola, F. 2016cm. *Peromyscus beatae*. The IUCN Red List of Threatened Species 2016:e.T136323A22364310.
Cassola, F. 2016cn. *Peromyscus gossypinus*. The IUCN Red List of Threatened Species 2016:e.T42653A115199668.
Cassola, F. 2016co. *Peromyscus keeni*. The IUCN Red List of Threatened Species 2016:e.T135164A115204632.
Cassola, F. 2016cp. *Peromyscus leucopus*. The IUCN Red List of Threatened Species 2016:e.T16669A115136270.
Cassola, F. 2016cq. *Peromyscus maniculatus*. The IUCN Red List of Threatened Species 2016:e.T16672A22360898.
Cassola, F. 2016cr. *Peromyscus polionotus*. The IUCN Red List of Threatened Species 2016:e.T42654A115199876.
Cassola, F. 2016es. *Peromyscus truei*. The IUCN Red List of Threatened Species 2016:e.T16694A115137578.
Cassola, F. 2016ct. *Phenacomys intermedius*. The IUCN Red List of Threatened Species 2016:e.T42636A115197827.
Cassola, F. 2016cu. *Phenacomys ungava*. The IUCN Red List of Threatened Species 2016:e.T42637A115198018.
Cassola, F. 2016cv. *Poliocitellus franklinii*. The IUCN Red List of Threatened Species 2016:e.T41787A22265037.
Cassola, F. 2016cw. *Reithrodontomys fulvescens*. The IUCN Red List of Threatened Species 2016:e.T19407A115150749.
Cassola, F. 2016cx. *Reithrodontomys humulis*. The IUCN Red List of Threatened Species 2016:e.T42678A115201061.
Cassola, F. 2016cy. *Reithrodontomys megalotis*. The IUCN Red List of Threatened Species 2016:e.T19410A115151098.
Cassola, F. 2016cz. *Scapanus orarius*. The IUCN Red List of Threatened Species 2016:e.T41474A115188698.
Cassola, F. 2016da. *Scapanus townsendii*. The IUCN Red List of Threatened Species 2016:e.T41475A22322352.
Cassola, F. 2016db. *Sciurus carolinensis*. The IUCN Red List of Threatened Species 2016:e.T42462A22245728.
Cassola, F. 2016dc. *Sciurus nayaritensis*. The IUCN Red List of Threatened Species 2016:e.T20015A115155124.
Cassola, F. 2016dd. *Sigmodon hispidus*. The IUCN Red List of Threatened Species 2016:e.T20213A115157685.
Cassola, F. 2016de. *Sigmodon toltecus*. The IUCN Red List of Threatened Species 2016:e.T136559A115209897.
Cassola, F. 2016df. *Sorex arcticus*. The IUCN Red List of Threatened Species 2016:e.T41385A115182930.
Cassola, F. 2016dg. *Sorex bendirii*. The IUCN Red List of Threatened Species 2016:e.T41389A115183051.
Cassola, F. 2016dh. *Sorex cinereus*. The IUCN Red List of Threatened Species 2016:e.T41392A115183208.
Cassola, F. 2016di. *Sorex fumeus*. The IUCN Red List of Threatened Species 2016:e.T41396A22312838.
Cassola, F. 2016dj. *Sorex haydeni*. The IUCN Red List of Threatened Species 2016:e.T41399A115183736.
Cassola, F. 2016dk *Sorex hoyi*. The IUCN Red List of Threatened Species 2016:e.T41400A115183871.
Cassola, F. 2016dl. *Sorex longirostris*. The IUCN Red List of Threatened Species 2016:e.T41401A115184004.
Cassola, F. 2016dm. *Sorex merriami*. The IUCN Red List of Threatened Species 2016:e.T41403A115184201.

Cassola, F. 2016dn. *Sorex nanus*. The IUCN Red List of Threatened Species 2016:e.T41406A115184594.
Cassola, F. 2016do. *Sorex palustris*. The IUCN Red List of Threatened Species 2016:e.T41410A115184897.
Cassola, F. 2016dp. *Sorex tenellus*. The IUCN Red List of Threatened Species 2016:e.T41419A22318690.
Cassola, F. 2016dq. *Sorex trowbridgii*. The IUCN Red List of Threatened Species 2016:e.T41421A115185589.
Cassola, F. 2016dr. *Sorex ugyunak*. The IUCN Red List of Threatened Species 2016:e.T41423A115185865.
Cassola, F. 2016ds. *Synaptomys cooperi*. The IUCN Red List of Threatened Species 2016:e.T42639A115198182.
Cassola, F. 2016dt. *Tamias striatus*. The IUCN Red List of Threatened Species 2016:e.T42583A115191543.
Cassola, F. 2016du. *Tamiasciurus douglasii*. The IUCN Red List of Threatened Species 2016:e.T42586A115192165.
Cassola, F. 2016dv. *Tamiasciurus hudsonicus*. The IUCN Red List of Threatened Species 2016:e.T42587A115192299.
Cassola, F. 2016dw. *Thomomys bulbivorus*. The IUCN Red List of Threatened Species 2016:e.T42594A22216513.
Cassola, F. 2016dx. *Thomomys mazama*. The IUCN Red List of Threatened Species 2016:e.T21810A115163696.
Cassola, F. 2016dy. *Thomomys monticola*. The IUCN Red List of Threatened Species 2016:e.T42596A22216069.
Cassola, F. 2016dz. *Thomomys talpoides*. The IUCN Red List of Threatened Species 2016:e.T42597A115193142.
Cassola, F. 2016ea. *Thomomys townsendii*. The IUCN Red List of Threatened Species 2016:e.T42598A115193369.
Cassola, F. 2016eb. *Urocitellus armatus*. The IUCN Red List of Threatened Species 2016:e.T42463A22264746.
Cassola, F. 2016ec. *Urocitellus beldingi*. The IUCN Red List of Threatened Species 2016:e.T42464A22264836.
Cassola, F. 2016ed. *Urocitellus columbianus*. The IUCN Red List of Threatened Species 2016:e.T42466A22265632.
Cassola, F. 2016ee. *Urocitellus parryii*. The IUCN Red List of Threatened Species 2016:e.T20488A22262403.
Cassola, F. 2016ef. *Urocitellus richardsonii*. The IUCN Red List of Threatened Species 2016:e.T42561A22262546.
Cassola, F. 2016eg. *Zapus hudsonius*. The IUCN Red List of Threatened Species 2016:e.T42613A115194664.
Cassola, F. 2016eh. *Zapus princeps*. The IUCN Red List of Threatened Species 2016:e.T42614A115195084.
Cassola, F. 2016ei. *Zapus trinotatus*. The IUCN Red List of Threatened Species 2016:e.T23192A115167834.
Cassola, F. 2017a. *Sciurus aberti*. The IUCN Red List of Threatened Species 2017:e.T42461A22245623.
Cassola, F. 2017b. *Sorex rohweri*. The IUCN Red List of Threatened Species 2017:e.T136282A22317740.
Cassola, F. 2017c. *Sorex sonomae*. The IUCN Red List of Threatened Species 2017:e.T41418A22318770.
Cassola, F. 2017d. *Synaptomys borealis*. The IUCN Red List of Threatened Species 2017:e.T42638A22377185.
Cassola, F. 2018. *Marmota caligata*. The IUCN Red List of Threatened Species 2018:e.T42456A122560084.
Cassola, F. 2019. *Sorex preblei*. The IUCN Red List of Threatened Species 2019:e.T41413A117935777.
Castañeda-Rico, S., L. León-Paniagua, E. Vázquez-Domínguez, and A. G. Navarro-Sigüenza. 2014. Evolutionary diversification and speciation in rodents of the Mexican lowlands: the *Peromyscus melanophrys* species group. Molecular Phylogenetics and Evolution 70:454–463.
Castleberry, S. B., M. T. Mengak, and W. M. Ford. 2006. *Neotoma magister*. Mammalian Species 789:1–5.
Castro-Arellano, I., and E. Vázquez. 2016a. *Pappogeomys bulleri*. The IUCN Red List of Threatened Species 2016:e.T92474664A22217245.
Castro-Arellano, I., and E. Vázquez. 2016b. *Peromyscus aztecus*. The IUCN Red List of Threatened Species 2016:e.T16651A22361906.
Castro-Arellano, I., and E. Vázquez. 2016c. *Peromyscus difficilis*. The IUCN Red List of Threatened Species 2016:e.T16658A115135855.
Castro-Arellano, I., and E. Vázquez. 2016d. *Peromyscus levipes*. The IUCN Red List of Threatened Species 2016:e.T16670A115136642.
Castro-Arellano, I., and E. Vázquez. 2016e. *Peromyscus megalops*. The IUCN Red List of Threatened Species 2016:e.T16674A115136745.
Castro-Arellano, I., and E. Vázquez. 2017. *Osgoodomys banderanus*. The IUCN Red List of Threatened Species 2017:e.T15629A22359053.
Castro-Arellano, I., and E. Vázquez. 2019. *Peromyscus furvus*. The IUCN Red List of Threatened Species 2019:e.T16661A22359896.
Castro-Arellano, I., H. Zarza, and R. A. Medellín. 2000. *Philander opossum*. Mammalian Species 638:1–8.
Castro-Arellano, I., R. Timm, and S. T. Álvarez-Castañeda. 2016a. *Heteromys irroratus*. The IUCN Red List of Threatened Species 2016:e.T12074A22225187.
Castro-Arellano, I., R. Timm, J. Matson, and N. Woodman. 2016b. *Notiosorex evotis*. The IUCN Red List of Threatened Species 2016:e.T136273A115205277.
Ceballos, G. 1990. Comparative natural history of small mammals from tropical forest in western Mexico. Journal of Mammalogy 71:263–266.
Ceballos, G., and R. A. Medellín. 1988. *Diclidurus albus*. Mammalian Species 316:1–4.
Ceballos, G., and A. Miranda. 2000. A field guide to the mammals from the Jalisco coast. Fundación Ecológica de Cuixmala-Universidad Nacional Autónoma de México. Distrito Federal, México.
Ceballos-G, G., and D. E. Wilson. 1985. *Cynomys mexicanus*. Mammalian Species 248:1–3.
Ceballos, G., H. Zarza, and M. A. Steele. 2002. *Xenomys nelsoni*. Mammalian Species 704:1–3.
Cerqueira, R. 1985. The distribution of *Didelphis* in the South America (Polyprotodontia, Didelphidae). Journal of Biogeography 12:135–145.
Cervantes, F. A. 1993. *Lepus flavigularis*. Mammalian Species 423:1–3.
Cervantes F. A., and L. Guevara. 2010. Rediscovery of the critically endangered Nelson's small-eared shrew (*Cryptotis nelsoni*), endemic to Volcán San Martín, Eastern México. Mammalian Biology 75:451–454.
Cervantes, F. A., and C. Lorenzo. 1997. *Sylvilagus insonus*. Mammalian Species 568:1–4.
Cervantes, F. A., C. Lorenzo, and R. S. Hoffmann. 1990. *Romerolagus diazi*. Mammalian Species 360:1–7.
Cervantes, F. A., V. J. Sosa, J. Martínez, R. M. González, and R. C. Dowler. 1993. *Pappogeomys tylorhinus*. Mammalian Species 433:1–4.
Cervantes, F. A., C. Lorenzo, J. Vargas, and T. Holmes. 1992. *Sylvilagus cunicularius*. Mammalian Species 412:1–4.
Cervantes, F. A., J. Martínez, and R. M. González. 1994. Karyotype of the Mexican tropical voles *Microtus quasiater* and *M. umbrousus* (Arvicolinae: Muridae). Acta Theriologica 39:373–377.
Cervantes, F. A., S. T. Álvarez-Castañeda, B. VillaRamírez, C. Lorenzo, A. Rojas, A. L. Colmenares, and J. Vargas. 1996. Natural history of the insular black Jackrabbit (*Lepus insularis*) from Espiritu Santo Island, Mexico. The Southwestern Naturalist 41:186–189.
Cervantes, F. A., J. P. Ramírez-Silva, A. Marín, and G. L. Portales. 1999a. Allozyme variation of cottontail rabbits (*Sylvilagus*) from Mexico. Zeitschrift für Saügetierkunde 64:356–362.
Cervantes, F. A., C. Lorenzo, and O. G. Ward. 1999b. Chromosomal relationships among spiny pocket mice, *Liomys* (Heteromydae), from Mexico. Journal of Mammalogy 80:823–832.
Cervantes, F. A., C. Lorenzo, V. Farías, and J. Vargas. 2008. *Lepus flavigularis*. The IUCN Red List of Threatened Species 2008:e.T11790A3306162.

Chambers, R. R., P. D. Sudman, and R. D. Bradley. 2009. A phylogenetic assessment of Pocket Gophers (*Geomys*): evidence from nuclear and mitochondrial genes. Journal of Mammalogy 90:537–547.

Chapman, F. M. 1894. Remarks on certain land mammals from Florida, with a list of the species known to occur in the state. Bulletin of the American Museum of Natural History 6:333–346.

Chapman, J. A. 1974. *Sylvilagus bachmani*. Mammalian Species 34:1–4.

Chapman, J. A. 1975a. *Sylvilagus transitionalis*. Mammalian Species 55:1–4.

Chapman, J. A. 1975b. *Sylvilagus nuttallii*. Mammalian Species 56:1–3.

Chapman, J. A. and G. Ceballos. 1990. Chapter 5: The Cottontails. Pp. 95–110, *in* Rabbits, hares and pikas: status survey and conservation action plan (Chapman, J. A., and J. C. Flux, eds). IUCN. Gland, Switzerland.

Chapman, J. A., and J. E. C. Flux.. 1990. Rabbits, hares and pikas. Status Survey and Conservation Action Plan. International Union for the Conservation of Nature and Natural Resources, Gland, Suiza.

Chapman, J. A., and G. A. Feldhamer. 1981. *Sylvilagus aquaticus*. Mammalian Species 151:1–4.

Chapman, J. A., and G. Willner. 1978. *Sylvilagus audubonii*. Mammalian Species 106:1–4.

Chapman, J. A., and G. Willner. 1981. *Sylvilagus palustris*. Mammalian Species 153:1–3.

Chapman, J. A., J. G. Hockman, and M. M. Ojeda. 1980. *Sylvilagus floridanus*. Mammalian Species 136:1–8.

Chapman, J. A., K. R. Dixon, W. Lopez-Forment, and D. E. Wilson. 1983. The New World jackrabbits and hares (genus Lepus).- I. Taxonomic history and populations status. Acta Zoologica Fenica 179:49–51.

Chapman J. A., K. L. Cramer, N. J. Dippenaar, and T. J. Robinson. 1992. Systematics and biogeography of the New England cottontail, *Sylvilagus transitionalis* (Bangs, 1895), with the description of a new species from the Appalachian Mountains. Proceedings of the Biological Society of Washington 105:841–866.

Chapskii, K. K. 1975. Obosnavanie dvukh novykh podvidov nastoyashchikh tyulenei semeistva Phocidae. Trudy Zoologischeskogo Instituta Akademii Nauk SSSR 53:282–333.

Chesser, R. K. 1983. Cranial variation among population of the black tailed prairie dog in New Mexico. Occasional Papers, the Museum Texas Tech University 84:1–13.

Christman, G. N. 1971. The mountain bison. American Western 8:44–47.

Choate, J. R. 1969. Taxonomic status of the shrew, *Notiosores* (*Xenosorex*) *phillipsii* Schaldach, 1966 (Mammalia: Insectivora). Proccedings of the Biological Society of Washington 82:469–476.

Choate, J. R. 1970. Systematics and zoogeography of Middle Americas shrews of the genus *Cryptotis*. University of Kansas Publications, Museum of Natural History 19:195–317.

Choate, J. R. 1973. *Cryptotis mexicana*. Mammalian Species 28:1–3.

Choate, J. R., and E. D. Fleharty. 1974. *Cryptotis goodwini*. Mammalian Species 44:1–3.

Choate, J. R., and S. L. Williams. 1978. Biogeographic interpretation of variation within and among populations of the prairie vole, *Microtus ochrogaster*. Occasional Papers, The Museum, Texas Tech University 49:1–25.

Choromanski-Norris, J., E. K. Fritzell, and A. B. Sargeant. 1989. Movements and habitat use of Franklin's ground squirrels in duck-nesting habitat. Journal of Wildlife Management 53:324–331.

Clare, E. L., A. M. Adams, A. Z. Maya-Simões, J. L. Eger, P. D. N. Hebert, and M. B. Fenton. 2013. Diversification and reproductive isolation: cryptic species in the only New World high-duty cycle bat, *Pteronotus parnellii*. BMC Ecology and Evolution 13:26. https://doi.org/10.1186/1471-2148-13-26.

Clark, J. L. 1964. The great arc of wild sheep. University of Oklahoma Press. Norman, U.S.A.

Clark, T. W., E. Anderson, C. Douglas, and M. Strickland. 1987. *Martes americana*. Mammalian Species 289:1–8.

Clark, T. W., R. S. Hoffmann, and C. F. Nadler. 1971. *Cynomys leucurus*. Mammalian Species 7:1–4.

Cláudio, V. C., R. L. Novaes, A. L. Gardner, M. R. Nogueira, D. E. Wilson, J. E. Maldonado, J. A. Oliveira, and R. Moratelli. 2023. Taxonomic re-evaluation of New World *Eptesicus* and *Histiotus* (Chiroptera: Vespertilionidae), with the description of a new genus. Zoologia (Curitiba) 40:e22029.

Clawson, R. G., J. A. Clawson, and T. L. Best. 1994a. *Tamias alpinus*. Mammalian Species 461:1–6.

Clawson, R. G., J. A. Clawson, and T. L. Best. 1994b. *Tamias quadrimaculatus*. Mammalian Species 469:1–6.

Cloutier, D., and D. W. Thomas. 1992. *Carollia perspicillata*. Mammalian Species 417:1–9.

Clutton-Brock, J., G. B. Corbert, and M. Hills. 1976. A review of the family Canidae, with a classification by numerical methods. Bulletin of the British Museum (Natural History). Zoology 29:117–199.

Coates-Estrada, R., and A. Estrada. 1986. Manual de identificación de campo de los mamíferos de la Estación de la Biología "Los Tuxtlas." Instituto de Biología, Universidad Nacional Autónoma de México, México. Distrito Federal, México.

Cockrum, E. L. 1961. The recent mammals of Arizona: their taxonomy and distribution. University of Arizona Press. Tucson, U.S.A.

Cockrum, E. L. and A. L. Gardner. 1960. Underwood's mastiff bat in Arizona. Journal of Mammalogy 41:510–511.

Cockrum, E. L., and G. Van R. Bradshaw. 1963. Notes on mammals from Sonora, Mexico. American Museum Novitates 2138:1–9.

Corbet, G. B. 1978. The mammals of the Palaearctic region: A taxonomic review. British Museum (Natural History). London, U.K.

Cole, F. R., and D. E. Wilson. 2006. *Leptonycteris yerbabuenae*. Mammalian Species 797:1–7.

Cole, F. R., and D. E. Wilson. 2009. *Urocitellus canus* (Rodentia: Sciuridae). Mammalian Species 834:1–8.

Cole, F. R., and D. E. Wilson. 2010. *Microtus miurus* (Rodentia: Cricetidae). Mammalian Species 42:75–89.

Colella, J. P., E. J. Johnson, and J. A. Cook. 2018. Reconciling molecules and morphology in North American *Martes,* Journal of Mammalogy 99:1323–1335.

Colella, J. P., R. E. Wilson, S. L. Talbot, and J. A. Cook. 2019. Implications of introgression for wildlife translocations: the case of North American martens. Conservation Genetics, 20:153–166.

Colella, J. P., L. M. Frederick, S. L. Talbot, and J. A. Cook. 2021. Extrinsically reinforced hybrid speciation within Holarctic ermine (*Mustela* spp.) produces an insular endemic. Diversity and Distributions 27:747–762.

Collins, A. C. 2008. The taxonomic status of spider monkeys in the twenty-first century. Pp. 50–78, *in* Spider Monkeys: behavior, ecology and evolution of the genus *Ateles* (Campbell, C. J. ed.). Cambridge, United Kingdom.

Collins, P. W. 1982. Origin and differentiation of the island fox: a study of evolution in insular populations. M. A. thesis, The University of California. Santa Barbara, U.S.A.
Collins, P. W. 1993. Taxonomic and biogeographic relationships of the island fox (*Urocyon littoralis*) and gray fox (*Urocyon cinereoargenteus*) from western North America. Pp. 351–390, *in* Proceedings of the third Channel Islands symposium: recent advances in California Islands research (Hochberg, F. G., ed.). Santa Barbara Museum of Natural History, Santa Barbara, U.S.A.
Collins, A., and J. Dubach. 2000. Phylogenetic relationships among spider monkeys (*Ateles*) haplotypes based on mitochondrial DNA variation. International Journal of Primatology 21:381–420.
Coltman, D. W., G. Stenson, M. O. Hammill, T. Haug, S. Davis, and T. L. Fulton. 2007. Panmictic population structure in the hooded seal (*Cystophora cristata*). Molecular Ecology 16:1639–1648.
Commissaris, L. R. 1960. Morphological and ecological differentiation of *Peromyscus merriami* from southern Arizona. Journal of Mammalogy 41:305–310.
Committee on Taxonomy. 2014. List of marine mammal species and subspecies. Available at: www.marinemammalscience.org, accessed November 2019.
Connior, M. B. 2011. *Geomys bursarius* (Rodentia: Geomyidae). Mammalian Species 43:104–117.
Connor, P. F. 1959. The bog lemming *Synaptomys cooperi* in southern New Jersey. Publication Museum Michigan State University, Biological Series 1:161–248.
Conroy, C. J., and J. A. Cook. 1999a. MtDNA evidence for repeated pulse of speciation within arvicoline and murid rodents. Journal Mammalian Evolution 6:221–245.
Conroy, C. J., and J. A. Cook. 1999b. *Microtus xanthognathus*. Mammalian Species 627:1–5.
Conroy, C. J., and J. A. Cook. 2000. Molecular systematics of a Holartic rodents (*Microtus*: Muridae). Journal of Mammalogy 81:344–359.
Conroy, C. J., Y. Hortelano, F. A. Cervantes, and J. A. Cook. 2001. The phylogenetic position of southern relictual species of *Microtus* (Muridae: Rodentia) in North America. Mammalian Biology 66:332–344.
Conway, M. C., and C. G. Schmitt. 1978. Record of the Arizona shrew *(Sorex arizonae)* from New Mexico. Journal of Mammalogy 59:631.
Cook, J. A. 2017. Southern bog lemming *Synaptomys cooperi*. Pp. 291–292, *in* Handbook of the mammals of the world, vol. 7 (Wilson, D. E., T. E. Lacher, Jr., and R. A. Mittermeier, eds.). Lynx Edicions, Barcelona, Spain.
Cooke, B. D., J. F. C. Flux, and N. Bonino. 2018. Introduced lagomorphs. Pp. 13–17, *in* Lagomorphs: Pikas, Rabbits, and Hares of the World (Smith, A. T., C. H. Johnston, P. Alves, and K. Hackländer, K., eds). Johns Hopkins University Press, Baltimore, U.S.A.
Coonan, T., K. Ralls., B. Hudgens., B. Cypher, and C. Boser. 2013. *Urocyon littoralis*. The IUCN Red List of Threatened Species 2013:e. T22781A13985603.
Corbet, G. B., and J. E. Hill. 1980. A world list of mammalian species. British Museum (Natural History). London, United Kingdom.
Cornejo-Latorre, C., P. Cortés-Calva, and S. T. Álvarez-Castañeda. 2017. The evolutionary history of the subgenus *Haplomylomys* (Cricetidae: *Peromyscus*). Journal of Mammalogy 98:1627–1640.
Cornely, J. E., and B. J. Verts. 1988. *Microtus townsendii*. Mammalian Species 325:1–9.
Cornely, J. E., and R. J. Baker. 1986. *Neotoma mexicana*. Mammalian Species 262:1–7.
Cornely, J. E., L. N. Carraway, and B. J. Verts. 1992. *Sorex preblei*. Mammalian Species 416:1–3.
Cortés-Calva, P., and S. T. Álvarez-Castañeda. 2001. *Peromyscus dickeyi*. Mammalian Species 659:1–2.
Cortés-Calva, P., S. T. Álvarez-Castañeda, and E. Yensen. 2001a. *Neotoma anthonyi*. Mammalian Species 663:1–3.
Cortés-Calva, P., E. Yensen, and S. T. Álvarez-Castañeda. 2001b. *Neotoma martinensis*. Mammalian Species 657:1–3.
Cortés-Calva, P., J. P. Gallo-Reynoso, J. Delgadillo-Rodríguez, C. Lorenzo, and S. T. Álvarez-Castañeda. 2013. The effect of feral dogs and other alien species on native mammals of Isla de Cedros, Mexico. Natural Areas Journal 33:466–473.
Cortés-Ortiz, L., E. Bermingham, C. Rico, E. Rodríguez-Luna, I. Sampaio, and M. Ruiz-García. 2003. Molecular systematics and biogeography of the Neotropical monkey genus, *Alouatta*. Molecular Phylogenetics and Evolution 26:64–81.
Cortés-Ortiz, L, T. F. Duda, D. Canales-Espinosa, F. García-Orduña, E. Rodríguez-Luna, and E. Bermingham. 2007. Hybridization in large-bodied New World primates. Genetics 176:2421–2425.
Cortés-Ortiz, L., L. Aguiar, I. Agostini, M. A. Kelaita, F. E. Silva, and J. C. Bicca-Marques. 2015. Hybridization in howler monkeys: Current understanding and future directions. Pp. 107–131, *in* Howler Monkeys: adaptive radiation, systematics, and morphology (Kowalewski, M. M., P. A. Garber, L. Cortés-Ortiz, B. Urbani, and D. Youlatos, eds.). Springer. New York, USA.
Cortes-Ortíz, L., D. Canales-Espinosa, F. M. Cornejo, D. Guzman-Caro, A. Link, P. Moscoso, P. Méndez-Carvajal, E. Palacios, V. Rodríguez, M. Rosales-Meda, D. Solano, K. Williams-Guillén, and S. de la Torre. 2020a. *Alouatta palliata*. The IUCN Red List of Threatened Species 2020:e.T39960A17925090.
Cortes-Ortíz, L., D. Canales-Espinosa, L. K. Marsh, R. A. Mittermeier, P. Méndez-Carvajal, M. Rosales-Meda, D. Solano, and K. Williams-Guillén. 2020b. *Ateles geoffroyi*. The IUCN Red List of Threatened Species 2020:e.T2279A17929000.
Cortes-Ortíz, L., M. Rosales-Meda, L. K. Marsh, and R. A. Mittermeier. 2020c. *Alouatta pigra*. The IUCN Red List of Threatened Species 2020:e. T914A17926000.
Costa, L. P., and J. L. Patton. 2006. Diversidade e limites geográficos e sistemáticos de marsupiais brasileiros. Pp. 321–341, *in* Os marsupiais do Brasil (Cáceres, N. C., and E. L. A. Monteiro-Filho, eds.). Campo Grande, Editora UFMS.
Cowan, I. M. 1936. Distribution and variation in deer (Genus *Odocoileus)* of the Pacific coastal region of North America. California Fish and Game 22:155–246.
Cowan, I. M. 1940. Distribution and variation in the native sheep of North America. American Midland Naturalist 24:505–580.
Cowan, I. M. 1956. What and where are the mule and black-tailed deer? Pp. 334–359, *in* The deer of North America (Taylor, W. P. ed.). The Stackpole Co., Harrisburg and The Wildlife Management Institution, Washington, U.S.A.
Cowan, I. M., and C. J. Guiguet. 1956. The mammals of British Columbia. British Columbia Provincial Museum of Natural History and Anthropology Handbook, Victoria, British Columbia 11:1–413.
Cowan, I. M., and C. J. Guiguet. 1965. The mammals of British Columbia. Third Edition. British Columbia Province Museum, Handbook 11:1–414.
Cowan, I. M., and W. McCrory. 1970. Variation in the mountain goat, *Oreamnos americanus* (Blainville). Journal of Mammalogy 51:60–73.

Craig, B. A., and J. E. Reynolds III. 2004. Determination of manatee population trends along the Atlantic coast of Florida using a Bayesian approach with temperature-adjusted aerial survey data. Marine Mammal Science 20:386–400.

Cramer, M. J., and G. N. Cameron. 2001. *Geomys texensis*. Mammalian Species 679:1–3.

Cramer, M. J., M. R. Willig, and C. Jones. 2001. *Trachops cirrhosus*. Mammalian Species 656:1–6.

Crase, F. T. 1973. New size records for the western gray squirrel. The Murrelet 54:20–21.

Cronin, M. A. 1992. Intraspecifi c mitochondrial DNA variation in North American cervids. Journal of Mammalogy 73:70–82.

Cronin, M. A., J. L. Bodkin, B. E. Ballachey, J. A. Estes, and J. C. Patton. 1996. Mitochondrial-DNA variation among subspecies and populations of Sea otters (*Enhydra lutris*). Journal of Mammology 77:546–557.

Csuti, B. A. 1979. Patterns of adaptation and variation in the Great Basin Kangaroo rat (*Dipodomys microps*). University of California Publications in Zoology 111:1–69.

Cuarón, A. D., and E. Vázquez. 2018. *Heteromys nelsoni*. The IUCN Red List of Threatened Species 2018:e.T10009A22223445.

Cuarón, A. D., and P. C. de Grammont. 2017. *Cryptotis goodwini*. The IUCN Red List of Threatened Species 2017:e.T48269679A123794650.

Cuarón, A. D., and P. C. de Grammont. 2018a. *Cryptotis magna*. The IUCN Red List of Threatened Species 2018:e.T5766A22285160.

Cuarón, A. D., and P. C. de Grammont. 2018b. *Sorex sclateri*. The IUCN Red List of Threatened Species 2018:e.T20394A22316927.

Cuarón, A. D., I. J. March, and P. M. Rockstroh. 1989. A second armadillo (*Cabassous centralis*) for the faunas of Guatemala and Mexico. Journal of Mammalogy 70:870–871.

Cuarón, A. D., P. C. de Grammont, and K. McFadden. 2016a. *Procyon pygmaeus*. The IUCN Red List of Threatened Species 2016:e. T18267A45201913.

Cuarón, A. D., P. C. de Grammont, N. Woodman, and J. Matson. 2016b. *Cryptotis mayensis*. The IUCN Red List of Threatened Species 2016:e. T136488A22284758.

Cuarón, A. D., F. Reid, J. F. González-Maya, and K. Helgen. 2016c. *Galictis vittata*. The IUCN Red List of Threatened Species 2016:e. T41640A45211961.

Cuarón, A. D., F. Reid, K. Helgen, and J. F. González-Maya. 2016d. *Eira barbara*. The IUCN Red List of Threatened Species 2016:e. T41644A45212151.

Cuarón, A. D., J. F. González-Maya, K. Helgen, F. Reid, J. Schipper, and J. W. Dragoo. 2016e. *Mephitis macroura*. The IUCN Red List of Threatened Species 2016:e.T41634A45211135.

Cuarón, A. D., K. Helgen, and F. Reid. 2016f. *Conepatus semistriatus*. The IUCN Red List of Threatened Species 2016:e.T41633A45210987.

Cuarón, A. D., K. Helgen, and F. Reid. 2016g. *Spilogale gracilis*. The IUCN Red List of Threatened Species 2016:e.T136797A45221721.

Cuarón, A. D., K. Helgen, F. Reid, J. Pino, and J. F. González-Maya. 2016h. *Nasua narica*. The IUCN Red List of Threatened Species 2016:e. T41683A45216060.

Cuarón, A. D., P. C. de Grammont, N. Woodman, and J. Matson. 2017. *Sorex saussurei*. The IUCN Red List of Threatened Species 2017:e. T41416A22317311.

Cuarón, A. D., P. C. de Grammont, and J. Matson. 2018. *Sorex stizodon*. The IUCN Red List of Threatened Species 2018:e.T20395A22316681.

Cudworth, N. L., and J. Koprowski. 2010. *Microtus californicus* (Rodentia: Cridetidae). Mammalian Species 42:230–243.

Cudworth, N. L., and M. B. Grenier. 2015. *Thomomys clusius* (Rodentia: Geomyidae). Mammalian Species 47:57–62.

Culver, M., W. E. Johnson, J. Pecon-Slattery, and S. J. O'Brien. 2000. Genomic ancestry of the American puma (*Puma concolor*). Journal of Heredity 91:186–197.

Currier, M. J. P. 1983. *Felis concolor*. Mammalian Species 200:1–7.

Cypher, B., and R. List. 2014. *Vulpes macrotis*. The IUCN Red List of Threatened Species 2014:e.T41587A62259374.

Czaplewski, N. J. 1983. *Idionycteris phyllotis*. Mammalian Species 208:1–4.

Czech, H. A., A. A. Bohlman, and W. B. Sutton. 2017. New Records for *Sorex hoyi* (American pygmy shrew) in Alabama. Southeastern Naturalist 16:464–472.

Czernay, S. 1987. Spiesshirsche und Pudus. A. Ziemsen Verlag. Wittenberg Lutherstadt, Germany.

Da Cunha-Tavares, V., and C. A. Mancina. 2008. *Phyllops falcatus* (Chiroptera: Phyllostomidae). Mammalian Species 811: 1–7.

Dalby, P. L., and H. A. Lillevick. 1969. Taxonomic analysis of electrophoretic blood serum patterns in the cotton rat, *Sigmodon*. Publications of the Museum Michigan State University, Biological Series 4:65–104.

Dalquest, W. W. 1947. Notes on the natural history of the bat, *Myotis yumanensis*, in California, with a description of a new race. The American Midland Naturalist 38:224–247.

Dalquest, W. W. 1948. Mammals of Washington. University of Kansas Museum of Natural History Publications 2:1–444.

Dalquest, W. W. 1950. Record of mammals from San Luis Potosi. Occasional Papers of the Museum of Zoology, Louisiana State University 23:1–15.

Dalquest, W. W. 1951. Six new mammals from the state of San Luis Potosi, Mexico. Journal of the Washington Academy of Sciences 41:361–364.

Dalquest, W. W. 1953. Mammals of the Mexican state of San Luis Potosi. Louisiana State University Press. Baton Rouge, U.S.A.

Dalquest, W. W., and D. R. Orcutt. 1942. The biology of the least shrew-mole, *Neurotrichus gibbsii minor*. The American Midland Naturalist 27:387–401.

Dalquest, W. W., and J. H. Roberts. 1951. Behavior of young grisons in captivity. The American Midland Naturalist 46:359–366.

Dalquest, W. W., H. J. Werner, and J. H. Roberts. 1952. The facial glands of a fruit eating bat, *Artibeus jamaicensis* Leach. Journal of Mammalogy 33:102–103.

Dannelid, E. 1989. Medial tines on the upper incisors and other dental features used as identification characters in European shrews of the genus Sorex (Mammalia, Soricidae). Z. Säugetierkunde 54:205–214.

Da Silva, A. G., Gaona, O., and Medellín, R. A. 2008. Diet and trophic structure in a community of fruit-eating bats in Lacandon Forest, Mexico. Journal of Mammalogy 89:43–49.

Dávalos, L. M. 2006. The geography of diversification in the mormoopids (Chiroptera: Mormoopidae). Biological Journal of the Linnean Society 88:101–118.

Davalos, L., J. Molinari, H. Mantilla-Meluk, C. Medina, J. Pineda, and B. Rodriguez. 2016. *Pteronotus personatus*. The IUCN Red List of Threatened Species 2016:e.T18709A115145223.

Davalos, L., J. Molinari, B. Miller, and B. Rodriguez. 2018. *Peropteryx kappleri*. The IUCN Red List of Threatened Species 2018:e. T16707A22100544.

Davalos, L., J. Molinari, H. Mantilla-Meluk, C. Medina, J. Pineda, and B. Rodriguez. 2019. *Mormoops megalophylla*. The IUCN Red List of Threatened Species 2019:e.T13878A22086060.

Davidow-Henry, B. R., J. K. Jones, Jr. and R. R. Hollander. 1989. *Cratogeomys castanops*. Mammalian Species 338:1–6.

Davis, B. L., S. L. Williams, and G. Lopez. 1971. Chromosomal studies of Geomys. Journal of Mammalogy 52:617–620.

Davis, C. S., I. Stirling, C. Strobeck, and D. W. Coltman. 2008. Population structure of ice-breeding seals. Molecular Biology 17:3078–3094.

Davis, W. B. 1937. Variations in Townsend pocket gophers. Journal of Mammalogy 18:145–158.

Davis, W. B. 1939. The recent mammals of Idaho. Caxton Printers. Caldwell, U.S.A.

Davis, W. B. 1940. Distribution and variation of pocket gophers (genus *Geomys*) in the southwestern United States. Bulletin of the Texas Agricultural Experiment Station 590:1–38.

Davis, W. B. 1942. The systematic status of four kangaroo rats. Journal of Mammalogy 23:328–333.

Davis, W. B. 1944a. Geographic variation in brown lemmings (Genus *Lemmus*). The Murrelet 25:19–25.

Davis, W. B. 1944b. Notes on Mexican mammals. Journal of Mammalogy 25:370–403.

Davis, W. B. 1958. Review of Mexican bats of the Artibeus "cinereus" complex. Proceedings of the Biological Society of Washington 71: 163–166.

Davis, W. B. 1959. Taxonomy of the eastern pipistrel. Journal of Mammalogy 40:521–531.

Davis, W. B. 1965. Review of the *Eptesicus brasiliensis* complex in Middle America with the description of a new subspecies from Costa Rica. Journal of Mammalogy 46:229–240.

Davis, W. B. 1966a. Review of South American bats of the genus *Eptesicus*. The Southwestern Naturalist 11:245–274.

Davis, W. B. 1966b. The mammals of Texas. Bulletin Texas Parks and Wildlife Departament 41:1–267.

Davis, W. B. 1968. Revision of the genus *Uroderma*. Journal of Mammalogy 49:676–698.

Davis, W. B. 1969. A review of the small fruit bats (genus *Artibeus)* of Middle America. Part I. The Southwestern Naturalist 14:15–29.

Davis, W. B. 1970. A review of the small fruit bats (genus *Artibeus)* of Middle America. Part II. The Southwestern Naturalist 14:389–402.

Davis, W. B. 1973a. Geographic variation in the fishing bat, Noctilio leporinus. Journal of Mammalogy 54:862–874.

Davis, W. B. 1973b. Geographic variation in the fishing bat, Noctilio leporinus. Journal of Mammalogy 54:862–874.

Davis, W. B. 1974. The mammals of Texas. Bulletin Texas Parks and Wildlife Deptament 41:1–294.

Davis, W. B. 1975. Individual and sexual variation in Vampyressa bidens. Journal of Mammalogy 56:262–265.

Davis, W. B. 1976. Geographic variation in the lesser noctilio, Noctilio albiventris (Chiroptera). Journal of Mammalogy 57:687–707.

Davis, W. B. 1980. New *Sturnira* (Chiroptera: Phyllostomidae) from Central and South America, with key to currently recognized species. Occasional Papers, Museum of Texas Tech University 70:1–5.

Davis, W. B. 1984. Review of the large fruit-eating bats of the "*Artibeus lituratus*" complex (Chiroptera: Phyllostomidae) in Middle America. Occasional Papers, Museum of Texas Tech University 93:1–16.

Davis, W. B., and D. C. Carter. 1962. Notes on Central American bats with description of a new subspecies of *Mormoops*. The Southwestern Naturalist 7:64–74.

Davis, W. B., and D. C. Carter. 1964. A new species of fruit eating bat (genus Artibeus) from Central America. Proceedings of the Biological Society of Washington 77:119–121.

Davis B. L., and R. J. Baker. 1974. Morphometrics, evolution, and cytogenetics of mainland bats of the genus Macrotus Gray (Chiroptera: Phyllostomatidae). Systematic Zoology 23:26–39.

Davis, J., and W. Z. Lidicker, Jr. 1975. The taxonomic status of the southern sea otter. Proceedings California Academy of Science 40:429–437.

Davis, W. B., and D. C. Carter. 1978. A review of the round-eared bats of the *Tonatia silvicola* complex, with description of three new taxa. Occasional Papers, Museum of Texas Tech University 53:1–13.

Davis, W. B., and A. L. Gardner. 2008. Genus *Eptesicus* Rafinesque, 1820. Pp. 440–450, *in* Mammals of South America. Vol. 1: Marsupials, xenarthrans, shrews, and bats (Gardner, A. L., ed.). University of Chicago Press. Chicago, U.S.A.

Davis, W. H., and C. L. Rippy. 1968. Distribution of *Myotis lucifugus* and *Myotis austroriparius* in the southeastern United States. Journal of Mammalogy 49: 113–117.

Davis, W. B., and D. J. Schmidly. 1994. The mammals of Texas. Texas Parks and Wildlife Press. Austin, U.S.A.

Davis, W. B., D. C. Carter, and R. H. Pine. 1964. Noteworthy records of Mexican and Central American bats. Journal of Mammalogy 45:375–387.

Dawson, N. G., J. P. Colella, M. P. Small, K. D. Stone, S. L. Talbot, and J. A. Cook. 2017. Historical biogeography sets the foundation for contemporary conservation of martens (genus *Martes*) in northwestern North America, Journal of Mammalogy 98:715–730.

de Grammont, P. C., and A. Cuarón. 2016a. *Neotoma goldmani*. The IUCN Red List of Threatened Species 2016:e.T14588A115122856.

de Grammont, P. C., and A. Cuarón. 2016b. *Notocitellus adocetus*. The IUCN Red List of Threatened Species 2016:e.T20477A22265744.

de Grammont, P. C., and A. Cuarón. 2016c. *Peromyscus yucatanicus*. The IUCN Red List of Threatened Species 2016:e.T16696A22362477.

de Grammont, P. C., and A. Cuarón. 2016d. *Sciurus alleni*. The IUCN Red List of Threatened Species 2016:e.T20004A22248517.

de Grammont, P. C., and A. D. Cuarón. 2018a. *Tamiasciurus mearnsi*. The IUCN Red List of Threatened Species 2018:e.T21378A22250725.

de Grammont, P. C., and A. D. Cuarón. 2018b. *Microtus oaxacensis*. The IUCN Red List of Threatened Species 2018:e.T13449A22346873.

de Grammont, P. C., and A. D. Cuarón. 2018c. *Microtus umbrosus*. The IUCN Red List of Threatened Species 2018:e.T42635A22346455.

de Grammont, P. C., and A. D. Cuarón. 2018d. *Sigmodon alleni*. The IUCN Red List of Threatened Species 2018:e.T115591937A22355693.

de Grammont, P. C., and J. Matson. 2018. *Sorex macrodon*. The IUCN Red List of Threatened Species 2018:e.T20392A22314529.

de Grammont, P. C., A. Cuarón, and E. Vázquez. 2016. *Sciurus colliaei*. The IUCN Red List of Threatened Species 2016:e.T20007A22248115.

de la Sancha, N., R. Pérez-Hernandez, L. P. Costa, D. Brito, and N. Cáceres. 2016. *Philander opossum*. The IUCN Red List of Threatened Species 2016:e.T40516A22176779.

de la Torre, L. 1956. The dental formula of the bats of the genus *Diaemus*. Proceedings of the Biological Socociety of Washington 69:191–192.

de la Torre, J. A., and R. A. Medellín. 2010. *Pteronotus personatus*. Mammalian Species 42:244–250.

de Oliveira, T. G. 1998a. *Herpailurus yagouaroundi*. Mammalian Species 578:1–6.

de Oliveira, T. G. 1998b. *Leopardus wiedii*. Mammalian Species 579:1–6.

de Oliveira, T., A. Paviolo, J. Schipper, R. Bianchi, E. Payan, and S. V. Carvajal. 2015. *Leopardus wiedii*. The IUCN Red List of Threatened Species 2015:e.T11511A50654216.

de Villa-Meza, A., R. Ávila-Flores, A. D. Cuarón, and D. Valenzuela-Galván. 2011. *Procyon pygmaeus* (Carnivora: Procyonidae). Mammalian Species 43:87–93.

Decher, J., and J. R. Choate. 1995. *Myotis grisescens*. Mammalian Species 510:1–7.

Decker, D. M., and W. C. Wozencraft. 1991. Phylogenetic analysis of recent procyonid genera. Journal of Mammalogy 72:42–55.

D'Elía, G., P. Fabre, and E. P. Lessa. 2019. Rodent systematics in an age of discovery: recent advances and prospects. Journal of Mammalogy 100:852–871.

Delgado, C., D. Tirira, M. Gómez-Laverde, J. Matson, and R. Samudio. 2016a. *Reithrodontomys mexicanus*. The IUCN Red List of Threatened Species 2016:e.T19411A115151358.

Delgado, C., M. Aguilera, R. Timm, and R. Samudio. 2016b. *Sigmodon hirsutus*. The IUCN Red List of Threatened Species 2016:e. T136426A115207583.

Delibes-Mateos, M., R. Villafuerte, B. D. Cooke, and P. C. Alves. 2018. *Oryctolagus cuniculus* (Linnaeus, 1758). Pp. 99–104, *in* Lagomorphs: Pikas, Rabbits, and Hares of the World (Smith, A. T., C. H. Johnston, P. Alves, and K. Hackländer, K., eds). Johns Hopkins University Press, Baltimore, U.S.A.

DeMaster, D. P., and I. Stirling. 1981. *Ursus maritimus*. Mammalian Species 145:1–7.

Demastes, J. W., T. A. Spradling, M. S. Hafner, D. J. Hafner, and D. L. Reed. 2002. Systematics and phylogeography in pocket gophers in the genera *Cratogeomys* and *Pappogeomys*. Molecular Phylogenetics and Evolution 22:144–154.

Demboski, J. R., and J. A. Cook. 2001. Phylogeography of the dusky shrew, *Sorex monticolus* (Insectivora, Soricidae): insight into deep and shallow history in northwestern North America. Molecular Ecology 10:1227–1240.

Demboski, J. R., and J. A. Cook. 2003. Phylogenetic diversification within the *Sorex cinereus* group (Soricidae). Journal of Mammalogy 84:144–158.

Desmastes, J. W., A. L. Butt, M. S. Hafner, and J. E. Light. 2003. Systematics of a rare species of pocket gopher *Pappogeomys alcorni*. Journal of Mammalogy 84:753–761.

Deutsch, C. J., J. P. Reid, R. K. Bonde, D. E. Easton, H. I. Kochman, and T. J. O'Shea. 2003. Seasonal movements, migratory behavior, and site fidelity of West Indian manatees along the Atlantic coast of the United States. Wildlife Monographs 51:1–77.

Deutsch, C. J., C. Self-Sullivan, and A. Mignucci-Giannoni. 2008. *Trichechus manatus*. The IUCN Red List of Threatened Species 2008:e. T22103A9356917.

DeWalt, T. S., E. G. Zimmerman, and J. V. Planz. 1993a. Mitochondrial-DNA phylogeny of species of the *boylii* and *truei* group of the genus *Peromyscus*. Journal of Mammalogy 74:352–362.

DeWalt, T. S., P. D. Sudman, M. S. Hafner, and S. K. Davis. 1993b. Phylogenetic relationships of pocket gophers (*Cratogeomys* and *Pappogeomys*) based on mitochondrial DNA cytochrome b sequences. Molecular Phylogenetics and Evolution 2:193–204.

Dice, L. R. 1937. Mammals of the San Carlos Mountains and vicinity. University of Michigan Studies, Scientific Series 12:245–268.

Dick, T. S. 1997. *Reithrodontomys humulis*. Mammalian Species 565:1–6.

Dickerson, B. R., R. R. Ream, S. N. Vignieri, and P. Bentzen. 2010. Population structure as revealed by mtDNA and microsatellites in northern fur seals, *Callorhinus ursinus*, throughout their range. PLoS ONE:e10671.

Dickey, D. R. 1928. Five new mammals of the genus *Peromyscus* from El Salvador. Proceedings of the Biological Society of Washington 41:1–6.

Diersing, V. E. 1980. Systematics and Evolution of the Pygmy Shrews (Subgenus *Microsorex*) of North America. Journal of Mammalogy 61:76–101.

Diersing, V. E. 1981. Systematic status of *Sylvilagus brasiliensis* and *S. insonus* from North America. Journal of Mammalogy 62:539–556.

Diersing, V. E. 2019. Taxonomic revision of the long-tailed shrew, *Sorex dispar* Batchelder, 1911, from the Appalachian Region of North America, with the description of a new subspecies. Journal of Mammalogy 100:1837–1846.

Diersing, V. E., and D. F. Hoffmeister. 1977. Revision of the shrews *Sorex merriami* and a description of a new species of the subgenus *Sorex*. Journal of Mammalogy 58:321–333.

Diersing, V. E., and D. F. Hoffmeister. 1981. Distribution and systematics of the Masked Shrew (*Sorex cinereus*) in Illinois. Natural History Miscellanea 213:1–11.

Dobson, G. E. 1878. Catalogue of the Chiroptera of the British Museum. British Museum. London, England.

Dobson, G. E. 1885. Notes on species of Chiroptera in the collection of the Genoa Civic Museum, with descriptions of new species. Annali del Museo Civico di Storia Naturale di Genova 22:16–19.

DOF (Diario Oficial de la Federación). 2010. NOM-059-SEMARNAT-2010, Protección ambiental-Especies nativas de México de flora y fauna silvestres-Categorías de riesgo y especificaciones para la inclusión, exclusión o cambio-Lista de especies en riesgo. Diario Oficial de la Federación, 30 de diciembre de 2010. Ciudad de México. México.

DOF (Diario Oficial de la Federación). 2019. Modificación del anexo normativo III. lista de especies en riesgo de la Norma Oficial Mexicana NOM-059-SEMARNAT-2010, Protección ambiental-Especies nativas de México de flora y fauna silvestres-Categorías de riesgo y especificaciones para la inclusión, exclusión o cambio-Lista de especies en riesgo. Diario Oficial de la Federación, 14 de noviembre de 2019. Ciudad de México, México.

Dokuchaev, N. E. 1994. Siberian shrew, Sorex minutissimus, founding Alaska. Zoologicheskii Zhurnal 73:254–256.

Dokuchaev, N. E. 1997. A new species of shrew (Soricidae, Insectivora) from Alaska. Journal of Mammalogy 78: 811–17.

Dolan, J. M., Jr. 1963. Beitrag zur systematischen Gliederung des Tribus Rupicaprini Simpson, 1945. Zeitschrift für zoologische Systematik und Evolutionsforschung 1:311–407.

Dolan, P. G. 1989. Systematics of Middle American mastiff bats of the genus *Molossus*. Special Publications of the Museum, Texas Tech University 29:1–71.

Dolan, P. G., and D. C. Carter. 1977. *Glaucomys volans*. Mammalian Species 78:1–6.

Dolan, P. G., and D. C. Carter. 1979. Distributional notes and records for Middle American Chiroptera. Journal of Mammalogy 60:644–649.

Dolan, P. G., and R. L. Honeycutt. 1978. Systematic and evolutionary implications of genic variation in the mastiff bat, *Eumops* (Chiroptera: Molossidae). Bat Research News 19:72.

Domínguez-Castellanos, Y., and J. Ortega. 2003. *Liomys spectabilis*. Mammalian Species 718:1–3.

Domning, D. 2016. *Hydrodamalis gigas*. The IUCN Red List of Threatened Species 2016:e.T10303A43792683.
Domning, D. P., and H. Furusawa. 1995. Summary of taxa and distribution of Sirenia in the North Pacific Ocean. *Island Arc* 3: 506–512.
Domning, D. P., J. Thomason, and D. G. Corbett. 2007. Steller's sea cow in the Aleutian Islands. Marine Mammal Science 23:976–983.
Donkin, R. A. 1985. The peccary-with observations on the introduction of pigs to the New World. Transactions of the American Philosophical Society 75:1–152.
Doroff, A., and A. Burdin. 2015. *Enhydra lutris*. The IUCN Red List of Threatened Species 2015:e.T7750A21939518.
Dowler, R. C., and M. D. Engstrom. 1988. Distributional records of mammals from the southwestern Yucatan peninsula of Mexico. Annals Carnegie Museum 57:159–166.
Dowler, R. C., and H. H. Genoways. 1978. *Liomys irroratus*. Mammalian Species 82:1–6.
Dragoo, J. W. 2009. Family Mephitidae (skunks). Pp. 532–563, *in* Handbook of the mammals of the world, vol. 1 Carnivores (Wilson, D. E., and R. A. Mittermeier, eds.). Lynx Edicions, CI and IUCN. Barcelona, Spain.
Dragoo J. W., and R. L. Honeycutt 1997. Systematics of Mustelid-like Carnivores. Journal of Mammalogy 78:426–443.
Dragoo, J. W., and S. R. Sheffield. 2009. *Conepatus leuconotus* (Carnivora: Mephitidae). Mammalian Species 827:1–8.
Dragoo, J. W., J. R. Choate, T. L. Yates, and T. P. O'Farrell. 1990. Evolutionary and taxonomic relationships among North American arid-land foxes. Journal of Mammalogy 71:318–332.
Dragoo, J. W, R. D. Bradley, R. L. Honeycutt, and J. W. Templeton. 1993. Phylogenetic relationships among the skunks: a molecular perspective. Journal of Mammalian Evolution 1:255–267.
Dragoo, J. W., R. L. Honeycutt, and D. J. Schmidly. 2003. Taxonomic status of the white-backed hog-nosed skunks, genus *Conepatus* (Carnivora: Mephitidae). Journal of Mammalogy 84:159–176.
Drew, R. E., J. G. Hallett, K. B. Aubry, K. W. Cullings, S. M. Koepf, and W. J. Zielinski. 2003. Conservation genetics of the fisher (*Martes pennanti*) based on mitochondrial DNA sequencing. Molecular Ecology 12:51–62.
Driscoll, C. A., M. Menotti-Raymond, A. L. Roca, K. Hupe, W. E. Johnson, E. Geffen, E. H. Harley, M. Delibes, D. Pontier, A. C. Kitchener, N. Yamaguchi, S. J. O'Brien, and D. W. Macdonald. 2007. The Near Eastern origin of cat domestication. Science 317:519–523.
Duarte, J. M. B., S. González, and J. E. Maldonado. 2008. The surprising evolutionary history of South American deer. Molecular Phylogenetics and Evolution 49:17–22.
Duckworth, J. W., N. S. Kumar, M. Anwarul Islam, H. Sagar Baral, and R. Timmins. 2015. Axis axis. The IUCN Red List of Threatened Species 2015:e.T41783A22158006.
Durish, N. D., K. E. Halcomb, C. W. Kilpatrick, and R. D. Bradley. 2004. Molecular systematics of the Peromyscus truei species group. Journal of Mammalalogy 85:1160–1169.
Durant, S. D. 1952. Mammals of Utah. University of Kansas Publications, Museum of Natural History 6:1–549.
Duplechin, R. M., and R. D. Bradley. 2014. *Peromyscus oaxacensis* Merrian, 1898 Oaxacan deermouse. Pp. 383–384, *in* Mammals of México (Ceballos, G., ed.). John Hopkins University Press. Baltimore, U.S.A.
Easterla, D. A. 1965. The spotted bat in Utah. Journal of Mammalogy 46:665–668.
Easterla, D. A. 1968a. First records of the pocketed free-tailed bat for Texas. Journal of Mammalogy 49:515–516.
Easterla, D. A. 1968b. Range extension of the southern bog lemming in Arkansas. The Southwestern Naturalist 13:364.
Easterla, D. A. 1968c. Hispid cotton rat north of the Missouri River. The Southwestern Naturalist 13:364–365.
Easterla, D. A. 1968d. First records of *Blarina brevicauda* minima in Missouri and Arkansas. The Southwestern Naturalist 13:448–449.
Easterla, D. A. 1970. First records of the spotted bat in Texas and notes on its natural history. The American Midland Naturalist 83:306–308.
Easterla, D. A. 1971. Notes on young and adults of the spotted bat, *Euderma maculatum.* Journal of Mammalogy 52:475–476.
Edelman, A. J. 2003. *Marmota olympus*. Mammalian Species 736:1–5.
Edelman, A. J. 2019. *Sylvilagus obscurus* (Lagomorpha: Leporidae). Mammalian Species 51:128–135.
Edwards, R. L. 1963. Observations on the small mammals of the southeastern shore of Hudson Bay. Canadian Field Naturalist 77:1–12.
Edwards, C. W., and R. D. Bradley. 2001. Molecular phylogenetics of the *Neotoma floridana* species group. Journal of Mammalogy 82:791–798.
Edwards, C. W., and R. D. Bradley. 2003. Molecular systematics of the genus *Neotoma.* Molecular Phylogenetics and Evolution 25:489–500.
Edwards, C. W., C. F. Fulhorst, and R. D. Bradley. 2001. Molecular phylogenetics of the *Neotoma albigula* species group: further evidence of a paraphyletic assemblage. Journal of Mammalogy 82:267–279.
Eger, J. L. 1974. A new subspecies of the bat *Eumops auripendulus* (Chiroptera: Molossidae), from Argentina and Eastern Brazil. Life Sciences Occasional Papers, Royal Ontario Museum 25:1–8.
Eger, J. L. 1977. Systematic of the genus *Eumops* (Chiroptera: Molossidae). Life Sciences Contributions, Royal Ontario Museum 110:1–69.
Eger, J. L. 1990. Patterns of geographic variation in the skull of Neartic ermine (*Mustela erminea*). Canadian Journal of Zoology 68:1241–1249.
Eger, J. L. 1995. Morphometric variation in the Nearctic collared lemming (*Dicrostonyx*). Journal of Zoology of London 235:143–161.
Eger, J. L. 2008. Family Molossidae P. Gervais, 1856. Pp. 399–439, *in* Mammals of South America. Vol. 1: Marsupials, xenarthrans, shrews, and bats (Gardner, A. L., ed.). University of Chicago Press. Chicago, U.S.A.
Egoscue, H. J. 1979. *Vulpes velox.* Mammalian Species 122:1–5.
Eisenberg, J. F. 1989. Mammals of the Neotropics: The northern Neotropics, Panama, Columbia, Venezuela, Guyana, Suriname, French Guiana. University of Chicago. Chicago, U. S. A.
Eisenberg, J. F., and K. H. Redford. 1999. Mammals of the Neotropics, 3. The central Neotropics. University of Chicago Press. Chicago, U.S.A.
Eizirik, E., T. Haag., A. S. Santos, F. M. Salzano, L. Silveira, F. C. C. Azevedo, and M. M. Furtado. 2008. Jaguar conservation genetics. Cat News Special Issue 4:31–34.
Elder, F. F. B., and M. R. Lee. 1985. The chromosomes of *Sigmodon ochrognathus* and *S. fulviventer* suggest a realignment of *Sigmodon* species groups. Journal of Mammalogy 66:511–518.
Elizalde-Arellano, C., J. C. López-Vidal, E. Quhart, J. I. Campos-Rodríguez, and R. Hernández-Arciga. 2010. Nuevos registros y extensiones de distribución de mamíferos para Guanajuato, México. Acta Zoológica Mexicana (n. s.) 26:73–98.
Elizalde-Arellano, C., J. C. López-Vidal, E. Uría-Galicia, H. Montellano-Rosales, J. Arroyo-Cabrales, and R. A. Medellín. 2008. Citologia vaginal y ciclo estral de *Diphylla ecaudata*. Pp. 253–268, *in* Avances en el estudio de los mamíferos de México (Lorenzo, C., E. Espinoza, and J. Ortega, eds.). Volumen II. Asociación Mexicana de Mastozoología, A. C. y el Colegio de la Frontera Sur. San Crístóbal de las Casas, México.

Elizalde-Arellano, C., J. C. López-Vidal, J. Arroyo-Cabrales, R. A. Medellín, and J. W. Laundré. 2007. Food sharing behavior in the hairy–legged vampire bat *Diphylla ecaudata*. Acta Chiropterologica 9:314–319.

Ellerman, J. R. 1940. The families and genera of living rodents. Volume I. Rodents other than Muridae. British Museum (Natural History), London, United Kingdom.

Ellerman, J. R. 1941. The families and genera of living rodents. British Museum (Natural History), London 2:1–690.

Ellerman, J. R., and T. C. S. Morrison-Scott. 1951. Checklist of Palaearctic and indian mammals 1758 to 1946. British Museum (Natural History), London, United Kingdom.

Ellerman, J. R., R. W. Hayman, and G. W. C. Holt. 1941. The families and genera of living rodents, with a list of names from (1758–1936). Brtish Museum, Natural history London 2:1–690.

Ellerman, J. R., and T. C. S. Morrison-Scott. 1966. Checklist of Palaearctic and Indian Mammals 1758 to 1946. Second ed. British Museum (Natural History), London, 810 pp.

Elliot, D. G. 1901. A synopsis of the mammals of North America and the adjacent seas. Field Columbian Museum Publications, Zoology Series 2:1–522.

Elliot, D. G. 1903. A list of a collection of Mexican mammals with descriptions of some apparently new forms. Field Columbian Museum Publications, Zoology Series 71:141–522.

Elliot, D. G. 1905. A checklist of mammals of the North American continent, the West Indes and the neighboring seas. Field Columbian Museum Publications Zoological series 6:1–192.

Elliott, C. L., and J. T. Flinders. 1991. *Spermophilus columbianus*. Mammalian Species 372:1–9.

Emmons, L. H. 1997. Mammals of the Rio Urucuti Basin, south central Chuquisaca, Bolivia. Pp. 30–33, *in* A rapid assessment of the humid forest of south central Chuquisaca, Bolivia (Schulenberg, T. S., and K. Awbrey, eds.). Conservation International, RAP Working Papers 8, Washington, U.S.A.

Emmons, L. H. 2005. A Revision of the Genera of Arboreal Echimyidae (Rodentia: echimyidae, Echimyinae), with descriptions of two new genera. Pp. 247–310, *in* Mammalian Diversification: From Chromosomes to Phylogeography (A Celebration of the Career of James L. Patton) (Lacey, E. A., and P. Myers). University of California Publications in Zoology.

Emmons, L. 2016a. *Cuniculus paca*. The IUCN Red List of Threatened Species 2016:e.T699A22197347.

Emmons, L. 2016b. *Dasyprocta punctata*. The IUCN Red List of Threatened Species 2016:e.T89497686A78319610.

Emmons, L. 2016c. *Erethizon dorsatum*. The IUCN Red List of Threatened Species 2016:e.T8004A22213161.

Emmons, L. H., and F. Feer. 1990. Neotropical rainforest mammals. A field guide. University of Chicago Press. Chicago, U.S.A.

Engstrom, M. D. 1984. Chromosomal, genic, and morphological variation in the *Oryzomys melanotis* species group. PhD. Dissertation, Texas A & M University, College Station, U.S.A.

Emmons, L., and F. Feer. 1997. Neotropical rainforest mammals: a field guide (second edition.) University of Chicago Press, Chicago

Engstrom, M. D., and D. E. Wilson. 1981. Systematics of *Antrozous dubiaquercus* (Chiroptera: Vespertilionidae), with comments on the status of *Bauerus* Van Gelder. Annals Carnegie Museum 50:371–383.

Engstrom, M. D., H. H. Genoways, and P. K. Tucker. 1987a. Morphological variation, karyology, and systematic relationships of *Heteromys gaumeri* (Rodentia: Heteromyidae). Pp. 289–303, *in* Studies in Neotropical mammalogy: essays in honor of Philip Hershkovitz (Patterson, B. D., and R. M. Timm, eds.). Fieldiana: Zoological, new series 39:1–506.

Engstrom, M. D., T. E. Lee, and D. E. Wilson. 1987b. *Bauerus dubiaquercus*. Mammalian Species 282:1–3.

Engstrom, M. D., O. Sánchez-Herrera, and G. Urbano-Vidales. 1992. Distribution, geographic variation, and systematic relationships within *Nelsonia* (Rodentia: Sigmodontinae). Proceedings of the Biological Society of Washington 105:867–881.

Erratum. 2014. Erratum and Nomenclature Committee Opinion on use of *Perognathus alticola* versus *P. alticolus*. Journal of Mammalogy 95:910.

Ernest, K. A., and M. A. Mares. 1987. *Spermophilus tereticaudus*. Mammalian Species 274:1–9.

Escobedo-Morales, L. A., L. León-Paniagua, J. Arroyo-Cabrales, and F. Greenaway. 2006. Distributional records for mammals from Chiapas, México. The Southwestern Naturalist 51:269–272.

Escobedo-Morales, L. A., S. Mandujano, L. E. Eguiarte, M. A. Rodrguez-Rodrguez, and J. E. Maldonado. 2016. First phylogenetic analysis of mesoamerican brocket deer *Mazama pandora* and *Mazama temama* (Cetartiodactyla: Cervidae) based on mitochondrial sequences: implications on neotropical deer evolution. Mammalian Biology: Zeitschrift für Säugetierkunde 81:303–313.

Escobedo-Morales, L. A., L. León-Paniagua, E. Martínez-Meyer, and S. Mandujano. 2023. Reevaluation of the status of the Central American brocket deer *Mazama temama* (Artiodactyla: Cervidae) subspecies based on morphological and environmental evidence. Journal of Mammalogy 104:333–346.

Eshelman, B. D., and G. N. Cameron. 1987. *Baiomys taylori*. Mammalian Species 285:1–7.

Eshelman, B. D., and C. S. Sonnemann. 2000. *Spermophilus armatus*. Mammalian Species 637:1–6.

Espinoza, J., C. Lorenzo, and E. Rios. 2011. Variación morfológica y morfométrica de *Heteromys desmarestianus* en Chiapas, México. Therya 2:139–154.

Esquivel, D. A., M. J. R. Pereira, J. D. Stuhler, D. M. Rossoni, P. M. Velazco, and F. M. Bianchi. 2022. Multiples lines of evidence unveil cryptic diversity in the *Lophostoma brasiliense* (Chiroptera: Phyllostomidae) complex. Systematics and Biodiversity, 20:2110172.

Estes, J. A. 1980. *Enhydra lutris*. Mammalian Species 133:1–8.

Esteva, M., A. F. Cervantes, S. V. Bryant, and J. A. Cook. 2010. Molecular phylogeny of long-tailed shrews (genus *Sorex*) from Mexico and Guatemala. Zootaxa 2615:47–65.

Ewer, R. F. 1973. The carnivores. Cornell University Press, Ithaca, New York.

Fabre, P.-H., J. L. Patton, and Y. L. R. Leite. 2016. Family Echimyidae (hutias, South American spiny-rats and coypu). Pp. 552–641, *in* Handbook of the Mammals of the World. Vol 6. Lagomorphs and Rodents I (Wilson, D. E., T. E. Lacher, Jr, and R. A. Mittermeier, eds.). Lynx Edicions. Barcelona, Spain.

Faries, K. M., T. V. Kristensen, J. Beringer, J. D. Clark, D. White, Jr, and L. S. Eggert. 2013. Origins and genetic structure of black bears in the Interior Highlands of North America. Journal of Mammalogy 94:369–377.

Fashing, N. J. 1973. Implications of karyotypic variation in the kangaroo rat, *Dipodomys heermanni*. Journal of Mammalogy 54:1018–1020.

Faulhaber, C. A., and A. T. Smith. 2008. *Sylvilagus palustris*. The IUCN Red List of Threatened Species 2008:e.T41303A10435830.

Fauteaux, D., G. Lupien, F. Fabianek, J. Gagnon, M. Séguy, and L. Imbeau. 2014. An illustrated key to the mandibles of small mammals of eastern Canada. Canadian Field Naturalist 128:25–37.
Fay, F. H. 1981. Walrus *Odobenus rosmarus* (Linneaus, 1758). Pp. 1–23, *in* Handbook of Marine Mammals. The walrus, sea lions, fur seals, and sea otter (Ridgway, S. H., and R. Harrison, eds.). Academic Press. San Diego, U.S.A.
Fay, F. H. 1982. Ecology and biology of the Pacific walrus, *Odobenus rosmarus divergens* Illiger. North American Fauna 74:1–279.
Fay, F. H. 1985. *Odobenus rosmarus*. Mammalian Species 238:1–7.
Feijó, A., B. Patterson, and P. Cordeiro-Estrela. 2018. Taxonomic revision of the long-nosed armadillos, Genus *Dasypus* Linnaeus, 1758 (Mammalia, Cingulata). PLoS ONE 13:e0195084.
Feijó, A., J. F. Vilela, J. Cheng, M. A. A. Schetino, R. T. F. Coimbra, C. R. Bonvicino, F. R. Santos, B. D. Patterson, and P. Cordeiro-Estrela. 2019. Phylogeny and molecular species delimitation of long-nosed armadillos (*Dasypus*: Cingulata) supports morphology-based taxonomy. Zoological Journal of the Linnean Society 186:813–825.
Fellers, G. M., W. Z. Lidicker Jr., A. Linzey, and NatureServe. 2016. *Aplodontia rufa*. The IUCN Red List of Threatened Species 2016:e. T1869A115057269.
Feldhamer, G. A. 1993. Habitat partitioning, body size, and timing of parturition in pygmy shrews and associated soricids. Journal of Mammalogy 74:403–411.
Felten, V. H. 1956. Fledermäuse (Mammalia, Chiroptera) aus El Salvador. Senckenbergiana Biologica 37:179–212.
Fenton, M. B., and R. M. R. Barclay. 1980. *Myotis lucifugus*. Mammalian Species 142:1–8.
Fernández, J. A. 2012. Phylogenetics and biogeography of the microendemic rodent *Xerospermophilus perotensis* (Perote ground squirrel) in the Oriental Basin of Mexico. Journal of Mammalogy 93:1431–1439.
Fernández, J. A. 2014. Mitochondrial phylogenetics of a rare Mexican endemic Nelson's woodrat, *Neotoma nelsoni* (Rodentia: Cricetidae), with comments on its biogeographic history. The Southwestern Naturalist 59:81–90.
Fernández, J. A., F. García-Campusano, and M. S. Hafner. 2010. *Peromyscus difficilis*. Mammalian Species 867:220–229.
Fernández, J. A., F. A. Cervantes, and M. S. Hafner. 2012. Molecular systematics and biogeography of the Mexican endemic kangaroo rat, *Dipodomys phillipsii* (Rodentia: Heteromyidae). Journal of Mammalogy 93:560–571.
Ferrell, C. S., and D. E. Wilson. 1991. *Platyrrhinus helleri*. Mammalian Species 373:1–5.
Fertl, D., A. J. Schiro, G. T. Regan, C. A. Beck, N. Adimey, L. Price-May, A. Amos, G. A. J. Worthy, and R. Crossland. 2005. Manatee occurrence in the northern Gulf of Mexico, west of Florida. Gulf and Caribbean Research 17:69–94.
Festa-Bianchet, M. 2020a. *Ovis canadensis*. The IUCN Red List of Threatened Species 2020:e.T15735A22146699.
Festa-Bianchet, M. 2020b. *Ovis dalli*. The IUCN Red List of Threatened Species 2020:e.T39250A22149895.
Festa-Bianchet, M. 2022. *Oreamnos americanus*. The IUCN Red List of Threatened Species 2022:e.T42680A211860282. Accessed on 11 August 2022.
Findley, J. S. 1955a. Taxonomy and distribution of some American shrews. University of Kansas Publications, Museum of Natural History 7:613–618.
Findley, J. S. 1955b. Speciation of the wandering shrew. University of Kansas Publications, Museum of Natural History 9:1–68.
Findley, J. S. 1960. Identity of the long-eared myotis of the Southwest and Mexico. Journal of Mammalogy 41:16–20.
Findley, J. S. 1967. A black population of the Goldman pocket mouse. The Southwestern Naturalist 12:191–192.
Findley, J. S. 1972. Phenetic relationships among bats of the genus *Myotis*. Systematic Zoology 21:31–52.
Findley, J. S. 1987. The natural history of New Mexican mammals. University of New Mexico Press, Albuquerque, U.S.A.
Findley, J. S., and C. Jones. 1960. Geographic variation in the yellow-nosed cotton rat. Journal of Mammalogy 41:462–469.
Findley, J. S., and C. Jones. 1967. Taxonomic relationships of bats of the species *Myotis fortidens*, *M. lucifugus*, and *M. occultus*. Journal of Mammalogy 48:429–444.
Findley, J. S., A. H. Harris, D. E. Wilson, and C. Jones. 1975. Mammals of New Mexico. University of New Mexico Press. Albuquerque, U.S.A.
Fisler, G. F. 1965. Adaptations and speciation in harvest mice of the marshes of San Francisco Bay. University of California Publications in Zoology 77:1–108.
Fish, F. E., B. R. Blood, and B. D. Clark. 1991. Hydrodynamics of the feet of fish-catching bats: influence of the wáter surface on drag and morphological design. Journal of Experimental Zoology 258:164–173.
Fitch, J. H., and K. A. Shump. 1979. *Myotis keenii*. Mammalian Species 121:1–3.
Fitch, J. H., K. A. Shump, A. U. Shump. 1981. *Myotis velifer*. Mammalian Species 149:1–5.
Fleharty, E. D. 1960. The status of the gray-necked chipmunk in New Mexico. Journal of Mammalogy 41:235–242.
Fleming, T. H. 1988. The short-tailed fruit bat: a study in plant animal interactions. The University of Chicago Press, Chicago.
Fleming, T. H., E. T. Hooper, and D. E. Wilson. 1972. Three Central American bat communities: structure, reproductive cycles, and movement patterns. Ecology 53:655–670.
Flyger, V., and J. E. Gates. 1982. Fox and gray squirrels. Pp. 209–229, *in* Wild mammals of North America (Chapman, J. A., and G. A. Feldhamer, eds.). Johns Hopkins University Press. Baltimore, U.S.A.
Flynn, J. J., M. A. Nedbal, J. W. Dragoo, and R. L. Honeycutt. 2000. Whence the Red Panda? Molecular Phylogenetics and Evolution 17:190–199,
Flynn, L. J., L. L. Jacobs, Y. Kimura, and E. H. Lindsay. 2019. Rodent Suborders. Fossil Imprint 75:292–298.
Folkow, L. P., E. S. Nordoy, and A. S. Blix. 2010. Remarkable development of diving performance and migrations of hooded seals (*Cystophora cristata*) during their first year of life. Polar Biology 33:433–441.
Fonseca, R. M., S. R. Hoofer, C. A. Porter, C. A. Cline, D. A. Parish, F. G. Hoffmann, and R. J. Baker 2007. Morphological and molecular variation within Little big-eared bats of the genus M*icronycteris* (Phyllostomidae: Micronycterinae) from San Lorenzo, Ecuador. Pp. 721–746, *in* The quintessential naturalist: Honoring the life and legacy of Oliver P. Pearson (Kelt, D. A., E. P. Lessa, J. Salazar-Bravo, and J. L. Patton, eds.). University of California Publications in Zoology 134:1–981.
Forsten, A., and P. M. Youngman. 1982. *Hydrodamalis gigas*. Mammalian Species 165:1–3.
Ford, L. S., and R. S. Hoffmann. 1988. *Potos flavus*. Mammalian Species 321:1–9.
Francis, C. M., and N. B. Simmons. 2022. On behalf of the G. B. T. W. G. of the IUCN SSC Bat Specialist Group. On the taxonomy of *Myotis lucifugus*.

Franzmann, A. W. 1978. Moose. Pp. 67–82, *in* Big game of North America: ecology and management (Schmidt, J. L. and D. L. Gilbert, eds.). Stackpole Books. Harrisburg, U.S.A.

Franzmann, A. W. 1981. *Alce alce*. Mammalian Species 154:1–7.

Franzmann, A. W., *et al.* 1978. Alaskan moose measurements and weights and measurement-weight relationships. Canadian Journal of Zoology 56:298–306.

Frase, B. A., and R. S. Hoffmann. 1980. *Marmota flaviventris*. Mammalian Species 135:1–8.

Freeman, P. W. 1981. A multivariate study of the family Molossidae (Mammalia: Chiroptera): morphology, ecology, evolution. Fieldiana Zoology (New Series) 7:1–173.

French, T. W. 1980. Sorex longirostris. Mammlian Species 148:1–3.

French, T. W., and G. L. Kirkland Jr., 1983. Taxonomy of the Gaspe shrew, *Sorex gaspensis*, and the rock shrew, *S. dispar*. Canadian Field-Naturalist 97:75–78.

Frey, J. K. 1999. Mogollon vole/*Microtus mogollonensis*. Pp. 634–635, *in* The Smithsonian book of North American Mammals (Wilson, D. E., and S. Ruff, eds.). Smithsonian Institution Press. Washington, U.S.A.

Frey, J. K., and F. A. Cervantes. 1997a. *Microtus umbrosus*. Mammalian Species 555:1–3.

Frey, J. K., and F. A. Cervantes. 1997b. *Microtus oaxacensis*. Mammalian Species 556:1–3.

Frey, J. K., and C. T. LaRue. 1993. Note on the distribution of the Mogollon vole (*Microtus mogollonensis*) in New Mexico and Arizona. The Southwestern Naturalist 38:176–178.

Frey, R., and T. Riede. 2013. The anatomy of vocal divergence in North American Elk and European red deer. Journal of Morphology 274:307–19.

Frey, J. K., and D. W. Moore. 1990. Status of Hayden's shrew (*Sorex haydeni*) in Kansas. The Southwestern Naturalist 35:84–86.

Frick, W. F., J. P. Hayes, and P. A. Heady, III. 2008. Island biogeography of bats in Baja California, Mexico: Patterns of bat species richness in a near-shore archipelago. Journal of Biogeography 35:353–364.

Frick, W. F., J. P. Hayes, P. A. Heady, III, and J. P. Haynes. 2009. Facultative nectar-feeding behavior in a gleaning insectivorous bat (*Antrozous pallidus*). Journal of Mammalogy 90:1157–1164.

Fritzell, E. K., and K. J. Haroldson. 1982. *Urocyon cinereoargenteus*. Mammalian Species 189:1–8.

Fujita, M. S., and T. H. Kunz. 1984. *Pipistrellus subflavus*. Mammalian Species 228:1–6.

Fulton, T. L., and C. Strobeck. 2010. Multiple markers and multiple individuals refine true seal phylogeny and bring molecules and morphology back in line. Proceedings of Biological Sciences (1684):1065–1070.

Fumagalli, L., P. Taberlet, D. T. Stewart, L. Gielly, J. Hausser, and P. Vogel. 1999. Molecular phylogeny and evolution of *Sorex* shrews (Soricidae: Insectivora) inferred from mitocondrial DNA sequence data. Molecular Phylogenetics and Evolution 11:222–235.

FWS. 2022. U.S. Fish and Wildlife Service. https://www.fws.gov/. Accessed on 10 September 2022.

Galbreath, K. E., D. J. Hafner, and K. Zamudio. 2009. When cold is better: climate-driven elevation shifts yield complex patterns of diversification and demography in an alpine specialist (American pika, Ochotona princeps). Evolution 63:2848–2863.

Galbreath, K. E., D. J. Hafner, K. Zamudio, and K. K. Agnew. In press. Isolation and introgression in the Intermountain West: incongruence at mitochondrial and nuclear loci illuminates a complex biogeographic history in the American pika (Ochotona princeps). Journal of Biogeography 37:344–362.

Galindo-Leal, C., and G. Zuleta. 1997. The distribution, habitat, and conservation status of the Pacific water shrew, *Sorex bendirii*, in British Columbia. Canadian-Field Naturalist 111:422–428.

Gallina, S., and H. Lopez Arevalo. 2016. *Odocoileus virginianus*. The IUCN Red List of Threatened Species 2016:e.T42394A22162580.

Gallo-Reynoso, J. P. 1994. Factors affecting the population status of the Guadalupe fur seal, *Arctocephalus townsendi* (Merriam, 1897), at Isla Guadalupe, Baja California, México. PhD thesis. University of California, Santa Cruz, U.S.A.

Gallo-Reynoso, J. P., and A. L. Figueroa-Carranza. 1996. Size and weight of Guadalupe fur seals. Marine Mammal Science 12:318–321.

Gallo-Reynoso, J. P., A. L. Figueroa-Carranza, I. D. Barba-Acuña, D. Borjes-Flores, and I. J. Pérez-Cossío. 2020. Steller Sea Lions (*Eumetopias jubatus*) Along the Western Coast of Mexico. Aquatic Mammals 46:411–416.

Gannon, W. L. 1988. *Zapus trinotatus*. Mammalian Species 315:1–5.

Gannon, W. L. 1999. *Zapus trinotatus*. Pp. 669–670, *in* The Smithsonian book of North American mammals (Wilson, D. E., and S. Ruff, eds.). Smithsonian Institution Press. Washington, U.S.A.

Gannon, W. L., and R. B. Forbes. 1995. *Tamias senex*. Mammalian Species 502:1–6.

Gannon, W. L., and T. E. Lawlor. 1989. Variation of the chip vocalization of three species of Townsend chipmunks (genus *Eutamias).* Journal of Mammalogy 70:740–753.

Gannon, M. R., M. R. Willig, and J. K. Jones, Jr. 1989. *Sturnira lilium*. Mammalian Species 333:1–5.

Gannon, W. L., R. B. Forbes, and D. E. Kain. 1993. *Tamias ochrogenys*. Mammalian Species 445:1–4.

Gentry, A., J. Clutton-Brock, and C. P. Groves. 2004. The naming of wild animal species and their domestic derivatives. Journal of Archaeological Science 31:645–651.

Garbino, G. S. T., R. Gregorin, I. P. Lima, L. Loureiro, L. M. Moras, R. Moratelli, M. R. Nogueira, A., C. Pavan, V. C. Tavares, and L. Peracchi. 2020. Updated checklist of Brazilian bats: versão 2020. Comitê da Lista de Morcegos do Brasil—CLMB. Sociedade Brasileira para o Estudo de Quirópteros (Sbeq). accessed on: 27 octuber 2020.

Garbino, G. S. T., G. Hernández-Canchola, L. León-Paniagua, and V. C. Tavares. 2024. A new Mexican endemic species of yellow-eared bat in the genus Vampyressa (Phyllostomidae, Stenodermatinae). Journal of Mammalogy 105:563–576.

García, M., C. Jordan, G. O'Farril, C. Poot, N. Meyer, N. Estrada, R. Leonardo, E. Naranjo, Á. Simons, A. Herrera, C. Urgilés, C. Schank, L. Boshoff, and M. Ruiz-Galeano. 2016. *Tapirus bairdii*. The IUCN Red List of Threatened Species 2016:e.T21471A45173340.

García-Estrada, C., A. Damon, C. Sánchez-Hernández, L. Soto-Pinto, and G. Ibarra-Núñez. 2006. Bat diversity in montane rainforest and shaded coffee under different management regimes in southeastern Chiapas, Mexico. Biological Conservation 132:351–361.

García-Perea, R. 1992. New data on the systematics of *Lynxes*. Cat News 16:15–16.

García-Rodríguez, R. 1998. Phylogeography of the West Indian manatee (*Trichechus manatus*): how many populations and how many taxa? Molecular Ecology 7:1137–1149.

Gardner, A. L. 1963. Nota acerca de la distribución de los murciélagos en México. Revista de la Sociedad Mexicana de Historia Natural 24:41–44.

Gardner, A. L. 1965. New bat records from the Mexican state of Durango. Proceedings of the Western Foundation Vertebrate Zoology 1:101–106.
Gardner, A. L. 1973. The systematic of the genus *Didelphis* (Marsupialia: Didelphidae) in North and Middle America. Special Publications Museum Texas Tech University 4:1–81.
Gardner, A. L. 1977. Feeding habits. Pp. 293–350, *in* Biology of bats of the New World family Phyllostomatidae, Part 2 (Baker, R. J., J. K. Jones, Jr., and D. C. Carter, eds.). Special Publications Museum, Texas Tech University 13:1–364.
Gardner, A. L. 1982. Virginia oposum. Pp. 3–36, *in* Wild mammals of North America (Chapman, J. A., and G. A. Feldhamer, eds.). Johns Hopkins Press. Baltimore, U.S.A.
Gardner, A. L. 1986. The taxonomic status of *Glossophaga morenoi* Martinez y Villa, 1938 (Mammalia: Chiroptera: Phyllostomidae). Proceedings of the Biological Society of Washington 99:489–492.
Gardner, A. L. 1993. Order Didelphimorphia. Pp. 15–23, *in* Mammal species of the world: a taxonomic and geographic reference. (Wilson, D. E., and D. M. Reeder, eds.). Second edition. Smithsonian Institution Press. Washington, U.S.A.
Gardner, A. L. 2005. Order Pilosa. Pp. 100–103, in Mammal species of the world: a taxonomic and geographic reference (Wilson, D. E., and D. M. Reeder, eds.). Third edition. Johns Hopkins University Press. Baltimore, U.S.A.
Gardner, A. L. 2005a. Order Didelphiomorpha. Pp. 3–18, *in* Mammal species of the world: a taxonomic and geographic reference (Wilson, D. E., and D. M. Reeder, eds.). Third edition. Johns Hopkins University Press. Baltimore, U.S.A.
Gardner, A. L. 2005b. Order Cingulata. Pp. 94–99, *in* Mammal species of the world: a taxonomic and geographic reference (Wilson, D. E., and D. M. Reeder, eds.). Third edition. Johns Hopkins University Press. Baltimore, U.S.A.
Gardner, A. L. 2005c. Order Pilosa. Pp. 100–103, *in* Mammal species of the world: a taxonomic and geographic reference (Wilson, D. E., and D. M. Reeder, eds.). Third edition. Johns Hopkins University Press. Baltimore, U.S.A.
Gardner, A. L. 2007a. Suborder Vermilingua Illiger, 1811. Pp. 168–177, *in* Mammals of South America. Vol. 1. Marsupials, xenarthrans, shrews, and bats (Gardner, A. L. ed.). University of Chicago Press. Chicago, U.S.A.
Gardner, A. L. 2007b. Tribe Sturnirini [Miller, 1907]. Pp. 363–376, *in* Mammals of South America. Vol. 1. Marsupials, xenarthrans, shrews, and bats (Gardner, A. L., ed.). University of Chicago Press. Chicago, U.S.A.
Gardner, A. L. 2008. Mammals of South America. Vol. 1: marsupials, xenarthrans, shrews, and bats. University of Chicago Press. Chicago, U.S.A.
Gardner, A. L., and D. C. Carter. 1972. A review of the Peruvian species of *Vampyrops* (Chiroptera: Phyllostomatidae). Journal of Mammalogy 53:72–82.
Gardner, A. L., and V. L. Naples. 2007. Genus *Tamandua* Gray, 1825. Pp. 173–177, *in* Mammals of South America. Vol. 1. Marsupials, xenarthrans, shrews, and bats (Gardner, A. L. ed.). University of Chicago Press, Chicago, U.S.A.
Gardner, A. L., and J. L. Patton. 1972. New species of *Philander* (Marsupialia: Didelphidae) and *Mimon* (Chiroptera: Phyllostomidae) from Peru. Occasional Papers of the Museum of Zoology, Louisiana State University 43:1–12.
Gardner, A. L., and J. Ramírez-Pulido. 2020. Type localities of Mexican land mammals, with comments on taxonomy and nomenclature. Special Publications of the Museum of Texas Tech University 73: 1–134.
Gardner, A. L., R. K. LaVal, and D. E. Wilson. 1970. The distributional status of some Costa Rican bats. Journal of Mammalogy 51:712–729.
Garibaldi, A. 2019. *Dicrostonyx unalascensis*. The IUCN Red List of Threatened Species 2019:e.T39974A22331447.
Garrison, T. E. and T. L. Best. 1990. *Dipodomys ordii*. Mammalian Species 353:1–10.
Garshelis, D. L., B. K. Scheick, D. L. Doan-Crider, J. J. Beecham, and M. E. Obbard. 2016. *Ursus americanus*. The IUCN Red List of Threatened Species 2016:e.T41687A114251609.
Gatica-Colmina, A., B. Navarrete-Laborde, A. Ortiz-González, and O. C. Rosas-Rosas. 2014. Nuevo registro de distribución del puerco espín del norte *Erethizon dorsatum* en Chihuahua, México. Acta Zoológica Mexicana 30:399–402.
Gaudin, T. J., and D. G. Branham. 1998. The Phylogeny of the Myrmecophagidae (Mammalia, Xenarthra, Vermilingua) and the Relationship of Eurotamandua to the Vermilingua. Journal of Mammalian Evolution 5:237–265.
Geist, V. 1971. Mountain sheep: a study in behavior and evolution. University Chicago Press. Chicago, U.S.A.
Geist, V. 1981. Behavior: adaptive strategies in mule deer. Lincoln, NE. Pp. 157–224, in Mule and black-tailed deer of North America (Wallmo, O. C., ed.). University of Nebraska.
Geist, V. 1998. Deer of the world: their evolution, behavior, and ecology. Stackpole Books. Mechanicsburg, U.S.A.
Geffen, E., A. Mercure, D. J. Girman, D. W. Macdonald, and R. K. Wayne. 1992. Phylogenetic relationships of the foxlike canids: mitochondrial DNA restriction fragment, site and cytochrome b sequence analyses. Journal of Zoology (London) 228:27–39.
Gelatt, T., and K. Sweeney. 2016. *Eumetopias jubatus*. The IUCN Red List of Threatened Species 2016:e.T8239A45225749.
Gelatt, T., R. Ream, and D. Johnson. 2015. *Callorhinus ursinus*. The IUCN Red List of Threatened Species 2015:e.T3590A45224953.
Genoways, H. H. 1971. A new species of spiny pocket mouse (genus *Liomys*) from Ialisco, Mexico. Occasional Papers of the Museum of Natural History, University of Kansas 5:1–7.
Genoways, H. H. 1973. Systematic and evolutionary relationships of spiny pocket mice, genus *Liomys*. Special Publications Museum Texas Tech University 5:1–368.
Genoways, H. H., and R. J. Baker. 1996. A new species of the genus *Rhogeessa,* with comments on geographic distribution and speciation in the genus. Pp. 83–87, *in* Contributions in mammalogy: a memorial volume honoring Dr. J. Knox Jones, Jr. (Genoways, H. H., and R. J. Baker, eds.). Museum of Texas Tech University, Lubbock, U.S.A.
Genoways, H. H., and E. C. Birney. 1974. *Neotoma alleni*. Mammalian Species 41:1–4.
Genoways, H. H., and J. H. Brown. 1993. Biology of the Heteromyidae. American Society of Mammalogists, Special Publication 10:1–719.
Genoways, H. H., and J. R. Choate. 1972. A multivariate analysis of systematic relationships among populations of the short-tailed shrews (genus *Blarina*) in the southeast. Systematic Zoology 21:106–116.
Genoways, H. H., and J. R. Choate. 1998. Natural history of the southern short-tailed shrew, *Blarina carolinensis*. Occasional Papers of the Museum of Southwestern Biology 8:1–43.
Genoways, H. H., and J. K. Jones, Jr. 1967. Notes on distribution and variation in the Mexican big-eared bat, *Plecotus phyllotis*. The Southwestern Naturalist 12:477–480.
Genoways, H. H., and J. K. Jones, Jr. 1969. Taxonomic status of certain long-eared bats (genus *Myotis*) from the southwestern United States and Mexico. The Southwestern Naturalist 14:1–13.

Genoways, H. H., and J. K. Jones, Jr. 1971. Systematics of southern banner-tailed kangaroo rats of the *Dipodomys phillipsii* group. Journal of Mammalogy 52:265–287.
Genoways, H. H., and J. K. Jones, Jr. 1972. Mammals from southwestern North Dakota. Occasional Papers Museum, Texas Tech University 6:1–36.
Genoways, H. H., and J. K. Jones, Jr. 1975. Annotated checklist of mammals of the Yucatan Peninsula, Mexico. IV. Carnivora, Sirenia, Perissodactyla, Artiodactyla. Occasional Papers The Museum, Texas Tech University 26:1–22.
Genoways, H. H., and S. L. Williams. 1979. Records of bats (Mammalia, Chiroptera) from Suriname. Annals of the Carnegie Museum 48:323–335.
Genoways, H. H., R. J. Baker, and J. E. Corney. 1977. Mammals of the Guadalupe Mountains National Park, Texas. Pp. 271–332, *in* Biological investigations of the Guadalupe Mountains National Park, Texas (Genoways, H. H., and R. J. Baker, eds.). National Park Service, Proceedings and Transactions Series Number 4:1–442.
Genoways, H. H., and S. L. Williams. 1984. Results of the Alcoa Foundation-Suriname expeditions. IX. Bats of the genus Tonatia (Mammalia: Phyllostomatidae) in Suriname. Annals of the Carnegie Museum 53:327–346
Genoways, H. H., and S. L. Williams. 1986. Results of the Alcoa Foundation-Surinam expeditions. XI. Bats of the genus Micronycteris (Mammalia: Chiroptera) in Surinam. Annals of the Carnegie Museum, 55:303–324.
Gentry, A. 2006. Mammal Species of the World. A taxonomic and geographic reference. 2005. D. E. Wilson & D.M. Reeder (Eds.) Ed. 3, 2 vols. Baltimore, Johns Hopkins University Press: A nomenclatural review. Bulletin of Zoological Nomenclature 63:215–219.
George, S. B. 1986. Evolution and historical biogeography of sorycine shrews. Systematic Zoology 35:153–162.
George, S. B. 1988. Systematics, historical biogeography, and evolution of the genus *Sorex*. Journal of Mammalogy 69:443–461.
George, S. B. 1989. *Sorex trowbridgiis*. Mammalian Species 337:1–5.
George, S. B. 1990. Unusual records of shrews in New Mexico. Southwestern Naturalist 35:464–465.
George, S. B., and J. D. Smith. 1991. Inter- and intraspecific variation among coastal and island populations of *Sorex monticolus* and *Sorex vagrans* in the Pacific Northwest. Pp. 75–91, *in* The biology of the Soricidae (Findley, J. S., and T. L. Jones, eds.). The Museum of Southwestern Biology, University of New Mexico, Albuquerque, U.S.A.
George, S. B., R. Choate, and H. H. Genoways. 1981. Distribution and taxonomic status of *Blarina hylophaga* Elliot (Insectivora: Soricidae). Annals of the Carnegie Museum 50:493–513.
George, S. B., J. R. Choate, and H. H. Genoways. 1986. *Blarina bravicauda*. Mammalian Species 261:1–9.
George, S. B., H. H. Genoways, J. R. Choate, and R. J. Baker. 1982. Karyotipic relationships within the short tail shrews genus Blarina. Journal of Mammalogy 63:639–645.
Giannatos, G., J. Herrero, and S. Lovari. 2007. *Capra hircus (Europe assessment).* The IUCN Red List of Threatened Species 2007:e. T136383A4283792.
Gibb, G. C., F. L. Condamine, M. Kuch, N. Moraes-Barros, M. Superina, H. N. Poinar, and F. Delsuc. 2016. Shotgun mitogenomics provides a reference phylogenetic framework and timescale for living xenarthrans. Molecular Biology and Evolution 33:621–642.
Gier, H. T. 1975. Ecology and social behavior of the coyote. Pp. 247–262, *in* The wild canids (Fox, M. W., ed.), Van Nostrand Reinhold. New York, U.S.A.
Gilbert, D. A., N. Lehman, S. J. O'Brien, and R. K. Wayne. 1990. Genetic fingerprinting reflects population differentiation in the California Channel Island fox. Nature 344:764–767.
Gilbert, C., A. Ropiquet, and A. Hassanin. 2006. Mitochondrial and nuclear phylogenies of Cervidae (Mammalia, Ruminantia): Systematics, morphology, and biogeography. Molecular Phylogenetic Evolution 40:101–17.
Gill, A. E. and E. Yensen. 1992. Biochemical differentiation in the Idaho ground squirrel, *Spermophilus brunneus* (Rodentia: Sciuridae). Great Basin Naturalist 52:155–159.
Gillihan, S. W., and K. R. Foresman. 2004. *Sorex vagrans*. Mammalian Species 744:1–5.
Girón, L. E., J. G. Owen, and M. E. Rodríguez. 2010. Van Gelder's Bat (*Bauerus dubiaquercus*) from El Salvador, Central America. The Southwestern Naturalist 55:585–587.
Glass, B. P. 1947. Geographic variation in *Perognathus hispidus*. Journal of Mammalogy 28:174–179.
Glass, B. P., and R. J. Baker. 1965. Vespertilio subulatus Say, 1823: proposed suppression under the plenary powers (Mammalia, Chiroptera) Z.N.(S.) 1701. Bulletin of Zoological Nomenclature 22:204–205.
Glass, B. P., and R. J. Baker. 1968. The status of the name *Myotis subulatus* Say. Proceedings of the Biological Society of Washington 81:257–260.
Glass, B. P., and C. M. Ward. 1959. Bats of the genus *Myotis* from Oklahoma. Journal of Mammalogy 40:194–201.
Global Invasive Species Database. 2022a. Species profile: *Capra hircus*. Downloaded from http://www.iucngisd.org/gisd/speciesname/Capra+hircus on 20–12–2022.
Global Invasive Species Database. 2022b. Species profile: *Axis axis*. Downloaded from http://www.iucngisd.org/gisd/speciesname/Axis+axis on 20–12–2022.
Global Invasive Species Database. 2022c. Species profile: *Dama dama*. Downloaded from http://www.iucngisd.org/gisd/speciesname/Dama+dama on 20–12–2022.
Global Invasive Species Database. 2022d. Species profile: *Sus scrofa*. Downloaded from http://www.iucngisd.org/gisd/species.php?sc=73 on 20–12–2022.
Global Invasive Species Database. 2022e. Species profile: *Canis lupus*. Downloaded from http://www.iucngisd.org/gisd/speciesname/Canis+lupus on 20–12–2022.
Godin, A. J. 1977. Wild mammals of New England. Johns Hopkins University Press, Baltimore, U.S.A.
Goldman, E. A. 1905. Twelve new wood rats of the genus *Neotoma*. Proceedings of the Biological Survey of Washington 18:27–34.
Goldman, E. A. 1910. Revision of the wood rat of the genus *Neotoma*. North American Fauna 31:1–124.
Goldman, E. A. 1911. Revision of the spiny pocket mice (genera *Heteromys* and *Liomys*) North American Fauna 34:1–70.
Goldman, E. A. 1915. Five new mammals from Mexico and Arizona. Proceedings of the Biological Society of Washington 28:133–137.
Goldman, E. A. 1918. The rice rats of North America (genus *Oryzomys*). North American Fauna 43:1–100.
Goldman, E. A. 1920. Mammals of Panama. Smithonian Miscellaneous Collections 69:1–306.
Goldman, E. A. 1927. A new woodrat from Arizona. Proceedings of the Biological Society of Washington 40:205–206.
Goldman, E. A. 1932. Revision of the wood rats of *Neotoma lepida* group. Journal of Mammalogy 13:59–67.

Goldman, E. A. 1937. The wolves of North America. Journal of Mammalogy 18:37–45.

Goldman, E. A. 1943. The races of the ocelot and margay in Middle America. Journal of Mammalogy 24:372–385.

Goldman, E. A. 1944. The wolves of North America. Pp. 387–636, *in* Classification of wolves. The American Wildlife Institution. Washington, U.S.A.

Goldman, E. A. 1945. A new pronghorn antelope from Sonora. Proceedings of the Biological Society of Washington 58:3–4.

Goldman, E. A. 1950. Raccoons of North and Middle America. North American Fauna 60:1–153.

Goldman, E. A. 1951. Biological investigations in Mexico. Smithsonian Miscellaneous Collection 115:1–476.

Goldman, E. A., and R. T. Moore. 1945. The biotic provinces of Mexico. Journal of Mammalogy 26:347–360.

Golley, F. B. 1962. Mammals of Georgia: a study of their distribution and functional role in the ecosystem. University Georgia Press. Athens, U.S.A.

Gompper, M. E. 1995. *Nasua narica*. Mammalian Species 487:1–10.

Gompper, M. E., and D. M. Decker. 1998. Nasua nasua. Mammalian Species. 580:1–9.

Gompper, M, and D. Jachowski. 2016. *Spilogale putorius*. The IUCN Red List of Threatened Species 2016:e.T41636A45211474.

Gongora, J., R. Reyna-Hurtado, H. Beck, A. Taber, M. Altrichter, and A. Keuroghlian. 2011. *Pecari tajacu*. The IUCN Red List of Threatened Species 2011:e.T41777A10562361.

Gonzalez, E., R. Barquez, and B. Miller. 2016a. *Lasiurus blossevillii*. The IUCN Red List of Threatened Species 2016:e.T88151055A22120040.

Gonzalez, E., R. Barquez, and J. Arroyo-Cabrales. 2016b. *Lasiurus cinereus*. The IUCN Red List of Threatened Species 2016:e.T11345A22120305.

González, C. A. L., D. E. Brown, and J. P. Gallo-Reynoso. 2003. The ocelot *Leopardus pardalis* in north-western Mexico: ecology, distribution and conservation status. Oryx 37:358–364.

González-Ruiz, N., E. Navarro-Frías, and T. Álvarez. 2000. Notas sobre algunos nuevos registros de murciélagos del Estado de México, México. Vertebrata Mexicana 9:1–6.

González, S., J. E. Maldonado, J. Ortega, A. C. Talarico, L. Bidegaray-Batista, J. E. Garcia, and J. M. B. Duarte. 2009. Identification of the endangered small red brocket deer (*Mazama bororo*) using noninvasive genetic techniques (Mammalia, Cervidae) Molecular Ecology Resources. 2009:754–758.

González-Cózatt, F. X., R. M. Vallejo, and F. A. Cervantes. 2007. Avances en el estudio de la sistemática de lagomorfos utilizando marcadores moleculares: filogenia del género *Sylvilagus* basada en secuencias del gen 16s. Pp. 31–46, *in* Tópicos de sistemática, biogeografia, ecología y conservación de mamíferos (Sánchez-Rojas, G., and A. Rojas-Martínez, eds.). Universidad Autónoma del Estado de Hidalgo. Pachuca, México.

González-Ruiz, N., and S. T. Álvarez-Castañeda. 2005. *Peromyscus bullatus*. Mammalian Species 770:1–3.

González-Ruiz, N., J. Ramírez-Pulido, and H. H. Genoways. 2006a. Geographic distribution, taxonomy, and conservation status of Nelson's woodrat (*Neotoma nelsoni*) in Mexico. The Southwestern Naturalist 51: 112–126.

González-Ruiz, N., J. Ramírez-Pulido, and J. Arroyo-Cabrales. 2011. A new species of mastiff bat (Molossidae: *Molossus*) from Mexico. Mammalian Biology 76:461–469.

González-Ruiz, N., S. T. Álvarez-Castañeda, A. G. Monroy-Gamboa, and J. Ramírez-Pulido. 2006b. *in litt. Neotoma nelsoni* [with keys of Genus *Neotoma*] (Rodentia: Cricetidae). Mammalian Species.

Goodwin, G. G. 1932. New records and sorne observations on Connecticut mammals. Journal of Mammalogy 13:36–40.

Goodwin, G. G. 1934. Mammals collected by A. W. Anthony in Guatemala, 1924–1928. Bulletin of the American Museum of Natural History 68:1–60.

Goodwin, G. G. 1940. Three new bats from Honduras and the first record of Enchisthenes hartii (Thomas) from North America. American Museum Novitates 1075:1.3

Goodwin, G. G. 1942a. A summary of recognizable species of *Tonatia*, with descriptions of two new species. Journal of Mammalogy 23:204–209.

Goodwin, G. G. 1942b. Mammals of Honduras. Bulletin of the American Museum of Natural History 79:107–195.

Goodwin, G. G. 1946. Mammals of Costa Rica. Bulletin of the American Museum of Natural History 87:271–474.

Goodwin, G. G. 1954. Southern records for Arctic mammals. Journal Mammalogy 35:258.

Goodwin, G. G. 1955. New tree-climbing rats from Mexico and Colombia. American Museum Novitates, American Museum of Natural History 1738:1–5.

Goodwin, G. G. 1958. Bats of the genus *Rhogeessa*. American Museum Novitates 1923:1–17.

Goodwin, G. G. 1959. Descriptions of some new mammals. American Museum Novitates 1967:1–8.

Goodwin, G. G. 1964. A new species and a new subspecies of *Peromyscus* from Oaxaca, Mexico. American Museum Novitates 2183:1–8.

Goodwin, G. G. 1966. A new species of vole (Genus *Microtus)* from Oaxaca, Mexico. American Museum Novitates 2243:1–4.

Goodwin, G. G. 1969. Mammals from the state of Oaxaca, Mexico, in the American Museum of Natural History. Bulletin of the American Museum of Natural History 141:1–270.

Goodwin, G. G., and A. M. Greenhall. 1961. A review of the bats of Trinidad and Tobago: descriptions, rabies infections, and ecology. Bulletin of the American Museum of Natural History 122:187–302.

Gordon, K. 1943. The natural history and behavior of the western chipmunk and the mantled ground squirrel. Oregon State College Studies in Zoology 5:1–104.

Gosselin, J., and F. Boily. 1994. Unusual southern occurrence of a juvenile bearded seal, *Erignathus barbatus*, in the St. Lawrence estuary, Canada. Marine Mammal Science 10:480–483.

Graham, R. W., and M. W. Graham. 1994. Late Quaternary distribution of *Martes* in America. Pp. 26–58, *in* Martens, sables, and fishers: biology and conservation (Buskirk, S. W., A. S. Harestad, M. G. Raphael, and R. A. Powell, eds.). Cornell University Press. Ithaca, New York.

Graham, R. 2001. "Comment on "Skeleton of extinct North American sea mink (*Mustela macrodon*)" by Mead et al.". Quaternary Research 56:419–421.

Green, J. S., and J. T. Flinders. 1980. *Brachylagus idahoensis*. Mammalian Species 125:1–4.

Green, M. M. 1943. The three pelages of the smoky shrew. Canadian Field Naturalist 57:96.

Greenbaum, I. F., and C. J. Phillips. 1974. Comparative anatomy and general histology of tongues of long-nosed bats (Leptonycteris sanborni and L. nivalis) with reference to infestation of oral mites. Journal of Mammalogy 55:489–504.

Greenbaum, I. F., R. L. Honeycutt, and S. E. Chirhart. 2019. Taxonomy and phylogenetics of the *Peromyscus maniculatus* species group. Pp. 559–576, *in* From field to laboratory: A memorial volume in honor of Robert J. Baker (Bradley, R. D., H. H. Genoways, D. J. Schmidly, and L. C. Bradley, eds.). Special Publications, Museum of Texas Tech University 71:1–911.

Greenhall, A. M., and W. A. Schutt. 1996. *Diaemus youngi*. Mammalian Species 533:1–7.

Greenhall, A. M., G. Joermann, and U. Schmidt. 1983. *Desmodus rotundus*. Mammalian Species 202:1–6.

Greenhall, A. M., U. Schmidt, and G. Joermann. 1984. *Diphylla ecaudata*. Mammalian Species 227:1–3.

Gregorin, R., G. L. Capusso, and V. R. Furtado. 2008. Geographic distribution and morphological variation in *Mimon bennettii* (Chiroptera, Phyllostomidae). Iheringia, Série Zoologia 98:404–411.

Grinnell, H. W. 1918. A synopsis of the bats of California. University of California Publications in Zoology 17:223–404.

Grinnell, J. 1922. A geographical study of the kangaroo rats of California. University California Publications in Zoology 24:1–124.

Grinnell, J. 1932. Habitat relations of the giant kangaroo rat. Journal of Mammalogy 13:305–320.

Grinnell, J. 1933. Mammals of California. Proceedings of the California Academy of Sciences 3:71–234.

Grinnell, J., and J. Dixon. 1918. Natural History of the ground squirrels of California. Monthly Bulletin of the State Commission of Horticulture 7:597–708.

Grinnell, J., and J. M. Linsdale. 1929. A new kangaroo rat from the upper Sacramento Valley, California. University of California Publications in Zoology 30:453–459.

Grinnell, J., J. Dixon, and J. M. Linsdale. 1937a. Fur-bearing mammals of California. University California Press, Berkeley 2:377–777.

Grinnell, J., and R. T. Orr. 1934. Systematic review of the *californicus* group of the rodent genus *Peromyscus*. Journal of Mammalogy 15:210–220.

Grinnell, J., and T. I. Storer. 1924. Animal life in the Yosemite: an account of the mammals, birds, reptiles, and amphibians in a cross-section of the Sierra Nevada. University of California Press. Berkeley, U.S.A.

Grinnell, J., J. B. Dixon, and J. M. Linsdale. 1937b. Fur-bearing mammals of California. University of California Press, Berkeley 1:1–1375.

Groves, C. P. 1981. Systematic relationships in the Bovini (Artiodactyla, Bovidae). Zeitschrift für Zoologisches Systematik und Evolutionsforschung 19:264–278.

Groves, C. 1982. Cranial and dental characteristics in the systematics of Old World Felidae. Carnivore 5:28–39.

Groves, C. P. 1986. The taxonomy, distribution and adaptations of recent equids. Pp. 11–65, *in* Equids in the Ancient World (Meadow, R. H., and H. P. Uepermann, eds.). Dr Ludwig Reichert Verlag. Wiesbaden, Germany.

Groves, C. P. 2005. Order Primates. Pp. 111–184, *in* Mammal species of the world. A taxonomic and geographic reference (Wilson, D. E., and D. A. M. Reeder, eds.). Third edition. The Johns Hopkins University Press. Baltimore, U.S.A.

Groves, C. 2006. The genus *Cervus* in eastern Eurasia. European Journal of Wildlife Research 52:14–22.

Groves, C., and P. Grubb. 2011. Ungulate taxonomy. The Johns Hopkins University Press. Baltimore, U.S.A.

Grubb, P. 1993. Order Artiodactyla. Pp. 377–414, *in* Mammal species of the world. A taxonomic and geographic reference (Wilson, D. E., and D. A. M. Reeder, eds.). Third edition. The Johns Hopkins University Press. Baltimore, U.S.A.

Grubb, P. 2005. Order Artiodactyla. Pp. 637–722, *in* Mammal species of the world. A taxonomic and geographic reference (Wilson, D. E., and D. A. M. Reeder, eds.). Third edition. The Johns Hopkins University Press. Baltimore, U.S.A.

Grzimek, B. 1975. Grzimek's Animal Life Encyclopaedia, Mammals, I-IV. Van Nostrand Reinhold. New York, U.S.A.

Guerrero, J. A., E. de Luna, and C. Sánchez-Hernández. 2003. Morphometrics in the quantification of character state identity for the assessment of primary homology: an analysis of character variation of the genus *Artibeus* (Chiroptera: Phyllostomidae). Biological Journal of the Linnean Society 80:45–55.

Guerrero, J. A., E. de Luna, and D. González. 2004. Taxonomic status of *Artibeus jamaicensis triomylus* inferred from molecular and morphometric data. Journal of Mammalogy 85:866–874.

Guerrero, J. A., J. Ortega, D. González, and J. E. Maldonado. 2008. Molecular phylogenetics and taxonomy of the fruit-eating bats of the genus *Artibeus* (Chiroptera: Phyllostomidae). Pp. 125–146, *in* Avances en el estudio de los mamíferos de México, Volumen II (Lorenzo, C., E. Espinoza, and J. Ortega eds.). Asociacin Mexicana de Mastozoologia, A. C. y el Colgio de la Forntera Sur. San Cristóbal de las Casas, México.

Guevara, L. 2017. They can dig it: semifossorial habits of the Mexican small-eared shrew (Mammalia: *Cryptotis mexicanus*). Revista Mexicana de Biodiversidad 88:1003–1005.

Guevara, L. 2023. A new species of small-eared shrew (Soricidae, Cryptotis) from El Triunfo Biosphere Reserve, Chiapas, Mexico. Journal of Mammalogy 104:546–561.

Guevara, L., and F. Cervantes. 2010. Rediscovery of the critically endangered Nelson's small-eared shrew (*Cryptotis nelsoni*), endemic to Volcán San Martín, Eastern México. Mammalian Biology 75:451–454.

Guevara, L., and F. Cervantes. 2013. Molecular systematics of small-eared shrews (Soricomorpha, Mammalia) within *Cryptotis mexicanus* species group from Mesoamérica. Acta Theriologica 59:233–242.

Guevara, L., and V. Sánchez-Cordero. 2018. New records of a critically endangered shrew from Mexican cloud forests (*Soricidae*, *Cryptotis nelsoni*) and prospects for future field research. Biodiversity Data Journal 6:e26667.

Guevara, L., V. Sánchez-Cordero, L. León-Paniagua, and N. Woodman. 2014a. A new species of small-eared shrew (Mammalia, Eulipotyphla, *Cryptotis*) from the Lacandona rain forest, Mexico. Journal of Mammalogy 95:739–753.

Guevara, L., C. Lorenzo, S. Ortega-García, and V. Sánchez-Cordero. 2014b. Noteworthy records of an endemic shrew from Mexico (Mammalia, Soricomorpha, *Crytotis griseoventris*), with comments on taxonomy. Mammalia 78:405–408.

Guevara, L., F. Cervantes, and V. Sánchez-Cordero. 2015. Riqueza, distribución y conservación de los topos y las musarañas (Mammalia, Eulipotyphla) de México. Therya 6:43–68.

Guevara-Chumacero, L. M., R. López-Wilchis, F. F. Pedroche, J. Juste, C. Ibáñez, and I. D. Barriga-Sosa. 2010. Molecular phylogeography of Pteronotus davyi (Chiroptera: Mormoopidae) in Mexico. Journal of Mammalogy 91:220–232.

Guggisberg, C. A. W. 1975. Wild cats of the world. Taplinger Publishing Company. New York, U.S.A.

Guilday, J. E. 1968. Grizzly bears from eastern North America. The American Midland Naturalist 79:247–250.

Gunn, A. 2016. *Rangifer tarandus*. The IUCN Red List of Threatened Species 2016:e.T29742A22167140.

Gunn, A., and M. Forchhammer. 2022. *Ovibos moschatus*. The IUCN Red List of Threatened Species 2022:e.T29684A22149286.

Gureev, A. A. 1979. Fauna SSSR. Mlekopitajuschie [Mammalia], Tom. 4, Vyp. 2. Nasekomojadnye (Mammalia, Insectivora). Nauka, Russia.

Gurnell, J. C. 1987. The natural history of squirrels. Facts on File. New York, U.S.A.

Gustavsson, I., and C. O. Sundt. 1968. Karyotypes in five species of deer (*Alces alces* L., *Capreolus capreolus* L., *Cervus elaphus* L., *Cervus nippon nippon* Temm. and *Dama dama* L.). Hereditas 60:233–247.

Gutiérrez, E. E., S. A. Jansa, and R. S. Voss. 2010. Molecular systematics of mouse opossums (Didelphidae: *Marmosa*): assessing species limits using mitochondrial DNA sequences, with comments on phylogenetic relationships and biogeography. American Museum Novitates 3692:1–22.
Gutiérrez, E. E., K. M. Helgen, M. M. McDonough, F. Bauer, M. T. Hawkins, L. A. Escobedo-Morales, B. D. Patterson, and J. E. Maldonado. 2017. A gene-tree test of the traditional taxonomy of American deer: the importance of voucher specimens, geographic data, and dense sampling. ZooKeys. 2017:87.
Gutiérrez-Costa, M. A., F. X. González-Cózatl, M. M. Ramírez-Martínez, L. I. Iñiguez-Dávalos, and D. S. Rogers. 2021. Molecular data suggest that *Heteromys irroratus bulleri* should be recognized as a species-level taxon. Therya 12:139–148.
Gwinn, R. N., G. H. Palmer, and J. L. Koprowski. 2011. *Sigmodon arizonae*. Mammalian Species 43:149–154.
Hacklander, K., and S. Schai-Braun. 2018. *Lepus europaeus* Pallas, 1778 European Hare. Pp. 187–190, *in* Lagomorphs: Pikas, Rabbits, and Hares of the World (Smith, A. T., C. H. Johnston, P. C. Alves, and K. Hackländer, eds). Johns Hopkins University Press. Baltimore, USA.
Hacklander, K., and S. Schai-Braun. 2019. *Lepus europaeus*. The IUCN Red List of Threatened Species 2019:e.T41280A45187424.
Hafner, D. J. 1984. Evolutionary relationships of the Nearctic Sciuridae. Pp. 3–23, *in* The biology of ground-dwelling squirrels: annual cycles, behavioral ecology, and sociality (Murie, J. O., and G. R. Michener, eds.). University of Nebraska Press. Lincoln, U.S.A.
Hafner, D. J. 1992. Speciation and persistence of a contact zone in Mojave Desert ground squirrels, subgenus *Xerospermophilus*. Journal of Mammalogy 73:770–778.
Hafner, D. 2019. *Dipodomys ornatus*. The IUCN Red List of Threatened Species 2019:e.T92464382A92464420.
Hafner, D. J., and K. N. Geluso. 1983. Systematic relationships and historical zoogeography of the desert pocket gopher, *Geomys arenarius*. Journal of Mammalogy 64:405–413.
Hafner, D. J., and A. T. Smith. 2010. Revision of the subspecies of the American pika, *Ochotona princep*s (Lagomorpha: Ochotonidae). Journal of Mammalogy 91:401–417.
Hafner, D. J., and T. L. Yates. 1983. Systematic status of the Mohave ground squirrel, *Spermophilus mohavensis* (subgenus *Xerospermophilus*). Journal of Mammalogy 64:397–404.
Hafner, D. J., J. C. Hafner, and M. S. Hafner. 1979. Systematic status of Kangaroo Mice, genus *Microdipodops*: morphometric, chromosomal, and protein analyses. Journal of Mammalogy 60:1–10.
Hafner, D. J., K. E. Petersen, and T. L. Yates. 1981. Evolutionary relationships of jumping mice (genus *Zapus*) of the southwestern United States. Journal of Mammalogy 62:501–512.
Hafner, D. J., B. R. Riddle, and S. T. Álvarez-Castañeda. 2001. Evolutionary relationships of white-footed mice (*Peromyscus*) on islands in the Sea of Cortez, México. Journal of Mammalogy 82:775–790.
Hafner, D. J., M. S. Hafner, G. L. Hasty, T. A. Spradling, and J. W. Demastes. 2008. Evolutionary relationships of pocket gophers (*Cratogeomys castanops* species group) of the Mexican Altiplano. Journal of Mammalogy 89:190–208.
Hafner, J. C. 1978. Evolutionary relationships of kangaroo mices, genus *Microdipodops*. Journal of Mammalogy 59:354–366.
Hafner, J. C., and M. S. Hafner. 1983. Evolutionary relationships of Heteromyid rodents. Great Basin Naturalist, Memories 7:3–29.
Hafner, J. C., J. E. Light, D. J. Hafner, M. S. Hafner, E. Reddington, D. S. Rogers, and B. R. Riddle. 2007. Basal clades and molecular systematic of heteromyid rodents. Journal of Mammalogy 88:1129–1145.
Hafner, M. S. 1982. A biochemical investigation of geomyoid systematics (Mammalia: Rodentia). Zeitschrift für Zoologische Systematik und Evolutionsforschung 20:118–130.
Hafner, M. S., J. C. Hafner, J. L. Patton, and M. F. Smith. 1987. Macrogeographic patterns of genetic differentiation in the Pocket Gopher *Thomomys umbrinus*. Systematic Zoology 36:18–34.
Hafner, D. J., and R. M. Sullivan. 1995. Historical and ecological biogeography of Nearctic pikas (Lagomorpha: Ochotonidae). Journal of Mammalogy 76:302–321.
Hafner, M. S., T. A. Spradling, J. E. Light, D. J. Hafner, and J. R. Demboski. 2004. Systematic revision of pocket gophers of the *Cratogeomys gymnurus* species group. Journal of Mammalogy 85:1170–1183.
Hafner, M. S., J. E. Light, D. J. Hafner, S. V. Brant, T. A. Spradling, and J. W. Demastes. 2005. Cryptic species in the Mexican pocket gopher *Cratogeomys merriami*. Journal of Mammalogy 86:1095–1108.
Hafner, M. S., D. J. Hafner, J. W. Demastes, G. L. Hasty, J. E. Light, and T. A. Spradling. 2009. Evolutionary relationships of pocket gophers of the genus *Pappogeomys* (Rodentia: Geomyidae). Journal of Mammalogy 90:47–56.
Hafner, M. S., A. R. Gates, V. Mathis, J. W. Demastes, and D. J. Hafner. 2011. Redescription of the pocket gopher *Thomomys atrovarius* from the Pacific coast of mainland Mexico. Journal of Mammalogy 92:1367–1382.
Hahn, W. L. 1905. A new bat from Mexico. Proceedings of the Biological Society of Washington 18:247–248.
Haiduk, M. W., C. Sánchez-Hernández, and R. J. Baker. 1988. Phylogenetic relationships of *Nyctomys* and *Xenomys* to other cricetine genera based on data from G-banded chromosomes. The Southwestern Naturalist 33:397–403.
Haiduk, M. W., J. W. Bickham, and D. J. Schmidly. 1979. Karyotypes of six species of *Oryzomys* from Mexico and Central America. Journal of Mammalogy 60:610–615.
Hall, A. 2002. Gray seal *Halichoerus grypus*. Pp. 522–524, *in* Encyclopedia of Marine Mammals (Perrin, W. F., B. Wursig, and J. G. M. Thewissen, eds.). Academic Press, San Diego, U.S.A.
Hall, E. R. 1941. Revision of the rodent genus *Microdipodops*. Field Mususem of Natural History, Zoological Series 27:233–277.
Hall, E. R. 1944. Four new ermines from the islands of southeastern Alaska. Proceedings of the Biological Society of Washington 57:35–42.
Hall, E. R. 1945. A revised classification of the American ermines with description of a new subspecies from the western Great Lakes Region. Journal of Mammalogy 26:175–182.
Hall, E. R. 1946. Mammals of Nevada. University of California Press. Berkeley, U.S.A.
Hall, E. R. 1951. American weasels. University of Kansas Publications of the Museum of Natural History 4:1–466
Hall, E. R. 1951a. American weasley. University of Kansas Publications, Museum of Natural History 4:1–466.
Hall, E. R. 1951b. Synopsis of the North American Lagomorpha. University of Kansas Publications, Museum of Natural History 5:119–202.
Hall, E. R. 1955. Handbook of the mammals of Kansas. University of Kansas Museum of Natural History, Miscellaneous Publications 7:1–303.

Hall, E. R. 1960. *Oryzomys couesi* only subspecifically differences from the marsh rice rat *Oryzomys palustris*. The Southwestern Naturalist 5:171–173.
Hall, J. S. 1962. A life history and taxonomic study of the Indiana bat, Myotis sodalis. Scince Publication, Reading Public Museum and Art Gallery 12:1–68.
Hall, E. R. 1981. The mammals of North America. Second edition. John Wiley and Sons. New York, U.S.A.
Hall, E. R. 1984. Geographic variation among brown and grizzly bears (*Ursus arctos*) in North America. Special Publications, Museum of Natural History, University of Kansas 13:1–16.
Hall, E. R., and E. L. Cockrum. 1953. A synopsis of the North American microtine rodents. University of Kansas Publications, Museum of Natural History 5:373–498.
Hall, E. R., and F. H. Dale. 1939. Geographic races of the kangaroo rat, *Dipodomys microps*. Occasional Papers of the Museum of Zoology, Louisiana State University 4:47–62.
Hall, E. R., and W. W. Dalquest 1950. Synopsis of the American Bat of the genus *Pipistrellus*. University of Kansas Publications, Museum of Natural History 1:591–602.
Hall, E. R., and J. K. Jones, Jr. 1961. North American yellow bats, "Dasypterus," and a list of the named kinds of the genus Lasiurus Gray. University Kansas Publiction. Museum of Natural History 14:73–98.
Hall, E. R., and W. W. Dalquest. 1963. The mammals of Veracruz. University of Kansas Publications, Museum of Natural History 14:165–362.
Hall, E. R, and W. W. Dalquest. 1963a. The mammals of Veracruz. University of Kansas Publications, Museum of Natural History 14:165–362.
Hall, E. R., and W.W. Dalquest. 1963b. The mammals of Veracruz. University of Kansas Publication, Museum of Natural History 14:165–362.
Hall, E. R., and D. F. Hoffmeister. 1942. Geographic variation in the canyon mouse, *Peromyscus maniculatus*. Journal of Mammalogy 23:51–65.
Hall, E. R., and H. H. Genoways. 1970. Taxonomy of the *Neotoma albigula*-group of woodrats in central Mexico. Journal of Mammalogy 51:504–516.
Hall, E. R., and R. M. Gilmore. 1932. New mammals from St. Lawrence Island, Bering Sea, Alaska. University of California Publications in Zoology 38:391–404.
Hall, E. R., and K. R. Kelson. 1959. The mammals of North America. The Ronald Press Co. New York, U.S.A.
Hall, E. R., and M. B. Ogilvie. 1960. Conspecificity of two pocket mice, *Perognathus goldmani* and *P. artus*. University of Kansas Publications, Museum of Natural History 9:513–518.
Hallett, J. G. 1978. *Parascalops breweri*. Mammalian Species 98:1–4.
Hallett, J. G. 1999. Hairy-tailed mole *Parascalops breweri*. Pp. 62–63, *in* The Smithsonian book of North American Mammals (Wilson, D. E., and S. Ruff, eds.). Smithsonian Institution Press. Washington, U.S.A.
Halls, L. K. 1978. White-tailed deer. Pp. 43–65, *in* Big game of North America: ecology and management (Schmidt, J. L., and D. L. Gilbert, eds.). Stackpole Books. Harrisburg, U.S.A.
Halls, L. K. 1984. White-tailed deer: ecology and management. Stackpole Books. Harrisburg, U.S.A.
Hamilton, M. J., R. K. Chesser, and T. L. Best. 1987. Genetic Variation in the Texas Kangaroo Rat, *Dipodomys elator* Merriam. Journal of Mammalogy 68:775–781.
Hamilton, W. J. 1940. The biology of the smoky shrew (*Sorex fumeus fumeus* Miller). Zoologica 25:473–492.
Hamilton, W. J. 1955. A new subspecies of Blarina brevicauda from Florida. Proceedings of the Biological Society of Washington 68:37–39.
Hammerson, G. A. 2016. *Sorex pacificus*. The IUCN Red List of Threatened Species 2016:e.T41409A22318109.
Hammerson, G. A., J. Matson, F. Reid, and N. Woodman. 2019. *Sorex neomexicanus*. The IUCN Red List of Threatened Species 2019:e. T136608A22319242.
Hanák, V., and V. Mazák. 1991. Enciclopedia de los Animales, Mamíferos de todo el Mundo. Susaeta. Madrid, España.
Handley, C. O. 1953. Three new lemmings (*Dicrostonyx*) from Arctic America. Journal of Washington Academic Science 43:197–200.
Handley, C. O., Jr. 1955. A new species of free-tailed bat (genus *Eumops*) from Brazil. Proceedings of the Biological Society of Washington 68:177–178.
Handley, C. O., Jr. 1959. A revision of American bats of the genera *Euderma* and *Plecotus*. Proceedings of the United States National Museum 110:95–246.
Handley, C. O., Jr. 1960. Description of new bats from Panama. Proceedings of the United States National Museum 3442:459–479.
Handley, C. O., Jr. 1966. Description of new bats (*Chiroderma* and *Artibeus*) from Mexico. Anales del Instituto de Biología, Universidad Nacional Autónoma de México 36:297–301.
Handley, C. O., Jr. 1976. Mammals of the Smithsonian Venezuela Project. Brigham Young University Sciences Bulletin, Biological Series 20:1–91.
Handley, C. O., Jr. 1980. Mammals of the Smithsonian Venezuelan Project. Brigham Young University, Science Bulletin, Biological Series 20:1–91.
Handley, C. O., Jr. 1991. The identity of *Phyllostoma planirostre* Spix, 1823 (Chiroptera: Stenodermatinae). Pp. 12–17, *in* Contributions to mammalogy in honor of Karl F. Koopman (Griffiths, T. A., and D. Klingener, eds.). Bulletin of the American Museum of Natural History 206:1–432.
Hansen, C. G. 1980. Physical characteristics. Pp. 52–63, *in* The desert bighorn: its life history, ecology, and management (Monson, G., and L. Sumner, eds.). University of Arizona Press. Tucson, U.S.A.
Hansen, R. M. 1954. Molt patterns in ground squirrels. Proceedings of the Utah Academy of Sciences, Arts and Letters 31:57–60.
Hanson, J. D., J. L. Indorf, V. J. Swier, and R. D. Bradley. 2010. Molecular divergence within the *Oryzomys palustris* complex: evidence for multiple species. Journal of Mammalogy 91:336–347.
Harding, L. E., and Smith, F. A. 2009. Mustela or Vison? Evidence for the taxonomic status of the American mink and a distinct biogeographic radiation of American weasels. Molecular Phylogenetics and Evolution 52:632–642.
Harington, C. R. 1966. A polar bear's life. Report on Polar Bears. Lectures presented at the Eighth Annual Meeting of the Washington Area Associates. Pp. 3–7. Arctic Institute of North America Research Paper 34. Calgary, Canada.
Harper, F. 1961. Land and fresh-water mammals of the Ungava Peninsula. University of Kansas Museum Natural History, Misclaneous Publications 27:1–178.
Harrington, C. R. 1966. Extralimital occurrences of walruses in the Canadian Artic. Journal of Mammalogy 47:506–513.
Harris, A. H. 1974. *Myotis yumanensis* in interior southwestern North America, with comments on *Myotis lucifugus*. Journal of Mammalogy 55:589– 607.

Harris, C. J. 1968. Otters; a study of the recent Lutrinae. Weinfield and Nicolson. London, United Kingdom.
Harris, A. H. 1999. Water shrew. Pp 38–39, in The Smithsonian Book of North American Mammals (Wilson, D. E., and S. Ruff, eds). Smithsonian Institution Press, Washington, D. C.
Harris, D. E., B. Lelli, and G. Jakush. 2002. Harp seal records from the southern Gulf of Maine: 1997–2001. Northeastern Naturalist 9:331–340.
Harris, W. P., Jr. 1937. Revision of *Sciurus variegatoides,* a species of Central American squirrel. Miscellaneous Publications of the Museum of Zoology, University of Michigan 38:1–39.
Harrison, D. L. 1975. *Macrophyllum macrophyllum*. Mammalian Species 62:1–3.
Harrison, R. G., S. M. Bogdanowicz, R. S. Hoffmann, E. Yensen, and P. W. Sherman. 2003. Phylogeny and evolutionary history of the ground squirrels (Rodentia: Marmotinae). Journal of Mammalian Evolution 10:249–276.
Hart, E. B. 1992. *Tamias dorsalis*. Mammalian Species 399:1–6.
Hart, E. B., M. C. Belk, E. Jordan, and M. W. Gonzalez. 2004. *Zapus princeps*. Mammalian Species 749:1–7.
Hartl, G. B., R. Göltenboth, M. Grillitsch, and R. Willing. 1988. On the biochemical systematics of the Bovini. Biochemical Systematics and Ecology 16:575–579.
Hartman, G. D., and T. L. Yates. 1985. *Scapanus orarius*. Mammalian Species 253:1–5.
Hartman, G. D. 1999. Star-nosed mole *Condylura cristata*. Pp. 65–67, *in* The Smithsonian book of North American Mammals (Wilson, D. E., and S. Ruff, eds.). Smithsonian Institution Press. Washington, U.S.A.
Hass, C. C. 2009. Competition and coexistence in sympatric bobcats and pumas. Journal of Zoology 278:174–180.
Hassanin, A., and E. J. P. Douzery. 2003. Molecular and morphological phylogenies of Ruminantia and the alternative position of the Moschidae. Systematic Biology 52:206–228.
Hassanin, A., F. Delsuc, A. Ropiquet, C. Hammer, B. Jansen Van Vuuren, C. Matthee, M. Ruiz-Garcia, F. Catzeflis, V. Areskoug, T. T. Nguyen, A. Couloux. 2012. Pattern and timing of diversification of Cetartiodactyla (Mammalia, Laurasiatheria), as revealed by a comprehensive analysis of mitochondrial genomes. Comptes Rendus Biologies 335:32–50.
Hatch, F. T., E. J. Ridley, and J. A. Mazrimas. 1971. Some *Dipodomys* species: ecologic and taxonomic features, estrous cycle, and breeding attempts. U.S. Atomic Energy Commission, Topical Report, Lawrence Livermore Laboratory UCRL-51140:1–25.
Hartman, D. S. 1971. Behavior and ecology of the Florida manatee, *Trichechus manatus latirostris* (Harlan), at Crystal River, Citrus County. Ph.D. dissertation. Cornell University, Cornell, U.S.A.
Hatt, R. T. 1932. The vertebral columns of ricochetal rodents. Bulletin of the American Museum of Natural History 63:599–738.
Hawes, M. L. 1975. Ecological adaptations in two species of shrews. Ph. D. dissertation. University of British Columbia. Vancouver, Canada.
Hayes, J. P. 1996. *Arborimus longicaudus*. Mammalian Species, 532:1–5.
Hayssen, V. 1991. *Dipodomys microps*. Mammalian Species 389:1–9.
Hayssen, V., F. Miranda, and B. Pasch. 2011. *Cyclopes didactylus* (Pilosa: Cyclopedidae). Mammalian Species 44(895):51–58.
Hayssen, V., J. Ortega, A. Morales-Leyva, and N. Martínez-Méndez. 2013. *Cabassous centralis* (Cingulata: Dasypodidae). Mammalian Species 45(898):12–17.
Hayward, B. J. 1970. The natural history of the cave bat, Myotis velifer. Western New Mexico University Research Science 1:1 74.
Heaney, L. R. 1983. *Sciurus granatensis* (ardilla roja, ardilla chisa, red-tailed squirrel). Pp. 489–490, *in* Costa Rican natural history (Janzen, D. H. ed.). The University of Chicago Press, Chicago, U.S.A.
Heaney, L. R., and R. M. Timm. 1983. Relationships of pocket gophers of the genus *Geomys* from the central and northern Great Plains. University of Kansas Publications, Museum of Natural History, Miscellaneous Publications 74:59.
Heckeberg, N. S. 2020. The systematics of the Cervidae: a total evidence approach. PeerJ, 8, e8114.
Heckeberg, N. S., D. Erpenbeck, G. Wörheide, and G. E. Rössner. 2016. Systematic relationships of five newly sequenced cervid species. PeerJ. 4:e2307.
Heffelfinger, J. R. 2000. Status of the name *Odocoileus hemionus crooki* (Mammalia: Cervidae). Proceedings of the Biological Society of Washington 113:319–333.
Helgen, K. 2016. *Conepatus leuconotus*. The IUCN Red List of Threatened Species 2016:e.T41632A45210809.
Helgen, K., and F. Reid. 2016a. *Martes americana*. The IUCN Red List of Threatened Species 2016:e.T41648A45212861.
Helgen, K., and F. Reid. 2016b. *Mephitis mephitis*. The IUCN Red List of Threatened Species 2016:e.T41635A45211301.
Helgen, K., and F. Reid. 2016c. *Mustela frenata*. The IUCN Red List of Threatened Species 2016:e.T41654A45213820.
Helgen, K., and F. Reid. 2016d. *Taxidea taxus*. The IUCN Red List of Threatened Species 2016:e.T41663A45215410.
Helgen, K., and F. Reid. 2018. *Martes pennanti*. The IUCN Red List of Threatened Species 2018:e.T41651A125236220.
Helgen, K. M., and D. E. Wilson. 2003. Taxonomic status and conservation relevance of the raccoons (*Procyon* spp.) of the West Indies. Journal of Zoology, London 259:69–76.
Helgen, K. M., and D. E. Wilson. 2005. A systematic and zoogeographic overview of the raccoons of Mexico and Central America. Pp. 221–236, *in* Contribuciones mastozoológicas en homenaje a Bernardo Villa (Sánchez-Cordero, V., and R. A. Medellín, eds.). Instituto de Biología e Instituto de Ecología, Universidad Nacional Autónoma de México y Comisión Nacional para el Conocimiento y Uso de la Biodiversidad. Distrito Federal, México.
Helgen, K., F. Reid, and R. Timm. 2016a. *Spilogale angustifrons*. The IUCN Red List of Threatened Species 2016:e.T136636A45221538.
Helgen, K., R. Kays, and J. Schipper. 2016b. *Potos flavus*. The IUCN Red List of Threatened Species 2016:e.T41679A45215631.
Helgen, K., A. D. Cuarón., J. Schipper, and J. F. González-Maya. 2016c. *Spilogale pygmaea*. The IUCN Red List of Threatened Species 2016:e. T41637A45211592.
Helgen, K. M., F. R. Cole, L. E. Helgen, and D. E. Wilson. 2009. Generic revision in the Holartic ground squirrel genus *Spermophilus*, Journal of Mammalogy 90:270–305.
Helm, J. D., III, C. Sánchez-Hernández, and R. H. Baker. 1974. Observaciones sobre los ratones de las marismas, *Peromyscus perfulvus* Osgood (Rodentia Cricetidae). Anales del Instituto de Biología, Universidad Nacional Autónoma de México, Serie Zoología 45:141–146.
Hemmer, H. 1978. The evolutionary systematics of living Felidae: Present status and current problems. Carnivore 1:71–79.
Hennings, D., and R. S. Hoffmann. 1977. A review of the taxonomy of the *Sorex vagrans* species complex from western North America. Occasional papers of the Museum of Natural History, the University of Kansas 68:1–35.

Hensley, A. P., and K. T. Wilkins. 1988. *Leptonycteris nivalis*. Mammalian Species 307:1–4.
Hensley, A. P., and K. T. Wilkins. 1988. Leptonycteris nivalis. Mammalian Species. 307:1–4.
Henttonen, H., B. Sheftel, M. Stubbe, R. Samiya, J. Ariunbold, V. Buuveibaatar, S. Dorjderem, Ts. Monkhzul, M. Otgonbaatar, and M. Tsogbadrakh. 2016. *Sorex minutissimus*. The IUCN Red List of Threatened Species 2016:e.T29666A115171049.
Heptner, V. G., K. K. Chapskii., V. A. Arsen'ev, and V. E. Sokolov. 1996. Mammals of the Soviet Union. Smithsonian Institution Libraries and National Science Foundation.
Herd, R. M. 1983. *Pteronotus parnellii*. Mammalian Species 209:1–5.
Herd, R. M., and M. B. Fenton. 1983. An electrophoretic, morphological, and ecological investigation of a putative hybrid zone between *Myotis lucifugus* and *Myotis yumanensis* (Chiroptera: Vespertilionidae). Canadian Journal of Zoology 61:2029–2050.
Hermanson, J. W. and T. J. O'Shea. 1983. *Antrozous pallidus*. Mammalian Species 213:1–8.
Hernández-Camacho, J. 1977. Notas para una monografia de *Potos flavus* (Mammalia: Carnivora) en Colombia. Caldasia 11:147–181.
Hernández-Camacho, J., and G. A. Cadena. 1978. Notas para la revisión del género *Lonchorhina* (Chiroptera, Phyllostomidae). Caldasia 12:199–251.
Hernández-Canchola, G., and L. León-Paniagua. 2020. *Sturnira parvidens* (Chiroptera: Phyllostomidae). Mammalian Species 52:57–70.
Hernández-Canchola, G., and L. León-Paniagua. 2021. About the specific status of *Baiomys musculus* and *B. brunneus*. Therya 12:291–301.
Hernández-Canchola, G., L. León-Paniagua, and J. A. Esselstyn. 2021a. Mitochondrial DNA indicates paraphyletic relationships of disjunct populations in the *Neotoma mexicana* species group. Therya 12:411–421.
Hernández-Canchola, G., J. Ortega, and L. León-Paniagua. 2021b. *Sturnira hondurensis* (Chiroptera: Phyllostomidae). Mammalian Species 53:23–34.
Hernández-Canchola, G., L. León-Paniagua, and J. A. Esselstyn. 2022. Mitochondrial DNA and other lines of evidence clarify species diversity in the *Peromyscus truei* species group (Cricetidae: Neotominae). Mammalia.
Hernández-Meza, B., Y. Domínguez-Castellanos, and J. Ortega. 2005. *Myotis keaysi*. Mammalian Species 785:1–3.
Hernández-Mijangos, L. A., R. Gálvez-Mejía, M. Díaz-Negrete, and C. M. Cruz-Durante. 2008. Nuevas localidades en la distribución de murciélagos filostóminos (Chiroptera: Phyllostomidae) en Chiapas, México. Revista Mexicana de Mastozoología 12:163–169.
Herrera, N. D., K. C. Bell, C. M. Callahan, E. Nordquist, B. A. J. Saver, J. Sullivan, J. R. Demobski, and J. M. Good. 2022. Genomic resolution of cryptic species diversity in chipmunks. Evolution 76:2004–2019.
Herron, M. D., T. A. Castoe, and C. L. Parkinson. 2004. Sciurid phylogeny and the paraphyly of Holarctic ground squirrels (*Spermophilus*). Molecular Phylogenetics and Evolution 31:1015–1030.
Hershkovitz, P. 1950. Mammals of northern Colombia. Preliminary report no 6: Rabbits (Leporidae), with notes on the classification and distribution of the South American forms. Proceedinsg of the United States National Museum 100:327–375.
Hershkovitz, P. 1951. Mammals from British Honduras, Mexico, Jamaica and Haiti. Fieldiana: Zoology 31:547–569.
Hershkovitz, P. 1954. Mammals of northern Colombia, preliminary report No. 7. Tapir (genus *Tapirus*) with a systematic review of the American species. Proceedings of the United States National Museum 103:165–496.
Hershkovitz, P. 1971. A new rice rat of the *Oryzomys palustris* group (Cricetinae, Muridae) from northwestern Colombia, with remarks on distribution. Journal of Mammalogy 52:700–709.
Hershkovitz, P. 1981. *Philander* and four-eyed opossums once again. Proceedings of the Biological Society of Washington 93:943–946.
Hershkovitz, P. 1997. Composition of the family Didelphidae Gray, 1821 (Didelphoidea: Marsupialia), with a review of the morphology and behavior of the included four-eyed pouched opossums of the genus *Philander* Tiedemann, 1808. Fieldiana, Zoology, New Series 86:1–103.
Hibbard, C. W. 1950. Mammals of the Rexroad formation from Fox Canyon, Meade County, Kansas. Contribution Museum of Paleontolgy, University of Michigan 8:113–192.
Hicks, S. A., and S. M. Carr. 1997. Are there two species of pine marten in North America? Genetic and evolutionary relationship within *Martes*. Pp. 15–28, *in* Martes: taxonomy, ecology, techniques, and management (Proulx, G., H. N. Bryant, and P. M. Woodard, eds.). Proceedings of the Second International Martes Symposium. Provincial Museum of Alberta. Edmonton, Canada.
Hidalgo-Mihart, M. G., L. Cantú-Salazar., A. González-Romero, and C. A. López-Gonzalez. 2004. Historical and present distribution of coyote (*Canis latrans*) in Mexico and Central America. Journal of Biogeography 31:2025–2038.
Higdon, J. W., O. R. P. Bininda-Emonds, R. M. D. Beck, and S. H. Ferguson. 2007. Phylogeny and divergence of the pinnipeds (Carnivora: Mammalia) assessed using a multigene dataset. BMC Evolutionary Biology 7:216.
Hildebrand, M. 1952. The integument in Canidae. Journal of Mammalogy 33:419–428.
Hildebrand, M. 1954. Comparative morphology of the body skeleton in recent Canidae. University California Publications in Zoology 52:399–470.
Hillman, C. N., and T. W. Clark. 1980. *Mustela nigripes*. Mammalian Species 126:1–3.
Hilton, C. D., and T. L. Best. 1993. *Tamias cinereicolli*s. Mammalian Species 436:1–5.
Hinesley, L. L. 1979. Systematics and distribution of two chromosome forms in the Southern Grasshopper Mouse, Genus *onychomys*. Journal of Mammalogy 60:117–128.
Hirshfeld, J. R., and W. G. Bradley.1977. Growth and development of two species of chipmunks: e*utamias panamintinus* and *E. palmeri*. Journal of Mammalogy 58:44–52.
Hoffmann, R. S. 1985. The correct name for the Paleartic brown, or Flat-skulled, shew is Sorex roboratus. Proceedings of the Biological society of Washington 98:17–28.
Hoffmann, R. S. 1987. A review of the systematics and distribution of Chinese red-toothed shrews (Mammalia: Soricinae). Acta Theriologica Sinica 7:100 139.
Hoffmann, F. G., and R. J. Baker 2001. Systematics of bats of the genus *Glossophaga* (Chiroptera: Phyllostomidae) and phylogeography in *G. soricina* based on the cytochrome-*b* gene. Journal of Mammalogy 82:1092–1101.
Hoffmann, F. G., and R. J. Baker 2003. Comparative phylogeography of short-tailed bats (*Carollia*: Phyllostomidae). Molecular Ecology 12:3403–3414.
Hoffmann, F. G., S. R. Hoofer, and R. J. Baker 2008. Molecular dating of the diversification of Phyllostominae bats based on nuclear and mitocondrial DNA sequences. Molecular Phylogenetics and Evolution 49:653–658.

Hoffmann, F. G., J. G. Owen, and R. J. Baker. 2003. mtDNA perspective of chromosomal diversification and hybridization in Peters' tent-making bat (*Uroderma bilobatum*: Phyllostomidae). Molecular Ecology 12:2981–2993.
Hoffmann, M., and C. Sillero-Zubiri. 2021. *Vulpes vulpes*. The IUCN Red List of Threatened Species 2021:e.T23062A193903628.
Hoffmann, R. S., and R. D. Fisher. 1978. Additional Distributional Records of Preble's shrew (*Sorex preblei*). Journal of Mammalogy 59:883–884.
Hoffmann, R. S., and J. G. Owen. 1980. Sorex tenellus and Sorex nanus. Mammalian Species 131:1–4.
Hoffmann, R. S., and J. W. Koeppl. 1985. Zoogeography. Pp. 85–115, *in* Biology of New World Microtus (Tamarin, R. H. ed.). American Society of Mammalogists. Shippensburg, Pennsylvania.
Hoffmann, R. S., and D. L. Pattie. 1968. A guide to Montana mammals: identification, habitat, distribution and abundance. University of Montana. Missoula, U.S.A.
Hoffmann, R. S., and R. S. Peterson. 1967. Systematics and zoogeography of *Sorex* in the Bering Strait area. Systematic Zoology 16:126–136.
Hoffmann, R. S. and A. T. Smith. 2005. Order Lagomorpha. Pp. 185–211, *in* Mammal Species of the World (Wilson, D. E., and D. M. Reeder, eds.). John Hopkins University Press. Baltimore, U.S.A.
Hoffmann, R. S., and R. J. Thorington. 2005. Family Sciuridae. Pp. 754–818, *in* Mammal Species of the World (Wilson, D. E., and D. M. Reeder, eds.). John Hopkins University Press. Baltimore, U.S.A.
Hoffmann, R. S., J. W. Koeppel, and C. F. Nadler. 1979. The relationships of the Amphiberingian marmots (Mammalia: Sciuridae). Occasional Papers of the Museum of Natural History, University of Kansas 83:1–56.
Hoffmann, R. S., P. L. Wright, and F. E. Newby. 1969. The distribution of some mammals in Montana. I. Mammals other than bats. Journal of Mammalogy 50:579–604.
Hoffmann, R. S., C. G. Anderson, R. W. Thorington, Jr., and L. R. Heaney. 1993. Family Sciuridae. Pp. 419–465, *in* Mammal Species of the World, a Taxonomic and Geographic Reference (Wilson, D., and D. M. Reeder, eds.), Smithsonian Institution Press. Washington, U.S.A.
Hoffmeister, D. F. 1951. A taxonomic and evolutionary study of the pinon mouse, *Peromyscus truei.* Illinois Biological Monographs 21:1–104.
Hoffmeister, D. F. 1956. Mammals of the Graham (Pinaleno) Mountains, Arizona. The American Midland Naturalist 55:257–288.
Hoffmeister, D. F. 1957. Review of the long-nosed bats of the genus Leptonycteris. Journal of Mammalogy 38:454–461.
Hoffmeister, D. F. 1969. The species problem in the *Thomomys bottae-Thomomys umbrinus* complex of gophers in Arizona. Miscellaneous Publications, Museum of Natural History, University of Kansas 51:75–91.
Hoffmeister, D. F. 1970. The seasonal distribution of bats in Arizona: a case for improving mammalian range maps. The Southwestern Naturalist 15:11–22.
Hoffmeister, D. F. 1971. Mammals of Grand Canyon. University of Illinois Press. Urbana, U.S.A.
Hoffmeister, D. F. 1974. The taxonomic status of *Perognathus penicillatus minimus* Burt. The Southwestern Naturalist 19:213–214.
Hoffmeister, D. F. 1977. Noteworthy range extensions of mammals in northem Mexico and Arizona. The Southwestern Naturalist 22:150–151.
Hoffmeister, D. F. 1981. *Peromyscus truei.* Mammalian Species 161:1–5.
Hoffmeister, D. F. 1986. Mammals of Arizona. The University of Arizona Press and the Arizona Game and Fish Department. Tucson, U.S.A.
Hoffmeister, D. E., and V. E. Diersing. 1978. Review of the tassel-eared squirrels of the subgenus *Otosciurus*. Journal of Mammalogy 59:402–413.
Hoffmeister, D. F., and L. S. Ellis. 1979. Geographic variation in *Eutamias quadrivitattus* with comments on the taxonomy of other Arizonan chipmunks. The Southwestern Naturalist 24:655–666.
Hoffmeister, R. G., and D. F. Hoffmeister. 1991. The hyoid in North American squirrels, Sciuridae, with remarks on associated musculature. Anales del Instituto de Bioloñia, Universidad Nacional Autónoma de México, Serie Zoología 62:219–234.
Hoffmeister, D. F., and P. H. Krutzsch. 1955. A new subspecies of *Myotis evotis* (H. Allen) from southeastern Arizona and Mexico. Chicago Academy of Sciences, Natural History Miscellaneous 151:1–4.
Hoffmeister, D. F., and M. R. Lee. 1963. Taxonomic review of cottontails, *Sylvilagus floridanus* and *Sylvilagus nuttallii*, in Arizona. The American Midland Naturalist 70:138–148.
Hoffmeister, D. F., and M. R. Lee. 1967. Revision of the pocket mice *Perognathus penicillatus*. Journal of Mammalogy 48:361–380.
Hoffmeister, D. F., and I. A. Nader. 1963. Distributional notes on Arizona mammals. Transactions of Illinois State Academy of Science 56:9293.
Hoffmeister, D. F., and L. de la Torre. 1960. A revision of the Wood Rat *Neotoma stephensi*. Journal of Mammalogy 41:476–491.
Hogan, K. M., M. C. Hedin, H. S. Koh, S. K. Davis, and I. F. Greenbaum. 1993. Systematics and taxonomic implications of karyotypic, electrophoretic, and mitochondrial-DNA variation in *Peromyscus* from the Pacific Northwest. Journal of Mammalogy 74:819–831.
Hoisington-Lopez, J. L., L. P. Waits, and J. Sullivan. 2012. Species limits and integrated taxonomy of the Idaho ground squirrel (*Urocitellus brunneus*): genetic and ecological differentiation. Journal of Mammalogy 93:589–604.
Holden, M. E., and G. G. Musser. 2005. Family Dipodidae. Pp. 871–893, *in* Mammal Species of the World (Wilson, D. E., and D. M. Reeder, eds.). The Johns Hopkins University Press. Baltimore, U.S.A.
Hollister, N. 1911. A systematic synopsis of the muskrats. North American Fauna 32:1–47.
Hollister, N. 1914. A systematic account of the grasshopper mice. Proceedings of the United States Natural History Museum 47:427–488.
Hollister, N. 1915. The genera and subgenera of raccoons and their allies. Proceedings United States National Museum 49:143–150.
Hollister, N. 1916. A systematic account of the prairie dogs. North American Fauna 40:1–37.
Holloway, G. I., and R. M. R. Barclay. 2001. *Myotis ciliolabrum.* Mammalian Species 670:1–5.
Homan, J. A., and H. H. Genoways. 1978. An analysis of hair structure and its phylogenetic implications among heteromyid rodents. Journal of Mammalogy 59:740–760.
Honacki, J. H., K. E. Kinman, and J. W. Koeppl (eds.). 1982. Mammal species of the world: a taxonomic and geographic reference. Allen Press, Inc. and The Association of Systematics Collections. Lawrence, U.S.A.
Honeycutt, R. L., and D. J. Schmidly. 1979. Chromosomal and morphological variation in the plains pocket gopher, *Geomys bursarius* in Texas and adjacent states. Occasional Papers, The Museum, Texas Tech University 58:1–54.
Honeycutt, R. L., and S. L. Williams, 1982. Genic differentiation in pocket gophers of the genus *Pappogeomys*, with comments on intergeneric relationships in the subfamily Geominae. Journal of Mammalogy 63:208–217.
Hood, C. S., and A. L. Gardner. 2008. Family Emballonuridae Gervais, 1856. Pp. 188–207, *in* Mammals of South America. Vol. 1. Marsupials, xenarthrans, shrews, and bats (Gardner, A. L., ed.). University of Chicago Press. Chicago, U.S.A.
Hood, C. S., and J. K. Jones, Jr. 1984. *Noctilio leporinus.* Mammalian Species 216:1–7.

Hood, C. S., and J. Pitocchelli. 1983. *Noctilio albiventris*. Mammalian Species 197:1–5.
Hood, C. S., L. W. Robbins, R. J. Baker, and H. S. Shellhammer. 1984. Chromosomal studies and evolutionary relationships of an endangered species, *Reithrodontomys raviventris*. Journal of Mammalogy 65:655–667.
Hoofer, S. R., and R. A. Van den Bussche. 2001. Phylogenetic relationships of plecotine bast and allies based on mitochondrial ribosomal sequences. Journal of Mammalogy 82:131–137.
Hoofer, S. R., and R. A. Van den Bussche. 2003. Molecular phylogenetics of the chiropteran family Vespertilionidae. Acta Chiropterologica 5:1–63.
Hoofer, S. R., R. A. Van Den Bussche, and I. Horáček. 2006. Generic status of the American pipistrelles (Vespertilionidae) with description of a new genus. Journal of Mammalogy 87:981–992.
Hoofer, S. R., S. A. Reeder, E. W. Hansen, and R. A. Van Den Bussche. 2003. Molecular phylogenetics and taxonomic review of noctilionoid and vespertilionoid bats (Chiroptera: Yangochiroptera). Journal of Mammalogy 84, 809–821.
Hoofer, S. R., S. Solari, P. A. Larsen, R. D. Bradley, and R. J. Baker. 2008. Phylogenetics of the Fruit-eating Bats (Phyllostomidae: Artibeina) inferred from mitochondrial dna sequences. Occasional papers, Museum of Texas Tech University 277:1–15
Hoogland, J. L. 1995. The black-tailed prairie dog: social life of a burrowing mammal. The University of Chicago Press. Chicago, U.S.A.
Hoogland, J. L. 1996. *Cynomys ludovicianus*. Mammalian Species 535:1–10.
Hooper, E. T. 1940a. A new race of pocket gopher of the species *Geomys lutescens* from Colorado. Occasional Papers of the Museum of Zoology, University of Michigan 420:1–3.
Hooper, E. T. 1940b. Geographical variation in bushy-tailed woodrats. University of California Publications in Zoology 42:407–424.
Hooper, E. T. 1944. San Francisco Bay as a factor influencing speciation in rodents. Miscellaneous Publications, Museum of Zoology, University of Michigan 59:1–89.
Hooper, E. T. 1947. Notes on Mexican mammals. Journal of Mammalogy 28:40–57.
Hooper, E. T. 1952. A systematic review of the harvest mice (genus *Reithrodontomys*) of Latin America. Miscellaneous Publications of the Museum of Zoology, University of Michigan 77:1–255.
Hooper, E. T. 1953. Notes on mammals of Tamaulipas, Mexico. Occasional Papers of the Museum of Zoology, University of Michigan 544:1–12.
Hooper, E. T. 1954. A synopsis of the cricetine rodents genus *Nelsonia*. Occasional Papers Museum Zoology, University of Michigan 550:1–12.
Hooper, E.T. 1955. Notes on mammals of western Mexico. Occasional Papers of the Museum of Zoology University of Michigan 565:1–26.
Hooper, E. T. 1957. Dental patterns in mice of the genus *Peromyscus*. Miscellaneous Publications, Museum of Zoology, University of Michigan 99:1–59.
Hooper, E. T. 1958. The male phallus in mice of the genus *Peromyscus*. Miscellaneous Publications, Museum of Zoology, University of Michigan 105:1–24.
Hooper, E. T. 1968. Classification. Pp. 27–74, *in* Biology of *Peromysus* (Rodentia) (King, J. A. ed.). Special Publications American Society of Mammalogists 2:1–593.
Hooper, E. T. 1972. A synopsis of the rodent genus *Scotinomys*. Occasional Papers of the Museum of Zoology, University of Michigan 665:1–32.
Hooper, E. T., and B. S. Hart. 1962. A synopsis of recent North American microtine rodents. Miscellaneous Publications of the Museum of Zoology, University of Michigan 120:1–68.
Hooper, E. T., and G. G. Musser. 1964. Notes on classification of the rodents of the genus *Peromyscus*. Occasional Papers of the Museum of Zoology, University of Michigan 635:1–13.
Hope, A. G., E. Waltari, N. E. Dokuchaev, S. Abramov, T. Dupal, A. Tsvetkova, H. Henttonen, S. O. MacDonald, and J. A. Cook. 2010. High-latitude diversification within Eurasian least shrews and Alaska tiny shrews (Soricidae). Journal of Mammalogy 91:1041–1057.
Hope, A. G., K. A. Speer, J. R. Demboski, S. L. Talbot, and J. A. Cook. 2012. A Climate for speciation: rapid spatial diversification within the *Sorex cinereus* complex of shrews. Molecular Phylogentic and Evolution 64:671–684.
Hope, A. G., N. Panter, J. A. Cook, S. L. Talbot, and D. W. Nagorsen. 2014. Multilocus phylogeography and systematic revision of North American water shrews (genus: *Sorex*). Journal of Mammalogy 95: 722–738.
Hope, A. G., J. L. Malaney, K. C. Bell, F. Salazar-Miralles, A. S. Chavez, B. R. Barber, and J. A. Cook. 2016. Revision of widespread red squirrels (genus: *Tamiasciurus*) highlights the complexity of speciation within North American forests. Molecular Phylogenetics and Evolution 100:170–182.
Hope, A. G., R. B. Stephens, S. D. Mueller, V. V. Tkach, and J. R. Demboski. 2020. Speciation of North American pygmy shrews (Eulipotyphla: Soricidae) supports spatial but not temporal congruence of diversification among boreal species. Biological Journal of the Linnean Society 129:41–60.
Hopton, M. E., and G. N. Cameron. 2001. *Geomys knoxjonesi*. Mammalian Species 672:1–3.
Houseal, T. W., I. F. Greenbaum, D. J. Schmidly, S. A. Smith, and K. M. Davis. 1987. Karyotypic variation in *Peromyscus boylii* from Mexico. Journal of Mammalogy 68:281–296.
Howell, A. H. 1936. Revision of the American Artic hare. Jurnal of Mammlogy 17:315–337.
Howard, W. E. 1949. A means to distinguish skulls of coyotes and domestic dogs. Journal of mammalogy 30:169–171.
Howell, A. B. 1926. Vole of the genus *Phenacomys*. North American Fauna 48:1–66.
Howell, A. B. 1927. Revision of the American lemming mice (Genus *Synaptomys*). North American Fauna 50:1–38.
Howell, A. H. 1906. Revision of the skunks of the genus *Spilogale*. North American Fauna 26:1–55.
Howell, A. H. 1909. Description of a new bat from Nickajack Cave, Tennessee. Proceedings of the Biological Society of Washington 22:45–47.
Howell, A. H. 1914a. Revision of the harvest mice (genus *Reithrodontomys*). North American Fauna 36:1–97.
Howell, A. H. 1914b. Ten new marmots from North America. Proceedings of the Biological Society of Washington, 27:1318.
Howell, A. H. 1915a. Descriptions of a new genus and seven new races of flying squirrels. Procceding of the Biological Society Washington 28:109–114.
Howell, A. H. 1915b. Revision of the American marmots. United States Department of Agriculture Bureau of Biological Survey, North American Fauna 37:1–80.
Howell, A. H. 1918. Revision of the flying squirrels. North American Fauna 44:1–64.
Howell, A. H. 1920. Description of a new race of the Florida water-rat (*Neofiber alleni*). Journal of Mammalogy 1:79–80.

Howell, A. H. 1922. Diagnoses of seven new chipmunks of the genus *Eutamias*, with a list of the American species. Journal of Mammalogy 3:178–185.
Howell, A. H. 1924. Revision of the American pikas. North American Fauna 47:1–57.
Howell, A. H. 1929. Revision of the American chipmunks (genera *Tamias* and *Eutamias*). North American Fauna 52:1–64.
Howell, A. H. 1938. Revision of the North American ground squirrels, with a classification of the North American Sciuridae. North American Fauna 56:1–256.
Howell, A. H. 1940. A new race of the harvest mouse (*Reithrodontomys*) from Virginia. Journal of Mammalogy 21:346.
Hrachovy, S. K., R. D. Bradley, and C. Jones. 1996. *Neotoma goldmani*. Mammalian Species 545:1–3.
Hsu, T. C., and M. L. Johnson. 1970. Cytological distinction between *Microtus montanus* and *Microtus canicaudus*. Journal of Mammalogy 51:824–826.
Huchon, D., F. M. Catzeflis, and E. J. P. Douzery. 2000. Variance of molecular datings, evolution of rodents and the phylogenetic affinities between Ctenodactylidae and Hystricognathi. Proceedings of the Royal Society B: Biological Sciences, 267:393–402.
Huckaby, D. G. 1980. Species limits in the *Peromyscus mexicanus* group (Mammalia: Rodentia: Muroidea). Contributions in Science, Natural History Museum of Los Angeles County 326:1–24.
Hückstädt, L. 2015. *Mirounga angustirostris*. The IUCN Red List of Threatened Species 2015:e.T13581A45227116.
Huey, L. M. 1925. Three new kangaroo rats of the genus *Dipodomys* from Lower California. Proccedings of the Biological Society of Washington 38:83–84.
Huey, L. M. 1951. The kangaroo rats (*Dipodomys*) of Baja California, Mexico. Transactions of the San Diego Society of Natural History 11:205–256.
Huey, L. M. 1926. The description of a new subspecies of *Perognathus* from Lower California with a short discussion of the taxonomic position of other peninsular members of this genus. Proceedings of the Biological Society of Washington 39:67–70.
Huey, L. M. 1960. Comments on the pocket mouse, *Perognathus fallax*, with descriptions of two new races from Baja California, Mexico. Transactions of the San Diego Society of Natural History 12:413–420.
Huey, L. M. 1964. The mammals of Baja California, Mexico. Transactions of the San Diego Society of Natural History 13:85–168.
Humphrey, S. R. 1974. Zoogeography of the nine-banded armadillo (*Dasypus novemcinctus*) in the United States. Bio-Science 24:457–462.
Humphrey, S. R., and T. H. Kunz. 1976. Ecology of a Pleistocene relict, the western big-eared bat (*Plecotus townsendii*), in the southern Great Plains. Journal of Mammalogy 57:470–494.
Hundertmark, K. 2016. *Alces alces*. The IUCN Red List of Threatened Species 2016:e.T56003281A22157381.
Hundertmark, K. J., and R. T. Bowyer. 2004. Genetics, evolution, and phylogeography of Moose. Alces: A Journal Devoted to the Biology and Management of Moose 40:103–122.
Hundertmark, K. J., G. F. Shields, I. G. Udina, R. T. Bowyer, A. A. Danilkin, and C. C. Schwartz. 2002. Mitochondrial phylogeography of moose (*Alces alces*): late Pleistocene divergence and population expansion. Molecular Phylogenetics and Evolution 22:375–387.
Hundertmark, K. J., R. T. Bowyer, G. F. Shields, and C. C. Schwartz. 2003. Mitochondrial phylogeography of moose (*Alces alces*) in North America. Journal of Mammalogy 84:718–728.
Hunt, J. L., L. A. McWilliams, T. L. Best, and K. G. Smith. 2003. *Eumops bonariensis*. Mammalian Species 733:1–5.
Hunt, J. L., J. E. Morris, and T. L. Best. 2004. *Nyctomys sumichrasti*. Mammalian Species 754:1–6.
Hunter, L. 2015. Wild Cats of the World. Bloomsbury Publishing. New York, U.S.A.
Hurtado, N., and V. Pacheco. 2014. Phylogenetic analysis of the genus *Mimon* Gray, 1847 (Mammalia, Chiroptera, Phyllostomidae) with description of a new genus. Therya 5:751–791.
Hurtado, N., and G. D'Elía. 2018. Taxonomy of the genus *Gardnerycteris* (Chiroptera: Phyllostomidae). Acta Chiropterologica 20:99–115.
Husar, S. L. 1978. *Trichechus manatus*. Mammalian Species 93:1–5.
Husson, A. M. 1962. The bats of Suriname. Zoologische Verhandelingen, Rijksmuseum van Natuurlijke Historie te Leiden 58:1–282.
Husson, A. M. 1978. The mammals of Suriname. Zoologische Monographieën van het Rijksmuseum van Natuurlijke Historie 2:1–569.
Hutterer, R. 2005. Order Soricomorpha. Pp. 220–311, *in* Mammal species of the world. A taxonomic and geographic reference (Wilson, D. E., and D. A. M. Reeder, eds.). Third edition. The Johns Hopkins University Press. Baltimore, U.S.A.
Hwang, Y. T., and S. Larivière. 2001. *Mephitis macroura*. Mammalian Species 686:1–3.
Indorf, J. L., and M. S. Gaines. 2013. Genetic divergence of insular marsh rice rats in subtropical Florida. Journal of Mammalogy 94:897–910.
Ingles, L. G. 1965. Mammals of the Pacific states. Stanford University Press. Stanford, U.S.A.
Intress, C., and T. L. Best. 1990. *Dipodomys panamintinus*. Mammalian Species 354:1–7.
Itoo, T., H. Katoh, K. Wada, K. Shimazaki, and K. Arai. 1977. Report of an ecological survey of the Steller sea lion in the coastal waters of Hokkaido, northern Japan, 1975–1976 winter. Geiken Tsushin 305:1–18.
Iudica, C. A. 2000. Systematic revision of the Neotropical fruit bats of the genus *Sturnira*: a molecular and morphological approach. Ph.D. dissertation. University of Florida. Gainsville, U.S.A.
Ivanitskaya, E. Y., A. I. Kozlovskii, N. V. Orlov, Y. M. Kovalskaya, and M. I. Baskevich. 1986. New data on karyotypes of common shrews (*Sorex*, Soricidae, Insectivora) in fauna of the USSR. Zoologicheskii Zhurnal 65:1228–1236.
Jackson, D. J., and J. A. Cook. 2020. A precarious future for distinctive peripheral populations of meadow voles (*Microtus pennsylvanicus*). Journal of Mammalogy 101:36–51.
Jackson, H. H. T. 1914. New moles of the genus *Scalopus*. Proceedings of the Biological Society of Washington 27:19–21.
Jackson, H. H. T. 1915. A review of the American moles. North American Fauna 38:1–100.
Jackson, H. H. T. 1925. The *Sorex arcticus* and *Sorex arcticus cinereus* of Kerr. Journal of Mammalogy 6:55–56.
Jackson, H. H. T. 1928. A taxonomical revision of the American long-tailed shrew (genera *Sorex* and *Microsorex*). North American Fauna 51:1–238.
Jackson, H. H. T. 1933. Five new shrew of the genus *Cryptotis* from Mexico and Guatemala. Proceedings of the Biological Society of Washington 46:79–82.
Jackson, H. H. T. 1947. A new shrew (genus Sorex) from Coahuila, Mexico. Proceedings of the Biological Society of Washington 60:131–132.
Jackson, H. H. T. 1951. Classification of the races of coyote. Pp. 227–341, *in* The clever coyote (Young, S. P., and H. H. T. Jackson, eds.). North American Wildlife Management Institute. Washington, U.S.A.

Jackson, H. H. T. 1961. Mammals of Wisconsin. The University of Wisconsin Press. Madison, U.S.A.

James, A. H., and D. K. James. 1984. *Dipodomys californicus* in Sierra Valley, Plumas County, California. California Fish and Game 70:58–64.

Jameson, E. W., and H. Peeters. 1988. California mammals. University of California Press. Berkeley, U.S.A.

Jameson, E. W., and H. Peeters. 2004. Mammals of California. University of California Press. Berkeley, U.S.A.

Janecek, L. L. 1990. Genetic variation in the *Peromyscus truei* group (Rodentia: Cricetidae). Journal of Mammalogy 71:301–308.

Jannett, F. J., Jr. 1976. Bacula of *Dipodomys ordii compactus* and *Dipodomys elator.* Journal of Mammalogy 57:382–387.

Jannett, F.J., Jr., and R. J. Oehlenschlager. 1994. Range extension and first Minnesota records of the Smokey Shrew *Sorex fumeus*. The American Midland Naturalist 131:364–365.

Jarrell, G. H., and S. O. MacDonald. 1989. Checklist to the mammals of Alaska. University of Alaska Museum. Fairbanks, U.S.A.

Jarrell, G. H., and K. Fredga. 1993. How many kinds of Lemmings? A taxonomic overview. Pp. 46–57, *in* The biology of Lemming (Stenseth, N. C., and R. A. Ims, eds.). Linnean Society Symposium series 15. Academic Press. London, United Kingdom.

Jenkins, S. H., and P. E. Busher. 1979. *Castor canadensis*. Mammalian Species 120:1–8.

Jenkins, S. H., and B. D. Eshelman. 1984. *Spermophilus beldingi*. Mammalian Species 221:1–8.

Jennings, J. B., T. L. Best, S. E. Burnett, and J. C. Rainey. 2002. *Molossus sinaloae*. Mammalian Species 691:1–5.

Jimenez, G. A. 1968. Nuevos registros de murciélago para Nuevo León, México. Anales del Instituto de Biología, Universidad Nacional Autónoma de México, serie Zoología 39:133–144.

Jiménez, J. J. 1972. Comparative post-natal growth in five species of the *Sigmodon* II. Cranial character relationships. Revista de Biología Tropical 20:5–27.

Jiménez-Guzmán, A. 1968. Nuevos registros de murciélagos para Nuevo León, México. Anales de Instuto de Biología, Universidad Nacional Autónoma de México, Serie Zoología 39:133–144.

Johnson, D. H. 1943. Systematic review of the chipmunks (genus *Eutamias*) of California. University of California Publications in Zoology 48:63–147.

Johnson, D. H. 1951. The water shrews of the Labrador Peninsula. Proceedings of the Biological Society of Washington 64:109–116.

Johnson, D. W., and D. M. Armstrong. 1987. *Peromyscus crinitus*. Mammalian Species 287:1–8.

Johnson, M. L. 1968. Application of blood protein electrophoretic studies to problems in mammalian taxonomy. Systematic Zoology 17:23–30.

Johnson, M. L., and S. B. Benson. 1960. Relationships of the pocket gophers of the *Thomomys mazama-talpoides* complex in the Pacific Nortthwest. The Murrelet 41:17–22.

Johnson, M. L., and C. W. Clanton. 1954. Natural history of *Sorex merriami* in Washington State. The Murrelet 35:1–4.

Johnson, M. L. and S. B. George. 1991. Species limits within the *Arborimus longicaudus* species-complex (Mammalia: Rodentia) with a description of a new species of California. Contributions in Science, Natural History Museum of Los Angeles County 429:1–16.

Johnson, M. L., and C. Maser. 1982. Generic relationships of *Phenacomys albipes*. Northwest Science 56:17–19.

Johnson, M. L., and B. T. Ostenson. 1959. Comments on the nomenclature of some mammals of the Pacific Northwest. Journal of Mammalogy 40:574–576.

Johnson, M. L., and T. L. Yates. 1980. A new Townsend's mole (*Scapanus townsendii*) from the State of Washington. Occasional Papers, The Museum, Texas Tech University 63:1–6.

Johnson, W. E., and R. K. Sealander. 1971. Protein variation and systematics in kangaroo rats (genus *Dipodomys*). Systematic Zoology 20:377–405.

Johnson, W. E., R. K. Sealander, M. H. Smith, and Y. J. Kim. 1972. Biochemical genetics of sibling species of the cotton rat (*Sigmodon*). Studies in Genetics VII. University of Texas Publications 7213:297– 305.

Johnson, W. E., E. Eizirik, J. Pecon-Slattery, W. J. Murphy, A. Antunes, E. Teeling, and S. J. O'Brien. 2006. The late Miocene radiation of modern Felidae: a genetic assessment. Science 311:73–77.

Jolley, T. W., R. L. Honeycutt, and R. D. Bradley. 2000. Phylogenetic relationships of pocket gophers (genus *Geomys*) based on the mitochondrial 12s rRNA gene. Journal of Mammalogy 81:1025–1034.

Jones, C. 1977. *Plecotus rafinesquii*. Mammalian Species 69:1–4.

Jones, C., and C. N. Baxter. 2004. *Thomomys bottae*. Mammalian Species 7421–14.

Jones, C., and N. J. Hildreth. 1989. *Neotoma stephensi*. Mammalian Species 328:1–3.

Jones, C., and J. N. Layne. 1993. *Podomys floridanus*. Mammalian Species 427:1–5.

Jones, C., and R. W. Manning. 1989. *Myotis austroriparius*. Mammalian Species 332:1–3.

Jones, C., and R. D. Suttkus. 1971. Wing loading in *Plecotus rafinesquii.* Journal of Mammalogy 52:458–460.

Jones, C., R. S. Hoffman, D. W. Rice, M. D. Engstrom, D. Bradley, D. J. Schmidly, C. Jones, and R. J. Baker. 1997. Revised checklist of North American mammals north of Mexico, 1997. Occasional Papers, The Museum, Texas Tech University 173:1–20.

Jones, J. K., Jr. 1964. Bats from western and southern Mexico. Transactions of Kansas Academy of Science 67:509–516.

Jones, J. K., Jr. 1966. Bats from Guatemala. University ok Kansas Publication, Museum of Natural History 16:439–472.

Jones, J. K., Jr. 1977. *Rhogeessa gracilis*. Mammalian Species 76:1–2.

Jones, J. K., Jr. 1978. A new bat of the genus *Artibeus* from the Lesser Antillean Island of St. Vincent. Occasional Papers, The Museum, Texas Tech University 51:1–6.

Jones, J. K., Jr. 1982. *Reithrodontomys spectabilis*. Mammalian Species 193:1.

Jones, J. K., Jr., and T. Álvarez. 1962. Taxonomic status of the free-tailed bat, *Tadarida yucatanica* Miller. University of Kansas Publications, Museum of Natural History 9:125–133.

Jones, J. K., Jr., and T. Álvarez. 1964. Additional records of mammals from the Mexican state of San Luis Potosí. Journal of Mammalogy 45:302–303.

Jones, J. K., Jr., and J. Arroyo-Cabrales. 1990. *Nyctinomops aurispinosus*. Mammalian Species 350:1–3.

Jones, J. K., Jr., and E. C. Birney. 1988. Handbook of mammals of the north-central states. The University of Minnesota Press. Minneapolis, U.S.A.

Jones, J. K., Jr., and D. C. Carter. 1976. Annotated checklist with keys to subfamilies and genera. Pp. 7–38, *in* Biology of bats of the New World family Phyllostomatidae, Part I (Baker, R. J., J. K. Jones, Jr., and D. C. Carter, eds.). Special Publications, The Museum, Texas Tech University 10:1–218.

Jones, J. K., Jr., and J. R. Choate. 1978. Distribution of two species of long-eared bats of the genus *Myotis* on the northern Great Plains. Prairie Naturalist 10:49–52.
Jones, J. K., Jr., and H. H. Genoways. 1967. Notes on the Oaxacan vole, *Microtus oaxacensis* Goodwin, 1966. Journal of Mammalogy 48:320–321.
Jones, J. K., and H. H. Genoways. 1975. *Dipodomys phillipsii*. Mammalian Species 51:1–3.
Jones, J. K., and H. H. Genoways. 1978. *Neotoma phenax*. Mammalian Species 108:1–3.
Jones, J. K., Jr., and J. A. Homan. 1974. *Hylonycteris underwoodi*. Mammalian Species 32:1–2.
Jones, J. K., Jr., and C. S. Hood. 1993. Synopsis of South American bats of the family Emballonuridae. Occasional Papers, The Museum, Texas Tech University, 155:1–32.
Jones, Jr., J. K., and T. E. Lawlor. 1965. Mammals from Isla Cozumel, Mexico, with de scription of a new species of harvest mouse. University of Kansas Publications, Museum of Natural History 16:409–419.
Jones, J. K., Jr., and R. R. Johnson. 1967. Sirenians. Pp. 366–373, *in* Recent mammals of the world-a synopsis of families (Anderson, S., and J. K. Jones eds.), Ronald Press Co. New York, U.S.A. Jones, J. K., Jr., and T. E. Lawlor. 1965. Mammals from Isla Cozumel, Mexico, with description of a new species of harvest Mouse. University of Kansas, Publications Museum of Natural History 16:411–419.
Jones, J. K., Jr., and R. W. Manning. 1992. Illustrated key to skulls of genera of North American land mammals. Texas Tech University Press. Lubbock, U.S.A.
Jones, C., and R. D. Suttkus. 1975a. Notes on the natural history of Plecotus rafinesquii. Occasional Papers of the Museum of Natural Science, Louisiana State University 47:1–7
Jones, C., and R. D. Suttkus. 1975b. Notes on the natural history of Plecotus rafinesquii. Occasional Papers of the Museum of Zoology, Louisiana State University 47:1–14.
Jones, J. K., Jr., T. Álvarez, and M. R. Lee. 1962. Noteworthy mammals of Sinaloa, Mexico. University of Kansas Publications, Museum of Natural History 14:145–159.
Jones, J. K., Jr., J. D. Smith, and T. Alvarez. 1965. Notes on bats from the cape regi6n of Baja California. Transactions of the San Diego Society of Natural History 14:53–56.
Jones, J. K., Jr., H. H. Genoways, and L. C. Watkins. 1970. Bats of the genus *Myotis* from western Mexico, with a key to the species. Transactions of the Kansas Academy of Sciences 73:409–418.
Jones, J. K., Jr., J. D. Smith, and R. W. Turner. 1971. Noteworthy records of bats from Nicaragua, with a checklist of the chiropteran fauna of the country. Occasional Papers Museum Natural History, University of Kansas 2:1–35.
Jones, J. K., Jr., J. R. Choate, and A. Cadena. 1972. Mammals from the Mexican state of Sinaloa. 2. Chiroptera. Occasional Papers Museum of Natural History, University of Kansas 6: 1–29.
Jones, J. K., Jr., J. D. Smith, and H. H. Genoways. 1973. Annotated checklist of mammals of the Yucatan Peninsula, Mexico: I. Chiroptera. Occasional Papers, The Museum, Texas Tech University 13:1–31.
Jones, J. K., Jr., H. H. Genoways, and T. E. Lawlor. 1974. Annotated checklist of mammals of the Yucatan Peninsula, Mexico II. Rodentia. Occasional Papers, The Museum, Texas Tech University 22:1–24.
Jones, J. K., Jr., D. C. Carter, H. H. Genoways, R. S. Hoffmann, and D. W. Rice. 1982. Revised checklist of North American mammals north of Mexico. Occasional Papers, The Museum, Texas Tech University 80:1–21.
Jones, J. K., Jr., D. C. Carter, and W. D. Webster. 1983a. Records of mammals from Hidalgo, Mexico. The Southwestern Naturalist 28:378–380.
Jones, J. K., Jr., D. M. Armstrong, R. S. Hoffman, and C. Jones. 1983b. Mammals of the northern Great Plains. University of Nebraska Press. Lincoln, U.S.A.
Jones, J. K., Jr., D. C. Carter, H. H. Genoways, R. S. Hoffmann, D. W. Rice, and C. Jones. 1986. Revised checklist of North American mammals north of Mexico. 1986. Occasional Papers, The Museum, Texas Tech University 107:1–22.
Jones, J. K., Jr., J. Arroyo-Cabrales, and R. D. Owen. 1988. Revised checklist of bats (Chiroptera) of Mexico and Central America. Occasional Papers, The Museum, Texas Tech University 120:1–34.
Jones, J. K., Jr., R. W. Manning, and J. R. Geotze. 1991. Noteworthy records of seven species of small mammals from westcentral Texas. Occasional Papers, The Museum, Texas Tech University 143:1–4.
Jones, J. K., Jr., R. S. Hoffmann, D. W. Rice, C. Jones, R. J. Baker, and M. D. Engstrom. 1992. Revised checklist of North American mammals north of Mexico, 1991. Occasional Papers, The Museum, Texas Tech University 146:1–23.
Juárez-López, R., M. Pérez-López, Y. Bravata-de la Cruz, A. J. de la Cruz, F. M. Contreras-Moreno, D. Thornton, and M. G. Hidalgo-Mihart. 2017. Range extension of the northern Naked-Tailed Armadillo (*Cabassous centralis*) in southern Mexico. Western North American Naturalist 77:398–403.
Judd, S. R., S. P. Cross, and S. Pathak. 1980. Non-Robertsonian chromosomal variation in *Microtus montanus*. Journal of Mammalogy 61:109–113.
Jung, T. S., B. G. Slough, D. W. Nagorsen, and P. M. Kukka. 2014. New records of the Ogilvie Mountains Collared Lemming (*Dicrostonyx nunatakensis*) in central Yukon. The Canadian Field-Naturalist 128:265–268.
Junge, J. A., and R. S. Hoffmann. 1979. An annotated key to the long-tailed shrews (genus Sorex) of the United States and Canada, with notes on Middle American Sorex. Occasional Papers, Museum Natural History, University of Kansas 94:1–48.
Junge, J. A., and R. S. Hoffman. 1981. An annotated key to the long-tailed shrews (genus *Sorex*) of the United States and Canada, with notes on the Middle American *Sorex*. Occasional Papers of the Museum of Natural History, the University of Kansas 94:1–48.
Junge, J. A., R. S. Hoffmann, and R. W. DeBry. 1983. Relationships within the Holarctic *Sorex arcticus-Sorex tundrensis* species complex. Acta Theriologica 28:339–350.
Kain, D. E. 1985. The systematic status of *Eutamias ochrogenys* and *Eutamias senex* (Rodentia: Sciuridae). M. A. thesis, Humboldt State University. Arcata, U.S.A.
Kalmbach, E. R. 1943. The armadillo: its relation to agricultura and game. Texas Game, Fish and Oyster Comm., Austin, Texas,
Kalcounis-Rueppell, M. C., and T. R. Spoon. 2009. *Peromyscus boylii* (Rodentia: Cricetidae). Mammalian Species 838:1–14.
Kays, R. 2018. *Canis latrans*. The IUCN Red List of Threatened Species 2018:e.T3745A163508579.
Kazanskaia, E., M. Kuznetsova, and D. Alexey. 2007. Phylogenetic reconstructions in the genus Capra (Bovidae, Artiodactyla) based on the mitochondrial DNA analysis. Genetika 43:245–53.

Keinath, D. A., H. R. Griscom, and M. D. Andersen. 2014. Habitat and distribution of the Wyoming pocket gopher (*Thomomys clusius*). Journal of Mammalogy 95:803–813.

Keith, J. O. 1965. The Abert squirrel and its dependence on Ponderosa pine. Ecology 46:150–163.

Kellnhauser, R. T. 1983. The acceptance of *Lontra* Gray for the New World river otters. Canadian Journal of Zoology 61:278–279.

Kellogg, R. 1956. What and where are the whitetails? Pp. 31–55, *in* The deer of North America (Taylor, W. P. ed.), The Stackpole Co. Harrisburg, U.S.A.

Kelly, P. A. 2018. *Sylvilagus bachmani.* Pp. 122–125, *in* Pikas, Rabbits and Hares of the World (Smith, A., C. H. Johnston, P. C. Alves, and K. Hackländer, eds.). The Johns Hopkins University Press/IUCN/SSC. Baltimore, U.S.A.

Kelly, M., D. Morin, and C. A. Lopez-Gonzalez. 2016. *Lynx rufus.* The IUCN Red List of Threatened Species 2016:e.T12521A50655874.

Kelly, P. A., C. Lorenzo, and S. T. Álvarez-Castañeda. 2019. *Sylvilagus bachmani.* The IUCN Red List of Threatened Species 2019:e. T41302A45192710.

Kelson, K. R. 1952. Comments on the taxonomy and geographic distribution of some North American woodrats (genus *Neotoma*). University of Kansas Publications, Museum of Natural History 5:233–242.

Kelt, D. A. 1988a. *Dipodomys californicus.* Mammalian Species 324:1–4.

Kelt, D. A. 1988b. *Dipodomys heermanni.* Mammalian Species 323:1–7.

Kenagy, G. J. 1973. Daily and seasonal patterns of activity and energetics in a heteromyid rodent community. Ecology 54:1201–1219.

Kenagy, G. J., and G. A. Bartholomew. 1985. Seasonal reproductive patterns in five coexisting California desert rodent species. Ecological Monographs 55:371–397.

Kennedy, M. L., M. L. Beck, and T. L. Best. 1980. Intraspecific Morphologic Variation in Ord's Kangaroo Rat, *Dipodomys ordii*, from Oklahoma. Journal of Mammalogy 61:311–319.

Kennedy, M. L., T. L. Best, and M. J. Harvey. 1984. Bats of Colima, Mexico. Mammalia 48:397–408.

Kennedy, M. L., P. K. Kennedy, M. A. Bogan, and J. L. Waits. 2002. Taxonomic assessment of the black bear (*Ursus americanus*) in the eastern United States. The Southwestern Naturalist 47:335–347.

Kenneth, T. W. 1987. *Lasiurus seminolus.* Mammalian Species 280:1–5.

Kenyon, K. W. 1960. A ringed seal from the Pribiof Islands, Alaska. Journal of Mammalogy 41:520–521.

Kenyon, K. W. 1969. The sea otter in the eastern Pacific Ocean. North American Fauna, 68:1–352.

Kenyon, K. W. 1977. Caribbean monk seal extinct. Journal of Mammalogy 58:97–98.

Keuling, O., and K. Leus. 2019. *Sus scrofa.* The IUCN Red List of Threatened Species 2019:e.T41775A44141833.

Keuroghlian, A., A. Desbiez, R. Reyna-Hurtado, M. Altrichter, H. Beck, A. Taber, and J. M. V. Fragoso. 2013. *Tayassu pecari.* The IUCN Red List of Threatened Species 2013:e.T41778A44051115.

Kitchner, A. C., *et al.* 2017. A revised taxonomy of the Felidae. The final report of the Cat Classification Task Force of the IUCN/SSS Cat Specialist Group. Cat News Special Issue 11. Stämpfli Publikationen AG, Bern, Switzerland.

Kilpatrick, C. W., and K. L. Crowell. 1985. Genic variation of the rock mole, *Microtus chrotorrhinus.* Journal of Mammalogy 66:94–101.

King, C. M. 1983a. *Mustela erminea.* Mammalian Species 195:1–8.

King, C. M. 1989. The natural history of weasels and stoats. Christopher Helm Publishers. London, United Kingdom.

King, J. E. 1954. The otariid seals of the Pacific coast of America. Bulletin of the British Museum (Natural History) Zoology 2:311–337.

King, J. E. 1956. The monk seals genus *Monachus.* Bulletin of the British Museum (Natural History), Zoology Series 3:203–256.

King, J. E. 1966. Relationships of the hooded and elephant seals (genera *Cystophora* and *Mirounga*). Journal of Zoology, London 148:385–398.

King, J. E. 1983b. Seals of the world. Second edition. British Museum (Natural History). London, United Kingdom.

King, S. R. B., L. Boyd, W. Zimmermann, and B. E. Kendall. 2015. *Equus ferus.* The IUCN Red List of Threatened Species 2015:e. T41763A97204950.

Kinlaw, A. 1995. *Spilogale putorius.* Mammalian Species 511:1–7.

Kilpatrick, C. W., N. Pradhan, and R. W. Norris. 2021. A re-examination of the molecular systematics and phylogeography of taxa of the *Peromyscus aztecus* species group, with comments on the distribution of *P. winkelmanni.* Therya 12:331–346.

Kirkland, G. L. 1977. A Re-examination of the subspecific status of the Maryland Shrew, *Sorex cinereus fontinalis* Hollister. Proceedings of the Pennsylvania Academy of Science 51:43–46.

Kirkland Jr., G. L., 1981. *Sorex dispar* and *Sorex gaspensis.* Mammlian Species 155:1–4.

Kirkland Jr., G. L., and F. J. J. Jannett. 1982. *Microtus chrotorrhinus.* Mammalian Species 180:1–5.

Kirkland Jr., G. L., and H. M. van Duesen. 1979. The shrews of the *Sorex dispar* group: *Sorex dispar* Batchelder and *Sorex gaspensis* Anthony and Goodwin. American Museum Novitates 2675:1–21.

Kirkland, G. L., and J. M. Levengood. 1987. First record of the Maryland Shrew (*Sorex fontinalis*) from West Virginia. Proceedings of the Pennsylvania Academy of Science 61:35–37.

Kirkland, G. L., and D. F. Schmidt. 1996. Sorex arcticus. Mammalian Species 524:1–5.

Kirkpatrick, R. D., A. M. Cartwright, J. C. Brier, and E. J. Spicka. 1975. Additional mammal records from Belize. Mammalia 39:330–331.

Kiser, W. M. 1995. *Eumops underwoodi.* Mammalian Species 516:1–4.

Kivett, V. K., J. O. Murie, and A. L. Steiner. 1976. A comparative study of scent-gland location and related behavior in some northwestern Nearctic ground squirrel species (Sciuridae): an evolutionary approach. Canadian Journal of Zoology 54:1294–1306.

Koepfli, K. P., and R. K. Wayne. 1998. Phylogenetic relationships of otters (Carnivora: Mustelidae) based on mitochondrial cytochrome b sequences. Journal of Zoology, London 246:401–416.

Koepfli, K. P., K. A. Deer, G. J. Slater, C. Begg, K. Begg, L. Grassman, M. Lucherini, G. Veron, and R. K. Wayne. 2008. Multigene phylogeny of the Mustelidae: Resolving relationships, tempo and biogeographic history of a mammalian adaptive radiation. BMC Biology 6:10.

Koop, B. F., R. J. Baker, and J. T. Mascarello. 1985. Cladistical analysis of chromosomal evolution withing the genus *Neotoma.* Occasional Papers, The Museum, Texas Tech University 96:1–9.

Koopman, K. F. 1971. The systematic and historical status of the Florida *Eumops* (Chiroptera, Molossidae). American Museum Novitates 2478: 1–6.

Koopman, K. F. 1978. Zoogeography of Peruvian bats with special emphasis on the role of the Andes. American Museum Novitates 2651:1–33.

Koopman, K. F. 1982. Biogeography of the bats of South America. Pp. 273–300, *in* Mammalian biology of South America (Mares, M. A., and H. H. Genoways, eds.). Special Publication, Pymatuning Laboratory of Ecology, University of Pittsburgh. Pittsburgh, U.S.A.

Koopman, K. F. 1984. Bats. Pp, 145–186, *in* Orders and families of Recent mammals of the world (Anderson, S., and J. K. Jones, Jr., eds.). John Wiley and Sons. New York, U.S.A.

Koopman, K. F. 1993. Order Chiroptera. Pp. 137–242, *in* Mammal species of the world (Wilson, D. E., and D. M. Reeder, eds.). Second ed. Smithsonian Institution Press. Washington, U.S.A.

Koopman, K. F. 1994. Chiroptera: systematics. Handbook of zoology: a natural history of the phyla of the animal kingdom. VIII. Mammalia. Walter de Gruyter. New York, U. S. A.

Koopman, K. F., M. K. Hecht, and E. Ledecky-Janecek. 1957. Notes on the mammals of the Bahamas with special reference to the bats. Journal of Mammalogy 38:164–174.

Koprowski, J. L. 1994a. *Sciurus carolinensis*. Mammalian Species 480:1–9.

Koprowski, J. L. 1994b. *Sciurus niger*. Mammalian Species 479:1–9.

Koprowski, J. L. 2017. *Ammospermophilus nelsoni*. The IUCN Red List of Threatened Species 2017:e.T1149A22251492.

Koprowski, J., Roth, L., N. Woodman, J. Matson, L. Emmons, and F. Reid. 2016. *Sciurus deppei*. The IUCN Red List of Threatened Species 2016:e.T20001A115154440.

Koprowski, J., Roth, L., F. Reid, N. Woodman, R. Timm, and L. Emmons. 2017. *Sciurus aureogaster*. The IUCN Red List of Threatened Species 2017:e.T20006A22248035.

Koprowski, J. L., E. A. Goldsteins, K. R. Bennett, and C. Pereira Mendes. 2016a. Family Sciuridae (tree, flying and ground squirrels, chipmunks, marmots and prairie dogs). Pp. 648–836, *in* Handbook of the Mammals of the world 6. Lagomorpha and Rodents. (Wilson, D. E., T. E. Lacher, Jr., and R. A. Mittermeier, eds.). Lynx Edicions, IUCN. Barcelona, Spain.

Koprowski, J. L., N. Ramos, B. S. Pasch, and C. A. Zugmeyer. 2006. Observations on the ecology of the endemic Mearns's squirrel (*Tamiasciurus mearnsi*). Southwestern Naturalist 51:426–430.

Koprowski, J. L., M. A. Steele, and N. Ramos-Lara. 2016b. *Tamiasciurus mearnsi* (Rodentia: Sciuridae). Mammalian Species 48:66–72.

Kovacs, K. M. 2009. Bearded seal *Erignathus barbatus*. Pp. 97–101, *in* Encyclopedia of Marine Mammals (Perrin, W. F., B. Wursig, and J. G. M. Thewissen, eds.). Academic Press. San Diego, U.S.A.

Kovacs, K. M. 2015. *Pagophilus groenlandicus*. The IUCN Red List of Threatened Species 2015:e.T41671A45231087.

Kovacs, K. M. 2016a. *Erignathus barbatus*. The IUCN Red List of Threatened Species 2016:e.T8010A45225428.

Kovacs, K. M. 2016b. *Cystophora cristata*. The IUCN Red List of Threatened Species 2016:e.T6204A45225150.

Kovacs, K. M., and D. M. Lavigne. 1986. *Crystophora cristata*. Mammalian Species 258:1–9.

Kovacs, K. M., A. Aguilar, D. Aurioles, V. Burkanov, C. Campagna, N. J. Gales, T. Gelatt, S. D. Goldsworth., S. J. Goodma., G. J. G. Hofmeyr, T. Härkönen, L. Lowry, L. Lydersen, J. Schipper, T. Sipilä, C. Southwell, D. Thompson, and F. Trillmich. 2012. Global threats to pinnipeds. Marine Mammal Science 28:414–436.

Krebs, C. J., and D. L. Murray. 2018. *Lepus americanus*. Pp. 163–165, *in* Pikas, Rabbits and Hares of the World (Smith, A., C. H. Johnston, P. C. Alves, and K. Hackländer, eds.). The Johns Hopkins University Press/IUCN/SSC. Baltimore, U.S.A.

Kretzoi, M. 1964. Über einige homonyme und synonyme Säugetiernamen. Vertebrata Hungarica 6:131–138.

Kretzoi, M. 1969. Skizze einer Arvicoliden-Phylogenie. Vertebrata Hungarica 12:111–121.

Krumbiegel, I. 1980. Die unterartliche Trennung des Bisons, *Bison bison* (Linné, 1788), und seine Rückzüchtung. Säugetierkundliche Mitteilungen 28:148–160.

Krutzsch, P. H. 1954. North American jumping Mice (Genus *Zapus*). University of Kansas Publications, Museum of Natural History 7:349–472.

Krystufek, B., and V. Vohralik. 2013. Taxonomic revision of the Palaearctic rodents (Rodentia). Part 2. Sciuridae: *Urocitellus, Marmota* and *Sciurotamias*. Lynx 44:27–138.

Kumirai, A., and J. K. Jones, Jr. 1990. *Nyctinomops femorosaccus*. Mammalian Species 349:1–5.

Kunz, T. H. 1982. *Lasionycteris noctivagans*. Mammalian Species 172:1–5.

Kunz, T. H., and R. A. Martin. 1982. *Plecotus townsendii*. Mammalian Species 175:1–6.

Kurose, N., A. V. Abramov, and R. Masuda. 2000. Intrageneric diversity of the cytochrome b gene and phylogeny of Eurasian species of the genus Mustela (Mustelidae, Carnivora). Zoological Science 17:673–679

Kurta, A. 1995. Mammals of the Great Lakes region. Revised edition. University of Michigan Press. Ann Arbor, U.S.A.

Kurta, A., and R. H. Baker. 1990. *Eptesicus fuscus*. Mammalian Species 356:1–10.

Kurta, A., and G. C. Lehr. 1995. *Lasiurus ega*. Mammalian Species 515:1–7.

Kurten, B., and E. Anderson. 1980. Pleistocene mammals of North America. Columbia University Press. New York, U.S.A.

Kuwayama, R., and T. Ozawa. 2000. Phylogenetic relationships among European red deer, wapiti, and sika deer inferred from mitochondrial DNA sequences. Molecular Phylogenetics and Evolution 15:115–123.

Kwiecinski, G. G. 1998. *Marmota monax*. Mammalian Species 591:1–8.

Kwiecinski, G. G. 2006. *Phyllostomus discolor*. Mammalian Species 801:1–11.

Kyle, C. J., C. Strobeck, R. D. Weir, H. Davis, and N. J. Newhouse. 2004. Genetic structure of sensitive and endangered northwestern badger populations (*Taxidea taxus taxus* and *T. t. jeffersonii*). Journal of Mammalogy 85:633–639.

Laabs, D. M., M. L. Allaback, and D. R. Mitchell. 2022. Distribution, morphology, and karyotype of San Joaquin pocket mice from the western Mojave Desert. California Fish and Wildlife Journal 108:e10.

Lacher, T. E., Jr. 2016a. Family Caviidae (Cavies, Capibaras, and Maras). Pp. 406–439, *in* Handbook of the mammals of the world 6. Lagomorpha and Rodentia I (Wilson, D. E., T. E. Lacher, Jr., and R. A. Mittermeier, eds.). Lynx Edicions. Barcelona, Spain.

Lacher, T. 2016b. *Geomys personatus*. The IUCN Red List of Threatened Species 2016:e.T9055A22218317.

Lacher, T., and S. T. Álvarez-Castañeda. 2016. *Neotoma albigula*. The IUCN Red List of Threatened Species 2016:e.T14582A115122696.

Lacher, T., and S. T. Álvarez-Castañeda. 2019. *Chaetodipus goldmani*. The IUCN Red List of Threatened Species 2019:e.T4332A22226182.

Lacher, T., R. Timm, and S. T. Álvarez-Castañeda. 2016a. *Dipodomys simulans*. The IUCN Red List of Threatened Species 2016:e. T136630A115210884.

Lacher, T., R. Timm, and S. T. Álvarez-Castañeda. 2016b. *Onychomys arenicola*. The IUCN Red List of Threatened Species 2016:e. T15337A115127153.

Lacher, T., R. Timm, and S. T. Álvarez-Castañeda. 2016c. *Onychomys torridus*. The IUCN Red List of Threatened Species 2016:e. T15339A115127523.

Lacher, T., R. Timm, and S. T. Álvarez-Castañeda. 2016d. *Otospermophilus variegatus*. The IUCN Red List of Threatened Species 2016:e. T20495A22263993.

Lacher, T., R. Timm, and S. T. Álvarez-Castañeda. 2016e. *Perognathus flavescens*. The IUCN Red List of Threatened Species 2016:e. T16634A115134864.

Lacher, T., R. Timm, and S. T. Álvarez-Castañeda. 2016f. *Perognathus flavus*. The IUCN Red List of Threatened Species 2016:e.T16635A115135061.

Lacher, T., R. Timm, and S. T. Álvarez-Castañeda. 2016g. *Perognathus merriami*. The IUCN Red List of Threatened Species 2016:e. T16637A115135383.

Lacher, T., R. Timm, and S. T. Álvarez-Castañeda. 2016h. *Peromyscus boylii*. The IUCN Red List of Threatened Species 2016:e.T16652A115135521.

Lacher, T., R. Timm, and S. T. Álvarez-Castañeda. 2016i. *Peromyscus gratus*. The IUCN Red List of Threatened Species 2016:e.T16663A115136134.

Lacher, T., R. Timm, and S. T. Álvarez-Castañeda. 2016j. *Peromyscus nasutus*. The IUCN Red List of Threatened Species 2016:e.T16682A115137189.

Lacher, T., R. Timm, and S. T. Álvarez-Castañeda. 2016k. *Peromyscus pectoralis*. The IUCN Red List of Threatened Species 2016:e. T16684A115137332.

Lacher, T., R. Timm, and S. T. Álvarez-Castañeda. 2016l. *Reithrodontomys montanus*. The IUCN Red List of Threatened Species 2016:e. T19413A115151509.

Lacher, T., R. Timm, and S. T. Álvarez-Castañeda. 2016m. *Sciurus griseus*. The IUCN Red List of Threatened Species 2016:e.T20011A115154716.

Lacher, T., R. Timm, and S. T. Álvarez-Castañeda. 2016n. *Sigmodon ochrognathus*. The IUCN Red List of Threatened Species 2016:e. T20217A115157990.

Lacher, T., R. Timm, and S. T. Álvarez-Castañeda. 2016o. *Thomomys bottae*. The IUCN Red List of Threatened Species 2016:e.T21799A115163311.

Lacher, T., S. T. Álvarez-Castañeda, and R. Timm. 2016p. *Thomomys umbrinus*. The IUCN Red List of Threatened Species 2016:e. T21800A115163507.

Lacher, T., R. Timm, and S. T. Álvarez-Castañeda. 2016q. *Xerospermophilus spilosoma*. The IUCN Red List of Threatened Species 2016:e. T42563A22262899.

Lacher, T., R. Timm, and S. T. Álvarez-Castañeda. 2016r. *Xerospermophilus tereticaudus*. The IUCN Red List of Threatened Species 2016:e. T20493A22264318.

Lacher, T., S. T. Álvarez-Castañeda, and R. Timm. 2016s. *Neotamias dorsalis*. The IUCN Red List of Threatened Species 2016:e.T42571A115190634.

Lacher, T., S. T. Álvarez-Castañeda, and R. Timm. 2016t. *Neotoma micropus*. The IUCN Red List of Threatened Species 2016:e.T14591A115123286.

Lacher, T., R. Timm, and S. T. Álvarez-Castañeda. 2017. *Neotoma lepida*. The IUCN Red List of Threatened Species 2017:e.T116988741A123797359.

Lacher, T., D. Hafner, and R. Timm. 2019a. *Geomys arenarius*. The IUCN Red List of Threatened Species 2019:e.T9054A22218401.

Lacher, T., R. Timm, and S. T. Álvarez-Castañeda. 2019b. *Peromyscus eremicus*. The IUCN Red List of Threatened Species 2019:e. T16659A143641683.

Lackey, J. A. 1967a. Biosystematics of *heermanni* group kangaroo rats in southern California. Transactions of the San Diego Society of Natural History 14:313–344.

Lackey, J. A. 1967b. Growth and development of *Dipodomys stephensi*. Journal of Mammalogy 48:624–632.

Lackey, J. A. 1991a. *Chaetodipus arenarius*. Mammalian Species 384:1–4.

Lackey, J. A. 1991b. *Chaetodipus spinatus*. Mammalian Species 385:1–4.

Lackey, J. A. 1996. *Chaetodipus fallax*. Mammalian Species 517:1–6.

Lackey, J. A., and T. L. Best. 1992. *Chaetodipus goldmani*. Mammalian Species 419:1–5.

Lackey, J. A., D. G. Huckaby, and B. G. Ormiston. 1985. *Peromyscus leucopus*. Mammalian Species 247:1–10.

Lacoste, K. N., and G. B. Stenson. 2000. Winter distribution of harp seals (*Phoca groenlandica*) off eastern Newfoundland and southern Labrador. Polar Biolgy 23:805–811.

Laidre, K. L., I. Stirling, L. F. Lowry, Ø. Wiig, M. P. Heide-Jørgensen, and S. H. Ferguson. 2008. Quantifying the sensitivity of arctic marine mammals to climateinduced habitat change. Ecological Applications 18:S97-S125.

Lamoreux, J. 2017. *Cratogeomys planiceps*. The IUCN Red List of Threatened Species 2017:e.T136249A22216740.

Lanier, H., and D. Hik. 2016. *Ochotona collaris*. The IUCN Red List of Threatened Species 2016:e.T41257A45182533.

Lanier, H. C. 2018. *Ochotona collaris*. Pp. 36–39, *in* Pikas, Rabbits and Hares of the World (Smith, A., C. H. Johnston, P. C. Alves, and K. Hackländer, eds.). The Johns Hopkins University Press/IUCN/SSC. Baltimore, U.S.A.

Lanier, H. C., and C. Nielsen. 2019. *Sylvilagus aquaticus*. The IUCN Red List of Threatened Species 2019:e.T41296A45190578.

Laerm, J. 1981. Systematic status of the Cumberland Island pocket gopher, *Geomys cumberlandius*. Brimleyana 6:141–161.

Larivière, S. 1999a. *Mustela vison*. Mammalian Species 608:1–9.

Larivière, S. 1999b. *Lontra longicaudis*. Mammalian Species 609:1–5.

Larivière, S. 2001. *Ursus americanus*. Mammalian Species 647:1–11.

Larivière, S. and M. Pasitschniak-Arts. 1996. *Vulpes vulpes*. Mammalian Species 537:1–11.

Larivière, S., and L. R. Walton. 1997. *Lynx rufus*. Mammalian Species 563:1–8.

Larivière, S., and L. R. Walton. 1998. *Lontra canadensis*. Mammalian Species 587:1–8.

Larrison, E. J., and D. R. Johnson. 1981. Mammals of Idaho. University Press of Idaho, Moscow, U.S.A.

Larry, G. M. 1978. *Chironectes minimus*. Mammalian Species 109:1–6.

Larsen, P. A., S. R. Hoofer, M. C. Bozeman, S. C. Pedersen, H. H. Genoways, C. J. Phillips, D. E. Pumo, and R. J. Baker. 2007. Phylogenetics and phylogeography of the *Artibeus jamaicensis* complex based on cytochrome-*b* DNA sequences. Journal of Mammalogy 88:712–727.

Larsen, P. A., M. R. Marchán-Rivadeneira, and R. J. Baker. 2010. Taxonomic status of Andersen's fruit-eating bat (*Artibeus jamaicensis aequatorialis*) and revised classification of *Artibeus* (Chiroptera: Phyllostomidae). Zootaxa 2648:45–60.

Lassieur, S., D. E. Wilson. 1989. *Lonchorhina aurita*. Mammalian Species 347:1–4.

Latch, E. K., J. R. Heffelfinger, J. A. Fike, and O. E. Rhodes. 2009. Species-wide phylogeography of North American mule deer (*Odocoileus hemionus*): cryptic glacial refugia and postglacial recolonization. Molecular Ecology 18:1730–1745.

Laundré, J. W. 2018. *Sylvilagus audubonii*. Pp. 120–122, *in* Pikas, Rabbits and Hares of the World (Smith, A., C. H. Johnston, P. C. Alves, and K. Hackländer, eds.). The Johns Hopkins University Press/IUCN/SSC. Baltimore, U.S.A.

Laundré, J., and T. W. Clark. 2003. Managing puma hunting in the western United States: Through a metapopulation approach. Animal Conservation 6:159–170.

LaVal, R. K. 1967. Records of bats from the southeastern United States. Journal of Mammalogy 48:645–648.

LaVal, R. K. 1970. Infraspecific relationships of bats of the species *Myotis austroriparius*. Journal of Mammalogy 51:542–552.

LaVal, R. K. 1973a. A revision of the Neotropical bats of the genus *Myotis*. Natural History Museum, Los Angeles County Science Bulletin 15:1–54.

LaVal, R. K. 1973b. Systematic of the genus *Rhogeessa* (Chiroptera: Vespertilionidae). Occasional Papers fo the Museum of Natural History, University of Kansas 19:1–47.

LaVal, R. K., and M. L. LaVal. 1980. Ecological studies and management of Missouri bats, with emphasis on cavedwelling species. Terrestrial series, Missouri Deptament of Conservation, Jefferson City 8:1–53.

Lavariega, M. C., N. Martín-Regalado, y R. M. Gómez-Ugalde. 2012. Mamíferos del centro-occidente de Oaxaca, México. Therya 3:349–370.

Lavigne, D. M. 2002. Harp seal Pagophilus groenlandicus. Pp. 560–562, in Encyclopedia of Marine Mammals (Perrin, W. F., B. Wursig, and J. G.,M. Thewissen, eds.). Academic Press, San Diego. California, USA.

Lavoie, M., A. Renard, and S. Larivière. 2019. *Lynx canadensis* (Carnivora: Felidae). Mammalian Species 51:136–154.

Lawlor, T. E. 1982a. The evolution of body size in mammals: evidence for insular populations in Mexico. American Naturalist 119:54–72.

Lawlor, T. E. 1982b. *Ototylomys phyllotis*. Mammalian Species 181:1–3.

Lawlor, T. E. 1969. A systematic study of the rodent genus *Ototylomys*. Journal of Mammalogy 50:28–42.

Lawlor, T. E. 1971a. Evolution of *Peromyscus* on Northern Island in the Gulf of California, Mexico. Transactions of the San Diego Society of Natural History 16:91–124.

Lawlor, T. E. 1971b. Distribution and relationships of six species of *Peromyscus* in Baja California and Sonora, Mexico. Occasional Papers of the Museum of Zoology, University of Michigan 661:1–22.

Lawlor, T. E. 1979. Handbook to the orders and families of living mammals. Mad River Press. Eureka, U.S.A.

Lawlor, T. E. 1983. The mammals. Pp. 265–284, *in* Island biogeography in the Sea of Cortez (Case, T. J. and M. L. Cody, eds.). University of California Press. Berkeley, U.S.A.

Lawrence, B., and W. H. Bossert. 1967. Multiple character analysis of *Canis lupus, latrans, andfamiliaris,* with a discussion of the relationships of *Canis niger.* American Zoology 7:223–232.

Layne, J. N. 1990. The Florida mouse. Pp. 1–21, *in* Burrow associates of the gopher tortoise (Dodd, Jr., C. K., R. E. Ashton, Jr., R. Franz, and E. Wester, eds.). Eighth Annual Meeting, Gopher Tortoise Council, Florida Museum of Natural History, University of Florida. Gainesville, U.S.A.

Lazcano-Barrero, M. A., and J. M. Packard. 1989. The occurrence of manatees (*Trichechus manatus*) in Tamaulipas, Mexico. Marine Mammal Science 5:202–205.

Lazell, J. D., Jr. 1981. Field and taxonomic studies of tropical American raccoons. National Geographic Society Research Reports 13:381–385.

Le Boeuf, B. J., K. W. Kenyon, and B. Villa-Ramírez. 1986. The Caribbean monk seal is extinct. Marine Mammal Science 2:70–72.

LeBoeuf, B. J., D. E. Crocker, D. P. Costa, S. B. Blackwell, P. M. Webb, et al. 2000. Foraging ecology of northern elephant seals. Ecological Monographs 70:353–382.

Lee, D. S., J. D. Funderburg, Jr., and M. K. Clark. 1982. A distributional survey of North Carolina mammals. Occasional Papers, North Carolina Biological Survey 1982 10:1–70.

Lee, H. S., and D. F. Hoffmeister. 2003. *Sorex arizonae*. Mammalian Species 732:1–3.

Lee, M. R., and F. B. Elder. 1977. Karyotypes of eight especies of Mexican rodents (Muridae). Journal of Mammalogy 58:479–487.

Lee, M. R., and D. F. Hoffmeister. 1963. Status of certain fox squirrels in Mexico and Arizona. Proceedings of the Biological Society of Washington 76:181–189.

Lee, M. R., and D. J. Schmidly. 1977. A new species of *Peromyscus* (Rodentia: Muridae) from Coahuila, Mexico. Journal of Mammalogy 58:263–268.

Lee, M. R., D. J. Schmidly, and C. H. Huheey. 1972. Chromosomal variation in certain populations of *Peromyscus boylii* and its systematic implications. Journal of Mammalogy 53:697–707.

Lee, T. E., Jr., B. R. Riddle and P. L. Lee. 1996. Speciation in the desert pocket mouse (*Chaetodipus penicillatus* Woodhouse). Journal of Mammalogy 77:58–68.

Lent, P. C. 1988. *Ovibos moschatus*. Mammalian Species 302:1–9.

Lee Jr, T. E., S. R. Hoofer, and R. A. Van Den Bussche. 2002. Molecular phylogenetics and taxonomic revision of the genus Tonatia (Chiroptera: Phyllostomidae). Journal of Mammalogy 83:49–57.

León, L., T. V. Monterrubio, and M. S. Hafner. 2001. *Cratogeomys neglectus*. Mammalian Species 685:1–4.

León-Paniagua, L., A. G. Navarro-Sigüenza, B. E. Hernández-Baños, and J. C. Morales. 2007. Diversification of the arboreal mice of the genus *Habromys* (Rodentia: Cricetidae: Neotominae) in the Mesoamerican highlands. Molecular Phylogenetics and Evolution 42:653–664.

León-Paniagua, L. 2017. Family Cricetidae. Pp. 321–334, *in* Handbook of mammals of the world, volume 7, Rodent II (Wilson D. E., T. E. Lacher, Jr., and R. A. Mittermeier, eds.). Lyx Edicions. Barcelona, Spain.

Léon-Tapia, M. A., J. A. Fernández, Y. Rico, F. A. Cervantes, and A. Espinosa de los Monteros. 2020. A new mouse of the *Peromyscus maniculatus* species complex (Cricetidae) from the highlands of central Mexico. Journal of Mammalogy 101:1117–1132.

Leopold, A. S. 1959. Wildlife of Mexico: the game birds and mammals. University of California Press, Berkeley, U.S.A.

Lessa, E., and J. L. Patton, 1989. Structural constraints, recurrent shapes, and allometry in pocket gophers (genus *Thomomys*). Biological Journal of the Linnean Society 36:349–363.

Levenson, H. R. 1990. Sexual size dimorphism in chipmunks. Journal of Mammalogy 71:161–170.

Levenson, H. R., R. S. Hoffman, C. F. Nadler, L. Deutsch, and S. D. Freeman. 1985. Systematics of the holartic chipmunks (*Tamias*). Journal of Mammalogy 66:219–242.

Levin, E. Y., and V. Flyger. 1971. Uroporphyrinogen III cosynthetase activity in the fox squirrel (*Sciurus niger*). Science 174:59–60.

Lewis, T. H. 1983. The anatomy and histology of the rudimentary eye of *Neurotrichus*. Northwest Science 57:8–15.

Lewis, S. E., and D. E. Wilson. 1987. *Vampyressa pusilla*. Mammalian Species 292:1–5.

Lewis-Oritt, N., C. A. Porter, and R. J. Baker. 2001. Molecular systematics of the family Mormoopidae (Chiroptera) based on cytochrome *b* and recombination activating gene 2 sequences. Molecular Phylogenetics and Evolution 20:426–436.

Li, X., A. C. Tzika, Y. Liu, K. V. Doninck, Q. Zhu, and M. C. Milinkovitch. 2010. Preliminary genetic status of the spotted seal *Phoca largha* in Liaodong Bay (China) based on microsatellite and mitochondrial DNA analyses. Trends in Evolutionary Biology 2:33–38.

Lidicker, W. Z., Jr. 1960a. An analysis of intraspecific variation in the kangaroo rat *Dipodomys merriami*. University California Publication in Zoology 67:125–217.

Lidicker, W. Z., Jr. 1960b. The baculum of *Dipodomys ornatus* and its implications for superspecific groupings of kangaroo rats. Journal of Mammalogy 41:495–499.

Lidicker, W. Z., Jr. 1960c. A new subspecies of the cliff cipmunk from Central Chihuahua. Proceedings of the Biological Society of Washington 73:267–274.

Lim, B. K. 1987. *Lepus townsendii*. Mammalian Species 288:1–6.

Lim, B. K., W. A. Pedro, and F. C. Passos. 2003. Differentiation and species status of the Neotropical yellow-eared bats *Vampyressa pusilla* and *V. thyone* (Phyllostomidae) with a molecular phylogeny and review of the genus. Acta Chiropterologica 5:15–29.

Lim, B. K., M. D. Engstrom, N. B. Simmons, and J. M. Dunlop. 2004. Phylogenetics and biogeography of least sac-winged bats *(Balantiopteryx)* based on morphological and molecular data. Mammalian Biology 69:225– 237.

Lim, B. 2015. *Balantiopteryx io*. The IUCN Red List of Threatened Species 2015:e.T2532A22030080.

Lim, B., and B. Miller. 2016. *Rhynchonycteris naso*. The IUCN Red List of Threatened Species 2016:e.T19714A22010818.

Lim, B. K., B. Miller, F. Reid, J. Arroyo-Cabrales, A. D. Cuarón, and P. C. de Grammont. 2016a. *Balantiopteryx plicata*. The IUCN Red List of Threatened Species 2016:e.T2533A22029659.

Lim, B. K., B. Miller, F. Reid, J. Arroyo-Cabrales, A. D. Cuarón, and P. C. de Grammont. 2016b. *Diclidurus albus*. The IUCN Red List of Threatened Species 2016:e.T6561A21986615.

Linares, O. J. 1986. Murciélagos de Venezuela. Cuadernos Lagoven. Caracas, Venezuela.

Lindqvist, C., L. Bachmann, L. W. Andersen, E. W. Born, U. Arason, K. M. Kovacs, C. Lydesren, A. V. Abramov, and Ø. Wiig. 2009. The Laptev Sea walrus *Odobenus rosmarus laptevi*: an enigma revisted. Zoologica Scripta 38:113–127.

Lindsay, S. L. 1981. Taxonomic and biogeographic relationships of Baja California chickarees (*Tamiasciurus*). Journal of Mammalogy 62:673–682.

Lindsay, S. L. 1982. Systematic relationships of parapatric tree squirrel species (*Tamiasciurus*) in the Pacific Northwest. Canadian Journal of Zoology 60:2149–2156.

Ling, J. K., and M. M. Bryden. 1992. *Mirounga leonina*. Mammalian Species 391:1–8.

Linzey, A. V. 1983. *Synaptomys cooperi*. Mammalian Species 210:1–5.

Linzey, A. V., and NatureServe. 2017a. *Thomomys clusius*. The IUCN Red List of Threatened Species 2017:e.T42595A22216284.

Linzey, A. V., and NatureServe. 2017b. *Thomomys idahoensis*. The IUCN Red List of Threatened Species 2017:e.T21809A22215570.

Linzey, A.V., R. Timm, S. T. Álvarez-Castañeda, I. Castro-Arellano, and T. Lacher. 2008a. *Sciurus arizonensis*. The IUCN Red List of Threatened Species 2008:e.T20005A9132802.

Linzey, A. V. and NatureServe (Hammerson, G., J. C. Whittaker, and Norris, S. J.) 2008b. *Neotoma magister*. The IUCN Red List of Threatened Species 2008:e.T14581A4446084.

Linzey, A. V., J. Matson, and S. Pérez. 2016a. *Neotoma mexicana*. The IUCN Red List of Threatened Species 2016:e.T14590A115123126.

Linzey, A. V., R. Timm, S. T. Álvarez-Castañeda, I. Castro-Arellano, and T. Lacher. 2016b. *Ammospermophilus leucurus*. The IUCN Red List of Threatened Species 2016:e.T42452A115189458.

Linzey, A. V., R. Timm, S. T. Álvarez-Castañeda, I. Castro-Arellano, and T. Lacher. 2016c. *Ictidomys mexicanus*. The IUCN Red List of Threatened Species 2016:e.T20487A22262744.

Linzey, A. V., R. Timm, S. T. Álvarez-Castañeda, I. Castro-Arellano, and T. Lacher. 2016d. *Chaetodipus baileyi*. The IUCN Red List of Threatened Species 2016:e.T4328A115068087.

Linzey, A. V., R. Timm, S. T. Álvarez-Castañeda, I. Castro-Arellano, and T. Lacher. 2016e. *Chaetodipus californicus*. The IUCN Red List of Threatened Species 2016:e.T4329A115068220.

Linzey, A. V., R. Timm, S. T. Álvarez-Castañeda, I. Castro-Arellano, and T. Lacher. 2016f. *Chaetodipus eremicus*. The IUCN Red List of Threatened Species 2016:e.T136606A115210279.

Linzey, A. V., R. Timm, S. T. Álvarez-Castañeda, I. Castro-Arellano, and T. Lacher. 2016g. *Chaetodipus hispidus*. The IUCN Red List of Threatened Species 2016:e.T4333A115068352.

Linzey, A. V., R. Timm, S. T. Álvarez-Castañeda, I. Castro-Arellano, and T. Lacher. 2016h. *Chaetodipus intermedius*. The IUCN Red List of Threatened Species 2016:e.T4334A115068570.

Linzey, A. V., R. Timm, S. T. Álvarez-Castañeda, I. Castro-Arellano, and T. Lacher. 2016i. *Chaetodipus nelsoni*. The IUCN Red List of Threatened Species 2016:e.T4335A115068711.

Linzey, A. V., R. Timm, S. T. Álvarez-Castañeda, I. Castro-Arellano, and T. Lacher. 2016j. *Chaetodipus penicillatus*. The IUCN Red List of Threatened Species 2016:e.T4336A115068852.

Linzey, A.V., R. Timm, S. T. Álvarez-Castañeda, I. Castro-Arellano, and T. Lacher. 2016k. *Chaetodipus spinatus*. The IUCN Red List of Threatened Species 2016:e.T4338A115068992.

Linzey, A. V., R. Timm, S. T. Álvarez-Castañeda, I. Castro-Arellano, and T. Lacher. 2016l. *Peromyscus merriami*. The IUCN Red List of Threatened Species 2016:e.T16680A115137090.

Linzey, A. V., R. Timm, L. Emmons, and F. Reid. 2016m. *Sciurus niger*. The IUCN Red List of Threatened Species 2016:e.T20016A115155257.

Linzey, A. V., R. Timm, N. Woodman, J. Matson, and R. Samudio. 2016n. *Oryzomys couesi*. The IUCN Red List of Threatened Species 2016:e. T15592A115128044.

Linzey, A. V., S. Shar, D. Lkhagvasuren, R. Juškaitis, B. Sheftel, H. Meinig, G. Amori, and H. Henttonen. 2016o. *Microtus oeconomus*. The IUCN Red List of Threatened Species 2016:e.T13451A115113894.

Linzey, A. V., R. Timm, S. T. Álvarez-Castañeda, I. Castro-Arellano, and T. Lacher. 2019. *Sciurus arizonensis*. The IUCN Red List of Threatened Species 2019:e.T20005A22247935.

Linzey, A. V., H. Henttonen, B. Sheftel, and N. Batsaikhan. 2020. *Myodes rutilus*. The IUCN Red List of Threatened Species 2020:e. T4975A164372228.

Lira-Torres, I., and M. Briones-Salas. 2011. Impacto de la ganadería extensiva y caceria de susbsistencia sobre abundancia relativa de mamíferos en la Selva Zoque, Oaxaca, México. Therya 2:217–244.

Lira-Torres, I., and M. Briones-Salas. 2012. Abundancia relativa y patrones de actividad de los mamíferos de los Chimalapas, Oaxaca, México. Acta Zoológica Mexicana (n. s.) 28:566–585.

Lissovsky, A. A., and E. V. Obolenskaya. 2011. The structure of craniometrical diversity of grey voles *Microtus* subgenus *Alexandromys*. Proceedings of the Zoological Institute RAS 315:461–477.

Lissovsky, A. A., T. V. Petrova, S. P. Yatsentyuk, F. N. Golenishchev, N. I. Putincev, I. V. Kartavtseva, I. N. Sheremetyeva, and N. I. Abramson. 2017. Multilocus phylogeny and taxonomy of East Asian voles *Alexandromys* (Rodentia, Arvicolinae). Zoologica Scripta 47:9–20.

List, R., O. R. W. Pergams, J. Pachecho, J. Cruzado, and G. Ceballos. 2010. Genetic divergence of *Microtus pennsylvanicus chihuahuensis* and conservation implications of marginal population extinctions. Journal of Mammalogy 91:1093–1101.

Litvaitis, J., and H. C. Lanier. 2019. *Sylvilagus transitionalis*. The IUCN Red List of Threatened Species 2019:e.T21212A45181534.

Litvaitis, M. K., W. Lee, J. A. Litvaitis, and T. D. Kocher. 1997. Variation in the mitochondrial DNA of the *Sylvilagus* complex occupying the northeastern United States. Canadian Journal of Zoology 75:595–605.

Livaitis, J. A., M. Livaitis, A. Kovach, and H. Kilpatrick. 2018. *Sylvilagus transitionalis*. Pp. 156–157, *in* Pikas, Rabbits and Hares of the World (Smith, A., C. H. Johnston, P. C. Alves, and K. Hackländer, eds.). The Johns Hopkins University Press/IUCN/SSC. Baltimore, U.S.A.

Long, C. A. 1965. The mammals of Wyoming. University of Kansas Publications, Museum of Natural History 14:493–758.

Long, C. A. 1972. Taxonomic revision of the North American Badger, *Taxidea taxus*. Journal of Mammalogy 53:725–759.

Long, C. A. 1973. *Taxidea taxus*. Mammalian Species 26:1–4.

Long, C. A. 1974. *Microsorex hoyi* and *Microsorex thompsoni*. Mammalian Species 33:1–4.

Long, C. A. 1999b. American badger, *Taxidea taxus*. Pp. 177–179, *in* The Smithsonian book of North American mammals (Wilson, D. E., and S. Ruff, eds.). Washington, U.S.A.

Long, C. A., and R. S. Hoffmann.1992. *Sorex preblei* from the Black Canyon, first record for Colorado. The Southwestern Naturalist 37:318–319.

Long, C. A., and C. F. Long. 1965. Dental abnormalities in North American badgers, genus *Taxidea*. Transactions of the Kansas Academy of Sciences 68:145–155.

Longhofer, L. K., and R. D. Bradley. 2006. Molecular systematics of the genus *Neotoma* based on DNA sequences from intron 2 of the Alcohol dehydrogenase gene. Journal of Mammalogy 87:961–970.

Lópes-Forment, W. 1981. Algunos aspectos ecológicos del murciélago Balantiopteryx plicata plicata Peters, 1867 (Chiroptera: Emballonuridae) en México. Anales del Instituto de Biología, Universidad Nacional Autónoma de México, serie zoología 50:673–699.

López-Cuamatzi, I. L., J. Ortega, S. M. Ospina-Garcés, G. Zúñiga, M. C. MacSwiney. 2024. Molecular and morphological data suggest a new species of big-eared bat (Vespertilionidae: Corynorhinus) endemic to northeastern Mexico. PLoS ONE 19(2): e0296275. https://doi.org/10.1371/journal.pone.0296275

López-González, C. 1998. Systematics and biogeography of the bats of Paraguay. Ph.D. dissertation, Texas Tech University, Lubbock, U.S.A.

López-González, C. 2003. Murciélagos (Chiroptera) del Estado de Durango, México: Composición, distribución y estado de conservación. Vertebrata Mexicana 13:15–23.

López-González, C., and D. F. García-Mendoza. 2006. Murciélagos de la Sierra Tarahumara, Chihuahua, México. Acta Zoológica Mexicana (n. s.) 22:109–135.

López-González, C., and S. J. Presley. 2001. Taxonomic status of *Molossus bondae* J. A. Allen, 1904 (Chiroptera: Molossidae), with description of a new subspecies. Journal of Mammalogy 82:760–774.

López-González, C., S. J. Presley, R. D. Owen, and M. R. Will. 2001. Taxonomic status of *Myotis* (Chiroptera: Vespertilionidae) in Paraguay. Journal of Mammalogy 82:138–160.

López-Wilchis, R., and J. López-Jardines. 1998. Los mamíferos de México depositados en colecciones de Estados Unidos y Canadá. Universidad Autónoma Metropolitana, Distrito Federal, México.

López-Wilchis, R., L. M. Guevara-Chumacero, N. P. Ángeles, J. Juste, C. Ibáñez, and I. D. L. A. Barriga-Sosa. 2012. Taxonomic status assessment of the Mexican populations of funnel-eared bats, genus *Natalus* (Chiroptera: Natalidae). Acta Chiropterologica 14:305–316.

López-Wilchis, R., J. W. Torres-Flores, J. Arroyo-Cabrales. 2020. *Natalus mexicanus* (Chiroptera: Natalidae). Mammalian Species 52:27–39.

Lorenzini, R., and L. Garofalo. 2015. Insights into the evolutionary history of *Cervus* (Cervidae, tribe Cervini) based on Bayesian analysis of mitochondrial marker sequences, with first indications for a new species. Journal of Zoological Systematics and Evolutionary Research 53:340–349.

Lorenzo, C., and S. T. Álvarez-Castañeda. 2011. *Sylvilagus mansuetus*. The IUCN Red List of Threatened Species 2011:e.T21210A9257211.

Lorenzo, C., and D. E. Brown. 2019. *Lepus alleni*. The IUCN Red List of Threatened Species 2019:e.T41272A45185265.

Lorenzo, C., and F. A. Cervantes. 1995. The G-banded karyotype of the Tapeti rabbit (*Sylvilagus brasiliensis*) from Chiapas, Mexico. Revista Sociedad Mexicana Historia Natural 46:1–5.

Lorenzo, C., and C. H. Johnston. 2019. *Lepus insularis*. The IUCN Red List of Threatened Species 2019:e.T11794A45177986.

Lorenzo, C., and H. C. Lanier. 2019a. *Sylvilagus cunicularius*. The IUCN Red List of Threatened Species 2019:e.T21211A45181292.

Lorenzo, C., and H. C. Lanier. 2019b. *Sylvilagus graysoni*. The IUCN Red List of Threatened Species 2019:e.T21206A45180643.

Lorenzo, C., P. Cortés-Calva, G. Ruiz-Campos, and S. T. Álvarez-Castañeda. 2013. Current distributional status of two subspecies of *Sylvilagus bachmani* in the Baja California Peninsula, Mexico. Western North American Naturalist 73:219–223.

Lorenzo, C., F. A. Cervantes, and M. A. Aguilar. 1993. The karyotypes of some Mexican cottontail rabbits of the genus *Sylvilagus*. Pp. 129–136, *in* Avances en el estudio de los mamíferos de México (Medellín, R. A., and G. Ceballos, eds.). Publicación especial. Asociación Mexicna de Mastozoología, A. C. Distrito Federal, México.

Lorenzo, C., T. Rioja, A. Carrillo, and F. A. Cervantes. 2008. Population fluctuations of *Lepues flavigularis* (Lagomorpha: Leporidae) at Tehuantepec Isthmus, Oaxaca, Mexico. Acta Zoológica Mexicana (n. s.) 24:207–220.

Lorenzo, C., A. Carrillo-Reyes, M. Gómez-Sánchez, A. Vélazquez, and E. Espinoza. 2011. Diet of the endangered Tehuantepec jackrabbit, *Lepus flavigularis*. Therya 2:67–76.

Lorenzo, C., D. E. Brown, S. Amirsultan, and M. García. 2014. Evolutionary history of the antelope jackrabbit, *Lepus alleni*. Journal of the Arizona-Nevada Academy of Sciences 45:70–75.

Lorenzo, C., T. Rioja-Paradela, and A. Carrillo-Reyes. 2015. State and knowledge and conservation of endangered and critical endangered lagomorphs worldwide. Therya 6:11–30.

Lorenzo, C., S. T. Álvarez-Castañeda, S. G. Pérez-Consuegra, and J. L. Patton. 2016. Revision of the Chiapan deer mouse, *Peromyscus zarhynchus*, with the description of a new species. Journal of Mammalogy 97:910–918.

Lorenzo, C., T. Rioja-Paradela, A. Carrillo-Reyes, M. de la Paz-Cuevas, and S. T. Álvarez-Castañeda. 2018a. *Sylvilagus mansuetus*. Pp. 145–147, *in* Pikas, Rabbits and Hares of the World (Smith, A., C. H. Johnston, P. C. Alves, and K. Hackländer, eds.). The Johns Hopkins University Press/IUCN/SSC. Baltimore, U.S.A.

Lorenzo, C., T. Rioja-Paradela, A. Carrillo-Reyes, M. de la Paz-Cuevas, F. A. Cervantes, and S. T. Álvarez-Castañeda. 2018b. *Lepus insularis*. Pp. 198–200, *in* Pikas, Rabbits and Hares of the World (Smith, A., C. H. Johnston, P. C. Alves, and K. Hackländer, eds.). The Johns Hopkins University Press/IUCN/SSC. Baltimore, U.S.A.

Lorenzo, C., J. Vázquez, L. Rodríguez-Martínez, A. Bautista, A. García-Méndez, and F. A. Cervantes. 2018c. *Sylvilagus cunicularius*. Pp. 131–135, *in* Pikas, Rabbits and Hares of the World (Smith, A., C. H. Johnston, P. C. Alves, and K. Hackländer, eds.). The Johns Hopkins University Press/IUCN/SSC. Baltimore, U.S.A.

Lorenzo, C., J. P. Ramírez-Silva, F. A. Cervantes, and R. Ferrera-Muro. 2018d. *Sylvilagus graysoni*. Pp. 142–144, *in* Pikas, Rabbits and Hares of the World (Smith, A., C. H. Johnston, P. C. Alves, and K. Hackländer, eds.). The Johns Hopkins University Press/IUCN/SSC. Baltimore, U.S.A.

Lorenzo, C., F. A. Cervantes, J. Vargas, and R. Ferrera-Muro. 2018e. *Sylvilagus insonus*. Pp. 144–145, *in* Pikas, Rabbits and Hares of the World (Smith, A., C. H. Johnston, P. C. Alves, and K. Hackländer, eds.). The Johns Hopkins University Press/IUCN/SSC. Baltimore, U.S.A.

Lorenzo, C., T. Rioja-Paradela, A. Carrillo-Reyes, E. C. Sáintiz-López, and J. Bolaños. 2018f. *Lepus flavigularis*. Pp. 191–193, *in* Pikas, Rabbits and Hares of the World (Smith, A., C. H. Johnston, P. C. Alves, and K. Hackländer, eds.). The Johns Hopkins University Press/IUCN/SSC. Baltimore, U.S.A.

Lorenzo, C., J. Bolaños-Citalán, D. Navarrete-Gutiérrez, J. A. Pérez-López, and L. Guevara. 2019b. In search of shrews of Chiapas: analysis of their distribution and conservation. Therya 10:121–129.

Lorenzo, C., J. E. Bolaños-Citalán, and O. G. Retana-Guiascón. 2020. Rediscovery of *Heteromys nelsoni* in its type locality after over a century. Mammalian 84:6–9.

Lotze, J., and S. Anderson 1979. *Procyon lotor*. Mammalian Species 119:1–8.

Loureiro, L. O., M. D. Engstrom, and B. K. Lim. 2020. Single nucleotide polymorphisms (SNPs) provide unprecedented resolution of species boundaries, phylogenetic relationships, and genetic diversity in the mastiff bats (*Molossus*). Molecular Phylogenetics and Evolution 143:106690.

Loughlin, T. R. 1997. Using the phylogeographic method to identify Steller sea lion stocks. Pp. 329–341, in Molecular genetics of marine mammals, incorporating the proceedings of a workshop on the analysis of genetic data to address problems of stock identity as related to management of marine mammals (Dizon, A., S. J. Chivers, and W. Perrin eds.). Society of Marine Mammalogy, Special Report No. 3.

Loughlin, T. R. 2009. Steller Sea Lion: Eumetopias jubatus. Pp 1107–1110, in Encyclopedia of Marine Mammals (Second Edition) (Perrin, W. F., B. Würsig, and J. G. M. Thewissen, eds.). Academic Press,

Loughlin, T. R., M. A. Perez, and R. L. Merrick. 1987. *Eumetopias jubatus*. Mammalian Species 283:1–7.

Loughlin, T. R., G. A. Antonelis, J. D. Baker, A. E. York, C. W. Fowler, R. L. DeLong, and H. W. Braham. 1994. Status of the northern fur seal population in the United States during 1992. Pp. 9–28, *in* Fur Seal Investigations 1992 (Sinclair, E. H., ed.). United States Department of Commerce, NOAA Technical Memorandum NMFS-AFSC 45:9–28.

Loughry, J., C. McDonough, and A. M. Abba. 2014. *Dasypus novemcinctus*. The IUCN Red List of Threatened Species 2014:e.T6290A47440785.

Lowery, G. H., Jr. 1974. The mammals of Louisiana and its adjacent waters. Louisiana State University Press, Baton Rouge, U.S.A.

Lowrey, C. 2016. *Neotamias palmeri*. The IUCN Red List of Threatened Species 2016:e.T21355A22267875.

Lowry, L. 2016a. *Histriophoca fasciata*. The IUCN Red List of Threatened Species 2016:e.T41670A45230946.

Lowry, L. 2016b. *Odobenus rosmarus*. The IUCN Red List of Threatened Species 2016:e.T15106A45228501.

Lowry, L. 2016c. *Phoca vitulina*. The IUCN Red List of Threatened Species 2016:e.T17013A45229114.

Lowry, L. 2016d. *Pusa hispida*. The IUCN Red List of Threatened Species 2016:e.T41672A45231341.

Lluch, B. C. 1965. Further notes on the biology of the manatee. Anales del Instituto Nacional de Investigaciones en Biologia Pesquera, México 1:405–419.

Luckett, W. P., and J.-L. Hartenberger. 1985 Evolutionary relationships among rodents: a multidisciplinary analysis. Plenum Press. New York, U.S.A.

Ludwing, D. R. 1984. *Microtus richardsoni*. Mammalian Species 223:1–6.

Lyon, M. W., Jr. 1903. Classification of the hares and their allies. Smithsonian Miscellaneous Collections 45:321–463.

Lyon, M. W., Jr. 1936. Mammals of Indiana. American Midland Naturalist 17:1–384.

MacCracken, J. G., D. W. Uresk, and R. M. Hansen. 1985. Habitat used by shrews in southeastern Montana. Northwest Science 59:24–27.

Macdonald, D. W., N. Yamaguchi, A. C. Kitchener, M. Daniels, K. Kilshaw, and D. Driscoll. 2010. The Scottish wildcat: On the way to cryptic extinction through hybridisation: past history, present problem, and future conservation. Pp. 471–491, *in* Biology and Conservation of Wild Felids (Macdonald D. W., and A. J. Loveridge, eds). Oxford University Press. Oxford, UK.

MacDonald, S. O., and J. A. Cook. 2009. Recent mammals of Alaska. University of Alaska Press. Fairbanks, U.S.A.

MacDonald, S. O., and C. Jones. 1987. *Ochotona collaris*. Mammalian Species 281:1–4.

MacClintock, D. 1970. Squirrels of North America. Van Nostrand Reinhold Company, New York, U.S.A.

Macêdo, R. H., and M. A. Mares. 1988. *Neotoma albigula*. Mammalian Species 310:1–7.

MacSwiney, G. M. C., B. Bolívar-Cimé, F. M. Clarke, and P. A. Racey. 2009a. Insectivorous bat activity at cenotes in the Yucatan Peninsula, Mexico. Acta Chiropterologica 11:139–147.

MacSwiney, G. M. C., S. Hernández-Betancourt, and R. Avila-Flores. 2009b. *Otonyctomys hatti*. Mammalian Species 825:1–5.

Mailliard, J. 1924. A new deer mouse (*Peromyscus slevini*) from the Gulf of California. Proceedings of the California Academy of Sciences 12:1219–1222.

Malaney, J. L., J. R. Demboski, and J. A. Cook. 2017. Integrative species delimitation of the widespread North American jumping mice (Zapodinae). Molecular Phylogenetics and Evolution 114:137–152.

Maldonado, J. E., C. Vilá, and R. K. Wayne. 2001. Tripartite genetic subdivisions in the Ornate Shrew (*Sorex ornatus*). Molecular Ecology 10:127–147.

Maldonado, J. E., F. Hertel, and C. Vilá. 2004. Discordant patterns of morphological variation in genetically divergent populations of Ornate Shrews (*Sorex ornatus*). Journal of Mammalogy 85:886–896.

Maldonado, J. E., S. Young, L. H. Simons, S. Stone, L. D. Parker, and J. Ortega. 2015. Conservation genetics and phylogeny of the Arizona shrew in the "Sky Islands" of the Southwestern United States. Therya 6:401–420.

Manning, R. W. 1993. Systematics and evolutionary relationships of the long-eared myotis, *Myotis evotis* (Chiroptera: Vespertilionidae). Special Publications, The Museum, Texas Tech University 37:1–58.

Manning, R. W., and J. K. Jones, Jr. 1988a. *Perognathus fasciatus*. Mammalian Species 303:1–4.

Manning, R. W., and J. K. Jones, Jr. 1988b. A new subspecies of fringed *Myotis*, *Myotis thysanodes*, from the northwestern coast of the United States. Occasional Papers of the Museum, Texas Tech University 123:1–6.

Manning, R. W., and J. K. Jones, Jr. 1989. *Myotis evotis*. Mammalian Species 329:1–5.

Manning, R. W., F. D. Yancey II, and C. Jones. 1996. Nongeographic variation and natural history of two sympatric species of pocket mice, *Chaetodipus nelsoni* and *Chaetodipus eremicus,* from Brewster County, Texas. Pp. 191–195, *in* Contributions in mammalogy: a memorial volume honoring Dr. J. Knox Jones, Jr. (Genoways, H. H., and R. J. Baker, eds.). Museum of Texas Tech University. Lubbock, U.S.A.

Manning, R. W., M. R. Heaney, M. Sagot, and R. J. Baker. 2014. Noteworthy record of Crawford's desert shrew (*Notiosorex crawfordi*) from southern Nevada. The Southwestern Naturalist 59:145–147.

Manning, T. H. 1971. Geographical variation in the polar bear *Ursus maritimus* Phipps. Canadian Wildlife Service. Ottawa, Canada.

Mantilla-Meluk, H. 2014. Defining Species and Species Boundaries in Uroderma (Chiroptera: Phyllostomidae) with a Description of a New Species. Occasional Papers, The Museum, Texas Tech University 325:1–25.

Mantilla-Meluk, H., and R. J. Baker. 2010. New species of *Anoura* (Chiroptera: Phyllostomidae) from Colombia, with systematic remarks and notes on the distribution of the *A. geoffroyi* complex. Occasional Papers, The Museum, Texas Tech University 292:1–24.

Mantilla-Meluk, H., and J. Muñoz-Garay. 2014. Biogeography and taxonomic status of *Myotis keaysi pilosatibialis* LaVal 1973 (Chiroptera: Vespertilionidae). Zootaxa 3793:60–70.

Mantooth, S. J., and T. L. Best. 2005a. *Chaetodipus eremicus*. Mammalian Species 768:1–3.

Mantooth, S. J., and T. L. Best. 2005b. *Chaetodipus penicillatus*. Mammalian Species 767:1–7.

Mantooth, S. J., C. Jones, and R. D. Bradley. 2000. Molecular systematics of *Dipodomys elator* (Rodentia: Heteromyidae) and its phylogeographic implications. Journal of Mammalogy 81:885–894.

Mantooth, S. J., D. J. Hafner, R. W. Bryson, and B. R. Riddle. 2013. Phylogeographic diversification of antelope squirrels (*Ammospermophilus*) across North American deserts. Biological Journal of the Linnean Society 109:949–967.

Manville, R. H., and P. G. Favour, Jr. 1960. Southern distribution of the Atlantic walrus. Journal of Mammalogy 41:499–503.

Manville, R. H., and S. P. Young. 1965. Distribution of Alaskan mammals. United States Bureau of Sport Fisheries and Wildlife, Circular 211:1–45.

Mares, M. A., R. A. Ojeda, and R. M. Barquez. 1989. Guide to the mammals of Salta Province, Argentina. University of Oklahoma Press. Norman, U.S.A.

Marchán–Rivadeneira, M. R., P. A. Larsen, C. J. Phillips, R. E. Strauss, and R. J. Baker. 2012. On the association between environmental gradients and skull size variation in the great fruit-eating bat, *Artibeus lituratus* (Chiroptera: Phyllostomidae). Biological Journal of the Linnean Society 105:623–634.

Marcot, J. D. 2007. Molecular phylogeny of terrestrial artiodactyls. conflicts and resolution. Pp. 4–18, *in* The Evolution of Artiodactyls (Prothero, D., and S. E. Foss, eds). Johns Hopkins University Press. Baltimore, U.S.A.

Marcy, A. E., S. Fendorf, J. L. Patton, and E. A. Hadly. 2013. Morphological adaptations for digging and climate-impacted soil properties define Pocket Gopher (*Thomomys* spp.) distributions. PLoS ONE 8(5):e64935.

Marshall, L. G. 1978. *Chironectes minimus*. Mammalian Species 109:1–6.

Marines-Macías, T., P. Colunga-Salas, and L. León-Paniagua. 2021. *Reithrodontomys microdon* (Rodentia: Cricetidae). Mammalian Species 54(1020):1–8.

Martin, G. M. 2016. *Marmosa mexicana*. The IUCN Red List of Threatened Species 2016:e.T40504A22173751.

Martin, G. M. 2017. *Tlacuatzin canescens*. The IUCN Red List of Threatened Species 2017:e.T12813A22177663.

Martin, C. O., and D. J. Schmidly. 1982. Taxonomic review of the pallid bat, *Antrozous pallidus* (Le Conte). Special Publications, The Museum, Texas Tech University 18:1–48.

Martínez, L., and B. Villa. 1938. Contribuciones al conocimiento de los murciélagos de México I. Anales del Instituto de Biología, Universidad Nacional Autónoma de México 9:339–360.

Martínez-Borrego, D., E. Arellano, F. X. González-Cózatl, and D. S. Rogers. 2020. *Reithrodontomys mexicanus* (Rodentia: Cricetidae). Mammalian Species 52:114–124.

Martínez-Borrego, D., E. Arellano, F. X. González-Cózatl, I. Castro-Arellano, L. León-Paniagua, and D. S. Rogers. 2022. Molecular systematics of the *Reithrodontomys tenuirostris* group (Rodentia: Cricetidae) highlighting the *Reithrodontomys microdon* species complex. Journal of Mammalogy 103:29–44.

Martínez-Chapital, S. T., G. D. Schnell, C. Sánchez-Hernández, and M. L. Romero-Almaraz. 2017. *Sigmodon mascotensis* (Rodentia: Cricetidae). Mammalian Species 49:109–118.

Masseti, M., and D. Mertzanidou. 2008. *Dama dama*. The IUCN Red List of Threatened Species 2008:e.T42188A10656554.

Mathis, V. L., M. S. Hafner, D. J. Hafner, and J. W. Demastes. 2013a. Resurrection and redescription of the pocket gopher *Thomomys sheldoni* from the Sierra Madre Occidental of Mexico. Journal of Mammalogy 94:544–560.

Mathis, V. L., M. S. Hafner, D. J. Hafner, and J. W. Demastes. 2013b. *Thomomys nayarensis*, a new species of pocket gopher from the Sierra del Nayar, Nayarit, Mexico. Journal of Mammalogy 94:983–994.

Mathis, V. L., M. S. Hafner, and D. J. Hafner. 2014. Evolution and phylogeography of the *Thomomys umbrinus* species complex (Rodentia: Geomyidae). Journal of Mammalogy 95:754–771.

Matocq, M. D. 2002a. Morphological and molecular analysis of a contact zone in the *Neotoma fuscipes* species complex. Journal of Mammalogy 83:866–883.

Matocq, M. D. 2002b. Phylogeographical structure and regional history of the dusky-footed woodrat, *Neotoma fuscipes*. Molecular Ecology 11: 229–242.

Matson, J. O. 1975. *Myotis planiceps*. Mammalian Species 60:1–2.

Matson, J. O. 1980. The status of banner-tailed kangaroo rats, genus *Dipodomys,* from central Mexico. Journal of Mammalogy 61:563–566.

Matson, J. 2018. *Sorex pribilofensis*. The IUCN Red List of Threatened Species 2018:e.T20391A22314622.

Matson, J. 2020. *Microtus guatemalensis*. The IUCN Red List of Threatened Species 2020:e.T13432A22350117.

Matson, J. O., and R. H. Baker. 1986. Mammals of Zacatecas. Special Publications, The Museum, Texas Tech University 24:1–88.

Matson, J., and P. C. de Grammont. 2018. *Sorex milleri*. The IUCN Red List of Threatened Species 2018:e.T20397A22316066.

Matson, J. O., and N. Ordóñez-Garza. 2017. The taxonomic status of Long-tailed shrews (Mammalia: genus *Sorex*) from Nuclear Central America. Zootaxa 4236:461–483.

Matson, J. O., and D. R. Patten. 1975. Notes on some bats from the state of Zacatecas, Mexico. Contributions in Science, Los Angeles County Museum of Natural History 263:1–12.

Matson, J., N. Woodman, and F. Reid. 2016a. *Sorex vagrans*. The IUCN Red List of Threatened Species 2016:e.T41425A115186125.

Matson, J., N. Woodman, I. Castro-Arellano, and P. C. de Grammont. 2016b. *Sorex monticola*. The IUCN Red List of Threatened Species 2016:e. T41405A194054902.

Matson, J., N. Woodman, I. Castro-Arellano, and P. C. de Grammont. 2016c. *Scalopus aquaticus*. The IUCN Red List of Threatened Species 2016:e.T41471A115188304.

Matson, J., N. Woodman, I. Castro-Arellano, and P. C. de Grammont. 2016d. *Scapanus latimanus*. The IUCN Red List of Threatened Species 2016:e.T41473A115188559.

Matson, J., P. C. de Grammont, and I. Castro-Arellano. 2017a. *Sorex emarginatus*. The IUCN Red List of Threatened Species 2017:e. T41395A22312750.

Matson, J., N. Woodman, I. Castro-Arellano, and P. C. de Grammont. 2017b. *Sorex mediopua*. The IUCN Red List of Threatened Species 2017:e. T136656A22318332.

Matson, J., N. Woodman, I. Castro-Arellano, and P. C. de Grammont. 2017c. *Sorex ventralis. The IUCN Red List of Threatened Species* 2017:e. T41426A22315839.

Matson, J., N. Woodman, I. Castro-Arellano, and P. C. de Grammont. 2017d. *Sorex veraecrucis*. The IUCN Red List of Threatened Species 2017:e. T136811A22319172.

Matson, J., N. Woodman, I. Castro-Arellano, and P. C. de Grammont. 2017e. *Sorex veraepacis*. The IUCN Red List of Threatened Species 2017:e. T41427A22315421.

Matson, J., A. D. Cuarón, and P. C. de Grammont. 2018. *Cryptotis nelsoni*. The IUCN Red List of Threatened Species 2018:e.T136389A22284939.

Matson, J., N. Woodman, I. Castro-Arellano, and P. C. de Grammont. 2019. *Sorex ixtlanensis*. The IUCN Red List of Threatened Species 2019:e. T136339A22319335.

Mascarello, J. T. 1978. Chromosomal, biochemical, mensural, penile, and cranial variation in desert woodrats (*Neotoma lepida*). Journal of Mammalogy 59:477–495.

Mattson, D. J. 1998. Diet and morphology of extant and recently extinct northern bears. Ursus 10:479–96.

Mattson, D. J., and T. Merrill. 2002. Extirpations of grizzly bears in the contiguous United States, 1850–2000. Conservation Biology 16:1123–1136.

Matyushkin, E. N. 1979. Rysi Golarktiki. Sbornik Trudov Zoologicheskovo Muzeya MGU 18:76–161.

Mayer, W. V. 1952. The hair of California mammals with keys to the dorsal guard hairs of California mammals. The American Midland Naturalist 48:480–512.

Mayer, J. J., and R. M. Wetzel. 1987. *Tayassu pecari*. Mammalian Species 293:1–7.

Mayer, J. J., and I. L. Brisbin, Jr. 2009. Wild Pigs: Biology, Damage, Control Techniques and Management Archived 24 October 2014 at the Wayback Machine, Savannah River National Laboratory Aiken, South Carolina, SRNL-RP-2009-00869.

Mazrimas, J. A., and F. T. Hatch. 1972. A possible relationship between satellite DNA and the evolution of kangaroo rat species (genus *Dipodomys*). Nature New Biology 240:102–105.

McAllister, J. A., and R. S. Hoffmann. 1988. *Phenacomys intermedius*. Mammalian Species 305:1–8.

McBee, K., and R. J. Baker. 1982. *Dasypus novemcinctus*. Mammalian Species 162:1–9.

Mccabe, R. E., and R. T. Mccabe. 1984. Of slings and arrows: an historical retrospection. Pp. 19–72, *in* White-tailed deer: ecology and management (L. K. Halls, ed.). Stackpole Books. Harrisburg, U.S.A.

McCarty, R. 1975. *Onychomys torridus*. Mammalian Species 59:1–5.

McCarty, R. 1978. *Onychomys leucogaster*. Mammalian Species 87:1–6.

McCay, T. S. 2001. *Blarina carolinensis*. Mammalian Species 673:1–7.

McCarthy, T. J. y Pérez, S. G. (2006). Land and freshwater mammals of Guatemala: faunal documentation and diversity. En E. Cano (Ed.), Biodiversidad de Guate mala (pp. 625–674). Guatemala, Guatemala: Universidad del Valle de Guatemala.

McCleery, R., C. Faulhaber, and A. Sovie. 2018. *Sylvilagus palustris*. Pp. 152–154, *in* Pikas, Rabbits and Hares of the World (Smith, A., C. H. Johnston, P. C. Alves, and K. Hackländer, eds.). The Johns Hopkins University Press/IUCN/SSC. Baltimore, U.S.A.

McClellan, D. A., and D. S. Rogers. 1997. *Peromyscus zarhynchus*. Mammalian Species 562:1–3.

McDaniel, V. R., M. J. Harvey, R. Tumlison, and K. N. Paige. 1982. Status of the small-footed bat, *Myotis leibii leibii,* in the southern Ozarks. Proceedings of the Arkansas Academy of Sciences 36:92–94.

McDonald, J. N. 1981. North American bison: their classification and evolution. University of California Press. Berkeley, U.S.A.

McDonald, R. A., A. V. Abramov, M. Stubbe, J. Herrero, T. Maran, A. Tikhonov, P. Cavallini, A. Kranz, G. Giannatos, B. Krytufek, and F. Reid. 2019. *Mustela nivalis*. The IUCN Red List of Threatened Species 2019:e.T70207409A147993366.

McDonough, M. M., L. K. Ammerman, R. M. Timm, H. H. Genoways, P. A. Larsen, and R. J. Baker. 2008. Speciation within bonneted bats (Genus *Eumops*): the complexity of morphological, mitochondrial, and nuclear data sets in systematics. Journal of Mammalogy 89:1306–1315.
McDonough, M. M., A. W. Ferguson, R. C. Dowler, M. E. Gompper, J. E. Maldonado. 2022. Phylogenomic systematics of the spotted skunks (Carnivora, Mephitidae, Spilogale): Additional species diversity and Pleistocene climate change as a major driver of diversification. Molecular Phylogenetics and Evolution 167:107266
McFadden, K. W., and S. Meiri. 2013. Dwarfism in insular carnivores: a case study of the Pygmy Raccoon. Journal of Zoology 289:213–221.
McGowan, C., W. S. Davidson, and L. A. Howes. 1999. Genetic analysis of an endangered Pine Marten (*Martes americana*) population from Newfoundland using randomly amplified polymorphic DNA markers. Canadian Journal of Zoology 77:661–666.
McGrew, J. C. 1979. *Vulpes macrotis*. Mammalian Species 123:1–6.
McKenna, M., and S. K. Bell. 1997. Classification of mammals above species level. Columbia University Press. New York, U.S.A.
McKnight, M. L. 2005. Phylogeny of the *Perognathus longimembris* species group based on mitochondrial cytochrome–*b*: how many species? Journal of Mammalogy 86:826–832.
McLaughlin, C. A. 1984. Protrogomorph, Sciuromorph, Castorimorph, Myomorph (Geomyoid, Anomaluroid, Pedetoid, and Ctenodactyloid) rodents. Pp. 267–288, *in* Orders and families of recent mammals of the World (Anderson, S., and J. K. Jones, Jr., eds.). John Wiley & Sons. New York, U.S.A.
McLean, B. S. 2018. *Urocitellus parryii* (Rodentia: Sciuridae). Mammalian Species 50:84–99.
McLean, B. S., D. J. Jackson, and J. A. Cook. 2016. Rapid divergence and gene flow at high latitudes shape the history of Holarctic ground squirrels (*Urocitellus*). Molecular Phylogenetics and Evolution 102:174–188.
McLellan, B. N., M. F. Proctor, D. Huber, and S. Michel. 2017. *Ursus arctos*. The IUCN Red List of Threatened Species 2017:e.T41688A121229971.
McLellan, L. J. 1984. A morphometric analysis of *Carollia* (Chiroptera, Phyllostomidae). American Museum Novitates 2791:1–35.
McLeod, B. A., T. R. Frasier, and Z. Lucas. 2014. Assessment of the extirpated Maritimes Walrus using morphological and ancient DNA analyses. PLoS ONE 9:e.99569.
McManus, J. J. 1942. *Didelphis virginiana*. Mammalian Species 40:1–6.
McManus, J. J. 1974. *Didelphis virginiana*. Mammalian Species 40:1–6.
McTaggart Cowan, I. 1938. Geographic distribution of color phases of the red fox and black bear in the Pacific Northwest. Journal of Mammalogy 19:202–206.
Mead, R. A. 1968. Reproduction in western forms of the spotted skunk (genus Spilogale). Journal of Mammalogy 49:373–390
Mead, J. I., A. E. Spiess, and K. D. Sobolik. 2000. Skeleton of extinct North American Sea Mink (*Mustela macrodon*). Quaternary Research 53:247–262.
Meade, L. 1992. New distributional records for selected species Kentucky mammals. Transactions of the Kentucky Academy of Science 53:127–132.
Meagher, M. 1986. *Bison bison*. Mammalian Species 266:1–8.
Mearns, E. A. 1890. Description of supposed new species and subspecies of mammals from Arizona. Bulletin of the American Museum of Natural History 2:277–307.
Mearns, E. A. 1896. Preliminary diagnoses of new mammals from the mexican border of the United States. Proceedings of the United States Natural History Museum 18:443–447.
Mearns, E. A. 1907. Mammals of the Mexican boundary of the United States. Pt. I, families Didelphidae to Muridae. Bulletin United States National Museum 56:1–530.
Mech, L. D. 1974. *Canis lupus*. Mammalian Species 37:1–6.
Medellín, R. A. 1988. Prey of *Chrotopterus auritus,* with notes on feeding behavior. Journal of Mammalogy 69:841–844.
Medellín, R. A. 1989. *Chrotopterus auritus*. Mammalian Species 343:1–5.
Medellín, R. 2016a. *Leptonycteris nivalis*. The IUCN Red List of Threatened Species 2016:e.T11697A22126172.
Medellín, R. 2016b. *Leptonycteris yerbabuenae*. The IUCN Red List of Threatened Species 2016:e.T136659A21988965.
Medellín, R. A., and H. T. Arita. 1989. *Tonatia evotis* and *Tonatia silvicola*. Mammalian Species 334:1–5.
Medellín, R. A., D. E. Wilson, and L. Navarro. 1985. *Micronycteris brachyotis*. Mammalian Species 251:1–4.
Medellín, R. A., G. Ceballos, and H. Zarza. 1989. *Spilogale pygmaea*. Mammalian Species 600:1–3.
Medellin, R. A., G. Ceballos, and H. Zarza. 1998. Spilogale pygmaea. Mammalian Species 600:1–3
Medellín, R. A., M. Equihua, and M. A. Amín. 2000. Bat diversity and abundance as indicator of disturbance in neotropical rainforest. Conservation Biology 14:1666–1675.
Medellín, R. A., G. Cancino, M. Apamar Clemente, and R. O. Guerrero. 1992. Noteworthy records of three mammals from Mexico. The Southwestern Naturalist 37:427–429.
Medellín, R. A., H. T. Arita, and O. Sanchez. 1997. Identificacion de los murcielagos de México, clave de campo. Asociacion Mexicana de Mastozoologia, A. C., Mexico, Publicaciones Especiales 2:1–83.
Medellín, R. A., A. L. Gardner, and J. M. Aranda. 1998. The taxonomic status of the Yucatán brown brocket, *Mazama pandora* (Mammalia: Cervidae). Proceedings of the Biological Society of Washington 111:1–14.
Medellín, R. A., M. Equihua, and M. Amín. 2000. Bat diversity and abundance as indicators of disturbance in Neotropical rainforests. Conservation Biology 14:1666–1675.
Medellín, R. A., H. T. Arita, and O. Sánchez. 2008. Identificación de los murciélagos de México, clave de campo, segunda edición. Instituto de Ecología, Universidad Nacional Autónoma de México – Comisión para el Conocimiento y Uso de la Biodiversidad. Distrito Federal, México.
Meiri, M., P. Kosintsev, K. Conroy, S. Meiri, I. Barnes, and A. Lister. 2018. Subspecies dynamics in space and time: a study of the red deer complex using ancient and modern DNA and morphology. Journal of Biogeography 45:367–380.
Meisfjord, J., and R. C. Sundt. 1996. Genetic variation between populations of the harp seal, *Phoca groenlandica*. Journal of Marine Science 53:89–95.
Mejenes-López, S. de M. A., M. Hernández-Bautista, J. Barragán-Torres, and J. Pacheco-Rodríguez. 2010. Los mamíferos en el estado de Hidalgo, México. Therya 1:161–188.

Mendoza, E., T. Fuller, H. A. Thomassen, W. Buermann, D. Ramírez-Mejía, and T. B. Smith. 2013. A preliminary assessment of the effectiveness of the Mesoamerican Biological Corridor for protecting potential Baird's tapir (Tapirus bairdii) habitat in southern Mexico. Integrative Zoology 8:35–47.

Mengak, M. T. 1987. Abundance and distribution of shrews in western South Carolina. Brimleyana 13:63–66.

Melo-Ferreira, J., F. A. Seixas, E. Cheng, L. S. Mills, and P. C. Alves. 2014. The hidden history of the snowshoe hare, *Lepus americanus*: extensive mitochondrial DNA introgression inferred from multilocus genetic variation. Molecular Ecology 23: 4617–4630.

Menu, H. 1984. Révision du statut de *Pipistrellus subflavus* (F. Cuvier, 1832). Proposition d'un taxon generique noveau: *Perimyotis nov. gen.* Mammalia 48:409–416.

Mercure, A., K. Ralls, K. P. Koepfli, and R. K. Wayne. 1993. Genetic subdivisions among small canids: mitochondrial DNA, differentiation of swift, kit, and arctic foxes. Evolution 47:1313–1328.

Merriam, C. H. 1884. Mammals of the Adirondack region, northeastern New York. Vol. I and II. Transactions Linnaean Society, New York. New York, U.S.A.

Merriam, C. H. 1889a. Preliminary revision of North American pocket mice (genera *Perognathus* et *Cricetodipus* auct.) with description of new species and subspecies and a key to the known forms. North American Fauna 1:1–29.

Merriam, C. H. 1889b. Revision of the North American Pocket Mice. North American Fauna 1:1–29.

Merriam, C. H. 1889c. Description of a new spermophile from southern California. North American Fauna 2:15–16.

Merriam, C. H. 1890a. Descriptions of three new kangaroo rats, with remarks on the identity of *Dipodomys ordii* of Woodhouse. North American Fauna 4:41–49.

Merriam, C. H. 1890b. Description of a new prairie dog from Wyoming. North American Fauna 4:33–35.

Merriam, C. H. 1890c. Description of twenty-six new species of North American mammals. North American Fauna 4:1–55.

Merriam, C. H. 1891. Results of a biological reconnaissance of south-central Idaho. North American Fauna 5:1–416.

Merriam, C. H. 1892a. Description of a new prairie dog (*Cynomys mexicanus*) from Mexico. Proceedings of the Biological Society of Washington 7:157–158.

Merriam, C. H. 1892b. Description of a new genus and species of murine rodent *(Xenomys nelsoni)* from the state of Colima, western Mexico. Proceedings of the Biological Society of Washington 7:159–163.

Merriam, C. H. 1893. Descriptions of eight new ground squirrels of the genera *Spermophilus* and *Tamias* from California, Texas, and Mexico. Proceedings of the Biological Society of Washington 8:129–138.

Merriam, C. H. 1894a. A new subfamily of murine rodents -the Neotominae- with description of a new genus and species and a synopsis of the known forms. Proceedings of the Academy Natural Sciences of Philadelphia 14:225–252.

Merriam, C. H. 1894b. Abstract of a study of the American wood rats, with descriptions of fourteen new species and subspecies of the genus *Neotoma*. Proceedings of the Biological Survey of Washington 9:117–128.

Merriam, C. H. 1894c. Preliminary descriptions of eleven new kangaroo rats of the genera *Dipodomys* and *Perodipus*. Proceedings of the Biological Society of Washington 9:109–116.

Merriam, C. H. 1895b. Revision of the shrews of the American genera *Blarina* and *Notiosorex*. North American Fauna 10:1–34.

Merriam, C. H. 1897. Notes on the chipmunks of the genus *Eutamias* occurring west of the east base of the Cascade-Sierra system, with descriptions of new forms. Proceedings of the Biological Society of Washington 11:189–212.

Merriam, C. H. 1900. Descriptions of twenty-six new mammals from Alaska and British North America. Proceedings of the Washington Academy of Sciences 2:13–30.

Merriam, C. H. 1902. Twenty new pocket mice *(Heteromys* and *Liomys)* from Mexico. Proceedings of the Biological Society of Washington 15:41–50.

Merriam, C. H. 1903. Four new mammals, including a new genus (*Teanopus*), from Mexico. Proceedings of the Biological Society of Washington 16:79–82.

Merriam, C. H. 1904a. Two new squirrels of the *aberti* group. Proceedings of the Biological Society of Washington 17:129–130.

Merriam, C. H. 1904b. New and little known kangaroo rats of the genus *Perodipus.* Proceedings of the Biological Society of Washington, 17:139–145.

Merriam, C. H. 1918. Review of the grizzly and big brown bears of North America (genus *Ursus).* North American Fauna 41:1136.

Merritt, J. F. 1978. *Peromyscus californicus*. Mammalian Species 85:1–6.

Merritt, J. F. 1981. *Clethrionomys gapperi*. Mammalian Species 146:1–9.

Michener, G. R., and J. W. Koeppl. 1985. *Spermophilus richardsonii.* Mammalian Species 243:1–8.

Mies, R., A. Kurta, and D. G. King. 1996. *Eptesicus furinalis*. Mammalian Species 526:1–7.

Miller, B., and B. Rodriguez. 2016a. *Myotis elegans*. The IUCN Red List of Threatened Species 2016:e.T14156A115121563.

Miller, B., and B. Rodriguez. 2016b. *Lasiurus intermedius*. The IUCN Red List of Threatened Species 2016:e.T11352A115101697.

Miller, B., F. Reid, J. Arroyo-Cabrales, A. D. Cuarón, and P. C. de Grammont. 2015a. *Carollia sowelli*. The IUCN Red List of Threatened Species 2015:e.T136268A22003903.

Miller, B., F. Reid, J. Arroyo-Cabrales, A. D. Cuarón, and P. C. de Grammont. 2015b. *Carollia subrufa*. The IUCN Red List of Threatened Species 2015:e.T3906A22133926.

Miller, B., F. Reid, J. Arroyo-Cabrales, A. D. Cuarón, and P. C. de Grammont. 2015c. *Trachops cirrhosus*. The IUCN Red List of Threatened Species 2015:e.T22029A22042903.

Miller, B., F. Reid, J. Arroyo-Cabrales, A. D. Cuarón, and P. C. de Grammont. 2015d. *Dermanura phaeotis*. The IUCN Red List of Threatened Species 2015:e.T83683287A21997769.

Miller, B., F. Reid, J. Arroyo-Cabrales, A. D. Cuarón, and P. C. de Grammont. 2016a. *Eumops underwoodi*. The IUCN Red List of Threatened Species 2016:e.T8248A22025754.

Miller, B., F. Reid, J. Arroyo-Cabrales, A. D. Cuarón, and P. C. de Grammont. 2016b. *Hylonycteris underwoodi*. The IUCN Red List of Threatened Species 2016:e.T10598A22036808.

Miller, B., F. Reid, J. Arroyo-Cabrales, A. D. Cuarón, and P. C. de Grammont. 2016c. *Glossophaga commissarisi*. The IUCN Red List of Threatened Species 2016:e.T9273A22108801.

Miller, B., F. Reid, J. Arroyo-Cabrales, A. D. Cuarón, and P. C. de Grammont. 2016d. *Glossophaga leachii*. The IUCN Red List of Threatened Species 2016:e.T9274A128959800.

Miller, B., F. Reid, J. Arroyo-Cabrales, A. D. Cuarón, and P. C. de Grammont. 2016e. *Artibeus jamaicensis*. The IUCN Red List of Threatened Species 2016:e.T88109731A21995883.

Miller, B., F. Reid, J. Arroyo-Cabrales, A. D. Cuarón, and P. C. de Grammont. 2016f. *Centurio senex*. The IUCN Red List of Threatened Species 2016:e.T4133A22009493.

Miller, B., F. Reid, J. Arroyo-Cabrales, A. D. Cuarón, and P. C. de Grammont. 2016g. *Rhogeessa tumida*. The IUCN Red List of Threatened Species 2016:e.T19685A22006890.

Miller, B., F. Reid, J. Arroyo-Cabrales, A. D. Cuarón, and P. C. de Grammont. 2016h. *Eptesicus fuscus*. The IUCN Red List of Threatened Species 2016:e.T7928A22118197.

Miller, B., F. Reid, J. Arroyo-Cabrales, A. D. Cuarón, and P. C. de Grammont. 2016i. *Molossus sinaloae*. The IUCN Red List of Threatened Species 2016:e.T13650A22106433.

Miller, F. L., R. H. Russell, and A. Gunn. 1977. Distributions, movements, and numbers of Peary caribou and muskoxen on western Queen Elizabeth Islands, Northwest Territories, 1972–74. Canadian Wildlife Service Report Series 40:1–55.

Miller, G. S., Jr. 1896. The genera and subgenera of voles and lemmings. North American Fauna 12:1–85.

Miller, G. S., Jr. 1897. Revision of the North American bats of the family Vespertilionidae. North American Fauna 13:1–135.

Miller, G. S., Jr. 1900. Note on *Micronycteris brachyotis* (Dobson) and *M. microtis* Miller. Proceedings of the Biological Society of Washington 13:154–155.

Miller G. S., Jr. 1902. Twenty new American bats. Proceedings of the Academy Natural Scieces Philadelphia 54:389–412.

Miller, G. S., Jr. 1907. The families and genera of bats. Bulletin of the United States National Museum 57:1–282.

Miller, G. S., Jr. 1912a. List of North American land mammals in the United States National Museum, 1911. Bulletin of the United States Natural Museum 79:1–455.

Miller, G. S. 1912b. Catalogue of the mammals of Western Europe (Europe exclusive of Russia) in the collection of the British Museum. William Clowes and Sons. London, United Kingdom.

Miller, G. S., Jr. 1914. The generic name of the collared peccaries. Proceedings of the Biological Society of Washington 27:215.

Miller, G. S., and R. Kellogg. 1955. List of North American Recent Mammals. Bulletin of the United States Natural Museum 205:1–954.

Miller, J. S., Jr. 1913. Revision of the bats of the genus *Glossophaga*. Proceeding of the United States National Museum 46:413–429.

Miller, J. S., Jr., and G. M. Allen. 1928. The American bats of the genus *Myotis* and *Pizonyx*. Bulletin of the United States Natural Museum 144:1–209.

Millions, D. G., and B. J. Swanson. 2007. Impact of natural and artificial barriers to dispersal on the population structure of bobcats. Journal of Wildlife Management 71:96–102.

Mills, L., and A. T. Smith. 2019. *Lepus americanus*. The IUCN Red List of Threatened Species 2019:e.T41273A45185466.

Milner, J., C. Jones, and J. K. Jones, Jr. 1990. *Nyctinomops macrotis*. Mammalian Species 351:1–4.

Miranda, F., D. A. Meritt, D. G. Tirira, and M. Arteaga. 2014. *Cyclopes didactylus*. The IUCN Red List of Threatened Species 2014:e.T6019A47440020.

Miranda, F. R., D. M. Casali, F. A. Perini, F. A. Machado, and F. R. Santos. 2018. Taxonomic review of the genus *Cyclopes* Gray, 1821 (Xenarthra: Pilosa), with the revalidation and description of new species. Zoological Journal of the Linnean Society 183:687–721.

Modi, W. S. 1986. Karyotypic differentiation among two sibling species pairs of New World microtine rodents. Journal of Mammalogy 67:159–165.

Modi, W. S. 1987. Phylogenetic analyses of chromosomal banding patterns among the Nearctic Arvicolidae (Mammalia: Rodentia). Systematic Zoology 36:109–136.

Modi, W. S. 1996. Phylogenetic history of LINE-1 among arvicolid rodents. Molecular Biology and Evolution 13:633–641.

Modi, W. S., and M. R. Lee. 1984. Systematic implications of chromosomal banding analyses of populations of *Peromyscus truei* (Rodentia: Muridae). Proceedings of the Biological Society of Washington 97:716–723.

Moehlman, P. D., F. Kebede, and H. Yohannes. 2015. *Equus africanus*. The IUCN Red List of Threatened Species 2015:e.T7949A45170994.

Moehrenschlager, A., and M. Sovada. 2016. *Vulpes velox*. The IUCN Red List of Threatened Species 2016:e.T23059A57629306.

Molinari, J., E. E. Guitiérrez, and B. K. Lim. 2023. Systematics and biogeography of *Anoura cultrata* (Mammalia, Chiroptera, Phyllostomidae): a morphometric, niche modeling, and genetic perspective, with a taxonomic reappraisal of the genus. Zootaxa 5297:151–188.

Montiel-Parra, G., A. L. Carlos-Delgado, R. Paredes-León, and T. M. Pérez. 2019. Epizoic arthropods of the Mexican Shrew, *Sorex oreopolus* (Mammalia: Soricidae). Therya 10:33–37.

Monk, R. R., and J. K. Jones, Jr. 1996. *Perognathus flavescens*. Mammalian Species 525:1–4.

Monroy-Gamboa, A. G. 2021. The ghost mammals from Mexico and their implications. Therya 12:477–486.

Monroy-Gamboa, A. G., S. T. Álvarez-Castañeda, and A. L. Trujano-Álvarez. Accepted. *Tylomys nudicaudus*. Mammalian Species.

Monson, G., and L. Sumner. 1980. The desert bighorn: its life history, ecology, and management. University of Arizona Press. Tucson, U. S. A.

Montgelard, C., F. M. Catzeflis, and E. Douzery. 1997. Phylogenetic relationships of artiodactyls and cetaceans as deduced from the comparison of cytochrome b and 12S rRNA mitochondrial sequences. Molecular Biological Evolution 14:550–559.

Moore, C. M., and P. W. Collins. 1995. *Urocyon littoralis*. Mammalian Species 489:1–7.

Moore, C. M., and L. L. Janacek. 1990. Genic relationships among North American *Microtus* (Mammalia: Rodentia). Annals of Carnegie Museum 59:249–259.

Moore, J. C. 1953. Distribution of marine mammals to Florida waters. The American Midland Naturalist 49:117–158.

Moore, J. C. 1960. The relationships of the gray squirrel; *Sciurus carolinensis*, to its nearest relatives. Proceedings of the Annual Conference of the Southeastern Association of Game and Fish Commissioners 13:356–363.

Moore, S. E., and E. I. Barrowclough. 1984. Incidental sighting of a ribbon seal (*Phoca fasciata*) in the western Beaufort Sea. Arctic 37:290.

Morales, J. C., and M. D. Engstrom. 1989. Morphological variation in the painted spiny pocket mouse, *Liomys pictus* (Family Heteromyidae), from Colima and southern Jalisco, México. Royal Ontario Museum, Life Sciences, Occasional Papers 38:1–16.

Morales, A. E., and B. C. Carstens. 2018. Evidence that Myotis lucifugus "Subspecies" are five nonsister species, despite gene flow. Systematic Biology 67:756–769

Morales-Malacara, J. B., and R. López-Wilchis. 1990. Epizoic fauna of *Plecotus mexicanus* (Chiroptera: Vespertilionidae) in Tlaxcala, Mexico. Journal of Medical Entomology 27:440–445.
Morales-Malacara, J. B., and J. Juste. 2002. Two new species of the genus Periglischrus (Acari: Mesostigmata: Spinturnicidae) on two bat species of the genus Tonatia (Chiroptera: Phyllostomidae) from southeastern Mexico, with additional data from Panama. Journal of Medical Entomology 2:398–311.
Moreno-Valdez, A., R. L. Honeycutt, and W. E. Grant. 2004. Colony dynamics of *Leptonycteris nivalis* (Mexican long-nosed bat) related to flowering *Agave* in Northern Mexico. Journal of Mammalogy 85:453–459.
Morales-Vela, B., and L. D. Olivera-Gómez. 1997. Distribución del manati (*Trichechus manatus*) en la costa norte y centro-norte del estado de Quintana Roo, México. Anales del Instituto de Biología, Serie Zoology 68:153–164.
Morales-Vela, B., J. A. Padilla-Saldivar, and A. A. Mignucci-Giannoni. 2003. Status of the manatee (*Trichechus manatus*) along the northern and western coasts of the Yucatán Peninsula, México. Caribbean Journal of Science 39:42–49.
Mowat, G., D. C. Heard, and C. J. Schwarz. 2013. Predicting grizzly bear density in western North America. PLoS ONE 8:e82757.
Müdespacher-Ziehl, C., R. Espiritu-Mora, M. Martínez-Coronel, and S. Gaona. 2005. Chromosomal studies of two populations of *Peromyscus difficilis felipensis* (Rodentia: Muridae). Cytologia 70:243–248.
Muñiz-Martínez, R., C. López-González, J. Arroyo-Cabrales, and M. Ortiz Gómez. 2003. Noteworthy records of free–tailed bats (Chiroptera: Molossidae) from Durango, México. The Southwestern Naturalist 48:138–144.
Murie, J. O. 1973. Population characteristics and phenology of a Franklin ground squirrel (*Spermophilus franklinii*) colony in Alberta, Canada. The American Midland Naturalist 90:334–340.
Murray J. L., and G. L. Gardner. 1997. *Leopardus pardalis*. Mammalian Species 548:1–10.
Musser, G. G. 1964. Notes on geographic dlstributlon, habitat, and taxonomy of some mexican mammals. Occasional Papers of the Museum of Zoology, University of Michigan 636:1–22.
Musser, G. G. 1968. A systematic study of the Mexican and Guatemalan gray squirrel, *Sciurus aureogaster* F. Cuvier (Rodentia: Sciuridae). Miscellaneous Publications of the Museum of Zoology, University of Michigan 137:1–112.
Musser, G. G. 1969. Notes on *Peromyscus* (Muridae) of Mexico and Central America. American Museum Novitates 2357:1–23.
Musser, G. G. 1970. Identity of the type-specimens of *Sciurus aureogaster* F. Cuvier and *Sciurus nigrescens* Bennett (Mammalia, Sciuridae). American Museum Novitates 2438:1–19.
Musser, G. G., and M. D. Carleton. 2005. Superfamily Muroidea. Pp. 894–1534, *in* Mammal species of the world. A taxonomic and geographic reference (Wilson, D. E., and D. A. M. Reeder, eds.), Third edition. The Johns Hopkins University Press. Baltimore, U.S.A.
Mycroft, E. E., A. B. A. Shafer, and D. T. Stewart. 2011. Cytochrome-b sequences variation in Water Shrews (*Sorex palustris*) from eastern and western North America. Northeastern Naturalist 18:497–508.
Nader, I. A. 1966. Roots of teeth as a generic character in the kangaroo rats, *Dipodomys*. Bulletin of the Biological Research Centre, Baghdad 2:62–69.
Nader, I. A. 1978. Kangaroo rats: intraspecific variation in *Dipodomys spectabilis* Merriam and *Dipodomys deserti* Stephens. Illinois Biological Monographs 49:1–116.
Nadler, C. F. 1966. Chromosomes and systematics of American ground squirrels of the subgenus *Spermophilus*. Journal of Mammalogy 47:579–596.
Nadler, C. F. 1968. The chromosomes of *Spermophilus townsendi* (Rodentia: Sciuridae) and report of a new subspecies. Cytogenetics 7: 144–157.
Nadler, C. F., and D. A. Sutton. 1967. Chromosomes of some squirrels (Mammalia-Sciuridae) from the genera *Sciurus* and *Glaucomys*. Experientia 23:249–251.
Nadler, C. F., R. S. Hoffmann, and K. R. Greer. 1971. Chromosomal divergence during evolution of ground squirrel populations (Rodentia: *Spermophilus*). Systematic Zoology 20:298–305.
Nadler, C. F., R. S. Hoffmann, and A. Woolf. 1973. G-Band patterns as chromosomal markers, and the interpretation of chromosomal evolution in wild sheep *(Ovis)*. Experientia 29:117–119.
Nadler, C. F., R. S. Hoffmann, N. N. Vorontsov, J. W. Koeppl, L. Deutsch, and R. I. Sukernik. 1982. Evolution in ground squirrels. II. Biochemical comparisons in Holarctic populations of *Spermophilus*. Zeitschrift für Säugetierkunde 47:198–214.
Nagorsen, D. W. 1987. *Marmota vancouverensis*. Mammalian Species 270:1–5.
Nagorsen, D. W. 1996. Opossums, shrews and moles of British Columbia. UBC Press, Vancouver, British Columbia, Canada.
Nagorsen, D. W. 2002. An identification manual to the small mammals of British Columbia. Ministry of Sustainable Resource Management, Ministry of Water, Land and Air Protection, and Royal British Columbia Museum. Victoria, Canada.
Nagorsen, D. W., N. Panter, and A. G. Hope. 2017. Are the western water shrew (*Sorex navigator*) and American water shrew (*Sorex palustris*) morphologically distinct? Canadian Journal of Zoology 95:727–736.
Nájera-Cortazar, L. A., S. T. Álvarez-Castañeda, and E. De Luna. 2015. An analysis of Myotis peninsularis (Vespertilionidae) blending morphometric and genetic datasets. Acta Chiropterologica 17:37–47.
Naranjo Piñera, E. J. 2007. Uso sustentable y conservación de ungulados silvestres en la Selva Lacandona, Chiapas, México. Pp. 183–196, in Tópicos en sistemática, biogeogra fía, ecología y conservación de mamíferos. México (Sánchez–Rojas., and A. Rojas Martínez, eds.). Universidad Autónoma del Estado de Hidalgo.
Naranjo, E. J., and E. Espinoza-Medinilla. 2001. Los mamíferos de la Reserva Ecológica Huitepec, Chiapas, México. Revista Mexicana de Mastozoología 5:58–67.
Naranjo Piñera, E. J., C. Lorenzo Monterrubio, and A. Horváth. 2005. La diversidad de mamíferos en Chiapas. Pp. 221–252, in Diversidad biológica en Chiapas (González–Espinoza, M., N. Ramírez–Marcial, and L. Ruíz–Montoya, coords.). ECOSUR, COCYTECH y Plaza y Valdés Editores.
Nash, D. J., and R. N. Seaman. 1977. *Sciurus aberti*. Mammalian Species 80:1–5.
Nason, E. S. 1948. Morphology of hair of eastern North American bats. American Midland Naturalist 39:345–361.
NatureServe. 2016a. *Urocitellus townsendii*. The IUCN Red List of Threatened Species 2016:e.T20476A112212554.
NatureServe. 2016b. *Urocitellus washingtoni*. The IUCN Red List of Threatened Species 2016:e.T20475A112211642.
NatureServe (Morefield, J.). 2016c. *Microdipodops pallidus*. The IUCN Red List of Threatened Species 2016:e.T42607A115193914.
NatureServe. 2017. *Sorex lyelli*. The IUCN Red List of Threatened Species 2017:e.T41402A22313470.

Navarrete, D., and J. Ortega. 2011. *Tamandua mexicana*. Mammalian Species 43:56–63.

Navarro L. D., and D. E. Wilson. 1982. *Vampyrum Spectrum*. Mammalian Species 184:1–4.

Navarro L. D., T. Jiménez, and J. Juárez. 1990. ·Los mamíferos de Quintana Roo. Pp. 371–450, *in* Diversidad biológica en la Reserva de la Biósfera de Sian Ka'an Quintana Roo, México (Navarro D., and J. G. Robinson, eds.). Centro de Investigaciones de Quintana Roo and Program of Studies in Tropical Conservation, University of Florida. Quintana Roo, México.

Naylor, L., and N. Roach. 2016. *Sorex jacksoni*. The IUCN Red List of Threatened Species 2016:e.T20390A22314744.

McNab, B. K. 1969. The economics of temperature regulation in Neotropical bats. Compo Biochemetri Physiology 31:227–268.

Neiswenter, S. A., and B. R. Riddle. 2010. Diversification of the *Perognathus flavus* species group in emerging arid grasslands of western North America, Journal of Mammalogy 91:348–362.

Neiswenter, S. A., and B. R. Riddle. 2011. Landscape and climatic effects on the evolutionary diversification of the *Perognathus fasciatus* species group, Journal of Mammalogy 92:982–993.

Neiswenter, S. A., D. J. Hafner, J. E. Light, G. D. Cepeda, K. C. Kinzer, L. F. Alexander, and B. R. Riddle. 2019. Phylogeography and taxonomic revision of Nelson's pocket mouse (*Chaetodipus nelsoni*). Journal of Mammalogy 100:1847–1864.

Nelson, E. W. 1898. Descriptions of new squirrels from Mexico and Central America. Proceedings of the Biological Sciety of Washington 12:145–156.

Nelson, E. W. 1899. Mammals of the Tres Marias Islands. North American Fauna 14:15–20.

Nelson, E. W. 1909. The rabbits of North America. North American Fauna 29:1–278.

Nelson, E. W. 1912. A new subspecies of pronghorn antelope from Lower California. Proceedings of the Biological Society of Washington 25:107–108.

Nelson, E. W. 1922. Lower California and its natural resources. National Academy of Sciences, First Memories 16:1–194.

Nelson, E. W., and E. A. Goldman. 1909. Eleven new mammals from Lower California. Proceedings of the Biological Society of Washington 22:23 –28.

Nelson, E. W., and E. A. Goldman. 1923. Six new pocket mice from Lower California and notes on the status of several described species. Proceedings of the Biological Society of Washington 42:103–112.

Nelson, E. W., and E. A. Goldman. 1929. Four new pocket gophers of the genus *Heterogeomys* from Mexico. Proceedings of the Biological Society of Washington 42:147–152.

Nelson, E. W., and E. A. Goldman. 1931. Three new Raccoons from Mexico and Salvador. Proceedings of the Biological Society of Washington 44:17–22.

Nelson, E. W., and E. A. Goldman. 1932. Two new cacomistles from Mexico, with remarks on the genus *Jentinkia*. Journal of the Washington Academy of Sciences 22:484–488.

Nelson, E. W., and E. A. Goldman. 1934a. Revision of the pocket gophers of the genus *Cratogeomys*. Proceedings of the Biological Society of Washington 47:135–154.

Nelson, E. W., and E. A. Goldman. 1934b. Pocket gophers of the genus *Thomomys* of the Mexican mainland and bordering territory. Journal of Mammalogy 15:105–124.

Nelson, T. W., and K. Shump, Jr. 1978. Cranial variation and size allometry in *Agouti paca* from Ecuador. Journal of Mammalogy 59:387–394.

Newman A., Bush M., Wildt D.E., Van Dam D., Frankenhuis M.T., Simmons L., Phillips L., O'Brien S.J. Biochemical genetic variation in eight endangered or threatened felid species. J. Mamm. 1985;66(2):256–267.

Nielsen C. K., and L. K. Berkman 2018a. *Sylvilagus aquaticus*. Pp. 117–200, *in* Pikas, Rabbits and Hares of the World (Smith, A., C. H. Johnston, P. C. Alves, and K. Hackländer, eds.). The Johns Hopkins University Press/IUCN/SSC. Baltimore, U.S.A.

Nielsen C. K., and L. K. Berkman. 2018b. *Sylvilagus floridanus*. Pp. 137–140, *in* Pikas, Rabbits and Hares of the World (Smith, A., C. H. Johnston, P. C. Alves, and K. Hackländer, eds.). The Johns Hopkins University Press/IUCN/SSC. Baltimore, U.S.A.

Nielsen, C., and H. C. Lanier. 2019. *Sylvilagus floridanus*. The IUCN Red List of Threatened Species 2019:e.T41299A45191626.

Nielsen, C., D. Thompson, M. Kelly, and C. A. Lopez-Gonzalez. 2015. *Puma concolor*. The IUCN Red List of Threatened Species 2015:e. T18868A97216466.

Novaes, R. L., G. Garbino, V. Cláudio, and R. Moratelli. 2018. Separation of monophyletic groups into distinct genera should consider phenotypic discontinuities: the case of Lasiurini (Chiroptera: Vespertilionidae). Zootaxa. 4379:439–440.

Nowak, R. M. 1991. Walker's mammals of the world. Fifth edition. The Johns Hopkins University Press. Baltimore, U.S.A.

Nowak, R. M. 1992a. The red wolf is not a hybrid. Conservation Biology 6:593–595.

Nowak, R. M. 1992b. Walker's mammals of the world. Fifth ed. The Johns Hopkins University Press, Baltimore, 2:643–1629.

Nowak, R. M. 1999. Walker's mammals of the world. Sixth edition. The Johns Hopkins University Press. Baltimore, U.S.A.

Nowak, R. M. 2005. Walker's Carnivores of the World. Johns Hopkins University Press, Baltimore, U.S.A. and London, UK.

Nowak, R. M., and J. L. Paradiso. 1983. Walker's mammals of the world. Fourth Edition. Johns Hopkins University Press, Baltimore, U.S.A.

Nowell, K., and P. Jackson. 1996. Wild cats. Status survey and conservation action plan. IUCN/SSC Cat Specialist Group. Cambridge. United Kingdom. The Burlington Press.

Nyakatura, K., and O. R. Bininda-Emonds. 2012. Updating the evolutionary history of Carnivora (Mammalia): a new species-level supertree complete with divergence time estimates. BMC Biology 10:12

Oaks, E. C., P. J. Young, G. L. Kirkland, and D. F. Schmidt. 1987. *Spermophilus variegatus*. Mammalian Species 272:1–8.

O'Corry-Crowe, G. M., and R. L. Westlake. 1997. Molecular investigations of spotted seals (*Phoca largha*) and harbor seals (*P. vitulina*), and their relationship in areas of sympatry. Pp. 291–304, *in* Molecular genetics of marine mammals (Dizon, A. E., S. J. Chivers, and W. F. Perrin, eds). The Society of Marine Mammalogy, Lawrence, U.S.A.

O'Corry-Crowe, G., and C. Bonin. 2009. The biogeography and population structure of spotted seals (*Phoca largha*) as revealed by mitochondrial and microsatellite DNA. Draft Report to the National Marine Mammal Laboratory. Harbor Branch Oceanographic Institute, Florida Atlantic University, Fort Pierce, U.S.A.

O'Farrell, M. J., and A. R. Blaustein. 1974a. *Microdipodops megacephalus*. Mammalian Species 46:1–3.

O'Farrell, M. J., and A. R. Blaustein. 1974b. *Microdipodops pallidus*. Mammalian Species 47:1–2.

O'Farrell, M. J., and E. H. Studier. 1980. *Myotis thysanodes*. Mammalian Species 137:1–5.

O'Gara, B. W. 1969. Horn casting by female pronghorns. Journal of Mammalogy 50:373–375.

O'Gara, B. W. 1978. *Antilocapra americana.* Mammalian Species 90:1–7.

O'Gara, P. W. 1990. The pronghorn (*Antilocapra americana*). Pp. 231– 264, *in* Horns, pronghorns, and antlers (Bubenik, G. A., and A. B. Bubenik, eds.). Springer-Verlag. New York, U.S.A.

Ohdachi, S. D., M. Hasegawa, M. A. Iwasa, P. Vogel, T. Oshida, L.-K. Lin, and H. Abe. 2006. Molecular phylogenetics of soricid shrews (Mammalia) based on mitochondrial cytochrome–*b* gene sequences: with special reference to the Soricinae. Journal of Zoology 270:177–191.

Ojasti, J., and O. J. Linares. 1971. Adiciones a la fauna de murciélagos de Venezuela con notas sobre las especies del género *Diclidurus*. Acta Biológica Venezolana 7:421–441.

Ojeda, R., C. Bidau, and L. Emmons. 2016. *Myocastor coypus*. The IUCN Red List of Threatened Species 2016:e.T14085A121734257.

Okhotina, M. V. 1983. A taxonomic revision of *Sorex arcticus* Kerr, 1972. (Soricidae, Insectivora). Zoologeskii Zhurnal 62:409–417.

Olguín-Monroy, H. C., L. León-Paniagua, U. M. Samper-Palacios, and V. Sánchez-Cordero. 2008. Mastofauna de la region de los Chimalapas, Oaxaca, México. Pp. 165–216, *in* Avances en el estudio de los mamíferos de México, Volumen II. (Lorenzo, C., E. Espinoza, and J. Ortega, eds.). Asociación Mexicana de Mastozoología, A. C. Distrito Federal, México.

Oliveira T. G. 1998a. *Herpailurus yagouaroundi*. Mammalian Species 578:1–6.

Oliveira T. G. 1998b. *Leopardus wiedii*. Mammalian Species 579:1–6.

Oliver, W. L. R. 1976. The management of yapoks (*Chironectes minimus*) at Jersey Zoo, with observations on their behavior. Annual Reptort Jersey Wildlife Preservation Trust 13:32–36.

Ordóñez-Garza, N., and R. D. Bradley. 2011. *Peromyscus schmidlyi* (Rodentia: Cricetidae). Mammalian Species 43:31–36.

Ordoñez-Garza, N., J. O. Matson, R. E. Strauss, R. D. Bradley, and J. Salazar-Bravo. 2010. Patterns of phenotypic and genetic variation in three species of endemic Mesoamerican *Peromyscus* (Rodentia: Cricetidae). Journal of Mammalogy 91:848–859.

Ordoñez-Garza, N., C. W. Thompson, M. K. Unkefer, C. W. Edwards, J. G. Owen, and R. D. Bradley. 2014. Systematics of the *Neotoma mexicana* species group (Mammalia: Rodentia: Cricetidae) in Mesoamerica: new molecular evidence on the status and relationships of *N. ferruginea* Tomes, 1862. Proceedings of the Biological Society of Washington 127:518–532.

Orr, R. T. 1934. Two new woodrat from Lower California, Mexico. Proceedings Biological Society Washington 47:109–112.

Orr, R. T. 1940. The rabbits of California. Occasional Papers California Academy of Sciences 19:1–227.

Ortega, J., and H. T. Arita. 1997. *Mimon bennettii*. Mammalian Species 549:1–4.

Ortega, J., and I. Alarcón-D. 2008. *Anoura geoffroyi*. Mammalian Species 818:1–7.

Ortega, J., and I. Castro-Arellano. 2001. *Artibeus jamaicensis*. Mammalian Species 662:1–9.

Ortega, J., M. Tschapka, T. P. González-Terrazas, G. Suzán, and R. A. Medellín. 2009. Phylogeography of *Musonycteris harrisoni* along the pacific coast of Mexico. Acta Chiropterologica 11:259–270.

Ortega Reyes, J., D. G. Tirira, M. Arteaga, and F. Miranda. 2014. *Tamandua mexicana*. The IUCN Red List of Threatened Species 2014:e. T21349A47442649.

Ortiz-Martínez T., V. Rico-Gray, and E. Martínez-Meyer. 2008. Predicted and verified distributions of *Ateles geoffroyi* and *Alouatta palliata* in Oaxaca, Mexico. Primates 49:186–19.

Ortega, J., B. Vite-de León, A. Tinajero-Espitia, and J. A. Romero-Meza. 2008. *Carollia subrufa*. Mammalian Species 823:1–4.

Osgood, W. H. 1900. Revision of the pocket mice of the genus *Perognathus*. North American Fauna 18:1–73.

Osgood, W. H. 1904. Thirty new mice of the genus *Peromyscus* from Mexico and Guatemala. Proceedings of the Biological Society of Washington 17:55–77.

Osgood, W. H. 1907. Four new pocket mice. Proceedings of the Biological Society of Washington 20:19–22.

Osgood, W. H. 1909. Revision of the mice of the genus *Peromyscus*. North American Fauna 28:33–252.

Osgood, W. H. 1913. Two new mouse opossums from Yucatan. Proceedings of the Biological Society of Washington 26:175–176.

Ostroff, A. C., and E. J. Finck. 2003. *Spermophilus franklinii*. Mammalian Species 724:1–5.

Owen J. G., and R. S. Hoffmann. 1983. *Sorex ornatus*. Mammalian Species 212:1–5.

Owen, J. G., D. J. Schmidly, and W. B. Davis. 1984. A morphometric analysis of three species of *Carollia* (*Chiroptera,* Glossophaginae) from Middle America. Mammalia 48:85–93.

Owen, J. G., R. J. Baker, and S. L. Williams. 1996. Karyotypic variation in spotted skunks (Carnivora: Mustelidae: *Spilogale*) from Texas, Mexico and El Salvador. Texas Journal of Science 48:119–122.

Packard, R. L. 1960. Speciation and evolution of the pygmy mice, genus *Baiomys*. University of Kansas Publications, Museum of Natural History 9:579–670.

Packard, R. L., and F. W. Judd. 1968. Comments on some mammals from western Texas. Journal of Mammalogy 49:535–538.

Packard, R. L., and J. B. Montgomery, Jr. 1978. *Baiomys musculus*. Mammalian Species 102:1–3.

Palmer, F. G. 1937. Geographic variation in the mole *Scapanus latimanus*. Journal of Mammalogy 18:280–314.

Paradiso, J. L. 1967. A review of the wrinkle-faced bats (*Centurio senex* Gray) with a description of a new subspecies. Mammalia 31:595–604.

Paradiso, J. L. 1969. Mammals of Maryland. North American Fauna 66:1–193.

Paradiso, J. L., and R. M. Nowak. 1972. *Canis rufus*. Mammalian Species 1972:1–4.

Pasitschniak-Arts, M. 1993. *Ursus arctos*. Mammalian Species 439:1–10.

Pasitschniak-Arts, M., and S. Larivière. 1995. *Gulo gulo*. Mammalian Species 499:1–10.

Patterson, B. D. 1980a. Montane mammalian biogeography in New Mexico. The Southwestern Naturalist 25:33–40.

Patterson, B. D. 1980b. A new subspecies of *Eutamias quadrivittatus* (Rodentia: Sciuridae) from the Organ Mountains, New Mexico. Journal of Mammalogy 61:455–464.

Patterson, B. D. 1984. Geographic variation and taxonomy of Colorado and hopi chipmunks (genus *Eutamias*). Journal of Mammalogy 65:442–456.

Patterson, B. D., and L. R. Heany. 1987. Preliminary analysis of geographic variation in red-tailed chipmunks (*Eutamias ruficaudus*). Journal of Mammalogy 68:782–791.

Patterson, B. D., and Norris, R. W. 2016. Towards a uniform nomenclature for ground squirrels: the status of the Holarctic chipmunks. Mammalia 80:241–251.

Patterson, B. D., H. E. Ramírez-Chaves, J. F. Vilela, A. E. R. Soares, F. Grewe. 2021. On the nomenclature of the American clade of weasels (Carnivora: Mustelidae). Journal of Animal Diversity 3:1–8.
Pattie, D. 1973. Sorex bendirii. Mammalian Species 27:1–2.
Patton, J. L. 1967a. Chromosome studies of certain pocket mice, genus *Perognathus* (Rodentia: Heteromyidae). Journal of Mammalogy 48:27–37.
Patton, J. L. 1967b. Chromosomes and evolutionary trends in the pocket mouse subgenus *Perognathus* (Rodentia: Heteromyidae). The Southwestern Naturalist 12:429–438.
Patton, J. L. 1969. Chromosome evolution in the pocket mouse, *Perognathus goldmani* Osgood. Evolution 23:645–662.
Patton, J. L. 1972. Patterns of geographic variation in karyotype in the Pocket Gopher, *Thomomys bottae* (Eydoux and Gervais). Evolution 26:574–586.
Patton, J. L. 1973. An analysis of natural hybridization between the pocket gophers *Thomomys bottae* and *Thomomys umbrinus*, in Arizona. Journal of Mammalogy 54:561–584.
Patton, J. L. 1993a. Family Geomyidae. Pp. 469–476, *in* Mammal species of the world, a taxonomic and geographic reference (Wilson, D. E., and D. M. Reeder, eds.). Second edition. Smithsonian Institution Press, Washington, U.S.A.
Patton, J. L. 1993b. Family Heteromidae. Pp. 477–486, *in* Mammal species of the world, a taxonomic and geographic reference (Wilson, D. E., and D. M. Reeder, eds.). Second edition. Smithsonian Institution Press. Washington, U.S.A.
Patton, J. L. 1999a. *Thomomys bulbivorus*. Pp. 468–469, *in* The Smithsonian Book of North American Mammals (Wilson, D. E., and S. Ruff, eds.). Smithsonian Institution Press. Washington, U.S.A.
Patton, J. L. 1999b. *Thomomys monticola*. Pp. 473–474, *in* The Smithsonian Book of North American Mammals (Wilson, D. E., and S. Ruff, eds.). Smithsonian Institution Press. Washington, U.S.A.
Patton, J. L. 1999c. *Thomomys talpoides*. Pp. 474–477, *in* The Smithsonian Book of North American Mammals (Wilson, D. E., and S. Ruff, eds.). Smithsonian Institution Press. Washington, U.S.A.
Patton, J. L. 2005a. Family Heteromyidae. Pp. 844–858, *in* Mammal species of the world. A taxonomic and geographic reference (Wilson, D. E., and D. A. Reeder, eds.). Third edition. The Johns Hopkins University Press. Baltimore, U.S.A.
Patton, J. L. 2005b. Family Geomyidae. Pp. 859–870, *in* Mammal species of the world. A taxonomic and geographic reference (Wilson, D. E., and D. A. Reeder, eds.). Third edition. The Johns Hopkins University Press. Baltimore, U.S.A.
Patton, J. L., and S. T. Álvarez-Castañeda. 1999. Family Heteromyidae. Pp. 351–443, *in* Mamíferos del Noroeste Mexicano (Álvarez-Castañeda, S. T., and J. L. Patton, eds.). Centro de Investigaciones Biológicas del Noroeste, S. C. La Paz, México.
Patton, J. L., and S. T. Álvarez-Castañeda. 2005. Phylogeography on the desert woodrat, *Neotoma lepida*, with comments on systematics and biogeographic history. Pp. 375–388, *in* Contribuciones mastozoologicas en honor de Bernanrdo Villa (Sánchez-Cordero, V., and R. Medellín, eds.). Instituto de Biología, Universidad Nacional Autónoma de México. Distrito Federal, México.
Patton, J. L., and S. T. Álvarez-Castañeda. 2017. *Neotoma insularis*. The IUCN Red List of Threatened Species 2017:e.T116989038A119112253.
Patton, J. L., and M. N. Da Silva. 1997. Definition of species of pouched four-eyed opossum (Didelphidae, *Philander*). Journal of Mammalogy 78:90–102.
Patton, J. L., and R. E. Dingman. 1968. Chromosome studies of pocket gophers, genus *Thomomys*. I. The specific status of *Thomomys umbrinus* (Richardson) in Arizona. Journal of Mammalogy 49:1–13.
Patton, J. L., and L. H. Emmons. 2015. Family Dasyproctidae Bonaparte, 1838. Pp. 733–762, *in* Mammals of South America Volume 2: Rodents (Patton, J. L., U. F. G. Pardiñas, and G. D'Elía, eds). University of Chicago Press. Chicago, U.S.A.
Patton, J. L., and R. N. Fisher. 2023. Taxonomic reassessment of the Little pocket mouse, *Perognathus longimembris* (Rodentia, Heteromyidae) of southern California and northern Baja California. Therya 14:131–160.
Patton, J. L., and A. L. Gardner. 1971. Parallel evolution of multiple sex-chromosome systems in the phyllostomatid bats, *Carollia* and *Choeroniscus*. Experientia 27:105–106.
Patton, J. L., and M. F. Smith. 1981. Molecular evolution in *Thomomys*: Phyletic systematics, paraphyly, and rates of evolution. Journal of Mammalogy 62:493–500.
Patton, J. L., and M. F. Smith. 1989. Genetic structure and the genetic and morphologic divergence among pocket gopher species (genus *Thomomys*). Pp. 284–304, *in* Speciation and its consequences (Otte, D., and J. A. Endler, eds.). Sinauer Associates Incorporated. Sunderland, U.S.A.
Patton, J. L., and M. F. Smith. 1990. The evolutionary dynamics of the Pocket Gopher *Thomomys bottae*, with emphasis on California populations. University of California Publications, zoology 123.
Patton, J. L., and M. F. Smith. 1991. Molecular evolution in *Thomomys*: phyletic systematics, paraphyly, and rates of evolution. Journal of Mammalogy 62:493–500.
Patton, J. L., and M. F. Smith. 1994. Paraphyly, polyphyly, and the nature of species boundaries in pocket gophers (Genus *Thomomys*). Systematic Biology 43:11–26.
Patton, J. L., H. MacArthur, and S. Y. Yang. 1976. Systematic relationships of the four-toed populations of *Dipodomys heermanni*. Journal of Mammalogy 57:159–163.
Patton, J. L., S. W. Sherwood, and S. Y. Yang. 1981. Biochemical systematics of chaetodipine pocket mice, genus *Perognathus*. Journal of Mammalogy 62:477–492.
Patton, J. L., M. F. Smith, R. D. Price, and R. A. Hellenthal. 1984. Genetics of hybridization between the pocket gophers *Thomomys bottae* and *Thomomys townsendii* in northeastern California. The Great Basin Naturalist 44:431–440.
Patton, J. L., D. G. Huckaby, and S. T. Álvarez-Castañeda. 2007. The evolutionary history and a systematic revision of woodrats of the *Neotoma lepida* Group. University of California Publications in Zoology 135:1–411.
Patton, J. L., D. F. Williams, P. A. Kelly, B. L. Cypher, and S. E. Phillips. 2019. Geographic variation and evolutionary history of *Dipodomys nitratoides* (Rodentia: Heteromyidae), a species in severe decline. Journal of Mammalogy 100:1546–1563.
Paulson, D. D. 1988a. *Chaetodipus baileyi*. Mammalian Species 297:1–5.
Paulson, D. D. 1988b. *Chaetodipus hispidus*. Mammalian Species 320:1–4.
Pavan, A. C., and G. Marroig. 2016. Integrating multiple evidences in taxonomy: species diversity and phylogeny of mustached bats (Mormoopidae: *Pteronotus*). Molecular Phylogenetics and Evolution 103:184–198.

Pavan, A. C., and Y. G. Marroig. 2017. Timing and patterns of diversification in the Neotropical bat genus *Pteronotus*. Mormoopidae. Molecular Phylogenetics and Evolution 108:61–69.
Pavan, A. C., and V. da C. Tavares. 2020. *Pteronotus gymnonotus* (Chiroptera: Mormoopidae). Mammalian Species 52:40–48.
Paviolo, A., P. Crawshaw, A. Caso, T. de Oliveira, C. A. Lopez-Gonzalez, M. Kelly, C. De Angelo, and E. Payan. 2015. *Leopardus pardalis*. The IUCN Red List of Threatened Species 2015:e.T11509A97212355.
Pavlinov, I. Y., and A. A. Lissovsky. 2012. The mammals of Russia: A Taxonomic and Geographic Reference. KMK Scientific Press Ltd. Moscow, Russia.
Pech-Canché, J. M., C. MacSwiney, and E. Estrella 2010. Importancia de los detectores ultrasónicos para mejorar los inventarios de murciélagos Neotropicales. Therya 1:227–234.
Pelton, M. R. 2003. Black bear (*Ursus americanus*). Pp. 547–555, *in* Wild mammals of North America: biology, management, and conservation (Feldhamer, G. A., B. C Thompson, and J. A. Chapman, eds). Johns Hopkins University Press. Baltimore, U.S.A.
Pembleton, E. F., and S. L. Williams. 1978. *Geomys pinetis*. Mammalian Species 86:1–3.
Peppers, L., and R. D. Bradley. 2000. Cryptic species in *Sigmodon hispidus*: evidence from DNA sequence. Journal of Mammalogy 81:332–343.
Peppers, L. L., D. S. Carroll, and R. D. Bradley. 2002. Molecular systematics of the genus *Sigmodon* (Rodentia: Muridae): evidence from the mitochondrial cytochrome *b* gene. Journal of Mammalogy 83:396–407.
Pérez, E. M. 1992. *Agouti paca*. Mammalian Species 404:1–7.
Perez, S., P. C. de Grammont, and A. D. Cuarón. 2017. *Myotis fortidens*. The IUCN Red List of Threatened Species 2017:e.T14161A22056846.
Pérez, S. G., M. R. Jolón, J. E. Mérida, and A. J. Andino-Madrid. 2019. First record of the shrew *Cryptotis lacandonensis* (Eulipotyphla: Soricidae) for Guatemala. Therya 10:187–193.
Pérez-Hernandez, R., D. Brito, T. Tarifa, N. Cáceres, D. Lew, and S. Solari. 2016a. *Chironectes minimus*. The IUCN Red List of Threatened Species 2016:e.T4671A22173467.
Pérez-Hernandez, R., D. Lew, and S. Solari. 2016b. *Didelphis virginiana*. The IUCN Red List of Threatened Species 2016:e.T40502A22176259.
Pérez-Irineo, G., and A. Santos-Moreno. 2012. Diversidad de mamíferos terrestres de talla grande y media de la selva caducifolia del noreste de Oaxaca, México. Revista Mexicana de Biodiversidad 83:164–169.
Perini, F. A., C. A. M. Russo, and C. G. Schrago. 2010. The evolution of South American endemic canids: a history of rapid diversification and morphological parallelism. Journal of Evolutionary Biology 23:311–322.
Perry, E. A., G. B. Stenson, S. E. Bartlett, W. S. Davidson, and S. M. Carr. 2000. DNA sequence analysis identifies genetically distinguishable populations of harp seals (*Phoca groenlandicus*) in the northwest and northeast Atlantic. Marine Biology 137:53–58.
Peters, S. L., B. K. Lim, and M. D. Engstrom. 2002. Systematics of dog-faced bats (*Cynomops*) based on molecular and morphometric data. Journal of Mammalogy 83:1097–1110.
Peterson, A. T., L. Canseco-Márquez, J. L. Contreras-Jiménez, G. Escalona-Segura, O. Flores-Villela, J. García-López, B. Hernández-Baños, C. A. Jiménez-Ruíz, L. León-Paniagua, S. Mendoza-Amaro, A. G. Navarro-Sigüenza, V. Sánchez-Cordero, and D. E. Willard. 2004. A preliminary biological survey of Cerro Piedra Larga, Oaxaca, Mexico: Birds, mammals, reptiles, amphibians, and plants. Anales del Instituto de Biología, Universidad Nacional Autónoma de México, Serie Zoología 75:439–466.
Peterson, R. L. 1955. North American moose. University of Toronto Press, Toronto, Canada.
Peterson, R. L. 1966. The mammals of eastern Canada. Oxford University Press. Toronto, Canada.
Peterson, R. S., and A. Symansky. 1963. First record of the Gaspé shrew from New Bnmswick. Journal of Mammalogy 44:278–279.
Peterson, R. S., C. L. Hubbs, R. L. Gentry, and R. L. Delong. 1968. The Guadalupe fur seal: habitat, behavior, population size, and field identification. Journal of Mammalogy 49:665–675.
Pfeiffer, C. J., and G. H. Gass. 1963. Note on the longevity and habits of captive *Cryptotis parva*. Journal of Mammalogy 44:427–428.
Phillips, C. J. 1971. The dentition of glossophagine bats: development, morphological characteristics, variation, pathology, and evolution. Miscellaneous Publications Museum of Natural History, University of Kansas 54:1–138.
Phillips, C. J., and J. K. Jones, Jr. 1971. A new species of the long-nosed bat, *Hylonycteris underwoodi*, from Mexico. Journal of Mammalogy 52:77–80.
Phillips, C. J., and J. K. Jones, Jr. 1969. Dental abnormalities in North American bats. 1. Emballonuridae, Noctilionidae, and Chilonycteridae. Transactions of the Kansas Academy of Sciences 71:509–520.
Phillips, C. J., G. W. Grimes, and G. L. Forman. 1977. Oral biology. Pp. 121–246, *in* Biology of bats of the New World family Phyllostomatidae, Part II (Baker, R. J., J. K. Jones, Jr., and D. C. Carter, eds.), Special Publications Museum, Texas Tech University 3:1–364.
Phillips, M. 2018. *Canis rufus* (errata version published in 2020). The IUCN Red List of Threatened Species 2018:e.T3747A163509841.
Phillips, C. D., J. W. Bickham, J. C. Patton, and T. S. Gelatt. 2009. Systematics of Steller sea lions (Eumetopias jubatus): subspecies recognition based on concordance of genetics and morphometrics. Museum of Texas Tech University Occasional Papers 283:1–15.
Phuong, M. A., M. C. W. Lim, D. R. Wait, K. C. Rowe, and C. Moritz. 2014. Delimiting species in the genus *Otospermophilus* (Rodentia: Sciuridae), using genetics, ecology, and morphology. Biological Journal of the Linnean Society 113:1136–1151.
Piaggio, A. J., and G. S. Spicer. 2001. Molecular phylogeny of the chipmunks inferred from mitochondrial cytochrome *b* and cytochrome oxidase II gene sequences. Molecular Phylogenetics and Evolution 20:335–350.
Piaggio, A. J., and S. L. Perkins. 2005. Molecular phylogeny of North American long-eared bats (Vespertilionidae: *Corynorhinus*); inter-and intraspecific relationships inferred from mitochondrial and nuclear DNA sequences. Molecular Phylogenetics and Evolution 37:762–775.
Piaggio, A. J., E. W. Valdez, M. A. Bogan, and G. S. Spicer. 2002. Systematics of *Myotis occultus* (Chiroptera: Vespertilionidae) inferred from sequences of two mitochondrial genes. Journal of Mammalogy 83:386–395.
Pine, R. H. 1972. The bats of the genus *Carollia*. Technical Monography, Texas A&M University, Texas Agriculture Experimental Station 8:1–125.
Pine, R. H. 1973. Anatomical and nomenclatural notes on opossums. Proceedings of the Biological Society of Washington 86:391–402.
Pine, R. H. 1993. A new species of Thyroptera Spix (Mammalia: Chiroptera: Thyropteridae) from the Amazon Basin of northeastern Perú. Mammalia 57:213–225.
Pine, R. H., D. C. Carter, and R. K. Laval. 1971. Status of *Bauerus* Van Gelder and its relationships to other nyctophiline bats. Journal of Mammalogy 52:663–669.
Pineda, J., and B. Rodriguez. 2015. *Eumops hansae*. The IUCN Red List of Threatened Species 2015:e.T8245A22026314.

Pino, J., R. Samudio Jr, J. F. González-Maya, and J. Schipper. 2020. *Bassariscus sumichrasti*. The IUCN Red List of Threatened Species 2020:e. T2613A166521324.
Pinto Sandoval, E. D., L. D. Rola, J. A. Morales-Donoso, S. Gallina, R. Reyna-Hurtado, and J. M. Barbanti Duarte. 2022. Integrative analysis of *Mazama temama* (Artiodactyla: Cervidae) and designation of a neotype for the species. Journal of Mammalogy 103:447–458.
Pitra, C., J. Fickel, E. Meijaard, and C. P. Groves. 2004. Evolution and phylogeny of Old World deer. Molecular Phylogenetics and Evolution 33:880–895.
Pizzimenti, J. J. 1975. Evolution of the prairie dog genus *Cynomys*. Occasional Papers Museum Natural History, University of Kansas 39:1–73.
Pizzimenti, J. J. 1976a. Genetic divergence and morphological convergence in the prairie dogs, *Cynomys gunnisoni* and *Cynomys leucurus*. I. Morphological and ecological analyses. Evolution 30:345–366.
Pizzimenti, J. J. 1976b. Genetic divergence and morphological convergence in the prairie dogs, *Cynomys gunnisoni* and *Cynomys leucurus*. II. Genetic analyses. Evolution 30:367–379.
Pizzimenti, J. J., and G. D. Collier. 1975. *Cynomys parvidens*. Mammalian Species 52:1–3.
Pizzimenti, J. J., and R. S. Hoffmann. 1973. *Cynomys gunnisoni*. Mammalian Species 25:1–4.
Pizzimenti, J. J., and C. F. Nadler. 1972. Chromosomes and serum proteins of the Utah prairie dog, *Cynomys parvidens* (Sciuridae). The Southwestern Naturalist 17:279–286.
Planz, J. V., E. G. Zimmerman, T. A. Spradling, and D. R. Akins. 1996. Molecular phylogeny of the *Neotoma floridana* species group. Journal of Mammalogy 77:519–535.
Plumpton, D., and J. K. Jones, Jr. 1992. *Rhynchonycteris naso*. Mammalian Species 413:1–5.
Pocock, R. I. 1920. On the external and cranial characters of the European badger (*Meles*) and of the American badger (*Taxidea*). Proceedings of the Zoological Society of London 1920:423–436.
Pocock, R. I. 1921. The external characters and classification of the Procyonidae. Proceedings of the Zoological Society of London 1921:389–422.
Pocock, R. I. 1928. The structure of the auditory bulla in the Procyonidae and the Ursidae with a note on the bulla of *Hyaena*. Proceedings of the Zoological Society of London 1928:963–974.
Pocock, R. I. 1941. The races of the ocelot and the margay. Field Museum of Natural History Zoological Series 27:319–369.
Poglayen-Neuwall, I. 1989. Procyonids. Pp. 450–468, *in* Grzimek's Encyclopedia of Mammals (Parker, S., ed.). McGraw-Hill. New York, U.S.A.
Poglayen-Neuwall, I., and D. E. Toweill. 1988. *Bassariscus astutus*. Mammalian Species 327:1–8.
Polaco, O. J., and R. Muñiz-Martínez. 1987. Los murciélagos de la costa de Michoacán, México. Anales de la Escuela Nacional de Ciencias Biológicas 31:63–89.
Polziehn, R. O., and C. A. Strobeck. 2002. Phylogenetic comparison of red deer and wapiti using mitochondrial DNA. Molecular Phylogenetics and Evolution 22:342–356.
Poole, K. G. 2003. A review of the Canada lynx, *Lynx canadensis*, in Canada. Canadian Field Naturalist 117:360–376.
Ports, M. A., and S. B. George. 1990. *Sorex preblei* in the northern Great Basin. Great Basin Naturalist 50:93–95.
Porter, C. A., S. R. Hoofer, R. A. Van Den Bussche, T. E. Lee, Jr., and R. J. Baker. 2003. Systematics of round–eared bats (Tonatia and Lophostoma) based on nuclear and mitochondrial DNA sequences. Journal of Mammalogy 84:791–808.
Porter, C. A., N. E. Beasley, N. Ordóñez-Garza, L. L. Lindsey, D. S. Rogers, N. Lewis-Rogers, J. W. Sites, and R. D. Bradley. 2017. A new species of big-eared climbing rat, genus *Ototylomys* (Cricetidae: Tylomyinae), from Chiapas, Mexico. Journal of Mammalogy 98:1310–1329.
Porter, C. A., S. R. Hoofer, C. A. Cline, F. G. Hoffmann, and R. J. Baker. 2007. Molecular phylogenetics of the phyllostomid bat genus *Micronycteris* with descriptions of two new subgenera. Journal of Mammalogy 88:1205–1215.
Powell, R. A. 1981. *Martes pennanti*. Mammalian Species 156:1–6.
Powell, R. A. 1993. The Fisher: Life History, Ecology, and Behavior. University of Minnesota Press. Minneapolis, U.S.A.
Powell, J. A., and G. B. Rathbun. 1984. Distribution, and abundance of manatees along the northern coast of the Gulf of Mexico. Northeast Gulf Science 7:1–97.
Presley, S. J. 2000. *Eira barbara*. Mammalian Species 636:1–6.
Prestrud, P. 1991. Adaptations by the arctic fox (*Alopex lagopus*) to the polar winter. Arctic 44:132–138.
Prothero, D. R., D. Domning, R. E. Fordyce, S. Foss, C. Janis, S. Lucas, K. L. Marriott, G. Metais, D. Naish, K. Padian, G. Rössner, N. Solounias, M. Spaulding, R. M. Stucky, J. Theodor, and M. Uhen. 2022. On the Unnecessary and Misleading Taxon "Cetartiodactyla". Journal of Mammal Evolution 29:93–97.
Puckett, E. E., P. D. Etter, E. A. Johnson, and L. S. Eggert. 2015. Phylogeographic analyses of American black bears (*Ursus americanus*) suggest four glacial refugia and complex patterns of postglacial admixture. Molecular Biology and Evolution 32:2338–2350.
Quakenbush, L. T. 1988. Spotted seal. Pp. 107–124, in Selected marine mammals of Alaska: species accounts with research and management recommendations (Lentfer, J. W., ed.). U.S. Marine Mammal Commission. Washington, USA.
Quakenbush, L., J. Citta, and J. Crawford. 2009. Biology of the spotted seal (*Phoca largha*) in Alaska from 1962 to 2008. Report to National Marine Fisheries Service. Fairbanks, Alaska.
Quigley, H., R. Foster, L. Petracca, E. Payan, R. Salom, and B. Harmsen. 2017. *Panthera onca*. The IUCN Red List of Threatened Species 2017:e. T15953A123791436.
Qumsiyeh, M. B., C. Sanchez-Hernandez, S. K. Da-Vis, J. C. Patton, and R. J. Baker. 1988. Chromosomal evolution in *Geomys* as revealed by G- and C-band analysis. The Southwestern Naturalist 33:1–13.
Rachlow, J., Becker, P. A., and L. Shipley. 2016. *Brachylagus idahoensis*. The IUCN Red List of Threatened Species 2016:e.T2963A45176206.
Rachlow, J. L., P. A. Becker, and L. A. Shipley. 2018. *Brachylagus idahoensis*. Pp. 87–90, *in* Pikas, rabbits and hares of the world (Smith, A., C. H. Johnston, P. C. Alves, and K. Hackländer, eds.). The Johns Hopkins University Press/IUCN/SSC. Baltimore, U.S.A.
Rafinesque, C. S. 1817. Descriptions of seven new genera of North American quadrupeds. The American Monthly Magazine and Critical Review 2:44–46.
Rainey, D. G., and R. H. Baker. 1955. The pigmy woodrat, *Neotoma goldmani*, its distribution and systematic position. University of Kansas Publications, Museum of Natural History 7:619–624.
Ralls, K., J. Ballou, and R. L. Brownell, Jr. 1983. Genetic diversity in California sea otters: theoretical considerations and management implications. Biological Conservation 25:209–232.

Raman, T. R. S. 2013. The Chital (*Axis axis* Erxleben). University Press.
Ramey, C. A., and D. J. Nash. 1976. Geographic variation in Abert's squirrel (*Sciurus aberti*). The Southwestern Naturalist 21:135–139.
Ramey II, R. R., H.-P. Liu, C. W. Epps, L. M. Carpenter, and J. D. Wehausen. 2005. Genetic relatedness of the Preble's meadow jumping mouse (*Zapus hudsonius preblei*) to nearby subspecies of *Z. hudsonius* as inferred from variation in cranial morphology, mitochondrial DNA and microsatellite DNA: implications for taxonomy and conservation. Animal Conservation 8:329–346.
Ramírez-Chaves, H. E., M. Alarcón Cifuentes, E. A. Noguera-Urbano, W. A. Pérez, M. M. Torres-Martínez, P. A. Ossa-López, F. A. Rivera-Páez, and D. M. Morales-Martínez. 2023. Systematics, morphometrics, and distribution of *Eptesicus fuscus miradorensis*, with notes on baculum morphology and natural history. Therya 14:299–311.
Ramírez-Pulido, J. 1969. Contribución al conocimiento de los mamíferos del Parque Nacional "Lagunas de Zempoala", Morelos, México. Anales del Instituto de Biología, Universidad Nacional Autónoma de México, Serie Zoología 40:253–290.
Ramírez-Pulido, J., and C. Sánchez-Hernández. 1971. *Tylomys nudicaudus* from the Mexican States of Puebla and Guerrero. Journal of Mammalogy 52:481.
Ramírez-Pulido, J., and C. Sánchez-Hernández. 1972. Regurgitaciones de lechuzas, procedentes del cañon del Zopilote, Guerrero, México. Revista de la Sociedad Mexicana de Historia Natural 33:107–112.
Ramírez-Pulido, J., R. López Wilchis, C. Müdespacher and I. E. Lira. 1983. Lista y bibliografia reciente de los mamíferos de México. Universidad Autónoma Metropolitana, Iztapalapa. Distrito Federal, México.
Ramírez-Pulido, J., M. C. Brition, A. Perdomo, and A. Castro. 1986. Guia de los mamiferos de México. Universidad Autonoma Metropolitana. Distrito Federal, Méxco.
Ramírez-Pulido, J., N. González-Ruiz, A. L. Gardner, and J. Arroyo-Cabrales. 2014. List of recent land mammals of Mexico, 2014. Museum of Texas Tech University Special Publications 63:1–69.
Randi, E., N. Mucci, M. Pierpaoli, and E. Douzery. 1998. New phylogenetic perspectives on the Cervidae (Artiodactyla) are provided by the mitochondrial cytochrome b gene. Proceedings of the Royal Society of London. Series B: Biological Sciences 265:793–801.
Rausch, R. L. 1953. On the status of some Artic mammals. Artic 6:91–148.
Rausch, R. L., and V. R. Rausch. 1968. On the biology and systematic position of *Microtus abbreviatus* Miller, a vole endemic to the St. Matthew Islands, Bering Sea. Zeitschrift für Säugetierkunde 33:65–99.
Rausch, R. L., and V. R. Rausch. 1972. Observations on chromosomes of *Dicrostonyx torquatus stevensoni* Nelson and chromosomal diversity in varying lemmings. Zeitschrift für Säugetierkunde 33:65–69.
Rausch, V. R., and R. L. Rausch. 1974. The chromosome complement of the yellow-cheeked vole *Microtus xanthognathus* (Leach). Canadian Journal Genetetic and Cytology 16:267–272.
Rausch, R. L., and V. R. Rausch. 1975. Relationships of the red-backed vole, *Clethrionomys rutilus* (Pallas) in North America: Karyotypes of the subspecies *dawsoni* and *albiventer*. Systematic Biology 24:163–170.
Rausch, R. L., and V. R. Rausch. 1997. Evidence for specific independence of the shrew (Mammalia: Soricidae) of St. Paul Island (Pribilov Islands, Bering Sea). Zeitschrift für Säugetierkunde 62:193–202.
Rausch, R. L., J. E. Feagin, and V. R. Rausch. 2007. *Sorex rohweri* sp. nov. (Mammalia, Soricidae) from northwestern North America. Mammalian Biology 72:93–105.
Reddell, J. R. 1968. The hairy-legged vampire, *Diphylla ecaudata,* in Texas. Journal of Mammalogy 49:769.
Redford, K. H., and J. F. Eisenberg. 1992. Mammals of the Neotropics. The southern cone: Chile, Argentina, Uruguay, Paraguay. The University of Chicago Press, Chicago, U.S.A.
Redondo, R. A. F., L. P. S. Brina, R. F. Silva, A. D. Ditchfield, and F. R. Santos. 2008. Molecular systematics of the genus *Artibeus* (Chiroptera: Phyllostomidae). Molecular Phylogenetics and Evolution 49:44–58.
Reducker, D. W., T. L. Yates, and I. F. Greenbaum. 1983. Evolutionary affinities among southwestern long-eared Myotis (Chiroptera: Vespertilionidae). Journal of Mammalogy 64:666–677.
Reed, C. A. 1951. Locomotion and appendicular anatomy in three soricoid insectivores. The American Midland Naturalist 45:513–671.
Reed, E. K. 1955. Bison beyond the Pecos. Texas Journal of Sciences 7:130–135.
Reed, K. M., and J. R. Choate. 1986. Natural history of the plains pocket mouse in agriculturally disturbed sandsage prairie. Prairie Naturalist 18:79–90.
Reich, L. M. 1981. *Microtus pennsylvanicus*. Mammalian Species 159:1–8.
Reid, F. A. 1997. A field guide to the mammals of Central America and southeast Mexico. Oxford University Press. New York, U.S.A.
Reid, F. A. 2006. A field guide to mammals of North America north of Mexico. Peterson Field Guides. Houghton Mifflin Company. New York, U.S.A.
Reid, F. A. 2009. A field guide to the mammals of Central America and Southeast Mexico. Oxford University Press. New York, U.S.A.
Reid, F. 2016a. *Sciurus variegatoides*. The IUCN Red List of Threatened Species 2016:e.T20024A22246448.
Reid, F. 2016b. *Sorex maritimensis*. The IUCN Red List of Threatened Species 2016:e.T136779A22312357.
Reid, F., and J. Pino. 2016. *Peromyscus mexicanus*. The IUCN Red List of Threatened Species 2016:e.T16681A22363818.
Reid, F., and R. Timm. 2016. *Ototylomys phyllotis*. The IUCN Red List of Threatened Species 2016:e.T15666A115129272.
Reid, F., and E. Vázquez. 2016a. *Baiomys musculus*. The IUCN Red List of Threatened Species 2016:e.T2465A115062118.
Reid, F., and E. Vázquez. 2016b. *Handleyomys rostratus*. The IUCN Red List of Threatened Species 2016:e.T15612A115128466.
Reid, F., and E. Vázquez. 2016c. *Heteromys gaumeri*. The IUCN Red List of Threatened Species 2016:e.T10007A22223646.
Reid, F., and E. Vázquez. 2016d. *Heteromys pictus*. The IUCN Red List of Threatened Species 2016:e.T12075A22225256.
Reid, F., and E. Vázquez. 2016e. *Reithrodontomys microdon*. The IUCN Red List of Threatened Species 2016:e.T19412A22386904.
Reid, F., E. Vázquez, and L. Emmons. 2008a. *Habromys lophurus*. The IUCN Red List of Threatened Species2008:e.T9610A13004630.
Reid, F., E. Vázquez, L. Emmons, and T. McCarthy. 2008b. *Handleyomys saturatior*. The IUCN Red List of Threatened Species 2008:e.T15613A4908339.
Reid, F., E. Vázquez, L. Emmons, and A. D. Cuarón. 2008c. *Rheomys thomasi*. The IUCN Red List of Threatened Species 2008:e.T19486A8909070.
Reid, F., K. Helgen, and A. Kranz. 2016a. *Mustela erminea*. The IUCN Red List of Threatened Species 2016:e.T29674A45203335.
Reid, F., J. Pino, and R. Samudio. 2016b. *Scotinomys teguina*. The IUCN Red List of Threatened Species 2016:e.T20052A22390040.

Reid, F., R. Samudio, and J. Pino. 2016c. *Reithrodontomys sumichrasti*. The IUCN Red List of Threatened Species 2016:e.T19417A115151678.
Reid, F., M. Schiaffini, and J. Schipper. 2016d. *Neovison vison*. The IUCN Red List of Threatened Species 2016:e.T41661A45214988.
Reid, F., J. Schipper, and R. Timm. 2016e. *Bassariscus astutus*. The IUCN Red List of Threatened Species 2016:e.T41680A45215881.
Reid, F., E. Vázquez, and L. Emmons. 2016f. *Reithrodontomys gracilis*. The IUCN Red List of Threatened Species 2016:e.T19408A115150933.
Reig, S. 1997. Biogeographic and evolutionary implications of size variation in North American least weasels (Mustela nivalis). Canadian Journal of Zoology 75:2036–2049.
Repenning, C. A. 1967. Subfamilies and genera of the Soricidae. Geological Survey Paper 565:1–74.
Repenning, C. A., R. S. Peterson, and C. L. Hubbs. 1971. Contributions to the systematics of the southern fur seals, with particular reference to the Juan Fernández and Guadalupe species. Pp. 1–34, *in* Antarctic Pinnipedia, Antarctic Research Series 18, (Burt, W. H., ed.). American Geophysical Union. New York, U.S.A.
Rezsutek, M., and G. N. Cameron. 1993. *Mormoops megalophylla*. Mammalian Species 448:1–5.
Rheingantz, M. L., and C. S. Trinca. 2015. *Lontra longicaudis*. The IUCN Red List of Threatened Species 2015:e.T12304A21937379.
Rheingantz, M. L., P. Rosas-Ribeiro, J. Gallo-Reynoso, V. C. Fonseca da Silva, R. Wallace, V. Utreras, and P. Hernández-Romero. 2021. *Lontra longicaudis*. The IUCN Red List of Threatened Species 2021:e.T12304A164577708.
Rhoads, S. N. 1894. Description of a New Genus and Species of Arvicoline Rodent from the United States. American Naturalist 28:182–185.
Rhymer, J. M., J. M. Barbay, and H. L. Givens. 2004. Taxonomic relationship between *Sorex dispar* and *S. gaspensis*: inferences from mitochondrial DNA sequences. Journal of Mammalogy 85:331–337.
Rice, D. W. 1998. Marine Mammals of the World: Systematics and Distribution. Society for Marine Mammalogy, Special Publication Number 4. Lawrence, U.S.A.
Rice, D. W., K. Kenyon, and B. D. Lluch 1965. Pinniped populations at Islas Guadalpe, San Benito, and Cedros, Baja California, in 1965. Transactions of the San Diego Society of Natural History 14:73–84.
Rickart, E. A. 1982. Annual cycles of activity and body composition in *Spermophilus townsendii mollis*. Canadian Journal of Zoology 60:3298–3306.
Rickart, E. A. 1986. Postnatal growth of the Piute ground squirrel (*Spermophilus mollis*). Journal of Mammalogy 67:412–416.
Rickart, E. A. 1987. *Spermophilus townsendii*. Mammalian Species 268:1–6.
Rickart, E. A., and P. B. Robertson. 1985. *Peromyscus melanocarpus*. Mammalian Species 241:1–3.
Rickart, E. A., and E. Yensen. 1991. *Spermophilus washingtoni*. Mammalian Species 371:1–5.
Rickart, E. A., L. R. Heaney, and R. S. Hoffman. 2004. First record of *Sorex tenellus* from the central Great Basin. The Southwestern Naturalist 49:132–134.
Riddle, B. R. 1999. Mearns's grasshopper mouse *Onychomys arenicola*. Pp. 588, *in* The Smithsonian book of the North American mammals (Wilson, D., and S. Ruff, eds.). Smithsonian Institution Press. Washington, U.S.A.
Riddle, B. R., D. J. Hafner, and L. F. Alexander. 2000a. Comparative phylogeography of Baileys' pocket mouse (*Chaetodipus baileyi*) and the *Peromyscus eremicus* species group: historical vicariance of the Baja California Peninsular desert. Molecular Phylogentic and Evolution 17:161—172.
Riddle, B. R., D. J. Hafner, and L. F. Alexander. 2000b. Phylogeography and systematics of *Peromyscus eremicus* species group and historical biogeography of North American warm regional deserts. Molecular Phylogentic and Evolution 17:145—160.
Riddle, B. R., T. Jezkova, M. E. Eckstut, V. Oláh-Hemmings, and L. N. Carraway. 2014. Cryptic divergence and revised species taxonomy within the Great Basin pocket mouse, *Perognathus parvus* (Peale, 1848), species group. Journal of Mammalogy 95:9–25.
Rideout, C. B., and R. S. Hoffmann. 1975. *Oreamnos americanus*. Mammalian Species 63:1–6.
Riechers Pérez, A. 2004. Análisis mastofaunístico de la zona sujeta a conservación ecológica Laguna Bélgica, Chiapas, México. Anales del Instituto de Biología, Serie Zoología 75:363–382.
Riley, G. A., and R. T. McBride. 1975. A survey of the red wolf (*Canis rufus*). Pp. 263–277, *in* The wild canids (Fox, M. W., ed.), Van Nostrand Reinhold. New York, U.S.A.
Riley S. P. D., E. E. Boydston, K. R. Crooks, and L. M. Lyren. 2010. Bobcats (*Lynx rufus*). Pp. 121–140, *in* Urban carnivores: ecology, conflict, and conservation (Gehrt, S. D., S. P. D. Riley, and B. L. Cypher, eds). The Johns Hopkins Unieristy Press. Baltimore, U.S.A.
Rios, E., and S. T. Álvarez-Castañeda. 2010. Phylogeography and systematics of the San Diego pocket mouse (*Chaetodipus fallax*). Journal of Mammalogy 91:293–301.
Rios, E., and S. T. Álvarez-Castañeda. 2011. *Peromyscus guardia*. Mammalian Species 43:172–176.
Rios, E., and S. T. Álvarez-Castañeda. 2013. Nomenclatural change of *Chaetodipus dalquesti*. Western North America Naturalist 73:399–400.
Rios, E., C. Lorenzo, and S. T. Álvarez-Castañeda. 2016. Genetic evaluation of a microendemic and threatened rodent (*Heteromys nelsoni*) from Southern Mexico. Mammalia 81:289–296.
Risch, D., C. W. Clarke, P. J. Cockeron, A. Elepfandt, K. M. Kovacs, C. Lydersen, I. Stirling, and S. M. Van Parijs. 2007. Vocalizations of male bearded seals, *Erignathus barbatus*: classification and geographical variation. Animal Behavior 73:747–762.
Rizo-Aguilar, A., C. Delfín-Alfonso, A. González-Romero, and J. A. Guerrero. 2016. Distribution and density of the zacatuche rabbit (*Romerolagus diazi*) at the Protected Natural Area "Corredor Biológico Chichinautzin". Therya 7:333–342.
Rivas-Camo, N. A., P. A. Sabido-Villanueva, C. R. Peralta-Muñoz, and R. A. Medellin. 2020. Cuba in Mexico: first record of *Phyllops falcatus* (Gray, 1839) (Chiroptera, Phyllostomidae) for Mexico and other new records of bats from Cozumel, Quintana Roo. ZooKeys 973:153–162.
Roach, N. 2016. *Microdipodops megacephalus*. The IUCN Red List of Threatened Species 2016:e.T42606A115193770.
Roach, N. 2017a. *Marmota vancouverensis*. The IUCN Red List of Threatened Species 2017:e.T12828A22259184.
Roach, N. 2017b. *Chaetodipus ammophilus*. The IUCN Red List of Threatened Species 2017:e.T96812094A22226639.
Roach, N. 2018a. *Geomys tropicalis*. The IUCN Red List of Threatened Species 2018:e.T9056A22218038.
Roach, N. 2018b. *Cynomys parvidens*. The IUCN Red List of Threatened Species 2018:e.T6090A22260975.
Roach, N. 2018c. *Dipodomys ingens*. The IUCN Red List of Threatened Species 2018:e.T6678A22227241.
Roach, N. 2018d. *Dipodomys nitratoides*. The IUCN Red List of Threatened Species 2018:e.T6683A22228395.
Roach, N. 2018e. *Dipodomys stephensi*. The IUCN Red List of Threatened Species 2018:e.T6682A22228640.
Roach, N. 2020. *Microtus breweri*. The IUCN Red List of Threatened Species 2020:e.T13417A22349291.
Roach, N., and L. Naylor. 2016a. *Cryptotis obscura*. The IUCN Red List of Threatened Species 2016:e.T136462A22286153.

Roach, N., and L. Naylor. 2016b. *Xerospermophilus mohavensis*. The IUCN Red List of Threatened Species 2016:e.T20474A22266305.

Roach, N., and L. Naylor. 2017. *Cryptotis griseoventris*. The IUCN Red List of Threatened Species 2017:e.T48269619A123794519.

Roberts, J. D., and G. Mills. 1983. Funny lookin' rats. Texas Parks and Wildlife 41:12–15.

Roberts, H. R., D. J. Schmidly, and R. D. Bradley. 1988. *Peromyscus spicilegus*. Mammalian Species 596:1–4.

Roberts, H. R., D. J. Schmidly, and R. D. Bradley. 2001. *Peromyscus simulus*. Mammalian Species 669:1–3.

Robertson, P. B., and E. A. Rickart. 1976. *Cryptotis magna*. Mammalian Species 61:1–2.

Robinson, J. W., and R. S. Hoffmann. 1975. Geographical and interspecific cranial vanation in big-eared ground squirrels (*Spermophilus*): a multivariate study. Systematic Zoology 24:79–88.

Robinson, T. J., F. F. B. Elder, and J. A. Chapman. 1983. Evolution of chromosomal variation in cottontails, genus *Sylvilagus* (Mammalia: Lagomorpha): *S. aquaticus, S. florindanus*, and *S. transitionalis*. Cytogenetics and Cell Genetics 35:216–222.

Robinson, T. J., F. F. B. Elder, and J. A. Chapman. 1984. Evolution of chromosomal variation in cottontails, genus *Sylvilagus* (Mammalia: Lagomorpha). II. *Sylvilagus audubonii*, *S. idahoensis*, *S. nuttallii* and *S. palustris*. Cytogenetics and Cell Genetics 38:282–289.

Robinson, P. W., D. P. Costa, D. E. Crocker, J. P. Gallo-Reynoso, C, D, Champagne, M. A. Fowler, et al. 2012. Foraging Behavior and Success of a Mesopelagic Predator in the Northeast Pacific Ocean: Insights from a Data-Rich Species, the Northern Elephant Seal. PLoS ONE 7(5): e36728.

Roest, A. I. 1973. Subspecies of the sea otter, Enhydra lutris. Contribution Science, Los Angeles County Museum 252:1–17.

Rodriguez, B., and J. Cajas. 2015. *Dermanura tolteca*. The IUCN Red List of Threatened Species 2015:e.T2140A21997479.

Rodriguez, B., and B. Miller. 2015. *Cynomops mexicanus*. The IUCN Red List of Threatened Species 2015:e.T136611A21987867.

Rodriguez, B., and W. Pineda. 2015. *Macrophyllum macrophyllum*. The IUCN Red List of Threatened Species 2015:e.T12615A22025883.

Rodriguez, R. M., and L. K. Ammerman, 2004. Mitochondrial DNA divergence does not reXect morphological diVerence between *Myotis californicus* and *Myotis ciliolabrum*. Journal of Mammalogy 85:842–851.

Roe, F. G. 1970. The North American buffalo. Second edition, University of Toronto Press, Toronto, Canada.

Røed, K. H., M. A. D. Ferguson, M. Çrête, and T. A. Bergerud. 1991. Genetic variation in transferrin as a predictor for differentiation and evolution of caribou from Eastern Canada. Rangifer 11:65–74.

Roehrs, Z. P., J. B. Lack, and R. A. Van Den Bussche. 2010. Tribal phylogenetic relationships within Vespertilioninae (Chiroptera: Vespertilionidae) based on mitochondrial and nuclear sequence data. Journal of Mammalogy 91:1073–1092.

Roemer, G., B. Cypher, and R. List. 2016. *Urocyon cinereoargenteus*. The IUCN Red List of Threatened Species 2016:e.T22780A46178068.

Roest, A. I. 1964. A ribbon seal from California. Journal of Mammalogy 45:416–420.

Rogers, D. S. 1989. Evolutionary implications of chromosomal variation among spiny pocket mice, genus *Heteromys* (Order Rodentia). The Southwestern Naturalist 34:85–100.

Rogers, D. S. 1990. Genetic evolution, historial biogeography, and systematic relationships amongs spiny pocket mice (subfamily Heteromynae). Journal of Mammalogy 71:668–685.

Rogers, D. S., and J. E. Rogers. 1992. *Heteromys nelsoni*. Mammalian Species 397:1–2.

Rogers, D. S., and J. A. Skoy. 2011. *Peromyscus furvus*. Mammalian Species 43:209–215.

Rogers, D. S., and M. W. González. 2010. Phylogenetic relationships among spiny pocket mice (*Heteromys*) inferred from mitochondrial and nuclear sequence data. Journal of Mammalogy 91:914–930.

Rogers, D. S., and D. J. Schmidly. 1982. Systematics of spiny pocket mice (genus *Heteromys)* of the *desmarestianus* species group from Mexico and northern Central America. Journal of Mammalogy 63:375–386.

Rogers, D. S., and V. L. Vance. 2005. Phylogenetics of Spiny Pocket Mice (Genus *Liomys*): analysis of cytochrome b based on multiple heuristic approaches. Journal of Mammalogy 86:1085–1094.

Rogers, D. S., R. N. Leite, and R. J. Reed. 2011. Molecular phylogenetics of an endangered species: the Tamaulipan woodrat (*Neotoma angustapalata*). Conservation Genetics 12:1035–1048.

Rogers, M. A. 1991a. Evolutionary differentiation within the northern great basin pocket gopher, *Thomomys townsendii*. I. Morphological variation. The Great Basin Naturalist 51:109–126.

Rogers, M. A. 1991b. Evolutionary differentiation within the northern great basin pocket gopher, *Thomomys townsendii*. II. Genetic variation and biogeographic considerations. The Great Basin Naturalist 51:127–152.

Romo-Vázquez, E., L. Le6n, and O. Sánchez. 2005. A new species of *Habromys* (Rodentia: Sigmodontinae) from México. Proceedings of the Biological Society of Washington 118:605–611.

Roots, E. H., and R. J. Baker. 1998. *Rhogeessa genoways*i. Mammalian Species 589:1–3.

Roots, E. H., and R. J. Baker. 2007. *Rhogeessa parvula*. Mammalian Species 804:1–4.

Rose, R. K. 1981. *Synaptomys* not extinct in the Dismal Swamp. Journal Mammalogy 62:844–845.

Rose, R. K., and A. V. Linzey. 2021. *Synaptomys cooperi* (Rodentia: Cricetidae). Mammalian Species 53:95–111.

Rossi, R. V., R. S. Voss, and D. P. Lunde. 2010. A revision of the didelphid marsupial genus *Marmosa*. Part 1. The species in Tate's "*mexicana*" and "*mitis*" sections and other closely related forms. Bulletin of the American Museum of Natural History 334:1–83.

Rostland, E. 1960. The geographic range of the historic bison in the southeast. Annals Association American Geographers 50:395–407.

Roth, E. L. 1976. A new species of pocket mouse (*Perognathus:* Heteromyidae) from the cape region of Baja California Sur, Mexico. Journal of Mammalogy 57:562–566.

Ruedas, L. A. 1986. Chromosomal variability in the New England cottontail, *Sylvilagus transitionalis* (Bangs)[sic], 1895 with evidence of recognition of a new species. Master Thesis. Fordham University. New York, U.S.A.

Ruedas, L. A. 1998. Systematics of *Sylvilagus* Gray, 1867 (Lagomorpha: Leporidae) from Southwestern North America. Journal of Mammalogy 79:1355–1378.

Ruedas, L. A. 2018a. *Sylvilagus cognatus*. Pp. 130–131, *in* Pikas, rabbits and hares of the world (Smith, A., C. H. Johnston, P. C. Alves, and K. Hackländer, eds.). The Johns Hopkins University Press/IUCN/SSC. Baltimore, U.S.A.

Ruedas, L. A. 2018b. *Sylvilagus gabbi*. Pp. 140–142, *in* Pikas, rabbits and hares of the world (Smith, A., C. H. Johnston, P. C. Alves, and K. Hackländer, eds.). The Johns Hopkins University Press/IUCN/SSC. Baltimore, U.S.A.

Ruedas, L. A., and R. C. Dowler. 2018. *Sylvilagus robustus*. Pp. 154–155, *in* Pikas, rabbits and hares of the world (Smith, A., C. H. Johnston, P. C. Alves, and K. Hackländer, eds.). The Johns Hopkins University Press/IUCN/SSC. Baltimore, U.S.A.

Ruedas, L. A., and J. Salazar-Bravo. 2007. Morphological and chromosomal taxonomic assessment of *Sylvilagus brasiliensis gabbi* (Leporidae). Mammalia 71:63–69.

Ruedas, L., and A. T. Smith. 2019a. *Sylvilagus gabbi*. The IUCN Red List of Threatened Species 2019:e.T87491157A87491160.

Ruedas, L. A., and A. T. Smith. 2019b. *Sylvilagus robustus*. The IUCN Red List of Threatened Species 2019:e.T41310A165116781.

Ruedas, L. A., R. C. Dowler, and E. Aita. 1989. Chromosomal variation in the New England cottontail, *Sylvilagus transitionalis*. Journal of Mammalogy 70:860–864.

Ruedas, L. A., *et al.* 2017. A prolegomenon to the systematics of South American cottontail rabbits (Mammalia, Lagomorpha, Leporidae: *Sylvilagus*): designation of a neotype for *S. brasiliensis* (Linnaeus, 1758), and restoration of *S. andinus* (Thomas, 1897) and *S. tapetillus* Thomas, 1913. Miscellaneous Publications Museum of Zoology, University of Michigan 205:1–67.

Ruedas, L. A., L. López, and J. M. Mora. 2023. A propaedeutic to the taxonomy of the Eastern cottontail rabbit (Lagomorpha: Leporidae: *Sylvilagus floridanus*) from Central America. Therya 14:99–119.

Ruedi, M., and F. Mayer. 2001. Molecular systematics of bats of the genus *Myotis* (Vespertilionidae) suggests deterministic ecomorphological convergences. Molecular Phylogenetics and Evolution 21:436–448.

Ruiz-Soberanes, J. A., and G. Gómez-Álvarez. 2010. Estudio mastofaunístico del Parque Nacional Malinche, Tlaxcala, México. Therya 1:97–110.

Ruschi, A. 1951. Morcegos do Estado do Espírito Santo: Descricão de *Diphylla ecaudata* Spix e algumas observacões a seu respeito. Boletin do Museu de Biologia. "Prof. Mello Leitão" (S. Teresa, Brasil) 3:1–7.

Russell, R. J. 1968a. Evolution and classification of the pocket gophers of the subfamily Geomyinae. University of Kansas Publications, Museum of Natural History 16:473–579.

Russell, R. J. 1968b. Revision of the pocket gophers of the genus *Pappogeomys*. University of Kansas Publications, Museum of Natural History 16:581–776.

Ryan, J. M. 1989. Comparative myology and phylogenetic systematics of the Heteromydae (Mammalia, Rodentia). Miscellaneous Publications, Museum of Zoology, University of Michigan 176:1–103.

Ryan, R. M. 1968. Mammals of Deep Canyon, Colorado-Desert, California. The Desert Museum, Palm Springs, California, U.S.A.

Rydell, J., H. T. Arita, M. Santos, and J. Granados 2002. Acoustic identification of insectivorous bats (order Chiroptera) of Yucatan, Mexico. Journal of the Zoological Society of London 257:27–36.

Sakahira, F., and M. Niimi. 2007. Ancient DNA Analysis of the Japanese Sea Lion (*Zalophus japonicus japonicus* Peters, 1866): Preliminary results using mitochondrial control-region sequences. Zoological Sciences 24:81–85.

Sampaio, E., B. Lim, and S. Peters. 2016a. *Diphylla ecaudata*. The IUCN Red List of Threatened Species 2016:e.T6628A22040157.

Sampaio, E., B. Lim, and S. Peters. 2016b. *Micronycteris schmidtorum*. The IUCN Red List of Threatened Species 2016:e.T13383A22124156.

Sampaio, E., B. Lim, S. Peters, B. Miller, A. D. Cuarón, and P. C. de Grammont. 2016c. *Lophostoma brasiliense*. The IUCN Red List of Threatened Species 2016:e.T21984A115164165.

Sampaio, E., B. Lim, and S. Peters. 2017. *Molossus coibensis*. The IUCN Red List of Threatened Species 2017:e.T102208365A22106904.

Samudio, R., J. Pino, and F. Reid. 2016. *Nyctomys sumichrasti*. The IUCN Red List of Threatened Species 2016:e.T14999A115124642.

Samudio, R., Jr. 2016. Family Cuniculidae (Pacas). Pp. 398–405, *in* Handbook of the mammals of the world 6. Lagomorpha and Rodentia I (Wilson, D. E., T. E. Lacher, Jr., and R. A. Mittermeier, eds.). Lynx edicions. Barcelona, Spain.

Sanborn, C. C. 1932. The bats of the genus *Eumops*. Journal of Mammalogy 13:347–357.

Sanborn, C. C. 1933. Bats of the genera Anoura and Lonchoglossa. Field Museum of Natural History, Zoology Series 20:23–28.

Sanborn, C. C. 1936. Records and measurements of Neotropical bats. Field Museum of Natural History, Zoological series 20:93–106.

Sanborn, C. C. 1937. American bats of the subfamily Emballonurinae. Field Museum of Natural History, Zoological series 20:321–354.

Sanborn, C. C. 1941. Descriptions and records of neotropical bats. Field Museum of Natural History, Zoology Series 27:371–387.

Sanborn, C. C. 1949. Bats of the genus *Micronycteris* and its sub-genera. Fieldiana Zoology 31:215–233.

Sánchez-Casas, N., and T. Álvarez. 2000. Palinofagia de los murciélagos del género *Glossophaga* (Mammalia: Chiroptera) en México. Acta Zoológica Mexicana (n. s.) 81:23–62.

Sánchez-H., O., G. Téllez-G., R. A. Medellín, and G. Urbano-V. 1986. New records of mammals from Quintana Roo. Mammalia 50:275–278.

Sánchez-Cordero, V., C. Bonilla, and E. Cisneros. 1993. Thomas' mastiff bat, *Promops centralis* (Molossidae) in Oaxaca, Mexico. Bat Reserch News 34:65.

Sánchez-Cordero, V., L. Guevara, C. Lorenzo, and S. Ortega-García. 2014. Noteworthy records of an endemic shrew from Mexico (Mammalia, Soricomorpha, *Cryptotis griseoventris*), with comments on taxonomy. Mammalia 78:405–408.

Sánchez-Hernández, C., M. de la L. Romero-Almaraz, and C. García-Estrada. 2005. Mamíferos. Pp. 283–303, *in* Biodiversidad del estado de Tabasco, México (Bueno, J., F. Álvarez, and S. Santiago, eds.). Instituto de Biología, UNAM–CONABIO. Distrito Federal, México.

Sánchez-Hernández, C., and G. Gavino-de la Torre. 1988. Registros de murciélagos para la Isla de Pefia, Nayarit, México. Anales del Instituto de Biologia, Universidad Nacional Autónoma de México, Serie Zoología 58:939–940.

Sánchez-Hernández, C., and M. de la L. Romero-Almaraz. 1995. Murciélagos de Tabasco y Campeche, una propuesta para su conservación. Cuadernos del Instituto de Biología, Universidad Nacional Autónoma de México 24:1–213.

Sánchez-Hernández, C., M. de la L. Romero-Almaraz, R. D. Owen, A. Núñez-Garduño, and R. López-Wilchis. 1999. Noteworthy records of mammals from Michoacan, Mexico. The Southwestern Naturalist 44:231–235.

Sánchez-Hernández, C., G. D. Schnell, and M. de la L. Romero-Almaraz. 2009. *Peromyscus perfulvus* (Rodentia: Cricetidae). Mammalian Species 833:1–8.

Sánchez-Hernández C., M. L. Romero-Almaraz, A. Taboada-Salgado, J. A. Almazán-Catalán, G. D. Schnell, and L. Sánchez-Vázquez. 2010. Five albino bats from Guerrero and Colima, Mexico. Chiroptera Neotropical 16:522–527.

Sánchez-Hernández, C., M. de la L. Romero-Almaraz, G. D. Schnell, M. L. Kennedy, T. L. Best, R. D. Owen, and C. López-González. 2002. Bats of Colima, Mexico: new records, geographic distribution, and reproductive condition. Occasional Papers Sam Noble, Oklahoma Museum of Natural History 12:1–23.

Sanchez-Herrera, O., G. Tellez-Giron, R. A. Medellín, and G. Urbano-Vidales. 1986. New records of mammals from Quintana Roo. Mammalia 50:275–278.

Sanchez-Rojas, G., and S. Gallina-Tessaro. 2016. *Odocoileus hemionus*. The IUCN Red List of Threatened Species 2016:e.T42393A22162113.

Santos-Moreno, A. 2000. Records of Bats from Oaxaca, Mexico. The Southwestern Naturalist 55:454–456

Santos-Moreno, A., and A. M. Alfaro-Espinoza. 2009. Mammalian prey of barn owl (*Tyto alba*) in southeastern Oaxaca, Mexico. Acta Zoológica Mexicana (n. s.) 25:143–149.

Santos-Moreno, A., M. Briones-Salas, G. González-Pérez, and T. de J. Ortiz. 2003. Noteworthy records of two rare mammals in Sierra Norte de Oaxaca, Mexico. The Southwestern Naturalist 48:312–313.

Santos-Moreno, A., S. García-Orozco, and E. E. Pérez-Cruz. 2010. Records of bats from Oaxaca, Mexico. The Southwestern Naturalist 55:454–456.

Santos-Moreno, A., C. T. Aldape-López, C. Benítez-Díaz, and M. Martínez Coronel. 2016. Ampliación del límite superior de distribución altitudinal de tres especies de mamíferos en Oaxaca, México. Revista Mexicana de Biodiversidad 87:267–269.

Schai-Braum, S., and K. Hackländer. 2018. *Lepus arcticus.* Pp. 165–168, *in* Pikas, rabbits and hares of the world (Smith, A., C. H. Johnston, P. C. Alves, and K. Hackländer, eds.). The Johns Hopkins University Press/IUCN/SSC. Baltimore, U.S.A.

Schaldach, W. J. 1960. *Xenomys nelson* Merriam, sus relaciones y sus hábitos. Revista de la Sociedad de Historia Natural 21:425–434.

Schaldach, W. J. 1964. Notas breves sobre algunos mamíferos del sur de México. Anales del Instituto de Biología, Universidad Nacional Autónoma de México 35:129–137.

Schaldach, W. J. 1966. New forms of mammals from southern Oaxaca, Mexico, with notes on some mammals of the coastal range. Säugetierkundliche Mitteilungen 14:286–297.

Schaldach, W. J., and C. A. McLaughlin. 1960. A new genus and species of glossophagine bat from Colima, Mexico. Contributions in Science, Los Angeles County Museum 37:1–8.

Schusterman, R. J. 1981. Steller sea lion Eumetopias jubatus (Schreber, 1776). Pp. 119–141, in Handbook of marine mammals. Vol. 1. (Ridgeway, S. H., and R. J. Harrison, eds). Academic Press, London and New York.

Scheel D. M., G. J. Slater, S-O. Kolokotronis, C. W. Potter, D. S. Rotstein, K. Tsangaras, A. D. Greenwood, and K. M. Helgen. 2014. Biogeography and taxonomy of extinct and endangered monk seals illuminated by ancient DNA and skull morphology. ZooKeys 409:1–33.

Scheffer, T. H. 1941. Ground squirrel studies in the Four-Rivers country, Washington. Journal of Mammalogy 22:270–279.

Scheffer, V. B. 1958. Seals, sea lions, and walruses, a review of the Pinnipedia. Stanford University Press. Stanford.

Scheick, B. K., and W. McCown. 2014. Geographic distribution of American black bears in North America. Ursus 25:24–33.

Scheuering, E. 2018. *Arborimus longicaudus*. The IUCN Red List of Threatened Species 2018:e.T42615A22389366.

Schmidly, D. J. 1972. Geographic variation in the white-ankled mouse, *Peromyscus pectoralis*. Southwestern Naturalist 17:113–138.

Schmidly, D. J. 1973. Geographic variation and taxonomy of *Peromyscus boylii* from Mexico and the southern United States. Journal of Mammalogy 54:111–130.

Schmidly, D. J. 1974. *Peromyscus pectoralis*. Mammalian Species 49:1–3.

Schmidly, D. J. 1977. The Mammals of Trans-Pecos Texas. Texas A & M University Press. College Station, U.S.A.

Schmidly, D. J. 1983. Texas Mammals east of the Balcones fault zone. Texas A&M University Press. College Station, U.S.A.

Schmidly, D. J. 2004. The Mammals of Texas. Revised Edition. University of Texas Press. Austin, U.S.A.

Schmidly, D. J. and F. S. Hendricks. 1976. Systematics of the southern races of Ord's kangaroo rat, *Dipodomys ordii.* Bulletin of the Southern California Academy of Sciences 75:225–237.

Schmidly, D. J., R. D. Bradley, and P. S. Cato. 1988. Morphometric differentiation and taxonomy of three chromosomally characterized groups of *Peromyscus boylii* from east-central Mexico. Journal of Mammalogy 69:460–480.

Schmidly, D. J., M. R. Lee, W. S. Modi, and E. G. Zimmerman. 1985. Systematics and notes on the biology of *Peromyscus hooperi.* Occasional Papers, the Museum, Texas Tech University 97:1–40.

Schmidt, C. A., and M. D. Engstrom. 1994. Genic variation and systematics of rice rats (*Oryzomys palustris* species group) in southern Texas and northeastern Tamaulipas, Mexico. Journal of Mammalogy 75:914–928.

Schmidt, C. A., M. D. Engstrom, and H. H. Genoways. 1989. *Heteromys gaumeri.* Mammalian Species 345:1–4.

Schmidt, U. 1978. Vampirfledermäuse. Die Neue Brehm-BÚcherei. Ziemsen Verlag, Wittenberg Lutherstadt, Germany.

Schnell, G. D., T. L. Best, and M. L. Kennedy. 1978. Interspecific morphologic variation in kangaroo rats (*Dipodomys*): degree of concordance with genic variation. Systematic Biology 27:34–48.

Schwartz, A. 1953. A systematic study of the water rat (*Neofiber alleni*). Occasional Papers Museum Zoology, University of Michigan 547:1–27.

Schwartz, A. 1956. A new subspecies of the Longtail Shrew (*Sorex dispar* Batchelder) from the Southern Appalachian Mountains. Journal of the Elisha Mitchell Scientific Society 72:24–30.

Schwartz, A., and E. P. Odum. 1957. The woodrats of the eastern United States. Journal of Mammalogy 38:197–206.

Schwartz, C. C., S. D. Miller, and M. A. Haroldson. 2003a. Grizzly bear. Pp. 556–586, *in* Wild mammals of North America: biology, management, and conservation (Feldhamer, G. A., B. C. Thompson, and J. A. Chapman, eds). Johns Hopkins University Press. Baltimore, U.S.A.

Schwartz, C. W., and E. R. Schwartz. 1981. The wild mammals of Missouri. Revised ed. University of Missouri Press and Missouri Department of Conservation. Columbia, U.S.A.

Schwartz, M. K., K. L. Pilgrim, K. S. Mckelvey, E. L. Lindquist, J. J. Claar, S. Loch, and L. F. Ruggiero. 2003b. Hybridization between Canada lynx and bobcats: Genetic results and management implications. Conservation Genetics 5:349.

Scott, F. W. 1987. First record of the long-tailed shrew, *Sorex dispar*, from Nova Scotia. Canadian Field-Naturalist 101:404–407.

Scott, F. W., and C. G. van Zyll de Jong. 1989. New Nova Scotia records of the Long-Tailed shrew, *Sorex dispar*, with comments on the taxonomic status of *Sorex dispar* and *Sorex gaspensis*. Le Naturaliste Canadien 116:145–154.

Sealander, J. A. 1979. A guide to Arkansas mammals. River Road Press. Conway, U.S.A.

Segura-Trujillo, C. A., and S. Navarro-Pérez. 2010. Escenario y problemática de conservación de los murciélagos (Chiroptera) cavernícolas del Complejo Volcánico de Colima, Jalisco-Colima, México. Therya 1:189–206.

Segura, V., F. Prevosti, and G. Cassini. 2013. Cranial ontogeny in the Puma lineage, *Puma concolor*, *Herpailurus yagouaroundi*, and *Acinonyx jubatus* (Carnivora: Felidae): a three-dimensional geometric morphometric approach. Zoological Journal of the Linnean Society 169:235–250.

Segura-Trujillo, C. A., A. L. Trujano-Álvarez and S. T. Álvarez-Castañeda. 2021. *Peromyscus ochraventer.* Mammalian Species, in press.

Sera, W. E., and C. N. Early. 2003. *Microtus montanus*. Mammalian Species 716:1–10.

Serfass, T. 2021. *Lontra canadensis*. The IUCN Red List of Threatened Species 2021:e.T12302A164577078.

Seton, E. T. 1929. Lives of game animals. Doubleday, Doran Co. Carden City, U. S. A.

Setzer, H. W. 1949. Subspeciation in the kangaroo rat, *Dipodomys ordii*. University of Kansas Publications, Museum of Natural History 1:473–573.
Severinghaus, W. D. 1977. Description of a new subspecies of prairie vole, *Microtus ochrogaster*. Proceedings of the Biological Society of Washington 90:49–54.
Severinghaus, W. D., and D. F. Hoffmeister. 1978. Qualitative cranial characters distinguishing *Sigmodon hispidus* and *Sigmodon arizonae* and the distribution of these two species in northern Mexico. Journal of Mammalogy 59:868–870.
Seymour, K. L. 1989. *Panthera onca*. Mammalian Species 340:1–9.
Shackleton, D. M. 1985. *Ovis canadensis*. Mammalian Species 230:1–9.
Shafer, A. B. A., and D. T. Stewart. 2007. Phylogenetic relationships among Nearctic shrews of the genus *Sorex* (Insectivora, Soricidae) inferred from combined cytochrome *b* and inter-SINE fingerprint data using Bayesian analysis. Molecular Phylogenetics and Evolution 44:192–203.
Shamel, H. H. 1931. Notes in the North American bats of the genus *Tadarida*. Proceedings of the United States National Museum 78:1–27.
Shaughnessy, P. D., and F. H. Fay. 1977. A review of the taxonomy and nomenclature of North Pacific harbour seals. Journal of Zoology (London) 182:385–419.
Shaw, W. T. 1934. The ability of the giant kangaroo rat as a harvester and storer of seeds. Journal of Mammalogy 15:275–286.
Sheeler-Gordon, L., and J. S. Smith. 2001. Survey of bat populations from Mexico and Paraguay for rabies. Journal of Wildlife Diseases 37:582–593.
Sheffield, S. R., and C. M. King. 1994. *Mustela nivalis*. Mammalian Species 454:1–10.
Sheffield, S. R., and H. H. Thomas. 1997. *Mustela frenata*. Mammalian Species 570:1–9.
Shellhammer, H. 1982. *Reithrodontomys raviventris*. Mammalian Species 169:1–3.
Shohfi, H. A., C. J. Conroy, A. R. Wilhelm, and J. L. Patton. 2006. New records of *Sorex preblei* and *S. tenellus* in California. The Southwestern Naturalist 51:108–111.
Shump, K. A., and R. H. Baker. 1978a. *Sigmodon alleni*. Mammalian Species 95:1–2.
Shump, K. A., and R. H. Baker. 1978b. *Sigmodon leucotis*. Mammalian Species 96:1–2.
Shump, K. A., and A. U. Shump. 1982a. *Lasiurus borealis*. Mammalian Species 183:1–6.
Shump, K. A., and A. U. Shump. 1982b. *Lasius cinereus*. Mammalian Species 185:1–5.
Silva-Caballero, A., and O. C. Rosas-Rosas. 2022. Rediscovery of the Tamaulipas white-sided jackrabbit (*Lepus altamirae*) after a century from its description. Therya Notes 3:1–3.
Simmons, N. B. 1996. A new species of *Micronycteris* (Chiroptera, Phyllostomidae) from northeastern Brazil: with comments on phylogenetic relationships. American Museum Novitates (3158):1–34.
Simmons, N. B. 1998. A reappraisal of interfamilial relationships of bats. Pp. 1–26, *in* Bat biology and conservation (Kunz, T. H., and P. A. Racey, eds.). Smithsonian Institution Press. Washington, U.S.A.
Simmons, N. B. 2005. Order Chiroptera. Pp. 312–529, *in* Mammal species of the world. A taxonomic and geographic reference (Wilson, D. E., and D. A. Reeder, eds.). Third edition. The Johns Hopkins University Press. Baltimore, U.S.A.
Simmons, N. B., and T. M. Conway. 2001. Phylogenetic relationships of mormoopid bats (Chiroptera: Mormoopidae) based on morphological data. Bulletin of the American Museum of Natural History 258:1–97.
Simmons, N. B., and C. O. Handley, Jr. 1998. A revision of *Centronycteris* Gray (Chiroptera: Emballonuridae) with notes on natural history. American Museum Novitates 3239:1–28.
Simmons, N. B., and R. S. Voss. 1998. The mammals of Paracou, French Guiana: a neotropical lowland rainforest fauna part I. Bats. Bulletin of the American Museum of Natural History 237:1–219.
Simons, L. H., and D. F. Hoffmeister. 2003. *Sorex arizonae*. Mammalian Species 732:1–3.
Simons, L. H., and W. E. Van Pelt. 1999. Occurrence of *Sorex arizonae* and other shrews (Insectivora) in southern Arizona. The Southwestern Naturalist 44:334–342.
Simpson, G. G. 1945. The principles of classification and a classification of mammals. Bulletin American Museum Natural History 85:1–350.
Simpson, M. R. 1993. *Myotis californicus*. Mammalian Species 428:1–4.
Singh, M., A. Kumar, and H. N. Kumara. 2020. *Macaca mulatta*. The IUCN Red List of Threatened Species e.T12554A17950825.
Skinner, M. F., and O. C. Kaisen. 1947. The fossil *Bison* of Alaska and preliminary revision of the genus. Bulletin American Museum Natural History 89:123–256.
Smith, A. T., and D. E. Brown. 2019a. *Sylvilagus audubonii*. The IUCN Red List of Threatened Species 2019:e.T41297A45190821.
Smith, A. T., and D. E. Brown. 2019b. *Sylvilagus nuttallii*. The IUCN Red List of Threatened Species 2019:e.T41300A45192243.
Smith, A.T., and E. Beever. 2016. *Ochotona princeps*. The IUCN Red List of Threatened Species 2016:e.T41267A45184315.
Smith, A.T., and C. H. Johnston. 2019a. *Lepus arcticus*. The IUCN Red List of Threatened Species 2019:e.T41274A45185887.
Smith, A.T., and C. H. Johnston. 2019b. *Lepus othus*. The IUCN Red List of Threatened Species 2019:e.T11795A45178124.
Smith, A. T., and A. R. Ruedas. 2016. *Sylvilagus cognatus*. The IUCN Red List of Threatened Species 2016:e.T41309A45193738.
Smith, A. T., and M. L. Weston. 1990. *Ochotona princeps*. Mammalian Species 352:1–8.
Smith, A. T., E. A. Beever, and C. Ray. 2018. *Ochotona prínceps*. Pp. 67–72, *in* Pikas, rabbits and hares of the world (Smith, A., C. H. Johnston, P. C. Alves, and K. Hackländer, eds.). The Johns Hopkins University Press/IUCN/SSC. Baltimore, U.S.A.
Smith, F. A. 1997. *Neotoma cinerea*. Mammalian Species 564:1–8.
Smith, H. C. 1988. The wandering shrew, *Sorex vagrans,* in Alberta. Canadian Field-Naturalist 102:254–256.
Smith, H. C. 1993. Alberta mammals: an atlas and guide. Provincial Museum of Alberta, Edmonton, Canada.
Smith, J. D. 1970. The Systematic status of the black howler monkey, *Aloutta pigra* Lawrence. Journal of Mammalogy 51:358–369.
Smith, J. D. 1972. Systematics of the chiropteran Family Mormoopidae. Miscellaneous Publications, University of Kansas Publications, Museum of Natural History 56:1–132.
Smith, A. T., N. A. Formozov, R. S. Hoffman, C. Zheng, and M. A. Erbajeva. 1990. The pikas. Pp 14–60, in Rabbits, hares and pikas. Status survey and conservation action plan, IUCN, Chapman, J. A., and J. E. C. Flux, eds). Gland, Switzerland.
Smith, J. E., J. D. Long, D. I. Russell, K. L. Newcomb, and V. D. Muñoz. 2016. *Otospermophilus beecheyi* (Rodentia: Sciuridae). Mammalian Species 48:91–108.
Smith, M. E., and M. C. Belk. 1996. *Sorex monticolus*. Mammalian Species 528:1–5.

Smith, M. F. 1999. *Thomomys townsendii*. Pp. 477–478, *in* The Smithsonian book of North American Mammals (Wilson, D. E., and S. Ruff). Smithsonian Institution Press. Washington, U.S.A.
Smith, W. P. 1991. *Odocoileus virginianus*. Mammalian Species 388:1–13.
Smolen, M. J. 1981. *Microtus pinetorum*. Mammalian Species 147:1–7.
Smolen, M. J., and B. L. Keller. 1987. *Microtus longicaudus*. Mammalian Species 271:1–7.
Snow, J. L., J. K. Jones, Jr., W. D. Webster. 1980. *Centurio senex*. Mammalian Species 138:1–3.
Snyder, D. P. 1982. *Tamias striatus*. Mammalian Species 168:1–8.
Sokolov, V. E., and V. N. Orlov. 1980. Opredelitel' mlekopitayushchickh Mongol'skoi Narodnoi Respubliki [Guide to the mammals of the Mongolian People's Republic]. Nauka, Moscow, Russia.
Solari, S. 2015a. *Chiroderma villosum*. The IUCN Red List of Threatened Species 2015:e.T4668A22037709.
Solari, S. 2015b. *Lonchorhina aurita*. The IUCN Red List of Threatened Species 2015:e.T12270A22039503.
Solari, S. 2015c. *Phylloderma stenops*. The IUCN Red List of Threatened Species 2015:e.T17168A22134036.
Solari, S. 2015d. *Saccopteryx bilineata*. The IUCN Red List of Threatened Species 2015:e.T19804A22004716.
Solari, S. 2015e. *Saccopteryx leptura*. The IUCN Red List of Threatened Species 2015:e.T19807A22005807.
Solari, S. 2015f. *Uroderma magnirostrum*. The IUCN Red List of Threatened Species 2015:e.T22783A22048094.
Solari, S. 2016. Molossus alvarezi. The IUCN Red List of Threatened Species 2016: e.T88087329A88087332.
Solari, S. 2016a. *Anoura geoffroyi*. The IUCN Red List of Threatened Species 2016:e.T88109511A88109515.
Solari, S. 2016b. *Dermanura azteca*. The IUCN Red List of Threatened Species 2016:e.T2123A22000362.
Solari, S. 2016c. *Dermanura watsoni*. The IUCN Red List of Threatened Species 2016:e.T99586593A21997358.
Solari, S. 2016d. *Eumops floridanus*. The IUCN Red List of Threatened Species 2016:e.T136433A21984011.
Solari, S. 2016e. *Pteronotus mesoamericanus*. The IUCN Red List of Threatened Species 2016:e.T88018392A88018395.
Solari, S. 2016f. *Pteronotus parnellii*. The IUCN Red List of Threatened Species 2016:e.T88017638A22077695.
Solari, S. 2016g. *Vampyrodes major*. The IUCN Red List of Threatened Species 2016:e.T88151984A88151987.
Solari, S. 2017a. *Eumops nanus*. The IUCN Red List of Threatened Species 2017:e.T87994060A87994063.
Solari, S. 2017b. *Rhogeessa bickhami*. The IUCN Red List of Threatened Species 2017:e.T88151726A88151729.
Solari, S. 2017c. *Sturnira hondurensis*. The IUCN Red List of Threatened Species 2017:e.T88154577A88154581.
Solari, S. 2018a. *Bauerus dubiaquercus*. The IUCN Red List of Threatened Species 2018:e.T1789A22129523.
Solari, S. 2018b. *Choeronycteris mexicana*. The IUCN Red List of Threatened Species 2018:e.T4776A22042479.
Solari, S. 2018c. *Enchisthenes hartii*. The IUCN Red List of Threatened Species 2018:e.T2130A21996891.
Solari, S. 2018d. *Glyphonycteris sylvestris*. The IUCN Red List of Threatened Species 2018:e.T13384A22123687.
Solari, S. 2018e. *Lampronycteris brachyotis*. The IUCN Red List of Threatened Species 2018:e.T13376A22131330.
Solari, S. 2018f. *Lichonycteris obscura*. The IUCN Red List of Threatened Species 2018:e.T88120245A22057648.
Solari, S. 2018g. *Lophostoma evotis*. The IUCN Red List of Threatened Species 2018:e.T21986A22041302.
Solari, S. 2018h. *Macrotus californicus*. The IUCN Red List of Threatened Species 2018:e.T12652A22031754.
Solari, S. 2018i. *Macrotus waterhousii*. The IUCN Red List of Threatened Species 2018:e.T12653A22032004.
Solari, S. 2018j. *Myotis grisescens*. The IUCN Red List of Threatened Species 2018:e.T14132A22051652.
Solari, S. 2018k. *Myotis leibii*. The IUCN Red List of Threatened Species 2018:e.T14172A22055716.
Solari, S. 2018l. *Myotis occultus*. The IUCN Red List of Threatened Species 2018:e.T136650A21990499.
Solari, S. 2018m. *Myotis septentrionalis*. The IUCN Red List of Threatened Species 2018:e.T14201A22064312.
Solari, S. 2018n. *Perimyotis subflavus*. The IUCN Red List of Threatened Species 2018:e.T17366A22123514.
Solari, S. 2018o. *Tonatia saurophila*. The IUCN Red List of Threatened Species 2018:e.T41530A22004890.
Solari, S. 2018p. *Vampyrum spectrum*. The IUCN Red List of Threatened Species 2018:e.T22843A22059426.
Solari, S. 2019a. *Baeodon alleni*. The IUCN Red List of Threatened Species 2019:e.T19679A21989577.
Solari, S. 2019b. *Baeodon gracilis*. The IUCN Red List of Threatened Species 2019:e.T19681A22007578.
Solari, S. 2019c. *Corynorhinus mexicanus*. The IUCN Red List of Threatened Species 2019:e.T17599A21976792.
Solari, S. 2019d. *Eumops ferox*. The IUCN Red List of Threatened Species 2019:e.T87994072A87994075.
Solari, S. 2019e. *Gardnerycteris crenulatum*. The IUCN Red List of Threatened Species 2019:e.T13560A88177260.
Solari, S. 2019f. *Lasionycteris noctivagans*. The IUCN Red List of Threatened Species 2019:e.T11339A22122128.
Solari, S. 2019g. *Lasiurus seminolus*. The IUCN Red List of Threatened Species 2019:e.T11353A22119113.
Solari, S. 2019h. *Myotis nigricans*. The IUCN Red List of Threatened Species 2019:e.T14185A22066939.
Solari, S. 2019i. *Myotis volans*. The IUCN Red List of Threatened Species 2019:e.T14210A22069325.
Solari, S. 2019j. *Myotis yumanensis*. The IUCN Red List of Threatened Species 2019:e.T14213A22068335.
Solari, S. 2019k. *Natalus mexicanus*. The IUCN Red List of Threatened Species 2019:e.T123984355A22011975.
Solari, S. 2019l. *Nycticeius humeralis*. The IUCN Red List of Threatened Species 2019:e.T14944A22015223.
Solari, S. 2019m. *Pipistrellus hesperus*. The IUCN Red List of Threatened Species 2019:e.T17341A22129352.
Solari, S. 2019n. *Pteronotus gymnonotus*. The IUCN Red List of Threatened Species 2019:e.T18706A22077065.
Solari, S. 2019o. *Rhogeessa aeneus*. The IUCN Red List of Threatened Species 2019:e.T136810A22043785.
Solari, S. 2019p. *Rhogeessa parvula*. The IUCN Red List of Threatened Species 2019:e.T19684A22007495.
Solari, S. 2019q. *Sturnira parvidens*. The IUCN Red List of Threatened Species 2019:e.T88154376A88154380.
Solari, S. 2019r. *Uroderma bilobatum*. The IUCN Red List of Threatened Species 2019:e.T22782A22048748.
Solari, S. 2019s. *Molossus aztecus*. The IUCN Red List of Threatened Species 2019:e.T13645A22107522.
Solari, S. 2019t. *Nyctinomops aurispinosus*. The IUCN Red List of Threatened Species 2019:e.T14993A22010682.
Solari, S. 2019u. *Promops centralis*. The IUCN Red List of Threatened Species 2019:e.T88087651A22036112.
Solari, S. 2021. *Myotis lucifugus*. The IUCN Red List of Threatened Species 2021:e.T14176A208031565.
Solari, S., V. Pacheco, and E. Vivar. 1999. New distribution of Peruvian bats. Revista Peruana de Biología 6:152–159.
Solari, S., and M. Camacho. 2019. *Micronycteris microtis*. The IUCN Red List of Threatened Species 2019:e.T136424A21985267.

Solari, S., and L. Davalos. 2019. *Pteronotus davyi*. The IUCN Red List of Threatened Species 2019:e.T18705A22077399.

Solari, S., and D. Lew. 2015. *Caluromys derbianus*. The IUCN Red List of Threatened Species 2015:e.T3650A22175821.

Solari, S., S. R. Hoofer, P. A. Larsen, A. D. Brown, R. J. Bull, J. A. Guerrero, J. Ortega, J. P. Carrera, R. D. Bradley, and R. J. Baker. 2009. Operational criteria for genetically defined species: analysis of the diversification of the small fruit–eating bats, *Dermanura* (Phyllostomidae: Stenodermatinae). Acta Chiropterologica 11:279–288.

Solari, S., C. Mancina, and L. Davalos. 2019. *Phyllops falcatus*. The IUCN Red List of Threatened Species 2019:e.T17176A22133485.

Soler-Frost, A., R. A. Medellín, and G. N. Cameron. 2003. *Pappogeomys bulleri*. Mammalian Species 717:1–3.

Solmsen, E.-H., and H. Schilemann. 2007. *Choeroniscus minor*. Mammalian Species 822:1–6.

Sosa, V. J., A. Hernández-Huerta, and J. A. Vargas-Contreras 2005. Los mamíferos. Pp. 522–537, *in* Historia Natural de la Reserva de la Biosfera El Cielo, Tamaulipas, México (Sánchez-Ramos, G., P. Reyes-Castillo, and R. Dirzo, eds.). Universidad Autónoma de Tamaulipas. Tamaulipas, México.

Spencer, A. W. 1966. Identification of the dwarf shrew, *Sorex nanus*. Journal of Colorado-Wyoming Academy of Sciences 5:89.

Spitsyn, V. M., I. N. Bolotov, A. V. Kondakov, A. L. Klass, I. A. Mizin, A. A. Tomilova, N. A. Zubrii, and M. Y. Gofarov. 2021. A new Norwegian Lemming subspecies from Novaya Zemlya, Arctic Russia. Ecologica Montenegrina 40:93–117.

Spradling, T. A., J. W. Demastes, D. J. Hafner, P. L. Milbach, F. A. Cervantes, and M. S. Hafner. 2016. Systematic revision of the pocket gopher genus *Orthogeomys*. Journal of Mammalogy 97:405–423.

Spaulding, M., M. A. O'Leary, and J. Gatesy. 2009. Relationships of Cetacea (Artiodactyla) among mammals: increased taxon sampling alters interpretations of key fossils and character evolution. PLoS One 4(9): e7062. https://doi.org/10.1371/journal.pone.0007062

Stadelmann, B., L. G. Herrera, J. Arroyo-Cabrales, J. J. Flores-Martínez, B. P. May, and M. Ruedi. 2004. Molecular systematics of the Fishing bat *Myotis* (*Pizonyx*) *vivesi*. Journal of Mammalogy 85:133–139.

Stadelmann, B., L. K. Lin, T. H. Kunz, and M. Ruedi. 2007. Molecular phylogeny of New World *Myotis* (Chiroptera, Vespertilionidae) inferred from mitochondrial and nuclear DNA genes. Molecular Phylogenetics and Evolution 43:32–48.

Stangl, F. B., Jr., B. F. Koop, and C. S. Hood. 1983. Occurrence of *Baiomys taylori* (Rodentia: Cricetidae) on the Texas High Plains. Occasional Papers Museum, Texas Tech University 85:1–4.

Stalling, D. T. 1990. *Microtus ochrogaster*. Mammalian Species 355:1–9.

Statham, M., J. Murdoch, J. Janecka, K. Aubry, C. Edwards, C. Soulsbury, O. Berry, Z. Wang, D. Harrison, M. Pearch, L. Tomsett, J. Chupasko, and B. Sacks. 2014. Range-wide multilocus phylogeography of the red fox reveals ancient continental divergence, minimal genomic exchange and distinct demographic histories. Molecular Ecology 23:4813–4830.

Steele, M. A. 1998. *Tamiasciurus hudsonicus*. Mammalian Species 586:1–9.

Steele, M. A. 1999. *Tamiasciurus douglasii*. Mammalian Species 630:1–8.

Steiner, C. C., and O. A. Ryder, 2011. Molecular phylogeny and evolution of the Perissodactyla. Zoological Journal of the Linnean Society 163:1289–1303.

Stephen, C. T. 1988. *Spermophilus saturatus*. Mammalian Species 322:1–4.

Stephen, L. W. 1982. *Geomys personatus*. Mammalian Species 170:1–5.

Stephen, L. W., J. Ramírez-Pulido, and R. J. Baker. 1985. *Peromyscus alstoni*. Mammalian Species 242:1–4.

Stephen, L. W., and R. J. Baker. 1974. *Geomys arenarius*. Mammalian Species 36:1–3.

Steppan, S. J., M. R. Akhverdyan, E. A. Lyapunova, D. G. Fraser, N. N. Vorontov, R. S. Hoffmann, and M. J. Braun. 1999. Molecular Phylogeny of the Marmots (Rodentia: Sciuridae): Tests of Evolutionary and Biogeographic Hypotheses. Systematic Biology 48:715–734.

Stewart, B. S., and H. R. Huber. 1993. *Mirounga angustirostris*. Mammalian Species 449:1–10.

Stewart, D. T., and A. J. Baker. 1994a. Patterns of sequence variation in the mitochondrial D-Loop region of shrews. Molecular Biology and Evolution 11:9–21.

Stewart, D. T., and A. J. Baker. 1994b. Evolution of mtDNA D-loop sequences and their use in phylogenetic studies of shrews in the subgenus *Otisorex* (*Sorex*: Soricidae: Insectivora). Molecular Phylogenetics and Evolution 3:38–46.

Stewart, D. T., and A. J. Baker. 1997. A phylogeny of some taxa of Masked Shrews (*Sorex cinereus*) based on mitochondrial DNA D-loop sequences. Journal of Mammalogy 78:361–376.

Stewart, D. T., T. B. Herman, and T. Teferi. 1989. Littoral feeding in a high-density insular population of *Sorex cinereus*. Canadian Journal of Zoology 67: 2074–2077.

Stewart, D. T., A. J. Baker, and S. P. Hindocha. 1993. Genetic differentiation and population-structure in *Sorex haydeni* and *S. cinereus*. Journal of Mammalogy 74:21–32.

Stewart, D. T., N. D. Perry, and L. Fumagalli. 2002. The maritime shrew, *Sorex maritimensis* (Insectivora: Soricidae): A newly recognized Canadian endemic. Canadian Journal of Zoology 80:94–99.

Stewart, D. T., M. McPherson, J. Robichaud, and L. Fumagalli. 2003. Are there two species of pygmy shrews (*Sorex*)? Revisiting the question using DNA sequence data. Canadian Field-Naturalist 117: 82–88.

Stewart, R. E. A., K. M. Kovacs, and M. Acquarone. 2014. Walrus of the North Atlantic. NAMMCO Scientific Publications 9:7–12.

Stock, A. D. 1971. Chromosome evolution in the genus *Dipodomys* and its phylogenetic implications. Mammalian Chromosomes Newsletter 12:122–128.

Stock, A. D. 1974. Chromosome evolution in the genus *Dipodomys* and its taxonomic and phylogenetic implications. Journal of Mammalogy 55:505–526.

Stone, W., and J. A. G. Rehn. 1903. On the terrestrial vertebrates of portions of southern New Mexico and western Texas. Proceedings of the Academy of Natural Sciences of Philadelphia 55:16–34.

Stone, K. D., R. W. Flynn, and J. A. Cook. 2002. Post-glacial colonization of northwestern North America by the forest associated American marten (*Martes americana*). Molecular Ecology 11:2049–2064.

Stone, K. D., and J. A. Cook. 2002. Molecular evolution of Holarctic martens (genus Martes, Mammalia: Carnivora: Mustelidae). Molecular Physlogical. Evolution 24:169–179.

Stoner, K. E., M. Quesada, V. Rosas-Guerrero, and J. A. Lobo. 2002. Effects of forest fragmentation on the Colima long-nosed bat (*Musonycteris harrisoni*) foraging in tropical dry forest of Jalisco, Mexico. Biotropica 34:462–467.

Straney, D. O., and J. L. Patton. 1980. Phylogenetic and environmental determinants of geographic variation of the pocket mouse *Perognathus goldmani* Osgood. Evolution 34:888–903.

Streubel, D. P., and J. P. Fitzgerald. 1978a. *Spermophilus spilosoma*. Mammalian Species 101:1–4.

Streubel, D. P., and J. P. Fitzgerald. 1978b. *Spermophilus tridecemlineatus*. Mammalian Species 103:1–5.

Strickland, M. A., C. W. Douglas, M. Novak, and N. P. Huzinger. 1982. Fisher *Martes pennanti*. Pp. 586–598, *in* Wild Mammals of North America: biology, management, and economics (Chapman, J. A., and G.A. Feldhamer, eds.). Johns Hopkins University Press. Baltimore, U.S.A.

Sudman, P. D., J. K. Wickliffe, P. Horner, M. J. Smolen, J. W. Bickham, and R. D. Bradley. 2006. Molecular systematics of pocket gophers of the genus *Geomys*. Journal of Mammalogy 87:668–676.

Sudman, P. D., J. R. Choate, and E. G. Zimmerman. 1987. Taxonomy of chromosomal races of *Geomys bursarius lutescens* Merriam. Journal of Mammalogy 68:526–543.

Sulentich, J. M., L. R. Williams, and G. N. Cameron. 1991. *Geomys breviceps*. Mammalian Species 383:1–4.

Sulkava, S. 1999. Sorex isodon Turov, 1924. Pp 50–51, in The Atlas of European Mammals (Mitchell-Jones, A. J., G. Amori, W. Bogdanowicz, B. Kryštufec, P. J. H. Reijnders, F. Spitzenberger, M. Stubbe, J. B. M. Thissen, V. Vohralík, and J. Zima, eds.). Poyser, London, UK.

Sullivan, E. G. 1956. Gray fox reproduction, denning, range, and weights in Alabama. Journal of Mammalogy 37:346–351.

Sullivan, J. K., and C. W. Kilpatrick. 1991. Biochemical systematics of the *Peromyscus aztecus* assemblage. Journal of Mammalogy 72:681–696.

Sullivan, J. K., J. A. Markert, and C. W. Kilpatrick. 1997. Phylogeography and molecular systematics of the *Peromyscus aztecus* species group (Rodentia: Muridae) inferred using parsimony and likelihood. Systematic Biology 46:426–440.

Sullivan, J. K., E. Arellano, and D. Rogers. 2000. Comparative phylogeography of Mesoamerica highland rodents: concerted versus independent response to past climatic fluctuations. The American Naturalist 155:755–768.

Sullivan, R. M., and T. L. Best. 1997a. Systematics and morphological variation in two chromosomal forms of the agile kangaroo rat (*Dipodomys agilis*). Journal of Mammalogy 78:775–797.

Sullivan, R. M., and T. L. Best. 1997b. Effects of environment on phenotypic variation and sexual dimorphism in *Dipodomys simulans* (Rodentia: Heteromyidae). Journal of Mammalogy 78:798–810.

Sullivan, R. M., D. J. Hafter, and T. L. Yates. 1986. Genetic of a contact zone between three chromosomal forms of the grasshopper mous of the genus (*Onychomys*): a reassessment. Journal of Mammalogy 67:640–659.

Sundt, R. C., G. Dahle, and G. Naevdal. 1994. Genetic-variation in the hooded seal, *Cystophora cristata*, based on enzyme polymorphism and multilocus DNA-fingerprinting. Hereditas 121:147–155.

Sunqist, M., and F. Sunquist. 2002. Wild cats of the world. University of Chicago Press. Chicago, U.S.A.

Sutton, D. A. 1982. The female genital bone of chipmunks, genus *Eutamias*. The Southwestern Naturalist 27:393–402.

Sutton, D. A. 1992. *Tamias amoenus*. Mammalian Species 390:1–8.

Sutton, D. A. 1993. *Tamias townsendii*. Mammalian Species 435:1–6.

Sutton, D. A., and C. F. Nadler. 1974. Systematic revision of three Townsend chipmunks (*Eutamias townsendii*). The Southwestern Naturalist 19:199–211.

Sutton, D. A., and B. D. Patterson. 2000. Geographic variation of the western chipmunks *Tamias senex* and *T. siskiyou*, with two new subspecies from California. Journal of Mammalogy 81:299–316.

Swanepoel, P., and H. H. Genoways. 1979. Morphometries. Pp 13–106, in Biology of the bats of the New Worldfamily Phylostomidae. Part III (Baker, R. J., J. K. Jones, Jr., and D. C. Carter, eds.). Special Publications, The Museum, Texas Tech University Press 16:1–441.

Swank, W.G., and J. G. Teer. 1989. Status of the Jaguar. Oryx 23:14–21.

Sweeney, J. R., J. M. Sweeney, and S. W. Sweeney. 2003. Feral hog, Sus scrofa. Pp. 1164–1179, *in* Wild mammals of North America: Biology, management, and conservation (Feldhammer, G. A, B. C. Thompson, and J. A. Chapman, eds.). The Johns Hopkins University Press. Baltimore, U.S.A.

Tamarin, R. H., and T. H. Kunz. 1974. *Microtus breweri*. Mammalian Species 45:1–3.

Tamsitt, J. R., and D. Valdivieso. 1963. Records and observations on Colombian bats. Journal of Mammalogy 44:168–180.

Tarver, J. E., M. dos Reis, S. Mirarab, R. J. Moran, S. Parker, J. E. O'Reilly, B. L. King, M. J. O'Connell, R. J. Asher, T. Warnow, K. J. Peterson, P. C. J. Donoghue, and D. Pisani. 2016. The interrelationships of placental mammals and the limits of phylogenetic inference. Genome Biology and Evolution 8:330–44.

Taranin, A. V., E. G. Ufimtseva, E. S. Belousov, D. V. Ternovsky, Y. G. Ternovskaya, and O. K. Baranov. 1991. Evolution of antigenic structure of immunoglobulin chains in mustelids (Carnivora, Mustelidae). Zoologicheskii Zhurnal 70:105–112.

Tate, C. M., J. F. Pagels, and C. O. Handley, Jr. 1980. Distribution and systematic relationship of two kinds of short-tailed shrews (Soricidae: *Blarina*) in south-central Virginia. Proceedings of the Biological Society of Washington 93:50–60.

Tate, G. H. H. 1933. A systematic revision of the marsupial genus *Marmosa,* with a discussion of the adaptive radiation of the murine opossums (*Marmosa*). Bulletin of the American Museum of Natural History 66:137–144.

Tate, G. H. H. 1939. The mammals of the Guiana region. Bulletin of the American Museum of Natural History 76:151–229.

Tate, G. H. H. 1942. Results of the Archbold expeditions. No. 47: review of the Vespertilioninae bats, with special attention to genera and species of the Archbold collections. Bulletin of the American Museum of Natural History 80:221–296.

Tate, G. H. H. 1947. Albino prairie dog. Journal of Mammalogy 28:62.

Taub, D.M., and P. T. Mehlman. 1989. Development of the Morgan Island rhesus monkey colony. Puerto Rico Health Sciences Journal 8:159–69.

Tavares, V., and S. Burneo. 2015. *Trinycteris nicefori*. The IUCN Red List of Threatened Species 2015:e.T13381A22123365.

Tavares, V., and H. Mantilla. 2015. *Thyroptera tricolor*. The IUCN Red List of Threatened Species 2015:e.T21879A97207863.

Tavares, V., and J. Molinari. 2015. *Choeroniscus godmani*. The IUCN Red List of Threatened Species 2015:e.T4772A22041805.

Tavares, V., A. Muñoz, and J. Arroyo-Cabrales. 2015. *Vampyressa thyone*. The IUCN Red List of Threatened Species 2015:e.T136671A21989318.

Taylor, C. L., and R. F. Wilkinson Jr. 1988. First record of *Sorex longirostris* (Soricidae) in Oklahoma. The Southwestern Naturalist 33:248.

Taylor, F. H. C., M. Fujinaga, and F. Wilke. 1955. Distribution and food habits of the fur seals of the North Pacific Ocean; Report of cooperative investigations by the Governments of Canada, Japan, and the United States of America, February-July 1952. United States Fish and Wildlife Service, Washington, U.S.A.

Taylor, W. P. 1916. The status of the beavers of western North America, with a consideration of the factors in their speciation. University of California Publications in Zoology 12:413–495.

Taylor, W. P. 1956. The deer of North America. The Stackpole Company. Harrisburg, U.S.A.

Teer, J. G., J. W. Thomas, and E. A. Walker. 1965. Ecology and management of white-tailed deer in the Llano Basin of Texas. Wildlife Monographs 15:1–62.

Tejedor, A. 2005. A new species of funnel–eared bat (Natalidae: *Natalus*) from Mexico. Journal of Mammalogy 86:1109–1120.

Tejedor, A. 2006. The type locality of *Natalus stramineus* (Chiroptera: Natalidae): implications for the taxonomy and biogeography of the genus *Natalus*. Acta Chiropterologica 8:361–380.

Tejedor, A. 2011. Systematics of funnel-eared bats (Chiroptera: Natalidae). Bulletin of the American Museum of Natural History 353:1–140.

Tellez, G., and J. Ortega. 1999. *Musonycteris harrisoni*. Mammalian Species 622:1–3.

Tener, J. S. 1965. Muskoxen in Canada: a biological and taxonomic review. Queen's Printer and Controller of Stationery. Ottawa, Canada.

Thaeler, C. S., Jr. 1968a. An analysis of three hybrid populations of pocket gophers (genus *Thomomys*). Evolution 22:543–555.

Thaeler, C. S., Jr. 1968b. An analysis of the distribution of pocket gopher species in northeastern California (genus *Thomomys*). University of California Publlications, Zoology 86:1–46.

Thaeler, C. S., Jr. 1972. Taxonomic Status of the Pocket Gophers, *Thomomys idahoensis* and *Thomomys pygmaeus* (Rodentia, Geomyidae). Journal of Mammalogy 53:417–428.

Thaeler, C. S., Jr. 1977. Taxonomic status of *Thomomys talpoides confinus*. Murrelet 58:49–50.

Thaeler, C. S., Jr. 1980. Chromosome numbers and systematics relations in the genus *Thomomys* (Rodentia: Geomyidae). Journal of Mammalogy 61:414–422.

Thaeler, C. S., Jr., and L. L. Hinesley. 1979. *Thomomys clusius*, a rediscovered species of pocket gopher. Journal of Mammalogy 60:480–488.

Thomas, D. C., and P. Everson. 1982. Geographic variation in caribou on the Canadian Arctic islands. Canadian Journal of Zoology 60:2442–2454.

Thomas, H. H., and T. L. Best. 1994. *Lepus insularis*. Mammalian Species 465:1–3.

Thomas, O. 1888. Catalogue of the Marsupialia and Monotremata in the collection of the British Museum (Natural History). British Museum. London, United Kingdom.

Thomas, O. 1903. New mammals from Chiriqui. Annals and Magazine of Natural History, including Zoology, Botany and Geology, London 7:376–382.

Thomson, C. E. 1982. *Myotis sodalis*. Mammalian Species 163:1–5.

Thorington, R. W., Jr., and R. S. Hoffmann. 2005. Family Sciuridae. Pp. 754–818, *in* Mammal species of the world: a taxonomic and geographic reference (Wilson, D. E., and D. M. Reeder, eds.). Third edition. Johns Hopkins University Press, Baltimore, U.S.A.

Thorington, Jr., R. W., J. L. Koprowski, M. A. Steele, and J. F. Whatton. 2012. Squirrels of the world. The John Hopkins University Press. Baltimore, U.S.A.

Thornton, W. A., and G. C. Creel. 1975. The taxonomic status of kit foxes. Texas Journal of Sciences 26:127–136.

Thornton, W. A., G. C. Creel, and R. E. Trimble. 1971. Hybridization in the fox genus *Vulpes* in west Texas. The Southwestern Naturalist 15:473–484.

Tiemann-Boege, I., C. W. Kilpatrick, D. J. Schmidly, and R. D. Bradley. 2000. Molecular phylogenetics of *Peromyscus boylii* species group (Rodentia: Muridae) based on mitichondrial cytochrome *b* sequences. Molecular Phylogenetics and Evolution 16:366–378.

Timm, R. M. 1985. *Artibeus phaeotis*. Mammalian Species 235:1–6.

Timm, R. 2016. *Onychomys leucogaster*. The IUCN Red List of Threatened Species 2016:e.T15338A115127288.

Timm, R. M., and H. H. Genoways. 2003. West Indian mammals from the Albert Schwartz Collection: Biological and historical information. Scientific Papers, Natural History Museum, University of Kansas 29:1–47.

Timm, R. M., and H. H. Genoways. 2004. The Florida bonneted bat, *Eumops floridanus* (Chiroptera: Molossidae): distribution, morphometrics, systematics, and ecology. Journal of Mammalogy 85:852–865.

Timm, R., and J. Matson. 2018. *Notiosorex villai*. The IUCN Red List of Threatened Species 2018:e.T136688A22293097.

Timm, R., S. T. Álvarez-Castañeda, I. Castro-Arellano, and T. Lacher. 2016a. *Ammospermophilus harrisii*. The IUCN Red List of Threatened Species 2016:e.T42399A115189204.

Timm, R., S. T. Álvarez-Castañeda, I. Castro-Arellano, and T. Lacher. 2016b. *Ammospermophilus interpres*. The IUCN Red List of Threatened Species 2016:e.T42451A115189324.

Timm, R., S. T. Álvarez-Castañeda, I. Castro-Arellano, and T. Lacher. 2016c. *Baiomys taylori*. The IUCN Red List of Threatened Species 2016:e. T2466A115062269.

Timm, R., S. T. Álvarez-Castañeda, J. Frey, and T. Lacher. 2016d. *Dipodomys spectabilis*. The IUCN Red List of Threatened Species 2019:e. T6693A22229212.

Timm, R., S. T. Álvarez-Castañeda, and T. Lacher. 2016e. *Dipodomys merriami*. The IUCN Red List of Threatened Species 2016:e. T92465716A115515430.

Timm, R., S. T. Álvarez-Castañeda, and T. Lacher. 2016f. *Neotoma leucodon*. The IUCN Red List of Threatened Species 2016:e.T136793A115212444.

Timm, R., S. T. Álvarez-Castañeda, and T. Lacher. 2016g. *Otospermophilus beecheyi*. The IUCN Red List of Threatened Species 2016:e. T20481A22263743.

Timm, R., S. T. Álvarez-Castañeda, and T. Lacher. 2016h. *Peromyscus fraterculus*. The IUCN Red List of Threatened Species 2016:e. T136412A115207364.

Timm, R., A. D. Cuarón, F. Reid, K. Helgen, and J. F. González-Maya. 2016i. *Procyon lotor*. The IUCN Red List of Threatened Species 2016:e. T41686A45216638.

Timm, R., J. Matson, D. Tirira, C. Boada, M. Weksler, R. P. Anderson, B. Rivas, C. Delgado, and M. Gómez-Laverde. 2016j. *Handleyomys alfaroi*. The IUCN Red List of Threatened Species 2016:e.T15585A115127897.

Timm, R., J. Matson, Woodman, N., and I. Castro-Arellano. 2016k. *Notiosorex crawfordi*. The IUCN Red List of Threatened Species 2016:e. T41456A115187458.

Timm, R., F. Reid, N. Woodman, T. McCarthy, and J. Matson. 2016l. *Otonyctomys hatti*. The IUCN Red List of Threatened Species 2016:e. T15664A115129163.

Timm, R., S. T. Álvarez-Castañeda, and T. Lacher. 2017a. *Oryzomys nelsoni*. The IUCN Red List of Threatened Species 2017:e.T15583A22388135.
Timm, R., J. Matson, N. Woodman, P. C. de Grammont, and I. Castro-Arellano. 2017b. *Notiosorex cockrumi*. The IUCN Red List of Threatened Species 2017:e.T136666A22293361.
Timm, R., S. T. Álvarez-Castañeda, and T. Lacher. 2018a. *Rheomys mexicanus*. The IUCN Red List of Threatened Species 2018:e.T19484A22354600.
Timm, R., T. Lacher, and S. T. Álvarez-Castañeda. 2018b. *Reithrodontomys bakeri*. The IUCN Red List of Threatened Species 2018:e.T136821A22385642.
Tirira, D. G., J. Díaz-N., M. Superina, and A. M. Abba. 2014. *Cabassous centralis*. The IUCN Red List of Threatened Species 2014:e.T3412A47437304.
Torres-Flores, J. W., and R. López-Wilchis. 2018. *Microtus quasiater* (Rodentia: Cricetidae). Mammalian Species 50:59–66.
Torres-Morales, L., D. F. García-Mendoza, C. López-González, and R. Muñiz-Martínez. 2010. Bats of northwestern Durango, Mexico: species richness at the interface of two biogeographic regions. The Southwestern Naturalist 55:347–362.
Torres-Morales, L., G. Rodríguez-Aguilar, S. A. Cabrera-Cruz, and R. Villegas-Patraca. 2008. Primer registro de Eumops nanus (Chiroptera: Molossidae) en Oaxaca, México Mastozoología. Neotropical 21:373–378
Trouessart, E. L. 1905. Catalogus mammalium tam viventium quam fossilium. Quinquennale supplementium (1899–1904). Cetacea, Edentata, Marsupialia, Allotheria, Monotremata. R. Friedlander & Sons. Berlin, Germany.
Trefethen, J. B. 1975. The wild sheep in modern North America. Winchester Press. New York, U.S.A.
Tremor, S., S. Vanderplank, and E. Mellink. 2019. The San Quintín Kangaroo rat is not extinct. Bulletin Southern California Academy of Sciences 118:71–75.
Treviño-Villareal, J.1991. The annual cycle of the Mexican prairie dog (*Cynomys mexicanus*). Occasional Papers of the Museum of Natural History, University of Kansas 139:1–27.
Trombulak, S. C. 1988. *Spermophilus saturatus*. Mammalian Species 322:1–4.
True, F. W. 1896. A revision of the American moles. Proceeding of the United States Natural Museum 19:1–111.
Trujano-Álvarez, A. L., and S. T. Álvarez-Castañeda. 2007. Taxonomic revision of *Thomomys bottae* in the Baja California Sur lowlands. Journal of Mammalogy 88:343–350.
Trujano-Álvarez, A. L., and S. T. Álvarez-Castañeda. 2010. *Peromyscus mexicanus*. Mammalian Species 42:111–118.
Trujano-Álvarez, A. L., and S. T. Álvarez-Castañeda. 2013. Phylogenetic structure among pocket gopher populations, genus *Thomomys* (Rodentia: Geomyidae), on the Baja California Peninsula. Zoological Journal of the Linean Society 168:873–891.
Tumlison, R. 1987. *Felis lynx*. Mammalian Species 269:1–8.
Tumlison, R. 1991. Bats of the genus *Plecotus* in Mexico: discrimination and distribution. Occasional Papers, Museum of Texas Tech University 140:1–19.
Tumlison, R. 1992. *Plecotus mexicanus*. Mammalian Species 401:1–3.
Tumlison, R., and M. E. Douglas. 1992. Parsimony analysis and the phylogeny of the plecotine bats (Chiroptera: Vespertilionidae). Journal of Mammalogy 73:276–285.
Tsytsulina, K., N. Formozov, B. Sheftel, M. Stubbe, R. Samiya, J. Ariunbold, and V. Buuveibaatar. 2016. *Sorex tundrensis*. The IUCN Red List of Threatened Species 2016:e.T41422A115185726.
UCN SSC Antelope Specialist Group. 2016. *Antilocapra americana*. The IUCN Red List of Threatened Species 2016:e.T1677A115056938.
Udina, I. G., A. A. Danilkin, and G. G. Boeskorov. 2002. Genetic Diversity of Moose (*Alces alces* L.) in Eurasia. Russian Journal of Genetics 38:951–957.
Vannatta, J. M., J. A. Gore, V. L. Mathis, and B. D. Carver. 2020. *Eumops floridanus* (Chiroptera: Molossidae). Mammalian Species 53:125–133.
Van Den Bussche, R. A., and R. J. Baker. 1993. Molecular phylogenetics of the New World bat genus *Phyllostomus* based on cytochrome-*b* DNA sequence variation. Journal of Mammalogy 74:793–802.
Van Den Bussche, R. A., R. J. Baker, H. A. Wichman, and M. J. Hamilton. 1993. Molecular phylogenetics of stenodermatine bat genera: Congruence of data from nuclear and mitochondrial DNA. Molecular Biology and Evolution 10:944–959.
Van den Bussche, R. A., J. L. Hudgeons, and R. J. Baker. 1998. Phylogenetic accuracy, stability, and congruence. Relationships within and among the New World bat genera *Artibeus*, *Dermanura*, and *Koopmania*. Pp. 59—71, *in* Bat biology and conservation (Thomas H., K., and P. A. Racey, eds.). Smithsonian Institution Press. Washington, U.S.A.
Van de Graff, K. M. 1975. Reproductive ecology of some Sonoran desert rodents. Ph.D. dissertation, Northern Arizona University. Flagstaff, U.S.A.
Van Gelder, R. G. 1959. A taxonomic revision of the spotted skunks (genus Spilogale). Bulletin of the American Museum of Natural History 117:229–392.
Van Gelder, R. G. 1959b. A taxonomic revision of the spotted skunks (genus *Spilogale*). Bulletin of the American Museum of Natural History 117:229–392.
Van Gelder, R. G. 1960. Results of the Puritan-American Museum of Natral History Expedition to western Mexico. 10, Marine mammals from the coast of Baja California and the Tres Marías Islands, México. American Museum Novitates 1992:1–27.
van Zyll de Jong, C. G. 1972. A systematic review of the Nearctic and Neotropical river otters (genus *Lutra,* Mustelidae, Carnivora). Life Sciences Contributions, Royal Ontario Museum 80:1–104.
van Zyll de Jong, C. G. 1979. Distribution and systematic relationships of long-eared Myotis in western Canada. Canadian Journal of Zoology. 57:987–994.
van Zyll de Jong, C. G. 1982. Relationships of amphiberingian shrews of the *Sorex cinereus* group. Canadian Journal of Zoology 60:1580–1587.
van Zyll de Jong, C. G. 1983a. Handbook of Canadian Mammals. National Museums of Canada, Ottawa, Canada.
van Zyll de Jong, C. G. 1983b. Handbook of Canadian mammals: bats. National Museums of Canada, Ottawa, Ontario 2:1–212.
van Zyll de Jong, C. G. 1983c. A morphometric analysis of North American shrews of the *Sorex arcticus* group, with special consideration of the taxonomic status of *S. a. maritimensis.* Le Naturaliste Canadien 110:373–378.
van Zyll de Jong, C. G. 1984. Taxonomic relationships of Nearctic small-footed bats of the *Myotis leibii* group (Chiroptera: Vespertilionidae). Canadian Journal of Zoology 62:2519–2526.
Van Zyll De Jong, C. G. 1985. Handbook of Canadian mammals. 2. Bats. Canadian Museum of Natural. Canada, 212 pp.

van Zyll de Jong, C. G. 1987. A phylogenetic study of the *Lutrinae* (Carnivora; Mustelidae) using morphological data. Canadian Journal of Zoology 65:2536–2544.
van Zyll de Jong, C. G., and G. L. Kirkland, Jr. 1989. A morphometric analysis of the *Sorex cinereus* group in central and Eastern North America. Journal of Mammalogy 70:110–122.
van Zyll de Jong, G. G. 1993. A morphometric analysis of cranial Variation in Holarctic. Canadiern Museum of Nature. Ottawa, Canada 57:77–93
van Zyll de Jong, C. G. 1999a. Tundra shrew, *Sorex tundrensis*. Pp. 44–45, *in* The Smithsonian book of North American Mammals (Wilson, D. E., and S. Ruff, eds.). Smithsonian Institution Press. Washington, U.S.A.
van Zyll de Jong, C. G. 1999b. Barren ground shrew, *Sorex uyunak*. Pp. 45, *in* The Smithsonian book of North American Mammals (Wilson, D. E., and S. Ruff, eds.). Smithsonian Institution Press. Washington, U.S.A.
Van Gelder, R. G. 1977. Mammalian hybrids and generic limits. American Museum Novitates 2635:1–25.
Van Gelder, R. G. 1978. A review of canid classification. American Museum Novitates 2646:1–25.
Van Staaden, M. J., and J. K. Jones, Jr. 1997. Comparative morphology of dorsal hair of New World bats of the Family Molossidae. Pp. 373–391, *in* Homenaje al Profesor Ticul Álvarez (Arroyo-Cabrales, J., and O. J. Polaco, coords.). Instituto Nacional de Antropología e Historia, Colección Científica. Distrito Federal, México.
Vargas, K., D. Brown, E. Wisely, and M. Culver. 2019. Reinstatement of the Tamaulipas white-sided jackrabbit, *Lepus altamirae*, based on DNA sequence data. Revista Mexicana de Biodiversidad 90:2–9.
Vargas-Contreras, J. A., G. Escalona-Segura, J. D. Cú-Vizcarra, J. Arroyo-Cabrales, and R. A. Medellín. 2008. Estructura y diversidad de los ensambles de murciélagos en el centro y sur de Campeche, México. Avances en el estudio de los mamíferos de México 2:551–577.
Vashon, J. 2016. *Lynx canadensis*. The IUCN Red List of Threatened Species 2016:e.T12518A101138963.
Vázquez, A., and F. Gopar-Merino. 2018. *Romerolagus diazi*. Pp. 198–200, *in* Pikas, Rabbits and Hares of the World (Smith, A., C. H. Johnston, P. C. Alves, and K. Hackländer, eds.). The Johns Hopkins University Press/IUCN/SSC. Baltimore, U.S.A.
Vázquez, E. 2016a. *Peromyscus guatemalensis*. The IUCN Red List of Threatened Species 2016:e.T16665A22361195.
Vázquez, E. 2016b. *Peromyscus hylocetes*. The IUCN Red List of Threatened Species 2016:e.T136416A22361620.
Vázquez, E. 2017a. *Habromys delicatulus*. The IUCN Red List of Threatened Species 2017:e.T136683A22376548.
Vázquez, E. 2017b. *Heterogeomys lanius*. The IUCN Red List of Threatened Species 2017:e.T42591A22215417.
Vázquez, E. 2017c. *Chaetodipus lineatus*. The IUCN Red List of Threatened Species 2017:e.T136361A22225840.
Vázquez, E. 2018a. *Habromys simulatus*. The IUCN Red List of Threatened Species 2018:e.T9611A22376731.
Vázquez, E. 2018b. *Handleyomys chapmani*. The IUCN Red List of Threatened Species 2018:e.T15591A22328808.
Vázquez, E. 2018c. *Xenomys nelsoni*. The IUCN Red List of Threatened Species 2018:e.T23115A22359234.
Vázquez, E. 2019a. *Handleyomys rhabdops*. The IUCN Red List of Threatened Species 2019:e.T15611A22328169.
Vázquez, E. 2019b. *Reithrodontomys tenuirostris*. The IUCN Red List of Threatened Species 2019:e.T19418A22385808.
Vázquez, E., and S. T. Álvarez-Castañeda. 2016. *Peromyscus melanophrys*. The IUCN Red List of Threatened Species 2016:e.T16677A115136842.
Vázquez, E., and F. Reid. 2016. *Peromyscus gymnotis*. The IUCN Red List of Threatened Species 2016:e.T16666A22361063.
Vázquez, E., L. Emmons, F. Reid, and A. D. Cuarón. 2008. *Dasyprocta mexicana*. The IUCN Red List of Threatened Species 2008:e.T6285A12596623.
Vázquez, E., L. Emmons, and T. McCarthy. 2016a. *Orthogeomys grandis*. The IUCN Red List of Threatened Species 2016:e.T42590A115193029.
Vázquez, E., L. Emmons, F. Reid, and A. D. Cuarón. 2016b. *Heteromys salvini*. The IUCN Red List of Threatened Species 2016:e.T12076A22225346.
Vázquez, E., L. Emmons, F. Reid, and A. D. Cuarón. 2016c. *Orthogeomys hispidus*. The IUCN Red List of Threatened Species 2016:e.T15549A115127668.
Vázquez, E., L. Emmons, F. Reid, and A. D. Cuarón. 2016d. *Sciurus yucatanensis*. The IUCN Red List of Threatened Species 2016:e.T20026A115156290.
Vázquez, E., F. Reid, and A. D. Cuarón. 2016e. *Coendou mexicanus*. The IUCN Red List of Threatened Species 2016:e.T20629A22214103.
Vázquez, E., L. Emmons, and T. McCarthy. 2017. *Tylomys nudicaudus*. The IUCN Red List of Threatened Species 2017:e.T22573A22340628.
Vázquez, E., P. C. de Grammont, and A. D. Cuarón. 2018. *Reithrodontomys spectabilis*. The IUCN Red List of Threatened Species 2018:e.T19416A22386261.
Vázquez, L. B., G. N. Cameron, and R. A. Medellín. 2001. *Peromyscus aztecus*. Mammalian Species 649:1–4.
Veal, R., and W. Caire. 1979. *Peromyscus eremicus*. Mammalian Species 118:1–6.
Velazco, P. M. 2005. Morphological phylogeny of the bat genus *Platyrrhinus* Saussure, 1860 (Chiroptera: Phyllostomidae) with the description of four new species. Fieldiana Zoology (n. s.) 105:1–54.
Velazco, P. M., and B. D. Patterson. 2013. Diversification of the yellow-shouldered bats, genus *Sturnira* (Chiroptera, Phyllostomidae), in the New World tropics. Molecular Phylogenetics and Evolution 68:683–698.
Velazco, P. M., and B. D. Patterson. 2014. Two new species of yellow-shouldered bats, genus *Sturnira* Gray, 1842 (Chiroptera, Phyllostomidae) from Costa Rica, Panama and western Ecuador. ZooKeys 402:43–66.
Velazco, P. M., A. L. Gardner, and B. D. Patterson. 2010. Systematics of the Platyrrhinus helleri species complex (Chiroptera: Phyllostomidae), with descriptions of two new species. Zoological Journal of the Linnean Society 159:785–812.
Velazco, P. M., and N. B. Simmons. 2011. Systematics and taxonomy of great striped-faced bats of the genus *Vampyrodes* Thomas, 1900 (Chiroptera: Phyllostomidae). American Museum Novitates 3710:1–35.
Velázquez, A. 1994. Distribution and population size of *Romerolagus diazi* on El Pelado volcano, Mexico. Journal of Mammalogy 75:743–749.
Velázquez, A., and J. A. Guerrero. 2019. *Romerolagus diazi*. The IUCN Red List of Threatened Species 2019:e.T19742A45180356.
Velázquez, A., F. A. Cervantes, and C. Galindo-Leal. 1993. The volcano rabbit *Romerolagus diazi*, a peculiar lagomorph. Lutra 36:62–70.
Verts, B. J., and L. N. Carraway. 1984. Keys to the mammals of Oregon. O.S.U. Book Stores, Inc. Corvallis, U.S.A.
Verts, B. J., and L. N. Carraway. 1987a. *Thomomys bulbivorus*. Mammalian Species 273:1–4.
Verts, B. J., and L. N. Carraway. 1987b. *Microtus canicaudus*. Mammalian Species 267:1–4.
Verts, B. J., and L. N. Carraway. 1995. *Phenacomys albipes*. Mammalian Species 494:1–5.
Verts, B. J., and L. N. Carraway. 1998. Land mammals of Oregon. University of California Press. Berkeley, U.S.A.
Verts, B. J., and L. N. Carraway. 1999. *Thomomys talpoides*. Mammalian Species 618:1–11.

Verts, B. J., and L. N. Carraway. 2000. *Thomomys mazama*. Mammalian Species 641:1–7.

Verts, B. J., and L. N. Carraway. 2001a. *Scapanus latimanus*. Mammalian Species 666:1–7.

Verts, B. J., and L. N. Carraway. 2001b. *Tamias minimus*. Mammalian Species 653:1–10.

Verts, B. J., and L. N. Carraway. 2002. *Neotoma lepida*. Mammalian Species 699:1–12.

Verts, B. J., and L. N. Carraway. 2003. *Thomomys townsendii*. Mammalian Species 719:1–6.

Verts, B. J., and G. L. Kirkland, Jr. 1988. *Perognathus parvus*. Mammalian Species 318:1–8.

Verts, B. J., L. N. Carraway, and A. Kinlaw. 2001. *Spilogale gracilis*. Mammalian Species 674:1–10.

Vianna, J. A., R. K. Bonde, S. Caballero, J. P. Giraldo, R. P. Lima, A. M. Clark, M. Marmontel, B. Morales-Vela, M. J. De Souza, L. Parr, M. A. Rodríguez-Lopez, A. A. Mignucci-Giannoni, J. A. Powell, and F. R. Santos. 2006. Phylogeography, phylogeny, and hybridization in trichechid sirenians: implications on manatee conservation. Molecular Ecology 15:433–447.

Vidal-López, R., and M. Martínez-Coronel. 2000. Nuevo molósido para la quiropterofauna de Chiapas, México. Vertebrata Mexicana 8:1–3

Villa-R., B. 1958. El mono araña (*Ateles geoffroyi*) encontrado en la costa de Jalisco y en la region central de Tamaulipas. Anales de Instituto de Biología, Universidad Nacional Autónoma de México, Serie Zoología 28:345–347.

Villa-R., B. 1960. *Tadarida yucatanica* in Tamaulipas. Journal of Mammalogy 41:314–319.

Villa, B. 1966. Los Murciélago de México. Universidad Nacional Autónoma de México, Instituto de Biología.

Villa-R., B. 1967. Los murciélagos de Mexico. Instituto de Biología, Universidad Nacional Autónoma de México. Distrito Federal, México.

Villa-R., B., and E. L. Cockrum. 1962. Migration in the guano bat *Tadarida brasilensis mexicana* (Saussure). Journal of Mammalogy 43:43–64.

Villa, B., and E. R. Hall. 1949. Subspeciation of the Pocket Gophers of Kansas. University Kansas publications, Museum of Natural History 1:219–234.

Villa-R., B., G. M. Ruiz, B. O. Bonilla, and B. V. Cornejo. 1967. Rabia en dos especies de rnurciélagos insectivoros género *Pteronotus,* en condiciones naturales, colectados en Jalisco, México. Anales de Instituto de Biología, Universidad Nacional Autónoma de México, Serie Zoología 38:9–16.

Villafuerte, R, and M. Delibes-Mateos. 2019. *Oryctolagus cuniculus*. The IUCN Red List of Threatened Species 2019:e.T41291A170619657.

Villalpando, J. A., and T. Álvarez. 2000. Murciélagos (Chiroptera) de Michoacán en la colección de mamíferos de la Escuela Nacional de Ciencias Biológicas del Instituto Politécnico Nacional, México. Anales de la Escuela Nacional de Ciencias Biológicas, México 46:119–188.

Villalpando-R., J. A., and J. Arroyo-Cabrales. 1996. Una nueva localidad para *Rhogeessa mira* (Chiroptera: Vespertilionidae) en la cuenca baja del río Balsas, Michoacán, México. Vertebrata Mexicana 2:9–11.

Voigt, D. R. 1987. Red fox. Pp. 379–392, *in* Wild furbearer management and conservation in North America (Nowak, M., J. A. Baker, M. E. Obbard, and B. Malloch, eds.). Ontario Ministry of Natural Resources. Ontario, Canada.

Volobouev, V. T., and C. G. van Zyll de Jong. 1988. The Karyotype of *Sorex articus maritimensis* (Insectivora, Soricidae) and its systematic implications. Canadian Journal of Zoology 66:1968–1972.

Vonhof, M. J. 2000. *Rhogeessa tumida*. Mammalian Species 633:1–3.

Voss, R. S. 1988. Systematics and ecology of *Ichthyomyine* rodents (Muroidea): patterns of morphological evolution in a small adaptive radiation. American Museum of Natural History 188:259–493.

Voss, R. S. 2011. Revision notes on Neotropical porcupines (Rodentia: eretizontidae) 3. An annoted checklist of the species of Coendou Lacépède, 1799. Annals Museum Novitates 3720:1–36.

Voss, R. S., and S. A. Jansa. 2003. Phylogenetic studies on didelphid marsupials II. Nonmolecular data and new IRBP sequences: separate and combined analyses of didelphine relationships with denser taxon sampling. Bulletin American Museum of Natural History 276:1—82.

Voss, R. S., M. Gómez-Laverde, and V. Pacheco. 2002. A new genus for *Aepeomys fuscatus* Allen, 1912, and *Oryzomys intectus* Thomas, 1921:enigmatic murid rodents from Andean cloud forests. American Museum Novitates 3373:1–42.

Voss, R. S., C. Hubbard, and S. A. Jansa. 2013. Phylogenetic relationships of the new world porcupines (Rodentia:eretizontidae): Implication for taxonomy, morphological evolution and biogeography. American Museum Novitates 3769:1–36.

Voss, R. S., E. E. Gutiérrez, S. Solari, R. V. Rossi, and S. A. Jansa. 2014. Phylogenetic relationships of mouse opossums (Didelphidae, *Marmosa*) with a revised subgeneric classification and notes on sympatric diversity. American Museum novitates 3817:1–27.

Voss, R. S., J. Díaz-Nieto, and S. A. Jansa. 2018. A Revision of *Philander* (Marsupialia: Didelphidae), Part 1: *P. quica*, *P. canus*, and a New Species from Amazonia. American Museum Novitates 3891:1–70.

Voss, R. S., D. W. Fleck, and S. A. Jansa. 2019. Mammalian diversity and Matses ethnomammalogy in Amazonian Peru Part 3: Marsupials (Didelphimorphia). Bulletin of the American Museum of Natural History 432:1–90.

Wade-Smith J., and B. J. Verts. 1982. *Mephitis mephitis*. Mammalian Species 173:1–7.

Wahle, R., E. Roth, and P. Horner. 2018. *Dipodomys elator*. The IUCN Red List of Threatened Species 2018:e.T6675A22227507.

Walker, E. P. 1975. Mammals of the world. Third Ed. Johns Hopkins University Press, Baltimore 1:1–646.

Walker, P. L. 1980. Archeological evidence for the recent extinction of three terrestrial mammals on San Miguel Island. Pp. 703–713, *in* The California Islands: proceedings of a multidisciplinary symposium (Power, D. M. ed.). Santa Barbara Museum Natural History, Santa Barbara, California, U.S.A.

Wallmo, O. C. 1981. Mule and black-tailed deer distribution and habitats. Pp. 1–25, in Mule and black-tailed deer of North America (Wallmo, 0. C., ed.). University Nebraska Press, Lincoln

Wang, H. G., R. D. Owen, C. Sánchez-Hernández, and M. de la L. Romero-Almaraz. 2003. Ecological characterization of bat species distributions in Michoacan, Mexico, using a geographic information system. Global Ecology and Biogeography 12:65–85.

Wang, K., J. A. Lenstra, L. Liu, Q. Hu, T. Ma, Q. Qiu, and J. Liu. 2018. Incomplete lineage sorting rather than hybridization explains the inconsistent phylogeny of the wisent. Communications Biology 1:1–9.

Warner, R. M. 1982. *Myotis auriculus*. Mammalian Species 191:1–3.

Warner, R. M., and N. J. Czaplewski. 1984. *Myotis volans*. Mammalian Species 224:1–4.

Warren, E. R.1936. The marmots of Colorado. Journal of Mammalogy 17:392–398.

Warren, E. R. 1942. The mammals of Colorado. 2nd ed. University Oklahoma Press, Norman, U.S.A.

Watkins, L. C. 1969. Observations on the distribution and natural history of the evening bat (*Nycticeius humeralis*) in northwestern Missouri and adjacent Iowa. Transactions of Kansas Academy of Science 72:330–336.

Watkins, L. C. 1972. *Nycticeius humeralis*. Mammalian Species 23:1–4.

Watkins, L. C. 1997. *Euderma maculatum*. Mammalian Species 77:1–4.

Watkins, L. C., J. K. Jones, Jr., and H. H. Genoways. 1972. Bats of Jalisco, Mexico. Special Publications, Museum of Texas Tech University 1:1–44.

Wayne, R. K., and S. M. Jenks. 1991. Mitonchondrial DNA analysis implying extensive hybridization of the endangered red wolf *Canis rufus*. Nature 351:565–568.

Wayne, R. K., S. B. George, D. Gilbert, and P. W. Collins. 1991a. The channel island fox (*Urocyon littoralis*) as a model of genetic change in small populations. Pp. 639–649, *in* The unity of evolutionary biology (Dudley, E. C., ed.). Proceedings of the fourth international congress of systematic and evolutionary biology. Dioscorides Press, Portland, Oregon 2:1–1048.

Wayne, R. K., S. George, D. Gilbert, P. Collins, and D. Girman. 1991b. A morphologic and genetic study of the island fox, *Urocyon littoralis*. Evolution 45:1849–1868.

Weber, M., and R. A. Medellín. 2010. Yucatán brown brocket deer *Mazama pandora* (Merriam 1901). Pp. 211–216, *in* (Duarte, J. M. B., and S. González, eds.), Neotropical Cervidology: Biology and Medicine of Latin American Deer. Funep, in collaboration with IUCN. Jaboticabal, Brazil and Gland, Switzerland.

Weber, M., P. C. de Grammont, and A. D. Cuarón. 2016. *Mazama pandora*. The IUCN Red List of Threatened Species 2016:e.T29622A22154219.

Webster, D. 1968. Comparative middle and inner ear structure of Heteromyidae. The Anatomical Record 160:447.

Webster, W. D. 1993. Systematics and evolution of the bats of the genus *Glossophaga*. Special Publications, Museum ofTexas Tech University 36:1–184.

Webster, W. D., and J. K. Jones, Jr. 1980. Taxonomic and nomenclatorial notes on the bats of the genus *Glossophaga* in North America, with description of a new species. Occasional Papers, Museum of Texas Tech University 71:1–12.

Webster, W. D., and J. K. Jones, Jr. 1982a. A new subspecies of *Glossophaga commissarisi* (Chiroptera: Phyllostomidae) from western Mexico. Occasional Papers, Museum of Texas Tech University 76: 1–6.

Webster, W. D., and J. K. Jones, Jr. 1982b. *Reithrodontomys megalotis*. Mammalian Species 167:1–5.

Webster, W. D., and J. K. Jones, Jr. 1982c. *Artibeus aztecus*. Mammalian Species 177:1–3.

Webster, W. D., and J. K. Jones, Jr. 1982d. *Artibeus toltecus*. Mammalian Species 178:1–3.

Webster, W. D., and J. K. Jones, Jr. 1983. *Artibeus hirsutus* and *Artibeus inopinatus*. Mammalian Species 199:1–3.

Webster, W. D., and J. K. Jones, Jr. 1984a. *Glossophaga leachii*. Mammalian Species 226:1–3.

Webster, W. D., and J. K. Jones, Jr. 1984b. A new subspecies of *Glossophaga mexicana* (Chiroptera: Phyllostomidae) from southern Mexico. Occasional Papers, Museum of Texas Tech University 91:1–5.

Webster, W. D., and J. K. Jones, Jr. 1985. *Glossophaga mexicana*. Mammalian Species 245:1–2.

Webster, W. D., and J. K. Jones, Jr. 1993. *Glossophaga commissarisi*. Mammalian Species 446:1–4.

Webster, W. D., J. K. Jones, Jr. and R. J. Baker. 1980. *Lasiurus intermedius*. Mammalian Species 132:1–3.

Webster W. D., L. W. Robbins, R. L. Robbins, and R. J. Baker. 1982. Comments on the status of *Mussonycteris harrisoni* (Chiroptera: Phyllostomidae). Occasional Papers, Museum of Texas Tech University 78:1–5.

Webster, W. D., N. D. Moncrief, J. R. Choate, and H. H. Genoways. 2011. Systematic revision of the northern short-tailed shrew, *Blarina brevicauda* (Say). Virginia Museum of Natural History Memoir 10:1–77.

Weksler, M. 2003. Phylogeny of Neotropical oryzomyine rodents (Muridae: Sigmodontinae) based on the nuclear IRBP exon. Molecular Phylogenetics and Evolution 29: 331–349.

Weksler, M., A. R. Percequillo, and R. S. Voss. 2006. Ten new genera of oryzomyine rodents (Cricetidae: Sigmodontinae). American Museum Novitates 3537:1–29.

Weksler, M., H. C. Lanier, and L. E. Olson. 2010. Eastern Beringian biogeography: historical and spatial genetic structure of singing voles in Alaska. Journal of biogeography 37:1414–1431.

Weksler, M., M. Aguilera, and F. Reid. 2016. *Oligoryzomys fulvescens*. The IUCN Red List of Threatened Species 2016:e.T15248A115126740.

Weller, T. J., S. A. Scott, T. J. Rodhouse, P. C. Ormsbee, and J. M. Zinck. 2007. Field identification of the cryptic vespertilionid bats, Myotis lucifugus and M. yumanensis. Acta Chiropterologica 9:133–147.

Wells-Gosling, N., and L. R. Heaney. 1984. *Glaucomys sabrinus*. Mammalian Species 229:1–8.

Weston, M. L. 1981. The *Ochonta alpina* complex: A statistical re-evaluation. Pp. 73–89, *in* Proceedings of the World lagomorph conference (Myers, K. K., and C. D. MacInnes, eds.). Guelp University Press, Guelph, Canada.

Wetterer, A. L., M. V. Rockman, and N. B. Simmons. 2000. Phylogeny of Phyllostomid bats (Mammalia: Chiroptera): data from diverse morphological systems, sex chromosomes, and restriction sites. Bulletin of the American Museum of Natural History (248):1–200.

Wetzel, R. M. 1955. Speciation and dispersal of the southern bog lemming, *Synaptomys cooperi* (Baird). Journal of Mammalogy 36:1–20.

Wetzel, R. M. 1975. The species of *Tamandua* Gray (Edentata, Myrmecophagidae). Proceedings of Biological Society of Washington 88:95–112.

Wetzel, R. M. 1977. The Chacoan peccary, *Catagonus wagneri* (Rusconi). Bulletin of the Carnegie Museum of Natural History 3:1–36.

Wetzel, R. M. 1980. Revision of the naked-tailed armadillos, genus *Cabassous* McMurtrie. Annals of Carnegie Museum 49:323–357.

Wetzel, R. M. 1982. Systematics, distribution, ecology, and conservation of South American edentates. Pp. 345–375, *in* Mammalian biology in South America (Mares, M. A., and H. H. Genoways, eds.). Pymatuning Symposia in Ecology. Vol. 6. Special Publication Series. Pymatuning Laboratory of Ecology, University of Pittsburgh. Pittsburgh, U.S.A.

Wetzel, R. M. 1985. The identification and distribution of Recent Xenarthra (= Edentata). Pp. 5–21, in The evolution and ecology of armadillos, sloths, and vermilinguas (Montgomery, G. G. ed.). Smithsonian Institution Press. Washington, U.S.A.

Wetzel, R. M. 1985a. The identification and distribution of Recent Xenarthra (= Edentata). Pp. 5–21, *in* The evolution and ecology of armadillos, sloths, and vermilinguas (Montgomery, G. G. ed.). Smithsonian Institution Press. Washington, U.S.A.

Wetzel, R. M. 1985b. Taxonomy and distribution of armadillos, Dasypodidae. Pp. 23–48, *in* The evolution and ecology of armadillos, sloths, and vermilinguas (Montgomery, G. G., ed.). Smithsonian Institution Press, Washington, U.S.A.

Wetzel, R. M., and E. Mondolfi. 1979. The subgenera and species of long-nosed armadillos, genus *Dasypus* L. Pp. 43–63, *in* Vertebrate ecology in the northern neotropics (Eisenberg, J. F., ed.). Smithsonian Institution Press, Washington, U.S.A.

Weyandt, S. E., and R. A. Van Den Bussche. 2007. Phylogeographic structuring and mammals: the case of the pallid (*Antrozous pallidus*). Journal of Biogeography 34:1233–1245.

Wharton, C. H. 1968. First records of *Microsorex hoyi* and *Sorex cinereus* from Georgia. Journal of Mammalogy 49:158–159.
Whitaker, J. O. Jr. 1972. *Zapus hudsonius*. Mammalian Species 11:1–7.
Whitaker, J. O. Jr. 1974. *Cryptotis parva*. Mammalian Species 43:1–8.
Whitaker, Jr. J. O. 1999a. Smoky Shrew (*Sorex fumeus*). Pp. 22–23, *in* The Smithsonian book of North American Mammals (Wilson, D. E., and S. Ruff, eds). Smithsonian Institution Press, Washington, U.S.A.
Whitaker, J. O., Jr. 1999b. Woodland jumping mouse *Napaeozapus insignis*. Pp. 665–666, *in* The Smithsonian book of North American Mammals (Wilson, D. E., and S. Ruff, eds.). Smithsonian Institution Press. Washington, U.S.A.
Whitaker, J. O., Jr. 1999c. Meadow jumping mouse *Zapus hudsonius*. Pp. 666–667, *in* The Smithsonian Book of North American Mammals (Wilson, D. E., and S. Ruff, eds.). Smithsonian Institution Press. Washington, U.S.A.
Whitaker, J. O. 2004. Sorex cinereus. Mammalian Species 743:1–9
Whitaker, J. O. Jr, and T. W. French. 1984. Foods of six species of sympatric shrews from New Brunswick. Canadian Journal of Zoology 62:622–626.
Whitaker, J. O. Jr., and W. J. Hamilton. 1998. Mammals of the eastern United States. Cornell University Press. Ithaca, U.S.A.
Whitaker, J. O. Jr., and NatureServe 2018. *Reithrodontomys raviventris*. The IUCN Red List of Threatened Species 2018:e.T19401A22385344.
Whitaker, J. O. Jr., and R. E. Wrigley. 1972. *Napeozapus insignis*. Mammalian Species 14:1–6.
Whittaker, J. C., G. Hammerson, L. Master, and S. J. Norris. 2016. *Sorex dispar*. The IUCN Red List of Threatened Species 2016:e. T41394A115183478.
White, J. A. 1953. Genera and subgenera of chipmunks. University of Kansas Publications, Museum of Natural History 5:543–561.
Whorley, J. R., S. T. Álvarez-Castañeda, and G. J. Kenagy. 2004. Genetic structure of desert ground squirrels over a 20-degree-latitude transect from Oregon through the Baja California peninsula. Molecular Ecology 13:2709–2720.
Wiig, Ø., and R. W. Lie. 1984. An analysis of the morphological relationships between the hooded seals (*Cystophora cristata*) of Newfoundland, the Denmark Strait and Jan Mayen. Journal of Zoology (London) 203:227–240.
Wiig, Ø., S. Amstrup, T. Atwood, K. Laidre, N. Lunn, M. Obbard, E. Regehr, and G. Thiemann. 2015. *Ursus maritimus*. The IUCN Red List of Threatened Species 2015:e.T22823A14871490.
Wiley, R. W. 1980. *Neotoma floridana*. Mammalian Species 139:1–7.
Wilkins, K. T. 1986. *Reithrodontomys montanus*. Mammalian Species 257:1–5.
Wilkins, K. T. 1987. *Lasiurus seminolus*. Mammalian Species 280:1–5.
Wilkins, K. T. 1989. *Tadarida brasiliensis*. Mammalian Species 331:1–10.
Wilkins, K. T., and D. J. Schmidly. 1979. Identification and distribution of three species of pocket mice (genus *Perognathus*) in Trans-Pecos Texas. The Southwestern Naturalist 24:17–32.
Williams, D. F. 1968. A new record of Myotis thysanodes from Washington. The Murrelet 49:26–27.
Williams, D. F. 1978a. Taxonomic and karyologic comments on small brown bats, genus *Eptesicus*, from South America. Annals of Carnegie Museum 47:361–383.
Williams, D. F. 1978b. Systematics and ecogeographic variation of the Apache pocket mouse (Rodentia: Heteromyidae). Bulletin of Carnegie Museum of Natural History 10:1–57.
Williams, D. F. 1979. Checklist of California mammals. Annals Carnegie Museum Natural History 48:425–433.
Williams, D. F. 1984. Habitat associations of some rare shrews (*Sorex*) from California. Journal of Mammalogy 65:325–328.
Williams, D. F., and J. S. Findley. 1979. Sexual and size dimorphism in vespertilionid bats. The American Midland Naturlist 102:113–216.
Williams, D. F., and H. H. Genoways. 1979. A systematic review of the olive-backed pocket mouse, *Perognathus fasciatus* (Rodentia: Heteromyidae). Annals Carnegie Museum 48:73102.
Williams, D. F., and K. S. Kilburn. 1991. *Dipodomys ingens*. Mammalian Species 377:1–7.
Williams, D. F., H. H. Genoways, and S. E. Braun. 1993. Taxonomy. Pp. 38–196, *in* Biology of the Heteromyidae (Genoways, H. H., and J. H. Brown, eds.). American Society of Mammalogists.
Williams, L. R. 1999. *Geomys attwateri*. Pp. 481–483, *in* The Smithsonian book of North American Mammals (Wilson, D. E., and S. Ruff, eds.). Smithsonian Institution Press. Washington, U.S.A.
Williams, L. R., and G. N. Cameron. 1991. *Geomys attwateri*. Mammalian Species 382:1–5.
Williams, S. L., and H. H. Genoways. 1981. Systematic review of the Texas pocket gopher, *Geomys personatus* (Mammalia: Rodentia). Annals of Carnegie Museum of Natural History 50:435–473.
Williams, S. L., and H. H. Genoways. 2008. Subfamily Phyllostominae Gray, 1825. Pp. 255–300, *in* Mammals of South America. Vol. 1. Marsupials, xenarthrans, shrews, and bats (Gardner, A. L., ed.). University of Chicago Press. Chicago, U.S.A.
Williams-Guillén, K., and E. I. Perfecto. 2010. Effects of agricultural intensification on the assemblage of leaf–nosed bats (Phyllostomidae) in a coffee landscape in Chiapas, Mexico. Biotropica 42:605–613.
Williamson, D. F. 2002. In the black. Status, management, and trade of the American black bear (*Ursus americanus*) in North America, World Wildlife Fund. Washington, U.S.A.
Willis, K. B., M. R. Willig, and J. K. Jones. 1990. Vampyrodes caraccioli. Mammalian Species 359:1–4
Willig, M. R. 1983. Composition, microgeographic variation, and sexual dimorphism in Caatingas and Cerrado bat communities from northeast Brazil. Bulletin of the Carnegie Museum of Natural History 23:1–131.
Willig, M. R. 1985. Reproductive activity of female bats from Northeast Brazil. Bat Research News 26:17–20.
Willner, G. R., G. A. Feldhamer, E. E. Zucker, and J. A. Chapman. 1980. *Ondatra zibethicus*. Mammalian Species 141:1–8.
Wilson, D. E. 1976. Cranial variation in polar bears. Pp. 447–453, *in* Bears –their biology and management (Pelton, R., J. W. Lentfer, and G. E. Folk, eds.). International Union for the Conservation of Nature 40:1–467.
Wilson, D. E. 1982. Wolverine *Gulo gulo*. Pp. 644–652, *in* Wild Mammals of North America (Chapman, J. A., and G. A. Feldhamer, eds.). The Johns Hopkins University Press. Baltimore, U.S.A.
Wilson, D. E. 1991. Mammals of the Tres Marias Islands. Bulletin of the American Museum of Natural History 206:214–250.
Wilson, D. E., and R. K. LaVal. 1974. *Myotis nigricans*. Mammalian Species 39:1–3.

Wilson, D. E., and D. A. Reeder. 1993. Mammal species of the World. A taxonomic and geographic reference. Second edition. The Smithsonian Institution Press. Washington, U.S.A.

Wilson, G. M., and J. R. Choate. 1997. Taxonomic status and biogeography of the southern bog lemming, *Synaptomys cooperi*, on the central Great Plains. Journal of Mammalogy 78:444–458.

Wilson, D. E., and J. S. Findley. 1977. Thyroptera tricolor, Mammalian Species 71:1–3

Wilson, D. E., and D. A. M. Reeder, 2005. Mammal species of the world. A taxonomic and geographic reference. Third edition. The Johns Hopkins University Press. Baltimore, U.S.A

Wilson, D. E., M. A. Bogan, R. L. Brownell, A. M. Burdin, and M. K. Maminov. 1991. Geographic Variation in Sea Otters, Enhydra lutris. Journal of Mammalogy 72:22–36

Whitaker, J. O., and W. J. Hamilton. 1998. Mammals of the Eastern United States. 3rd ed. Cornell University Press. Ithaca, U.S.A.

Wolf, J. B.W., D. Taut, and F. Trillmich. 2007. Galapagos and Californian sea lions are separate species: genetic analysis of the genus *Zalophus* and its implications for conservation management. Frontiers in Zoology 4:1–13.

Wolfe, J. L. 1982. *Oryzomys palustris*. Mammalian Species 176:1–6.

Wolfe, J. L., and A. V. Linzey. 1977. *Peromyscus gossypinus*. Mammalian Species 70:1–5.

Wood, A. E. 1935. Evolution and relationships of the Heteromyid rodents with new form from the Tertiary of western North America. Annals of Carnegie Museum 24:73–262.

Wood, G. W., and R. H. Barrett. 1979. Status of wild pigs in the United States. Wildlife Society Bulletin 7:237–246.

Woodburne, M. O. 1968. The cranial myology and osteology of *Dicotyles tajacu*, the collared peccary, and its bearing on classification. Memoirs of the Southern California Academy of Sciences 7:1–48.

Woodhouse, S. W. 1852. Description of a new species of Pouched Rat of the genus *Geomys*, Raf. Proceedings of the Academy of Natural Sciences of Philadelphia 6:201–202.

Woodman, N. 2005. Evolution and Biogeography of the Mexican small-eared shrew of the *Cryptotis mexicana* group (Insectivora: Soricidae). Pp. 523–534, *in* Contribuciones mastozoologicas en Homenaje a Bernardo Villa (Sánchez-Cordero, V., and R. A. Medellin, eds.). Instituto de Biología e Instituto de Ecología, Universidad Nacional Autónoma de México y Comisión Nacional para el Conocimiento y Uso de la Biodiversidad. Distrito Federal, México.

Woodman, N. 2010. Two new species of shrews (Soricidae) from the western highlands of Guatemala. Journal of Mammalogy 91:566–579.

Woodman, N. 2012. Taxonomic status and relationships of *Sorex obscurus parvidens* Jackson, 1921, from California. Journal of Mammalogy 93:826–838.

Woodman, N. 2018. American Recent Eulipotyphla: Nesophontids, Solenodons, Moles, and Shrews in the New World. Smithsonian Contributions to Zoology, 650:1–107.

Woodman, N. 2019. *Cryptotis tropicalis*. The IUCN Red List of Threatened Species 2019:e.T136757A22286227.

Woodman, N., and A. W. Ferguson. 2021. The relevance of a type locality: the case of *Mephitis interrupta* Rafinesque, 1820 (Carnivora: Mephitidae). Journal of Mammalogy 102: 1583–1591.

Woodman, N., and R. B. Stephens. 2009. At the foot of the shrew: manus morphology distinguishing closely related *Cryptotis goodwini* and *Crytotis griseoventris* (Mammaliae Soricidae) in Central America. Biological Journal of the Linnean Society 99:118–134.

Woodman, N., and R. M. Timm. 1999. Geographyc variation and evolutionary relationships among broad-clawed shrews of *Cryptotis goldmani*-group (Mammalia: Insectivore: Soricidae). Fieldiana: Zoology: New Series 91:1–35.

Woodman, N., and R. M. Timm. 2000. Taxonomy and evolutionary relationships of Phillips' small-eared shrew, *Cryptotis phillipsii* (Schaldach, 1966), from Oaxaca, Mexico (Mammalia: Insectivora: Soricidae). Proceedings of the Biological Society of Washington 113:339–355.

Woodman, N., J. O. Matson, J. J. McCarthy, R. P. Eckerlin, W. Bulmer, and N. Ordóñez-Garza. 2012. Distributional records of shrews (Mammalia, Soricomorpha, Soricidae) from northern Central America with the first records of *Sorex* from Honduras. Annals of Carnegie Museum 80:207–237.

Woodman, N., S. T. Álvarez Castañeda, I. Castro-Arellano, and P. C. de Grammont. 2016a. *Megasorex gigas*. The IUCN Red List of Threatened Species 2016:e.T41454A22319710.

Woodman, N., J. Matson, A. D. Cuarón, and P. C. de Grammont. 2016b. *Cryptotis merriami*. The IUCN Red List of Threatened Species 2016:e. T136398A115207240.

Woodman, N., J. Matson, A. D. Cuarón, and P. C. de Grammont. 2016c. *Cryptotis parva*. The IUCN Red List of Threatened Species 2016:e. T41377A115182514.

Woodman, N., J. Matson, and I. Castro-Arellano. 2016d. *Sorex arizonae*. The IUCN Red List of Threatened Species 2016:e.T20396A115158374.

Woodman, N., J. Matson, A. D. Cuarón, and P. C. de Grammont. 2019. *Cryptotis alticola*. The IUCN Red List of Threatened Species 2019:e. T136789A22284844.

Woods, C. A. 1973. *Erethizon dorsatum*. Mammalian Species 29:1–6.

Woods, C. A. 1992. Endangered: Florida saltmarsh vole (*Microtus pennsylvanicus dukecampbelli*). Pp. 131–139, *in* Rare and endangered biota of Florida: vol. 1. Mammals (Humphrey, S. R., ed.). University Press of Florida. Gainesville, Florida.

Woods, C. A., and C. W. Kilpatrick. 2005. Infraorder Hystricognathi. Pp. 1538–1599, *in* Mammal Species of the World (Wilson, D. E., and D. M. Reeder, eds). Third edition. The Johns Hopkins University Press, Baltimore, U.S.A.

Woods, C. A., W. Post, and C. W. Kilpatrick. 1982. *Microtus pennsylvanicus* (Rodentia: Muridae) in Florida: a Pleistocene relict in a coastal salt-marsh. Bulletin of the Florida State Museum Biological Science 28:25–52.

Woodward, S. M. 1994. Identification of *Sorex monticolus* and *Sorex vagrans* from British Columbia. Northwest Science 68:277–283.

Worley, K., C. Strobeck, S. Arthur, J. Carey, H. Schwantje, A. Veitch, and D. W. Coltman. 2004. Population genetic structure of North American thinhorn sheep (*Ovis dalli*). Molecular Ecology 13:2545–2556.

Wozencraft, W. C. 1989. Classification of the recent carnivora. Pp. 569–594, *in* Carnivore Behaviour, Ecology and Evolution (Gittleman, J. L., ed.). Chapman, and Hall. London, United Kingdom.

Wozencraft, W. C. 1993. Order Carnivora, Family Mustelidae. Pp. 309–325, *in* Mammal species of the world. A taxonomic and geographic reference (Wilson, D. E., and D. A. M. Reeder, eds.). Second edition. The Johns Hopkins University Press. Baltimore, U.S.A.

Wozencraft, W. C. 2005. Order Carnivora. Pp. 512–628, *in* Mammal species of the world. A taxonomic and geographic reference (Wilson, D. E., and D. A. M. Reeder, eds.). Third edition. The Johns Hopkins University Press. Baltimore, U.S.A.

Wright, D. B. 1989. Phylogenetic relationships of *Catagonus wagneri*: Sister taxa from the Tertiary of North America. Pp. 281–308, *in* Advances in Neotropical mammalogy (Redford, K. H., and J. F. Eisenberg, eds). Sandhill Crane Press, Gainesville, U.S.A.

Wright, P. L. 1953. Intergradation between *Martes americana* and *Martes caurina* in Western Montana. Journal of Mammalogy 34:70–87.

Wynen, L. P., S. D. Goldsworthy, S. J. Insley, M. Adams, J. W. Bickham, J. Francis, J. P. Gallo, A. R. Hoelzel, P. Majluf, R.W.G. White, and R. Slade. 2001. Phylogenetic relationships within the eared seals (Otariidae: Carnivora): Implications for the historical biogeography of the family. Molecular Phylogenetics and Evolution 21:270–284.

Yamaguchi, N., A. Kitchener, C. Driscoll, and B. Nussberger. 2015. *Felis silvestris*. The IUCN Red List of Threatened Species 2015:e. T60354712A50652361.

Yancey, F. D., II. 1997. The mammals of Big Bend Ranch State Park. Museum of Texas Tech University, Lubbock.

Yancey, F. D., J. R. Goetze, and C. Jones. 1998a. *Saccopteryx bilineata*. Mammalian Species 581:1–5.

Yancey, F. D., J. R. Goetze, and C. Jones. 1998b. *Saccopteryx leptura*. Mammalian Species 582:1–3.

Yates, T. L. 1978. The systematics and evolution of North American moles (Insectivora: Talpidae). Ph. D. thesis. Texas Tech University. Lubbock, U.S.A.

Yates, T. L. 1999a. American shrew mole *Neurotrichus gibbsii*. Pp. 56–57, *in* The Smithsonian book of North American Mammals (Wilson, D. E., and S. Ruff, eds.). Smithsonian Institution Press. Washington, U.S.A.

Yates, T. L. 1999b. Broad-footed mole *Scapanus latimanus*. Pp. 57–58, *in* The Smithsonian book of North American Mammals (Wilson, D. E., and S. Ruff, eds.). Smithsonian Institution Press. Washington, U.S.A.

Yates, T. L. 1999c. Townsend's mole *Scapanus townsendii*. Pp. 60–61, *in* The Smithsonian book of North American Mammals (Wilson, D. E., and S. Ruff, eds.). Smithsonian Institution Press. Washington, U.S.A.

Yates, T. L. 1999d. Eastern mole *Scalopus aquaticus*. Pp. 63–65, *in* The Smithsonian book of North American Mammals (Wilson, D. E., and S. Ruff, eds.). Smithsonian Institution Press. Washington, U.S.A.

Yates, T. L. H. H. Genoways, and J. K. Jones, Jr. 1979. Rabbits (genus Sylvilagus) of Nicaragua. Mammalia 43:113–124.

Yates, T. L., and I. F. Greenbaum. 1982. Biochemical Systematics of North American Moles (Insectivora: Talpidae). Journal of Mammalogy 63:368–374.

Yates, T. L., and R. J. Pedersen. 1982. Moles. Pp. 37–51, *in* Wild mammals of North America: biology, management, and economics (Chapman, J. A., and G. A. Feldhamer, eds.). The Johns Hopkins University Press, Baltimore.

Yates, T. L., and J. Salazar-Bravo. 2005. A revision of *Scapanus latimanus*, with the revalidation of a species of Mexican mole. Pp. 489–506, *in* Contribuciones mastozoológicas en homenaje a Bernardo Villa (Sánchez-Cordero, V., and R. A. Medellín, eds.). Instituto de Biología e Instituto de Ecología, Universidad Nacional Autónoma de México y Comisión Nacional para el Conocimiento y Uso de la Biodiversidad. Distrito Federal, México.

Yates, T. L., and D. J. Schmidly. 1975. Karyotype of the Eastern mole (*Scalopus aquaticus*), with comments on the karyology of the family Talpidae. Journal of Mammalogy 56:902–905.

Yates, T. L., and D. J. Schmidly. 1977. Systematics of *Scalopus aquaticus* (Linnaeus) in Texas and adjacent states. Occasinal Papers, Museum of Texas Tech University 45:1–36.

Yates, T. L., and D. J. Schmidly. 1978. *Scalopus aquaticus*. Mammalian Species 105:1–4.

Yates, T. L., R. J. Baker, and R. K. Barnett. 1979a. Phylogenetic analysis of karyological variation in three species of peromyscine rodents. Systematic Zoology 28:40–48.

Yates, T. L. H. H. Genoways, and J. K. Jones, Jr. 1979b. Rabbits (genus *Sylvilagus*) of Nicaragua. Mammalia 43:113–124.

Yee, D. A. 2000. *Peropteryx macrotis*. Mammalian Species 643:1–4.

Yensen, E. 2000. *Urocitellus brunneus*. The IUCN Red List of Threatened Species 2000:e.T20497A9196134.

Yensen, E. 2018. *Urocitellus endemicus*. The IUCN Red List of Threatened Species 2018:e.T20498A117636227.

Yensen, E. 2019. *Urocitellus mollis*. The IUCN Red List of Threatened Species 2019:e.T116989381A116989399.

Yensen, E., and NatureServe. 2017a. *Urocitellus canus*. The IUCN Red List of Threatened Species 2017:e.T42465A22265551.

Yensen, E., and NatureServe. 2017b. *Urocitellus elegans*. The IUCN Red List of Threatened Species 2017:e.T42467A22265347.

Yensen, E., and P. W. Sherman. 1997. *Spermophilus brunneus*. Mammalian Species 560:1–5.

Yensen, E., and T. Tarifa. 2003. *Galictis vittata*. Mammalian Species 727:1–8.

Yensen, E., and M. Valdés-Alarcón. 1999. Family Sciuridae. Pp. 239–320, *in* Mamíferos del noroeste de México (Álvarez-Castañeda, S. T. and J. L. Patton, eds.). Centro de Investigaciones Biológicas del Noroeste, S. C. La Paz, México.

Young, C. J. and J. K. Jones, Jr. 1982. *Spermophilus mexicanus*. Mammalian Species 164:1–4.

Young, C. J. and J. K. Jones, Jr. 1983. *Peromyscus yucatanicus*. Mammalian Species 196:1–3.

Young, C. J., and J. K. Jones, Jr. 1984. *Reithrodontomys gracilis*. Mammalian Species 218:1–3.

Young, S. P., and E. A. Goldman. 1944. The wolves of North America. The American Wildlife Institution. Washington, U.S.A.

Young, S. P., and E. A. Goldman. 1946. The puma: Mysterious American cat. American Wildlife Institute. Washington, U.S.A.

Young, S. P., and H. H. T. Jackson. 1951. The clever coyote. American Wildlife Management Institute. Washington, U.S.A.

Youngman, P. M. 1967. A new subspecies of varying lemming, *Dicrostonyx torquatus* (Pallas), from Yukon Territory (Mammalia, Rodentia). Proceedings of the Biological Society of Washington 80:31–34.

Youngman, P. M. 1975. Mammals of the Yukon Territory. National Museum Canada, Publication Zoology 10:1–192.

Youngman, P. M. 1982. Distribution and systematics of the European mink, *Mustela lutreola* Linnaeus, 1761. Acta Zoologica Fennica 166:1–48.

Zagorodnyuk, I. V. 1990. Kariotipicheskaya izmenchuivost'I systematika serykh polevok (Rodentia, Arvicolini). Soobshchenic 1. Vidovoi sostav I khromosomnye chisla. Vestnik Zoologii 2:26–37.

Zarza, H., G. Ceballos, and M. A. Steele. 2003. *Marmosa canescens*. Mammalian Species 725:1–4.

Zamora-Gutierrez, V., and J. Ortega. 2020. *Lichonycteris obscura* (Chiroptera: Phyllostomidae). Mammalian Species 52:165–172.

Zegers, D. A. 1984. *Spermophilus elegans*. Mammalian Species 214:1–7.

Zeppelin, T. K., and A. J. Orr. 2010. Stable isotope and scat analyses indicate diet and habitat partitioning in northern fur seals Callorhinus ursinus across the eastern Pacific. Marine Ecology Progress Series 409:241–253.
Zhang, W. Q., and M. H. Zhang. 2012. Phylogeny and evolution of Cervidae based on complete mitochondrial genomes. Genetics and Molecular Research 11:628–635.
Ziegler, A. C., F. G. Howarth, and N. B. Simmons. 2016. A second endemic land mammal for the Hawaiian Islands: a new genus and species of fossil bat (Chiroptera: Vespertilionidae). American Museum Novitates (3854):1–52.
Zima, J., L. Lukacova, and M. Macholan. 1998. Chromosomal evolution in shrews. Pp. 175–218, in Evolution of shrews (Wojcik, J. M., and M. Wolsan, eds.). Białowieża, Poland: Mammal Research Institute, Polish Academy of Science, Białostockie Zakłady Graficzne.
Zimmerman, E. G. 1970. Karyology, systematics, and chromosomal evolution in the rodent genus, *Sigmodon*. Publications of the Museum, Michigan State University, Biological Series 4:385–454.
Zimmerman, E. G. 1999. *Geomys bursarius*. Pp. 485–486, *in* The Smithsonian book of North American Mammals (Wilson, D. E., and S. Ruff). Smithsonian Institution Press. Washington, U.S.A.
Zimmerman, E. G., and M. R. Lee. 1968. Variation in the chromosomes of the cotton rat, *Sigmodon hispidus*. Chromosoma 24:243–250.
Zimmerman, E. G., B. J. Hart, and C. W. Kilpatrick. 1975. Biochemical genetics of the *boylii* and *truei* groups of the genus *Peromyscus* (Rodentia). Comparative Biochemistry and Physiology 52B:541–545.
Zimmerman, E. G., C. W. Kilpatrick, and B. J. Hart. 1978. The genetics of speciation in the rodents genus *Peromyscus*. Evolution 32:565–579.

Index

S. T. Álvarez-Castañeda, *Mammals of North America - Volume 1*, https://doi.org/10.1007/978-3-031-41661-3

C

D

H

O

P

U

MIX
Papier aus verantwortungsvollen Quellen
Paper from responsible sources
FSC® C105338

If you have any concerns about our products,
you can contact us on
ProductSafety@springernature.com

In case Publisher is established outside the EU,
the EU authorized representative is:
Springer Nature Customer Service Center GmbH
Europaplatz 3, 69115 Heidelberg, Germany

Printed by Libri Plureos GmbH
in Hamburg, Germany